Discover Sociology

Fifth Edition

The fifth edition of Discover Sociology *is dedicated to the young adults who faced the myriad challenges of 2020 and ACTED by caring for family and friends, volunteering to support others, learning about issues, speaking out for justice, and voting. Gen Z will be an amazing generation!*

As always, I am grateful for the support of my family: my patient and wonderful husband, Joe, and my two children, Niklavs and Anna, who have grown into thoughtful and amazing young adults.

—DSE

I am grateful to my husband, Joe, who has always urged me to pursue my interests and dreams; to Dr. Lynn White, whose work ethic and long-ago mentoring have continued to inspire and motivate me; and to my parents, George and Janet, who helped me to develop a curious mind well before I discovered sociology.

—SLW

Discover Sociology

Fifth Edition

Daina S. Eglitis

The George Washington University

William J. Chambliss

The George Washington University

Susan L. Wortmann

Nebraska Wesleyan University

Los Angeles | London | New Delhi
Singapore | Washington DC | Melbourne

FOR INFORMATION:

SAGE Publications, Inc.
2455 Teller Road
Thousand Oaks, California 91320
E-mail: order@sagepub.com

SAGE Publications Ltd.
1 Oliver's Yard
55 City Road
London EC1Y 1SP
United Kingdom

SAGE Publications India Pvt. Ltd.
B 1/I 1 Mohan Cooperative Industrial Area
Mathura Road, New Delhi 110 044
India

SAGE Publications Asia-Pacific Pte. Ltd.
18 Cross Street #10-10/11/12
China Square Central
Singapore 048423

Acquisitions Editor: Jeff Lasser
Content Development Editor: Tara Slagle
Editorial Assistant: Tiara Beatty
Production Editor: Tracy Buyan
Copy Editor: Talia Greenberg
Typesetter: diacriTech
Proofreader: Eleni Maria Georgiou
Cover Designer: Scott Van Atta
Marketing Manager: Jennifer Jones

Printed in Canada

Library of Congress Cataloging-in-Publication Data

Names: Eglitis, Daina Stukuls, author. | Chambliss, William J., author. | Wortmann, Susan L., author.
Title: Discover sociology / Daina S. Eglitis, The George Washington University, William J. Chambliss, The George Washington University, Susan L. Wortmann, Nebraska Wesleyan University.
Description: Fifth Edition. | Thousand Oaks : SAGE Publications, Inc., 2021. | Revised edition of Discover sociology, [2020] | Includes bibliographical references and index.
Identifiers: LCCN 2021007611 | ISBN 9781071820087 (paperback) | ISBN 9781071820070 (epub) | ISBN 9781071839782 (other)
Subjects: LCSH: Sociology.
Classification: LCC HM585 .C4473 2021 | DDC 301--dc23
LC record available at https://lccn.loc.gov/2021007611

This book is printed on acid-free paper.

21 22 23 24 25 10 9 8 7 6 5 4 3 2 1

BRIEF CONTENTS

DETAILED CONTENTS

DETAILED CONTENTS

PREFACE

The German physicist Albert Einstein wrote that "the important thing is not to stop questioning. Curiosity has its own reason for existing." Indeed, *curiosity* is the bedrock of all scientific inquiry because curiosity underlies the motivation and passion to seek answers to challenging questions. But curiosity is not enough. To be a component of good sociology, curiosity must be disciplined: Answers must be sought within the scientific tradition of gathering data through systematic observations and then shared with careful empirical and theoretical explanation of the findings. A key goal of *Discover Sociology* is to pique students' curiosity about the social world—and then give them the academic tools to study that world, analyze it, and even change it.

There are many introductory sociology textbooks, some of which are very good. We believe that our contribution to the marketplace of sociological texts and ideas is a book that engages the sociology student's curious mind—and then offers them the theoretical, conceptual, and empirical tools to analyze and understand the issues that affect our world, both local and global.

We have written this book in a way that we hope will encourage students to keep reading, not only because of assigned pages but also because, with the encouragement of the instructor and the text, they have a desire to know more! We also endeavor to show the discipline of sociology as a source of critical skills valued in the job market and in graduate and professional education. We are delighted that previous editions of *Discover Sociology* have been well received, and it is our goal in this edition to continue to engage students with timely and interesting openers, the newest available data on important social phenomena, and carefully constructed theoretical and empirical discussions. We hope that this edition will also expand the reach of *Discover Sociology* with an innovative technological platform, Vantage, that is instructor and student friendly.

CHAPTER OPENERS THAT SPEAK TO STUDENTS

In this book, you will find chapters that begin with openers drawn from contemporary issues and events and that endeavor to speak to readers and to the kinds of experiences or concerns they have as students, in families, and at workplaces. The beginning of each chapter also features "What Do You Think?" questions to engage students' curiosity and give a preview of interesting issues that will be covered in the chapter.

SOCIOLOGY IS A SCIENTIFIC DISCIPLINE

Every chapter in the book integrates empirical research from sociology, highlighting the point that sociology is about the *scientific understanding* of the social world—rigorous research can illuminate the sociological roots of diverse phenomena and institutions, ranging from poverty and deviance to capitalism and the nuclear family. Research may also result in conflicting or ambiguous conclusions. Students learn that social life is complex and that sociological research is an ongoing effort to explain why things are as they are—and how they might change.

KEY THEMES

Each chapter's material highlights key themes in this book:

The sociological imagination, a foundational concept in the discipline, is interwoven throughout the book. We seek to illuminate for student readers the nexus between private troubles and public issues and to show with clear examples and explanations how individual lives are shaped by social forces—but also how, in turn, individuals can act to shape the world around them.

Power is a key theme in sociology—and in this text. Sociologists want to know how power is acquired, distributed, reproduced, and exercised in social relationships and institutions. The unequal distribution of economic, social, and political power is an important topic of sociological inquiry, and this text offers many examples that probe manifestations of and explanations for power and resource disparities.

This text emphasizes the importance of being a *critical consumer of information.* We are surrounded by sources of data that stream into our lives from the Internet, television and newspapers, peers and colleagues, friends and family, and academic studies. Many of our chapters feature informative examinations of statistical information on phenomena like unemployment, poverty, and the gender wage gap. These are intended to help students understand the sources and assumptions behind these figures and to recognize what they illuminate, and what they obscure about our social world.

Social media increasingly shapes our lives, our beliefs, and our actions. Social science is only beginning to grasp the significance of these dramatic developments. Across chapters, this text endeavors to provide a sociological perspective on social media's functions, contributions, and consequences.

Global issues are interwoven throughout this text. The book seeks to help students develop a fuller understanding of the place of their lives, their communities, and their country in an interconnected, interdependent, and multicultural international environment—and to enable them to see how other countries around the world are experiencing societal changes and challenges.

These key themes—the sociological imagination, power and inequality, the critical consumption of information, social media and its effects, and a global perspective—are woven throughout this text. Like past editions, this edition invites students to learn about the social world. In this edition, we also ask them to reflect on the dramatic events of the year that kicked off the 2020s, to consider where we find ourselves, and to apply sociological insights to consider what's next.

NEW IN THIS EDITION

An important goal of the fifth edition of *Discover Sociology* is to retain the best features of prior editions while responding to the ideas and requests of reviewers and faculty for expanded coverage of issues such as intersectionality, rising economic inequality, popular culture, and the concerns of our contemporary population of college students in the United States. This edition features updated social indicators, bringing in the latest data available from the U.S. Census Bureau, the Bureau of Labor Statistics, the Centers for Disease Control and Prevention, and the Pew Research Center, among others, to ensure that discussions and figures remain timely.

What's Next?

This edition of *Discover Sociology* takes a new approach to wrapping up individual chapters. Rather than ending chapters with a standard conclusion that summarizes content, we reflect on the significant and often dramatic events of the beginning of the 2020s, including the global COVID-19 pandemic, acute economic crisis and displacement, a contentious and arguably unprecedented election year, and mass movements demanding justice and social change. We consider them from a sociological perspective, raising questions about where we find ourselves and *what's next.* We hope that this new feature will encourage instructors and students to talk about the sociological significance of the events of this tumultuous time and to consider how they may shape the years to come.

A Greater Focus on Intersectionality in Sociological Study

In recent decades, sociologists have increasingly sought to identify ways in which achieved and ascribed characteristics such as gender, race, ethnicity, age, class, religion, sexual orientation, and sexuality intersect with one another in social practices and institutions and, significantly, how these intersections affect access to, for instance, education, occupational status, and political voice. Put another way,

> When it comes to social inequality, people's lives and the organization of power in a given society are better understood as being shaped not by a single axis of social division, be it race or gender or class, but by many axes that work together and influence each other. (P. H. Collins & Bilge, 2016, p. 2)

Instructors have asked for fuller and more sustained attention to intersectionality as a key feature in the sociological analysis of inequality. In this book, we have endeavored to respond with new discussion, research, and examples to show student readers the significance of an intersectional approach.

Connecting Sociology and Career Success

As an instructor of introductory sociology, you are probably asked by students, "What can I do with a sociology degree?" This is an important question for students and instructors. This book offers a unique feature that speaks directly and specifically to this question.

In *Discover Sociology*, all of the chapters, beginning with Chapter 2, feature an essay that accomplishes two major tasks. First, every essay highlights specific skills students learn as sociology majors and describes those skills in ways that provide students a vocabulary they can use in the job market. Second, each essay profiles a graduate with a degree in sociology who is putting his or her skills to work in an interesting occupation or workplace. Graduates share, in their own words, what they learned from sociology and how it has contributed to their skills, knowledge, and career.

The fifth edition of *Discover Sociology* offers readers an updated array of career profiles: Twelve fully new profiles of sociology majors provide students with insights into how sociology can prepare them for a spectrum of diverse and fulfilling careers.

It is important to note that this feature is not only for sociology majors! Sociology is often among the general education courses completed by students across a variety of disciplines, and it can help all students develop important skills—such as critical thinking, data literacy, and written communication—that they will need in the workplace.

Clearly, for students across disciplines, there is value in understanding and naming the skills that they gain when they study sociology. We encourage all students to take advantage of this valuable feature.

Photos and Graphics

The photographs in this edition have been carefully selected to help students put images together with ideas, events, and phenomena. A good photo can engage a student's curiosity and offer a visual vehicle for remembering the material under discussion. This has been our goal in choosing the photos included here. The book also features clear and visually appealing graphics, including tables, figures, and maps, to attract students' attention and enhance learning.

Glossaries for Learning

This book provides glossaries in each chapter that provide students easy access to definitions of key concepts, phenomena, and institutions. Additionally, key terms are bolded in the text, and a comprehensive glossary is available at the end of the book.

Chapter Review

Every chapter ends with a summary of key learning points and a set of discussion questions to review what students have learned and to foster critical thinking about the materials.

ACKNOWLEDGMENTS

We are grateful to the terrific editors and marketing staff at SAGE, including Jeff Lasser, Tara Slagle, Tiara Beatty, Rob Bloom, Jennifer Jones, and Christina Fohl. Thank you as well to this edition's Vantage team, Deb Hartwell, Claudine Bellanton, Melissa Seserko, Isabelle Cheng, Jonathan Hritz, Ashlee Blunk, and Kelly DeRosa. It is a privilege to work with this creative, smart, and supportive group. Thank you as well to SAGE's amazing and hardworking sales staff, including a long-time friend, Jerry Higgins. We are also indebted to colleagues and graduate students who have helped over four editions with the materials that went into the book. Among those who contributed ideas and assistance are the Department of Sociology at GWU's Michelle Kelso, Steve Tuch, Greg Squires, Ivy Ken, Antwan Jones, Fran Buntman, Hiromi Ishizawa, Emily Morrison, and Xolela Mangcu, and the Department of Sociology at NWU's Joan Gilbreth and David Iaquinta. We would like to extend special thanks to the excellent research assistants who have supported this book: for this edition, Holden Nix, Kimberly Krane, and Jeremy Wortmann; for work on Chapter 16, E. Hanah Nore; for the third edition and second core edition, Marwa Moaz; for the original core edition, Srushti Upadhyay; for the fourth edition, Ertrell Harris; for work on Chapter 6 of editions three and four, Anna Eglitis; for the second edition, Ann Horwitz and Chris Moloney; and for the first edition, Chris Moloney, Jee Jee Kim, Claire Cook, Scott Grether, Ken Leon, Ceylan Engin, and Adam Bethke. This project could not have been brought to completion without the valuable help and skills of all of the amazing people named.

Daina would like to extend her gratitude to her family, particularly her husband, Joe; mother, Silvija; and children, Niklavs and Anna. They continue to be an important source of inspiration and information for this project. Sue would like to extend her deepest appreciation to her husband, Joe; her children, Jeremy and Alex; and their spouses, Lindsay and Allison. They have gifted her with their ideas, encouragement, and humor.

Finally, we thank all of the reviewers listed subsequently, who contributed to *Discover Sociology* with excellent suggestions, creative insights, and helpful critiques.

REVIEWERS FOR THE FIFTH EDITION

Matthew Green, College of DuPage
Daniel Kavish, Southwestern Oklahoma State University
Derek Roberts, Monroe County Community College
James Shockey, University of Arizona
Megan Smith, University of North Carolina, Charlotte
Sumi Srinivason, Cuyahoga Community College

REVIEWERS FOR THE CORE CONCEPTS, SECOND EDITION

Marianne Cutler, East Stroudsburg University of Pennsylvania
Michelle L. Johnson, Muskegon Community College
Ryan Jerome LeCount, Hamline University
Christina Mendoza, Chabot College
Susan B. Murray, San Jose State University
Jamie Oslawski-Lopez, Indiana University Kokomo
Max Probst, Bucks County Community College
Elizabeth Robinson, Pacific Oaks College
Frank A. Salamone, Westchester Community College

REVIEWERS FOR THE FOURTH EDITION

Jessica Bishop-Royse, DePaul University
Scott Coahran, Merced College
Heather Downs, Jacksonville University
Candan Duran-Aydintug, University of Colorado Denver

M. Faye Hanson-Evans, University of Texas Arlington
Kia Heise, California State University–Los Angeles
Ting Jiang, Metropolitan State University of Denver
Robert S. Mackin, Texas A&M University
Brian Monahan, Baldwin Wallace University
Naghme Morlock, Gonzaga University
Marvin Pippert, Young Harris College
Milanika Tuner, independent researcher
Alicia M. Walker, Missouri State University
Kristi D. Wood-Turner, West Virginia University

REVIEWERS FOR THE CORE CONCEPTS, FIRST EDITION

Michael Bourgoin, CUNY, Queens College
Gerri Brown, Copiah-Lincoln Community College
Marianne Cutler, East Stroudsburg University of Pennsylvania
Kellie J. Hagewen, College of Southern Nevada
Mark Killian, Whitworth University
Rosalind Kopfstein, Western Connecticut State University
Ryan Jerome LeCount, Hamline University
Ho Hon Leung, SUNY Oneonta
Sherry N. Mong, Capital University
Kaitlyne A. Motl, University of Kentucky
Susan B. Murray, San Jose State University
Carolyn Pevey, Germanna Community College
Thomas Piñeros Shields, University of Massachusetts at Lowell
Karen Platts, Bucks County Community College
Frank A. Salamone, Westchester Community College
Kamesha Spates, Kent State University
Jennifer Valentine, Tidewater Community College
Abraham Waya, Boston University
Lia Chervenak Wiley, The University of Akron

REVIEWERS FOR THE THIRD EDITION

Laura Chambers Atkins, Jacksonville University
Marian Colello, Strayer University
Leslie Elrod, University of Cincinnati
Matthew Green, College of DuPage
Othello Harris, Miami University
Belinda Hartnett, Strayer University
Rick Jones, Marquette University
Lauren Kempton, University of New Haven
Veena S. Kulkarni, Arkansas State University
Elaine Leeder, Sonoma State University
Olena Leipnik, Sam Houston State University
Peter LeNeyee, Strayer University
Robert Sean Mackin, Texas A&M University
Aurelien Mauxion, Columbia College
Debra M. McCoy, Strayer University
Virginia Merlini, Strayer University
Allan Mooney, Strayer University
Andrew J. Prelong, University of Northern Colorado
Angela Primm-Bethea, Strayer University
Terri Slonaker, San Antonio College
Lia Chervenak Wiley, The University of Akron
Susan L. Wortmann, Nebraska Wesleyan University

REVIEWERS FOR THE SECOND EDITION

Dianne Berger-Hill, Old Dominion University
Alison J. Bianchi, University of Iowa
Michael Bourgoin, Queens College, The City University of New York
Paul E. Calarco Jr., Hudson Valley Community College
Nicolette Caperello, Sierra College
Susan E. Claxton, Georgia Highlands College
Sonya R. De Lisle, Tacoma Community College
Heather A. Downs, Jacksonville University
Leslie Elrod, University of Cincinnati
S. Michael Gaddis, University of Michigan
Cherly Gary-Furdge, North Central Texas College
Louis Gesualdi, St. John's University
Todd Goodsell, University of Utah
Matthew Green, College of DuPage
Ashley N. Hadden, Western Kentucky University
Othello Harris, Miami University
Michael M. Harrod, Central Washington University
Sarah Jacobson, Harrisburg Area Community College
Kimberly Lancaster, Coastal Carolina Community College
Katherine Lawson, Chaffey Community College
Jason J. Leiker, Utah State University
Kim MacInnis, Bridgewater State University
Barret Michalec, University of Delaware
Amanda Miller, University of Indianapolis
Christine Mowery, Virginia Commonwealth University
Scott M. Myers, Montana State University
Frank A. Salamone, Iona College

Bonita A. Sessing-Matcha, Hudson Valley Community College
Richard States, Allegany College of Maryland
Myron T. Strong, Community College of Baltimore County
Heather Laine Talley, Western Carolina University
P. J. Verrecchia, York College of Pennsylvania
Jerrol David Weatherly, Coastal Carolina Community College
Debra L. Welkley, California State University, Sacramento
Luis Zanartu, Sacramento City College

REVIEWERS FOR THE FIRST EDITION

Kristian P. Alexander, Zayed University
Lori J. Anderson, Tarleton State University
Shannon Kay Andrews, University of Tennessee at Chattanooga
Joyce Apsel, New York University
Gabriel Aquino, Westfield State College
Janet Armitage, St. Mary's University
Dionne Mathis Banks, University of Florida
Michael S. Barton, University at Albany
Jeffrey W. Basham, College of the Sequoias
Paul J. Becker, University of Dayton
Alison J. Bianchi, University of Iowa
Kimberly Boyd, Piedmont Virginia Community College
Mariana Branda, College of the Canyons
Jennifer Brennom, Kirkwood Community College
Denise Bump, Keystone College
Nicolette Caperello, Sierra College
Michael J. Carter, California State University, Northridge
Vivian L. Carter, Tuskegee University
Shaheen A. Chowdhury, College of DuPage
Jacqueline Clark, Ripon College
Susan Eidson Claxton, Georgia Highlands College
Debbie Coats, Maryville University
Angela M. Collins, Ozarks Technical Community College
Scott N. Contor, Idaho State University
Denise A. Copelton, The College at Brockport, State University of New York
Carol J. Corkern, Franklin University
Jennifer Crew Solomon, Winthrop University
William F. Daddio, Georgetown University
Jeffrey S. Debies-Carl, University of New Haven
Melanie Deffendall, Delgado Community College
Marc Jung-Whan de Jong, State University of New York, Fashion Institute of Technology
David R. Dickens, University of Nevada, Las Vegas
Keri Diggins, Scottsdale Community College
Amy M. Donley, University of Central Florida
Amanda Donovan, Bristol Community College
Heather A. Downs, Jacksonville University
Daniel D. Doyle, Bay College
Dorothy E. Everts, University of Arkansas–Monticello
Gary Feinberg, St. Thomas University
Bernie Fitzpatrick, Western Connecticut State University
Tonya K. Frevert, University of North Carolina at Charlotte
S. Michael Gaddis, University of North Carolina at Chapel Hill
Robert Garot, John Jay College of Criminal Justice
Todd A. Garrard, University of Texas at San Antonio
Cherly Gary-Furdge, North Central Texas College
Marci Gerulis-Darcy, Metropolitan State University
Louis Gesualdi, St. John's University
Jennifer E. Givens, University of Utah
John Glass, Collin College
Malcolm Gold, Malone University
Thomas B. Gold, University of California, Berkeley
Matthew Green, College of DuPage
Johnnie M. Griffin, Jackson State University
Randolph M. Grinc, Caldwell College
Greg Haase, Western State College of Colorado
Dean H. Harper, University of Rochester
Anne S. Hastings, University of North Carolina at Chapel Hill
Anthony L. Haynor, Seton Hall University
Roneiko Henderson-Beasley, Tidewater Community College
Marta T. Henriksen, Central New Mexico Community College
Klaus Heyer, Nunez Community College
Jeremy D. Hickman, University of Kentucky
Bonniejean Alford Hinde, College of DuPage
Joy Crissey Honea, Montana State University Billings
Caazena P. Hunter, University of North Texas
John Iceland, Pennsylvania State University
Robert B. Jenkot, Coastal Carolina University
Wesley G. Jennings, University of South Florida
Audra Kallimanis, Wake Technical Community College
Ali Kamali, Missouri Western State University
Leona Kanter, Mercer University
Earl A. Kennedy, North Carolina State University
Lloyd Klein, York College, The City University of New York
Julie A. Kmec, Washington State University
Todd M. Krohn, University of Georgia
Veena S. Kulkarni, Arkansas State University
Karen F. Lahm, Wright State University
Amy G. Langenkamp, Georgia State University
Barbara LaPilusa, Montgomery College

Jason LaTouche, Tarleton State University
Ke Liang, Baruch College, The City University of New York
Carol S. Lindquist, Bemidji State University
Travis Linnemann, Kansas State University
Stephen Lippmann, Miami University
David G. LoConto, Jacksonville State University
Rebecca M. Loew, Middlesex Community College
Jeanne M. Lorentzen, Northern Michigan University
Betsy Lucal, Indiana University South Bend
George N. Lundskow, Grand Valley State University
Crystal V. Lupo, Auburn University
Brian M. Lynch, Quinebaug Valley Community College
Kim A. MacInnis, Bridgewater State University
Mahgoub El-Tigani Mahmoud, Tennessee State University
Lori Maida, The State University of New York
Hosik Min, Norwich University
Madeline H. Moran, Lehman College, The City University of New York
Amanda Moras, Sacred Heart University
Rebecca Nees, Middle Georgia College
Christopher Oliver, University of Kentucky
Sophia M. Ortiz, San Antonio College
Kathleen N. Overmiller, Marshall University
Josh Packard, Midwestern State University
Marla A. Perry, Nashville State Community College
Daniel Poole, Salt Lake Community College
Shana L. Porteen, Finlandia University
Eric Primm, University of Pikeville
Jeffrey Ratcliffe, Drexel University
Jo Reger, Oakland University
Daniel Roddick, Rio Hondo College
David Rohall, Western Illinois University
Olga I. Rowe, Oregon State University
Josephine A. Ruggiero, Providence College
Frank A. Salamone, Iona College
Stephen J. Scanlan, Ohio University
Michael D. Schulman, North Carolina State University
Maren T. Scull, University of Colorado Denver
Shane Sharp, Northern Illinois University
Mark Sherry, The University of Toledo
Amber M. Shimel, Liberty University
Vicki Smith, University of California, Davis
Dan Steward, University of Illinois at Urbana–Champaign
Myron T. Strong, Community College of Baltimore County
Richard Sullivan, Illinois State University
Sara C. Sutler-Cohen, Bellevue College
Joyce Tang, Queens College, The City University of New York
Debra K. Taylor, Metropolitan Community College–Maple Woods
Richard Tewksbury, University of Louisville
Kevin A. Tholin, Indiana University–South Bend
Brian Thomas, Saginaw Valley State University
Lorna Timmerman, Indiana University East
Cynthia Tooley-Heddlesten, Metropolitan Community Colleges–Blue River
Okori Uneke, Winston-Salem State University
Paula Barfield Unger, McLennan Community College
P. J. Verrecchia, York College of Pennsylvania
Joseph M. Verschaeve, Grand Valley State University
Edward Walker, University of California, Los Angeles
Tom Ward, New Mexico Highlands University
Lisa Munson Weinberg, Florida State University
Casey Welch, Flagler College
Shonda Whetstone, Blinn College
S. Rowan Wolf, Portland Community College
Loreen Wolfer, University of Scranton
Jason Wollschleger, Whitworth University
Kassia R. Wosick, New Mexico State University

ABOUT THE AUTHORS

Daina S. Eglitis (PhD, University of Michigan–Ann Arbor) is an associate professor of sociology and international affairs and the director of undergraduate studies in the Department of Sociology at The George Washington University (GWU). Her scholarly interests include class and social stratification, historical sociology, contemporary theory, gender, and culture. She is the author of *Imagining the Nation: History, Modernity, and Revolution in Latvia* (Penn State Press, 2002), as well as numerous articles on collective memory and history, social inequality, and demographic change in Eastern Europe. She has held two Fulbright awards in Latvia and is a past recipient of research fellowships and awards from the U.S. Holocaust Memorial Museum, the American Council of Learned Societies, the National Council for Eurasian and East European Research, the International Research and Exchanges Board, and the Woodrow Wilson International Center for Scholars. Dr. Eglitis is the author of "The Uses of Global Poverty: How Economic Inequality Benefits the West," an article widely used by undergraduate students. At GWU, she teaches courses on contemporary sociological theory, class and inequality, and introductory sociology, among others. She presents and writes on the topic of teaching and learning and is the author of the *Teaching Sociology* articles "Performing Theory: Dramatic Learning in the Theory Classroom" (2010) and "Social Issues and Problem-Based Learning in Sociology: Opportunities and Challenges in the Undergraduate Classroom" (2016). Outside the classroom, Dr. Eglitis is an avid reader of fiction (recent discoveries include *The Book of Night Women, The Thing Around Your Neck*, and *Where the Crawdads Sing*) and loves to travel to new places.

Susan L. Wortmann (PhD, University of Nebraska–Lincoln) is a professor of sociology and gender studies at Nebraska Wesleyan University (NWU). Her scholarly interests include classical sociological theory, gender, religion, and health. She instructs classes in these subjects and introduction to sociology. She is the author of *Society: A User's Guide*, a collection of introductory sociology readings. She has published articles on meaning and religious participation, sociological theory, and project evaluation, and has contributed to Blackwell's *Encyclopedia of Sociology* and the *Encyclopedia of Consumer Culture*. Dr. Wortmann also practices applied sociology in her community, using sociological methods, theories, and concepts in collaboration with community partners to address social issues. These include developing a survivor-centered response to campus sexual assault; increasing diabetes and mental health access for Karen, Yezidi, Sudanese, and Vietnamese refugees and immigrants; and creating and evaluating programming for at-risk youth. She is a recipient of community, colleague, and student-nominated awards and honors. Like Dr. Eglitis, when she is not teaching, Dr. Wortmann is an avid fiction reader and traveler. Her favorite travels have involved mutual discovery with students: how face-to-face contact fostered mutual understanding in a rural village in Izmir Province, Turkey; the reality of resilience beyond struggle on the Pine Ridge Indian Reservation, South Dakota; and the importance of instructor–student trust while exploring abandoned underground tunnels in Havana, Cuba.

William J. Chambliss (PhD, Indiana University) was a professor of sociology at The George Washington University from 1986 to 2014. During his long and distinguished career, he wrote and edited close to two dozen books and produced numerous articles for professional journals in sociology, criminology, and law. The integration of the study of crime with the creation and implementation of criminal law was a central theme in his writings and research. His articles on the historical development of vagrancy laws, the legal process as it affects different social classes and racial groups, and his efforts to introduce the study of state-organized crimes into the mainstream of social science research are among the most recognized achievements of his career. Dr. Chambliss was the recipient of numerous awards and honors, including a doctorate of laws *honoris causa*, University of Guelph, Guelph, Ontario, Canada, 1999; the 2009 Lifetime Achievement Award, Sociology of Law, American Sociological Association; the 2009 Lifetime Achievement Award, Law and Society, Society for the Study of Social Problems; the 2001 Edwin H. Sutherland Award, American Society of Criminology; the 1995 Major Achievement Award, American Society of Criminology; the 1986 Distinguished Leadership in Criminal Justice, Bruce Smith Sr. Award, Academy of Criminal Justice Sciences; and the 1985 Lifetime Achievement Award, Criminology, American Sociological Association. Professor Chambliss also served as president of the American Society of Criminology and the Society for the Study of Social Problems.

©Michael H/Getty Images

1 DISCOVER SOCIOLOGY

WHAT DO YOU THINK?

1. Can societies be studied scientifically? What does the scientific study of societies entail?
2. What is a theory? What role do theories play in sociology?
3. In your opinion, what social issues or problems are most interesting or important today? What questions about those issues or problems would you like to study?

LEARNING OBJECTIVES

1.1 Describe the sociological imagination.

1.2 Understand the significance of critical thinking in the study of sociology.

1.3 Trace the historical development of sociological thought.

1.4 Identify key theoretical paradigms in the discipline of sociology.

1.5 Identify the three main themes of this book.

SOCIOLOGY AND THE CURIOUS MIND

A goal of this book is to take you on a sociological journey. But let's begin with a basic question: *What is sociology?* First of all, sociology is a discipline of and for curious minds. Sociologists are deeply committed to answering the question, "Why?" Why are some people desperately poor and others fabulously wealthy? Why does racial segregation in housing and public education exist, and why does it persist more than half a century after civil rights laws were enacted in the United States? What accounts for the decline of marriage among the poor and the working class—as well as among the millennial generation? Why is the proportion of women entering and completing college rising while men's enrollment has fallen? Why, despite this, do men as a group still earn higher incomes than do women as a group? And how is it that social media is simultaneously praised as a vehicle of transformational activism and criticized as a cause of social alienation and civic disengagement? Take a moment to think about some *why* questions you have about society and social life: As you look around you, hear the news, and interact with other people, what strikes you as fascinating but perhaps difficult to understand? What are you curious about?

Scientific: A way of learning about the world that combines logically constructed theory and systematic observation.

Sociology will take you on a journey to understanding and generating new knowledge about human behavior, social relations, and social institutions on a larger scale.

©Westend61/Getty Images

Sociology is an academic discipline that takes a scientific approach to answering the kinds of questions our curious minds imagine. When we say that sociology is **scientific**, we mean that it is *a way of learning about the world that combines logically constructed theory and systematic observation*. The goal of sociological study and research is to base answers to questions, like those we just posed, on careful examination of the roots of social phenomena, such as poverty, segregation, and the wage gap. Sociologists do this with *research methods*—surveys, interviews, observations, and archival research, among others—that yield data that can be tested, challenged, and revised. In this text, you will see how sociology is done—and you will learn how to do sociology, yourself.

Concisely stated, **sociology** is *the scientific study of human social relations, groups, and societies*. Unlike *physical and natural sciences*, such as physics, geology, chemistry, and biology, sociology is one of several *social sciences* engaged in the scientific study of human beings and the social worlds they consciously create and inhabit. The purpose of sociology is to understand and generate new knowledge about human behavior, social relations, and social institutions on a larger scale. The sociologist adheres to the principle of **social embeddedness**: *the idea that economic, political, and other forms of human behavior are fundamentally shaped by social relations*. Thus, sociologists pursue studies on a wide range of issues occurring within, between, and among families, communities, states, nations, and the world. Other social sciences, some of which you may be studying, include anthropology, economics, political science, and psychology.

Sociology seeks to construct a body of scientific and rigorous knowledge about social relations, groups, and societies. A new area of interest is the way social media is changing the way we interact with our social environment and with one another.

©FreshSplash/Getty Images

Sociology: The scientific study of human social relations, groups, and societies.

Social embeddedness: The idea that economic, political, and other forms of human behavior are fundamentally shaped by social relations.

Sociology is a field in which students have the opportunity to build strong core knowledge about the social world with a broad spectrum of important skills, ranging from gathering and analyzing information to identifying and addressing social problems to effective written and oral communication. Throughout this book, we draw your attention to important skills you can gain through the study of sociology and the kinds of jobs and fields in which these skills can be put to work.

Doing sociology requires that you build a foundation for your knowledge and understanding of the social world. Some key foundations of sociology are the *sociological imagination* and *critical thinking*. We turn to these next.

THE SOCIOLOGICAL IMAGINATION

As we go about our daily lives, it is easy to overlook the fact that large-scale economic, political, and cultural forces shape even the most personal aspects of our lives. When parents divorce, for example, we tend to focus on individual explanations: A father was devoted more to his work than to his family; a mother may have felt trapped in an unhappy marriage but stuck with it for the sake of young children. Yet while personal issues are inevitable parts of a breakup, they can't tell the whole story. When many U.S. marriages end in divorce, forces larger than incompatible personalities or marital discord are at play. But what are those greater social forces, exactly?

As sociologist C. Wright Mills (1916–1962) suggested half a century ago, uncovering the relationship between what he called *personal troubles* and *public issues* calls for a **sociological imagination** (Mills, 1959/2000b). The sociological imagination is *the ability to grasp the relationship between individual lives and the larger social forces that shape them*—that is, to see where biography and history intersect.

Sociological imagination: The ability to grasp the relationship between individual lives and the larger social forces that help to shape them.

In a country such as the United States, where individualism is part of the national heritage, people tend to believe that each person creates his or her life's path and largely disregards the social context in which this happens. When we cannot get a job, fail to earn enough to support a family, or experience marital separation, for example, we tend to see it as a personal trouble. We do not necessarily see it as a public issue. The sociological imagination, however, invites us to make the connection and to step away from the vantage point of a single life experience to see how powerful social forces—for instance, changes in social norms, racial or gender discrimination, large shifts in the economy, or the beginning or end of a military conflict—shape the obstacles and opportunities that contribute to the unfolding of our life's story. Among Mills's (1959/2000b) most often cited examples is the following:

> When, in a city of 100,000, only one man is unemployed, that is his personal trouble, and for its relief we properly look to the character of the man, his skills, and his immediate opportunities. But when in a nation of 50 million employees, 15 million men are unemployed, that is an issue, and we may not hope to find its solution within the range of opportunities open to any one individual. The very structure of opportunities has collapsed. Both the correct statement of the problem and the range of possible solutions require us to consider the economic and political institutions of the society, and not merely the personal situation and character of a scatter of individuals. (p. 9)

Workers with a college degree fared better than those without a degree in the COVID-19 recession because they were more likely to work in jobs that can be done remotely. While about 62% of those with a college degree had that option, just 9% of those without a high school degree did (Kochhar, 2020).

©FreshSplash/Getty Images

To apply the idea to contemporary economic conditions, we might look at recent college graduates. The COVID-19 pandemic hit the U.S. population and job market with force beginning in March of 2020. For nearly a decade, the job market for recent graduates had been relatively robust: Unemployment was low and while challenges, including underemployment and high levels of student debt, were present, the picture for those completing higher education at both the undergraduate and graduate levels was generally good. In early 2019, the rate of unemployment of young college graduates (ages 21–24) was about 6.4% for men and 4.1% for women. For college graduates 25 years and older, the rate was lower still: In April of 2019, it was just 2.1% (Bureau of Labor Statistics, 2019). In April of 2020, however, it had risen precipitously, reaching 8.4% (U.S. Bureau of Labor Statistics, 2020). For new graduates, the job prospects were worse. According to a poll taken that month, even among students who had jobs lined up after graduation, about three quarters had their positions cancelled, delayed, or moved to remote work (Bahler, 2020). The sociological imagination invites us to recognize that employment or unemployment, like many other phenomena, are an outcome not only of individual choices, actions, and luck, but of larger social, economic, and political forces. Biography and history are, as the sociological imagination recognizes, closely intertwined.

Understanding this relationship is particularly critical for people in the United States, who often regard individuals as solely responsible for their social, educational, and economic successes and failures. For instance, it is easy to fault the poor for their poverty, assuming they only need to work harder and pull themselves up by their bootstraps. We may neglect the powerful role of social forces such as racial and ethnic discrimination, the outsourcing or automation of manufacturing jobs that used to employ those without a college education, or the dire state of public education in many economically distressed rural and urban areas. The sociological imagination implores us to seek the intersection between private troubles, such as a family's poverty, and public issues, such as lack of access to good schooling and jobs paying a living wage, to develop a more informed and comprehensive understanding of the social world and social issues.

Agency: The ability of individuals and groups to exercise free will and to make social changes on a small or large scale.

Structure: Patterned social arrangements that have effects on agency and are, in turn, affected by agency.

It is useful, when we talk about the sociological imagination, to bring in the concepts of *agency* and *structure*. Sociologists often talk about social actions—individual and group behavior—in these terms. **Agency** can be understood as *the ability of individuals and groups to exercise free will and to make social changes on a small or large scale.* **Structure** is a complex term but may be defined as *patterned social arrangements that have effects on agency and are, in turn, affected by agency.* Structure may enable or constrain social action. For example, sociologists talk about the class structure, which is composed of social groups that hold varying amounts of resources, such as money, political voice, and social status. They also identify normative structures—for instance, they might analyze patterns of social norms regarding "appropriate" gender behaviors in different cultural contexts.

Sociologists take a strong interest in the relationship between structure and agency. Consider that, on the one hand, we all have the ability to make choices—we have free will, and we can opt for one path over another. On the other hand, the structures that surround us impose obstacles on us or afford us opportunities to exercise agency: We can make choices, but they may be enabled or constrained by structure. For instance, in the early 1900s, we could surely have found bright young women in the nascent U.S. middle class who wanted to study law, medicine, or philosophy. The social norms of the time, however, held that young women of this status were better off marrying and caring for a husband, home, and children. There were also legal constraints to women's entry into higher education and the paid labor force. Women of color, poor women, and immigrant women were even less likely to have opportunities for education. So, although the women in our example might have individually aspired to an education and career, their dreams were constrained by powerful normative, educational, and legal structures, as well as economic constraints, that actively excluded them from the public sphere.

Homelessness is one of many public issues that sociologists seek to understand more fully. Homelessness may be chronic or short-term. It is often rooted in a lack of low-cost housing, particularly in urban areas where the cost of living is high.

©urbazon/Getty Images

A growing number of high school graduates are enrolling in college immediately after high school. Those from higher-income schools, however, are more likely to do so than their peers at low-income schools: 69% of graduates from higher-income schools matriculate immediately compared to 55% from low-income schools (National Student Clearinghouse Research Center, 2019). What sociological factors might explain this gap?

©Paul Bradbury/Getty Images

Consider as well the relationship between the class structure and individual agency as a way of thinking about social mobility in U.S. society. If, for instance, a young man whose parents are well educated and whose family is economically comfortable wishes to go to college and study to be an architect, engineer, or lawyer, his position in the class structure (or the position of his family) is *enabling*—that is, it raises the probability that he will be able to make this choice and realize it. If, however, a young man from a poor family with no college background embraces these same dreams, his position in the class structure is likely to be *constraining*: Not only does his family have insufficient economic means to pay for college, but he also may be studying in an underfunded or underperforming high school that cannot provide the advanced courses and other resources he needs to prepare for college. A lack of college role models also may be a factor. This does not mean that the first young man will inevitably go to college and realize his hopes and the second will not; it does, however, suggest that structural conditions favor the first college aspirant over the second.

To understand why some students go to college and others do not, sociologists would say that we cannot rely on individual choice or will (agency) alone—structures, whether subtly or quite obviously, exercise an influence on social behavior and outcomes. At the same time, we should not see structures

FIGURE 1.1 ■ Structure and Agency

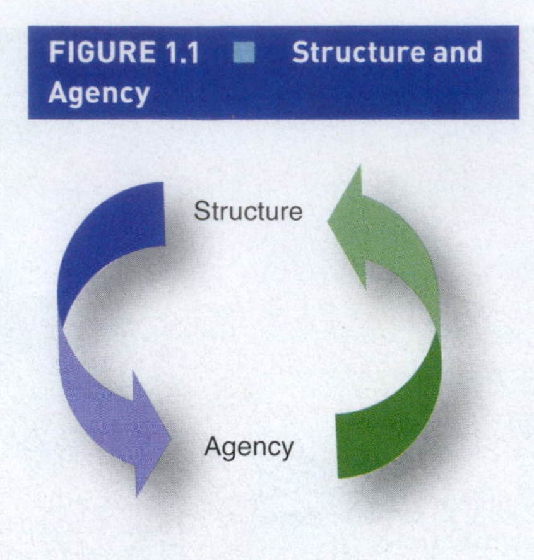

as telling the whole story of social behavior because history shows the power of human agency in making change, even in the face of obstacles. Agency itself can transform structures. For example, think about the ways women's historical activism helped to transform gender norms, educational practices, and legal protections for women today. Sociologists weigh both agency and structure and study how the two intersect and interact. For the most part, sociologists understand the relationship as *reciprocal*—that is, it goes in both directions, as structure affects agency and agency, in turn, can change the dimensions of a structure (Figure 1.1).

CRITICAL THINKING

Critical thinking: The ability to evaluate claims about truth by using reason and evidence.

Taking a sociological perspective requires more than an ability to use the sociological imagination. It also entails **critical thinking**, *the ability to evaluate claims about truth by using reason and evidence.* In everyday life, we often accept things as true because they are familiar, feel right, or are consistent with our beliefs. Critical thinking takes a different approach—recognizing weak arguments, rejecting statements not supported by empirical evidence, and questioning our assumptions. One of the founders of modern sociology, Max Weber, captured the spirit of critical thinking in two words when he said that a key task of sociological inquiry is to acknowledge "inconvenient facts."

Critical thinking requires us to be open-minded, but it does not mean that we must accept all arguments as equally valid. Those supported by logic and backed by evidence are clearly preferable to those that are not. For instance, we may passionately agree with Thomas Jefferson's famous statement, "That government is best that governs least." Nevertheless, as sociologists we must also ask, "What evidence backs up the claim that less government is better under all circumstances?" To think critically, it is useful to follow six simple rules (adapted from Wade & Tavris, 1997):

1. **Be willing to ask any question, no matter how difficult.** The belief in small government is a cherished U.S. ideal. But sociologists who study the role of government in modern society must be willing to ask whether there are circumstances under which more—not less—government is better. Government's role in areas such as homeland security, education, and health care has grown in recent decades—what are the positive and negative aspects of this growth?
2. **Think logically, and be clear.** Logic and clarity require us to define concepts in a way that allows us to study them. "Big government" is a vague concept that must be made more precise and measurable before it provides for useful research. Are we speaking of federal, state, or local government, or all of these? Is "big" measured by the cost of government services, the number of agencies or offices within the government, the number of people working for it, or something else? What did Jefferson mean by "best," and what would that "best" government look like? Who would have the power to define this notion?
3. **Back up your arguments with evidence.** Founding Father Thomas Jefferson is a formidable person to quote, but quoting him does not prove that smaller government is better in the 21st century. To find evidence, we need to seek out studies of contemporary societies to see whether there is a relationship between a population's well-being and the size of government in terms of agencies or public employees, or the breadth of services it provides. Because studies may offer contradictory evidence, we also need to be able to assess the strengths and weaknesses of arguments on different sides of the issue.
4. **Think about the assumptions and biases—including your own—that underlie all studies.** You may insist that government has a key role to play in modern society. On the other

hand, you may believe with equal passion that big government is one root of the problems in the United States. Critical thinking requires that we recognize our beliefs and biases. Otherwise, we might unconsciously seek out only evidence that supports our argument, ignoring evidence to the contrary. Passion has a role to play in research: It can motivate us to devote long hours to studying an issue. But passion should not play a role when we are weighing evidence and drawing conclusions.

5. **Avoid anecdotal evidence.** It is tempting to draw a general conclusion from a single experience or anecdote, but that experience may illustrate the exception rather than the rule. For example, you may know someone who just yesterday received a letter mailed 2 years ago, but that is not evidence that the U.S. Postal Service is inefficient or does not fulfill its mandates. To determine whether this government agency is working well, you would have to study its entire mail delivery system and its record of work over time.

6. **Be willing to admit when you are wrong or uncertain about your results, and when results themselves are mixed.** Sometimes, we expect to find support for an argument only to find that things are not so clear. For example, consider the position of a sociologist who advocates small government and learns that Japan and Singapore initially became economic powerhouses because their governments played leading roles in promoting growth or a sociologist who champions an expanded role for government but learns from the downturn of the 1990s in the Asian economies that some societal needs can be better met by private enterprise. Empirical evidence may contradict our beliefs: We learn from recognizing erroneous assumptions and having a mind open to new information.

Critical thinking also means becoming a critical consumer of the information that surrounds us—news, social media, surveys, texts, magazines, and scientific studies. To be a good sociologist, it is important to look beyond the commonsense understanding of social life and develop a critical perspective. Being a critical consumer of information entails paying attention to the sources of information we encounter, asking questions about how data were gathered, and keeping our minds open to information and evidence that challenge our beliefs. In this text, for example, we look at commonly cited data, including figures on poverty, unemployment, and crime, and inquire into how those phenomena are measured and what the resulting figures illuminate—and what they obscure—about those social issues.

THE DEVELOPMENT OF SOCIOLOGICAL THINKING

Humans have been asking questions about the nature of social life as long as people have lived in societies. Aristotle and Plato wrote extensively about social relationships more than 2,000 years ago. Ibn Khaldun, an Arab scholar writing in the 14th century, advanced several sociological concepts we recognize today, including ideas about social conflict and cohesion. Yet modern sociological concepts and research methods did not emerge until the 19th century, after the industrial revolution, and then largely in those European nations undergoing dramatic societal changes, such as industrialization and urbanization.

The Birth of Sociology: Science, Progress, Industrialization, and Urbanization

We can trace sociology's roots to four interrelated historical developments that gave birth to the modern world: *the scientific revolution, the Enlightenment, industrialization*, and *urbanization*. Since these developments initially occurred in Europe, it is not surprising that sociological perspectives and ideas evolved there during the 19th century. By the end of the 19th century, sociology had taken root in North America as well; some time later, it gained a foothold in Central and South America, Africa, and Asia. Sociology throughout the world initially bore the stamp of its European and North American origins, although recent decades have brought a greater diversity of perspectives to the discipline.

The Scientific Revolution

The rise of modern natural and physical sciences, beginning in Europe in the 16th century, offered scholars a more advanced understanding of the physical world. The success of natural science contributed to the belief that science could be fruitfully applied to human affairs, thereby enabling people to improve society or even perfect it. Auguste Comte (1798–1857) coined the term *sociology* to characterize what he believed would be a new "social physics"—that is, the scientific study of society.

The Enlightenment

Inspired in part by the success of the physical sciences, French philosophers in the 18th century, such as Voltaire (1694–1778), Montesquieu (1689–1755), Diderot (1713–1784), and Rousseau (1712–1778), promised that humankind could attain lofty heights by applying scientific understanding to human affairs. Enlightenment ideals such as equality, liberty, and fundamental human rights found a home in the emerging social sciences, particularly sociology. Émile Durkheim (1858–1917), considered by many to be the first modern sociologist, argued that sociological understanding would create a more egalitarian, peaceful society, in which individuals would be free to realize their full potential. Many of sociology's founders shared the hope that a fairer and more just society would be achieved through the scientific understanding of society.

The Industrial Revolution

The industrial revolution, which began in England in the mid- to late 18th century and soon spread to other countries, dramatically changed European societies. Traditional agricultural economies and the small-scale production of handicrafts in the home gave way to more efficient, profit-driven manufacturing based in factories. For instance, in 1801 in the English city of Leeds, there were about 20 factories manufacturing a variety of goods. By 1838, Leeds was home to 106 woolen mills alone, employing 10,000 people.

Industrialization brought new workers to cities, and urban populations grew dramatically in a short period of time. Sociologists such as Émile Durkheim theorized the normative effects of moving from small, traditional communities to diverse, unfamiliar, populous cities.

Small towns, including Leeds, were transformed into bustling cities, showcasing extremes of wealth and poverty, as well as opportunity and struggle. In the face of rapid social change and growing inequality, sociologists sought to gain a social scientific perspective on what was happening and how it had come about. German theorist and revolutionary Karl Marx (1818–1883), who had an important impact on later sociological theory concerning modern societies and economies, predicted that industrialization would make life increasingly intolerable for the masses. He believed that private property ownership by the wealthy allowed for the exploitation of working people and that its elimination would bring about a future of equality for all.

Urbanization: The Population Shift Toward Cities

Industrialization fostered the growth of cities as people streamed from rural fields to urban factories in search of work. By the end of the 19th century, more than 20 million people lived in English cities. The population of London alone exceeded 7 million by 1910.

Norms: Accepted social behaviors and beliefs.

Anomie: A state of normlessness that occurs when people lose touch with the shared rules and values that give order and meaning to their lives.

Early industrial cities were often fetid places, characterized by pollution and dirt, crime, and crowded housing tenements. In Europe, some sociologists lamented the passing of communal village life and its replacement by a savage and alienating urban existence. Durkheim, for example, worried about the breakdown of stabilizing beliefs and values in modern urban society. He argued that whereas traditional communities were held together by shared culture and **norms**, or *accepted social behaviors and beliefs*, modern industrial communities were threatened by **anomie**, or a *state of normlessness that* *occurs when people lose sight of the shared rules and values that give order and meaning to their lives.* In a state of anomie, individuals often feel confused and anxious because they do not know how to interact with each other and their environment. Durkheim raised the question of what would hold societies and communities together as they shifted from homogeneity and shared cultures and values to heterogeneous masses of diverse cultures, norms, and occupations.

Nineteenth-Century Founders

Despite its largely European origins, early sociology sought to develop universal understandings that would apply to other peoples, times, and places. The discipline's principal acknowledged founders—Auguste Comte, Harriet Martineau, Émile Durkheim, Karl Marx, and Max Weber—left their marks on sociology in different ways.

Social statics: The way society is held together.

Social dynamics: The laws that govern social change.

Positivism: An approach to research that is based on scientific evidence.

Auguste Comte

Auguste Comte (1798–1857), a French social theorist, is credited with founding modern sociology, naming it, and establishing it as the scientific study of social relationships. The twin pillars of Comte's sociology were the study of **social statics**, *the way society is held together*, and the analysis of **social dynamics**, *the laws that govern social change*. Comte believed that social science could be used effectively to manage the social change resulting from modern industrial society, but always with a strong respect for traditions and history.

Comte proclaimed that his new science of society was **positivism**. This meant that it was to be *based on facts alone*, which should be determined scientifically and allowed to speak for themselves. Comte argued that this purely factual approach was the proper method for sociology. He argued that all sciences—and all societies—go through three stages. The first stage is a theological one, in which key ways of understanding the world are framed in terms of superstition, imagination, and religion. The second stage is a metaphysical one, characterized by abstract speculation but framed by the basic belief that society is the product of natural rather than of supernatural forces. The third and last stage is one in which knowledge is based on scientific reasoning from the "facts." Comte saw himself as leading sociology toward its final, positivist stage.

As a founding figure in the social sciences, Auguste Comte is associated with positivism, the belief that the study of society must be anchored in facts and the scientific method.

Comte left a lasting mark on modern sociology. The scientific study of social life continues to be the goal of sociological research. His belief that social institutions have a strong impact on individual behavior—that is, that our actions are the products of personal choices and the surrounding social context—remains at the heart of sociology.

Harriet Martineau

Harriet Martineau (1802–1876) was an English sociologist who, despite deafness and other physical challenges, became a prominent social and historical writer. Her greatest handicap was being a woman in male-dominated intellectual circles that failed to value female voices. Today, she is frequently recognized as the first major woman sociologist.

Deeply influenced by Comte's work, Martineau translated his six-volume treatise on politics into English. Her editing helped make Comte's esoteric prose accessible to the English-speaking world, ensuring his standing as a leading figure in sociology. Martineau was also a distinguished scholar in her own right. She wrote dozens of books; more than 1,000 newspaper columns; and 25 novels, including a three-volume study, *Society in America* (Martineau, 1837), based on observations of the United States that she made during a tour of the country.

Martineau, like Comte, sought to identify basic laws that govern society. She derived three of her four laws from other theorists. The fourth law, however, was her own and reflected her progressive

Harriet Martineau translated into English the work of sociologist Auguste Comte, who dismissed women's intellect, saying, "Biological philosophy teaches us that . . . radical differences, physical and moral, distinguish the sexes . . . biological analysis presents the female sex . . . as constitutionally in a state of perpetual infancy, in comparison with the other" (Kandal, 1988, p. 75). The widespread exclusion of the voices of women and people of color from early sociology represents one of its key limitations.

Émile Durkheim pioneered some of sociology's early research on such topics as social solidarity and suicide. His work continues to inform sociological study and understanding of social bonds and the consequences of their unraveling.

(today we might say *feminist*) principles: For a society to evolve, it must ensure social justice for women and other oppressed groups. In her study of U.S. society, Martineau treated slavery and women's experience of dependence in marriage as indicators of the limits of the moral development of the United States. In her view, the United States was unable to achieve its full social potential while it was morally stunted by persistent injustices, such as slavery and women's inequality. The question of whether the provision of social justice is critical to societal development remains a relevant and compelling one today.

Émile Durkheim

Auguste Comte founded and named the discipline of sociology, but French scholar Émile Durkheim (1858–1917) set the field on its present course. Durkheim established the early subject matter of sociology, laid out rules for conducting research, and developed an important theory of social change.

Social facts: Qualities of groups that are external to individual members yet constrain their thinking and behavior.

For Durkheim, sociology's subject matter was **social facts**, *qualities of groups that are external to individual members yet constrain their thinking and behavior*. Durkheim argued that such social facts as religious beliefs and social duties are external—that is, they are part of the social context and are larger than our individual lives. They also have the power to shape our behavior. You may feel compelled to act in certain ways in different contexts—in the classroom, on a date, at a religious ceremony—even if you are not always aware of such social pressures.

Durkheim also argued that only social facts can explain other social facts. For example, there is no scientific evidence that men have an innate knack for business compared with women, but in 2020, women headed only 37 of the *Fortune* 500 companies. A Durkheimian approach would highlight women's experience in society—where historically they have been socialized into domestic roles, restricted to certain noncommercial professions, or discriminated against in hiring—and the fact that the social networks that foster mobility in the corporate world today are still primarily male to help explain why men dominate the upper ranks of the business world.

Durkheim's principal concern was explaining the impact of modern society on **social solidarity**, *the bonds that unite the members of a social group*. In his view, in traditional society, these bonds are based on similarity—people speak the same language, share the same customs and beliefs, and do similar work tasks. He called this *mechanical solidarity*. In modern industrial society, however, bonds based on similarity break down. Everyone has a different job to perform in the industrial division of labor, and modern societies are more likely to be socially diverse. Nevertheless, workers in different occupational positions are dependent on one another for things such as safety, education, and the provision of food and other goods essential to survival. The people filling these positions may not be alike in culture, beliefs, or language, but their dependence on one another contributes to social cohesion. Borrowing from biology, Durkheim called this *organic solidarity*, suggesting that modern society functions as an interdependent organic whole, like a human body.

Social solidarity: The bonds that unite the members of a social group.

Yet organic solidarity, Durkheim argued, is not as strong as mechanical solidarity. People no longer necessarily share the same norms and values. The consequence, according to Durkheim, is anomie. In this weakened condition, the social order may disintegrate and pathological behavior increase (Durkheim, 1922/1973a).

Collective conscience: The common beliefs and values that bind a society together.

Consider whether the United States, a modern and diverse society, is held together primarily by organic solidarity or whether the hallmark of mechanical solidarity, a **collective conscience**—*the common beliefs and values that bind a society together*—is in evidence. Do public demonstrations of patriotism on nationally significant anniversaries such as September 11 and July 4 indicate mechanical solidarity built on a collective sense of shared values, norms, and practices? Or do the deeply divisive politics of recent years suggest that social bonds are based more fully on practical interdependence?

Class conflict: Competition between social classes over the distribution of wealth, power, and other valued resources in society.

Proletariat: The working class; wage workers.

Karl Marx

The extensive writings of Karl Marx (1818–1883) influenced the development of economics and political science as well as sociology. They also shaped world politics and inspired communist revolutions in Russia (later the Soviet Union), China, and Cuba, among others.

Marx's central idea was deceptively simple: Almost all societies throughout history have been divided into economic classes, with one class prospering at the expense of others. All human history, Marx believed, should be understood as the product of **class conflict**, *competition between social classes over the distribution of wealth, power, and other valued resources in society* (Marx & Engels, 1998).

In the period of early industrialization in which he lived, Marx condemned capitalism's exploitation of *working people*, the **proletariat**, by the *ownership class*, the **bourgeoisie**. As we will see in later chapters, Marx's views on conflict and inequality are still influential in contemporary sociological thinking, even among sociologists who do not share his views on society.

Karl Marx was a scholar and critic of early capitalism. His theories of class struggle and alienation have influenced modern sociologists' thinking on the dynamics of social class.

Marx focused his attention on the emerging capitalist industrial society (Marx, 1867/1992a, 1885/1992b, 1894/1992c). Unlike his contemporaries in sociology, however, Marx saw capitalism as a transitional stage to a final period in human history in which economic classes and the unequal distribution of rewards and opportunities linked to class inequality would disappear and be replaced by a utopia of equality.

Many of Marx's predictions have come to pass. Indeed, his critical analysis of the dynamics of capitalism proved insightful. Among other things, Marx argued that capitalism would lead to accelerating technological change, the replacement of workers by machines, and the growth of monopoly capitalism.

Bourgeoisie: The capitalist (or property-owning) class.

Marx presciently predicted that ownership of the **means of production**, *the sites and technology that produce the goods* (and sometimes services) *we need and use*, would come to be concentrated in fewer and fewer hands. As a result, he believed, a growing wave of people would be thrust down into the proletariat, which owns only its own labor power. In modern society, large corporations have progressively swallowed up or pushed out smaller businesses; where small lumberyards and pharmacies used to serve

Means of production: The sites and technology that produce the goods we need and use.

many communities, corporate giants, such as Home Depot, CVS, and Best Buy, have moved in, putting locally owned establishments out of business.

In many U.S. towns, small-business owners have joined forces to protest the construction of "big box" stores such as Walmart (now the largest private employer in the United States), arguing that these enormous establishments, although they offer cheap goods, wreak havoc on local retailers and bring only the meager economic benefit of masses of entry-level, low-wage jobs. From a Marxist perspective, we might say that the local retailers, in resisting the incursion of the big-box stores into their communities, are fighting their own proletarianization. Even physicians, many of whom used to own their own means of production in the form of private medical practices, have increasingly been driven by economic necessity into working for large health maintenance organizations (HMOs), where they are salaried employees.

Verstehen: The German word for interpretive understanding; Weber's proposed methodology for explaining social relationships by having the sociologist imagine how subjects might perceive a situation.

Unlike Comte and Durkheim, Marx thought social change would be revolutionary, not evolutionary, and would be the product of oppressed workers rising up against a capitalist system that exploits the many to benefit the few.

Max Weber

Formal rationality: A context in which people's pursuit of goals is shaped by rules, regulations, and larger social structures.

Bureaucracies: Formal organizations characterized by written rules, hierarchical authority, and paid staff, intended to promote organizational efficiency.

Max Weber (1864–1920), a German sociologist who wrote at the beginning of the 20th century, left a substantial academic legacy. Among his contributions are an analysis of how Protestantism fostered the rise of capitalism in Europe (Weber, 1904–1905/2002) and insights into the emergence of modern bureaucracy (Weber, 1946/1919). Weber, like other founders of sociology, took up various political causes, condemning injustice wherever he found it. Although pessimistic about capitalism, he did not believe, as did Marx, that some alternative utopian form of society would arise. Nor did he see sociologists enjoying privileged insights into the social world that would qualify them to wisely counsel rulers and industrialists, as Comte (and, to some extent, Durkheim) had envisioned.

Weber believed that an adequate explanation of the social world begins with the individual and incorporates the meaning of what people say and do. Although he argued that research should be scientific and value free, Weber also believed that to explain what people do, we must use a method he termed **Verstehen**, *the German word for interpretive understanding. This methodology*, rarely used by sociologists today, *sought to explain social relationships by having the sociologist/observer imagine how the subjects being studied might have perceived and interpreted the situation*. Studying social life, Weber felt, is not the same as studying plants or chemical reactions because human beings act on the basis of meanings and motives.

Max Weber made significant contributions to the understanding of how capitalism developed in Western countries and its relationship to religious beliefs. His work on formal rationality and bureaucracy continues to influence sociologists' study of modern society.

Weber's theories of social and economic organization have also been highly influential (Weber, 1921/2012). Weber argued that the modern Western world showed an ever-increasing reliance on logic, efficiency, rules, and reason. According to him, modern societies are characterized by the development and growing influence of **formal rationality**, a context in which people's pursuit of goals is increasingly shaped by rules, regulations, and larger social structures. One of Weber's most widely known illustrations of formal rationality comes from his study of **bureaucracies**, *formal organizations characterized by written rules, hierarchical authority, and paid staff, intended to promote organizational efficiency*. Bureaucracies, for Weber, epitomized formally rational systems: On the one hand, they offer clear, knowable rules and regulations for the efficient pursuit of particular ends, such as obtaining a passport or getting financial aid for higher education. On the other hand, he feared, the bureaucratization of modern society would also progressively strip people of their humanity and creativity and result in an iron cage of rationalized structures with irrational consequences.

Weber's ideas about bureaucracy were remarkably prescient in their characterization of our bureaucratic (and formally rationalized) modern world. Today, we are confronted regularly with both the incredible efficiency and baffling irrationality of modern bureaucratic structures. Within moments of entering into an efficiently concluded contract with a wireless phone service provider, we can become consumers of a cornucopia of technological opportunities, with the ability to chat on the phone with or receive text messages from a friend located almost anywhere in the world, post photographs or watch videos online, and pass the time on social-media platforms. Should we later be confused by a bill and need to speak to a company representative, however, we may be shuttled through endless repetitions of an automated response system that never seems to offer us the option of speaking with another human being. Today, Weber's theorized irrationality of rationality is alive and well.

Significant Founding Ideas in U.S. Sociology

Sociology was born in Europe, but it took firm root in U.S. soil, where it was influenced by turn-of-the-century industrialization and urbanization, as well as racial strife and discrimination. Strikes by organized labor, corruption in government, an explosion of European immigration, racial segregation, and the growth of city slums all helped mold early sociological thought in the United States. By the late 1800s, numerous universities in the United States were offering sociology courses. The first faculties of sociology were established at the University of Kansas (1889), the University of Chicago (1892), and Atlanta University (1897). Next, we look at a handful of sociologists who have had an important influence on modern sociological thinking in the United States. Throughout the book, we will learn about more U.S. sociologists who have shaped our perspectives today.

Robert Ezra Park

The Sociology Department at the University of Chicago, which gave us what is often known as the "Chicago School" of sociology, dominated the new discipline in the United States at the start of the 20th century. Chicago sociologist Robert Ezra Park (1864–1944) pioneered the study of urban sociology and race relations. Once a muckraking journalist, Park was an equally colorful academic, reportedly coming to class in disheveled clothes and with shaving soap still in his ears. But his students were devoted to him, and his work was widely recognized. His 1921 textbook, *An Introduction to the Science of Sociology*, coauthored with his Chicago colleague Ernest Burgess, helped shape the discipline. The Chicago School studied a broad spectrum of social phenomena, from hoboes and flophouses (inexpensive dormitory-style housing) to movie houses, dance halls, and slums, and from youth gangs and mobs to residents of Chicago's ritzy Gold Coast.

Park was a champion of racial integration, having once served as personal secretary to the African American educator Booker T. Washington. Yet racial discrimination was evident in the treatment of Black sociologists, including W. E. B. Du Bois, a contemporary of many of the sociologists working in the Chicago School.

W. E. B. Du Bois, the first African American to receive a PhD from Harvard, wrote 20 books and more than 100 scholarly articles on race and race relations. Today, many of his works are classics in the study of African American lives and race relations in the United States at the turn of the 20th century.

W. E. B. Du Bois

A prominent Black sociologist and civil rights leader at the African American Atlanta University, W. E. B. Du Bois (1868–1963) developed ideas that were considered too radical to find broad acceptance in the sociological community. At a time when the U.S. Supreme Court had ruled that segregated "separate but equal" facilities for Blacks and whites were constitutional and when lynching of Black Americans had reached an all-time high, Du Bois condemned the deep-seated racism of white society. Today, his writings on race relations and the lives of Black Americans are classics in the field.

Du Bois sought to show that racism was widespread in U.S. society. He was also critical of Black Americans who had "made it" and then turned their backs on those who had not. One of his most enduring ideas is that in U.S. society, African Americans are never able to escape a fundamental awareness of race. They experience a **double consciousness**—*an awareness of themselves as both Americans*

Double consciousness: Among African Americans, an awareness of themselves as both American and Black, and never free of racial stigma.

and Black, and never free of racial stigma. He wrote, "The Negro is sort of a seventh son . . . gifted with second-sight . . . this sense of always looking at one's self through the eyes of others" (Du Bois, 2008, p. 12). Today, as in Du Bois's time, physical traits such as skin color continue to shape people's perceptions and interactions in significant and complex ways.

Charlotte Perkins Gilman

Charlotte Perkins Gilman (1860–1935) was a well-known novelist, feminist, and sociologist of her time. Because of her family's early personal and economic struggles, she had only a few years of formal schooling in childhood, although she would later enroll at the Rhode Island School of Design. She read widely, however, and she was influenced by her paternal aunts, who included suffragist Isabella Beecher Stowe and writer Harriet Beecher Stowe, author of *Uncle Tom's Cabin*, an antislavery novel written in 1852.

Gilman's (1892) most prominent publication was her semiautobiographical short story, "The Yellow Wallpaper," which follows the decline of a married woman shut away in a room (with repellent yellow wallpaper) by her husband, ostensibly for the sake of her health. Gilman used the story to highlight the consequences of women's lack of autonomy in marriage. She continued to build this early feminist thesis in the book *Women and Economics: A Study of the Economic Relation Between Men and Women as a Factor in Social Evolution* (Gilman, 1898/2006)

> The labor of women in the house, certainly, enables men to produce more wealth than they otherwise could; and in this way women are economic factors in society. But so are horses. The labor of horses enables men to produce more wealth than they otherwise could. The horse is an economic factor in society. But the horse is not economically independent, nor is the woman. (p. 7)

Gilman was critical of women's enforced dependence in marriage and society. Her work represents an early and notable effort to look at sex roles in the family not as static, natural, and inevitable, as many saw them at the time, but as dynamic social constructions with the potential to change and to bring greater autonomy to women in the home and society.

Robert K. Merton

After World War II, sociology began to apply sophisticated quantitative models to the study of social processes. There was also a growing interest in the grand theories of the European founders. At Columbia University, Robert K. Merton (1910–2003) undertook wide-ranging studies that helped further establish sociology as a scientific discipline. Merton is best known for his theory of deviance (Merton, 1938), his work on the sociology of science (Merton, 1996), and his iteration of the distinction between manifest and latent functions as a means for more fully understanding the relationships between and roles of sociological phenomena and institutions in communities and society (Merton, 1968). He emphasized the development of theories in what he called the *middle range*—midway between the grand theories of Weber, Marx, and Durkheim and quantitative studies of specific social problems.

C. Wright Mills

Columbia University sociologist C. Wright Mills (1916–1962) is best known in the discipline for describing the *sociological imagination*, the imperative in sociology to seek the nexus between private troubles and public issues. In his short career, Mills was prolific. He renewed interest in Max Weber by translating many of his works into English and applying his ideas to the contemporary United States. But Mills, who also drew on Marx, identified himself as a "plain Marxist." His concept of the sociological imagination can be traced, in part, to Marx's famous statement that "man makes history, but not under circumstances of his own choosing," meaning that even though we are agents of free will, the social context has a profound impact on the obstacles or opportunities in our lives.

Mills synthesized Weberian and Marxian traditions, applying sociological thinking to the most pressing problems of the day, particularly inequality. He advocated an activist sociology with a sense of social responsibility. Like many sociologists, he was willing to turn a critical eye on "common knowledge," including the belief that the United States is a democracy that represents the interests of all people. In a provocative study, he examined the workings of the "power elite," a small group of wealthy businessmen, military leaders, and politicians who Mills believed ran the country largely in their own interests (Mills, 1956/2000a).

Women in Early Sociology

Why did so few women social scientists find a place among sociology's founders? After all, the American (1775–1783) and French (1789–1799) revolutions elevated such lofty ideals as freedom, liberty, and equality. Yet long after these historical events, women and minorities were still excluded from public life in Europe and North America. Democracy—which gives people the right to participate in their governance—was firmly established as a principle for nearly a century and a half in the United States before women achieved the right to vote in 1920. In France, it took even longer—until 1945.

The sociological imagination involves viewing seemingly personal issues through a sociological lens and recognizing the relationship between biography and history. C. Wright Mills is best known for coining this catchy and popular term.

©Fritz Goro/The LIFE Picture Collection/Getty Images

Sociology as a discipline emerged during the first modern flourishing of feminism in the 19th century. Yet women and people of non-European heritage were systematically excluded from influential positions in the European universities where sociology and other modern social sciences originated. In 1861, feminist scholar Julie Daubié won a prize from the Lyon Academy for her essay, "Poor Women in the Nineteenth Century." For many years, she had been denied entry to higher education, but eventually she because the first woman in the country to earn a Baccalaureate degree (Kandal, 1988). Between 1840 and 1960, almost no women held senior academic positions in the sociology departments of any European or U.S. universities, with the exception of exclusively women's colleges.

Several female scholars managed to overcome these obstacles to make significant contributions to sociological inquiry. For example, in 1792, the British scholar Mary Wollstonecraft published *A Vindication of the Rights of Women*, arguing that scientific progress could not occur unless women were allowed to become men's equals by means of universal education. In France in 1843, Flora Tristan called for equal rights for women workers, "the last remaining slaves in France." Also in France, Aline Valette published *Socialism and Sexualism* in 1893, nearly three quarters of a century before the term *sexism* found its way into spoken English (Kandal, 1988).

An important figure in early U.S. sociology is Jane Addams (1860–1935). Addams is best known as the founder of Hull House, a settlement house for the poor, sick, and aged that became a center for political activists and social reformers. Less well known is the fact that under Addams's guidance, the residents of Hull House engaged in important research on social problems in Chicago. *Hull-House Maps and Papers*, published in 1895, pioneered the study of Chicago neighborhoods, helping to shape the research direction of the Chicago School of sociology. Following Addams's lead, Chicago sociologists mapped the city's neighborhoods, studied their residents, and helped create the field of community studies. Despite her prolific work—she authored 11 books and hundreds of articles and received the Nobel Peace Prize for her dedication to social reform in 1931—she never secured a full-time position at the University of Chicago, and the school refused to award her an honorary degree.

Underappreciated during her time, Jane Addams was a prominent scholar and early contributor to sociology. She is also known for her political activism and commitment to social reform.

©Hulton Archive/Getty Images

As Harriet Martineau, Jane Addams, Julie Daubié, and others experienced, early female sociologists were not accorded the same status as their male counterparts. Only recently have many of their writings been rediscovered and their contributions acknowledged in sociology.

WHAT IS SOCIOLOGICAL THEORY?

Often, multiple sociologists look at the same events, phenomena, or institutions and draw different conclusions. How can this be? One reason is that they may approach their analyses from different theoretical perspectives. In this section, we explore the key theoretical paradigms in sociology and look at how they are used as tools for the analysis of society.

Sociological theories: Logical, rigorous frameworks for the interpretation of social life that make particular assumptions and ask particular questions about the social world.

Sociological theories are *logical, rigorous frameworks for the interpretation of social life that make particular assumptions and ask particular questions about the social world.* The word *theory* is rooted in the Greek word *theoria*, which means "a viewing." An apt metaphor for a theory is a pair of glasses. You can view a social phenomenon such as socioeconomic inequality; poverty, deviance, or consumer culture; or an institution such as capitalism or the family by using different theories as lenses.

The word *theory* is rooted in the Greek word theoria, which means "a viewing." Sociological theories provide us different lenses for viewing social phenomena and institutions. They are analytical tools that illuminate some aspects of those phenomena and institutions while obscuring others.

©AerialPerspective Images/Getty Images

As you will see in the next section, in the discipline of sociology, several major categories of theories seek to examine and explain social phenomena and institutions. Imagine the various sociological theories as different pairs of glasses, each with colored lenses that change the way you see an image: You may look at the same institution or phenomenon as you put on each pair, but it will appear differently, depending on the glasses you are wearing. Keep in mind that sociological theories are not "truths" about the social world. They are logical, rigorous analytical tools that we can use to inquire about, interpret, and make educated predictions about the world around us. From the vantage point of any sociological theory, some aspects of a phenomenon or an institution are illuminated while others are obscured. In the end, theories are more or less useful depending on how well *empirical data*—that is, knowledge gathered by researchers through scientific methods—support their analytical conclusions. Next, we outline the basic theoretical perspectives that we will be using in this text.

The three dominant theoretical perspectives in sociology are *structural functionalism, social conflict theory*, and *symbolic interactionism*. We will outline their basic characteristics and revisit them again throughout the book. Symbolic interactionism shares with the functionalist and social conflict paradigms an interest in interpreting and understanding social life. Nevertheless, the first two are **macrolevel paradigms**, *concerned with large-scale patterns and institutions*. Symbolic interactionism is a **microlevel paradigm**—it is *concerned with small-group social relations and interactions*.

Macrolevel paradigms: Theories of the social world that are concerned with large-scale patterns and institutions.

Microlevel paradigm: A theory of the social world that is concerned with small-group social relations and interactions.

Structural functionalism, social conflict theory, and symbolic interactionism form the basic foundation of contemporary sociological theorizing (Table 1.1). Throughout this book, we will introduce variations on these theories, as well as new and evolving theoretical ideas in sociology.

The Functionalist Paradigm

Structural functionalism: A theory that seeks to explain social organization and change in terms of the roles performed by different social structures, phenomena, and institutions; also known as functionalism.

Structural functionalism (or *functionalism*—the term we use in this book) *seeks to explain social organization and change in terms of the roles performed by different social structures, phenomena, and institutions.* Functionalism characterizes society as made up of many interdependent parts—an analogy often cited is the human body. Each part serves a different function, but all parts work together to ensure the equilibrium and health of the entity as a whole. Society, too, is composed of a spectrum of different parts with a variety of different functions, such as the government, the family, religious and educational institutions, and the media. According to the theory, together, these parts contribute to the smooth functioning and equilibrium of society.

The key question posed by the functionalist perspective is, What function does a particular institution, phenomenon, or social group serve for the maintenance of society? That is, what contribution

TABLE 1.1 ■ The Three Principal Sociological Paradigms

	THEORETICAL PERSPECTIVE AND FOUNDING THEORIST(S)		
	STRUCTURAL FUNCTIONALISM (Émile Durkheim)	**SOCIAL CONFLICT (Karl Marx)**	**SYMBOLIC INTERACTIONISM (Max Weber, George Herbert Mead)**
Assumptions about self and society	Society is a system of interdependent, interrelated parts, like an organism, with groups and institutions contributing to the stability and equilibrium of the whole social system.	Society consists of conflicting interests, but only some groups have the power and resources to realize their interests. Some groups benefit from the social order at the expense of other groups.	The self is a social creation; social interaction occurs by means of symbols such as words, gestures, and adornments; shared meanings are important to successful social interactions.
Key focus and questions	Macrosociology: What keeps society operating smoothly? What functions do different societal institutions and phenomena serve for society as a whole?	Macrosociology: What are the sources of conflict in society? Who benefits and who loses from the existing social order? How can inequalities be overcome?	Microsociology: How do individuals experience themselves, one another, and society as a whole? How do they interpret the meanings of particular social interactions?

does a given institution, phenomenon, or social group make to the equilibrium, stability, and functioning of the whole? Note the underlying assumption of functionalism: Any existing institution or phenomenon does serve a function; if it served no function, it would evolve out of existence. Consequently, the central task of the functionalist sociologist is to discover what function an institution or a phenomenon—for instance, the traditional family, capitalism, social stratification, or deviance—serves in the maintenance of the social order.

Émile Durkheim is credited with developing the early foundations of functionalism. Among other ideas, Durkheim observed that all known societies have some degree of deviant behavior, such as crime. The notion that deviance is functional for societies may seem counterintuitive: Ordinarily, we do not think of deviance as beneficial or necessary to society. Durkheim, however, reasoned that since deviance is universal, it must serve a social function—if it did not serve a function, it would cease to exist. Durkheim concluded that one function of deviance—specifically, of society's labeling of some acts as deviant—is to remind members of society what is considered normal or moral; when a society punishes deviant behavior, it reaffirms people's beliefs in what is right and good.

From the structural functionalist perspective, if a phenomenon or institution exists and persists, it must serve a function. Talcott Parsons posited that traditional sex roles in the American family were positively functional because they ensured that the positions of husbands and wives were complementary rather than competitive. This analysis failed to recognize the inequality inherent in women's economic dependence on men in this relationship.

Talcott Parsons (1902–1979) expanded functionalist analysis by looking at individual social institutions, such as government, the economy, and the family, and how they contribute to the functioning of society as a whole (Parsons, 1967, 2007). For example, he wrote that traditional sex roles for men and women contribute to stability on both the microsociological level (that of the family, in this instance) and the macrosociological level (society). Parsons argued that traditional socialization produces "instrumental–rational and work-oriented–males and "expressive," or sensitive, nurturing, and emotional females. Instrumental males, he reasoned, are well suited for the competitive world of work, whereas their expressive female counterparts are appropriately prepared to care for the family. According to Parsons, these roles, which he saw as a product of socialization rather than nature, are complementary and positively functional, leading men and women to inhabit different spheres of the social world. Complementary rather than competing roles contribute to solidarity in a marriage by reducing competition between husband and wife. Critics have rejected this idea as a justification of inequality.

As this example suggests, functionalism is conservative in that it tends to accept rather than question the status quo; it holds that any institution or phenomenon exists because it is functional for society, rather than asking whether it might benefit one group to the detriment of other groups. One of functionalism's key weaknesses is a failure to recognize inequalities in the distribution of power and resources and how those affect social relationships.

Merton attempted to refine the functionalist paradigm by demonstrating that not all social structures work to maintain or strengthen the social organism, as Durkheim and other early functionalists seemed to suggest. According to Merton, a social institution or phenomenon can have both positive functions and problematic dysfunctions. Merton broadened the functionalist idea by suggesting that **manifest functions** are the *obvious and intended functions of a phenomenon or institution.* **Latent functions**, by contrast, are *functions that are not recognized or expected.* He used the famous example of the Hopi rain dance, positing that although the manifest function of the dance was to bring rain, a no less important latent function was to reaffirm social bonds in the community through a shared ritual. Consider another example: A manifest function of war is usually to vanquish an enemy, perhaps to defend a territory or to claim it. Latent functions of war—those that are not the overt purpose but may still have powerful effects—may include increased patriotism in countries engaged in the war, a rise in the profits of companies manufacturing military equipment or contracting workers to the military, and changes in national budgetary priorities.

Manifest functions: The obvious and intended functions of a phenomenon or institution.

Latent functions: Functions of a phenomenon or institution that are not recognized or expected.

The Social Conflict Paradigm

Social conflict paradigm: A theory that seeks to explain social organization and change in terms of the conflict that is built into social relations; also known as conflict theory.

In contrast to functionalism, the **social conflict paradigm** (which we refer to in this book as *conflict theory*) *seeks to explain social organization and change in terms of the conflict that is built into social relationships.* Conflict theory is rooted in the ideas about class and power put forth by Marx. Although Durkheim's structural functionalist lens asked how different parts of society contribute to stability, Marx asked about the roots of conflict. Conflict theorists pose the questions, Who benefits from the way social institutions and relationships are structured, and who loses? The social conflict paradigm focuses on what divides people rather than on what unites them. It presumes that group interests drive relationships and that various groups in society (for instance, social classes, ethnic and racial groups, and women and men) will act in their own interests. Conflict theory thus assumes not that interests are shared but that they may be different and irreconcilable and, importantly, that only some groups have the power and resources to realize their interests. As a result, conflict is—sooner or later—inevitable.

The manifest function of a vehicle is to transport a person efficiently from Point A to Point B. One latent function is to say something about the status of the driver.

©REUTERS/Alessandro Bianchi

From Marx's perspective, the bourgeoisie benefits directly from the capitalist social order. If, as Marx suggests, the capitalist class has an interest in maximizing productivity and profit and minimizing costs (including the cost of labor in the form of workers' wages) and the working class has an interest in earning more and working less, then the interests of the two classes are difficult to reconcile. The more powerful group in society generally has the upper hand in furthering its interests.

After Marx, the body of conflict theory expanded tremendously. In the 20th century and today, theorists have extended the reach of the perspective to consider how control of culture and the rise of technology (rather than just control of the means of production) underpin class domination (Adorno, 1975; Horkheimer, 1947), as well as how the expanded middle class can be accommodated in a Marxist perspective (Wright, 1998). Many key ideas in feminist theory take a conflict-oriented perspective, although the focus shifts from social class to gender power and conflict (Connell & Messerschmidt, 2005), as well as ways in which race is implicated in relations of power (Collins, 1990).

The social conflict perspective posits that groups in society often have differing, irreconcilable interests. However, only some groups have the power and resources to realize their interests. For example, low-wage workers have a clear interest in earning more, but this may conflict with the interest of those who employ them in minimizing pay of workers in favor of higher profits. Which of these two groups has greater power to realize its interest?

Recall Durkheim's functionalist analysis of crime and deviance. According to this perspective, society defines crime to reaffirm people's beliefs about what is right and to dissuade them from deviating. A conflict theorist might argue that dominant groups in society define the behaviors labeled *criminal* or *deviant* because they have the power to do so. For example, street crimes such as robbery and carjacking are defined and punished as criminal behavior. They are also represented in reality television programs, movies, and other cultural products as images of criminal deviance. On the other hand, corporate or white-collar crime, which may cause the loss of money or even lives, is less likely to be clearly defined, represented, and punished as criminal. From a conflict perspective, white-collar crime is more likely to be committed by members of the upper class (for instance, business or political leaders or financiers) and is less likely to be punished harshly compared with street crime, which is associated with the lower-income classes, although white-collar crime may have even greater economic and health consequences. A social conflict theorist would draw our attention to the fact that the decision makers who pass our laws, particularly at the federal level, are mostly members of the upper class and govern in the interests of capitalism and their own socioeconomic peers.

A key weakness of the social conflict paradigm is that it overlooks the forces of stability, equilibrium, and consensus in society. The assumption that groups have conflicting (even irreconcilable) interests and that those interests are realized by those with power at the expense of those with less power fails to account for forces of cohesion and stability in societies.

Symbolic Interactionism

Symbolic interactionism argues that *both the individual self and society as a whole are the products of social interactions based on language and other symbols.* The term *symbolic interactionism* was coined by U.S. sociologist Herbert Blumer (1900–1987) in 1937, but the approach originated in the lectures of George Herbert Mead (1863–1931), a University of Chicago philosopher allied with the Chicago School of sociology. The symbolic interactionist paradigm argues that people acquire their sense of who they are only through interaction with others. They do this by means of **symbols**, *representations of things that are not immediately present to our senses.* Symbols include such things as words, gestures,

Symbolic interactionism: A microsociological perspective that posits that both the individual self and society as a whole are the products of social interactions based on language and other symbols.

Symbols: Representations of things that are not immediately present to our senses.

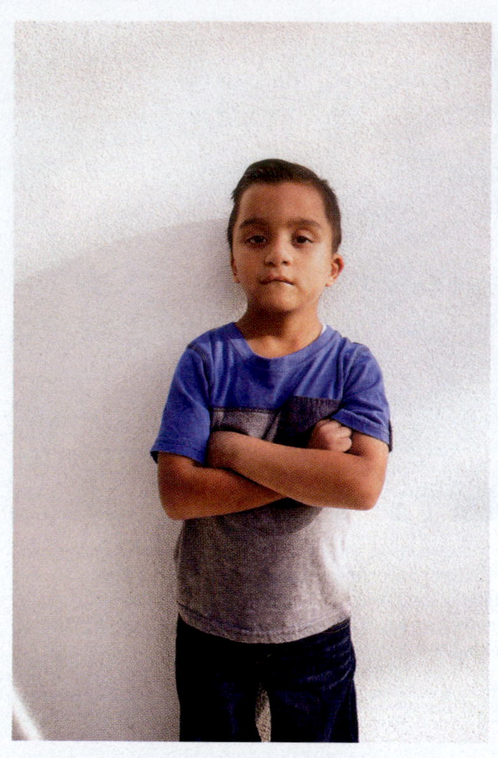

Symbolic interactionists take an interest in how people's self-perceptions are shaped by perceptions and actions of others. Consider how a child's self-perception might differ depending on whether he is labeled a "bad boy" or a "good boy who made a mistake and did a bad thing." Symbolic interactionists argue that labels matter. What do you think?

©The Good Brigade/Getty Images

emoticons, and tattoos, among others. Symbolic interactionists take an interest in how we develop social selves. Among other things, symbolic interactionists study the process of socialization and how we learn the culture of our own societies and internalize the roles, norms, and practices of those societies.

Recall our earlier discussions of the theoretical interpretations of deviance and crime. A symbolic interactionist might focus on the ways in which people label some acts or groups as *deviant* (a symbolic act that uses language), the sociological factors that make such a label stick, and the meanings underlying such a label. If you are accused of committing a crime you did not commit, how will the label of *criminal* affect the way others see you? How will it affect the way you see yourself, and will you begin to act differently as a result? Can being labeled *deviant* be a self-fulfilling prophecy? For the symbolic interactionist, sociological inquiry is the study of how people interact and how they create and interpret symbols in the social world.

Although symbolic interactionist perspectives draw our attention to important microlevel processes in society, they may miss the larger structural context of those processes, such as examining who has the power to make laws defining what or who is criminally deviant. For this reason, many sociologists use both macro- and microlevel perspectives when analyzing social phenomena such as deviance.

The three paradigms we described lead to diverse images of society, research questions, and conclusions about the patterns and nature of social life. Each "pair of glasses" can provide a different perspective on the social world. Throughout this text, the three major theoretical paradigms—and some new ones we will encounter in later chapters—will help us understand key issues and themes of sociology.

PRINCIPAL THEMES IN THIS BOOK

We began this chapter with a list of *why* questions with which sociologists are concerned and about which any one of us might be curious. Behind these questions, we find several major themes, which are also some of the main themes in this book. Three important focal points for sociology—and for us—are (1) power and inequality and the ways in which the unequal distribution of social, economic, and political resources shape opportunities, obstacles, and relationships; (2) the societal changes occurring as a result of globalization and the growing social diversity of modern communities and societies; and (3) the powerful impact of technological change on modern lives, institutions, and states.

Power and Inequality

Power: The ability to mobilize resources and achieve goals despite the resistance of others.

As we consider broad social topics such as gender, race, social class, and sexual orientation and their effects on social relationships and resources, we will be asking who has **power**—*the ability to mobilize resources and achieve goals despite the resistance of others*—and who does not. We will also ask about variables that influence the uneven distribution of power and how some groups use power to create advantages for themselves (and disadvantages for others) and how disadvantaged groups mobilize to challenge the powerful.

Inequality: Differences in wealth, power, political voice, educational opportunities, and other valued resources.

Power is often distributed unequally and can be used by those who possess it to marginalize other social groups. **Inequality** refers to *differences in wealth, power, political voice, educational opportunities, and other valued resources*. The existence of inequality not only raises moral and ethical questions about fairness, but it also can tear at the very fabric of societies, fostering social alienation and instability. Furthermore, it may have negative effects on local and national economies. Notably, economic inequality is increasing both within and between many countries around the globe, a fact that makes understanding the roots and consequences of this phenomenon—that is, asking the *why* questions—ever more important.

In recent decades, sociologists have increasingly focused on recognizing ways in which social statuses such as gender, sexuality, race, ethnicity, class, religion, and others overlap and intersect with one another in our daily lives and, importantly, how these overlapping characteristics affect individual and group access to resources such as power, political voice, health, well-being, education, and justice. Put another way, "When it comes to social inequality, people's lives and the organization of power in a given society are better understood as being shaped not by a single axis of social division, be it race or gender or class, but by many axes that work together and influence each other" (Collins & Bilge, 2016, p. 2). **Intersectionality**, a term coined by legal scholar Kimberlé Crenshaw, captures the idea that people's individual and group experiences are shaped by overlapping oppression and privilege built into the structure of our societies. It also enables us to think critically and in a more complex way about social realities. To enlist an intersectional lens, we must carefully consider differences and similarities between groups of people and how these intersect.

Intersectionality: The confluence of social statuses that shape people's lives, access to resources and power, justice, health, and well-being.

Globalization and Diversity

Globalization is *the process by which people all over the planet have become increasingly interconnected economically, politically, socially, culturally, and environmentally.* Globalization is not new. It began nearly 200,000 years ago, when humans first spread from their African cradle into Europe and Asia. For thousands of years, humans have traveled, traded goods, and exchanged ideas over much of the globe, using seaways or land routes such as the famed Silk Road, a stretch of land that links China and Europe. But the rate of globalization took a giant leap forward with the industrial revolution, which accelerated the growth of global trade. It made another dramatic jump with the advent of the Information Age, drawing together individuals, cultures, and countries into a common global web of information exchange. In this book, we consider a spectrum of manifestations, functions, and consequences of globalization in areas such as the economy, culture, and the environment.

Globalization: The process by which people worldwide become increasingly connected economically, politically, socially, culturally, and environmentally.

Growing contacts between people and cultures have made us increasingly aware of social diversity as a feature of modern societies. **Social diversity** is *the social and cultural mixture of different groups in society and the societal recognition of difference as significant.* The spread of culture through the globalization of media and the rise of migration has created a world in which almost no place is isolated. As a result, many nations today, including the United States, are characterized by a high degree of social diversity.

Social diversity: The social and cultural mixture of different groups in society and the societal recognition of difference as significant.

Social diversity brings a unique set of sociological challenges. People everywhere have a tendency toward **ethnocentrism**, a *worldview whereby one judges other cultures by the standards of one's own culture and regards one's own way of life as normal—and often superior to others.* From a sociological perspective, no group can be said to be more human than any other. Yet history abounds with examples of people lashing out at others whose religion, language, customs, race, or sexual orientation differed from their own.

Ethnocentrism: A worldview whereby one judges other cultures by the standards of one's own culture and regards one's own way of life as normal and often superior to others.

Technology and Society

Technology is the practical application of knowledge to transform natural resources for human use. The first human technology was probably the use of rocks and other blunt instruments as weapons, enabling humans to hunt large animals for food. Agriculture—planting crops such as rice or corn in hopes of reaping a yearly harvest—represents another technological advance, one superior to simple foraging in the wild for nuts and berries. The use of modern machinery, which ushered in the industrial revolution, represents still another technological leap, multiplying the productivity of human efforts.

Today, we are in the midst of another revolutionary period of technological change: the information revolution. Thanks to the microchip, the Internet, and mobile technology, an increasing number of people around the world now have instant access to a mass of information that was unimaginable just 20 years ago. The information revolution is creating postindustrial economies based far more heavily on the production of knowledge than on the production of goods, as well as new ways of communicating that have the potential to draw people around the world together—or tear them apart. No less importantly, revolutions in robotics and artificial intelligence promise to alter the world of work in ways that we are only beginning to recognize.

Together, these three themes provide the foundation for this text. Our goal is to develop a rigorous sociological examination of power and inequality, globalization and diversity, and technology and society to help you better understand the social world from its roots to its contemporary manifestations to its possible futures.

SOCIOLOGISTS ASK, WHAT'S NEXT?

In this book, we will be wrapping up each chapter with a brief reflection on *what's next*. That is, we want to invite you to consider how the events of our time may shape our individual and shared future. In this book we illustrate that a sociological perspective can help us see and understand the world, our lives, and the lives of those around us with new eyes. It highlights the ways that we both influence and are powerfully influenced by the social world around us: Society shapes us, and we, in turn, shape society. There are significant changes and challenges ahead in areas including technology, politics, the economy and labor market, public health, and social justice, among others. Sociology can give us important tools for understanding what may lie ahead—and to consider how social change is made.

Why are the issues and questions posed by sociology incredibly compelling for all of us to understand? One reason is that, as we will see throughout this book, many of the social issues sociologists study—crime, marriage, gender and sexuality, race and ethnicity, unemployment, poverty, consumption, discrimination, and many others—are related to one another in ways we may not immediately see. A sociological perspective helps us to make connections between diverse social phenomena. When we understand these connections, we are better able to understand social issues, to address social problems, and to make (or vote for) policy choices that benefit society.

In the coming chapters we highlight some significant social changes and phenomena that have had a broad impact. For instance, the COVID-19 pandemic, which emerged in mainland China in December 2019 and quickly spread across the globe, was not just an epidemiological and medical phenomenon; it was also social and political. It impacted our norms of interaction, social practices, and behaviors, and caused change in social institutions, including schools, families, and the economy. Recent years have also seen mass public mobilizations around the pursuit of criminal justice reform and social justice, not only in the United States, where the disproportionate killing of Black Americans by law enforcement officers has been the object of social protest, but around the world. These are just two of the many dramatic changes and challenges of our time that led so many of us to ask, What's next?

WHAT CAN I DO WITH A SOCIOLOGY DEGREE?

An Introduction

Have you ever wondered what you can do with a sociology degree, or how you can take the skills you'll learn in this major and use them in your career? This book can help you answer that question: Near the end of each chapter, we feature a short essay that links your study of sociology to potential career fields. In the "What Can I Do With a Sociology Degree?" feature, we highlight the professional skills and core knowledge that the study of sociology helps you develop. This set of skills and competencies, which range from critical thinking and written communication skills to aptitude in qualitative and quantitative research to the understanding of diversity, prepares you for the workforce, as well as for graduate and professional school.

In every chapter that follows, this feature describes a specific skill that you can develop through the study of sociology. Each chapter also profiles a sociology graduate who shares what he or she learned through the study of sociology and how that particular skill has been valuable in his or her job. A short U.S. Bureau of Labor Statistics overview of the occupational field in which the graduate is working, its educational requirements, median income, and expected growth potential is also included.

Although this feature highlights sociology graduates, it also speaks to students taking sociology who are majoring in other disciplines—being aware of your skills and having the ability to articulate them precisely and clearly is important, no matter what your chosen field of study or career path.

SUMMARY

- **Sociology** is the **scientific** study of human social relationships, groups, and societies. Its central task is to ask what the dimensions of the social world are, how they influence our behavior, and how we, in turn, shape and change them.
- Sociology adheres to the principle of **social embeddedness**, the idea that economic, political, and other forms of human behavior are fundamentally shaped by social relationships. Sociologists seek to study, through scientific means, the social worlds that human beings consciously create.
- The **sociological imagination** is the ability to grasp the relationship between our individual lives and the larger social forces that help to shape them. It helps us see the connections between our private lives and public issues.
- **Critical thinking** is the ability to evaluate claims about truth by using reason and evidence. Often, we accept things as true because they are familiar, seem to mesh with our own experiences, and sound right. Critical thinking instead asks us to recognize poor arguments, reject statements not supported by evidence, and even question our own assumptions.
- Sociology's roots can be traced to the scientific revolution, the Enlightenment, industrialization and the birth of modern capitalism, and the urbanization of populations. Sociology emerged in part as a tool to enable people to understand the dramatic changes taking place in modern societies.
- Sociology generally traces its classical roots to Auguste Comte, Émile Durkheim, Max Weber, and Karl Marx. Early work in sociology reflected the concerns of the men who founded the discipline.
- In the United States, scholars at the University of Chicago focused on reforming social problems stemming from industrialization and urbanization. Women and people of color worked on the margins of the discipline because of persistent discrimination.
- Sociologists base their study of the social world on different theoretical perspectives that shape theory and guide research, often resulting in different conclusions. The major sociological paradigms are **structural functionalism**, the **social conflict paradigm**, and **symbolic interactionism**.
- Major themes in sociology include the distribution of **power** and growing inequality, **globalization** and its accompanying social changes, the growth of **social diversity**, and the way advances in technology have changed communication, commerce, and communities.
- The early founders of sociology believed that scientific knowledge could lead to shared social progress. Some modern sociologists question whether such shared scientific understanding is indeed possible.

KEY TERMS

scientific (p. 2)
social conflict paradigm (p. 18)
social diversity (p. 21)
social dynamics (p. 9)
social embeddedness (p. 3)
social facts (p. 10)
social solidarity (p. 11)
social statics (p. 9)
sociological imagination (p. 3)
sociological theories (p. 16)
sociology (p. 3)
structural functionalism (p. 16)
structure (p. 4)
symbolic interactionism (p. 19)
symbols (p. 19)
Verstehen (p. 12)

DISCUSSION QUESTIONS

1. Think about Mills's concept of the sociological imagination and its ambition to draw together what Mills called *private troubles* and *public issues*. Think of a private trouble that sociologists might classify as also being a public issue. Share your example with your classmates.
2. In the chapter, we asked why women's voices were marginal in early sociological thought. What factors explain the dearth of women's voices? What about the lack of minority voices? What effects do you think these factors may have had on the development of the discipline?
3. What is critical thinking? What does it mean to be a critical thinker in our approach to understanding society and social issues or problems?
4. Recall the three key theoretical paradigms discussed in this chapter—structural functionalism, conflict theory, and symbolic interactionism. Discuss the ways these diverse "glasses" analyze deviance, its labeling, and its punishment in society. Try applying a similar analysis to another social phenomenon, such as class inequality or traditional gender roles.
5. Identify the three main themes of this book and explain their importance. Which most interests you and why?

©iStockphoto.com/choness

2 DISCOVER SOCIOLOGICAL RESEARCH

WHAT DO YOU THINK?

1. What kinds of research questions could one pose to guide studies of sociological phenomena such as long-term poverty, cyberbullying, family or veteran homelessness, Generation Z consumption habits, or the high dropout rate in some high schools?
2. What factors may affect the honesty of people's responses to survey or interview questions?
3. What makes a sociological research project ethical or unethical?

LEARNING OBJECTIVES

2.1 Describe the scientific method and distinguish between qualitative and quantitative research.

2.2 Describe the components of a scientific theory and how a scientific theory is tested.

2.3 Identify key methods employed in sociological research and explain when it is appropriate to use them.

2.4 Understand the basic steps in building a sociological research project.

NO ROOF OVERHEAD: RESEARCHING EVICTION IN AMERICA

In *Evicted: Poverty and Profit in the American City*, sociologist Matthew Desmond (2016a) writes that

> [M]illions of Americans are evicted every year because they can't make rent. . . . In 2013, 1 in 8 poor renting families nationwide were unable to pay all of their rent, and a similar number thought it would be likely they would be evicted soon. (pp. 4–5)

Desmond argues that eviction is not only a consequence of poverty but also a cause because the lack of a stable home undermines the ability of the poor to get and keep a job and to establish children in good schools, and it can lead to stress, depression, and even suicide. As a *New York Times* book review of *Evicted* poignantly notes, "Living in extreme poverty in the United States means waging an almost gladiatorial battle for creature comforts that luckier people take for granted. And of all those comforts, perhaps the most important is a stable, dignified home" (Senior, 2016, para. 4).

Desmond builds his research around a powerful, on-the-ground ethnographic account of the lives of eight Milwaukee families caught in a web of destitution and despair as they try to navigate the private rental market in that city. They include Arleen and her two young sons, fighting to find safe haven as Arleen struggles with money, depression, and the behavioral troubles of her boys. Desmond also offers an account of the multigenerational Hinkston family, including young teenager Ruby, whose efforts at the public library to construct a bright, pleasant virtual home with a free online computer game are a grim contrast to her own living conditions in low-rent housing, which are characterized by instability, cockroaches, and chronically clogged plumbing.

©ERIC BARADAT/AFP via Getty Images

Desmond points out that their stories are not isolated accounts; rather, in 2013, 67% of poor renting families received no housing assistance—the demand for housing help far outpaces the availability of subsidized apartments and housing vouchers. This leaves families to seek what they hope will be permanent shelter in a private low-rent housing market that is rife with dismal, dirty, and even dangerous living conditions. Significantly, even the worst housing may stretch the resources of many families beyond their means: The majority of poor families spend over half their income on rent, whereas about one quarter spend over 70% (Desmond, 2016a, p. 4). An unanticipated expense, a dispute with a landlord, or the loss of a job can easily put families on the street. The lack of resources and an eviction record can keep them there for a significant period of time.

In 2020, the pain of eviction reached an even greater proportion of Americans: Among the most significant economic consequences of the COVID-19 pandemic was a soaring rate of evictions as tenants who lost jobs and income as a result of the economic slowdown ran short of funds to pay their rent. At the end of September 2020, the Eviction Lab, which hosts a public website, had logged over 50,600 evictions in 17 monitored cities since the start of the pandemic (https://evictionlab.org/eviction-tracking/).

Desmond's work is a good example of qualitative sociological research, and he recognizes its significance to academic and policy debates. By using a scientific approach and rigorous field research, Desmond casts light on the little-examined but significant problem of evictions. He recognizes the struggles of those who are most at risk of eviction—low-income minority women: "Women living in black neighborhoods in Milwaukee represent 9.6% of the population, but 30% of evictions" (Desmond, 2015, pp. 3–4). Importantly, he also sees that there is profit to be made from the misery of others, and he documents the multitude of ways in which landlords exploit the low-end market for their benefit, skimping on repairs, failing to provide even basic appliances (apparently a legal action), and keeping even low rents high enough that if tenants fail to pay, the landlord can evict them, keep the deposit, and move on to a new renter (Desmond, 2016a). As a sociologist, Desmond has described and defined his problem, examined its causes and consequences, and provided policy prescriptions to address it.

In this chapter, we examine the ways sociologists like Matthew Desmond study the social world. First, we distinguish between sociological understanding and common sense. Then we discuss the key steps in the research process itself. We examine how sociologists test their theories using a variety of research methods, and finally, we consider the ethical implications of doing research on human subjects.

SOCIOLOGY AND COMMON SENSE

Using science means using a unique way of seeing to investigate the world around us. The essence of the **scientific method** is straightforward: It is *a process of gathering empirical (scientific and specific) data, creating theories, and rigorously testing theories.* In sociological research, theories and empirical data exist in a dynamic relationship (Figure 2.1). Some sociological research begins from general theories, which offer "big picture" ideas: **Deductive reasoning** *starts from broad theories about the social world but proceeds to break them down into more specific and testable hypotheses.* Sociological **hypotheses** are *ideas about the world, derived from theories, that describe possible relationships between social phenomena.* Some research begins from the ground up: **Inductive reasoning** *starts from specific data, such as interviews, observations, or field notes, that may focus on a single community or event and endeavors to identify larger patterns from which to derive more general theories.*

Sociologists employ the scientific method in both quantitative and qualitative research. **Quantitative research**, which is often done through methods such as large-scale surveys, *gathers data that can be quantified and offers insight into broad patterns of social behavior* (for example, the percentage of U.S. adults who use corporal punishment such as spanking with their children) *and social attitudes* (for example, the percentage of U.S. adults who approve of corporal punishment) without necessarily exploring the meaning of or reasons for the identified phenomena. **Qualitative research**, such as that conducted by Matthew Desmond, *is characterized by data that cannot be quantified (or converted into numbers), focusing instead on generating in-depth knowledge of social life, institutions, and processes* (for example, why parents in particular demographic groups are more or less likely to use spanking as a method of punishment). It relies on the gathering of data through methods such as focus groups, participant and nonparticipant observation, interviews, content analysis, and archival research. Generally, population samples in qualitative research are small because they focus on in-depth understanding.

Scientific method: A process of gathering empirical (scientific and specific) data, creating theories, and rigorously testing theories.

Deductive reasoning: Starts from broad theories about the social world but proceeds to break them down into more specific and testable hypotheses.

Hypotheses: Ideas about the world, derived from theories, that describe possible relationships between social phenomena.

Inductive reasoning: Starts from specific data, such as interviews, observations, or field notes, that may focus on a single community or event and endeavors to identify larger patterns from which to derive more general theories.

Quantitative research: Research that gathers data that can be quantified and offers insight into broad patterns of social behavior and social attitudes.

Qualitative research: Research that is characterized by data that cannot be quantified (or converted into numbers), focusing instead on generating in-depth knowledge of social life, institutions, and processes.

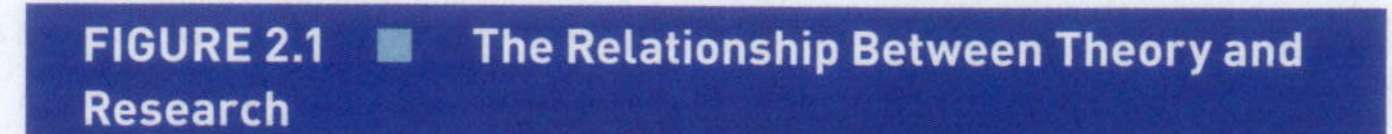
FIGURE 2.1 ■ The Relationship Between Theory and Research

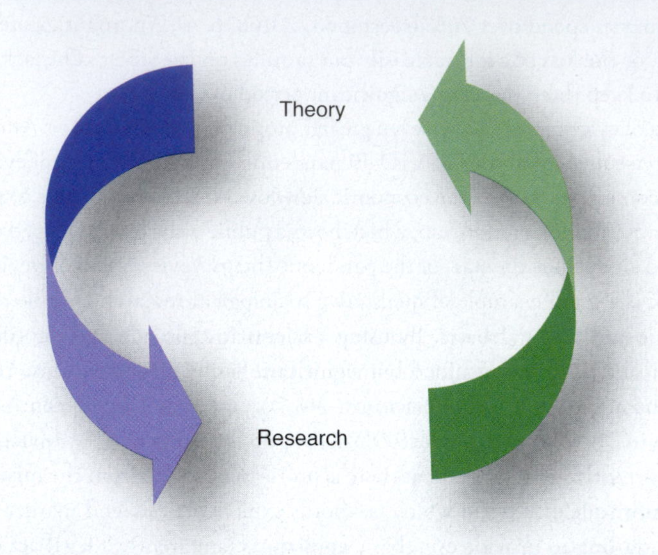

TABLE 2.1 ■ Annual Percentage of Drug Use Among 8th, 10th, and 12th Graders, 2019

	MARIJUANA	COCAINE	CRACK COCAINE	LSD	HEROIN
8th	11.8	0.7	0.4	1.6	0.3
10th	28.8	1.5	0.6	2.3	0.3
12th	35.7	2.2	1.0	3.6	0.4

Source: National Institute on Drug Abuse. (2019). *Monitoring the future study: Trends in prevalence of various drugs for 8th graders, 10th graders, and 12th graders; 2016–2019 (in percent).* https://www.drugabuse.gov/drug-topics/trends-statistics/monitoring-future/monitoring-future-study-trends-in-prevalence-various-drugs

Personal experience and common sense about the world are often fine starting points for sociological research. Researchers are driven by a passion to understand the world around them, and curiosity that emerges from personal experience can be a powerful motivator. Personal experience and even common sense can, however, mislead us. In the 14th century, common sense suggested to people that the Earth was flat; after all, it *looks* flat. Today, influenced by stereotypes and media portrayals of youthful deviance, many people believe that drug use is common among high schoolers. But common sense misleads—or is far more nuanced. The Earth is not flat (as you know!), and, while the rate of marijuana use has risen among young people, the use of hard drugs is very low (Table 2.1).

Consider the following ideas, which many believe to be true, although all are false.

Common Wisdom

I know women who earn more than their husbands or boyfriends. The gender wage gap is no longer an issue in the United States.

Sociological Research

Data show that men as a group earn more than women as a group. For example, in mid-2020, men had a weekly median income of $1,087, compared with $913 for women for all full-time occupations (U.S. Bureau of Labor Statistics, 2020). There is some statistical variation, but data suggest that women as a group earn 84 cents to a dollar that men earn. These figures compare all men and all women who work full time and year-round. Reasons for the gap include worker characteristics (such as experience, education, and ability or willingness to negotiate salary), job characteristics (such as hours required), devaluation of women's work by society, and pay discrimination against female workers (American Association of University Women, 2016; Cabeza et al., 2011; Reskin & Padavic, 2002). Although some women, of course, earn more than some men, the overall pattern of men outearning women remains in place today. This topic is discussed in greater detail in Chapter 10.

Common Wisdom

Homeless people lack adequate shelter because they do not work.

Sociological Research

Finding safe, permanent housing is a challenge for many Americans, even those who work for pay. Low wages and poor benefits in the service industry, where many less-educated people work, as well as a shortage of adequate housing options for low-income families, can make finding permanent shelter a challenge and eviction a chronic risk (Desmond, 2016a). The precarity of housing for millions of families was highlighted in 2020 when the COVID-19 pandemic resulted in massive job losses that rendered many Americans unable to meet their rental and mortgage obligations.

There is also a population of the long-term homeless who cannot work: "Nearly all of the long-term homeless have tenuous family ties and some kind of disability, whether it is a drug or alcohol addiction, a mental illness, or a physical handicap" (Culhane, 2010, para. 4). Alas, this is a group that would benefit from housing in facilities that can treat their ailments so they can attain self-sufficiency. The important sociological subjects of poverty and access (or lack thereof) to resources like safe housing are discussed more fully in Chapter 7.

Common Wisdom

Education is the great equalizer. All children in the United States have the opportunity to get a good education. Low academic achievement is an individual failure.

Sociological Research

Public education is free and available to all in the United States, but the quality of education varies dramatically. Consider the fact that in many states and localities, a major source of public school funding is local property taxes, which constitute an average of about 45% of funding (state and federal allocations make up the rest) (National Public Radio, 2010). As such, communities with high property values have richer sources of funding from which to draw educational resources, whereas poor communities—even those with high tax rates—have more limited resource pools.

As well, high levels of racial segregation persist in U.S. schools. A U.S. Government Accountability Office report found that the proportion of schools that are highly segregated by race and class—that is, where more than 75% of children get free or reduced-price lunch and more than 75% are Black or Hispanic—is rising, climbing from 9% to 16% of schools between 2001 and 2014. It is also significant that students in the high-poverty and majority-Black or Hispanic schools were less likely to have access to the range of math and science courses available to their peers in better-off schools and to be subject to harsher disciplinary measures (U.S. Government Accountability Office, 2016). Research also shows a relationship between academic performance and class and racial segregation: Students who are not isolated in poor, racially segregated schools perform better on a variety of academic measures than those who are (Condron, 2009; Logan et al., 2012).

The problem of low academic achievement is complex, and no single variable can explain it. At the same time, the magnitude and persistence of this problem suggest that we are looking at a phenomenon that is a public issue rather than a personal trouble. We discuss issues of class, race, and educational attainment further in Chapter 12.

Even deeply held and widely shared beliefs about society and social groups may be inaccurate—or more nuanced and complex than they appear on the surface. Until it is tested, common sense is merely conjecture. Careful research allows us to test our beliefs to gauge whether they are valid or merely anecdotal. From a sociological standpoint, empirical evidence is granted greater weight than common sense. By basing their decisions on scientific evidence rather than on personal beliefs or common wisdom, researchers and students can draw informed conclusions and policy makers can ensure that policies and programs are data driven and maximally effective.

RESEARCH AND THE SCIENTIFIC METHOD

Scientific theories: Explanations of how and why scientific observations are as they are.

Scientific theories are *explanations of how and why scientific observations are as they are.* A good scientific theory has the following characteristics:

- *It is logically consistent.* One part of the theory does not contradict another part.
- *It can be disproved.* If the findings contradict the theory, then we can deduce that the theory is wrong. Although we can say that testing has failed to disprove the theory, we cannot assume the theory is true if testing confirms it. Theories are always subject to further testing, which may point to needed revisions, highlight limitations, or strengthen conclusions.

Concepts: Ideas that summarize a set of phenomena.

Theories are made up of **concepts**, *ideas that summarize a set of phenomena.* Concepts are the building blocks of research and prepare a solid foundation for sociological work. Some key concepts in sociology are *social stratification, social class, power, inequality,* and *diversity,* which we introduced in the opening chapter.

Operational definition: Describes the concept in such a way that it can be observed and measured.

To gather data and create viable theories, we need to define concepts in ways that are precise and measurable. A study of social class, for example, would need to begin with a working definition of that term. An **operational definition** of a concept *describes the concept in such a way that it can be observed and measured.* Many sociologists define *social class* in terms of dimensions such as income, wealth, education, occupation, and consumption patterns. Each of these aspects of class has the potential to be measurable. We may construct operational definitions in terms of *qualities* or *quantities* (Babbie, 1998; Neuman, 2000). In terms of qualities, we might say that the *upper-middle class* is composed of working professionals who have completed advanced degrees, even though there may be a broad income spread between those with a master's degree in fine arts and those with a master's degree in business administration. This definition is based on an assumption of *class* as a social position that derives from educational attainment. Alternatively, by using quantity as a key measure, we might operationally define *upper class* as households with an annual income greater than $200,000 and *lower class* as households with an annual income of less than $30,000. This definition takes income as the preeminent determinant of class position, irrespective of education or other variables.

Some research on bullying relies on self-reports, whereas other data come from peer reports. Research (Bronson & Carson, 2019) suggests that more than twice as many students (11%) were labeled bullies in peer reports than in self-reports (5%), highlighting the fact that definitions of what constitutes bullying may differ, and any method of data collection has limitations.

Consider a social issue of contemporary interest—bullying. Imagine that you want to conduct a research study of bullying to determine how many female middle schoolers have experienced bullying in the past academic year. You would need to begin with a clear definition of *bullying* that operationalizes the term. That is, to measure how many girls have experienced bullying, you would need to articulate what constitutes bullying. Would you include physical bullying? If so, how many instances of being pushed or punched would constitute bullying? What kinds of verbal behaviors would be considered bullying? Would you include the newer phenomenon of cyberbullying? To study a phenomenon such as bullying, it is not enough to assume that we know it when we see it. Empirical research relies on the careful and specific definition of terms and the recognition of how definitions and methods affect research outcomes.

Relationships Between Variables

In studying social relationships, sociologists identify key *variables.* A **variable** is *a concept that can take on two or more possible values.* For instance, marital status can be married or unmarried, work status can be employed or unemployed, and geographic location can be urban, suburban, or rural. We can measure variables both *quantitatively* and *qualitatively.* **Quantitative variables** include *factors that can be counted,* such as rates of employment or unemployment, marriage rates, crime victimization rates, and drug use frequency. **Qualitative variables** are *variables that express qualities and do not have numerical values.* Qualitative variables might include attitudinal characteristics such as a parent's preference for a private or public school, or a commuter's preference for riding public transportation or driving to work.

Variable: A concept or its empirical measure that can take on two or more possible values.

Qualitative variables: Variables that express qualities and do not have numerical values.

Quantitative variables: Factors that can be counted.

Sociological research often tries to establish a relationship between two or more variables. Suppose you want to find out whether more education is associated with higher earnings. After asking people about their years of schooling and their annual incomes, both of which are quantitative variables, you could estimate the degree of *correlation* between the two. **Correlation**—literally, "co-relationship"—is *the degree to which two or more variables are associated with one another.* Correlating the two variables *years of education* and *annual income* demonstrates that the greater the education, the higher the income (Figure 2.2). Do you see the exception to that relationship? How might you explain it?

Correlation: The degree to which two or more variables are associated with one another.

When two variables are correlated, we are often tempted to infer a **causal relationship**, *a relationship between two variables in which one variable is the cause of the other.* Nevertheless, even though two variables are correlated, we cannot assume that one causes the other. For example, ice cream sales rise during the summer, as does the homicide rate. These two events are correlated in the sense that both increase during the hottest months. Yet, because the rise in ice cream sales does not *cause* rates of homicide to increase (nor, clearly, does the rise in homicide rates cause a spike in ice cream consumption), these two phenomena do not have a causal relationship. Correlation does not equal causation.

Causal relationship: A relationship between two variables in which one variable is the cause of the other.

Sometimes an observed correlation between two variables is the result of a **spurious relationship**—*a correlation between two or more variables caused by another factor that is not being measured rather than a causal link between the variables themselves.* In the example given earlier, the common factor missed in the relationship is the temperature. When it's hot, more people want to eat ice cream. Studies also show that rising temperatures are linked to an increase in violent crimes—although after a certain

Spurious relationship: A correlation between two or more variables caused by another factor that is not being measured rather than a causal link between the variables themselves.

FIGURE 2.2 ■ Correlation Between Education and Median Weekly Earnings in the United States, 2019

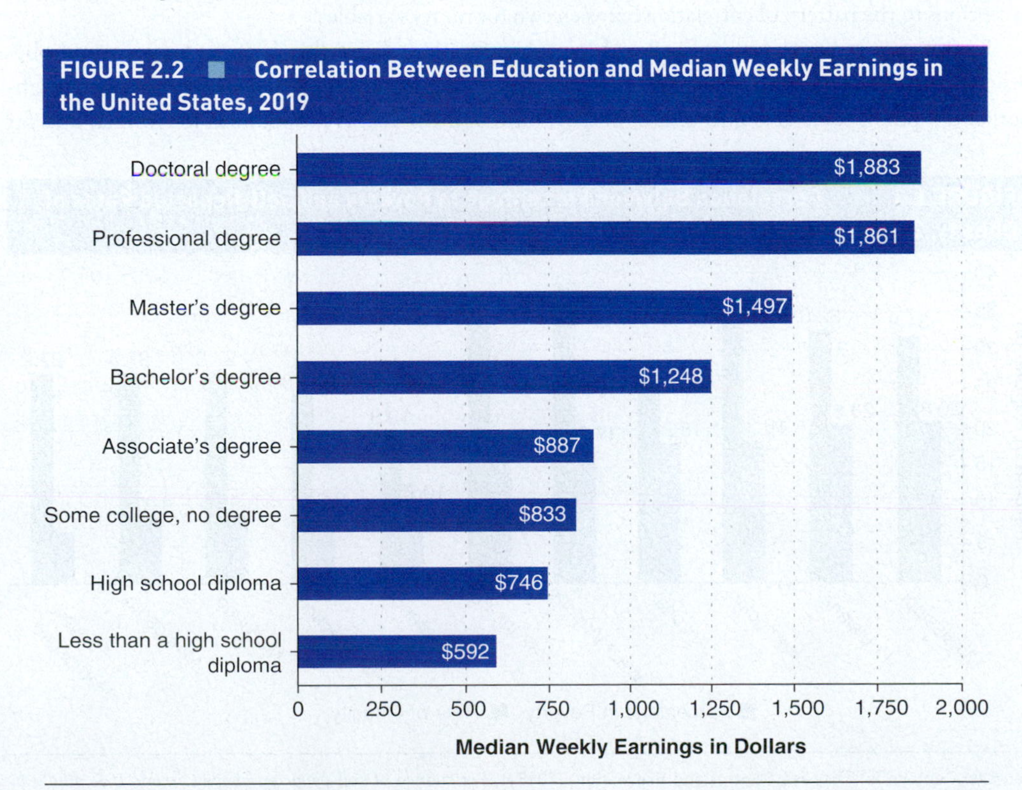

Source: Bureau of Labor Statistics. (2019). *Unemployment rates and earnings by educational attainment. Employment projections.* U.S. Government Printing Office.

temperature threshold (about 90 degrees), crimes wane again (Gamble & Hess, 2012). Among the reasons more violent crimes are committed in the warm summer months is the fact that people spend more time outdoors in social interactions, which can lead to confrontations.

Let's take a look at another example: Imagine that a college newspaper publishes a study concluding that coffee drinking causes poor test grades. The story is based on a survey of students at the college that found that those who reported drinking a lot of coffee the night before an exam scored lower than did their peers who had consumed little or no coffee. Having studied sociology, you wonder whether this relationship might be spurious. What is the "something else" that is not being measured here? Could it be that students who did not study in the days and weeks prior to the test and stayed up late the night before cramming—probably consuming a lot of coffee as they fought sleep—earned lower test grades than did their peers who studied earlier and got adequate sleep the night before the test? The overlooked variable, then, is the amount of studying students did in the weeks preceding the exam, and we are likely to find a positive correlation and evidence of causation in looking at time spent studying and grade outcomes.

Sociologists attempt to develop theories systematically by offering clear operational definitions, collecting unbiased data, and identifying evidence-based relationships between variables. Sociological research methods usually yield credible and useful data, but we must always critically analyze the results to ensure their validity and reliability and to check that hypothesized relationships are not spurious.

Testing Theories and Hypotheses

Once we have defined concepts and variables with which to work, we can endeavor to test a theory by positing a hypothesis. Hypotheses enable scientists to check the accuracy of their theories. For example, data show that some positive correlation exists between obesity and poverty rates at the state level: Mississippi, West Virginia, Kentucky, and Louisiana, which are among the poorest states in the country, are also among the states with the highest obesity rates (Figure 2.3). As well, four of the ten wealthiest states in the United States are among those with the lowest obesity rates. A **positive correlation** is *a relationship showing that as one variable rises or falls, the other does as well.* As noted earlier, sociologists are quick to point out that correlation does not equal causation. Researchers are interested in creating and testing hypotheses to explain cases of positive correlation—they are also interested in explaining exceptions to the pattern of correlation between two (or more) variables.

Positive correlation: A relationship showing that as one variable rises or falls, the other does as well.

In fact, researchers have explored and hypothesized the relationship between poverty and obesity. Among the conclusions they have drawn is that living in poverty—and living in a poor neighborhood—puts people at higher risk of obesity, although the risk is pronounced for women and far

FIGURE 2.3 ■ Correlation Between Percentage in Poverty and Self-Reported Rates of Obesity, 2018

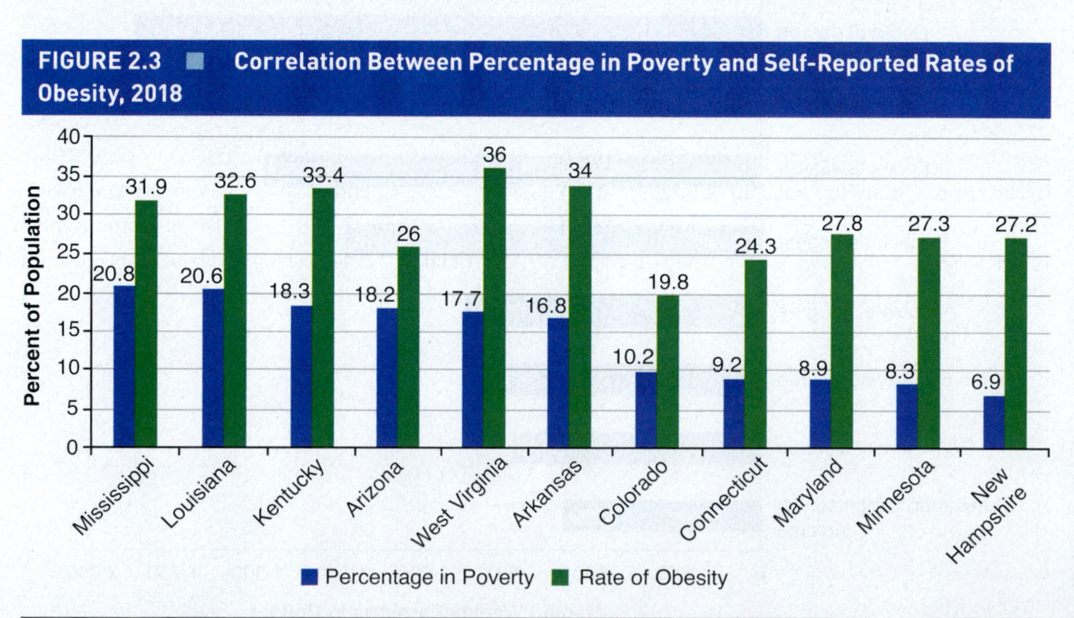

Sources: Centers for Disease Control and Prevention. (2018e). *Prevalence of self-reported obesity among U.S. adults by race/ethnicity, state and territory, BRFSS, 2018.* https://www.cdc.gov/obesity/data/prevalence-maps.html; United States Census Bureau. (2018). *Poverty: 2017 and 2018.* https://www.census.gov/content/dam/Census/library/publications/2019/acs/acsbr18-02.pdf

less clear for men (Hedwig, 2011; Smith, 2009). Factors that researchers have identified as contributing to a causal path between poverty and obesity include a lack of access to healthy food choices and safe, accessible spaces for physical exercise. A low-income area in which at least one third of residents live more than 1 mile away (or 10 miles away in rural areas) from a supermarket or large grocery store is referred to by the U.S. Department of Agriculture as a "food desert" (Block & Subramanian, 2015). Food deserts make it difficult for people living in poverty to access healthy food, especially if they do not have the time or affordable means to travel to a full-service supermarket. A deficit of time to cook healthy foods and to exercise, a lack of funds to purchase high-quality foods, and the stress induced by poverty are all contributing factors as well. Although the data cannot lead us to conclude decisively that poverty is a *cause* of obesity, research can help us to gather evidence that supports or refutes a hypothesis about the relationship between these two variables.

Physical exercise is a key to good health. Lack of access to safe places to walk and play can contribute to problems of overweight and obesity for individuals and communities.

In the case of a **negative correlation**, *one variable increases as the other decreases.* As we discuss later in Chapter 11, which focuses on the family and society, researchers have found a negative correlation between male unemployment and rates of marriage. That is, as rates of male unemployment in a community rise, rates of marriage in the community fall. Observing this relationship, sociologists have conducted research to test explanations for this relationship (Edin & Kefalas, 2005; Wilson, 2010). What kind of hypothesis would you offer to test this relationship?

Negative correlation: A relationship showing that as one variable increases, the other decreases.

Keep in mind that we can never prove theories to be decisively right—we can only prove them wrong. Proving a theory right would require the scientific testing of absolutely every possible hypothesis based on that theory—a fundamental impossibility. In fact, good theories are constructed in a way that makes it logically possible to prove them wrong. This is Karl Popper's (1959) famous **principle of falsification, or falsifiability,** which holds that *to be scientific, a theory must lead to testable hypotheses that can be disproved if they are wrong.*

Principle of falsification (or falsifiability): The principle, advanced by philosopher Karl Popper, that to be scientific, a theory must lead to testable hypotheses that can be disproved if they are wrong.

Validity and Reliability

For theories and hypotheses to be testable, both the concepts used to construct them and the measurements used to test them must be accurate. When our observations adequately reflect the real world, our findings have **validity**—that is, *the degree to which concepts and their measurements accurately represent what they claim to represent.* For example, suppose you want to know whether violent or property crime in the United States has gone up or down. For years, sociologists depended on police reports to measure crime: Official crime statistics in the United States generally come from the Uniform Crime Report (UCR). Researchers could assess the validity of these official statistics, which are based on reported crimes, only if subsequent surveys were administered nationally to victims of crime. If the victim tallies matched those of the police reports, then researchers could say the police reports were a valid measure of crime in the United States. The National Crime Victimization Survey (NCVS) enables researchers to assess validity because it offers information on crimes that have not been reported to authorities and would thus not appear in the UCR.

Validity: The degree to which concepts and their measurements accurately represent what they claim to represent.

Sociologists are also concerned with the reliability of their findings. **Reliability** is *the extent to which researchers' findings are consistent with the findings of different studies of the same thing or with the findings of the same study over time.* Sociological research may suffer from problems of validity and reliability because of **bias**, *a characteristic of results that systematically misrepresent the true nature of what is being studied.* Bias can creep into research as a result of the use of inappropriate measurement instruments.

Reliability: The extent to which researchers' findings are consistent with the findings of different studies of the same thing or with the findings of the same study over time.

Bias: A characteristic of results that systematically misrepresent the true nature of what is being studied.

For example, suppose the administrator of a city wants to know whether homelessness has risen in recent years. She operationally defines *the homeless* as those who sleep in the street or in shelters and dispatches her team of researchers to city shelters to count the number of people occupying shelter beds

TABLE 2.2 ■ How Truthful Are Survey Respondents? (in Percentages)

	THREAT OF VALIDATION		NO THREAT OF VALIDATION	
Survey Question	***Anonymous***	***Named***	***Anonymous***	***Named***
Ever smoked?	63.5	72.9	60.5	67.8
Smoked in the last month?	34.5	39.5	25.9	21.8
Smoked in the last week?	26.0	25.5	14.4	17.6

Source: Adams, J., Parkinson, L., Sanson-Fisher, R. W., & Walsh, R. A. (2008). Enhancing self-report of adolescent smoking: The effects of bogus pipeline and anonymity. *Addictive Behaviors, 33*(10), 1291–1296.

or sleeping on street corners or park benches. A sociologist reviewing the research team's results might question the administrator's definition of what it means to be homeless and, by extension, her findings. Are the homeless solely those spending nights in shelters or on the streets? What about those who stay with friends or in a hotel after eviction, or camp out in their cars? In this instance, a sociologist might suggest that the city's measure is biased because it misrepresents (and undercounts) the homeless population by failing to define the concept in a way that captures the broad manifestations of homelessness.

Bias can also occur in research when respondents do not tell the truth (see Table 2.2). An example of this is a study in which respondents were asked whether they used illegal drugs or had driven while impaired. All were asked the same questions, but some were wired to a machine they were told was a lie detector. The subjects who thought their truthfulness was being monitored by a lie detector reported higher rates of illegal drug use than did subjects who did not. Based on the assumption that actual drug use would be about the same for both groups, the researchers concluded that the subjects who were not connected to the device were under-reporting their actual illegal drug use and that simply asking people about drug use would lead to biased findings because respondents would not tell the truth.

Social desirability bias: A response bias based on the tendency of respondents to answer a question in a way that they perceive will be favorably received.

The reticence to truthfully self-report drug use and impaired driving may also be related to a phenomenon that researchers call **social desirability bias**, which is *a response bias based on the tendency of respondents to answer a question in a way that they perceive will be favorably received.* That is, many respondents want to present themselves positively to the interviewer. Social desirability bias is most likely to be a problem in studies that examine, for instance, participation in physical exercise (Brenner & Delamater, 2014) or "cyberloafing" at work (Akbulut et al., 2017). Because exercising and being productive at work are widely perceived as positive activities, there is greater reticence to report behavior that does not adhere to perceived norms. We discuss other ways in which social desirability bias may affect research findings further along in this chapter.

Objectivity in Scientific Research

Even if sociologists develop theories based on good operational definitions and collect valid and reliable data, like all human beings, they have passions and biases that may color their research. For example, criminologists (many of whom, until recently, were male) long ignored the criminality of women because they assumed that women were not disposed toward criminal behavior. Researchers therefore did not have an accurate picture of women and crime until this bias was recognized and rectified.

Objectivity: The ability to represent the object of study accurately.

Value neutrality: The characteristic of being free of personal beliefs and opinions that would influence the course of research.

Personal values and beliefs may affect a researcher's **objectivity** or *ability to represent the object of study accurately.* In the 19th century, sociologist Max Weber argued that for scientific research to be objective, it has to have **value neutrality**—*the characteristic of being free of the influence of personal beliefs and opinions that would influence the course of research.* The sociologist should acknowledge personal biases and assumptions, make them explicit, and prevent them from getting in the way of observation and reporting.

How can we best achieve objectivity? First, recall Karl Popper's principle of falsification, which proposes that the goal of research is not to prove our ideas correct but to find out whether they are wrong. To accomplish this, researchers must be willing to accept that the data they collect might contradict their most passionate convictions. Research should deepen human understanding, not prove a particular point of view.

A second way we can ensure objectivity is to invite others to draw their own conclusions about the validity of our data through **replication**, *the repetition of a previous study using a different sample or population to verify or refute the original findings.* For research to be replicated, the original study must spell out in detail the research methods employed. If potential replicators cannot conduct their studies exactly as the original study was performed, they might accidentally introduce unwanted variables. To ensure the most accurate replication of their work, researchers should archive original materials such as questionnaires and field notes and allow replicators access to them.

Replication: The repetition of a previous study using a different sample or population to verify or refute the original findings.

Popper (1959) describes scientific discovery as an ongoing process of confrontation and refutation. Sociologists usually subject their work to this process by publishing their results in scholarly journals. Submitted research undergoes a rigorous process of peer review, in which other experts in the field of study examine the work before the results are finalized and published. Once research has been published in a reputable journal such as the *American Sociological Review* or the *Journal of Health and Social Behavior,* other scholars read it with a critical eye. The study may then be replicated in different settings.

DOING SOCIOLOGICAL RESEARCH

Sociological research requires careful preparation and a clear plan that guides the work. The purpose of a sociological research project may be to obtain preliminary knowledge that will help formulate a theory or to evaluate an existing theory about society and social life. As part of the strategy, the researcher selects from a variety of **research methods**—*specific techniques for systematically gathering data.* In the following sections, we look at a range of research methods and examine their advantages and disadvantages. We also discuss how you might prepare a sociological research project of your own.

Research methods: Specific techniques for systematically gathering data.

Sociological Research Methods

Sociologists employ a variety of methods to learn about the social world (Table 2.3). Since each has strengths and weaknesses, a good research strategy may be to use several different methods. If they all yield similar findings, the researcher is more likely to have confidence in the results. The principal methods are the survey, fieldwork (either participant observation or detached observation), experimentation, working with existing information, and participatory research.

Survey Research

A **survey** uses *a questionnaire or interviews administered to a group of people in-person or by telephone or e-mail to determine their characteristics, opinions, and behaviors.* Surveys are versatile, and sociologists often use them to test theories or to gather data. Some survey instruments, such as National Opinion Research Center questionnaires, consist of closed-ended questions that respondents answer by choosing from among the responses presented. Others, such as the University of Chicago's Social Opportunity Survey, consist of open-ended questions that permit respondents to answer in their own words.

Survey: A research method that uses a questionnaire or interviews administered to a group of people in-person or by telephone or e-mail to determine their characteristics, opinions, and behaviors.

An example of survey research conducted for data collection is the largest survey in the nation, the U.S. Census, which is conducted every 10 years. The census is not designed to test any particular theory. Rather, it gathers voluminous data about U.S. residents that researchers, including sociologists, use to test and develop a variety of theories. In this text, you will find U.S. Census data in many chapters.

Usually, a survey is conducted on *a small number of people,* a **sample**, selected to represent a **population**, *the whole group of people to be studied.* The first step in designing a survey is to identify the population of interest. Imagine that you are doing a study of sociological factors that affect grades in college. Who would you survey? Members of a certain age group only? People in the airline industry? Pet owners? To conduct a study well, we need to identify clearly the survey population that will most

Sample: A small number of people; a portion of the larger population selected to represent the whole.

Population: The whole group of people studied.

TABLE 2.3 ■ Key Sociological Research Methods

RESEARCH METHOD	APPROPRIATE CIRCUMSTANCES
Survey research	When basic information about a large population is desired. Sociologists usually conduct survey research by selecting samples that are representative of the entire populations of interest.
Fieldwork	When detailed information is sought, and when surveys are impractical for getting the information desired (for example, in studying youth gangs or gamblers). Fieldwork usually relies on small samples, especially compared to surveys.
Detached observation	When researchers desire to stay removed from the people being studied and must gather data in a way that minimizes impact on the subjects. Detached observations are often supplemented with face-to-face interviews.
Participant observation	When first-hand knowledge of the subjects' direct experience is desired, including a deeper understanding of their lives.
Experimentation	When it is possible to create experimental and control groups that are matched on relevant variables but provided with different experiences in the experiment.
Use of existing information	When direct acquisition of data is either not feasible or not desirable because the event studied occurred in the past or because gathering the data would be too costly or too difficult.
Participatory research	When a primary goal is training people to gain political or economic power and acquire the necessary skills to do the research themselves.

effectively help us answer the research question. In your study, you would most likely choose to survey students now in college because they offer the best opportunity to correlate grades with circumstances and behaviors.

Once we have identified a population of interest, we need to select a sample, as we are unlikely to have the time or money to talk to all the members of a given population, especially if it is a large one. Other things being equal, larger samples better represent the population than do smaller ones. Nevertheless, with proper sampling techniques, sociologists can use small (and therefore inexpensive) samples to represent large populations. For instance, a well-chosen sample of 1,000 U.S. consumers can be used to represent 100,000 U.S. consumers with a fair degree of accuracy, enabling surveys to make predictions about economic behavior with reasonable confidence. Sampling is also used for looking at social phenomena such as marriages and online dating in a population: A recent paper, based on a sample of 4,200 U.S. adults, suggests that those with Internet access at home are more likely to have partners, even controlling for other factors (Rosenfeld & Thomas, 2012).

Ideally, a sample should reflect the composition of the population we are studying. For instance, if you want to be able to use your research data about college students to generalize about the entire college student population of the United States, you would need to collect proportional samples from two-year colleges, four-year colleges, large universities, community colleges, online schools, and so on. It would not be adequate to survey only students at online colleges or only female students at private four-year schools.

Random sampling: Sampling in which everyone in the population of interest has an equal chance of being chosen for the study.

To avoid bias in surveys, sociologists may use **random sampling**, whereby *everyone in the population of interest has an equal chance of being chosen for the study.* Typically, they make or obtain a list of everyone in the population of interest. Then they draw names or phone numbers, for instance, by chance until the desired sample size is reached (today, such work can be done by computers). Large-scale random sample surveys permit researchers to draw conclusions about large numbers of people on the basis of small numbers of respondents. For our survey of college students, we could (theoretically) take all U.S. college students as our starting point and sample randomly from that group. We might also choose to use a stratified sample: In **stratified sampling**, *researchers divide a population into a series of subgroups* (for instance, students at four-year public universities, students at two-year colleges, students at online schools, etc.) *and take random samples from within each group.* This can be used to ensure representation from all subgroups (such as college students at different types of schools) in the final research sample.

Stratified sampling: Dividing a population into a series of subgroups and taking random samples from within each group.

Researchers may assemble survey respondents through other sampling means. For example, they may use convenience sampling or snowball sampling. Imagine you are doing a survey of college students to learn what factors students consider when they choose a major. You may opt for a *convenience sample* of students at your school: This could include students in your classes; friends from clubs or organizations on campus; or, if you live on campus, people in your dormitory. The term *convenience sample* suggests that the selection is driven by convenience rather than by systematic sampling.

You might use *snowball sampling* if you know a lot of students in your major but not in other majors. In such a case, if you wanted a wider sample, you could ask a few people you do know in other majors to refer classmates from those majors. From those classmates, you could expand your reach still further into other majors. Your sample then expands like a snowball, building from a core group outward through recruitment. Researchers sometimes rely on snowball sampling when they are trying to access a group that is insular or difficult to reach, such as sex workers or drug addicts.

Nonrandom samples such as those gathered through convenience or snowball sampling can be suggestive of findings, but they are rarely generalizable by themselves and must be used with care.

In constructing surveys, sociologists are also concerned with ensuring that the questions and their possible responses will capture the respondents' points of view. The wording of questions is an important factor; poor wording can produce misleading results, as the following example illustrates. In 1993, an American Jewish Committee/Roper poll was taken to examine public attitudes and beliefs about the Holocaust. To the astonishment of many, the results indicated that fully 22% of survey respondents expressed a belief that the Holocaust had never happened. Not immediately noticed was the fact that the survey contained some very awkward wording, including the question "Does it seem possible or does it seem impossible to you that the Nazi extermination of the Jews never happened?" Can you see why such a question might produce a questionable result? The question's compound structure and double-negative wording almost certainly confused many respondents.

The American Jewish Committee released a second survey with different wording: "Does it seem possible to you that the Nazi extermination of the Jews never happened, or do you feel certain that it happened?" The results of the second poll were quite different. Only about 1% of respondents thought it was possible the Holocaust never happened, while 8% were unsure (Kagay, 1994). Despite the follow-up poll that corrected the mistaken perception of the previous poll's results, the new poll was not as methodologically rigorous as it could have been; a single survey question should ask for only one type of response. The American Jewish Committee's second survey contained a question that attempted to gauge two different responses simultaneously.

A weakness of surveys is that they may reveal what people *say* rather than what they *do*. Responses are sometimes self-serving, intended to make the interviewee look good in the eyes of the researcher. As noted earlier, social desirability bias is a response bias based on the tendency of respondents to answer a question in a way that they perceive will be favorably received. An example of this can be found in measures on voter turnout. Because voting is a socially desirable behavior, research suggests that self-reported voting behavior may not match up with actual voter turnout: That is, there is a tendency for people to say they voted in an election even if they did not (Presser, 1990). The respondent's bias toward choosing a response that they believe will be perceived as socially acceptable by the interviewer may also affect survey findings on political candidates or social issues. For example, in research conducted before the U.S. Supreme Court legalized same-sex marriage nationally, Powell (2013) found that there was a gap between public support expressed in pre-election surveys for local or state ballot initiatives legalizing same-sex marriage and actual voting-day support. He determined that "other things equal, election day opposition to same-sex marriage is between 5% and 7% greater than found in pre-election polls" (p. 1065). The wish to avoid stigma by voicing a position perceived to be socially acceptable to the interviewer may have an effect on survey responses to socially sensitive issues.

Question order can also affect survey findings in part because respondents have a desire to be consistent in their responses, making for a "consistency effect" (Schuman & Presser, 1981). A study (Wilson et al., 2008) on the issue of question order noted that "public opinion polls show that the public expresses greater support for gender-targeted AA [affirmative action] than race-targeted AA, but no research has addressed the extent to which expressed support for one group influences expressed support toward the other" (p. 514). The authors set out to determine if asking respondents about one or the other affirmative-action target group would affect their stated attitudes about the other. In fact, they found that question order affected responses. Specifically, respondents who were asked about affirmative action for women *first* were more likely to favor it than to oppose it: That is, about 63% supported affirmative action for women and 29% rejected it. When respondents were asked about affirmative action for women *after* being asked about such programs for racial minorities, support dropped: 57% supported affirmative action for women and 34% rejected it. Similarly, a greater percentage of respondents expressed support for racially targeted affirmative action when the question was asked after a question about affirmative action for women (57%) than when it was asked first (50%). The authors write that "results suggest that for the American public as a whole, support for one type of AA program is indeed affected by whether that program is considered by itself or in the context of both types of AA programs" (p. 518).

Constructing a survey that accurately represents attitudes and practices is challenging. A well-constructed survey, however, can overcome these problems. Awareness of pitfalls is important. Furthermore, assuring respondents of anonymity, assigning interviewers with whom respondents feel comfortable, and building in questions that ask for the same information in different ways can reduce bias in survey research.

Fieldwork

Fieldwork: A research method that uses in-depth and often extended study to describe and analyze a group or community; also called *ethnography.*

Fieldwork is *a research method that uses in-depth and often extended study to describe and analyze a group or community.* Sometimes called *ethnography,* it takes the researcher into the field, where he or she directly observes—and sometimes interacts with—subjects in their social environment.

Social scientists, including sociologists and anthropologists, have employed fieldwork to study everything from hoboes and working-class gangs in the 1930s (Anderson, 1940; Whyte, 1943) to prostitution and drug use among inner-city women (Maher, 1997) to Vietnam veterans motorcycling across the country to the Vietnam Veterans Memorial in Washington, D.C. (Michalowski & Dubisch, 2001), to poor families experiencing eviction (Desmond, 2016a).

Most fieldwork combines several different methods of gathering information. These include interviews, detached observation, and participant observation.

Interview: A detailed conversation designed to obtain in-depth information about a person and their activities.

An **interview** is *a detailed conversation designed to obtain in-depth information about a person and their activities.* When used in surveys, interview questions may be either open-ended or closed-ended. They may also be formal or informal. In fieldwork, the questions are usually open-ended to allow respondents to answer in their own words. Sometimes the interviewer prepares a detailed set of questions; at other times, the best approach is simply to have a list of relevant topics to cover.

Leading questions: Questions that tend to elicit particular responses.

Good researchers guard against influencing respondents' answers. In particular, they avoid the use of **leading questions**—*questions that tend to elicit particular responses.* Imagine a question on attitudes toward the marine environment that reads "Do you believe tuna fishing with broad nets, which leads to the violent deaths of dolphins, should be regulated?" The bias in this question is obvious—the stated association of broad nets with violent dolphin deaths creates a bias in favor of a *yes* answer. Accurate data depend on good questions that do not lead respondents to answer in particular ways.

Sociologists may use snowball sampling in their research. Snowball sampling involves using a core group of known respondents as sources to contact new respondents, expanding the core group outward like a snowball.
©iStockphoto.com/Nadezhda1906

Sometimes a study requires that researchers in the field keep a distance from the people they are studying and simply observe without getting involved. The people being observed may or may not know they are being observed. This approach is called *detached observation.* In his study of two delinquent gangs (the Saints and the Roughnecks), sociologist William J. Chambliss spent many hours observing gang members without being involved in what they were doing. With the gang members' permission, he sat in his car with the window rolled down so he could hear them talk and watch their behavior while they hung out on a street corner. At other times, he would observe them playing pool while he played at a nearby table. Chambliss sometimes followed gang members in his car as they drove around in theirs and sat near enough to them in bars and cafés to hear their conversations. Through his observations at a distance, he was able to gather detailed information on the kinds of delinquencies the gang members engaged in. He was also able to unravel some of the social processes that led to their behavior and observe other people's reactions to it (Chambliss, 1973).

Detached observation is particularly useful when the researcher has reason to believe other forms of fieldwork might influence the behavior of the people to be observed. It is also helpful for checking the validity of what the researcher has been told in interviews. A great deal of sociological information about illegal behavior has been gathered through detached observation.

One problem with detached observation is that the information gathered is likely to be incomplete. Without talking to people, we are unable to check our impressions against their experiences. For this reason, detached observation is usually supplemented by in-depth interviews. In his study of the delinquent gang members, Chambliss (1973, 2001) periodically interviewed them to complement his findings and check the accuracy of his detached observations.

Another type of fieldwork is *participant observation,* a mixture of active participation and detached observation. Participant observation can sometimes be dangerous. Chambliss's (1988b) research on organized crime and police corruption in Seattle, Washington, exposed him to threats from the police and organized crime network members who feared he would reveal their criminal activities. Desmond's (2016a) work on eviction included participant observation; he spent significant amounts of time with the Milwaukee residents he studied, seeking to carefully document their voices and experiences.

Experimentation

Experiments are *research techniques for investigating cause and effect under controlled conditions.* We construct experiments to measure the effects of **independent or experimental variables** (*variables the researcher changes intentionally*) on **dependent variables** (*variables that change as a result of changes in other variables*). To put it another way, researchers modify one controllable variable (such as diet or exposure to violent movie scenes) to see what happens to another variable (such as willingness to socialize, or the display of aggression). Some variables, such as sex, ethnicity, and height, do not change in response to stimuli and thus do not make useful dependent variables.

Experiments: Research techniques for investigating cause and effect under controlled conditions.

Independent or experimental variables: Variables the researcher changes intentionally.

Dependent variables: Variables that change as a result of changes in other variables.

In a typical experiment, researchers select participants who share characteristics such as age, education, social class, or experiences that are relevant to the experiment. The participants are then randomly assigned to two groups. The first, called the *experimental group,* is exposed to the independent variable—the variable the researchers hypothesize will affect the subjects' behavior. The second group is assigned to the *control group.* These subjects are not exposed to the independent variable—they receive no special attention. The researchers then measure both groups for the dependent variable. For example, if a neuroscientist wanted to conduct an experiment on whether listening to classical music affects performance on a math exam, he or she might have an experimental group listen to Mozart, Bach, or Chopin for an hour before taking a test. The control group would take the same test but would not listen to any music beforehand. In this example, exposure to classical music is the independent variable, and the quantifiable results of the math test are the dependent variable.

To study the relationship between violent video game play and aggression, researchers examined the sustained violent video game play and aggressive behavior of 1,296 Chinese adolescents ranging in age from 12 to 19 (Shao & Wang, 2019). Their results showed a strong correlation between playing violent video games and being more likely to engage in or approve of violence. Research in this field is complicated, however: Another study found that the level of competitiveness in a video game, and not the violence itself, had the greatest influence on aggressive behavior (Adachi & Willoughby, 2011). More research on this topic may help differentiate between the effects of variables and avoid conclusions based on spurious relationships.

When looking at the relationship between violent video games and violent behavior, researchers must account for many variables. What variables would you choose to study and why?

©Godong /UIG via Getty Images

Working With Existing Information

Sociologists frequently work with existing information and data gathered by other researchers. Why would researchers choose to reinterpret existing data? Perhaps they want to do a secondary analysis of statistical data collected by an agency such as the U.S. Census Bureau, which makes its materials available to researchers studying issues ranging broadly from education to poverty to racial residential segregation. Or they may want to work with archival data to examine the cultural products—posters, films, pamphlets, and such—used by an authoritarian regime in a given period to legitimate its power or disseminate it by a social movement such as the civil rights movement to spread its message to the masses.

Statistical data: Quantitative information obtained from government agencies, businesses, research studies, and other entities that collect data for their own or others' use.

Statistical data include *quantitative information obtained from government agencies, businesses, research studies, and other entities that collect data for their own or others' use.* The U.S. Bureau of Justice Statistics, for example, maintains a rich storehouse of information on several criminal justice social indicators, such as prison populations, incidents of crime, and criminal justice expenditures. Many other government agencies routinely conduct surveys of commerce, manufacturing, agriculture, labor, and housing. International organizations such as the United Nations and the World Bank collect annual data on the health, education, population, and economies of nearly all countries in the world. Many businesses publish annual reports that yield basic statistical information about their financial performance.

Document analysis: The examination of written materials or cultural products: previous studies, newspaper reports, court records, campaign posters, digital reports, films, pamphlets, and other forms of text or images produced by individuals, government agencies, private organizations, or others.

Document analysis is *the examination of written materials or cultural products: previous studies, newspaper reports, court records, campaign posters, digital reports, films, pamphlets, and other forms of text or images produced by individuals, government agencies, private organizations, and others.* Nevertheless, because such documents are not always compiled with accuracy in mind, good researchers exercise caution in using them. People who keep records are often aware that others will see the records and take pains to avoid including anything unflattering. The diaries and memoirs of politicians are good examples of documents that are invaluable sources of data but that must be interpreted with great caution. The expert researcher looks at such materials with a critical eye, double-checking with other sources for accuracy where possible.

This type of research may include historical research, which entails the analysis of historical documents. Often, such research is comparative, examining historical events in several different countries for similarities and differences. Unlike historians, sociologists usually identify patterns common to different times and places; historians tend to focus on particular times and places and are less likely to draw broad generalizations from their research. An early master of the sociological approach to historical research was Max Weber (1919/1946, 1921/1979), who contributed to our understanding of (among many other things) the differences between religious traditions in the West and those in East Asia.

Content analysis is the systematic examination of forms of documented communication. A researcher can take a content analysis approach by coding and analyzing patterns in cultural products such as music, laws, tweets, blogs, and works of art. An exciting aspect of social science research is that your object of curiosity can become a research question. In 2009, sociologists conducted a content analysis of 403 gangsta rap songs to assess whether rap's reputation of being misogynistic (hostile to women) was justified (Weitzer & Kubrin, 2009). The analysis found that although only about a fifth of the songs in the sample contained lyrics that were notable for their "objectification, exploitation, and victimization" of women (p. 25), most portrayals of women were still gender stereotypical and disempowering.

Participatory Research

Although sociologists usually try to avoid having an impact on the people they study, one research method is employed specifically to foster change. *Participatory research* supports an organization or community trying to improve its situation when it lacks the necessary economic or political power to do so by itself. The researcher fully participates by training the members to conduct research on their own while working with them to enhance their power (Freire, 1972; Whyte, 1991). Such research might be part of, for instance, empowering a community to act against the threat of HIV/AIDS, as has been done in places such as San Francisco, California, and Nairobi, Kenya. Participatory research is an effective way of conducting an empirical study while also furthering a community or organizational goal that will benefit from the results of the study.

DOING SOCIOLOGY: A STUDENT'S GUIDE TO RESEARCH

Sociological research seldom follows a formula that indicates exactly how to proceed. Sociologists often have to feel their way as they go, responding to the challenges that arise during research and adapting new methods to fit the circumstances. Thus, the stages of research can vary, even when sociologists agree about the basic sequence. At the same time, for student sociologists, it is useful to understand the key building blocks of good sociological research. As you read through the following descriptions of the stages (Figure 2.4), think about a topic of interest to you and how you might use that as the basis for an original research project.

Frame Your Research Question

"Good research," Thomas Dewey observed, "scratches where it itches." Sociological research begins with the formulation of a question or questions to be answered. Society offers an endless spectrum of compelling issues to study: Does exposure to violent video games affect the incidence of aggressive behavior in adolescents? Does religious faith affect voting behavior? Is family income a good predictor of performance on standardized college entrance tests such as the ACT or SAT? Beyond the descriptive

FIGURE 2.4 ■ Sociological Research Formula

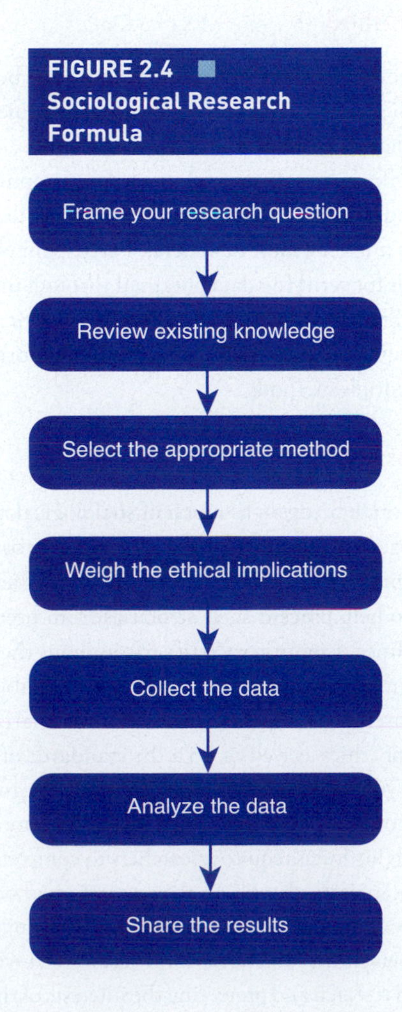

aspects of social phenomena, sociologists are also interested in how relationships between the variables they examine can be explained.

Formulating a research question precisely and carefully is one of the most important steps toward ensuring a successful research project. Research questions come from many sources. Some arise from problems that form the foundation of sociology, including an interest in socioeconomic inequalities and their causes and effects or the desire to understand how power is exercised in social relationships. Sociologists are also mindful that solid empirical data are important to public policies on issues of concern such as poverty, occupational mobility, and domestic violence.

Keep in mind that you also need to define your terms. Recall our discussion of operationalizing concepts. For example, if you are studying middle school bullying, you need to make explicit your definition of bullying and how that will be measured. The same holds true if you are studying a topic such as illiteracy or aggressive behavior.

Review Existing Knowledge

Once you identify the question you want to ask, you need to conduct a review of the existing literature on your topic. The literature may include published studies, unpublished papers, books, dissertations, government documents, newspapers and other periodicals, and increasingly, data disseminated on the Internet. The key focus of the literature review, however, is usually published and peer-reviewed research studies. Your purpose in conducting the literature review is to learn about studies that have already been done on your topic of interest so that you can set your research in the context of existing studies. You will also use the literature review to highlight how your research will contribute to this body of knowledge.

Select the Appropriate Method

Now you are ready to think about how your research question can best be answered. Which of the research methods described earlier will give the best results for the project and is most feasible for your research circumstances, experience, and budget?

If you wish to obtain basic information from a relatively large population in a short period of time, then a survey is the best method to use. If you want to obtain detailed information about a smaller group of people, then interviews might be most beneficial. Participant observation and detached observation are ideal research methods for verifying data obtained through interviews or, for the latter, when the presence of a researcher might alter the research results. Document analysis and historical research are good choices for projects focused on inaccessible subjects and historical sociology. Remember, sociological researchers often use multiple methods.

Weigh the Ethical Implications

Research conducted on other human beings—as much of sociological research is—poses certain ethical problems. An outpouring of outrage after the discovery of gruesome experiments conducted by the Nazis during World War II prompted the adoption of the Nuremberg Code, a collection of ethical research guidelines developed to help prevent such atrocities from ever happening again (Table 2.4). In addition to these basic guidelines, scientific societies throughout the world have adopted their own codes of ethics to safeguard against the misuse and abuse of human subjects.

Before you begin your research, it is important that you familiarize yourself with the American Sociological Association's code of ethics as well as with the standards of your school, and carefully follow both. Ask yourself whether your research will cause the subjects any emotional or physical harm. How will you guarantee their anonymity? Does the research violate any of your own ethical principles?

Most universities and research institutes require researchers to complete particular forms before undertaking experiments using human subjects, describing the research methods to be used and the groups of subjects who will take part. Depending on the type of research, a researcher may need to obtain written agreement from the subjects for their participation. Approval of research involving human subjects is granted with an eye to both fostering good research and protecting the interests of those partaking in the study.

TABLE 2.4 ■ The Nuremberg Code

	DIRECTIVES FOR HUMAN EXPERIMENTATION
1.	The voluntary consent of the human subject is absolutely essential.
2.	The experiment should be such as to yield fruitful results for the good of society.
3.	The experiment should be so designed and based on the results of animal experimentation and a knowledge of the natural history of the disease.
4.	The experiment should be so conducted as to avoid all unnecessary physical and mental suffering and injury.
5.	No experiment should be conducted where there is an *a priori* reason to believe that death or disabling injury will occur.
6.	The degree of risk to be taken should never exceed that determined by the humanitarian importance of the problem to be solved by the experiment.
7.	Proper preparations should be made and adequate facilities provided to protect the experimental subject against even remote possibilities of injury, disability, or death.
8.	The experiment should be conducted only by scientifically qualified persons.
9.	During the course of the experiment, the human subject should be at liberty to bring the experiment to an end.
10.	During the course of the experiment, the scientist in charge must be prepared to terminate the experiment at any stage if he has probable cause to believe, in the exercise of the good faith, superior skill, and careful judgment required of him, that a continuation of the experiment is likely to result in injury, disability, or death to the experimental subject.

Source: The Nuremberg Code. *United States Holocaust Memorial Museum.* https://www.ushmm.org/information/exhibitions/online-exhibitions/special-focus/doctors-trial/nuremberg-code

Collect and Analyze the Data

Collecting data is the heart of research. It is time-consuming but exciting. During this phase, you will gather the information that will allow you to contribute to the sociological understanding of your topic. If your data set is qualitative (for example, open-ended responses to interview questions or observations of people), you will proceed by carefully reviewing and organizing your field notes, documents, and other sources of information. If your data set is quantitative (for example, completed closed-ended surveys), you will proceed by entering data into spreadsheets, comparing results, and analyzing your findings using statistical software.

Your analysis should offer answers to the research questions with which you began the study. Be mindful in interpreting your data, and avoid conclusions that are speculative or not warranted by the actual research results. Do your data support or contradict your initial hypothesis? Or are they simply inconclusive? Report *all* of your results. Do your findings have implications for larger theories in the discipline? Do they suggest the need for further study of another dimension of the issue at hand? Good research need not have results that unequivocally support your hypothesis. A finding that refutes the hypothesis can be instructive as well.

Share the Results

However fascinating your research may be to you, its benefits are amplified when you take advantage of opportunities to share it with others. You can share your findings with the sociological community by publishing the results in academic journals. Before submitting research for publication, you must learn which journals cover your topic areas and review those journals' standards for publication. Some colleges and universities sponsor undergraduate journals that offer opportunities for students to publish original research.

During the Nuremberg Trials, which brought key figures of the Nazi Party of Germany to justice, the practices of some Nazi medical personnel were found to be unethical and even criminal. The Nuremberg Code, which emerged from these trials, established principles for any type of human experimentation.

Other outlets for publication include books, popular magazines, newspapers, video documentaries, and websites. Another way to communicate your findings is to give a presentation at a professional meeting. Many professional meetings are held each year; at least one will offer a panel suited to your topic. In some cases, high-quality undergraduate papers are selected for presentation. If your paper is one, relevant experts at the meeting will likely help you interpret your findings further.

WHAT'S NEXT? SOCIOLOGY AND THE WORLD ONLINE

In this chapter, you learned about the process of research in sociology. As you will see throughout this text, sociologists have endeavored to study a wide variety of subjects, from early efforts such as Émile Durkheim's classical study of suicide to contemporary studies of the relationship between video game play and aggression. The 2020s will open the door to new—and largely unprecedented—areas of research for sociologists. Consider the topic of young people spending time online. Over the last decade or so, survey research has documented a marked rise in the time children and teens spend engaged with social media, watching videos, playing games, or doing other activities online. About 45% of teens say they are online almost constantly (Pew Research Center, 2018a).

Researchers have also noted some disturbing trends with respect to esteem, depression and anxiety, and loneliness, which appear to be somewhat more acute among young people who spend a lot of time online. A study from the *Canadian Journal of Psychiatry* showed that the more time kids spent on digital screens, the more severe were their symptoms of depression and anxiety (Boers et al., 2020). While we know that correlation is not causation, data are suggestive of a relationship that invites further examination.

So what's new in the 2020s? The COVID-19 pandemic that saw the mass retreat of children and teens from daycares, K–12 schools, and colleges and universities in early 2020 has also seen the rapid rise in hours these young people are spending online, whether for education, interaction, or entertainment. Early data show that children ages 6–12 are spending at least 50% more time in front of screens daily, and traffic to kids' apps and digital services has increased by nearly 70% in the United States (Cheng & Wilkinson, 2020).

Before the pandemic, many scientists, medical professionals, and teachers were concerned about the time young people were spending online: Recommendations to get up, get out, and get off one's phone were more a norm than an anomaly. In the acute phase of the pandemic, this recommendation fell to the wayside and, as we just discussed, hours spent on a computer or smartphone grew precipitously. What will this mean for a generation of young people already steeped in technology? What will be the benefits? What about the costs? Consider what kinds of research questions sociologists may want to ask in the future and what methods will help to answer them. What would you like to know?

WHAT CAN I DO WITH A SOCIOLOGY DEGREE?

Quantitative Research Skills

Sociologists use quantitative research skills to conduct systematic empirical investigations of social phenomena using statistical methods. Quantitative research encompasses those studies in which data are expressed in terms of numbers. Important sources of quantitative data include surveys and observations. The objective of quantitative research in sociology is to gather rigorous data and to use numerical data to characterize the dimensions of an issue or the extent of a problem (this could include, for instance, the collection of statistical data on rates of obesity and poverty in neighborhoods or states and the calculation of the correlation of the two phenomena). Data may be used to develop or test hypotheses about the roots of a sociological phenomenon or problem.

Knowledge of quantitative methods is a valuable skill in today's job market. Learning to collect and analyze quantitative data, which is an important part of a sociological education, prepares you to do a wide variety of job tasks, including survey development, questionnaire design, market research, brand health tracking, and financial quantitative modeling and analysis.

Amber Henderson, Survey Statistician, U.S. Census Bureau

The George Washington University, MA in Sociology

I work in the Center for Survey Measurement as a statistician at the U.S. Census Bureau. The goal of the Census Bureau is to provide timely, accurate, and quality data while minimizing the various sources of survey error. When fielding a survey, it must go through all of the phases of what we call the survey life cycle. This includes tasks such as project planning, data collection, data analyses, and reporting. During my first year at Census, I used statistical software packages to manipulate, edit, and analyze data for surveys on education. Statistical software is a valuable tool for those who work with data. I used it frequently to run basic descriptive statistics and to check the data for error. For example, if a respondent gave a date of birth that indicated they were 12 years of age and listed his or her marital status as married, I would flag these data points for potential inconsistencies.

In my current role at Census, I do a lot more survey research where I specialize in structured cognitive interviewing and develop survey questions. The core sociology courses I took both during undergraduate and graduate school prepared me for my career at Census. I use a lot of what I learned in my courses on sociological research methods and data analysis to choose the best research method and work effectively and accurately with the Census Bureau's survey data. People often look puzzled when they learn you want to study sociology, but what they do not realize is that it's a multidimensional field. Sociology and my professors taught me both the qualitative and quantitative skills I needed to land my dream job. I wouldn't change a thing!

Career Data: Statisticians and Mathematicians

- 2019 Median Pay: $92,030 per year
- $44.35 per hour
- Typical Entry-Level Education: Master's degree
- Job Growth, 2019–2029: 33% (much faster than average)

Source: Bureau of Labor Statistics. (2019). *Occupational outlook handbook.*

SUMMARY

- Unlike commonsense beliefs, sociological understanding puts our biases, assumptions, and conclusions to the test.
- As a science, sociology combines logically constructed theory and systematic observation to explain human social relations.
- **Inductive reasoning** generalizes from specific observations; **deductive reasoning** consists of logically deducing the empirical implications of a particular theory or set of ideas.
- A good theory is logically consistent, testable, and valid. The **principle of falsification** holds that if theories are to be scientific, they must be formulated in such a way that they can be disproved if wrong.
- Sociological **concepts** must be operationally defined to yield measurable or observable variables. Often, sociologists operationally define **variables** so they can measure these in quantifiable values and assess **validity** and **reliability** to eliminate **bias** in their research.
- Quantitative analysis permits us to measure correlations between variables and identify **causal relationships**. Researchers must be careful not to infer causation from correlation.

- Qualitative analysis is often better suited than **quantitative research** to producing a deep understanding of how the people being studied view the social world. On the other hand, it is sometimes difficult to measure the reliability and validity of **qualitative research**.
- Sociologists seek **objectivity** when conducting their research. One way to help ensure objectivity is through the **replication** of research.
- Research strategies are carefully thought-out plans that guide the gathering of information about the social world. They also suggest the choice of appropriate **research methods**.
- Research methods in sociology include **survey** research (which often relies on random sampling), **fieldwork** (including participant observation and detached observation), **experiments**, working with existing information, and participatory research.
- Sociological research typically follows seven steps: framing the research question, reviewing the existing knowledge, selecting appropriate methods, weighing the ethical implications of the research, collecting data, analyzing data, and sharing the results.
- To be ethical, researchers must be sure their research protects the privacy of subjects and does not cause them unwarranted stress. Scientific societies throughout the world have adopted codes of ethics to safeguard against the misuse and abuse of human subjects.

KEY TERMS

bias (p. 33)
causal relationship (p. 31)
concepts (p. 30)
correlation (p. 31)
deductive reasoning (p. 27)
dependent variables (p. 39)
document analysis (p. 40)
experiments (p. 39)
fieldwork (p. 38)
hypotheses (p. 27)
independent or experimental variables (p. 39)
inductive reasoning (p. 27)
interview (p. 38)
leading questions (p. 38)
negative correlation (p. 33)
objectivity (p. 34)
operational definition (p. 30)
population (p. 35)
positive correlation (p. 32)
principle of falsification (or falsifiability) (p. 33)
qualitative research (p. 27)
qualitative variables (p. 31)
quantitative research (p. 27)
quantitative variables (p. 31)
random sampling (p. 36)
reliability (p. 33)
replication (p. 35)
research methods (p. 35)
sample (p. 35)
scientific method (p. 27)
scientific theories (p. 30)
social desirability bias (p. 34)
spurious relationship (p. 31)
statistical data (p. 40)
stratified sampling (p. 36)
survey (p. 35)
validity (p. 33)
value neutrality (p. 34)
variable (p. 31)

DISCUSSION QUESTIONS

1. Think about a topic of contemporary relevance that interests you (for example, poverty, juvenile delinquency, teen births, or racial neighborhood segregation). Using what you learned in this chapter, create a simple research question about the topic. Match your research question to an appropriate research method. Share your ideas with classmates.
2. What is the difference between quantitative and qualitative research? Give an example of each from the chapter. In what kinds of cases might one choose one or the other research method to effectively address an issue of interest?

3. Sociologists often use interviews and surveys as methods for collecting data. What are potential problems with these methods of which researchers need to be aware? What steps can researchers take to ensure that the data they are collecting are of good quality?

4. Imagine that your school has recently documented a dramatic rise in plagiarism reported by teachers. Your sociology class has been invited to study this issue. Consider what you learned in this chapter about survey research and design a project to assess the problem.

5. As discussed in the chapter, researchers have used experiments to test the relationship between playing violent video games and aggressive behavior. Results of such experiments are mixed, as some suggest a significant correlation between sustained play and a willingness to engage in or approve of aggressive behavior, while others do not show a clear relationship, or attribute the relationship to another variable—the level of competitiveness of the games. How would you design an experiment to test this relationship? What ethical challenges might experiments testing this relationship face? How would you ensure that your experiment conforms to the ethics established for human subjects research?

CULTURE AND MASS MEDIA

WHAT DO YOU THINK?

1. Do we act in ways that are inconsistent with our cultural beliefs and values? Why or why not?
2. Does graphic violence in films, television, music, and video games have an effect on youth attitudes and behaviors? What about adults?
3. Do we share a "global" culture? If so, what are its key characteristics?

LEARNING OBJECTIVES

3.1 Define the component parts of culture, including beliefs, values, norms, and taboos.

3.2 Recognize the importance of language in representing and creating culture.

3.3 Discuss the relationship between culture and mass media and the debate over mass culture and violence.

3.4 Explain how sociologists theorize the relationship between culture and social class.

3.5 Apply functionalist and conflict perspectives to the phenomenon of global culture.

MONSTERS AS CULTURAL SYMBOLS

Americans are intensely fascinated with all types of monsters—zombies, vampires, even scary clowns like the iconic dancing Pennywise in Stephen King's *It*. Our fascination is deeply reflected in American culture: the shows we watch, the books we read, the video games we play, and the way we celebrate our holidays. For instance, many of us (up to 17.3 million at one point) have watched Rick and his compatriots in *The Walking Dead* spend a decade fighting for survival against the feared "walkers" (zombies) who hunt them (Bukszpan, 2019). Millions have been entertained by Stephen King's monsters and supernatural beings—as of 2016, King had written over 50 such books and sold over 350 million copies (Heller, 2016). And he has published over 13 new titles since then! More than half of Americans said in 2019 that they celebrate Halloween, spending over $8.8 billion on Halloween-related goods, with over 2 million adults dressing as vampires and 1.4 million costuming as zombies (National Retail Federation, 2019). Our fascination with monsters is such that it has even spawned an interdisciplinary academic field called "monster studies."

Why would we begin a chapter on culture with monster stories? It turns out that our monsters are inherently cultural products with something to say about the societies from which they arise. For instance, in the words of a recent article that quotes a foundational text of monster studies: "[M]onsters [are] a product of culture, functioning as signs and symbols for social problems, and [demand] to be acknowledged on account of their mere presence" (Erle & Hendry, 2020, p. 3, citing Cohen). When Mary Shelley published *Frankenstein* in 1818, she was writing during a time of great social and industrial progress. Shelley not only birthed Dr. Frankenstein and his monster, she also revealed cultural ideas and fears about technology, science, and progress itself: Who exactly was the monster in the story, the creation or the creator?

Zombie popularity, according to Kyle W. Bishop (2010), has risen after traumatic societal calamities: In the United States, this included such events as the terrorist attacks on New York City and Washington, D.C., on September 11, 2001; the disease fears generated by deadly outbreaks of viruses such as SARS; and even Hurricane Katrina. Zombie stories can resonate with a public that is anxious about the threat of societal disaster, whether natural or human-made. Zombies evoke a fear response, although the object of fear is not necessarily the zombie itself: "Because the aftereffects of war, terrorism, and natural disasters so closely resemble the scenarios of zombie cinema . . . [these films have] all the more power to shock and terrify a population that has become otherwise jaded by more traditional horror films" (Bishop, 2009, p. 18).

The annual Halloween festival with monsters hanging from buildings and a spooky bones parade through New Cathedral Street in Manchester, UK, in October 2019.

©Barbara Cook/Alamy Live News

For social theorist Michel Foucault, monsters also are a reflection of our norms and values, a way to differentiate between who we would consider as "normal, abnormal, and sexually deviant" (Foucault, 2005, cited in Erle & Hendry, 2020, p. 3). Monsters draw the boundaries between what is human and what is "other," which can also reflect our fears of other cultures, and those from different nations, ethnicities, and sexualities. In short, to sociologists, all cultural products are more than entertainment—they are a mirror of society. Popular culture in the form of films, television, or video games may capture our utopian dreams, but it is also a net that catches and reflects pervasive societal fears and anxieties. Monsters may provide us cultural catharsis, but when we apply our sociological imagination to the grotesque and frightening, we might also find new topics for research and enhanced understanding.

In this chapter, we consider culture and media (a key vehicle of culture), and we illustrate how culture both constructs and reflects society in the United States and around the globe. We begin our discussion by exploring the basic concept of *culture*, its material and nonmaterial components, and its ideal and real aspects. We then look at contemporary issues of language in a changing world. We also examine how media messages reflect and affect behaviors and attitudes. We investigate the sociological question of whether culture and taste are linked to class identity and social reproduction. Finally, we examine the evolving relationship between global and local cultures, in particular, the influence of U.S. mass media on the world.

CULTURE: CONCEPTS AND APPLICATIONS

What do you think of when you see the word *culture*? Perhaps you envision songs, foods, dances, or festivals. Or maybe you think of life in a different region of the world where the language and styles of dress differ from yours. Culture, from a sociological perspective, is all of this and more. Sociologists define **culture** as *the beliefs, norms, behaviors, rituals, languages, and collective identities and memories developed by members of a particular group that bring meaning to their social worlds.* They point out that culture is essential to our life experiences; it helps us form our ideas of who we are and influences how we think, to whom we speak, and many other facets of our identities and interactions.

Culture: The beliefs, norms, behaviors, languages, and products common to the members of a particular group that bring meaning to their social worlds.

Culture also shapes and permeates our lives from the ways in which we are born, live, and die, to the homes we live in, the objects we use daily, and the languages we enlist in conversation, poetry, stories, and music. It also creates collective memories that large groups of people share. As humans, we create culture while simultaneously stepping into a culture that has already been created for us. And culture *makes us human* in apparent and subtle ways. Importantly, it makes our lives *meaningful* (ASA statement on culture, 2020, available at https://www.asanet.org/topics/culture). In this section we provide the sociological tools to help you uncover the components of culture, to explore some facts about familiar and not-so-familiar cultural groups, and to consider the impact of culture in your life and the lives of others.

Material and Nonmaterial Culture

Like many students, you may be reading this chapter on your smartphone while pausing to check social media, send a text, and shop or make an appointment. Today, smartphones are an important part of what sociologists refer to as **material culture** or *the physical objects that are created, embraced, or consumed by society that help shape people's lives.*

Material culture: The physical objects that are created, embraced, or consumed by society that help shape people's lives.

Material culture emerges from the physical environment we inhabit. Take, for instance, the types of shelters that characterize a community. In seaside communities, it is common to see elevated homes built on stilts or piles to protect against flooding. The materials used to construct homes have historically been those available in the immediate environment—wood, thatch, or mud. Importantly, the global trade in timber, marble, granite, and modern housing materials has transformed the relationship between place and shelter in many countries, allowing people to create structures that amalgamate the products of many cultures.

Nonmaterial culture: The abstract creations of human cultures, including language, ideas about behavior, and social practices.

To understand culture sociologically, you also must understand its **nonmaterial culture**, or *the abstract creations of humans, including language, ideas about behavior, and social practices.* Nonmaterial culture encompasses aspects of the social experience, such as beliefs, norms, values, language, family forms, and institutions. It also reflects the natural environment in which a culture has evolved.

Nonmaterial and material culture are intertwined, with nonmaterial culture often providing the meaning to material culture objects. Returning to our previous example, we should point out that homes not only reflect available material objects for building, but they also reflect our values, norms, and beliefs. So, *why and how* we build—size, design, and location—is inherently cultural. Take the typical American house. In 1973, according to Kilman (2016), the average size of a new home in the United States was 1,660 square feet. But, in 2018, the average size had grown to over 2,386 square feet (U.S. Census Bureau, 2018). Beginning in the early 2000s, though, some Americans had rejected the environmental impact of these increasingly bigger homes, joining a "tiny house movement" of small homes ranging from about 400 to 700 square feet in size, or nearly four times smaller than the big homes being built. The tiny home movement, then, is not so much about material culture as it is evidence of how nonmaterial culture shapes it: "The main assumption of the tiny house movement is that homeowners can reduce the environmental impact and increase affordability by reducing their spatial footprint" (Ford & Gomez-Lanier, 2017, p. 394).

Tiny homes range from 400 to 700 square feet. Tiny houses are based on a global movement meant to decrease the environmental impact of housing. While initially designed to decrease consumption and to be affordable, advertising for tiny homes shows an increasing emphasis on words such as *chic*, *cute*, and *luxurious*.

When you think sociologically about your own living arrangements, how would you say your home reflects both material and nonmaterial culture? And, returning to our earlier example, how does an item that you use consistently—your smartphone—also reflect both? Do you think that nonmaterial culture influences material culture, or vice versa?

TABLE 3.1 ■ Cultural Concepts and Characteristics

CONCEPT	CHARACTERISTICS
Beliefs	Particular ideas that people accept as true
Values	General ideas about what is good, right, or just in a culture
Norms	Culturally shared rules governing social behavior (oughts and shoulds)
Folkways	Conventions (or weak norms), the violation of which is not very serious
Mores	Strongly held norms, the violation of which is very offensive
Taboos	Very strongly held norms, the violation of which is highly offensive and even unthinkable
Laws	Norms that have been codified

Beliefs

Beliefs: Particular ideas that people accept as true.

What do you believe and why? Sociologists broadly define **beliefs** as *particular ideas that people accept as true.* Our beliefs can be based on many phenomena: for example, personal experience, faith, superstition, science, tradition, or our social groups. Importantly, beliefs do not need to be objectively true, but, to paraphrase the words of sociologists W. I. Thomas and Thomas (1928), beliefs may be understood as *real* when they are real in their consequences. For example, during the witch hunts in early colonial America, the shared belief in the existence of witches led to rituals of accusation, persecution, and execution in communities such as Salem, Massachusetts. From 1692 through 1693, more than 200 people were accused of practicing witchcraft; of these, 20 were executed, 19 by hanging and 1 by being pressed to death between heavy stones.

Moving forward nearly 300 years to the late 20th century, in the 1980s a "Satanic Panic" arose in the United States, a set of beliefs driven by mass media alleging that organized satanic ritual abuse was occurring in preschools and other sites. This concern soon extended to allegedly satanically connected companies, groups, and popular culture. A scared public even linked cartoons like *The Smurfs* and *Transformers* and *Thundercats* to satanism (Romano, 2016). This shared set of beliefs inspired numerous accusations, massive law enforcement training, criminal investigations, therapeutic interventions, international conventions, an entire cultural industry of panic, and a number of accusations, arrests, and even jail sentences (Beck, 2015; Romano, 2016). Today, similar beliefs about forces have arisen in Internet-generated panics around international child pedophile rings. In July 2020, for instance, the retail company Wayfair was hit with untrue rumors originating on the Internet site Reddit that, under the cover of selling expensive cupboards, Wayfair was instead actually engaged in child trafficking (Evon, 2020).

Importantly, beliefs, like other aspects of culture, are dynamic rather than static: When belief in the existence of witchcraft waned, so did the witch hunts. In 1711, a bill was passed that restored "the rights and good names" of those who had been accused, and in 1957, the state of Massachusetts issued a formal apology for the events of the past (Blumberg, 2007). So, too, the "Satanic Panics" waned in the mid-1990s, as state and federal investigations found little evidence to support the allegations of systematic abuse (Romano, 2016). And Snopes, an Internet site that checks the veracity of emergent stories, quickly tamped down the Wayfair cupboard story. These examples show an extreme set of beliefs around good and evil that come to take on a life of their own. Many beliefs, however, are benign. Thinking about your experience, can you identify other beliefs in society that have been real in their consequences? What was the basis for these beliefs?

Values

Values: The abstract and general standards in society that define ideal principles, such as those governing notions of right and wrong.

Have you ever been asked to share your personal values or to identify the core values of a group to which you belong? To sociologists, **values** are a significant nonmaterial component of culture. We define them as *the abstract and general standards in society that define ideal principles, such as those governing notions of right and wrong.* Values may unite or divide us. We often attach values to societal institutions such as the state (national or patriotic values), the community (local values), and families (familial values). Because we use values to legitimate and justify our behavior as members of a country or community or as individuals, we also tend to staunchly defend them (Kluckhohn & Strodtbeck, 1961).

Structural functionalists including Talcott Parsons (1951) propose that values play a critical role in social integration, bringing people together around shared ideals. Do Americans, then, have a specific set of shared values that unite them?

Sociologists often study what people say they value through survey research. According to a classic study by Robin M. Williams Jr. (1970), "American values" include personal achievement, hard work, material comfort, and individuality. U.S. adults value science and technology, efficiency and practicality, morality and humanitarianism, equality, and "the American way of life." Some of these values are crystalized in what has been called the "American Dream," a general belief that "hard work, financial security, and career success" will result in "each new generation being better off than the one before it" (Lopez et al., 2018). A 2017 Pew Research Center survey examined the question of what respondents value as part of the American Dream: Interestingly, an "essential" part of the American Dream cited by 77% of respondents was "freedom of choice in how to live" (Figure 3.1; Smith, 2017).

FIGURE 3.1 ■ Views About the American Dream, 2017

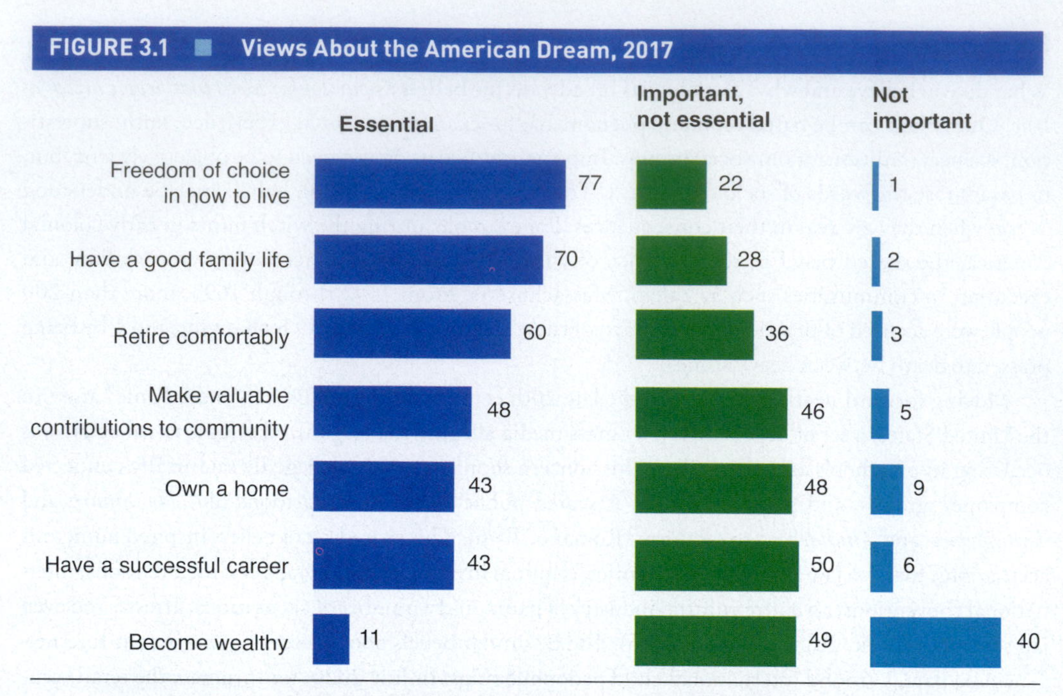

Source: Smith, S. (2017, October 31). *Most think the "American dream" is within reach for them.* Pew Research Center. http://www.pewresearch.org/fact-tank/2017/10/31/most-think-the-american-dream-is-within-reach-for-them/ft_17-10-31_americandream_definitions/

Conflict theorists, on the other hand, point out that power often resides behind seemingly universal phenomena. So, a conflict theorist might also point out that although Americans hold shared values that transcend race, ethnicity, gender, class, and other demographic characteristics, there are also stark and growing differences that emerge around them. Scholars, while recognizing the widespread cultural adoption of the American Dream, have also critiqued its achievability, pointing out that, cultural values aside, intersectional structural issues such as racism and class inequalities block access to achieving it (Ehrenreich, 2001; Kendi, 2020).

Looking intersectionally at American values allows us to explore more closely how race and ethnicity, and the life experiences associated with race, might impact beliefs in such a dream (Lopez et al., 2018). For instance, a 2018 Pew study of Latinos shows that they are *more* likely than other Americans to believe that they can get ahead with hard work and determination, but that they also reported experiencing it as hard to achieve (Figure 3.2; Lopez et al., 2018). Finally, a 2019 Pew survey showed some racial differences emerging in attitudes about the possibilities in America and the American Dream: Whites were becoming less optimistic, seeing an America in decline, while Blacks and Hispanics were more becoming more optimistic (Figure 3.3; Parker et al., 2019). How might a sociologist interpret these differences over time and between racial groups?

FIGURE 3.2 ■ Latino/a Views About the American Dream, 2018

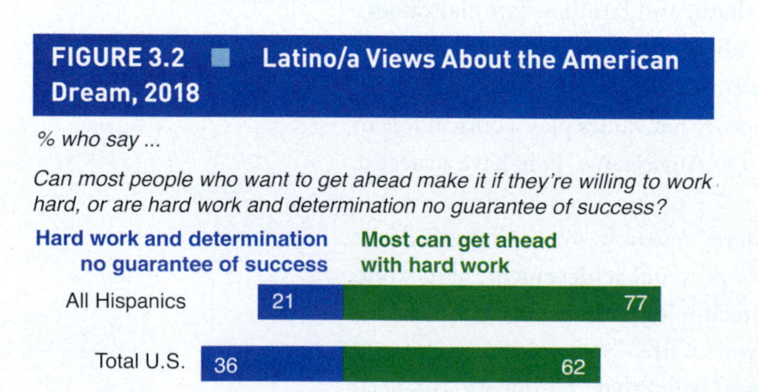

Source: Lopez, M. H., Gonzalez-Barrera, A., & Krogstad, J. M. (2018, September 11). *Latinos are more likely to believe in the American dream, but most say it is hard to achieve.* Pew Research Center. https://www.pewresearch.org/fact-tank/2018/09/11/latinos-are-more-likely-to-believe-in-the-american-dream-but-most-say-it-is-hard-to-achieve/

In the 1980s a political "culture war" emerged around issues ranging from social welfare to traditional family values (Hunter, 1992). This political division has continued to grow and divide Americans. For instance, according to Parker (2012), between 1987 and 2012, there was a dramatic split between the share of Republicans, Democrats, and independents who agreed that "the government should take care of people who can't take care of themselves" (see Figure 3.4). A 2017 study suggests one possible explanation for this split: Republican respondents are likely to attribute poverty to a lack of effort, while their Democratic counterparts are more likely to attribute it to circumstances beyond a person's control (see Figure 3.5; Smith, 2017).

By the spring and summer of 2020 sharp partisan splits had extended to the way people viewed public health. While the infection rate and death toll of the COVID-19 virus continued to climb, a Pew survey conducted in July 2020 showed that the vast majority of Democrats (85%) and those who leaned Democrat viewed COVID-19 as a major threat to public health. Conversely, only 46% of those who identified as Republican or Republican-leaning viewed COVID-19 as a major threat (Tyson, 2020, para. 1).

Norms

In any culture, a set of ideas exists about what is right and wrong, just and unjust. Émile Durkheim referred to such ideas as "morality," an invisible but real social fact that he saw as "coercive on human behavior" (Durkheim, 1973a). In the 1970s sociologist Robert Nisbet, drawing on similar ideas, said, "The moral order of society is a kind of tissue of 'ought's': negative ones which forbid certain actions and positive ones which [require certain] actions" (p. 226). Today, sociologists refer to these ideas of "ought" and "ought not" as norms, *the accepted social behaviors and beliefs* that spring from culture and guide how we choose to behave: where to stand relative to others in an elevator, how long to hold someone's gaze in conversation, how to celebrate or grieve, why to marry, and how to resolve disagreements or conflicts. Some norms are enshrined in laws; others are inscribed in our psyches and consciences.

To provide an example of elements of material and nonmaterial culture we have thus far explored, we turn to weddings. These ceremonies reflect and create culture, and they are widely reflected in popular culture. Weddings are also significant economically: The term *wedding industrial complex* (Ingraham, 1999) has been used to describe a massive industry in the United States that generates over $72 billion in revenue and employs over a million people (Schmidt, 2017). This comes as little surprise when we consider that in 2019, the estimated average amount spent on a wedding was just over $24,000 (Wedding Report, 2019). Many weddings involve significant consumerism. Advertisers often enlist wedding images to promote products from cosmetics to furniture to major purchases such as cars, homes, and luxurious vacations.

FIGURE 3.3 ■ Pessimism About the American Dream, 2019

% saying ...

When the public looks to the future of the U.S. over the next 30 years, they see ...

A country declining in stature on the world stage

A widening gap between the haves and the have-nots

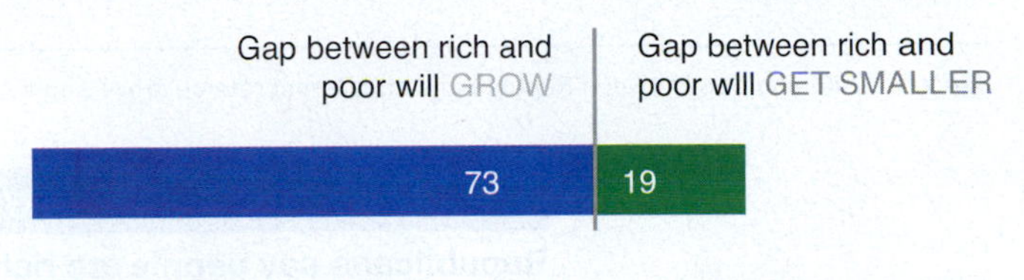

Growing political polarization

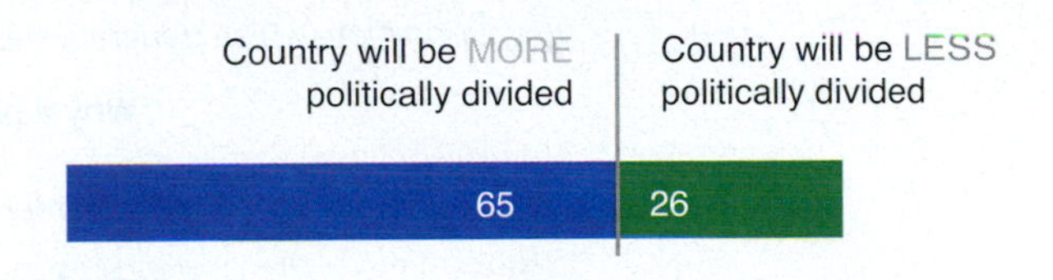

And they are worried that the country's political leaders are not up to the challenge

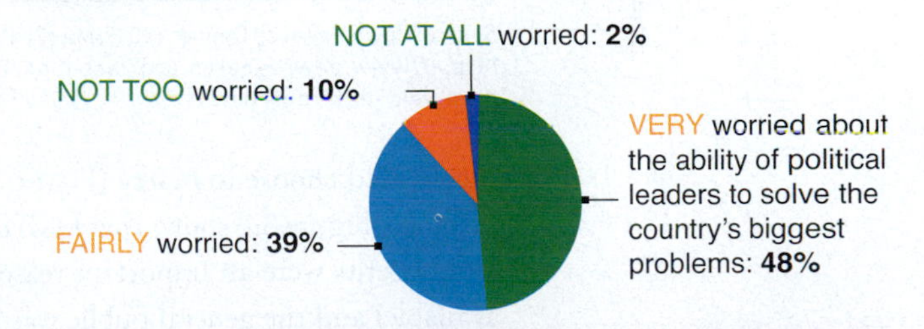

Source: Parker, K., Morin, R., & Horowitz, J. (2019, March 19). *Looking to the future, public sees an America in decline on many fronts*. Pew Research Center. https://www.pewsocialtrends.org/2019/03/21/public-sees-an-america-in-decline-on-many-fronts/

Weddings are also a major part of popular culture. Popular movies and television shows often feature weddings (for example, *Crazy Rich Asians* [2018], *Four Weddings and a Funeral* [2019], and *Love, Wedding, Repeat* [2020]) and fodder for reality shows. For example, *Say Yes to the Dress* enthralls viewers with the drama of choosing a wedding gown and *Four Weddings* pits four brides against one another to pull off the "perfect wedding," while *90 Day Fiancé* follows long-distance couples who must decide whether or not to wed before the foreign partner's visa expires. Mass media images can also reflect or influence public attitudes. Before the U.S. Supreme Court legalized same-sex marriage in 2015, public attitudes had already shifted to be much more accepting. According to one poll, watching television programs such as *Modern Family*, which prominently features a gay couple, made some viewers more supportive (Appelo, 2012).

Weddings reflect beliefs and values. For example, many American couples, regardless of sexuality, share the beliefs in love, companionship, and making a lifelong commitment as the top reasons why

FIGURE 3.4 ■ Political Party Affiliation and Support for Social Welfare

Percent who agree that the government should . . .

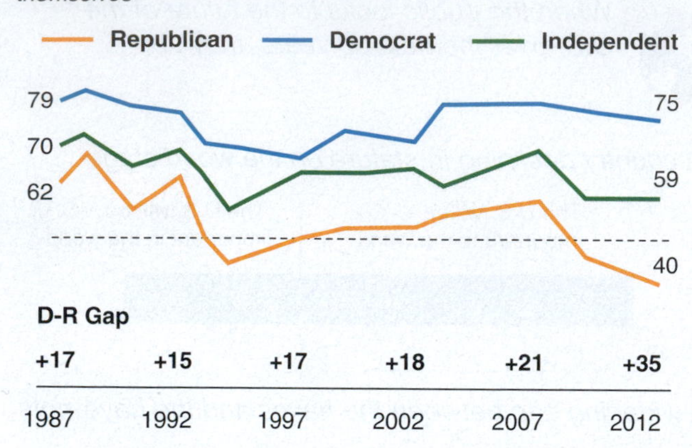

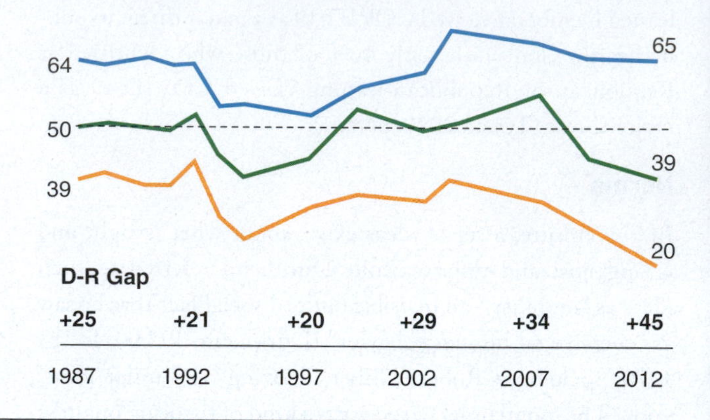

Source: "Widening Gap Between Republicans and Democrats on Why People Are Rich and Poor." Pew Research Center, Washington, DC, May 2, 2017.

FIGURE 3.5 ■ Political Party Affiliation and Attitude Toward Wealth and Poverty

Republicans say people are rich – and poor – because of their own efforts; Democrats more likely to point to a person's circumstances and advantages

In your opinion, which generally has more to do with...

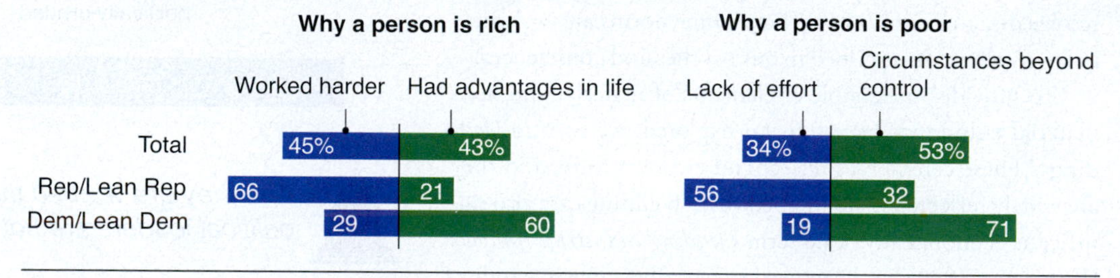

Source: Pew Research Center. (2017, May 2). *Widening gap between Republicans and Democrats on why people are rich and poor.* https://www.pewresearch.org/fact-tank/2017/05/02/why-people-are-rich-and-poor-republicans-and-democrats-have-very-different-views/

they would choose to marry (Figure 3.6; Masci et al., 2019). Importantly, differences also exist—for instance, Figure 3.6 shows that LGBTQ+ Americans in 2013 were twice as likely to say that legal rights and benefits were an important reason to marry (the latest survey for which data of this nature were available) and the general public was almost twice as likely to say that having children were (Masci et al., 2019). Consider, however, that while this is the latest survey from Pew that regards attitudes by sexuality on marriage, indeed, this survey actually predated the Supreme Court's 2015 ruling allowing gays and lesbians to legally marry. Do you think that this decision may make a difference in LGBTQ+ responses to reasons to marry? Why or why not?

It is also important to note that the top three values that we identified are values that are not universal worldwide—for instance, arranged marriages orchestrated by parents or matchmakers are still very common in many countries (and occur within the United States as well). We can also connect weddings with folkways, mores, taboos, and laws.

Folkways: Fairly weak norms that are passed down from the past, the violation of which is generally not considered serious within a particular culture.

Sociologist William Graham Sumner (1906–1959) distinguished among several different kinds of norms (folkways and mores), each of which can be applied to weddings. **Folkways** are *fairly weak norms that are passed down from the past, the violation of which is generally not considered serious within a particular culture.*

A historical folkway of many U.S. wedding rituals is the "giving away" of the bride: The father of the bride symbolically "gives" his daughter to the groom, signaling a change in the woman's identity

FIGURE 3.6 ■ Contrasting LGBT and General Public Responses to Why Get Married?

	WHY GET MARRIED? % SAYING EACH IS A...		
	VERY IMPORTANT REASON	SOMEWHAT IMPORTANT REASON	NOT AN IMPORTANT REASON
Love			
LGBT	84	12	4
General public	88	9	2
Companionship			
LGBT	71	24	5
General public	76	19	3
Making a lifelong commitment			
LGBT	70	24	5
General public	81	14	4
For legal rights and benefits			
LGBT	40	36	17
General public	23	38	37
Financial stability			
LGBT	35	44	31
General public	28	48	22
Having children			
LGBT	28	41	31
General public	49	30	19
Having a relationship recognized in a religious ceremony			
LGBT	17	29	53
General public	30	33	35

Source: Masci, D., Brown, A., & Kiley, J. (2019, June 24). *5 facts about same-sex marriage.* Pew Research Center. https://www.pewresearch.org/fact-tank/2019/06/24/same-sex-marriage/

Mores: Strongly held norms, the violation of which seriously offends the standards of acceptable conduct of most people within a particular culture.

from daughter to wife. Some couples today reject this ritual as patriarchal because it recalls earlier historical periods when a woman was treated as property given—literally—to her new husband by her previous keeper, her father. Some couples also now choose to walk down the aisle together to signal an equality of roles and positions. Although the sight of a couple going to the altar together might raise a few eyebrows among more traditional guests, this violation of the "normal" way of doing things does not constitute a serious cultural transgression. Because culture is dynamic, it may in time become a folkway, itself.

Mores (pronounced "MOR-ays") are *strongly held norms, the violation of which seriously offends the standards of acceptable conduct of most people within a particular culture.* In a typical American wedding, the officiant plays an important role in directing the events, and the parties enacting the ritual are expected to respond conventionally. For instance, when the officiant asks the guests whether anyone objects to the union, the standard of conduct is for no one to object. If an objector surfaces (more often in television programs and films than in real life), the response of the guests is shock and dismay: The ritual has been disrupted and the scene violated.

The marriage of Prince Harry and Meghan Markle in May 2018 captured worldwide attention. Prince Harry and Duchess Meghan also broke with long-standing royal norms by "stepping back" from public duties as working royals.

Taboos: Powerful mores, the violation of which is considered serious and even unthinkable within a particular culture.

Laws: Codified norms or rules of behavior.

Ideal culture: The values, norms, and behaviors that people in a society profess to embrace.

Real culture: The values, norms, and behaviors that people in a society actually embrace and exhibit.

Taboos are *powerful mores, the violation of which is considered serious and even unthinkable within a particular culture.* The label of taboo is commonly reserved for behavior that is extremely offensive: Incest, for example, is a nearly universal taboo. There may not be any taboos associated with the wedding ritual itself in the United States, but there are some relating to marital relationships. For instance, while in some U.S. states it is not illegal to marry a first cousin, in most modern communities, doing so violates a taboo against intermarriage in families.

Laws are *codified norms or rules of behavior.* Laws formalize and institutionalize society's norms. Continuing with our example, marriages are also legal acts requiring an authorized officiant and, for most states, a registered license. Laws regulate who can and cannot marry. In the 20th century two important Supreme Court decisions around race and sexuality formalized changing marital norms. For instance, antimiscegenation laws institutionalized racial discrimination and segregation by prohibiting interracial marriage in many parts of the United States until in 1967 the Supreme Court ruled in *Loving v. Virginia* that these marriage bans violated the 14th Amendment's Equal Protection and Due Process Clauses. Until very recently, marriage was legally open only to heterosexual adults. Over time, however, the normative climate shifted, and a majority of Americans expressed support for same-sex marriage. In June 2015, the Supreme Court of the United States ruled in *Obergefell v. Hodges* that state-level bans on same-sex marriage also violated the 14th Amendment's Equal Protection and Due Process Clauses. Today, marriage is legally open to heterosexual and gay couples, although there have been instances of county clerks in some states refusing to grant marriage licenses to same-sex couples because they claim it violates their beliefs.

Before the U.S. Supreme Court legalized same-sex marriage in 2015, public attitudes had already shifted to be much more accepting. According to one poll, watching television programs such as *Modern Family*, which prominently featured a gay couple, made some viewers more supportive (Appelo, 2012).

©AF archive/Alamy Stock Photo

What beliefs, values, and rituals do you see reflected in the weddings you have attended? The weddings you have watched on television? What norms are upheld in weddings that you have attended? What, if any, norms were violated? How are weddings associated with consumption?

Ideal and Real Culture in U.S. Society

Beauty is only skin deep. Don't judge a book by its cover. All that glitters is not gold. These three messages, often reflected in U.S. culture, represent a commitment of sorts that society will value our inner qualities more than our outward appearances. They are also examples of **ideal culture**, *the values, norms, and behaviors that people in a given society profess to embrace*, even though the actions of the society may often contradict them. The importance of a certain type of beauty in American culture is promoted from childhood onward in the stories told by our parents, teachers, and the media.

Sociologists often use the research method of content analysis to understand messages contained in material culture. Their careful analysis of the images and language of shows have shown that iconic Disney stories such as *Snow White, Cinderella*, and *Sleeping Beauty* connect beauty with morality and goodness (and, white characters) and unattractiveness with malice, jealousy, and other negative traits. The link between unattractive (or unconventional) appearance and unattractive behavior is unmistakable, especially in female figures. Consider other characters many American children are exposed to early in life, such as nasty Cruella de Vil in *101 Dalmatians*, Maleficent in *Sleeping Beauty*, and the angry octopod Ursula in *The Little Mermaid*.

The ugly stepsisters, Anastasia and Drizella, from the story of Cinderella are only two of many children's story characters who combine an unattractive appearance with flawed personalities. How do we reconcile the idea that "beauty is only skin deep" with the images of popular culture?

©Moviestore collection Ltd/Alamy Stock Photo

On the other hand, **real culture** consists of *the values, norms, and behaviors that people in a given society actually embrace and exhibit.* In the United States, for instance, interviews, surveys, and participant observation have shown that conventional attractiveness offers consistent advantages in the workplace (Hamermesh,

2011; Marlowe et al., 1996; Shahani-Denning, 2003; Tews et al., 2009), the courtroom (Ahola et al., 2009; DeSantis & Kayson, 1997; Gunnell & Ceci, 2010; Taylor & Butcher, 2007), and the classroom (Chia et al., 1998; Poteet, 2007).

On television (another medium that disseminates important cultural stories), physical beauty and social status are powerfully linked, and characters are often young, white, heterosexual, wealthy, and able-bodied. Since most of us do not fit all of those categories, its viewers may experience feelings of inadequacy as they follow their favorite celebrities on Instagram, watch popular series on television or Netflix, or leaf through ads in popular magazines.

Mulan, an action movie, was released directly to Disney+ to stream in September 2020. After accusations of "whitewashing" characters and an online petition to feature Asians in Asian roles that generated thousands of signatures, Disney selected Chinese actor Liu Yifei to play Mulan (https://www.bbc.com/news/world-asia-china-42177082).

©David M. Benett/Dave Benett/WireImage

Using our sociological imagination, we can deduce that the weight concerns many people—particularly women—experience as personal troubles are in fact linked to public issues. For instance, worrying about (and even obsessing over) weight is a widely shared phenomenon. Millions of women diet regularly, and some manifest extreme attention to weight in the form of eating disorders (National Institute of Mental Health, 2020). In addition to dieting, social media allows us another venue to manipulate our body presentation. Photo editing applications allow us to capture the perfect angle and alter our image to fit society's body ideal in only a few clicks, something that used to take professionals hours to do—or that could not be done at all (Cosslett, 2016). According to one report, up to 90% of teen girls have edited pictures to appear thinner or even to avoid cyberbullying (Bingham, 2015). One researcher also found that some girls had images of Victoria's Secret models as their phone screensaver as "thinspirations," a motivational term often used in pro-anorexia/bulimia circles that promotes thinness and glorifies certain aspects of thinness, such as visible ribs and thigh gaps (Cosslett, 2014, 2016).

As individuals, we experience the consequences of an artificially created ideal as a personal trouble—unhappiness about our appearance—but the deliberate construction and dissemination of an unattainable ideal for the purpose of generating profits is surely a public issue. Reflecting a conflict perspective, psychologist Sharlene Hesse-Biber (1997) has suggested that to understand the eating disorders and disordered eating so common among U.S. women, we ought to ask not "'What can women do to meet the ideal?' but, instead, 'Who benefits from women's excessive concern with thinness?'" (p. 32).

Cultural inconsistency: A contradiction between the goals of ideal culture and the practices of real culture.

These examples illustrate a **cultural inconsistency**, *a contradiction between the goals of ideal culture and the practices of real culture*, in our society's treatment of conventional attractiveness. Cultural inconsistencies are common. For instance, take the cultural ideal of "honesty as the best policy." Childhood stories embrace honesty. Think of *Pinocchio*: Were you warned as a child not to lie because it might cause your nose to grow? What about children's stories like *Edwurd Fudwupper Fibbed Big* (the young Edwurd's lies become so consequential that a military response becomes necessary), or *The Berenstain Bears and the Truth* (the Berenstain children learn that they must tell their mother that they broke the lamp while she was gone), or *The Empty Pot* (Ping, a young boy, is rewarded for his honesty)—all about the importance of telling the truth? Despite these messages, most people do lie, with the average person lying at least once a day (Serota & Levine, 2015). Indeed, the 45th president of the United States, Donald J. Trump, was tracked by the *Washington Post* (2020) to have made over 20,055 false or misleading claims over the course of 1,276 days in office.

You might wonder why, if the culture emphasizes honesty, do people lie? Sociologists, like the symbolic interactionist Erving Goffman (1959), tied lying to the need to protect or project a certain image of ourselves. Goffman called this *misrepresentation*, arguing that all of us, as social actors, engage in this practice because we are concerned with defining a situation— whether it be a date or a job interview or a meeting with a professor or boss—in a manner favorable to ourselves. Job seekers, for instance, may pad their résumés to leave the impression on potential employers that they are qualified or worthy, even though this practice could result in disqualifying them for a job were they to be found out. Indeed, a 2018 CareerBuilder survey found that nearly 75% of employers had detected lying on a résumé! Common lies included misrepresentations of educational credentials, skill sets, dates of employment, and prior job responsibilities. According to the same survey, "39 percent of hiring managers said they spend less than a minute looking at a résumé and 23 percent spend less than 30 seconds" (CareerBuilder, 2015, 2018), suggesting that some dishonesty probably goes unnoticed.

The workplace is not the only site of consequential lies. Many studies suggest that academic dishonesty such as cheating on exams or plagiarizing is common in both secondary and higher educational institutions. For example, a 2012 study of 23,000 high school students revealed that about half had cheated on a test in the past year. Just under a third also said that they had used the Internet to plagiarize assigned work (Josephson Institute Center for Youth Ethics, 2012). Notably, a 2009 study suggests that about half of teens age 17 and younger believe cheating is necessary for success (Josephson Institute of Ethics, 2009).

Turning to American higher education, Barthel, in a 2016 *Atlantic* article, calls cheating "omnipresent" (Barthel, 2016, para. 1). She cites Harvard's investigation of 125 students for improper collaboration on a final exam, the University of Georgia's 2014 identification of 603 possible incidents of cheating, and Dartmouth's 2015 suspension of 64 students for cheating in an ethics class (Barthel, 2016, para. 1). A more recent study of 734 engineering students from four higher education institutions showed that student attitudes and behaviors around cheating varied by the setting (proctored vs. nonproctored). Students were more likely to cheat in an unproctored setting, to think that cheating in proctored settings is more unacceptable than in nonproctored, and to place the onus to decrease cheating on the instructor and the college. As one student put it: "Anyone would do anything they can get away with if they are desperate enough and if it means succeeding. If you are an instructor and you give an exam, then expect at least some level of dishonesty" (Dyer et al., 2020, p. 21).

In 2019, American higher education was rocked by cheating scandals related not to students, but instead to their parents who were accused of a number of behaviors such as:

> having a stand-in take a college entrance exam, photo-doctoring to paste a student's head onto the body of an athlete, bribing college sports coaches and paying up to $75,000 for falsified exam results, all in the name of getting their kids admission through what Singer called "a side door" to schools like Yale University, the University of Texas at Austin and the University of Southern California (Nadworny & Kamenetz, 2019, para. 3).

By mid-2020, 55 people, 38 of them parents, including celebrities Felicity Huffman and Lori Loughlin, had been charged with "conspiring with consultant, William 'Rick' Singer, to use bribery and fraud to secure their children's admission to top schools" (Raymond, 2020, para. 4). Huffman paid $15,000 to improve her daughter's SAT scores while Loughlin and her husband, Mossimo Giannulli, pleaded guilty to "conspiring to fraudulently secure admission for their daughters to the University of Southern California" (Hughes & Haynes, 2020). One parent, Karen Littlefair, paid $9,000 to "have someone take online courses for her son so he could graduate from Georgetown University" (Raymond, 2020, para. 2).

How does your educational experience compare? Do you encounter high levels of plagiarism and cheating? If so, do you think these are primarily a result of access to opportunities enabled by the Internet and smart watches—or are they a product of trends in the culture that diminish the importance of integrity? How might you design a sociological study to examine these questions at your own school?

Felicity Huffman, escorted by husband, William H. Macy, makes her way to the entrance of the John Joseph Moakley United States Courthouse on September 13, 2019, in Boston, where she was sentenced for her role in the College Admissions scandal. Huffman, sentenced to 14 days in jail, only served 11. She was also required to "pay a fine of $30,000 and perform 250 hours of community service" (Fieldstadt & Kaplan, 2019, para. 3).

Ethnocentrism

Foods are cultural products. Have you ever rejected an unfamiliar food because you had never eaten anything like it? Perhaps you found it unappealing or even thought it disgusting. Because we tend to perceive our own cultural products and ways of doing things as natural and normal, we may even consider items as basic as our food as the standard by which to judge all foods. In so doing, though, we engage in ethnocentrism, *the act of using a worldview to judge other cultures and cultural products by the standards of our own and regarding our own way of life as normal and superior.*

A number of aspects of our life compared to others can invoke ethnocentrism—the clothes we wear and the way we ornament our bodies, the expressions we use, the rituals we enlist, the ways we live. Other societies' rituals of death, for example, can look astonishingly different from our own. Take, for instance, this description of an ancient burial practice from Russia's North Caucasus:

> Scythian-Sarmatian burials were horrible but spectacular. A royal would be buried in a *kurgan* [burial mound] alongside piles of gold, weapons, horses, and, Herodotus writes, "various members of his household: one of his concubines, his butler, his cook, his groom, his steward, and his chamberlain—all of them strangled." A year later, 50 fine horses and 50 young men would be strangled, gutted, stuffed with chaff, sewn up, then impaled and stuck around the *kurgan* to mount a ghoulish guard for their departed king. (Smith, 2001, pp. 33–34)

Let's interpret this historical fragment using two different cultural perspectives. From an **etic perspective**—that is, *the perspective of the outside observer*—the burial ritual looks bizarre and shockingly cruel. Nevertheless, to understand it fully and avoid a potentially ethnocentric perspective, we need to call on an **emic perspective**, *the perspective of the insider*, and ask, "What did people in this period believe about the royals? What did they believe about the departed and the experience of death itself? What did they believe about the utility of material riches in the afterlife and the rewards the afterlife would confer on the royals and those loyal to them?" What death rituals does your society (or even your group) have that might appear exotic or strange to outsiders even though you see them as normal?

Etic perspective: The perspective of the outside observer.

Emic perspective: The perspective of the insider, the one belonging to the cultural group in question.

Putting aside the ethnocentric perspective allows us to embrace **cultural relativism**, *a worldview whereby we understand the practices of another society sociologically, in terms of that society's norms and values and not our own.* In this way, we can come closer to understanding cultural beliefs and practices such as those that surround the end of life. Whether the body of the departed is viewed or hidden, buried or burned, feasted with or feasted for, danced around or sung about, a culturally relativist perspective allows sociologists to examine the roots of these practices. As you read the following paragraphs by anthropologist Herbert Miner (1956) describing the Nacirema and their rituals, try to practice cultural relativism.

Cultural relativism: A worldview whereby the practices of a society are understood sociologically in terms of that society's norms and values, and not the norms and values of another society.

> Nacirema culture is characterized by a highly developed market economy which has evolved in a rich natural habitat. Although much of the people's time is devoted to economic pursuits, a large part of the fruits of these labors and a considerable portion of the day are spent in ritual activity. The focus of this activity is the human body, the appearance and health of which loom as a dominant concern in the ethos of the people. . . .
>
> The fundamental belief underlying the whole system appears to be that the human body is ugly and that its tendency is to debility and disease. Incarcerated in such a body, man's only hope is to avert these characteristics through the use of powerful influences of ritual and ceremony. Every household has one or more shrines devoted to this purpose. The more powerful individuals in this society have several shrines in their houses. . . .
>
> The focal point of the shrine is a box or chest which is built into the wall. In this chest are kept many charms and magical potions without which no native believes he could live. These preparations are secured from a variety of specialized practitioners. However, the medicine men do not provide the curative potions for their clients but decide what the ingredients should be and then write them down in an ancient and secret language. This writing is understood only by the medicine men and by the herbalists who, for another gift, provide the required charm. (pp. 503–504)

What looks strange here, and why? Did you already figure out that *Nacirema* is *American* spelled backward? Miner invites his readers to see American rituals linked to the body and health not as natural but as *part of a culture.* Can you think of other norms or practices in the United States that we could view from this perspective? What about the all-American game of baseball, the high school graduation ceremony, the language of texting, or the cultural obsession with celebrities or automobiles?

Subcultures

Subcultures: Cultures that exist together with a dominant culture but differ from it in some important respects.

Sociologists do not presume that in any given country—or even community—there is only one culture. While one culture may dominate, significant cultural identities exist in addition to, or sometimes in opposition to, the dominant one. These are called **subcultures**, *cultures that exist together with a dominant culture but differ from it in some important respects.*

Some subcultures may embrace most of the values and norms of the dominant culture while simultaneously choosing to preserve the values, rituals, and languages of their (or their parents' or grandparents') cultures of origin. Members of ethnic subcultures such as Armenian Americans and Cuban Americans may follow political events in their heritage countries or prefer their children to marry within their groups. It is comfort in the subculture rather than rejection of the dominant culture that supports the vitality of many ethnic subcultures.

In a few cases, however, ethnic and other subcultures do reject the dominant culture surrounding them. The Amish choose to elevate tradition over modernity in areas such as transportation (many still use horse-drawn buggies), occupations (they rely on simple farming), and family life (women are seen as subordinate to men), and they lead a retreatist lifestyle in which their community is intentionally separated from the dominant culture.

Sociologists sometimes also use the term *counterculture* to designate subcultural groups whose norms, values, and practices deviate from those of the dominant culture. The hippies of the 1960s, for example, are commonly cited as a counterculture to mainstream middle America, although many of those who participated in hippie culture aged into fairly conventional middle-class lives.

Even though there are exceptions, most subcultures in the United States are permeated by the dominant culture, and the

In the Tibetan sky burial, the body is left on a mountaintop exposed to the elements. This once-common practice of "giving alms to the birds" represented belief in rebirth and the idea that the body is an unneeded, empty shell.

©Imaginechina Limited / Alamy Stock Photo

In Indonesia, mass cremations take place where bodies are placed in sarcophagi of various sizes with animal representations.

©SONNY TUMBELAKA/AFP/Getty Images

In New Orleans, a casket is paraded through the street. Death and burial rituals are components of culture.

©Bob Sacha/Corbis Documentary/Getty Images

influence runs both ways. To return to our earlier example of foods—what, for example, is an "all-American" meal? Your answer may be a hamburger and fries. But what about other U.S. staples, such as Chinese takeout, Italian pastas and pizzas, and Mexican burritos? Mainstream culture has also absorbed the influence of the United States' multicultural heritage: Salsa music, created by Cuban and Puerto Rican American musicians in 1960s New York, is widely popular, and world music, a genre that reflects a range of influences from the African continent to Brazil, has a broad U.S. following. Some contemporary pop music, as performed by artists such as Lady Gaga, incorporates elements of British glam, U.S. hip-hop, and central European dance. The influence is apparent in sports as well: Soccer, now often the youth game of choice in U.S. suburbs, was popularized by players and fans from South America and Europe. Mixed martial arts, a combat sport popularized by the U.S. organization Ultimate Fighting Championship, incorporates elements of Greco-Roman wrestling, Japanese karate, Brazilian jiu-jitsu, and Muay Thai (from Thailand).

CULTURE AND LANGUAGE

As you go about your day, you consistently encounter symbols filled with meanings that you recognize and understand. (Recall that symbols are cultural representations of social realities or, as we put it in Chapter 1, representations of things that are not immediately present to our senses.) For instance, as you read this sentence, you are drawing on symbols that we, your authors, take to have meaning, and whose meaning we are trying to convey to you. You will later produce a series of symbols for your instructor to show if you have successfully understood us. Symbols may take the form of letters or words, images, rituals, or actions. **Language**, the most significant of symbols, is a *symbolic system composed of verbal, nonverbal, and written representations that are vehicles for conveying meaning.* Language, as Ng and Deng (2017) put it, is the vehicle through which

Language: A symbolic system composed of verbal, nonverbal, and written representations that are vehicles for conveying meaning.

> humans express and communicate their private thoughts and feelings as well as enact various social functions. The social functions include co-constructing social reality between and among individuals, performing and coordinating social actions such as conversing, arguing, cheating, and telling people what they should or should not do. Language is also a public marker of ethnolinguistic, national, or religious identity, so strong that people are willing to go to war for its defense, just as they would defend other markers of social identity, such as their national flag. (p. 2)

To fully appreciate the importance of language as a cultural marker, let's take a moment to consider the worldwide magnitude of language similarity and difference (Figure 3.7). Ethnologue Languages of the World (2021) reveals that worldwide there are over 7,117 languages spoken today, but over half of the world's population speaks just *23* of these languages.

In 2019, according to Ethnologue (as shown in Figure 3.7), 1.27 billion people worldwide spoke English. At the same time, English "is a global language with official or special status in at least 75 countries" (British Council, n.d., cited in Ng & Deng, 2017, p. 6). Mandarin was the second most spoken language, with over 1.12 billion speakers. More than 637 million people speak Hindi, the primary official language of India, as a first language, and over 538 million speak Spanish. By way of contrast, the world's 3,500 least widely spoken languages share only 8.25 million speakers. Aka, another language of India, has between 1,000 and 2,000 native speakers. The Mexican language of Seri has between 650 and 1,000 speakers, while only five fluent speakers remain for Native American Euchee (or Yuchi) language (http://www.ourmothertongues.org/language/Euchee/5).

Attendees of the 2019 San Diego Comic-Con cosplay (short for costume play) as their favorite characters from comic books and movies. Cosplayers are a subculture organized around looking like a character from popular culture.

©Rich Polk/Getty Images for IMDb

In the 1930s, Edward Sapir and Benjamin Whorf developed the *Sapir–Whorf hypothesis*, which posits that our understandings and actions emerge from language—that is, the words and concepts of our own languages structure our perceptions of the social world. Take, for instance, the simple power of a

FIGURE 3.7 ■ Most Common Languages Spoken Worldwide, 2019

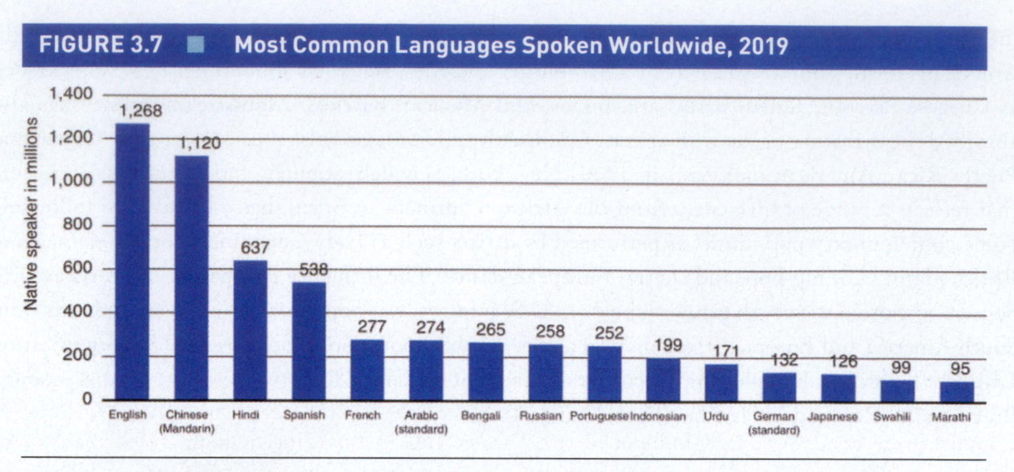

Source: Statista, based on data from Ethnologue.com.

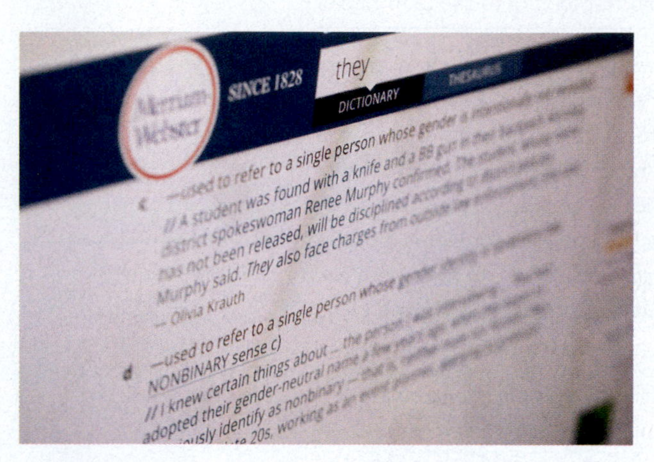

In 2019, the *Merriam-Webster Dictionary* chose as its word of the year the nonbinary pronoun "they."

©AP Photo/Jenny Kane

personal pronoun (he, she, they) in English to convey (or not) a speaker's gender (Baron, 2020).

Language is also closely tied to cultural objects and practices. The Aka language, for example, has more than 26 words to describe beads, a rich vocabulary suited for a culture in which beads not only are decorative objects but also convey status and facilitate market transactions. In the Seri language, to inquire where someone is from you ask, "Where is your placenta buried?" This question references a historical cultural practice of burying a newborn's afterbirth by covering it with sand, rocks, and ashes (Rymer, 2012). Losing language means losing culture. About 40% of the world's languages are endangered (Ethnologue Languages of the World, 2021), or as an article in *National Geographic* put it more visually, "One language dies every 14 days" (Rymer, 2012). As languages like Aka and Seri die out, usually replaced by dominant tongues such as Spanish, English, Chinese, Arabic, and Russian, we lose the opportunity to more fully understand the historical and contemporary human experience and the natural world. For instance, the fact that some small languages have no words linked to specific numbers and instead use only relative designations such as *few* or *many* opens the possibility that our number system may be a product of culture rather than of innate cognition, as many believe. Or consider that the Seri culture, based in the Sonoran Desert, has names for animal species that describe behaviors that natural scientists are only beginning to document (Rymer, 2012).

To summarize: Language is a cultural vehicle that enables communication, illuminates beliefs and practices, roots a community in its environment, and contributes to the cultural richness of our world. While a large variety of languages exist worldwide, only a few languages, including English, are the primary language spoken by most people. Each language lost represents the erasure of cultural history, knowledge, and human diversity (Living Tongues Institute for Endangered Languages, n.d.).

Language, Social Integration, and Social Conflict

Given the worldwide variety of languages amidst the reality that many fewer are official or actually spoken, does language serve to bring us together or to separate us? Functionalists see language as conveying shared values and norms that maintain social bonds both between individuals and between people and society and provides a sense of belonging in a complex, even alienating, social world (Parsons & Smelser, 1956; Smelser, 1962). On the other hand, conflict theorists see language as a medium that "reveals power, reflects power, maintains existing dominance, unites and divides a nation, and creates influence" (Ng & Deng, 2017, p. 13).

Consider that in the United States over 500 different languages (including sign language) coexist. English is the most commonly spoken language, with Spanish being the second most spoken, but the United States has no official, federal language. However, 32 states list English as an official language (Misachi, 2018).

Struggles over language in the United States date to pre-colonization. But in the 20th and 21st centuries, English-only movements have arisen in response to native populations, the arrival of diverse immigrant groups, and the increased use of Spanish in the United States. For instance, "reformers" in the late 19th and early 20th centuries advocated for Indian boarding schools to educate Native American children to be "patriotic and productive members of society" (Marr, n.d., para. 1). Children were forcibly separated from their parents, stripped of Native names and Native culture, instead given English names and Victorian-era clothes, had their hair cut, and were allowed to speak only in English. As Marr describes it, "Believing that Indian ways were inferior to those of whites, [reformers, like Captain Richard Henry Pratt, founder of the Carlisle Indian Boarding School,] subscribed to the principle, 'kill the Indian and save the man'" (Marr, n.d.). Another English-only battle raged around immigration in the early 20th century. President Theodore Roosevelt, responding to the high rate of immigration, wrote,

> We have room for but one language in this country, and that is the English language, for we intend to see that the crucible turns our people out as Americans . . . and not as dwellers in a polyglot boarding house.

Today, English-only battles still occur in the United States, with supporters urging the passage of legislation to make English the only official language of the United States.

In truth, many people do not just speak one, pure language. They may, instead, "code switch" between different languages or between different types of dialects. Code switching is often applied to the phenomenon of expressing oneself differently around people of different racial and ethnic groups. For example, a 2019 Pew Research study found that almost half of Black college graduates and 42% of Hispanic college graduates say they feel the need to code switch at least sometimes around people of another race or ethnicity (see Figure 3.8; Dunn, 2019). Professionals may also "code switch" between the language of their disciplines and the popular vernacular—for instance, your "sore throat" may be "pharyngitis" to your doctor, but your doctor will likely concur verbally to you that you indeed have a redness indicative of a "sore throat" so that you can understand each other.

FIGURE 3.8 ■ Nearly Half of Black College Grads and 42% of Hispanic College Grads Feel They Need to Change How They Talk Around People of Other Races or Ethnicities

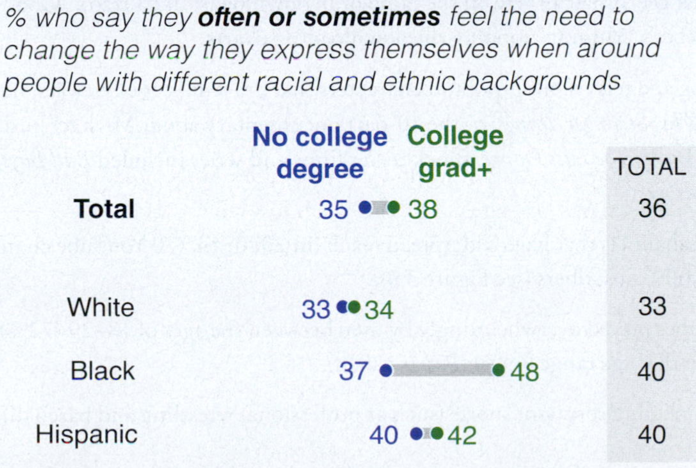

% who say they ***often or sometimes*** *feel the need to change the way they express themselves when around people with different racial and ethnic backgrounds*

	No college degree	College grad+	TOTAL
Total	35	38	36
White	33	34	33
Black	37	48	40
Hispanic	40	42	40

Source: Dunn, A. (2019, September 24). *Younger, college-educated Black Americans are most likely to feel need to "code-switch."* Pew Research Center. https://www.pewresearch.org/fact-tank/2019/09/24/younger-college-educated-black-americans-are-most-likely-to-feel-need-to-code-switch/

Multiculturalism: A commitment to respecting cultural differences rather than trying to submerge them into a larger, dominant culture.

Many people embrace cultural diversity and emphasize the value of **multiculturalism**, *a commitment to respecting cultural differences rather than trying to submerge them into a larger, dominant culture.* Multiculturalism recognizes that the country is as likely to be enriched by its differences as it is to be divided by them. In a globalizing world, knowledge of other cultures and proficiency in languages (multilingualism) other than English are important. In a far-reaching study of over 100 articles from multiple countries, Fox et al. (2019) find significant benefits to multilingualism, including individual health and lifestyle advantages such as improved cognitive abilities over the life course and increased creativity, and increased economic avenues such as employability, and social potential such as increased ability to communicate and understand other cultures.

CULTURE AND MASS MEDIA

Cultural stereotypes might lead us to believe that people who attend the symphony, are knowledgeable about classic literature and fine wines, or possess a set of distinctive manners are "cultured." This implies a value judgment: Being cultured is better than being uncultured. From a sociological perspective, however, we all are cultured. That is to say, we all participate in and identify with a culture or cultures. Sociologists, then, distinguish instead between high culture and popular culture as distinctly different types of cultures accessed by different publics.

High culture: Music, theater, literature, and other cultural products that are held in particularly high regard in society.

High culture consists of *music, theater, literature, and other cultural products that are held in particularly high regard in society.* What is considered high culture in a society often reflects a particular set of distinctive "tastes" that, according to sociologist Pierre Bourdieu (1930–2002), are often associated with social class and can change over time. For instance, William Shakespeare's plays were popular with the English masses when they were staged in open public theaters during his lifetime and lobster was a meal of the poor in colonial America. Today, both activities tend to be cost-prohibitive for widespread consumption, and Shakespeare typically draws an audience that has been educated to appreciate his works.

Popular culture: The entertainment, culinary, and athletic tastes shared by the masses.

Popular culture encompasses *the entertainment, culinary, and athletic tastes shared by the masses.* It is widely available, less costly to consume, and its use transcends social class. Because it is an object of mass consumption, popular culture plays a key role in shaping values, attitudes, and consumption in society. It is an important topic of sociological study because, as we noted in our opening story, it not only *shapes* but also *is shaped by* society. Examples of popular culture and its omnipresence abound:

- Foods and beverages from chain stores and restaurants like Starbucks lattes, McDonald's french fries, and Popeye's chicken sandwiches.
- Music that gets broad airplay on the radio or in downloads, like Lizzo's "Good as Hell" or Justin Bieber's "Yummy" topping the downloading charts.
- Streaming and television shows and characters that draw masses of viewers, like *Thirteen Reasons Why; Stranger Things;* or the 10-part documentary about Michael Jordan and the Chicago Bulls, *The Last Dance.* Blockbuster films and series included *Bad Boys for Life*, and *Black Panther.*
- YouTube channels that have widespread reach (in 2020, 43,770 YouTube channels exist, with over 250,000 subscribers (see Figure 3.9).
- Video games played overwhelmingly by men between the ages of 18–29 (72%) but also by women in that age range (49%) (Perrin, 2018).
- Sports, including spectator sports (such as professional wrestling and baseball).

Mass media: Media of public communication intended to reach and influence a mass audience.

Popular culture is spread widely through **mass media**, *a media of public communication intended to reach and influence a mass audience.* Mass media comes in many forms: television, streaming content, books, songs, etc. The mass media permeates our lives, but its rise is more recent than we may realize. That is, prior to the development of printing presses and the spread of literacy, most communication was oral and local. Importantly, as theorist Jürgen Habermas (1962/1989) points out, the concomitant rise of mass media and European coffee houses and salons constituting a "public sphere" contributed to the rise of a more informed, democratic society in industrialized 18th century Europe. Newly literate populations met in these public sphere venues to discuss daily events and, informed by mass circulating written forms of information, to exchange informed opinions. These public discussions and the understandings that emerged from them helped shape public attitudes and became the foundation for informed political debates.

In the 20th century, the technologies of mass media spread from radio to television and the Internet. Indeed, by 2020, Americans were spending over 463 minutes, or over 7.5 hours per day, with electronic mass media (Watson, 2020). Sociologically speaking, mass media is of interest to both functionalists and conflict theorists. On the one hand, as we indicated earlier, mass media can be a powerful and effective means for conveying information and contributing to the development of an informed citizenry: Mass-circulation newspapers such as *The New York Times* and *The Wall Street Journal*, television networks such as CNN and the BBC, and radio news stations such as National Public Radio inform us about important issues. Importantly, the readership of news in the form of print papers continues to decline, with an estimated print and digital newspaper circulation of 28.6 million people (Pew Research Center, 2019b).

This widespread media penetration, and the concomitant social changes it engendered, inspired many theorists to think critically about its social impact.

Theorists like Marshall McLuhan (1964) suggested that "the medium is the message"— meaning that the medium itself has an influence on how the message is received and perceived. In other words, in looking at only a particular message, we may miss the power of the messenger itself and how that transforms social life. Television, for instance, is fundamentally different from print in how it communicates information. McLuhan also asserted that television constructed a *global village* in which people around the world, who did not and never would know one another, could be engaged with the same news event. For example, it was reported by FIFA, the world's governing body of soccer, that in the summer of 2019, more than a billion people (about a seventh of the world's population) tuned in for the Women's World Cup held in France (Glass, 2019).

FIGURE 3.9 ■ A Week in the Life of Popular YouTube Channels

43,770 channels with at least **250,000** subscribers

56% posted a video during the first week of 2019

243,254 videos

Just **17%** were in English

Over **48,000** total hours

Over **14 billion** views in one week

Source: Van Kessel, P., Toor, S., & Smith, A. (2019, July 25). *A week in the life of popular YouTube channels.* Pew Research Center. https://www.pewresearch.org/internet/2019/07/25/a-week-in-the-life-of-popular-youtube-channels

On the other hand, critical theorists of the Frankfurt School Theodor Adorno, Max Horkheimer, and Herbert Marcuse (Horkheimer, 1947; Horkheimer & Adorno 1944/2007; Marcuse, 1964) argued that mass media promoted disengagement, distraction, and consumption. They made a distinction between *culture*, which, they suggested, retains the potential to be a vehicle for creativity, critique, and social change, and the *culture industry*, which they saw as a capitalist industry manufacturing homogenized, predictable, and banal cultural products that pacify and sell rather than inform and provoke. Indeed, they judged the culture industry as one that promises an "escape from reality but it really offers an escape from the last thought of resisting that reality" (Horkheimer & Adorno, 1944/2007, p. 116). To the Frankfurt School, the culture industry told consumers that happiness was linked to buying similar products, watching similar shows, and thinking similar thoughts. They thought that the magic (or deception) of the culture industry was all the more powerful for the fact that its coercion and "unfreedom" were pleasant to the masses (Marcuse, 1964).

The Frankfurt School would likely not be surprised by the increasing amount of advertising featured directly in lyrics and images of popular music. A report in *The Atlantic Monthly* points out that more brands are paying for product placement in pop songs. The article quotes Adam Kluger, chief executive officer (CEO) of the Kluger Agency, which specializes in "lyrical product placements," as saying that a brand placement in a hit single "can easily offset the entire production and marketing budget" for the song (Brennan, 2015, p. 40). Brand references in music are growing. According to William

Team USA soccer players, including Megan Rapinoe, celebrate after winning the 2019 Women's World Cup.

©PHILIPPE DESMAZES/AFP via Getty Images

Brennan's calculation, these references appeared 109 times in the top 30 *Billboard* songs in 2012, compared with 47 times in 2002—and zero times in 1962. The popular song "I Am the One," by DJ Khalid and featuring Justin Bieber, Chance the Rapper, Lil Wayne, and Quavo, references two fashion brands (Chanel and Gucci). The song also talks about Netflix. Similarly, in "Closer," the Chainsmokers (a DJ duo) sing to their love interest, asking her to meet "in the backseat of [her] Rover." Product placement is also present in music videos. For example, Ariana Grande's 2016 "Focus on Me" video features the young singer with her Samsung Galaxy Note phone, dancing in a galaxy-themed background. The appearance of brand-name goods and companies in popular music, whether purchased by an advertiser or not, is common. A *National Public Radio* report noted that an examination of the top 20 songs of the 3 years up to 2017 determined that about 212 different brands had been mentioned in songs. The recent rap song "Bad and Boujee," by Migos and Lil Uzi Vert, features 19 brand names (Lonsdorf, 2017).

Similarly, Jürgen Habermas (1962/1989), who we encountered earlier, feared that potential for the development of an active public sphere has been largely quashed by the culture industry and the rise of a corporatized media that have substituted mass entertainment based largely on consumerism for meaningful debate, elevating sound bites over sound arguments.

Another theorist, Douglas Kellner, concurred that modern technology and media—and television, in particular—constitute a threat to human freedom of thought and action in the realm of social change. Kellner suggested that the television industry "has the crucial ideological functions of legitimating the capitalist mode of production and delegitimating its opponents" (1990, p. 9). That is, mainstream television appears to offer a broad spectrum of opinions, but in fact, it systematically excludes opinions that seem to question the fundamental values of capitalism (for example, the right to accumulate unlimited wealth and power) or to critique not individual politicians, parties, and policies but the system within which they operate. Because television is such a pervasive force in our lives, the boundaries it draws around debates on capitalism, social change, and genuine democracy are significant.

Finally, French theorist Jean Baudrillard's (1981/1994, 1988) concepts of simulacra and hyperreality also allow us to think about popular culture, advertising, and imaginary worlds that come to take on a life of their own. Baudrillard identified the rise of **simulacra** (copies of objects and things that have no reality in the first place) resulting from advanced technologies of industrial and post-industrial societies such as photography and video capabilities as well as mass reproduction. He believed that in the late 20th and early 21st centuries mass reproduction of images and realities had led to a culture where language and images were separated completely, and reproductions of reality are more real than the reality they intend to produce. He called this **hyperreality**, identifying a phenomenon like the completely made-up worlds of Disneyland, Disneyworld, and Las Vegas. To Baudrillard, these destinations have no reality outside of their façade, rely on mass media to promote them, create popular culture, and encourage only empty consumption.

Simulacra: Copies of objects and things that have no reality in the first place.

Hyperreality: Mass reproduction of images and realities leading to a culture where language and images are separated completely, and reproductions of reality are more real than the reality they intend to produce.

What do you think? Does mass media encourage us to be passive, making us indifferent and vulnerable to manipulation? Has it created a series of false worlds that signify nothing except consumption? Or does mass media actually allow us to expand human creativity, freedom, and action?

In the section that follows, we turn our attention to another dimension of culture: the controversial relationship between culture, mass media, and the negative but pervasive phenomenon of sexual violence against women.

Culture, Media, and Violence

Consider these numbers: In the United States, approximately 1 in 5 women and 1 in 38 men have experienced rape in their lifetime (Centers for Disease Control and Prevention [CDC], 2020l). In 2018 alone, according to the National Crime Victimization Survey, 734,630 rapes, attempted rapes, or sexual assaults occurred in the United States (Morgan & Oudekerk, 2019). In short, sexual violence is quite common in American society. And women are more likely than men to experience it. How do we sociologically explain sexual violence's prevalence?

An individualist explanation of this phenomenon might focus on the deviant individuals who commit sexual assaults and the outcomes of particular and individual circumstances. However, a sociologist would note that given the magnitude of the problem, the examination of individual cases alone, while important, is sociologically inadequate. In short, a sociologist would advocate for us to enlist what C. Wright Mills called the sociological imagination, conceptualizing sexual violence in American society as a personal trouble *and* a public issue. For instance, a sociologist might ask these questions: Why are there so many deviant individuals? What can we learn by looking instead at the patterns of sexual assault as a way to find widespread causes and consequences?

As it applies to our learning in this chapter, sociologists and other researchers have identified one contributing factor as the existence of a **rape culture**, *a social culture that provides an environment conducive to rape and that blames victims for their victimization* (Boswell & Spade, 1996; Maxwell, 2014, cited in Zaleski et al., 2016; Sanday, 1990). A rape culture is characterized by widespread normalization of sexual violence and rape in both material and nonmaterial cultural elements such as laws, beliefs, language, and media. For instance, feminist theorist and legal scholar Catharine MacKinnon (1982) argued that historical legislative and judicial processes regarding rape are male-created and male-centered. Consider, for instance, that all 50 states did not have laws against marital rape until 1993, based, at least in part, on the notion that a woman could not be raped by her husband because sexual consent was taken as implied in the marital contract. These laws changed only because of grassroots organizing by groups of women (Pauly, 2019). At the same time, however, the issue of marital rape has not been completely abolished—17 states still have restrictions against convicting a spouse from raping a partner "who is unconscious, drugged, or otherwise incapacitated" (Pauly, 2019).

Rape culture: A social culture that provides an environment conducive to rape that blames victims for their victimization.

Some sociological research suggests that popular culture promotes rape culture by normalizing violence. This is not to argue that culture is a direct cause of sexual violence (or other kinds of violence) but rather to suggest that popular culture renders violence part of the social scenery by making its appearance so common in films, video games, and music videos that it evolves from being shocking to being utterly ordinary. It thus desensitizes us to this violence, *and* makes it seem like the norm (Katz & Jhally, 2000a, 2000b; Newsom, 2015). How does this process occur?

Some scholars argue that popular media embraces *violent masculinity*—a form of masculinity that associates *being a man* with being aggressive and merciless. Popular action films often highlight violent male protagonists, such as Keanu Reeves's character in *John Wick: Chapter 3* (2019). Outside of the realm of fiction, dominant male sports such as football and hockey elevate physical violence as entertainment and venerate the toughest players on the field or the ice. While popular culture features many images of male violence against other men, violent images may also normalize violence against women. A variety of genres of music including heavy metal, rap, hip-hop, and country western have long been associated with misogynistic lyrics and videos, although they are hardly alone in their objectification of women (Weitzer & Kubrin, 2009). Many commercial films also feature rough—even violent—treatment of women, offered as entertainment. The most gratuitous violence in films and shows such as *Hush* (2016), *The Cutting Room* (2015), *Game of Thrones* (2011–2019), and *Westworld* (2016–2020) is reserved for female victims.

Popular culture's most predictable normalization of violence against women occurs in pornography, a multibillion-dollar-a-year industry in the United States. Fictionalized portrayals of sexual activity range from a compliant and always-willing partner to violent rape simulations in which consent is clearly refused.

Although researchers do not propose that lyrics or images disseminated by mass media cause sexual violence *directly*, some suggest that popular culture's persistent use of sex-starved, compliant, and easily victimized female characters sends messages that forced sex is no big deal, that women really want to be raped, and that some invite rape by their appearance. In a study of 400 male and female high school students, Cassidy and Hurrell (1995, cited in Workman & Freeburg, 1999) determined that respondents who heard a vignette about a rape scenario and then viewed a picture of the victim (in reality, a

Many popular sports, including hockey, football, and boxing, feature violence as a key component of the entertainment. What makes the physical violence of these sports enthralling for audiences?

©Mike Ehrmann/Getty Images

model for the research) dressed in provocative clothing were more likely than those who saw her dressed in conservative clothing, or who saw no picture at all, to judge her responsible for her assailant's behavior and to say his behavior was justified and not really rape. More recent studies have reproduced findings that myths about rape are widely used to explain and even justify sexual violence (Hammond et al., 2011). A more recent study by Zaleski et al. (2016) explored the presence of rape culture on social media by analyzing 4,239 comment threads following a news story of sexual assault on Facebook and in news articles. They found that victims were commonly blamed for being sexually assaulted, especially when those accused of rape were celebrities.

Are cultural products that normalize violence (for example, music and movies) a sociological *antecedent* of real violence? Thinking intersectionally, do you believe that different groups of people receive different messages about violence as a norm or violence as deviance? Also, consider the points in this chapter's opening vignette about entertainment reflecting culture. Do you think violence in music and movies indicates a culture of violence? Why or why not?

CULTURE, CLASS, AND INEQUALITY

Social class reproduction: The way in which class status is reproduced from generation to generation, with parents passing on a class position to their offspring.

Cultural capital: Wealth in the form of knowledge, ideas, verbal skills, and ways of thinking and behaving.

Does social class influence what music we listen to, what foods we eat, and what clothes we wear? If so, why and how? For sociologists, these questions are interesting and explorable. Sociologist Pierre Bourdieu (1930–2002), for instance, was especially interested in how culture was involved in **social class reproduction**, or, *the way in which class status is reproduced from generation to generation.* While Karl Marx argued that the key to power in a capitalist system is economic capital, particularly possession of the means of production, Bourdieu (1984) indicated that two additional types of capital (social and cultural) are also involved in social power. Important to this chapter is the concept of **cultural capital**, *wealth that takes the form of knowledge, ideas, verbal skills, and ways of thinking and behaving.* Bourdieu extends this idea by suggesting that cultural capital is a source of power. For instance, take a moment to consider the potential differences in cultural capital between children from upper-class versus children from working-class backgrounds by thinking through how a young child might enter the educational system.

Children of the upper and middle classes come into the education system—the key path to success in modern industrial societies—with a set of language and academic skills, beliefs, and models of success and failure that fit into and are validated by mainstream schools. Their parents have likely bought many books for them and they may have their own iPad and other technologies to learn language. Children from less-privileged backgrounds enter with a smaller amount of validated cultural capital; their skills, knowledge base, and styles of speaking are not those that schools conventionally recognize and reward. For example, while a child from a working-class immigrant family may know how to care for her younger siblings, prepare a good meal, and translate for non–English-fluent parents, her parents (similar to many first-generation immigrants) may have worked multiple jobs and may not have had the skills to read to her or the time or money to expose her to enriching activities. By contrast, her middle-class peers are more likely to have grown up with parents who regularly read to them, took them to shows and museums, and quizzed them on multiplication problems. Although both children come to school with knowledge and skills, the cultural capital of the middle-class child can be more readily "traded" for academic success—and eventual economic gains.

In short, schools serve as locations where the cultural capital of the better-off classes is exchanged for educational success and credentials. This difference in scholastic achievement then translates into economic capital as high achievers assume prestigious, well-paid positions in the workplace. Those who do not have the cultural capital to trade for academic success are often tracked into jobs in society's lower tiers. Class is reproduced as cultural capital begets academic achievement, which begets economic capital, which again begets cultural capital for the next generation.

Clearly, however, the structure of institutional opportunities, while unequal, cannot alone account for broad reproduction of social class across generations. Individuals, after all, make choices about education, occupations, and the like. They have free will—or, as sociologists put it, *agency*—which is understood as the capacity of individuals to make choices and to act independently. Bourdieu (1977) argues that agency must be understood in the context of structure. To this end, he introduces the concept of **habitus**, *the internalization of objective probabilities and the subsequent expression of those probabilities as choice.* Put another way, people come to want that which their own experiences and those of the people who surround them suggest they can realistically have—and they act accordingly.

Habitus: The internalization of objective probabilities and subsequent expression of those probabilities as choice.

Consider the following hypothetical example of habitus in practice. In a poor rural community where few people go to college, fewer can afford it, and the payoff of higher education is not obvious because there are few immediate role models with such experience, Bourdieu would argue that an individual's choice not to prioritize getting into college reflects both agency *and* structure. That is, she makes the choice not to prepare herself for college or to apply to college, but going to college would likely not have been possible for her anyway as a result of her economic circumstances and perhaps as a result of an inadequate education in an underfunded school. By contrast, the habitus of a young upper-middle-class person makes the choice of going to college almost unquestionable. Nearly everyone around her has gone or is going to college, the benefits of a college education are broadly discussed, and she is socialized from her early years to understand that college will follow high school—alternatives are rarely considered. Furthermore, a college education is accessible—she is prepared for college work in a well-funded public school or a private school, and family income, loans, or scholarships will contribute to making higher education a reality. Bourdieu thus suggests that social class reproduction appears on its face to be grounded in individual choices and merit, but fundamental structural inequalities that underlie class reproduction often go unrecognized (or, as Bourdieu puts it, "misrecognized"), a fact that benefits the well-off.

CULTURE AND GLOBALIZATION

We live in an increasingly interconnected world. Given the information in this chapter, you might wonder how culture impacts and is impacted by **globalization**, *the process by which people worldwide become increasingly connected economically, politically, culturally, and environmentally.* Sociologists and anthropologists have debated, for instance, whether this cultural contact results in widespread cultural homogenization (likeness) or cultural heterogeneity (difference). Or, some ask, does it produce something in between? They have also asked: Is globalization, on balance, positive or negative for countries, communities, and corporate entities? We will begin by exploring how conflict and functionalist perspectives might offer us different ways of seeing contemporary **global culture**, *a type of homogenized culture—some would say U.S. culture—that has spread across the world in the form of Hollywood films, fast-food restaurants, and popular music heard in virtually every country.*

Globalization: The process by which people worldwide become increasingly connected economically, politically, socially, culturally, and environmentally.

Global culture: A type of culture—some would say U.S. culture—that has spread across the world in the form of Hollywood films, fast-food restaurants, and popular music heard in virtually every country.

Let's consider the idea of global culture. We often see images of such—a global landscape dotted in every corner with the Golden Arches and Colonel Sanders and Coke beckoning the masses to consume fried foods and drink soda. Language is an important part of this culture—returning to our earlier discussion of language, you might recall that English is the most common language spoken around the world. Indeed, Pew shows that in 2017, 91% of students in Europe were taking English in their classrooms, with variation among European countries (Devlin, 2020).

And, considering popular culture, we see the effects of globalization—and of Americanization, in particular—in cultural representations such as U.S. pop music and videos. We might feel that this is flattening or Americanizing culture: familiar songs of Beyoncé, Ed Sheeran, Drake, Justin Bieber, Pitbull, Bad Bunny, and Lil Uzi Vert broadcast on radio stations from Bangladesh to Bulgaria to Belize, while rebroadcasts of such popular U.S. soap operas as *The Bold and the Beautiful* provide a picture of ostensibly "average" U.S. lives on the world's television screens. And we should consider the amount of money that follows these products: For example, about 70% of studio revenue in Hollywood is generated in overseas markets; that is, many films make far more money abroad than in the United States. Action-oriented films in particular garner large audiences in markets such as China and Russia (Brook, 2014): For example, *Avengers: Endgame*, the top-grossing film of 2019, earned over $2.7 billion, with $1.9 billion coming from international markets (Watson, 2019). *Black Panther*, the highest-grossing film in 2018 in the United States, earned $700.6 million domestically and over $646 million internationally (Nash, n.d.[a], n.d.[b]).

Journalist Thomas Friedman has suggested that, although most countries cannot resist the forces of globalization, it is not inevitably homogenizing. His concept of *glocalization* highlights the idea of cultural hybrids born of a pastiche of both local and global influences. In *The Lexus and the Olive Tree: Understanding Globalization*, (Friedman, 2000) writes that

> [T]he most important filter is the ability to "glocalize." I define healthy glocalization as the ability of a culture, when it encounters other strong cultures, to absorb influences that naturally fit into and can enrich that culture, to resist those things that are truly alien and to compartmentalize those things that, while different, can nevertheless be enjoyed and celebrated as different. (p. 295)

Many U.S. films earn more money abroad than they do in the United States. Action films such as *Avengers: Endgame* are particularly popular with moviegoers around the globe. What makes these films appealing to a global audience?

©BFA/ Alamy Stock Photo

Social-cultural anthropologist Arjun Appadurai (1996, 2014) argues that a landscape of global flows exists across the world, and that these flows ultimately produce new forms of culture. To Appadurai, these landscapes include flows of people (ethnoscapes), flows of technologies (technoscapes), flows of money (financescapes), flows of mass media (mediascapes), and flows of ideas (ideoscapes). These lie outside of the nation-state and, indeed, they are interpreted differently by different individual entities (like multinational corporations or religious groups) and publics. That is to say, globalization is also subjective, and its effects are mediated locally—people incorporate global flows into their cultures.

A functionalist examining the development and spread of a broad global culture might begin by asking, "What is its function?" They could deduce that globalization spreads not only material culture in the form of food and music but also nonmaterial culture in the form of values and norms. They might conclude that globalized norms and values can strengthen social solidarity by allowing people to share symbols (such as language) and ideas and images that serve to increase communication. Therefore, globalization serves the integrative function of creating some semblance of a common culture that can foster mutual understanding and a foundation for dialogue with the potential to reduce conflict between states and societies.

In its more than half-century of operation, McDonald's has become one of the most recognized icons of U.S. life and culture; Ronald McDonald is said to be the most recognized figure in the world after Santa Claus. McDonald's serves 68 million customers every day in over 36,000 restaurants in 120 countries around the globe.

©iStockphoto.com/paulprescott72

Recall from Chapter 1 that functionalism assumes that the social world's many parts are interdependent. Indeed, globalization highlights both the cultural and the economic interdependence of countries and communities. The 2002 book *Global Hollywood* by Toby Miller et al. describes what its authors call a *new international division of cultural labor*, a system of cultural production that crosses the globe, making the creation of culture an international rather than a national phenomenon (though profits still flow primarily into the core of the filmmaking industry in Hollywood). Mirrlees (2013) emphasizes that films, TV shows, and the like are also produced by a blend of technological and human labor, debunking the notion of a "solitary author" and highlighting instead the "thousands of cultural workers" who produce cultural products.

A conflict theorist might also describe how the globalization of cheap fast food can cripple small independent eateries that serve indigenous (and arguably healthier) cuisine. An influx of global corporations inhibits some local people from owning their own means of production and providing employment to others. The demise of local restaurants, cafés, and food stalls represents a loss of the cuisines and thus the unique cultures of indigenous peoples. It also forces working people to depend on large corporations for their livelihoods, depriving them of economic independence.

Although functionalism and conflict theory offer different interpretations of globalization, both offer valuable insights. Globalization may bring people together through common entertainment, eating experiences, and communication technologies, and, at the same time, it may represent a threat—real or perceived—to local cultures and economies as indigenous producers are marginalized and the sounds and styles of different cultures are replaced by a single mold set by Western entertainment marketers.

In *The Globalization of Nothing*, sociologist George Ritzer (2007) proposes a view of globalization that integrates what he calls *grobalization*, the product of "the imperialistic ambitions of nations, corporations, organizations, and the like and their desire . . . to impose themselves on various geographic areas" (p. 15). Ritzer adds that the "main interest of the entities involved in grobalization is in seeing their power, influence, and in many cases, profits grow (hence the term *grobalization*) throughout the world" (p. 16). The concept of *grobalization* draws from classical sociological theorists such as Karl Marx and Max Weber. For instance, where Marx theorized capitalism's imperative of economic imperialism, Ritzer offers contemporary examples of grobalization's economic and cultural imperialism, exporting not only brand-name products but also the values of consumerism and the practical vehicles of mass consumption, such as credit cards.

How will the world's cultures shift in the decades to come? Will they globalize or remain localized? Will they glocalize or grobalize? Clearly, the material culture of the West, particularly of the United States, is powerful: It is pushed into other parts of the world by markets and merchants, but it is also pulled in by people eager to hitch their stars to the modern Western world. Local identities and cultures continue to shape people's views and actions, but there is little reason to believe that McDonald's, KFC, and Coca-Cola will drop out of the global marketplace. The dominance of U.S. films, music, and other cultural products is also likely to remain a feature of the world cultural stage.

WHAT'S NEXT? CHALLENGING CULTURAL SYMBOLS

In this chapter we introduced the topic of culture, a powerful, complex, and vital component of individual and community identities. We asserted that our shared realities are significantly shaped by both material objects and the nonmaterial social facts that they represent: our norms, values, and beliefs. We illustrated the importance of language, both as a carrier and creator of culture. We addressed the importance of mass media in promoting and creating culture, highlighting the rape culture and its negative impact.

We end this chapter at a time of significant cultural change. Leading up to and through 2020, American newspapers and 24-hour news cycles prominently featured images of conflict around flags, statues, names of sports teams, and symbolic kneeling at sporting events. For the majority of citizens, these cultural representations and practices were increasingly viewed as symbols of racial discrimination, violence, and stereotyping. Meanwhile, the resolve of some groups who saw themselves as holding on to cultural representations associated with tradition grew stronger.

In the summer of 2020, spurred by the record millions of Americans who took to the streets to support #Black Lives Matter, a number of corporate entities began to change their symbols. For example, NASCAR banned the Confederate flag at its races (Liasson, 2020); Quaker Oats removed Aunt Jemima from its pancake syrup (O'Kane, 2020); the NFL announced that the Washington Redskins football team was "retiring" its name and developing a new one (Booker, 2020); and NFL Commissioner Roger Goodell apologized to athletes for that organization's previous stance against kneeling, reversing course by "encouraging all to speak out and peacefully protest" (Romo, 2020, para. 1).

Importantly, 100 years earlier, sociologist W. E. B. Du Bois, seeing racial inequality as a global, economic, *and* cultural phenomenon, clearly identified *language* as a global cultural marker of racism:

> This theory of human culture and its aims has worked itself through warp and woof of our daily thought with a thoroughness that few recognize. Everything great, good, efficient, fair,

and honorable is "white"; everything mean, bad, blundering, cheating, and dishonorable is "yellow"; bad taste is "brown"; and the devil is "black." (Du Bois, 1920/2003, p. 69)

- *What sociological methods would you use to study if and how racism is embedded in culture?*
- *Given the emergence of large-scale social movements such as #MeToo and #BlackLivesMatter, how might popular culture begin to reflect changing attitudes in both material and nonmaterial culture?*
- *We have discussed the major impact of mass media in culture. What role do corporations play in influencing culture? Religious groups? Schools?*

So, what's next in the continuing struggle around the racism embedded in culture? Some questions follow for you to consider. As sociologists, we close with two important observations: (1) Humans create the cultures that have the power to unite and to divide us. (2) Humans have the power to change and challenge culture. Cultural forces of unity and division, particularly around racial divides, are likely to be an ongoing sociological issue in our lifetimes. The role of individuals, groups, mass media, educators, and even corporations will be an important area for study and theorizing in the next several decades.

WHAT CAN I DO WITH A SOCIOLOGY DEGREE?

Critical Thinking

Critical thinking entails the evaluation of claims about society, politics, the economy, culture, the environment, or any other area of knowledge with the application of reason and evidence. *Critical thinking skills* are very broad, but they have several key elements, including the ability to rigorously evaluate data, carefully and systematically analyze a problem or situation, and draw conclusions that recognize strengths and weaknesses of an argument or position. They also include the inclination and knowledge needed to be a critical consumer of information, thoughtfully questioning rather than simply accepting arguments or solutions at face value.

Critical thinking in sociology is not about *criticizing* but about developing a nuanced understanding of phenomena, institutions, and practices. Every chapter in this book seeks to help you sharpen your critical thinking by enabling you to see more deeply into the social processes and structures that affect our lives and society and to raise questions about social phenomena we may take for granted. As a sociologist in training, you will develop a more comprehensive understanding of your social world and will learn to ask critical questions about why things are as they are and to seek out and find evidence-based answers to those questions.

Neej Patel, Saint Louis University, BS in Sociology and Public Health

Michigan State University College Human Medicine, MD Candidate

With my eyes set on applying to medical school, I chose to pursue sociology because it most perfectly combined my interests in science, people, and culture. What I didn't know when making the choice was that it would color the way that I think about everything from disparities in our communities to everyday conversations with patients about treatment options. With a focus on health and medicine, my undergraduate degree in Sociology set me up for success beyond a traditional classroom.

As a medical student and future physician I value the understanding of societal disparities rooted in social injustice towards the underserved. Specifically, I have gained the research tools to qualitatively and quantitatively analyze information related to race, ethnicity, income, education, social status and other demographic features. In this way I am now better able to communicate the lived experiences of our communities to others. Additionally, it was my connection to sociology, which fueled my interest to serve internationally in the slum communities of Ahmedabad, India. In this way, the discipline has allowed me to better connect with the most vulnerable populations and make actionable change in our society.

Currently, I am privileged to be able to learn the trade of medicine, while simultaneously exploring the art of care. Health is innately socialized in our communities and lives and I believe that my sociology background helps me see the bigger picture behind each patient. The field of sociology has equipped me with tools to understand the human in front of me, both within and outside of the medical context. Sociology has broadened my understanding of what it means to care for others by assessing their past, listening to their present, and working together towards a better future.

Career Data: Physicians and Surgeons

- 2019 Median Pay: $208,000 per year
- $100 per hour
- Typical Entry-Level Education: Doctoral or professional degree
- Job Outlook, 2019–2029: 4% (as fast as average)

Source: Bureau of Labor Statistics. (2019). Occupational outlook handbook.

SUMMARY

- **Culture** consists of the beliefs, norms, behaviors, and products common to members of a particular social group. **Language** is an important component of cultures. The Sapir–Whorf hypothesis points to language's role in structuring perceptions and actions. Culture is a key topic of sociological study because, as human beings, we have the capacity to develop it through the creation of artifacts such as songs, foods, and values. Culture also influences our social development: We are products of our cultural beliefs, behaviors, and biases.
- Sociologists and others who study culture generally distinguish between material and nonmaterial culture. **Material culture** encompasses physical artifacts—the objects created, embraced, and consumed by a given society. **Nonmaterial culture** is generally abstract and includes culturally accepted ideas about living and behaving. The two are intertwined because nonmaterial culture often gives particular meanings to the objects of material culture.
- Norms are the common rules of a culture that govern people's actions. **Folkways** are fairly weak norms, the violation of which is tolerable. **Mores** are strongly held norms; violating them is subject to social or legal sanction. **Taboos** are the most closely held mores; violating them is socially unthinkable. **Laws** codify some, although not all, of society's norms.
- **Beliefs** are particular ideas that people accept as true, although they need not be objectively true. Beliefs can be based on faith, superstition, science, tradition, or experience.
- **Values** are the general, abstract standards of a society and define basic, often idealized principles. We identify national values, community values, institutional values, and individual values. Values may be sources of cohesion or of conflict.
- **Ideal culture** consists of the norms and values that the people of a society profess to embrace. **Real culture** consists of the real values, norms, and practices of people in a society.
- Ethnocentrism is the habit of regarding one's own way of life as superior and judging other cultures by the standards of one's own.
- Sociologists entreat us to embrace **cultural relativism**, a perspective that allows us to understand the practices of other societies in terms of those societies' norms and values rather than our own.
- Multiple cultures may exist and thrive within any country or community. Some of these are **subcultures**, which exist together with the dominant culture, differing in some important respects from it.
- **High culture** is an exclusive culture often limited in its accessibility and audience. High culture is widely associated with the upper class, which both defines and embraces its content. **Popular**

culture encompasses entertainment, culinary, and athletic tastes that are broadly shared. As mass culture, popular culture is more fully associated with the middle and working classes. Theorists have identified mass culture as both leading to increased literacy and early democratization and, conversely, of creating passive audiences, mesmerized by images, who are urged to consumption.

- **Rape culture** is a social culture that provides an environment conducive to rape and blames the victim for their victimization. Sociologists argue that mass media that marginalizes and normalizes sexual assault and violence against women is a component of modern culture that may contribute to the high number of rapes and attempted rapes in the United States.
- Social class is a primary vehicle for the reproduction of **cultural capital**, a form of cultural knowledge that also provides access to social power.
- **Global culture** has spread across the world in the form of Hollywood films, fast-food restaurants, and popular music heard in almost every country. While some view this as homogenizing, others believe globalization allows for the emergence of new forms of culture and individualistic expressions.

KEY TERMS

beliefs (p. 53)
cultural capital (p. 70)
cultural inconsistency (p. 59)
cultural relativism (p. 61)
culture (p. 51)
emic perspective (p. 61)
etic perspective (p. 61)
folkways (p. 56)
global culture (p. 71)
globalization (p. 71)
habitus (p. 71)
high culture (p. 66)
hyperreality (p. 68)
ideal culture (p. 58)
language (p. 63)
laws (p. 58)
mass media (p. 66)
material culture (p. 51)
mores (p. 57)
multiculturalism (p. 66)
nonmaterial culture (p. 52)
popular culture (p. 66)
rape culture (p. 69)
real culture (p. 58)
social class reproduction (p. 70)
simulacra (p. 68)
subcultures (p. 62)
taboos (p. 58)
values (p. 53)

DISCUSSION QUESTIONS

1. This chapter discusses tensions between ideal and real cultures in attitudes and practices linked to conventional attractiveness and honesty. Can you think of other cases where ideal and real cultures collide?
2. How is language a source of both social integration and social conflict? Give one example of each from this chapter and one example of each from your own experience.
3. Following the ideas of the critical theorists in sociology, this chapter suggests that mass media may play a paradoxical role in society, offering both the information needed to bring about an informed citizenry while also producing mass entertainment that distracts and disengages individuals from important debates. Which of these functions do you think is more powerful?
4. What is cultural capital? What, according to Bourdieu, is its significance in society? How does one acquire valued cultural capital, and how is it linked to the reproduction of social class?
5. Compare and contrast the views on globalization and culture, including ideas of cultural homogeneity and cultural heterogeneity. Which do you think best reflects the impact of globalization on culture and the impact of culture on globalization?

©Klaus Vedfelt/DigitalVision/Getty Images

4 SOCIALIZATION AND SOCIAL INTERACTION

WHAT DO YOU THINK?

1. Are our personalities already formed at birth?
2. Does the influence of social media on children and teens outweigh the influence of other socialization agents, such as families, schools, and sports?
3. Have you ever presented yourself differently to create a particular impression? Why, and how did you do this? What do you think this says about social interaction?

LEARNING OBJECTIVES

4.1 Describe how sociologists theorize the birth of the social self.

4.2 Explain the significance of agents of socialization in the development of the self.

4.3 Discuss socialization across the life course.

4.4 Define the concept of *total institutions*.

4.5 Theorize social interaction through the sociological perspectives of dramaturgy, ethnomethodology, and conversational analysis.

MY ROBOT, MY FRIEND

Do you think that human beings can have authentic emotional relationships with robots?

In popular culture, humans have long engaged in friendly, professional, or practical interactions with robots. Consider the long-running *Star Wars* franchise: Among the central relationships in the now four-decade-old original film are the bonds between Luke Skywalker and his robotic sidekicks, C3PO and R2-D2. For instance, the 2018 *Star Wars* film, *Solo: A Star Wars Story*, highlighted a strong emotional connection between the space pirate Lando Calrissian and his spunky female companion and copilot, a robot named L3-37. When L3-37 is destroyed in a shootout, Lando mourns the loss of his friend. Today, humanoid robots are making their debut in our workplaces and social spaces. What will this mean for social interaction in the future? What are the risks and benefits of human relationships with robots?

This humanoid robot, unveiled at the International Robot Exhibition in Tokyo, Japan, could one day provide someone with companionship.

©Andia/UIG via Getty Images

A writer for *The Atlantic Monthly*, noting the increasing interaction with humans and robots, muses: "The human–machine relationship is rapidly evolving as a result. Humanity, and what it means to be a human, will be defined in part by the machines people design" (LaFrance, 2016, para. 4). Some companies already market "companion robots" that perform simple interactions such as reminding older adults to take medications. Researchers say these same robots will soon be able to lift and dress humans, further moving them into more intimate human spaces. Consider the possibility of "humanized" robots providing important opportunities for interaction to those who might currently lack it, such as those with disabilities or the elderly. Do you think their presence will fill the gap of insufficient human interaction?

Sociologist and physician Nicholas Christakis and colleagues point out that artificial intelligence might positively change our human interactions. Several of their studies indicate that when people interact with robots or bots that make mistakes and own up to them, the humans relate *better* to each other. Christakis further cites the research of political scientist Kevin Munger, who, through the use of bots to intervene in racism online, "showed that, under certain circumstances, a bot that simply reminded the perpetrators that their target was a human being, one whose feelings might get hurt, could cause that person's use of racist speech to decline for more than a month" (Christakis, 2019, para. 8).

Others, though, caution about the danger of developing emotional attachments to machines that cannot reciprocate but may be able to manipulate us to consume or alter our relationships with others. As Simon (2017) puts it:

> As AI gets smarter and smarter, it will be easier to trick people—especially children and the elderly—into thinking the relationship is reciprocal. And such a bond is a powerful thing. Imagine an unscrupulous toy maker inventing a doll so sophisticated that it appears animate to a child. Now imagine the toy maker exploiting that bond by having the doll tell the kid to buy a personality upgrade for $50. (para. 15)

How do you think robots will influence our interactions? Can robots be our friends, companions, and colleagues? Will they allow us to form stronger social bonds and positively influence our interactions? Can they help address some social challenges like isolation and loneliness? How could they create or intensify social inequalities? In your opinion, do the potential benefits outweigh the costs?

In this chapter, we explore the process of becoming human through socialization and the vital role of social interaction. We begin by looking into the "nature versus nurture" debate, and the sociological, theoretical, and research-based contributions to this issue. We describe some of the typical "agents" that help shape our social selves and our behavioral choices over the life course. We explore the radically different type of socialization that occurs in total institutions. Finally, we turn to social interaction, discovering ways in which sociologists conceptualize our presentation of self and our group interactions. Throughout, we also try to show how social power and social diversity provide important variations for different social groups. And, as our opening vignette points out, socialization and social interaction are subject to new insights as new technologies emerge to influence the way in which we interact not only with other humans, but also with robots and other forms of artificial intelligence.

THE BIRTH OF THE SOCIAL SELF

Socialization is *the process by which people learn the culture of their society.* It is a lifelong and active process in which individuals construct their sense of who they are, how to think, and how to act as members of their culture. Socialization is our primary way of reproducing culture, including the norms, values, beliefs, and practices that represent "normal" ways of being in the world. Socialization takes place every day, usually without our thinking about it: when we speak, when others react to us, when we observe others' behavior—whether in person or on a screen—and in almost every other human interaction. The principal agents of socialization (including parents, teachers, religious institutions, peers, television, and social media) exert enormous influence on us.

Socialization: The process by which people learn the culture of their society.

Social scientists have debated the relative influence of genetic inheritance ("nature") and cultural and social experiences ("nurture") in shaping people's lives (Coleman & Hong, 2008; Ridgeway & Correll, 2004). If inborn biological predispositions explain differences in behaviors and interests between, say, sixth-grade boys and girls or between a professional thief and the police officer who apprehends her, then understanding socialization will do little to help us understand those differences. On the other hand, if biology cannot adequately explain differences in attitudes, characters, and behaviors, then it is imperative that we examine the effects of socialization. Social scientists have found little support for the idea that personalities and behaviors are rooted exclusively in human nature. Indeed, little human behavior is purely natural. For example, humans have a biological capacity for language, but language is learned and develops only through interaction. The weight of socialization in the development of language, reasoning, and social skills is dramatically illustrated in cases of children raised in isolation. If a biologically inherited mechanism alone triggered language, it would do so even in people who grow up deprived of human contact. If socialization plays a key role, however, then such people would not only have difficulty learning to use language, but they would also lack a capacity to play the social roles to which most of us are accustomed. Few now argue that behavior is entirely determined by either socialization or biology, seeing an interaction between the two. What they disagree about, however, is which is more *important* in shaping an individual's personality, philosophy of life, and social actions. In this text, we lean toward socialization because we believe the evidence points in that direction.

One of the most extensively documented cases of social isolation occurred more than 200 years ago. In 1800, a "wild boy," later named Victor, was seen by hunters in the forests of Aveyron, a rural area of France (Shattuck, 1980). Victor had been living alone in the woods for most of his 12 or so years and could not speak, and although he stood erect, he ran using both arms and legs like an animal. Victor was taken into the home of Jean-Marc-Gaspard Itard, a young medical doctor who, for the next 10 years, tried to teach him the social and intellectual skills expected of a child his age. According to Itard's careful records, Victor managed to learn a few words, but he never spoke in complete sentences. Although he eventually learned to use the toilet, he continued to evidence "wild" behavior, including public masturbation. Despite the efforts of Itard and others, Victor was incapable of learning more than rudimentary social and intellectual skills; he died in Paris in 1828.

Other studies of the effects of isolation have centered on children raised by their parents, but in nearly total isolation. For 12 years, from the time she was one and a half years old, "Genie" (a pseudonym) saw only her father, mother, and brother, and only when one of them came to feed her. Genie's father did not allow his wife or Genie to leave the house or have any visitors. Genie was either strapped to a child's potty-chair or placed in a sleeping bag that limited her movements. Genie rarely heard any conversation. If she made noises, her father beat her (Curtiss, 1977; Rymer, 1993).

When Genie was 13, her mother took her and fled the house. Genie was unable to cry, control her bowels, eat solid food, or talk. Because of her tight confinement, she had not even learned to focus her eyes beyond 12 feet. She was constantly salivating and spitting, and she had little controlled use of her arms or legs (Rymer, 1993). Gradually, Genie learned some of the social behavior expected of a child. For example, she learned to wear clothing and use the toilet. Nevertheless, although intelligence tests did not indicate reasoning disability, even after 5 years of concentrated effort on the part of a foster mother, social workers, and medical doctors, Genie never learned to speak beyond the level of a 4-year-old, and she did not interact with others. Although she responded positively to those who treated her with sympathy, Genie's social behavior remained severely underdeveloped for the rest of her life (Rymer, 1993).

Genie's and Victor's experiences underscore the significance of socialization, especially during childhood. Their cases show that even biologically rooted capacities do not develop into recognizable human ways of acting and thinking, unless the individual interacts with other humans in a social environment. Children raised in isolation fail to develop complex language, abstract thinking, notions of cooperation and sharing, or even a sense of themselves as social beings. In other words, they do not develop the hallmarks of what we know as humanity (Ridley, 1998).

Theories by sociologists and other social scientists explain the role of socialization in the development of social selves. What these theories recognize is that whatever the contribution of biology, ultimately, people as social beings are made, not born. Below, we explore four approaches to understanding socialization: behaviorism, symbolic interactionism, developmental stage theories, and psychoanalytic theories.

Behaviorism and Social Learning Theory

Behaviorism: A psychological perspective that emphasizes the effect of rewards and punishments on human behavior.

Behaviorism is *a psychological perspective that emphasizes the effect of rewards and punishments on human behavior.* It arose during the late 19th century to challenge the then-popular belief that human behavior results primarily from biological instincts and drives (Baldwin & Baldwin, 1986,

1988; Dishion et al., 1999). Early behaviorist researchers, such as Ivan Pavlov (1849–1936) and John Watson (1878–1958) and, later, B. F. Skinner (1904–1990), demonstrated that even behavior thought to be purely instinctual (such as a dog salivating when it sees food) may be produced or extinguished through the application of rewards and punishments. Thus, a pigeon will learn to press a bar if that triggers the release of food (Skinner, 1938, 1953; Watson, 1924). Behaviorists concluded that both animal and human behavior can be learned, and neither is purely instinctive.

When they turned to human beings, behaviorists focused on **social learning**, *the way people adapt their behavior in response to social rewards and punishments* (Baldwin & Baldwin, 1986; Bandura, 1977; Bandura & Walters, 1963). They were particularly interested in the satisfaction people get from imitating others. Social learning theory thus combines the reward-and-punishment effects identified by behaviorists with the idea that we model the behavior of others; that is, we observe the way people respond to others' behavior.

Social learning: A perspective that emphasizes the way people adapt their behavior in response to social rewards and punishments.

Social learning theory would predict, for example, that if a boy gets high fives from his friends for talking back to his teacher—a form of encouragement rather than of punishment—he is likely to repeat this behavior. What's more, other boys may imitate him. Social learning researchers have developed formulas for predicting how rewards and punishments affect behavior. For instance, rewards given repeatedly may become less effective when the individual becomes satiated: If you have just eaten a huge piece of cake, you are less likely to feel rewarded by the prospect of another.

Social behaviorism is not widely embraced today as a rigorous perspective on human behavior. One reason is that only in carefully controlled laboratory environments is it easy to demonstrate the power of rewards and punishments. In real social situations, the theory is of limited value as a predictor. For example, whether a girl who is teased ("punished") for engaging in a "masculine" pursuit such as football or wrestling will lose interest in the sport depends on many other variables, such as the support of family and friends and her own enjoyment of the activity. The simple application of rewards and punishments is hardly sufficient to explain why people repeat some behaviors and not others.

In addition, behaviorist theories violate Karl Popper's principle of falsification (discussed in Chapter 2). Since what was previously rewarding may lose effectiveness if the person is satiated, if a reward does not work, we can always attribute its failure to satiation. Therefore, no matter the outcome of the experiment, the theory has to be true; it cannot be proved false. For these reasons, sociologists find behaviorism an inadequate theory of socialization. To explain how people become socialized, they highlight theories within symbolic interaction.

Socialization as Symbolic Interaction

Recall that Chapter 1 introduced *symbolic interactionism* as a theory that posits our self and our societies emerge from the language and other symbols we share in social interaction. Symbolic interactionism is especially useful in explaining how individuals develop a social identity and a capacity for social interaction (Blumer, 1969, 1970; Hutcheon, 1999; Mead, 1934, 1938).

An early contribution to symbolic interactionism was Charles Horton Cooley's (1864–1929) concept of the **looking-glass self**, the *self-image that results from our interpretation of other people's views of us*. For example, children who are frequently told they are capable and smart may tend to see themselves as such and act accordingly. On the other hand, children who are repeatedly told they are incapable or not smart may lose pride in themselves and act the part. According to Cooley (1902/1964), we are constantly forming ideas about how others perceive and judge us, and the resulting *self-image*—the way we view ourselves—is, in turn, the basis of our social interaction with others.

Looking-glass self: The concept developed by Charles Horton Cooley that our self-image results from how we interpret other people's views of us.

Cooley, recognizing that not everyone we encounter is equally important in shaping our self-image, introduced the notion of primary and secondary groups. **Primary groups** are *small groups characterized by intense emotional ties, face-to-face interaction, intimacy, and a strong, enduring sense of commitment*. Families, close friends, and lovers are all examples of primary groups likely to shape our self-image. **Secondary groups**, on the other hand, come together for reasons that are *functional or fleeting rather than emotional or enduring*. These groups may be based on interests or economic exchange: They could include employees of a workplace, members of a running club, or even students in a sociology class. We could also think of the vast number of "friends" that we acquire on social media platforms as secondary

Primary groups: Small groups characterized by intense emotional ties, face-to-face interaction, intimacy, and a strong, enduring sense of commitment.

Secondary groups: Groups that are impersonal and characterized by functional or fleeting relationships.

Society has unwritten but widely understood rules for standing with strangers in an elevator. We learn these conventional practices from interactions and observations. What are these unwritten rules? What constitutes a violation, and how are these rules enforced? How did rules of social distance in elevators change during COVID-19?

©iStockphoto.com/monkeybusinessimages

groups. For instance, nearly 7 of every 10 Americans use Facebook, and their "friends" there can number in the hundreds (*Social Media Fact Sheet*, 2019). Secondary groups typically have less influence in shaping our self-image than do primary groups, although, arguably, social media has unleashed the modern power of "likes" that can have a profound effect on one's self-image. Both primary and secondary groups act on us throughout our lives; our self-image is not set at some early stage but continues to develop throughout adulthood (Barber, 1992; Berns, 1989). Both primary and secondary groups also serve as **reference groups**, *groups that provide standards for judging our attitudes or behaviors.* When you consider your friends' reactions to your clothes, hairstyle, or new smartphone, you are using your peers as a reference in shaping your decisions.

Reference groups: Groups that provide standards for judging our attitudes or behaviors.

I: According to George Herbert Mead, the part of the self that creatively responds to a social situation. It is the impulse to act; it is creative, innovative, unthinking, and largely unpredictable.

Me: According to George Herbert Mead, the part of the self through which we see ourselves as others see us.

Role-taking: The ability to take the roles of others in interaction.

Significant others: According to George Herbert Mead, the specific people who are important in children's lives and whose views have the greatest impact on the children's self-evaluations.

George Herbert Mead (1863–1931), widely regarded as the founder of symbolic interactionism, explored the ways in which self and society shape one another. Mead thought the self was made up of two parts: the "I" and the "me." The **I** is *our impulse to respond to a social situation; it is creative, innovative, unthinking, and largely unpredictable.* The **me** is *our social self, the part of the self through which we see ourselves as others see us, and which urges us to conform to social norms and practices.* (Note the similarity between Mead's *me* and Cooley's *looking-glass self.*) Predictably, the I and the me are often in tension. When the *I* initiates a spontaneous act, the me raises society's response: *How will others regard me if I act this way?* Mead believed that theme is often capable of controlling the I. For instance, a common battle between your I and me during the early days of the 2020 COVID-19 pandemic may have gone something like this:

I: It is a beautiful, sunny day. I want to go out and do things with friends like I used to. I'm going to text my friends, sneak out of the house, and get together with them.

Me: My state has a stay-at-home order to enforce social distancing. If I go out with friends, we will be violating that order, and I know that we are told that social distancing is really important for everyone to remain well. I'm also likely to worry my parents.

Mead further argued that we develop a sense of self through **role-taking**, *the ability to take the roles of others in interaction.* For example, a young girl playing soccer may pretend to be a coach; in the process, she learns to see herself (as well as other players) from a coach's perspective. Mead proposed that childhood socialization relies on an expanding ability to take on such roles, moving from the extreme self-centeredness of the infant to an adult ability to take the standpoint of society as a whole. He outlined four principal stages in socialization that reflect this progression: the preparatory, play, game, and adult stages. The attainment of each stage results in an increasingly mature social self.

1. During the *preparatory stage*, children younger than 3 years old relate to the world as though they are the center of the universe. They do not engage in true role-taking but respond primarily to things in their immediate environments, such as their mothers' breasts, the colors of toys, or the sounds of voices.
2. Children 3 or 4 years of age enter the *play stage*, during which they learn to take the attitudes and roles of the people with whom they interact. **Significant others** are *specific people important in children's lives whose views have the greatest impact on the children's self-evaluations.* By role-playing at being mothers or fathers, for example, children come to see themselves as their parents see them. Nevertheless, according to Mead, they have not yet acquired the complex sense of self that lets them see themselves through the eyes of *many* different people or society.
3. The *game* stage begins when children are about 5 and learn to take the roles of multiple others. This stage is directly comparable to the learning required to play a variety of games. To be

an effective basketball player, for example, you must be able to see yourself from the perspective of teammates, the other team, and the coach and play accordingly. You must know the rules of the game. To be successful in society, we all must gain the ability to see ourselves as others see us, to understand societal "rules," and to act accordingly. The game stage, therefore, signals the development of a self that is aware of societal positions and perspectives.

Young children learn to see the world from the perspective of others in part through play, which allows them to take the role of another person. Which stage of George Herbert Mead's socialization theory do these children exemplify?

©JGI/Jamie Grill/Getty Images

4. Game playing takes the child to the final *adult stage*. At this stage, young people begin to internalize the **generalized other**, *the abstract sense of society's norms and values by which people evaluate themselves*. They act on a set of socially normative principles that may or may not serve their self-interest—for example, voluntarily joining the military to fight in a war that might injure or kill them because patriotic people are expected to defend their country or choosing to return a wallet full of cash to its owner because taking something that belongs to someone else is wrong. By the adult stage, a person is capable of understanding abstract and complex cultural symbols, such as love and hate, success and failure, friendship, patriotism, and morality.

Generalized other: The abstract sense of society's norms and values by which people evaluate themselves.

George Herbert Mead (1934) was optimistic that our introduction to a multitude of "generalized others" through socialization would lead to a more democratic, "universal society" where the needs of the individual and the stability of society would be in balance. He thought that societies would be able to incorporate a variety of cultures and their generalized others and that individual people would adapt their behavior in terms of others' cultures. The end result, for Mead, would be a citizenry aware and appreciative of the value of other cultures and a type of "universal discourse" that was enriched by individual, local, national, and global interactions. Do you think Mead's dream of a highly multicultural, globalized world is realistic? What processes of globalization might work toward "universal discourse"? What processes might work against it?

Stages of Development: Piaget and Kohlberg

Like Mead and Cooley, the Swiss social psychologist Jean Piaget (1896–1980) believed humans are socialized in stages. Piaget devoted a lifetime to researching how young children develop the ability to think abstractly and make moral judgments (Piaget, 1926, 1928, 1930, 1932). His theory of **cognitive development**, based largely on studies of Swiss children at play (including his own), argues that *an individual's ability to make logical decisions increases as the person grows older*. Piaget noted that infants are highly **egocentric**, *experiencing the world as if it were centered entirely on them*. In stages over time, socialization allows children to learn to use language and symbols, to think abstractly and logically, and to see things from different perspectives.

Cognitive development: The theory, developed by Jean Piaget, that an individual's ability to make logical decisions increases as the person grows older.

Egocentric: Experiencing the world as if it were centered entirely on oneself.

Piaget also developed a theory of moral development, which holds that as they grow, people learn to act according to abstract ideas about justice or fairness. This theory parallels his idea of cognitive development, since both describe overcoming egocentrism and acquiring the ability to take other points of view. Eventually, children come to develop abstract notions of fairness, learning that rules should be judged relative to the circumstances. For example, even if the rules say "Three strikes and you're out," an exception might be made for a child who has never played the game or who is physically challenged.

Lawrence Kohlberg (1927–1987) extended Piaget's ideas about moral development. In his best-known study, subjects were told the story of the fictitious "Heinz," who was unable to afford a drug that might prevent his wife from dying of cancer. As the story unfolds, Heinz breaks into the druggist's shop and steals the medication. Kohlberg asked his subjects what they would have done, emphasizing that there is no right or wrong answer. Using experiments such as this, Kohlberg (1969, 1983, 1984) proposed three principal stages (and several substages) of moral development:

1. The *pre-conventional stage*, during which people seek simply to achieve personal gain or avoid punishment. People at this stage might support Heinz's decision to steal on the grounds that it would be too difficult to get the medicine by other means or oppose it on the grounds that Heinz might get caught and go to jail. Children are typically socialized into this rudimentary form of morality between ages 7 and 10.

2. The *conventional stage*, during which the individual is socialized into society's norms and values and would feel shame or guilt about violating them. People at this stage might support Heinz's decision to steal on the grounds that society would judge him as callous if he let his wife die. Or, they might oppose his decision because people would call Heinz a thief if he were caught. Children are socialized into this more developed form of morality at about age 10, and most people remain in this stage throughout their adult lives.

3. The *post-conventional stage*, during which the individual invokes general, abstract notions of right and wrong. Even though Heinz has broken the law, his transgression has to be weighed against the moral cost of sacrificing his wife's life. People at the highest levels of post-conventional morality will go beyond social convention entirely, appealing to a higher set of abstract principles.

Some scholars have argued that Kohlberg's theory reflects a strong male bias because it derives from male rather than from female experience. Foremost among Kohlberg's critics was Carol Gilligan (1982; Gilligan et al., 1989), who argued that men may be socialized to base moral judgment on abstract principles of fairness and justice, but women are socialized to base theirs on empathy, compassion, and caring. She showed that women scored lower on Kohlberg's measure of moral development because they valued how other family members were affected by Heinz's decision more than abstract considerations of justice. Because it assumes that abstract thinking represents a higher stage of development, Gilligan suggests, Kohlberg's measure is biased in favor of male socialization.

Research testing Gilligan's ideas has found that men and women alike adhere to *both* care-based and justice-based forms of moral reasoning (Gump et al., 2000; Jaffee & Hyde, 2000). Differences between the sexes in these kinds of reasoning are, in fact, small or nonexistent. For example, studies of federal employees (Peek, 1999), a sample of men and women using the Internet (Anderson, 2000), and a sample of Mexican American and Anglo American students (Gump et al., 2000) all found no significant difference between men and women in the degree to which they employ care-based and justice-based styles of moral reasoning. Indeed, a recent study of nurses' ethical decision making in geriatric rehabilitation units (Juujärvi et al., 2019) showed that the female nurses saw their jobs as *requiring* both types of moral reasoning even though these types could be conflictual. Conflicts emerged, for instance, when nurses encountered situations that required a care-based "ethic of care" while also adhering to the justice-based "rules of care delivery." Nurses needed to adhere to the rules while also providing care to patients who wanted to go home but were not ready, patients who wanted to stay in the hospital but the hospital team saw them as ready to go home, or patients who wanted to return home but their family saw them as not yet ready. To "anticipate and prevent controversies," the nurses enlisted both care and justice:

> . . . by careful listening and showing an understanding of their difficult situations. They explain the official admission criteria for different forms of care that are regarded as impartial rules constituting fairness in society. Ethical decision-making is ultimately justified by arguments of impartiality, enabled through individual assessment and evidence-based knowledge. (Juujärvi et al., 2019, p. 193)

Given the results of these and other studies, it is likely that people enlist both types of reasoning in different situations. Consider a time in your life when you had to enlist moral reasoning. Did you use a care of justice perspective, or a little bit of both? Do you think gender played a role in your decision making? Why or why not? What other social factors and identities might influence how we reason?

Biological Needs Versus Social Constraints: Freud

Sigmund Freud (1856–1939), an Austrian psychiatrist, had a major impact on the study of socialization as well as on the disciplines of psychology and psychiatry. Freud (1905, 1929, 1933) founded the field of **psychoanalysis**, *a psychological perspective that emphasizes the complex reasoning processes of the conscious and unconscious mind.* He stressed the role of the unconscious mind in shaping human behavior and theorized that early-childhood socialization is essential in molding the adult personality by age 5 or 6. In addition, Freud sought to demonstrate that in order to thrive, a society must socialize its members to curb their instinctive needs and desires.

Psychoanalysis: A psychological perspective that emphasizes the complex reasoning processes of the conscious and unconscious mind.

Id: According to Sigmund Freud, the part of the mind that is the repository of basic biological drives and needs.

Ego: According to Sigmund Freud, the part of the mind that is the "self," the core of what is regarded as a person's unique personality.

Superego: According to Sigmund Freud, the part of the mind that consists of the values and norms of society insofar as they are internalized, or taken in, by the individual.

According to Freud, the human mind has three components: the id, the ego, and the superego (Figure 4.1). The **id** is *the repository of basic biological drives and needs,* which Freud believed to be primarily bound up in sexual energy. (*Id* is Latin for "it," reflecting Freud's belief that this aspect of the human personality is not even truly human.) The **ego** (Latin for "I") is *the "self," the core of what we regard as a person's unique personality.* The **superego** *consists of the values and norms of society insofar as they are internalized, or taken in, by the individual.* The concept of the superego is similar to the notion of a conscience.

Freud believed that babies are all id. Left to their own devices, they will seek instant gratification of their biological needs for food, physical contact, and nurturing. Therefore, according to Freud, to be socialized, they must eventually learn to suppress such gratification. The child's superego consists of cultural *shoulds* and *should nots*. It struggles constantly with the biological impulses of the id. The child's emerging ego serves as a mediator between their id and their superego. In Freud's view, a child

FIGURE 4.1 ■ The Id, Ego, and Superego, as Conceived by Freud

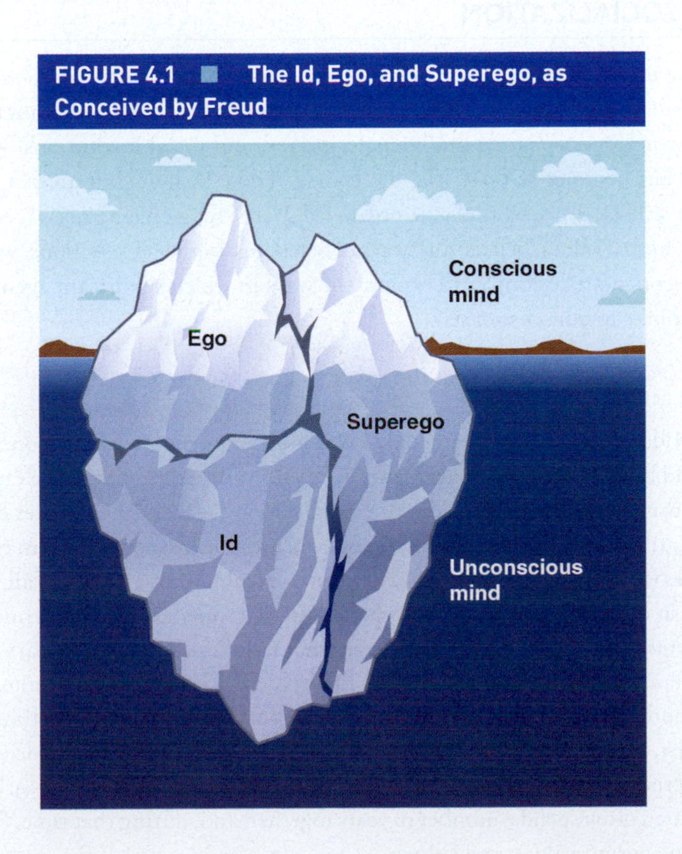

TABLE 4.1 ■ Comparison of Mead's and Freud's Theories of Socialization

MEAD'S STAGES	FREUD'S PSYCHOANALYTIC THEORY
Preparatory: The individual takes highly limited roles, viewing the world through his or her own eyes.	**Id:** The repository of basic biological drives and needs, which seeks instant gratification.
Play: The individual takes the roles of significant others, *one at a time*.	**Ego:** The self that, once developed, balances the forces of the id and superego. The ego is necessary in the socialization process for the individual to become a well-adjusted adult.
Game: The individual views the world through the eyes of multiple others, simultaneously.	**Superego:** The values and norms of society. May conflict with the id.
Maturity: The individual is able to take the attitude of the **generalized other** and can view the world through the eyes of society as a whole.	

Source: Adapted from Mead, G. H. (1934). *Mind, self, and society.* University of Chicago Press.

will grow up to be a well-socialized adult to the extent that their ego succeeds in bending their id's biological desires to meet the social demands of their superego.

Since Freud claimed that personality is set early in life, he viewed change as difficult for adults, especially if psychological troubles originate in experiences too painful to remember or face. Individuals must become fully aware of their repressed or unconscious memories and unacceptable impulses if they ever hope to change (Freud, 1933). Freud's psychoanalytic therapy focused on accessing deeply buried feelings to help patients alter current behaviors and feelings. Table 4.1 compares Mead's and Freud's views. Importantly, whereas Mead saw socialization as a lifelong process relying on many socialization agents, for Freud, it stopped at a young age.

AGENTS OF SOCIALIZATION

Who do you most credit for helping you develop your sense of self? A sociologist would say that this question cannot be adequately answered by listing just one person. Sociologists point to the importance of a number of *agents of socialization* (individuals, groups, and social institutions) who contribute to the development of a sense of self over the life course. These include, but are not limited to, families, schools and teachers, peers, workplaces, organized sports, religion, and mass media. Each of these agents reinforces norms, values, beliefs, and ways of behaving. In the next sections, we highlight sociological studies, theories, and examples to explore the ways in which significant agents of socialization (Figure 4.2) contribute to our sense of self.

Families

Sociologists often identify families as being the most important agent of socialization. Because U.S. culture is diverse, and because family arrangements differ, it is difficult to describe a "typical" American family or who within it performs the majority of socialization (Glazer, 1997; Stokes & Chevan, 1996). For instance, in Latinx families, grandparents, aunts, uncles, cousins, and in-laws may share child-rearing responsibilities (Supple et al., 2006). On the other hand, almost one fourth of all U.S. children live in households with only one parent and no other adults (Kramer, 2019). Furthermore, child-rearing practices and norms, values, and cultural messages that families promote may vary by many factors, for example, the family's race-ethnicity (Umaña-Taylor & Hill, 2020), socioeconomic status (Lareau, 2011), and neighborhood (Banerjee, 2014). Still, some important patterns of socialization are typical of most families. First, families are the first place that children typically encounter society. Second, families often supply crucial physical and emotional support to children during their earliest years. Third, families and children often spend a number of years together, and, during that time, families play a key role in transmitting norms, values, and culture.

FIGURE 4.2 ■ Agents of Socialization

Source: Supple et al. (2006).

Our families are typically the first to teach us about safety and danger. For example, early commonly given messages might include: "Don't touch the stove because it is hot," "Don't talk to strangers," or "Look both ways before crossing the street." These important messages, however, can vary between and even within families. For ethnic–racial minority families, socialization might include vital strategies to protect oneself against ethnic-racial discrimination and violence such as "how to interact with police officers, monitoring their children's friends, and carefully selecting school or extracurricular environments" (Elliot & Aseltine, 2013, cited in Nomaguchi & Milkie, 2020, p. 206). Furthermore, who gives the talk (mothers or fathers) and how they conceptualize children's risk shows that parents may perceive that girls and boys bear different risk and socialize their children differently. In her in-depth interviews with 30 Black mothers, for instance, Malone Gonzalez (2019) found that when mothers gave their children the "making it home talk" (a talk about strategies to survive police violence and make it home alive) the mothers were more likely to perceive their sons as being at higher risk and apply this talk more specifically to them.

Our families are also typically the first to teach us socially defined roles and behaviors associated with being a father, mother, sister, brother, uncle, aunt, or grandparent, and how children should think and act within these roles. But family are also "gendered institutions" (Risman, 1998) who also teach us messages about gender. Families may gender-stereotype household labor, socializing sons to do "outdoor" tasks such as yardwork and grilling, and daughters to do "indoor" work such as cooking and cleaning. Interestingly, research shows that while gender stereotyping may be less prevalent in same-sex and transgender families, the extent to which LGBTQ+ families challenge stereotyping depends on factors such as personal preference, number of hours worked outside of the home, relationship dynamics, income inequality between partners, and even personal traits of the parents (Reczek, 2020).

Our families also exist in a class-based society where one's occupation is an important determinant of one's position in the social class structure. Since many working-class jobs demand conformity while middle- and upper-middle-class jobs are more likely to offer independence, how social class impacts parents' child-rearing practices may be a key factor in explaining differences in family socialization of children (Kohn, 1989; Lareau, 2002, 2011). For instance, parents whose jobs require them to be subservient to authority and to follow orders without raising questions typically stress obedience and respect for authority at home, while parents whose work gives them freedom to make their own decisions and be creative are likely to socialize their children into norms of creativity and spontaneity (Lareau 2002, 2011).

Importantly, the way parents relate to children affects almost every aspect of the child's behavior, including the ability to resolve conflicts through the use of reason instead of violence and the propensity for emotional stability or distress. The likelihood that young people will be victims of homicide, commit suicide, engage in acts of aggression against other people, use drugs, complete their secondary education, or have an unwanted pregnancy also is greatly influenced by childhood experiences in the family (Campbell & Muncer, 1998; McLoyd & Smith, 2002; Muncer & Campbell, 2000). For example, children who are regularly spanked or otherwise physically punished may internalize the idea that violence is an acceptable means of achieving goals. They are more likely than peers who are not spanked to engage in aggressive delinquent behavior, to have low self-esteem, suffer depression, and do poorly in school (Straus et al., 1997).

Family patterns are changing rapidly in the United States, partly because of declining marriage rates and low overall fertility rates. Such changes can affect socialization practices and outcomes. Consider that children raised by a single parent may experience economic hardship that determines where they go to school or with whom they socialize outside the home. Being an only child is also a different experience from being raised in a family of multiple children, where both attention and roles may differ. Children raised in blended families (the result of remarriage) may have stepparents and stepsiblings whose norms, values, and behaviors are unfamiliar. Same-sex-couple families may challenge, or reinforce, conventional modes of socialization, particularly with respect to gender socialization. Migrant parents and their children may experience significant challenges to parental socialization. A study of migrant Mayan parents who were separated from their children between the United States and Guatemala shows that parents, unable to enforce "*mandados*" (directives) from a distance, instead attempted to "maintain a presence" in children's lives by providing children "*consejos*" (advice) (Hershberg, 2018).

School and Teachers

Schools are significant agents of socialization where most children and adolescents spend a considerable amount of time over many years (although, if we look globally, 248 million, or one sixth of all children in 2018, were not in school (UNESCO, 2019). Think of all of the educational systems that students in the United States typically move through—daycares, preschools, pre-kindergartens, kindergartens, elementary and secondary schools, and colleges. Compared to 100 years ago, U.S. children also spend more hours each day and more days each year in school (although they spend less time in school than their peers in Europe and Asia). Most states, for instance, require a school year of 170–185 days in length (National Center for Education Statistics, 2018).

Schools help youth prepare for adult society by teaching reading, writing, math, and other subjects. But schools are also expected to teach values and norms such as patriotism, competitiveness, morality, and respect for authority, as well as basic social skills. Some sociologists call this the **hidden curriculum**, *the unspoken classroom socialization into the norms, values, and roles of a society that schools provide along with the "official" curriculum.* The hidden curriculum can also teach students lessons about gender, race, sexuality, and social class. For instance, students may learn lessons about stereotypical gender roles taught through teachers' differing expectations of boys and girls, with, for instance, boys pushed to pursue higher math courses while girls are encouraged to embrace language and literature (Sadker et al., 2003). They may learn "lessons" that reinforce intersecting class status and racial inequalities, with predominantly white middle- and upper-class children having access to classes and schools with advanced subjects, advanced technology, and outstanding teachers, and predominantly children of color and poor children provided a smaller selection of less academically challenging or vocational classes and limited access to advanced teaching technologies and highly trained educators (Bowles & Gintis, 1976; Kozol, 2005). Simultaneously, the hidden curriculum may also include what is *not* taught: For example, if an English class typically relies on reading material with white main characters, this may teach students of color that their cultures are not appreciated or that people of their ethnic group cannot be heroes. The absence of curriculum representing one's group is clearly illustrated in The Human Rights Campaign's study in 2018 of 12,000 LGBTQ+ youth who reported that only 13% of them heard positive messages about being LGBTQ+ in school. As one respondent put it: "At my school LGBT topics aren't discussed. Ever. I wish they were, but they are usually avoided" (p. 8).

Hidden curriculum: The classroom socialization into the norms, values, and roles of a society that schools provide along with the "official" curriculum.

The hidden curriculum is also present in higher education curriculum, books, classrooms, and student life activities. For example, a study of long-term college outdoor educators identified a number of ways that they noticed a gender-based hidden curriculum operating in their field including "prioritizing . . . values and traits perceived to be predominantly male, linguistic sexism, assumptions about outdoor identity, outdoor career messages, gender insensitive facilitation and teaching, and the centering of White men in the field's history" (Warren et al., 2019, p. 143).

Thinking back over grade school and high school, can you identify any examples of a hidden curriculum in your or your classmates' experiences? What about in college?

Peers

Peers are people of the same age and, often, of the same social status and interests. Peer socialization begins very early—usually during the first several years of life, when a child starts to play with other children outside of the family. It continues throughout the life course. When children enter school peer socialization intensifies, and it becomes especially compelling during adolescence, when conformity to the norms and values of friends is particularly important (Harris, 2009; Ponton, 2000; Sebald, 2000). Judith Rich Harris (2009) argues, for example, that after the first few years of life, a child's friends' opinions outweigh those of their parents. Harris (2009) suggests that to manage these predominant peer group influences, parents must try to ensure that their children have the "right" friends. Importantly, technology and social media have extended the reach and number of "friends," as 95% of American youth ages 13–17 own or have access to cell phones and 45% say they are online "on a near constant basis" (Anderson & Jiang, 2018, para. 3). This ability to interact with much larger groups of which one is not specifically a part extends the traditional notion of peer groups and will continue to provide new avenues for sociologists to explore. As an illustration, a recent study shows that when individuals are exposed to peers' aggressive texts about someone outside their peer group it can increase both text-based and in-person aggression (Vollet et al., 2020).

Young children learn to see the world from the perspective of others in part through play, which allows them to take the role of another person. Which stage of George Herbert Mead's socialization theory do these children exemplify?

©Kate Geraghty/Sydney Morning Herald/Fairfax Media via Getty Images via Getty Images

Sociologically speaking, taken as a whole, adolescents constitute and create a subculture that also plays an extremely important part in their socialization. Researchers have associated these characteristics with adolescent subculture (Hine, 2000; Sebald, 2000):

1. A set of norms not shared with the adult or childhood cultures and governing interaction, statuses, and roles.
2. An *argot* (the special vocabulary of a particular group) that is not shared with nonadolescents and is often frowned upon by adults and school officials. Think about the jargon used by young people who text—many adults can read it only with difficulty!
3. Various underground media and preferred media programs, music, the Internet, and social media.
4. Unique fads and fashions in dress and hairstyles that often lead to conflict with parents and other adult authorities over their appropriateness.
5. A set of "heroes, villains, and fools." Sometimes adults are the "villains and fools," while the adults' "villains and fools" are heroes in the adolescent subculture.

6. A more open attitude than that found in the general culture toward experimentation with drugs and, at times, violence (fighting, for example).

Keep in mind that this list characterizing adolescent subculture was based on research from several decades ago. While some may still hold true, other characteristics may no longer be relevant.

American adolescents have typically spent more time with peers than with their families as a result of school, athletic activities, and other social and academic commitments. However, the amount of time spent in face-to-face interaction time with peers has declined from previous generations according to a study of a nationally representative sample of over 8 million U.S. adolescents between 1976–2017. The study also found that adolescent feelings of loneliness had "increased sharply after 2011," with "adolescents low in in-person social interaction and high in social media use reporting the most loneliness" (Twenge et al., 2019).

Anticipatory socialization: Adoption of the behaviors or standards of a group one emulates or hopes to join.

Sociologists use the term **anticipatory socialization** to describe the process of *adopting the behaviors or standards of a group one emulates or hopes to join.* People engage in anticipatory socialization when they model their behavior and attitudes on future expectations rather than present experience. Consider the behavior of teens who seek to be part of a group like a campus fraternity or sorority or medical interns who seek to become doctors—both sets of individuals will likely begin to dress and act like members of these groups before they actually are officially inducted or receive their degrees. You may be like one of many American college students today who have participated in volunteer experiences, internships, practicums, and clinicals to explore and prepare for a future career. If you have participated in one of these activities, in what ways could you describe your experience using the concept of anticipatory socialization?

Youth peer groups are also important because they influence adult behavioral patterns, including the development of self-esteem and self-image, career choices, ambition, and deviant behavior (Cohen, 1955; Hine, 2000; Sebald, 2000). But, as we show in the next several sections, peer socialization continues to shape us throughout our lives—through sports, religious groups, work settings, on social media, and in other settings.

Organized Sports

Organized sports are a fundamental part of the lives of millions of children in the United States: By one estimate, 21.5 million children and teens ages 6 to 17 participate in at least one organized sport (Kelley & Carchia, 2013). About 40% of those between ages 6 and 12 are involved in a team sport (Rosenwald, 2016) and the average time per week U.S. kids spend on sports is just under 12 hours (The Aspen Institute Project Play, 2019). If it is the case, as psychologist Erik Erikson (1950) posited, that in middle childhood children develop a sense of "industry or inferiority," then it is surely the case that in a sports-obsessed country such as the United States, one avenue for generating this sense of self is through participation in sports.

Team membership and mastery of sports skills are widely recognized as valuable; they are presumed to provide a number of social benefits such as building character, contributing to hard work, instilling competitiveness, and developing the ability to perform in stressful situations and under the gaze of others (Friedman, 2013). Sports also provide opportunities for parent–child socialization through shared moments of interaction such as in "the car ride home" after the sporting event.

Sociologically speaking, sports also transmit messages about gender, race, and sexuality. For instance, some research suggests that sports benefit girls through lower rates of teen sexual activity and pregnancy (Sabo et al., 1999) and higher rates of college attendance, labor force participation, and entry into male-dominated occupations (Stevenson, 2010). Some studies have also found improved academic performance relative to nonparticipants for all athletes, although they have shown some variation in this effect by race and gender (Eccles & Barber, 1999; Miller et al., 2005). At the same time, it is important to point out that sports socialization is clearly mediated by social class, as the affordability of and access to sports clearly vary. As a study from The Aspen Institute found:

> Kids from lower-income homes face increasing participation barriers. In 2018, 22% of kids ages 6 to 12 in households with incomes under $25,000 played sports on a regular basis, compared to 43% of kids from homes earning $100,000 or more. Kids from the lowest-income homes are more than three times as likely to be physically inactive. (The Aspen Institute Project Play, 2019, p. 1)

Sports socialization has also been associated with reinforcement of gender stereotypes (Jakubowska & Byczkowska-Owczarek, 2018) and homophobia (Denison & Kitchen, 2015). In a study of more than 1,400 teenagers, Osborne and Wagner (2007) found that boys who participated in "core" sports (football, basketball, baseball, and/or soccer) were three times more likely than their nonparticipant peers to express homophobic attitudes. In a country in which sports and sports figures are widely venerated and participation (particularly for boys) is labeled as "masculine," there may also be negative effects for boys who are not athletic or who do not enjoy sports.

Looking back at your own socialization, what norms, values, beliefs, and messages about gender and sexuality do you credit to sports?

Religion

Sociologists Émile Durkheim (2008/1912) and Talcott Parsons (1970) both noted that religion plays a role in socialization. While Durkheim linked religion to a strong sense of collectivity and social solidarity, Parsons pointed out that religion acts as an agent of socialization, teaching fundamental values and beliefs that contribute to shared cultural norms. Different religions function in similar ways, providing a sense of what is right and wrong and directions and structure for conduct and organization of followers' lives. Some groups provide abstract teachings about morality, service, or self-discipline, directing believers to, for example, serve their fellow human beings or to avoid the sin of vanity. Others teach specific rules about behaviors. For instance, the Amish faith entreats young men to remain clean-shaven prior to marriage, but married men must grow beards. Sikh men of India wear turbans that cover their hair, which they do not cut. Hasidic Jewish men are expected to grow beards and many grow long sidelocks, while married Hasidic women cover their hair in public (Freeman & Posner, n.d.).

In the United States, religion continues to be a significant part of many people's lives. About 65% of U.S. adults indicate they are members of a religious group (Pew Forum on Religion and Public Life, 2019b), and nearly 36% attend religious services once a week (Pew Forum on Religion and Public Life, 2015a). Even among those who declare themselves unattached to any particular religion, the overwhelming majority of Americans (90%) believe in some type of higher power (Fahmy, 2018). At the same time, the number of people who are religiously unaffiliated (those who identify as "atheist, agnostic, or 'nothing in particular'") has risen from 17% in 2009 to 26% of the population in 2019 (Pew Forum on Religion and Public Life, 2019b, para. 1).

Family and religion often teach and reinforce to children similar socialization messages such as how to conduct themselves in society and what is right or wrong. Our parents' religiosity or irreligiosity has an impact on whether or not we choose a religious identity as adults.

©Roberto Machado Noa/LightRocket via Getty Images

Parents and extended family members are significant agents in connecting (or not connecting) children to a religious group, and this socialization may have long-lasting effects. Notably, of those who say they attend religious services at least once or twice a month, a majority (69%) agree that, among other reasons, they attend "so children will have a moral foundation" (Pew Forum on Religion and Public Life, 2018). Parental religiosity has been shown to influence the continued religiosity of children even through their adulthood, while, conversely, parental irreligiosity has been shown to provide a form of "irreligious socialization" that influences adult irreligiosity. As Merino (2021) puts it:

> Variation in exposure to religion during childhood is an important factor among individuals raised with no religion. . . . Those with religiously unaffiliated parents as children are

significantly less likely to express a religious preference as adults, while those who would sometimes attend religious services as children are significantly more likely to do so. (p. 12)

Like other agents of socialization and social control, religion directs its followers to choose certain paths and behaviors and not others. This is not to say that religion compels us to behave in a certain way, but rather that socialization often leads us to control our own behavior because we fear social ostracism or other negative consequences. When you think intersectionally, a term we introduced in Chapter 1, what are some of the messages about gender, race, social class, or sexuality that you or someone you know could credit to religious socialization? Do changes in religion influence changes in societal norms or behaviors, or do changes in societal norms and behaviors foster changes in religion?

Mass Media and Social Media

Among the most influential agents of socialization in modern societies are technology and the mass media, a category that includes newspapers, magazines, movies, radio, television, the Internet, and social media such as Facebook, Twitter, Snapchat, and Instagram.

Television is a long-standing influential agent of socialization: According to Nielsen ratings, the typical American spends 4 hours and 27 minutes in front of a TV screen per day; this includes regularly broadcast television programs and DVR/time-shifted TV viewing (Nielsen, 2017). When we add in time spent watching streaming content from Netflix, Hulu, or other services—60% of Americans subscribe to at least one of these types of services (Nielsen, 2020)—it is clear that, overall, screen time is rising in the United States. Americans now own four digital devices on average, and, as shown in Figure 4.3, the typical U.S. consumer spends 75 hours a week using content across devices (Nielsen, 2018).

Child psychologists, sociologists, and parents' groups pay special attention to the impact of TV and other media violence on youth. Media studies during the past 20 years have largely come to a common conclusion: Media violence has the clear potential to socialize children, teenagers, and even adults into a greater acceptance of real-life violence. This is true regardless of gender or race. Much media violence is directed against women, and a large body of research supports the conclusion that media violence promotes tolerance among men for sexual violence, including rape (Rodenhizer & Edwards, 2017; Ward, 2016; Ybarra et al., 2014). The argument is not that viewing violent shows is a direct cause

FIGURE 4.3 ■ Average Number of Hours per Week of Screen Time

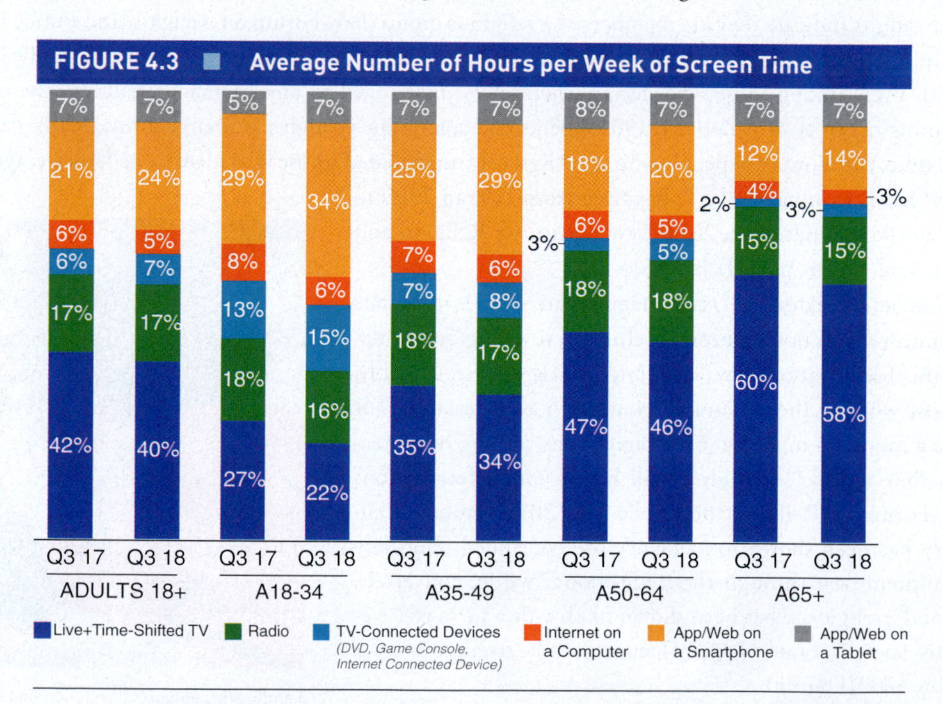

Source: Nielsen.com (2018). The Nielsen Total Audience Report Q3 2018.

Note: Data are not available for smartphone usage for 2- to 17-year-olds, but some studies have estimated that children 2 to 11 years old may spend as much as 14 hours a week on smartphones, and teens 12 to 17 years old may spend nearly 27 hours per week on their smartphones (Houghton et al., 2015).

of violence; rather, viewers may become desensitized to the sight of violence. Still, given that most people who are exposed to violence in the media do not become violent, the part played by the media as an agent of socialization is probably less important than the contribution made by other agents, such as family and peers.

"Selfies" exemplify the ways technology and social media combine to help us accomplish a presentation of self. How would Erving Goffman analyze the interaction and presentation of selves in this picture?

©Andy Cross/The Denver Post via Getty Images

The media play a role in socialization by creating fads and fashions such as how people should look, what they should wear, what kinds of friendships they should have, and acceptable gender expression. Children's cartoons, advertisements, video games, and streaming content often stereotype gender, sexualities, and people of different races and ethnicities. Teenage girls, for example, are depicted as boy crazy and obsessed with their looks; teenage boys are shown as active, independent, and sexually and physically aggressive (Kahlenberg & Hein, 2010; Maher et al., 2008). Females' roles also portray mostly familial or romantic ideals, whereas males fulfill work-related roles (Lauzen et al., 2008). These stereotypes have been found to influence children's gender perceptions (Aubrey & Harrison, 2004; Gerding & Signorielli, 2014). Additionally, gender stereotypes influence beliefs across the spectrum of sexual orientation, with gay teens embracing stereotypes in ways comparable to their heterosexual peers (Bishop et al., 2014). Finally, an interesting study of 246 racially diverse video gamers between the ages of 18 and 25 (a group the researchers call "emerging adulthood") who played games associated with violent masculinity ("self-described as containing violence and beliefs that masculinity should entail aggression, toughness, dominance, or restrictive emotionality") found that these players consistently held traditional gendered masculinity norms and values *regardless* of their gender (Blackburn & Scharrer, 2019). That is to say, playing video games associated with violent masculinity appears to have the possibility to "cultivate attitudes" that link the game's masculinity stories to the attitudes that male and female players adopt about how gender should be performed (p. 319).

Social media has a significant and likely increasing impact on socialization that may also vary by race and ethnicity, social class, sexuality, and gender. While in this chapter's "peer socialization" section we pointed to the findings of a 2018 Pew survey that showed 45% of teens say that they are online "almost constantly," we note that the number of teens who answered affirmatively nearly *doubled* in under 5 years—growing from just under a quarter of teens (24%) in 2014–2015. That survey also shows that Hispanic teens were more likely than whites to report using the Internet almost constantly (54% vs. 41%) and girls were more likely than boys to be near constant users (50% to 39%, respectively; Anderson & Jiang, 2018). How does your own social media usage compare with these results? Why do you think that is?

A variety of studies point out the mixed effects of online interactions. When online interactions are mixed with off-line, face-to-face interactions, use of the Internet and social media can foster new personal relationships and build communities (Valentine, 2006; Wellman & Hampton, 1999). The types of friendships adolescents create and maintain through social media reflect the friendships they have off-line (Mazur & Richards, 2011). Additionally, online spaces can provide both safe and negative spaces for LGBTQ+ youth—a 2012 Human Rights Campaign Youth Survey, for instance, finds that while 73% of over 10,000 LGBTQ+ youth between ages 13 and 18 said they could be more honest about themselves online, 70% had also encountered negative messaging about being LGBTQ+ online. Since online interaction is often anonymous and occurs from the safety of familiar places, people with characteristics society tends to stigmatize or who engage in behaviors considered deviant or dangerous may enter virtual communities where such differences are not perceived or punished, interests are shared, and friendships emerge (Crowe & Hoskins, 2019; McKenna & Bargh, 1998). For instance, "Jazz," a digital member of the online subculture of Ana and Mia girls (those who advocate anorexia and bulimia), explains her "secret" community in familial terms:

> Ana is like our special secret you know. No one understands us in RL (Real life) but online I have so many sisters to share my life with. Only they get Ana, only a sister can comprehend the daily struggle that Ana asks of us. (Crowe & Hoskins, 2019, p. 5)

As we discuss elsewhere in the text, researchers have also linked high levels of social media use to declines in communication within households, shrinking social circles, and increased depression and loneliness (Dokoupil, 2012a, 2012b; Kraut et al., 1998; Twenge et al., 2019). Extreme cases can develop into Internet addiction, a relatively recent phenomenon characterized by a search for social stimulation and escape from real-life problems (Armstrong et al., 2000). Although the Internet can be a valuable learning tool for children, it can also damage their development by decreasing the time they spend in face-to-face interactions and exposing them to inappropriate information and images (Bremer & Rauch, 1998; Lewin, 2011b; Livingstone & Brake, 2010).

Another form of negative socialization is *cyberbullying*—taunting, teasing, or verbal attacks through e-mail, text, or social networking sites with the intent to hurt the victim (Van DeBosch & Van Cleemput, 2008). Cyberbullying is a problem of acute concern to social workers, child psychologists, and school administrators (Slovak & Singer, 2011; Watts et al., 2017). Children and adolescents who are bullied in real life can be simultaneously cyberbullies *and* victims of cyberbullying (Dilmac, 2009; Smith et al., 2008; Tyman et al., 2010). Victims take to the Internet to get revenge, often through anonymous attacks, but this perpetuates the bullying cycle online and in real life. One study found that hurtful cyberteasing between adolescents in romantic relationships can escalate into real-life shouting, throwing of objects, or hitting (Madlock & Westerman, 2011).

Interestingly, teens, who are the most avid users of social media, are not in agreement on its larger effects. A recent study found that while 45% of teens indicated that the effects of social media were neither positive nor negative, 31% responded that the effects were mostly positive, citing connections with family and friends as the most positive aspects. At the same time, 24% categorized effects as mostly negative, citing bullying and rumor spreading as the worst aspects (Anderson & Jiang, 2018). How would you characterize the effects of social media on society? On your peers? On your own life? If you were going to study the impact of social media, which type of media would you choose? How would you study that usage sociologically?

Work

In 2019, over 130 million Americans were working full-time and over 26 million were working part-time (Bureau of Labor Statistics, 2020). For many teens and adults, then, the workplace is an important site of socialization where many of us spend a significant amount of our time. Workplace norms and expectations can differ from those we experience in primary groups such as the family and peer groups.

Employment often socializes us in two areas: our formal job roles and our roles as members of a collective (sharing the same employer). To be successful employees we must be taught what Haveman and Wetts (2019) call the *formal* and *informal* organization of our worksite:

> *Formal* features include the configuration of offices and positions, the officially designated linkages between them (the "organization chart"), and written job descriptions, rules, and procedures. *Informal* features include the actual (as opposed to official) communication and influence channels (who really talks to whom, not just who is supposed to talk to whom, who sways decision making), actual behavior (what people do every day, not what job descriptions say they should do), and informal norms and practices (what is expected and valued). (p. 2)

Through the process of workplace socialization, we, as employees, may also internalize the values and norms of our employer. Indeed, we may even come to identify with the employer: Notice that employees who are speaking about their workplaces will often refer to them rather intimately, saying, for instance, not that "*Company X* is hiring a new sales manager," but rather that "*we* are hiring a new sales manager."

When you reflect on workplace socialization, what have you learned about the formal and informal organization of your worksite? Have you noticed others, or even yourself referring to your workplace as a site with which you personally identify? What other norms and values have you learned at your worksite?

Interestingly, even "occupations" outside the bounds of legality are governed by rules and roles learned through socialization. Harry King, a professional thief studied by William J. Chambliss, learned not only how to break into buildings and open safes but also how to conform to the culture of the professional thief. A professional thief never "rats" on a partner, for example, or steals from mom-and-pop stores. In addition, King acquired a unique language that enabled him to talk to other thieves while in the company of nonthieves ("Square Johns"), police officers, and prison guards (King & Chambliss, 1984).

Meyrowitz (1985) writes that "old people are respected [in media portrayals] to the extent that they can behave like young people" (p. 153). Betty White, a highly recognized actor holding the Guinness Book of World Records' "longest career by a female entertainer" (over 80 years!), often plays humorous roles that appeal to younger crowds.

©Vincent Sandoval/WireImage/Getty Images

SOCIALIZATION AND AGING

Most theories of socialization focus on infancy, childhood, and adolescence, but socialization in various forms does not stop once people become adults. As you have seen, we experience ongoing socialization in many areas. Take the example of the workforce where older adults are now working much longer than they were 20 years ago. According to Pew Research, in 2019, 19.8% of U.S. adults ages 65 and older were working full or part-time, a number that doubled since 2000 (Desilver, 2019). This upward trend is driven by a variety of factors, such as increased life expectancy, improved health, higher education, changes to Social Security, and the need for more retirement savings (Toossi & Torpey, 2017). Continued employment for older adults allows their continued socialization to new ways of thinking, being, and acting, while also providing younger coworkers an opportunity for the socialization of older peers.

Anticipatory socialization is likely to be experienced by aspiring retirees as they envision their futures. They may read different publications, engage in different activities, and pay more attention to how friends experience retirement and aging. They may begin to be more aware of ageism (prejudice and discrimination based on age), noticing, for example, the clear underrepresentation and stereotyping of older adults in mass media. Take, for example, a study of the 100 top-grossing films of 2016, which clearly highlighted the confluence of age-related stereotyping and its intersectionality with gender, race, sexuality, and disability. It showed, for instance, the rarity in which older women, especially women of color, appeared in films. (While 57% of the films had older lead or supporting characters, 39% had *no* older female characters, 88% had no Black older women, 95% had no Asian older women, and *100% of the movies had no Latina older women characters!*) Ageist comments about older characters were found in 25% of the films, with over half of the films having "comments about health, including mental well-being, memory, and hearing," which were "surprising, given that few seniors faced health issues in the stories told" (Smith et al., 2018, p. 12). Finally, only three LGBTQ+ characters existed in all 100 films, and, of those, all three were gay males. Thus, no lesbians or transgender characters, either in leading or support roles, were featured in the top-grossing 100 films of 2016 examined by researchers.

Aging brings change, restructuring of experiences, and accompanying resocialization. Health and well-being challenges may include being taught new ways of behaving around nutrition and physical activity. Social networks may change, shrink, or even broaden as older people move, lose spouses, partners, and close friends (Abramson, 2015). New peers may emerge. Very old people may spend time in the hospital or in a nursing home, which requires being socialized into a total institution (discussed in the next section). All of these can change the structure of people's lives, influence their perceptions, and cause them to modify their behaviors. As Jaber Gubrium's (1975/1995) classic study of the nursing home "Murray Manor" reveals, one element of resocialization of Murray Manor residents was learning the new rules about something as basic as how to spend one's time ("passing time"). Older adults are mistakenly perceived to be more likely than younger adults to disengage from society. Although some do, research suggests that most older people remain active as long as they are healthy (Rubin, 2006). In fact, the notion that seniors are disengaged is belied by the fact that many adults become more politically active in their older years. In the 2016 U.S. presidential election, for instance, the majority of poll workers were over 60, and people over 65 constituted over a quarter of American voters (Barthel & Stocking, 2020).

Older adults are increasingly likely to be engaged digitally (Figure 4.4). Facebook is especially popular, used by well over half of people (68%) from ages 50 to 64 and nearly half of people ages 65 and older (Perrin & Anderson, 2019). Interestingly, age is not the only factor that influences technology use: Although younger seniors are more likely to use technology, use is also significantly affected by income and education. Overall, seniors with higher incomes and levels of education use technology more. For example, 81% of seniors whose household income is above $75,000 own smartphones, compared to 27% of those with earnings less than $30,000 (Anderson & Perrin, 2017). Technology offers seniors a spectrum of ways to stay connected to family and friends and to meet new friends: Some research suggests that adults older than 60 are the fastest-growing segment of the online dating market. Accordingly, the Internet offers a variety of dating sites targeted specifically to older Americans who, according to researchers Wendy K. Watson and Claude Stelle,

> appear to market themselves differently on online dating sites than younger adults. Gone is the focus on appearance and status . . . the senior population appears to be more interested in honest self-representation and being compatible rather than discussing areas such as sexual prowess and nightlife. (Bowling Green State University, 2012, para. 4)

Clearly, socialization is a lifelong process, with different agents of socialization emerging as more salient at various points of our life. Our early primary socialization lays a foundation for our social selves, which continue to develop through processes of secondary socialization, including our interactions with technology, media, education, and work. But can we be resocialized? That is, can our social

FIGURE 4.4 ■ Use of Different Online Platforms by Demographic Groups

% OF U.S. ADULTS WHO SAY THEY EVER USE THE FOLLOWING ONLINE PLATFORMS OR MESSAGING APPS

	YOU TUBE	FACE BOOK	INSTAGRAM	PINTEREST	LINKEDIN	SNAPCHAT	TWITTER	WHATSAPP	REDDIT
U.S. adults	73%	69%	37%	28%	27%	24%	22%	20%	11%
Men	78	63	31	15	29	24	24	21	15
Women	68	75	43	42	24	24	21	19	8
White	71	70	33	33	28	22	21	13	12
Black	77	70	40	27	24	28	24	24	4
Hispanic	78	69	51	22	16	29	25	42	14
Ages 18–29	91	79	67	34	28	62	38	23	22
18–24	90	76	75	38	17	73	44	20	21
25–29	93	84	57	28	44	47	31	28	23
30–49	87	79	47	35	37	25	26	31	14
50–64	70	68	23	27	24	9	17	16	6
65+	38	46	8	15	11	3	7	3	1

Source: Perrin, A., & Anderson, M. (2019, April 10). *Share of U.S. adults using social media, including Facebook, is mostly unchanged since 2018*. Pew Research Center. https://pewrsr.ch/2VxJuJ3

selves be torn down and reconstituted in new forms that conform to the norms, roles, and rules of entirely different social settings? We explore this question in the following section.

Total institutions: Institutions that isolate individuals from the rest of society to achieve administrative control over most aspects of their lives.

Resocialization: The process of altering an individual's behavior through control of his or her environment, for example, within a total institution.

TOTAL INSTITUTIONS AND RESOCIALIZATION

Although individuals typically play an active role in their own socialization, in one setting—the total institution—they experience little choice. **Total institutions** are *institutions that isolate individuals from the rest of society to achieve administrative control over most aspects of their lives.* Examples include prisons, youth and immigrant detention facilities, the military, hospitals (especially mental hospitals), religious orders such as monasteries and convents, and live-in drug and alcohol treatment centers. Administrative control is achieved through rules that govern all aspects of daily life, from dress to schedules to interpersonal interactions. The residents of total institutions are subject to inflexible routines rigidly enforced by staff supervision (Goffman, 1961; Malacrida, 2005).

These Marines are part of a total institution in which they are subject to regimentation and control of their daily activities by an authoritative body. They are expected to exhibit obedience to authority and to elevate the collective over the individual good.

©Scott Olson/Getty Images

A major purpose of total institutions is **resocialization**, *the process of altering an individual's behavior through control of his or her environment.* Goffman (1961) referred to this as the "mortification of self," or the process of degrading and, over time, transforming the self of the individual subject to the discipline of the total institution. The first step is to break down the sense of self. In a total institution, every aspect of life is managed and monitored. The individual is stripped of identification with the outside world. Institutional haircuts, uniforms, language, round-the-clock inspections, and abuse, such as the harassment of new recruits to a military school, contribute to breaking down the individual's sense of self. In extreme situations, such as in concentration camps, psychological and physical torture may also be used.

Once the institutionalized person is "broken," the institution begins rebuilding the personality. Desirable behaviors are rewarded with small privileges and undesirable behaviors are severely punished. For instance, a prisoner who exhibits model behavior may be rewarded with privileges such as options for work duty. Conversely, a prisoner who exhibits undesirable behavior may lose privileges such as family visitation or be more severely punished by "administrative segregation" or prolonged isolation in restrictive housing or one's cell. Since the goal of the total institution is to change attitudes as well as behaviors, even a hint that the resident continues to harbor undesirable ideas may provoke a range of disciplinary actions.

How effective are total institutions in resocializing individuals? The answer depends partly on the methods used, partly on the individual, and partly on peer pressure. In the most extreme total institutions, Nazi concentration camps, some inmates came to identify with their guards and torturers, even helping them keep other prisoners under control. Most, however, resisted resocialization until their death or release (Bettelheim, 1979). Even when an institution is initially successful at resocialization, individuals who return to their original social environments often revert to earlier behavior. This reversal confirms that socialization is an ongoing process, continuing throughout a person's lifetime as a result of changing patterns of social interaction.

SOCIAL INTERACTION

Socialization at every stage of life occurs primarily through *social interaction*—human communication guided by the combination of spoken words, gestures, body language, and other symbols and combined with ordinary, taken-for-granted rules that enable people to live, work, and socialize together (Ridgeway & Smith-Lovin, 1999). Sociologists look behind the everyday aspects of social interaction to understand how it unfolds and to identify the social norms and language that make it possible.

Social interaction usually requires that we conform to social norms and conventions. According to Scheff (1966), people who violate the norms of interaction are seen as abnormal or potentially dangerous. A person in a crowded elevator who persists in engaging strangers in loud conversations, for example, and disheveled homeless people who shuffle down the street muttering to themselves will evoke anxiety, if not scorn. Norms govern a wide range of interactive behaviors. For example, making eye contact when speaking to someone is valued in mainstream U.S. culture. By contrast, among the Navajo and the Australian Aborigines, as well as in many East Asian cultures, direct eye contact is considered disrespectful, especially with a person of greater authority. While American men are socialized to avoid displays of intimacy with other men, such as walking arm in arm, Nigerian men who are close friends or relatives hold hands when walking together. In Italy, Spain, Greece, and some Middle Eastern countries, men commonly throw their arms around each other's shoulders, hug, and even kiss. What reaction do you think is likely to occur to American men who violate the norm of not displaying intimacy?

Norms also govern how close we stand to friends and strangers in making conversation. American anthropologist Edward T. Hall studied *the human use of space*, a field he termed "proxemics." Hall described four distinct zones of space (Figure 4.5) and hypothesized who is (and is not) permitted to enter the zones: For instance, the "intimate zone" is only for one's close friends, relatives, and significant others, while the "personal space zone" could be used by friends, colleagues, or classmates. Violations of space, he suggested, could lead to anger, discomfort, or anxiety (Hall, 1966). Hall also showed that different cultures varied in their norms around space. In North American and Northern European cultures, for instance, people avoid standing closer than a couple of feet from one another unless they are on intimate terms. A recent study also underscores Hall's observation. In the study, researchers asked 8,943 respondents in 42 countries to indicate on a graph showing two figures how they identified appropriate personal distances between themselves, their intimates, and strangers. Across cultures,

FIGURE 4.5 ■ Hall's Interpersonal Spaces

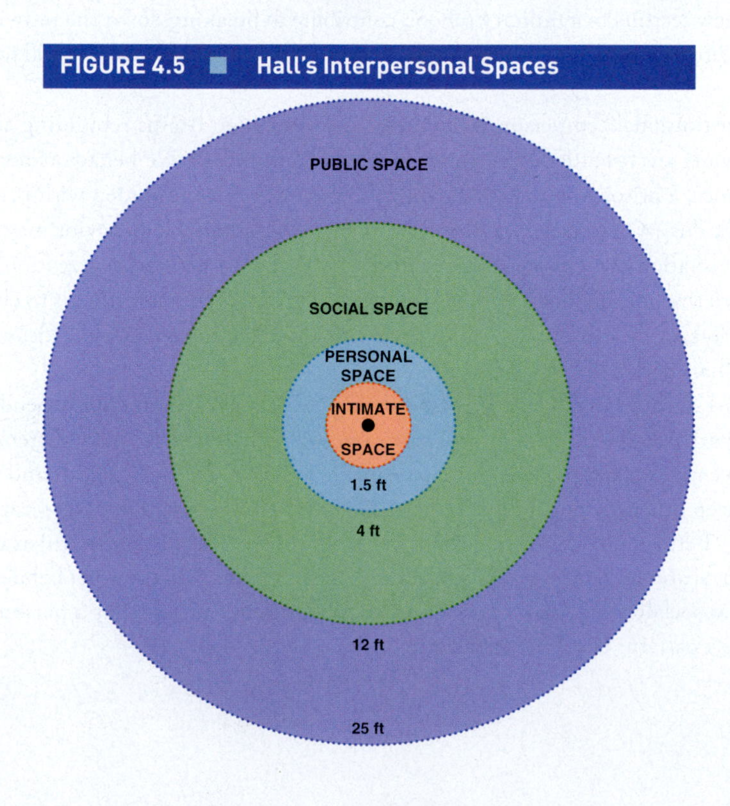

some commonalties emerged: Women preferred more distance from strangers than did men, and older respondents wanted more space than did younger ones. But there were also important differences: Respondents from countries with warmer climates tolerated close proximity more than did those from colder climates, a determination that aligns with hypotheses about "contact cultures" (South America, the Middle East, Southern Europe) and "noncontact cultures" (Northern Europe, North America, Asia) (Hall, 1966). At the same time, there were some interesting variations within the "contact" or "noncontact" cultures: For instance, while Romanians preferred significant distance from strangers, friends were welcome to move close, but their neighbors the Hungarians wanted to keep both strangers and friends at some distance. Argentinians and other South Americans needed less personal space than did those from several other cultures, including those of Asian countries and the Middle Eastern country of Saudi Arabia (Sorokowska et al., 2017).

Studies of Social Interaction

What aspects of daily life and interaction do you think would make an interesting research project? What methods would you use to study these aspects? Symbolic interactionists can study a diverse range of interactions. They often use qualitative methods such as participant observation, interviews, and focus groups to illuminate taken-for-granted meanings, strategies, and discourses. Their studies, a sampling of which are featured below, reveal a rich variety of important aspects and patterns of everyday life.

- The strategies homeless youth use to manage and alleviate stigma, including creating friendships or attempting to pass as nonhomeless, as well as acting aggressive and fighting back (Roschelle & Kaufman, 2004).
- How women in abusive relationships come to redefine their situations (Hattery, 2001; Johnson & Ferraro, 1984) and how they might both draw from *and* reject discourses that label them as "victims" (Leisenring, 2006).
- The ways in which a sense of "corporate social responsibility" is promoted and learned by corporate executives in the work environment (Shamir, 2011).
- The way online gamers coordinate their individual actions with one another and through the user interface to succeed at games such as *World of Warcraft* (Williams & Kirschner, 2012).
- The meanings that new medical students give to their white coats, and how they use their coats to invoke social status (Vinson, 2019).
- The stigma and shaming that parents experience in the child support system and how interactions may result in parents being labeled as "responsible fathers" or "deadbeat dads" (Brittany, 2019).

We now turn our attention to two different approaches addressing social interaction: Erving Goffman's dramaturgical analysis and conversation analysts' efforts to study the way people manage routine talk.

The Dramaturgical Approach: Erving Goffman

Erving Goffman (1959, 1961, 1963a, 1967, 1972), a major figure in the study of social interaction, developed a set of theoretical ideas that make it possible to observe and describe social interaction. Goffman used what he termed the **dramaturgical approach**, *the study of social interaction as if it were governed by the practices of theatrical performance.*

According to Goffman, people in their everyday lives are concerned, similar to actors on a stage, with the **presentation of self**, *the creation of impressions in the minds of others to define and control social situations.* For instance, to serve many customers simultaneously, a waiter must take charge with a "presentation of self" that is polite but firm, ensuring that customers do not take too much time ordering.

Dramaturgical approach: Developed by Erving Goffman, the study of social interaction as if it were governed by the practices of theatrical performance.

Presentation of self: The creation of impressions in the minds of others to define and control social situations.

To assert control, a waiter might simply say, "I'll give you a few minutes to decide what you want" and walk away.

Continuing the metaphor of a theatrical performance, Goffman divides spheres of interaction into two stages, front stage and backstage. In the *front stage*, people are social actors engaged in a process of impression management through the use of props, costumes, gestures, and language. A professor lecturing to her class, a young couple on their first date, and a job applicant in an interview are all governed by existing social norms, so the professor will not arrive in her nightgown; nor will the prospective employee greet his interviewer with a high-five rather than with a handshake. Just as actors in a play must stick to their scripts, so too, suggests Goffman, do we as social actors risk consequences (such as failed interactions) if we diverge from the normative script. As people interact, they monitor themselves and each other, looking for clues that reveal the impressions they are making on others. This ongoing effort at what Goffman called "impression management" results in a continual realignment of the individuals' "performances" as the "actors" refit their roles using dress, objects, voice, and gestures in a joint enterprise.

Goffman offers insights into the techniques we as social actors have in our repertoire. Among them are the following:

- *Dramatic realization* is the actor's effort to mobilize his or her behavior to draw attention to a characteristic of the role he or she is assuming. For instance, a baseball umpire needs to embody authority to both teams and fans so he makes his calls loudly and with bold gestures.
- *Idealization* is an actor's effort to embody in his or her behaviors the officially accredited norms and values of a community or society. People with fewer economic resources might purchase fake designer apparel and accessories (clothes, bags, shoes, or watches) to appear monied.
- *Misrepresentation* is part of every actor's repertoire, ranging from kind deception (telling a friend she looks great when she doesn't) to self-interested untruth (telling a professor a paper was lost in a computer crash when it was never written) to bald-faced prevarication (lying to conceal an affair). The actor wants to maintain a desired impression in the eyes of the audience: The friend would like to be perceived as kind and supportive, the student as conscientious and hardworking, and the spouse as loyal and loving.
- *Mystification* is largely reserved for those with status and power and serves to maintain distance from the audience to keep people in awe. Corporate leaders keep their offices on a separate floor and don't mix with employees, while celebrities may avoid interviews and allow their on-screen roles to define them as savvy and smart.

We also engage in impression management in teams, a group of two or more people ("actors") cooperating to create a definition of the situation favorable to them. For example, sports team members work together, although some may be more skilled than others, to convey a definition of themselves as a highly competent and competitive group. Or the members of a family may work together to convey to their dinner guests that they are content and happy by acting cooperatively and smiling at one another during the group interaction.

The film *The Wizard of Oz* offers a good example of Goffman's idea of mystification. Although the wizard is really, in his own words, "just a man," he maintains his status in Oz by hiding behind a curtain and using a booming voice and fiery mask to convey the impression of awesome power.

The example of the family gives us an opportunity to explore Goffman's concept of the *backstage*, where actors let down their masks and relax or even practice their impression management. Before the dinner party, the home is a backstage. One parent is angry at the other for getting cheap rather than expensive wine; one sibling refuses to speak to the parent who grounded her; and the other won't stop texting long enough to set the table. Then the doorbell rings. Like magic, the home becomes the front stage as the adults smilingly welcome their guests and the kids begin to carry out trays of snacks and drinks.

The guests may or may not sense some tension in the home, but they play along with the scenario so as not to create discomfort. When the party ends, the home reverts to the backstage, and each actor can relax his or her performance.

Like George Herbert Mead, Erving Goffman sees the social self as an outcome of society and social interactions. Goffman, however, characterizes the social self not as a *possession*—a dynamic but still essentially real self—but rather as a *product* of a given social interaction, which can change as we seek to manage impressions for different audiences. Would you say that Mead or Goffman offers a better characterization of us as social actors?

Ethnomethodology and Conversation Analysis

Routine, day-to-day social interactions are the building blocks of social institutions and ultimately of society itself. **Ethnomethodology** is used to study *the body of common-sense knowledge and procedures by which ordinary members of a society make sense of their social circumstances and interactions. Ethno* refers to "folk" or ordinary people; *methodology* refers to the methods they use to govern interaction—which are as distinct as the methods used by sociologists to study them. Ethnomethodology was created through Harold Garfinkel's work in the early 1960s. Garfinkel (1963, 1985) sought to understand exactly what goes on in social interactions after observing that our interpretation of social interaction depends on the context. For example, if a child on a playground grabs another child's ball and runs with it, a teacher may see this as a sign of the child's aggressiveness, while peers may see it as a display of courage. Social interaction and communication are not possible unless most people have learned to assign similar meanings to the same interactions. By studying the specific contexts of concrete social interactions, Garfinkel sought to understand how interacting people come to share similar interpretations of their interactions.

Ethnomethodology: A sociological method used to study the body of common-sense knowledge and procedures by which ordinary members of a society make sense of their social circumstances and interaction.

Garfinkel also believed that in all cultures, people expect others to talk in a way that is coherent and understandable. If this does not happen, they will become anxious and upset. He also argued that the need to make sense of conversations is even more fundamental to social life than cultural norms. Without ways of arriving at meaningful understandings, it is impossible to achieve communication, a fundamental part of culture. Procedures that determine how we make sense of conversations are so important to social interaction, that another field developed from ethnomethodology to focus specifically on talk itself: *conversation analysis*.

Conversation analysis investigates *the way participants in social interaction recognize and produce coherent conversation* (Schegloff, 1990, 1991). In this context, *conversation* includes about any form of verbal communication—from phenomena as routine as small talk to as fraught as police crisis negotiators dealing with a suicidal person, and from events as procedural and scripted as congressional hearings and court proceedings (Heritage & Greatbatch, 1991; Hopper, 1991; Stokoe & Sikveland, 2019; Zimmerman, 1984, 1992). Conversation analysis illustrates that social interaction is not simply a random succession of words and phrases but, instead, a reciprocal process typically occurring in a sequential, patterned manner. To ensure that conversation occurs smoothly, we employ strategies such as *turn-taking*, a strategy allowing us to understand when to take the initiative to speak and when to respond. A person's turn ends once the other conversant indicates they have understood the message. So, when you answer "Fine" to the question "How are you?" you show that you have understood the question and are ready to move ahead with your "real" conversation. Incidentally, "small talk" is not meaningless. Instead, it carries the expectation that our conversational partner will respond. As Elizabeth Stokoe, a British conversation analysist of many different types of talk, explains in an interview in *The Guardian*:

Conversation analysis: The study of how participants in social interaction recognize and produce coherent conversation.

> *Interviewer:* So, all of these things like "Hello, how are you?", "Are you having a good day?" and so on are not meaningless?
>
> *Stokoe:* They are certainly not meaningless if they are absent. For example, when I moved to a new house, the first thing my new neighbor said to me was: "Your gutter is leaking." My response was: "Hello." Now that I've fixed the gutter they say: "How are you?" (Tucker, 2018)

Zoom and other software technologies were widely enlisted by business, education, religion, and health institutions as a substitute for face-to-face communication during COVID-19. Conversational analysis can shed light on communication software phenomena like turn-taking, interrupted conversations, and conversational repair.

©OLIVIER DOULIERY/AFP via Getty Images

Conversations can break down for many reasons. For instance, if you answer "What do you mean?" or "Green" to the question "How are you?" your conversation is likely to break down. Conversation analysts have identified several techniques commonly used to repair such breakdowns. For example, if you begin speaking, but realize midsentence that the other person is already speaking, you can repair this awkward situation by pausing until the original speaker finishes his or her turn, and then restart your turn.

New communication software technologies like Zoom and Microsoft Teams enable individuals and groups of people to be together in conversation from a distance. But they may also change conversational flow, as the typical biological and conversational cues for interaction are altered (Sklar, 2020). Have you noticed differences of turn-taking on these technologies compared to face-to-face communication? What strategies have you or others used to repair conversation when it is interrupted because of the technology enlisted?

Research has also emphasized that conversations are impacted by the larger social structure and power relations that follow us into conversation (Wilson, 1991). For instance, dispatchers may hold power over the caller in emergency phone calls (Whalen et al., 1990; Zimmerman, 1984, 1992), questioners may hold power over testifiers in governmental hearings (Molotch & Boden, 1985), men may hold power over women in male–female interactions (Campbell et al., 1992 Fishman, 1978; West, 1979; West & Zimmerman, 1977, 1983; Zimmerman & West, 1975, 1980), and whites may hold power over minority groups (Hamlet, 2020). Even at the most basic and personal level—a private conversation between two people—social structures exercise a potentially powerful influence. For instance, as Hoops (2020) shows in his conversation analysis of the 2017 film *The Big Sick*, Pakistani Kumail Manji's white friends, in attempting to "identify with his experience" of being disowned by his entire family, do not understand that this will mean losing not only his primary social group but lifelong connections with his cultural identity. They actually end up "marginalizing what Kumail is going through":

> After the climactic fight with his parents, his [Kumail's] roommate Chris enters the room and says, "I was just listening at the door. My mom kicked me out for dealing weed at sixteen, so I get it man," suggesting these two experiences are equivalent, unaware of the effect his anecdote functions to marginalize what Kumail is going through. (Hoops, p. 189)

Ethnomethodology and conversation analysis continue to be relevant to sociologists, constituting one of the 52 special interest sections on the American Sociological Association's website (https://www.asanet.org/asa-communities/asa-sections).

WHAT'S NEXT? SOCIALIZATION, SOCIAL INTERACTION, AND THE INTERNET

We began this chapter questioning the potential role of robots to mediate human interaction in the future. We illustrated the powerful forces of socialization and human interaction. We noted that these forces pattern our daily lives, contour our lifelong experiences, mold our own sense of self, and, perhaps most crucially, ultimately make us human. As you now know, sociologists also identify the importance of "agents" of socialization in this process. We end this chapter with a more mundane, but widely pervasive, force in our daily lives: the Internet.

In 2018, almost half of American teens said they "almost constantly" accessed the Internet, and in 2019, 80% of Americans said they accessed it at least once a day (Anderson & Jiang, 2018; Perrin

& Kumar, 2019). These statistics, it is important to note, were gathered before the 2020 COVID-19 pandemic, when, as Koeze and Popper (2020) put it: "The virus changed the way we internet." We believe this is true, and we foresee the Internet to have a continuing, significant impact on social interaction and socialization. For instance, at the time of this writing, the requirements of social distancing necessitated the use of the Internet and all manner of apps to connect us with workplaces, schools, and religious groups. People also sought out video chat apps like Houseparty and Nextdoor to connect with friends (Koeze & Popper, 2020).

Pre-2020 sociologists were already interested in the potential for the Internet to allow us to experiment with new identities and to seek new forms of and forums for social interaction. We have shown some of the ways in which this may result in positive experiences for underrepresented groups, and also result in continuations of face-to-face behaviors such as bullying. Sociologists will continue to ask how our presentation of self is transformed when we create social selves in the anonymous space of social media. Given the current events, sociologists may need to rethink some of their ideas about the influence of agents such as parents and schools; perhaps these may recede in importance—or grow. You may have noted that sociological research sometimes lags behind quickly changing realities. This is because sociological studies require data gathering, analysis, and peer review, all of which take time. We are interested in the questions you have going forward. For instance:

- *What research question(s) would you ask to learn more about the impact of the Internet on the socialization process? What research methods would you use to answer those questions?*
- *What change have you noticed in interactions with people (peers, especially) on the Internet in the past 3 to 5 years? To what do you attribute these changes?*

WHAT CAN I DO WITH A SOCIOLOGY DEGREE?

Applying Insights About Interaction: Interpersonal Skills

Interpersonal skills are those that allow for constructive, effective communication and relationships between two or more people. The development of interpersonal skills may seem increasingly irrelevant in a world of telework and where more and more of our interactions occur via impersonal devices and even anonymous social media posts. But unless you are a lone worker in a remote outpost, your job will inevitably require some amount of interpersonal interaction. Sociology not only provides a conceptual framework for understanding the nature of social interactions (Chapters 4 and 5), but it also nurtures the development of important interpersonal skills essential to the workplace. The sociological imagination (Chapter 1) gives us perspective—on ourselves and others. Understanding that our locations in the social structure impact who we are, what we have, and how we think is an important step in building more effective relationships with individuals, groups, and communities. It helps to clarify our own biases and assumptions.

The study of sociology also increases *perspective taking*, or the ability to put yourself in someone else's shoes. Sociology can help illuminate cultural differences in communication and interaction, the understanding of which can prevent misunderstandings or misconceptions. For example, rather than assuming that a quiet student is unengaged or a parent who doesn't come to Parent–Teacher Association meetings is uninvolved, a teacher may question whether cultural differences regarding the teacher–student role or access to transportation or translation services is preventing involvement. This can help increase empathy and cultural competence and decrease prejudice, stereotyping, and discrimination. Sociology seeks to ask questions rather than make assumptions about people's motivations and behaviors. Understanding diverse perspectives is the first step in being able to work effectively across differences. By helping to illuminate sources of difference and conflict, and encouraging dialogue across those differences (of race, class, religion, age, gender, and sexuality), sociology can foster important communication and conflict resolution skills that facilitate not just interpersonal but also intergroup relationships. These are all essential skills when working with others to accomplish a task and build community in the workplace.

Devin Frawley, Social Services Assistant at Consulate Health Care

Florida Gulf Coast University, BA in Sociology

My sister was diagnosed with bipolar depression at the age of 16. At that time, I was studying at Florida Gulf Coast University in Fort Myers, Florida. I spent most of my freshman year of college driving to and from Orlando (where my family lived) to visit my sister in numerous hospitalizations. I watched as my sister's mental illness consumed the lives of each and every one of my family members, and I watched as my sister fell further and further into a place so dark, we didn't know if light could be reached. My parents finally admitted her to a rehabilitation center, which saved my sister's life and ignited my passion. I was taking introduction to sociology classes at the time, as an undeclared major, and started connecting themes within the courses to issues involving mental illness, which is a widespread problem in our country. I wrote a couple of papers on the subject and even did my course project on social issues related to mental illness, using my sociological imagination to make the connection between private troubles and public issues. By the end of the semester, my sister was healing rapidly, and I had officially declared myself a sociology major.

I now hold the position of social services assistant at a subacute rehabilitation center. My job duties include attaining resources for residents as well as residents' family members. I was asked a question during my interview: "Why a career in social services at a rehabilitation center?" I answered the question honestly, telling my supervisor this story and how, ultimately, I felt I could better benefit the lives of others due to my experience as one of those "others" myself. I used my interpersonal skills to acquire the position by explaining that my sense of place, location, and life experiences have prepared me for this role. My supervisor later told me I had an outstanding interview and that although I didn't have as much experience as the other candidates, she could tell I was more driven and motivated by this field of work than the other candidates. I credit landing the position to an important aspect of my degree in sociology: the acquisition of interpersonal skills.

Career Data: Social and Human Service Assistant

- 2019 Median Pay: $35,060 per year
- $16.85 per hour
- Typical Entry-Level Education: High school degree or equivalent
- Job Outlook: 13% (much faster than average)

Source: Bureau of Labor Statistics. (2019). *Occupational outlook handbook.*

SUMMARY

- **Socialization** is a lifelong, active process by which people learn the cultures of their societies and construct a sense of who they are.
- What we often think of as "human nature" is, in fact, learned through socialization. Sociologists argue that human behavior is not determined biologically, although biology plays some role; rather, human behavior develops primarily through social interaction.
- Although some theories emphasize the early years, sociologists generally argue that socialization takes place throughout the life course. The theories of Sigmund Freud and Jean Piaget emphasize the early years, while those of George Herbert Mead, Lawrence Kohlberg, and Judith Harris give more consideration to the whole life course (although Mead's **role-taking** theory focuses on the earlier stages of the life course). According to Mead, children acquire a sense of self through symbolic interaction, including the role-taking that eventually enables the adult to take the standpoint of society.
- Kohlberg built on Piaget's ideas to argue that a person's sense of morality develops through different stages, from that in which people strictly seek personal gain or seek to avoid punishment to the stage in which they base moral decisions on abstract principles.
- Our immediate families provide the earliest and typically foremost source of socialization, but also significant are school, work, peers, religion, sports, and mass media.

- Agents of socialization provide patterned guidance for the development of our norms, values, beliefs, and behaviors, but the messages we receive and the behaviors we learn through agents of socialization may differ depending on our social class, race, and gender. Additionally, we may actively seek to learn and emulate new norms, values, and beliefs through the process of **anticipatory socialization.**
- In **total institutions**, such as prisons, the military, and hospitals, individuals are isolated so that society can achieve administrative control over their lives. By enforcing rules that govern all aspects of daily life, from dress to schedules to interpersonal interactions, total institutions can open the way to **resocialization**, breaking down the person's sense of self and rebuilding their personality.
- According to Erving Goffman's **dramaturgical approach**, we are all actors concerned with the **presentation of self** in social interaction. People perform their social roles on the "front stage" and are able to avoid performing on the "backstage."
- Socialization occurs at every stage of the life course through social interaction. This interaction requires understanding cultural norms that vary across global cultures. Symbolic interactionists use a variety of qualitative methods to study social interaction.
- **Ethnomethodology** is a method of analysis that examines the body of common-sense knowledge and procedures by which ordinary members of a society make sense of their social circumstances and interaction.
- **Conversation analysis**, which builds on ethnomethodology, is the study of the way participants in social interaction recognize and produce coherent conversation.

KEY TERMS

anticipatory socialization (p. 90)
behaviorism (p. 80)
cognitive development (p. 83)
conversation analysis (p. 101)
dramaturgical approach (p. 99)
ego (p. 85)
egocentric (p. 83)
ethnomethodology (p. 101)
generalized other (p. 83)
hidden curriculum (p. 88)
I (p. 82)
id (p. 85)
looking-glass self (p. 81)
me (p. 82)
presentation of self (p. 99)
primary groups (p. 81)
psychoanalysis (p. 85)
secondary groups (p. 81)
socialization (p. 79)
social learning (p. 81)
reference groups (p. 82)
resocialization (p. 97)
role-taking (p. 82)
significant others (p. 82)
superego (p. 85)
total institutions (p. 97)

DISCUSSION QUESTIONS

1. Define agents of socialization and describe the agents of socialization that sociologists identify as particularly important. Which of these would you say have the most profound effects on the construction of our social selves? Make a case to support your choices.
2. Sports are an important part of Americans' lives—many enjoy playing them while others follow their favorite sports teams closely in the media. How are sports an agent of socialization? What roles, norms, or values are conveyed through this agent of socialization? Does this vary by sport?
3. What role does the way people react to you play in the development of your personality and your self-image? How can the reactions of others influence whether or not you develop skills as an athlete or a student or a musician, for example?

4. What are the key characteristics of total institutions such as prisons and mental institutions? How does socialization in a total institution differ from ordinary socialization? How is it similar?

5. Recall Goffman's ideas about social interaction and the presentation of self. How have social media sites such as Facebook, Snapchat, and Instagram affected the presentation of self? Have there been changes to what Goffman saw as our front and back stages?

6. Differentiate between conversation analysis and ethnomethodology. If you could research some aspect of your life using one of these methods what would you examine? Why?

©John Minchillo-Pool/Getty Images

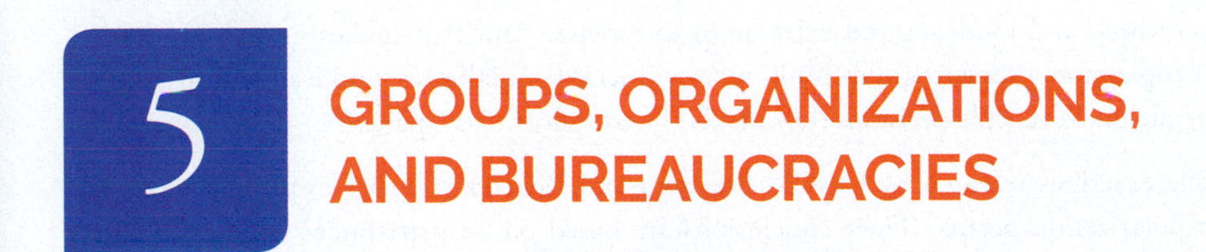

5 GROUPS, ORGANIZATIONS, AND BUREAUCRACIES

WHAT DO YOU THINK?

1. Do most people conform to the expectations of the groups to which they belong? What explains conformity? What explains dissent?
2. What kinds of groups have been spawned by social media in the last decade? How are social media–based groups the same as and different from in-person groups?
3. Why do many people think of bureaucracies as inefficient and annoying? What would be the alternative?

LEARNING OBJECTIVES

5.1 Describe primary and secondary groups and their effects.

5.2 Discuss the power of groups in terms of their composition, leadership, and conformity.

5.3 Explain the sociological concepts of economic, cultural, and social capital.

5.4 Describe three types of formal organizations.

5.5 Apply a sociological lens to modern bureaucracies.

GROUP POLARIZATION: WHEN DOES DIFFERENCE BECOME EXTREMISM?

There is satisfaction in spending time with people who share one's views about the world. Whether the subject is politics and political parties, professional sports, higher education, public health and public policy, or social protest, we may feel more comfortable with those who we perceive as thinking like we do.

A recent article in the magazine *The Atlantic Monthly* offers a perspective on the potential consequences of eschewing company that embraces a diversity of viewpoints:

> But for all the benefits of agreement, solidarity, and spending time with like-minded people, there is compelling evidence of a big cost: the likeminded make us more confident that we know everything and more set and extreme in our views. And that makes groups of like-minded people more prone to groupthink, more vulnerable to fallacies, and less circumspect and moderate in irreversible decisions they make. (Friedersdorf, 2017, para. 8)

The article describes research by Cass R. Sunstein and Reid Hastie that examines why—and how—group polarization occurs. Their conclusions are based on an experiment conducted in the Colorado cities of Boulder, which tends to lean Democratic, and Colorado Springs, which tends to lean Republican. About 60 participants were gathered and put into smaller groups with politically like-minded peers. The research is described in a blog written by the researchers:

> Citizens expressed their views in three ways: anonymously, before deliberation began; in small groups, which deliberated and tried to reach verdicts; and anonymously, after deliberation concluded. Our key question was this: What would be the effect of deliberation on people's views?
>
> Here are our three major findings. (1) Liberals, in Boulder, became distinctly more liberal on all three issues. Conservatives, in Colorado Springs, became distinctly more conservative on all three issues. The result of deliberation was to produce extremism—even though deliberation consisted of a brief (15 minute) exchange of facts and opinions! (University of Chicago Law School Faculty Blog, 2006, paras. 3–4)

The authors add that differences between the groups grew more pronounced and similarities within the groups also increased.

Why did group polarization increase after people interacted with like-minded peers? Sunstein and Hastie offered some hypotheses. First, participants were exposed to new information in the form of arguments they may not have heard or thought of before, but those arguments functioned to confirm their beliefs further, not to challenge them. Second, participants wanted to be received favorably by the group and would adjust their perspectives, if only slightly, to edge closer to the dominant perspective. Third, if, as the researchers contend, less confidence in one's position has a moderating effect on beliefs, then more agreement increases confidence in one's position—which, in turn, increases extremism (Friedersdorf, 2017; Sunstein & Hastie, 2014).

Consider the sociological significance of the Colorado experiment. In the age of social media—and particularly during the COVID-19 pandemic, when many of us were not able to spend substantial time outside of our own homes—people are spending significant time online, doing schoolwork or paid work, playing games, and watching videos. Many are also voraciously consuming news and opinions about the politics and events of the day and interacting virtually with like-minded people, usually on sites that comport with their political views. What does Sunstein and Hastie's finding suggest about potential consequences of pervasive interaction with the like-minded and limited, if any, interaction with diverse perspectives? Can this help to explain political and social polarization in the United States? Is this a problem for the country—and can it be addressed? What do you think?

We begin this chapter with an overview of the nature of social groups, looking at primary and secondary groups and their effects on our lives. We also examine the power of groups in fostering integration and enforcing conformity, among other key functions. We then turn to a discussion of the importance of capital in social group formation and action, followed by an exploration of the place of organizations in society. Next, we address a topic about which sociologist Max Weber wrote extensively and with which we all have some experience—bureaucracies. We end the chapter with a consideration of the modern roles of governmental and nongovernmental organizations in the pursuit of social change.

THE NATURE OF GROUPS

The male elephant is a solitary creature, spending much of its life wandering alone, interacting with other elephants only when it is time to mate or if another male intrudes on its territory. Female elephants, by contrast, live their lives in groups. Human beings are similar to the female elephant: We are social animals who live our entire lives in the company of others. Each of us is born into an emotionally and biologically connected group we know as the family. As we mature, we become increasingly interconnected with other people, some our own age and others not, at school, on sports teams, and through various other social interactions and increasingly via social media. We consolidate and accumulate friends, teammates, and classmates—different groups with whom we interact on a regular basis. Eventually, we get jobs and engage with coworkers and other people we encounter in the course of our work. Along the way, we may form and maintain friendship groups, either in person or virtually, that share our interests in particular activities or lifestyles, such as bowling, model airplanes, video games, or music. Sometimes we are part of groups that gather for special events, such as watching a college football game or attending a political demonstration.

Secondary groups may evolve into primary groups for some members. For example, when young people who play together on a sports team begin to socialize away from the field or court, they may form bonds of friendship that come to constitute a primary group.

©Mariah Wild/Universal Orlando via Getty Images

A moment's reflection on these types of groups reveals that they differ in many important ways, particularly in the degree of intimacy and social support their members experience. Sociologists distinguish between *primary* and *secondary* groups. *Primary groups* are characterized by intense emotional ties, intimacy, and identification with membership in the group. *Secondary groups* are large, impersonal groups with minimal emotional and intimate ties (Cooley, 1909).

Primary groups are of significance because they exert a long-lasting influence on the development of our social selves (Cooley, 1964). Charles Horton Cooley (1864–1929), who first introduced the distinction between primary and secondary groups, argued that people belong to primary groups mainly because these groups satisfy personal needs of belonging and fulfillment. People become part of secondary groups such as business organizations, schools, work groups, athletic clubs, and governmental bodies to achieve specific goals: to earn a living, to get a college degree, to compete in sports, and so on. Social media has blurred the lines of this distinction: Arguably, while belonging to, for instance, a professional development group online can foster the achievement of academic or occupational goals, many people also turn to social media groups to find a sense of belonging among others who, for instance, share the same political views, enjoy particular video games or musical genres, or struggle with similar physical or mental challenges. (For a summary of the characteristics of primary and secondary groups, see Table 5.1.)

THE POWER OF GROUPS

As you learned in Chapter 4, we often judge ourselves by how we think we appear to others, which Cooley termed the *looking-glass self*. Groups as well as individuals provide the standards by which we make these self-evaluations. Robert K. Merton (1968), following Herbert Hyman (1942), elaborated on the concept of the *reference group* as a measure by which we evaluate ourselves. Importantly, a reference group provides a standard for judging our own attitudes or behaviors.

For most of us, the family is the reference group with the greatest impact in shaping our early view of ourselves. As we mature, and particularly during adolescence, peers replace or at least compete with the family as the reference group through which we define ourselves. Today, thanks to the growth of social media, many people establish virtual reference groups and intimate primary groups with people they have never seen face to face.

Reference groups may be primary, such as the family, or secondary, such as a group of soldiers in the same branch of service in the military. They may even be fictional. One of the chief functions of advertising, for example, is to create sets of imaginary reference groups that will influence consumers' buying habits. We are invited to purchase a particular vehicle or fragrance, for instance, in order to join an ostensibly exclusive group of sophisticated, sexy consumers of that item. Reference groups can have powerful effects on our purchasing decisions, as well as other social actions.

Does Size Matter?

Another significant way in which groups differ has to do with their size. German sociologist Georg Simmel (1858–1918) was one of the first to call attention to the influence of group size on people's behavior.

TABLE 5.1 ■ The Characteristics of Primary and Secondary Groups

CHARACTERISTIC	PRIMARY GROUP	SECONDARY GROUP
Social distance of relationships	Low: face-to-face	High: indirect, remote
Intimacy	High: "fusion of personalities"	Low: relatively impersonal
Importance in forming the social self	Fundamental: earliest complete experience of social unity	Secondary: occurs later in life, no all-embracing experience of social unity
Degree of mutual identification with others	High: "we"	Low: "they"
Degree of permanence	High: changes but slowly over time	Low: likely to change over time, sometimes significantly
Examples	Family; children's playgroups; neighborhood and community groups; sports teams; clubs, fraternities, and sororities	Secondary schools, colleges, and universities; businesses and other workplaces; government agencies; bureaucracies of all sorts

Since Simmel's time, small-group researchers have conducted laboratory experiments to discover how group size affects both the quality of interaction in the group and the group's effectiveness in accomplishing certain tasks (Levine & Crowther, 2008; Lucas & Lovaglia, 1998).

The simplest group, which Simmel (1955) called a **dyad**, *consists of two persons.* Simmel reasoned that dyads, which offer both intimacy and conflict, are likely to be simultaneously intense and unstable. To survive, they require the full attention and cooperation of both parties. Dyads are typically the sources of our most elementary social bonds, often constituting the groups in which we are most likely to share our deepest secrets. The commitment two people make through marriage is one way to form a dyadic group. But dyads can also be very fragile. If one person withdraws from the dyad, it vanishes. That is why, as Simmel believed, a variety of cultural and legal norms arise to support dyadic groups, including marriage, in societies where such groups are regarded as an important source of social stability.

FIGURE 5.1 ■ A Dyad and a Triad

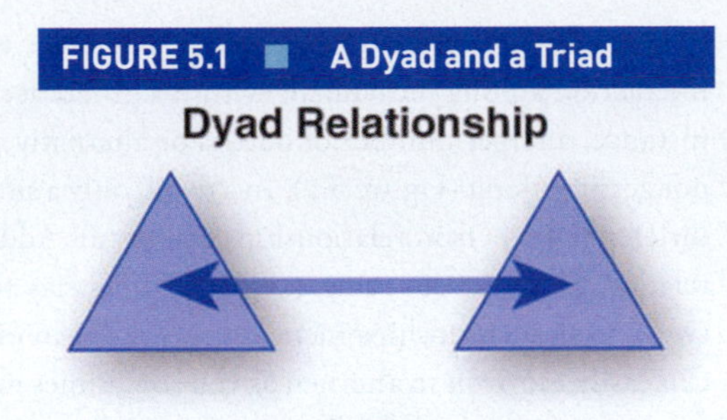

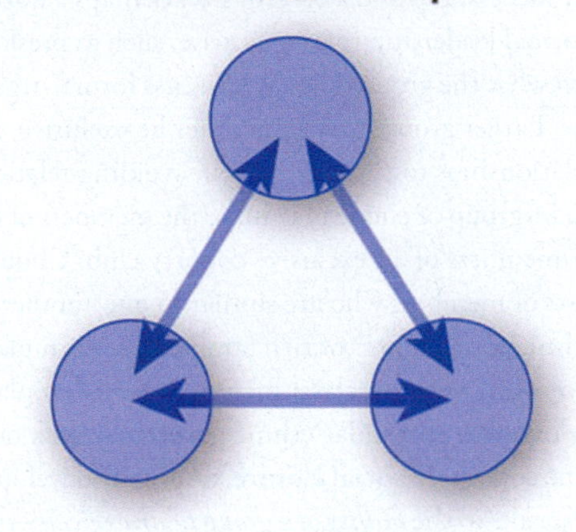

Dyad: A group consisting of two persons.

Adding one other person to a dyad changes the group relationship, producing what Simmel termed a **triad**. Triads are apt to be more stable than dyads, since the presence of a third person relieves some of the pressure on the other two members to always get along and maintain the energy of the relationship. One person can temporarily withdraw his or her attention from the relationship without necessarily threatening it. In addition, if two members have a disagreement, the third can play the role of mediator, as when you try to patch up a falling-out between two friends or classmates (see Figure 5.1).

Triad: A group consisting of three persons.

Alliance (or coalition): A subgroup that forms between group members, enabling them to dominate the group in their own interest.

On the other hand, however, an **alliance (or coalition)** *may form between two members of a triad, enabling them to "gang up" on the third member, thereby destabilizing the group.* Alliances are most likely to form when no member is clearly dominant and all three are competing for the same thing—for example, when three friends are given a pair of tickets to a concert and have to decide which two will go. Larger groups share some of the characteristics of triads. For instance, in fictional stories such as *Lost, The Walking Dead*, or *Lord of the Flies*, alliances form within groups of disaster survivors as individuals forge special relationships with one another to get access to greater power or resources for survival.

Theoretically, in forming an alliance, a triad member is most likely to choose the weaker of the two other members, if there is one. But why would this be the case if picking a stronger member would strengthen the alliance? Choosing a weaker member enables the member seeking to form the alliance to exercise more power and control within the alliance. However, in some revolutionary coalitions, the two weaker members may form an alliance to overthrow the stronger one (Goldstone, 2001; Grusky, Bonacich, & Webster, 1995).

In the popular television program *The Walking Dead*, survivors of a zombie apocalypse cohabit in small groups and pool their resources to flee from or fight the zombies. Across seasons of the show, alliances and leaders have often shifted in response to group needs and individual ambitions.

©HRAUN via Getty Images

Going from a dyad to a triad illustrates an important sociological principle first identified by Simmel: *As group size increases, the intensity of relationships within the group decreases while overall group stability increases.* There are exceptions to every principle, however. Intensity of interaction among individuals within a group decreases as the size of the group increases because, for instance, a larger number of outlets or alternative arenas for interaction exist for individuals who are not getting along (Figure 5.2). In a dyad, only a single relationship is possible; however, in a triad, three different two-person relationships can occur. Adding a fourth person leads to six possible two-person relationships, not counting subgroups that may form. In a 10-person group, the number of possible two-person relationships increases to 45! When one relationship doesn't work out to your liking, you can easily move on to another, as you sometimes may do at large gatherings.

Larger groups tend to be more stable than smaller ones because the withdrawal of some members does not threaten the survival of the entire group. For example, sports teams do not cease to exist simply because of the loss of one player, even though that player might have been important to the team's overall success. Beyond a certain size, perhaps a dozen people, groups may also develop a formal structure. Formal leadership roles may arise, such as president or secretary, and official rules may develop to govern what the group does. We discuss formal organizations later in this chapter.

Larger groups can sometimes be exclusive, since it is easier for their members to limit their social relationships to the group itself, avoiding relationships with nonmembers. This sense of being part of an in-group or clique may unite the members of high school sports teams, campus Greek organizations, or members of an exclusive country club. Cliquishness is especially likely to occur when a group consists of members who are similar to one another in social characteristics like age, gender, class, race, or ethnicity. Members of rich families, for example, may sometimes be reluctant to fraternize with people from the working class; men may prefer to play basketball only with other men; and students who belong to a particular ethnic group may seek out each other's company in the dormitory or cafeteria. The concept of **social closure**, originally developed by Max Weber, is especially relevant here, insofar as it speaks to the *ability of a group to strategically and consciously exclude outsiders or those deemed "undesirable" from participating in the group or enjoying the group's resources* (Murphy, 1988; Parkin, 1979).

Social closure: The ability of a group to strategically and consciously exclude outsiders or those deemed "undesirable" from participating in the group or enjoying the group's resources.

Groups do not always exclude outsiders, however (Blau, 1977; Stolle, 1998). For example, if your social group or club is made up of members from different social classes, racial or ethnic groups, or generations, you are more likely to appreciate diversity thanks to your firsthand experience. This experience with difference may perhaps lead you to be more inclusive of others not like yourself in other

FIGURE 5.2 ■ A Complex Network of Relationships

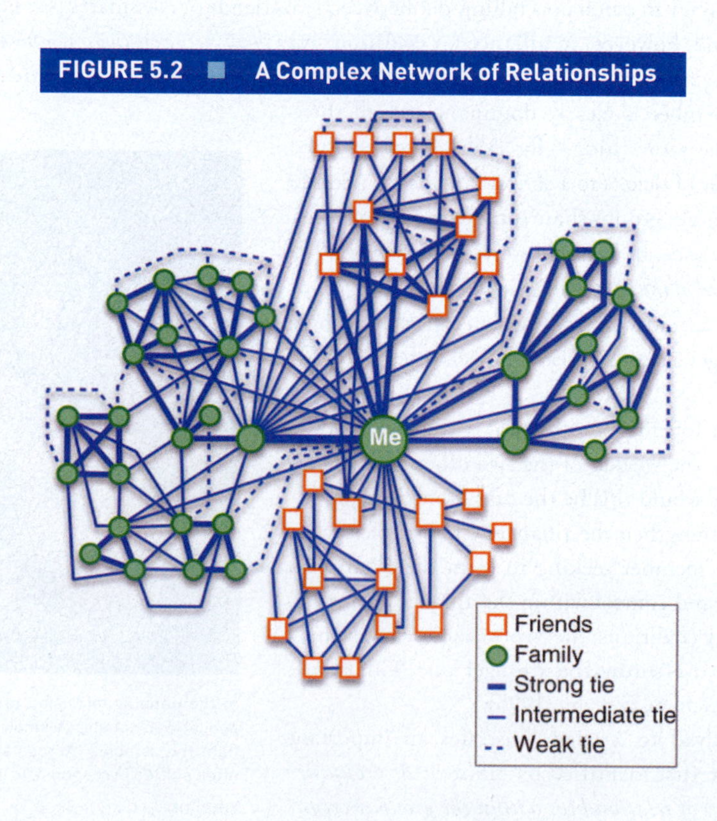

aspects of your life; for example, in bringing together a group to work on a project. As we saw in the opening story, groups with a greater diversity of viewpoints are also less likely to fall into extremism and overconfidence in the rightness of their positions.

Nelson Mandela played an influential role in leading South Africa out of apartheid in the late 1980s and early 1990s. Prior to assuming the presidency of the postapartheid South African government, Mandela, writing from a prison cell, inspired many South Africans, as well as people in other parts of the world, to form anti-apartheid coalitions and groups.

©Pool BOUVET/DE KEERLE/Gamma-Rapho via Getty Images

Types of Group Leadership

A leader is a person able to influence the behavior of other members of a group. All groups tend to have leaders, even if the leaders do not have formal titles. Leaders come in a variety of forms: autocratic, charismatic, democratic, laissez-faire, bureaucratic, and so on. Some leaders are especially effective in motivating members of their groups or organizations, inspiring them to achievements they might not ordinarily accomplish. Such a **transformational leader** goes beyond the merely routine, *instilling in group members a sense of mission or higher purpose and thereby changing (transforming) the nature of the group itself* (Burns, 1978; Kanter, 1983; Mehra, Dixon, Brass, & Robertson, 2006).

Transformational leaders leave their marks on their organizations and can be vital inspirations for social change in the world. Nelson Mandela, the first Black African president of postapartheid South Africa, had spent 27 years in prison, having been convicted of treason against the white-dominated government. Nonetheless, his moral and political position was so strong that upon his release, he immediately assumed leadership of the African National Congress (ANC), leading that political group to the pinnacle of power in South Africa and then assuming the office of president of the country.

Most leaders are not as visionary as Mandela, however. A leader who simply "gets the job done" is a **transactional leader**, *concerned with accomplishing the group's tasks, getting group members to do their jobs, and making certain the group achieves its goals.* Transactional leadership is routine leadership. For example, the teacher who effectively gets through the lesson plan each day but does not necessarily transform the classroom into a place where students explore new ways of thinking and behaving that can change their educational lives is exercising transactional leadership.

For leaders to be effective, they must somehow get others to follow them. How do they do that? At one end of the spectrum, a leader might coerce people into compliance and subordination; at the other end, people may willingly consent to a leader's exercise of power. The basic sociological notion of *power,* the ability to mobilize resources and achieve a goal despite the resistance of others, captures the first point (Emerson, 1962; Hall, 2003; Weber 1921/2012). The related notion of **legitimate authority**, *power exercised over those who recognize it as deserved or earned,* captures the second (Blau, 1964). For example, prison guards often rely on the use of force to ensure compliance with their orders, whereas professors must depend on their legitimate authority if they hope to keep their students attentive and orderly.

Sociologists have typically found authority to be more interesting than the exercise of raw force. After all, it is not surprising that people will follow orders when coerced to do so by force. But why do they go along with authority when they are *not* overtly compelled to do so?

Part of the answer is that people often regard authority as legitimate when it seems to accompany the leadership position. A professor would appear to possess the right to expect students to listen attentively and behave respectfully. Similarly, a commanding officer occupies a position that enables him or her to exercise power over soldiers. Power stemming from an official leadership position is termed **positional power**; that *power depends on the leader's role in the group*

Transformational leader: A leader who is able to instill in group members a sense of mission or higher purpose, thereby changing (transforming) the nature of the group itself.

Transactional leader: A leader who is concerned with accomplishing the group's tasks, getting group members to do their jobs, and making certain that the group achieves its goals.

Legitimate authority: A type of power that is recognized as deserved or earned.

Positional power: Power that depends on the leader's role in the group.

Malala Yousafzai is one example of a young woman who has led through personal power: While she has not held a formal position of power, she has succeeded in mobilizing large numbers of supporters to work toward girls' education across the globe, including through her foundation, the Malala Fund.

©CHRISTOPHE PETIT TESSON/AFP via Getty Images

Greta Thunberg is another young woman who has led through personal power. She has successfully mobilized large numbers of supporters to work toward addressing the threat of climate change, in addition to speaking at the UN Climate Action Summit in 2019.

©JOHN THYS/AFP via Getty Images

Personal power: Power that depends on the ability to persuade rather than the ability to command.

(Chiang, 2009; Hersey, Blanchard, & Natemeyer, 1987; Raven & Kruglianski, 1975).

At the same time, some leaders derive their power from their unique ability to inspire others. Power that derives from the leader's personality is termed **personal power**; *it depends on the ability to persuade rather than the ability to command* (van Dijke & Poppe, 2006). Consider the cases of Malala Yousafzai and Greta Thunberg. Malala, as she is often known to those who admire her achievements and courage, was born in Mingora, Pakistan, in 1997. Even as a teen, she was a fierce advocate for young women's education, a position that put her in opposition to conservative religious and political groups in her country. At the age of 15, Malala was seriously wounded in an attempt on her life by the Taliban, an extreme Islamist group that rejects education and rights for girls and women. When she recovered, she returned to her campaign for girls' global education. She earned a worldwide following for her persistence and willingness to speak out for her cause, becoming, in the process, the youngest recipient of the Nobel Peace Prize in history. Greta Thunberg is another very young woman who has gained widespread recognition, including being selected as *Time* magazine's Person of the Year in 2019, with her advocacy for action on climate change. She began as a single protester waging a "school strike for climate" on Fridays to draw attention to the issue. She began her strikes in 2018. By September 20, 2019, her entreaties for action helped inspire the largest climate demonstration in history when over 4 million people, many of them young people like herself, joined a worldwide strike. Neither Malala nor Greta have positional power, but they have nurtured enormous followings with personal power derived from courage, persistence, and the embrace of public issues that fostered interest and action.

In most situations, the effective exercise of personal power, rather than positional power, is more likely to result in highly motivated and satisfied group members. When group members are confused or ill prepared to undertake a particular task, however, they seem to prefer the more command-oriented style associated with positional leadership (Hersey et al., 1987; Mizruchi & Potts, 1998; Patterson, 1989; Podsakoff & Schriesheim, 1985; Schaefer, 2011).

Conformity to Groups

Following group norms such as getting tattoos or piercings or wearing trendy footwear seems relatively harmless. At the same time, conformity to group pressure can lead to destructive behavior such as drug abuse or serious crimes. For this reason, sociologists and social psychologists have long sought to understand why most people tend to go along with others—and under what circumstances they do not.

FIGURE 5.3 ■ The Asch Experiments: A Study in Conformity

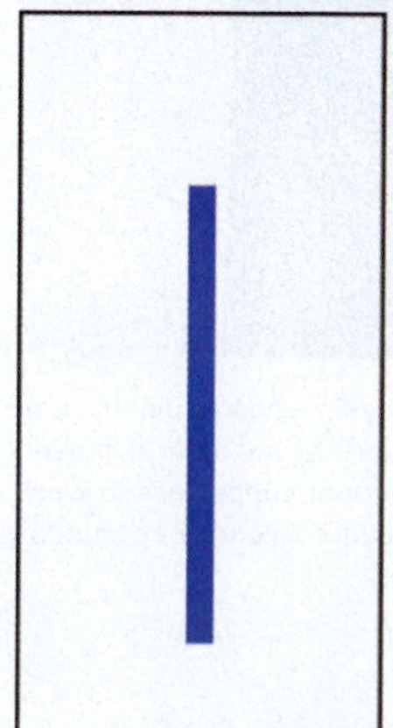

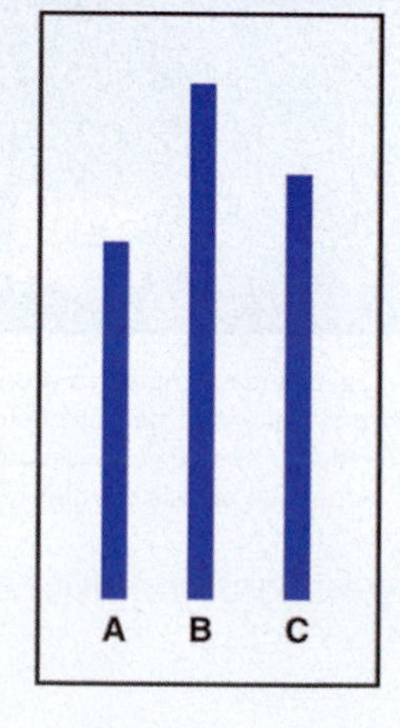

Some of the earliest studies of conformity to group pressures were conducted by psychologist Solomon Asch more than 60 years ago. In one of his classic experiments, Asch (1952) told the group of young male undergraduates in the study that they were taking a "vision test." Subjects were instructed to identify which of three lines of different length most closely matched a fourth (Figure 5.3). The differences between the line lengths were obvious; subjects had no difficulty making the correct match. Asch then arranged a version of the experiment in which the lines to be matched were presented in a group setting, with each person calling out the answer one at a time.

In the second version of the experiment, all but one of the subjects were accomplices of Asch's, who intentionally attempted to deceive the outsider in the group by saying that two lines that were clearly unequal in length were identical. The experiment was conducted with 18 trials, 12 of which were "clinical"—that is, in a dozen of the trials, Asch's accomplices called out the incorrect answer to determine if the subject would follow. Asch's experiment found that about 32% of subjects conformed on the experimental trials; across 12 trials,

75% of subjects conformed at least once. In the experiments where no intentionally misleading answers were given, the subjects offered incorrect responses less than 1% of the time. Asch's experiments demonstrated that many people are willing to discount their own perceptions rather than contradict group consensus.

Significantly, while Asch showed the power of conformity, his experiment also recognized the potential power of dissent. When even one of the experimenter's seven confederates rejected the group consensus, the subject was far more likely to assert that the line lengths were unequal. A later study (Allen & Levine, 1968) affirmed this point, showing that the presence of even a single dissenter could dramatically reduce conformity. This point recalls our opening story and reaffirms the value of a diversity of perspectives and opinions in a group.

Obedience to Authority

Another widely cited study of human behavior was conducted by Stanley Milgram (1963). One of his specific research questions concerned what allowed ordinary German citizens to go along with and even participate in the mass killing of Jews, Roma (also known as *Gypsies*), homosexuals, the disabled, and others who were judged socially undesirable by the Nazis during World War II. Obedience is a kind of conformity, an outcome of conformity to expected norms, rules, or roles. Milgram thus desired to find the boundaries of obedience, to identify how far a person would be willing to go if an authority figure encouraged him or her to complete a given task, even if that task was harmful to another person. His study produced some chilling answers.

In Milgram's experiment, male volunteers were told by an actor dressed in a white lab coat (an authoritative prop) to read aloud pairs of words from a list someone in another room was to memorize and repeat. Whenever the "learner" (who was, unbeknownst to the volunteer subject, an accomplice in the research) made a mistake, the subject was instructed to give the learner an electric shock by flipping a switch on an authentic-looking shock generator (which was actually fake). With each mistake, the voltage of the purported shock was to be increased, until it eventually reached the highest levels, visibly labeled on the machine "450 volts—danger, severe shock."

In reality, the learner never received any electric shocks, but he reacted audibly and physically as if he had, emitting cries that grew louder and more pained, pounding on the table, and moving about in his chair (the cries were prerecorded and played back). Meanwhile, the "scientist" ordered the subject to proceed with the experiment and continue administering shocks, saying things such as, "The experiment requires that you continue" even when the learner expressed concern about his "bad heart."

In what is likely the most widely known variation of the study, about 65% of participants in the study obeyed the commands to keep going, administering what they believed to be electric shocks up to the maximum voltage until nothing but an eerie silence came from the other room (Milgram, 1963; Romm, 2015). What happened here? How could ordinary, basically good people so easily conform to orders that turned them into potential accomplices to injury or death?

The answer, Milgram found, was deceptively simple: Ordinary people will conform to orders given by someone in a position of power or authority, *particularly when the authority is understood to take responsibility for the action*. They will do so even when those orders result in harm to other human beings. Many ordinary Germans who participated in the mass execution of Jews in Nazi concentration camps allegedly did so on the ground that they were "just following orders." Milgram's research, though ethically questionable, produced sobering findings for those who believe that only "other people" would bow to authority.

Milgram's experiments have produced consternation and introspection, but they have also produced critique. Apart from the ethical problems of an experiment that caused its subjects (the "teachers") psychological distress—and that cannot be repeated

Most people would say that they are not capable of committing horrendous acts, yet Stanley Milgram's famous experiment illustrated how obedience to authority can lead people to commit actions that result in harm to others.

due to ethical issues—some critics have called attention to the strict categories of "obedience" and "disobedience" that Milgram used to categorize the actions of his subjects. Matthew Hollander, who reviewed audio recordings of 117 participants in the Milgram studies, points out that these classifications miss some key dynamics of the situation. Specifically, Hollander suggests that both "disobedient" and "obedient" subjects engaged in protest, using phrases like "I can't do this anymore." The difference, he says, is that some subjects were more successful at following through at resistance (Hollander, 2015). Obedience, then, was not unquestioned and uncritical, but stressful and problematic.

As noted earlier, Milgram's experiments have not been reproduced, though a variety of other events give us an opportunity to reflect on the phenomenon of obedience to authority, and the obstacles to resistance.

The 2012 film *Compliance,* which is based on real-life incidents, depicts the events that unfold when a prank phone caller, an unidentified male, calls a fast-food establishment in Kentucky and, pretending to be a police officer, enlists the aid of the store manager to help him crack an ostensibly important case. Once the manager agrees to help the "officer," the prankster tells the manager to perform increasingly invasive acts against a female employee. Obeying a figure believed to be a legitimate (though unseen) authority, in several of the incidents on which the movie is based, restaurant managers actually strip-searched female employees (Kavner, 2012; Wolfson, 2005).

Another example of obedience to authority even in the face of dangerous consequences took place at the Edgewood Arsenal in Maryland, the site of top-secret military experiments involving more than 7,000 U.S. soldiers from the 1950s through the 1970s. Many soldiers volunteered for duty at Edgewood unaware of, or even deceived about, exactly what would be asked of them. Once at Edgewood, they were informed that if they refused to participate in any required duties, they could face jail time for insubordination or receive an unsatisfactory review in their personnel files, and during the Vietnam era, some were reportedly threatened with being sent to war. The soldiers were experimented on repeatedly, often exposed to a variety of dangerous chemical and biological toxins, including sarin gas, VX gas, LSD, tranquilizers, and barbiturates, some of which produced extended and untreated hallucinations (Martin, 2012; Young and Martin 2012). Military authority is built on a chain of command that requires obedience, and the Edgewood Arsenal experiments showcase a situation with greater obstacles to resistance than the Milgram experiments or the events depicted in *Compliance.* At the same time, they underscore a key characteristic of obedience across a variety of scenarios: We as individuals are likely to comply with any demands made by persons in positions of authority whether out of fear of the repercussions associated with failure to comply, belief in the legitimacy of demands made by authorities, or trepidation to resist even when we sense danger to ourselves or others.

Groupthink

Common wisdom may suggest that we put our heads together to solve a problem, but pressures to go along with the crowd sometimes result in poor decisions rather than creative new solutions to problems. You have probably had the experience of feeling uneasy about voicing your opinion while in a group struggling with a difficult decision. Irving Janis (1972, 1989; Janis & Mann, 1977) coined the term **groupthink** to describe *what happens when members of a group ignore ways of thinking and plans of action that go against the group consensus.* Not only does groupthink frequently embarrass potential dissenters into conforming, but it can also produce a shift in perceptions so that group members rule out alternative possibilities without seriously considering them. Groupthink may facilitate a group's reaching a quick consensus, but the consensus may also be ill chosen.

Groupthink: A process by which the members of a group ignore ways of thinking and plans of action that go against the group consensus.

Janis undertook historical research to see whether groupthink had characterized U.S. foreign policy decisions, including the infamous Bay of Pigs invasion of Cuba in 1961. Newly elected president John F. Kennedy inherited from the preceding administration a plan to provide U.S. supplies and air cover while an invasion force of exiled Cubans parachuted into Cuba's Bay of Pigs to liberate the country from Fidel Castro's communist government. A number of Kennedy's top advisers were certain the plan was fatally flawed but refrained from countering the emerging consensus. As it happened, the invasion was a disaster. The ill-prepared exiles were immediately defeated, Kennedy suffered public embarrassment, and the Cold War standoff between the Soviet Union and the United States deepened.

How could Kennedy's advisers, people of strong will and independent judgment educated at elite universities, have failed to voice their concerns adequately? Janis identified a number of possible reasons. For one, they were hesitant to disagree with the president, lest they lose his favor. Nor did they want to diminish group harmony in a crisis situation where teamwork was important. The cohesion of the group was of key concern. In addition, there was little time for them to consult outside experts who might have offered radically different perspectives. All these circumstances contributed to a single-minded pursuit of the president's initial ideas rather than an effort to generate effective alternatives.

Consider as well the case of Penn State University student Timothy Piazza. In 2017, Timothy was a sophomore student pledging the Beta Theta Psi fraternity. After a night of heavy drinking with his fraternity brothers, the young man tumbled down the steps of the frat house, resulting in a traumatic brain injury. According to a news account of the incident, "As he drifted in and out of consciousness, about 20 brothers failed to dial 911 or get outside help from the Penn State University campus—waiting 12 harrowing hours before one of them finally called emergency responders" (Ortiz & Lubell, 2017, para. 2). Two days later, Timothy was dead. Why did no one call for help sooner, when immediate medical intervention might have saved him?

One possibility is that *groupthink* played a role. According to Professor Alan Reifman of Texas Tech University,

> Leaders of a group become committed to a course of action—in this case, making feckless attempts to revive Mr. Piazza without calling the authorities—and follow through on it with great single-mindedness. In groupthink, many of the group members appear to be on the "same page" in executing the plan, but if there is any dissent, it is suppressed. (Ortiz & Lubell, 2017, para. 9)

Indeed, in this instance, one young man entreated the others to allow him to take Timothy to the hospital, but he was dismissed.

When decisions are based primarily on how other group members will react, rather than on ethical, professional, legal, or medical considerations, groupthink can—as these examples show—lead to devastating outcomes.

Think about your own experiences working with groups, whether at work, on a class project, or in a campus organization. Have you ever "gone along to get along" or felt pressured to choose a particular path of action in spite of your own reservations? Or, conversely, have you ever chosen to refuse to conform in spite of the pressure? What factors affected your decision in either case?

ECONOMIC, CULTURAL, AND SOCIAL CAPITAL

One of the most important additions to the sociological study of groups is the contribution of the French school of thought known as **structuralism**, or *the idea that an overarching structure exists within which culture and other aspects of society must be understood.* A leading proponent, the French sociologist Pierre Bourdieu, provides an analytical framework that extends our understanding of the way group relationships and memberships shape our lives. Bourdieu argues that several forms of *capital*—that is, social currency—stem from our association with different groups. These forms of capital are of importance in the reproduction of socioeconomic status in society.

Structuralism: The idea that an overarching structure exists within which culture and other aspects of society must be understood.

Economic capital, the most basic form, consists of *money and material that can be used to access valued goods and services.* Depending on the social class you are born into and the progress of your education and career, you will have more or less access to economic capital and more or less ability to take advantage of this form of capital. Another form is *cultural capital,* or your interpersonal skills, habits, manners, linguistic styles, tastes, and lifestyles. For instance, in some social circles, having refined table manners, being knowledgeable about particular sports or other leisure activities, having cultural interests defined as tasteful, or speaking with a distinctive accent place a person in a social class that enhances his or her access to jobs, social activities, and friendship groups. Cultural capital may also be enhanced by credentials that confirm one's socially elite status, such as a degree from a prestigious university or membership in an exclusive club or organization.

Economic capital: Money and material that can be used to access valued goods and services.

We often hear the phrase, "It's not what you know; it's who you know." Indeed, history shows that social networks are important. The Bush family has had a disproportionate impact on American political life. Members of the Bush family have twice been elected to the presidency and have held the governorships of Florida and Texas. Members of the family continue to be prominent in public life.

©jean-Louis Atlan/Sygma via Getty Images

Friendship groups and other social contacts also provide **social capital**, *the personal connections and networks that enable people to accomplish their goals and extend their influence* (Bourdieu, 1984; Coleman, 1990; Putnam, 2000). College students who join fraternities and sororities expect that their "brothers" or "sisters" will help them get through the often challenging social and academic experiences of college. Other political, cultural, or social groups on campus offer comparable connections and opportunities. Many new (as well as more seasoned) employees and prospective employees join LinkedIn, a social media site that offers possibilities for people to expand their professional social networks, a key part of nurturing social capital.

Social capital: The personal connections and networks that enable people to accomplish their goals and extend their influence.

While social capital is strongly influenced by socioeconomic class status, it may also be related to gender, to race, and intersectionally to both gender and race (McDonald & Day, 2010; McDonald, Lin, & Ao, 2009). In a study of social networks and their relationship to people's information about job leads, sociologists Matt Huffman and Lisa Torres (2002) found that women benefited from being part of networks that included more men than women; those who had more women in their social networks had a diminished probability of hearing about good job leads. Interestingly, the predominance of men or women in a man's social network made no discernible difference. The researchers suggested that perhaps the women were less likely to learn about job leads, and notably, when they knew of leads, they were more likely to pass them along to men than to other women. Similarly, McDonald and Mair (2010) and Trimble and Kmec (2011) explored issues of networking in relation to women's career opportunities over their lifetimes and also the extent to which networks aid women in attaining jobs. Both teams of researchers found that social capital in the form of networks of relations has very distinct and important effects for women. The advent of professional networking sites online offers sociologists the opportunity to expand this research to see if and how gender affects social networks and their professional benefits, as research from the field of psychology suggests that job networking sites (including LinkedIn) play an important role in the job acquisition process (Bohnert & Ross, 2010).

Economic, cultural, and social capital confer benefits on individuals, at least in part, through membership in particular social groups. Characteristics such as class, race, ethnicity, and gender, among others, can have effects on the capital one has. Membership in organizations such as fraternities, exclusive golf clubs, or college alumni associations can offer important network access. These are some examples of the kinds of organizations that shape our lives and society, sometimes to our benefit, sometimes to our disadvantage. In the next section, we look at organizations and their societal functions through the sociological lens.

ORGANIZATIONS

People frequently band together to pursue activities they could not readily accomplish by themselves. A principal means for accomplishing such cooperative actions is the **organization**, *a group with an identifiable membership that engages in concerted collective actions to achieve a common purpose* (Aldrich & Marsden, 1988). An organization can be a small primary group, but it is more likely to be a larger, secondary one: Universities, churches, armies, and business corporations are all examples of organizations. Organizations are a central feature of all societies, and their study is a core concern of sociology today.

Organization: A group with an identifiable membership that engages in concerted collective actions to achieve a common purpose.

Organizations tend to be highly formal in modern industrial and postindustrial societies. A **formal organization** is *rationally designed to achieve particular objectives, often by means of explicit rules, regulations, and procedures.* Examples include a state or county's department of motor vehicles or the federal Internal Revenue Service. As Max Weber (1919/1946) first recognized almost 100 years ago, modern societies are increasingly dependent on formal organizations. One reason is that formality is often a

Formal organization: An organization that is rationally designed to achieve particular objectives, often by means of explicit rules, regulations, and procedures.

requirement for legal standing. For a college or university to be legally accredited, for example, it must satisfy explicit written standards governing everything from faculty hiring to fire safety. Today, formal organizations are the dominant form of organization across the globe.

Types of Formal Organizations

Thousands of different kinds of formal organizations serve every imaginable purpose. Sociology seeks to simplify this diversity by identifying the principal types. Amitai Etzioni (1975) grouped organizations into three main types, based on the reasons people join them: utilitarian, coercive, and normative. In practice, of course, many organizations, especially utilitarian and normative organizations, include elements of more than one type.

Utilitarian organizations are *those that people join primarily because of some material benefit they expect to receive in return for membership.* For example, you probably enrolled in college not only because you want to expand your knowledge and skills but also because you know that a college degree will help you get a better job and earn more money later in life. In exchange, you have paid tuition and fees, devoted countless hours to studying, and agreed to submit to the rules that govern your school, your major, and your courses. Many of the organizations people join are utilitarian, particularly those in which they earn a living, such as corporations, factories, and banks.

Utilitarian organization: A group with an identifiable membership that engages in concerted collective actions to achieve a common purpose.

Coercive organizations are *those in which members are forced to give unquestioned obedience to authority.* People are often forced to join coercive organizations because they have been either sentenced to punishment (prisons) or remanded for mandatory treatment (mental hospitals or drug treatment centers). Coercive organizations may use force or the threat of force, and sometimes confinement, to ensure compliance with rules and regulations. Guards, locked doors, barred windows, and monitoring are all features of jails, prisons, and mental hospitals. Sometimes people join coercive organizations voluntarily, but once they are members, they may not have the option of leaving as they desire. An example of such an organization is the military: While enlisting is voluntary in the United States, once a person joins, he or she is subject to discipline and the demand for submission to authority in a rigidly hierarchical structure. Coercive organizations are examples of total institutions, about which you read in Chapter 4. By encompassing all aspects of people's lives, total institutions can radically alter people's thinking and behavior.

Coercive organizations: Organizations in which members are forced to give unquestioned obedience to authority.

Normative organizations, or **voluntary associations**, are those *that people join of their own will to pursue morally worthwhile goals without expectation of material reward.* Belonging to such organizations may offer social prestige or moral or personal satisfaction. (Of course, such organizations may also serve utilitarian purposes, such as a charitable group you join partly to hand out your business card and boost your chances for monetary gain.)

Normative organizations (or voluntary associations): Organizations that people join of their own will to pursue morally worthwhile goals without expectation of material reward.

The United States is a nation of normative organization joiners. Individuals affiliate with volunteer faith-related groups such as the YMCA, Hillel, and the Women's Missionary Society of the African Methodist Episcopal Church; charitable organizations such as the Red Cross; social clubs and professional organizations; politically oriented groups such as the National Association for the Advancement of Colored People (NAACP); and self-help groups such as Alcoholics Anonymous and Overeaters Anonymous. According to the National Center for Charitable Statistics (2016), there are more than a million public charities; 105,000 private foundations; and 368,000 other types of nonprofit organizations, such as fraternal organizations and civic leagues. Many of these organizations provide their members with a sense of connectedness while enabling them to accomplish personal and moral goals.

Normative organizations may also erect barriers based on social class, race, ethnicity, and gender. Those traditionally excluded from such organizations, including women, Hispanics/Latinxs, Native Americans, African Americans, and other people of color, have, in response, formed their own voluntary associations. Although it may seem that these, too, are exclusionary, such groups have a different basis for their creation—the effort to remedy social inequality. Social justice, as a result, is often their primary concern.

Next, we shift our gaze from voluntary and coercive organizations to a phenomenon that is familiar to most of us—the bureaucracy. While we have some control over our membership in many organizations, we are all—as U.S. residents, taxpayers, students, or recipients of mortgage or college loans, among others—subject to the reach of modern bureaucracy.

BUREAUCRACIES

The authority structure of most large organizations today is bureaucratic. In this section, we will look at the modern bureaucracy—the way it operates and some of its shortcomings. We will also see how bureaucratic structures have been modified or reformed to offer an alternative type of organization.

Max Weber (1919/1946) was the first sociologist to examine the characteristics of bureaucracy in detail. As noted in Chapter 1, Weber defined a *bureaucracy* as a type of formal organization based on written procedural rules, arranged into a clear hierarchy of authority, and staffed by full-time paid officials. Although Weber showed that bureaucracies could be found in many different societies throughout history, he argued that they became a dominant form of social organization only in modern society, where they came to touch key aspects of our daily lives. In particular, Weber suggested that bureaucracies are a highly rational form of organization because they were devised to achieve organizational goals with the greatest degree of efficiency—that is, to optimize the achievement of a task.

Note that when Weber characterized bureaucracies as *rational,* he did not assume that they would always be *reasonable.* By *rational,* he meant that they were organized based on knowable rules and regulations that laid out a particular path to a goal rather than on general or abstract principles or ideologies. As we know from our own contacts with bureaucratic structures—whether they involve long waits on the phone to speak to a human being rather than a computer or the confusing pursuit of the correct person to whom one must turn in a critical student loan application—they are not, in fact, always reasonable.

To better understand the modern bureaucracy, Weber (1919/1946) identified what he referred to as the *ideal type* of this form of organization, describing the characteristics that would be found if the quintessential bureaucracy existed (Figure 5.4). While Weber recognized that no actual bureaucracy necessarily possesses all of the characteristics he identifies, he argued that, by clearly articulating them, he was describing a standard against which actual bureaucracies could be judged and understood.

Written Rules and Regulations

The routine operation of the bureaucracy is governed by written rules and regulations, the purpose of which is to ensure that universal standards govern all aspects of bureaucratic behavior. Typically, rules govern everything from the hiring of employees to the reporting of an absence due to illness. They are usually spelled out in an organizational manual or handbook, now often available to employees on a human resources website that describes in detail the requirements of each organizational position. While these rules and regulations can be lengthy and complex, they are—in theory—knowable, and the expectation is that those who work in and seek the services of a given organization will adhere to them—sometimes even if they don't seem to make sense!

- **Specialized offices:** Positions in a bureaucracy are organized into "offices" that create a division of labor within the organization. The duties of each office, such as bookkeeping or paying invoices, are described in the organizational manual. Each office specializes in one particular bureaucratic function to the exclusion of all others. Such specialization is one of the reasons why bureaucratic organization is said to be efficient; bureaucratic officials are supposed to become experts at their particular tasks, efficient cogs in a vast machine. The efficiency is, ideally, beneficial for the organization and its clients. If you are seeking to clear up a problem with your tuition bill, you will not visit the admissions office because you know the expert advice you seek is to be found in the student accounts office.

- **Hierarchy:** A bureaucracy is organized according to the vertical principle of hierarchy, so that each office has authority over one or more lower-level offices, and each in turn is responsible to a higher-level office. At the top, the leader of the organization stands alone; in the well-known words of then–U.S. president Harry S. Truman, "The buck stops here." The organizational chart of a bureaucracy therefore generally looks similar to a pyramid. Again, efficiency is achieved through the knowable hierarchy of power that governs the organization.

FIGURE 5.4 ■ The Ideal Typical Bureaucracy

- **Impersonality in record keeping:** Within a bureaucracy, communications are likely to be formal and impersonal. Written forms—paperwork or the electronic equivalent—substitute for more personalized human contact, because bureaucracies must maintain written records or databases of all important actions. Modern computer technology has vastly increased the ability of organizations to maintain and access records. In some ways, this is an advantage—for example, when it allows you to register for classes via smartphone instead of standing in line for hours waiting to fill out forms. On the other hand, you may regret the loss of human contact and the inflexibility of the process, however efficient it may be. This impersonality also, ideally, has the effect of ensuring that all clients are treated equally and efficiently rather than capriciously; in reality, however, people with substantial economic, social, or cultural capital often have the easiest time navigating bureaucracies.
- **Technically competent administrative staff:** A bureaucracy generally seeks to employ a qualified professional staff. Anyone who by training and expertise is able to perform the duties of a particular position in an office of the organization is deemed eligible to fill the position. Work in the bureaucracy is a full-time job, ideally providing a career path for the bureaucrat, who must demonstrate the training and expertise necessary to fill each successive position. In its ideal form, the system is a meritocracy—that is, positions are filled on the basis of merit or qualifications, typically demonstrated by performance on competitive exams, rather than on applicants' knowing the "right" people. In practice, however—as is true of the other characteristics listed previously—an actual bureaucracy is unlikely to meet this standard fully. In fact, getting hired into the organization and advancing in it are likely to be influenced not strictly by objective criteria such as education and experience but also by such social variables as age, gender, race, and social connections.

What would you say are the ideal-typical characteristics of a bureaucracy? Does your description comport with that put forth by Max Weber?

©Myung J. Chun/Los Angeles Times via Getty Images

A Critical Evaluation

Bureaucracies popularly evoke images of paper pushers and annoying red tape. In their studies of bureaucracy, sociologists, too, have had much to say about this form of organization, with mixed conclusions. Max Weber recognized that bureaucracies can, indeed, provide organizational efficiency in getting the job done. In contrast to earlier organizational forms, many of which filled positions through nepotism, bribes, or other non–merit-based forms of promotion and were founded to serve the needs of their leaders or small elite groups, modern bureaucracies have many redeeming qualities in spite of the frustrations they cause.

At the same time, Weber argued that a bureaucracy may create what he termed an *iron cage*—a prison of rules and regulations from which there is little escape (DiMaggio & Powell, 1983; Weber, 1904–1905/2002). The iron cage, which Weber memorably described as having the potential to be a "polar night of icy darkness," is a metaphor. We become "caged" in bureaucratic structures when we build them to serve us (as rules and regulations would ideally do), but they ultimately come to trap us by denying our humanity, creativity, and autonomy.

As you think about this metaphor of the iron cage, consider encounters with bureaucratic structures that you have had: If you've ever had the feeling that solving a personal or family problem concerning tuition, taxes, or immigration would require speaking to a human being with the power to make a decision or to see that your case is an exception in some way—but no such human was available!—then you can see what Weber meant. We make rules and regulations to keep order and to have a set of knowable guidelines for action and decisions, but what happens when the rules and regulations and their enforcement become the *ends* of an organization rather than a *means* to an end? Then we are in the iron cage.

Sociologists have identified a number of specific problems that plague bureaucracies, many of which may be familiar and could be thought of as representing *irrationalities of rationality*:

- **Waste and incompetence:** As long as administrators appear to be doing their jobs—filing forms, keeping records, responding to memos, and otherwise keeping busy—nobody really wants to question whether the organization as a whole is performing effectively or efficiently. Secure in their positions, bureaucrats may become inefficient, incompetent, and often indifferent to the clients they are supposed to serve.
- **Trained incapacity:** We have all seen bureaucrats who "go by the book," even when a situation clearly calls for fresh thinking. Thorstein Veblen (1899), a U.S. sociologist and contemporary of Weber's, termed this tendency *trained incapacity,* a learned inability to exercise independent thought. However intelligent they may otherwise be, such bureaucrats make poor judgments when it comes to decisions not covered by the rule book. They become so obsessed with following the rules and regulations that they lose the ability and flexibility to respond to new situations.
- **Goal displacement:** Bureaucracies may lose sight of the original goals they were created to accomplish. Large corporations such as General Motors and government organizations such as the Department of Homeland Security employ thousands of middle-level employees whose job it is to handle the paperwork required in manufacturing automobiles or computers or in protecting the country. Perhaps understandably, such people may, over time, become preoccupied with getting their own jobs done and, driven by the need to ensure the continuation of particular practices or programs linked to their positions, eventually lose touch with the larger goals of the organization. This shift in focus adds to costs, lowers efficiency, and may prove detrimental to corporations that compete in a global economy and governments seeking to accomplish goals and stay within tight budgets.

Although Weber presented a sometimes-chilling picture of bureaucracies operating as vast, inhuman machines, we all recognize that, in practice, there is often a human face behind the counter. In fact, much important work done in bureaucratic organizations is achieved through informal channels and personal ties and connections rather than through official channels, as sociologist Peter Blau showed in his research (Blau & Meyer, 1987). For example, a student who wishes to register late for a class may avoid having to get half a dozen signatures if he or she knows the professor or a staff person in the registrar's office. However, because of the shortcomings of bureaucratic forms of organization, some theorists have argued for the development of alternative organizational forms. We discuss some of these after looking at the relationship between bureaucracy and democracy.

Bureaucracy and Democracy

Max Weber argued that bureaucracies were an inevitable component of modern society, with its large-scale organizations, complex institutional structure, and concern with rationality and efficiency. Yet many observers have viewed bureaucracy as a stifling, irrational force that dominates our lives and threatens representative government. In *Les Employés* (1841/1985), French novelist Honoré de Balzac, who popularized the term *bureaucracy,* called it "the giant power wielded by pigmies" and a "fussy and meddlesome" government. Do bureaucracies inevitably lead to a loss of freedom and erosion of democracy? Are there more humanistic alternatives to bureaucracies that allow freer, more fulfilling participation in the organization? Let's look briefly at the views of sociologist Robert Michels on the incompatibility between democracy and bureaucracy, then see what some people have done to try to reform this organizational structure.

Michels (1876–1936), another contemporary of Weber's, argued that bureaucracy and democracy are fundamentally at odds. He observed that the Socialist Party in Germany, originally created to democratically represent the interests of workers, had become an oligarchy, a form of organization in which a small number of people exert great power. For him, this was an example of what he termed the **iron law of oligarchy**, *an inevitable tendency for a large-scale bureaucratic organization to become ruled undemocratically by a handful of people.* (*Oligarchy* is defined as the rule of a small group over many people.) Following Weber, Michels argued that in a large-scale bureaucratic organization, the closer you are to the top, the greater the concentration of power. People typically get to the top because they are ambitious, hard-driving, and effective in managing the people below or because they have economic, cultural, and social capital to trade for proximity to power. Once there, leaders increase their social capital through specialized access to information, resources, and influential people—access that reinforces their power. They also often appoint subordinates who are loyal supporters and thus further enhance their position. Such leaders may come to regard the bureaucracy as a means to meet their own needs or those of their social group. The democratic purposes of an organization may become subordinate to the needs of the dominant group.

Iron law of oligarchy: Robert Michels's theory that there is an inevitable tendency for a large-scale bureaucratic organization to become ruled undemocratically by a handful of people.

Since all modern societies require large-scale organizations to survive, Michels believed that democracies—or, in some cases, organizations—may sow the seeds of their own destruction by breeding bureaucracies that eventually grow into undemocratic oligarchies. Following Michels, one can make a case that institutions such as the U.S. Congress show tendencies to act in the interests of political parties, powerful members, or wealthy donors rather than the interests of constituents. For example, throughout the summer of 2020, as the COVID-19 pandemic was causing significant distress in the U.S. labor market and economy, the U.S. Congress failed to pass a relief package that would have extended unemployment benefits to jobless Americans and relief for states and cities struggling with budget shortfalls. While real differences and disagreements existed between the Republican and Democratic Parties and representatives, politicians were also keen to cater to their perceived constituencies in the months before the 2020 election. The question of whether legislators were representing their constituents or the interests of donors, corporations, and other interest groups when they opted for frequent recesses over focused negotiations is one that Michels might have posed in 2020.

In response to what they feel is the stifling effect of bureaucratic organizations, some people have sought alternative forms of organizations designed to allow greater freedom and more fulfilling participation. For example, as part of the sweeping countercultural spirit of the late 1960s and early 1970s, many youthful activists joined collectives, small organizations that operate by cooperation and consensus. Food cooperatives, employee-run newspapers and health clinics, and "free schools" sprang up as organizations that sought to operate by consensus rather than by bureaucracy. Members of these organizations shunned hierarchy, avoided a division of labor based on expertise, and happily sacrificed efficiency in favor of more humanistic relationships.

The founders of these organizations believed they were reviving more personal organizational arrangements that could better enable society to reach certain goals. Although these organizations initially met with some success and left a legacy, they also confronted a larger society in which more conventional forms of organization effectively shut them out. Members of such organizations as the food cooperatives and employee-run newspapers and health clinics of the 1960s and 1970s favored the values of cooperation and service over the more competitive and materialistic values of the larger society. In the exuberance of that period, members of collectives believed they were forging a radically new kind of antibureaucratic organization.

In her examination of early collectives, sociologist Joyce Rothschild-Whitt (1979) studied several that self-consciously rejected bureaucracy in favor of more cooperative forms. In one health clinic, for example, all jobs were shared (to the extent legally possible) by all members: Doctors would periodically answer telephones and clean the facility, while nurses and paramedical staff would conduct examinations and interview patients. While the doctors were paid somewhat more than the other staff members, the differences were not large and were the subject of negotiation by everyone who worked at the clinic.

As long as the collectives remained small, they were able to maintain their founders' values. On the other hand, vastly reduced pay differentials between professional and nonprofessional staff, job sharing, and collective decision making often made it difficult for the collectives to compete for employees with organizations that shared none of these values (Rothschild-Whitt, 1979). Doctors, for example, could make much more money in conventional medical practice, without being expected to answer telephones or sweep the floor. Over time, the original cooperative values tended to erode, and many of the new organizations came to resemble conventional organizations in the larger society. Still, more than three decades after Rothschild-Whitt studied them, a number of these original groups still exist. Although they may have lost some of their collective zeal, they still operate more cooperatively than most traditional organizations.

One foray away from hierarchically and bureaucratically organized entities has been made by the online retailer Zappos. In early 2014, the company announced that it planned to introduce a "holacracy," replacing traditional management structures with "self-governing 'circles.'" The goal of holacracy, according to a media account of the practice, is to "organize a company around the work that needs to be done instead of around the people who do it." Hence, a holacracy is devoid of job titles; instead, employees are integrated into multiple circles of cooperative workers. A few other companies are experimenting with holacracy as well (McGregor, 2014). Do workers perform well in contexts of dispersed or ambiguous authority? Results at Zappos, according to *Fortune* magazine, have been mixed, and after 3 years of the new management style, the company fell off the 100 Best Companies to Work For list, where it had occupied a place for the past 8 years. The change has been described as fostering both chaos and new ideas, and is slated to continue (Reingold, 2016).

The Global Organization

Organizations from multinational corporations to charitable foundations span the globe and increasingly contribute to what some sociologists believe is a "homogenization" of the world's countries (McNeely, 1995; Neyazi, 2010; Scott & Meyer, 1994; Thomas, Meyer, Ramirez, & Boli, 1987). You can listen to the same music, employ the same Internet search engine, see the same films, and eat the same meals (if you wish) in Bangalore and Baku as you do in Berlin and Boston.

Global organizations are not new. The Hanseatic League, a business alliance between German merchants and cities, dominated trade in the North and Baltic Seas from the mid-12th to the mid-18th centuries. The British East India Company virtually owned India and controlled the vast bulk of trade throughout the Far East for several centuries. In 1919, following World War I, the League of Nations was formed, uniting the most economically and militarily powerful nations of the world in an effort to ensure peace and put an end to war. When Germany withdrew and began expanding its borders throughout Europe, however, the League dissolved.

After World War II, a new effort at international governance was made in the form of the United Nations (UN), begun in 1945. The United Nations is still important and active today: Its power is limited, but its influence has grown. It not only mediates disputes between nations, but it is also ever present in international activities ranging from fighting hunger and HIV/AIDS to mobilizing peacekeeping troops and intervening to address conflicts and their consequences.

International organizations exist in two major forms: those established by national governments and those established by private organizations. We consider each separately below.

International Governmental Organizations

The first type of global organization is the **international governmental organization (IGO)** *established by treaties between governments to facilitate and regulate trade between the member countries, promote national security* (both the League of Nations and the United Nations were created after highly destructive world wars), *protect social welfare or human rights, or ensure environmental protection.*

International governmental organization (IGO): An international organization established by treaties between governments to facilitate and regulate trade between the member countries, promote national security, protect social welfare and human rights, or ensure environmental protection.

Some of the most powerful IGOs today were created to unify national economies into large trading blocs. One of the most complex IGOs is the European Union (EU), whose rules now govern 28 countries in Europe; five additional countries have applied for EU entry. The European Union was formed to create a single European economy in which businesses could operate freely across borders in search of markets and labor and workers could move freely in search of jobs without having to go through customs or show passports at border crossings. EU member states have common economic policies, and 18 of them share a single currency (the euro). Not all Europeans welcome economic unity, however, since it means their countries must surrender some of their economic power to the EU as a whole. Being economically united by a single currency also means the economic problems felt by one country are distributed among all the other countries to some degree. Thus, when economic crisis hit Europe in 2008, the severe economic woes of Greece, Portugal, and Spain (among others) caused serious problems for stronger EU economies, like that of Germany.

IGOs can also wield considerable military power, provided that their member countries are willing to do so. The North Atlantic Treaty Organization (NATO) and the United Nations, for example, have sent troops from some of their participating nations into war zones in Iraq and Afghanistan in recent years. Yet, because nations ultimately control the use of their own military forces, there are limits to the authority of even the most powerful military IGOs, whose strength derives from the voluntary participation of their member nations.

International nongovernmental organization (INGO): An international organization established by agreements between the individuals or private organizations making up the membership and existing to fulfill an explicit mission.

IGOs often reflect inequalities in power among their members. For example, the UN Security Council is responsible for maintaining international peace and security and is therefore the most powerful organization within the United Nations. Its five permanent members include the United Kingdom, the United States, China, France, and Russia, which gives these countries significant clout over the Security Council's actions. The remaining 10 Security Council member countries are elected by the UN General Assembly for 2-year terms and therefore have less lasting power than the permanent members.

International Nongovernmental Organizations

The second type of global organization is the **international nongovernmental organization (INGO)**, *established by agreements between the individuals or private organizations making up the membership and existing to fulfill an explicit mission*. Examples include Doctors Without Borders (Médecins sans Frontiers), the International Sociological Association, the World Wildlife Federation, and the International Red Cross. Global business organizations (GBOs) represent a subtype within the broader category of INGOs. The concept of the GBO captures the fluid and highly interconnected nature of our modern globalized labor market in which employees often interact and communicate with people from other nations and cultures, facilitated by technology and online networks.

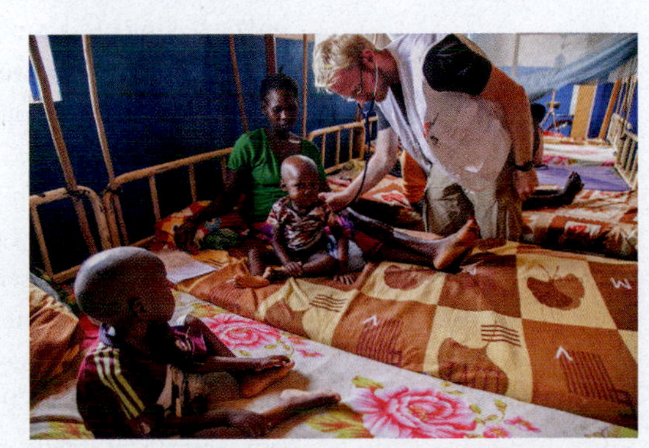

Médecins sans Frontières (MSF, or Doctors Without Borders) works around the world to serve underserved or conflict-torn countries and communities. In this photo, a doctor from MSF examines children suffering from malnutrition in a hospital in Bambari, Central African Republic.

The international nongovernmental organization called the Committee to Protect Journalists (CPJ) is dedicated to protecting press freedom worldwide and to supporting the safety of journalists who report the news. Jamal Khashoggi, a Saudi Arabian national who wrote for the U.S.-based *Washington Post*, was murdered in Istanbul, Turkey, in October 2018. The CPJ has sought to draw attention to his killing, which it believes was carried out with the consent of the Saudi government, which limits press freedom.

©YASIN AKGUL/AFP via Getty Images

Amnesty International is an INGO comprised of over 7 million people, including 2 million members and more than 5 million activists and supporters across more than 150 countries and territories, including over a million in the United States (Amnesty International, n.d.). One of its goals is to help secure the freedom of people imprisoned because of their political beliefs or actions, especially those in immediate danger of torture or execution.

Amnesty International has helped thousands of individual prisoners since it was founded in Britain in 1961. It functions as a global pressure group made up of ordinary citizens. Anyone can join, pay nominal annual dues, and become part of a global "urgent action network" that is regularly mobilized to send government officials faxes, letters, and e-mail and social media entreaties on behalf of prisoners. Amnesty International also sends delegations to countries where government abuses are rampant. The reports these delegations write draw worldwide media attention to the world's prisons and political prisoners.

NGOs such as Amnesty International highlight the power that ordinary citizens have when they band together toward the achievement of a defined goal. If you could create an NGO or INGO to draw attention to a social or political issue, what issue would you choose? How would you persuade others to support your cause?

Although they are far more numerous than IGOs and have achieved some successes, INGOs have less power over state actions and policies, since legal power (including enforcement) ultimately lies with governmental organizations and treaties. At the same time, their influence in individual countries can be considerable.

WHAT'S NEXT? GROUPS, ORGANIZATIONS, AND BUREAUCRACIES

In this chapter, we examined the topic of groups and significance of groups in fostering a sense of belonging in our social environment. But while we may get comfort from group membership, whether that is a primary or secondary group, it is valuable to recognize that groups may exist in tension with one another. Both research and anecdotal evidence point to a growing polarization in U.S. society. One particularly salient fissure in the United States today is political. Arguably, people's sense of belonging, whether to a politically progressive left-wing group that orients toward Democratic politics or to a conservative right-wing group that orients toward support for Republicans, is sharpened by having not only a feeling of like-mindedness with other members but by highlighting the contrast with the other group. This is a topic we explored in our opening story.

The year 2020 brought significant new challenges and changes to U.S. society in the form of the COVID-19 pandemic, the rise of massive movements for social change that are demanding an end to racism in American institutions and practices, the growing visibility of white nationalist groups espousing ideologies of racial exclusion and hate, and a highly contentious national election. Political polarization, never absent, has become a defining feature of the landscape.

Another key feature of the U.S. landscape is the ubiquity and power of social media. Social media is widely blamed for exacerbating political polarization. Is this trend likely to continue? And if, as Sunstein and Hastie (2014) suggest, limiting our interactions largely to the like-minded fosters extremism, can social media also contribute to reducing divisions and opening the minds of users to new ideas? Or should we seek an antidote to extremism elsewhere? What do you think?

WHAT CAN I DO WITH A SOCIOLOGY DEGREE?

Leadership Skills and Teamwork

Sociology students not only examine theories and research about groups and organizations, they often have multiple opportunities both in and out of class to develop leadership and teamwork skills that are essential in the workplace. In fact, many of the skills discussed in other sections in this book are essential to leadership and teamwork, especially problem solving, understanding diversity, interpersonal skills, and community resource skills.

Effective leaders must be able to harness organizational and community resources as well as human capital to meet goals. They must be able to identify and address challenging group dynamics to ensure that people work together effectively. Leadership of a team requires the ability to work with team members and stakeholders of diverse backgrounds. Sociology students gain a critical understanding of issues of power and inequality and intergroup relationships, especially the ways in which social identities of race, class, gender, sexual orientation, and so forth influence interaction and participation. Sociology students learn to ask critical questions about who is participating and who has the power to make decisions. The interpersonal competencies discussed elsewhere, particularly increased self-awareness and perspective taking and communication and conflict resolution skills, are essential to effective leadership and teamwork in diverse workplaces. Sociology students are increasingly exposed to experiential learning opportunities that encourage the development of these skills firsthand through student leadership programs, internships, and service learning in various organizations.

Steven Rice, Human Resources Generalist, Charlotte, NC

Jacksonville University, B.S. Sociology

Working in Human Resources (HR), one must be equipped with handling diversity in the workplace. One goal of HR is to ensure that employees have a safe work environment. Sociology provided me with a deep dive into social factors that can come about in the workplace. Such factors as: gender, race, ethnicity, age, and education play important parts in HR. While studying sociology, I would find myself being faced with the challenge to open my mind to other's thoughts and specify guaranteeing that minority voices are heard. This has now carried over into my HR career. For example, if HR wanted to plan an employee event to increase employee engagement, I would revert to some sociology charged questioning. A few questions I might ask would be, "can this event be enjoyed by all ages of the company?" or "does this event take in account all cultures?" For example, a Christmas party might become a "Winter Celebration." Overall, in HR, you always want to keep everyone's different social backgrounds in mind.

Another key factor of HR is helping develop policies and educating employees on processes. Sociology helped me keen in on my communication skills as well as analyzing data, both of which are effective tools to have when developing policies and educating employees. For a company to be successful, there must be processes in place and HR is there to help implement said policies. When I have a training with employees, I feel as if I have the advantage of sociology. I understand that everyone is different, and I try to be as inclusive as possible and make all feel valued.

Even though it took some time to find the right fit for a major, from being undecided to studying psychology, I finally landed in the world of sociology. I am beyond grateful that I ended up with sociology as my major and even more thrilled that it has been beneficial in my HR career. I would recommend that everyone take at least one sociology class, you may not keep it as a major, but you will certainly learn some incredible things.

Career Data: Human Resource Managers

- 2019 Median Pay: $116,720 per year
- $56.11 per hour
- Typical Entry-Level Education: Bachelor's degree
- Job Outlook, 2019–2029: 6% (faster than average)

Source: Bureau of Labor Statistics, *Occupational outlook handbook*, 2020.

SUMMARY

- The importance of social groups in our lives is one of the salient features of the modern world. Social groups are collections of people who share a sense of common identity and regularly interact with one another based on shared expectations. There are many conceptual ways to distinguish social groups sociologically in order to better understand them.
- Among the most important characteristics of a group is whether or not it serves as a reference group—that is, a group that provides standards by which we judge ourselves in terms of how we think we appear to others, what sociologist Charles Horton Cooley termed the *looking-glass self*.
- Group size is another variable that is an important factor in group dynamics. Although their intensity may diminish, larger groups tend to be more stable than smaller groups of two (**dyads**) or three (**triads**) people. While even small groups can develop a formal group structure, larger groups develop a formal structure.
- Formal structures include some people in leadership roles—that is, those group members who are able to influence the behavior of the other members. The most common form of leadership is **transactional**—that is, routine leadership concerned with getting the job done. Less common is **transformational leadership**, which is concerned with changing the very nature of the group itself.
- Leadership roles imply that the role occupant is accorded some power, the ability to mobilize resources and get things done despite resistance. Power derives from two principal sources: the personality of the leader (**personal power**) and the position that the leader occupies (**positional power**). Max Weber highlighted the importance of charisma as a source of leadership as well as leadership deriving from traditional authority (a queen inherits a throne, for example).
- In general, people are highly susceptible to group pressure. Many people will conform to group norms or obey orders from an authority figure, even when there are potentially negative consequences for others or even for themselves.
- Important aspects of groups are the networks that are formed between groups and among the people in them. Networks constitute broad sources of relationships, direct and indirect, including connections that may be extremely important in business and politics. Women, people of color, and people with lower incomes typically have less access to the most influential economic and political networks than do upper-class white males in U.S. society.
- As a consequence of unequal access to powerful social networks, there is an unequal division of social capital in society. **Social capital** is the knowledge and connections that enable people to cooperate with one another for mutual benefit and to extend their influence. Some social scientists have argued that social capital has declined in the United States during the last quarter century—a process they worry indicates a decline in Americans' commitment to civic engagement.
- **Formal organizations** are organizations that are rationally designed to achieve their objectives by means of rules, regulations, and procedures. They may be **utilitarian**, **coercive**, or **normative**, depending on the reasons for joining. One of the most common types of formal organizations in modern society is the bureaucracy. Bureaucracies are characterized by written rules and regulations, specialized offices, a hierarchical structure, impersonality in record keeping, and professional administrative staff.
- The **iron law of oligarchy** holds that large-scale organizations tend to concentrate power in the hands of a few people. As a result, even supposedly democratic organizations tend to become undemocratic when they become large.

- A number of organizational alternatives to bureaucracies exist. These include collectives, which emphasize cooperation, consensus, and humanistic relations. Networked organizations, which increase flexibility by reducing hierarchy, are similar to collectives in their organization.
- Two important forms of global organizations are **international governmental organizations (IGOs)** and **international nongovernmental organizations (INGOs)**. Both kinds of organizations play increasingly important roles in the world today, and IGOs—particularly the United Nations—may become key organizational actors as the pace of globalization increases.

KEY TERMS

alliance (or coalition) (p. 111)
coercive organizations (p. 119)
dyad (p. 111)
economic capital (p. 117)
formal organization (p. 118)
groupthink (p. 116)
international governmental organization (IGO) (p. 125)
international nongovernmental organization (INGO) (p. 125)
iron law of oligarchy (p. 123)
legitimate authority (p. 113)
normative organizations (or voluntary associations) (p. 119)
organization (p. 118)
personal power (p. 114)
positional power (p. 113)
social capital (p. 118)
social closure (p. 112)
structuralism (p. 117)
transactional leader (p. 113)
transformational leader (p. 113)
triad (p. 111)
utilitarian organizations (p. 119)

DISCUSSION QUESTIONS

1. Can you think of a time when a group to which you belonged was making a decision you thought was wrong—ethically, legally, or otherwise—but you went along anyway? How do your experiences confirm or refute Janis's characterization of groupthink and its effects?
2. List the primary and secondary groups of which you are a member, then make another list of the primary and secondary groups to which you belonged 5 years ago. Which groups in these two periods were most important for shaping (a) your view of yourself, (b) your political beliefs, (c) your goals in life, and (d) your friendships?
3. Think of a time when you chose to go along to get along with a group decision, even when you were inclined to think or behave differently. Think of a time when you opted to dissent, choosing a path different from that pursued by your group or organization. How would you account for the different decisions? How might sociologists explain them?
4. What did Stanley Milgram seek to test in his human experiments at Yale University? What did he find? Do you think that a similar study today would find the same results? Why or why not?
5. Max Weber suggested that bureaucracy, while intended to maximize efficiency in tasks and organizations, could also be highly irrational. He coined the term *the iron cage* to talk about the web of rules and regulations he feared would ensnare modern societies and individuals. On the one hand, societies create organizations that impose rules and regulations to maintain social order and foster the smooth working of institutions such as the state and the economy. On the other hand, members of society may often feel trapped and dehumanized by these organizations. Explain this paradox using an example of your own encounters with the "iron cage" of bureaucracy.

©Peter Titmuss/Alamy Stock Photo

DEVIANCE AND SOCIAL CONTROL

WHAT DO YOU THINK?

1. Is everyone deviant at least some of the time? Does this make deviance normal?
2. How might deviance be linked to social inequality?
3. What methods are most effective for controlling deviance? Does all deviance need to be controlled?

LEARNING OBJECTIVES

6.1 Define *deviance* from a sociological perspective.

6.2 Describe key theories used in sociology to explain deviance.

6.3 Identify types of deviance studied by sociologists.

6.4 Discuss how deviance is controlled and punished in the United States today.

DEVIANCE IN THE BLOOD?

In 2015, Elizabeth Holmes was riding high. At the age of 26, she was the founder and CEO of Theranos, a health technology company valued at $9 billion. Named one of *Time Magazine*'s "100 most influential people in the world," she was lauded by former U.S. secretary of state (and Theranos board member) Henry Kissinger as a "tech visionary" (Kissinger, 2015). Holmes climbed to fame over the course of 7 years. At age 19, she had dropped out of Stanford, drawing on family connections to quickly raise nearly $6 million to develop *Therapatch*, a "microchip sensing project that would analyze the blood and make 'a process control decision' about how much drug to deliver, communicating those findings wirelessly to a patient's doctor" (Carreyrou, 2019, p. 15).

Unfortunately, Holmes's machine never worked the way it was said to. By 2016, *Forbes Magazine* called her one of the "World's 19 Most Disappointing Leaders." By 2018, Theranos was dissolved, and investors had lost nearly $1 billion (Carreyrou, 2019). Among them were prominent figures such as Secretary of Education Betsy DeVos and media mogul Rupert Murdoch (Carreyrou, 2019, p. 301). Holmes was indicted in 2018, and by 2020 she had been charged with a dozen felony fraud counts. According to Baron (2020):

Holmes and former company president Sunny Balwani are charged with allegedly bilking investors out of hundreds of millions of dollars, and defrauding doctors and patients, with false claims that the company's machines could conduct a full range of tests using just a few drops of blood. The two have denied the allegations, with lawyers for Holmes arguing in a December court filing that the government's case was "unconstitutionally vague" and lacked specific claims of misrepresentation.

In September 2020, awaiting a trial that had been postponed because of COVID-19, Holmes claimed that she had "a mental condition that affected her issue of guilt" (Baron, 2020). In earlier chapters, we have pointed out that sociologist C. Wright Mills encouraged us to look beyond individual troubles to social issues. Why, then, draw on one story? Sociologically speaking, Holmes's individual story highlights many of the social patterns and theories we will explore around deviance and crime in this chapter. For instance, Holmes was indicted in 2018 of accusations that are associated with white-collar crime. During that time, 172.3 U.S. citizens per 10 million were convicted of white-collar crimes. So, was her alleged crime deviant? Holmes is currently undergoing psychological evaluation because of her claims of a mental condition. How might a sociologist interpret her actions? How might her age, gender, and social class fit into answering that question? Based on patterns we uncover in this chapter, what types of punishment might she receive?

From these questions, you might get a sense that the sociological study of deviance is broad. It explores theories, behaviors, and conditions, of which only a small part fall under the category of crime, including homicides or robberies. We begin this chapter, then, by looking at how *deviance* is defined. We then examine different perspectives that sociologists employ to understand and explain deviant behavior. Finally, we consider the ways in which U.S. society exercises social control over groups and behaviors defined as deviant, including both criminal and noncriminal behavior.

Elizabeth Holmes, former CEO of Theranos

©Yichuan Cao/NurPhoto via Getty Images

WHAT IS DEVIANCE?

Deviance *is any attitude, behavior, or condition that violates cultural norms or societal laws and results in disapproval, hostility, or sanction if it becomes known.* By contrast, a **crime** *is any act defined in the law as punishable by fines, imprisonment, or both.*

Sociologists suggest that what is labeled as deviant depends on cultural norms and changes over time and place. Nonbinary dressing, once limited largely to underground drag queen culture, is slowly becoming more accepted in U.S. society. Gender-fluid garb can now be seen on the red carpet (as worn by Billy Porter at the 2019 Academy Awards), and influencers like Bretman Rock have millions of followers on social media.

Dan Macmedia/Getty Images

Several important aspects of our definition of deviance deserve greater elaboration. First, *deviance* is a broad term that may encompass crimes but often refers to noncriminal attitudes, practices, beliefs, and conditions. Second, deviance can be attached to any social group or individual. Third, what is considered deviant can include things that are not consciously chosen, such as medical or physical conditions and mental or physical illnesses. Fourth, deviance is relative and subjective—definitions of deviance vary from place to place, across time and space, and among groups within society. Finally, deviance is a label applied to the attitudes, practices, or conditions of other people or groups. As such, moral, social, and legal judgments play a role in decisions regarding what is or is not deviant.

Is all deviance criminal? In fact, most of the attitudes, practices, and conditions considered deviant by society at any given time are not criminal. For example, while having extensive tattoos or piercings may be considered deviant, it is not criminal. The follow-up question—Are all criminal acts deviant?—has a more complicated answer. Although the label *crime* applies to acts that are widely agreed to be deviant in nature (for example, murder, robbery, rape, the sexual exploitation of children, and arson), such consensus is lacking regarding other kinds of crimes. Use of illegal drugs, some types of gambling, vagrancy, and adult prostitution are a few examples of crimes that lack societal consensus about their deviance. Once an act is labeled *criminal*, formal sanctions can be applied to control it and the people who engage in it. Importantly, not every act labeled a crime is considered deviant by all of society's members; nor is every form of deviance criminalized.

Deviance: Any attitude, behavior, belief, or condition that violates cultural norms or societal laws and results in disapproval, hostility, or sanction if it becomes known.

Crime: Any act defined in the law as punishable by fines, imprisonment, or both.

The diversity of opinion surrounding deviance and criminality stems from the fact that most societies today are pluralistic. **Pluralistic societies** are *made up of many diverse groups with different norms and values*, which may or may not change over time. In a pluralistic society, what is deviant for one social group may be acceptable or normal in another, and even long-held beliefs and practices are sometimes subject to transformation over time both within and between groups.

Pluralistic societies: Societies made up of many diverse groups with different norms and values.

Consider examples of deviance surrounding marriage, punishment, and drug use. For instance, is it deviant to be married to more than one person at the same time? In the 1800s, members of the Church of Jesus Christ of Latter-day Saints (Mormons) practiced polygamy—specifically, men could have multiple wives—but by the end of the 19th century, the church officially condemned that practice. While polygamy remains illegal today, polygamous families have been featured in documentaries and their own television shows, most famously the reality show *Sister Wives*, which aired on TLC in 2010 and has run for 10 seasons. It featured the Brown family, consisting of husband Kody Brown and his four wives, Meri, Janelle, Christine, and Robyn, and their children.

Or, consider societal views on who should be punished and how severe that punishment should be by age. Is it acceptable to execute a teenager for a **capital offense**, a *crime punishable by death?* In

Capital offense: Crimes punishable by death.

2005, after centuries of supporting such a practice, the U.S. Supreme Court ruled it unconstitutional to execute anyone for a crime they committed before the age of 18. Prior to this ruling, 22 individuals had been executed for crimes they committed while younger than age 18 (Death Penalty Information Center, n.d.).

Finally, what is your opinion on the legalization and use of marijuana? Over the past five decades, public opinion around marijuana use has changed dramatically. Today, as Figure 6.1 illustrates, two thirds of Americans favor the legalization of marijuana. In 1969, however, well over two thirds thought it should be illegal (Geiger & Gramlich, 2019). Furthermore, public attitudes changed rapidly even in the past two decades, where, in 2000, 63% of Americans said marijuana use should be illegal (Geiger & Gramlich, 2019). By mid-2020, however, only 11 states had legalized marijuana for recreational use, while 35 allowed residents to use it for medicinal purposes only (Weed News Marijuana legality map, 2020, https://www.weednews.co/marijuana-legality-states-map/). Consider, too, that in the remaining states it is classed a Schedule 1 drug, which is characterized by the United States Drug Enforcement Administration with drugs like heroin as having "no currently accepted medical use and a high potential for abuse" (United States Drug Enforcement Administration, 2020, para. 4). Thus, in the case of marijuana, deviance and criminality can vary dramatically even between state borders.

In each of these cases, norms pertaining to the acceptability of practices or punishments shifted over time, creating a new normal. What other types of behaviors or practices can you think of that were once considered deviant but are no longer so? Conversely, can you think of any behaviors or practices not considered deviant that have been known to have been defined this way? Why do you think this might be?

FIGURE 6.1 ■ Changing Public Opinion Around Marijuana Use, 1969–2019

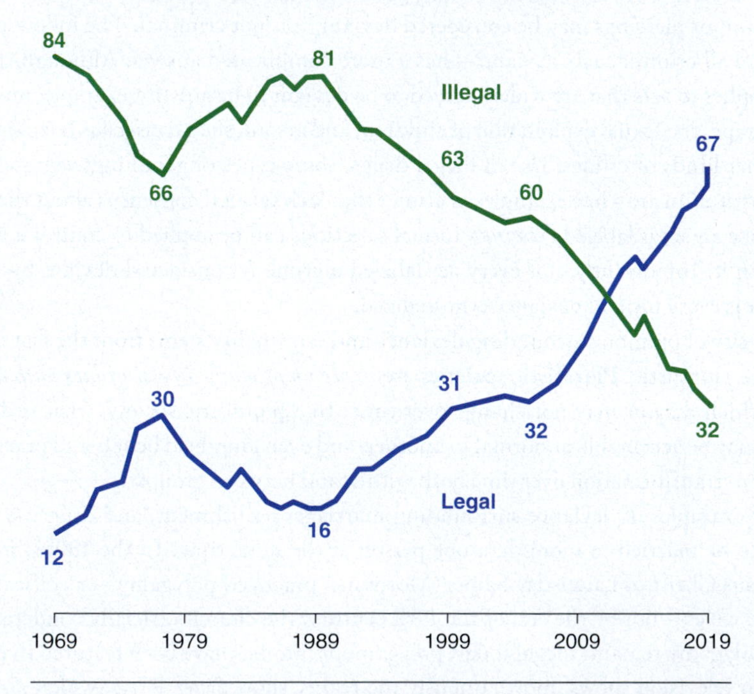

Source: Geiger, A. W., & Gramlich, J. (2019, November 22). *6 Facts about marijuana.* Pew Research Center. https://www.pewresearch.org/fact-tank/2019/11/22/facts-about-marijuana/

HOW DO SOCIOLOGISTS EXPLAIN DEVIANCE?

How do sociologists explain deviance? Next, we look at a range of theoretical perspectives divided broadly into explanatory and interactionist categories. Theories that try to explain *why* deviance does (or does not) occur, including biological, functionalist, and conflict perspectives, differ from interactionist theories, which seek to understand *how* deviance is defined, constructed, and enacted through social processes such as labeling. They also differ from critical theories that focus specifically on deviance, power, and social inequalities such as gender, race, and their intersections.

Biosocial Perspectives

Early social scientists were convinced that deviant behavior—from alcoholism to theft to murder—was caused by individual biological or anatomical abnormalities (Hooton, 1939). Some early researchers claimed that *skull configurations of deviant individuals differed from those of nondeviants*, a theory known as **phrenology**. For instance, Cesare Lombroso (1896) claimed that deviants were "atavisms," or *throwbacks to primitive early humans*. He thought that they had body types and physical features that differed from those who did not commit crimes, and he spent considerable time trying to measure and describe these differences in men (Sheldon, 1949). Lombroso, with William Ferraro, believed that women offenders were biologically more primitive than men. In *The Female Offender* (1895) they examined incarcerated women's physical characteristics, describing what they saw as possible physical abnormalities or features that they believed led to criminal behavior. As Mallicoat puts it:

Phrenology: A theory that the skull configurations of deviant individuals differ from those of nondeviants.

> They were convinced that women who engaged in crime would be less sensitive to pain, less compassionate, generally jealous, and full of revenge—in short, criminal women possessed all of the worst characteristics of the female gender while embodying the criminal tendencies of the male. (Mallicoat, 2019, p. 243)

These early biological theories tended to reflect the racial, gender, and class prejudices of the day and were used to punish racial minorities and the poor. They have since been disproven. Research continues, however, in examining the possibility of the influence of biology on individual crime and deviance. Nevertheless, most biological theories today do not attribute deviance to biology alone.

Contemporary researchers look to the interaction between biological and environmental factors (Denno, 1990; Kanazawa & Still, 2000; Mednick et al., 1987; Muncie, 2019; Wright & Cullen, 2012).

Advances in medical technology, especially increased use of magnetic resonance imaging (MRI and fMRI [functional MRI]), have enabled researchers to uncover patterns of brain function, physiology, and response unique to some deviant or criminal individuals (Giedd, 2004). On the other hand, Muncie (2019) points out that while researchers have examined many biochemical factors ("hormone imbalances; testosterone, vitamin, adrenalin and blood sugar levels; allergies; slow brain-wave activity; lead pollution; epilepsy; and the operation of the autonomic nervous system") no research has found a "direct causal relationship" between deviance and these factors.

Biosocial approaches have influenced social policy making in deviance and crime in important ways. As Walsh and Hemmens (2019) point out, these approaches played a "pivotal role" in the U.S. Supreme Court's decision to outlaw the juvenile death penalty (addressed earlier). As they put it: "In writing the majority opinion in *Roper v. Simmons* (2005), Justice Anthony Kennedy noted the neurobiological evidence for the physical immaturity of the adolescent brain, which was brought to the Court's attention by the American Medical and Psychological Associations" (Walsh & Hemmens, 2008, cited in Walsh & Hemmens, 2019, p. 331).

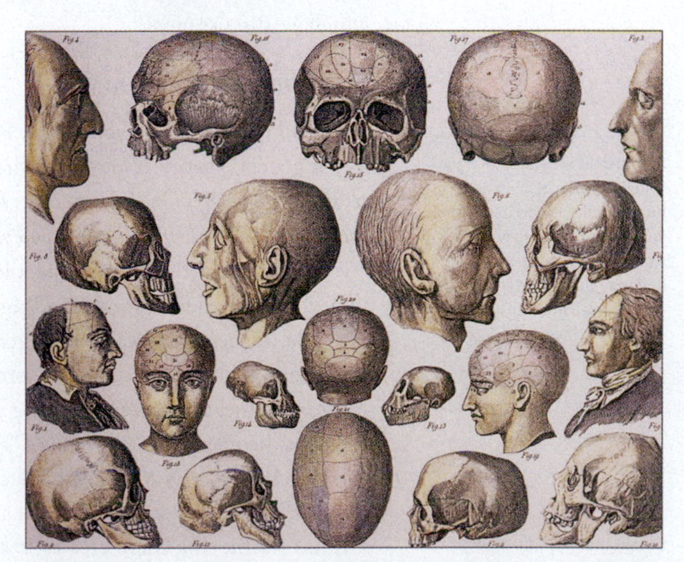

Cesare Lombroso, an early criminologist, theorized that criminals were throwbacks to primitive humans. Although his theory has been disproven, the search for biological causes of criminality continues.

Functionalist Perspectives

Functionalist approaches to deviance emphasize that society is characterized by a high degree of consensus on norms and values. Deviance, therefore, calls into question the collective rule or rules. In general, then, functionalism regards widespread deviance as an abnormality that society seeks to eliminate, much as an organism seeks to rid itself of a parasite. At the same time, it sees a certain amount of deviant behavior as normal and useful—or functional—for society. It suggests that deviance—or the labeling of some behaviors as deviant—contributes to social solidarity by enhancing members' sense of the boundary between right and wrong (Durkheim, 1893/1997).

Deviance and Social Solidarity

Émile Durkheim (1858–1917) is associated with foundational statements about the role of crime and deviance in society. Durkheim did not see deviance as a potentially abnormal part of society (Muncie, 2019). He hypothesized that deviant behavior serves a positive function by drawing moral boundaries that establish what behavior is acceptable or not within a community.

Durkheim also argued that we can describe a society lacking large-scale consensus on what is right and wrong as being in a state of *anomie*, a condition of confusion that occurs when people lose sight of the shared rules and values that give order and meaning to their lives. In his famous study *Suicide* (1897/1951), Durkheim sought to show that anomie is a principal cause of suicide, itself a deviant act. He gathered extensive data on suicide in France and Italy and found that these data supported the theory that societies characterized by high levels of anomie also have high levels of suicide. Moreover, he argued that his research demonstrated that suicide rates vary depending on the level of *social solidarity*, the social bonds that unite members of a group. He discovered, for example, that single men had higher rates of suicide than married men, Protestants a higher rate than Catholics, and men a higher rate than women. He suggested that the higher rates were correlated with lower levels of social solidarity in the groups to which people were attached.

Durkheim's research methods—the statistics as well as the sampling procedures—were primitive, compared to current methods. Since he first published his research, however, hundreds of studies have looked at suicide differences between men and women, between industrialized and developing countries, and even among the homeless. Most of these empirical studies have found considerable support for Durkheim's anomie theory (Cutright & Fernquist, 2000; Diaz, 1999; Kubrin, 2005; Lester, 2000; Simpson & Conklin, 1989; Wasserman, 1999). Durkheim's theory influenced influential contemporary theories of deviance, including those of Robert K. Merton, Richard Cloward, and Lloyd Ohlin, which we discuss next.

Structural Strain Theory

Structural strain: In Merton's reformulation of Durkheim's functionalist theory, a form of anomie that occurs when a gap exists between the culturally defined goals of a society and the means available in society to achieve those goals.

In the 1930s, American sociologist Robert K. Merton (1968) adapted Durkheim's concept of anomie into a general theory of deviance that he called "structural strain." According to Merton, **structural strain** is *a form of anomie that occurs when a gap exists between the culturally defined goals of a society and the means available in society to achieve those goals.*

Merton argued that most people in a society share a common understanding of the goals they should pursue and an understanding of the culturally legitimate means for achieving those goals. For example, success (as measured in terms of wealth, consumption, and prestige) is a widely regarded, important goal in U.S. society. Moreover, people largely agree on the legitimate means for achieving success—education, an enterprising spirit, and hard work, among others.

Strain theory: The theory that when there is a discrepancy between the cultural goals for success and the means available to achieve those goals, rates of deviance will be high.

Most people pursue the goal of "success" by following established social norms. Merton referred to this as *conformity.* Nevertheless, success is not always attainable through conventional means or conformity. When this occurs, Merton argued, the resulting contradiction between societal goals and the means of achieving them creates *strain*, which may result in four different types of deviant behavior. His **strain theory** suggests that *when there is a discrepancy between the cultural goals for success and the means available to achieve those goals, rates of deviance will be high.* Reactions to the discrepancy will lead to the types of responses depicted in the first column in Table 6.1. Since Merton's original formulation of strain theory, other researchers have expanded on his work. For example, several researchers have

TABLE 6.1 ■ Merton's Structural Strain Theory

TYPE OF RESPONSE	CULTURAL GOALS	LEGITIMACY OF MEANS
Conformity*	Accept	Accept
Innovation	Accept	Reject
Ritualism	Reject	Accept
Retreatism	Reject	Reject
Rebellion	Reject/substitute	Reject/substitute

Source: Data from Merton, R. K. (1968, orig. 1938). *Social theory and social structure* (pp. 230–246). Free Press.

*Conformity is not considered a deviant response, but rather the "norm."

looked at how strain may differ by gender. Kaufman (2009) explored the relation between general strains and gender, finding that serious strains may affect men and women differently and influence their inclination to engage in deviance. Women, Kaufman found, are especially likely to engage in deviance in response to depression. On the other hand, Moon and Morash (2017) found that the intersection of age and gender impacts how youth experience strain and, as a consequence, the different types of juvenile delinquency in which they might engage (violent and property crimes for boys and status offenses for girls).

Differential Opportunity Theory

Merton's theory helps us understand that structural conditions can lead to high rates of deviance, including crime. However, it neglects the fact that not everyone has the same access to deviant solutions. This is the point made by Richard Cloward and Lloyd Ohlin (1960), who developed **differential opportunity theory** as an extension of Merton's strain theory.

Differential opportunity theory: The theory that people differ not only in their motivations to engage in deviant acts but also in their opportunities to do so.

According to Cloward and Ohlin, *people differ not only in their motivations to engage in deviant acts but also in their opportunities to do so.* These opportunities include three elements: (1) a specific situation that might lead to deviance—for example, a particularly opportunistic time or space; (2) the potential that the deviant behavior will result in some sort of reward; and (3) the lack of guardians who act as barriers to the behavior (Hollin, 2019). Deviance, then, is more likely to occur in a community, place, or time when the opportunities for it exist and when it provides a monetary, social, political, or individual reward.

Take the case of drug dealing: It requires a demand for illicit drugs, access to supplies of those drugs through producers, and lack of oversight in some way. Without these, there is little opportunity for individuals to become drug dealers. Similarly, unless you have access to funds you can secretly convert to your own use, you are unlikely to consider embezzlement as an option, much less carry it out.

Social Control Theory

Social control theorists adopt functionalists' claim that a society's norms and values are the starting point for understanding deviance, and that deviance is to be expected in society. But they differ in that they emphasize the *development of relationships* that keep people from deviance and "bind them to the social order" (Walsh & Hemmens, 2019, p. 199). These social relationships are varied—parents, peers, schools, religion, etc. (Rankin & Kernsmith, 2019).

For instance, Gottfredson and Hirschi (2004) **social control theory** *explains that the probability of delinquency or deviance among youth is rooted in social control.* Gottfredson and Hirschi differ from Durkheim and Merton, however, regarding the importance of a general state of anomie in creating deviance, arguing instead that a person's acceptance or rejection of societal norms depends on that individual's life experiences. They assert that deviance arises from **social bonds**, or *individuals' connections to others*, especially institutions, rather than from anomie. Forming strong social bonds with people and institutions that disapprove of deviance, they argue, keeps people from engaging in deviant

Social control theory: The theory that explains that the probability of delinquency or deviance among children and teenagers is rooted in social control.

Social bonds: Individuals' connections to others.

behaviors. Conversely, people who do not form strong social bonds will engage in deviant acts because they have nothing to lose by acting on their impulses and do not fear the consequences of their actions.

Furthermore, control theorists argue that most deviant acts are spontaneous. For example, a group of teenagers, seeing a homeless person sleeping on a bench, suddenly decide to take the person's backpack, or a man learns of a house whose owners are on vacation and decides to burglarize it. While some people will succumb to such temptations, those who do not, according to Gottfredson and Hirschi, have a greater willingness to conform. This willingness, in turn, comes from associating with people who are committed to conventional social roles and morality.

Bernie Madoff on his way to court for 11 counts of fraud in 2009. A prominent citizen, Madoff was found guilty of defrauding investors of over $17 billion and sentenced to 150 years in prison. In what ways does his behavior fit Cloward and Ohlin's differential opportunity theory?

AFP PHOTO/Timothy A. Clary

Some evidence supports this theory. For example, delinquency is somewhat less common among youth who have strong family attachments, perform well in school, and feel they have something to lose by appearing deviant in the eyes of others (Gottfredson & Hirschi, 1990/2004; Hirschi, 1969). On the other hand, critics point out some ways in which control theories can fail to address deviant behavior. For instance, as it concerns connections to others, youth who engage in deviant behavior may have strong attachments, but these attachments may be to peers who encourage their deviance (Rankin & Kernsmith, 2019).

The success of many white-collar criminals suggests that they have spent their lives conforming to societal norms, yet they also commit criminal acts that cost U.S. taxpayers, investors, and pension holders billions of dollars (McLean & Elkind, 2003). In short, we could scarcely argue that white-collar criminals like the infamous Bernie Madoff, the former executive chair of the NASDAQ stock market and socially prominent financier who pleaded guilty to massive financial fraud in 2009, did *not* have strong social bonds to society. In fact, these bonds are what allowed him to find so many people to defraud. Thus, although social control theory may prove useful in certain instances, it has limitations.

Conflict Perspectives

Recall that conflict perspectives posit that groups in society have different interests and unequal access to resources with which to realize those interests. In contrast to the functionalist perspective, these perspectives do not assume shared norms and values. Rather, they presume that groups with power will use that power to maintain control in society and keep other groups at a disadvantage. As we will see next, conflict theory can be fruitfully used in the study of deviance.

Subcultures and Deviance

Subcultural theories: Theories that explain deviance in terms of the conflicting interests of different segments of the population.

More than three quarters of a century ago, Thorsten Sellin (1938) pointed out that the cultural diversity of modern societies results in conflicts between social groups over what kinds of behavior are right and wrong. Sellin argued that deviance is best explained through **subcultural theories**, which *identify the conflicting interests of different segments of the population*, whether it be over culture (as in Sellin's case) or, more generally, over particular rituals or behaviors. For example, immigrants and refugees may bring norms and values with them from their original cultures that conflict with the norms and values of the adopted country. Thus, they may be perceived—and sometimes punished—as deviant by the dominant culture. Some practices of migrant communities might breach U.S. conventions: For instance, some Southeast Asian families (among others) still choose to follow traditional customs of arranged marriages for adult children. Although not the norm in the United States, it is not illegal. Indeed, the popular Netflix 2020 show *Indian Matchmaking*, while not designed for the "Western gaze," has brought the work of Mumbai-based matchmaker Sima Taparia and her "marriage and then love marriages" into many homes (Villarreal, 2020). Other customary practices, however, violate U.S. criminal laws. For instance, the practice of female genital circumcision by some immigrants from North Africa and the Middle East is a violation of U.S. law. As well, even though physical violence against a woman is understood in some communities as a husband's prerogative in marriage, immigrants to the United States who practice domestic violence are subject to arrest and prosecution.

It is not only cultural differences between immigrants and the host country that create subcultures of perceived or real deviance. Sociologists also analyze a very wide range of subcultures perceived as deviant—from juvenile gangs to white racist groups, from kink to Bronie culture, and from a host of other groups (Chambliss & King, 1984; Cohen, 1955; Crome, 2015; Etter, 1998; Hamm, 2002; Meyer & Chen, 2019).

The My Little Pony fans who are adult and mostly male call themselves Bronies. This group of fans has developed a community in online forums and at an annual BronyCon fan convention (pictured here) based around values espoused in My Little Pony.

Photo by ANDREW CABALLERO-REYNOLDS/AFP via Getty Images

Class-Dominant Theory

Class-dominant theories *propose that what is labeled deviant or criminal—and therefore who gets punished—is determined by the interests of the dominant class* (Quinney, 1970). For example, since labor is central to the functioning of capitalism, those who do not work will be labeled as deviant in capitalist societies (Spitzer, 1975). In a similar vein, since private property is a key foundation of capitalism, those who engage in acts against property, such as stealing or vandalism, will be labeled as criminal. And because profits are realized through buying and selling things in the capitalist marketplace, unregulated market activities (such as selling drugs on the street, making alcohol without a license, or even operating a catering service out of one's home without proper licensing) will also be defined as deviant and criminal.

Class-dominant theories: Theories that propose that what is labeled deviant or criminal—and therefore who gets punished—is determined by the interests of the dominant class.

Critics of class-dominant theory point out that laws against the interests of the ruling class do get passed. Laws prohibiting insider trading on the stock market, governing the labor practices of corporations, and giving workers the right to strike and form trade unions were all signed over the strident opposition of big business interests (Chambliss, 2001). It is important to incorporate these facts to account for both the power *and* limitations of capitalists in capitalist society.

Structural Contradiction Theory

Rather than seeing the ruling class as all-powerful in determining what is deviant or criminal, **structural contradiction theory** argues that *conflicts generated by fundamental contradictions in the structure of society produce laws defining certain acts as deviant or criminal* (Chambliss, 1988a; Chambliss & Hass, 2011; Chambliss & Zatz, 1994). For instance, there is a fundamental structural contradiction in capitalist economies between the need to maximize profits (which keeps wages down) and the need to maximize consumption (which requires high wages). Consider a U.S. business that, to maximize profits, keeps wages and salaries down, perhaps by moving its factories to a part of the world where wages are low. As jobs move overseas, the availability of jobs to unskilled and semiskilled workers in the United States declines. The loss of jobs produces downward pressure on wages and a loss of purchasing power. Yet capitalism depends on people buying the things that are produced— corporations cannot profit unless they sell their products.

Structural contradiction theory: The theory that conflicts generated by fundamental contradictions in the structure of society produce laws defining certain acts as deviant or criminal.

Trapped in the contradiction between norms valuing consumerism and an economic system that can make consumption of desired material goods and services difficult or even impossible for many, some (but not all) people will resolve the conflict by resorting to deviant and criminal acts such as cheating on income taxes and writing bad checks, or profiting from illegal markets by selling drugs or committing theft. Of course, it is not only lower-income people who deviate to increase their consumption of material goods. Everyone who wants to enjoy a higher standard of living is a candidate for deviant or criminal behavior, according to structural contradiction theory. The head of a giant corporation may be as tempted to violate criminal laws to increase company profits (and personal income) as is the 13-year-old from a poor family who snatches a pair of sunglasses from the store.

Structural contradiction theory holds that societies with the greatest gaps between what people earn and what they are normatively enticed to consume will have the highest levels of deviance. Since industrial societies differ substantially in this regard, we can compare them to test this theory. Societies such as Finland, Denmark, Sweden, and Norway, for instance, provide a "social safety net" that guarantees all citizens a basic standard of living. Therefore, poorer residents in Scandinavian countries are able to come much closer to what their societies have established as a "normal" standard of living. The fact that rates of assault, robbery, and homicide are anywhere from 3 to 35 times higher in the United States than in these countries (depending on which country is compared) is what structural contradiction theory would predict (Archer & Gartner, 1984).

Globalization is another example of the way structural contradictions lead to changes in deviant behavior. The ability to trade worldwide increases the wealth of populations and nations that are able to take advantage of global markets. Nevertheless, it also increases opportunities for criminal activities such as money laundering, stealing patents and copyrights, and trafficking in people, arms, and drugs. Cybercrime is also a global phenomenon. During the COVID-19 outbreak, for instance, particularly lucrative both domestically and internationally were products and "remedies" that were COVID-19 related. Consequently, the FBI reported a number of national and international COVID-19–related crimes that it referred to as "pandemic scamming," including testing and treatment scams, international spying around vaccines, extortion, cryptocurrency scams, fraudulent unemployment insurance claims around COVID-19, and many others (FBI, 2020b). Just 3 months after the start of the pandemic in the United States, the FBI had already received 3,600 complaints of such COVID-19–related online scams (https://www.justice.gov/opa/pr/department-justice-announces-disruption-hundreds-online-covid-19-related-scams).

Critical Perspectives

Critical perspectives of deviance and crime emerge from inequalities in society. Critical theorists question the status quo, showing its unequal gender and racialized dimensions. They highlight many contemporary issues of social inequalities tied to recent movements like #MeToo and #BlackLivesMatter. In this section we explore two critical theories of deviance—feminist theories and critical race theory.

Feminist perspectives on deviance: Perspectives that suggest that studies of deviance have been biased because almost all the research has been done by, and about, males, largely ignoring female perspectives on deviant behavior as well as analyses of differences in the types and causes of female deviance.

Feminist Theories

The sociological study of deviance was historically dominated by white male criminologists. As a result, theories and research overlooked or stereotyped women, crime, and deviance. **Feminist perspectives on deviance**, emerging from insights of feminism in the mid-1960s and 1970s, ensured that women and girls became their own focus of study (Campbell, 1984; Chesney-Lind, 1989, 2004; Chesney-Lind & Morash, 2013). By the 21st century, feminist theories on deviance and crime spanned a wide range of perspectives, with a special concentration on patriarchy, masculinities, femininities, and intersectionality (Chesney-Lind & Morash, 2013; Cook, 2016; Mallicoat, 2019). These studies highlighted a number of previously underexplored topics such as the importance of examining crime and deviance through a lens of race, class, gender, and sexuality; exploring gendered victimization as a starting point to theorizing differences in offending; understanding the conditions of women in prisons; and applying feminist insights to occupational gender segregation in criminal justice fields.

Feminist research has revealed interesting insights into girls' behavior and experiences in gangs. Girls sometimes form gangs in neighborhoods where male street gangs are prevalent. These girl gangs are often auxiliaries of male gangs engaged in selling drugs and committing petty crimes. Joining a gang may require an initiation ritual that involves violence, including rape.

An example of the importance of feminist theorizing is in the area of women's higher rate of victimization around crimes like sexual assault, domestic violence and abuse, and rape and connections with early victimization and subsequent offending (Chesney-Lind & Morash, 2013; Mallicoat, 2019). For example, studies show that before becoming involved in the juvenile justice system, many girls labeled delinquent were runaways escaping sexual and physical abuse. In a 2014 study, 31% of girls reported a personal experience of sexual violence in the home, 41% reported being physically abused, and 84% reported experiencing family violence. Girls reported having been sexually abused at a rate 4.4 times higher than boys (Levintova, 2015). This research supports the hypothesis that many girls labeled delinquent have been driven out of their homes by abusive parents or relatives.

Critical Race Theories

Critical race theories (CRTs) are *theories that focus on the relationship between race, power, and racism.* They emerged in the 1970s in legal studies as a critique of the role of law in institutionalizing racial discrimination. Kimberlé Crenshaw et al. identified the two goals of CRT:

> The first is to understand how a regime of white supremacy and its subordination of people of color has been created and maintained in America, and, in particular, to examine the relationship between that social structure and professed ideals such as "the rule of law" and "equal protection." The second is a desire not merely to understand the vexed bond between law and racial power but to change it. (Crenshaw et al., 1995, cited in West, 1995, p. xiii)

Critical race theories are particularly important in their emphasis on the contradictions between ideals of justice and realities of practice. They are also aimed at social action. As you will see in the sections that follow, including a racial lens is important in understanding sociological factors in justice (and injustice).

Critical race theories (CRTs): Theories that focus on the relationship between race, power, and racism.

Interactionist Perspectives

Interactionist perspectives provide a language and framework for looking at how deviance is constructed, including how individuals are connected to the social structure. Interactionist approaches also explain why some people are labeled deviant and behave in deviant ways while others do not. Like many interactionist approaches, labeling theory and differential association theory point out that we see ourselves through the eyes of others, and our resulting sense of ourselves conditions how we behave.

Labeling Theory

Labeling theory holds that *deviant behavior is a product of the labels people attach to certain types of behavior* (Asencio & Burke, 2011; Lemert, 1951; Tannenbaum, 1938). From this perspective, deviance is seen as socially constructed. That is, labeling theory holds that deviance is the product of interactions wherein the response of some people to certain types of behaviors produces a label of *deviant* or *not deviant.* Edwin Lemert's "stutterers" 1951) and Bill Chambliss's "saints and roughnecks" (1973) highlight the power of labeling (or escaping the label).

Labeling theory: A symbolic interactionist approach holding that deviant behavior is a product of the labels people attach to certain types of behavior.

Lemert observed the interactions of a group of people in the northwestern United States with an unusually high incidence of stuttering. He concluded that stuttering was common in the group partly because its members were stigmatized and labeled as stutterers (primary deviance). These stutterers then began to view themselves through this label and increasingly acted in accordance with it—which included a greater amount of stuttering than otherwise would have been the case (secondary deviance). Lemert argued that the labeling process has two steps: primary deviance and secondary deviance. **Primary deviance** occurs *at the moment an activity is labeled as deviant by others.* **Secondary deviance** occurs *when a person labeled deviant accepts the label as part of his or her identity and, as a result, begins to act in conformity with the label.* Also connected to primary and secondary deviance is the concept known as the **self-fulfilling prophecy**, or a behavior that responds to the situation that then becomes true (Merton, 1948). *In other words, as a consequence of labeling, people develop deviant self-concepts, living up to the label imposed on them.*

Primary deviance: The first step in the labeling of deviance. It occurs at the moment an activity is labeled deviant.

Secondary deviance: The second step in the labeling of deviance. It occurs when a person labeled deviant accepts the label as part of his or her identity and, as a result, begins to act in conformity with the label.

Self-fulfilling prophecy: As a consequence of labeling, people develop deviant self-concepts, living up to the label imposed on them.

Chambliss's (2001) observations of two groups of high school youth whom he termed the "Saints and the Roughnecks" also support labeling theory. In spite of engaging in similar kinds of crime and mischief, the working-class teens (the "Roughnecks") he studied in one community were labeled deviant while the middle-class boys (the "Saints") he observed were not. Chambliss sought to understand this difference. He noted that first, the actions of the Roughnecks were far more visible than those of the Saints. With their access to cars, the Saints were able to remove themselves from the sight of the community, but the Roughnecks congregated in a public area where they could be seen by teachers or the police. Second, the demeanors of the gang members differed. Although the Saints showed remorse and respect, the Roughnecks offered a barely veiled contempt for authority. This resulted in different responses to their misdeeds. Third, adults in the community showed bias toward the Saints, who were presumed to be "good boys sowing wild oats" rather than "bad boys." Chambliss concluded that labels matter. Those who were labeled as *bad* largely lived up (or down) to expectations. Those whose youthful transgressions were not transformed into labels lived up to more positive expectations and became successful adults.

These studies and others suggest that the ways we label individuals and groups have important effects. Think about your experiences of people who may have been labeled as deviant in high school or at work. Can you identify how and why people were labeled and how these labels did or did not stick?

Differential Association Theory

Differential association theory: The theory that deviant and criminal behavior is learned and results from regular exposure to attitudes favorable to acting in ways that are deviant or criminal.

Do we *learn* to be deviant from those around us? **Differential association theory** asserts that social relationships help to influence deviant behavior. This theory purports that *deviant and criminal behavior is learned and results from regular exposure to attitudes favorable to acting in ways that are deviant or criminal* (Burgess & Akers, 1966; Church et al., 2012;Sutherland, 1929). For example, a corporate executive who embezzles company funds may have learned the norms and values appropriate to this type of criminal activity by associating with others already engaged in it. Similarly, youth living in areas where selling and using drugs are common practices will be more likely than their peers not exposed to that subculture to develop attitudes favorable toward using and selling drugs. Conversely, populations with different subcultures or attitudes toward particular forms of deviance (such as illicit drug usage) may not experience the same rates of crime.

According to Sutherland's (1929) differential association theory, the more we associate with people whose behavior is deviant, the greater the likelihood that our behavior will also be deviant. Sutherland, therefore, linked deviance with such factors as the frequency and intensity of our associations with other people, how long they last, and how early in our lives they occur.

Much has changed since Sutherland's theory. Today, we interact extensively through technologies such as the Internet and smartphones, and this theory needs modification to consider new methods of interacting today in promoting both deviant and conforming behavior. Furthermore, feminist researchers have shown that gender may play an important role in mediating or exacerbating the impact of peers, with some studies indicating that gender is an important consideration: Boys may be likelier to be influenced by delinquent peers than girls (Mallicoat, 2019).

In these sections we have explored a range of different theories, each of which provides us with considerable insight into the social processes that lead to deviance and crime in society. While each of these theories has some empirical support, and we might conclude that each of these theories makes sense, each theory also is incomplete and cannot explain all the behaviors we classify as deviant or criminal.

As we have pointed out in earlier chapters, the debate and interaction between theories is essential to the development of scientific knowledge (Popper, 1959). Thus, even though there are many unique ways to examine the same topic without necessarily reaching one perfect answer with any of them, each of these theoretical orientations can be useful in explaining certain facets of deviance and crime. Researchers must decide which theory to use and why, and for others to determine whether the use of the theory was a success or a failure.

TYPES OF DEVIANCE

Deviance comes in many varieties, from the relatively benign to the extremely harmful. In this section, we explore two types of deviance that sociologists explore: everyday deviance and deviance of the powerful.

Everyday Deviance

Many acts could fall under the label of *everyday deviance*, or behaviors and actions in which people engage at some point (or even regularly) that are still considered deviant. Examples include plagiarism among high school or college students, shoplifting, underage alcohol consumption, viewing Internet pornography, smoking, eating meat, or calling in sick to work or school when you actually feel fine. Everyday deviance can be explained by a number of the theories we addressed earlier. The pluralistic nature of U.S. society influences the variability of everyday deviance.

For instance, taking a subcultural perspective, we may find that smoking is more acceptable among some societal subgroups (for instance, casino-goers) than it is among others (such as fitness instructors). So, too, with eating meat: The owner of a barbecue restaurant and a vegan are likely to hold starkly different views regarding the deviance of eating meat or subsisting on a vegan diet. How would theories like differential opportunity theory and social bonds theory explain calling in sick to work when you feel fine? Or, how do you think theories like class dominant theory and feminist theory would explain the widespread incidence of Internet pornography? Our point here is that deviance in its various forms has many potential explanations, which can be strengthened through the combination of different theoretical perspectives and studies to test the strength of the theories.

Deviance of the Powerful

The most powerful people in public life engage in many of the same types of deviance as ordinary men and women (McLean & Elkind, 2003; McLean & Nocera, 2010; Reiman & Leighton, 2012). Sociologically, deviance knows no class bounds. The crimes of the famous and powerful are ubiquitous and wide ranging, from the fraudulent reporting of corporate profits to the misleading of investors to bribery, corruption, misuse of public trust, and violence. The public and the mass media are often interested in deviance committed by athletes, celebrities, and prominent political and religious figures, and examples abound.

For instance, according to *Sports Illustrated*, between January 2012 and September 2014, 33 National Football League players were arrested on charges of domestic violence, battery, assault, and murder; nearly half of the charges involved violence against women (*Sports Illustrated*, 2014). Major League Baseball player Alex Rodriguez of the New York Yankees was suspended for the 2014 season for using banned substances and lying about it. In 2015, rapper Vanilla Ice was arrested for burglary and grand theft for stealing bicycles and a couch, among other items, from a foreclosed house. That same year, actor Bill Cosby, accused by more than 50 women of assault, was charged with aggravated indecent assault. In the spring of 2018, Cosby was convicted of three counts of sexual assault and sentenced to 3 to 10 years in prison.

As we earlier noted, COVID-19 provided additional opportunities for "celebrity" offender deviance. For instance, a 31-year-old NFL player was charged for "his alleged participation in a scheme to file fraudulent loan applications seeking more than $24 million in forgivable Paycheck Protection Program (PPP) loans guaranteed by the Small Business Administration (SBA) under the Coronavirus Aid, Relief, and Economic Security (CARES) Act" (https://www.justice.gov/opa/pr/nfl-player-charged-role-24-million-covid-relief-fraud-scheme). Among his charges are allegations of using the funds to purchase luxury goods including clothes and jewelry, gamble, and provide money to family members.

The response of the public to the deviance of political and religious leaders is often particularly pointed, given the trust, responsibility, and power vested in those individuals. In 2016, former Speaker of the U.S. House of Representatives Dennis Hastert pleaded guilty for illegal actions linked to a $3.5 million payment to silence allegations of sexual misconduct with a student when he was a high school teacher and coach decades ago. Prosecutors say he molested at least four high-school-age boys (Davey & Smith, 2016). In March 2018, Representative Blake Farenthold, a Republican from Texas, resigned his seat in Congress, bowing to pressure put on him after an investigation revealed that he had used $84,000 in taxpayers' money to settle a sexual harassment claim with a former congressional employee (*Politico*, 2018). In 2020, Jerry Falwell Jr. of Liberty University resigned after serving 13 years there as president following an emerging family sex scandal. Liberty, a Baptist college founded by Falwell's father in 1971, had grown to be a major evangelical institution, with an undergraduate enrollment of over 45,000 students coming to a sprawling campus of 7,000 acres. Falwell's alleged behavior broke with the norms and code of conduct that students, faculty, and staff of the university espoused (Graham, 2020).

Empirical data show that the powerful are more likely than those without it to escape punishment for deviance. On the other hand, public figures who are caught in acts of deviance are often subject to acute media attention and broad disdain, particularly when their acts violate public trust and waste public resources. What other public personalities have been punished for deviance? What unwritten rules did they violate? Were their punishments commensurate with their deviance? Why or why not?

BEHIND THE NUMBERS: COUNTING THE INCIDENCE AND TYPES OF CRIMES

You go to a restaurant with a friend, hanging your leather jacket by the door. As you prepare to leave, you realize that your jacket is gone. Do you report it to the restaurant management? Do you report it to the police? If you report it, will it count in official crime statistics?

How does the United States measure violent and property crime?

Two federal organizations, the FBI and the Bureau of Justice, collect data on crime. Property and violent crime are tracked in two specific ways (see Table 6.2):

- The official, limited records such as the FBI's Uniform Crime Reports (UCR)
- Individual and household survey data collected via the National Crime Victimization Survey (NCVS)

TABLE 6.2 ■ The Uniform Crime Report (UCR) Versus the National Crime Victimization Survey (NCVS)

	UNIFORM CRIME REPORT (UCR)	NATIONAL CRIME VICTIMIZATION SURVEY (NCVS)
Who administers it?	FBI	U.S. Census Bureau under the Department of Commerce
What crimes are reported?	Murder and nonnegligent manslaughter, forcible rape, robbery, aggravated assault, burglary, larceny–theft, motor vehicle theft, and arson.	Rape, sexual assault, robbery, aggravated and simple assault, theft, household burglary, and motor vehicle theft.
Who reports the crimes?	Law enforcement agencies around the country. (Participation in the program is optional.)	A randomly selected sample of the general U.S. population. (Because it is based on a sample, figures are extrapolated to the U.S. population.)
How are the crimes reported?	Through law enforcement reports that document citizens' reports to police of crimes and police-reported arrests.	Through an annual survey of 90,000 households that reveals the "frequency, characteristics, and consequences of criminal victimization in the United States." Every household is interviewed twice during the year and asked about a spectrum of crimes. All crimes are self-reported.

How do the UCR and NCVS compare on larceny and theft statistics?

Figure 6.2 compares the number of larcenies reported by the UCR and the number of thefts reported by the NCVS in 2013. We see that the NCVS figure of 10,050 per 100,000 population is about 5.3 times higher than the UCR figure of 1,899 per 100,000 population. Although the comparison is imperfect, it suggests that there is a gap between the incidence of crimes and their reporting to police.

FIGURE 6.2 ■ Comparison of UCR and NCVS Larceny and Theft Statistics

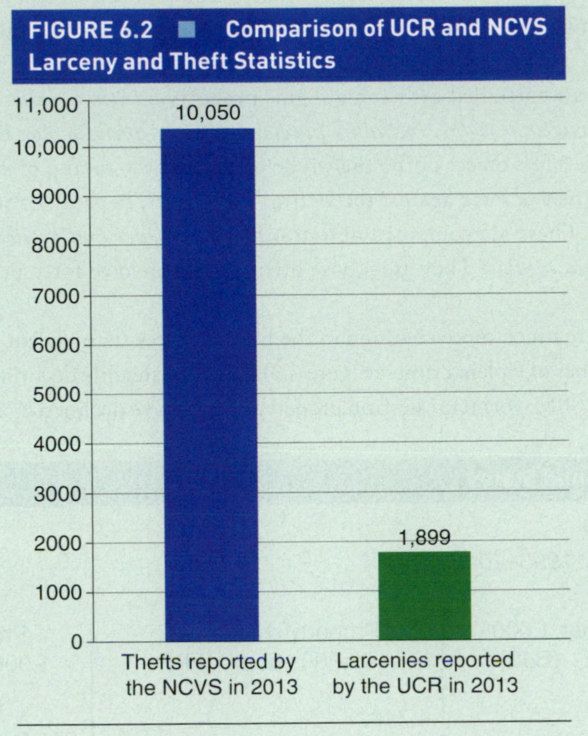

Source: Gramlich, J. (2019, November 20). *What the data says (and doesn't say) about crime in the United States.* Pew Research Center. https://www.pewresearch.org/fact-tank/2020/11/20/facts-about-crime-in-the-u-s/

Analysis

So: If you reported your missing coat, it will likely register in the Uniform Crime Report (UCR), where it will be recorded as a larceny, whether or not the perpetrator is caught, or you get your jacket back. If you did not report it, it will not appear in the UCR. If you happen to be a part of the national sample of the NCVS, it will be recorded as a theft in those figures. If you are not part of the sample, your loss will not be recognized in the figures.

Both the UCR and the NCVS provide valuable information about crime in the United States, but neither provide conclusive figures.

Crime

Our discussion thus far has been concerned primarily with deviance in a general sense—those attitudes, behaviors, and conditions that are widespread but generally not condemned by all or seen as especially serious. In this section, we discuss *crime*—acts that are sometimes considered deviant and are defined under the law as punishable by fines, imprisonment, or both. Law enforcement agencies across the United States take rigorous steps to record formally how much crime occurs, to prosecute criminal offenders, and to control and prevent crime. In so doing, however, they are also scrutinized by the public and the media, with recent reporting highlighting the fact that some commit their own crimes in the process of controlling and preventing crime.

In 2018, former pharmaceutical executive Martin Shkreli, who was the object of sharp public criticism after he raised the price of a vital anti-infective drug by 5,000%, received a jail sentence of seven years.

AP Photo/Seth Wenig

Violent and Property Crimes

Measuring the incidence of crime in the United States is challenging. For one thing, most crimes are not reported to the police (Gramlich, 2019) for a variety of reasons, including:

- Distrust of law enforcement
- Thinking reporting is too much of a bother

Recall that the FBI's Uniform Crime Report (UCR) counts both violent and property crimes by relying on official crime reports and arrest records. What does this report show us about these crimes in the United States?

Property crimes: Crimes that involve the violation of individuals' ownership rights, including burglary, larceny/theft, motor vehicle theft, and arson.

Violent crimes: Crimes that involve force or threat of force, including murder and negligent manslaughter, rape, robbery, and aggravated assault.

First, the UCR, as we detailed earlier, highlights **property crimes** as *crimes that involve the violation of individuals' ownership rights, including burglary, larceny–theft, motor vehicle theft, and arson.* According to the FBI, "The object of the theft-type offenses is the taking of money or property, but there is no force or threat of force against the victims" (FBI, 2017b, para. 1). Second, the UCR highlights **violent crimes**. These are composed of four offenses: *murder and negligent manslaughter, rape, robbery, and aggravated assault.* They are "those offenses that involve force or threat of force" (FBI, 2017a, para. 1).

Property crimes are much more common in the United States than violent crimes, although their number, like the number of violent crimes (Figure 6.3), has been steadily declining (Figure 6.4). Recent data show that since 2008, violent crimes and property crimes have declined by 23% (Gelb & Denney,

FIGURE 6.3 ■ Violent Crime Rate in the United States

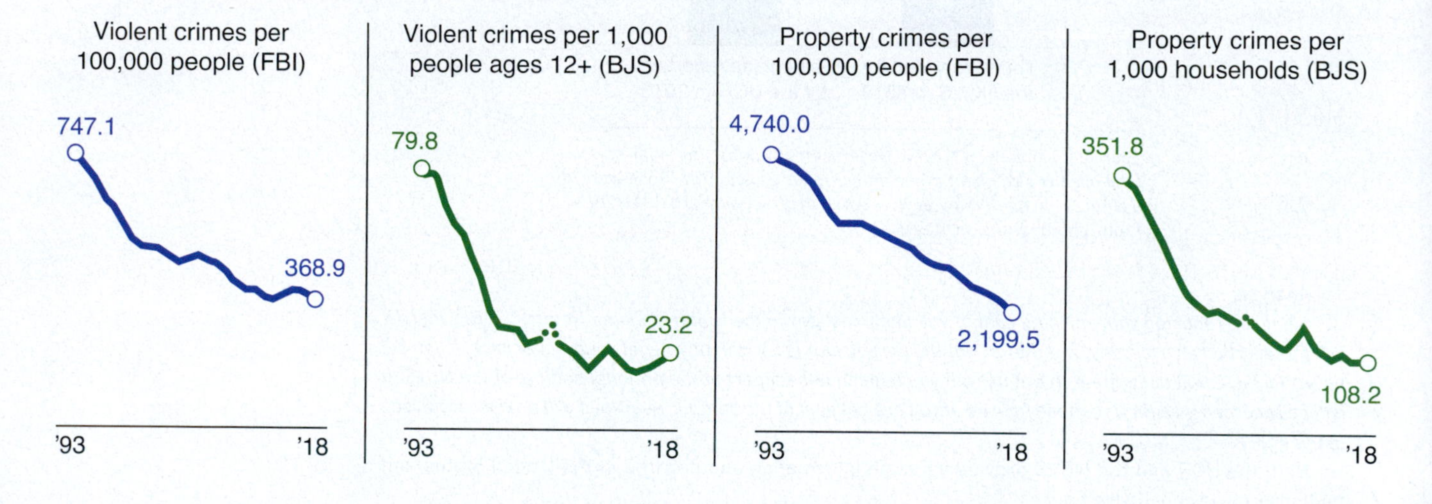

Source: Pew Research, Bureau of Justice Statistics, FBI.

Note: FBI figures include reported crimes only. Bureau of Justice Statistics (BJS) figures include unreported and reported crimes. 2006 BJS estimates are not comparable with other years due to methodological changes.

FIGURE 6.4 ■ Property Crime Rate in the United States

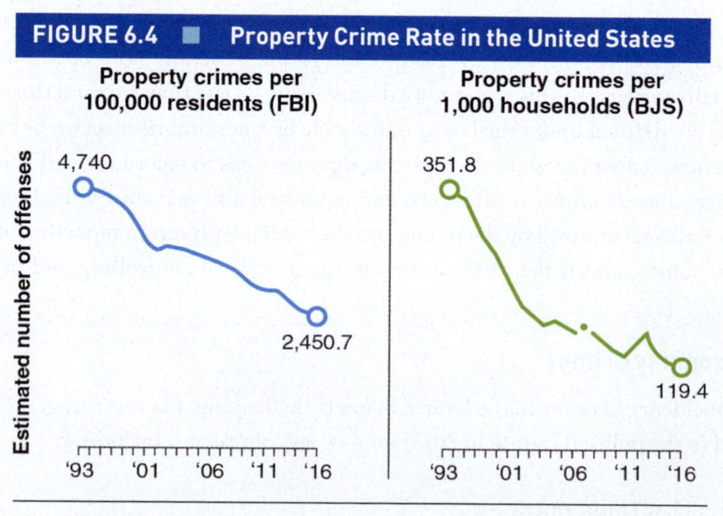

Source: Pew Research and Bureau of Justice Statistics and FBI.

2018). In 2019, the FBI reported that all types of crime had continued to decrease from 2018 (FBI, 2020a), with property crimes decreasing in all four regions of the United States.

The most common property crimes are larceny/theft, with possible charges ranging from misdemeanor to felony depending on their seriousness (Walsh & Hemmens, 2019). Age can play a role in the frequency of some types of crimes: Property crimes like arson, for instance, are usually committed by juveniles. Violent crimes are often what people think of when they think of crime. And, while violent crime is decreasing in the United States, it is still a common worry for many Americans. For instance, a Pew Research 2018 poll found that 57% of teens are worried that a school shooting could happen at their school (Graf, 2018).

One interesting and notable feature of property and violent crimes in the United States is this: Fewer than half of these are reported *and* fewer than half of the reported crimes are solved (Gramlich, 2017). Auto thefts are the crimes most likely to be reported, and murders are the crimes most likely to be solved (Gramlich, 2017).

Variations of serious deviance, including violent and property crimes, can be analyzed from a variety of perspectives. For instance, how might opportunity and societal strain perspectives differently explain stealing from one's employer or breaking into a neighbor's house to take food and other goods?

Organized Crime

Sociologists define **organized crime** as *crime committed by criminal groups that provide illegal goods and services.* The most prominent of these, according to the FBI (n.d.), are illegal gambling; human, migrant, drug, and sex trafficking; cybercrime; black marketeering, loan-sharking, and money laundering. Organized crime exists across nations (transnationally) and is increasingly facilitated by technology. Its costs to societies are extremely high, with numbers ranging from billions to trillions of dollars annually.

Organized crime: Crime committed by criminal groups that provide illegal goods and services.

To meet the demand for illegal goods and services, criminal organizations have flourished in U.S. urban areas since the 1800s (Woodiwiss, 2000). Over the years, they have recruited members and leadership from more impoverished groups, such as new immigrants in big cities, who may have great aspirations but limited means of achieving them (Block & Weaver, 2004). Consequently, organized crime has been dominated by the most recent arrivals to urban areas: Irish, Jewish, and Italian mobs in the 19th and early 20th centuries and Asian, African, South American, and Russian mobs through the mid-20th century (Albanese, 1989; Finckenauer & Waring, 1996; Hess, 1973).

Depictions of organized crime in movies and television shows such as *The Sopranos, The Godfather, Scarface,* and *Goodfellas* have popularized the erroneous impression that there is an international organization of criminals (the Mafia) dominated by Italian Americans. The reality is that organized crime consists of thousands of different groups throughout the United States and the world. No single ethnic group or organization has control over most or even a major share of these activities, which include weapons and drug smuggling (Block & Weaver, 2004; Chambliss, 1988a). Also included is **human trafficking**, which the United Nations defines as *the recruitment, transport, transfer, harboring or receipt of a person by such means as threat or use of force or other forms of coercion, abduction, fraud or deception for the purpose of exploitation* (United Nations, n.d.). On the other hand, the most serious organized crime threat today, according to the FBI, CIA, and IRS, is the Russian "Mafiya," an organized and well-educated transnational group strongly emerging after the fall of the Soviet Union (Walsh & Hemmens, 2019).

Human trafficking: The recruitment, transport, transfer, harboring, or receipt of a person by such means as threat or use of force or other forms of coercion, abduction, fraud, or deception for the purpose of exploitation.

White-Collar Crime

When most people think of crime, they think of the violent and property crimes just discussed—and they think of crimes committed by those in lower socioeconomic groups. But the forms of crime perpetrated by individuals and groups who possess great power, authority, and influence are also often deeply harmful to society. **White-collar crime** is *crime committed by people of high social status in connection with their work* (Sutherland, 1949/1983). There are two principal types: crimes committed for the benefit of the individual who commits them, and crimes committed for the benefit of the organization for which the individual works.

White-collar crime: Crime committed by people of high social status in connection with their work.

Among the many white-collar crimes that benefit the individual are the theft of money by accountants who alter their employers' or clients' books and the overcharging of clients by lawyers. More costly types of white-collar crime occur when corporations and their employees engage in criminal conduct either through *commission* (by doing something criminal) or *omission* (by failing to prevent something criminal or harmful from occurring).

White-collar crimes receive considerable media attention. In 2020, these crimes also made the news because the number of prosecutions for them had dropped to a 30-year low under the Donald J. Trump administration and because individuals, not businesses, were the only ones being prosecuted (Arends, 2020; Trac Reports, 2020).

Here is a sample of major white-collar criminal cases spanning over a decade:

- In 2008, at the height of the U.S. recession, Bernard "Bernie" Madoff, a former Wall Street executive, was arrested and charged with managing an intricate criminal scheme that stole at least $50 billion from corporate and individual investors. Madoff pleaded guilty to 11 criminal counts in Manhattan's federal district courthouse and was sentenced to 150 years in prison (Rosoff et al., 2010).
- In 2010, financial giant JPMorgan Chase was fined $48.6 million by British financial regulators for "failing to keep clients' funds separate from those of the firm." The error went undetected for more than 7 years and placed billions of dollars of clients' funds at risk of being lost (Werdigier, 2010).
- In 2014, General Motors Corporation, which received close to $50 billion in bailout funds from American taxpayers to stay financially solvent, was implicated in a scandal for failing to fix a defective ignition switch in its Chevrolet Cobalt line of vehicles (Isidore, 2012; Wald, 2014). The company allegedly chose not to make the fix, which caused vehicles to lose power while running and may have resulted in more than a dozen deaths, because doing so would have added to the cost of each car (Lienert & Thompson, 2014). The National Highway Transportation Safety Administration, which is charged with regulating and investigating complaints about motor vehicle safety, knew of the deaths linked to the vehicles with defective switches as early as 2007 but did not act (Wald, 2014).
- In 2014, JPMorgan was fined again, this time more than $461 million, by the Financial Crimes Enforcement Network (FinCEN) of the U.S. Department of the Treasury for violating the Bank Secrecy Act by failing to report the suspicious activities of Bernie Madoff (Financial Crimes Enforcement Network, 2014).
- In 2015, the Consumer Financial Protection Bureau (2015) fined Citibank $700 million for the deceptive marketing of credit card add-on products.
- In 2017, the car manufacturer Volkswagen AG (VW) agreed to plead guilty to three criminal felony counts and to pay a penalty of $2.8 billion for selling 590,000 diesel fuel vehicles in the United States that employed a "defeat device" to cheat on mandated emissions tests and then seeking to cover up the crime (U.S. Department of Justice Office of Public Affairs, 2017).

Police Corruption and Police Brutality

Police officers are frontline enforcers of laws to limit the amount of crime that occurs in society. When violent or property crimes are committed, people rely on the police for protection and investigation. Policing can be dangerous, and officers need to protect themselves and their colleagues. Members of the law enforcement community, however, are not immune to deviance, including corruption and brutality. They are, at the same time, in a unique position relative to deviance. As Martin and Kposowa put it, "They are a small group of people who are given permission to take another's life without being penalized" (2019, p. 8).

Police corruption and violence are not new, but they have increasingly been visible due to technologies such as phone videos, police body cameras, and media coverage (Martin & Kposowa, 2019). In 2016, for example, three New York Police Department commanders were arrested for taking extensive gifts (including hotel rooms, jewelry, and basketball tickets) from businesspeople seeking "illicit favors from the police" (Rashbaum & Goldstein, 2016).

Facebook CEO Mark Zuckerberg testifies virtually to the House Judiciary Subcommittee on Antitrust, Commercial and Administrative Law on Online Platforms and Market Power. In 2020, he was criticized for appearing to buy companies that were in competition with him—Instagram and WhatsApp.

Mandel Ngan-Pool/Getty Images

In 2020, #BlackLivesMatter saw millions of Americans and citizens abroad marching on the streets to protest police brutality against Black Americans. The sustained outpouring of marches and protests were spurred by the May 25 murder of George Floyd by a Minneapolis police officer. Floyd died as a result of the actions of police officer Derek Chauvin, who knelt on Floyd's neck for almost 8 minutes during an arrest that was caught on video camera (Willis et al., 2020).

Critical race theorists would point out that police brutality has a long and racialized history in the United States. The responses of southern police officers and sheriffs, such as Bull Connor in Birmingham, Alabama, to the 1960s civil rights marches provide well-documented examples. The videotaped beating of Rodney King in the 1990s by four white members of the Los Angeles Police Department was a vivid example of the treatment many minority city residents suffered at the hands and batons of the police. Public attention has particularly focused on the killing of unarmed Black men by police: Of 987 police shootings documented in 2017, in 68 documented cases, the victim was unarmed (*Washington Post*, "Fatal Force," 2017). According to a *Washington Post* analysis of 2016 data, "unarmed black men were seven times as likely as unarmed whites to die from police gunfire" (Somashekhar et al., 2015, para. 8). In one case, a University of Cincinnati police officer pulled a Black driver over for having no front license plate. The incident ended with the officer shooting Samuel Du Bose in the head. Because the killing was captured on a body camera, the officer was arrested and charged with murder (Blow, 2015), though in most recent cases, officers have not been charged with a crime. By August 30, 2020, police had fatally shot 661 people; 123 of them were Black (Statistical, 2020). In light of the multitude of killings of unarmed Black men, the question of how race affects officers' reaction to a suspect is important to both understanding and addressing the problem.

Both corruption and brutality represent important forms of criminal deviance committed by representatives of state authority. How can we make sense of these forms of deviance?

We might enlist Sutherland's differential association theory to consider both police corruption and the use of excessive force by law enforcement officers. Some studies and first-person accounts demonstrate that police officers are often exposed to various forms of deviance once they become members of the force (Kappeler et al., 1998; Maas, 1997). Exposure to attitudes favorable to the commission of deviance and crime, especially in light of the intensity and duration of the relationships police officers form with one another, may lead some to engage in those same types of deviant or criminal behaviors.

We can also view the police as having a distinct culture and, thus, see police brutality and corruption in terms of the subcultural expectations that accompany police work. Some members of the police subculture may see certain behaviors, actions, and perspectives, especially regarding the use of force, as a necessary part of accomplishing the demands of police work.

In their work on police brutality, Martin and Kposowa (2019) find that media stories covering police shootings of African American men do show that racial stereotypes and blaming-the-victim narratives are a secondary theme to emerge in the justifications of such shootings. They find the most common justification, however, is dehumanization of the victim through stereotyping. Thus, classical theories of deviance can also be enhanced by critical theories that examine the importance of race and racial inequality as well as its intersections with social class, gender, and sexuality.

Protesters in many U.S. cities have sought to draw attention to police brutality. Following the killing of George Floyd by a Minneapolis police officer a record number took to the streets in large and small communities.

Kent Nishimura/*Los Angeles Times* via Getty Images

State Crimes

Finally, we turn to perhaps another significant form of crime among the powerful: state crime. Although police brutality and corruption are especially egregious examples of crimes occurring among those with power, state crimes rank above even them in terms of the seriousness and potential harm that may result from their commission.

State crimes *consist of criminal or other harmful acts of commission or omission perpetrated by state officials in the pursuit of their jobs as representatives of the government*. Needless to say, governments do not normally keep statistics on their own criminal behavior. Nonetheless, we do know from various contemporary and historical examples that such crimes are not uncommon (Chambliss et al., 2010; Moloney & Chambliss, 2014; Rothe, 2009). Contemporary examples of state crime cover a spectrum of deviance and take place across the globe. Some examples include the routine torture of detainees at the U.S. military prison in Guantánamo Bay, Cuba; the secret transportation and torture of battlefield detainees in foreign prisons; and even the violation of international laws leading up to the 2004 invasion of Iraq and Syria's Russian-backed war crimes against its own citizens in 2019 and 2020 (Grey, 2006; Human Rights Watch, n.d.; Kramer & Michalowski, 2005; Paglen & Thompson, 2006; Ratner & Ray, 2004).

State crimes: Criminal or other harmful acts of commission or omission perpetrated by state officials in the pursuit of their jobs as representatives of the government.

In one study, the Chinese government was found to have taken a role in the trafficking of human body parts. The study's author found that the organs of executed prisoners were harvested by government-approved doctors and sold to corporations and other entities for use in organ transplantation and cosmetic surgery, often at substantial profits, without the consent of the prisoners' families (Lenning, 2007). In 2016, some members of Russia's Olympic team and the entire Paralympic team were barred from the Rio Olympic Games due to a far-reaching doping conspiracy. And in the 2018 Winter Olympic Games, athletes from Russia were forced to compete as the Olympic Athletes from Russia (OAR) team and were barred from carrying or wearing their country's flag as punishment for the country's history of doping in athletics.

SOCIAL CONTROL OF DEVIANCE

Social control: The attempts by certain people or groups in society.

Social power: The ability to exercise social control.

The persistence of deviant behavior in society leads inevitably to a variety of measures designed to control it. **Social control** is defined as the *attempts by certain people or groups in society to control the behaviors of other individuals and groups to increase the likelihood that they will conform to established norms or laws*. Thus, deviance, the definition of which tends to require some sort of moral judgment, also attracts attempts at social control, usually exercised by those people or groups who possess **social power**, or *the ability to exercise social control* to regulate the behaviors of other individuals and groups to increase the likelihood that they will conform to established norms or laws.

Informal social control: The unofficial means through which deviance and deviant behaviors are discouraged in society; most often occurs among ordinary people during their everyday interactions.

Informal social control is *the unofficial means through which deviance and deviant behaviors are discouraged in society; it most often occurs among ordinary people during their everyday interactions*. It ranges from frowning at someone's sexist assertion to threatening to take away a child's cell phone to coerce conformity to the parent's wishes. Informal social control mechanisms explain why people don't spit on the floor in a restaurant, do choose to pay back a friend who lent them $20, or say "thank you" in response to a favor. These behaviors and responses are governed not by formal laws but by informal expectations of which we are all aware and that lead us to make certain choices. Much of the time, these informal social controls lead us into conformity with societal or group norms and away from deviance. Informal controls are thus responsible for keeping most forms of noncriminal deviance in check.

Socialization, discussed in earlier chapters, plays a significant role in the success of informal social control. When parents seek to get their children to conform to the values and norms of their society, they teach them to do one thing and not another. Peer groups of workers, students, and friends also implement informal social control through means such as embarrassment, gossip, and criticism that work to control behavior and thus deviance. Bonds to institutions and people enact various informal social controls on our behavior. Such bonds have been shown to be crucial in explaining why some people engage in deviance and others do not (Laub & Sampson, 2003; Sampson & Laub, 1990).

Formal social control is defined as *official attempts to discourage certain behaviors and visibly punish others.* In the modern world, formal social control is most often exercised by societal institutions associated with the state, including the police, prosecutors, courts, and prisons. The goal of all these institutions is to suppress, reduce, and punish those individuals or groups who engage in criminal forms of deviance. Theft, assault, vandalism, cheating on income taxes, fraudulent reporting of corporate earnings, and insider trading on the stock market—all have been deemed crimes and represent forms of criminal deviance. As such, they are subject to formal social control.

Formal social control: Official attempts to discourage certain behaviors and visibly punish others; most often exercised by the state.

For an act to be criminal, several elements must be present. First, a specific law must prohibit the act and a punishment of either prison or a fine (or both) must be specified for violation of the law. Most important, the act must be intended, and the person committing the act must be capable of having the necessary intent. Someone judged to be mentally ill, which U.S. criminal law defines as the person "not knowing right from wrong" at the time of the act, cannot have the required legal intent and therefore cannot be held criminally liable for committing an illegal act. Nevertheless, the insanity defense is not accepted in all states: Idaho, Montana, Kansas, and Utah do not recognize insanity as a defense. In other state courts and federal courts, the burden of proof regarding a defendant's mental state is on the defendant. The U.S. Supreme Court refused to hear a case from Idaho in which a defendant claimed he had a constitutional right to claim insanity as a defense (Barnes, 2012).

Mainstream media and "official" depictions of deviance often overlook or ignore some of the more serious manifestations of deviance in the culture. So, too, do local communities. For example, in his research on the Saints and the Roughnecks, Chambliss (2001) found that middle-class boys were much more likely to have their deviant behavior written off as simply "sowing their wild oats," even though the behavior was dangerous and costly to society. On the other hand, the community was quick to judge, and apply deviant labels to, a lower-income group of boys. Such findings suggest a need for sociologists to delve more deeply into stereotypes of gangs and delinquency as phenomena associated almost exclusively with poor urban youth. Similarly, sociological research and theory remind us that focusing on the deviance of the poor and minorities blinds us to an understanding of deviance among the rich and powerful. The findings of sociologists who study deviant behavior generally and criminal deviance in particular suggest that social control policies such as imprisonment have limited effects.

In the sections that follow, we discuss some key issues associated with formal social control in U.S. society, including discipline in schools, the phenomenon of mass imprisonment, and the death penalty.

School-to-prison pipeline: The policies and practices that push students, particularly at-risk youth, out of schools and into the juvenile and criminal justice system.

Schools and Discipline: Is There a School-to-Prison Pipeline?

Were you or someone you know ever suspended or expelled from school? In recent years, many U.S. public schools have tightened disciplinary codes and punishments for infraction of them. As a result, the number of in and out of school suspensions and expulsions has climbed, even for children in primary school. For instance, children as young as preschoolers have been suspended for kicking off their shoes and crying. Where this is categorized as a disturbance to the classroom, the child may be subject to disciplinary action. Moreover, these punishments disproportionately impact African American and Latinx students (Peguero et al., 2018). All of this has led to the observation that children are diverging onto two roads: Some are tracked to careers and college, while others are tracked to prison.

Individuals without a high school degree are more likely to serve time in prison. Most state and federal prisons, like the one shown here, offer GED completion programs to the nearly two thirds of inmates who do not have a high school diploma.

The **school-to-prison pipeline** refers to *the policies and practices that push school children, particularly at-risk minority youth, out of classrooms and into the juvenile and criminal justice systems* (American Civil Liberties Union, n.d.; Potter et al., 2017).

Tightened discipline in school and increased safety measures implemented in the 1990s and beyond have not only affected high school students but has trickled down to preschools, increasing the probability of student infractions being harshly punished (Potter et al., 2017).

Zero-tolerance policies: School or district policies that set predetermined punishments for certain misbehaviors and punish the same way, no matter the severity or the context of the behavior.

What factors can help us understand the school-to-prison pipeline? First, following the 1999 Columbine, Colorado, shooting, when two high school students killed 13 students and injured 24 in a rampage and then shot themselves, some schools enacted **zero-tolerance policies**—*school or district policies that set predetermined punishments (usually suspension or expulsion) for certain misbehaviors and punish the same way, no matter the severity or the context of the behavior.* These policies, while potentially helpful, also caused problems. They punish children for a variety of infractions, including minor ones: Punishable offenses may range from carrying aspirin (which is considered a drug), to skipping class, to threatening a teacher, to concealing a weapon. Zero-tolerance policies led to a rise in detentions, suspensions, and expulsions (Potter et al., 2017; U.S. Department of Education Office for Civil Rights, 2014). The consequences of these are many. For instance, students are not learning in the classroom. They may experience stigma and avoid school after such punishment, which may, in turn, lead to more disciplinary behavior (Potter et al., 2017).

Second, school officials have increased police presence in the schools with the intention of protecting students. This, however, has resulted in more student arrests for infractions that would have previously been handled by the school. Together with zero-tolerance policies, this has resulted in more severe punishment and more student contact with the justice system (American Civil Liberties Union, n.d.). Students as young as 10 have been treated like criminals for minor offenses such as talking back or cutting class. For instance, in a case in Queens, New York, a 12-year-old girl was taken away in handcuffs after doodling on her desk with an erasable marker (Herbert, 2010). In-school police presence increases the risk of students being pushed into the juvenile justice system, which in turn puts them at higher risk for dropping out, as well as for future encounters with the criminal justice system.

Third, some observers have suggested that the increased emphasis on standardized testing results as a measure of school success may be introducing a perverse incentive to push out students who are less likely than their peers to perform well on these tests (Advancement Project, 2010). For example, a National Bureau of Economic Research paper used 4 years of data from Florida at the time that the state introduced a high-stakes testing regime. The author found that "while schools always tend to assign harsher punishments to low-performing students than high-performing students . . . this gap grows substantially throughout the testing window. Moreover, this testing window–related gap is only observed for students in testing grades" (Figlio, 2005, pp. 4–5).

Some policies that underpin the school-to-prison pipeline appear to be easing; for instance, the New York City Council, seeking greater transparency in the discipline process, now calls for detailed police reports on which students are arrested and why ("Criminalizing Children at School," 2013). Nevertheless, the disproportionately high suspension and expulsion rate for students of color continues to be a key concern. Black students are three times more likely than white students to be both suspended and expelled. The police presence in schools has exacerbated the racial divide: Although Black students only represent 16% of student enrollment, they represent 31% of school-related arrests (U.S. Department of Education Office for Civil Rights, 2014). The divide has called into question the equality of the school system and the justice system. How can schools balance the need to ensure student safety while providing support for the success of all their students? What do you think?

Imprisonment in the United States

To control serious, criminal forms of deviance, a society typically arrests and prosecutes those who have committed violent crimes or property crimes. Interestingly, however, of the roughly 11 million arrests in the United States in 2015, the highest number of arrests were for drug abuse violations (estimated at 1,488,707 arrests), larceny–theft (estimated at 1,160,390), and driving under the influence (estimated at 1,089,171) (Federal Bureau of Investigation, 2015). In 2016, about 47% of federal prisoners had been sentenced for drug offenses (Carson, 2018).

A contemporary indication of the persistence of formal social control is the very high number of people imprisoned (Figure 6.5). The United States still imprisons a vastly higher percentage of its

FIGURE 6.5 ■ U.S. State and Federal Prison Population, 1925–2016

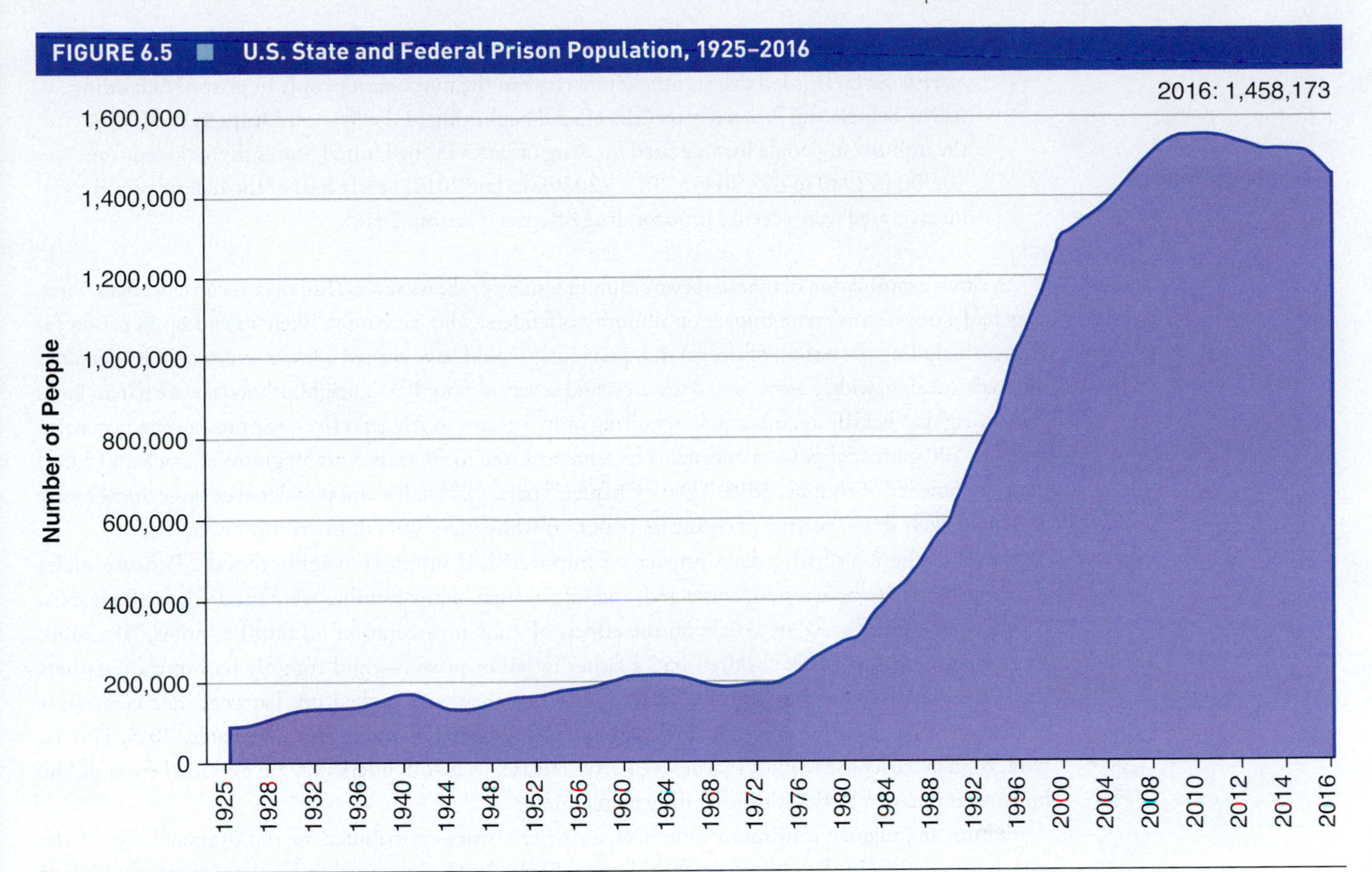

Source: Reprinted with permission from The Sentencing Project.

population than any other industrial society. While its prison population has declined from its peak of 7,339,600 incarcerated in 2008, in 2018, a total of 6,410,000—or, about 1 in 40 U.S. adults—was under some form of correctional system supervision in either prison or jail, on probation, or on parole (Maruschak & Minton, 2020).

Sociologists note that U.S. incarceration and correctional supervision *increased* significantly in the 1980s, at a time when both violent and property crime rates were *declining*. What accounts for that unprecedented rise amidst the decline of other types of crime? Among the key reasons, researchers cite the following:

1. **Mandatory minimum sentences**: Federal and state legislators in the 1980s passed *legislation stipulating that a person found guilty of a particular crime must be sentenced to a minimum number of years in prison*. This reduced judges' ability to use their discretion in sentencing. It led to a substantial increase in the average prison term.

2. **"Three strikes" laws**: Some *state and federal laws sentence an individual to life in prison who has been found guilty of committing three felonies or serious crimes punishable by a minimum of 1 year in prison*. "Three strikes" laws became popular after several high-profile murders were committed by former prisoners, raising concerns that their release was leading to more crimes in the community. Three strikes laws resulted in a significant increase in the number of people incarcerated. For instance, *The New York Times* reported in 2013 that 9,000 offenders were serving life in prison in California, comprising many whose third strike was a nonserious, nonviolent offense. In 2011, the U.S. Supreme Court ordered the state to reduce its prison population by 137.5% of capacity (*New York Times*, 2013).

3. The **"war on drugs"**: Mandatory minimums are intertwined with the "war on drugs," launched in the 1980s under the Ronald Reagan administration. To some degree, panic over the dangers of crack cocaine fueled *efforts by the U.S. state and federal governments to*

Mandatory minimum sentences: Legislation stipulating that a person found guilty of a particular crime must be sentenced to a minimum numbers of years in prison.

"Three strikes" laws: State and federal laws that sentence an individual to life in prison who has been found guilty of committing three felonies or serious crimes punishable by a minimum of a year in prison.

"War on drugs": Actions taken by U.S. state and federal governments to curb the illegal drug trade and reduce drug use by punishing drug possession, use, and trafficking more harshly.

curb the illegal drug trade and reduce drug use by punishing drug possession, use, and trafficking more harshly. This led to a significant increase in the number of people in prison. According to The Sentencing Project, after "the official beginning of the War on Drugs in the 1980s, the number of people incarcerated for drug offenses in the United States skyrocketed from 40,900 in 1980 to 452,964 in 2017" (2020). In late 2016, nearly half of the federal prisoners incarcerated were serving time for drug offenses (Carson, 2018).

A closer examination of mandatory minimum sentences shows several important consequences. First, they had a disproportionate impact on minority offenders, who were more likely to end up in prison for crimes (including possession of drugs) that previously would have earned a lesser sentence. The penalties for crack cocaine, widely associated with users and sellers in poor Black neighborhoods, were harsh: Laws "were weighted heavily against crack, requiring only 5 grams to trigger a five-year prison sentence, compared to 500 grams for powder cocaine. The sentences rose to 10 years with 50 grams of crack and 5,000 grams of powder" (Schuppe, 2016, "'Dark Chapter,'" para. 1). Penalties for powdered cocaine, more costly and more likely to be the drug of choice for better-off white users, were dramatically lower.

Second, the laws had a deep impact on impoverished minority neighborhoods. Because males were most likely to be arrested, convicted, and imprisoned, many families were suddenly without sons, brothers, and fathers. As an article on the effects of mass incarceration on families notes, "By 2000, more than 1 million black children had a father in jail or prison— and roughly half of those fathers were living in the same household as their kids when they were locked up. Paternal incarceration is associated with behavior problems and delinquency, especially among boys" (Coates, 2015, Part II, para. 8). The costs to families accrue even after release, as ex-offenders may be excluded from public housing and have a difficult time finding employment.

Third, mandatory minimum sentences for drug crimes contributed to the dramatic rise of the prison population in the United States (Figure 6.5). In 2007, the rate of imprisonment in the United States peaked at 767 people per 100,000 (Coates, 2015). More recently, it has dropped to 450 per 100,000, although this continues to be far more than other modern democratic states around the globe (Carson, 2015, 2018).

Over the past several decades, changes in criminal laws and criminal sentencing have resulted in much stricter forms of social control in relationship to certain types of crime. In turn, this led to a huge increase in the U.S. prison population, although the imprisonment rate has turned downward in recent years: At year's end in 2016, the United States imprisoned 450 persons per 100,000 residents of all ages and 582 persons per 100,000 residents age 18 or older. Both statistics represent the lowest rate of imprisonment in more than a decade and continue decreases that began in 2007 and 2008 (Carson, 2018).

Formal mechanisms of social control administered through the criminal justice system have not been applied equally or proportionally to all groups in society. Those most likely to be imprisoned and punished for engaging in criminal deviance are disproportionately people of color, as shown in Figure 6.6. Blacks and Hispanics are arrested and imprisoned at much higher rates than whites, despite the

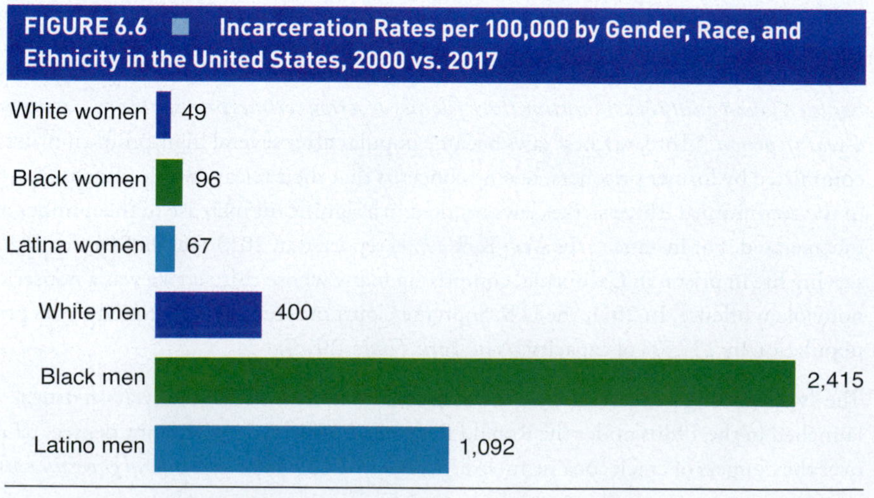

FIGURE 6.6 ■ Incarceration Rates per 100,000 by Gender, Race, and Ethnicity in the United States, 2000 vs. 2017

Source: Reprinted with permission from The Sentencing Project.

fact that whites make up a much larger proportion of the total U.S. population (Glaze & Herberman, 2013; The Sentencing Project, 2020). Statistics show that Black men have a 32% chance of serving time in prison at some point during their lives; for Latino men, the chance is 17% and for white men, 6% (Sentencing Project, n.d.). Black men are incarcerated at a rate of about 1,824 per 100,000; for Hispanic men, the rate is more than 820 per 100,000. White men, by contrast, have an incarceration rate of 312 per 100,000 (Carson & Anderson, 2016). If current trends continue, Black males between the ages of 19 and 34 will experience an even greater overrepresentation in the prison population (Figure 6.6).

Women constitute about 10% of the prison population in the United States (World Prison Studies, 2020). Women's incarceration rates have increased dramatically over the past four decades, largely due to tough crime policies and mandatory minimum drug-sentencing policies (Tripodi et al., 2019). Housed in a variety of local, state, and federal facilities, 95% of incarcerated women will reenter society (Turanovic et al., 2015), and nearly three quarters will be rearrested (Alper et al., 2018). Women from racial and ethnic minorities are disproportionately imprisoned, with African American women being incarcerated in state prisons at nearly double the rate of white women (Bronson & Carson, 2019).

People of color are more likely than whites to be arrested and imprisoned for several reasons, all of which relate to the extension of formal social control over criminal deviance (Mann & Zatz, 1998). First, impoverished inner-city residents are disproportionately nonwhite, and the inner city is where the "war on drugs" was most avidly waged (Chambliss, 2001). Second, the work of policing generally focuses on poor neighborhoods, where crowded living conditions force many activities onto the streets. Illegal activities are therefore much more likely to attract police attention in poor neighborhoods than in dispersed suburban neighborhoods. Finally, racism in practices of prosecution and sentencing also accounts for greater arrest and imprisonment rates of people of color. Even though many more whites than nonwhites are arrested for crimes, people of color are more likely to be imprisoned for their offenses (Austin et al., 1992; The Sentencing Project, 2020).

In addition to concerns over disparities in rates of imprisonment, conditions of health and safety in prison are of interest to sociologists. Because of their close proximity, inmates are susceptible to contracting certain illnesses at greater rates than are individuals outside of corrections facilities. For instance, by August 2020, COVID-19 had had a major impact on the health of those imprisoned. In the United States, according to the Death Penalty Information Center, more death row prisoners had died from COVID-19 than had been executed in a total of 20 years, and "more California death-row prisoners ha[d] been killed by the virus than ha[d] been executed in the state since 1993" (Death Penalty Information Center, 2020a).

The Stigma of Imprisonment

People who leave prison are said to have "done their time" or "paid their debt to society." Arguably, however, an array of laws and policies essentially ensure that ex-offenders continue to be stigmatized and punished by limiting their access to political voice, housing, and employment. According to the U.S. Department of Justice (n.d.), about 650,000 ex-offenders are released from prison every year. Notably, an estimated two thirds will reoffend within 3 years. A careful consideration of the roadblocks faced by former prisoners when they are released may, arguably, help society to understand and reduce the problem of recidivism.

Consider that upon release from prison, people need shelter, food, and clothing. Many ex-offenders face significant barriers to housing after release. Many released from incarceration do not know where they are going to live. Approximately one third expect to go to homeless shelters upon release. Ex-offenders have been, in many areas, banned from public housing, although studies show that stable housing can reduce recidivism. Notably as well, in some instances, entire families have been evicted from their homes in public housing after taking in a family member returning from prison. In response, several cities have begun to address housing policies: Accordingly, "local public housing authorities can no longer use arrest records as 'the sole basis for denying admission, terminating assistance or evicting tenants'" (Carpenter, 2015, para. 3). In 2016, the U.S. Department of Housing and Urban Development released new guidelines that also make it more difficult for landlords and home sellers to discriminate against those with criminal backgrounds (Abdullah, 2016).

Crucial to successful reentry is the securing of jobs. Prisons and jails often provide insufficient programming for inmates to increase job skills, leaving them subject to similar or worse economic conditions upon release. For instance, in her ethnographic research of six midwestern women's prisons, Wortmann (2020) found that none of these offered women college degree completion programs, and few offered training toward living wage occupations that could be accessed by a majority of their prisoners. She also found that prison jobs available for women were highly gendered—such as sewing, cooking, and laundry.

Most important, ex-offenders face the often daunting challenge of finding employment. Data from the New York State Division of Parole show that only 36% of able-bodied parolees who had been out of prison for 30 days or more were employed in 2014. As an article on the problem notes,

> Former inmates often face enormous challenges finding work after they've been released: not only have many of them been out of the workforce for years, but often their criminal record prevents them from even getting their foot in the door in the first place. (Vega, 2015, para. 4)

A national "ban the box" movement has arisen in some areas, with the goal of eliminating the box that job applicants are sometimes required to check on their applications if they have been convicted of a crime, most often a felony. A secondary goal of ban the box is to reduce racial disparities in employment discrimination as Black men, even without a criminal record, are more likely to be turned down for employment (Doleac & Hansen, 2018; Pager, 2003). Recently, 14 states and several dozen cities passed laws requiring employers to postpone background checks until the later stages of the hiring process to reduce discrimination against ex-offenders (Appelbaum, 2015). Ensuring access to job opportunities in the legal economy is a key way to reintegrate former prisoners into society. An interesting study by Doleac and Hansen (2018) showed that an unintended consequence of banning the box might be resulting employer discrimination against Black and Hispanic men, with Blacks being particularly likely to be targeted.

Importantly, ex-offenders who have served time for a felony are denied a political voice in many states in this country. Although a poll released by YouGov showed that about 54% of Americans believe that ex-felons should have their right to vote restored once they have completed their sentences (Moore, 2016), only two states (Maine and Vermont) have laws that allow persons with felony convictions never to lose their right to vote and only 16 states have automatic restoration after release. In 11 states, "those with a felony lose their voting rights indefinitely for some crimes or require a governor's pardon in order for voting rights to be restored, or face an additional waiting period after completion of sentence" (National Conference of State Legislatures, 2020).

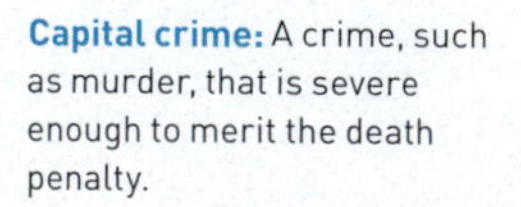

Capital crime: A crime, such as murder, that is severe enough to merit the death penalty.

The U.S. justice system releases several hundred thousand men and women from prisons and back into society every year. Arguably, social stigma, conditions in prisons themselves, laws and policies that erect obstacles to stable housing and employment, as well as participation in society through access to the ballot, play a role in determining the probability an ex-prisoner will follow one or the other path.

The Death Penalty in the United States

A **capital crime**, sometimes called a *capital offense*, is a *crime, such as murder, which is severe enough to merit the death penalty (capital punishment).* Forty-one offenses are listed by the federal government as being punishable by death. Crimes that fall under this category include espionage, treason, and aircraft hijacking, as well as murder in particular circumstances.

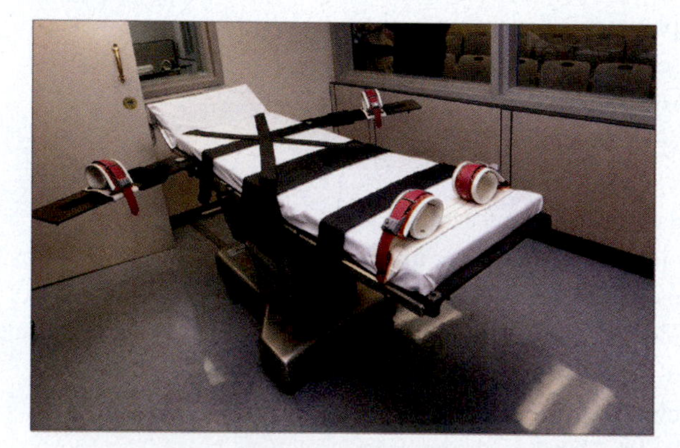

Since 1976, most executions in the United States (1,344) have been done by lethal injection, a method that replaced the electric chair in 30 states (Death Penalty Information Center, 2020b).

©AP Photo/Sue Orocki, File

The death penalty is a controversial practice in the United States. Few modern democratic states use it. In fact, by the end of 2019, according to Amnesty International, 106 countries had abolished the death penalty in law for all crimes, and 142 countries had abolished the death penalty in law or practice. It was most frequently enlisted (in order) in these countries: China, Iran, Saudi Arabia, Iraq, and Egypt (Amnesty International, 2020). According to Pew Research, men are more likely than women to support the death penalty (61% vs. 46%) and whites are more likely to support it than are Blacks and Hispanics (59% vs. 36% vs. 47%) (Masci, 2018, para. 6).

The death penalty varies state by state in the United States: Some states have and use the death penalty, others do not impose a death penalty, and still others have the death penalty but operate under a governor-imposed moratorium on the practice. Specifically, 28 states enlist capital punishment, with Texas and Georgia executing the most prisoners. In 2019, 22 people were executed in seven states; nine of these were in Texas. Twenty-two states and the District of Columbia have abolished the death penalty. Five states have enlisted a governor-imposed moratorium; that is, the governor has the discretion to halt the use of the death penalty while in office. For instance, California's governor, Gavin Newsom, in March 2019 signed an executive order that created a moratorium on the death penalty for the 737 prisoners on death row in California (Figure 6.7; Death Penalty Information Center, 2020b).

Sociologists point out the importance to view the death penalty intersectionally. For instance, rates vary by gender: Men are much likelier to be on death row than women. In 2019, 57 women were on death row as compared to 2,546 men (Death Penalty Information Center, 2020b). Rates also vary by race: While more white than Black men are on death row, death row inmates are disproportionately Black (Figure 6.8; NAACP Legal Defense Fund). Although more white prisoners were executed in the past year, the imposition of the death penalty has historically been biased against Black defendants, who, as we just mentioned, are disproportionately found on death row (Figure 6.8). For example, a 2007 Yale University study found that African American defendants received the death penalty at three times the rate of white defendants when the victim was white (cited in Amnesty International, 2012). In 2020, the Death Penalty Information Center released a report on the role racial segregation and discrimination had played in the death penalty, describing a continuing "pervasive discrimination" in capital punishment. As Robert Dunham, the report's editor, explains:

FIGURE 6.7 ■ Death Penalty in the United States

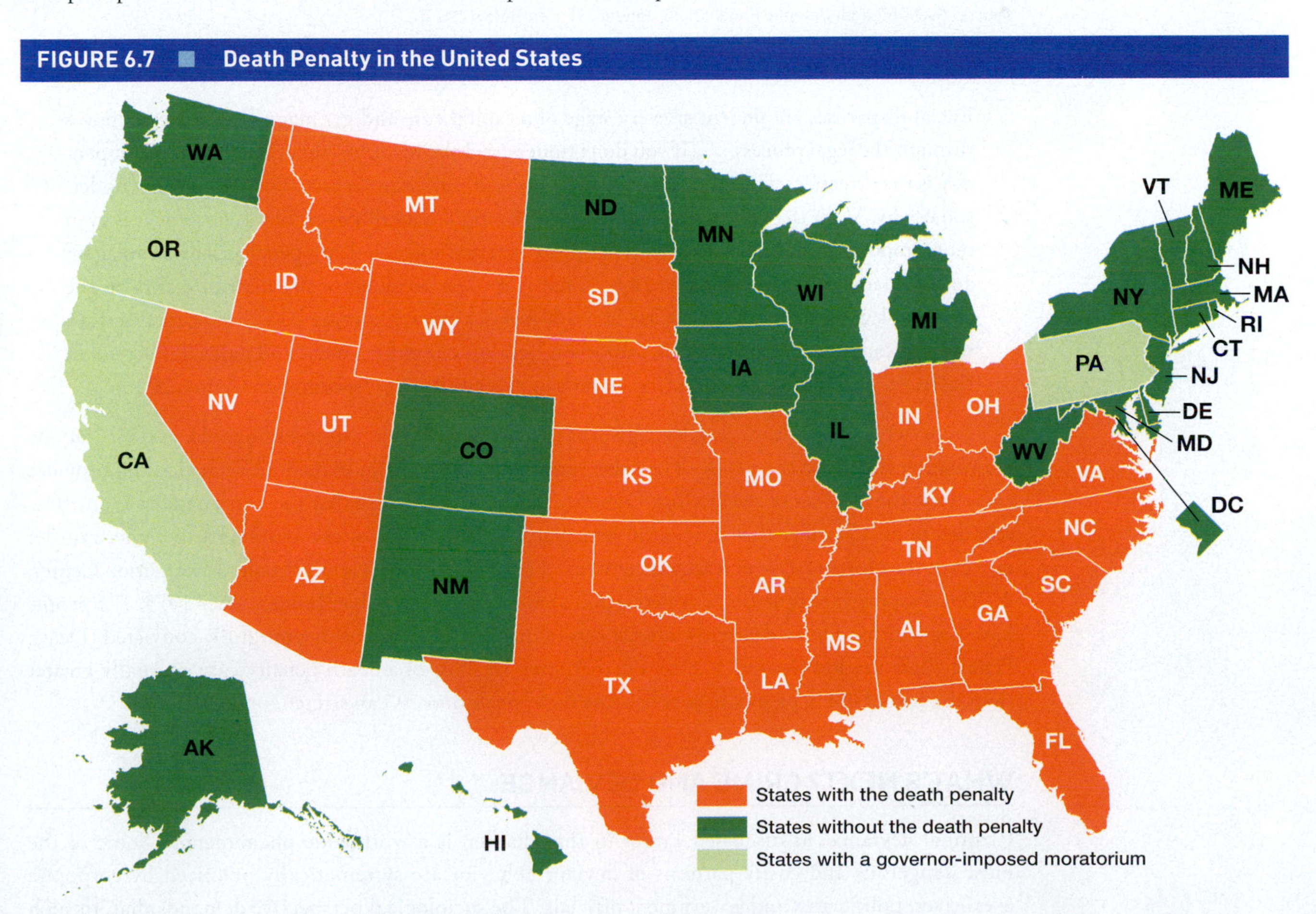

Source: ProCon.org

FIGURE 6.8 ■ Death Row Inmates by Race, 2020

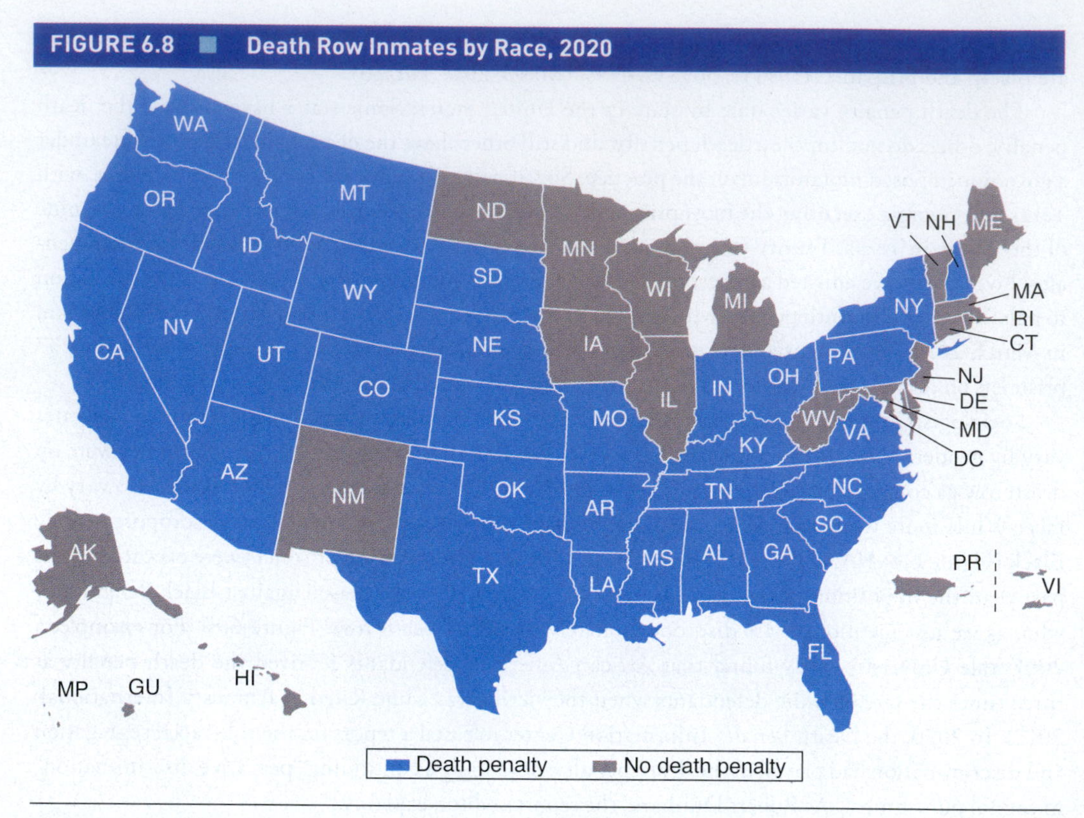

Source: NAACP Legal Defense Fund. (2020, January 1). Death Row U.S.A.

Note: Percentages may not equal 100% due to rounding.

> Racial disparities are present at every stage of a capital case and get magnified as a case moves through the legal process. . . . If you don't understand the history—that the modern death penalty is the direct descendant of slavery, lynching, and Jim Crow–segregation—you won't understand why. With the continuing police and white vigilante killings of Black citizens, it is even more important now to focus attention on the outsized role the death penalty plays as an agent and validator of racial discrimination. What is broken or intentionally discriminatory in the criminal legal system is visibly worse in death-penalty cases. Exposing how the system discriminates in capital cases can shine an important light on law enforcement and judicial practices in vital need of abolition, restructuring, or reform. (Death Penalty Information Center, 2020b)

Proponents of the death penalty argue that it may deter future crime and uphold a safe society by eliminating threats to the public. They also suggest that it is a form of retributive justice and provides closure for the families of the victims. Opponents challenge the idea that the death penalty is an effective deterrent: Consider that the South, which accounts for 80% of executions, has a higher murder rate than the Northeast, which accounts for just .5% of executions (Death Penalty Information Center, 2020b). Proponents argue that erroneous punishments are irreversible: Indeed, since 1973, 171 people have been released from death row after it was found that they had been wrongfully convicted (Death Penalty Information Center, 2020b). Furthermore, the cost of a death penalty case is usually greater than the cost of life in prison, in part because the appeals process can stretch for years.

WHAT'S NEXT? CRIME AND DEVIANCE

Criminal deviance, as discussed earlier in this chapter, is a worldwide phenomenon. Some of the most dangerous and costly patterns of deviant behavior are systematically practiced by corporate executives, politicians, and government officials. The sociological perspective demands that, in such cases, we ask questions about power and who has the power to define deviance and to enforce their definitions. Also important are globalization and digital networks, which have increased the potential for transnational crime networks to gain wealth and political power (Block & Weaver, 2004; FBI, n.d.; Naylor, 2002). At the top of the list of crimes facilitated by processes linked to globalization are money laundering and the smuggling and illicit selling of drugs, weapons, and the trafficking of human beings. So, what's next?

One important area of theorizing, research, policy, and intervention will continue to be human trafficking, a major international criminal enterprise. Worldwide, human trafficking is the second-largest criminal activity and fastest-growing criminal enterprise (Mallicoat, 2019, p. 188). The U.S. Department of State's Advisory Council on Human Trafficking identifies these five groups as especially vulnerable: those trafficked for labor; LGBTQI; indigenous populations; people with special needs, disabilities, and who are elderly; and boys and men who, they note, comprise a significant number of those trafficked but lack services. They point out that the challenges for addressing these issues are:

> [L]imited public awareness, lack of data, limited service providers available to address their needs, fewer prosecutions of cases involving victims from these populations, and distinctly inadequate use of grant funding for these population. (United States Advisory Council on Human Trafficking, 2020)

Based on what you have learned in this chapter, what unique role do you think sociologists can play in helping to understand and address human trafficking? What types of studies do you think would be most effective in studying the scope and issues involved in human trafficking? Which theoretical perspectives best address this issue and why?

WHAT CAN I DO WITH A SOCIOLOGY DEGREE?

Understanding and Fostering Social Change

Social change comes about as a result of shifts in the social order of society. Although changes may be evolutionary or revolutionary in pace, change is inevitable. Understanding social change and the factors that underlie its dynamics is key to bringing about positive change, whether at the micro or the macro level. Sociologists study factors that bring about large-scale social change—for instance, shifts in population growth or health, technological innovations, economic and labor market changes, the mobilization of civil society, or the rise of a charismatic leader—and seek to understand barriers to normative or structural change. They are also interested in factors that affect change or resistance to change in smaller groups and communities. Skills in the areas of leadership, communications, strategic thinking, motivation and mobilization, and advocacy can evolve from knowledge gained in the study of social change. Students interested in social change may also take advantage of internships or practicums in community or political organizations involved in fostering positive change. Supervised practice and the opportunity for reflection on your work nurture skills in the area of social change.

Careers in social change may focus on specific areas, including the environment, labor, human rights, free speech, legal reform, social justice, conflict resolution, poverty, health care, gender equity, economic justice, and corporate ethics. They may be careers in public service (such as in federal, state, or local government) or in the private sector (with advocacy organizations or in research-focused organizations, for instance). An understanding of social change and the development of skills associated with fostering positive social change are important in a wide variety of occupational fields.

Louis Mariel, Field Organizer with Bernie 2020 & People's Action

University of Nevada, Reno, BA in Sociology & Political Science

As an undergraduate freshman with no major and little experience in the "real world" of employment and politics, I went into Sociology to gain some grounding in the nuances of human society. I came out with a yearning passion to engage the world on critical issues of power dynamics and marginalization, undergirded by a profound cognizance of how society is shaped by material forces. Sociology taught me many things: a fondness for reading, research and analysis skills, a love of data, and genuine empathy regarding why we act as we do. Above all else though, Sociology gave me the tool of critical thinking, a mindset to cut through the whirlwind of events that kept happening around me.

As a junior with two years of Sociology classes under my belt, I felt emboldened like never before to create change in my community. With friends and classmates, I started and led a club on campus dedicated to direct action and community-guided education. We read and discussed together, fed the local homeless folks, raised money for community gardens, and developed ourselves in relation to the world we lived in. The authors I read, from Weber to Gramsci, Durkheim to Du Bois, gave me a unique lens to understand the value of this work. After I graduated, I found my way into the Bernie 2020 campaign as a field organizer, tasked with building a movement by and for the working class. Day in and day out, I talked to my community about historic rates of income inequality, public health as tied to policy, systemic racism, and how climate change developed alongside industrial society and class relations. Now I do the same with a national progressive network, People's Action. I don't think I'll ever put down organizing work again.

As a working professional with a degree in Sociology, I know that the ideas we examined were not abstracted theoreticals to be debated in a distant ivory tower. They were critical elements of my life, and of yours, of your family and job and school and relationships, which affect you from the moment you're born. Every current event and groundbreaking story around us is shaped by the norms that contain it, the organizations that advance it, and the social conditions which necessitated it. Take Sociology—you will unlock a whole world of new perspectives which will stick with you for the rest of your life.

Career Data: Political Scientists

- 2019 Median Pay: $122,220 per year
- $58.76 per hour
- Typical Entry-Level Education: Master's degree
- Job Outlook, 2019–2029: 6% (fast as average)

Source: Bureau of Labor Statistics. (2020). *Occupational outlook handbook.*

SUMMARY

- Notions of what constitutes **deviance** vary considerably and are relative to the norms and values of particular cultures as well as the labels applied by certain groups or individuals to specific behaviors, actions, practices, and conditions. Even **crimes**, which are particular forms of especially serious deviance, are defined differently from place to place and over time, and their definitions also depend on social and political factors.
- In **pluralistic societies** such as the United States, it is difficult to establish universally accepted notions of deviance.
- Most sociologists do not believe there is a direct causal link between biology and deviance. Whatever the role of biology, deviant behaviors are culturally defined and socially learned.
- Functionalist theorists explain deviance in terms of the functions it performs for society. Émile Durkheim argued that some degree of deviance serves to reaffirm society's normative boundaries. Robert K. Merton argued that deviance reflects **structural strain** between the culturally defined goals of a society and the means society provides for achieving those goals. **Differential opportunity theory** emphasizes access to deviance as a major source of deviance. **Social control theory** focuses on the presence of interpersonal bonds as a means of keeping deviance in check.
- Conflict theories explain deviance in terms of the conflicts between different groups, classes, or subcultures in society. **Class-dominant theories** of deviance emphasize how wealthy and powerful groups are able to define as deviant any behavior that runs counter to their interests. **Structural contradiction theory** argues that conflicts are inherent in social structure; it sees the sorts of structural strains identified by Merton as being built into society itself. **Feminist perspectives** argue for the importance of examining gender (and other social identities) as

important factors in deviance and crime. They point out that many women labeled as deviant, such as delinquent girls, are, in fact, escaping gendered sexual and physical abuse. **Critical race theories** address the importance of examining race and racism in understanding who is labeled as deviant or criminal and how racism is used in the legal and criminal justice systems to unequally impact people of color.

- Symbolic interactionist theorists argue that deviance, like all forms of human behavior, results from the ways in which we come to see ourselves through the eyes of others. One version of symbolic interactionism is **labeling theory**, which argues that deviance results mainly from the labels others attach to our behavior.
- **Violent crimes** are the most heavily publicized, but the most common are **property crimes** and victimless crimes. Although there is a public perception in the United States that crime is increasing, all crime rates have decreased.
- Although crime is often stereotyped as concentrated among poor racial minorities, crimes are committed by all groups of people. **White-collar crime** and **state crime** are two examples of crime committed by people in positions of wealth and power. They exact enormous financial and personal costs from society.
- Deviance and criminal deviance are both controlled socially through mechanisms of **informal social control**, such as socialization, and **formal social control**, such as arrests and imprisonment.
- Means of formal social control include the growth of **zero-tolerance policies** and policing in schools, **mandatory minimum sentences** for certain crimes and mass incarceration, and the use of the death penalty.
- A disproportionate number of people of color are arrested and incarcerated in the United States. This phenomenon is linked to a variety of factors that include the concentration of the **"war on drugs"** on poor, minority neighborhoods and racism in the arrest, prosecution, and sentencing of those who engage in deviant behavior.

KEY TERMS

capital crime (p. 156)
capital offense (p. 133)
class-dominant theories (p. 139)
crime (p. 133)
Critical race theories (CRTs) (p. 141)
deviance (p. 133)
differential association theory (p. 142)
differential opportunity theory (p. 137)
feminist perspectives on deviance (p. 140)
formal social control (p. 151)
human trafficking (p. 147)
informal social control (p. 150)
labeling theory (p. 141)
mandatory minimum sentences (p. 153)
organized crime (p. 147)
phrenology (p. 135)
pluralistic societies (p. 133)
primary deviance (p. 141)
property crimes (p. 146)
school-to-prison pipeline (p. 151)
secondary deviance (p. 141)
self-fulfilling prophecy (p. 141)
social bonds (p. 137)
social control (p. 150)
social control theory (p. 137)
social power (p. 150)
state crimes (p. 150)
strain theory (p. 136)
structural contradiction theory (p. 139)
structural strain (p. 136)
subcultural theories (p. 138)
"three strikes" laws (p. 153)
violent crimes (p. 146)
"war on drugs" (p. 153)
white-collar crime (p. 147)
zero-tolerance policies (p. 152)

DISCUSSION QUESTIONS

1. Discuss the differences in definitions of crime and deviance. Explain how each are socially constructed and the implications of that social construction for social interaction and social structure.

2. Labeling theories in the area of criminology suggest that labeling particular groups as deviant can set in motion a self-fulfilling prophecy. That is, people may become that which is expected of them—including becoming deviant or even criminally deviant. Can you think of other social settings where labeling theory might be applied?

3. Why, according to sociologists, are the crimes of the powerful (politicians, businesspeople, and other elites) less likely to be severely punished than those of the poor, even when those crimes have mortal consequences?

4. Broadly describe patterns of incarceration in the United States. Why, according to the chapter, did the rate of imprisonment rise in the United States in the 1980s? Why are a disproportionate number of inmates people of color? What challenges do they face upon reentry?

©iStockphoto.com/peeterv

7 SOCIAL CLASS AND INEQUALITY IN THE UNITED STATES

WHAT DO YOU THINK?

1. What is income? What is wealth? Why is it important to distinguish between the two when studying class status and inequality?
2. What factors explain rising income and wealth inequality in the United States?
3. What explains the existence and persistence of widespread poverty in the United States?

LEARNING OBJECTIVES

7.1 Identify characteristics of stratification in traditional and modern societies.

7.2 Describe components of social class, including income, wealth, occupation, status, and political voice.

7.3 Describe dimensions of socioeconomic inequality in the United States.

7.4 Describe and discuss the problems of household and neighborhood poverty.

7.5 Analyze the existence and persistence of stratification and poverty from sociological perspectives.

HOMELESS AND ELDERLY

In 1935, the U.S. Congress passed a measure, signed into law by President Franklin D. Roosevelt, that created Social Security. Social Security was intended to provide income support for retiring Americans, based on their working years, that would keep them economically secure. Today, Social Security, which is funded by payroll taxes, continues to provide monthly income for millions of seniors. The Medicare program, enacted in 1965, which is funded with taxpayer dollars, provides a basic health care safety net to older adults as well, giving them access to preventive and acute medical care. It is not a coincidence that Americans age 65 and over have poverty rates lower than those of children and lower than some groups of younger adults.

At the same time, many older Americans struggle with basic needs. Among those needs is access to safe and stable housing. By one estimate, the number of elderly people experiencing homelessness will rise threefold, as Baby Boomers, who have comprised the most significant share of the homeless in the United States, age (Santos, 2020). A recent report suggests that the economic conditions that existed for those born at the tail end of the Baby Boom (that is, between 1955 and 1964) were less favorable than those that awaited those born earlier: Specifically, they were more likely to encounter an excess supply of workers in the labor market, which pushed down wages and reduced opportunities, and significant competition in the housing market, which drove up prices. While many members of this age cohort found their economic footing, especially if they were able to earn a college degree, others struggled throughout their adult lives (Culhane et al., 2019). Baby Boomers also comprise a disproportionate share of the military veteran population, a population that has faced high levels of homelessness (Longshore, 2017). Economic deprivation in the Baby Boom generation has, however, been largely obscured by a powerful image of that generation as the most prosperous in U.S. history (Santos, 2020).

The Baby Boomer generation is at a heightened risk for homelessness. What are particular challenges that older adults who are homeless may face?

A recent report notes that there are challenges ahead not only for those older adults experiencing homelessness, but also for the institutions that serve them. For example, according to the report, "Older homeless adults have medical ages that far exceed their biological ages," and they experience conditions like cognitive decline and poor mobility at rates similar to those who are 20 years older. These conditions are likely to lead to higher costs for medical and nursing home care as struggling adults age (Culhane et al., 2019).

The COVID-19 pandemic, by some accounts, exacerbated the vulnerabilities of older adults, even those who are housed. Those who are near retirement but lost a job during the pandemic may be less likely than younger workers to find a new job in a labor market where they face age discrimination (Santos, 2020). This creates additional economic pressure that could push them out of stable and permanent housing.

Economic insecurity, in the form of homelessness, hunger, lack of access to education and health care, among others, is a chronic and continuing challenge in the United States today. It is, however, often masked by the images of prosperity and plenty that populate popular culture and social media. How pervasive is economic deprivation? Who is prospering—and who is falling behind? What does the U.S. class hierarchy look like today? Those are questions we seek to address in this chapter.

We begin this chapter with an examination of forms of stratification in traditional and modern societies, followed by a discussion of the characteristics of caste, social class, and stratification. Next, we look at important quantitative and qualitative dimensions of inequality and both household and neighborhood poverty in the United States. Finally, we turn to a discussion of theoretical perspectives on class and inequality.

STRATIFICATION IN TRADITIONAL AND MODERN SOCIETIES

In the United States today, there is substantial **social inequality**—a *high degree of disparity in income, wealth, power, prestige, and other resources.* Sociologists capture the disparities between social groups conceptually with an image from geology: They suggest that society, similar to the Earth's surface, is made up of different layers. **Social stratification** is thus *the systematic ranking of different groups of people in a hierarchy of inequality.* Sociologists seek to outline the quantitative dimensions and the qualitative manifestations of social stratification in the United States and around the globe, but—even more important—they endeavor to identify the social roots of stratification.

Social inequality: A high degree of disparity in income, wealth, power, prestige, and other resources.

Social stratification: The systematic ranking of different groups of people in a hierarchy of inequality.

Stratification systems are considered *closed* or *open,* depending on how much mobility between layers is available to groups and individuals within a society. Caste societies (closed) and class societies (open) represent two important examples of systems of stratification.

Caste Societies

In a **caste society,** *social positions are closed, so that individuals remain at the social level of their birth throughout life.* Social mobility is virtually impossible because social status is the outcome of *ascribed* rather than of *achieved* characteristics. Ascribed characteristics are those that are assigned from birth and cannot typically be changed, including age, birth order, caste, race, and ethnicity. Achieved characteristics are those that can be earned or chosen.

Caste society: A system in which social positions are closed, so that individuals remain at the social level of their birth throughout life.

Historically, castes have been present in some agricultural societies, such as rural India and South Africa prior to the end of white rule in 1992. In the United States before the end of the Civil War in 1865, slavery imposed a racial caste system because enslavement was usually a permanent condition (except for those enslaved persons who escaped or were freed by their owners). In the eyes of the law, the slave was a form of property without personal rights. Some argue that institutionalized racial inequality and limits on social mobility for African Americans remained fixtures of the U.S. landscape even after the end of slavery (Alexander, 2010; Dollard, 1957; Immerwahr, 2007). Indeed, enforced separation of Blacks and whites was supported by federal, state, and local laws on education, family formation, public spaces, and housing as late as the 1960s.

Caste systems are far less common in countries and communities today than they were in centuries past. For example, India is now home to a rising middle class, but it has long been described as a caste-based society because of its historical categorization of the population into four basic castes (or *Varnas*): priests, warriors, traders, and workmen. These categories, which can be further divided, are based on the country's majority religion, Hinduism. At the bottom of this caste hierarchy one finds the *Dalits* or *untouchables*, the lowest caste.

Since the 1950s, India has passed laws to integrate the lowest caste members into positions of greater economic and political power. Some norms have also changed, permitting members of different castes to intermarry. Although members of the lowest castes still lag in educational attainment compared with higher-caste groups, India today is moving closer to a class system.

Class Societies

Class society: A system in which social mobility allows an individual to change their socioeconomic position.

In a **class society,** *social mobility allows an individual to change their socioeconomic position.* Class societies exist in modern economic systems and are defined by several characteristics. First, they are *economically based*, at least in theory—the hierarchy of social positions is determined largely by economic status (whether earned or inherited) rather than by religion or tradition. Second, class systems are *relatively fluid:* Boundaries between classes are violable and can be crossed. In fact, in contrast to caste systems, in class systems, social mobility is looked at favorably. It has long been an American aspiration for parents to hope that their children will live better than they do. Finally, class status is understood to be *achieved rather than ascribed:* Status is, ideally, not related to a person's position at birth, religion, race, or other inherited categories, but to the individual's merit and achievements in education, entrepreneurship, and work.

As we will see in this chapter, these ideal-typical characteristics of class societies do not necessarily correspond to historical or contemporary reality, and class status can be profoundly affected by factors such as race, gender, and class of birth.

Social categories: Categories of people who share common characteristics without necessarily interacting or identifying with one another.

Achieved status: Social position linked to an individual's acquisition of socially valued credentials or skills.

Ascribed status: Social position linked to characteristics that are socially significant but cannot generally be altered (such as race or sex).

SOCIOLOGICAL BUILDING BLOCKS OF SOCIAL CLASS

Nearly all socially stratified systems share three characteristics. First, rankings apply to **social categories**—*categories of people who share common characteristics without necessarily interacting or identifying with one another.* In many societies, women may be ranked differently than men, wealthy people differently than the poor, and highly educated people differently than those with little schooling. Individuals may be able to change their rank (through education, for instance), but the categories themselves continue to exist as part of the social hierarchy.

Second, people's opportunities and experiences are shaped by how their social categories are ranked. Ranking may be linked to **achieved status**, *social position linked to a person's acquisition of socially valued credentials or skills*, or **ascribed status**, *social position linked to characteristics that are socially significant but cannot generally be altered (such as race or sex).* Although anyone can exercise individual agency, membership in a social category may influence whether an individual's path forward (and upward) is characterized by obstacles or opportunities.

Across the United States, increases in rent have dramatically outpaced increases in wages, contributing to high rates of eviction in many poor neighborhoods. By one estimate, about a fifth of Black women renters have faced eviction, often more than once (Desmond, 2016a, 2016b).

The third characteristic of a socially stratified system is that the hierarchical positioning of social categories tends to change slowly over time. Members of groups that enjoy prestigious and preferential rankings in the social order tend to remain at the top, although the expansion of opportunities may change the composition of groups over time.

Societal stratification has evolved through historical stages. The earliest human societies, based on hunting and gathering, had little social stratification; there were few resources to divide, so differences within communities were not very pronounced, at least materially. Advances in agriculture produced considerably more wealth and a consequent rise in social stratification. The hierarchy in agricultural societies increasingly came to resemble a pyramid, with a large number of poor people at the bottom and successively smaller numbers in the upper tiers of better-off members.

Modern capitalist societies are, predictably, even more complex: Some sociologists suggest that the shape of class stratification resembles a teardrop (Figure 7.1), with a large number of people in the middle ranks, a slightly smaller number of people at the bottom, and very few people at the top.

Before we continue, let's look at what sociologists mean when they use the term *class*. **Class** refers to *a person's economic position in society, which is usually associated with income, wealth, and occupation (and sometimes associated with political voice).* Class position at birth strongly influences a person's **life chances**, *the opportunities and obstacles the person encounters in education, social life, work, and other areas critical to social mobility.* **Social mobility** is *the upward or downward status movement of individuals or groups over time.* Many middle-class Americans have experienced downward mobility in recent decades. Upward social mobility may be experienced by those who earn educational credentials or have social networks they can tap. A college degree is one important step toward upward mobility for many people.

Class: A person's economic position in society, which is usually associated with income, wealth, and occupation (and sometimes associated with political voice).

Life chances: The opportunities and obstacles a person encounters in education, social life, work, and other areas critical to social mobility.

Social mobility: The upward or downward status movement of individuals or groups over time.

The class system in the United States is complex, as class is composed of multiple variables. We may, however, identify some general descriptive categories. Our descriptions follow the class categories used by Gilbert and Kahl, as shown in Figure 7.1 (Gilbert, 2011). At the bottom of the economic ladder, one finds what economist Gunnar Myrdal (1963), writing in the 1960s, called the *underclass*: "a class of unemployed, unemployables, and underemployed who are more and more hopelessly set apart from the nation at large" (p. 10). The term has been used by sociologists such as Erik Olin Wright (1994) and William Julius Wilson (1978), whose work on the "black underclass" described that group as "a massive population at the very bottom of the social ladder plagued by poor education and low-paying, unstable jobs" (p. 1).

People who perform manual labor or work in low-wage sectors such as food service and retail jobs are generally understood to be working class, although some sociologists distinguish those in the *working class* from the *working poor*. Households in both categories cluster below the median household income in the United States and are characterized by breadwinners whose education beyond high school is limited or nonexistent. People in both categories depend largely on hourly wages, even though the working poor have lower incomes and little or no wealth; although they are employed, their wages fail to lift them above the poverty line, and many struggle to meet even basic needs. Author David Shipler (2005) suggests that they are "invisible," as U.S. mainstream culture does not equate work with poverty.

Those who provide skilled services of some kind (whether legal advice, electrical wiring, nursing, or accounting services) and work for someone else are considered—and usually consider themselves—middle class. Lawyers, teachers, social workers, plumbers, auto sales representatives, and store managers are all widely considered to be middle class, although there may be significant income, wealth, and educational differences among them, leading some observers to distinguish between the (middle) *middle class* and the *upper middle class*. As most Americans describe themselves in surveys as "middle class," establishing quantitative categories is challenging. In fact, in 2010, the White House Task Force on the Middle Class, led by Vice President Joe Biden, opted for a descriptive rather than a statistical

FIGURE 7.1 ■ **Class in the United States (Gilbert-Kahl Model)**

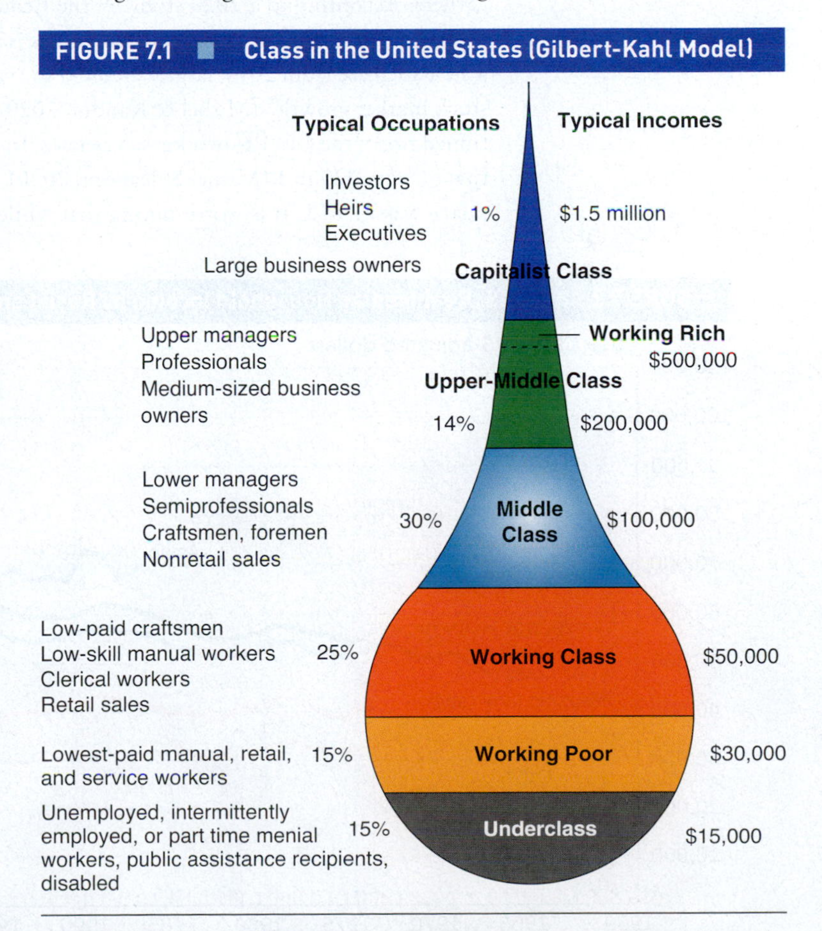

Source: Gilbert, D. L. (2020). *The American class structure in an age of growing inequality.* SAGE.

definition of the middle class, suggesting that its members are "defined by their aspirations more than their income. [It is assumed that] middle class families aspire to homeownership, a car, college education for their children, health and retirement security and occasional family vacations" (U.S. Department of Commerce, Economics and Statistics Administration, 2010).

Those who own or exercise substantial financial control over large businesses, financial institutions, or factories are generally considered to be part of the upper class, a category Gilbert and Kahl term the *capitalist class* (Gilbert, 2011). This is the smallest of the categories and consists of those whose wealth and income, whether gained through work, investment, or inheritance, are dramatically greater than those of the rest of the population.

Below, we look more closely at some key components of social class position: income, wealth, occupation, status, and political voice.

Income

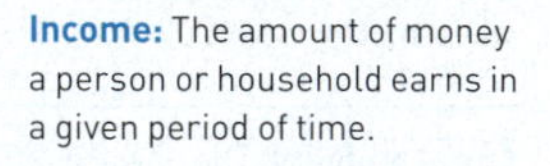

Income: The amount of money a person or household earns in a given period of time.

Income is *the amount of money a person or household earns in a given period of time.* Income is earned most commonly at a job and less commonly through investments. Household income also includes government transfers such as Social Security payments, veterans' benefits, or disability checks. Income typically goes to pay for food, clothing, shelter, health care, and other costs of daily living. It has a fluid quality in that it flows into a household in the form of pay-period checks and then flows out again as the mortgage or rent is paid, groceries are purchased, and other daily expenses are met.

U.S. household incomes have largely stagnated over the past decades, a topic we cover later in the chapter. Effects of the recent economic crisis have not been felt evenly, but they have been experienced by all U.S. ethnic and racial groups (Figure 7.2).

Income gains in the United States, however, have been disproportionately concentrated among top earners. According to a 2020 study by the Economic Policy Institute, average chief executive officer (CEO) pay in the nation's top 350 companies is around $21.3 million per year. That figure represents a 14% increase from 2018, largely credited to "vested stock awards and exercised stock options tied to stock market growth" (Mishel & Kandra, 2020). The increase in average CEO salary marks the continued rise of the CEO-to-worker salary ratio. In 2018, the CEO-to-worker ratio was 293 to 1. By 2019, that grew to 320 to 1 (Mishel & Kandra, 2020). In 1965, by comparison, the ratio of CEO-to-worker salary was 21 to 1. It is worth noting that while the COVID-19 pandemic has left millions without

FIGURE 7.2 ■ **U.S. Real Median Household Income by Racial and Ethnic Group, 1969–2019**

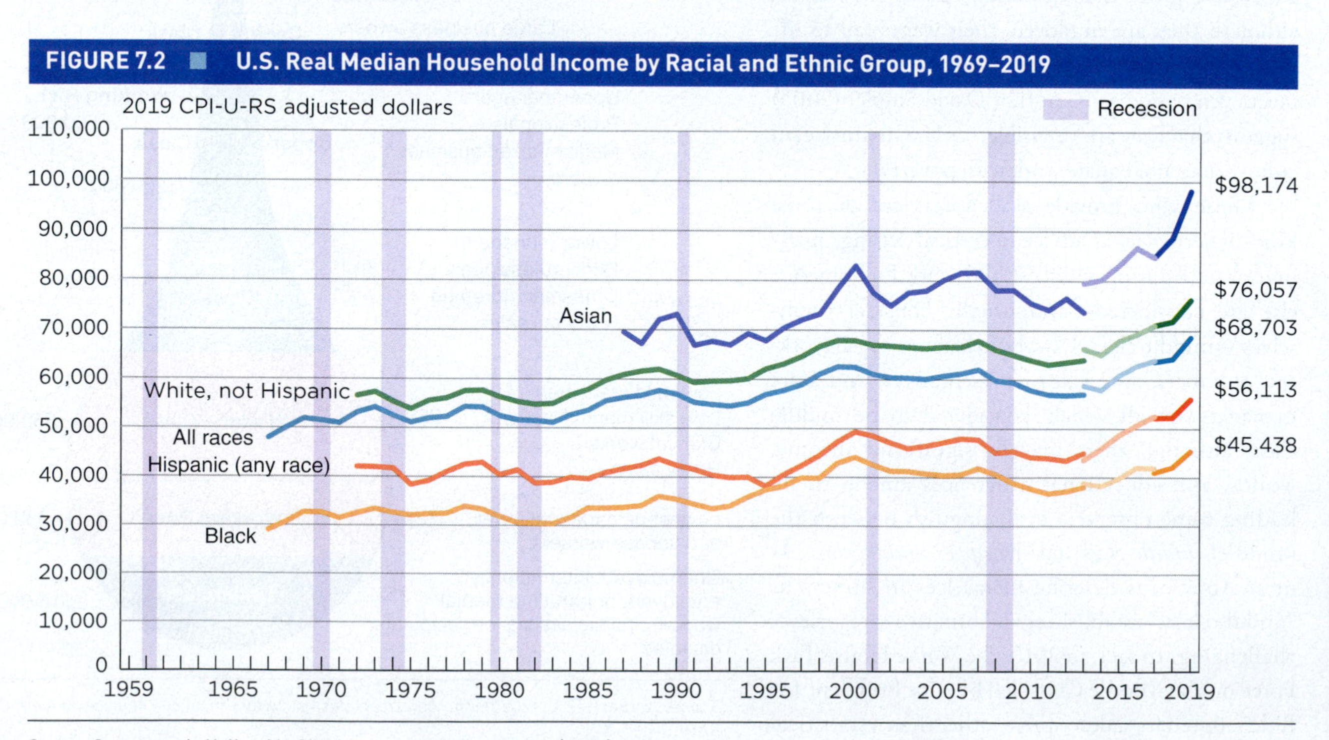

Source: Semenga, J., Kollar, M., Shrider, E. A., & Creamer, J. F. (2020). *Table A-1. Income and poverty in the United States: 2019.* https://www.census.gov/content/dam/Census/library/publications/2020/demo/p60-270.pdf

work, it is possible that we could see yet another increase in average CEO salary. This is due to the fact that CEO salary is tied to the stock market, which has recovered throughout 2020 and at times has approached record highs in spite of the crisis in the labor market (McGregor, 2020).

Wealth (or net worth): The value of everything a person owns minus the value of everything he or she owes.

Wealth

Wealth (or net worth) differs from income in that it is *the value of everything a person owns minus the value of everything he or she owes.* Wealth becomes a more important source of status as people rise on the income ladder.

For most people in the United States who possess any measurable wealth, the key source of wealth is home equity, which is essentially the difference between the market value of a home and what is owed on the mortgage. This form of wealth is *illiquid* (as opposed to *liquid*); illiquid assets are those that are logistically difficult to transform into cash because the process is lengthy and complicated. So, a family needing money to finance car repairs, meet educational expenses, or even ride out a period of unemployment cannot readily transform its illiquid wealth into cash.

For most working- and middle-class Americans, homeownership is their primary source of wealth: About two thirds of the median household's wealth is comprised of housing equity. Changes in home values due to large-scale economic shifts can significantly affect household wealth.

©iStockphoto.com/jhorrocks

Economists and sociologists treat **net financial assets** as *a measure of wealth that excludes illiquid personal assets, such as the home and vehicles.* Examples of net financial assets are stocks, bonds, cash, and other forms of investment assets. These are the principal sources of wealth used by the rich to secure their position in the economic hierarchy and, through reinvestment and other financial vehicles, to accumulate still more wealth.

Wealth, unlike income, is built up over a lifetime and may be passed down to the next generation. It is used to create new opportunities rather than merely to cover routine expenditures. Income buys shoes, coffee, and car repairs; wealth buys a high-quality education, business ventures, and access to travel and leisure that are out of reach of most, as well as financial security and the creation of new wealth. Those who possess wealth have a decided edge at getting ahead in the stratification system. In the United States, wealth is largely concentrated at the very top of the economic ladder.

How would you define the U.S. middle class? Should it be defined by income, wealth, or occupation? Should it be defined by aspirations and achievements?

©Diversity Photos/Getty Images

Occupation

Net financial assets: A measure of wealth that excludes illiquid personal assets, such as the home and vehicles.

Occupation: A person's main vocation or paid employment.

An **occupation** is *a person's main vocation.* In the modern world, this generally refers to *paid employment.* Occupation is an important determinant of social class because it is the main source of income in modern societies. The U.S. Bureau of Labor Statistics tracks 840 detailed occupational categories in the United States. Sociologists have used various classifications to reduce these to a far smaller number of categories. For example, jobs are described as *blue collar* if they are based primarily on manual labor (factory workers, agricultural laborers, truck drivers, and miners) and *white collar* if they require mainly analytical skills or formal education (doctors, lawyers, and business managers). The term *pink collar* is sometimes used to describe semiskilled, low-paid service jobs that are primarily held by women (waitresses, sales clerks, and receptionists).

In the 1990s, some writers adopted the term *gold collar* to categorize the jobs of young professionals who commanded huge salaries and high occupational positions very early in their professional careers thanks to the technology bubble and economic boom of the 1990s (Wonacott, 2002). After the bubble burst, gold-collar workers were more often found in the financial sector, earning very substantial salaries and benefits. The economic recession that commenced in 2007 put a damper on growth in salaries and benefits of gold-collar workers, but they have risen again in recent years.

Status

Status: The prestige associated with a social position.

Status refers to *the prestige associated with a social position.* It varies based on factors such as family background and occupation. A considerable amount of social science research has gone into classifying occupations according to the degrees of status or prestige they hold in public opinion.

We might expect white-collar jobs to rank more highly in prestige than blue-collar jobs, but do they? Doctors and scientists are indeed at the top of the prestige scale—but so are less highly paid professionals such as nurses and firefighters (who top the poll results discussed in this section, with 80% of respondents indicating that firefighters have "very great prestige"). Also in the top ranks are military officers and emergency medical technicians (with 78% and 72% conferring "very great prestige" on them, respectively). At the bottom are politicians, stockbrokers, accountants, and real estate agents (only 32% of respondents indicated "very great prestige" for real estate agents). It seems occupations that require working with ideas (scientist, engineer) or providing professional services that contribute to the public welfare (emergency medical technician, doctor, firefighter) have the highest prestige, and perhaps surprisingly, the U.S. public does not always link prestige to income (Mekouar, 2016).

Prestige rankings of specific occupations have been relatively stable over time, although changes do occur. For instance, since 1977, scientists have gained 16 points (after falling by 9 points in 2009) and journalists have increased by 30 points to 47%. What factors might explain these shifts? Have societal changes taken place that might contribute to our understanding of why occupations such as these rise and fall on the prestige scale?

Political Voice

Political power: The ability to exercise influence on political institutions and/or political actors to realize personal or group interests.

Political power is *the ability to exercise influence on political institutions and/or political actors to realize personal or group interests.* It involves the mobilization of resources (such as money or technology or political support of a desired constituency) and the successful achievement of political goals (such as the passage of legislation favorable to a particular group).

Sociological analyses of power have revealed a pyramid-shaped stratification system in the United States as well as in most advanced industrial societies, including those of Western Europe. At the top are a handful of political figures, businesspeople, and other leaders with substantial power over political decision making and the national economy. Moving down the pyramid, we encounter more people—and less power (Domhoff, 2009).

Sociologist C. Wright Mills began to write as early as the 1950s about the existence of a "power elite," which he defined as a group comprising elites from the executive branch of government, the military, and the corporate community who share social ties, a common worldview born of socialization in prestigious schools and clubs, and professional links that create revolving doors between positions in these three areas (Mills, 2000a).

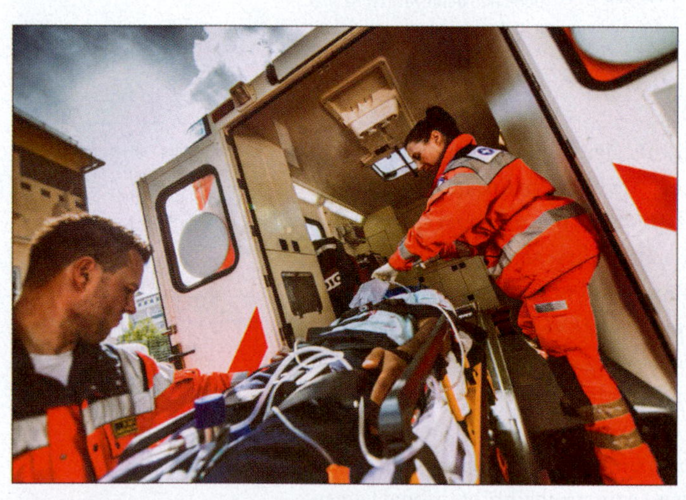

In the 2016 Harris poll, emergency medical technicians/paramedics were ranked among the top 10 most prestigious professions. According to the Bureau of Labor Statistics, in 2019, the average salary of these first responders was less than $35,400 annually. What explains the discrepancy between the social status and economic valuation of this occupation?

©iStockphoto.com/simonkr

In contrast to the pluralist perspective on U.S. democracy, which suggests that political power is fluid and passes, over time, among a spectrum of groups and interests that compete in the political arena, Mills offered a critical perspective. He described a concentration of political power in the hands of a small elite. According to Mills, even though power over local issues remains largely in the hands of elected legislatures and interest groups, decision-making power over issues of war and peace, global economic interests, and other matters of international and national consequence remain with the power elite. The power of the masses is little more than an illusion in Mills's view; the masses are composed of "entirely private" individuals wrapped up in personal concerns and largely disconnected from the political process.

In recent elections, the U.S. middle class has been at the center of political discourse, but are decision makers addressing its fundamental economic concerns, including stagnating wages and challenges such as the steep cost of higher education? Or do the interests of the wealthy guide policy making? What do you think? In the following, we look more closely at trends in inequality in the United States.

CLASS AND INEQUALITY IN THE UNITED STATES: DIMENSIONS AND TRENDS

The United States prides itself on being a nation of equals. Indeed, except for the period of the Great Depression of the 1930s, inequality declined throughout much of the 20th century, reaching its lowest levels during the 1960s and early 1970s. But during the past three decades, inequality has been on the rise again. The rich have gotten much richer, middle-class incomes have stagnated, and a growing number of poor are struggling to make ends meet.

Income Inequality

Sociologist Richard Sennett (1998) writes,

> Europeans from [Alexis de] Tocqueville on have tended to take the face value for reality; some have deduced we Americans are indeed a classless society, at least in our manners and beliefs—a democracy of consumers; others, like Simone de Beauvoir, have maintained we are hopelessly confused about our real differences. (p. 64)

Was Tocqueville right or Beauvoir? Are we classless or confused? What are the dimensions of our differences? Let us look at what statistics tell us.

Every year, the U.S. Census Bureau calculates how income is distributed across the population of earners. All households are ranked by annual income and then categorized into *quintiles*, or fifths. The U.S. Census Bureau calculates how much of the *aggregate income*, or total income, generated in the United States each quintile gets. In other words, imagine all legally earned and reported income thrown into a big pot—that is, the aggregate income. The Census Bureau wants to know (and we do, too!) how much of this income goes to each quintile of earners. In a society with equal distribution across quintiles, each fifth of earners would get about one fifth of the income in the pot. Conversely, in a society with complete inequality across quintiles, the top would get everything, and the bottom quintiles would be left empty-handed. The United States, like all other countries, falls between these two hypothetical extremes.

In Figure 7.3, we see how aggregate income in the United States is divided among quintiles of earners. When we look at the pie, we see that income earners at different levels take in disparate proportions of the income total. Those in the bottom quintile take in just over 3% of the aggregate income, while those in the top quintile get more than half; that means the top 20% of earners bring in as much as all in the bottom 80% combined. No less significant is the fact that the top 5% take in about 23% of the total income—more than the bottom 40% combined (Semenga et al., 2020).

Data compiled by economists Emmanuel Saez and Thomas Piketty, with a formula that uses pretax income (as do the census figures) but includes capital gains, suggest an even more stratified picture. According to Saez and Piketty's calculations, about 50% of pretax income goes to the top 10% of earners: Notably, the income is not composed primarily of wages but of capital income such as capital gains and dividends (Tankersley, 2016). Within this well-off decile (or tenth) of earners, there is a still more dramatic division of income, because the top 1% of earners takes about a fifth of the aggregate income (Saez, 2010). Clearly, gains have been concentrated at the top of the income ladder. As economist Joseph Stiglitz (2012) points out, the fraction of the aggregate income taken by the upper 1% has doubled since 1980, while the fraction that goes to the upper 0.1% has nearly tripled over that period.

When we study issues such as income inequality, we benefit from understanding the data we gather in their historical context. Figure 7.3 presents a snapshot of one moment in time, but what about decades past? The economic prosperity of the middle to late 1990s brought some benefit to most American workers: The median U.S. income rose faster at the end of the 1990s than it had since the

FIGURE 7.3 ■ Shares of Aggregate U.S. Income by Quintile, 2019

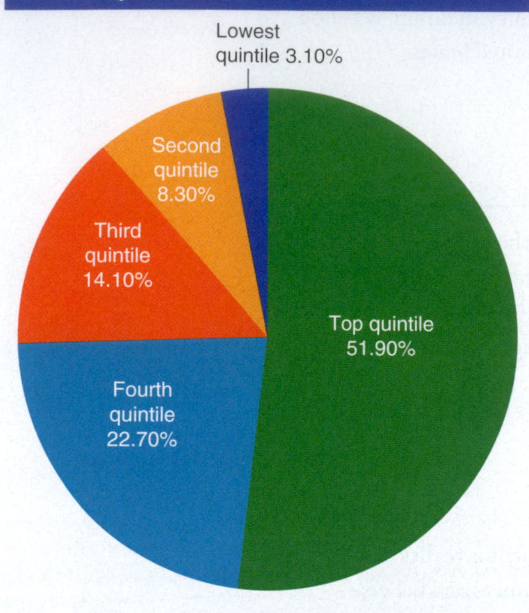

Source: Semega, J., Fontenot, K., & Kollar, M. (2018). *Table A-2. Income and poverty in the United States.* https://www.census.gov/library/publications/2018/demo/p60-263.html; Semega, J., Kollar, M., Shrider, E. A., & Creamer, J. F. (2020). *Table A-4. Income and poverty in the United States.* https://www.census.gov/data/tables/2020/demo/income-poverty/p60-270.html

Note: Values add up to 99%

FIGURE 7.4 ■ Changes in Income Inequality in the United States, 1968–2018

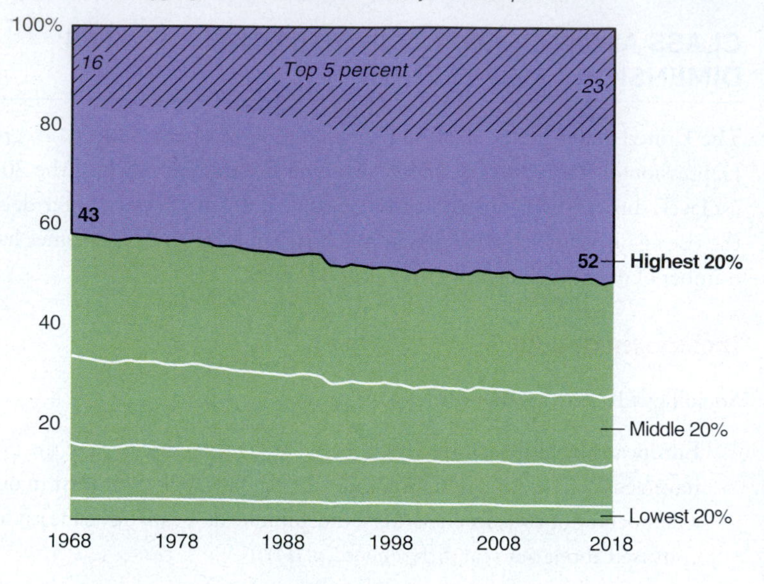

Source: Pew Research Center. (2020, February 7). *6 facts about economic inequality in the U.S.* https://www.pewresearch.org/fact-tank/2020/02/07/6-facts-about-economic-inequality-in-the-u-s/#:~:text=The%20share%20of%20American%20adults,from%2025%25%20to%2029%25.

period from the late 1940s to the middle 1950s. Saez (2010) calculates that the real annual growth of income among the bottom 99% of earners grew 2.7% in the period he terms the "Clinton Expansion" (1993–2000), but it dropped during the 2000–2002 recession (by 3.3%) and again during the 2007–2008 period (by 6.9%). On the whole, the period from 1993 to 2008 saw a real annual growth of only 0.75% for the incomes of the bottom 99%. Over the same period, the top 1% of earners experienced a real annual growth of almost 4% (although this group also experienced significant losses in the recessionary periods).

From about World War II until the middle 1970s, the top 10% earned less than a third of the national income pool (Pearlstein, 2010). In the years since, however, the incomes of people at or near the top have risen far faster than those of earners at the bottom or middle of the income scale (Figure 7.4). The stagnation of wages is illustrated by the poor growth of average wages of young high school and college graduates (Figure 7.5). The economic position of college graduates is significantly better than that of high school graduates, but wage growth has been slow for both groups and young college graduates face a rising burden of student debt that puts a further drag on their economic position (Gould et al., 2019).

Underemployed: Working in jobs that do not make full use of one's skills or working part time when one would like to be working full time.

Even for those with a college degree in hand, the labor market can be challenging: The Economic Policy Institute estimates that more than 11% of men and nearly 10% of women graduates are **underemployed** (Gould et al., 2019). That is, they are *working in jobs that do not make full use of their skills* (for instance, a college graduate working in a job that does not require a college degree, or where most employees in an equivalent position do not possess a college degree) *or working part time when they would like to be working full time.* Notably, Figure 7.6 shows only those graduates who are working part time when they would like to be working full time. A publication by the Federal Reserve Bank of New York estimates that in September 2020, 43% of new college graduates were working in jobs that did not require a college degree (Federal Reserve Bank of New York, 2020).

FIGURE 7.5 ■ Average Hourly Wages of U.S. College and High School Graduates Ages 21–24, 1989–2017

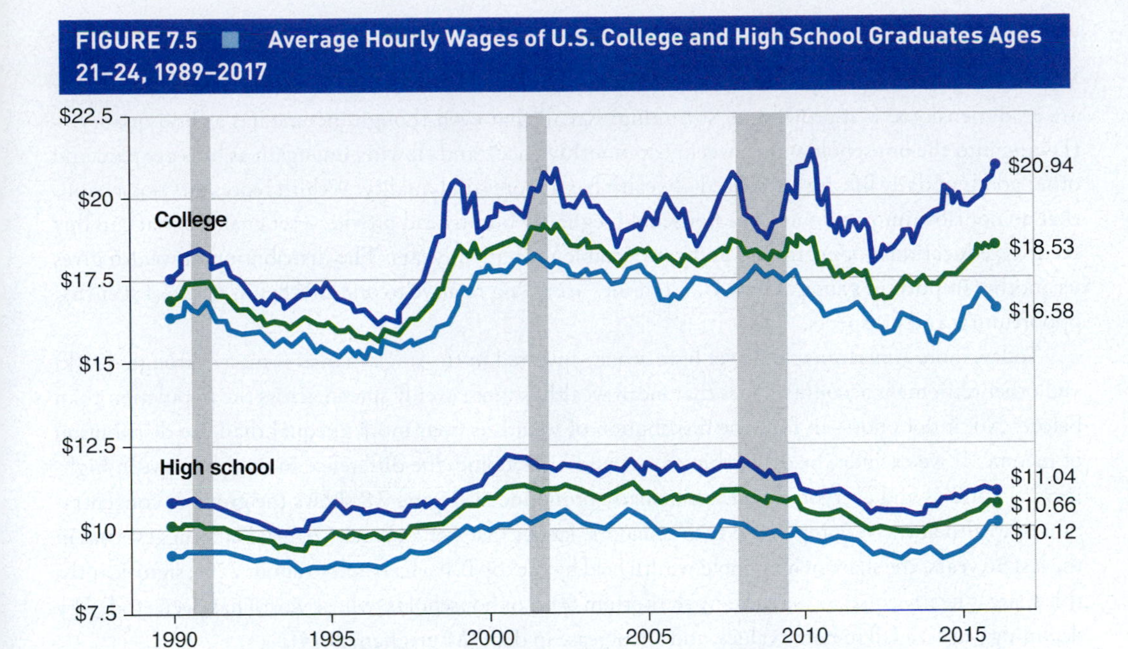

Source: Kroeger, T., & Gould, E. (2017, May 4). *The class of 2017. Figure O.* Economic Policy Institute. Reprinted with permission.

FIGURE 7.6 ■ Underemployment by Gender and Race/Ethnicity, 2000 and 2019

Underemployment of young college graduates (ages 21–24) not enrolled in further schooling, by gender and race/ethnicity, 2000 and 2019

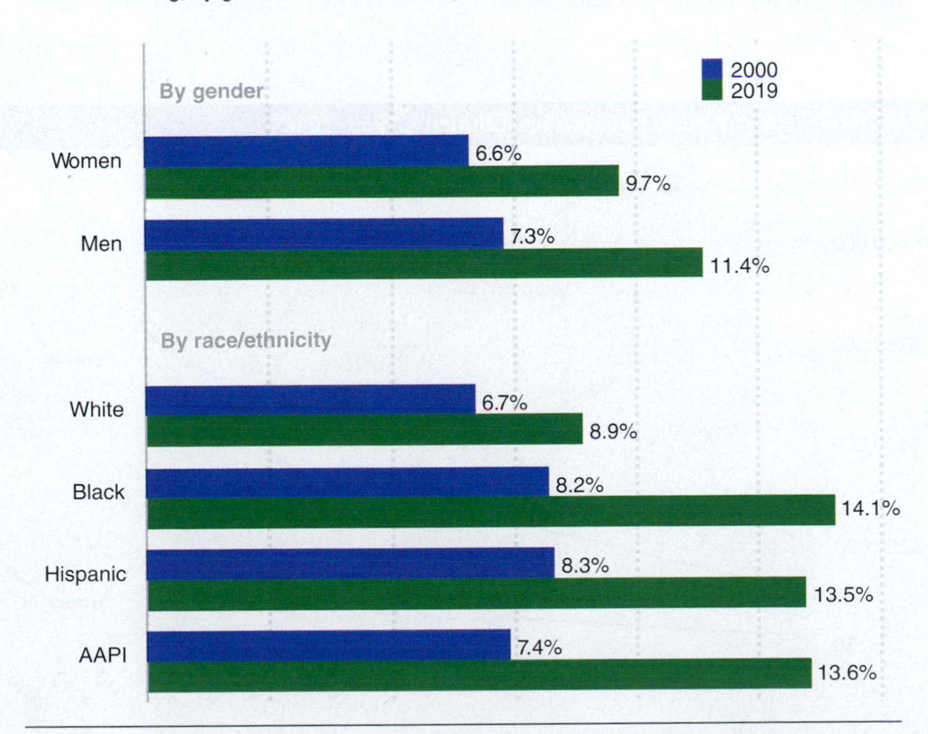

Source: Gould, E., Mokhiber, Z., & Wolfe, J. (2019, May 14). *The class of 2019.* Economic Policy Institute. https://www.epi.org/publication/class-of-2019-college-edition/

Note: AAPI stands for Asian American/Pacific Islander. Data for 2000 and 2018 use an average of January 1998–December 2000 and March 2015–February 2018, respectively.

Wealth Inequality

We see the growth of inequality in the distribution of income, but what about the wealth gap? What are its dimensions? Is it growing or shrinking? Recall that even though income has a fluid quality—flowing into the household with a weekly or monthly check and flowing out again as bills are paid and other goods of daily life are purchased—wealth has a more solid quality. Wealth represents possessions that do not flow into and out of the household regularly but instead provide a set of assets that can buy security, educational opportunity, and comfortable retirement years. The distribution of wealth gives us another important gauge of how U.S. families are doing relative to one another in terms of security, opportunity, and prospects.

Today, more Americans than ever have money invested in the stock market—many through 401(k) and other retirement accounts. Does that mean wealth is more evenly spread across the population than before? No, it does not—in fact, the distribution of wealth is even more unequal than the distribution of income. If we exclude the ownership of cars and one's home, the difference in wealth between high-income families and everyone else is particularly pronounced. Figure 7.8 shows the growing concentration of wealth at the very top of the U.S. economic ladder (Saez & Zucman, 2014). Data suggest that in the last 30 years, the share of household wealth held by the top 0.1% has risen to about 22%; significantly, this share is nearly equal to that held by the bottom 90% of households, whose worth has been eroded by declining wages, a fall in home values, and an increase in debt (Monaghan, 2014).

Minority groups hold far fewer net financial assets than whites. The wealth held by minority households has historically lagged; for instance, in 1990, Black households held about 1% of total U.S. wealth (Conley, 1999). This percentage rose markedly in the economic boom years of the 1990s and continued to expand into the 2000s. Black household wealth reached an average of just over $12,000 in 2005. This climb, however, was reversed by the housing crisis and the Great Recession, which saw a fall in household wealth among minorities. In 2013, the U.S. median net worth of a household was just over $81,400 and differences by race and ethnicity were stark: Although white median net worth was $134,000, Black household wealth was just $11,000 and Hispanic household wealth was about $14,000 (Jones, 2017; Kochhar & Fry, 2014). Mean (rather than median) wealth differences are even more stark (Urban Institute, 2017; Figure 7.7).

FIGURE 7.7 ■ Wealth Gap by Race/Ethnicity, 1963–2016

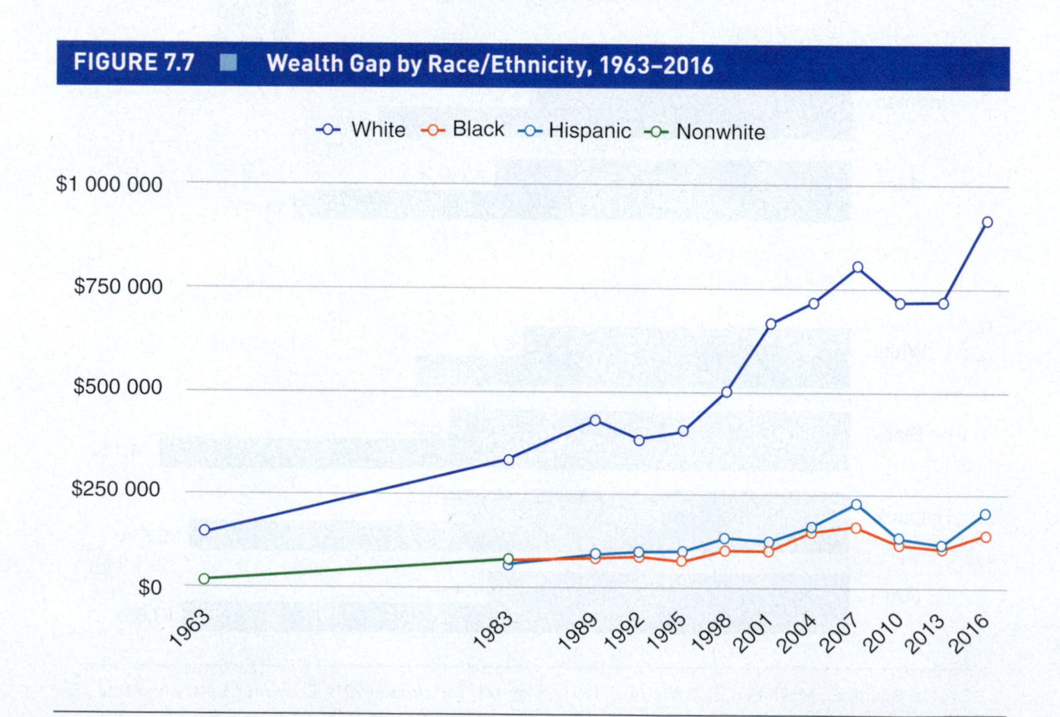

Source: Urban Institute. (2017). *Nine charts about wealth inequality in America.* https://apps.urban.org/features/wealth-inequality-charts

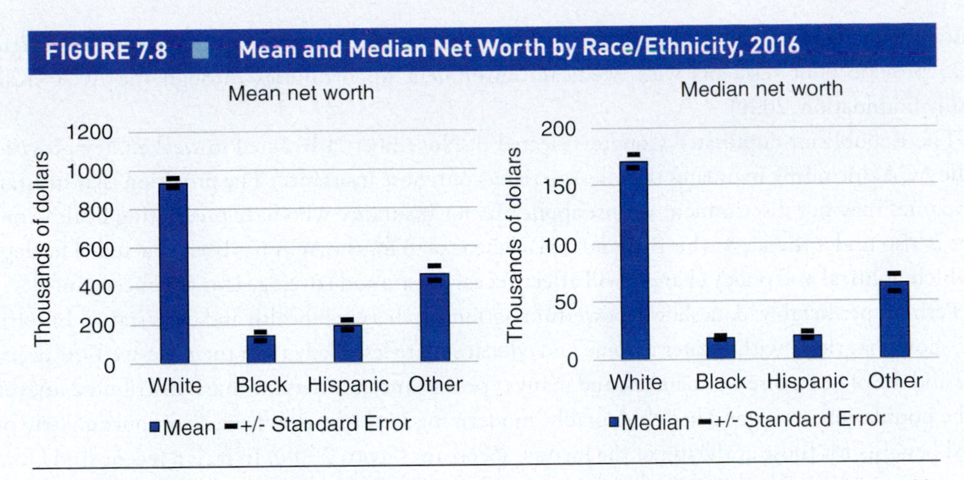

FIGURE 7.8 ■ Mean and Median Net Worth by Race/Ethnicity, 2016

Source: Dettling, L. J., Hsu, J. W., Jacobs, L., Moore, K. B., & Thompson, J. P. (2017). *Recent trends in wealth-holding by race and ethnicity: Evidence from the survey of consumer finances.* https://www.federalreserve.gov/econres/notes/feds-notes/recent-trends-in-wealth-holding-by-race-and-ethnicity-evidence-from-the-survey-of-consumer-finances-20170927.htm

Many U.S. families have zero or negative net worth: In 2016, this included 9% of white families, 19% of Black families, and 13% of Hispanic families (see Figure 7.8; Dettling et al., 2017). A recent report on U.S. wealth examined the distribution of wealth by quintile: The findings show that about 90% of wealth is held by the top quintile; the bottom two quintiles hold zero or negative wealth (Wolff, 2017).

Other Gaps: Inequalities in Health Care, Health, and Access to Consumer Goods

Along with the gap in income and wealth, there is a critical gap in employer benefits, including health insurance. From the 1980s to the 1990s, health care coverage for workers in the bottom quintile of earners fell more dramatically than for any other segment of workers: From a rate of 41% coverage, it dropped to 32% in the late 1990s (Reich, 2001). In 2011, about 25% of those living in households earning less than $25,000 a year were uninsured, along with more than 21% of those in households earning $25,000–$49,999 (U.S. Census Bureau, 2012). Altogether, more than 15% (48.6 million) of the U.S. population was without health insurance, including 7 million children younger than 18 years of age (U.S. Census Bureau, 2012).

Many city neighborhoods lack access to large, well-stocked supermarkets with competitive prices. Residents must often choose between overpriced (and often poor-quality) goods and a long trip to a suburban market. The low rates of private vehicle ownership among the urban poor can make shopping for healthy food a burden.

Many of the jobs created in the 1980s and 1990s were positions in the service sector, which includes retail sales and food service. Although the *quantity* of jobs created in this period helped push down the unemployment rate, the *quality* of jobs created for those with less education was not on par with the quality of many of the jobs lost as U.S. manufacturing became automated or moved overseas. Many service sector jobs pay wages at or just above minimum wage and have been increasingly unlikely to offer employer benefits.

One goal of President Barack Obama's Patient Protection and Affordable Care Act (known simply as the Affordable Care Act, or ACA), signed into law in 2010, was to expand insurance coverage for the working poor. In the first year the law was in effect, more than 8 million people signed up for health insurance plans under the Affordable Care Act, and 57% of those people had been uninsured before enrolling in ACA-compliant plans. Moreover, those who enrolled in ACA-compliant plans reported slightly worse health than nonenrollers, and most said that they would not have sought insurance if the law had not taken effect (Hamel et al., 2014).

The rolls of insured Americans have grown with expanded access through Medicaid, a government health insurance program that primarily serves the poor, many of whom do not have employer-provided insurance and cannot afford to purchase insurance. At this time, 39 states (including Washington, D.C.) provide poor residents with Medicaid enrollment opportunities through the ACA (Kaiser Family Foundation, 2020).

The Republican-dominated Congress elected in November 2016 acted to weaken key provisions of the ACA, including mandates that all Americans purchase insurance. The provision that insurance companies may not discriminate against applicants for insurance who have preexisting health conditions is also under threat. At this time, no replacement plan for the ACA has been enacted. The degree to which political and policy changes will affect access to care and coverage is as yet uncertain.

Perhaps predictably, data show a powerful relationship between health and class status. Empirical data show that those with greater income and education are less likely than their less-well-off peers to have and die of heart disease, diabetes, and many types of cancer. Just as income is distributed unevenly in the population, so is good health. Notably, modern medical advances have disproportionately provided benefits for those at the top of the income spectrum (Scott, 2005). In fact, a recent study found that greater income is associated with greater longevity. The gap in life expectancy between the richest and poorest Americans is about 15 years for men and 10 years for women (Cutler et al., 2016).

Children in disadvantaged families are more likely than their better-off peers to have poor physical and mental health. According to the Kaiser Family Foundation (2008), the rate of hospitalization for asthma for Black children, who are more likely than their peers to be poor, is four to five times higher than that for white children. The problem is not only the lack of health insurance in families—although this factor is important—but also the lack of physical activity that may result when children don't have safe places to play and exercise and when their families are unable to provide healthy foods because both money and access to such foods are limited. No less important, poverty can take a toll on mental health and mental capacity in the young: A recent study found that children who grow up poor are at risk of reduced short-term spatial memory as well as persistent feelings of powerlessness, and they are more likely to exhibit antisocial behaviors such as bullying (Dallas, 2017). Significantly, research suggests that families living in poverty have higher levels of the stress hormone, cortisol, and lower levels of the happiness hormone, serotonin (Johnston, 2016).

The problem of poor health may be related to another disadvantage experienced by those on the lower economic rungs: Many poor urban neighborhoods lack access to high-quality goods at competitive prices. Most middle-class shoppers purchase food at large chain grocery stores that stock items like fresh fruit and vegetables and meat at competitive prices. In contrast, inhabitants of poor neighborhoods are likely to shop at small stores that have less stock and higher prices because large grocery chains choose not to locate in poor areas. If they want to shop at big grocery stores, poor residents may need to travel great distances, a substantial challenge for those who do not own cars and a costlier proposition for those who do. As one study noted, "Lower-income shoppers must travel further and/or have fewer shopping options than do higher-income shoppers" (Hatzenbuehler et al., 2012, p. 54).

Food deserts: Areas (often urban neighborhoods or rural towns) characterized by poor access to healthy and affordable food.

Some researchers refer to *areas (often urban neighborhoods or rural towns) characterized by poor access to healthy and affordable food* as **food deserts** (DeChoudhury et al., 2016). A *USA Today* article describes the situation of Louisville retiree Jessie Caldwell, who regularly makes an hour-long bus trip to get fresh vegetables or meat:

> For her and many others, it's often tempting to go to a more convenient mini-market or grab some fast food. "The corner stores just sell a lot of potato chips, pop and ice cream," she said. "But people are going to eat what's available." (Kenning & Halladay, 2008, para. 8)

Although we do not always think of access to stores with competitive prices and fresh goods as an issue of class inequality, lack of such access affects people's quality of life, conferring advantage on the already advantaged and disadvantage on those who struggle to make ends meet. Writer Barbara Ehrenreich (2001) highlights this point:

> There are no secret economies that nourish the poor; on the contrary, there are a host of special costs. If you can't put up the two months' rent you need to secure an apartment, you end up paying through the nose for a room by the week. If you have only a room, with a hot plate

at best, you can't save by cooking up huge lentil stews that can be frozen for the week ahead. You eat fast food and hot dogs and Styrofoam cups of soup that can be microwaved in a convenience store. (p. 27)

WHY HAS INEQUALITY GROWN?

There is a significant split between the fortunes of those who are well educated and those who do not or cannot attend college. The demand for labor over the past several decades has been differentiated on the basis of education and skills—workers with more education are more highly valued, while those with little education are becoming less valuable. These effects are among the results of the transition to a postindustrial economy in the United States.

The nation's earlier industrial economy was founded heavily on manufacturing. U.S. factories produced a substantial proportion of the goods Americans used—cars, washing machines, textiles, and the like—and a big part of the economy depended on this production for its prosperity. This is no longer the case. In the postindustrial economy of today, the United States manufactures a smaller proportion of the goods Americans consume and fewer goods overall. Many manufacturing jobs have either been automated or gone abroad, drawn to the low-cost labor in developing countries. New manufacturing jobs created in the United States offer lower wages overall than did their predecessors in the unionized factories of the industrial Midwest (we discuss this issue in greater detail in Chapter 15). The modern U.S. economy has produced larger numbers of jobs in the production of knowledge and information and the provision of services.

One group that has grown is made up of professionals who engage in what former secretary of labor Robert Reich (1991) has called *symbolic analysis*, or "problem-solving, problem-identifying, and strategic-brokering activities" (p. 111). These occupational categories—law, engineering, business, technology, and the like—typically pay well and offer some job security, but they also require a high level of skill and at least a college education. Even well-educated middle-class workers, however, have been touched by automation and outsourcing. As well, a rising proportion of middle-level jobs have converted from relatively stable and secure long-term positions to contractual work.

The fastest-growing sector of the postindustrial economy beginning in the 1980s was the service sector, which includes jobs in food service, retail sales, health care (for instance, home health aides and nurse's aides), janitorial and housecleaning services, and security. By providing jobs to those with less education, the service sector has, in a sense, moved into the void left by the manufacturing sector of the industrial economy over the past few decades. The service sector, which offers lower pay scales and fewer benefits, does not typically provide the kinds of jobs that offer a solid road to the middle class. Another difference is that manufacturing, especially in the automotive and steel industries, was overwhelmingly a male bastion, while service jobs favor women. The "advantage" enjoyed by less-educated women over their male counterparts does not, however, translate into substantial economic gains for women or their families. Wage gains for women overall have been more fully driven by gains made by college-educated women.

In the period following the official end of the Great Recession, the bulk of new jobs created by the economy were low-wage positions, many of them service sector positions in areas such as hospitality, tourism, and retail (National Employment Law Project, 2014). By 2014, however, the economy was on track to create more high-wage jobs, and the job market of "good jobs"—those with above-median wages, insurance, and retirement benefits—was growing. Notably, however, the "good jobs" being created were almost exclusively available only to those with higher education (Carnevale et al., 2016).

The stratification of the U.S. labor force into a low-wage service sector, heavily tilted toward jobs in hospitality, retail, and security, and a well-paid knowledge and technology sector with clear orientation toward the highly educated, appears to be continuing unabated. The narrative of a "disappearing middle class" has become a common theme in both social science and mainstream political discourse. Do you see such a trend in your own community? How does this narrative coexist with Americans' long-existing tendency to self-identify as middle-class, almost regardless of income?

Poverty in the United States

> There is a familiar America. It is celebrated in speeches and advertised on television and in the magazines. It has the highest mass standard of living the world has ever known.
>
> In the 1950s, this America worried about itself, yet even its anxieties were products of abundance. There was introspection about Madison Avenue and tail fins; there was discussion of the emotional suffering taking place in the suburbs. In all this, there was an implicit assumption that the basic grinding economic problems had been solved in the United States. . . .
>
> While this discussion was being carried on, there existed another America. In it dwelt between 40,000,000 and 50,000,000 citizens of this land. They were poor. They still are.
>
> To be sure, the other America is not impoverished in the same sense as those poor nations where millions cling to hunger as a defense against starvation. This country has escaped such extremes. That does not change the fact that tens of millions of Americans are, at this very moment, maimed in body and spirit, existing at levels beneath those necessary for human decency. If these people are not starving, they are hungry, and sometimes fat with hunger, for that is what cheap foods do. They are without adequate housing and education and medical care. (Harrington, 1963, pp. 1–2)

These words, first published in 1963, helped to open the eyes of many to the plight of the U.S. poor, who were virtually invisible to a postwar middle class comfortably ensconced in suburbia. Michael Harrington's classic book, *The Other America: Poverty in the United States*, also caught the interest of President John F. Kennedy's administration and, later, the Johnson administration, which inaugurated the War on Poverty in 1964.

When President Lyndon B. Johnson began his War on Poverty, around 36 million U.S. citizens lived in poverty. Within a decade, the number had dropped sharply, to around 23 million. But then, beginning in the early 1970s, poverty again began to climb, reaching a high of 39 million people in 1993 before receding. Poverty climbed again in the period of economic crisis, but it has begun to decline slowly in its aftermath (Figure 7.9).

We pause on the topic of "official poverty" because it is important for us to be critical consumers of information. We are surrounded by statistics, subject to a barrage of information about the proportion of the population who support the president or reject the health care initiatives of a political party, about the numbers of teen pregnancies and births, about the percentages who are unemployed or in poverty.

FIGURE 7.9 ■ Poverty Levels in the United States, 1959–2019

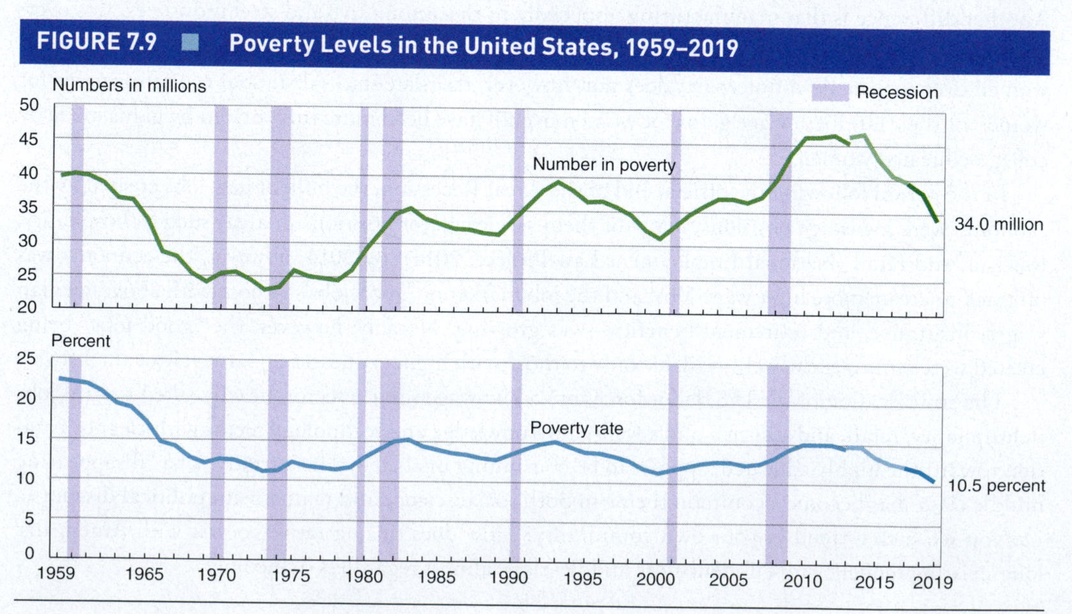

Source: Semenga, Fontent, & Kollar (2017, 2018). US Census Bureau.

TABLE 7.1 ■ Poverty Rates of Selected U.S. Groups, 2019

Category	Percentage
All people	10.5%
Whites	9.1%
Blacks	18.8%
Hispanics	15.7%
Asians	7.3%
Under 18 years of age	14.4%
65 years of age and older	8.9%

Source: Semega, J., Kollar, M., Shrider, E. A., & Creamer, J. F. (2020). *Income and poverty in the United States.* https://www.census.gov/content/dam/Census/library/publications/2020/demo/p60-270.pdf

These statistics illuminate the social world around us and offer us a sense of what we as a nation are thinking or earning or debating. On the other hand, statistics—including social indicators such as the poverty numbers (Table 7.1)—may also obscure some important issues. To use indicators such as the poverty numbers wisely, we should know where they come from and what their limitations are.

What is poverty from the perspective of the U.S. government? How were these numbers generated? The **official poverty line** is *the dollar amount set by the government as the minimum necessary to meet the basic needs of a family.* In 2019, the U.S. government used the following thresholds:

Official poverty line: The dollar amount set by the government as the minimum necessary to meet the basic needs of a family.

- One person, younger than age 65: $13,300
- One person, age 65 or older: $12,261
- Three persons (one adult, two children): $20,598
- Four persons (two adults, two children): $25,926
- Five persons (two adults, three children): $30,510
- From the federal government's perspective, those whose pretax income falls beneath the threshold are officially poor; those whose pretax income is above the line (whether by $10 or $10 million) are nonpoor.

How does the U.S. government measure poverty? In the early 1960s, an economist at the Social Security Administration, Mollie Orshansky, used a 1955 U.S. Department of Agriculture study to establish a poverty line. She learned from the study that about one third of household income went to food, so she calculated the cost of a "thrifty food basket" and tripled it to take into account other family needs such as transportation and housing. Then she adjusted the figure again for the size and composition of the family and the age of the head of household. The result was the poverty threshold, which is illustrated in Figure 7.10.

Orshansky's formula represented the first systematic federal attempt to count the poor, and it has been in use for more than half a century. But its age makes it a problematic indicator for the 21st century. Some critics argue that it may underestimate the number of those struggling with material deprivation. Consider the following points:

FIGURE 7.10 ■ The Poverty Threshold Calculation

Source: Bishaw, A., & Macartney, S. (2010). *Poverty: 2008 and 2009. American community survey briefs.* U.S. Census Bureau.

The multiplier of three was used because food was estimated to constitute one third of a family budget in the 1960s. Food is a smaller part of budgets today (about one fifth), and housing and transport are much bigger ones. Using a higher multiplier would raise the official poverty line and, consequently, increase the number of households classified as poor.

The formula makes no adjustment for where people live, even though costs of living vary tremendously by region. Although a family of three may be able to make ends meet on a pretax income of about $19,000 in South Dakota or Nebraska, it is doubtful that the housing costs of such areas as Boston, San Francisco, New York City, and Washington, D.C., would permit our hypothetical family to survive in basic decency.

On the other hand, some critics have suggested that the poverty rate overestimates the problem because it does not account for noncash benefits that some poor families receive, including food stamps and public housing vouchers. Adding the value of those (although they cannot be converted into cash) would increase the income of some families, possibly raising them above the poverty line.

We might thus conclude that even though the official poverty statistics give us some sense of the problem of poverty and poverty trends over time, they must be read with a critical eye.

Notably, official poverty figures offer us a picture of what the Census Bureau calls the *annual poverty rate*. This figure captures the number of households whose total income over the 12 months of the year fell below the poverty threshold, but it does not illuminate how many families may have dropped into or climbed out of poverty and how many dwell there over a longer period. A Census Bureau report points out that even though the official poverty figure was around 12.7% in the period 2009 to 2011, about one third of U.S. households fell below the poverty threshold for at least 2 months. At the same time, only 3.5% remained poor for the full 3-year period under study (Edwards, 2014). Although few households languish at the very bottom of the economic ladder for years, it is significant that nearly 10 times more households experienced periods of poverty.

Inequality and poverty in the United States are serious issues that demand both analysis and attention. Even though inequality is part of any modern capitalist state, the steep rise of inequality in recent decades presents a challenge to societal mobility and perhaps, ultimately, stability.

FIGURE 7.11 ■ Neighborhood Poverty Maps

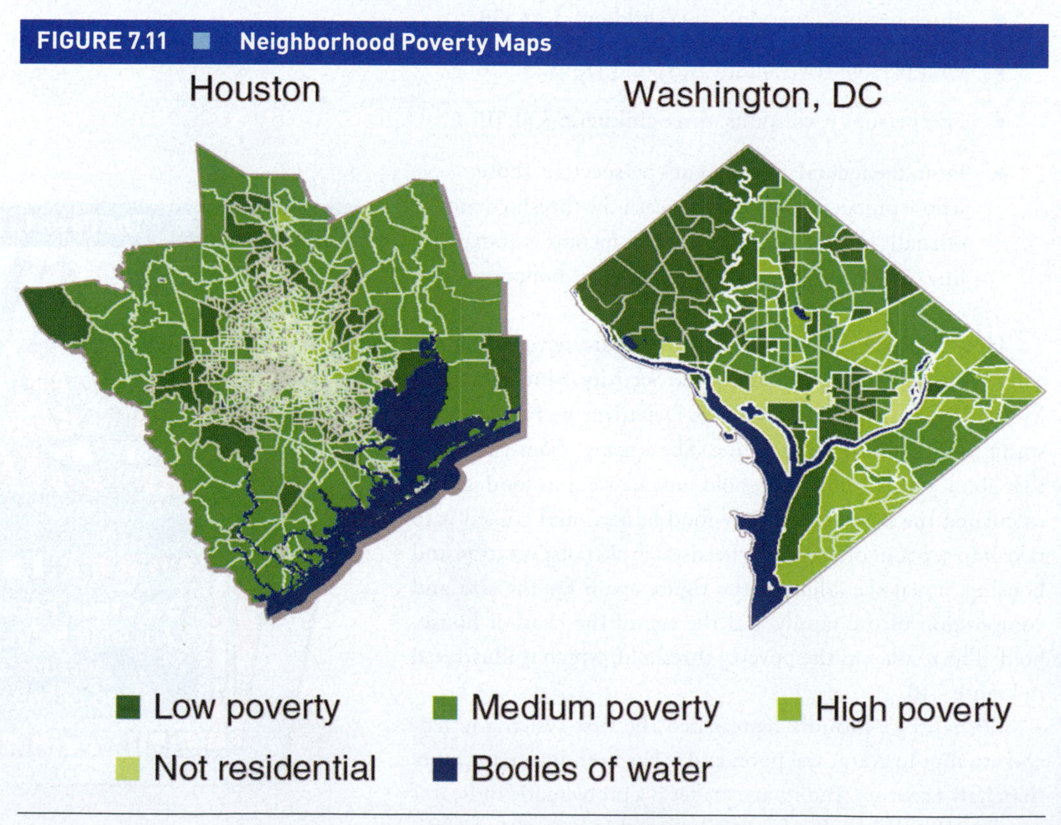

Source: The PEW Charitable Trusts. (2016, January 28). Neighborhood poverty and household financial security. www.pewtrusts.org/en/research-and-analysis/issue-briefs/2016/01/neighborhood-poverty-and-household-financial-security

The Problem of Neighborhood Poverty

Living in an impoverished neighborhood has significant consequences for both poor and nonpoor households. Diminished opportunities for work, education, consumption, and recreation affect entire neighborhoods.

©iStockphoto.com/Peeter Viisimaa

In this section, we discuss the issue of concentrated poverty, looking specifically at measures, causes, and consequences of high levels of neighborhood (or area) poverty. We thus distinguish between *household poverty*, which an individual or family may experience while living in a mixed-income neighborhood, and *neighborhood-level poverty*. Notably, research suggests that being poor in a poor neighborhood has more negative social, economic, and educational effects than household poverty in a more economically heterogeneous context (Wilson, 2010). Neighborhood poverty affects those households that are poor, but it also affects those in the neighborhood who are not officially poor.

A Census Bureau report shows that a growing proportion of Americans reside in *poverty areas*, defined in the report as census tracts featuring 20% or more households in poverty (Bishaw, 2014). (Census tracts are areas with between 1,200 and 8,000 residents; most tracts fall in the 4,000 range.) Poverty areas are further characterized by a lack of "good job opportunities, trusting neighbors, and strong community institutions" (Benzow & Fikri, 2020). In 2000, just over 18% of U.S. inhabitants lived in poverty areas; by 2010, nearly a quarter did. Poverty areas can be rural, suburban, or urban: Just over half are in central cities, another 28% are in the suburbs, and about 20% are outside metropolitan areas.

There has been significant growth in high-poverty neighborhoods in some metropolitan areas: There were an estimated 6,547 such neighborhoods in 2018, nearly double the amount in 1980. Geographically, both the Midwest and South each contain around 25% of the country's high-poverty neighborhoods, while the Northeast and the West account for 20% and 18%, respectively. As of 2018, "41% of the country's Black poor and 31% of its Hispanic poor lived in a high-poverty neighborhood, compared to 13% of its white poor" (Benzow & Fikri, 2020).

Female-headed households are more likely than other family types to live in poverty areas: In 2010, more than 38% of female-headed households resided in areas with more than 20% poverty.

Sociologists study the development of poverty areas, particularly in urban neighborhoods (Wilson, 1996, 2010). The rise of the suburbs in the post–World War II period fostered the out-migration of many city residents, particularly members of the white middle class, and was accompanied by a shift of public resources to new neighborhoods outside cities. Public housing built in U.S. cities around the same period was intended to offer affordable domiciles for the poor, but it also contributed to the development of concentrated poverty, as policies foresaw income limitations that foreclosed the possibility of maintaining mixed-income neighborhoods. Racially discriminatory policies and practices, including limitations on Black access to mortgages or to homes in white neighborhoods, made Blacks far more vulnerable than their white counterparts to becoming trapped in poor neighborhoods. Even today, Black and Hispanic households are more likely to reside in poverty areas (Bishaw, 2014).

As many families moved to the suburbs in the 1950s and 1960s, jobs eventually followed, contributing to a "spatial mismatch" between jobs in the suburbs and potential workers in urban areas (Wilson, 2010). The decline of manufacturing in the 1970s and the decades following also had a profound effect on some urban areas, such as Chicago and Detroit, which were deeply reliant on heavy industry for employment.

As noted earlier, area poverty compounds the negative effects of household poverty and presents challenges to all residents of economically disadvantaged neighborhoods. Research has shown, for instance, that nonpoor Black children are more likely than their white counterparts to reside in poor neighborhoods and to experience limitations to social mobility; no less important, children are more likely than adults to be living in high-poverty areas (Jargowsky, 2015). Poor areas are more likely than better-off or even mixed-income neighborhoods to experience high levels of crime, to have low-quality housing and education, and to offer few job opportunities to residents (Federal Reserve System &

Brookings Institution, 2008; Jargowsky, 2015). Among the challenges to poor neighborhoods is that individual households that have enough resources to leave may choose to do so, contributing to even less circulating capital and an increasing withdrawal of businesses, fewer employed residents, weaker social and economic networks, and more empty buildings (Wilson, 1996). Cities such as Detroit and Cleveland have, in fact, lost thousands of residents in recent decades.

Significantly, children who live in high-poverty neighborhoods are more likely to experience food insecurity than children in low-poverty neighborhoods (Morrissey et al., 2016). They are also less prepared for school than their peers (Wolf et al., 2017). Girls who live in poverty-stricken neighborhoods are also at higher risk of experiencing sexual harassment, exploitation, pressure, and sexual violence. Poor neighborhoods may, over time, develop what the researchers have called "coercive sexual environments or CSEs," in which sexual violence becomes "normal" (Zweig et al., 2015).

The revival of economically devastated neighborhoods is an important public policy challenge. How can the fortunes of poor—particularly very poor—neighborhoods be reversed? How does such a process begin and what does it entail? What do you think?

In the following section, we examine the issues of stratification and poverty from the functionalist and conflict perspectives. As you read, consider how these perspectives can be used as lenses for understanding phenomena such as income and wealth inequality, household and neighborhood poverty, and other issues discussed earlier.

WHY DO STRATIFICATION AND POVERTY EXIST AND PERSIST IN CLASS SOCIETIES?

We find stratification in virtually all societies, a fact that the functionalist and social conflict perspectives seek to explain. The functionalist perspective highlights the ways in which stratification is functional for society as a whole. Social conflict theorists, in contrast, argue that inequality weakens society as a whole and exists because it benefits those in the upper economic, social, and political spheres. We take a closer look at each of these theoretical perspectives next.

The Functionalist Explanation

Functionalism is rooted in part in the writings of sociologist Émile Durkheim (1893/1997), who suggested that we can best understand economic positions as performing interdependent functions for society as a whole. Using this perspective, we can think of social classes as equivalent to the different organs in the human body: Just as the heart, lungs, and kidneys serve different yet indispensable functions for human survival, so do the different positions in the class hierarchy.

In the middle of the 20th century, Kingsley Davis and Moore (1945) built on these foundations to offer a detailed functionalist analysis of social stratification. They argued that in all societies some positions—the most functionally important positions—require more skill, talent, and training than others. These positions are thus difficult to fill—that is, they may suffer a "scarcity of personnel." To ensure that they get filled, societies may offer valued rewards such as money, prestige, and leisure to induce the best and brightest to make sacrifices, such as getting a higher education, and to do these important jobs conscientiously and competently. According to Davis and Moore, social inequality is an "unconsciously evolved device by which societies ensure that the most important positions are conscientiously filled by the most qualified persons" (p. 243).

Meritocracy: A society in which personal success is based on talent and individual effort.

An implication of this perspective is that U.S. society is a **meritocracy**, *a society in which personal success is based on talent and individual effort.* That means your position in the system of stratification depends primarily on your talents and efforts: Each person gets more or less what he or she deserves or has earned, and society benefits because the most functionally important positions are occupied by the most qualified individuals. Stratification is then ultimately functional for society because the differential distribution of rewards ensures that highly valued positions are filled by well-prepared and motivated people. After all, Davis and Moore might say, we all benefit when we get economic information from good economists, drive across bridges designed by well-trained engineers, and cure our ills with pharmaceuticals developed by capable medical scientists.

Clearly, the idea that the promise of higher pay and prestige motivates people to work hard has some truth. Yet it is difficult to argue that the actual differences in rewards across positions are necessarily suitable ways of measuring the positions' relative worth to society (Tumin, 1953, 1963, 1985; Wrong, 1959). Is an NBA (National Basketball Association) point guard really worth more than a teacher or a nurse, for instance? Is a hedge fund manager that much more important than a scientist (particularly given that both positions require extensive education)?

Moreover, when people acquire socially important, higher-status positions by virtue of their skills and efforts, they are then often able to pass along their economic privilege, and the educational opportunities and social connections that go with it, to their children, even if their children are not particularly bright, motivated, or qualified. As Melvin Tumin (1953) points out in his critique of Davis and Moore, stratification may *limit* the discovery of talent in society rather than *ensure* it by creating a situation in which those who are born to privilege are given fuller opportunities and avenues to realize occupational success while others are limited by poor schooling, little money, and lack of networks upon which to call. Such a result would surely be dysfunctional for society rather than positively functional.

How would functionalism account for the fact that people are often discriminated against because of their skin color, sex, and other characteristics determined at birth that have nothing to do with their talents or motivations, resulting in an enormous waste of society's human skills and talents? Can you see other strengths or weaknesses to Davis and Moore's perspective on stratification?

In a twist on the functionalist perspective, sociologist Herbert Gans (1972) poses this provocative question: How is poverty positively functional in U.S. society? Gans begins with a bit of functionalist logic, namely, that if a social phenomenon exists and persists, it must serve a function or else it would evolve out of existence. But he does not assume that poverty is functional for everyone. So, *for whom* is it functional? Gans suggests that eliminating poverty would be costly to the better-off. Thus, poverty is functional for the nonpoor but not functional for the poor—or even for society as a whole.

Among the "benefits" to the nonpoor of the existence of a stratum of poor people, Gans includes the following:

Poverty ensures there will be low-wage laborers prepared—or driven by circumstances—to do society's dirty work. These are the jobs no one else wants because they are demeaning, dirty, and sometimes dangerous. A large pool of laborers desperate for jobs also pushes down wages, a benefit to employers.

Poverty creates a spectrum of jobs for people who help the poor (social welfare workers), protect society from those poor people who transgress the boundaries of the law (prison guards), or profit from the poor (owners of welfare motels and cheap grocery shops). Even esteemed sociologist Herbert Gans has built an academic career on analyzing poverty.

Poverty provides a market for goods and services that would otherwise go unused. Day-old bread, wilting fruits and vegetables, and old automobiles are not generally purchased by the better-off. The services of second-rate doctors and lawyers, among others, are also peddled to the poor when no one else wants them.

Beyond economics, the poor also serve cultural functions. They provide scapegoats for society's problems and help guarantee some status for those who are not poor. They also give the upper crust of society a socially valued reason for holding and attending lavish charity events.

Gans's (1972) point is stark. He notes that the functions served by the poor have *functional alternatives*—that is, they could be fulfilled by means other than poverty. Nevertheless, he suggests, those who are better off in society are not motivated to fight poverty comprehensively because its existence is demonstrably functional for them. Although he is not arguing that

Herbert Gans suggests that poverty ensures a pool of workers "unable to be unwilling" to do difficult and dirty jobs for low pay. Such jobs could also be filled in the absence of poverty through better pay and benefits. But, says Gans, this would be costly and, thus, dysfunctional to the nonpoor.

anyone is in favor of poverty (which is difficult to imagine), he is suggesting that "phenomena like poverty can be eliminated only when they become dysfunctional for the affluent or powerful, or when the powerless can obtain enough power to change society" (p. 288). Do you agree with his argument? Why or why not?

The Social Conflict Explanation

Social conflict theory draws heavily from the work of Karl Marx. As we saw in the opening chapter, Marx divided society into two broad classes: workers and capitalists, or *proletarians* and *bourgeoisie.* The workers do not own the factories and machinery needed to produce wealth in capitalist societies—they possess nothing of real value except their labor power. The capitalists own the necessary equipment—the *means of production*—but require the labor power of the workers to run it.

These economic classes are unequal in their access to resources and power, and their interests are opposed. Capitalists seek to keep labor costs as low as possible to produce goods cheaply and make a profit. Workers seek to be paid adequate wages and to secure safe, decent working conditions and hours. At the same time, the two groups are interdependent: The capitalists need the labor of the workers, and the workers depend on the wages they earn (regardless of how meager) to survive.

Although more than a century has passed since Marx formulated his theory, a struggle between workers and owners (or, in our time, between workers and owners, managers, and even stockholders, who all depend on a company's profits) still exists. Conflict is often based on the irreconcilability of these competing interests. A study found that collective action lawsuits alleging wage and hour violations have skyrocketed, increasing 400% in the past 11 years. Among companies such as Bank of America, Walmart, and Starbucks, Taco Bell has been one of the latest to be sued for allegedly forcing employees to work overtime without pay (Eichler, 2012).

The source of inequality, then, lies in the fact that the bourgeoisie own the means of production and can use their assets to make more money and secure their position in society. Most workers do not own substantial economic assets aside from their own labor power, which they use to earn a living. Although successful lawsuits for lost wages show that workers have avenues for asserting their rights against employers, the conflict perspective contrasts these small victories with the far more significant power and control exercised by large economic actors in modern society.

In short, the conflict perspective suggests that significant and persistent stratification exists because those who have power use it to create economic, political, and social conditions that favor them and their children, even if these conditions are detrimental to the lower classes. Inequality thus is not functional, as Davis and Moore argued. Rather, it is dysfunctional, because it keeps power concentrated in the hands of the few rather than creating conditions of meritocracy that would give equal opportunity to all.

Similar to the functionalist perspective, the conflict perspective has analytical weaknesses. It overlooks cooperative aspects of modern capitalist businesses, some of which have begun to take a more democratic approach to management, offering workers the opportunity to participate in decision-making processes in the workplace. Modern workplaces in the technology sector, for instance, thrive when decision making and the production of ideas come from various levels rather than only from the top down.

Like modernization theory, the dependency and world systems theoretical perspectives illuminate some aspects of the case while obscuring others. Variables such as the exploitative power of Western oil companies are a key part of understanding the failure of Nigeria to develop in a way that benefits the broader population, but conflict-oriented perspectives pay little attention to the agency of poor states and, in particular, their governing bodies in setting a solid foundation for development.

WHAT'S NEXT FOR ECONOMIC INEQUALITY?

What will economic inequality look like as we enter the third decade of the 21st century? It is too soon to decisively answer that question, but we do know that challenges, including growing gaps in income and wealth inequality in the United States, have been amplified by the public health challenge of COVID-19, which began in early 2020, and the dramatic economic downturn that accompanied the pandemic. They have also been sharpened by a contentious political environment that stood in the way of informed and effective policy making to reduce the negative effects of disease and downturn.

Economic inequality in the United States is closely tied to other types of inequality that we discuss in this text, and the consideration of *what's next* entails the recognition of how economic inequality might be affected by changes in areas like gender inequality and educational inequality. Recent data suggest that a significant number of women left the workforce during the pandemic. While some of this was driven by job losses in sectors where women are heavily concentrated, such as hospitality, some mothers were compelled to leave the workforce because their children could not attend school or daycare and were learning at home due to COVID-19 restrictions. By one estimate, in the month of September 2020, about 800,000 women dropped out of the labor force (Scott, 2020). This affected women's earning power directly, but will also likely be reflected in future data on household income and poverty.

We know that income is closely tied to educational attainment. Recent decades have seen a trend of higher college enrollment among students from low-income families, but the pandemic threatens to set back access to this important vehicle of social mobility. During the COVID-19 pandemic, many low-income college students struggled with challenges like poor Internet access during remote learning, lack of safe housing while dormitories were closed, and the loss of financial aid as colleges cut back on help to manage their own budget losses (Levin, 2020).

How will economic inequality—measured in terms of income, wealth, and social mobility, among others—look in the middle of the 2020s? Where will the effects of these crises be most visible in our data? Where will they be most visible in our lived experiences and those of the people around us?

WHAT CAN I DO WITH A SOCIOLOGY DEGREE?

Making an Evidence-Based Argument

We live in a sound-bite society where news and information are circulated in tweets and texts, and memes go viral faster than colds in a pre-K class. We regularly hear people make claims about their lives or society based on their personal experiences or things they've heard or read in passing. Ask yourself how many times you've forwarded a story without checking its validity, only to realize later that it was not true or only partially true. In a fast-moving social world, we may feel we lack the time or inclination (or even the skills) to determine whether the information we accept and share is fully reliable. Sociology invites us to become critical consumers of information, a skill that is of vital importance in both civil life and the job market.

In Chapters 1 and 2, you learned how sociology entreats us to go beyond "common wisdom" about social issues such as homelessness and the wage gap, and to study society scientifically. Unlike those conversations, in which someone makes an argument based on anecdotal evidence or a small and unrepresentative sample, sociology highlights the necessity of *evidence-based arguments*. Sociology students learn how to collect and analyze both quantitative and qualitative data, and evaluate that data for scientific rigor. Being able to make an evidence-based argument is an important skill in any occupation. Budget reports, employee evaluations, scientific and social scientific research, grant proposals, program and policy evaluations, and arguing a case in court all require evidence-based arguments. The ability to communicate effectively is a requirement of most jobs and necessitates showing or proving your points, not only stating your thoughts or opinions. Sociology students learn not only how to make evidence-based arguments but to evaluate the claims of others.

Tessa Constantine, Research Analyst at Building Movement Project, New York, NY

BA, Sociology, Carthage College
MA, Sociology, Columbia University

I chose to go into sociology because it seemed to hold similar ideals and ethics to my own. Now, having completed two degrees in the discipline, sociology has not only taught me to articulate my ideals, but to scientifically explore them. Having just graduated, I am starting as a research associate at a nonprofit looking at diversity gaps in leadership structures. The ability to tackle institutional issues surrounding race, gender, sexuality, and other identities is something I have always wanted to do but didn't imagine I would be doing so soon. Specifically, I have been able to work to improve these issues through the quantitative and qualitative data analysis that is central to my current position. These skills allow me to see the story the data tell and build a compelling, fact-based argument around them to disseminate to other organizations.

One of the wonderful aspects of the discipline is its versatility and ability to impact a multitude of fields. Working on social justice at a nonprofit has allowed me to dive headfirst into the issues I care about while having the capacity to relate and understand them in relation to a host of other social topics. Further, sociology has provided me with the skills to quantify these observations and disseminate the information for a broader audience. I believe the presentation of research is one of the most direct ways to stimulate change and challenge the status quo, and I am excited to start doing so full time.

Career Data: Survey Researcher

- 2019 Median Pay: $59,170 per year
- $28.45 per hour
- Typical Entry-Level Education: Master's degree
- Job Growth Outlook, 2019–2029: –4% (decline)

Source: Bureau of Labor Statistics. (2019). *Occupational outlook handbook*.

SUMMARY

- Class societies are more open than caste societies. In a **caste society**, a person's position in the hierarchy is determined by ascribed characteristics such as race or birth status. In a **class society**, a person's position is determined by what he or she achieves, and mobility is looked upon favorably. Nevertheless, barriers to mobility similar to those in caste societies still exist in class societies.
- **Class** refers to a person's economic role in society, associated with income, wealth, and the type of work he or she does. Class position strongly influences an individual's **life chances**—the opportunities and obstacles he or she encounters in areas such as education, social life, and work. Important components of class position are **occupation**, **income**, and **wealth**.
- We can measure inequality in the United States by looking at disparities in income, wealth, health, and access to credit and goods. All these indicators show that inequality in the United States is substantial and growing.
- Since the early 1970s, the gap between the rich and the poor has grown, as has the gap between the rich and everyone else. Some of this growth is attributable to the transformation of the U.S. economy from industrial to postindustrial, which has helped the best educated and hurt the least educated.
- Poverty is a significant problem in the United States: In 2015, 13.5% of the population was officially poor. The formula used to measure poverty gives us a sense of the problem, but it has limitations of which we should be aware.

- Researchers distinguish between household (or individual) poverty and neighborhood poverty. Studies suggest that living in a poor neighborhood amplifies the effects of poverty and also poses challenges (including limited mobility) for nonpoor residents.
- Functionalist theorists argue that inequality exists and persists because it is positively functional for society. According to this perspective, inequality is necessary to motivate the best people to assume the most important occupational positions.
- Conflict theorists argue that the privileged classes benefit from inequality and that inequality inhibits the discovery of talented people rather than fostering it. This perspective suggests that classes with differential access to power and resources are in conflict and that the interests of the well-off are most likely to be realized.

KEY TERMS

achieved status (p. 166)
ascribed status (p. 166)
caste society (p. 165)
class (p. 167)
class society (p. 166)
food deserts (p. 176)
income (p. 168)
life chances (p. 167)
meritocracy (p. 182)
net financial assets (p. 169)
occupation (p. 169)
official poverty line (p. 179)
political power (p. 170)
social categories (p. 166)
social inequality (p. 165)
social mobility (p. 167)
social stratification (p. 165)
status (p. 170)
underemployed (p. 172)
wealth (or net worth) (p. 169)

DISCUSSION QUESTIONS

1. What is the difference between *wealth* and *income*? Why is it sociologically important to make a distinction between the two? Which is greater in the United States today, the income gap or the wealth gap?
2. Herbert Gans talks about the "uses of poverty" for the nonpoor. Recall some of his points presented in the chapter and then add some of your own. Would you agree with the argument Gans makes about the existence and persistence of poverty? Why or why not?
3. What is the difference between individual or household poverty and neighborhood poverty? Why is the distinction important? How does being poor in a poor neighborhood amplify the effects of economic disadvantage?

BULENT KILIC/Getty Images

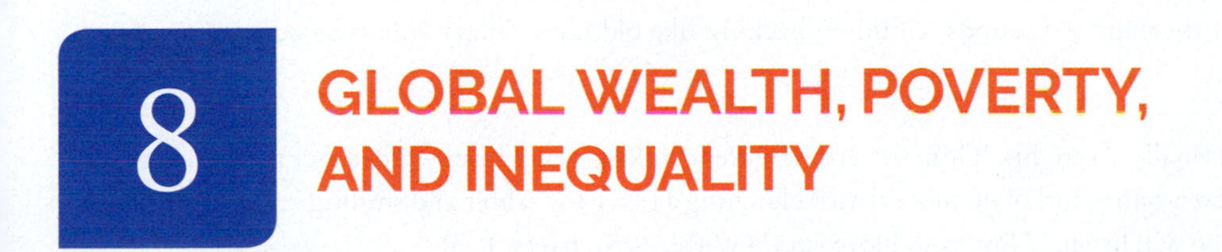

8 GLOBAL WEALTH, POVERTY, AND INEQUALITY

WHAT DO YOU THINK?

1. Why are some countries prosperous while others are poor?
2. How has the spread of modern technologies (such as mobile phones) had an impact on economic development across the globe?
3. What is the relationship between armed conflict and poverty of families and countries around the world?

LEARNING OBJECTIVES

8.1 Describe quantitative and qualitative dimensions of global wealth, poverty, and inequality.

8.2 Explain the relationship between armed conflict and poverty.

8.3 Discuss the role of technology in economic development across the globe.

8.4 Apply theoretical perspectives to analyze the existence and persistence of global inequality.

8.5 Describe characteristics of the global elite.

MIGRANTS IN LIMBO

The Greek resort island of Lesbos has long been a tranquil tourist haven in the turquoise Aegean Sea. In recent years, however, the small island has been transformed into a destination for war refugees fleeing broken homelands such as Syria and Iraq and economic migrants from Africa and Asia seeking better lives. A recent article in the *Washington Post* describes a scene from a migrant camp in Lesbos:

> The first thing you notice is the smell: the stench from open-pit latrines mingling with the odor of thousands of unwashed bodies and the acrid tang of olive trees being burned for warmth.
>
> Then there are the sounds: Children hacking like old men. Angry shouts as people joust for food.
>
> And, finally, the sights: Thin, shivering figures drinking water from washed-out motor oil jugs. A brown-haired girl of no more than 3 clutching a fuzzy toy rabbit and smiling as she repeats to all who will listen, "I love you. I love you." (Witte, 2018, paras. 1–3)

Several years have passed since the massive waves of migrants started arriving in Greece. The numbers have dropped off considerably since 2015 and 2016, when as many as 10,000 arrived in a single day on Lesbos (Witte, 2018). But large numbers of migrants still languish in decrepit camps on Lesbos, trapped in limbo between abandoned homelands and a safe, stable future home.

©MANOLIS LAGOUTARIS/Getty Images

Lesbos was a destination of convenience more than choice: It is a frontline port of call into Europe. While it is only 5 miles from the Turkish coast from which many migrants launched on smugglers' overfilled boats, problems such as poor weather conditions and boats that proved not to be seaworthy have led to hundreds of lost lives. The risks of the journey, which for many included threats of assault, kidnapping, robbery, and drowning, did not, alas, end when they reached the Greek island (Faiola, 2015).

The conditions at camps such as those situated on Lesbos are, according to aid workers, residents, and journalists, often very poor. Sanitation, food provision, medical care, and shelter are largely inadequate. By some accounts, however, the "appallingly bad conditions are no accident, but rather the result of a deliberate European strategy to keep people away" (Witte, 2018, para. 6). Indeed, the Western welcome mat has been pulled, and several thousand camp residents have waited as long as 2 years to move forward in their journey. Many will never move forward, as most European countries, as well as the United States, are permitting declining numbers of new migrants. Migration from poor countries in the Global South to wealthier states in the Global North is a long-standing phenomenon, but it has intensified and become more visible in recent years. Today, refugees desperately fleeing war-torn homelands mix with economic migrants in a mass movement of humanity pursuing physical safety and economic security.

The COVID-19 pandemic, which hit Europe in early 2020, exacerbated the poor living conditions for residents of these camps, and in September, the largest migrant camp, Moria, was devastated by massive fires. The fires were started by camp residents expressing "dissatisfaction" with the coronavirus-related lockdown measures put in place within the refugee camp, home to an estimated 13,000 people. For refugees exposed to extreme psychological stress and material deprivation in the camp, the lockdown became the final straw. Moria was destroyed by these fires. The tragedy has left families combing through the wreckage with nowhere to turn and even fewer resources (Labropoulou et al., 2020).

Across the globe, countries, communities, and households are stratified: While some struggle to meet the most basic needs, others enjoy comfortable lives, broad opportunities, and modern amenities. We begin this chapter with a look at some of the dimensions of global inequality, examining factors such as per capita income, literacy, education, sanitation, and health in order to understand more about the hierarchy of countries in our global system. We then add in another variable that exacerbates economic struggles of countries and communities: armed conflict. Next, we consider one area—mobile technology—that may be helping to shrink economic disparities globally. This is followed by a look at theoretical perspectives that seek to understand why these deep global disparities exist and persist. We also examine the phenomenon of an increasingly wealthy global elite whose influence crosses national boundaries. We end with a brief consideration of the question of why sociologists take an interest in global inequality.

DIMENSIONS OF GLOBAL INEQUALITY AND POVERTY

Many of the world's people are poor. According to a recent report from The World Bank, 10% of the world's population—about 734 million people—live on the equivalent of less than $1.90 a day. Nearly all of these economically marginal people reside in the developing world, with over half living in sub-Saharan Africa (The World Bank, 2018). Notably, the figure has been falling for years—even as the world population grows—but the appearance of the COVID-19 virus in 2020 may turn back progress: The World Bank estimates that an additional 40 to 60 million people will fall into extreme poverty by the end of the year as a result of the virus (The World Bank, 2020). Meanwhile, according to the global nonprofit organization Oxfam, the wealthiest 1% of the population holds more than twice as much wealth as 6.9 billion of the world's people (Oxfam, 2020).

We can look at inequality in terms of individuals or households, but we can also compare the economic positions of countries, recognizing a global class system with prosperous states, poor states, and a wide swath of countries in between. In this chapter, we look at some of the dimensions of **global inequality**, which can be defined as *the systematic disparities in income, wealth, health, education, access to technology, opportunity, and power among countries, communities, and households around the world.* While the focus is largely on differences among countries, we will see that these are only one part of a broader picture of global inequality.

Global inequality: The systematic disparities in income, wealth, health, education, access to technology, opportunity, and power among countries, communities, and households around the world.

We follow The World Bank in categorizing countries using four economic categories: high income, upper-middle income, lower-middle income, and low income (Figure 8.1; Serajuddin & Hamadeh, 2020). In 2020, The World Bank defined these classifications quantitatively using the following gross national income (GNI) per capita limits:

FIGURE 8.1 ■ High-Income, Upper-Middle-Income, Lower-Middle-Income, and Low-Income Countries, 2020

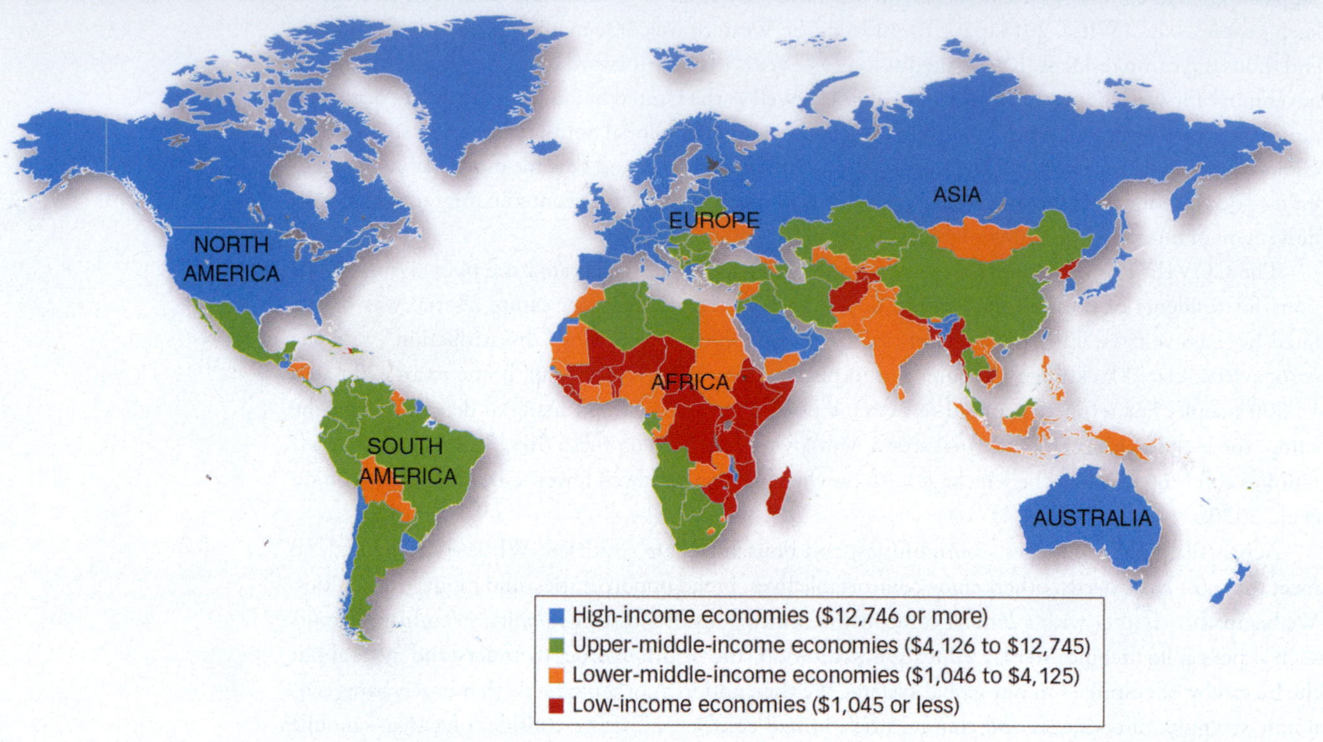

- High-income economies ($12,746 or more)
- Upper-middle-income economies ($4,126 to $12,745)
- Lower-middle-income economies ($1,046 to $4,125)
- Low-income economies ($1,045 or less)

Source: Reprinted with permission from The World Bank. (2020). *World Bank Country and Lending Groups.*

- Low-income economies: $1,045 or less
- Lower-middle-income economies: $1,046 to $4,125
- Upper-middle-income economies: $4,126 to $12,745
- High-income economies: $12,746 or more

Qualitatively, we can describe the *high-income countries* as those that are highly industrialized, characterized by the presence of mass education, and both urbanized and technologically advanced. Among the high-income countries, we find nations such as the United States, Canada, Japan, Germany, Norway, Estonia, and Australia. High-income countries are home to about 15% of the global population.

More than 70% of the world's population lives in *middle-income countries* (the lower- and upper-middle categories combined), which include a wide variety of nations; among them are former Soviet states such as Armenia and Belarus; South and Central American states such as Brazil and Belize; Middle Eastern countries such as Lebanon and Iran; Asian states such as Indonesia, India, and China; and African countries such as Morocco and Senegal. Many of these countries are on a path to economic diversification and development, though most also started down the road to urbanization and industrialization much later than the high-income countries and still lag in instituting mass education. Some middle-income countries, such as those in the Middle East and Africa, are home to vast natural resources, though the conversion of those resources to shared prosperity has, for reasons that theorists and observers debate, not been widespread.

Similar to high-income countries, *low-income countries* constitute a relatively small proportion of the global total. Many are agricultural states with rapidly growing populations. While urbanization is a growing phenomenon, cities in these countries often lack the jobs and services that rural migrants seek, and both rural and urban dwellers struggle with hunger and malnutrition, economic and educational deprivation, and preventable diseases. Low-income countries may have small and wealthy groups of elites, but they lack stable middle-class populations. Low-income countries include South Asian states such as Bangladesh and Cambodia as well as Africa's poorest countries, such as Somalia and the Central African Republic.

In the remainder of this section, we describe some ways in which inequalities are manifested around the globe. Later in the chapter, we consider the key question of *why* this inequality exists and persists and examine various theoretical perspectives on the issue.

We begin with a look at **gross national income–purchasing power parity per capita (GNI-PPP)**, *a comparative economic measure that uses international dollars to indicate the amount of goods and services someone could buy in the United States with a given amount of money.* At one end of the class spectrum, we find countries with very high GNI-PPP, such as the United States ($63,780), Canada ($49,430), Norway ($70,530), Germany ($55,980), and Japan ($43,010). In the middle are countries ranging from Botswana ($17,460) and Estonia ($35,340) to Turkey ($27,710) and Brazil ($14,530). At the bottom are countries whose GNI-PPP can be as low as that of Nicaragua ($5,700), Senegal ($3,330), or Gambia ($2,160; Population Reference Bureau, 2020). While GNI-PPP cannot tell us a great deal about the resources of individual families in the given countries, it gives us some insight into the economic resources available to the state and society from a macro perspective, and it offers a comparative measure for looking at stratification in the global system.

Gross national income–purchasing power parity per capita (GNI-PPP): A comparative economic measure that uses international dollars to indicate the amount of goods and services someone could buy in the United States with a given amount of money.

Hunger, Mortality, and Fertility in Poor Countries

As we saw in our discussion of social stratification in the United States, one important indicator of inequality is health (Table 8.1). A key aspect of good health is adequate food, in terms of both sufficient calories and basic nutrition. While the world has the capacity to produce enough food for all its inhabitants, the Food and Agriculture Organization of the United Nations (2017) estimates that 815 million people are chronically undernourished, and over 21 million children under age 5 are affected by *wasting*, or being too thin for one's height. At the turn of the millennium, the United Nations set as a goal

TABLE 8.1 ■ Global Inequality Indicators, 2020

	GNI-PPP	TOTAL FERTILITY RATE	INFANT MORTALITY RATE PER 1,000 LIVE BIRTHS	PERCENTAGE UNDERNOURISHED (2016–2018 data)
World	$16,885	2.3	31	10.8% (2018)
By Level of Development				
More Developed	$46,188	1.6	4	n/a
Less Developed	$10,814	2.5	34	n/a
Least Developed	$2,923	4.1	49	23.6%
By Specific Country				
United States	$63,780	1.7	5.7	<2.5%
France	$47,490	1.8	3.6	<2.5%
Mexico	$19,870	2.1	11	3.6%
China	$15,320	1.5	9	8.5%
Jordan	$10,050	2.7	17	12.2%
Yemen	n/a	3.7	43	38.9%
Niger	$870	7.1	69	16.5%

Sources: Population Reference Bureau. (2020). *2020 world population data sheet;* Food and Agriculture Organization of the United Nations. (2019). *Source on undernourishment.*

the substantial reduction of hunger around the globe. In fact, data suggest that there have been marked improvements in access to adequate food supplies in many of the world's regions, most notably in those that have experienced rapid economic growth, including parts of Southeast Asia and Central and South America. At the same time, hunger has increased in other areas, including sub-Saharan Africa.

An important cause of hunger at the household level is poverty; many of those who lack sufficient food do not have the economic resources to acquire it. Subsistence farmers in developing countries, many of whom survive from season to season on their own small-scale crop yields, are vulnerable to weather events and natural disasters that can push their families into destitution and starvation. Hunger at a community level is more complex. While entire communities may suffer poverty and malnutrition, large-scale hunger is often the outcome of political decisions or armed conflicts. For example, the government of Syria under Bashar al-Assad has persistently impeded the delivery of food supplies to civilians in areas held by his opponents in the Syrian civil war that has raged since 2011. Hundreds of thousands have been killed and millions more have been displaced, fleeing Syria in search of refuge (as we saw in the opening story). Many of those left behind are the victims of hunger, which some (including the Secretary-General of the United Nations, Ban Ki-Moon) claim are being used by al-Assad as a weapon of war (Melvin et al., 2016). Regardless of the causes of hunger, the costs of undernourishment are serious and often lasting: By one estimate from the United Nations Educational, Scientific and Cultural Organization, fully 21.3% of all children under age 5 are *stunted*—their growth progression is impaired by a lack of access to adequate nutrition (UNESCO, 2019).

Infant mortality rate: The number of deaths of infants under age 1 per 1,000 live births per year.

In evaluating global health, we can also compare across countries the **infant mortality rate**—*the number of deaths of infants under age 1 per 1,000 live births per year.* This figure gives some insight into the health status of populations, and of women and children in particular, because infant mortality rates are lowest in states that offer access to safe pre- and antenatal care and sanitary childbirth facilities as well as good nutrition during pregnancy. Consider the vast differences in the infant mortality rates among categories of countries: As of mid-2020, the most-developed countries had an infant mortality rate of 4 per 1,000 live births. By contrast, the less-developed countries had an infant mortality rate of 34 per 1,000 live births, and the least-developed countries posted an infant mortality rate of 49 per 1,000 live births. Across specific countries, rates vary from lows in countries such as Sweden (1.8) and Austria (2.7) to highs such as those in Haiti (59), Pakistan (62), and Sierra Leone (75; Population Reference Bureau, 2020). Demographic rates like the infant mortality rate tend to shift slowly rather than precipitously over time. The global COVID-19 pandemic that began in early 2020, however, may have effects that will become visible in the data in future years. Although early indicators in the pandemic suggest that infants and children are less vulnerable than older adults, the loss of an adult caregiver in the first year of life would substantially raise young children's risk of early death.

Total fertility rate (TFR): The average number of children a woman in a given country will have in her lifetime if age-specific fertility rates hold throughout her childbearing years (ages 15–49).

Global health indicators such as infant and child mortality rates are linked not only to income differences *between countries* but also to income stratification *within countries.* Data suggest that those countries with a highly unequal distribution of income also experience highly variable health outcomes. For instance, in Cambodia (which is deeply stratified by income), among the top fifth of income earners, the infant mortality rate is 23 per 1,000 live births, while for those in the bottom fifth, the rate is 77 per 1,000 live births (Population Reference Bureau, 2013).

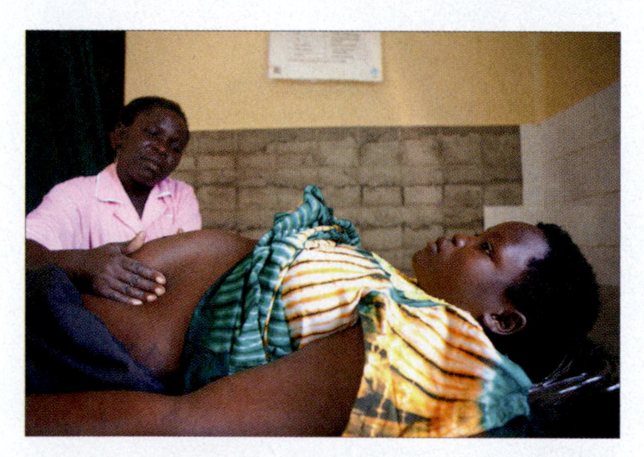

Total fertility rates have fallen across the developing world as more families gain access to information on family planning and to safe, effective contraception. In many of the countries of Western and Eastern Africa, however, there is a significant unmet need for contraception: 24% of partnered women indicate they would like to stop or delay childbearing but are not using any contraceptive method (United Nations, 2015).

©Jake Lyell/Water Aid/Alamy

Our discussion of health may also be linked to the issue of fertility. Demographers measure **total fertility rate (TFR)**, which is *the average number of children a woman in a given country will have in her lifetime if age-specific fertility rates hold throughout her childbearing years (ages 15–49).* We can use this measure to look at childbearing over space and time. It is notable that many of the world's poorest countries have the highest fertility rates. For example, while the TFR is 1.5 in Norway, 1.6 in Germany, 1.3 in Japan, and 1.7 in the United States, rates in the least-developed countries remain high. In 2020, some of the world's highest TFRs were found in Chad (5.9), Mozambique (4.9), Nigeria

(5.3), and Afghanistan (4.5). Rates of fertility within many countries also vary by economic status; for instance, in the African country of Uganda, women in the top fifth of the income hierarchy have a TFR of 4.0, while their sisters in the lowest fifth have a TFR that is nearly double (7.9; Population Reference Bureau, 2020).

What sociological factors help explain differences in fertility? One factor is the link between infant and child mortality and fertility: In regions or countries where early child survival is threatened by disease, poverty, or other risks, families may choose to have more children in order to ensure that some survive into adulthood to contribute to the household and care for elderly parents, particularly in countries without social welfare supports for retirees. Second, it has been said that children are a poor man's riches—indeed, in many agricultural economies, many hands are needed to do work, and children are active contributors to a family's economic well-being. Economic modernization correlates historically with drops in fertility (see Chapter 17 for a fuller discussion of this topic). As well, where a lack of access to maternal and child health care is common, there may also be little access to safe, effective contraceptives that would enable women to control their fertility. A recent United Nations (2015) report shows a significant unmet need for contraception among women: The figure reaches as high as 24% in sub-Saharan Africa and is 22% overall in the least-developed countries.

Safe Sanitation

Access to safe, hygienic sanitation facilities has emerged as an issue of public health concern for many poor communities but also as an issue of dignity and security for girls and women. Specifically, female inhabitants of communities that lack accessible, safe toilet facilities face serious vulnerabilities when they need to meet normal bodily needs. Consider the following story from June 2014: In a small Indian village in the northern state of Uttar Pradesh, two girls, ages 12 and 14, went together one evening to relieve themselves in the wild bamboo fields several minutes from their home, which has no bathroom facilities. In the darkness, the girls were brutally attacked, raped, and hanged from a mango tree, allegedly by three brothers (Banerjee, 2014).

The majority of India's 1.3 billion inhabitants do not have access to private toilets or latrines. According to the Population Reference Bureau (2013), only 60% of urban Indians and 24% of rural Indians were using "improved sanitation facilities" in 2011; while this measure focuses on the "hygienic separation of sewage from human contact" rather than safety or privacy, it points to a presence or lack of access to basic toilet facilities. In Uttar Pradesh, which is the country's largest state, about 64% of Indians have no indoor plumbing (McCarthy, 2014). Although some villages have public bathrooms, one press account notes that "many women avoid using them because they are usually in a state of disrepair and because men often hang around and harass the women." The same report quotes a local Indian police official's estimate that more than 60% of rapes occur in similar circumstances (Banerjee, 2014). A 2012 study reported that

> approximately 30% of women from the underprivileged sections of Indian society experience violent sexual assaults every year because lack of sanitation facilities forces them to go long distances to find secluded spots or public facilities to meet their bodily needs. (Bhatia, 2013)

The problem is not limited to India. Similar stories of violation and violence have been documented from Nepal to South Africa, where private toilet facilities are the exception rather than the rule (Bhatia, 2013). In the west African country of Senegal, for instance, only 21% of the population has access to what The World Bank terms safely managed sanitation services, defined as "safe disposal of human waste and access to handwashing with soap and water." Figures on access vary from a full 100% in most developed countries to 36% in Djibouti, 41% in Iraq, and 17% in Colombia. Figures in rural areas are lower still: In remote areas of Djibouti, for instance, the figure is only 19% (The World Bank, 2017).

According to the United Nations, about 2.5 billion people still lack access to basic, safe sanitation facilities, a figure linked not only (as noted previously) to violence toward and degradation of women but also to high rates of diarrheal diseases among children.

In 2018, about 258 million adolescents and children worldwide were out of school, and the number is rising.

©Thomas Koehler/Photothek via Getty Images

In some developing countries such as Bangladesh (pictured), India, and the Ivory Coast, child labor remains a serious human rights issue. Child advocates underscore the right to care, safety, and education for all young people.

©Nurun Nahar Nargish/Majority World/UIG via Getty Images

Education Matters

In most developed countries, nearly all young people complete primary school, and most move on to high school. In less-developed states, access to education is more limited, and the opportunity to go to school may be affected by a spectrum of factors. In some countries and communities, girls are discouraged or even prevented from attending school by economic or cultural factors. In others, school fees present obstacles to poor families who cannot afford to enroll their children. Sometimes schools and teachers are themselves not available because of the presence of armed conflict or the absence of communities that could sustain them.

A study by UNESCO found that in 2018, about 258 million adolescents and children worldwide were out of school. While many children do not attend school because of armed conflicts in their countries or regions, many of those not in school are, according to UNESCO, unlikely ever to attend school. School attendance is highly variable by gender, as 9 million girls of primary school age will never attend school, compared to about 3 million boys (UNESCO, 2019). Even where schools are available in poor countries, the quality of education is lacking: UNESCO (2014b) reports that about 250 million children are without basic literacy and numeracy skills, although about half of them have completed at least 4 years of school. Inadequate teacher preparation may combine with overcrowding (the African state of Malawi reported an average of 130 children in a Grade 1 classroom), lack of textbooks, and far distances from home to school to render efforts to educate children ineffective. The lack of safe transportation, running water, and toilet facilities may also discourage children, particularly girls, from attending school regularly (Rueckert, 2018).

Uneducated or poorly educated children pass into adulthood without basic skills. Literacy and numeracy skills not achieved in the years of primary school are rarely achievable in adulthood in developing countries, which have not established a tradition of adult education. UNESCO (2017) estimates that in 2016, there were 750 million fully illiterate adults worldwide, about two thirds of whom were women, though data show steady improvement across regions (Table 8.2).

Education improves the lives of communities and families in a multitude of ways. UNESCO (2014b) estimates that, on the global level, a year of school can equal a 10% boost in income. Education benefits both those who work for wages (by improving skills) and those who farm (by increasing access to knowledge about effective, efficient farming methods). Education also helps workers to avoid exploitation and better advocate for their interests. Apart from opening up broader avenues for economic advancement, better education is also linked to positive health outcomes. This relationship is particularly strong for women's education and child health outcomes. Research has documented a positive correlation between maternal education (even at the primary level) and decreased risk of child mortality (Glewwe, 1999; LeVine et al., 2012). For instance, a study on Nigeria found that better reading skills among mothers were linked to lower rates of child mortality (Smith-Greenaway, 2013). Other work suggests that greater maternal education translates into a greater probability that a woman's children will be educated (UNESCO, 2014b).

Significant strides have been made in many countries and regions in recent decades in educating young people (and women in particular), but much remains to be done. Even today, an estimated 102

TABLE 8.2 ■ Selected Literacy Rates Among Adults Ages 15 Years and Older

COUNTRY	LITERACY RATE: WOMEN	LITERACY RATE: TOTAL
Afghanistan	30%	43%
Brazil	93%	93%
Uganda	71%	77%
Italy	99%	99%
Mexico	95%	95%
China	95%	97%
Nigeria	53%	62%
Vietnam	94%	95%

Source: The World Bank. (2018). https://data.worldbank.org/indicator/SE.ADT.LITR.ZS

million young people around the globe ages 15 to 24 are unable to read or write a sentence (UNESCO, 2017).

ARMED CONFLICT AND POVERTY

War kills. War also destroys homes and communities, displaces populations, and strips people of the basic means to feed, clothe, and care for their families. In understanding global poverty and inequality, it is important to consider the relationship between armed conflict and the economic distress of families and countries. We look at two aspects of this relationship next. First, we consider the consequences of conflict and poverty for girls and women in the Middle Eastern country of Yemen. Second, we look at war refugees and the echoing economic effects in countries around the globe.

Child Brides in a Time of Crisis

According to the Population Reference Bureau (2015), early marriage for girls is on the decline. While many young women in the developing world are still marrying by age 18, there are far fewer child brides now than in the past. For example, between 1992 and 2012, the proportion of girls married by age 15 in the country of Niger fell from 47% to 28%. In Bangladesh, between 1993/1994 and 2011, the proportion dropped from 47% to 29%. In Ethiopia, the figure in 2011 was 16%, a small decline from the 2000 figure of 19%. Declines follow other trends in the empowerment of women and girls, including rising numbers of girls in school and greater opportunities for women to contribute to the paid workforce or to undertake entrepreneurial projects in their communities.

According to United Nations Children's Fund (UNICEF, 2019), child marriage is on the decline: Over 25 million child marriages have been prevented within the past decade, equating to a 15% decrease. Today, there are more than 650 million living women who were married as children, and 31% of young women and 4% of young men were married before their 18th birthday. It is estimated that 12 million girls under age 18 are married each year. At this rate, it is estimated that more than 150 million additional girls under age 18 will be married by 2030.

Positive trends in empowerment of women and girls, however, are often fragile. Armed conflict is among the factors that can derail progress. In Yemen, early marriage has long been a tradition. According to a recent *Washington Post* report, however, "before the civil war began last year, international and local activists had made progress toward ending the practice. They were campaigning for a law setting 18 as the minimum age for marriage and for girls to remain in school" (Raghavan, 2016, p. A1). The war, however, has set back these efforts. As of 2020, no law that regulates the minimum age for marriage in Yemen and the absence of any such law and the ongoing war have perpetuated

Yemeni women show their support for a proposed law banning marriage for girls under the age of 17. Because the country has no minimum age of marriage, girls as young as 8 may become wives before they even reach adolescence.

this practice: "In 2017 the UN's Office for Coordination of Humanitarian Affairs (OCHA) reported that 52% of Yemeni girls and women had gotten married before the age of 18" (Mahdi, 2020).

Yemen is the Middle East's poorest country: The per capita GNI is only $2,380 (Population Reference Bureau, 2018). The spread of armed conflict has deprived many families who had little to begin with of their homes and possessions. In this situation, some families have sought to acquire resources and pay debts—as well as reduce family expenses—by giving young daughters in marriage for a bride price.

The costs of early marriage for girls are manifold. At 164 per 100,000 births, Yemen has one of the world's highest maternal mortality figures, a statistic that can be tied to premature marriage and motherhood (WHO, UNICEF, UNFPA, World Bank Group, & the United Nations Population Division, 2019). Girls as young as 12 have died in childbirth in Yemen (Raghavan, 2016). Young wives are also unlikely to attend school. Yemen's ratio of school-age girls in school—only 40 of 100—is one of the lowest in the region (Population Reference Bureau, 2015).

Using our sociological imaginations, we can see that personal troubles—family poverty, young girls being forced to marry—are more fully illuminated when we see them in the context of public issues such as armed conflict and mass displacement of families and communities.

Refugees and Refuges

As we saw in the opening story, millions of refugees and migrants have fled war and poverty in the Middle East, Africa, and Asia. Many have sought to make their way to Western countries such as Germany, France, and Sweden; about a million Syrian refugees currently reside there (Connor, 2018). At the same time, it is important to recognize that in most cases, refugees and migrants from war-ravaged and destitute countries are fleeing to and finding refuge in other struggling states.

Consider the war in Syria, which has pitted the sitting authoritarian ruler, Bashar al-Assad, against both anti-Assad rebel groups and radical Islamist groups, including the Islamic State (or ISIS), who are fighting for power. According to a Pew Research report,

> Nearly 13 million Syrians are displaced after seven years of conflict in their country—a total that amounts to about six-in-ten of Syria's pre-conflict population. No nation in recent decades has had such a large percentage of its population displaced. (Connor, 2018, para. 1)

About half of the displaced Syrians remain in their own country; most of the rest reside in Turkey, Iraq, Jordan, and Lebanon. The mass influx of refugees has strained already scarce economic resources in these countries.

A similar situation exists in Africa. Armed conflicts in the Congo, Somalia, Sudan, and South Sudan have left about 6.6 million people displaced in sub-Saharan Africa (Mohamed & Chughtai, 2019). Most of these refugees either remain in their own countries or find shelter in neighboring states (see Figure 8.2), which include some of the globe's poorest countries (Pecanha & Wallace, 2015). For instance, the country of Chad is home to an estimated 360,000 refugees from Sudan, many of whom have been in the country for over a decade. Chad itself is very poor: It ranks 184th of 187 countries in the United Nations Human Development Index. It also strains beneath the weight of regional tensions, including political chaos in Libya and terrorist threats from the Nigerian-based terror group Boko Haram. Chad's economic struggles are shared by both Chadian citizens and the refugees within their borders (Boyce & Hollingsworth, 2015).

As of 2019, Uganda was home to the most African refugees: 1.4 million (Mohamed & Chughtai, 2019). Most are from South Sudan. As home to over 270,000 refugees, the Bidi Bidi camp in northern Uganda is the world's largest refugee camp. Refugees are drawn by Uganda's generous laws, which

FIGURE 8.2 ■ Refugees and Displaced Persons in Africa, Selected Countries

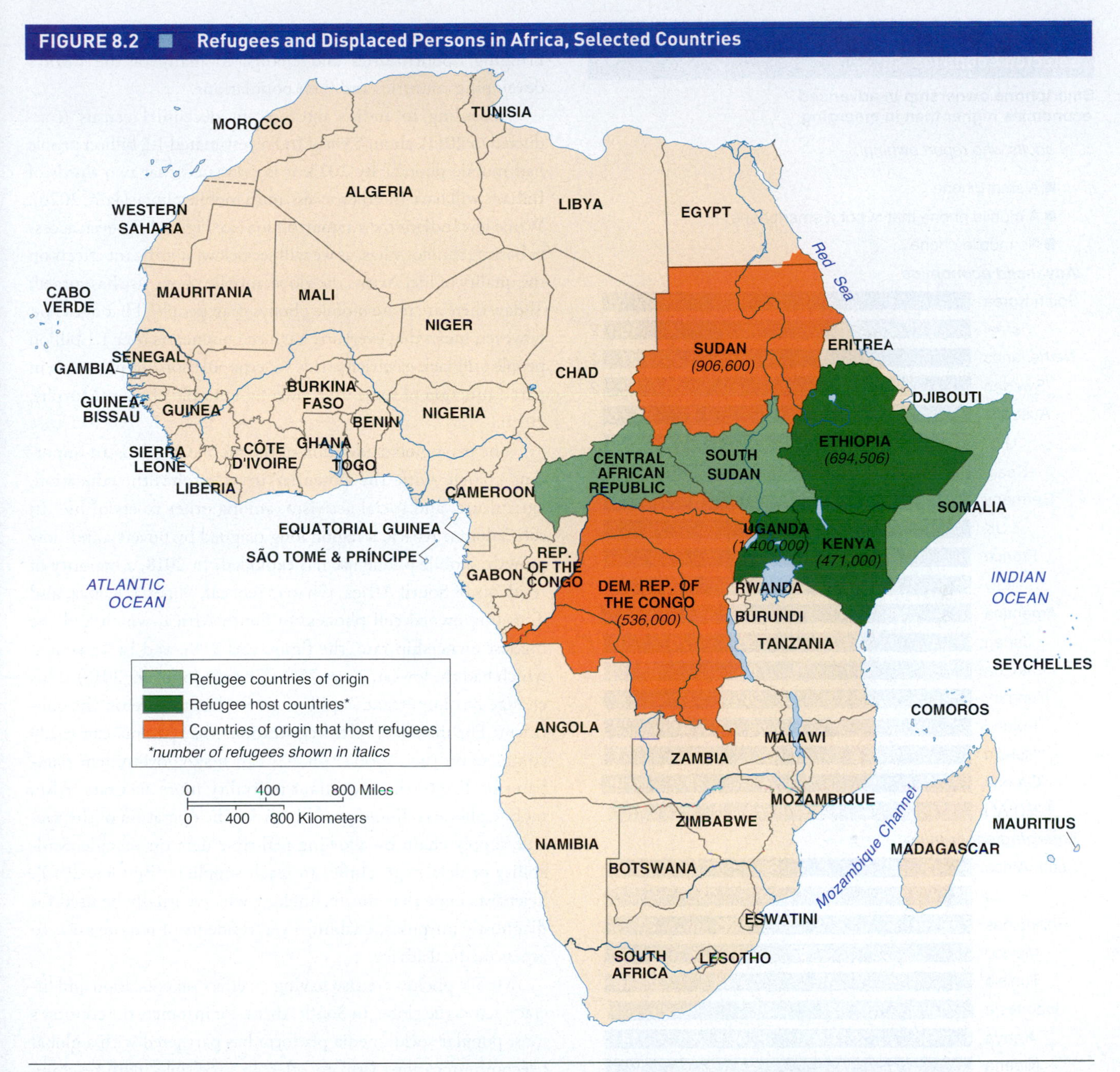

Source: Based on Mohammed, H., & Chughtai, A. (2019, February 9).

permit refugees to work and access public services such as education (Quartz Africa, 2017). The influx, however, has strained the resources of a country that is already among the world's poorest, with a GNI-PPP of $1,780 and its own very young population: Those age 15 and under comprise fully 47% of the population (Population Reference Bureau, 2020).

TECHNOLOGY: THE GREAT EQUALIZER?

As we have seen, global inequality is manifested today in a broad spectrum of ways, ranging from a lack of income and food to little access to good education and sanitation. At the same time, new technologies have emerged whose adoption has not been limited to the well-off. Access to the Internet is growing across the globe: In 2017, just about half of the world's population was using the Internet—up from just over 20% in 2007 (The World Bank, 2019).

FIGURE 8.3 ■ Smartphone Ownership Growth, Selected Countries

Smartphone ownership in advanced economies higher than in emerging

% of adults who report owning ...

- A smartphone
- A mobile phone that is not a smartphone
- No mobile phone

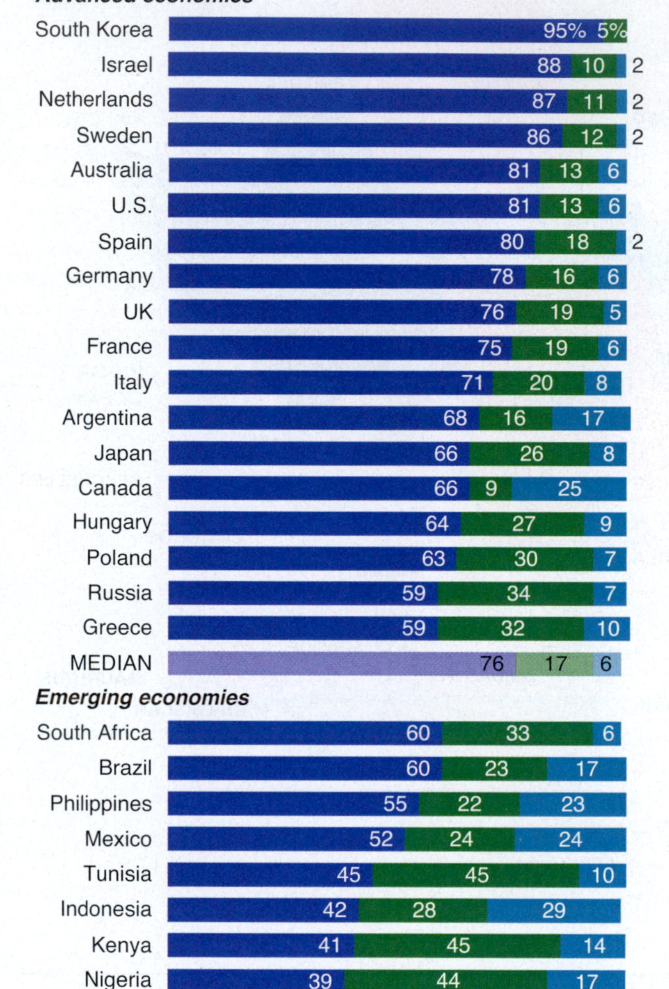

Source: Silver, L. (2019, February 5). *Smartphone ownership is growing around the world, but not always equally.* Pew Research. https://www.pewresearch.org/global/2019/02/05/smartphone-ownership-is-growing-rapidly-around-the-world-but-not-always-equally/

Note: Some totals do not exactly equal 100 due to rounding.

Consider as well the mobile phone, a technology that is bringing opportunities and change to many of the world's developing countries and their populations.

According to India's most recent decennial census (conducted in 2011), about 53% of India's estimated 1.2 billion people had mobile phones. By 2023, it is estimated that two thirds of Indians will have Internet access and a mobile phone (Jain, 2020). While few Indians own a smartphone (see Figure 8.3), even access to basic technology has, as we will see below, significant effects on the quality of life. Across the globe, mobile phone use has surged: Today, there are more mobile phones than people! That does not, however, mean that everyone has a cell phone: As over 1.1 billion people still lack electricity, this "overpopulation" of phones is in part a function of some users owning multiple phones (Murphy, 2019).

The mass global adoption of the mobile phone is an important change with the potential to affect health, education, agriculture, and social activism, among other facets of life. In sub-Saharan Africa, a region long plagued by poverty and slow growth, mobile phone use has exploded: In 2018, a majority of residents of South Africa, Ghana, Senegal, Nigeria, Kenya, and Tanzania owned cell phones. In South Africa, which had the highest ownership rate, the figure was 91%, and in Tanzania, which had the lowest, it was 75% (Silver & Johnson, 2018). This change has important implications for health care on the continent: For instance, birth registrations that parents can easily complete on their mobile phones can make government databases on live births and infant mortality more accurate. Also, mobile phone technology can improve the operation of the vaccine supply chain by allowing real-time data on vaccine availability or deficits in clinics to reach suppliers (Berkley, 2013). Scientists hope that this technology will eventually be used for diagnostic purposes, enabling even residents of remote areas to access medical advice.

Mobile phones are also having an effect on education and literacy across the globe. In South Africa, for instance, the country's most popular social media platform has partnered with a global telecommunications firm to offer an accessible math-teaching tool to users, and it is hoped that such technology can eventually be used to offer a wide variety of lessons to students (Ogunlesi & Busari, 2012). No less important, access to mobile phones is providing access to books for both new and advanced readers. As a 2014 UNESCO report notes, "Many people from Lagos to La Paz to Lahore . . . do not read for one reason: they don't have books" (West & Chew, 2014, p. 13). The report goes on to observe that "a well-respected study of 16 sub-Saharan African countries found that most primary schools have few or no books, and in many countries these low levels are not improving" (West & Chew, 2014, pp. 13–14). This lack of books may compromise reading acquisition and have longer-term academic consequences. As well, mobile phone technology offers opportunities to read—whether for pleasure, formal education, children, or news—to those who may lack the resources to obtain texts for reading in other ways. If 6 billion of the world's 7 billion people can now access working mobile phones, a new world of literacy may be at their fingertips.

Even agriculture in the developing world, much of which is still concentrated in small family farms, is being affected by mobile technology. In countries where agriculture is still a primary sector of the economy, technology has the potential to affect both micro- and macro-level economic indicators. According to a CNN report on mobile phones in Africa, "By serving as platforms for sharing weather information, market prices, and micro-insurance schemes, mobile phones are allowing Africa's farmers to make better decisions, translating into higher-earning potentials." Farmers with cattle, for instance, can use the mobile app iCow to track cows' gestation periods and to learn additional information about breeding and nutrition (Ogunlesi & Busari, 2012).

Mobile phone technology has begun to reach even the remotest parts of the globe, bringing new opportunities for contact and commerce as well as activism, education, and health care.

©Tom Gilks / Alamy Stock Photo

Finally, technology has brought to the developing world a new platform for social activism. Citizen mobilization against crime, corruption, violence, and other social ills has been fostered by mobile technology that rapidly passes information across communities. Mobile phones have played important roles in many events, from Kenya in 2008, where citizens used phones to report violent incidents in the country's disputed elections, to the citizen uprisings against authoritarian regimes in North African countries such as Tunisia in 2011 to Nigeria in 2014, where the viral Twitter campaign #BringBackOurGirls brought global attention to the plight of more than 300 girls kidnapped by the radical Islamist organization Boko Haram.

Can mobile phone technology contribute to bringing greater prosperity—and equality—to the planet? How might mass adoption of mobile phones contribute to addressing some of the inequalities described in this chapter?

THEORETICAL PERSPECTIVES ON GLOBAL INEQUALITY

In this section, we analyze global inequality from several theoretical perspectives, raising the questions of what explains global inequality and why it exists and persists. Later, we take a critical look at the theories.

Modernization theory: A market-oriented development theory that envisions development as evolutionary and guided by modern institutions, practices, and cultures.

Modernization theory is *a market-oriented development theory* associated with the work of Walt Rostow (1961) and others. In contrast to many perspectives on stratification, the modernization perspective asks not why some countries are poor but why some countries are rich. In asking this question, it makes the assumption that the historical norm in states has been poverty; that is, the populations of most countries at most times have subsisted rather than prospered. The answer it proposes is that affluent states have modern institutions, markets, and worldviews; by *modern*, Rostow meant those that emulate the democratic and capitalist states of the West. He argued that economically underdeveloped states can progress if they adopt Western institutions, markets, and worldviews. Rostow used the analogy of an airplane taking flight to illustrate his key ideas about the stages of development:

- **The traditional stage:** In this pre-Newtonian (that is, prescientific) stage, societies are present- and past-time oriented, looking back into history for models of economic and political behavior rather than looking forward and seeking new models. They embrace tradition over innovation. Economic development is limited by low rates of savings and investment and by a work orientation that elevates subsistence over ambition and prosperity. The airplane is grounded and has not yet begun its journey to affluence. Today, few such countries exist. One might look at individual communities within developing countries to find these traditional orientations.

In some developing countries, however, traditional beliefs about women's roles hinder their educational attainment and access to the labor market. Arguably, this is a cultural norm that also stifles national development, as it keeps a segment of the population that could potentially constitute half the workforce (women) from contributing its talents and skills.

- **The takeoff stage:** In this stage, societies are moving away from traditional cultural norms, practices, and institutions and are embracing economic development with a sense of purpose and increasing practices of savings and investment. The plane rises as the weight of tradition is cast overboard in favor of modernity.

Rostow, an originator of modernization theory, was an adviser to President John F. Kennedy, whose administration was responsible for the development of the Peace Corps. The Peace Corps, which has for decades sent young U.S. workers abroad to spread innovations in agriculture and technology, to teach English, and to train leaders in developing countries in methods of modern governance, could be seen as a vehicle for moving countries from the traditional to the takeoff stage. Today, some African countries with modernity-oriented leadership, such as Liberia, might be categorized as members of the takeoff group.

- **In flight with technological progress and cultural modernity:** In this next stage, as the plane moves forward, technology is spreading to areas such as agriculture and industry, innovation is increasing, and resistance to change is declining. Many people are adopting modern cultural values, and governance increasingly reflects the rule of law. Advanced countries facilitate these processes by offering advice and money.

Progress may take the form of industrialization, which drives greater urbanization as rural dwellers leave poor agricultural areas to seek their fortunes in cities. It may also be accompanied by lower fertility, driven by the increased use of contraception as opportunities for women grow in education and the labor market. India might be considered a modern example of this stage, as it has a growing educated middle class, rising urbanization and industrialization, and (for better or worse) soaring consumer ambitions.

- **The stage of high mass consumption and high living standards:** In this stage, there is a greater emphasis on the satisfaction of consumer desires, as new affluence expands the ranks of those with disposable income. This is the stage that advanced countries such as those of Western and Northern Europe, the United States, Israel, and Japan have reached.

As these stages suggest, modernization theory assumes we can understand a given state's level of development by looking at its political, economic, and social institutions and its cultural orientation. That is, the theory uses a country's *internal variables* as key measuring sticks. In contrast, two later theories take a conflict perspective, focusing on countries' conflicting interests, unequal resources, and exploitative relationships, though they emphasize different aspects of inequality.

Dependency theory: The theory that the poverty of some countries is a consequence of their exploitation by wealthy states, which control the global capitalist system.

Just as Marx posited a fundamentally exploitative relationship between the bourgeoisie and the proletariat, so too does **dependency theory** (Emmanuel, 1972; Frank, 1966, 1979; Ghosh, 2001), which argues that *the poverty of some countries is a consequence of their exploitation by wealthy states, which control the global capitalist system.* While exploitation originated in colonial relationships, when powerful Western states such as Britain, the Netherlands, and Belgium dominated countries such as the Congo, South Africa, and India, it continues through the modern vehicle of multinational corporations that reap great profits from the cheap labor and raw materials of poor countries while local populations draw only bare subsistence from their human and natural resources.

Dependency theory draws its name from the idea that prices on the global market for human and natural resources held by poor states are intentionally kept low to benefit high-income states, so low-income states cannot fully develop industrially, technologically, or economically. Thus, these states remain in a *dependency relationship* with the well-off states that buy and exploit their labor and raw materials. Whereas modernization theory implies that high-income states want to encourage the full development of low-income countries, dependency theory suggests there is a direct relationship between the affluence of one and the poverty of the other.

World systems theory: The theory that the global capitalist economic system has long been shaped by a few powerful economic actors, who have constructed it in a way that favors their class interests.

World systems theory shares some of these basic ideas. Immanuel Wallerstein (1974, 1974/2011a, 1980/2011b, 1989/2011c, 2011d), one of the pioneers of the theory, argues that *the global capitalist economic system has long been shaped by a few powerful economic actors, who have constructed it in a way that favors their class interests.* He suggests that the world economy is populated by three key categories of countries:

- **Core countries:** The core countries are economically advanced, technologically sophisticated, and home to well-educated, skilled populations. They control the vast majority of the world's wealth and reap the greatest benefits from the world economic order, including trade and production practices. They include the United States, Canada, the states of Western and Northern Europe, and Japan, among others.
- **Peripheral countries:** The peripheral states have low national incomes and low levels of technological and industrial development; many still depend on agriculture. They have been exploited by the core states for their cheap labor (and, historically, slave labor) and for cheap raw materials that are exported to advanced countries and made into finished goods that bring far greater profit to core companies and consumers. Peripheral countries include parts of Central and Latin America, Asia, and many of sub-Saharan Africa's states. Some of those in Africa provide the critical mineral components of modern electronics for which consumers pay top dollar, such as smartphones and iPads, but they still suffer dire poverty.
- **Semiperipheral states:** The semiperiphery shares some characteristics with both the core and peripheral states, occupying an intermediate and sometimes stabilizing position between them. Semiperipheral states such as China, India, and Brazil may be exploited by core states, but they may in turn have the capacity to exploit the resources of peripheral states. For example, China, which has advanced industrial capacity and a growing middle class of consumers, has begun to foster economic relationships with African countries that can offer oil resources for the populous and economically growing state.

World systems theory sees the world as dynamic rather than static, with peripheral and semiperipheral states seeking to rise in the ranks and core countries attempting to hold fast to global power. The key unit of analysis in world systems theory is less about individual countries (as it is in modernization theory) than it is about relationships between countries and regions of the world. Similar to dependency theory, world systems theory sees relationships between states, such as those between core and periphery states, as fundamentally exploitative; that is, some countries benefit to the detriment of others.

Next, we use these perspectives to examine the case of Nigeria, a developing African country. Application of the perspectives will help us to assess their utility as analytical tools for understanding development and global inequality.

Applying the Theories: The Case of Nigerian Oil Wealth

A *National Geographic* story on the Niger Delta begins like this:

> Oil fouls everything in southern Nigeria. It spills from the pipelines, poisoning soil and water. It stains the hands of politicians and generals, who siphon off its profits. It taints the ambitions of the young, who will try anything to scoop up a share of the liquid riches—fire a gun, sabotage a pipeline, kidnap a foreigner.
>
> Nigeria had all the makings of an uplifting tale: [A] poor African nation blessed with enormous sudden wealth. . . . By the mid-1970s, Nigeria had joined OPEC (Organization of Petroleum Exporting Countries), and the government's budget bulged with petrodollars. (O'Neill, 2007)

Using the case of Nigeria and its vast oil reserves in the southern Niger Delta, we can evaluate the theories we have described and compare how well they illuminate the case of Nigeria, a country with a per capita GNI-PPP of only $5,680 (Population Reference Bureau, 2018).

Recall that the modernization perspective highlights internal variables such as the lack of modern state, economic, and legal institutions and inadequately modern cultures to explain why some countries have lagged in development. A modernization theorist would thus point to the rampant culture of corruption and lack of rule of law that have characterized countries such as Somalia and North Korea, which Transparency International (2011), a corruption watchdog agency, has ranked as the most corrupt countries in the world. Nigeria is also near the top of the agency's list. Clearly, there are links between state corruption, the lack of an effective legal and civic structure, and the dire poverty in and around Port Harcourt. Though it is the capital of Nigeria's oil-rich Rivers state, Port Harcourt has "no electricity, no clean water, no medicine, [and] no schools" (O'Neill, 2007). But does the modernization perspective miss some key aspects of the problem of global poverty?

Critics argue that in attending almost exclusively to internal variables, the modernization perspective fails to recognize external obstacles to development and the ways in which well-off states benefit from the inferior economic position of poor states. According to *National Geographic*, in the wake of independence from colonial Britain, few observers expected that Nigeria would become a global oil source (Figure 8.4). In the decades that followed, however, five multinational oil companies—Royal Dutch Shell, Total, Italy's Agip, Exxon-Mobil, and Chevron—transformed the Rivers state. "The imprint: 4,500 miles . . . of pipelines, 159 oil fields, and 275 flow stations" (O'Neill, 2007). This massive oil infrastructure continues to leave a significant environmental footprint in the area. According to Amnesty International (2015),

> Royal Dutch Shell and the Italian multinational oil giant ENI have admitted to more than 550 oil spills in the Niger Delta last year, according to an Amnesty International analysis of the companies' latest figures. By contrast, on average, there were only 10 spills a year across the whole of Europe between 1971 and 2011. (para. 1)

The United Nations Development Programme and the International Crisis Group point to decades of problematic economic strategies employed by oil companies, which have taken advantage of weak environmental controls, offered little compensation for land and few employment opportunities to local communities, engaged in corrupt deals for oil, and used private security forces to commit violence against those who resisted their efforts to control the oil fields of the Niger Delta (Brock & Cocks, 2012; O'Neill, 2007).

From the dependency and world systems perspectives, a relationship of fundamental exploitation exists between Nigeria and high-income countries, including the United States and Britain, for which

Port Harcourt, located in Nigeria's Rivers state, is a key exporter of crude oil. Despite the region's valuable natural resources, many of its citizens still face extreme hardship (including poverty) and environmental threats.

©REUTERS/Akintunde Akinleye (left); ©Friedrich Stark / Alamy Stock Photo (right)

FIGURE 8.4 ■ Where the Oil Is in Africa

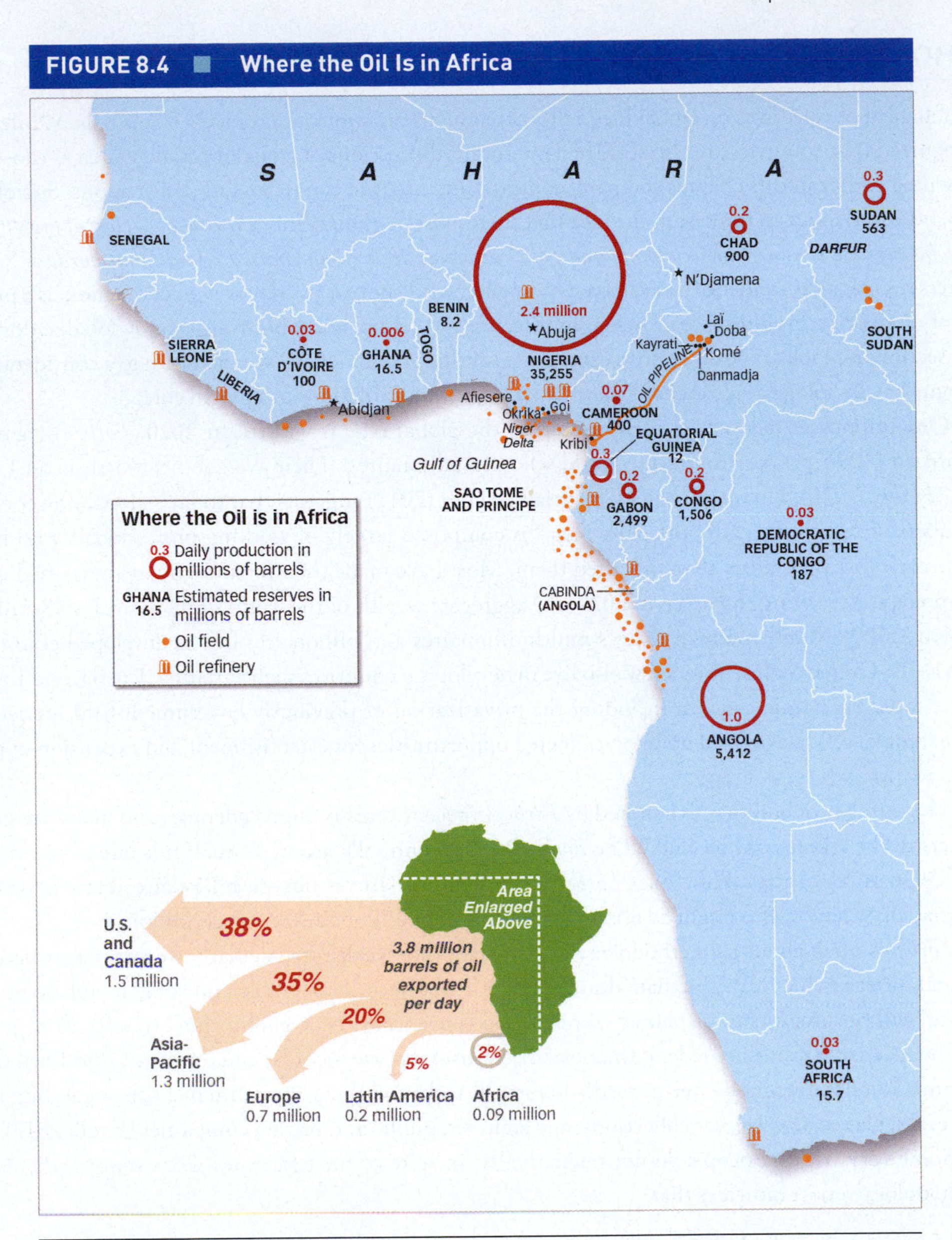

Source: Adapted from Mappery.com

Nigeria provides a critically important resource. If the United States is a core state, Nigeria appears from this perspective a peripheral state supplying oil, the basic raw building block of modern economies, without seeing the benefit of its own natural wealth. Semiperipheral states aggressively seeking to develop their own economies and wealth are also part of the picture: "China, India, and South Korea, all energy-hungry, have begun buying stakes in Nigeria's offshore [oil] blocks" (O'Neill, 2007). The dependency theory perspective suggests that developed and rapidly developing states benefit from lax government oversight of environmental pollution, low-wage pools of local labor, and corruptible officials willing to bend rules to accommodate corporate wishes.

Similar to modernization theory, the dependency and world systems theoretical perspectives illuminate some aspects of the case while obscuring others. Variables such as the exploitative power of Western oil companies are a key part of understanding the failure of Nigeria to develop in a way that benefits the broader population, but conflict-oriented perspectives pay little attention to the agency of poor states and, in particular, their governing bodies in setting a solid foundation for development.

WHO ARE THE GLOBAL ELITE?

Global elite: A transglobal class of professionals who exercise considerable economic and political power that is not limited by national borders.

Sociological perspectives on global inequality often focus on *countries* as objects of analysis. While it is recognized that countries are also stratified internally, discussions of global inequality such as those we presented earlier in this chapter compare, contrast, and analyze countries and their regions. Sociology also, however, takes an interest in the idea that there exists a **global elite**, a *transglobal class of professionals who exercise considerable economic and political power that is not limited by national borders.* Some writers suggest that an identifiable global power elite, while not an entirely new phenomenon, is a product of modernity, brought into growing significance by technological innovation and globalization. In this section, we look at the descriptive dimensions of this phenomenon. We then turn to a consideration of some key sociological ideas about the characteristics and functions of a global elite.

One important measure of membership in the global elite is wealth. In 2020, *Forbes* magazine identified 2,095 people around the world who are billionaires: Their average net worth is $4.1 billion (*Forbes*, 2020). Financial writer Chrystia Freeland (2012) suggests that today's global power elite, members of which she calls "the plutocrats," is composed largely of working professionals who have made their fortunes rather than inherited them. Most have made their fortunes in business, media, or technology. According to *Forbes*, in 2020, the aggregate wealth of the world's billionaires is $8 trillion (Dolan, 2020). While many of today's multimillionaires and billionaires live in developed countries such as the United States, quite a few also live in developing countries such as China, Russia, and India, where rapid economic growth, including the privatization of previously government-held industries (for example, oil) has opened up unprecedented opportunities for establishment and expansion of personal wealth.

The number of billionaires counted by *Forbes* in recent years is unprecedented, and many are newcomers to the "three comma club." The *Forbes* list is an annually awaited ritual: It is one of the magazine's signature features. How does *Forbes* know who is—and is not—a billionaire? How is wealth calculated? What or who might be missing in the count of billionaires and their billions?

Forbes is open about its methodology and the fact that wealth (particularly wealth that is tied up in stock prices, which can fluctuate dramatically) is not necessarily stable and some members of the group "will become richer or poorer within weeks—even days—of publication" (Kroll, 2018, para. 8). The magazine examines federal financial disclosures, course records, and Securities and Exchange Commission documents, as well as media reports. It seeks to value a spectrum of assets, including real estate, art, planes, jewelry, car collections, and stakes in public and private companies (Kroll, 2013).

Some very wealthy people do not make the list in spite of their fortunes. For example, the 2014 methodology report indicates that

> [*Forbes* does] not include royal family members or dictators who derive their fortunes entirely as a result of their position of power, nor do we include royalty who, often with large families, control the riches in trust for their nation. Over the years *Forbes* has valued the fortunes of these wealthy despots, dictators and royals but have listed them separately as they do not truly reflect individual, entrepreneurial wealth that could be passed down to a younger generation or truly given away. (Kroll & Dolan, 2014)

Consider the case of Russian president Vladimir Putin. According to a 2018 media report, "Vladimir Putin earned 38.5 million rubles (roughly $673,000) between 2011 and 2016, according to information publicly released by the Central Election Commission of the Russian Federation . . . giving the Russian president an average yearly salary of about $112,000" (Taylor, 2018, para. 1). However, the same article suggests that Putin is likely a multibillionaire and a second article asked if Putin may secretly be the richest man in the world because of his investments in highly valued Russian companies in the oil sector and ownership of one—or many—lavish palaces (Wile, 2017).

There are billionaires who do not make the list because they work hard to maintain their financial and personal privacy. After all, great wealth may invite unwanted attention from those seeking support or donations or from those wishing to exploit or steal another's fortune. There are significant sums of global cash stashed away in countries such as Switzerland, the Cayman Islands, Belize, and the

Bahamas, where they can be hidden from the gaze of national tax collectors (Shapiro, 2013). Wealth can also be held under alternative names and shell companies in complex financial arrangements that enable one to keep wealth under wraps.

There is significant interest in wealth and the wealthy. Great riches confer celebrity, and some billionaires seek attention while others avoid it. All, arguably, seek to keep as much of their money as possible. Consequently, counting the wealth of the billionaire club is a challenge.

In the context of a growing global list of billionaires, being a millionaire looks more commonplace than extraordinary, though millionaires are certainly few in number in the context of the global population of over 7.8 billion (as of mid-2020). The United States is home to a substantial proportion of the world's multimillionaires: In 2018, it had about 47,127 individuals worth $30 million or more. By comparison, Japan had 18,534; Brazil had 3,754; and Kenya had 125 (Knight Frank Research, 2019). According to the Boston Consulting Group (which tracks global assets), most of the wealth has grown due to a rising stock market and increases in asset prices rather than from newly created businesses (Frank, 2015).

While there is unprecedented national diversity on the current roster of the rich and the superrich, there has also been a marked rise in the gap between the world's economic elites and everyone else. According to a report by the development organization Oxfam, nearly half of the world's wealth is owned by 1% of the globe's population (Hardoon et al., 2016).

Seeking to outline a definition of the *global elite*, David Rothkopf (2008) suggests that

> the distribution of power has clearly shifted, not just away from the United States and Europe, but away from nations. . . . Had [sociologist C. Wright] Mills been writing today, he would have turned his attention from the national elite in America to a new and more important phenomenon: the rise of a global power elite, a superclass that plays a similar role in the hierarchy of the global era to the role that the U.S. power elite played in that country's first decade as a superpower. (p. 9)

Rothkopf emphasizes the idea that the global elite is powerful not only because it is rich, but because it is influential in political decision making, global markets and industries, technological innovation, the production of cultural or intellectual ideas, and even world religions. The global elite and its decisions have impacts on the lives of thousands or even millions of citizens, consumers, workers, and worshippers. The global power elite, Rothkopf argues, includes corporate executives, presidents and prime ministers of powerful states, technological innovators, those who control flows of global resources such as oil, media moguls, some military elites, and a handful of well-known and active cultural and religious figures. Their power is multiplied by the fact that they are deeply networked, sharing links to other members of the elite through both personal and professional ties. Their exercise of power is not always direct, but their influence is palpable in politics, economics, and media, among other areas.

Sociologist Leslie Sklair (2002) conceptualizes the notion of a modern global elite in his writings on "transnational capitalism." His work highlights the position that nations and borders are of declining significance in capitalist globalization. He argues that important objects of analysis in a globalizing world are what he terms *transnational practices*—"practices that cross state borders but do not originate with state actors, agencies, or institutions" (p. 10). He suggests that understanding the modern economic order requires recognizing how power has become transnational rather than limited by national borders. As part of his examination of transnational practices, Sklair theorizes the rise of a *transnational capitalist class* that is composed not of capitalists (that is, the bourgeoisie) in the classical Marxian sense but of four categories of members: (1) a *corporate fraction* drawn from transnational corporations, (2) a *state fraction* composed of global political elites, (3) a *technical fraction* representing globalizing professionals, and (4) a *consumerist fraction* made up of executives of marketing and media.

While Sklair's theorized class would appear to encompass a broader swath of members than Rothkopf's (2008) global elite, his characterization of the transnational capitalist class reflects an idea shared with Rothkopf that global economic integration and the mobility of capital have fostered the birth of a transglobal class. The members of this class have the following characteristics:

- They share global (not only local) interests and perspectives as well as consumer lifestyle choices.
- They seek to exercise control or influence over key political, economic, and cultural–ideological processes on a global level.
- They hail from different national backgrounds, but they see themselves as citizens of the world rather than citizens of particular states.

Freeland (2012) echoes the last point in her journalistic account of the global elite, noting that members often feel they have more in common with their fellow elites than with their countrymen. She points out that this, however, may not be a new development: She quotes an early theorist of capitalism, Adam Smith, who wrote in 1776, "The proprietor of land is necessarily a citizen of the particular country in which his estate lies. The proprietor of stock is properly a citizen of the world, and is not necessarily attached to a particular country" (p. 67).

The transnational capitalist class is a global elite that exercises political, economic, and cultural–ideological power and acts to organize "conditions under which its interests and the interests of the global system . . . can be furthered" (Sklair, 2002, p. 99). Sklair also theorizes a particular function for the transnational capitalist class: It is a vehicle by which *transnational corporations,* modern and powerful economic entities, expand and legitimize the consumerist culture and ideology that are needed to sustain the global system of capitalist production.

In an interesting variation on the conceptualization of globalized classes, Zygmunt Bauman (1998) writes of a *space war.* Bauman's analysis points not toward the cosmos but toward modern citizens' relationship to physical spaces and places, including countries. Bauman posits that the "winners" of globalization (and the space war) are those who are mobile, who can move through space to create value and meaning—he calls them *tourists* (though he does not mean that in the strictly conventional sense). They can move across the globe, enabled by transportation and communication technologies and their economic and professional resources. They exist, suggests Bauman, in time rather than in space because space is not constraining to them. What Bauman is describing is a globalized category of the world population whose mobility is enabled by education, economic resources, and social networks. For example, business professionals, high-level government bureaucrats and representatives, and cultural elites and celebrities have the means to seek out both personal pleasures and professional opportunities globally.

By contrast, the "losers" of modernity are those who are tied to places and spaces that have been devastated by globalization. Lacking resources, they are rooted in place and denied geographic and economic mobility. In thinking about this category, one might consider diverse groups, ranging from former automobile plant workers stripped of their livelihood by globalization and stuck in economically devastated cities, to poor urban dwellers in the developing world whose lives revolve around difficult and dangerous low-wage jobs that they cannot afford to lose. Among the losers of a globalizing world are also, Bauman points out, those who are on the move, but they differ from the mobile global elite; rather, they are what he terms *vagabonds,* moving across the globe as refugees or poor economic migrants, unable to live in their own countries and unwanted elsewhere.

Bauman does not specify the precise qualities of his loosely defined global classes, but his work points to the idea that globalization has created different experiences for different groups that are not easily characterized through reference to national boundaries. This perspective may lead us to consider global inequality as a modern phenomenon that crosses borders and shapes new opportunities as well as obstacles.

WHAT'S NEXT? GLOBAL INEQUALITY AND COVID-19

In this chapter, we have examined the phenomenon of global inequality. Inequality manifests in a variety of ways, including access to education, basic hygiene, health care, safe housing, and technology. The most common measures of inequality, however, are disparities in economic resources: income and wealth. Data from the period following 2010, some of which we have looked at in this chapter, suggest two key trends: First, the very rich continue to get richer, sometimes much richer. Second, extreme poverty has been on the decline in much of the world. What effects will the COVID-19 pandemic that shook global health, politics, economies, and societies in 2020 have on these trends?

The data from mid-2020 seem unequivocal. The poor around the globe will lose ground. As noted earlier, The World Bank estimates that about 40–60 million people will fall into extreme poverty by the end of 2020. While many of them reside in developing countries like Nigeria, India, and Ecuador, which have struggled to contain and respond to the virus, the United States may also experience a precipitous rise in poverty after years of falling rates. As unemployment rates in the summer of 2020 reached levels unseen since the Great Depression of the 1930s, a growing number of Americans have struggled to meet basic needs for food, shelter, and health care (Kochhar, 2020).

At the other end of the economic spectrum, the world's billionaires, particularly those linked to the technology sector, have seen their net worth skyrocket. By one estimate, on a July day in 2020, Amazon founder Jeff Bezos's wealth reached $189 billion after a single-day surge of $13 billion (Huddleston, 2020). Other billionaires around the world have also seen their wealth grow. They include U.S. billionaires Mark Zuckerberg of Facebook and entrepreneur Elon Musk; China's Pony Ma, who owns a substantial stake in the online gaming company Tencent; and India's Mukesh Ambani, a telecom and technology investor (Mitter, 2020).

What is the outlook for these trends in the post-COVID era? Will economic distress rise around the world? Will this foster social movements for change, disorder, or withdrawal and alienation from institutions? What will the continued expansion of wealth for a tiny global elite mean in this context? Can a profoundly unequal world become more equal in the decades ahead?

WHAT CAN I DO WITH A SOCIOLOGY DEGREE?

Active Understanding of Diversity

The development of an understanding of diversity is central to sociological study. As a sociology student, you will gain knowledge related to the histories, practices, and perspectives of diverse groups and will develop intercultural competence. You will also study and apply theories that lend themselves to the analysis of diversity and ways in which it can underpin societal harmony as well as conflict. The close understanding of diversity can inform important research about societal challenges and potential solutions. Active reflection on this knowledge supports the development of skills to effectively work through and with differences that may divide communities and individuals by race, ethnicity, gender, sexuality, religion, and class, among others. A key part of developing an active understanding of diversity is learning to see the world from the perspectives of others. As a major in the social scientific discipline of sociology, you will be exposed to a breadth of theoretical and empirical work that illuminates the social world from a variety of perspectives and helps you to see how the positions of different groups in the social structure may affect perceptions, practices, and opportunities. You will be well prepared for the diverse and dynamic workplace of the present and the future.

Eduardo Evaristo, English Language Instructional Assistant at Moreno Valley Unified School District

BA, Sociology, University of California, Berkeley

Growing up in a low-income gang- and drug-infested neighborhood, my chances to break out of a too-common cycle of neighborhood and family poverty and violence seemed minimal. I witnessed friends and relatives fall prey to this toxic environment, which was fostered by anger, drugs, and disadvantage. Lost in a system that often criminalizes young men of color, I was set to become another number, with my life choices limited by the available opportunities: engaging in gang-related activities or a life of hard labor.

It was not until I discovered sociology that I was able to conceptualize what I had experienced, allowing me to see that the social conditions of my life were the products of generational inequalities. With this new awareness, I saw that my life experiences could easily repeat themselves, forcing another child to undergo the same trajectory, but I was not just going to stand by and let it happen. With a sociology degree, I believe I have the tools to both understand and lessen the varying effects of neighborhoods' inequalities.

Navigating through the challenges of college was a task of its own, but today, I serve as a bridge between two worlds. As an English language instructional assistant in Moreno Valley Unified School District in California, I work to cultivate a new cycle for a better tomorrow for underprivileged students. By working with them to refine speaking, writing, and reading skills all school year, I help foster better educational outcomes for English learners. My presence and my work with my students also help them to dream beyond the disadvantages into which they were born. Studying sociology has helped me to see more clearly the structural obstacles in the classroom and outside that may confine my students to limited life opportunities. My awareness of the challenges they face only increases my determination not to give up on them and not to lose them to threats like the school-to-prison pipeline.

Through studying sociology, I learned that the cycles of poverty and violence can be shattered when education brings along opportunities to impoverished youth. One becomes cognizant about the social conditions and challenges of these students' lives and also gains the ability to help identify and bring empowerment to dismantle generational cycles of disadvantage.

Career Data: Kindergarten and Elementary School Teachers

- 2019 Median Pay: $59,420 per year
- Typical Entry-Level Education: Bachelor's degree
- Job Growth, 2019–2024: 4% (as fast as average)

Source: U.S. Bureau of Labor Statistics (2020).

SUMMARY

- **Global inequality** can be described in terms of disparities in income, wealth, health, education, and access to safe, hygienic sanitation, among other things. Global gaps in equality between high-income, middle-income, and low-income countries remain substantial, even as some countries are effectively addressing problems such as malnutrition.
- There is an important relationship between armed conflict and poverty at the household, community, and country levels. Conflict exacerbates the problems of poverty and raises the risks of hunger and exploitation as well as the loss of life and property.
- While global inequalities remain substantial, access to mobile technology is improving, as nearly 6 billion of the world's 7 billion people now have access to working mobile phones.
- **Modernization theory** posits that global underdevelopment exists in states that cling to traditional cultures and fail to build modern state and market institutions.
- **Dependency theory** and **world systems theory** highlight external variables that point out how high-income states benefit from the economic marginality of low-income states.
- Sociologists describe and analyze the phenomenon of a **global elite**, a transglobal class of professionals who exercise considerable economic and political power that is not limited by national borders. The global power elite, while not an entirely new phenomenon, is a product of modernity, brought into growing significance by technological innovation and globalization.
- Sklair theorizes a transnational capitalist class that is composed of elites with economic, political, and cultural ideological influence; he asserts that the members of this class organize the global order in a manner that realizes their own interests and contributes to the expansion and legitimation of a global consumerist ideology. Bauman looks at the development of population categories as defined by their relationships to space, in particular their mobility or lack of mobility.

KEY TERMS

dependency theory (p. 202)
global elite (p. 206)
global inequality (p. 191)
gross national income–purchasing power parity per capita (GNI-PPP) (p. 193)
infant mortality rate (p. 194)
modernization theory (p. 201)
total fertility rate (TFR) (p. 194)
world systems theory (p. 203)

DISCUSSION QUESTIONS

1. Why do many of the world's poorest countries have the highest fertility rates? What sociological factors can be used to explain the correlation?
2. How is armed conflict both a consequence and cause of economic scarcity?
3. Can the mass adoption of modern technologies (such as mobile phones) have an impact on poverty in developing countries? What does the chapter suggest? What other effects can you envision?
4. How do modernization theory, dependency theory, and world systems theory explain the existence and persistence of inequality between countries and regions? What are the strengths of these perspectives as analytical tools? What are their weaknesses?
5. What is meant by the term *global elite*? Who are the members of the global elite, and how do they differ from the upper-class elites described in an earlier chapter?

©John Moore/Getty Images

RACE AND ETHNICITY

WHAT DO YOU THINK?

1. What makes a group a minority? Does this term have both qualitative and quantitative dimensions?
2. Why does racial residential segregation exist and persist in many U.S. cities? What are the consequences of racial residential segregation?
3. How is the ethnic and racial composition of the United States changing?

LEARNING OBJECTIVES

9.1 Describe how sociologists understand race and ethnicity.

9.2 Describe different types of minority and dominant group relations in history and today.

9.3 Discuss theoretical approaches to the concepts of *ethnicity, racism*, and *minority status*.

9.4 Define *prejudice, stereotyping*, and *discrimination*.

9.5 Identify major racial and ethnic groups in the United States.

9.6 Define *genocide* and its relationship to national, ethnic, racial, or religious group membership.

In the summer of 2020 the team of NASCAR driver Darrell "Bubba" Wallace found what appeared to be a noose hanging in his No. 43 car's team garage. Wallace, the only Black driver in NASCAR's top tier, received nearly immediate support from dozens of NASCAR drivers, who several days later escorted his car to the front of the Talladega field. Meanwhile, the FBI investigated, ultimately concluding that Wallace was not the target of a hate crime because what the FBI called a "rope fashioned like a noose" had been in the garage long before his team was assigned it (Maxouris, 2020). NASCAR president Steve Phelps, identifying the noose as a "symbol of hate" in a "highly charged and emotional time," said that "our ultimate conclusion for this investigation is to ensure that this never happens again, that no one walks by a noose without recognizing the potential damage it can do" (Barrett, 2020). While the FBI did not classify the incident as a hate crime against Wallace, the presence of a noose in a southern garage raised the specter of the past practice of lynching as a tool of repression and violence against African Americans.

By one estimate, between the years 1882 and 1968, over 4,700 lynchings took place in the United States. *Lynching* is the extrajudicial killing, usually premeditated, of someone suspected of a transgression, though lynchings are also used to intimidate a larger group of people. Most victims of lynching in the United States were Black, though some were whites who were believed by the murderous mobs to be helping or associating with Black residents.

The National Memorial for Peace and Justice in Montogmery, Alabama, reminds us of the social significance of race on the lives of individuals.

©AP Photo/Emilio Morenatti

The majority of lynchings in U.S. history took place in the South. The end of slavery radically transformed the South and created strong resentments against African Americans. Dramatic changes disrupted a long-standing economic order based on forced labor and a social order based on dehumanization. White supremacy in the South embraced a status quo that privileged whites had no tolerance for change. Lynching functioned in this environment as a violent form of social control and a way for whites to reassert their power and position over Blacks (Tolnay & Beck, 1995). Mobs involved in lynching rarely suffered legal or other consequences for their violent actions, which is one reason why they continued for decades.

Notably, it took an African American woman, journalist Ida B. Wells, to bring the evidence of lynching to a worldwide audience. Wells, who herself was born enslaved, bravely traveled to communities where she tallied cases, wrote newspaper articles, published pamphlets, and in 1895, published a tabulated account of the number of lynchings occurring in the United States: *A Red Record: Tabulated Statistics and Alleged Causes of Lynchings in the United States.* Her methods were both qualitative and quantitative:

> Wells visited places where people had been hanged, shot, beaten, burned alive, drowned or mutilated. She examined photos of victims hanging from trees as mobs looked on, pored over local newspaper accounts, took sworn statements from eyewitnesses and, on occasion, even hired private investigators. (Smith, 2018, para. 12)

Wells brought the attention of an international audience and provided strategies for Blacks to fight back. In her 1892 book, *Southern Horrors: Lynch Law in All Its Phases*, she noted:

> I have shown how he may employ the boycott, emigration, and the press, and I feel that by combination of all these agencies can effectually stamp out lynch law, the last relic of barbarism and slavery. The gods help them who help themselves. (p. 48)

Today, the National Memorial for Peace and Justice in Montgomery, Alabama, commemorates the victims of lynching:

> At the center is a grim cloister, a walkway with 800 steel columns, all hanging from a roof. Etched on each column is the name of an American county and the people who were lynched there, most listed by name, many simply as "unknown." (Robertson, 2018, para. 3)

The memorial recognizes the dramatic violence that took many innocent lives. It offers an unflinching look at slavery, lynching, segregation, and today's mass incarceration, but it also offers pathways for the future: The museum exhibit closes with a voter registration booth, information on volunteering, and guidelines for discussing lynching in U.S. history with youth.

Race and ethnicity—and their significance in society—are key issues in sociology. We begin this chapter with a discussion of the sociological definitions of race and ethnicity. We then consider patterns of minority–majority group relations. Next, we look at theoretical perspectives on ethnicity, racism, and minority group status. This leads to a discussion of prejudice, stereotyping, and individual and institutional discrimination. We then briefly highlight the experiences of diverse racial and ethnic groups in the United States and how group membership may shape people's political, economic, and social status. Finally, we discuss genocide as a race-based atrocity that continues to claim new victims in the 21st century.

THE SOCIAL CONSTRUCTION OF RACE AND ETHNICITY

Sociologists Thomas and Thomas (1928) observed that "if [people] define situations as real, they are real in their consequences" (pp. 571–572). The wisdom of their observation is powerfully demonstrated by the way societies construct definitions of race and ethnicity and then respond as though the definitions represent objective realities.

Race

One of the most dynamic areas of scientific research in recent years has been the Human Genome Project. Among its compelling findings is the discovery that, genetically, all human beings are nearly identical. Less than 0.01% of the total gene pool contributes to racial differences (as manifested in physical characteristics), whereas thousands of genes contribute to traits that include intelligence, artistic and athletic talent, and social skills (Angier, 2000; Cavalli-Sforza et al., 1994). Based on this research, many scientists agree that "race is a *social concept*, not a *scientific* one... we all evolved in the last 100,000 years from the same small number of tribes that migrated out of Africa and colonized the world" (Angier, 2000, para. 6).

Race: A group of people who share a set of characteristics (usually physical characteristics) deemed by society to be socially significant.

Sociologists define a **race** as *a group of people who share a set of characteristics (usually physical characteristics) deemed by society to be socially significant*. Notice that this definition suggests that physical characteristics are not the only—or even necessarily the most important—way of defining *races*. For many years, Catholics and Protestants in Ireland defined one another as separate races, and the United States long considered Jews a separate racial category from Europeans, even though distinctive physical characteristics between these groups are in the eye of the beholder rather than objectively verifiable (Schaefer, 2009).

Race and ethnicity are *socially significant* categories, and as such sociologists find them important areas of study. For instance, following the observation of the Thomases, they can conclude that because race is defined as real, it is real in its consequences. Racial differentiation has historically been linked to power: Racial categories have facilitated the treatment (or maltreatment) of others based on membership in given racial groups. Even though sociologists generally agree that races and ethnic groups cannot be objectively, biologically differentiated, they nonetheless gather and analyze statistics and qualitative studies of groups by race and ethnicity. These help to illustrate the historical and contemporary consequences of group membership—from educational attainment to health and well-being to outcomes to poverty and experiences with the criminal justice system. This book features such data throughout its pages.

Ethnicity

Ethnicity: Characteristics of groups associated with national origins, languages, and cultural and religious practices.

Like race, the socially defined category **ethnicity** also has significant consequences for people's lives. Ethnicity refers to *characteristics of groups associated with national origins, languages, and cultural and religious practices*. Although ethnicity can be based on cultural self-identification (that is, one may choose to embrace one's Irish or Brazilian roots and traditions), an acknowledgment or degree of acceptance by a larger group is often necessary. For example, a third-generation Italian American may choose to self-identify as Italian. If she cannot speak Italian, however, others who identify as Italian may not see her as authentically Italian. The sociological significance of belonging to a particular racial or ethnic group is that society may treat group members differently, judging them or giving them favorable or unfavorable treatment based on membership and perceived affiliation.

Minorities

Minorities: Less powerful groups that are dominated politically, economically, and socially by a more powerful group and, often, discriminated against on the basis of characteristics deemed by the majority to be socially significant.

Any racial, ethnic, religious, or other group can constitute a minority in a society. **Minorities** are *less powerful groups that are dominated politically, economically, and socially by a more powerful group and, often, discriminated against on the basis of characteristics deemed by the majority to be socially significant*. Minorities are usually distinguished by physical and cultural attributes that make them recognizable to the dominant group. They are generally fewer in number than the dominant population, though this is not invariably the case. For instance, under *apartheid*, South Africa's system of legalized discrimination that mandated segregation, the ruling whites constituted less than 15% of the population while the Black population constituted well over three quarters of it (Chimere-dan, 1992).

South Africa's apartheid, a legalized system of racial discrimination and enforced segregation, allowed a minority of whites to sit atop a hierarchy of power while the majority of Black South Africans remained at the bottom. It deeply regulated every aspect of daily life including living arrangements, income, marriage, and sexuality. While officially ending in 1994 after many years of struggle, almost 30 years later, the effects still linger (Jones, 2019).

An important aspect of minority group status is that the group has less access than the majority to power and resources valued by society. Thus, sociologically, women in the United States—and in societies worldwide—may be considered a minority due to their lack of political and economic resources relative to men, despite the fact that they typically outnumber men. For example, women own less than 20% of the world's land (Villa, 2017). In short, minority status is determined by access to power, status, and resources, not by the numerical size of a group.

In the United States today, African Americans, Hispanics/Latinxs, Asians, and Native Americans are less numerous than the white majority and are considered minority groups. As racial and ethnic minority groups grow in number in the United States, how do you think that will change their minority group status?

MINORITY AND DOMINANT GROUP RELATIONS

Modern societies are characterized by racial and ethnic heterogeneity, as well as divisions between dominant and minority groups. The coexistence of racial and ethnic groups can be a source of social conflict. Among minorities' most frequently used methods of resolving such conflict are social movements designed to challenge and change existing social relations. In extreme cases, resolution may be sought through revolution or rebellion. Dominant populations respond to such social movements with a variety of social and political policies, ranging from expulsion and segregation to assimilation of minorities and the acceptance of cultural pluralism. We next discuss this spectrum of relationships between majority and minority groups.

Expulsion

Native Americans populated broad swaths of North America when the first European settlers arrived. White settlers and prospectors, inspired by the ideology of *Manifest Destiny*, the belief that they were divinely destined to settle North America, expanded westward, forcing Native Americans from long-held tribal lands. Wars broke out, and Native Americans were driven under military arms to march, sometimes thousands of miles, to areas designated by the government as Native American reservations. Sociologically, *the process of forcibly removing people from a particular area* is called **expulsion**. Today, over 300 reservations continue to exist in the United States (Elliott, 2016). Many are located in remote areas where jobs are scarce, housing is limited, and health care is lacking.

Expulsion: The process of forcibly removing a population from a particular area.

Expulsion continues to be a significant experience for many across the globe who are forced from their homes due to civil wars and ethnic and sectarian conflicts. According to recent figures from the United Nations High Commissioner for Refugees (2020a), at the end of 2019, over 37,000 people around the world daily were forced to flee their homes. This totals 45.7 million people internally displaced, with 79.5 million of them **refugees**, or people who have fled war, violence, conflict, or persecution and have crossed an international border to find safety in another country (United Nations High Commissioner for Refugees, 2020b). Approximately 3 million refugees today live in refugee camps where resources, food, and health care as well as educational opportunities are often sparse. Added to lack of resources, food, and health care, the COVID-19 epidemic also created increasing mental health challenges. As Filippo Grandi, the director of the UN High Commissioner for Refugees, put it in the fall of 2020:

Refugees: People who have fled war, violence, conflict, or persecution and have crossed an international border to find safety in another country.

Segregation: The practice of separating people spatially or socially on the basis of race or ethnicity.

> Many refugees tell us they see their futures crumbling. The issues that drove them from their countries remain unresolved and they can't return home. Many who have survived in exile by eking out a living in the informal economy have lost their jobs. They are also anxious about their health and that of their families, not knowing when the pandemic will end and how they can really protect themselves. They see a lack of solutions and lack hope in the future. (United Nations High Commissioner for Refugees, 2020c)

Thousands of asylum seekers await COVID-19 testing to gain entry to a temporary refugee camp on Greece's Lesbos island after an October 2020 fire destroyed their previous refugee camp of 25,000. Lesbos has received refugees since 1990, but an influx of Syrian refugees from Turkey and continued refugees from the Middle East and Africa have strained the small island, creating tensions with the local population.

©Nicolas Economou/NurPhoto via Getty Images

Segregation

Segregation is *the practice of separating people spatially or socially on the basis of race or ethnicity*. In South Africa, as we have earlier noted, apartheid was state policy until 1994. The release from prison of anti-apartheid activist Nelson Mandela in 1990 and the end of a ban on previously forbidden

In the Jim Crow South, segregation infiltrated virtually every aspect of everyday living, including building entrances, water fountains, and waiting lines. Blacks could not eat in the same restaurants as whites. When they rode the same buses, Blacks were required by law to sit at the back, and on trains, they occupied different cars than white riders. Professional offices required separate waiting rooms for Blacks and whites, and courts required different Bibles for swearing-in witnesses.

political groups began the slow process of desegregation. Apartheid was an extreme form of segregation that included prohibitions not only on where members of different racial and ethnic groups could live but also on where they could travel and at what hours of the day they could be in different parts of particular cities. Although no longer policy, the legacy of apartheid lives on in many South African communities, where the white, Black African, and "Coloured" populations still live separately and unequally. Today, the *rainbow generation,* or "the first cohort of modern black, biracial and other ethnic groups who would not live under the legal and political system crafted by South Africa's white minority," has come of age. In an article documenting their experiences, Rachel Jones (2019) interviews Sibonisile Tshabalala, a 25-year-old recent college graduate with a degree in engineering. Tshabalala describes a system that has improved generationally but still bears the inequality of the past:

> Sibonisile Tshabalala... says she gets a stipend from her contract job with a Johannesburg company. But it's far less that what some white college graduates are paid for staff positions. In some ways, Tshabalala says, post-apartheid is still a struggle. "When I look at my white peers, I realize that as hard as I have worked to gain an education, we did not start from the same playing field. My parents, my grandparents, my great grandparents, they suffered. And the consequences are still coming to me as a young black South African." (paras. 12–13)

Restrictive covenants: Contractual agreements that restrict the use of land, ostensibly in order to preserve the value of adjacent land or a neighborhood.

Segregation was also a feature of life in the United States. Before the 1960s, segregation on the basis of race was a legally upheld, common practice. This practice was maintained, for instance, through **restrictive covenants**, *contractual agreements that restrict the use of land, ostensibly in order to preserve the value of adjacent land or a neighborhood.* These covenants were used to prevent white homeowners from selling their homes to nonwhite buyers (Massey & Denton, 1993). The University of Minnesota's *Mapping Prejudice Program*, a website designed to map disparities in housing segregation that continue to make Minneapolis among the most segregated of U.S. cities today, documents the first "racially restrictive deed" that appeared in Minneapolis in 1910 in this way:

> In Minneapolis, the first racially restrictive deed appeared in 1910, when Henry and Leonora Scott sold a property on 35th Avenue South to Nels Anderson. The deed conveyed in that transaction contained what would become a common restriction, stipulating that the "premises shall not at any time be conveyed, mortgaged or leased to any person or persons of Chinese, Japanese, Moorish, Turkish, Negro, Mongolian or African blood or descent."

The civil rights movement of the 1960s succeeded in securing the passage of federal laws that outlawed segregation such as the Fair Housing Act of 1968. Despite this, high levels of racial residential segregation remain a reality in U.S. cities. Why is this the case today?

First, it is legal and commonplace for housing markets to be segregated by income. Since racial minorities as a group have lower household incomes than do whites as a group, some cannot afford to live in predominantly white neighborhoods, where housing costs may be higher. Second, laws outlawing the consideration of race in home rental and sale practices are not always followed by real estate agents, landlords, and lenders, who may steer minorities away from predominantly white neighborhoods (Squires, 2003; Squires et al., 2002). Third, white residents have often moved out of neighborhoods when increasing numbers of minority residents moved in (Woldoff, 2011). As Williams, who cites a National Fair Housing report puts it, the result of this "white flight" is that "approximately half of all Black persons and 40 percent of all Latinos live in neighborhoods without a White presence.... The average White person lives in a neighborhood that is nearly 80 percent White" (Williams, 2018, para. 29).

Notably, like Minneapolis, many other racially, ethnically diverse cities are, nonetheless, highly segregated. Consider also Chicago: The city's population is about 33% Black, 32% white, and 29%

Hispanic (Silver, 2015). At the neighborhood level, however, segregation by race and ethnicity is dramatic. For example, Chicago's south side is largely Black: Neighborhoods such as Washington Park are about 97% African American. By contrast, in the city's north, Lincoln Park neighborhoods are typically at least 80% white. To the west, the Cicero area is about 90% Hispanic (Block et al., 2015; Figure 9.1). Racial residential segregation remains a significant phenomenon, particularly in the larger, older cities of the Northeast and Midwest that are home to older housing stock and relatively large populations of poor minorities. Where integration has taken place, it has been limited mostly to areas with "relatively affluent and well-educated minority populations, low levels of anti-Black and anti-Latino sentiment, low rates of immigration, and permissive regimes of density zoning" (Rugh & Massey, 2014, p. 221).

Segregation has significant consequences for minority groups. Apart from denying them residential choice, it may compel them to live in poorer neighborhoods that offer less access to high-quality schools, jobs, and medical facilities (Kozol, 1995; Massey & Denton, 1993). As we noted in Chapter 7, the primary source of wealth for most working- and middle-class Americans is home equity. Lower average property values in minority neighborhoods translate to lower wealth in the community and

FIGURE 9.1 ■ Concentration of Whites, Blacks, and Hispanics in Chicago, 2010

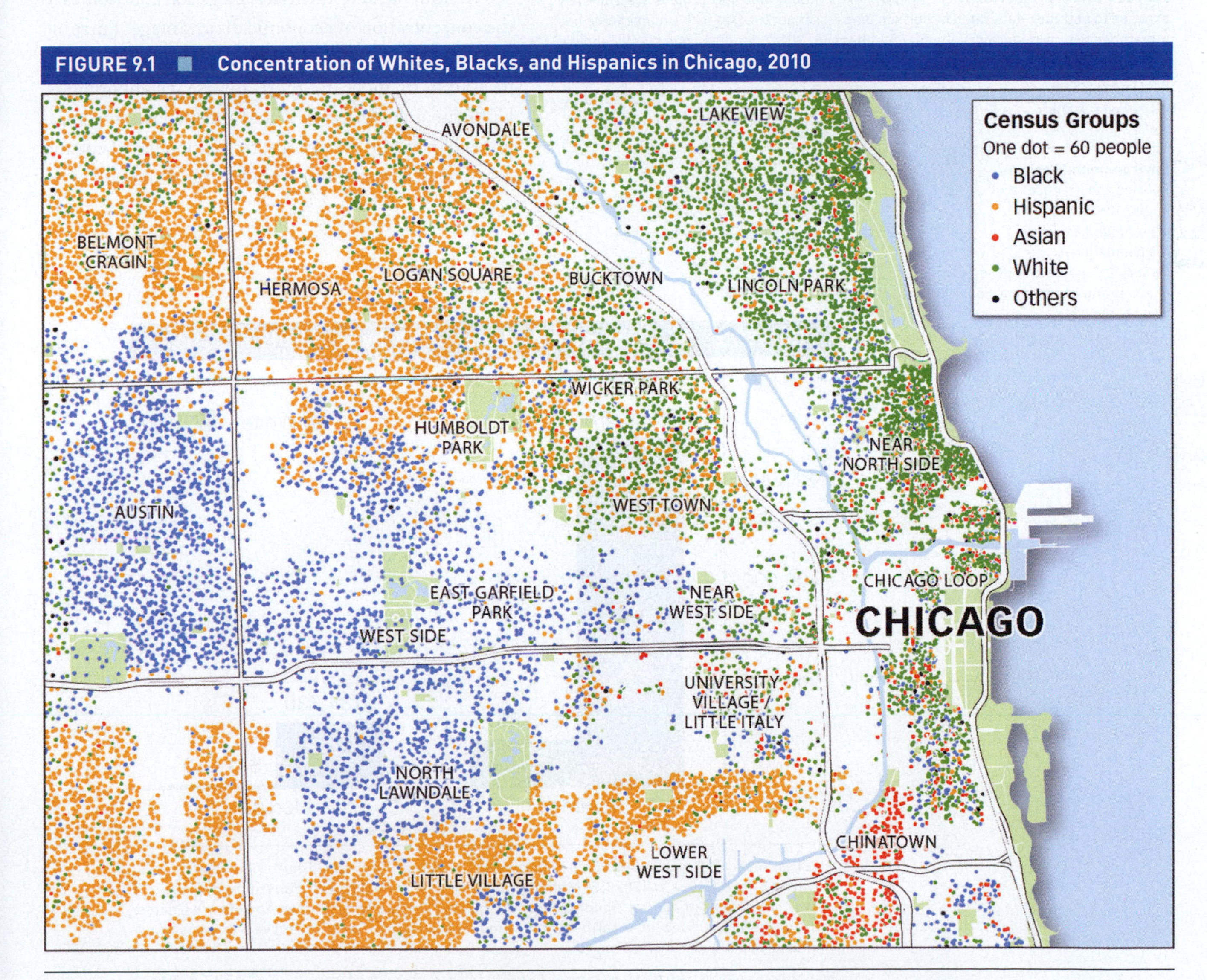

Source: From Mapping segregation. (2015, July 8). Copyright 2015. *The New York Times*. All rights reserved. Used by permission.

Families living in this West Port Arthur, Texas, public housing project are routinely exposed to pollution generated by surrounding oil refineries. One in five households has a member with a respiratory illness (Stephenson, 2014). Studies consistently show a correlation between living near toxic waste sites and poverty and race (Badger, 2014; Ihab et al., 2018).

Environmental racism: Policies and practices that impact the environment of communities where the population is disproportionately composed of ethnic and racial minorities.

in individual households. These effects can be seen in the racial wealth gap: In 2019, white median household net worth was $188,200; by contrast, Black median household wealth was around $24,100 and Hispanic household wealth was $36,100 (Figure 9.2; Bhutta et al., 2020). Furthermore, poor and segregated neighborhoods and regions are also more likely to be home to hazardous-waste facilities and other sources of pollution. Scholars have termed this **environmental racism**, or *policies and practices that impact the environment of communities where the population is disproportionately composed of ethnic and racial minorities.* This problem plagues minority communities from urban Los Angeles to rural Louisiana, leading to health inequalities that we further explore in Chapter 16.

In sum, racial residential segregation contributes to the concentration of economic disadvantage. Pursuing greater opportunities, members of minority groups may seek to move to more diverse neighborhoods. Unfortunately, historically and today, powerful social forces of past and present discrimination have kept and continue to keep many from achieving this goal.

FIGURE 9.2 ■ Median and Average Wealth, by Race, 2019

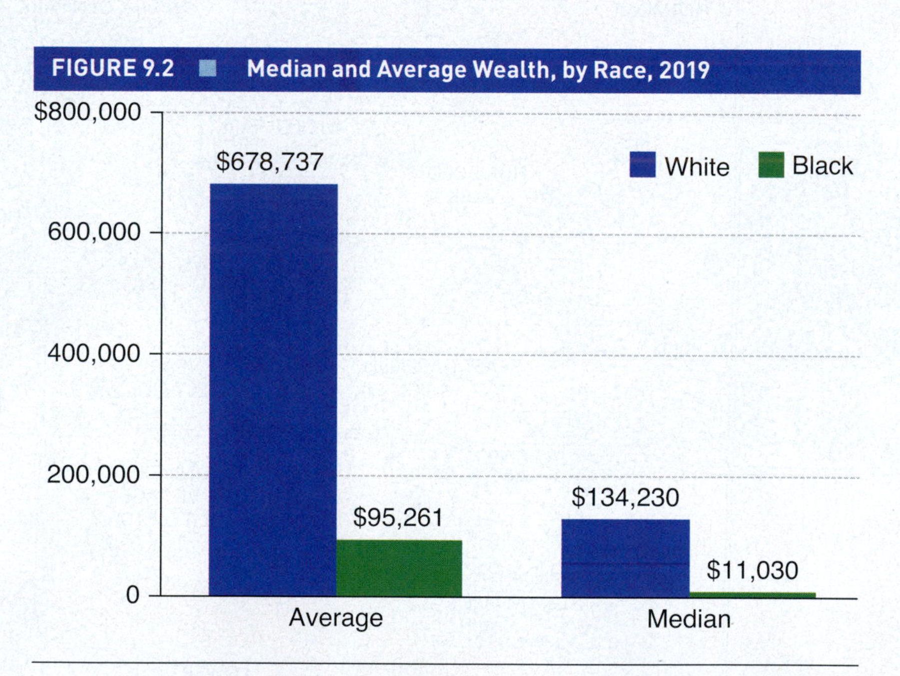

Sources: Federal Reserve Board, 2019 Survey of Consumer Finances; Bhutta, N., Chang, A. C., Dettling, L. J., & Hsu, J. W. (2020, September 28). *Disparities in wealth by race and ethnicity in the 2019 survey of consumer finances.* Federal Reserve. https://www.federalreserve.gov/econres/notes/feds-notes/disparities-in-wealth-by-race-and-ethnicity-in-the-2019-survey-of-consumer-finances-20200928.htm

Note: According to the Federal Reserve, wealth is defined as the difference between families' gross assets and their liabilities.

BEHIND THE NUMBERS: COUNTING—AND NOT COUNTING—HATE CRIMES

What is a hate crime? The Federal Bureau of Investigation (FBI) defines a *hate crime* as a "criminal offense against a person or property motivated in whole or in part by an offender's bias against a race, religion, disability, sexual orientation, ethnicity, gender, or gender identity."

Under FBI guidelines, an incident should be reported as a suspected hate crime if a "reasonable and prudent" person would conclude a crime was motivated by bias. Among the criteria for evaluation is whether an incident coincided with a significant holiday or date, specifically citing the celebration of Martin Luther King, Jr. day. A suspect need not be identified to meet the threshold for reporting (Cassidy, 2016, para. 12).

Who tracks hate crimes? The FBI tracks hate crimes based on the Uniform Crime Report. The Bureau of Justice Statistics (BJS) reports hate crimes from the National Crime Victimization Survey. (See Chapter 6's *Behind the Numbers* for differences between these systems.) The nonprofit group *Southern Poverty Law Center* also monitors hate crimes.

Why track hate crimes? Full reporting of hate crimes recognizes the importance of victims and their experiences. Effective law enforcement, policy making, resource allocation, and public awareness depend on reliable numbers.

What types of hate crimes are most commonly reported? In 2018, the FBI reported 7,120 separate incidents of hate crimes (Figure 9.3). Crimes based on race and ethnicity were most frequently reported, followed by crimes of religious bias and sexual orientation. Most crimes against persons were cases of intimidation or simple assault, while most property crimes were vandalism and destruction or damage of property.

FIGURE 9.3 ■ Hate Crime Breakdown, 2018

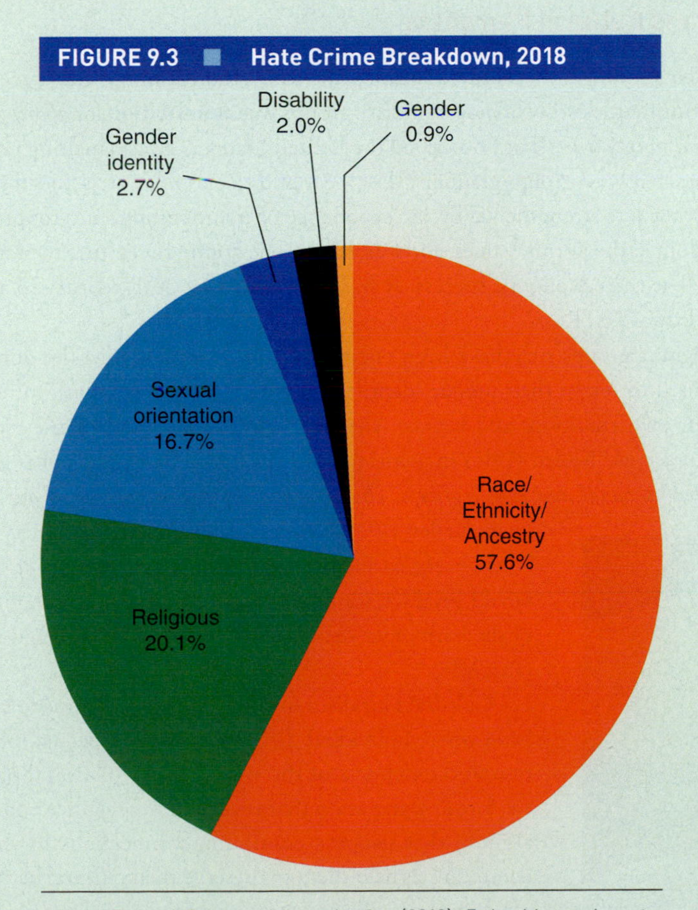

Source: Federal Bureau of Investigation (2019). *Federal hate crime statistics 2019.* https://www.fbi.gov/news/pressrel/press-releases/fbi-releases-2019-hate-crime-statistics

Are hate crime statistics accurate?

Both FBI and BJS likely underreport hate crimes (Potok, 2015; Shanmugasundaram, 2018). Why?

1. *Statistics gathering is not uniform:*
   - Law enforcement agencies are not obligated to report hate crimes. An Associated Press report points out that more than 2,700 city police and county sheriff's departments across the country (about 17% nationwide) have not submitted a single hate crime report to the FBI in the past 6 years (Cassidy, 2016, para. 4).
   - BJS statistics rely on self-reporting from 95,000 households across the United States. Thus, their numbers only show hate crimes as reported by these individuals (Shanmugasundaram, 2018).
2. *Victims may choose not to report:*
   - because they fear that the police cannot or will not help (Langton et al., 2013).
   - because they fear backlash from the community if crimes are reported.
3. *Investigators may not find a motive or may not have special training to identify hate crimes:*
   - Investigators must determine the existence of bias to label a crime a hate crime.
   - Motivation may not always be easy to discern, and only 12 states require training for those who investigate hate crimes (Shanmugasundaram, 2018).

Thinking sociologically, why would some law enforcement agencies be more likely to report hate crimes than others? What are the consequences of having different types of reporting systems? As a sociologist, what advice would you provide lawmakers about collecting hate crime data?

Assimilation and Cultural Pluralism

Assimilation: The absorption of a minority group into the dominant culture.

Cultural pluralism: The coexistence of different racial and ethnic groups as characterized by the acceptance of one another's differences.

Throughout much of the 20th century, sociologists who studied minorities, race, and ethnicity assumed that the ultimate destiny of most minority groups was **assimilation**, or *absorption of a minority group into the dominant culture*. They described the United States as a vast "melting pot" in which significant differences between groups gradually disappear and the population is boiled into a single cultural soup. This view was strengthened by the experience of many immigrant groups from European countries, who adopted the norms, values, and folkways of the dominant culture to increase their social acceptance as well as their economic success. It was also touted by 20th century sociologists such as Robert Park and Edward A. Ross.

Although migrant groups may have been encouraged to assimilate into the dominant majority group, many wanted to retain their unique identities. Native American nations, for example, sought to keep alive their own traditions and beliefs. Another form of racial and ethnic group interaction, introduced by Horace M. Kallen (1915) in a critique of Edward A. Ross, was **cultural pluralism**, *the coexistence of different racial and ethnic groups, characterized by the acceptance of one another's differences*. Kallen (1915) saw a culturally pluralist society as an orchestra where every nationality was an instrument with its unique role that would produce something entirely new and "the range and variety of the harmonies may become wider and richer and more beautiful" (Part VI).

In Emporia, Kansas, a woman and her daughter celebrate the Mexican Día de los Muertos (Day of the Dead). Which form of minority group interaction is best illustrated with such a celebration?

Cultural pluralism in the United States came to be seen as a salad bowl where individual differences were respected for their contribution to the richness of the whole society. Cultural pluralists criticize the forced segregation that results from prejudice and discrimination, arguing that people's continuing connections to their own ethnic communities help them to preserve their cultural heritages and, at the same time, provide networks of mutual social and economic support.

What examples of assimilation or cultural pluralism do you see in your community? What role, if any, do you think social media has to play in cultural pluralism? Assimilation?

THEORETICAL APPROACHES TO ETHNICITY, RACISM, AND MINORITY STATUS

Throughout this text, we seek to highlight the ways in which key theoretical perspectives in sociology can illuminate social phenomena, including socioeconomic stratification, poverty, and deviance. In this section, we consider ethnicity, racism, and minority group status through functionalist, conflict, and symbolic interactionist lenses.

The Functionalist Perspective

One of functionalism's key assumptions is that a social phenomenon exists and persists because it serves a positive function in a community or society, contributing to order and harmony. Beginning with classical sociologists Auguste Comte and Émile Durkheim and extending to contemporary functionalist theorizing, the basic functionalist assumption is that solidarity characterizes social groupings. Durkheim believed that social groups held together by mechanical solidarity and based on homogeneity in, among other things, language and culture are more culturally durable than are those based on organic solidarity, which involves interdependence (for instance, economic interdependence). This may help us understand the cohesion of many ethnic groups in the United States. Whether Armenian Americans, Egyptian Americans, Cuban Americans, or some other group, people gravitate toward those who are similar to them, a process rooted in shared pasts and practices.

At the same time, Durkheim was deeply disturbed by the anti-Semitism he saw in France in his day, embodied in the famous trial against a Jewish army captain named Alfred Dreyfuss, who had been court-martialed for treason. Writing of the "Dreyfuss Affair" in 1899, Durkheim noted that the anti-Semitism inherent in the charges against Dreyfuss constituted a "moral sickness" in French society. In short, Durkheim seemed to view ethnic discrimination, at least in France at the time, as a form of *anomie*, or lack of sufficient moral regulation in the society (Goldberg, 2008). Durkheim believed that this form of **scapegoating**, *or assigning misplaced blame to a person or group based on some alleged characteristic or action,* served a function. In this case, by scapegoating someone by their ethnicity, it allowed the disaffected in society to receive comfort by having a sense of someone on "whom to blame for [their] economic troubles and moral distress," which then made it seem that "everything was already better" (Durkheim, 1899, cited in Goldberg, 2008, p. 303).

Scapegoating: Assigning misplaced blame to a person or group based on an alleged characteristic or action.

Overall, however, it is challenging to apply the functionalist perspective to racism as a phenomenon. Racism is not positively functional for a community or society because, by definition, it marginalizes some members of the group. It can, however, be positively functional for some groups (and individuals) while being detrimental for others. By asking, "How is this functional for some groups?" we move toward a key conflict question: "Who benefits from this phenomenon or institution—and who loses?" In the next section, we examine who benefits from racism.

Conflict Perspectives

Consider that **racism**, *the idea that one racial or ethnic group is inherently superior to another*, offers a justification for racial inequality and associated forms of stratification such as socioeconomic inequality. If a powerful group defines itself as "better" than another group, then the unequal treatment and distribution of resources can be rationalized as acceptable or even natural. Various conflict perspectives underscore the importance of economic, social, and political power inherent in racism.

Racism: The idea that one racial or ethnic group is inherently superior to another; often results in institutionalized relationships between dominant and minority groups that create a structure of economic, social, and political inequality based on socially constructed racial or ethnic categories.

For instance, conflict theorists point to the confluence of racial, legal, political, economic, and gender dimensions of slavery in the United States. It was possible for whites in the South (and many in the North) to justify the maltreatment of Blacks because they did not see them as fully human. Senator John Calhoun, who represented South Carolina and died just over a decade before the Civil War, wrote in a letter that "the African is incapable of self-care and sinks into lunacy under the burden of freedom. It is a mercy to him to give him the guardianship and protection from mental death" (quoted in Silva, 2001, p. 71).

In legal terms, racism was clear in the Three-Fifths Compromise, which emerged at the Constitutional Convention in Philadelphia in 1787. Delegates from the North and South agreed that

Enslaved persons were leased by their owners to mine sandstone from local quarries, which the enslaved then used to build the iconic structures of U.S. democracy—the U.S. Capitol building and the White House—in Washington, D.C. Which theoretical perspective would best describe this reality?

©iStockphoto.com/lucky-photographer

for purposes of taxation and political representation, each enslaved person would be counted as three fifths of a person. Abolitionists wanted the compromise because they wanted to count only free people. Slaveholders and their supporters wanted to count enslaved people, whose presence would add to their states' population counts and thus their representation in Congress. Those who held slaves but did not permit them to be free or to vote still gained politically from their presence.

In economic terms, slavery benefited plantation owners and their families, who reaped the financial benefits of a population of the unpaid workers who performed demanding and difficult work and could be bought and sold, exploited and abused. Many scholars have argued that capitalism and economic development in the early United States would not have been so robust without the country's reliance on an enslaved labor force, which contributed most fully to the development of a growing agricultural economy in the South. The North, too, was home to numerous beneficiaries and proponents of slavery.

In social terms, slavery was also a gendered institution where Black women's sexual reproduction was necessary to make new bodies to enslave and their other labors were necessary for maintaining their families as well as white families. Indeed, as Cooper Owens shows in her award-winning book *Medical Bondage, Race, Gender, and the Origins of Medical Gynecology* (2018), early gynecological practice and research were built on the crude "operations" performed on Black enslaved women. These involved experiments and operations without anesthesia and with little regard for the women's well-being.

After the end of slavery, new forms of legalized racism continued, for example, in Jim Crow laws, which followed in the decade after slavery and lasted until the middle of the 20th century. These laws legally mandated segregation of public facilities in the South and fundamentally limited Blacks' ability to exercise their rights. The schools, accommodations, and opportunities afforded to Blacks were invariably inferior to those offered to whites. Local and state governments, therefore, could expend fewer funds on their Black populations, and white populations benefited from reduced competition in higher education and the labor market. Racism made Jim Crow laws both possible and widely acceptable because it offered a justification for their existence and persistence.

Although racism is clearly of no benefit to its victims and has negative effects on society as a whole, the conflict perspective entreats us to recognize the ways in which it has created wealth and social and political power for groups in society at the expense of minority groups. These benefits help explain the existence and persistence of racism over time.

The Symbolic Interactionist Perspective

Sociologist Louis Wirth (1945) noted that minority groups share particular traits. First, membership in a minority group is essentially involuntary—that is, someone is socially classified as a member of a group that is discriminated against and is not, in most instances, free to opt out. Second, as we discussed previously, minority status is a question not of numbers (minorities may outnumber the dominant group), but rather of control of valued resources. Third, minorities do not share the full privileges of mobility or opportunity enjoyed by the dominant group. Finally, membership in the minority group conditions the treatment of group members by others in society. Specifically, Wirth states, societal minorities are "treated as members of a category, irrespective of their individual merits" (p. 349).

Symbolic interactionists examine the consequences of racism in interaction. Wirth's definition has similarities to symbolic interactionist Erving Goffman's concept of a **stigma**, *an attribute that is deeply discrediting to an individual or a group because it overshadows other attributes and merits the individual or group may possess.* Goffman applied stigma to a number of consequential socially constructed differences, race being one of them, that influence interaction. Tyler (2018) and others (Howarth, 2006) have elaborated on Goffman's idea of stigma to show it not only as a "technology of dehumanization"

Stigma: An attribute that is deeply discrediting to an individual or a group because it overshadows other attributes and merits the individual or group may possess.

but also as a "form of power" that minority groups have resisted (p. 18). In other words, the social "structural" reality of racism is present in interactions, and manifests as *stigma*, but so too is resistance of that racism and stigma on the part of those who experience it. Symbolic interaction, then, can shed light on both the power to stigmatize and the resistance to it. As Howarth (2006) notes, this is not to say that people who are stigmatized should "think themselves out of stigma," but that "stigma needs to be seen as collectively constructed, institutionalized and resisted in the systems of difference, privilege and inequality that constitute the social structures and institutionalized practices of any society" (p. 12).

In 1968, Black Memphis sanitation workers went on strike carrying placards stating "I Am a Man" in bold red letters. This phrase became an embodiment of the civil rights movement, as it showed that workers were not only fighting for better wages or working conditions but actively fighting against racial stigma and the recognition of their humanity.

©Bettmann/Bettmann/Getty Images

W. E. B. Du Bois's concept of *double consciousness* (2008/1903) is also important to the symbolic interactionist perspective, overlapping with ideas of stigma. Du Bois describes the social psychological impact of being Black in a racist society that sees and judges skin color first as having a "double" effect on Black individuals, as both insiders and outsiders: It provides a unique vantage point to see one's society, but it also creates unease, a feeling of being a "problem," feelings that produce a split identity. Combining stigma and double consciousness may give us some insight into important dimensions of how racism, while potentially functional for groups in power, is felt, resisted, and potentially reframed by the groups that experience it.

In the next section, we turn to issues of prejudice and discrimination, considering how our judgments about someone's race or ethnicity—or about the racial or ethnic identity of an entire group—are manifested in practice.

PREJUDICE, STEREOTYPING, AND DISCRIMINATION

Prejudice is *a belief about an individual or a group that is not subject to change on the basis of evidence.* Prejudices are thus inflexible attitudes toward others. Sociologist Zygmunt Bauman (2001), writing about the Holocaust, eloquently captures this idea: "Man *is* before he *acts*; nothing he does may change what he is. That is, roughly, the philosophical essence of racism" (p. 60). Recall Wirth's point about the characteristics of a social minority: Membership in the disadvantaged group matters more than individual merit.

Prejudice: A belief about an individual or a group that is not subject to change on the basis of evidence.

When prejudices are strongly held, no amount of evidence is likely to change the belief. Among neo-Nazis, for example, prejudice against Jews runs deep. Some neo-Nazis deny the occurrence of the Holocaust, despite clearly authentic firsthand accounts, films, and photographs of Nazi Germany's concentration camps, where millions of Jews, Roma, Soviet prisoners of war, LGBTQ+ individuals, and others were killed during World War II. This Holocaust denial had been perpetrated on social media to the extent that Facebook CEO Mark Zuckerberg, in October 2020, reversed his social media policy, banning all content that "denies or distorts the Holocaust" (Bond, 2020, para. 1). Zuckerberg and his company pointed to an "alarming level" of ignorance about the Holocaust in the general population, citing a survey of 18- to 39-year-olds where "almost a quarter of people in US aged 18–39 said they believed the Holocaust was either a myth, had been exaggerated or were not sure about the genocide" (Bond, para. 3).

Another social phenomenon closely linked to prejudice is **stereotyping**—*the generalization of a set of characteristics to all members of a group.* Stereotyping offers a way for human beings to organize and categorize the social world—but these attributions are often deeply flawed. Consider the fact that media are common purveyors of ethnic and racial stereotypes. U.S. action movies often rely on them heavily: Italian Mafiosi, tough African American street gangs, violent Asian gangsters or martial artists, and Middle Eastern terrorists. Or consider a children's staple, Disney films. After years of criticism,

Stereotyping: The generalization of a set of characteristics to all members of a group.

October 2020 saw a major venue of its images, Disney+, listing warning labels on streaming content of classics such as *Peter Pan, Dumbo, The Aristocrats*, and others:

> This program includes negative depictions and/or mistreatment of people or cultures. These stereotypes were wrong then and are wrong now. Rather than remove this content, we want to acknowledge its harmful impact, learn from it and spark conversation to create a more inclusive future together. (quoted in Oxner, 2020, para. 6)

From a functionalist perspective, we may argue that even though stereotypes are often flawed, they are also functional for some groups—although dysfunctional for others. Consider that one of the social forces contributing to racism is the desire of one group to exploit another. Research on early contacts between Europeans and Africans suggests that negative stereotypes of Africans developed *after* Europeans discovered the economic value of exploiting African labor and the natural resources of the African continent. That is, white Europeans stereotyped Black Africans as inferior when such ideas suited the Europeans' need to justify enslavement and economic exploitation of this population. From a conflict perspective, we can note that stereotypes, in the case of both Facebook and Disney, may be challenged when economic interests, such as consumer boycotts, may have the power to sway those in powerful positions to discontinue their promotion. How do you think a symbolic interactionist might interpret stereotyping?

Discrimination: The unequal treatment of individuals on the basis of their membership in a group.

Discrimination is *the unequal treatment of individuals on the basis of their membership in a group.* Discrimination is often targeted and intentional, but it may also be unintentional—in either case, it denies groups and individuals equal opportunities and blocks access to valued resources.

Individual discrimination: Overt and intentional unequal treatment, often based on prejudicial beliefs.

Institutionalized discrimination: Discrimination enshrined in law, public policy, or common practice; it is unequal treatment that has become a part of the routine operation of such major social institutions as businesses, schools, hospitals, and the government.

Sociologists also distinguish between individual and institutional discrimination. **Individual discrimination** is *overt and intentional unequal treatment, often based on prejudicial beliefs.* If the manager of an apartment complex refuses to rent a place to someone on the basis of his or her skin color or an employer chooses not to hire a qualified applicant because he or she is foreign-born, that is individual discrimination. **Institutionalized discrimination** is *discrimination enshrined in law, public policy, or common practice—it is unequal treatment that has become part of the routine operation of such major social institutions as businesses, schools, hospitals, and the government.* Institutionalized discrimination is particularly pernicious because, although it may be clearly targeted and intentional, it also may be the outcome of customary practices and bureaucratic decisions that result in discriminatory outcomes.

Discrimination against African Americans and women was written into the Constitution of the United States, in effect barring members of both groups from voting and holding public office. In 1866, Congress passed the first civil rights act, allowing Black men the right to vote, hold public office, use public accommodations, and serve on juries. In 1883, however, the U.S. Supreme Court declared the 1866 law unconstitutional, and states passed laws restricting where minorities could live, go to school, receive accommodations, and such. These laws were upheld by the U.S. Supreme Court in the case of *Plessy v. Ferguson* in 1896. It took more than 75 years for the Court to reverse itself.

In the 1960s, the Supreme Court held that laws institutionalizing discrimination were unconstitutional. Open forms of discrimination, such as signs and advertisements that said, "Whites only" or "Jews need not apply," were deemed illegal. Research and experience suggest, however, that discrimination often continues in more subtle and complex forms. Consider, once again, the case of housing and discrimination.

Institutionalized discrimination affects opportunities for housing and, by extension, opportunities for building individual and community wealth through homeownership. We noted earlier in the chapter that African Americans often still live in segregated neighborhoods. Discrimination is part of the reason; for example, Blacks have been historically less likely to secure mortgages because of it. Paired testing studies in which researchers have sent Black and white applicants with nearly identical financial profiles to apply for mortgages have determined that whites as a group continue to be advantaged in their treatment by lending institutions (Silverman, 2005). Among other benefits, whites enjoy higher rates of approval and better loan conditions (Turner et al., 2009).

The turn of the millennium, however, ushered in a shift in mortgage lending. Banks had plenty of money to lend for home purchases, and among those they targeted were minorities, even those with low incomes and poor credit. The lending bonanza was no boon for these groups, however, because many of the loans made were subprime. (Subprime loans carry a higher risk that the borrower will default, and the terms are more stringent to compensate for this risk.) Subprime loans were five times more common in predominantly Black neighborhoods than in white ones (Pettit & Reuben, 2010), and many borrowers did not fully understand the terms of their loans, such as "balloon" interest rates that rise dramatically over the life of a loan. One consequence was a massive wave of foreclosures beginning in 2008. Although minority and poor communities were not the only ones affected by the subprime loan fiasco, they were disproportionately harmed and have been slow to regain economic ground. According to a recent investigation,

> Nationwide, home values in predominantly African American neighborhoods have been the least likely to recover.... Across the 300 largest U.S. metropolitan areas, homes in 4 out of 10 zip codes where blacks are the largest population group are worth less than they were in 2004. That's twice the rate for mostly white zip codes across the country. (Badger, 2016, para. 6)

In the sections that follow, we discuss other contemporary manifestations and consequences of individual and institutionalized discrimination in the United States, highlighting the criminal justice system, health, and the Internet housing market.

Criminal Justice, Politics, and Power

Racial disparities are evident in the U.S. criminal justice system in all aspects, including victimization, arrest, and length and severity of punishment. Michelle Alexander (2010) argues that the U.S. criminal justice system effectively functions as a modern incarnation of the Jim Crow laws. Today, she suggests, the expansion of the prison population to include nonviolent offenders, particularly people of color, has contributed to the development of a population denied opportunities to have a political voice, as well as education, work, and housing.

The Sentencing Project (2020b) highlights racial disparities in the criminal justice system. For instance, it shows how youth of color are disproportionately siphoned into the criminal justice system at an early age. Native American youth are three times more likely than white youth to be housed in juvenile detention centers; African American youth in 2011 were more than 269 times more likely to be arrested for violating curfew laws than were white youth and Latinx youth. And data on Latinx youth, despite reforms issued by the Juvenile Justice Delinquency Prevention Act of 2018 that mandated uniform data collection, are still, according to the Anna E. Casey Foundation, being undercounted in the system as either Black or white, and these inconsistencies "exist in all 50 states" (Alianza for Youth Justice and UCLA's Latino Policy and Politics Initiative, 2020, p. 25). As we have noted in our *Behind the Numbers* features throughout this book, undercounting populations, no matter what the subject, limits our understanding of the scope of the issue and ways to address it. What challenges do you see with an undercounting of Latinx youth in the justice system?

In terms of incarceration, while overall numbers have fallen significantly for all racial groups since 2006, Blacks and Hispanics remain overrepresented in the nation's prisons (Figure 9.4). Gramlich (2020) notes that the rate of Black male imprisonment is especially high in the age group of 35 to 39, where "about one-in-twenty black men were in state or federal prison in 2018" (para. 5).

One of the key rights of U.S. citizenship is the right to vote—that is, the right to have a voice in the country's political process. As we noted in Chapter 6, many former prisoners have been legally disenfranchised because of state laws that prohibit them from voting. In the mid-20th century, more than 70% of those incarcerated in the United States were white; by the end of the 20th century, those numbers had reversed, and most prisoners were nonwhite (Wacquant, 2002; Western & Pettit, 2010). This shift has had a profound effect on African Americans' political voice: 48 states prohibit inmates from voting while they are incarcerated; 14 states and the District of Columbia restore voting rights after incarceration; 4 states restore it after incarceration and parole; and 18 states restore it after incarceration, parole, and probation have been completed. Alabama, Arizona, Delaware, Florida, Iowa, Kentucky, Mississippi, Nebraska, Nevada, Tennessee, Virginia, and Wyoming foresee the possibility of a lifetime voting ban on anyone who has been convicted of a felony (Uggen et al., 2016).

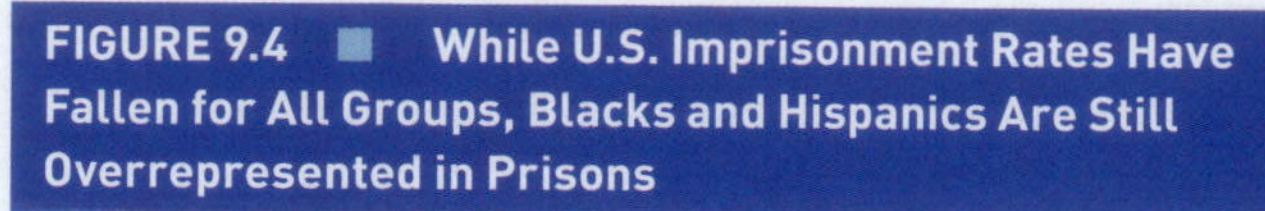

FIGURE 9.4 ■ While U.S. Imprisonment Rates Have Fallen for All Groups, Blacks and Hispanics Are Still Overrepresented in Prisons

Blacks, Hispanics make up larger shares of prisoners than of U.S. population.

U.S. adult population and U.S. prison population by race and Hispanic origin, 2018

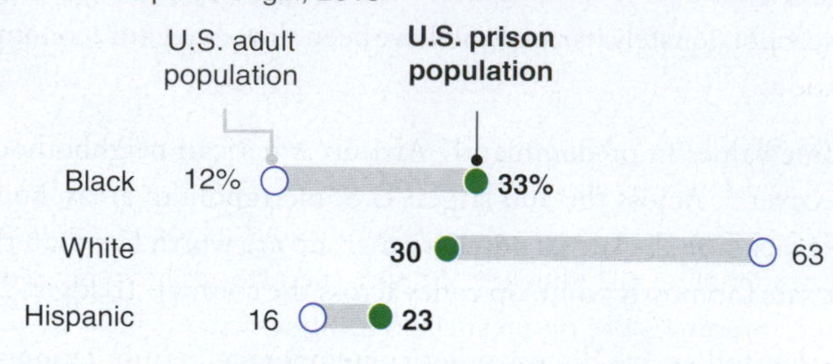

Source: Gramlich, J. (2020, May 6). *Black imprisonment rate in the United States has fallen by a third since 2006.* https://www.pewresearch.org/fact-tank/2020/05/06/share-of-black-white-hispanic-americans-in-prison-2018-vs-2006/

This disenfranchisement has a significant racialized impact, particularly on African American men. In October 2020, 1 in 16 African Americans of voting age was disenfranchised (a rate of approximately 6.2% compared to about 2% for Hispanics and 1.7% for other racial groups; The Sentencing Project, 2020a). Importantly, these rates also vary among states. As The Sentencing Project's 2020 report on disenfranchisement puts it:

> African American disenfranchisement rates vary significantly by state. In seven states—Alabama, Florida, Kentucky, Mississippi, Tennessee, Virginia, and Wyoming—more than one in seven African Americans is disenfranchised. (The Sentencing Project, 2020a)

Consequences of Prejudice and Discrimination: Race and Health

Sociologists studying discrimination take a significant interest in the health of minority populations. For example, researchers have argued that racism and other disadvantages suffered by Black women in the United States contribute to the much higher level of negative birth outcomes, including low birth weight and infant mortality and maternal mortality (Geronimus, 1992). We explore these issues more closely in Chapter 16.

In 2020, one emerging major health disparity was the disproportionate impact of COVID-19 on communities of color, most especially African Americans, Latinxs, and Native Americans. An October 13, 2020, analysis of mortality data from COVID-19 in the United States concluded:

> If they had died of COVID-19 at the same actual rate as White Americans, about 21,800 Black, 11,400 Latino, 750 Indigenous and 65 Pacific Islander Americans would still be alive. (APM Research Lab Staff, 2020https://www.apmresearchlab.org/covid/deaths-by-race)

Social epidemiology: The study of communities and their social statuses, practices, and problems with the aim of understanding patterns of health and disease.

Why such racial differences in mortality rates? To see such medical statistics through the prism of social science is a goal of **social epidemiology**, which is *the study of communities and their social statuses, practices, and problems with the aim of understanding patterns of health and disease.* According to the Centers for Disease Control and Prevention (CDC, 2020), a number of factors (most of which we have associated with institutional discrimination) could make communities of color more suspectable. Its list includes "Discrimination, healthcare access and utilization, occupation, educational and income and wealth gaps, and housing" (CDC, 2020).

Consider, for instance, the fact that the continuing segregation of neighborhoods means that many communities of color lack ready access to medical care, including COVID-19 testing, doctors, and hospitals. Also consider that statistically speaking, African American and Latinx workers do not have the same opportunity to work from home as do their white and Asian peers. As Moody (2020) puts it: "16.2% of Hispanic workers and 19.7% of black Americans are able to work from home, while about 30% of whites and 37% of Asian-Americans can" (para. 2). Additionally, these groups are less likely to have access to adequate personal protective equipment (PPE), paid leave, and other protections such as health insurance (Beaman & Taylor, 2020).

In an interesting study of the negative consequences of race, racism, and health, professor of sociology and psychiatry Jonathan Metzl turns a lens on the white population. His 2019 book *Dying of Whiteness* documents what he learned in traveling throughout the United States to try to understand why white people, especially white men, were beginning to see a decrease in their life expectancies. Metzl interviews a number of people, showing through their stories how the "defense of white 'ways of life' or concerns about minorities or poor people hoarding resources" have led to decreased life expectancies for white populations. These early deaths have been fueled, in part, by increased death by suicide and gun violence and decreased investment in public health. For instance, the majority of suicide deaths in the United States occur from guns and kill white men. Metzl advocates collaboration across groups through shared interactions and community investment. He critiques politically divisive rhetoric that "promises greatness" but actually results in the "biology of demise" (p. 281).

Race and Ethnicity in Hollywood

The Academy Awards have long failed to recognize achievements in film among people of color, prompting the hashtags #OscarsSoWhite (a hashtag created in 2015 in response to the almost exclusive nomination of white actors) and #WhitewashedOUT (a hashtag created to respond to the phenomenon of white actors playing Asian characters instead of Asians). Between 2014 and 2016, no African American actors were nominated for the top awards at the Oscars—Best Actor, Best Actress, or Best Director—which led to boycotts and a large decline in viewership (Kirk, 2019). In response to the marginalization of actors of color, some celebrities (such as Will Smith and Jada Pinkett Smith) also boycotted the 2016 Oscars. The Academy subsequently released a statement recognizing inequities and committing to increase diversity by doubling the number of women and people of color in its membership by 2020 (Wagner, 2016).

Three years later, the 2019 Oscars were notable for their gender and racial diversity. Out of the 52 presenters, there were 29 nonwhite presenters (Ng, 2019) and the Academy handed out a record number of trophies to 6 African Americans and 15 female artists and technicians (Ng, 2019). Rami Malek, an Egyptian American actor, received the Best Actor award in 2019 for his role in *Bohemian Rhapsody*. Spike Lee noted on camera that he would not have won an Oscar for Best Adapted Screenplay had it not been for the #OscarsSoWhite campaign and for Cheryl Boone Isaacs, the former Academy president who oversaw efforts to diversify its membership (Groom & Dobuzinskis, 2019). Other notable wins during the 2019 awards ceremony were Regina King winning Best Supporting Actress for *If Beale Street Could Talk*, and *Black Panther* racking up another two wins. Costume designer Ruth E. Carter and production designer Hannah Beachler were the first African American women to win for their respective categories.

However, 1 year later, Black nominations of Oscar winners hit a 3-year low, with only five Black people nominated for Oscars in 2020 even though a number of top films that year featured Black actors. On the other hand, the South Korean film *Parasite*, a subtitled film, took home four Oscars and became the first international film to win Best Picture in Oscar history (Jancelewicz, 2020).

So, what can we expect in coming years?

To answer this question, we might consider the notion of institutional discrimination—for instance, what practices are in place, and how have positions been filled? Actor George Clooney has suggested that the problem is not primarily who the Academy is choosing to nominate but rather "how many options are available to minorities in film, particularly in quality films" (Setoodeh, 2016, para. 2). Indeed, people of color are often placed in stereotypical roles such as gangsters, single parents, singers,

Parasite, the 2019 multi-award-winning film, featured the Kim and Park families in a satirical horror tale of South Korean social class inequality that ends in the tragedy of murder. It was the first foreign language film ever to win a Best Picture Oscar.

©BFA /Alamy Stock Photo

dancers, maids, sex workers, enslaved persons, or plains Indians. Their characters are more likely than others to be violent, hypersexualized, or poor; fewer are presented as leaders, intellectuals, professionals, or characters with important stories to tell. No less notably, minority characters are less likely to be prominent in films: A study done at the Annenberg School for Communication and Journalism at the University of Southern California found that in 600 popular films from 2007 to 2013, white actors populated about 75% of speaking roles. Nearly a fifth of films had no Black speaking characters at all (Smith et al., 2014). Additionally, as we noted previously, Hollywood has engaged in a long-standing practice of casting white actors to play minority characters. For instance, in the 2017 film *Ghost in the Shell*, Scarlett Johansson played a Japanese woman, and Ben Affleck played the role of Tony Mendez, a Hispanic character, in the 2012 film *Argo*. In the 2013 film *The Lone Ranger*, actor Johnny Depp was cast as the Lone Ranger's American Indian sidekick, Tonto.

Consider also, who is allowed to tell the stories, and who directs them? The seventh annual *UCLA Hollywood Diversity Report* shows that in 2019, for instance, people of color constituted only 9% of studio heads, 13.9% of film writers, and 15.1% of film directors (Hunt & Ramón, 2020). This report also shows steady progress being made on the diversification of characters, with lead actors who were not white nearly tripling since 2011—rising from 11% then to 27.6% in 2019.

Finally, it is important to note that the change promised in 2016 is beginning to be made, but more progress is necessary to reach a Hollywood that truly represents the diversity of audiences in the United States. As Hunt and Ramón (2020) note in their report:

> By 2018, Academy membership was 31 percent female and 16 percent minority, and the large class of 842 new members the following year was half women and fully 29 percent people of color, diversifying Academy membership further. (p. 41)

What types of changes, if any, do you think need to be made in films, television shows, and other cultural entertainment to present diverse stories? Who should make those changes? What are the benefits? To whom? In the next section, we look at some of the major groups that make up the U.S. population and see how their numbers are contributing to the changing composition of the country.

RACIAL AND ETHNIC GROUPS IN THE UNITED STATES

The area now called the United States was once occupied by hundreds of Native American nations. Later came Europeans, most of whom were English, although early settlers also included the French, Dutch, and Spanish, among others. Africans who were forcibly enslaved and brought to the United States added a substantial minority to the European population. Hundreds of thousands of Europeans continued to arrive from Norway, Sweden, Germany, and Russia to start farms and work in the factories of early industrial America. In the mid-1800s, Chinese immigrant males were brought in to provide the heavy labor for building railroads and mining gold and silver. Then, in 1882, the Chinese Exclusion Act suspended all immigration of Chinese immigrants for 10 years, and subsequent legislation limited the amount of immigration for people from the Middle East and Asia. Between 1890 and 1930, nearly 28 million people immigrated to the United States. After the 1950s, immigrants came mostly from Latin America and Asia.

Today, the U.S. population is increasingly diverse (see Figure 9.5). According to the United States Bureau of the Census, in 2019, whites composed about 60% of the population; Blacks about 13%; Hispanics 18.5%; Asians about 6%; American Indians and Alaska Natives about 1%; and people self-identifying as mixed race about 3%. If current trends hold, nonwhites will constitute a numeric majority nationwide by about 2040.

The United States has more immigrants than any other country in the world, according to Budiman (2020a, para. 1). The rapid growth of minority populations reflects growth from both new births and migration. Migration is the primary engine of population growth in the United States. Today, 44.8 million people in the United States are immigrants, and they account for almost 14% of the population (Budiman, 2020a). Well over three quarters of immigrants are in the country legally (Budiman, 2020a). Demographers estimate that by the middle of the 21st century, fully one fifth of the U.S. population will be immigrants. In a population estimated to reach more than 438 million by 2050, 67 million will be immigrants, and about 50 million will be the children and grandchildren of immigrants (Pew Research Center, 2008). Overall, the trend toward a more racially and ethnically diverse country is clear: At the time of the 2010 Census, the majority of children under 2 years old were nonwhite.

In the following sections, we look in more depth at the racial and ethnic groups that have come together to constitute U.S. culture and society today.

American Indians lived in the territorial United States long before the political state was established. They suffered the loss of territory and population as a result of U.S. government policies. Even today, American Indian communities face unique challenges.

American Indians

There is no agreement on how many people were living in North America when the Europeans arrived around the beginning of the 16th century, but anthropologists' best guess is about 20 million. Thousands of thriving societies existed throughout North and South America in this period. Encounter between European explorers and Native Americans soon became conquest, and despite major resistance, the Native nations were eventually defeated, and their lands confiscated by European settlers.

Between 1500 and 1800, the North American Indian population is thought to have been reduced from more than 20 million to fewer than 600,000 (Haggerty, 1991). American Indians, like conquered

FIGURE 9.5 ■ Racial/Ethnic Composition of the U.S. Population, 2019

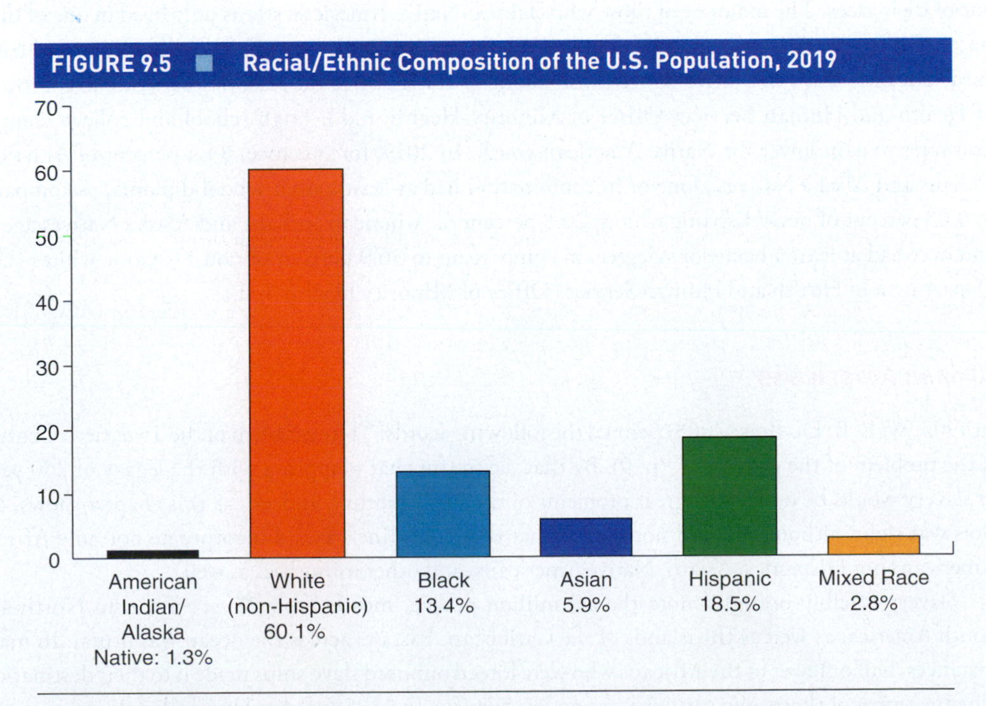

Source: U.S. Census Bureau. (2019, June 19). *Population estimates.* Quick Facts. https://www.census.gov/quickfacts/fact/table/US/PST045219

and oppressed peoples everywhere, did not passively accept the European invasion. The "Indian Wars" are among the bloodiest in U.S. history, but resistance was not enough. First, white settlers had access to superior weaponry that Native Americans could not effectively counter. Second, whites' control of political, economic, and environmental policies gave them a powerful advantage. Consider, for instance, the U.S. government's decision to permit the unregulated slaughter of bison (buffalo) in the post–Civil War era. From about 1866, when millions of bison roamed the American West and the Great Plains, to the 1890s, this key Native American resource was decimated until only a few thousand remained (Moloney, 2012; Smits, 1994).

Since the 1960s, some Native Americans have focused on efforts to force the U.S. government to honor treaties and improve their living conditions, particularly on reservations. In 1969, the group Indians of All Tribes (IAT), citing the 1868 Treaty of Fort Laramie, which states that all unused or abandoned federal land must be returned to the Indians, seized the abandoned island of Alcatraz, a former prison island in the San Francisco Bay, and issued a proclamation that contained the following (deliberately ironic) declarations:

> We, the native Americans, re-claim the land known as Alcatraz Island in the name of all American Indians by right of discovery....We feel that this so-called Alcatraz Island is more than suitable for an Indian Reservation, as determined by the white man's own standards.

The IAT went on to note the island's "suitability" by pointing out that, among other things, it had no fresh running water, no industries to provide employment, no health care facilities, and no oil or mineral rights for inhabitants. This comparison was intended to point to the devastating conditions on most reservations. The IAT occupied Alcatraz for more than 18 months. The occupation ended when some IAT members left due to poor living conditions, as the federal government shut off all power and water to the island, and others were removed when the government took control of the island in June 1971.

Native American people continue to face a number of challenges including issues of land rights, voting rights, sovereignty and legal issues, sacred practice rights, and challenges of domestic violence (Indianlaw.org, n.d.). Additionally, conditions on reservations can vary dramatically depending on location, job accessibility, health care accessibility, and financial stability of the tribe. In 2018, 2.8 million people identified as Native American, and an additional 2.9 identified Native American as one of their races. The majority of those who claimed Native American status only lived in one of three states: California, Arizona, and Oklahoma (USA Facts, 2020). Today, 573 federally recognized tribes exist, but only 22% of Native Americans/American Indians live on reservations (U.S. Department of Health and Human Services Office of Minority Health, n.d.). High school and college completion rates remain lower for Native American youth. In 2019, for instance, 84.4 percent of American Indians and Alaska Natives alone or in combination had at least a high school diploma, as compared to 93.3 percent of non-Hispanic whites; 20.8 percent of American Indians and Alaska Natives age 25 and over had at least a bachelor's degree, in comparison to 36.9 percent of non-Hispanic whites (U.S. Department of Health and Human Services Office of Minority Health, n.d.).

African Americans

In 1903, W. E. B. Du Bois (2008) penned the following words: "The problem of the Twentieth Century is the problem of the color line" (p. 9). By that, he meant that grappling with the legacy of 250 years of slavery would be one of the great problems of the 20th century. Indeed, as this chapter shows, Du Bois was right, although he did not foresee that the "color line" would incorporate not only African Americans but Hispanics, Asians, Native Americans, and other minorities as well.

Slavery forcibly brought more than 9 million people, mostly from West Africa, to North and South America, as well as the islands of the Caribbean. Passage across the ocean was brutal: In many instances, half or fewer of the Africans who were forced onboard slave ships made it to their destination. The treatment of those who survived was no less horrific. In 1808, a federal law ended the importation of enslaved people to the United States. As a result, the internal slave trade grew, and more enslaved people were torn from their families and sold to work the plantations that were expanding westward.

Slave revolts were frequent throughout the South. More than 250 Black uprisings against slavery were recorded from 1700 to 1865, and many others were not recorded (Greenberg, 1996; Williams-Meyers, 1996).

Immediately following the end of slavery, during the Reconstruction period, African Americans sought to establish political and economic equality. The exploitation of African Americans did not end with slavery: The sharecropping system, under which Blacks farmed someone else's land and paid the owner with a fraction of the crop, kept former enslaved peoples tied to landowners who often charged excessive sums for rent, seed, and tools. This meant that sharecroppers, rather than earning money, ended up owing money to the landowner. Some progress was made, however, and former enslaved men gained the right to vote, and in some jurisdictions, they constituted the majority of registered voters. Black legislators were elected in every southern state. Between 1870 and 1901, 22 Blacks served in the U.S. Congress, while hundreds of others served in state legislatures, on city councils, and as elected and appointed officials throughout the South (Holt, 1977). Their success was short-lived. White southern legislators, still a majority, passed Jim Crow laws, which excluded Blacks from voting and using public transportation. In time, every aspect of life was constrained for Blacks, including access to hospitals, schools, restaurants, churches, jobs, recreation sites, and cemeteries. By 1901, because literacy was required to vote and most Blacks had been effectively denied even a basic education, the registration of Black voters had dwindled to a mere handful (Morrison, 1987).

The North's higher degree of industrialization offered a promise of economic opportunity, even though social and political discrimination existed there as well. Between 1940 and 1970, more than 5 million African Americans left the rural South for what they hoped would be a better life in the North (Lemann, 1991). Whether they stayed in the South or moved to the North, however, many Blacks were still being denied fundamental rights nearly 100 years after slavery was abolished.

By the 1950s, the social landscape was changing. Having fought against a racist regime in Nazi Germany in the 1940s, the United States found it far more difficult to justify continued discrimination against its own minority populations. Legal mandates for equality, however, did not easily translate to actual practices, and the 1960s saw a wave of protest from African Americans eager to claim greater rights and opportunities. The civil rights movement was committed to nonviolent forms of protest. Participants in the movement used strikes, boycotts, voter registration drives, sit-ins, and freedom rides in their efforts to achieve racial equality. In 1964, following what was at the time the largest civil rights demonstration in the nation's history (a march on Washington, D.C.), the federal government passed the first in a series of civil rights laws that made it illegal to discriminate on the basis of race, sex, religion, physical disability, or ethnic origin.

All African Americans have not benefited from legislation aimed at abolishing discrimination (Wilson, 2010). In particular, the poorest Blacks—those living in disadvantaged urban and rural areas—lack the opportunities that could make a difference in their lives. Many poor Black Americans suffer from having little cultural and social capital; inadequate schools, lack of skills and training, and poor local job opportunities raise the risk of persistent poverty. Today, nearly half of African American workers are employed in unskilled or semiskilled service jobs—as health care aides, janitors, and food service workers, for example—compared with only a quarter of whites. Few of these occupations offer the opportunity to earn a living wage.

Over time, many African Americans have been broadly successful in improving their economic circumstances, particularly as they have gained access to education. The economic gains of African Americans, however, have been more tenuous than those of most other groups. The Great Recession beginning in 2007 also substantially eroded Black wealth and income in America. Other economic downturns, such as the impact of COVID-19 and subsequent losses of jobs and livelihoods, has also had a disproportionate impact on the economic stability of African Americans and other communities of color.

Although African Americans have had mixed success in gaining economic power, as a group, they have seen important gains in political power. In the early 1960s, there were only 103 African Americans holding public office in the United States. By 1970, about 715 Blacks held elected office at the city and

In 2020, the nation mourned the deaths of two long-standing Black members of Congress and civil rights leaders, Elijah Cummings and John Lewis. Here, John Lewis waits outside the Capitol for the memorial service of Elijah Cummings with other members of the Congressional Black Caucus. Lewis, badly beaten by police during the "Bloody Sunday" march on the Selma bridge, was among the many instrumental in the passage of the Voting Rights Act of 1965.

county levels. By the 1990s, however, the figure had surpassed 5,000, and in recent decades, major cities—including New York, Los Angeles, Chicago, and Washington, D.C.—have elected Black mayors, and Black representation on city and town councils and state bodies has grown dramatically. At the federal level, there have been fewer gains, although clearly, President Barack Obama's election in November 2008 and reelection in 2012 represented a profound political breakthrough for Blacks. As of 2020, Congress had 57 Black members, its highest number historically.

Latinxs

The category of Latinx (or *Hispanic*, the term used by the U.S. Census Bureau) includes people whose heritages lie in the many different cultures of Latin America. According to a 2020 Pew Research Center estimate, 60.6 million Hispanics reside in the United States, making them the largest minority group. They constitute about 18% of the U.S. population. They are the second fastest-growing racial or ethnic group after Asian Americans because of both high immigration rates and high birthrates (Krogstad & Noe-Bustamante, 2020).

Today, about half of Hispanic adults were born in the United States, and about half are foreign-born. Many trace their ancestry to a time when the southwestern states were part of Mexico. During the mid-19th century, the United States sought to purchase from Mexico what is now Texas and California. Mexico refused to sell, leading to the 1846–1848 U.S.–Mexican War. Following its victory, the U.S. government forced Mexico to sell two fifths of its territory, enabling the United States to acquire all the land that now makes up the southwestern states, including California, Texas, New Mexico, and Arizona. Along with this vast territory came the people who inhabited it, including tens of thousands of Mexicans who were forced by white settlers to forfeit their property and who suffered discrimination at the hands of the whites (Valenzuela, 1992).

Latinxs and Hispanics are often referred to in a way that implies uniformity across members of this group, but these categories encompass many different ethnic groups with different experiences. They also have different ethnic roots ranging from Central America to South America to the Caribbean. The very terms *Latinx* and *Hispanic* are used interchangeably by some groups, or exclusively by others. For instance, the Census Bureau enlists the term *Hispanic*, while the Pew Research Center enlists both *Hispanic* and *Latino*. More important, people select different names for themselves, leading the Pew Research Center to conclude: "Who is Hispanic? Anyone who says they are. And nobody who says they aren't" (Lopez et al., 2020, para. 3; see Figure 9.6 for differences in identification). Like African Americans, Latinxs have experienced prejudice and discrimination in the United States. We next feature the diverse experiences of two prominent Latinx groups: Mexican Americans and Cuban Americans.

Mexican Americans

Mexicans are the largest group of Latinxs in the United States, accounting for over 60% of the population (Krogstad & Noe-Bustamante, 2020). A large proportion of the Mexican American population lives in California, New Mexico, Texas, and Arizona, although migration to the American South has grown substantially. The immigration of Mexican Americans has long reflected the immediate labor needs of the U.S. economy (Barrera, 1979; Muller & Espenshade, 1985). During the 1930s, state and local governments forcibly sent hundreds of thousands of Mexican immigrants back to Mexico, but when the United States experienced a labor shortage during World War II, immigration was again encouraged. After the war, the *bracero* (manual laborer) program enabled 4 million Mexicans to work as temporary farm laborers in the United States, often under exploitative conditions. The program was ended in 1964, but by the 1980s, immigration was on the upsurge, with millions of people fleeing poverty and political turmoil in Latin America by illegally entering the United States.

FIGURE 9.6 ■ U.S. Hispanic Population

Hispanics have mixed views on how they describe their identity

% of Hispanics saying they describe themselves most often as...

Country of origin/heritage 50
Hispanic/Latino 23
American 23

Do you prefer the term "Hispanic" or "Latino"?

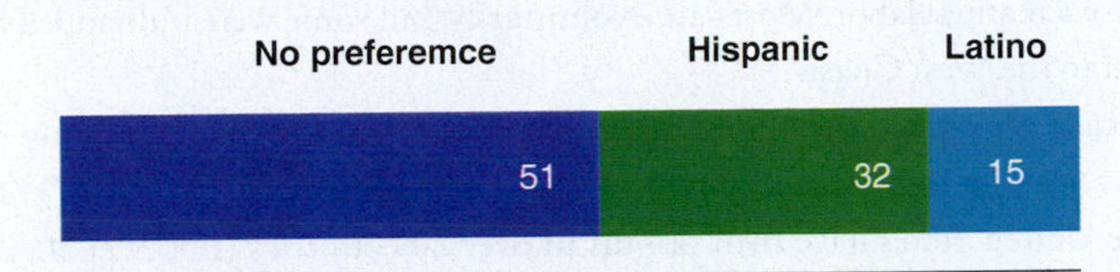

Source: Lopez, M. H., Krogstad, J. M., & Passel, J. S. (2020, September 15). *Who is hispanic?* https://www.pewresearch.org/fact-tank/2020/09/15/who-is-hispanic/

The debate over how to deal with undocumented immigrants seeking U.S. jobs has been discussed heatedly in American politics, although the numbers of such immigrants appear to be falling. In fact, data suggest that a host of factors—including the dangers of crossing the U.S.–Mexico border due to drug violence on the Mexican side, harsher immigration laws in states from Georgia to Arizona, and the dearth of employment in the wake of the recession—have pushed illegal immigration down (Massey, 2011). The vast majority of Mexicans in the United States (80%) are legal residents, and most (70%) are born in the United States (Krogstad & Noe-Bustamante, 2020).

Cuban Americans

When Fidel Castro came to power in Cuba in 1959, Cubans who opposed his communist regime sought to emigrate to the United States. These included some of Castro's political opponents, but most were middle-class Cubans whose standard of living was declining under Castro's economic policies and the U.S. economic embargo of Cuba. About half a million entered the United States, mostly through Florida, which is only 90 miles away from Cuba. After that first wave of immigration, Cubans came to the United States in several other waves, following policy changes in both countries.

The Cuban community has enjoyed both economic and, perhaps as a result, political success in Florida (Ferment, 1989; Stepick & Grenier, 1993). In contrast to many immigrants, who tend to come from the poorer strata of their native countries, Cuban immigrants were often highly educated professionals and businesspeople before they fled to the United States. They brought with them a considerable reserve of training, skill, and sometimes wealth.

The anticommunist stance of the new migrants was also a good fit with U.S. foreign policy and the effort to crush communist politics in the Western Hemisphere. The U.S. government provided financial assistance such as small-business loans to Cuban immigrants. It also offered a spectrum of language classes, job-training programs, and recertification classes for physicians, architects, nurses, teachers, and lawyers who sought to reestablish their professional credentials in the United States. About three quarters of all Cubans arriving before 1974 received some form of government benefits, the highest rate of any minority community.

In 2017, Cubans were the third-largest Hispanic population living in the United States, with 2.3 million calling it home (Noe-Bustamante et al., 2019). Although Cuban migrants have tended to have higher levels of education than Mexican immigrants do, their status in the United States has also been strongly influenced by the political and economic (that is, the structural) needs of the U.S. government. U.S. economic needs have determined whether the government encouraged or discouraged the migration of manual labor power from Mexico, while political considerations fostered a support network for migrants who endorsed the U.S. government's effort to end the Castro regime in Cuba.

Asian Americans

Asian immigrants were instrumental in the development of industrialization in the United States. In the mid-1800s, construction began on the transcontinental railroad that would link California with the industrializing East. At the same time, gold and silver were discovered in the American West. Labor was desperately needed to mine these natural resources and to work on building the railroad. At the time, China was suffering a severe drought. American entrepreneurs seeking labor quickly saw the potential fit of supply and demand, and hundreds of thousands of Chinese were brought to the United States as inexpensive manual labor. Most came voluntarily, but some were kidnapped by U.S. ship captains and brought to the West Coast.

Asian Americans are culturally and linguistically diverse, and the wide-ranging characteristics of this population cannot be fully captured by the term *Asian American* (Figure 9.7). The over 20 million Asians in the United States have their origins in over 20 countries (López et al., 2017). The Asian American population grew more rapidly between 2000 and 2015 than any other racial and ethnic group in the United States—from 11.9 million to 20.4 million (López et al., 2017). The largest numbers of Asian Americans are from India, China, and the Philippines, but those from Japan, Korea, and Vietnam number over 1 million people each. The number of Asian Americans is growing, primarily because of immigration. (Asian American birthrates, following the pattern in Asian countries, are low.) Asian Americans are expected to become the largest minority group in the United States in 50 years, surpassing the Hispanic population (López et al., 2017, para. 8).

FIGURE 9.7 ■ Countries of Origin of the Asian Population in the United States, 2015

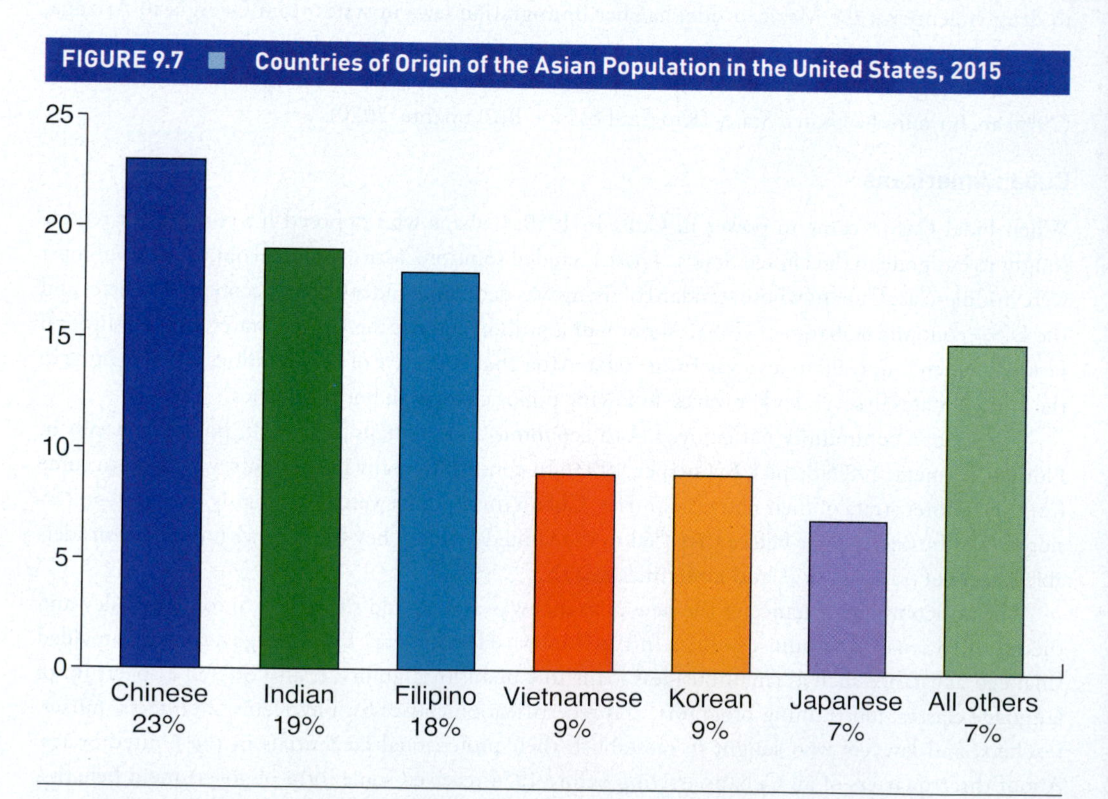

Source: Data from Budiman, A., Cilluffo, A., & Ruiz, N. G. (2019, May 22). *Key facts about Asian origin groups in the United States.* https://www.pewresearch.org/fact-tank/2019/05/22/key-facts-about-asian-origin-groups-in-the-u-s/

Note: Chinese includes those identifying as Taiwanese. Figures may not add up to 100% due to rounding.

Asian Americans, similar to members of other minority groups, have experienced prejudice and discrimination. For example, during World War II, many Japanese Americans were forcibly placed in internment camps, and their property was confiscated (Harth, 2001). The U.S. government feared their allegiances were with Japan—an enemy in World War II—rather than the United States. Although the teaching of German was banned in some schools, drastic measures were not taken against German or Italian Americans, although Germany and Italy were allied with Japan during the war.

In spite of obstacles, the Asian American population has been successful in making economic gains. Asian Americans as a whole have the highest median household income of any minority group, as well as the lowest rates of divorce, teenage pregnancy, and unemployment. Their economic success reflects the fact that, similar to Cuban Americans, recent Asian immigrants have come from higher socioeconomic backgrounds, bringing greater financial resources and more human capital (Zhou, 2009). Family networks and kinship obligations also play an important role. In many Asian American communities, informal community-based lending organizations provide capital for businesses, families and friends support one another as customers and employees, and profits are reinvested in the community (Ferment, 1989; Gilbertson & Gurak, 1993; Kasarda, 1993).

At the same time, there are substantial differences within the Asian American population. Nearly half of all Asian Indian Americans, for example, are professionals, compared with less than a quarter of Korean Americans, who are much more likely to run small businesses and factories. Income inequality is, according to Budiman et al. (2019), more rapidly growing for the "Asian American" group than for any other racial or ethnic group (para. 8). While the median income of Asian households in 2018 was almost $26,000 over all other households in the United States ($87,243 for Asian Americans and $61,937 for all other Americans), these numbers mask considerable inequality within the group. For instance, in 2016, Asians at the top 10% of the income ladder made over 10 times more than those at the bottom: $133,529 versus $12,478 (Kocchar & Cilluffo, 2018). Poverty and hardship are more prevalent among those from Southeast Asia, particularly immigrants from Cambodia and Laos, which are deeply impoverished countries with less-educated populations.

Arab Americans

Over 3.5 million Arab Americans live in the United States. Hailing from over 22 countries, they are a religiously and culturally diverse population. Contrary to popular perceptions, most members of today's Arab American population were born in the United States. Immigrants from Arab countries began coming to the United States in the 1880s. The last major wave of Arab migration occurred in the decade following World War II (PBS, 2011). The majority trace their roots back to Lebanon, Syria, Palestine, Egypt, and Iraq (Arab American Institute, 2012).

Arab Americans face a unique set of challenges and obstacles. Immediately following the September 11, 2001, terrorist attacks, Arab Americans as a group suffered stigma, and acts of discrimination against them increased sharply. The terror attacks provided an avenue for legitimating the views of racists and xenophobes (Salaita, 2005), but institutionalized prejudice was also obvious in the increase in "random" screenings and searches that targeted them.

Because the U.S. Census Bureau does not measure Arab Americans as a distinct group (an issue we broach later), official information on them is difficult to obtain. According to statistics from the Arab American Institute Foundation (2011), Arab Americans, as a demographic group, have been notably successful in these two areas: 45% have at least a bachelor's degree or higher (compared with 28% for the general population), and in 2008 their mean individual income was 27% higher than the national average.

©Andree Kehn/Sun Journal via AP

White Americans

White people have always made up a high percentage of the foreign-born population in the United States. The stereotype of minorities as people of color leaves out significant pockets of white ethnic minorities. In the early days of mass migration to the United States, the dominant group was white Protestants from Europe. Whites from Europe who were not Protestant—particularly Catholics and Jews—were defined as ethnic minorities, and their status was clearly marginal in U.S. society.

In mid-19th-century Boston, for example, Irish Catholics made up a large portion of the city's poor and were stereotyped as drunken, criminal, and generally immoral (Handlin, 1991). The term *paddy wagon*, used to describe a police van that picks up the drunk and disorderly, came from the notion that the Irish were such vans' most common occupants. (*Paddy* was a derogatory term for an Irishman.) Over time, however, white ethnic groups assimilated with relative ease because, for the most part, they looked much like the dominant population (Prell, 1999). To further their assimilation, some families changed their names and encouraged their children to speak English only. Today, most of these white ethnic groups—Irish, Italians, Russians, and others—have integrated fully into U.S. society (Figure 9.8). They are, on the whole, born wealthier, have more social capital, and encounter fewer barriers to mobility than do other racial and ethnic groups.

If dark skin has historically been an obstacle in U.S. society, light skin has been an advantage. That advantage, however, may go unrecognized. As Peggy McIntosh (1990) writes in her thought-provoking article "White Privilege: Unpacking the Invisible Knapsack":

> As a white person, I realized I had been taught about racism as something that puts others at a disadvantage, but had been taught not to see one of its corollary aspects, white privilege, which puts me at an advantage.... I have come to see white privilege as an invisible package of unearned assets that I can count on cashing in each day, but about which I was "meant" to remain oblivious. (p. 31)

McIntosh goes on to outline these "assets," including the following: When whites are told about their national heritage or the origins of civilization, they are shown that white people shaped it. Posters, picture books, greeting cards, toys, and dolls all overwhelmingly feature white images. As a white

FIGURE 9.8 ■ **Legal Migration to the United States by Region of Origin, 1820–2014**

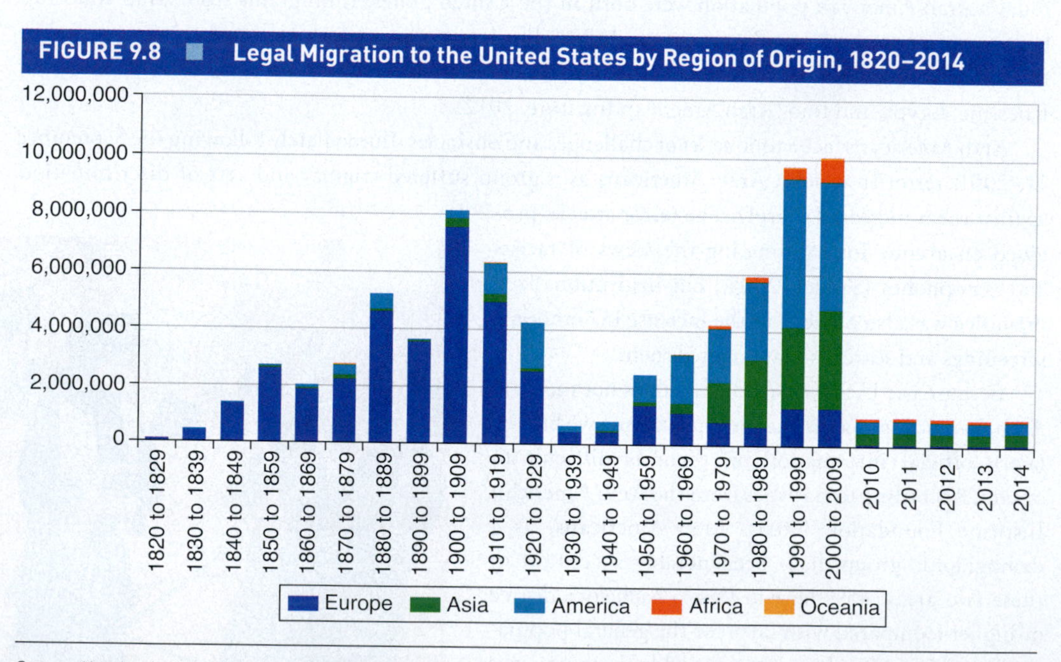

Source: Yearbook of immigration statistics: 2015 legal permanent residents. U.S. Department of Homeland Security.

person, she need not worry that her coworkers at an affirmative action employer will suspect she got her job only because of her race. If she needed any medical or legal help, her race alone would not be considered a risk factor or hindrance in obtaining such services. These are only a few among the 50 unearned privileges of white skin that McIntosh (1990) identifies. Her argument is essentially that obliviousness to the advantages of white skin "maintain[s] the myth of meritocracy, the myth that democratic choice is equally available to all" (p. 36). Recognition of skin privilege, in contrast, would be a challenge to the structure of power, which prefers to ignore whatever does not support the story of achievement as an outcome of merit.

Is the recognition of racial or ethnic advantages as important as the recognition of racial or ethnic disadvantages? Can this help to level the field of opportunity for all Americans? What do you think?

Multiracial Americans

Naming oneself or being named by someone else is an important sociological action. It may be an exercise of power or an act of recognition. Sociologist Pierre Bourdieu (1991) has pointed out that language is a medium of power, a vehicle that may confer status or function as a means of devaluation and exclusion. Historically, minority groups have been denied opportunities to name themselves. In 2000, for the first time, the U.S. Census Bureau's questionnaire permitted respondents to select more than one race in describing themselves. Also in 2000, about 2.4% of Americans, or 6.8 million people, identified themselves as multiracial. In the 2010 Census, that figure grew to 9 million people (2.9% of Americans). Interestingly, the self-reported multiracial population grew much more dramatically than the single-race population: The former rose by 32%, while the latter rose by just over 9% (Jones & Bullock, 2012).

In 2020, the Census registered a change in the way it recorded race and ethnicity (see Figure 9.9 for the 2020 race questions). As it describes this change:

> Individuals who identify as White, Black/African American, and/or American Indian or Alaska Native will be asked to specifically identify their racial origins. As outlined in the planned Census 2020 question, Black/African American individuals, for example, will be asked to print their specific origin (e.g. African American, Jamaican, Haitian, Nigerian, Ethiopian, Somali, etc.). Write in options will still be available for "Other Asian" and "Some other race" categories.

The rise in the numbers of U.S. residents describing themselves as belonging to more than one race should perhaps not come as any surprise. On the one hand, more children are being born into interracial partnerships: in 2016, 10.2% of heterosexual married couples were interracial, a number that has been rising each year (Rico et al., 2018). On the other hand, because this is a self-reported status, more people may be choosing to identify themselves as multiracial as the availability of the category becomes more widely known and as multiracial status becomes more broadly accepted.

Notably, being biracial or multiracial is not a new phenomenon. It is, in some respects, a newly named phenomenon that is also of growing interest to researchers. The politics of multiracialism can be complicated, with competing interests at play. For example, leading up to the 2000 Census, many mixed-race individuals and advocacy groups favored the addition of an entirely separate *multiple race* option on the Census questionnaire, as opposed to allowing respondents to check multiple boxes from among the established racial categories. They argued that the addition of such an option would allow for fluidity in the way that individuals identify themselves, freeing them from the constraints of categories imposed on them by the state. Others, including some prominent civil rights leaders, countered that identification according to the established racial categories is necessary to ensure that civil rights legislation is enforced and that inequalities in housing, income, education, and so on are documented accurately (Roquemore & Brunsma, 2008). On the one hand, the lived experience of multiracial Americans and their right to self-identify must be respected; on the other hand, the Census Bureau must fulfill its obligation to provide legislators with accurate information on inequality. Social science research tells us that individuals' racial identification plays an important role in how others interact with them, their social networks, and even their life chances (Khanna, 2010). Thus, the question of

FIGURE 9.9 ■ Race, Ethnicity, and Hispanic Origin Questions, U.S. Bureau of the Census Form, 2020

NOTE: Please answer BOTH Question 5 about Hispanic origin and Question 6 about race. For this Census, Hispanic origins are not races.

5. Is this person of Hispanic, Latino, or Spanish origin?

☐ No, not of Hispanic, Latino, or Spanish origin
☐ Yes, Mexican, Mexican Am., Chicano
☐ Yes, Puerto Rican
☐ Yes, Cuban
☐ Yes, another Hispanic, Latino, or Spanish origin - *Print origin, for example, Argentinean, Colombian, Dominican, Nicaraguan, Salvadoran, Spaniard, and so on.*

6. What is this person's race? *Mark* ☒ *one or more boxes.*

☐ White
☐ Black, African Am., or Negro
☐ American Indian or Alaska Native — *Print name of enrolled or principal tribe.*

☐ Asian Indian ☐ Japanese ☐ Native Hawaiian
☐ Chinese ☐ Korean ☐ Guamanian or Chamorro
☐ Filipino ☐ Vietnamese ☐ Samoan
☐ Other Asian — *Print race, for example, Hmong, Laotian, Thai, Pakistani, Cambodian, and so on.*
☐ Other Pacific Islander — *Print race, for example, Fijian, Tongan, and so on.*

☐ Some other race – Print race.

Is this person of Hispanic, Latino, or Spanish Origin?

☐ **No**, not of Hispanic, Latino, or Spanish origin
☐ Yes, Mexican, Mexican Am., Chicano
☐ Yes, Puerto Rican
☐ Yes, Cuban
☐ Yes, another Hispanic, Latino, or Spanish origin - *Print, for example, Salvadoran, Dominican, Colombian, Guatemalan, Spaniard, Ecuadorian, etc.*

What is this person's race?

Mark ☒ *one or more boxes* **AND** *print origins.*

☐ White - *Print, for example, German, Irish, English, Italian, Lebanese, Egyptian, etc.*

☐ Black or African Am - *Print, for example, African, American, Jamaican, Haitian, Nigerian, Ethiopian, Somali, etc.*

☐ American Indian or Alaska Native - *Print name for enrolied or principal tribe(s) for example, Navajo Nation, Blackfeet Tribe, Mayan, Aziec Native Village of Barrow Inupiat Traditional Government, Name Eskimo Community, etc.*

☐ Chinese ☐ Vietnamese ☐ Native Hawaiian
☐ Filipino ☐ Korean ☐ Samoan
☐ Asian Indian ☐ Japanese ☐ Chamorro
☐ Other Asian — *Print, for example, Pakistani, Cambodian, Hmong, etc.*
☐ Other Pacific Islander — *Print, for example, Tongan, Fijian, Marshallese, etc.*

☐ Some other race – *Print race or origin.*

Source: U.S. Census Bureau. (2020). *Race/ethnicity and the 2020 census*. https://www.census2020now.org/faces-blog/same-sex-households-2020-census-r3976#:~:text=The%20OMB%20definitions%20of%20race%20and%20ethnicity%20in,consistent%2C%20Census%202020%20will%20feature%20an%20important%20change

how a person self-identifies—and how others perceive them—has real consequences in that individual's everyday life.

We earlier introduced the experience of Arab Americans. One challenge that the Census poses for this group *is* the problem of inclusion. As an article by Adely (2020) explains, Arab Americans are not included as a possible group on the Census. Instead, people from the Middle East and North Africa are labeled as "white," a labeling that means that this diverse group of people does not have its own tallying for important markers such as income, education, housing, health, etc. As Rania Mustafa, the executive director of the Palestinian American Community Center, puts it:

> "Arabs have a minority experience and being identified as white takes away from that experience and simplifies it a lot," she said. "We don't have the white privilege and we don't have the white experience. We are a minority." She advises people to check off "other" in the race category and to fill in their country of origin. "It's a valid way to fill out the census and helps the community to be counted," she said. (cited by Adely, 2020, paras. 13–14)

Interestingly, according to Jurjevich (2019), the Census Bureau had suggested in 2017 that the Office of Management and Budget (OMB) adopt definitional changes including "a new race category for individuals identifying as Middle East and North African (MENA)." They note that "The OMB did not respond by December 31, 2017 to these recommendations, effectively forcing the Bureau to retain the race/ethnicity definitions used in Census 2010" (Jurjevich, 2019).

RACE AND ETHNICITY IN A GLOBAL PERSPECTIVE

Although our focus to this point has been on the United States, issues of racial and ethnic identities and inequalities are global in scope. Around the world, unstable minority and majority relationships, fears of rising migration from developing countries, and even ethnic and racial hatreds that explode into brutal violence highlight again how important it is for us to understand the causes and dynamics of categories that, while socially constructed, are powerfully real in their consequences.

European countries such as France, Germany, and Denmark have struggled in recent years to deal with challenges of immigration, discrimination, and prejudice. Migration to Northern and Western Europe from developing states, particularly those in North Africa and the Middle East, has risen, driven by a combination of *push factors* (social forces such as dire economic conditions and inadequate educational opportunities that encourage migrants to leave their home countries) and *pull factors* (forces that draw migrants to a given country, such as labor shortages that can be filled by low- or high-skilled migrants or an existing ethnic network that can offer new migrants support). In a short period of time, the large numbers of job seekers from developing countries in search of better lives have transformed the populations of many European countries from largely ethnically homogeneous to multicultural—a change that has caused some political and social tensions.

Tensions in Europe revolve, at least in part, around the fact that a substantial proportion of migrants from developing states are Muslim, and some have brought with them religious practices, such as veiling, that are not perceived to be a good fit with largely secular European societies. By some estimates, Muslims will account for about a fifth of the population of the European Union by 2050 (Mudde, 2011). In 2011, France, in which about 10% of the population is Muslim, became the first country in Europe to ban the full-face veil, also known as the *burqa*, which is the customary dress for some Muslim women. Similar legislation has been introduced—but not passed—in Belgium and the Netherlands. Earlier, when France banned all overt religious symbols from schools in 2004, civil defiance and violence followed. The Parisian suburbs burned as young French Muslims rioted, demanding opportunity and acceptance in a society that has traditionally narrowly defined who is "French."

In some cases, ethnic and racial tensions have had even more explosive effects. Earlier in this chapter, we talked about four forms that majority–minority relations can take: expulsion, segregation, assimilation, and cultural pluralism. Here, we add a fifth form: genocide.

Genocide: The Mass Destruction of Societies

Genocide: The mass, systematic destruction of a people or a nation.

Genocide can be defined as *the mass, systematic destruction of a people or a nation*. According to the Convention on the Prevention and Punishment of the Crime of Genocide, adopted by the United Nations General Assembly in 1948, when genocide was declared a crime under international law, it included "any of the following acts committed with intent to destroy, in whole or in part, a national, ethnical, racial or religious group, as such":

a. Killing members of the group;

b. Causing serious bodily or mental harm to members of the group;

c. Deliberately inflicting on the group conditions of life calculated to bring about its physical destruction in whole or in part;

d. Imposing measures intended to prevent births within the group;

e. Forcibly transferring children of the group to another group;

The mass killing of people because of their membership in a particular group is ages old. Hans Van Wees (2010) notes that ancient Christian and Hebrew religious texts contain evidence of early religiously motivated mass killings. Roger W. Smith (2002) points out that the Assyrians, Greeks, Romans, and Mongols, among others, engaged in genocide, but their perspective on mass killing was shaped by their time. He notes that rulers boasted about the thousands they killed and the peoples they eliminated. Genocide as both a term and a defined crime is much more recent.

The word *genocide* was coined in the 1940s by Polish émigré Raphael Lemkin, who left his native country in the late 1930s, several years before Adolf Hitler's Nazi army overran it. A lawyer and an academic, Lemkin was also a crusader for human rights, and aware of Nazi atrocities in Europe, he sought to inspire U.S. decision makers to act against the mass killing of Jews and other targeted groups. In a 1941 radio broadcast, British prime minister Winston Churchill decried the "barbaric fury" of the Nazis and declared, "We are in the presence of a crime without a name." Lemkin, who heard the broadcast, undertook to name the crime and, through this naming, to bring an end to it (Power, 2002).

Sociologist Pierre Bourdieu (1991) points to the significance of language in structuring the perception that social agents (such as governments or populations) have of the world around them. The very act of naming a phenomenon confers on it a social reality. Lemkin sought to use the act of naming as a vehicle for action against Nazi atrocities and against future threats to the physical, political, cultural, and economic existence of collectivities (Shaw, 2010). Lemkin settled on *genocide*, a hybrid of the Greek *geno*, meaning *tribe* or *race*, and the Latin *cide* (from *caedere*), or *killing*.

The 1948 UN Convention on the Prevention and Punishment of the Crime of Genocide created a potentially powerful new political reality, although it narrowed the definition from the broad articulation championed by Lemkin (Shaw, 2010). Theoretically, state sovereignty could no longer shield a country from the consequences of committing genocide because the nations that signed the UN convention were bound by it to prevent, suppress, and punish such crimes. At the same time, the notion of an obligation to intervene is loose and open to interpretation, as history has shown. Humanitarian organizations such as Genocide Watch and institutions such as the U.S. Holocaust Memorial Museum have sounded the alarm on recent atrocities, often labeling them genocide even as many governments have resisted using the term.

Genocides in the second half of the 20th century include slaughters carried out by the Khmer Rouge in Cambodia in the 1970s, Saddam Hussein's 1987–1988 campaign against ethnic Kurds in Iraq, and Serb atrocities in the former Yugoslav states of Bosnia-Herzegovina and Kosovo in the late 1990s (Figure 9.10). The 21st century has already seen the genocidal destruction of Black Africans, including the Darfuris, Abyei, and Nuba, by Arab militias linked to the government in Sudan.

One of the most discussed cases of genocide in the late 20th century took place in Rwanda, where, in 1994, no fewer than 800,000 ethnic Tutsis were killed by ethnic Hutus in the space of 100 days. Many killings were done by machete, and women and girls were targeted for sexual violence. By most

FIGURE 9.10 ■ Select Genocides Around the World, 1914–Present

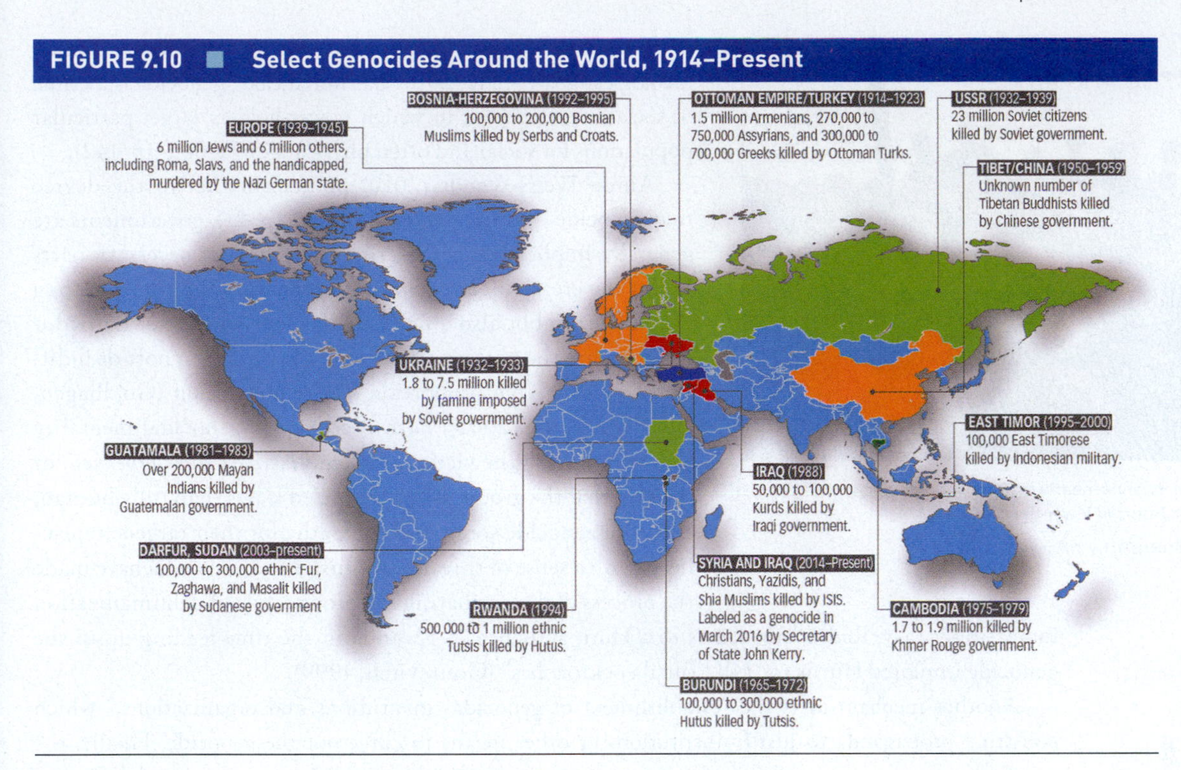

Source: Adapted from Online Resources from The Choices Program, Brown University.

Note: The term *genocide* has not been legally applied to all mass killings listed above. The number of victims has also been disputed for each of these killings.

accounts, Tutsis and Hutus are not so different from one another; they are members of the same ethnic group, live in the same areas, and share a common language. One cultural difference of considerable importance, however, is that for the Tutsis, the cow is sacred and cannot be killed or used for food. The Hutus, who have a long tradition of farming, were forced to leave large areas of land open for cattle grazing, thus limiting the fertile ground available for farming. As the cattle herds multiplied, so did tensions and conflict. Notably, for many years, the Tutsis, while fewer in number (representing approximately 14% of the population), dominated the Hutus politically and economically (Kapuscinski, 2001).

Rwanda was long a colony of Belgium, but in the 1950s, Tutsis began to demand independence, as did many other colonized peoples in Africa. The Belgians incited the Hutus to rebel against the Tutsi ruling class. The rebellion was unsuccessful, but the seeds for genocide were planted in fertile soil, and sporadic ethnic violence erupted in the decades that followed. Belgium finally relinquished power and granted Rwanda independence in 1962. From the early 1970s, the government was headed by a moderate Hutu leader, Major General Juvénal Habyarimana (Rwanda, 2008). In the early 1990s, General Habyarimana agreed to a power-sharing arrangement with Tutsis in Rwanda, angering some nationalist Hutus, who did not want to see a division of power. In 1994, the general was assassinated (his private jet was shot down), and radical Hutus took the opportunity to turn on both Tutsi countrymen and moderate Hutus, murdering at least 800,000 people in a few short months.

In the next section, we consider the question of how genocide—such as the mass killing of Rwanda's Tutsi population in 1994—looks through a sociological lens.

What Explains Genocide?

Taking a sociological perspective, we might ask, Who benefits from genocide? Why do countries or leaders make a choice to pursue or allow genocidal actions? Genocide is the product of conscious decisions made by those who stand to benefit from it and who anticipate minimal costs for their actions.

Chalk and Jonassohn (1990) suggest that those who pursue genocide do so to eliminate a real or perceived threat, to spread fear among enemies, to eliminate a real or potential threat to another group,

In the 1994 genocide in Rwanda, nearly a million ethnic Tutsis were systematically killed by ethnic Hutus in fewer than 100 days.

to gain wealth or power for a dominant group, and/or to realize an ideological goal. Shaw (2010) has noted that "genocide is a crime of social classification, in which power-holders target particular populations for social and often physical destruction" (p. 142).

Anton Weiss-Wendt (2010) points out that to the degree that genocide "requires premeditation" (p. 81), governments are generally implicated in the crime. Among the necessary parts is also the creation of a "genocidal mentality," not only among the leadership but also among a sufficient number of individuals to ensure the participation—or at least the support or indifference—of a large proportion of the population (Goldhagen, 1997; Markusen, 2002). Part of creating a genocidal mentality is dehumanizing the victims. By emphasizing their *otherness*, or the idea that the group to be victimized is less than fully human, agents of genocide stamp out sympathy for their targets, replacing it with a sense of threat or disgust. Modern media have made the process of disseminating the propaganda of dehumanization increasingly easy. Radio broadcasts to a Hutu audience in Rwanda in the time leading up to the genocide implored Hutus to "kill [Tutsi] cockroaches" (Gourevitch, 1999).

Another mechanism is the "establishment of genocidal institutions and organizations," which construct propaganda to justify institutions or other means to carry out the genocide. Finally, the recruitment and training of perpetrators are important, as is the establishment of methods of group destruction (Markusen, 2002). In Rwanda, it is notable that some Hutu Power groups had, in the early 1990s, begun to arm civilians with hand weapons such as machetes, ostensibly as self-defense against Tutsi rebels. Many of these were later used in the genocide. As Hutus and Tutsis lived side by side in many cities and rural areas, identification of victims was rapid and easy.

Choosing genocide as a policy incurs costs as well as benefits. Clearly, outside forces can impose costs on a genocidal regime; we have seen that signatories to the 1948 UN convention are theoretically obligated to intervene. But historically, intervention has often been too little and too late. Why, in the face of overwhelming evidence of genocide (as in Rwanda or Bosnia-Herzegovina or Sudan), does the international community not step in? One answer is that these states also make choices based on perceived costs and benefits. Intervention in the affairs of another state can be costly, and many leaders avoid it. In a historical overview of U.S. responses to genocide, for instance, Samantha Power (2002) argues that "American leaders did not act because they did not want to. They believed that genocide was wrong, but they were not prepared to invest the military, financial, diplomatic, or domestic political capital needed to stop it" (p. 508).

Genocide is an extreme and brutal manifestation of majority–minority relations. It is the effort to destroy indiscriminately an entire racial or ethnic group. Is genocide a continuing pattern for states or regimes wishing to destroy groups labeled by powerful leaders as inferior or dangerous? How should outside states—and concerned individuals—respond when conflict and intolerance evolve into genocide? What do you think?

WHAT'S NEXT? RACE AND ETHNICITY IN THE UNITED STATES

The United States is today at its most racially diverse in its over 244 years (Budiman, 2020b). And, as we have noted, racial and ethnic diversity will continue to grow. While race and ethnicity were strongly politicized under the Donald J. Trump administration, a survey in the October leading up to a contentious 2020 presidential election showed that the majority of Americans viewed the increasing racial diversity in the United States more *positively* than negatively. Indeed, they had grown more positive toward this diversity than they were in 2016, shortly before Trump was elected. The October survey also showed, perhaps unsurprisingly, that younger generations (Gen X and Millennials) held more positive attitudes toward racial diversity than had the generations before them (Budiman, 2020b).

On the other hand, race and ethnicity were certainly on the minds of some voters. In August 2020, 52% of registered voters (over half) said that race and ethnic inequality were important to their vote (Pew Research Center, 2020). In June and July, America saw what *The New York Times* labeled "the largest social movement" as thousands took to the streets to support #BlackLivesMatter and to protest racial inequality and police brutality. Finally, in early 2020 and scattered throughout the news that year, outlets featured stories about hate groups that were "prepping for a race war," with the Southern Poverty Law Center (2020) noting a 55% increase in white nationalist hate groups from 2017 to 2019. So, what's next? Thinking sociologically, what types of changes do you foresee in racial and ethnic relations in the next 10 years? In 20 years? How will individuals change and challenge institutional discrimination? What are the most important areas to tackle? Who will be part of the change? How would you study these sociologically?

WHAT CAN I DO WITH A SOCIOLOGY DEGREE?

Advocating for Social Justice

As you learned in Chapter 1, sociology emerged in the midst of the large-scale social changes of the 19th century. While many early sociologists such as Comte, Marx, and Durkheim sought to understand, predict, and even manage social change, others, including Park, Du Bois, Martineau and Addams, were not only social theorists and researchers but also social activists. They were outraged by social inequality and used their sociological knowledge to advocate for social justice. For Mills, social responsibility was an integral part of sociology.

Many of us have a desire for a more just society and world. The capacity to actively advocate for social justice, whether personally or professionally (as a career), requires knowledge and skills that can be gained by studying sociology. First and foremost, it requires that we understand the complex nature of injustice. We live in an individualistic culture, where success and failure are often seen as proportional to one's effort and talent. While *agency* (the power to exercise free will and affect change) is an important concept in sociology, as a field, sociology seeks to provide a more multifaceted understanding of social life. By examining how the unequal distribution of opportunities, resources, and power shapes life chances, sociology students gain a more sophisticated understanding of the nature of social inequality. They learn the critical difference between individual discrimination and discrimination that is built into the policies and practices of social institutions such as education and the criminal justice systems. Despite the centrality of social structure, sociology does not discount agency. Sociology students gain a framework for understanding the nature of social movements and examine forces that may foster or impede social change.

Many sociology students have opportunities to not only study social justice but also to engage in social justice work through group organizations on campus and service-learning, internships, and other experiential learning opportunities in off-campus community agencies. Anyone can work for social justice on an individual level by being aware of their own prejudices and privileges and by taking action in their personal and professional lives—for example, by confronting acts of prejudice and discrimination, being an advocate and ally for oppressed groups, and voting for candidates and causes that support a social justice mission. You can practice your own career in a way that contributes to social justice (for example, by doing pro bono work and using the power you may have to hire people) or you may choose to follow a career that has an explicit focus on social justice. Some colleges and universities now offer graduate degrees in social justice fields (such as social justice education). Many of the other career skills outlined in this book, including those concerning interpersonal skills (Chapter 4), diversity (Chapter 8), team work (Chapter 5), ethical decision making (Chapter 10), community resources and services (Chapter 13), and social change (Chapter 18), are necessary in social justice work.

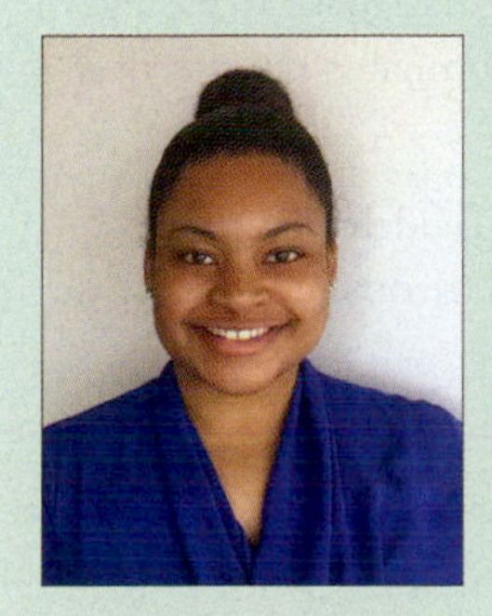

Kelsey Edwards, Program Assistant at Equal Justice Works

The George Washington University, BA in Sociology and Human Services & Social Justice, Minor in Sustainability

Through my sociology degree, I was able to study and research societal issues and problems affecting our population and develop solutions on how to address those issues to combat injustice. My sociology major has helped me understand the world around me, including the institutional structures and barriers that make it harder for people to succeed and cause an unleveled playing field. As a sociology major, I focused mainly on how the intersection of race, gender, and class affects one's ability to succeed in our society. Knowing the structural and institutional barriers in our society gives me a different perspective when meeting new people and hearing their background and testimonies.

After graduating from The George Washington University with a degree in sociology, I started working as a program assistant at Equal Justice Works. In my role, I work with lawyers and law students to seek social justice for clients in need, including veterans, the elderly, those experiencing homelessness, survivors of natural disasters, and those who are unemployed and underemployed. We provide resources and opportunities to individuals who are from underprivileged and marginalized backgrounds. I gained many practical skills as a sociology major, including research, writing, and critical thinking that has helped me grow as a working professional. When applying for grants and funding, we use research-based evidence, including quantitative and qualitative research, to show foundations and grant makers how our current and proposed programs can make a difference in the lives of the people we serve. Additionally, the major has given me a deeper understanding of the complex systems that are contributing to individual experiences of the clients our organization serves as well as all individuals experiencing injustice within our society.

My future aspirations are to work in race-based policy making to ensure that your race/ethnicity and where you live do not determine whether you live.

Regardless of one's ethnicity, socioeconomic status, or level of education, we all deserve equal access to opportunities and resources. With my background in sociology, I have the right skills, tools, and understanding of historical context, which allows me to successfully advocate for systemic change.

Career Data: Paralegals/Legal Assistants

- 2019 Median Pay: $51,740 per year
- $24.87 per hour
- Typical Entry-Level Education: Associate degree
- Job Outlook, 2019–2029: 10% (Much faster than average)

Source: Bureau of Labor Statistics. (2019). *Occupational outlook handbook.*

SUMMARY

- **Race** is not a biological category but a social construct. Its societal significance derives from the fact that people in a particular culture believe, falsely, that there are biologically distinguishable races and then act on the basis of this belief. The perceived differences among races are often distorted and lead to **prejudice** and **discrimination**.
- Many societies include different ethnic groups with varied histories, cultures, and practices.
- **Minorities**, typically because of their race or **ethnicity**, may experience prejudice and discrimination. Different types of minority–dominant group relations include **expulsion**, **assimilation**, **segregation**, and **cultural pluralism**.
- Prejudice usually relies on **stereotyping** and **scapegoating**. During difficult economic times, prejudice may increase as people seek someone to blame for their predicament.
- The civil rights struggle that began in the United States more than 60 years ago led to passage of civil rights and affirmative action legislation that has reduced but not eliminated the effects of prejudice and discrimination against minorities.
- Although discrimination is against the law in the United States, it is still widely practiced. **Institutionalized discrimination**, in particular, results in the unequal treatment of minorities in employment, housing, education, and other areas.
- The United States is a multiethnic, multiracial society. Minority groups—including American Indians, African Americans, Latinxs, Asian Americans, and Arab Americans—make up close to 40% of the population. This figure will continue to rise as a result of new births and immigration, changing the demographic face of the country.

- Historically, the vast majority of immigrants to the United States have come from Europe, but in recent years, the pattern has changed; currently, most immigrants come from Latin America and Asia.
- With a few exceptions, minority groups are disadvantaged in the United States relative to the majority-white population in terms of income, numbers living in poverty, and the quality of education and health care, as well as political voice. Although the disadvantages of dark skin are recognized, the privileges of light skin have not been widely acknowledged.
- **Genocide** is the mass and systematic destruction of a people or a nation. The 1948 United Nations Convention on the Prevention and Punishment of the Crime of Genocide obligates signatories to act against genocide.

KEY TERMS

assimilation (p. 222)
cultural pluralism (p. 222)
discrimination (p. 226)
environmental racism (p. 220)
ethnicity (p. 216)
expulsion (p. 217)
genocide (p. 242)
individual discrimination (p. 226)
institutionalized discrimination (p. 226)
minorities (p. 216)
prejudice (p. 225)
race (p. 216)
racism (p. 223)
refugees (p. 217)
restrictive covenants (p. 218)
scapegoating (p. 223)
segregation (p. 217)
social epidemiology (p. 228)
stereotyping (p. 225)
stigma (p. 224)

DISCUSSION QUESTIONS

1. If genetic differences between people who have different physical characteristics (such as skin color) are minor, why do we, as a society, continue to use race as a socially significant category?
2. As noted in the chapter, racial residential segregation remains a key problem in the United States. What sociological factors explain its persistence in an era when housing discrimination is illegal?
3. Compare and contrast functionalist, conflict, and symbolic interactionist theories on race and racism. Which do you think best helps you understand current race and ethnic relations in the United States? Why?
4. What links have researchers identified between race and health outcomes in the United States? What sociological factors explain relatively poorer health among minorities than among whites? How might public policy be used to address this problem?
5. Do you think immigration will continue to grow in coming decades in the United States? What kinds of factors might influence immigration trends?
6. What factors make communities vulnerable to genocide? How should other countries respond when genocide seems imminent or is already under way?

©WeAre Digital Vision

10 GENDER, SEXUALITY, AND SOCIETY

WHAT DO YOU THINK?

1. How are gender and sexuality social as well as personal?
2. Why do men as a group continue to earn more money than women as a group?
3. What are key issues of feminism in the United States in the 21st century?

LEARNING OBJECTIVES

10.1 Explain key concepts sociologists use to study sex, gender, and sexuality.

10.2 Explain how gender is learned, performed, and structured into families, schools, peers, and media.

10.3 Identify social inequalities of gender and sexuality in families, education, and work settings.

10.4 Identify the contributions of feminisms and masculinity studies to the study of gender and sexuality.

10.5 Contrast global examples of inequality and empowerment around gender.

FIGHTING FOR WHO I AM, WHO I LOVE, AND WHERE I STARTED

Sharice Davids made history in November 2018 when she was elected to represent Kansas in the 116th United States House of Representatives. Davids, a member of the Ho-Chunk Nation, was one of only two Native American women ever to have been elected to Congress. (The second—Deb Haaland, a Laguna Pueblo from New Mexico—was also elected in 2018.) Davids was also part of what the magazine *The Advocate* called a "rainbow wave," a historic 116th Congress where Americans elected 10 "out" lesbian, gay, bisexual, transgender, and queer (LGBTQ+) new members (Reynolds, 2019).

Davids's story is significantly intertwined with this chapter's topics and concepts. For example, the United States' democracy relies on the ideal of the equal political voice and representation of its diverse citizenry. But men have historically had more of each. Women, on the other hand, fought for over 100 years to achieve the voting rights granted by the 1920 ratification of the 19th Amendment. In October 2020, men continued to be heavily overrepresented in Congress, filling 409 of the 535 seats in Congress (CAWP, 2019). The 117th Congress will be increasingly diverse. By November 19, 2020, at least 51 women of color were elected to Congress, and at least 141 women won congressional seats, constituting an estimated 24% of congressional leaders (Kambhampati & Luna, 2020). Is the unequal representation of men and women in Congress an anomaly? Or is it a pattern in other social institutions such as our families, education, and work worlds? We explore these questions in this chapter.

Davids's biography also shows that gender coexists and *intersects* with other social identities in powerful ways. For instance, a closer historical look at suffrage by race and ethnicity shows that Native American women (and men) were barred from voting until 1924 because they were not considered citizens until the passage of the *Snyder Act*. Davids's social identities and lived experiences (collectively, what sociologists would call her *standpoint*) are clearly intertwined in her campaign message: fighting from "where she started" for "who she loved" in the effort to help "those who's ever been left out or left behind." In short, Davids not only breaks political barriers *just* for women, or *just* for Native American women, or *just* for Ho-Chunks, or *just* for lesbians, but instead simultaneously breaks barriers for all four.

Throughout this chapter, we explore the complexity of gender and sexuality for individuals and societies. We also explore sociological insights into how people learn, interpret, enact, challenge, and change societal messages and structures.

In the coming sections, we turn our discussion to the key concepts of sex, gender, and sexuality, and how sociologists use them in sociological study. Realizing these concepts are both social *and* personal, we encourage you to think of how our gendered (and sexed) selves emerge. We focus on agents of socialization—including the family, schools, peers, and media—and examining gender as a process, a performance, and an identity that intersects with our other social identities. We also examine how gender and sexuality shape our lives unequally. We discuss gendered norms, roles, practices, and patterned structures, focusing on gender and family, higher education, and the wage gap to understand how gender influences our experiences in key societal institutions. We provide an overview of sociological thinking about gender, feminisms, and masculinities. We end by highlighting several important global inequalities around gender and sexuality and empowerment.

In 2018, Sharice Davids became one of the two Native American women elected to serve in the U.S. House of Representatives. That year saw what *The Advocate* magazine called a "rainbow wave" of women, LGBTQ+, and racial-minority winning candidates.

©Kansas City Star/Getty Images

SOCIOLOGICALLY UNDERSTANDING SEX, GENDER, AND SEXUALITY

If you google the words *sex, gender, transgender*, or *sexuality* you will find a wide array of definitions, topics, ideas, examples, and terms all shaped by society and culture. Thus, sociologists actively engage in the study of all of these, using the methods and theories that you have discovered throughout this book to offer unique insights into lives, practices, and identities. We want to provide definitions and examples to help you use and apply these terms to your life. As you read this chapter, you might notice that some definitions may vary from informal usage of them. You may have noticed, for instance, that some people use "sex" and "gender" interchangeably. Sociologists, on the other hand, typically use the term *sex* to refer to biological distinctions and the term *gender* to refer to social distinctions. We provide more information on each of these in the sections that follow.

Sex: The anatomical and other biological characteristics that differ between males and females and that originate in human genes.

Intersex: Those born with characteristics associated with sex (e.g., genitals, gonads, and chromosome patterns) that do not fit typical binary notions of male or female bodies.

Sex and Gender

Sociologists associate **sex** with the *anatomical and other biological differences between males and females that originate in human genes*. Many biologically based sex differences, such as differences in genitalia, are usually present at birth. Others, triggered by male or female hormones, develop later—for example, menstruation, breast development, and differences in muscle mass, facial hair, height, and vocal characteristics. Importantly, though, we are not saying sex is entirely biological, or that biology is completely male or female, or that anatomy at birth is also clear-cut or determinative. In short, what has been labeled as sex is also socially constructed (Fausto-Sterling, 2000).

Consider the experience of those who are "**intersex**," or *born with characteristics associated with sexed notions of male or female bodies (including genitals, gonads, and chromosome patterns) that do not fit typical binary patterns* (United Nations Fact sheet, n.d.). Between .05% and 1.7% of the population is estimated to be intersex, a number that is similar to the percentage of people who have red hair (United Nations Fact Sheet, n.d., para. 3). Doctors, parents, and societies have rarely tolerated uncertainties in their infant's sex categorization. Accordingly, intersex children have been subjected to medicalization including cosmetic surgeries and medications intended to render them categorizable in a binary (either/or) sex system as male or female to reassure their parents that their children are categorizable. These surgeries and medications have long-lasting repercussions for those who have not had a say in their own bodies. Advocates have challenged these practices, and

Activists calling intersex medical procedures "medically sanctioned violence and torture" were instrumental in challenging these practices. In 2020, the Ann & Robert H. Lurie Children's Hospital of Chicago became the first hospital nationwide to formally apologize for performing cosmetic genital surgery on intersex infants (Neus, 2020).

©Courtney Pedroza/Chicago Tribune/Tribune News Service via Getty Images

Sociologists point out that how people display gender is socially important, but this presentation varies across cultures. For the Surma people of the Omo Valley in Ethiopia, ear labrets are worn by women and men, though lip plates, which stretch the lower lip away from the mouth, are only worn by women. What gender presentations or "displays" exist in your society?

©iStockphoto.com/Guenter Guni

the American Academy of Family Physicians in 2018 issued a statement opposing unnecessary genital surgery on intersex children.

Gender, to sociologists, encompasses *the norms, roles, and behavioral characteristics associated in a given society with being a man or a woman*. Gender is less about being biologically male or female than about conforming to one's societal norms of *masculinity* (being a man) and *femininity* (being a woman). These norms vary socially across time, space, place, and culture. Like sex, many societal ideas of masculinity and femininity are often *binary* (seeing some characteristics as the exclusive property of men and others the exclusive property of women). For instance, masculinity is often associated with toughness, while femininity is associated with nurturance. People are expected to live up to such gender norms and to present the appropriate gendered behaviors, actions, and ideas. Boys are often expected to present a stoic front, even in the face of upsetting events, and told that "real men don't cry." Sociologists refer to these behaviors as *gender displays* or as one's *gender presentation*. In reality, gender is complex, and people often behave in ways that are not associated with binary gender notions—in reality, many men and boys *do* cry.

Gender: The norms, roles, and behavioral characteristics associated in a given society with being a man or a woman.

No matter the norms, gender diversity exists both across and within societies. For instance, North American Indigenous communities have a history of *two-spirited* people, or those who " *identify with both masculinity and femininity*." They were culturally accepted, and many held ceremonial or spiritual significance in their communities (Savage, 2020, para. 4). In India, *hijras*, people who are intersex or transgender, historically "held special roles in India's royal courts." Today, while officially recognized by India's Supreme Court as a third gender in 2014, they also experience both social discrimination (housing segregation, lack of work, violence) and a place of social distinction—seen to have a link to divine power, they are often enlisted in ceremonies such as births and weddings (Pal, 2019).

Gender identity: One's sense of being a woman, man, or outside of the gender binary.

Cisgender: Those whose gender identity matches their sex assigned at birth.

Transgender: An umbrella term used to describe those whose gender identity, expression, or behavior differs from their assigned sex at birth or is outside the gender binary.

One's gender identity is personal and social. **Gender identity** is *one's sense of being a woman, man, girl, or boy, or being outside of the gender binary*. Cisgender and transgender are two ways to describe gender identity. **Cisgender** denotes people whose *gender identity matches their sex assigned at birth*. **Transgender** is an "umbrella term" used to describe a diverse group of people *whose gender identity, expression, or behavior differs from their assigned sex at birth or is outside the gender binary* (GLAAD, 2020). Notably, not everyone who is transgender describes their gender as binary male or female. As the *National Center for Transgender Equality* puts it: "Some transgender people identify as neither male nor female, or as a combination of male and female" (https://transequality.org/issues/resources/frequently-asked-questions-about-transgender-people). The terms *transgender* and *cisgender* both describe a relationship between gender identity and sex/gender assigned at birth. Figure 10.1 compares these terms: the left flowchart uses arrows to draw a progression from baby born, to assigned female at birth, to identifying as any gender other than as a woman, to transgender person; the right flowchart uses arrows from baby born, to assigned female at birth, to identifying as a woman, to cisgender woman. While this visual shows a general way to understand both terms, how a person self-identifies is the way they should be described.

How many people today identify as transgender? According to Flores et al. (2016), about 1.4 million people in the United States (or 0.6% of the population) so identify, a number that nearly doubled since their 2011 study. Flores also points out that the most common group to identify as transgender are between the ages of 18–24 (Flores et al., 2016). Indeed, a 2020 survey conducted by Harris Polls and GLAAD (an LGBTQ+ organization) found that 12% of Americans ages 18–34 identify as transgender or gender nonconforming.

Population estimates for transgender people may vary between surveys for a variety of social and methodological issues. For instance, surveys may not allow a nonbinary gender box to check for respondents, or respondents might not feel safe identifying their gender. Thinking sociologically, can you

FIGURE 10.1 ■ Transgender and Cisgender

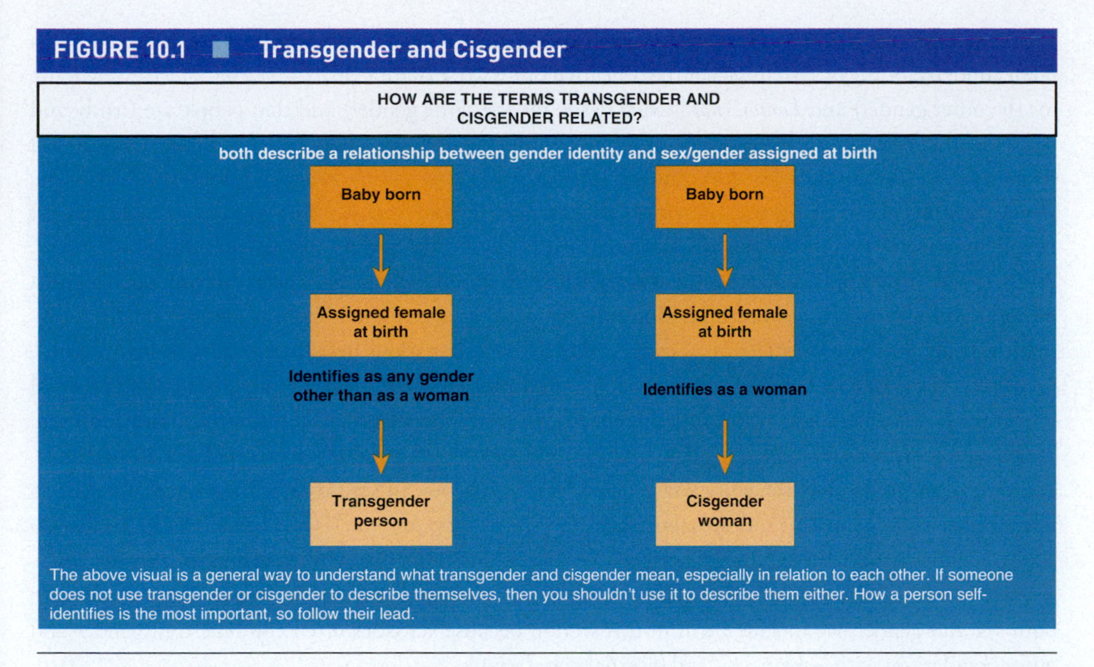

Source: Géza Kelemen, LuLu (2021), Safe Zone Program, LGBTQ+ Resources, George Mason University, https://lgbtq.gmu.edu/safe-zone/gender-pronouns/

identify other reasons that might cause variation in numbers in surveys that seek to understand gender identity?

Today's Gen Zers (those born after 1996) are most likely to think of gender in nonbinary terms. They are more likely to know someone who goes by a gender-neutral pronoun and, according to a recent Pew Research Center survey, 6 in 10 Gen Zers think that online profiles that ask about gender should include options other than "man or woman." This stands in contrast to their Baby Boomer and Silent Generation peers, who opposed such an option by 60% (Parker & Igielnik, 2020).

Patriarchy: Any set of social relationships in which men dominate women.

Sexuality: A term used to encompass sexual identity, attraction, and relationships.

Finally, in a society that still operates significantly on the gender binary, gender is associated with social power. Some feminist sociologists point out that gender has been used to provide men as a group more resources and social power than women. They associate gender with **patriarchy**, *a set of social relationships in which men dominate women*. For instance, women as a group lag significantly behind men as a group on a number of indicators worldwide including political representation, income, wealth, and others, leading the World Economic Forum to note that gender parity will not be achieved worldwide for 100 years! (Zahidi, 2019). In this chapter we explore such social relations as those embedded in laws, organizations, and institutions. For instance, in our opening story, we showed how the institution of politics and the associated political power in the United States are still disproportionately held by men. We also note that social change is slowly occurring: In November 2020 Kamala Harris became both the first woman and the first woman of color to be United States Vice President Elect.

Janet Mock, a writer, director, producer, and transgender activist, tells the story of "Ball Culture" in her FX series show, *Pose*. Mock, who transitioned in the 1980s, notes that she tells stories she craved as a young trans person: "Part of my public work is talking about my life experiences and what I've gone through" (Mock, 2019, quoted in Gross, 2019).

©Sthanlee B. Mirador/Sipa USA/Alamy Live News

Sexuality

Sexuality, both a private and public matter, is of interest to sociologists. Sexuality *encompasses sexual identity, sexual attraction, and sexual relationships.* It overlaps with many practices and institutions we discuss in this text. While related to both sex and gender, one's sexuality is personal, it may or may not align conventionally with one's sex and gender, and it may change over one's lifetime.

Queer theory: A theory that describes how societies develop ideas about sexuality and gender, rejecting claims that gender identities and sexuality are fixed or stable.

As with gender, many terms associated with sexuality in contemporary society are organized around a *binary* understanding of sexuality—or the belief that only two sexualities exist—*heterosexual* (sexual desire for the other gender) and *homosexual* (sexual desire for the same gender) and that people are firmly and permanently placed in one category or another. An important theory that challenged binary notions of sexuality is the interdisciplinary **queer theory**, *which describes how societies develop ideas about sexuality and gender, rejecting claims that gender identities and sexuality are fixed or stable.* The word *queer*, used against people who were thought to be gay, was reclaimed by LGBTQ+ groups as "self-affirming" (Perlman, 2019). Queer theory also challenges the ways in which gays and lesbians have been discriminated against by highlighting *homophobia* (discrimination and prejudice based on a fear of homosexuality) and *heterosexism* (the belief that heterosexuality is above all other sexual orientations in culture and social institutions). Queer theorists point out how social institutions and norms embrace, elevate, and reinforce heterosexual relationships. (Numerous examples are present in culture such as laws around marriage and adoption, media representations, even greeting cards; what other examples can you think of?) In Table 10.1 we present some terms to describe sexuality and gender identity. As you can see, some of these terms are used to refer exclusively to sexuality and sexual orientation (*lesbian, gay, bisexual*) while others can be applied to gender identity or sexuality and sexual orientation (*queer*) and others are applied to gender identity (*transgender*). While transgender relates to gender rather than sexuality and queer, or questioning can relate to both sex and gender, we include them in this section because statistics often combine transgender and queer populations with lesbian, gay, and bisexual populations.

How many people identify as LGBTQ+? According to Gates (2011), there are over 13 million LGBT people in the United States (1.4 million of those are transgender adults and 150,000 are transgender youth). According to the General Social Survey, about 10 million Americans identified as LGBT in 2016 (Gates, 2017). According to Gallup, the LGBTQ+ population constituted 4.5% of the population in 2017, a percentage that had grown since Gallup, in 2012, began asking the question: "Do you, personally, identify as lesbian, gay, bisexual or transgender?" (Newport, 2018). Gates also points out that while men and women are as likely to identify as gay/lesbian as bisexual, they are up to three times more likely to say that they have same-sex attraction or to have engaged in same-sex behavior than they are to *identify* as LGBT (Gates, 2011).

As we pointed out with transgender people, population estimates for LGBTQ+ people may vary between surveys for a variety of social and methodological issues. For instance, respondents might not feel safe identifying their sexual orientation, or surveys may not ask for this information, or different sampling measures may yield different results. Thinking sociologically, can you identify other reasons that might cause variation in numbers in surveys that seek to understand sexual orientation?

As sexual *minority* populations, LGBTQ+ have experienced stereotyping, prejudice, and discrimination. For instance, they are four times more likely to experience violent victimization than are non-LGBTQ+ populations (Flores et al., 2020). They also lack equal protection under the law in cases such as job discrimination in many states, with "only 20 states and D.C. hav[ing] laws that explicitly prohibit discrimination based on sexual orientation and gender identity" (National LGBTQ Workers Center, n.d.). However, attitudes worldwide have become generally much more accepting over the past several decades. For example, global responses varied widely to the question *Should homosexuality be accepted in society*? between 2002 and in 2017. In the United States in 2002, 51% said "yes," but in 2019, 72% said "yes." Globally, the wealthier the country, in general, the higher the level of acceptance: Sweden, Germany, Canada, and the Netherlands were among the most accepting, while Nigeria, Tunisia, and Kenya where the least. Even among these countries, acceptance had grown, however. Only 1 in 100 Kenyans in 2002 said homosexuality should be accepted in society. That number had grown to 14% in 2019 (Poushter & Kent, 2020).

LEARNING AND PERFORMING GENDERED SELVES IN A GENDERED WORLD

Norms, values, and practices around gender and sexuality vary, sometimes dramatically, among societies, cultures, and groups. They can diverge within one's own experiences, too, as Jennifer Finney Boylan illustrates in her books *She's Not There: A Life in Two Genders* (2003) and *Stuck in the Middle With You: A Memoir of Parenting in Three Genders* (2014). Boylan is an author, professor, and activist

TABLE 10.1 ■ Commonly Used Terms Associated With Sexuality (and Gender)

Asexual: Those who do not experience sexual attraction to any gender.
Bisexual: Those who have physical, romantic, and/or emotional attractions to both genders.
Gay: Those who have enduring physical, romantic, and/or emotional attractions to people of the same gender.
Lesbian: A woman who has enduring physical, romantic, and/or emotional attractions to other women.
LGBTQ+: Lesbian, Gay, Bisexual, Transgender, Queer or Questioning and + can stand for ally or other groups. Notably, transgender relates to gender rather than sexuality. Queer or questioning and + can relate to both sex and gender.
Pansexual: Those who have sexual attractions to people regardless of their gender.
Questioning: Those who are questioning their sexuality or gender identity.
Queer theory: A theory that challenges binary ideas of gender and sexuality.
Queer: Those who fall outside of gender and sexuality norms.
Sexual orientation: A person's sexual identity based on their enduring physical, romantic, or emotional attraction. Sexual orientation is often classified heterosexual, bisexual, homosexual, and asexual.

Source: Definitions informed by *GLAAD media guide* (10th ed.). https://www.glaad.org/reference

who transitioned from male to female in her 40s. Boylan remained married to her wife of many years, and they continued to raise their two sons together. Some parts of Boylan's life changed fairly markedly (such as her outward gender display) while others remained remarkably similar (such as parenting her sons).

Across the world's diverse cultures, men are generally expected to behave in culturally defined "masculine" ways and women in "feminine" ways. Few people fully conform to these stereotypes—most exhibit a blend of gendered behaviors. Still, in many cultures, people believe gender stereotypes represent fundamental, real sex differences, and label those who diverge too far from social scripts of gender deviant. Insights about gender and gender identity continue to emerge. We highlight three ways that sociologists have thought about gender and gender identity: gender socialization, gender performance, and gender as a structure.

In one view, **gender socialization**—*the process of learning social norms and values around gender*—has an important impact on gender identity. Influenced by the *agents of socialization* (families, peers, teachers, and media) who promote gendered ideas and behaviors, this view implies that people internalize gender as a social identity. While gender is not seen as natural, it is a fundamental socially influenced aspect of the social self.

Gender socialization: The process of learning social norms and values around gender.

In another view, gender is less an identity than a *performance*. As we saw in Chapter 4, sociologist Erving Goffman (1959) argued that our social selves are the *product* of social interactions and people, and groups tailor their presentation of self in a way most favorable to the given situation to ensure a believable performance. Building on some of Goffman's ideas, West and Zimmerman (1987) suggest that gender is something we *do* rather than a fixed identity. Thus, gender is seen as an *activity* that creates differences between men and women. These differences, although not biological, appear natural because they are so consistently enacted. They note that gender is also associated with power differences—membership in a gender category brings differential access to power and resources, affecting interpersonal relationships and social status.

Finally, gender can be seen as a *structure*—not only personally experienced or performed, but also present in "processes, practices, images, ideologies, and distributions of power in the various sectors of social life" (Acker, 1992, p. 567). Acker points out that institutions in the United States are organized around gender. As she puts it:

> The law, politics, religion, the academy, the state, and the economy . . . are institutions historically developed by men, currently dominated by men, and symbolically interpreted from the

standpoint of men in leading positions, both in the present and historically. These institutions have been defined by the absence of women. (Acker, 1992, p. 567)

In the next sections we apply these three ways of thinking about gender to families, schools, peers, and media. We also explore some of the ways that sociologists have shown gender to be taught, performed, negotiated, and sometimes subverted (Moon et al., 2019).

The Roots of Gender: Families and Gender Role Socialization

As we first illustrated in Chapter 4, children first learn a great deal about how to behave from their families. Their socialization includes gendered messages about how to act, leading some sociologists to call the home a "factory of gendered personalities" (Fenstermaker Berk, 1985; Goldscheider & Waite, 1991).

Until the middle of the 20th century, some sociologists argued that what they called *sex role differences* were positively functional for social harmony, order, and stability and that these were first learned in the family. Sociologist Talcott Parsons, for instance, offered a functionalist theory of **gender roles**—*activities, attitudes, and behaviors considered appropriately feminine or masculine* in the home and the workforce (Parsons, 1954; Parsons & Bales, 1955). Parsons argued that in a modern capitalist society, women raise children and maintain the family through emotional and physical nurturance—what he called *expressive roles*; men do so by earning the family income through their wage labor and rational guidance—what he called *instrumental roles*. Parsons thought this a necessary and functional gendered arrangement—preventing competition in the family and giving people clear social statuses.

Gender roles: The actions, attitudes, and behaviors that are considered appropriately masculine or feminine in a particular culture.

Many critiques have been leveled against Parson's theory, including its class, race, and gender biases that viewed all families as traditionally organized, capable to make ends meet without dual earners, and functional for individuals who were asked to fit such roles. Critics also question that the gendered division of labor is functional for society where over half of its inhabitants are women. However, sociologists continue to enlist Parson's insight that families do contribute strongly to children's gender role socialization. For instance, even before a child is born, many parents hold gendered beliefs about how their infants are supposed to look (as we have shown in the example of intersex children) and behave.

As one study found, mothers spoke differently to their babies *in utero* depending on whether they knew the babies were boys or girls (Smith, 2005). And you may have noticed that before children are born, their parents often decorate gendered nursery rooms or buy gender-typed toys and clothing. Others hold *gender reveal parties* or events where the expected sex of the baby at birth is "revealed," typically through pink or blue objects baked into cakes, contained in balloon confetti, revealed with colored silly string, displayed with fireworks, or even lit up on a Ferris wheel flashing blue or pink. Gieseler (2018) notes that this new ritual is not only one of parents, but also joins communities of people in "gendering" the baby publicly, through social media, and in encouraging consumerism.

A survey of Australian parents of 3- to 6-year-old children found that most thought same-gender-typed or gender-neutral toys were more desirable for their children than cross-gender-typed toys (Kollmayer et al., 2018). Do you think these parents in the United States would answer similarly? What is the significance, if any, of steering children to play with certain types of toys?

Parents also teach gender roles over a child's life course (sometimes deliberately and other times unwittingly) by the way they divide the labor in their homes, their chosen occupations, and the amount of time they spend with their children. Halpern and Perry-Jenkins (2016), for example, found that although a mother may have more egalitarian views about gender, their child will have more traditional gender role attitudes about women if the mother completes more traditional feminine tasks, such as doing most of the housework or having a stereotypical feminine job. Another study found that "spending more time with fathers in childhood predicted daughters attaining *less* and sons acquiring *more* gender-typed occupations in young adulthood" (Lawson et al., 2015, p. 26).

Parents also communicate gender roles by selecting books that present gender in covers, illustrations, images, and words. Take the example of a study of children's picture books from 1930–2017 that featured "anthropomorphized inanimate objects" as their main characters (for instance, a car named Bob). The researchers found that the vast majority of the 103 books they analyzed—three quarters—featured only male characters on their front covers. Moreover, male characters were more likely to be portrayed as leaders and heroes, likeable, and praised for their accomplishments (Berry & Wilkins, 2017, p. 13). And these male characters were also more likely to be portrayed with a face than were similar female characters—only 4% of male characters were faceless compared to 44% of female characters. Concluding that the absence of faces barred female characters from having voices and emotions, the authors noted: "Children learn about the world through children's literature; when children's books reinforce gender divisions in society, children come to see these divisions as normal" (p. 5).

Importantly, parents do not always have agency in the choices they make around gender socialization. That is to say that although many parents now hold more gender egalitarian views, gendered structures such as the marketplace may make available predominantly gender-typed items such as toys, clothing, and room accessories. For instance, a recent study shows children's rooms are still highly gendered: Girls' rooms have more dolls, floral accents, and dress-up clothes, while boys' rooms have more play guns and tools, sports equipment, and vehicles (MacPhee & Prendergast, 2019).

Thinking intersectionality, we can also note that socioeconomic status, race, and sexuality can influence parents' gendered expectations of their children, their children's behaviors, what they tell their children about gender, and the options they provide to their children. Consider these studies that show the following:

- *Parental social class can influence gender expectations.* One study found that families with a higher socioeconomic status, including parents with greater education and higher-paying jobs, have more equal gender expectations than do those with lower socioeconomic status (Samari & Coleman-Minahan, 2018).
- *Parental race can influence gender role expectations.* African American parents are more flexible in gender role expectations of their children, according to Skinner and McHale (2018). Their study indicated that spending extra time with male peers led African American boys to adopt more traditional gender role ideas while spending more time with parents led to boys exhibiting less stereotypically gendered behaviors.
- *Parental sexuality and gender identity can influence gender socialization.* Several qualitative studies of LGBTQ+ parents found differences in gender socialization practices. For instance, Oakley et al. (2017) found their LGBTQ+ parents said they both taught their children awareness of diverse families and enlisted proactive parenting strategies to help them respond to the stigma they might face because of their family diversity. And Kate (2016) found that LGBTQ+ parents enlisted the "gender buffet"—providing children a variety of options when it came to play, toys, clothing, or ideas. She also showed how these parents sometimes drew on their own gendered socialization where they felt their parents "mis(recognized) their gender expression or sexual identities" so that they would not "constrain their children's gender."
- *Parental socialization can also reinforce norms around sexuality, particularly heterosexuality.* Kate (2016) identifies *heteronormativity* (the belief that heterosexuality is the norm) as a central component of gender socialization as parents often "tend to link gender conformity and heterosexuality in their interpretations of and responses to children's behavior, and, conversely, they associate gender nonconformity with future homosexuality."

Gendered Playgrounds, Gendered Interactions: Schools and Peers

Schools are places where gender is taught and performed. They can also be seen as gendered structures. Their physical structure, the books from which students learn, and the classroom and extracurricular opportunities available to them can promote gendered messages and opportunities. The actions and judgments of our teachers, coaches, and peers also play a significant role in the way they learn and display gender.

Schools

Have you ever considered that spaces (buildings, rooms, outside areas) might be gendered? Think about elementary school. Boys' and girls' bathrooms might be the first spaces to come to mind, but what about playgrounds, locker rooms, hallways, places to eat and hang out, and classrooms? To understand the possibility of gendered space, researchers in a creative recent study provided iPads to sixth grade boys and girls, asking them to photograph six different types of spaces at their school (Ceridwen et al., 2019):

1. a place you really like/your favorite place
2. a place you hang out a lot
3. a place you avoid or don't like
4. a place you've never been to (in the school)
5. a place you used to love but no longer go
6. (for girls) a place that boys use that you don't/(for boys) a place that girls use that you don't

Researchers then analyzed the photos, also conducting focus groups with the children to understand the spots they chose. The girls and boys did, indeed, identify certain spaces as gendered. Boys' pictures and focus groups showed them to have access to large, open, and central spaces. Girls, on the other hand, took pictures of smaller spaces. They felt that they did not control the basketball court and soccer fields, for example, and that they were not able to compete with boys in these spaces. This, in turn, influenced their attitudes, or as one girl put it, "We're not very good." Still, the girls expressed a longing to be able to access those spaces. Conversely, boys did not have spaces where they were excluded from active play. In short, this study indicated that gendered space existed, and girls believed it barred them from some places and activities, particularly around sports. Our first point, then, is that gender has an often invisible space dimension that can have a real impact. One way in which school is a gendered institution, then, is by virtue of its spaces.

Also, schools provide both informal and formal learning, both of which can have gendered dimensions. Informally speaking, researchers have noted that schools are an important site of a "hidden curriculum," the unspoken socialization to norms, values, and roles—including gender roles (Basow, 2004; Margolis, 2001). For example, there is evidence that from an early age, shyness is discouraged in boys at school because it is viewed as violating the masculine norm of assertiveness (Doey et al., 2013). Teachers, administrators, and other adults in the school not only provide instruction, their occupational roles might also offer early lessons to students about their future career prospects. For example, pre-K and elementary teaching is still a largely gendered profession—that is to say, women staff the majority of these jobs (which are typically low-paying) while men staff administrative roles such as principals and superintendents (Figure 10.2).

The consequences of this gender dynamic are felt by both instructors and students. Research consistently shows that when jobs are gendered as primarily the domain of women, they pay less. In fact, according to an MIT study on teachers' wages, in most states teachers' salaries do not provide a living wage (Wong, 2019). Furthermore, the fact that elementary school teaching has been gender-typed as predominantly a "feminine" profession also disadvantages men who would like to step into teaching roles in these areas, because stereotypes that cause people to question men's "intentions of working with young children" or to think that only men with lower abilities would seek elementary education can keep men from such jobs or make them feel uncomfortable. Wong cites the experience of two male elementary school teachers whose responses also show instances of gender performance in that they say they have to display "just the right amount of maleness" and engage in strategies such as ensuring that another adult is always in the classroom and forging strong relationships with parents. As one interviewee puts it: "It's tiresome, and so a lot of the male elementary teachers say after a while, 'This is just too draining.'" (para. 14).

Formally, the curriculum still contains solid messages about gender. Books on history, for instance, are still heavily populated by male characters, and women, people of color, and sexual minorities are often shown in special features outside of the main text, where they can be overlooked. In 1997, David

FIGURE 10.2 ■ Percentage of Female Teachers in Educational Occupations, 2019

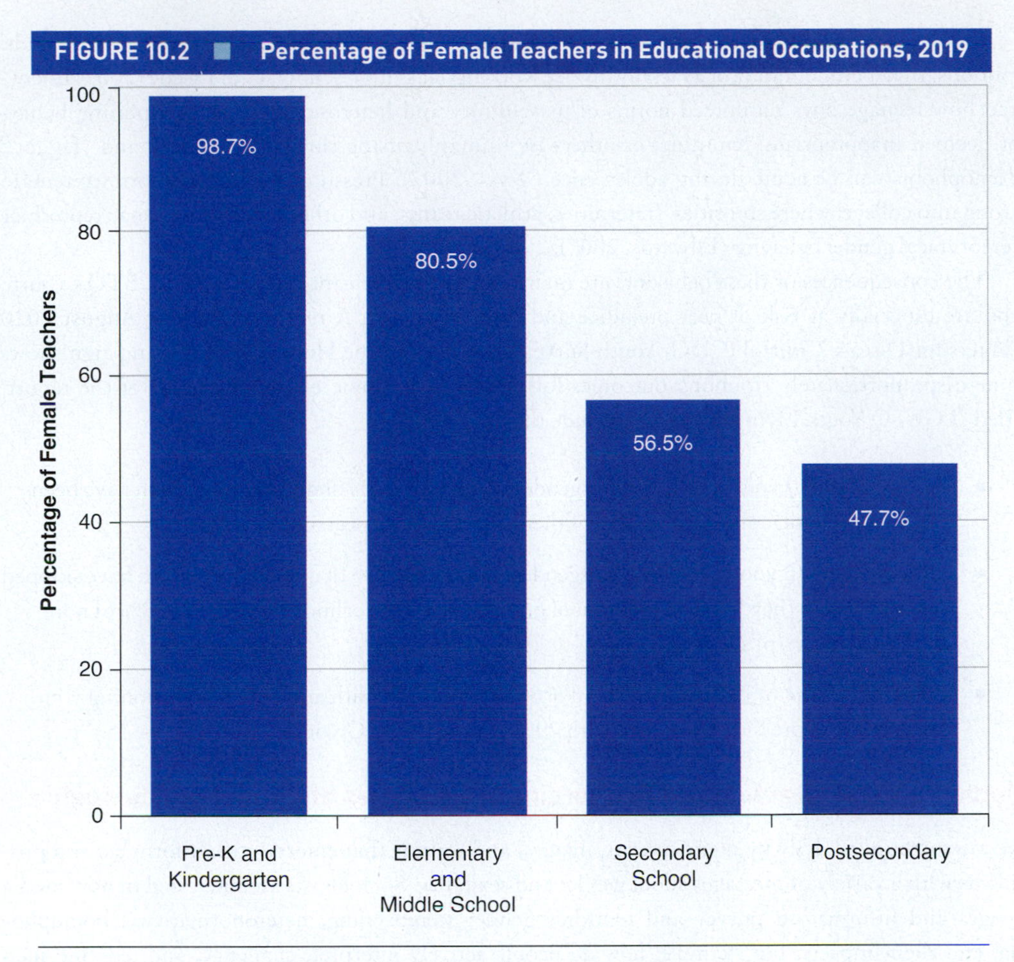

Source: U.S. Bureau of Labor Statistics. (2019). https://www.bls.gov/cps/cpsaat11.htm

Sadker and Myra Sadker famously argued in their book *Failing at Fairness* that in most history texts, students see few women. Their study of one 819-page text yielded less than a full page of references devoted to women. When the authors challenged school seniors to, in 5 minutes, name 20 famous U.S. women, past or present (not including sports or entertainment figures), most students were able to come up with only a handful, but had little trouble naming famous men. More recently, a study found that only 23% of people mentioned in introductory economic books, real or fictional, were women (Stevenson & Zlotnik, 2018). Even the most women-friendly texts featured women only 25% of the time, and their mentions often involved food, fashion, and household tasks.

Textbooks sometimes lag behind changes in society. How would you study gender and sexual diversity in textbooks? In behaviors on the playground? What other areas of study do you think are important to consider when looking at the institution of education?

Peers

Children quickly learn and display stereotypical gender roles in peer groups (Aina & Cameron, 2011; Martin & Fabes, 2001). This behavior is particularly acute among boys, who police peers and stigmatize perceived feminine traits. Playing in same-gender peer groups, interestingly enough, may have a different effect than playing more often in mixed groups. Martin and Fabes (2001) point to a "social dosage effect," such that play among young girls who stay with all-girl groups tends to reflect norms of encouragement and support, while young boys in same-gender groups develop a tendency toward more aggressive and competitive play. Research on elementary school playgroups has found that girls often gather in smaller groups, playing games that include imitation and the taking of turns. Boys, on the other hand, play in larger groups at rule-governed games that occur over larger physical distances than girls occupy (Richards, 2012; Thorne, 1993).

During late childhood and adolescence, children are especially concerned about their friends' opinions. In her ethnography of a racially diverse working-class high school, C. J. Pascoe (2007) uncovered how teenage boys reinforced norms of masculinity and heterosexuality by disciplining behavior deemed inappropriate, feminine, or otherwise unmanly, using slurs such as "fag" and "faggot." Homophobia can be acute during adolescence (Wyss, 2007). Pressures for peer conformity remain strong into college, where sororities, fraternities, athletic teams, and other social groups may reproduce stereotypical gender behavior (Edwards, 2007).

The consequences of these behaviors are many, but one significant impact is on LGBTQ+ youth, who are especially at risk of peer prejudice and discrimination. A recent analysis of August 2020 Centers for Disease Control (CDC) Youth Surveillance data by the Human Rights Campaign shows some disproportionately troubling outcomes for these youth. Some of the highlights of the report, titled "LGBTQ Youth Living in Crisis," include the following:

- 31% of LGBTQ youth, 43% of transgender youth and 40% of questioning youth have been bullied at school, compared to 16% of their non-LGBTQ peers; (p. 2)
- 24% of LGBTQ youth, 35% of transgender youth and 41% of questioning youth have skipped school because they felt unsafe at school or on their way to school, compared to 8% of non-LGBTQ youth; (p. 2)
- Over half (54%) of LGBTQ youth, 61% of transgender youth and 61% of questioning youth are battling depression compared with 29% of non-LGBTQ youth. (p. 3)

Media Power: Reflecting and Reinforcing Stereotypes of Gender and Sexuality

We are surrounded daily by many screens, images, and sounds that entertain us, inform us, and provide us with a variety of messages about gender and sexuality. Sociologists are interested in how media images and information purvey and reinforce gender stereotyping, heteronormativity, homophobia, and their impacts. For example, how do people actively interpret, challenge, and redefine such messages?

Television and Movies

Studies spanning five decades have identified television programming and advertising as major purveyors of gender stereotyping (see, for instance, Barner, 1999; Condry, 1989; Gauntlett, 2008). Youth are frequent consumers of media, with children ages 7 and under spending about 2 hours and 20 minutes every day engaging in screen time. TV and video viewing make up the largest share of that time at 72% (Common Sense Media, n.d.). For children 8 to 12 years of age, the average time is about 4 hours and 35 minutes every day (Bhattacharjee, 2017). During the COVID-19 pandemic in 2020, researchers noted the increasing time that children spent on television and social media, with Brown (2020) suggesting that may also result in increased viewing of stereotyped gender images:

An annual "Accelerating Acceptance" survey found a significant decrease in overall comfort with LGBTQ+ people between 2016 and 2018. Especially notable was the decrease in the support of 18- to 34-year-olds who identified as "allies" (GLAAD Accelerating Acceptance, 2020). Thinking sociologically, how would you discover the causes and consequences for the declining the level of allyship and support?

©SAUL LOEB/AFP via Getty Images

> On average, children in elementary school watch four and a half hours of television a day: At this rate of exposure, children see approximately 78,069 examples of "sexy girl" role models just in children's programming alone every year. And with schools, playgrounds, and after-school activities grounded, children are likely to consume much more media this year. (Brown, 2020)

Movie portrayals of gender and sexuality are also of interest to sociologists. Consider Disney movies. A linguistic analysis of several decades of Disney princess films uncovered a striking finding: In Disney's older "classic" princess films, female characters were more likely to speak than female characters in newer Disney films. For example, in *Cinderella* (1950), women spoke about 60% of lines. In *The Little Mermaid* (1989), female characters spoke 32% of the time, and in *Mulan* (1998), 23% of the time. In the hit film *Frozen* (2013), two young princesses take center stage, but female characters have only 41% of the spoken lines as opposed to male characters (Guo, 2016). Considering this information, are you surprised that a Disney film of the 1950s actually featured *more* speaking time for women than movies made over 60 years later?

A more recent Annenberg Foundation study looked at the top 100 popular films in 2018 (Smith et al., 2019). Of these, only 39 out of 100 presented a woman in the leading or co-leading role, and only 24 had an LGBTQ+ character (and of those, over 63% were white). Additionally, these statistics show that diverse characters were extremely underrepresented:

- 99 films had no American Indian or Alaska Native women characters
- 97 films had no native Hawaiian or Pacific Islander women characters
- 92 had no Middle Eastern or North African women characters
- 51 had no multiracial women characters
- 83 had no women characters who had a disability

Finally, while transgender people's stories and experiences are emerging in society, media representation is sparse. For instance, a study of the eight major films from the top eight major film companies between 2015–2019 showed no transgender characters in any of these movies. Indeed, as the authors of the study note: "The one nonbinary character in the year before that (2016) existed solely as a punchline" (GLAAD Media Institute, 2020, p. 4).

What are the consequences, if any, of the underrepresentation of women, LGBTQ+ people, racial and ethnic groups, and those with disabilities in major films?

Music, Gaming, Fantasy Sports

Music lyrics and videos, video games, and fantasy sports have long been associated with gender and sexuality stereotyping. We explore several current studies here.

In a review of the literature on popular music lyrics, Rasmussen and Densley (2017) cite research showing that a variety of music genres have undergone changes over the past few decades in the way that lyrics are gendered. For instance, they reveal that rap music became more violent between 1979 and 1997; that lyrics in pop music became more "socially disconnected, angry, negative, and antisocial"; and that the primary magazine devoted to music, *Rolling Stone*, increasingly featured sexualized women on its covers between 1969 and 2009 (p. 5).

Their research explores country western music, a genre listened to by about 76 million Americans weekly (Country Music Association, 2019). In a three-decade analysis of 750 country songs from 1990 to 2014, they update previous research that had shown that lyrics that gender stereotyped women had actually *decreased* between 1955 and 2005. Rasmussen and Densley's 2017 study finds that newer country music (in the 2010s) instead "tends to objectify women more and portray them as empowered less than in previous decades," and that the songs of male artists, not females, were primarily responsible for objectifying lyrics (p. 12).

The Canadian comedy series *Schitt's Creek* takes place in a town with no homophobia; it has gender-fluid characters and a pansexual main character, David (far right), played by Dan Levy. Levy, who is also the writer of the show, intentionally created a town that was not homophobic; nor were characters judged for gender fluidity. David's pansexuality was also directly addressed in the show (Gilchrist, 2020).

In short, they support the research cited earlier that lyrics actually changed in a more stereotyped direction.

One unanswered question of research on sexist images and lyrics is this: Do these images and lyrics have an impact on people's ideas and behaviors? If so, what is it? How do you think researchers might study this? What are some challenges they might encounter in connecting attitudes and images?

Playing video games, as we pointed out in Chapter 3, is also an area of gender socialization and potential gender stereotyping. By some estimates, fully 91% of youth aged 2–17 play video games. Gaming is very popular among teens—especially teenage boys, 90% of whom say they play video games. At the same time, video games are also played by three quarters (75%) of teenage girls (Perrin, 2018). As we noted in Chapter 3, Perrin (2018) shows that video game playing remains high for men aged 18–24 years, with almost three quarters (72%) of them saying they play video games. At the same time, just under half (49%) of women in that age group say they do, revealing a significant gender difference in video gaming activity as people age. Substantial majorities of men aged 18–24 also own a video game console (68%), compared to 46% of women (Perrin, 2018).

Interestingly, among adults who play video games, the perception of how games portray gender is mixed. For instance, Duggan (2015) found that 26% of all players and 35% of frequent players (those who labeled themselves *gamers*) do not think women are portrayed poorly in most games. Another 16% of players and 24% of gamers agreed that women are portrayed poorly in most games. This is significant because content studies suggest that popular video games consistently convey stereotypical gender images. Downs and Smith (2010), in an analysis of 60 video games with a total of 489 characters, determined that women were likely to be presented in hypersexualized ways that included being partially nude or wearing revealing attire. Reflecting common findings that men are *action* figures while women are largely *passive*, Haninger and Thompson (2004) discovered that in the 81 teen-rated games they sampled, fully 72 had playable male characters, while only 42 had playable female characters.

Sexuality, too, may be portrayed stereotypically in video games. But Ruberg (2018) argues that this does not mean that players tacitly accept these stereotypes. In fact, she argues that it is possible to "queer" video games. Ruberg does not look at the lack of representations of LGBTQ+ characters or games, or at how to improve these issues; rather, Ruberg suggests that all games have subversive tendencies and that "all games are fair territory for exploring queerly" (p. 210). Ruberg also is the lead organizer of an annual Queerness and Games Conference (QGCON) and has begun a new academic discipline blending game studies and queer studies called Queer Game Studies (http://ourglasslake.com/queer-game-studies-101/).

Finally, we turn to an increasingly popular pastime, fantasy sports. Played by over a fifth of Americans over age 18 in 2019 (Fantasy Sports and Gaming Association [FSGA], 2019), the game also has some strong gender (and age) dimensions. For instance, over 80% of players (and people wagering bets) are men, and most of them are young—over half are between the ages of 18 and 34.

Interested in the small percentage of women who play fantasy sports, Kissane and Winslow (2016) conducted a qualitative study of women players, interviewing them directly. They also examined secondary data from a survey conducted through related message boards and websites, on Facebook, and on Twitter that asked, "Do you ever feel like you are treated or perceived differently by other people involved in fantasy sports because you are a woman?" (p. 824). They found that women fantasy sports players saw this domain as a male space where they felt "highly visible." For instance, they enlisted several strategies when they encountered gendered stereotypes about women's lack of sports knowledge. The first, *exercising mediated agency*, involved asking for men players' "help" to improve their experience and play. The second, *conflicted agency*, was to accept and use gender stereotypes of women either to their advantage or to act unlike other women so that they didn't have stereotypes applied to them. On the other hand, several respondents noted how "underestimation" by male players worked to their advantage. As one participant said,

> They don't expect a lot out of that one woman who is in the league, and it's a good opportunity to kind of show off a little bit and be like, "Hey, you're underestimating me, and you shouldn't. I know exactly what I'm doing here and probably more than some of you." (p. 834)

GENDER INEQUALITIES: FAMILY, SCHOOL, AND WORK

Anthropological studies have found that inequalities in almost all known past and present societies favor men over women (Huber, 2006). Women are occasionally equal to men economically, politically, or socially, but in no known society do they have greater control over economic and political resources, exercise greater power and authority, or enjoy more prestige than men (Chafetz, 1984). While 143 of 195 countries guarantee gender equality in their constitutions, in practice, their laws, policies, practices, norms, beliefs, and stereotypes interfere (United Nations, 2015). Consider these statistics:

- 77 countries criminalize same-sex relationships;
- 66% of the 781 million adults and 126 million youth worldwide who are illiterate are women and girls;
- Women earn 10 to 30% less than men in equivalent jobs;
- Only 22% of political positions worldwide are filled by women, leaving 78% to be filled by men. (United Nations, 2015)

In the section that follows we explore inequalities in family life, education, and work in the United States.

Gender and Family Life

What types of tasks did your parent or parents daily do to keep your household going? What about your brothers or sisters? How long did each of these tasks take? Did some tasks take longer than others? Were there any patterns to who did what and why?

Sociologists ask these questions, consistently finding that significant gender inequality (including who is expected to do certain chores, how much time these chores take, and how often these chores occur) is involved in the reproduction of the household. That is to say that among heterosexual couples who are married or cohabiting, daily domestic tasks—childcare, cooking, cleaning, laundry, and shopping for necessities of living—can entail long hours of unpaid work that is gender stereotyped as "women's work." Arlie Hochschild (2012) called this phenomenon the **second shift**—*the unpaid housework women typically do after they come home from their paid employment.*

Second shift: The unpaid housework that women typically do after they come home from their paid employment.

In the United States, women still do the larger daily share of these tasks although the hours have fallen over time (Carlson & Hans, 2020). For instance, an older study by the University of Michigan found that, in 1976, the total amount of housework done by women averaged 26 hours per week, a number that had dropped to about 17 hours 30 years later. Men's housework contribution grew during that time period from 6 hours in 1976 to 13 hours in 2008 (Achen & Stafford, 2005; Reaney & Goldsmith, 2008).

Notably, men's domestic work was more likely to be nonroutine domestic tasks, such as making home repairs, preparing a barbeque, or taking the children on outings, activities which all took less time over the course of the week. Researchers have also found people increasingly have developed more household task sharing over time (Gibbs, 2009). Additionally, the type of couple or household arrangement may matter. Lesbian couples, for instance, are more likely to share household labor equally based on the partners ability to do the task or the quality of the task required (Brewster, 2017). Cohabiting heterosexual couples may be more gender egalitarian, as well. An older study of cohabiting heterosexual couples found that women spent less time doing housework than married women—even when researchers account for numbers of children and hours of paid work (Davis et al., 2007).

Interestingly, some research suggests that place of residence—in this case, the state where couples live—has an influence on the household division of labor. Ruppanner and Maume (2016) found that in states where women have more labor market power (measured by, among others, the number of women who are in the paid labor force, are college educated, and work in management), married men are spending more time on domestic tasks. Mothers in these states, whether or not they work, are spending less time on housework. Conversely, in states with lower female labor market power and greater cultural traditionalism (measured by, for instance, higher church attendance and higher marriage and fertility rates), married men spent less time on housework and mothers devoted more hours to home chores.

Childcare is one major issue for families, and, as we noted, that work has historically been gender-typed as the domain of women. It is also important to note that families are not self-reliant. They depend on community and government structures to ensure the health and stability of families. Today, research by the Institute for Women's Policy Research and others finds that providing childcare often places families in tenuous gendered positions for which there are little supports, particularly when children are small. Another related concern tied to gender is paid parental leave. As Figure 10.3 shows, of 41 countries, the United States is the only one that does not provide paid parental leave for new parents.

FIGURE 10.3 ■ U.S. Ranks Last in Government-Mandated Parental Leave

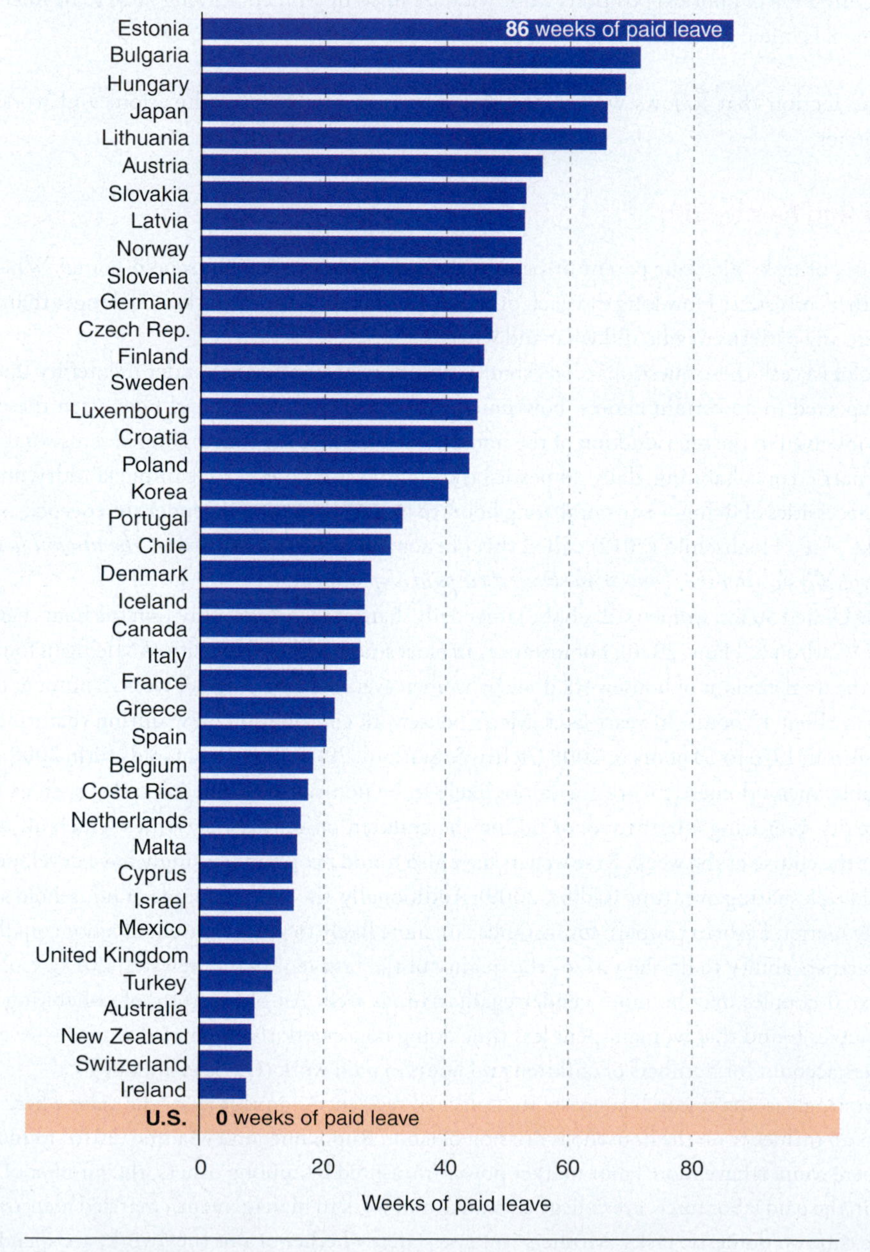

Source: Livingston, G., & Thomas, D. (2019, December 16). Among 41 countries, only U.S. lacks paid parental leave. Retrieved from https://www.pewresearch.org/fact-tank/2019/12/16/u-s-lacks-mandated-paid-parental-leave/

The majority of moms are working, and dads are also increasingly likely to take on more childcare responsibilities, perhaps less by a sense of gender egalitarianism than economic survival. As Livingston and Thomas (2019) put it:

> The share of moms who are working either full or part time in the United States has increased over the past half-century from 51% to 72%, and almost half of two-parent families now include two full-time working parents. At the same time, fathers—virtually all of whom are working—are taking on more child care responsibilities, as fatherhood has grown to encompass far more than just bringing home the bacon. (para. 1)

Consider, for instance, families who have children and want to continue their higher education while their children are young. Will they be able to have a place to take their children that is near them, on campus? This option, according to Green and Robeson (2020), is becoming *less* rather than more available. They note that the availability of on-site childcare centers at public and community colleges decreased between 2003–2015, and that today, less than half of all postsecondary institutions provide on-campus childcare (Green & Robeson, 2020).

Finally, during a time of major family disruption brought about by COVID-19, a *Council on Contemporary Families* study of heterosexual couples compared men and women's perceptions about who was doing the housework and childcare tasks before and during the pandemic. They conclude that COVID-19 both "exacerbated and reduced gender inequalities in the division of domestic labor." As they put it:

For women who continued to shoulder domestic work during the pandemic, housework and childcare responsibilities have become much more arduous. Not only are these women spending more time in classic tasks, but homeschooling has also been added to their plates. Nonetheless, though roughly half of women are doing most of the housework and childcare right now, according to our estimates another half of women are not. Among a sizeable number of families, the burden of domestic responsibilities has become more equal as fathers have increased their contributions to housework and childcare. (Carlson et al., 2020, para. 13)

Gender and Higher Education

Today, higher education is highly correlated with income, and who goes to college and who graduates is undergoing historical change (Fry, 2019). In the 18th- to mid-20th century, women in the United States were discouraged from pursuing higher education. In the late 19th century, beliefs about women's capacity to succeed at *both* education and reproduction led to barriers to access and acceptance. Some believed that the "ovaries—not the brain—were the most important organ in a woman's body" (Brumberg, 1997). As Brumberg writes,

> The most persuasive spokesperson for this point of view was Dr. Edward Clarke, a highly regarded professor at Harvard Medical School, whose popular book *Sex in Education; Or, A Fair Chance for the Girls* (1873) was a powerful statement of the ideology of "ovarian determinism." In a series of case studies drawn from his clinical practice, Clarke described adolescent women whose menstrual cycles, reproductive capacity, and general health were all ruined, in his opinion, by inattention to their special monthly demands [menstruation]. Clarke argued against higher education because he believed women's bodies were more complicated than men's; this difference meant that young girls needed time and ease to develop, free from the drain of intellectual activity. (p. 8)

Although the idea of a "brain–womb conflict" faded as the United States entered the 20th century, other beliefs persisted. Ordinary families with resources to support college study were likelier to invest them in their sons, who were expected to be their future families' primary breadwinners. And women encountered structural and ideological barriers. For example, until the passage of Title IX of the Education Amendments Act in 1972, some U.S. colleges and universities, particularly at the graduate and professional school level, limited or prohibited female enrollment. Title IX, often associated with increasing equity in women's access to collegiate athletic opportunities, has provided for gender equity policies in every educational program receiving federal funding. In recent years, Title IX has also increased emphasis on accountability for sexual harassment and assault on college campuses. In the 1940s, women were said to be going to college to obtain an MRS degree, a phrase to refer to the marriage or engagement of a woman pursuing higher education. Apparently, the MRS degree language is still alive and well. In 2020, the online *College Magazine*, "a national guide to daily campus life," featured an article entitled "Top 10 Schools to Find a Husband" listing "guy to girl" ratios to enhance the chances of getting an MRS degree (Droke, 2017).

In 1975, 29% of men and about 24% of women ages 18–24 were enrolled in a degree-granting postsecondary institution. Enrollment rates have grown for both men and women, but more women as a group have enrolled in college each year since 2000 than have men. This trend holds true, albeit to different degrees, across all major demographic groups (U.S. Department of Education, National Center for Education Statistics, 2020b). The data in Figure 10.4 show that for every group except Black women, college enrollment rates have grown since 2000. What other patterns do you see in these data?

Consequently, women's representation among bachelor's graduates has soared. It has also risen considerably in professional degree achievement. For example, in 1972, women earned only 7% of law degrees and 9% of medical degrees in the United States (National Organization for Women, n.d.). According to the American Bar Association Commission on the Status of Women Report (American Bar Association, 2019), in 2016, women earned 50% of the law degrees conferred, and they made up just over 51% of the total JD enrollment that year. In 2019, for the first time, 50.5% of medical school students were women (Association of American Medical Colleges, 2019). As a result, as shown in Figure 10.5, women now comprise a majority of the college-educated labor force, with 29.5 million women holding at least a bachelor's degree, compared to 29.3 million men (Fry, 2019).

Why are women now more likely than men to be enrolled in college? Researchers suggest several reasons. First, a high school degree is a prerequisite to college, and regardless of racial group, more men than women leave high school without diplomas (U.S. Department of Education, National Center for Education Statistics, 2020a). In high school, as well as college, struggles have been attributed not to differences in "cognitive abilities," which are comparable in boys and girls, but rather to differences in "'non-cognitive skills' among boys, including the inability to pay attention in class, to work with

FIGURE 10.4 ■ College Enrollment Rates by Gender and Race, 2000 and 2018

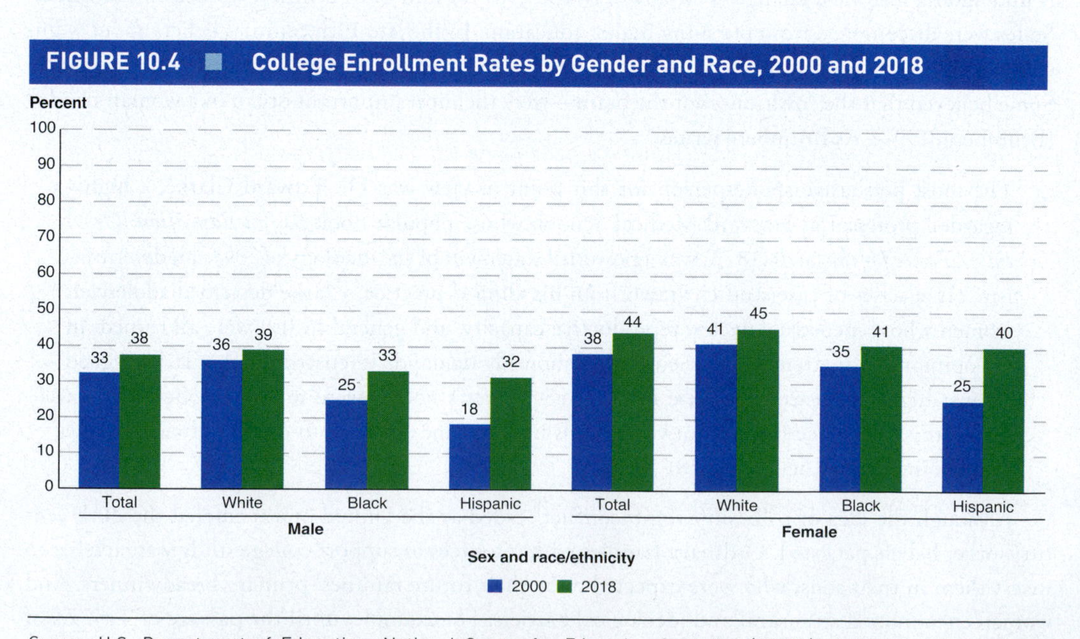

Source: U.S. Department of Education, National Center for Education Statistics (2020b). *College enrollment rates.* National Center for Education Statistics. https://nces.ed.gov/programs/coe/indicator_cpb.asp

FIGURE 10.5 ■ Women Are Now Half of the U.S. College-Educated Labor Force

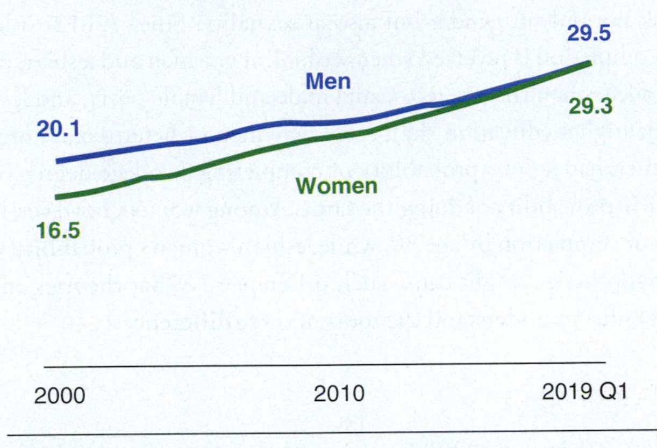

Source: Fry, R. (2019, June 20). *U.S. women near milestone in the college-educated labor force.* https://pewrsr.ch/2ZEVQB3

others, to organize and keep track of homework or class materials, and to seek help from others" (Jacob, 2002, p. 4). Previous research has shown that boys are also more likely than girls to have behavioral or disciplinary problems that lead to dropping out before completion of high school (Stearns & Glennie, 2006).

Second, young women as a group have higher grades than young men overall, even in science and math (O'Dea et al., 2018), and this factor is predictive of college matriculation and success.

Third, women perceive college as bringing greater returns. A Pew Research Center survey found that women are more likely than men to say college was "very useful" in increasing their knowledge and helping them grow intellectually (81% compared to 67%) as well as helping them grow and mature as a person (73% compared to 64%). Perceptions about the necessity of a college education for getting ahead in life were also split by gender: 77% of respondents indicated this was true for women, while 68% felt it was true for men (Wang & Parker, 2011).

Indeed, the benefits of a college education have grown for women (Goldin et al., 2006), and they may be more motivated to use college as a stepping-stone to a desired job. Olivieri (2014) argues that this means significant inroads into some historically male-dominated occupations: According to the author's calculations, of all white women who were either staying home or employed in 1960, only 8% were in male-dominated occupations. In 2010, fully 29% worked in these occupations, which included physicians, lawyers, managers, and scientists. By contrast, the proportion of men in female-dominated occupations, such as teaching and nursing, has remained low. And this occupational gender segregation, which we cover in the next section, remains significant in some fields. For instance, in 2019, women accounted for only 25% of college-educated workers in computer occupations and 15% of college-educated workers in engineering occupations (Fry, 2019).

Olivieri concludes that sexism is at the foundation of this phenomenon: Even though women are occupying a greater share of historically male jobs that require a college education, sexist notions about masculinity and women's jobs are keeping men from pursuing other occupations—including in growing fields such as nursing—that need a college credential. Rather, men may be choosing other historically masculine fields such as construction and manufacturing, even though those sectors have seen declines in recent decades.

For most American college students, the ability to secure loans has become a prerequisite to entering college. The fourth and final piece of the puzzle is that men are more likely than women to leave college without finishing a degree. Student debt may play a role in this process. For some, mounting debt during their undergraduate years leads to a rethinking of the benefit of the degree versus dropping out to enter the workforce. Data suggest that men may be more averse than women to accruing debt: Men drop out with lower levels of debt than women do, but they are also more likely to leave before graduating (Dwyer et al., 2013). Additionally, due to the wage gap that we detail in the following section, men can often more easily find a well-paying job without a college degree.

As we have noted, there are identifiable factors that help us understand the growing gap between men and women in college attendance and completion. Notably, however, the picture becomes more complex when we look not only at gender but also at sexuality. Fine's (2012) study suggests that the probability of college completion is reversed when we look at gay men and lesbian women: Gay men are more likely to finish college than their heterosexual male and female peers, and lesbian women are less likely to complete their higher education than either gay men or heterosexual men and women. One study found that gay men had a 44% probability of completing a college degree by age 30, while heterosexual men had a 28% probability of doing the same. Among women, however, heterosexual women had a 34% probability of completion by age 30, while lesbian women's probability was only 24% (Fine, 2012). What do you hypothesize might cause such differences? What theories and research methods might sociologists best enlist to understand the roots of these differences?

Gender and the Wage Gap

Gender wage gap: The difference between the earnings of women who work full-time year-round as a group and those of men who work full-time year-round as a group.

What do the years 2059, 2133, and 2220 have in common? According to the Institute for Women's Policy Research (IWPR, 2020a), these are years when the **gender wage gap**—or *the difference between the earnings of women who work full-time year-round as a group and those of men who work full-time year-round as a group*—will end.

The year 2059, for instance, is the year when an average of all data will cancel out the wage gap. But this general figure will not show the racial inequalities that will likely still exist. That is, as the IWPR reports, Black women's wages today lag behind white men's to the extent that gender wage gap parity for them may come as late as 2133 and Latina women's wage parity with white men may take until 2220! (IWPR, 2020b; Lacarte & Hayes, 2020). Why, given their rising achievements in education, do women's earnings in most occupational categories continue to lag behind those of men?

One important piece of the answer is the historical and institutional context of the United States' labor market, women's labor force participation, and how this has waxed and waned with structural changes such as war, legal advances, and technological change.

At the end of the 19th century, only one in five women age 16 or older was paid for her work; most paid female employees were young, unmarried, or poor (often all three) and held very low-wage jobs. During World War II, women's labor was needed on the domestic front and their labor force participation rose by 50%, or 6.7 million women (Rose, 2018). While that number declined after the war, by 1950 (5 years after the end of World War II) the proportion of American women working outside the home for pay had risen again. The enormous economic expansion that continued through the 1970s drew even more women into the paid workforce, attracted by higher pay and supported by the passage of laws such as the *Equal Pay Act* of 1963, which made it illegal to pay women a lower wage than men for doing the same job, and the *Pregnancy Discrimination Act* of 1978, which made it illegal for employers to discriminate against pregnant women.

By 1980, more than half of U.S. women were in the paid workforce. According to the Bureau of Labor Statistics, women's labor force participation peaked in 1999 at 60%. In 2018 (the last date for which there are current statistics), their labor force participation rate was 57.1%. That same year men participated at 69%. Incidentally, men's labor force participation rates were at their peak in 1948, when 86.8% of men were in the labor force (Bureau of Labor Statistics Reports, 2019).

Another important dimension to understanding changes in women's labor force participation are women's educational gains that opened up more professional and well-paid positions. For instance, in 1970, only 11% of women between the ages of 25–64 held a bachelor's degree. In 2018, that number had risen to 44% (Bureau of Labor Statistics Reports, 2019).

A continuing aspect of the gender wage gap is the fact that *men and women are still* (and more than we might think) *concentrated in different occupations.* These occupations pay differently, with those jobs gender-typed as "female" paying less (Hegewisch & Tesfaselassie, 2019). Researchers label this phenomenon **occupational gender segregation**. When we identify the 20 most common occupations for women and for men, only three appear on both lists: "first-line supervisors of retail sales workers; managers, all other; and retail salespersons" (IWPR, 2015). According to a study in *The Economist* (2019), "26 out of the 30 highest-paying jobs in the US are male-dominated. In comparison, 23 out of the 30 lowest-paying jobs in the US are female-dominated" (para. 4). According to a 2017 Pew Research Center study, while most men (69%) and most women (68%) say that gender has not made a difference in their job success, women report higher rates of gender discrimination in fields where men constitute the majority of workers. By contrast, women who work in mostly female workplaces are much less likely to say that sexual harassment is a problem (32% vs. 49% for those who work in mostly male workplaces) (Parker, 2018, para. 5).

Occupational gender segregation: The concentration of men and women in different occupations.

Recent data from the Bureau of Labor Statistics (Figure 10.6) show that gendered pay differences continue to exist across all men as a group and all women as a group and also within racial and ethnic groups. Why do gender occupational segregation and the gender pay gap exist and persist? To answer this question, we borrow some terms from economics. For the purposes of our analysis, **labor supply factors** *highlight reasons why women or men may prefer particular occupations*, preparing for, pursuing, and accepting these positions in the labor force. **Labor demand factors** *highlight the needs and preferences of the employer.*

Labor supply factors: Factors that highlight reasons why women or men may prefer particular occupations.

Labor demand factors: Factors that highlight the needs and preferences of the employer.

At the beginning of the 20th century, median earnings for women working in full-time, year-round jobs were only half as much as those for men. In 1963, they were still only about three fifths as much, but by 1999, the gap had narrowed, and women earned just over 72% of the median male wage. Why does this gender gap persist?

These gains were partly the result of postwar "baby boom" women, many of them educated and skilled, entering the workforce and moving into higher-paying jobs as they gained experience.

Labor supply factors underscore the importance of people's choices and decisions in career paths. Several decades ago, many high school– or college-educated working women were likely to work in one of three occupational categories: secretarial work, nursing, and teaching (below the college level). Why were women opting for these occupations? One factor was socialization; women were encouraged to choose feminine occupations, and many did. Another factor was choices women made based on their families' needs: For instance, a schoolteacher's daytime hours and summers off were a good fit with her children's schedules. Today, women are far less limited by either imagination or structural obstacles, as we have seen in the educational and occupational statistics. At the same time, the top 10 jobs most commonly occupied by women today still include traditionally feminine jobs filled by women since large numbers entered the workforce in the 1960s and 1970s (see Figure 10.7). We can also use labor

FIGURE 10.6 ■ Median Weekly Earnings of Full-Time Workers, by Race/Ethnicity and Gender, Second Quarter of 2020

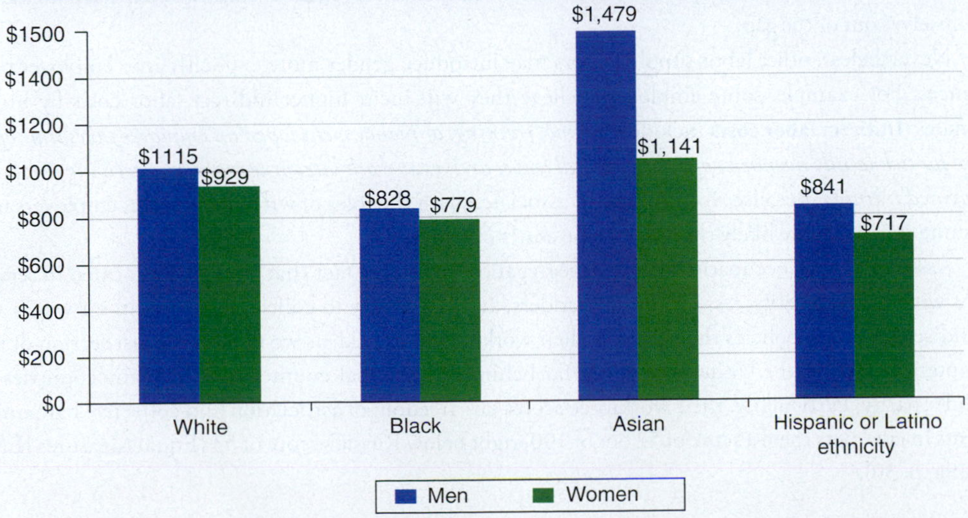

	Men	Women
White	$1115	$929
Black	$828	$779
Asian	$1,479	$1,141
Hispanic or Latino ethnicity	$841	$717

Source: Bureau of Labor Statistics (2020, July 17).

FIGURE 10.7 ■ Women's and Men's Earnings by Occupation in Gender-Segregated Fields

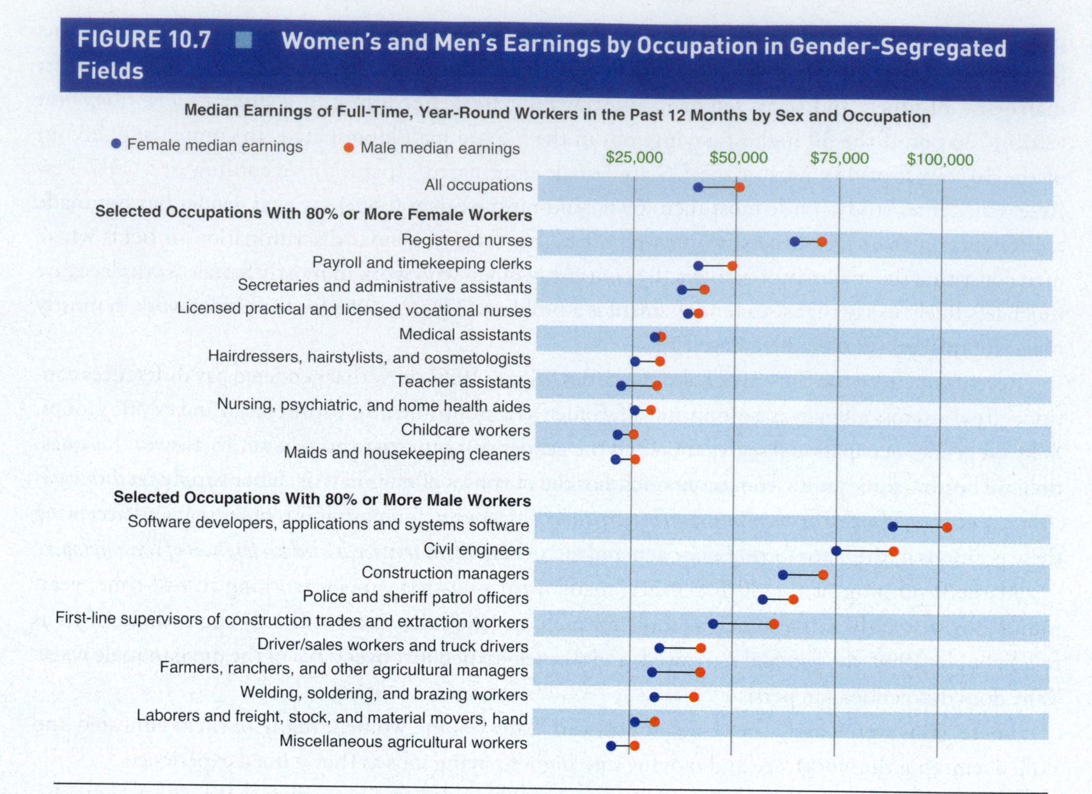

Source: U.S. Bureau of the Census. (2018). *Women's earnings by occupation.* https://www.census.gov/content/dam/Census/library/visualizations/2018/comm/womens-occupations.pdf

supply factors to talk about men's workforce preferences. Men are more broadly spread throughout the U.S. Census Bureau's occupational categories than are women, though in many categories, such as engineer and pilot, they make up a substantial share of all workers.

Human capital: The skills, knowledge, and credentials a person possesses that make him or her valuable in a particular workplace.

Labor demand factors highlight what employers need and prefer—employees with **human capital**, *the skills, knowledge, and credentials a person possesses that make him or her valuable in a particular workplace.* A landscaper seeking a partner will want to hire someone with skills in landscaping, an office manager will seek an administrative assistant who is tech savvy and organized, an accounting firm will want a well-trained certified accountant, and a legal firm a well-trained lawyer who has passed the bar. These preferences are not gendered but instead focus on skills, knowledge, and credentials. It is, however, notable that a pay gap exists at every educational level; in a sense, women cannot educate themselves out of the gap.

Indirect labor costs: The time, training, or money spent when an employee takes time off to care for sick family members, opts for parental leave, arrives at work late, or leaves a position after receiving employer-provided training.

Nevertheless, other labor supply factors may introduce gender more explicitly into employer preferences. For example, some employers believe they will incur higher indirect labor costs by hiring females. **Indirect labor costs** include *the time, training, or money spent when an employee takes time off to care for sick family members, opts for parental leave, arrives at work late, or leaves after receiving employer-provided training.* Because women are still associated with the roles of wife and mother, employers may assume they are more likely than men to be costly employees.

Associated with occupational gender segregation is also the fact that women are located in service jobs with lower benefits, less promotion ladders, and less ability to collectively bargain for wages and nondiscrimination policies that apply to their workplaces. Indeed, as we note in the last section of this chapter, women in the United States are far behind their global counterparts in many countries on gender parity. Particularly, their workforce scores on "freedom of association and collective bargaining rights in law" nets them a score of 52 out of 100, right below Russia's score of 53 (Equal Measures [EM], 2019a, p. 30).

Stereotypes may also condition employers' views, especially when jobs are perceived to be feminine or masculine. Just as a preschool or childcare center might be wary of hiring a man to work with small children or infants (because women are perceived to be more nurturing), a construction firm might hesitate to hire a woman to head a team of workers (because men are widely perceived to be more comfortable under male leadership).

Looking at labor supply and labor demand factors can help us to sort out how men and women have become concentrated in different occupations. Gender occupational segregation is significant because jobs dominated by men have historically paid more than jobs dominated by women. That is, those jobs have higher pay scales, which means that men's earnings tend to start higher and end higher than women's earnings.

What do you notice about the occupations featured in Figure 10.7 in terms of their occupational segregation? Wages? Gendered wages?

Even *within* occupational categories, however, men commonly earn more than their female counterparts. Consider a 2016 study of earnings among doctors in academia—in this case, physicians who teach at U.S. public medical schools. Research on over 10,000 faculty members (65% male, 35% female) at 24 schools found a gender earnings gap: Without accounting for rank and other differences such as specialty and years of residency, male doctors were found to average $257,000 annual salary, while women averaged $206,000. Importantly, even controlling for factors such as rank, age, specialty, publications, years of residency, and research funding, the researchers found a gap of close to $20,000 (Anupam et al., 2016). Differences exist across the occupational spectrum. Data show that, for example, in retail sales—a robust field of employment for both men and women—women earn about 70% of what men earn.

Among managers and waitstaff, women earn about 82% of men's earnings. Even in fields where women make up a significant share of all employees, men outearn their female peers: For instance, among registered nurses, women earn 90% of men's earnings (IWPR, 2015).

What explains these differences within occupations? First, although the Equal Pay Act of 1963 made it illegal to pay men and women different wages for the same work, differences have been documented in fields from journalism to construction to academia. For example, a study of several major media outlets, including *The Wall Street Journal* and *Barron's*, which are owned by Dow Jones, found that "'male reporters' at the company typically make 11 percent more than female reporters. . . . Male 'senior special writers,' a high-level distinction, also outearn their female counterparts by 11 percent" (Paquette, 2016, para. 15). *Pay secrecy*—not knowing what colleagues earn, and being hesitant to ask (or being forbidden to ask)—raises the risk that differences exist. According to the Institute for Women's Policy Research: "About half of all workers (51 percent of women and 47 percent of men) report that the discussion of wage and salary information is either discouraged or prohibited and/or could lead to punishment" (IWPR, 2014, para. 1).

Second, in some occupations, men and women concentrate in different specialties, with men tending to occupy the most lucrative sectors. For instance, male physicians are more likely than females to specialize in cardiology, which pays better than areas where women tend to concentrate, such as pediatrics and obstetrics. Similarly, women real estate agents are more likely to sell residential properties, while men are more likely to sell commercial properties, which bring in higher profits and commissions. Even among restaurant servers, men tend to concentrate in high-end restaurants, while women dominate in diners and chain restaurants.

Third, economist Claudia Goldin (2014) suggests that a key factor in understanding the wage gap both within occupations and more broadly is *temporal flexibility*. She argues that the persistence of the wage gap lies in how jobs are structured and remunerated: Many women are economically disadvantaged by their need or desire for flexibility in work hours. Women may work a comparable number of hours to men, but they are less likely to work odd hours or be willing or able to be available at any hour. Some writers have described this as a "caregiving penalty" (Slaughter, 2015), as the lack of women workers' flexibility is usually linked to their obligations to children or aging parents. Goldin suggests that "the gender gap in pay would be considerably reduced and might vanish altogether if firms did not have an incentive to disproportionately reward individuals who labored long hours and worked particular hours" (Goldin, 2014, p. 1091).

Glass escalator: The nearly invisible promotional boost that men gain in female-dominated occupations.

Glass ceiling: An artificial boundary that allows women to see the next occupational or salary level even as structural obstacles keep them from reaching it.

Researchers have offered various metaphors to explain differences in gendered promotion opportunities. For instance, men who work in traditionally women's fields, such as social work, nursing, and elementary school teaching, have been found to benefit from a **glass escalator**, a *nearly invisible promotional boost that men gain in female-dominated occupations.* Williams (1995) found that bosses often presumed men in these occupations wanted to move up, automatically moving them to promotional tracks even when they were ambivalent about leaving their current positions. For example, a teacher was assumed to want an administrative position or a nurse to want to be a head nurse. However, a later study by Wingfield (2008) of African American and white male nurses showed that race made a difference, as African American men did not experience the same glass escalator effect.

Another metaphor, the **glass ceiling**, is employed to describe *an artificial boundary that allows women to see the next occupational or salary level even as structural obstacles keep them from reaching it.* In 2006, a study of 1,200 executives in eight countries, including the United States, Austria, and Australia, found that a substantial proportion of women (70%) and a majority of men (57%) agreed that a glass ceiling prevents women from moving ahead in the business hierarchy (Clark, 2006). They may be correct; in 2019, women made up less than 5% of all CEOs (chief executive officers) in the S&P 500 (Gurdus, 2019). Furthermore, in 2020, only 38 women were CEOs of *Fortune* 500 companies (Benveniste, 2020).

More recently, Alegria (2019), in a study of women in the tech industry, enlisted the metaphor of a *step-stool* that encourages white women to move into middle management without necessarily removing the glass ceiling for them to move to executive management. Alegria also notes that this is generally true for white women only, not women of color who moved to management only through deliberate effort and with unquestionable qualifications" (2019).

The wage gap is influenced by gender, as well as race and ethnicity, but data available from sources like the U.S. Bureau of Labor Statistics do not capture another important aspect of this phenomenon: the significance of sexuality. This failure to include sexuality in the wage gap phenomenon is due largely to the lack of a holistically intersectional lens.

Studies suggest that lesbian, gay, and transgender workers experience high rates of discrimination in hiring, firing, and wages. As a recent report points out, "These . . . workers lack the necessary legal protections currently afforded to other categories of individuals that would help to combat and correct pay inequities that exist on the basis of sexual orientation and gender identity" (Burns, 2019, para. 2). Transgender workers, for instance, face wage disparities, in particular transgender women: One study determined that wages of these workers fell by a third after their transition (cited in Burns, 2019).

Finally, the *National Partnership for Women and Families* (2020) provides an interesting set of figures for your consideration. It notes that the average working woman would have enough money to purchase one of the following without the annual gender wage gap:

- Over 13 extra months of childcare
- 1 additional year of tuition and fees at a 4-year public university
- 7 months of premiums on health insurance
- 6 months of mortgage and utility payments
- 9 additional months of rent
- 65 weeks of food

In this section, we have reviewed some key aspects of the gender wage gap. Although it is a persistent problem in the United States, women have made significant strides toward closing the gap: Legal protection against discrimination and high rates of women's college completion are among the factors that have contributed to improvements in women's economic status. Importantly, women's improvements vary significantly by race, as we showed earlier in this section. Notably, a part of the declining gap can also be attributed to men's worsening labor market position, a topic we will cover in greater detail in Chapter 15. Wages of men without a college education have been on the decline in recent decades, as well-paying jobs in sectors such as manufacturing have been automated or moved to lower-wage areas abroad. A decreasing pay gap, then, is attributable to improvements for women—but also diminished economic prospects for some men.

FEMINISMS, FEMINIST SOCIOLOGY, AND MASCULINITIES

The fact that the world looks fundamentally different today is, to a large degree, thanks to the feminist movements of both the distant and the recent past. **Feminism** is *the belief that social equality should exist between the sexes*; the term also refers to *social movements aimed at achieving that goal.* Feminism is also an umbrella term for many different theories of *praxis*, or theories tied both to analysis and action. Overall, feminism seeks to explain, expose, and eliminate **sexism**, *the belief that one sex is innately superior to the other and is therefore justified in having a dominant social position.* In the United States and most other societies, sexism takes many forms, from stereotyping to prejudice and discrimination to sexual harassment and sexual violence. Feminists analyze why sexism exists and how it can be eliminated. Today's feminism is also *intersectional*, or as we described in Chapter 1, it sees gender as one component of other interlocking and inseparable identities including race, sexuality, social class, disability, citizenship status, and other socially significant categories that can affect how people experience their daily lives and impact the inequality they experience. In this section, we provide a brief overview of feminism, and three historical "waves" of feminist activity in the United States.

Feminism: The belief that social equality should exist between the sexes; also, the social movements aimed at achieving that goal.

Sexism: The belief that one sex is innately superior to the other and is therefore justified in having a dominant social position.

The *first wave* feminism that emerged in the United States was connected with *abolitionism*, the campaign to end slavery in the 1830s; this movement gave birth to the struggle by women to achieve basic rights, including the right to vote and own land. The emergence of the feminist movement in the 19th century is associated with the 1848 convention in Seneca Falls, New York, to pursue women's expanded rights organized by Elizabeth Cady Stanton and Lucretia Mott. Although their efforts were a landmark in women's history, the results they sought were achieved through consistent struggle through 1920, when women finally gained the vote in the United States with the passage of the 19th Amendment to the United States Constitution. (As we earlier noted, Native American women were not enfranchised until 1924, and xenophobia and racism kept many women of color from voting at the time of passage; Junn et al., 2020.) Feminist activism was limited to a small group of women and their male supporters in an environment that saw sex differences as natural. During the first wave, sociologist Charlotte Perkins Gilman wrote prolifically about women and gender issues outside of the vote including domestic work, economic inequality, health, domestic violence, and the constraints of femininity. In her book *Women and Economics* (1898/2006), Gilman described heterosexual marriage as what she saw as an "unnatural" *sexuoeconomic relation.* As she put it:

> We are the only species in which the female depends upon the male for food, the only animal species in which the sex-relation is also an economic relation." (*Women and Economics*, 1895/1998, p. 5)

To Gilman, women were expected to attract men who would marry them. Women would then be financially dependent on men, and in turn, to serve as caregivers for their husbands and children. Women's gender socialization included significant pressure to find husbands, who in turn felt obligated to support their wives. Thus, the sexual relationship between men and women also became an economic one, with negative effects for the relationship as well as women's autonomy.

The second wave of feminism is generally associated with the equal treatment of women in the home, workplace, and under the law. The 1963 publication of Betty Friedan's *The Feminine Mystique*, which argued that rigid stereotypes of femininity distorted women's real-life experiences and contributed to their unhappiness, helped initiate this wave, with social theorizing and activism that were much broader in scope (Bernard, 1981, 1982; Friedan, 1963, 1981). Women's experiences in the civil rights and anti–Vietnam War and LGBTQ+ social movements also helped shape a growing feminist consciousness.

Women and men were to be viewed not as fundamentally different but as similar; given equal opportunity, women would show themselves the equals of men in all respects. This revival of feminist thinking strongly appealed to the growing number of college-educated and professional women drawn to work and public life during the 1960s (Buechler, 1990), and an explosion of feminist thinking and activism followed.

A variety of different strands of feminisms emerged during these two waves with different emphases and understandings of the roots of inequality and its solutions.

Liberal feminism: The belief that women's inequality is primarily the result of imperfect institutions, which can be corrected by reforms that do not fundamentally alter society itself.

Liberal feminism, reflected in the work of Betty Friedan and today in organizations like the National Organization for Women (NOW), holds that *women's inequality is primarily the result of imperfect institutions, which can be corrected by reforms that do not fundamentally alter society itself.* To eliminate this inequality, liberal feminists have fought to elect women to the U.S. House and Senate, to enact legislation to ensure equal pay for equal work, and to protect women's rights to make choices about their fertility and their family lives. Today, for instance, the National Organization for Women features as one of its key issues winning a constitutional guarantee of equality for women through the Equal Rights Amendments, which was passed by Congress in 1972 and is only one state short of the required three quarters to become part of the United States Constitution (http://now.org/now-and-the-equal-rights-amendment/).

Socialist feminism: The belief that women's inequality results from the combination of capitalistic economic relations and male domination (patriarchy), arguing that both must be fundamentally transformed before women can achieve equality.

As its name implies, **socialist feminism** is rooted in the socialist tradition. It is deeply critical of capitalist institutions and practices and regards *women's inequality as the result of the combination of capitalistic economic relations and male domination (patriarchy), arguing that both must be fundamentally transformed before women can achieve equality* (Chafetz, 1997). This viewpoint originated in the writings of Karl Marx and Friedrich Engels, who argued that inequality, including that of women, is an inevitable feature of capitalism. Engels (1884/1942), for example, sought to demonstrate that the family unit was historically based on the exploitation and male "ownership" of women (the practice of a father "giving away" his daughter in marriage is rooted in the symbolic "giving" of a young woman from one keeper to a new one). Socialist feminism in the United States emerged in the 1960s, when liberal feminists became frustrated by the pace of social reform and sought to address more fundamental sources of women's oppression (Hartmann, 1984; Jaggar, 1983; MacKinnon, 1982; Rowbotham, 1973).

Radical feminism: The belief that women's inequality underlies all other forms of inequality, including economic inequality.

Some feminists also grew frustrated with the civil rights and antiwar organizations of the 1960s and 1970s, which were headed by male leaders who often treated women as second-class citizens. Mindful that full equality for women has yet to be achieved in any existing political or economic system, **radical feminism** argues that *women's inequality underlies all other forms of inequality.* Radical feminists focus on patriarchy, pointing to gender inequality in the economy, religion, and other institutions to argue that relations between the sexes must be radically transformed before women can hope to achieve true equality. They argue that if patriarchal norms and values go unchallenged, many women will accept them as normal, even natural. Thus, although men should also work to end male domination, it is only by joining with other women that women can empower themselves. Radical feminists advocated all-women efforts to provide shelters for battered women, rape crisis intervention, and other issues that affect women directly (Barry, 1979; Dworkin, 1981, 1987, 1989; Faludi, 1991; Firestone, 1971; Griffin, 1978, 1979, 1981; Millett, 1970). Together with other feminists, grassroots organizers, and people involved in advocacy and support, they successfully fought for the Violence Against Women Act of 1994 (VAWA), which provided communities with federal money to address domestic violence, sexual assault, transitional housing for survivors, shelters, and programming, as well as many other areas of assistance around these issues.

Multicultural feminism: The belief that inequality must be understood—and ended—for all women, regardless of race, class, nationality, age, sexual orientation, physical ability, or other characteristics.

Multicultural feminism aims to *understand and end inequality for all women, regardless of race, class, nationality, citizenship, age, sexual orientation, physical ability, or other characteristics* (Smith, 1990). Multicultural feminists seek to build coalitions among women, creating international and global organizations, networks, and programs to achieve women's equality. They acknowledge that much of the contemporary women's movement originated among heterosexual, white, and middle- or upper-class women in Europe and North America and that, as a result, its central ideas reflect these women's perspectives. These perspectives are being challenged and changed by feminists of color from Africa, Asia, and Latin America, as well as gay, bisexual, queer, and trans women, contributing to an enriched multicultural feminist understanding (Andersen & Collins, 1992; Anzaldúa, 1990; Chafetz, 1997; Narayan & Harding, 2000; Zinn et al., 1986).

Third-wave feminism emerged in the early 1990s as a response to some of the perceived shortcomings of second-wave feminism but also as a product of changing societal norms and opportunities—and the Internet. Although third-wave feminists have paid significant attention to issues such as gendered violence and reproductive rights, they also argue that any issue a feminist finds important can and should be discussed. Choice is a central tenet, as well as the use of social media including blogs, Twitter, and Instagram:

> [T]his feminism looks different, in many ways, than that of earlier generations. . . . [It] is shaped less by a shared struggle against oppression than by a collective embrace of individual freedoms, concerned less with targeting narrowly defined enemies than with broadening feminism's reach through inclusiveness, and held together not by a handful of national organizations and charismatic leaders but by the invisible bonds of the Internet and social media. (Sheinin et al., 2016, p. A1)

Is a fourth wave in progress? Pierce (2015), writing on the *National Organization of Women* (NOW) blog site, notes changes over time in feminisms such as the inclusion of transgender women, emphasis on body positivity and self-care, intersectionality, and people-first language. Undoubtedly, movements like #MeToo and women's marches are continuations of feminisms at work today.

Do you identify as a feminist? Why or why not? In 2009, only about a quarter of U.S. women described themselves as *feminist*, but more than two thirds believed the women's movement had made their lives better—including 75% of women under age 35. About half believed there was still a need for a strong women's movement (Alfano, 2009). Interestingly, a much more recent poll shows that 6 in 10 women say that feminism describes them "somewhat well" or "very well," but their responses vary by age, education, and political party (Barroso, 2020). Before you look at their responses in Figure 10.8, what types of people do you hypothesize would be more likely to say that feminism describes them well?

FIGURE 10.8 ■ Percentage of Women Who Say *Feminist* Describes Them Well, 2020

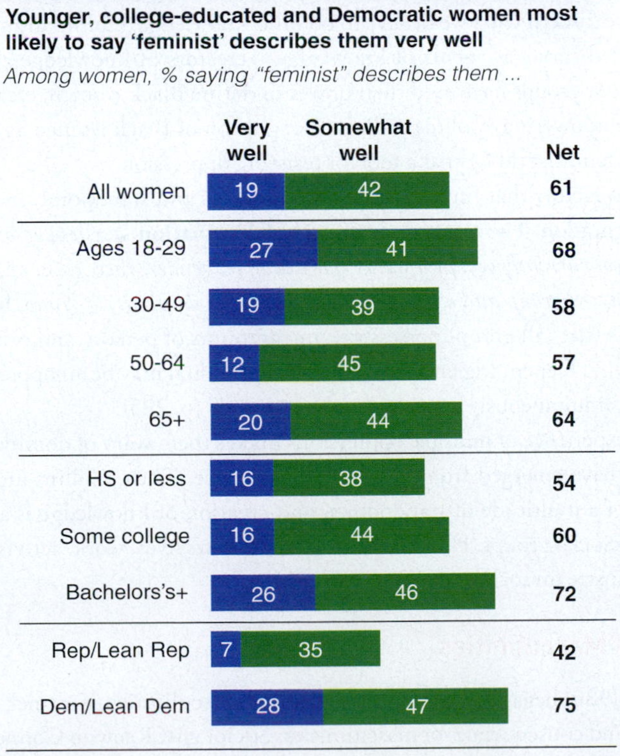

Younger, college-educated and Democratic women most likely to say 'feminist' describes them very well

Among women, % saying "feminist" describes them ...

	Very well	Somewhat well	Net
All women	19	42	61
Ages 18-29	27	41	68
30-49	19	39	58
50-64	12	45	57
65+	20	44	64
HS or less	16	38	54
Some college	16	44	60
Bachelors's+	26	46	72
Rep/Lean Rep	7	35	42
Dem/Lean Dem	28	47	75

Source: Barroso, A. (2020, July 7). *61% of American women say "feminist" describes them well; many see feminish as empowering, polarizing.* https://www.pewresearch.org/fact-tank/2020/07/07/61-of-u-s-women-say-feminist-describes-them-well-many-see-feminism-as-empowering-polarizing/#more-369224

Feminist Perspectives on Doing Sociology

Many sociologists today contribute to sociological knowledge around gender and sexuality, incorporating feminist and intersectional approaches. Two sociologists, Dorothy Smith and Patricia Hill Collins, offer valuable perspectives on what it means to *do* sociology from feminist perspectives, explicitly recognizing women as both subjects and creators of new knowledge.

Standpoint theory: A perspective that says the knowledge we create is conditioned by where we stand or by our subjective social position in our daily lives.

Dorothy Smith (1987, 1990, 2005) made important contributions to **standpoint theory**, which points out that *the knowledge we create is conditioned by our social position in our daily worlds*, or as Smith called them, our *"every day, every night" worlds.* Smith also points out that while our everyday practices shape and reaffirm institutional practices, much of our social worlds have been shaped from the perspectives of educated and often economically privileged white males. Smith thus suggests that our base of knowledge is incomplete because much of what we know—or think we know—about the social world has come from a limited number of perspectives able to shape texts and practices. Smith describes a *bifurcation of consciousness*, or a break that occurs when our individual experiences are different from and contradictory to the dominant point of view to which we must adhere. She has advocated a sociology that begins with the body and in the home, with our daily lives, as a corrective to narratives to such dominant discourses.

Standpoint theory also challenges the sociological (and general scientific) idea that researchers can be, in Max Weber's words, "value-free." Smith argues that standpoint does matter, since we do not so much *discover* knowledge as we *create* it from data we gather and interpret from our own standpoint. Recall Victorian doctor Edward Clarke's influential book about the brain–womb conflict, positing that higher education could damage young women's reproductive capacity. Could such "knowledge" have emerged from the research of a female physician of that time? Would Karl Marx or Talcott Parsons's theories look the same from a woman's perspective? What do you think?

Standpoint epistemology: A philosophical perspective that argues that what we can know is affected by the position we occupy in society.

Patricia Hill Collins (1990) has integrated elements of this idea into her articulation of "Black feminist thought." She offers **standpoint epistemology**, *a philosophical perspective that argues that what we can know is affected by the position we occupy in society.* Epistemology is the study of how we know what we know and how we discern what we believe to be valid knowledge. Collins argues that Black women have long been denied status as agents of knowledge—creators of knowledge about their own lives and experiences. Other groups have used their power to define Black women, creating a picture that is incomplete and disempowering. Collins calls for recognition of Black women as agents of knowledge and the use of Black feminist thought as a tool for resisting oppression.

Matrix of domination: A system of social positions in which any individual may concurrently occupy a status (for example, gender, race, class, or sexual orientation) as a member of a dominated group and a status as a member of a dominating group.

Collins points to factors that fundamentally affect status and standpoint, including gender, race, class, and sexual orientation. The concept of a **matrix of domination**, *a system of social positions in which any individual may concurrently occupy a status (for example, gender, race, class, or sexual orientation) as a member of a dominated group and a status as a member of a dominating group*, highlights this point. Collins (1990) writes that "all groups possess varying amounts of penalty and privilege in one historically created system. . . . Depending on the context, an individual may be an oppressor, a member of an oppressed group, or simultaneously oppressor and oppressed" (p. 225).

Black women's experience of multiple oppressions makes them wary of dominant frames of knowledge, few of which have emerged from their own experience. Thus, Collins argues, comprehensive knowledge is born of a multitude of standpoints, and creation of knowledge is a form of power that should extend across social groups. Her focus on Black women's lives, work, activism, and struggles is a vital addition to mainstream sociology (Wingfield, 2019).

The Sociology of Masculinities

Scholars in the mid-1980s began to develop a sociology of masculinities (Kimmel, 1986) to address the social construction and consequences of masculinities. Sociologist Raewyn Connell (2005) noted that *masculinities* are not the same as *men*, adding that masculinities concern in particular the position of men in society and the organization of gender. Among men, status and power are not evenly distributed and may diverge along lines of class, race, and sexuality, among others. Hence, the focus is not on recognizing and analyzing *masculinity* (in the singular) so much as it is on examining the variety of cultural, social, and institutional influences that "make men."

While scholars were identifying the plurality of masculinities in society, they also pointed out that a certain "*hegemonic*" type of masculinity may dominate. Kimmel (1996), for instance, drew on David and Brannon's (1976) idea that in U.S. culture there are four "basic rules of manhood":

- No "sissy stuff"—avoid any hint of femininity.
- Be a "big deal"—acquire wealth, power, and status.
- Be a "sturdy oak"—never show your emotions.
- "Give 'em hell"—exude a sense of daring and aggressiveness.

These "rules" reflect the concept of **hegemonic masculinity**, the culturally normative idea of male behavior, which often emphasizes strength, control, and aggression (Connell & Messerschmidt, 2005). Men's sports, such as football, reflect these ideals where toughness and aggression also have resulted in playing beyond pain to the detriment of male athletes' health. In an investigative book on brain injuries among football players and the reticence of the National Football League—as well as many players—to acknowledge the risks and injuries to athletes, the authors profile a player, Gary Plummer, who tells them that

Hegemonic masculinity: The culturally normative idea of male behavior that often emphasizes strength, control, and aggression.

> I had been playing football since I was eight years old, and there is nothing more revered in football than being a tough guy. . . . The coaches have euphemisms. They'll say: "You know that guy has to learn the difference between pain and injury." What he's saying is the guy's a pussy and he needs to get tough or he's not going to be on the team." (quoted in Fainaru-Wada & Fainaru, 2014, pp. 79–80)

Kimmel (1996, 2013) argues that some ideas about masculinity are so deeply ingrained in culture that when we discuss social problems such as teen violence (particularly shootings) in U.S. schools, we forget that we are talking almost entirely about the behavior of men. Consider the issue of mass shootings in the United States more generally: A *Washington Post* investigation of "mass shootings" (defined as shootings in which four or more people were killed by a lone shooter, though in three instances, there were two shooters) in the United States over the past 50 years found that of the 129 shooters, most were ages 20–49 and all but three were male (Berkowitz et al., 2016). A *Mother Jones* compilation of mass shootings between 2013–2019 found that of the 53 shootings that occurred during this time, all but three involved men (Reese, 2019). As journalist James Hamblin (2016) wrote in the aftermath of the killing of 49 victims at Pulse, a gay nightclub in Orlando, Florida, in June 2016, "That makes masculinity a more common feature than any of the elements that tend to dominate discourse—religion, race, nationality, political affiliation, or any history of mental illness" (para. 3).

Why are men more likely to be mass shooters—or to commit homicide (which is a crime committed by men over 90% of the time; Ford, 2015)? Educator Jackson Katz points to the rarity, and importance, of this question by offering a hypothetical scenario: "Imagine if 61 out of 62 mass killings were done by women? Would that be seen as merely incidental and relegated to the margins of discourse? . . . No. It would be the first thing people talked about" (quoted in Murphy, 2012, para. 18).

Some writers suggest that "men's studies"—the study of masculinities—should become as much a part of the college curriculum as "women's studies." In many respects, men are already well integrated into the curriculum: As Kimmel writes, "Every course that isn't in 'women's studies' is de facto a course in men's studies. Except we call it history, political science, literature, chemistry" (2000, pp. 5–6). At the same time, this fails to give scholars and students insight into ways in which masculinities shape boys, men, institutions, and societies. In an interview with *The New York Times*, Kimmel suggested that studying masculinity would entail a cross-disciplinary examination of

The National Football League long denied the dangers of repeated head trauma to players' long-term health. A masculine ethic of toughness in the face of injury helped to maintain this position among coaches and players.

> [W]hat makes men men, and how are we teaching boys to fill those roles? It would look at the effects of race and sexuality on masculine identity and the influence of the media and pop culture. It would also allow scholars to take seemingly unrelated phenomena—male suicide and the fact that men are less likely to talk about their feelings, say, or the financial collapse and the male tendency for risk-taking—and try to connect the dots. (Bennett, 2015, para. 25)

Finally, in a 2018 interview, Kimmel also shared some thoughts on how the study of masculinities incorporates *patriarchy*:

> Our analysis of patriarchy is not simply men's power over women; it's also some men's power over other men. Patriarchy's always been a dual system of power, and unless we acknowledge that second one, we won't get an idea of why so many men feel like they're complete losers in the gender game, and they're not at all privileged, and they'll resist any effort toward gender equality. I think we can make them allies. (Kimmel & Wade, 2018, para. 14)

Today, journals such as the *Journal of Men and Masculinities* offer a space for a number of scholars to explore emerging topics: diversity among men (race, ethnicity. sexuality, age, and class), men's movements, fatherhood, media representations of masculinity, sport, violence, and many others. Only by studying the social construction of masculinity can we understand that men are made and not born. And that, sociologists of masculinity believe, is an important step toward achieving gender equality in both attitudes and practices.

In these sections we have covered a range of perspectives and insights offered by scholars of feminisms and masculinity. What areas do you think are most important to continue to study and theorize? What areas are missing?

GENDER AND SEXUALITY IN A GLOBAL PERSPECTIVE

In many places across the globe, if you are reading these words, you are likely a man; millions of women are denied education and cannot read. Worldwide, girls and women lack literacy and other forms of education, are at greater risk of being denied medical care, are refused the right to own or inherit property, and are trafficked into the sex trade. They are less likely than boys and men to go to school, and to eat or get medical care when family resources are scarce. They are likely to have lower wages than men. They have also suffered additional challenges under the COVID-19 pandemic, from increased violence in their homes to caregiving struggles and challenges of working, such as transportation shutdowns.

On balance, then, what does gender equality currently look like around the world? Is it improving? How can we know? The 2019 Sustainable Development Goals Gender Equality Index (SDGGEI) provides some answers. This index, enlisted by the organization Equal Measures (EM), draws on 14 of 17 sustainable development goals surrounding gender equity in 129 countries to "examine 51 issues ranging from health, gender-based violence, climate change, decent work, and others" (EM, 2019b). Of the 129 countries surveyed, no country has fully achieved gender equality. As you might expect, some countries have higher levels of gender equality than others (for instance, Denmark receives the highest ranking, the United States ranks no. 28, and Chad ranks no. 129).

Figure 10.9 shows the progress of select countries on five dimensions of gender equity: "access to contraception, girls' education, political leadership, workplace equality laws, and safety." As you can see, some countries have slowed in their movements toward gender equality while some with lower levels of gender equality have made advances more rapidly in recent years. Why do you think that progress in some countries has stalled while in others has improved?

Women and Families

We earlier discussed gendered inequalities in American household labor. Women worldwide also bear the unequal burden of unpaid domestic care work also known as the second shift. According to the United Nations, women worldwide devote 1 to 3 hours more a day to housework than do men; 2 to 10

FIGURE 10.9 ■ Overall Gender Equality of Select Countries Since 2000

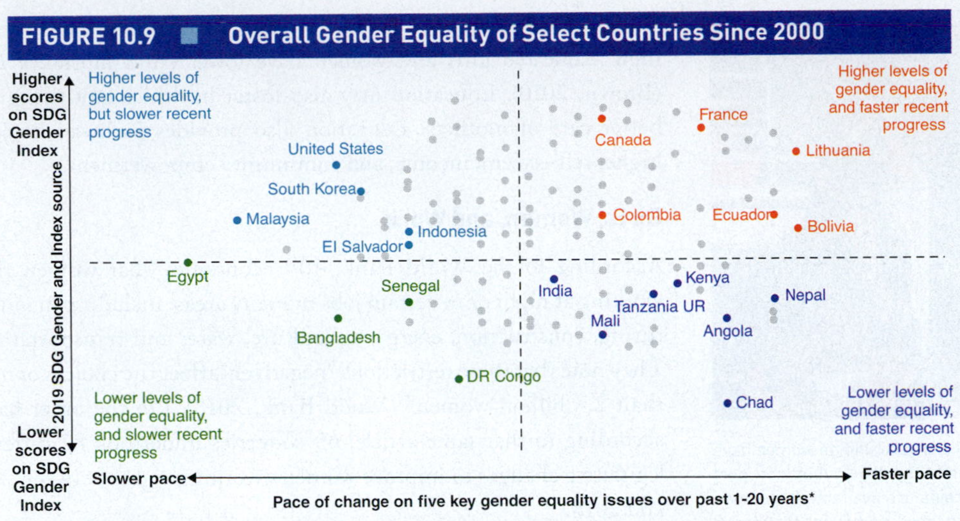

*For results for all countries, please visit the Gender Advocates Data Hub: data.em2030.org

Source: Equal Measures 2030. (2019b). *Bending the curve towards gender equality by 2030.* https://www.equalmeasures2030.org/products/bending-the-curve-towards-gender-equality-by-2030/

times the amount of time a day to care (for children, elderly, and the sick), and 1 to 4 hours less a day to market activities (United Nations Women, n.d.).

Take the example of Cuba. Many Cubans live in multigenerational households where women are often expected to do the disproportionate share of childcare, eldercare, and traditional cooking and cleaning while balancing work and maintaining the household. According to Cuba's 2016 National Survey for Gender Equality, women ages 15–74 engaged in 36 hours of unpaid domestic care work weekly; men engaged in 22 hours (Gonzalez, 2020). As Caridad Acosta, a retired woman who now cares full time for her mother, explained of the all-encompassing nature of her caregiving experience:

> I have taken care of my father, my aunts . . . everyone who gets sick. My husband had an ischemic stroke and lived for six more years and was in a coma for a month just before he passed away. I spent five years of that time dividing my time between my husband and my job in a workers' canteen. (quoted in Gonzalez, 2020)

Disproportionate caregiving expectations are an important roadblock to gender equity. They have a number of impacts, including increasing mental and physical health burdens on women and decreasing their ability to participate in the paid labor market.

Girls and Education

According to the United Nations (2020), 132 million girls worldwide are not enrolled in school. Gender parity for girls and boys in primary education has only been achieved in 66% of countries. Furthermore, only 25% have achieved gender parity in upper secondary education (UNICEF, 2020). Many challenges can keep girls from getting an education such as norms against girls' education, access, work, children, and unsafe or unavailable transportation. For instance, Srivastava (2019) notes that girls in India have dropped out of school because of lack of safe transportation, an issue that several Indian states have addressed by providing bikes to girls to travel to school.

In terms of gender equity, keeping girls in school in many countries helps to spread economic development. Educated women are also healthier and can pass these health gains to their families. In West African Burkina Faso's Sahel region, for instance, girls are more likely to drop out of school at an early age. They are more likely to get married and pregnant when they are between the ages of 16 and 19. As a result, they may average no more than 4.3 years of education. However, according to Power et al. (2019), if reasons related to school, work, and care work can be eliminated for girls in this region, 30% more will stay in school. Furthermore, relieving economic needs allows another 7% to stay in school.

In Mumbai, India, public transportation is crucial. Indian women have experienced high rates of harassment and assault on public transport and, as a result, separate passenger trains are available for women to travel to work. Indian women workers faced significant challenges under the COVID-19 pandemic when these trains suspended services during "nonpeak" hours.

REUTERS/Niharika Kulkarni

In many countries, maternal and infant mortality rates remain high. Educated girls and women have lower child mortality rates (Brown, 2010). Education may also foster healthier mothering and better care of mothers. Education also provides pathways to jobs, higher self-esteem, income, and community empowerment.

Girls, Women, and Work

According to the World Bank, 104 economies "bar women from working at night or in certain jobs in many areas, including manufacturing, construction, energy, agriculture, water and transportation." They note that these restrictions "negatively affect the choices of more than 2.7 billion women" (World Bank, 2018). On the other hand, according to that same article, 65 countries undertook 87 different legislative changes to improve women's working rights between 2016 and 2018.

Economically vulnerable women are also socially vulnerable. Impoverished girls worldwide, especially in countries with marked poverty, are particularly vulnerable to sexual exploitation and trafficking. According to the annual *Trafficking in Persons Report* published by the U.S. Department of State (2012b), it is difficult to pin down the extent of this "modern-day slavery." The International Labour Organization, however, estimates that almost 21 million people are victims of forced labor: about 11.4 million women and girls and 9.5 million men and boys (International Labour Organization, n.d.). Of those exploited by individuals or enterprises, 4.5 million are victims of forced sexual exploitation (International Labour Organization, n.d.).

In India, sex trafficking is pervasive (Kara, 2009). A British Broadcasting Corporation (Patel, 2013) examination of Indian brothels told the story of one woman:

> Guddi was only 11 years old when her family was persuaded by a neighbor to send her to the city of Mumbai hundreds of miles away from her poverty-stricken village in the eastern state of West Bengal.
>
> They promised her a well-paid job as a housemaid to help feed her family.
>
> Instead, she ended up at one of Asia's largest red-light districts to become a sex worker.
>
> Trafficked by her neighbor, she arrived at a brothel. She was raped by a customer and spent the next three months in hospital.
>
> Guddi's sad and harrowing story is similar to many of the estimated 20,000 sex workers in Kamathipura, established over 150 years ago during colonial rule as one of Mumbai's "comfort zones" for British soldiers. (paras. 1–5)

Kristof and WuDunn (2009) estimate that there are 2 to 3 million prostitutes in India.

The size of the sex trade in some developing states is rooted in several factors. First, in societies such as India, Pakistan, and other regional states, societal norms dictate that young couples wait until marriage to consummate a relationship. "Respectable" middle-class girls are expected to save their virginity for their husbands. For young men, then, access to prostitutes offers a penalty-free way to gain sexual pleasure and experience before marriage. Second, the girls and young women in the brothels are usually poor, illiterate villagers with no power or voice and few advocates or protectors (Kara, 2009). Police are not only unlikely to help them but may participate in their exploitation as well.

The diminished status conferred by deep poverty and being female imposes a profound double burden on girls and women. It is no coincidence that in a global environment that so often marginalizes women and girls, the trade in their bodies and lives is vast, widespread, and often ignored by authorities.

Gender and Pandemic: The Gendered Impact of COVID-19

Worldwide, COVID-19 had disproportionate impact on women and girls, bringing together many issues discussed in this section. Women domestically and internationally, for instance, work in many occupational fields where caregiving is on the front lines (for instance, health care, teaching, daycare, and eldercare). Like women in the United States, the burden of educating youth during COVID-19 fell disproportionately to them, causing them to be exposed to the virus, to work long second shifts when they returned home, and to leave their jobs to address family needs such as education.

During pandemics, rates of domestic violence have been noted to climb, and COVID-19 was no exception to an increase in domestic violence reported in some countries, especially when lockdowns may have resulted in people living together 24 hours a day (Faiola & Herrero, 2020). In a poignant excerpt in *The Washington Post*, "Zoila," a 24-year-old woman from El Salvador, told of her experience during lockdown:

> Zoila fell fast for the soft-spoken day laborer, moving in with him last year just two weeks after their first date. But after El Salvador imposed a strict coronavirus lockdown, she says, the man she thought she knew became an inescapable menace.
>
> "The quarantine changed everything," she said.
>
> Shut inside their one-room house in rural El Salvador, he began drinking heavily. Soon, she says, he was regularly violating the coronavirus curfew and seeing other women openly. He would return home at odd hours, wake her and demand meals. Drunk, he would taunt Zoila, 24 years old and pregnant, calling her worthless and threatening violence.
>
> Then one morning, she says, he grabbed her by the throat, slammed her against the wall and attempted to rape her. When she resisted, she said, the punching began, stopping only when fluid began trailing down her leg. Zoila screamed, fearing a miscarriage.
>
> "I remember that day, and I just want to cry," said Zoila, who gave birth to a daughter in June.
>
> "I was pregnant," Zoila said. "During what was supposed to be a time of joy for me, I felt only pain." (quoted in Faiola & Herrero, 2020)

In October 2020, the United Nations warned that women were especially hard hit by COVID's impact on "informal economies":

> Women have been hard hit by the collapse of informal economies, and with so many schools closed, children are increasingly at risk of online sexual exploitation and the worst forms of child labor. . . . Around the world, many migrant workers remain stranded with no way home and no social benefits where they are. Low-income workers have been hit by rising unemployment. (United Nations Human Rights, 2020)

It also warned that COVID-19 was likely to make addressing and assisting victims of human trafficking, a phenomenon impacting women, girls, and sexual minority groups, more difficult.

Change Happens: Women's Empowerment

A number of nongovernmental organizations (NGOs) and not-for-profit organizations operate globally to provide the tools for women to work within their communities to combat educational access, violence, poverty, gender discrimination, and a range of other issues. For instance, the United Nations explains that it works with a variety of partners to offer programs that:

> promote women's ability to secure decent jobs, accumulate assets, and influence institutions and public policies determining growth and development. One critical area of focus involves advocacy to measure women's unpaid care work, and to take actions so women and men can more readily combine it with paid employment. (United Nations Women, n.d., para. 4)

According to the nonprofit organization Oxfam, more than 700,000 women around the world had participated in Saving for Change by 2015. The women in the groups make small weekly deposits into a common fund and lend to one another from the fund at a 10% monthly interest rate. More women today are using cell phones to save and track money, which has led to greater savings.

©Rebecca Blackwell/Oxfam America

The nonprofit organization Oxfam America (http://www.oxfama-merica.org) is one example of such an organization. It features a program called Saving for Change, which emphasizes "savings-led microfinance." Participating groups of women save and pool their money, agree on guidelines for investing or lending in their communities, and organize their resources to serve local needs. The groups enhance not only the women's economic capital but also their social capital, building ties that support them in times of economic or other crises. Although Oxfam funds coaches who help the women get started, Saving for Change groups are not financed or managed from the outside; they are fully autonomous and run by the women themselves.

Take the example of women's experience in West Africa's Mali. Mali has high rates of early marriage and few legal protections for women. But Mali, a site where the Saving for Change program is in operation, also has a growing practice of savings-led microfinance, empowering women to save, earn, lend, and invest. According to Oxfam, "Households in villages [in Mali] with savings groups experienced an 8 percent increase in food security and saved 31 percent more on average" (Kramer, 2013). With coaching, women gain financial literacy and empowerment, despite their lack of schooling. More mature savings groups are serving the global market for local commodities such as shea butter, a popular cosmetic ingredient. Other countries where women are participating in savings-led microfinance programs include Senegal, Cambodia, and El Salvador (Oxfam, 2015).

Some fear that women's empowerment can foster backlash, manifested as violence or social repercussions. Some patriarchal societies may not be ready to see women take the initiative to address sexual exploitation, bring attention to crimes against women, or grow economically independent of men, and the victimization of women who step out of traditionally subjugated roles is a risk.

On the other hand, greater independence—social or economic—may allow women to leave violent relationships or challenge norms that marginalize them. The effects of women's economic empowerment can also go beyond their own lives, improving prospects for their children and communities. Studies suggest that when women earn and control economic resources, family money is more likely to go toward needs such as food, medicine, and housing (Kristof & WuDunn, 2009). Maternal and child mortality are reduced with women's empowerment, as are the poverty, marginality, and illiteracy that may lead desperate girls and women to the global sex trade. Mobilization of human and intellectual capital is a critical part of domestic development for a country. It is not a coincidence that countries offering opportunity and mobility to women prosper economically; where fully half the population is deprived of rights, education, and access to the labor market, the consequences are ultimately borne by the whole society and state.

WHAT'S NEXT? EQUALITY, GENDER, AND SEXUALITY

As we have shown throughout this chapter, gender, sex, and sexuality matter—from a teacher's encouragement to study different academic subjects, to choosing who one can love, to who does the chores within the household, to the cumulative impact of pay and promotions on the job and over one's life course. Worldwide, one's gender can determine whether someone can attend school, work in a paid job, earn a living wage, and be represented politically. And one's gender and sexuality can also result in discrimination, harassment, and violence. For instance, 76 countries worldwide still criminalize sexual acts with persons of the same sex (Amnesty International, 2020). Even organizations like the United Nations, who have made concerted efforts to tackle gender equality, which they have labeled their "greatest single rights challenge to human rights around the world" (Nichols, 2020), had to note that women in their own organization had a "perfect right to feel that they are not represented" when, at the October 2020 United Nations General Assembly, only nine women were representative of 190 countries (Nichols, 2020).

The majority of Americans in 2020 believe work remains to be done on gender equality (Horowitz & Igielnik, 2020). They are, however, split on what shape equality in the future should take. The authors note that "about three in ten men say women's gains have come at their expense." At the same time, men and women worldwide in 2020 are increasingly more likely to be optimistic about the future of gender equality (Horowitz & Fettererolf, 2020). In a survey of people from 34 countries, the majority of respondents leaned to gender egalitarianism in heterosexual households, saying "a marriage where both the husband and wife have jobs and take care of the house and children is a more satisfying way of life than one where the husband provides for the family and the wife takes care of the house and children" (Horowitz & Fettererolf, 2020). Concurrently, as we have shown in this chapter, attitudes toward LGBTQ+ remain sharply divided around the world (Poushter & Kent, 2020) while the numbers of people who identify as LGBTQ+ grow. At the same time, as we have shown in several sections of this chapter, COVID-19 may have complicated movement toward equality as it hampered programs, crippled economies, and impacted individual lives in many ways.

So, what's next? What do you predict will happen in terms of gender and sexual equality in the next 10 years? Why do you think that? How might different views of equality through the lens of feminisms, queer theory, and theories of masculinity contribute insights to pathways toward equality? What challenges do you predict around achieving gender equality from studying this chapter? Sexual equality? How might sociological research play a role in these issues?

WHAT CAN I DO WITH A SOCIOLOGY DEGREE?

Ethical Decision Making

As you learned in Chapter 6, ideas of right and wrong are socially constructed. Ethical standards are socially, culturally, and even historically specific as societies struggle to adopt new ethical standards that keep pace with the speed of discovery and technological innovation. While many occupations have written codes of ethics to guide employees, such as the Hippocratic Oath that instructs medical professionals to do no harm, most of us are on our own to make personal and professional decisions. Ethical decisionmaking requires understanding the impact of those outcomes on others, be they individuals, groups, communities, society, or even the world.

Sociology can help you develop the skills to make decisions that are both effective and ethical. Sociology demands a more complex understanding of social life and tries to get at the root of social issues, for example, by examining the links between housing discrimination and educational segregation or the limited effects of social control policies in the criminal justice system. Sociology also looks at the ways in which access and outcomes are affected by social categories of class, race, gender, sexual orientation, age, and religion. Sociology also asks us to think about our own theoretical perspectives and social locations. It illuminates how different assumptions about how society operates lead to very different explanations for the causes and therefore solutions to social problems. By understanding these differences, you can unmask your own and other's potential biases and better understand the impact of decisions. To make ethical decisions, you have to understand how different groups of people will be impacted and you need to understand root causes. Both are at the core of sociology.

Leah Hubbard, Associate at Estolano LeSar Perez Advisors

Loyola Marymount University, BA in Sociology and Music

University of Southern California, Master's of Public Affairs

ELP Advisors is a mission-driven consulting firm that exists to build better communities through strategic vision. We work with public agencies, foundations, businesses, nonprofits, and other stakeholders to provide innovative approaches to complex policy issues. When taking on a new project or providing strategic guidance to a client, we must understand and dissect the broad, complex issues of community and economic development, including markets, local and national policy, and social inequalities as well as consider the impact of potential outcomes on various groups of stakeholders. This helps us effectively address client challenges with integrity and equity.

Sociology provides the perfect opportunity to get a first look at these complex social systems at work. My theoretical classes gave me the ability to think critically through policy and planning recommendations and to consider their unintended consequences. On the other hand, my quantitative courses gave me the skills to translate data into narratives that can be used to inform and generate solutions. Consulting requires that I be a problem solver who is not only a creative thinker but an ethical decision maker who understands the roots of inequality and biases.

Career Data: Urban and Regional Planners

- 2019 Median Pay: $74,350 per year
- $35.75 per hour
- Typical Entry-Level Education: Master's degree
- Job Outlook, 2019–2029: 11% (Much faster than average)

Source: Bureau of Labor Statistics. (2019). *Occupational outlook handbook.*

SUMMARY

- Sociologists differentiate between **sex** and **gender** using the term *sex* to refer to biological differences between males and females and *gender* to refer to social differences between men and women and boys and girls. Sex and gender are often identified as binary differences, but today sociologists and other researchers recognize a continuum.
- **Gender roles** are the activities, attitudes, and behaviors considered appropriately masculine or feminine in a particular culture. Children begin to learn culturally appropriate masculine and feminine gender identities as soon as they are born, and these roles are reinforced and renegotiated throughout life. Gender roles are learned through social interaction with others. Early family influences, peer pressure, the mass media, and the "hidden curriculum" in schools are especially important sources of gender socialization.
- Women do more housework than men in all industrial societies, even when they engage in full-time paid employment outside the home. Women are also more vulnerable to childcare obligations and are disadvantaged by lack of daycares and family leave policies.
- Women typically work in lower-paying occupations than men and are paid less than men for similar jobs. While all women as a group suffer from the impact of the wage gap, women of color are paid lower wages than white women. Women in all groups have made gains but are still less likely to be promoted in most positions than are their male peers.
- **Liberal feminism** argues that women's inequality is primarily the result of imperfect institutions. **Socialist feminism** emphasizes that the combination of capitalistic economic relations and male domination creates inequality. **Radical feminism** focuses on **patriarchy** as the source of domination. Finally, **multicultural feminism** emphasizes ending inequality for all women, regardless of race, class, nationality, age, sexual orientation, or other characteristics. Third-wave feminism highlights women's agency, often enlisting social media and the Internet to inform, connect, and activate people around feminist issues and concerns.
- Both Dorothy Smith and Patricia Hill Collins point out the importance of individual women's standpoints in constructing a more robust discipline of sociology that incorporates more diverse experiences.

- Some sociologists advocate for the creation of a study of masculinities to understand more fully the sociological influences on the perceptions and practices of men and boys.
- Globally, being born female is still a risk and women's equality with men is still not achieved. Women are disadvantaged in access to power, education, occupations, and safety from violence. At the same time, women are taking the initiative in many developing areas to improve their own lives and those of their communities and families, assisted by national agencies, community groups, and nonprofit organizations. COVID-19 has created a number of barriers to equality for women and sexual minorities.

KEY TERMS

cisgender (p. 252)
feminism (p. 273)
gender (p. 252)
gender identity (p. 252)
gender roles (p. 256)
gender socialization (p. 255)
gender wage gap (p. 268)
glass ceiling (p. 272)
glass escalator (p. 272)
hegemonic masculinity (p. 277)
human capital (p. 270)
indirect labor costs (p. 270)
intersex (p. 251)
labor demand factors (p. 269)
labor supply factors (p. 269)
liberal feminism (p. 274)
matrix of domination (p. 276)
multicultural feminism (p. 274)
occupational gender segregation (p. 269)
patriarchy (p. 253)
queer theory (p. 254)
radical feminism (p. 274)
second shift (p. 263)
sex (p. 251)
sexism (p. 273)
sexuality (p. 253)
socialist feminism (p. 274)
standpoint epistemology (p. 276)
standpoint theory (p. 276)
transgender (p. 252)

DISCUSSION QUESTIONS

1. How are gender, sex, and sexuality sociologically related? How is each socially constructed? To answer this question well you should be sure to accurately use terms and concepts from your reading.
2. Explain how an individual is socialized into the gender they identify with at the following life stages: early childhood, preteen years, adolescence. For each stage also provide an example of how the individual may perform gender.
3. Throughout the chapter, we learned about gender inequalities in institutions including the family, education, and the workplace. Think about another institution, such as religion, politics, or criminal justice. What kinds of research questions could we create to study gender inequality in those institutions?
4. There have been several waves of feminism, and women have made significant gains legally, socially, and politically. Is feminism still needed in our society? What would be the key characteristics of a feminism that meets today's societal challenges?
5. How do inequalities globally faced by women contrast with the inequalities experienced in the United States? What are some of the ways in which gender inequalities are being addressed and who is addressing them?

©adamkaz/Getty Images

11 FAMILIES AND SOCIETY

WHAT DO YOU THINK?

1. What is a family? Who should have the power to define what a family is?
2. Why are U.S. young adults today less likely to marry than previous generations of young adults?
3. Do styles and priorities of parenting differ across the socioeconomic class spectrum?

LEARNING OBJECTIVES

11.1 Explain key concepts sociologists use to study families.

11.2 Apply theoretical perspectives to the study of marriage and the family.

11.3 Describe trends in family formation and family life in the United States, including marriage, divorce, childcare arrangements, and domestic violence, and describe family patterns in immigrant, Native American, and deaf families.

11.4 Explain what sociological research suggests about ways in which social class influences family formation and parenting practices.

11.5 Understand the relationship between globalization and family life, particularly for U.S. families and women from less developed countries.

ASSORTATIVE MATING AND RISING INEQUALITY IN THE UNITED STATES

While marriage has been on the decline in the United States, it remains the case that most Americans will marry at some point in their lives. How do people choose a mate? The question seems simple: The top responses in a 2019 survey showed that most people marry for love and companionship. Others responded that they were motivated by a desire for children or married because it made financial sense (Horowitz et al., 2019). So why do sociologists take an interest in the processes of dating, mating, and marriage, which seem driven by individual attractions and wishes? Sociologists take an interest because marriage decisions in societies often follow patterns that suggest notable commonalties in motivations and outcomes—including societal outcomes—of marriage.

Consider the phenomenon of assortative mating. Assortative mating means that people who are similar in education or income (or both) tend to marry each other. The phenomenon is also known as *homogamy:* a marriage of those who are in some respect the same. Assortative mating, research suggests, has been on the ascent in the United States for several decades.

The work of economist Branko Milanovic shows that over just about 50 years, marriage between those in the upper- and lower-income brackets has grown increasingly improbable. For example, Milanovic looked at high-income U.S. men between ages 20 and 35 in 1970 and asked what proportion of them married women in their own income group versus marrying women from a low-income group. He found that the ratio was about 1:1; that is, for every high-income female partner there was also one low-income female partner entering marriage. Today, says Milanovic, the ratio is 3:1. So high-income men in typical marriages are three times more likely to marry women of their own income group (Garcia & Vanek Smith, 2020).

The rise in assortative mating is a trend, according to Milanovic, that reflects positive social changes, but it also has economic consequences. On the one hand, the freedom to choose one's partner has grown, as has women's earning power, which is more likely to put them in a high-income bracket. On the other hand, assortative mating also exacerbates inequality (Greenwood et al., 2014; Milanovic, 2019). Milanovic points out that

> [I]f you have a rich guy marrying a poor woman and, likewise, a rich woman marrying a poor guy, these two couples will have the same income. But if you have the opposite situation—rich

men and rich women . . . mating and poor woman and poor . . . man mating—then, of course, the gap between the two couples is going to be exacerbated by the marriage. (Garcia & Vanek Smith, 2020)

Notably, a Pew Research study on marriage and cohabitation found that the most commonly cited obstacles to marriage were that respondents felt their partner was not financially ready for marriage (29%) and that respondents indicated that they themselves were not financially ready for marriage (27%). While these results can be read a number of ways and middle- or higher-income couples too may feel financially unprepared for marriage, research suggests that marriage continues to decline among adults who have lower educational attainment—and likely lower incomes. College-educated adults are more likely to be or to have been married than those with a high school education or less (Horowitz et al., 2019). The stratifying effects of assortative marriage may be exacerbated by a rising marriage gap.

Milanovic argues that as more high-income couples, whether of different sexes or of the same sex, pair off, there may be a homogeneity of interests and experiences that decisively separates them from other classes in society. This means that those who can afford private schools and other private amenities may lose connection to and interest in supporting public goods, like public parks, schools, health care, and the like, which are not part of their individual or shared experience. This may have broader societal implications as the distance between the rich and everyone else grows and continues to be reinforced by assortative mating and shifts in marital patterns that link to income and education (Milanovic, 2019).

In this chapter, we focus on the U.S. family in the past and today. We begin by introducing key concepts used in the sociological study of families and discuss the idea of the family as an institution. Then, we review functionalist and feminist perspectives on families and, in particular, on traditional sex roles in marital relationships. We devote a broad section of the chapter to an overview of modern U.S. families, looking at trends in marriage and divorce as well as family life in a sampling of subcultures—immigrant, Native American, and deaf families. Sociologists take a strong interest in issues of social class and its roots and effects, so next, we explore practices of child-rearing and differences across class, the decline of marriage in the poor and working classes, and work and family life in the middle class. Finally, we explore the relationship between globalization and family in the United States and beyond.

HOW DO SOCIOLOGISTS STUDY THE FAMILY?

Families come in a spectrum of forms, but they share important qualities. A **family**, at the most basic level, is two or more individuals who identify themselves as being related to one another, usually by blood, marriage, or adoption, and who share intimate relationships and dependency. The family is a key social institution. Although families and their structures vary, the family as an institution is an organized system of social relationships that both reflects societal norms and expectations and meets important societal needs. It plays a role in society as a site for the reproduction of community and citizenry, socialization and transmission of culture, and the care of the young and old. Families, as micro units in the social order, also serve as sites for the allocation of social roles, such as breadwinner and caregiver, and contribute to the economy as consumers.

Many families are formed through marriage. Sociologists define **marriage** as a culturally normative relationship, usually between two individuals, that provides a framework for economic cooperation, emotional intimacy, and sexual relations. Marriages may be legitimated by legal or religious authorities or, in some instances, by the norms of the prevailing culture. Although marriage has historically united partners of different sexes, same-sex marriages have become increasingly common in the United States and other countries, although their legal recognition is still incomplete.

Most societies have clear and widely accepted norms regarding the institution and practice of marriage that have varied across time and space. Two common patterns are **monogamy**, *a form of marriage in which a person may have only one spouse at a time*, and **polygamy**, *a form of marriage in which a person may have more than one spouse at a time*. Within the latter category are **polygyny**, *a form of marriage in which a man may have multiple wives*, and **polyandry**, *a form of marriage in which a woman may have multiple husbands*. In feudal Europe and Asia, monogamy prevailed, although in some parts of

Family: Two or more individuals who identify themselves as being related to one another, usually by blood, marriage, or adoption, and who share intimate relationships and dependency.

Marriage: A culturally approved relationship, usually between two individuals, that provides a framework for economic cooperation, emotional intimacy, and sexual relations.

Monogamy: A form of marriage in which a person may have only one spouse at a time.

Polygamy: A form of marriage in which a person may have more than one spouse at a time.

Polygyny: A form of marriage in which a man may have multiple wives.

Polyandry: A form of marriage in which a woman may have multiple husbands.

Asia, wealthy men supported concubines (similar to mistresses; Goody, 1983). George Peter Murdock's (1949) classic anthropological study of 862 preindustrial societies found that 16% had norms supportive of monogamy, 80% had norms that underpinned the practice of polygyny, and 4% permitted polyandry.

The polygynist practice of a man taking multiple wives is unusual (and not legally recognized) in the United States, but according to researchers at Brigham Young University, an estimated 30,000 to 50,000 U.S. residents practice polygamy. Many are members of breakaway sects of the Mormon Church.

Serial monogamy: The practice of having more than one wife or husband—but only one at a time.

Some sociologists have suggested that because divorce and remarriage are so common in postindustrial countries such as the United States, our marriage pattern might be labeled **serial monogamy**, *the practice of having more than one wife or husband—but only one at a time*. Most modern societies are strongly committed to monogamy and the selection of a lifelong mate—at least in principle, if not always in practice. Later in this chapter, we will explore a trend noted in our opener: More women and men in the United States, as well as other countries across the globe, are foregoing marriage altogether, turning the tables on an institution that has long been considered normative in the life course—and the social order.

Endogamous: A characteristic of marriages in which partners are limited to members of the same social group or caste.

Antimiscegenation laws: Laws prohibiting interracial sexual relations and marriage.

In many societies, marriages tend to be **endogamous**—*limited to partners who are members of the same social group or caste*. Sexual or marital partnerships outside the group may be cause for a range of sanctions, from family disapproval to social ostracism to legal consequences. Consider that in the United States, **antimiscegenation laws**—*laws prohibiting interracial sexual relations and marriage*—were ruled unconstitutional only in 1967. Until then, some states defined miscegenation as a felony, prohibiting residents from marrying outside their racial groups. Today, such laws are history. In fact, a rising number of new marriages are between spouses of different races or ethnicities: In 1967, intermarriages constituted only 3% of new marriages; today, the figure is 17% (Figure 11.2). Interestingly, the rate of intermarriage is dramatically different across U.S. racial and ethnic groups: As Figure 11.1 shows, Asian Americans and Hispanics are far more likely to marry outside of the group than are white or Black Americans (Livingston & Brown, 2017).

FIGURE 11.1 ■ Racial and Ethnic Intermarriage in the United States, 2017

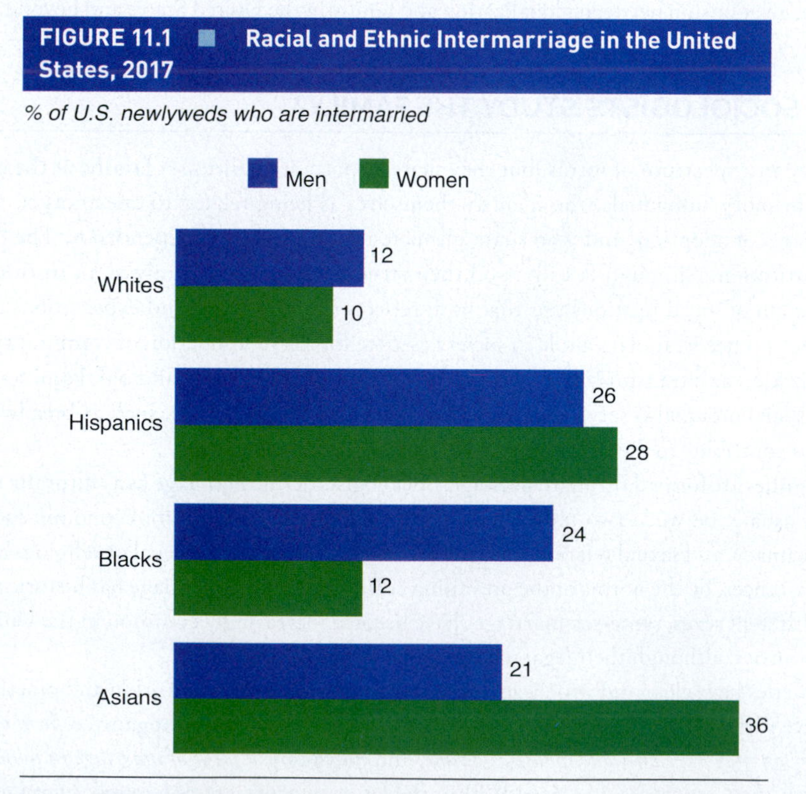

Source: Pew Research Center. (2017a). *Intermarriage in the U.S. 50 years after* Loving v. Virginia. https://www.pewsocialtrends.org/2017/05/18/intermarriage-in-the-u-s-50-years-after-loving-v-virginia/

FIGURE 11.2 ■ Since 1967, a Steady Increase in U.S. Intermarriage

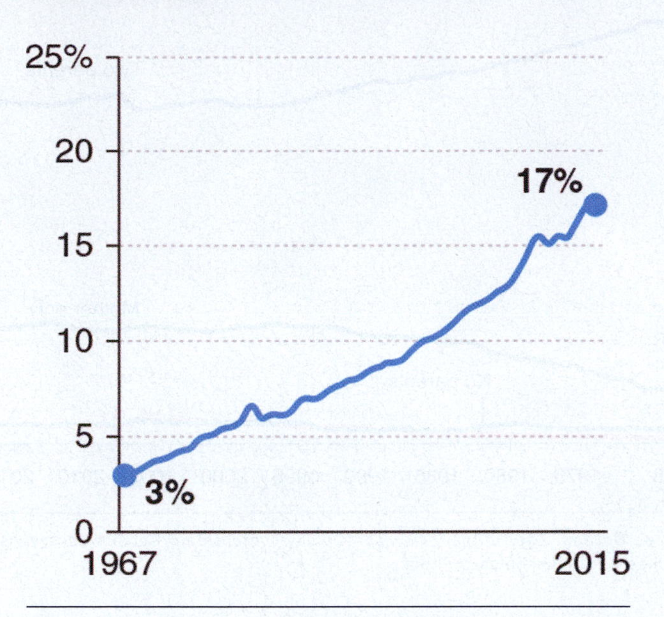

Source: Pew Research Center. (2017b, May 15). *Since 1967, a steady increase in U.S. intermarriage.* http://www.pewsocialtrends.org/2017/05/18/intermarriage-in-the-u-s-50-years-after-loving-v-virginia/

Families and the Work of Raising Children

The role of parent or primary caregiver in the United States and much of Europe has traditionally been assumed by biological parents (and occasionally stepparents), but this is one of many possible family formations in which adults have raised children in different times and places. Consider the Baganda tribe of Central Africa, in which the biological father's brother was traditionally responsible for raising the children (Queen, Habenstein, & Adams, 1961). The Nayars of southern India offer another variation, assigning responsibility to the mother's eldest brother (Renjini, 2000; Schneider & Gough, 1974). In Trinidad and other Caribbean communities, extended family members have often assumed the care of children whose parents have migrated north to seek work (in the United States, in most instances) (Ho, 1993).

A substantial number of children in the United States also live in **extended families**, social groups consisting of one or more parents, children, and other kin, often spanning several generations, living in the same household. An extended family may include grandparents, aunts and uncles, cousins, and other close relatives. In Northern and Western Europe, Canada, the United States, and Australia, most children live in **nuclear families**—families characterized by one or two parents living with their biological, dependent children in a household with no other kin—while extended families are more common in Eastern and Southern Europe, Africa, Asia, and Central and Latin America. In the United States, the extended-family form is most common among those with lower incomes, in rural areas, and among recent migrants and minorities.

Extended families: Social groups consisting of one or more parents, children, and other kin, often spanning several generations, living in the same household.

Nuclear families: Families characterized by one or two parents living with their biological, dependent children in a household with no other kin.

For close and extended family members to function as caregivers is neither new nor unusual. In fact, a growing number of children in the United States live with one or a pair of grandparents, although the proportion who live with neither parent is still only 4%. In 2019, about 70% of children lived with two parents, and just over 20% lived with only their mothers, while just over 4% resided with only their father and 2% lived with grandparents or a grandparent. Figures show that changes in the past half century have been steady but slow (U.S. Census Bureau, 2017; see Figure 11.3).

FIGURE 11.3 ■ Living Arrangements of Children, 1960–Present

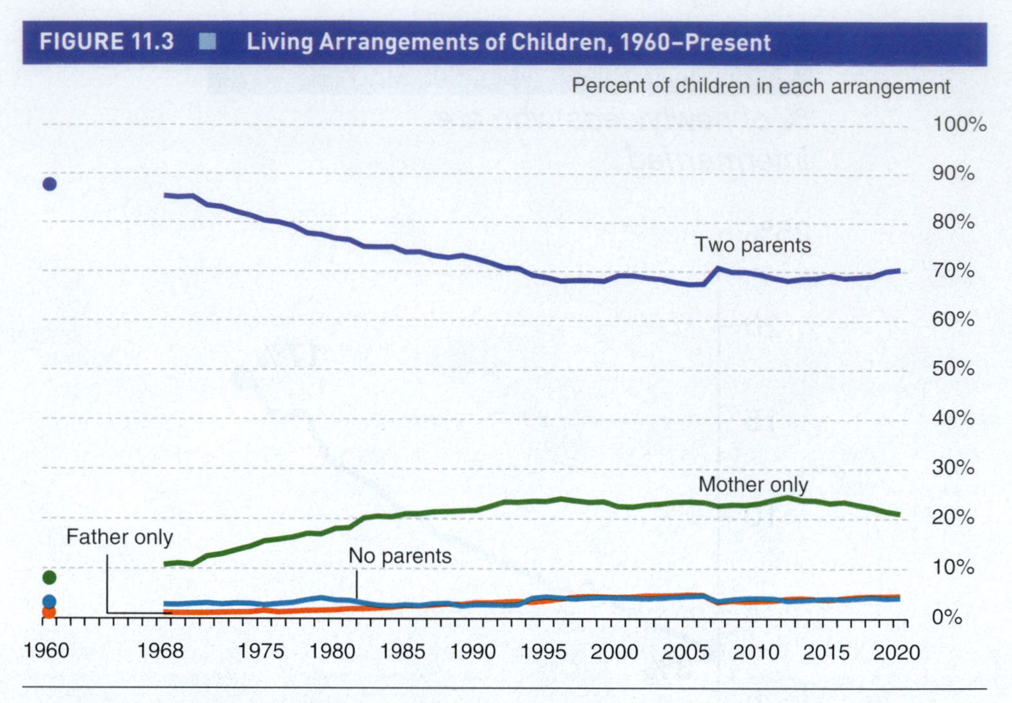

Source: U.S. Census Bureau, 2019. https://www.census.gov/content/dam/Census/library/visualizations/time-series/demo/families-and-households/ch-1.pdf

THEORETICAL PERSPECTIVES ON FAMILIES

When sociologists study families, their perspectives are shaped by their overall theoretical orientations toward society. Thus, as in the study of other institutions, it is helpful to distinguish between the functionalist and conflict perspectives, although, as we will see, there are some important variations and additions to these classic categories.

The Functionalist Perspective

Recall that functionalism asks the following question: What positive functions does a given institution or phenomenon serve in society? Based on the foundational assumption that if something exists and persists, it must serve a function, functionalist theory has highlighted, in particular, the economic, social, and cultural functions of the family. Arguably, the shift from an agricultural to industrial to postindustrial economy in the United States has made the family's economic purpose less central than its reproductive and socializing functions, although microlevel consumption decisions made by families continue to drive a national economy that is deeply dependent on consumer activity.

In his work on sex roles in the U.S. kinship system, sociologist Talcott Parsons (1954) theorized that men and women play different but complementary roles in families. In the "factory of personalities"—in other words, the family—socialization produces males and females prepared to pursue and fulfill different roles in the family and society. Parsons posited that women were socially prepared for the *expressive* role of mothers and wives, while men were prepared for *instrumental* roles in the public sphere, working and earning money to support the family. These complementary roles, he suggested, were positively functional, as they ensured cooperation rather than competition for status or position. Distinct sex roles also clarified the social status of the family, which was derived from the male's social position.

Aside from his belief that the family served the function of primary socialization—that is, the process of learning and internalizing social roles and norms (such as those relating to gender)—Parsons suggested that the nuclear family of his time functioned to support adult family members emotionally, a phenomenon he called *personality stabilization* (Parsons & Bales, 1955). In industrial societies, in which the nuclear family unit was often disconnected from the extended kin networks that characterized earlier eras, this stabilization function was particularly vital.

Writing in the 1950s, Parsons worried that disruption of the roles he observed in families could cause dysfunctions for the family and society. Indeed, there is some correlation between women's assumption of autonomous roles outside the home and the rise of divorce. Correlation is, of course, not causation. Possible explanations for the link include the advent of no-fault divorce laws, decline in the normative stigma related to divorce, and women's greater economic independence, which has enabled them to leave unhappy marriages that might earlier have been sustained by their dependence on spouses' wages.

Critics see Parsons's work as reinforcing and legitimating traditional roles. The functionalist perspective—and Parsons's application of it—has been criticized for neglecting power differentials inherent in a relationship where one party (the wife) is economically dependent on the other (the husband). In a capitalist system, power tends to accrue to those who hold economic resources. Functionalists also neglect family dysfunctions, including ways in which the nuclear family—central to modern society yet, in many respects, isolated from support systems such as kin networks—may perpetuate gender inequality and even violence.

The Feminist Approach: A Conflict Perspective . . . and Beyond

You can probably anticipate that in looking at the family, the conflict perspective will ask how it might produce and reproduce inequality. Feminist theories about the family have reflected a conflict orientation in its efforts to understand the family as a potential site of both positive support and unequal power. From the 1970s, a period following intense activity in the women's movement and a rise in the number of women taking jobs outside the home, feminist perspectives became central to sociological debates on the family.

Although early theorizing about the family highlighted its structure and roles as well as its evolution from the agricultural to the industrial era, feminist theorizing in the late 20th century turned its attention to women's experiences of domestic life and their status in the family and social world. Feminists endeavored to critique the **sexual division of labor** in modern societies. The sexual division of labor is *the phenomenon of dividing production functions by gender* (men produce, women reproduce) *and designating different spheres of activity: the "private" to women and the "public" to men*. Even though theorists, including Parsons, saw this division as fundamentally functional, feminists challenged a social order that gave males privileged access to the sphere offering capitalism's most prized rewards, including social status, opportunities for advancement, and economic independence.

Sexual division of labor: The phenomenon of dividing production functions by gender and designating different spheres of activity, the "private" to women and the "public" to men.

His and Her Marriage

An important sociological analysis that captures some of liberal feminism's key concerns is Jessie Bernard's (1982) *The Future of Marriage*. (See Chapter 10 for a fuller discussion of varieties of modern feminism, including liberal feminism.) Bernard confronts the issue of equality in marriage, positing that a husband and wife experience different marriages. In her analysis of marriage as a cultural system comprising beliefs and ideals, an institutional arrangement of norms and roles, and a complicated individual-level interactional and intimate experience, Bernard identifies *his* and *her* marriage experiences:

His marriage is one in which he may define himself as burdened and constrained (following societal norms that indicate this is what he *should* be experiencing) while, at the same time, experiencing authority, independence, and a right to the sexual, domestic, and emotional services of his wife.

Her marriage is one in which she may seek to define herself as fulfilled through her achievement of marriage (following societal norms that indicate this is what she *should* be experiencing) while, at the same time, experiencing associated female dependence and subjugation.

Bernard understood these gender-differentiated experiences as rooted in the cultural and institutional foundations of marriage in the era she studied. Marriage functioned, from this perspective, to allocate social roles and expectations—but not to women's advantage. In a good example of the sociological imagination, Bernard saw a connection between the personal experiences of individual men and women and the norms, roles, and expectations that create the context in which their relationship is lived.

Bernard's analysis pointed to data showing that married women, ostensibly fulfilled by marriage and family life, and unmarried men, ostensibly privileged by freedom, scored highest on stress

According to the Census Bureau, about one fifth of U.S. households comprise married couples with children. In 1950, about 43% of households fit this description. Some contemporary television shows both parody and reproduce traditional family images and gender roles.

©Lambert/Getty Images

indicators, while their unmarried female and married male counterparts scored lowest. Although this was true when Bernard was writing several decades ago, recent social indicators show a mix of patterns. Some are similar to those she identified. For instance, an article in the *Harvard Men's Health Watch Newsletter* reported,

> A major survey of 127,545 American adults found that married men are healthier than men who were never married or whose marriages ended in divorce or widowhood. Men who have marital partners also live longer than men without spouses; men who marry after age 25 get more protection than those who tie the knot at a younger age, and the longer a man stays married, the greater his survival advantage over his unmarried peers. In terms of mental health, married men also have a lower risk of depression and a higher likelihood of satisfaction with life in retirement than in their unmarried peers. (Harvard Medical School, 2019)

Other studies paint a different picture. For instance, an examination of a spectrum of marriage studies determined that married women were less likely to experience depression than their unmarried counterparts. Researchers controlled for such factors as the possibility that less depressed people were more likely to get married (which would confound results) and found that self-selection was not an issue. That is, marriage did seem to have positive health effects for women (Wood et al., 2007).

In fact, the issue is more complex than either Bernard's work or recent studies can embrace in a single narrative. Consider other variables at play here. Many of the health benefits connected with marriage are more pronounced for married men than for married women (Shmerling, 2016). Additionally, marriage as an institution does not appear to confer health benefits; rather, it is the *quality* of marriage that matters. Indeed, when studies measure it, marital satisfaction is a much stronger predictor of happiness than just being married, and being in a toxic relationship is decidedly bad for happiness (Simon-Thomas, 2019). According to one study, people in happy marriages rate their health better as they age (Proulx & Snyder-Rivas, 2013). Solid and low-conflict marriages are healthy, and unstable, high-conflict marriages are not. The never married are better off than those in high-conflict marriages (Parker-Pope, 2010).

Bernard's work gives us an opportunity to look at marriage as a *gendered institution*—that is, one in which gender fundamentally affects the experience of marriage. Although her analysis, which is nearly four decades old, cannot fully capture the reality of today's marriages, her recognition that men and women may experience marriage in different ways remains an important insight.

The feminist perspective and other conflict-oriented perspectives offer a valuable addition to functionalist theorizing. Nevertheless, their focus on the divisive and unequal aspects of family forms and norms may overlook the valuable functions of caring, socializing, and organizing that families have long performed and continue to perform in society. Indeed, both of these macrolevel approaches may have difficulty capturing the complexities of any family's lived experiences, particularly as they evolve and change over the years. Nonetheless, they offer a useful way of thinking about families and family members, their place in the larger social world, and the way they influence and are influenced by societal institutions and cultures.

The Psychodynamic Feminist Perspective

Sociologist Nancy Chodorow (1999) asks, "Why do women mother?" She suggests that to explain women's choice to *mother*, a verb that describes a commitment to the care and nurturing of children, and men's choice to *not mother* (that is, to assume a more distant role from child-rearing), we must look at personality development and relational psychology. Although mothering is rooted in biology, Chodorow argues that biology cannot fully explain mothering because fathers or other kin can perform key mothering functions as well.

While fathers are more involved with the day-to-day care of their children than in generations past, mothers continue to play the role of primary caregiver in most families. Would you predict that this division of roles will remain static for heterosexual couples? Why, or why not?

©Yasser Chalid/Moment/Getty Images

Drawing from Sigmund Freud's object relations perspective, Chodorow argues that an infant of either sex forms his or her initial bond with the mother, who satisfies all of the infant's basic needs. Later, the mother pushes a son away emotionally, whereas she maintains the bond with a daughter. Through such early socialization, daughters come to identify more fully with their mothers than with their fathers; boys, on the other hand, develop masculine personalities, but those draw from societal models of masculinity (or, sometimes, hypermasculinity) rather than predominantly from their fathers, who take a far less prominent role in child-rearing than do mothers. Chodorow suggests that masculinity in boys may thus develop, in part, as a negation and marginalization of qualities associated with femininity, which is rejected for both social and psychological reasons.

Women, reared by mothers who nurture close and critical bonds, are rendered "relational" through this process, seeking close bonds and defining themselves through relationships (Anna's mom, Joe's wife, and so on). Men, by contrast, define themselves more autonomously and have a harder time forming close bonds. Again, the roots of this difficulty are social (society defines men as autonomous and independent) and psychological (the pain of an early break in the mother–son bond results in fear or avoidance of these deep bonds). So why do women mother? Because men in heterosexual relationships are not socially or psychologically well prepared for close relational bonding, women choose to mother to reproduce this intimate connection with a child.

Although these processes play out primarily on the micro level of the family and relationships, Chodorow also recognizes macrolevel effects. She suggests that because of a lack of available male role models at home, the masculine personality develops, in part, as a negation of the feminine personality. Significantly, she argues, this devaluation of the feminine is institutionalized in society: That is, because men still largely dominate key institutions in society, traits associated with masculinity are more highly valued in areas like politics, business, and the law. Chodorow's work on sex roles and socialization in the family offers a unique marriage of Freud and feminism that is both challenging and compelling, asking us to consider the effects that psychological processes in early childhood have on social institutions.

U.S. FAMILIES YESTERDAY AND TODAY

The traditional nuclear family often appears in popular media and political debates as a nostalgic embodiment of values and practices to which U.S. families should return. Historian Stephanie Coontz (2000, 2005), who has written about the history of U.S. families, points out that the highly venerated traditional nuclear family model is, in fact, a fairly recent development.

Consider that in the preindustrial era, when the U.S. economy was primarily agricultural, families were key social and economic units. Households often included multiple generations and sometimes boarders or farmworkers. Families were typically large, and children were valued for their contributions to a family's economic viability, participating along with the other members in productive activities. Marriages tended to endure; divorce was neither normative nor especially easy to secure. At the same time, average life expectancy was about 45 years (Rubin, 1996). As life spans increased, divorce also became more common, replacing death as the factor most likely to end a marriage.

The period of early industrialization shifted these patterns somewhat, not least because it was accompanied by urbanization, which brought workers and their families to cities for work. The family's economic function declined; some children worked in factories, but the passage of child labor laws and the rise of mass public schooling made this increasingly uncommon (although, according to one source, at the end of the 19th century, a quarter of textile workers in the American South were children, whose cheap labor was a boon to employers; Wertheimer, 1977). Over time, children became more of an economic cost than a wage-earning benefit; in a related development, families became smaller and began to evolve toward the nuclear family model.

The basic nuclear family model, with a mother working in the private sphere of the home while focused on child-rearing and a father working in the public sphere for pay, evolved among middle-class families in the late 19th century. It was far less common among the working class at this time; working-class women, in fact, often toiled in the homes of the new middle class as housekeepers and nursemaids.

Coontz (2000) points out that, as the popular imagination suggests, the mother-as-homemaker and father-as-breadwinner model of the nuclear family is most characteristic of the widely idealized era of the 1950s. The post–World War II era witnessed a range of interconnected social phenomena, including suburbanization supported by federal government initiatives to build a network of highways and encourage home ownership; a boom in economic growth and wages that brought greater consumption power, along with technologies that made the home more comfortable and convenient; and a "baby boom," as a wave of pregnancies delayed by the years of war came to term.

Although prosperity and technology brought new opportunities to many, mass suburbanization largely left behind minorities, including Black Americans, who were not given full access to the government's subsidized mortgages (including mortgages subsidized through the G.I. Bill, which was theoretically available to all returning veterans of World War II) and were often left behind in segregated, devalued neighborhoods. As the jobs followed white workers to the suburbs, the economic condition of many Black families and their neighborhoods deteriorated.

Furthermore, it is not clear that all was well in the prosperous suburbs, either. As we noted in the section on feminist theoretical perspectives, some sociological observers detected a streak of discontent that ran through the idealized nuclear family. Betty Friedan's (1963) book *The Feminine Mystique* highlighted "the problem that has no name," a broad discontent born of women's exclusion from or marginalization in the workplace and the disconnect between their low status and opportunities and society's expectation that marriage and children were the ultimate feminine fulfillment. Coontz (2000) points out that tranquilizers, one of many medical innovations of the era, were largely consumed by women—and in considerable quantities, at least 1.15 million pounds in 1959 alone.

Marriage and Divorce in the United States

The traditional nuclear family with the man as breadwinner and the woman as caregiver is still in existence, although it has changed in many respects over recent decades—in 1970 it was about 40%, but by 2012 it had fallen to just 20% of U.S. households (Pew Research Center, 2012). At the same time, even though commentators often lament the "decline of the family," most children in the United States still live in two-parent households (see Figure 11.3). More children are living with single parents than in the past, but more adults are also living in nonfamily households, consisting of either a single householder or unrelated individuals. Today, about a quarter of adults live alone.

One reason for the growth of single-person households is the rising median age at first marriage, which in 2019 was 30 for men and 28 for women (Figure 11.4). Consider the fact that for men, the median age at first marriage was 22.8 in 1960, 26.1 in 1980, and 26.8 in 2000; for women, it was 20.3 in 1960, 23.9 in 1980, and 25.1 in 2000 (U.S. Census Bureau, 2017). The steep rise in the median age at first marriage suggests that many people are not marrying until their 30s—or even later. Most U.S. adults indicate a desire to marry, and most will at some point in their lives; more than 2.1 million married in 2018, and the U.S. marriage rate of 6.5 per 1,000 population exceeds that of many other economically advanced countries, including the states of Western and Northern Europe (U.S. Centers for Disease Control and Prevention, 2019).

At the same time, rates of marriage in the United States have declined, and far fewer adults today are married than in generations past (Figure 11.5). Consider that of the roughly 73 million men and women who compose the millennial generation, about 44% are married. By comparison, about 61% of the baby boomer generation was married at the same point in their lives (Barroso et al., 2020).

Marriage is only one way in which young people create families today. According to a Pew Research Center analysis, 12% of never-married young adults (25–34) are living with a partner (2020). Parenthood also has become increasingly separated from marriage, and a contemporary family may

FIGURE 11.4 ■ Median Age at First Marriage in the United States, 1890–2019

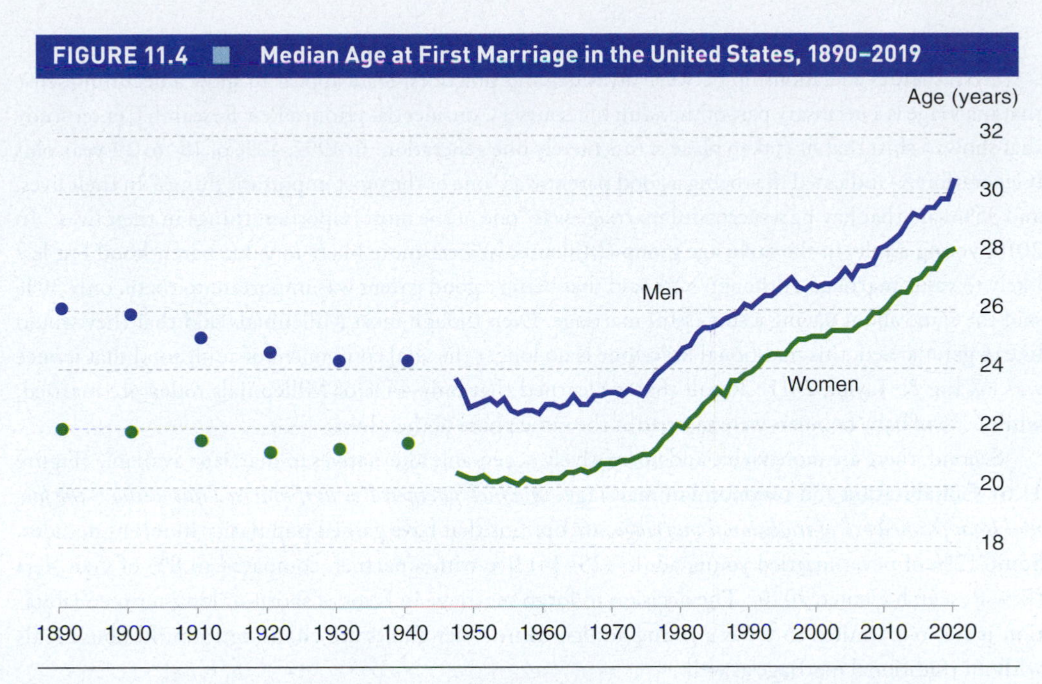

Source: U.S. Census Bureau. Decennial censuses, 1890 to 1940, and current population survey, annual social and economic supplements, 1947 to 2019.

FIGURE 11.5 ■ Marital Status in the United States (Adults Ages 15 and Over), 1960–2017

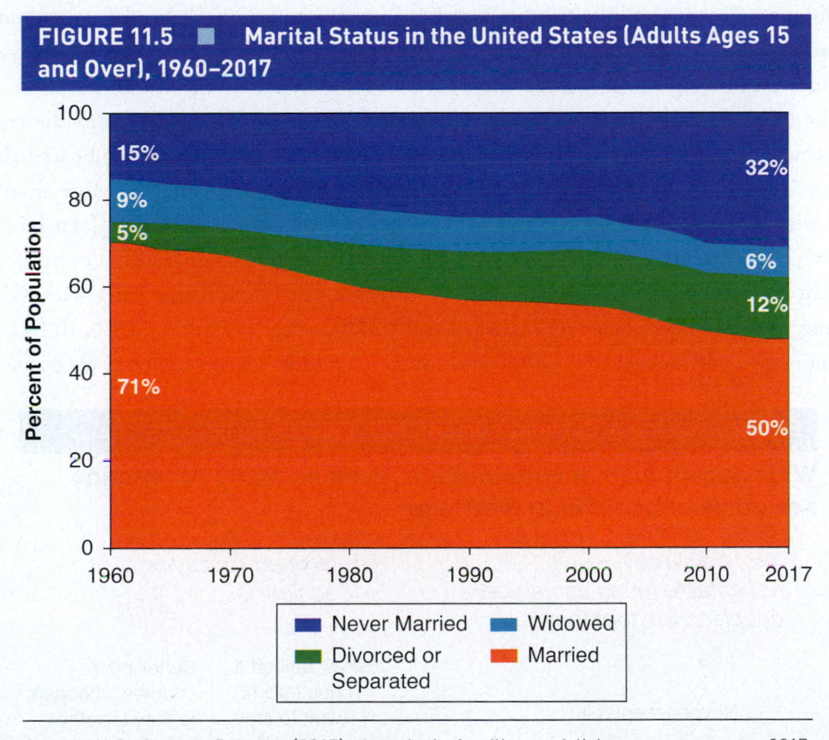

Source: U.S. Census Bureau. (2017). America's families and living arrangements: 2017: Adults (A table series): Table A1.

well consist of a never-married mother (or, less commonly, a father) raising a child or children. Today, 33% of millennial mothers have never been married; by contrast, that figure in 1960 was 23%.

At the same time, most young adults do not reject marriage as an institution. According to Pew, "only 4% of never-married adults ages 25 to 34 say they don't want to get married. A majority of them either want to marry (61%) or are not sure (34%)," although the collective attitude toward marriage is far less traditional than that of older generations: Only about a third of those in the 18 to 29 age group agreed that it was "very important" that a couple marries if they plan to spend their lives together, while about half of those 50 to 64 agreed, and fully 65% of those 65 and over agreed (Wang & Parker, 2014).

What, then, explains the decline of marriage in the millennial generation?

First, there is a relationship between attitudes and practices. Data appear to show a declining sense that marriage is a necessary part of the adult life course. Consider data from a Pew Research Center study that shows a shift that has taken place across merely one generation: In 1997, 42% of 18- to 29-year-olds (Generation X) indicated that being a good parent was "one of the most important things" in their lives, and 35% said that having a successful marriage was "one of the most important things in their lives." In 2010, young adults in the same age group (Millennials) were more likely to value parenthood but less likely to value marriage: Although 52% said that being a good parent was important to them, only 30% said the same about having a successful marriage. Even though most Millennials said that they would like to get married, this traditional milestone is no longer the vital component of adulthood that it once was (Wang & Taylor, 2011): Recall that we learned that only 44% of Millennials today are married, while 61% of baby boomers were married at the same phase of their lives.

Cohabitation: Living together as a couple without being legally married.

Common-law marriage: A type of relationship in which partners live as if married but without the formal legal framework of traditional marriage.

Second, there are more viable and normatively acceptable alternatives to marriage available (Figure 11.6). **Cohabitation** and **common-law marriage**, *in which partners live as if married but without the formal legal framework of traditional marriage*, are options that have gained popularity in recent decades. Some 12% of never-married young adults (25–34) live with a partner, compared to 8% of Gen Xers (Pew Research Center, 2020). The decision to forgo marriage in favor of short- or long-term cohabitation is not one limited to today's young adults: More older adults are choosing to build households without traditional marriage as well.

Third, economic circumstances, including the rising burden of student debt, are having an influence on decisions about family formation. Research shows some correlation between debt and the ability or willingness of young adults to start families (Smock et al., 2005). Students graduating from higher education in the past decade are the first U.S. generation to finance so much of their education with interest-bearing loans. A 2002 survey from Nellie Mae (a nonprofit corporation and, until recently, the largest private source of student loans) offers some early insights into the relationship between debt and delayed family formation: In the survey, 14% of borrowers indicated that "loans delayed marriage," a rise from 9% in 1987, when the debt burden was smaller. More than one fifth responded that they had "delayed having children because of student loan debt," an increase from 12% in 1987. More recently, an IHS Global Insight report highlighted the fact that even though other types of debt have declined, student loan debt continues to rise—and it correlates with a discernible trend among young adults of delaying marriage and childbearing (Dwoskin, 2012). In fact, financial concerns more generally appear to have an effect on young adults' decisions about marriage: In a Pew

FIGURE 11.6 ■ Cohabitation and Marriage Today

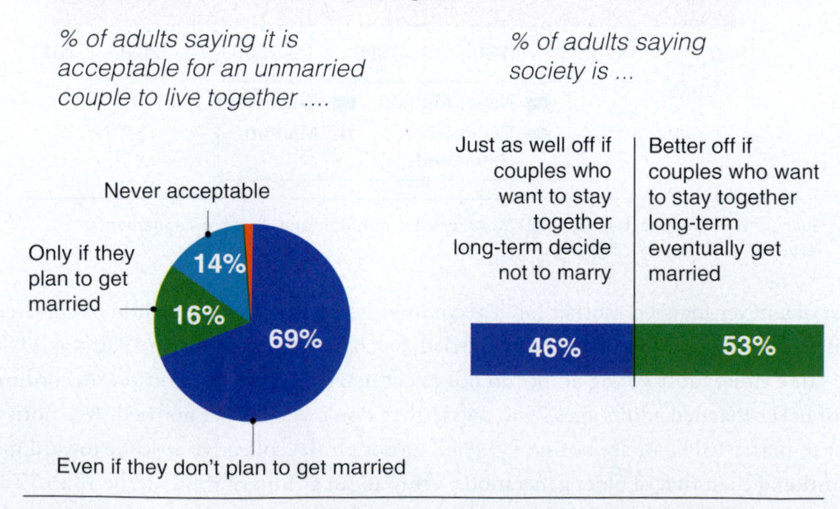

Source: Pew Research Center. (2019, November). *Key findings on marriage and cohabitation in the U.S.* https://www.pewresearch.org/fact-tank/2019/11/06/key-findings-on-marriage-and-cohabitation-in-the-u-s/

FIGURE 11.7 ■ Reasons Adults Give for Not Being Married

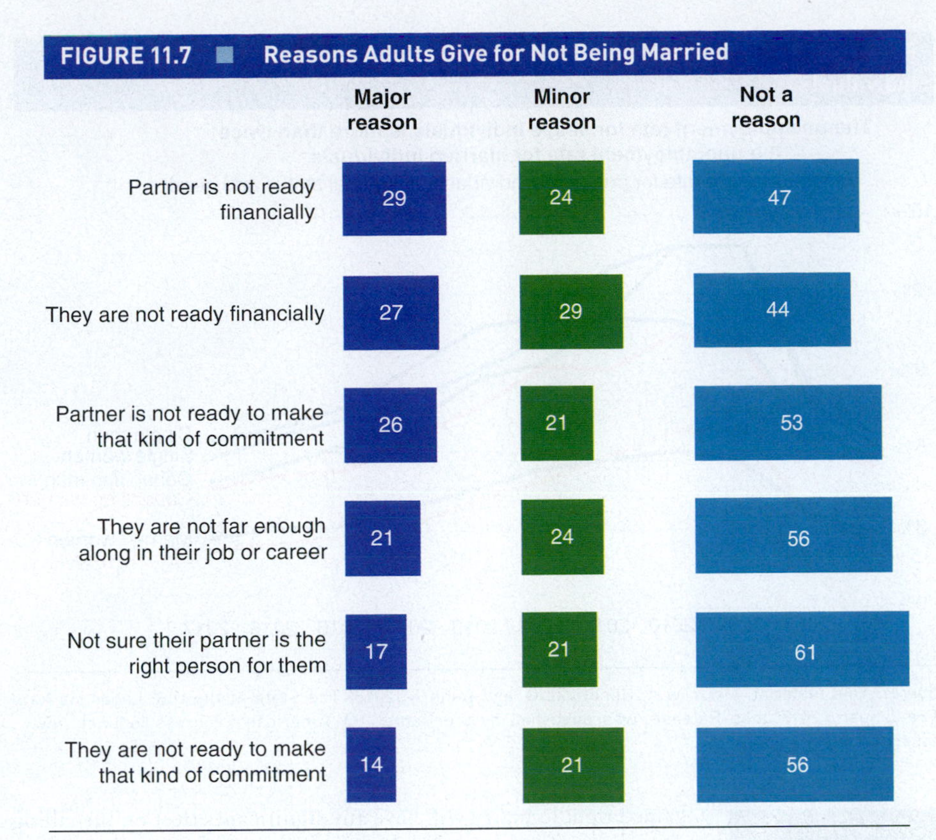

Source: Horowitz, J., Graf, N., & Livingston, G. (2019, November 6). *Marriage and cohabitation in the U.S.* Pew Research Center. https://www.pewsocialtrends.org/2019/11/06/marriage-and-cohabitation-in-the-u-s/

Research Center (2014a) survey, a third of young adults indicated that economic reasons were a key obstacle to getting married (Figure 11.7).

Fourth, the marriage market has shifted. If it is the case that nearly a third of young adults indicate that they have not yet met someone they would like to marry, some of that can be explained with the application of the sociological imagination. That is, rather than looking at this on an individual level, we may want to ask what sociological factors may underpin difficulty in meeting a suitable partner. For instance, even though there are more unmarried young adult men than unmarried young adult women (implying a robust pool of partners), a closer look at the marital pool shows that the unemployment rate for unmarried men is higher than for unmarried women (see Figure 11.8). To the degree that being employed is a variable that makes a man "marriageable," the data suggest that the marriage market may be weaker than it appears.

Interestingly, as younger generations of heterosexual adults have drifted away from marriage as a normative part of the life course, many of their gay and lesbian peers have been fighting for the opportunity to marry. Just a few years ago—2014—33 states prohibited same-sex marriage while only 17 states and the District of Columbia permitted it (Ahuja et al., 2014). States that rejected same-sex marriage tended to cite the same language as the federal law prohibiting same-sex marriage, the Defense of Marriage Act (DOMA): "The word 'marriage' means only a legal union between one man and one woman as husband and wife." DOMA was signed into law by President Bill Clinton in 1996. It also included the provision that states were not obligated to recognize same-sex marriages conducted in states or cities that permit them.

A dramatic shift took place in June of 2015, when the U.S. Supreme Court held in *Obergefell v. Hodges* that states must allow same-sex couples to marry and that they must recognize same-sex marriages from other states. The 5–4 decision indicated that a fundamental right to marry is guaranteed to same-sex couples by the Due Process Clause and the Equal Protection Clause of the Fourteenth Amendment to the Constitution. According to a Gallup poll reported in the *Washington Post* prior to the *Obergefell* decision, about 390,000 married same-sex couples resided in the United States. Another 1.2 million adults were living in same-sex domestic partnerships (Schwarz, 2015). Whether the newly

FIGURE 11.8 ■ Unemployment Rate for Single, Cohabiting, and Married Individuals

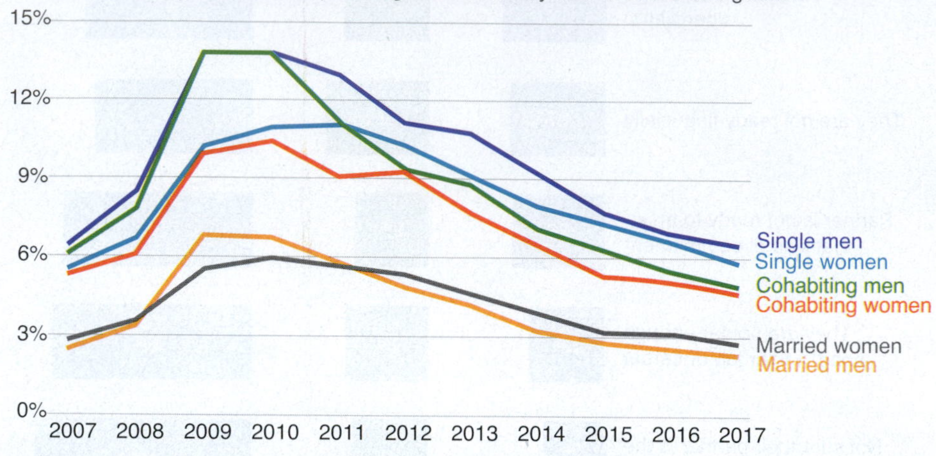

Source: This material, McGrew A, Hendricks G, and Bahn K (2018) The State of the U.S. Labor Market: Pre-January 2018 Jobs Release, was published by the Center for American Progress (online), www.americanprogress.org

Marriages between partners of a different race or ethnicity are on the rise in the United States. About 10% of all marriages are intermarriages, and in 2015, fully 17% of newlyweds married a partner of a different race or ethnicity (Livingston & Brown, 2017).

©Robert Alexander/Archive Photos Getty Images

The proportion of LGBT adults married to a same-sex partner is growing in the United States after the Supreme Court's legalization of same-sex marriage in the 2015 *Obergefell v. Hodges* decision.

©Hinterhaus Productions/DigitalVision/Getty Images

gained right to marry will have any significant effect on the falling rate of marriage in the United States remains to be seen, but recent data show that 2 years after the *Obergefell* decision, just over 10% of lesbian, gay, bisexual, and transsexual (LGBT) adults are now married to a same-sex partner (Jones, 2017).

With the decline of marriage, the United States has also experienced a decline in divorce. After rising through the 1960s and 1970s, the rate of divorce has leveled off. There is a relationship between these two phenomena, since a smaller number of marriages reduces the pool of people who can divorce. The rate overall is still high, however, and the United States has one of the highest divorce rates in the world, with the rate for second and later marriages exceeding that for first marriages.

Why is the U.S. divorce rate, while declining, persistently high? Historian Stephanie Coontz (2005) argues that divorce is, in part, driven by our powerful attachment to the belief that marriage is the outcome of romantic love. Although historically, many societies accepted marriage primarily as part of an economic or social contract—and some still do—modern U.S. adults are smitten with love. Yet the powerful early feelings and passion that characterize many relationships are destined to wane over time. In a social context that elevates romantic love and passion in films, music, and books, we may have less tolerance for the more measured emotions inherent in most long-term marriages. Could our strong focus on romantic love be a driver of both marriage *and* divorce? What do you think?

Who's Minding the Children?

In a reversal of a longtime (nearly four-decade) trend in the United States, increasing numbers of mothers are staying home to care for children: In 2016, 27% of mothers reported that they did not work outside the home (Figure 11.9). About two thirds of today's stay-at-home mothers are part of "traditional" families; that is, they care for children in the home while their

FIGURE 11.9 ■ Percentage of Stay-at-Home Mothers and Fathers in the United States, 1989–2016

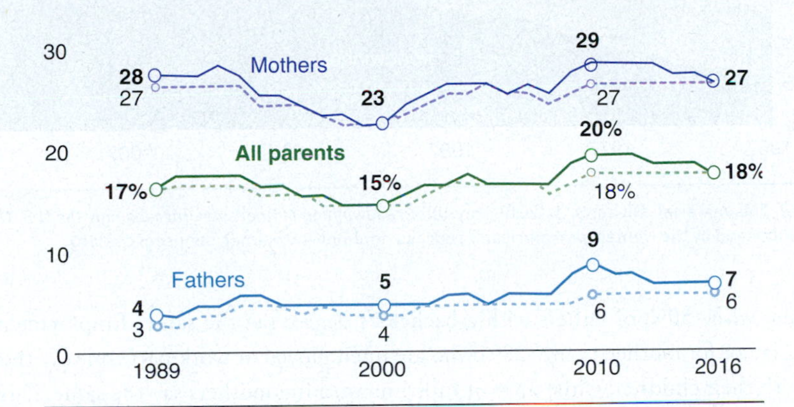

Source: Livingston, G. (2018, September 24). *Stay-at-home moms and dads account for about one-in-five U.S. parents.* https://www.pewresearch.org/fact-tank/2018/09/24/stay-at-home-moms-and-dads-account-for-about-one-in-five-u-s-parents/
Note: Based on mothers ages 18–69 with their own child(ren) younger than 18 in the household. Mothers are categorized based on employment status in the year prior to the survey. "Other" stay-at-home mothers are those who are married with a non-working or absent husband.

husbands work for pay. The rest include single and cohabiting women, as well as women whose husbands are unemployed. The shift toward fewer mothers working outside the home is driven by a variety of social, cultural, and economic factors. Mothers remain far more likely than fathers to remain home to care for the family (Pew Research Center, 2018). Other factors include the growing percentage of immigrant women who are mothers, and stagnating wages that have led some women to conclude that the costs of outside childcare outweigh the benefits of working for pay (Cohn, Livingston, & Wang, 2014).

At the same time, in the slow growth of another trend, more fathers are assuming the primary childcare role in the home. According to one study, about 17% of stay-at-home parents are fathers, up from 10% in 1989. Interestingly, this trend appears to be driven by labor market and health issues more than by changes in social or cultural norms that support more active fathering. Data show that even though 78% of stay-at-home mothers are motivated by a desire to care for the family, only 24% of men offer the same explanation: 40% of stay-at-home fathers indicate that they are at home because they are either ill or disabled (Pew Research Center, 2018). Notably, public attitudes about men as primary caregivers are, in spite of myriad changes in family life, still only nominally supportive. Livingston (2014) reports on a Pew Research Center poll in which about 76% of respondents said that they believed children are "just as well off" if their father works, but only 34% said that children are "just as well off" if their mother works, and 51% said that children are "better off" if the mother stays home; only 8% responded that the children are "better off" with the father at home.

Although the number of stay-at-home fathers is rising, 63% of fathers say that they do not spend enough time with their children. By contrast, about 35% of mothers say the same. While education is a factor in determining how fathers self-report their time spent with their children, it is not a factor for mothers: 69% of fathers who do not have a bachelor's degree say they spend too little time with

FIGURE 11.10 ■ Share of All Mothers Who Are Breadwinners or Co-Breadwinners, 1967–2017

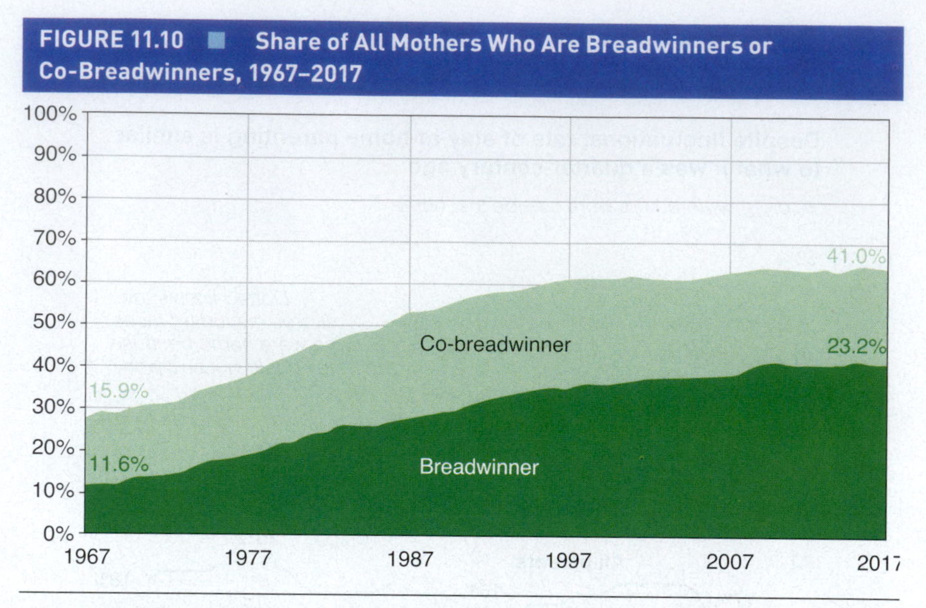

Source: This material, Glynn, S. J. (2019, May 10). *Breadwinning Mothers Are Increasingly the U.S. Norm*, was published by the Center for American Progress. (online) www.americanprogress.org

their children, while 50% of fathers with a bachelor's degree say the same. Employment is a better determining factor for mothers: Only 28% who are unemployed or work part time say they spend too little time with their children, while 43% of full-time working mothers say the same. Parents of both sexes share the sentiment that time spent at work is the biggest obstacle to spending time with children (Livingston, 2018). This is particularly salient in a time when a growing share of women are primary or co-breadwinners in the family (Figure 11.10).

Notably, the second leading reason that parents give for spending too little time with children differs by sex, with mothers saying the reason is because they have other family or household obligations and fathers responding that it is because their children do not live with them. Of all fathers, 17% live apart from all of their children, a percentage that varies based on education and race. Only 8% of fathers with a bachelor's degree or more live apart from their children, while 28% of those without a bachelor's degree say the same. As for race, more Black fathers report living apart from some or all of their children (47%), while 26% of Hispanic and 17% of white fathers say the same (Livingston, 2018).

With mothers and fathers of young children in the labor force, childcare has progressed further into paid means. Data from the Center for American Progress show that 55% of children age 5 and younger whose mothers work for pay are cared for in childcare centers or preschools (Malik, 2019).

In nearly half of two-parent households, both parents work full time. About 60% of mothers and 52% of fathers say it is difficult to find a balance between family and work demands (Pew Research Center, 2015a).

Choosing who will care for a young child is a highly personal decision for a family, but there are some discernible patterns in childcare arrangements by socioeconomic status. For example, poorer fathers are more likely to be stay-at-home parents than are their better-off male peers (Livingston, 2014). As well, working mothers who hold college degrees are most likely to use center-based childcare, which is often costly (ChildStats, 2013). How would you explain these patterns? What other patterns of childcare could you hypothesize? Next, we turn to an examination of family patterns in immigrant, Native American, and deaf families in the United States.

Immigration and Family Patterns

The United States has more foreign-born residents than any other country in the world. Given its low fertility rate, a substantial

FIGURE 11.11 ■ Top-10 Largest U.S. Immigration Groups, 2018

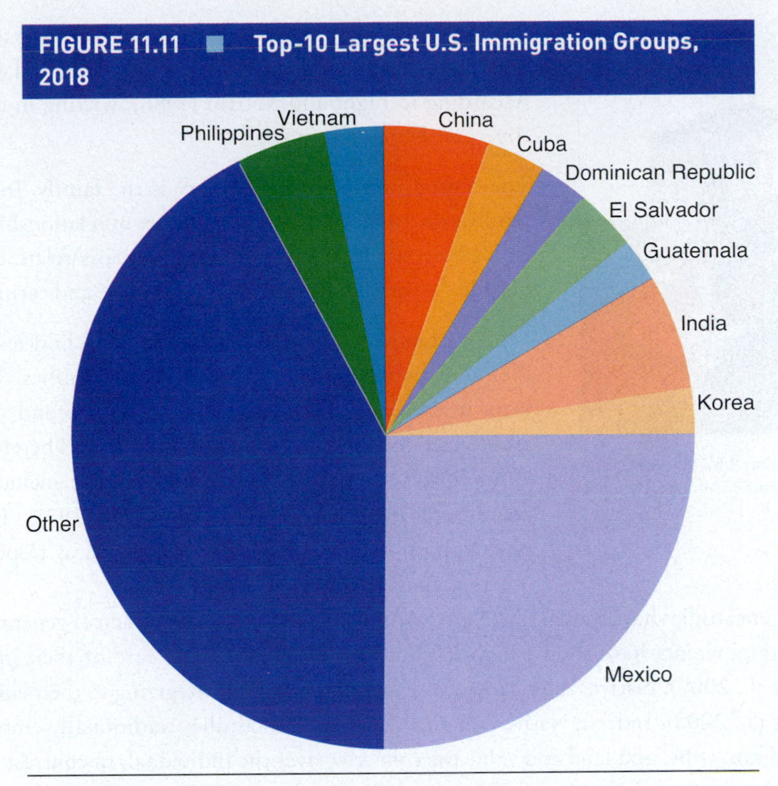

Source: Originally published in Migration Policy Institute's Migration Data Hub (www.migrationpolicy.org/programs/migration-data-hub).

proportion of the country's population growth is the result of immigration (Figure 11.11). With the proportion of foreign-born residents at about 13%, it is not surprising that immigrants and the cultures they carry with them have important effects on family patterns in the United States.

Predictably, more recent immigration correlates with a group's stronger ties to its homeland culture (Moore & Pinderhughes, 2001). Many scholars are now focusing on the *transnational* nature of immigrant families. Embodying transnationalism may mean living, working, worshipping, and being politically active in one nation while still maintaining strong political, social, religious, and/or cultural ties to other nations (Levitt, 2004). Family members may send money to relatives in their countries of origin, keep in nearly constant contact with those they left behind through modern communication technologies, and travel back and forth between countries frequently to maintain close emotional ties.

Interestingly, reflecting the influence of home cultures on migrants, first-generation migrants often have birthrates above those of the native-born U.S. population. One study found that immigrant women in the United States were slightly more likely to have given birth in the last 12 months (6%) than U.S.-born women in the same age group (5%) (Batalova, 2020). In the second generation, however, the rate typically declines to the average U.S. rate—that is, approximately 2.1 children per woman (Hill & Johnson, 2002). Nevertheless, some studies have found that the trend toward decreasing fertility in first- and second-generation immigrant women may actually be reversed in the third generation for some groups (Parrado & Morgan, 2008). Historically, fertility rates have decreased for immigrant families the longer they reside in the United States. It is important, however, to note that factors such as country of origin, education, religion, and cultural attitudes add to the complexity of understanding these trends.

Next, we narrow our focus to a pair of specific examples of family subcultures in the United States—those of Native Americans and deaf families.

America's First Nations: Native American Families

Family patterns among Native Americans are highly diverse. There are nearly 500 different nations, although more than half of American Indians identify as coming from six of these (U.S. Census

The 2010 U.S. Census counted just over half a million (557,185) American Indian and Alaska Native families in the United States. About 57% were reported to be married-couple families.

©Lawrence Migdale/Science Source

Bureau, 2017). American Indians often live in extended-family households that include uncles, aunts, cousins, and grandparents. According to Light and Martin (1986), writing in the *Journal of American Indian Education*,

> The central unit of Indian society is the family. Indian . . . families do not have the rigid structure of relationships found in Western white culture. Instead, Indians relate to people outside the immediate family in supportive and caring ways.

An example of family involvement in child-rearing outside the immediate parents is found in Sioux families. This involvement begins early in a child's life, when a second set of parents are selected for the newborn (Sandoz, 1961). Therefore, the total family involved in child-rearing and support includes unrelated members of the Indian community (Ryan, 1981). This community support and protection can be viewed as responsibility for others' actions (Light & Martin, 1986).

More recent studies have shown that Native American family ties extend across generations. Native American grandparents have shown high levels of involvement in the care of their grandchildren (Mutchler et al., 2007). Native Americans also report high levels of caregiving to their elderly relatives (McGuire et al., 2008). Indeed, Native American culture and families traditionally emphasize sensitivity toward kin, tribe, and land and value the collective over the individual, in contrast to the mainstream U.S. emphasis on individual fulfillment and achievement (Light & Martin, 1986; Newcomb, 2008). Research on Native Americans of the Southwest suggests that many members of these groups see their children as belonging to the entire American Indian nation: to the land, the sky, the tribe, and its history, customs, and traditions (Nicholas, 2009).

Some controversies over family life taking place within Native American communities mirror those taking place in U.S. society more generally. For instance, beginning in 2008, some tribes began to permit same-sex marriage; in that year, the Coquille tribe of Oregon was the first to pass a law defining marriage or domestic partnership as a "formal and express civil contract entered into by two persons, regardless of their sex." The sovereign Navajo Nation, together with tribes such as the Kickapoo and Chicksaw, has resisted this step, even after the legalization of same-sex marriage by the U.S. Supreme Court in 2015: According to the Navajo Nation's marriage law, "Marriage between persons of the same sex is void and prohibited." These sovereign tribes' laws are not affected by the Supreme Court ruling because the tribes were never parties to the U.S. Constitution (Drew, 2015). Interestingly, some research shows that gay partnerships were historically accepted in many tribal cultures, where the term *two spirits* was used to categorize LGBT members (Kronk, 2013).

Deaf Culture and Family Life

Census Bureau data show that about one fifth of the U.S. population has some kind of disability. How does disability affect family life? In this subsection, we examine the case of people who are deaf and the choices and challenges that family life brings for them. According to Gallaudet University (2012) figures, in 2012, approximately 13% of the U.S. population had hearing problems that may range from being fully deaf to being hard of hearing. Fewer than 1 in 1,000 were deaf before the age of 18; more than half became deaf at some point after childhood (Gallaudet Research Institute, 2005). Family life poses unique challenges for many people who are deaf, and as a consequence, some prefer to practice endogamy, marrying others who are deaf and therefore share a common experience.

There has been a movement within the deaf community to redefine the meaning of deafness to denote not a form of disability but a positive culture. Some people who are deaf see themselves as similar to an ethnic group: sharing a common language (American Sign Language, or ASL), possessing a strong sense of cultural identity, and taking pride in their heritage. Identifying as an ethnic group rather than as a disability group, some people who are deaf believe that cochlear implant surgery, a procedure through which some deaf people can become hearing, is problematic, especially when it is performed on children who cannot consent. Often, deaf people who undergo this surgery are still unable to attain mastery of any oral language (Lane, 2005). The National Association of the Deaf (2000) takes a cautionary stance on cochlear implants, advising hearing parents of deaf children to conduct thorough research, create a support system, and, most important, communicate with their children before undertaking the transition. Julie Mitchiner takes pride in being deaf. She writes,

> Growing up with deaf parents and attending deaf schools, I have a strong sense of pride of being deaf and being part of the Deaf community. I do not look at myself as disabled. I often say if I were given a choice to hear or stay deaf, I'd choose to stay deaf. It is who I am. My family, my friends, and my community have taught me that being deaf is part of our culture and is a way of life. (Mitchiner & Sass-Lehrer, 2011, p. 3)

Gallaudet University in Washington, D.C., is unique in serving specifically people who are deaf and hearing impaired by offering bilingual instruction in English and ASL. The rising proportion of students who are hearing or come from mainstream schools and do not know ASL has led to debates over the centrality of deaf culture at the school.

Many people who are deaf succeed in the hearing world, but they may confront daunting problems (Heppner, 1992). Often, they are not able to speak in a way that hearing people fully understand, and most hearing people do not know ASL. It is not surprising that an estimated 85% of people who are deaf choose to marry others who share their own language and culture (Cichowski & Nance, 2004) or that many deaf parents are wary when their deaf children form relationships with hearing people. When a deaf couple has a hearing child, the family must make difficult choices as they negotiate not only the ordinary challenges of child-rearing but also the raising of a child who may be "functionally hearing" but "culturally deaf" (Bishop & Hicks, 2009; Preston, 1994).

Families typically confer their own cultural status on their children, but this may not be true for 9 of 10 deaf children born to hearing parents. On the one hand, hearing parents want the same sorts of things for their deaf children as any parents want for their children: happiness, fulfillment, and successful lives as adults. Family dynamics between hearing parents and deaf children are also influenced by the way the parents perceive the condition of the child, namely, whether deafness is seen as a physiological (hearing) and medical problem to be rectified and "normalized," or as an integral part of the child that relates to a sense of identity and belonging (Avrahami-Winaver et al., 2020). These children often face challenges, such as parental disconnectedness, emotional distance, difficulties in developing an identity as a person who is Deaf/Hard of Hearing, and a lack of exposure to deaf culture. Many hearing parents of deaf children would like their children to mainstream into the hearing world as well as possible, in spite of the challenges.

On the other hand, some in the deaf community argue that the deaf children of hearing parents can never fully belong to the hearing world. Many in the deaf community believe hearing parents should send their deaf children to residential schools for the deaf, where they will be fully accepted, learn deaf culture, and be with people who share their experience of deafness (Dolnick, 1993; Lane, 1992; Sparrow, 2005).

The situations of deaf parents raising a hearing child and hearing parents raising a deaf child raise interesting and fundamental questions about what happens when family members are also members of different cultures and how the obstacles of difference within a microunit such as the family are negotiated.

Families in Crisis

Domestic (or family) violence is *physical or sexual abuse committed by one family member against another.* It may be perpetrated by adults against their children, by one spouse against another, by one sibling against another, or by adult children against their elderly parents. As little as three to four decades ago, domestic violence was rarely studied. Many people regarded violence in the home as a private matter, an attitude that was reflected in lawmaking as well, which provided few sanctions for violence that did not reach the level of severe injury or death. Today, domestic violence is understood to be a serious public issue, as researchers have come to realize that it is sadly commonplace.

Domestic (or family) violence: Physical or sexual abuse committed by one family member against another.

Accurate data on family violence are difficult to obtain for a variety of reasons. Abused partners or children are reluctant to call attention to the fact that they are abused. Police do not want to mediate or make arrests in family conflicts, and even today, the courts are hesitant to intervene in what are often perceived as family matters (Tolan et al., 2005). Some good estimates of the prevalence of this crime are available, however. According to the U.S. Bureau of Justice Statistics, between 2003 and 2013, domestic violence made up about a fifth of all violent crime in the United States. The most common form of domestic violence is *intimate partner violence* (IPV), that is, violence that involves current or former spouses or nonmarried partners (Truman & Morgan, 2014).

The COVID-19 pandemic has exacerbated the problem of domestic violence. Physicians at a large hospital in Boston saw a near-doubling of the proportion of domestic abuse cases that resulted in physical injury in 2020 in comparison with previous years. The injuries were also more severe, prompting concerns that victims had delayed seeking care even as the violence against them escalated (Healy, 2020). Additionally, the proportion of men and women whose abuse was physical—rather than verbal or emotional—was 80% higher in 2020 than in the previous 3 years put together. With victims isolated for extended periods of time with their abusers and few other social contacts, more opportunities for abuse have presented themselves. Stressors due to the pandemic may have also contributed to an increase in abuse, including the high unemployment rate, loss of savings, and uncertainty of the future.

In children, the toll of domestic violence—and its undercount—could be even worse due to the pandemic. Most cases of physical abuse against children are typically identified by teachers and school administrators; however, due to school closures and remote learning replacements, teachers lack the ability to easily identify such abuse. Thus, most cases of physical injury at the hands of a caregiver are more unlikely to be identified unless they are serious enough to prompt a visit to a doctor (Dvorak, 2020).

The National Intimate Partner and Sexual Violence Survey, an ongoing survey developed and administered by the Centers for Disease Control and Prevention (CDC), found in 2019 that 1 in 4 women and 1 in 10 men surveyed had been victims of IPV in their lifetimes. As defined by the CDC, IPV includes physical violence, rape, and stalking by a former or current partner or spouse. About 1 in 5 women and 1 in 7 men surveyed had experienced severe physical violence at the hands of their partners. Although these data are for experiences over the life course, even the data from a single year reveal a serious epidemic of IPV. According to the survey, more than 42 million people experienced IPV in 2015 (Centers for Disease Control and Prevention, 2018).

Child abuse—sexual and/or physical assaults on childtren by adult members of their families—is also common in our society. According to the U.S. Department of Health and Human Services (2020), approximately 3.5 million children were the subject of a child abuse investigation in 2018. On any given day, an estimated five children die as the result of abuse (U.S. Department of Health and Human Services, 2020), although the main form of abuse is neglect (U.S. Department of Health and Human Services, 2020). Boys and girls are equally likely to be physically abused, but girls are more likely to be sexually abused as well. According to Childhelp National Child Abuse Hotline (2020), an organization dedicated to the prevention of child abuse, the cycle of abuse is difficult to break; one study suggests that about 30% of abused and neglected children will later abuse their own children.

Elder abuse is the victimization of elderly persons by family members or other caregivers. In a National Center on Elder Abuse study, 10% of elderly U.S. adults (those 65 or older) surveyed reported experiencing either emotional, physical, or sexual abuse or potential neglect (National Center on Elder Abuse, 2020). Similar to child abuse, elder abuse is likely underreported because victims are often in a subordinate position in the family and unable to access help outside the home. Elder abuse also shares with child abuse some of its forms, including neglect—the failure of caregivers to provide for basic needs such as nutritious food and hygienic conditions—and physical abuse. Some aspects of elder abuse differ from abuse of other kinds of victims, however. For example, elder abuse may take the form of financial exploitation or outright theft of property. Those who care for elderly relatives may feel entitled to the resources the seniors possess—or they may simply take advantage of the older persons' vulnerabilities.

As sociologists examining a problem that is both a private trouble and a public issue, we need to ask, "Why does domestic violence exist and persist in family life?" The acceptance of a husband's "right" to subject his wife to physical discipline has roots in Anglo American culture. British common law permitted a man to strike his wife and children with a stick as a form of punishment, provided the stick was no thicker than his thumb; the phrase *rule of thumb* originates in this practice. Through the end of the 19th century in the United States, men could legally beat their wives (Renzetti & Curran, 1992). Research has found that domestic violence is most likely to be prevalent in societies in which family relationships are characterized by high emotional intensity and attachment, there is a pattern of male dominance and sexual inequality, a high value is placed on the privacy of family life, and violence is common in other institutional spheres, such as entertainment or popular culture (Straus & Gelles, 1990; Straus et al., 1988).

FIGURE 11.12 ■ The Opioid Epidemic

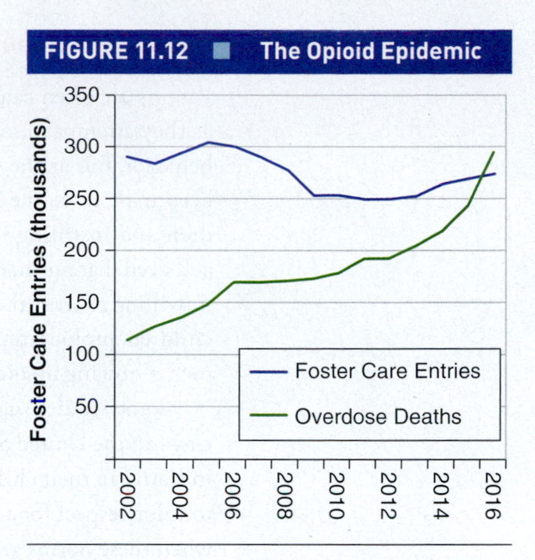

Source: Radel, L., Baldwin, M., Crouse, G., Ghertner, R., & Waters, A. (2018, March 7). *ASPE research brief: Substance use, the opioid epidemic, and the child welfare system: Key findings from a mixed methods study.* U.S. Department of Health and Human Services. https://aspe.hhs.gov/system/files/pdf/258836/SubstanceUseChildWelfareOverview.pdf

People have become more aware of child and spousal abuse and its spectrum of consequences in recent years, largely because of efforts by the women's movement to bring them into the open. Shelters for battered women and children enable victims to be protected from violence. Although still limited in number, such refuges have enabled women to get counseling while terminating abusive relationships in relative safety (Haj-yahia & Cohen, 2009). Family violence is clearly a dysfunctional social phenomenon. It is both a personal trouble and a public issue. A fuller understanding of its roots and consequences can contribute to both a better-informed national conversation about the problem and more robust efforts to address it.

Domestic violence is a persistent source of family crisis in society. A more recent phenomenon is the dramatic rise in opioid addiction in some U.S. regions and the effect this is having on families. According to recent data, on any given day in 2016, about 437,500 children were in foster care, and in the same year, over 273,000 entered foster care for the first time. **Foster care** is *a situation in which a child is cared for by people who are not his or her parents for either a brief or extended period of time.* Foster care may be used when a parent is determined by authorities to no longer be capable of safely caring for the child. A foster parent may be a relative but may also be a nonrelative who has been vetted by authorities. While not all children enter foster care because of a parent's drug addiction, the opioid crisis has been a catalyst in the rise of children needing foster homes. A judge in Indiana (which, together with other parts of the Midwest, has been hard-hit by the opioid crisis) recently noted that the intake of children into foster care has more than doubled in the past 2 years in his area (Simon, 2017). A recent *New York Times* report notes, "In Montana, the number of children in foster care has doubled since 2010. In Georgia, it has increased by 80 percent, and in West Virginia, by 45 percent." It continues,

Foster care: A situation in which a child is cared for by people who are not his or her parents for either a brief or extended period of time.

> The data points to drug abuse as a primary reason, and experts have identified opioids in particular. Neglect remains the main reason children enter foster care. But from 2015 to 2016, the increase in the number of children who came into foster care as a result of parental drug abuse was far greater than the increases in the 14 other categories, like housing instability, according to data from the federal Adoption and Foster Care Analysis and Reporting System. (Lachman, 2017, paras. 2–3)

Authorities in areas ravaged by the opioid crisis are struggling to find enough foster homes to care for children removed from threatening situations (Lachman, 2017; Figure 11.12). Widespread opioid addiction, born of prescription drugs such as OxyContin and Vicodin, given by health care providers for pain, has birthed a crisis of heroin and fentanyl use and abuse that has devastated families in many communities not only through death and injury but also through the dissolution of families (Quinones, 2015).

SOCIOECONOMIC CLASS AND FAMILY IN THE UNITED STATES

An array of family differences are linked to social-class differences. In this section, we consider research showing that social class may have an effect on child-rearing practices, as well as on family formation through marriage.

Social Class and Child-Rearing

Parents are often caught in a dilemma: They must instill some degree of conformity in their children as they attempt to socialize them into the norms and values that will be socially appropriate for adult behavior, but at the same time, they must foster a degree of independence—after all, children must eventually leave the nest and survive in the adult world. Given the tension between protecting children and instilling independence, it should not be surprising that in most families, neither is fully achieved. Parents may understand that they need to help build their children's independence yet still be unwilling to trust their children's judgment, especially during adolescence. They may hang on to their children, prolonging dependence past the point in which children are ready to make decisions on their own. Growing up includes some degree of conflict no matter what approach parents use.

Some studies suggest that parental attitudes toward children's independence may differ by social class. In the United States, middle- and upper-class families tend to value self-direction and individual initiative in their children. Working-class parents, by contrast, have been observed by some researchers to value respect for authority, obedience, and a higher degree of conformity and to rely on punishment when these norms are violated. Sociologist Melvin Kohn (1989), who spent many years studying class differences in child-rearing, attributes these differences to the parents' work experiences: Middle- and upper-class jobs often require individual initiative and innovation, while working-class jobs tend to emphasize conformity.

Annette Lareau (2002) goes beyond Kohn's focus on work experiences to argue that social class (which we experience in a multitude of ways) has an impact on family life, in particular on the styles of child-rearing in which parents engage. In reporting her study, in which 88 white and Black American families were interviewed and 12 were closely observed at home, Lareau writes,

> It is the interweaving of life experiences and resources, including parents' economic resources, occupational conditions, and educational backgrounds, that appears to be most important in leading middle-class parents to engage in concerted cultivation and working-class and poor parents to engage in the accomplishment of natural growth. (pp. 771–772)

These two concepts—concerted cultivation and accomplishment of natural growth—form a key foundation for Lareau's argument. She defines *concerted cultivation* as a style of parenting associated most fully with the middle class and characterized by an emphasis on negotiation, discussion, questioning of authority, and cultivation of talents and skills through, among other things, participation in organized activities. Lareau explains the *accomplishment of natural growth* as a parenting style associated with working-class and poor families. Directives rather than negotiation and explanation, a focus on obedience, and an inclination to care for children's basic needs characterize this style, in which parents leave children to play and grow in a largely unstructured environment. Notably, Lareau identifies a tension between obedience and trust in this style, suggesting it is characterized by something close to distrustful consent born of frustration with authority and dominant institutions but a sense of powerlessness in their presence. From this, she suggests, children take away an emerging sense of constraint. Consider the following observation by Edin and Kefalas (2005):

> While poor mothers see keeping a child housed, fed, clothed, and safe as noteworthy accomplishments, their middle-class counterparts often feel they must earn their parenting stripes by faithfully cheering at soccer league games, chaperoning boy scout camping trips and attending ballet recitals or martial arts competitions. (p. 141)

What are the effects of these differing styles of child-rearing, which Lareau argues are associated with class status? Examining the outcomes for children in her study as they approached adulthood, Lareau (2002) concludes that the accomplishment of natural-growth style not only tends to cultivate early independence but also leads young people toward jobs that require respect for authority and obedience to directives, such as those associated with the working class. In contrast, the concerted-cultivation approach to child raising tended to lead young people both to hold a sense of entitlement and to pursue careers that require a broad vocabulary and ease in negotiating with people in authority. Following up with the families she studied, Lareau found that all the middle-class children had completed high school and that most were attending college. Many of the children of low-income families had left high school, and few were in college. "In sum," she writes, "differences in family life lie not only in the advantages parents obtain for their children, but also in the skills they transmit to children for negotiating their own life paths" (p. 749).

These studies strongly suggest that family life—and practices of child-rearing, in particular—contribute to the reproduction of class status. Although structural factors, including obstacles related to educational and economic resources, are an important part of the picture, Lareau suggests that the orientations and skills developed in childhood, which become part of a young adult's human capital as he or she negotiates the path through school and toward a job, are also relevant to understanding socioeconomic outcomes and the reproduction of class status.

Economy, Culture, and Family Formation

Class status is linked with changing patterns of family life in another important way. Some sociologists believe that macrolevel economic changes, in particular the rise of a postindustrial economy and associated labor market, have had a powerful effect on microlevel practices of family formation (Edin & Kefalas, 2005; Wilson, 1996, 2010). In this section, we examine the sociological roots of the decline of marriage and the rise of nonmarital births in poor and working-class Black, white, and Latinx communities.

In 1965, Daniel Patrick Moynihan, a former professor of sociology who was then the assistant U.S. secretary of labor and would later become a U.S. senator from New York, published a controversial study of the African American family titled *The Negro Family: The Case for National Action*, which later came to be called the *Moynihan Report*. In it, Moynihan argued that lower-class Black family life was often dysfunctional, as reflected in high rates of family dissolution, single parenting, and the dominance of female-headed families.

Moynihan saw the breakdown of Black families, together with continued racial inequality, as leading to a new crisis in race relations. He identified the legacy of slavery, which intentionally broke up Black families, as one of the roots of the problem. He also argued that many rural Blacks had failed to adapt adequately to urban environments. Moynihan concluded that these patterns in family life were, at least in part, responsible for the failure of many low-income Black Americans to make it into the economic mainstream in the United States. He saw Black poverty and inner-city violence, which exploded in the urban riots of the 1960s, as partly the result of Black family breakdown.

The *Moynihan Report* was widely criticized as racist and sexist, and Moynihan was accused of blaming the victims of racism and poverty (primarily Black female heads of household) for their disadvantages. Many critics ignored Moynihan's attention to structural as well as cultural factors in his analysis, however. Although he wrote that "at the center of the tangle of pathology is the weakness of the family structure," he also implicated social phenomena such as unemployment, poverty, and racial segregation in the decline of families and the rise of dysfunctions. Still, critics focused largely on his cultural analysis, and Moynihan's findings were disregarded while sociologists avoided examining the connections among race, family characteristics, and poverty for at least two decades.

In 1987, sociologist William Julius Wilson published *The Truly Disadvantaged: The Inner City, the Underclass, and Public Policy*, in which he revisited the issues Moynihan had raised in the 1965 report. In the two intervening decades, much had changed—and much had stayed the same. The family patterns that Moynihan had identified as problematic, including nonmarital births and high levels of family dissolution, had become more pronounced among poor and working-class Black Americans. At the same time, social problems such as joblessness in the inner city had grown more acute, as deindustrialization and the movement of jobs to the suburbs dramatically reduced the number of positions available for less educated and low-skilled workers. These changes, Wilson noted, had important consequences for family formation.

Almost 10 years later, Wilson (1996) argued that falling rates of marriage and rising numbers of nonmarital births were, at least in part, rooted in the declining numbers of "marriageable men" in inner-city neighborhoods. He posited that the ratio of unmarried Black women to single and "marriageable" Black men of similar age was skewed by male joblessness, high rates of incarceration, and high death rates for young Black men. The loss of jobs in the inner city had a powerfully negative effect on male employment opportunities. The consequences were felt not only in the economic fortunes of communities but also—and no less importantly—in families, where women were choosing motherhood but options for marriage were diminished by the uneven ratio of single women to marriageable men. Although rates of nonmarital births had previously been high in Black communities, where extended families have traditionally been available to support mothers and children (Gerstel & Gallagher, 1994), Wilson saw new urban circumstances as central to the rise of nonmarital births among Black Americans from one quarter in 1965 (when Moynihan published his report) to about 70% by the middle 1990s, where it remains today (Table 11.1). By then, about half of Black American families were headed by a woman (Wilson, 2010).

TABLE 11.1 ■ Nonmarital Birthrate by Race/Ethnicity in the United States, 2020

White	Black	Hispanic
28.2%	69.9%	51.8%

When Wilson revisited this issue in 2010, he found that research on the relationship between male employment and rates of marriage and single parenthood offered mixed findings:

> Joblessness among black men is a significant factor in their delayed entry into marriage and in the decreasing rates of marriage after a child has been born, and this relationship has been exacerbated by sharp increases in incarceration that in turn lead to continued joblessness. Nevertheless, much of the decline in marriages in the inner city, including marriages that occur after a child has been born, remains unexplained when only structural factors are examined. (p. 108)

William Julius Wilson (2010) also cited cultural factors in the fragmentation of the poor Black family. Although sociologists have been reluctant to use culture as an explanation, not least due to fear of the backlash generated by the *Moynihan Report*, Wilson writes that structure and culture interact to create normative contexts for behavior. He points out, for instance, that

> [B]oth inner-city black males and females believe that since most marriages will eventually break up and no longer represent meaningful relationships, it is better to avoid the entanglements of wedlock altogether. . . . Single mothers who perceive the fathers of their children as unreliable or as having limited financial means will often—rationally—choose single parenthood. (p. 125)

In this social context, the stigma of unmarried parenthood is minimal, and behaviors rooted in structure become culturally normative.

Examining the phenomenon of low marriage rates and high nonmarital births in some communities, sociologists Edin and Kefalas (2005) looked at data on low-income white, Black, and Puerto Rican single mothers in Camden, New Jersey, and Philadelphia, Pennsylvania. Black communities have historically had higher nonmarital birthrates and lower rates of marriage than white and Latinx communities, but in low-income white and Latinx communities, comparable trends have taken root.

Edin and Kefalas found that the women in their study placed a high value on both motherhood and marriage. They saw motherhood as a key role and central achievement in their lives, and few made serious efforts to delay motherhood. If anything, they saw early motherhood as something that had forced them to mature and kept them from getting into trouble. At the same time, they held a utopian view of marriage that may, ironically, have put it out of their reach. Many dreamed of achieving financial security and owning a home before marrying—but their poverty made this a challenge. More problematic, perhaps, was that they did not consider the men in their lives—often the fathers of their children—to be good partners, their marriageability undermined by low education, joblessness, poor economic prospects, criminal records, violence, drug and alcohol abuse, and infidelity.

Notably, then, even though they highly valued marriage, the women had few realistic opportunities to achieve stability and independence first and little hope of finding stable partners. Motherhood, however, was an achievable dream, an opportunity to occupy an important social role and to achieve success as a parent, where other paths of opportunity were often blocked by poor structural circumstances.

Statistics show that children in single-mother homes are among those most likely to be born and to grow up impoverished. They are also more likely than their better-off peers to repeat the patterns of their parents and to remain in poverty. Here, some interesting sociological questions emerge: Is single parenthood a cause (not the only one) of poverty, or is living in poverty a sociological root of single parenthood? Or perhaps both are true? What are the implications of the answers to these questions for the design of public policies that address poverty?

Poor parents often struggle to provide for their children. For parents of young children, a common problem is diaper need. Disposable diapers, while taken for granted among better-off parents, are costly, and economically disadvantaged families may not be able to take advantage of discounts available for bulk purchases.

The relationships among class, poverty, and family patterns are complex, but Wilson and Edin and Kefalas offer sociological lenses for understanding some of the structural and cultural roots of low marriage rates and high nonmarital birthrates in many poor U.S. communities. As William Julius Wilson (2010) notes in his book *More Than Just Race: Being Black and Poor in the Inner City*, "How families are formed among America's poorest citizens is an area that cries out for further research" (p. 129). What other factors should sociologists examine? How would you conduct such a study? What kinds of questions would you ask?

Parenting in Poverty

Parents across the economic spectrum seek to raise healthy and happy children and to meet the many challenges of shepherding boys and girls from infancy to adulthood. But parenting in poverty presents a range of obstacles and fears that well-off families are far less likely to face.

For one thing, parenting in poverty is expensive. As a blog post in the *Washington Post* noted, for parenting families in the middle quintile of the household income spectrum, diapers consume just under 3% of their income; for those in the upper two quintiles, the figure is closer to 1% to 2%. By contrast, nearly 14% of the household income of the poorest quintile goes to diapers (Badger & Eilperin, 2016). What explains this dramatic difference? Consider the fact that today many middle- or upper-income families buy their disposable diapers in bulk from shopping clubs such as Sam's Club or through the mail from Amazon, which offers free shipping for subscribed members. This can considerably reduce the cost and increase the convenience of buying a good that has become a basic necessity of modern life.

For low-income families, these avenues are often foreclosed. Poor parents may not be able to afford a costly membership to a shopper's club—or even a car to travel to a distant warehouse shop where these items could be bought. They may not have space at home to store bulk purchases of diapers or other household goods. Those with little money may not be able to purchase the larger bags of diapers that offer a better value; instead, they are forced by economic circumstances to buy in small quantities. As the authors of the blog write,

> Cheap diapers are hard to come by for the families that have the least to spend on them. Mora, 27, scans coupons and travels for bargains. She tells her children "no" at the grocery store, when she has to choose between the kiwis they want and the diapers 2-year-old Nathan needs. Then, sometimes, when she finds a good deal, it comes in the wrong sizes. "Sometimes you have to decide between 'Okay, this box has 120 diapers, and this is the size that he doesn't use. But if I get the size that he's using, it's just 70 diapers, and I have 50 diapers more. So what should I do?'" she says, knowing that a too-small size might chafe her son's skin. "You just have to make things happen." (Badger & Eilperin, 2016, para. 3)

The challenges do not end at the doorstep of an individual household. Parenting in poverty often means raising children in economically disadvantaged neighborhoods, which present a further set of challenges. A recent Pew Research Center report titled "Parenting in America" indicates that "higher-income parents are nearly twice as likely as lower-income parents to rate their neighborhood as an 'excellent' or 'very good' place to raise kids (78% to 42%)" (p. 3). Fully 38% of families with incomes under $30,000 described their neighborhood as only a "fair or poor" place to raise children. Among families with incomes over $75,000, the figure was only 7%. The differences do not stop there: Nearly half of the poorest families expressed a fear that their child could get shot, and over half worried that their child could get beat up or attacked. Although these concerns were also expressed by better-off families, 22% of whom worried about shooting and 38% of whom were concerned about other physical violence, the figures suggest a lower perceived degree of threat (Pew Research Center, 2015a).

There are, to be sure, commonalities that many parents share, regardless of income. Comparable numbers of parents expressed fears of a son or daughter struggling with anxiety or depression or having problems with drugs or alcohol (Pew Research Center, 2015a). At the same time, families with greater economic means are likely to have fuller access to resources to address these problems, again highlighting some of the challenges to poor parents seeking, like their better-off peers, to make a good life for their children.

Family Life in the Middle Class

Today, parents in the U.S. middle class, particularly its upper fraction, devote an unprecedented amount of resources to child-rearing. Although some of the rising financial outlays are for basic needs and childcare, parents are also committing time and money to enrichment activities intended to give their offspring advantages in competition, education, and the future labor market. More economically advantaged parents are actively engaged in building what Hilary Levey Friedman (2013) calls "competitive kid capital." Friedman's study of 95 families with elementary school–age children who were involved in competitive after-school activities, such as chess, dance, and soccer, found that many parents

> saw their kids' participation in competitive afterschool activities as a way to develop certain values and skills: the importance of winning; the ability to bounce back from a loss to win in the future; to succeed in stressful situations; and to perform under the gaze of others. (p. 31)

Interestingly, Friedman points out that parents of upper-middle-class girls are more likely to enroll their girls in soccer or chess than in dance, pursuing an "aggressive femininity" that they perceive to offer their daughters a future labor market advantage.

Middle-class family life is often characterized by a strong commitment to constructive and active child-rearing (as we saw in Lareau's study and the research described previously) and to the parents' pursuit of careers. These competing commitments often leave parents without the time to do everything or to do it well and feeling rushed and stressed instead of satisfied.

Sociologist Arlie Russell Hochschild (2001b) has come to some interesting and perhaps surprising conclusions about this modern dilemma. Hochschild conducted a series of interviews with employees at a well-known *Fortune* 500 firm that had gone to some lengths to be family friendly, offering flextime, the option of part-time work, parental leave, job sharing, and even a course titled "Work–Life Balance for Two-Career Couples." But she noticed that the family-friendly measures didn't make much of a difference. Most employees said they put "family first," but they also felt strained to the limit, and almost none cut back on work time. Few took advantage of parental leave or the option of part-time employment.

Why do people say they want to strike a better balance between work and the rest of their lives and yet do nothing about it when they have the opportunity? Some reasons are practical; several employees in Hochschild's study feared that taking advantage of liberal work policies would count against them in their careers, while others simply needed the money—they couldn't afford to work less. Yet Hochschild identified a more surprising reason many people worked long hours: They liked being at work better than being at home. Previous research had shown that many men regard work as a haven, and Hochschild found that a notable number of working women now feel the same way. Despite the stress of long hours and guilt about being away from their families, they are reluctant to cut back on their commitment to paid work.

Hochschild found that both men and women often derive support, companionship, security, pride, and a sense of being valued when they are working. In the absence of family time and kin and community support at home, some parents sought and found—and sometimes preferred—a sense of competence and achievement in the workplace. In about a fifth of the families Hochschild studied, work (rather than home) was the site at which the parents derived the most satisfaction.

Many studies since have found that workplaces with more family-friendly policies have higher levels of workplace satisfaction and productivity as well as lower levels of stress (Bilal et al., 2010; Frye & Breaugh, 2004). Still, Hochschild's (2001b) research suggests that many people derive satisfaction from being workaholics. As she concludes, "Working families are both prisoners and architects of the time bind in which they find themselves" (p. 249).

GLOBALIZATION AND FAMILIES

The impact of globalization on families depends, to a substantial degree, on social class and country or region of residence. In this section, we look at ways in which different families experience globalization and its myriad costs and benefits.

Consider the economic and labor market impact of globalization on U.S. families. On the one hand, globalization has contributed to an increase in many employers' demands for men and women with high degrees of skill and formal training, particularly in technical fields. On the other hand, low-skilled U.S. workers have been priced out of many sectors of the global job market by the fact that lower-wage labor is readily available elsewhere around the globe. Globalization can produce national economic gains even as it diminishes the prospects of some categories of workers.

About 70% of the individuals who make up the U.S. labor force do not hold 4-year college degrees. As we see elsewhere in this text, these are the workers who have been hit hardest by global economic change. Writing about domestic manufacturing industries, journalist Louis Uchitelle (2007) observes,

> As customers defected, sales plummeted and failed to bounce back. Nowhere was that more apparent than in the auto industry's struggle with [lower-cost] Japanese imports. But nearly every manufacturer was hit, and the steep recession in 1981 and 1982 compounded the damage. The old world has never returned. (p. 8)

Many working-class families have found themselves confronting flat or declining incomes as a result of competition with a global workforce. Household incomes at the bottom of the economic spectrum have declined most dramatically since the 1970s, when globalization began to transform the domestic economy. This decline has had a multitude of effects on U.S. families. Recall our discussion in this chapter of the diminished pool of "marriageable" males. Here, we see some of the effects of declining job opportunities and wages in manufacturing, which used to offer gainful employment to less educated men.

The need for a family to have two incomes to make ends meet is one of the reasons for the dramatic increase in the number of women working outside the home. The movement of women into the paid workforce has, in turn, provided some women with a degree of economic independence and an opportunity to rethink the meaning of marriage. As women join the paid workforce, some postpone marriage until they are older. Couples choose cohabiting as an alternative to marriage, and when they do decide to have children, their families are likely to be smaller. Some may never marry; as we learned earlier, marriage is continuing to decline among young adults.

Globalization means greater mobility for families and more fluid ways of organizing work and life. The benefits of globalization enjoyed by some U.S. families, however, are accompanied by the losses suffered by others.

International Families and the Global Woman

Macrolevel processes of globalization affect U.S. families in a variety of ways. In this section, we examine the dual phenomena of *international families* and the *global woman* to emphasize some of the microlevel effects of globalization on women, particularly women from the developing world.

International families: Families that result from globalization.

Anthropologist Christine Ho (1993) has examined what she terms **international families**—*families that result from globalization*. Focusing on mothers who emigrate from the Caribbean to the United States, Ho documents how they often rely on child minding, an arrangement in which extended family members and even friends cooperate in raising the women's children while they pursue work elsewhere, often thousands of miles away. This practice adds a global dimension to cooperative child-rearing practices that are a long-standing feature of Caribbean culture.

Ho suggests in her profiles of these female global citizens—most of whom work in lower-wage sectors of the economy, including clerical work and childcare—that such global-family arrangements enable Caribbean immigrants to avoid becoming fully Americanized: International families and child minding provide a strong sense of continuity with their Caribbean homeland culture. Ho predicts that Caribbean immigrants will retain their native culture by regularly receiving what she characterizes as "bicultural booster shots" through the shuttling of family members between the United States and the Caribbean. At the same time, Ho notes, this process contributes to the Americanization of the Caribbean region, which may eventually give rise to an ever more global culture.

Ehrenreich and Hochschild (2002) have turned their attention to what they call the "global woman." Like Ho, they examine the female migrant leaving home and family to seek work in the wealthy "first world." Unlike Ho, however, they take a pointedly critical view of this phenomenon, suggesting that these female workers, many of them employed as nannies or housekeepers (or even prostitutes), are filling a "care deficit" in the wealthier countries, where many female professionals have pursued opportunities outside the home. In doing so, the migrants create a new deficit at home, leaving their own children and communities behind:

> Third World migrant women achieve their success only by assuming the cast-off domestic roles of middle- and high-income women in the First World—roles that have been previously rejected, of course, by men. And their "commute" entails a cost we have yet to fully comprehend. (Ehrenreich & Hochschild, 2002, p. 3)

Ehrenreich and Hochschild (2002) argue that Western global power, previously manifested in the extraction of natural resources and agricultural goods, has evolved to embrace an extraction of women's labor and love, which is transferred to the well-off at a cost to poorer countries, communities, and—most acutely, perhaps—families:

> The lifestyles of the First World are made possible by a global transfer of the services associated with a wife's traditional role—child care, homemaking, and sex—from poor countries to rich ones. To generalize and perhaps oversimplify: in an earlier phase of imperialism, northern countries extracted natural resources and agricultural products . . . from lands they conquered and colonized. Today, while still relying on Third World countries for agricultural and industrial labor, the wealthy countries also seek to extract something harder to measure and quantify, something that can look very much like love. (p. 4)

Arlie Hochschild defines the nanny chain thus: "An older daughter from a poor family in a third world country cares for her siblings, while her mother works as a nanny caring for the children of a nanny migrating to a first world country, who, in turn, cares for the child of a family in a rich country" (Hochschild, 2001a, para. 3).

©Terese Loeb Kreuzer/Alamy Stock Photo

Global women have been increasingly in demand in high-income countries to provide health and social care for children, elderly people, and the disabled. As of 2014, there are approximately 53 million domestic workers worldwide, the majority of whom are female temporary migrant workers (International Labour Organization, 2014). The demand for live-in caregivers and domestic workers is driven by aging populations, the aforementioned feminization of international migration, diminishing state provision of care services, and the incorporation of women into the labor force without policies in place to reconcile the demands of family life (such as childcare) and paid employment (Salami et al., 2017).

Although the women from the developing world are, for the most part, agents in their own choice to migrate to countries of the developed world in search of work (unless they are trafficked or tricked into migration), Ehrenreich and Hochschild point to powerful social forces that figure into this choice. On the one hand, many women encounter the *push* factor of poverty: the choice of facing destitution at home or leaving families behind to earn what are, for them, substantial wages abroad. On the other hand, there is the *pull* factor of opportunities abroad: their services are welcomed and needed, and they may gain human as well as economic capital. Even though these women make choices, their decisions may be driven by strong economic pressures and carry substantial noneconomic costs.

WHAT'S NEXT FOR FAMILIES AND FAMILY LIFE?

We opened this chapter with a story about assortative mating and a discussion of research that suggests it has exacerbated income inequality in U.S. society. Social forces, as we learned in this chapter, shape families and family life in important ways. What will be the effect on families of the dramatic economic and public health situations fostered by the global COVID-19 pandemic? The coming months and years will bring answers to a variety of questions.

First, will the economic crisis wrought by COVID-19 lead to a rise in divorce? Evidence from past economic crises suggests some possible outcomes. The Great Recession of 2008–2009 saw a decline in divorce, which may seem counterintuitive since economic challenges can put substantial stress on a marriage. However, the recovery that followed also saw a rise in divorce rates, suggesting that divorce, which can be costly, may be postponed by couples during periods of economic turmoil when a spouse may lose a job or health insurance or a couple cannot afford the legal process. A study conducted in 2012 of the divorce rate among a sample of about 2.8 million U.S. women in 2008–2011 showed a sharp decline between 2008 and 2009, and an increase after 2009 (Cohen, 2014; see Figure 11.13).

As the COVID-19 pandemic develops and its effects reverberate throughout U.S. society, will we see a repeat of this pattern?

Second, will there be noticeable effects on the birthrate? The 2008–2009 recession brought a decline in birthrates, which appear to have been more pronounced in the states that were hardest hit by unemployment increases (Cherlin et al., 2013). While there has been some speculation about a coming "baby boom" linked to weeks or months that couples spent quarantined together, there is more evidence to suggest we will see a "baby bust" (Yuhas, 2020). The 2008–2009 recession cost millions of people their jobs and homes, which are foundations for raising a family; even after data showed the recession had ended, for many young people, stable jobs were hard to find and owning a home was a distant dream. As 22 million Americans have lost their jobs since March of 2020, a similar effect on U.S. birthrates is a likely outcome.

Third, what will be the effects on longer-term household formation after a period in which the proportion of young adults living with their parents grew? According to Pew Research, in September 2020, the proportion of young adults (ages 18–29) living with their parents was the highest it had been since the Great Depression of the 1930s (Fry et al., 2020). How will this affect marriage and cohabitation rates in this generation? How will it affect their establishment of independent households? What about future fertility?

Have you observed short- or longer-term effects of the economic and public health crises of 2020? What questions do you have about the future of families and family life?

FIGURE 11.13 ■ Divorce Rates Among Women in the United States, 2008–2011

	2008	2009	2010	2011
Divorced women	1,309,921	1,219,656	1,250,086	1,251,239
Divorce per 1,000 married women	20.9	19.5	19.8	19.8

WHAT CAN I DO WITH A SOCIOLOGY DEGREE?

Problem Solving

Problem solving is a fundamental skill in social scientific disciplines such as sociology and in a wide variety of contemporary occupational fields. Managing and addressing complex problems by identifying their dimensions, researching their roots, and using the knowledge to craft well reasoned responses is a skill set that is developed through careful study, training in research and analysis, and practice. Problem solving is, in many respects, comprised of other key skills we discuss in this feature, including data and information literacy, critical thinking, quantitative and qualitative research competency, and understanding of diversity. At the same time, it is a skill that has its own characteristics as a product of sociological training. Sociological research data, which form the foundation of what sociologists do and teach, cannot solve problems; rather, research data contribute to the informed understanding of the dimensions of a problem. Data are also used to hypothesize the roots of a problem. Once the roots of a problem are identified, they can be addressed through, for instance, policy or community interventions. Research can be used to follow up on whether and how solutions worked and to rework hypotheses based on new information.

Researching the roots of a problem can involve a spectrum of different approaches, and a sociologist often needs to try more than one approach to generate a comprehensive picture of the problem. Social life is complex, and most serious social problems are not amenable to simple solutions. At the same time, the probability of successfully addressing a problem is appreciably greater when one has used careful research to understand its causes.

Katie Marquette, Community Relations Manager, Providence Health & Services Alaska

BA, Sociology, University of Alaska Anchorage

The problems encountered in different occupational fields vary, but the need for people who are skilled in breaking down a problem, defining it, analyzing it, crafting solutions based on good data, and effectively communicating identified paths of action is common across many areas.

Sociology teaches you to think critically about how policies and environmental factors impact certain populations and cultures. I studied sociology because I wanted to learn why some populations were less advantaged than others and what social determinants would change their life outcomes for the better.

In the nearly 10 years I have worked in the communications and public relations field, I have frequently utilized my knowledge of sociological theories and concepts. As a professional communicator, it is my job to break down complex topics and policies for the general public to understand. Knowing how those policies will impact different populations and cultures helps me create a message that is more succinct and relatable to my target audience.

My desire to work in health care is also tied to my interest in sociology. Access to affordable health care services is so critical in our society and affects issues like public safety and crime, the economy, and job creation—just to name a few. You cannot go to school, take care of your family, or go to work if you are not healthy.

In my current job, I work with local nonprofits to provide financial sponsorship on behalf of Providence Health & Services Alaska for programs that align with our organizational mission to serve the most poor and vulnerable. This includes support for homeless youth shelters, soup kitchens, women's shelters, and many other important programs in order to improve health outcomes for Alaska families.

In sociology, you learn about social stratification and how factors like wealth and income influence a person's ability to succeed in life. Communicating about the importance of affordable health care and access to critical social services directly relates to this discussion of social inequalities.

Career Data: Public Relations Specialists

- 2019 Median Pay: $61,150
- Typical Entry-Level Education: Bachelor's degree
- Job Growth Outlook, 2019–2029: 7% (faster than average)

Source: U.S. Bureau of Labor Statistics (2020).

SUMMARY

- The meaning of **family** is socially constructed within a particular culture. In the United States, as in other modern societies, the meaning and practices of family life have been changing.
- **Marriage**, found in some form in all societies, can take several different forms, from the most common—**monogamy**—to many variations of **polygamy**, in which a person has multiple spouses simultaneously.
- The functionalist perspective highlights the family's functionality in terms of social stability and order, emphasizing such activities as sex role allocation and child socialization.
- Feminist perspectives on the family are more conflict oriented, highlighting the **sexual division of labor** in society and its stratifying effects. Feminist perspectives also examine the different experiences of men and women in marriage and the way social expectations and roles affect those experiences. The psychodynamic feminist perspective takes a sociopsychological approach, emphasizing the impact of early mothering on the later assumption of gender roles.
- In U.S. society today, the composition of families and the roles within families are shifting. The age at first marriage has risen across the board, and rates of marriage have declined, particularly among the less educated. Same-sex marriage was recognized as legal in the entire United States in 2015. Young adults are less likely than prior generations to marry, although most still value parenthood. Nonmarital births account for more than 40% of all births. Divorce rates have leveled off but remain at a high level.
- Socioeconomic class status affects child-rearing practices and family formation patterns. Lower rates of marriage and high rates of nonmarital births are present in the working class and among the poor. Middle-class family life is often structured around the needs of children.
- In the United States, the effects of globalization include changes in household income and employment opportunities. Women from developing countries often leave their homes and children to work for families in the developing world.

KEY TERMS

antimiscegenation laws (p. 290)
cohabitation (p. 298)
common-law marriage (p. 298)
domestic (or family) violence (p. 305)
endogamous (p. 290)
extended families (p. 291)
family (p. 289)
foster care (p. 307)
international families (p. 314)
marriage (p. 289)
monogamy (p. 289)
nuclear families (p. 291)
polyandry (p. 289)
polygamy (p. 289)
polygyny (p. 289)
serial monogamy (p. 290)
sexual division of labor (p. 293)

DISCUSSION QUESTIONS

1. Why do people get married? Why do people *not* marry? Think about individual and sociological reasons. Link your answers to the discussion of marriage trends and the experience of marriage discussed in this chapter.
2. Recent data show some changes in the childcare practices of U.S. families. What do trends show? How do sociological factors help to explain the changes?

3. How does the case of deaf families with hearing children show the opportunities and challenges of family life characterized by different cultures? Can this case be compared with immigrant families with children? What similarities and differences can you identify?
4. Lareau's research suggests that middle- and working-class families have different child-rearing styles. How does she describe these styles? Why might the differences be sociologically significant? Does the essay on parenting in poverty help to shed light on differences?
5. Who is the "global woman"? What are the costs and benefits to women and families of a global labor market for care work?

©iStockphoto.com/PeopleImages

12 EDUCATION AND SOCIETY

WHAT DO YOU THINK?

1. Does your family's social class matter when it comes to your educational attainment?
2. Why does racial segregation exist and persist in many U.S. public schools?
3. Why do students drop out of college before completing a degree?

LEARNING OBJECTIVES

12.1 Describe the historical and contemporary role of education in society.

12.2 Apply conflict, functionalist, and symbolic interactionist theoretical perspectives to the institution of education.

12.3 Explain how education may function both to reduce and reproduce social inequalities.

12.4 Discuss key issues in U.S. higher education, including the relationship between education and income potential, the debate over college internships, and the college dropout phenomenon.

12.5 Learn about the importance of higher education globally and the growth of U.S. student contact with others from around the world.

HUNGRY AND HOMELESS COLLEGE STUDENTS?

During college have you or someone you know gone without food because there was no money to buy it? Have you or someone you know ever had to couch surf because you lacked housing? Were resources available on campus to help?

Food insecurity: A lack of consistent access to enough food for an active, healthy life.

Food insecurity, defined by the U.S. Department of Agriculture as *lack of consistent access to enough food for an active, healthy life*, can result in skipping meals to save money, trying to grab extra food at the cafeteria to have another meal for the day or the weekend, or experiencing the fatigue and anxiety that accompany hunger pangs. Housing insecurity, while not officially defined, can range from not having enough money to cover rent to being without a place to stay. In three recent national surveys, a number of U.S. college students revealed that issues of hunger and homelessness are part of their college experience.

For instance, food insecurity was experienced by well over a third of the 167,000 college students who responded to the 2020 #RealCollege Campus Basic Needs Security study and 43% of students in the Spring 2020 American College Health Association National College Health Assessment. Other surveys, like the 2020 *Student Experience in the Research University* (SERU) COVID survey of 30,697 undergraduates found lower rates of food insecurity (20%). But they identified important race and class differences in who was more likely to experience it. For instance, while less than 20% of white students (19%) noted food insecurity, it almost doubled for Black students (37%). And a majority (58%) of students with low incomes said they experienced it.

Housing insecurity was true of just under half (46%) of the respondents to #RealCollege study. And respondents to the SERU COVID survey who were working class were 16 times likelier than wealthy students to say that they had not been able to cover the cost of their housing. Seventeen percent of those students said they experienced homelessness, dealing with it by "temporarily staying with a relative or friend, or couch surfing" (Baker-Smith et al., 2020, p. 13).

Why are so many students confronting the challenges of school while also facing the threat of hunger and unstable living? Thinking sociologically, we can identify three factors. First, more first-generation students than ever are entering college. As a significant proportion of these students come from lower-income families, they may not be able to rely on family resources when they run short of money or meals. Furthermore, student meal plans are not necessarily structured in a way that meets the needs of those who are food insecure. College cafeterias often have limited hours that do not correspond to the schedules of students who are busy with school and work. Many schools do not permit students to roll over unused funds or meals to a future semester. This prompted students at Spelman and Morehouse Colleges in Georgia to go on a successful hunger strike to protest a school policy that prevents students from sharing unused meal vouchers with classmates who might need them (Mitchell, 2017). After less than a week, both institutions committed to offer 14,000 meals per year for students facing food insecurity (Carter, 2017).

Second, college costs continue to rise: Tuition is only one of many costs that students face. School fees, lab fees, books, housing, and food may add up to more than students' savings, loans, and scholarships can cover. Research suggests that students of color, LGBTQ+ students, and young people coming out of foster care into college are at particular risk of food insecurity and homelessness (Goldrick-Rab, 2020).

Third, the effects of COVID-19 on students and employment and the economy have continued to have an unequal effect on students by social class, including the loss of family income, increases in living and technology expenses, and loss of their own jobs or canceled internship experiences. Soria and Horgos (2020) note that students with low incomes and from working-class families were more likely to have had all of these experiences.

In total, these challenges can be significant barriers for students. Research shows a correlation between food insecurity, lower grade point averages (Maroto et al., 2015), and lower graduation rates (Goldrick-Rab, 2020) and higher risks for dropping out (Yavorski, 2017).

Student-led campaigns such as Swipe Out Hunger, which advocates for policies that would permit students to donate their meal "swipes" to students facing food insecurity, and the growing number of schools opening food pantries to assist students, are helping to address this largely hidden problem, but more remains to be done. Sociologists, as well as public policy analysts, social workers, and many others, have joined with students to provide solutions like subsidized housing, better federal benefit programs such as Supplemental Nutrition Assistance Program (SNAP), and many others (Goldrich-Rab, 2020).

In this chapter, we explore selected sociological issues in education, including key challenges that face schools and students at all levels. We begin with a discussion of the roots of mass public education in the United States and the development of the "credential society" currently driving rising enrollments in higher education. We take a critical look at education, using the functionalist, conflict, and symbolic interactionist perspectives to consider the American educational system. We then turn to the issue of education and inequality, examining education as a key to understanding how inequality is both reduced and reproduced in society. We also consider three selected issues in higher education: the relationship between higher education and income, the rise of internships, and the problem of college noncompletion. Finally, we look at education in a global perspective, considering how the U.S. higher educational system compares in outcomes to peer countries and taking a closer look at international interaction through study abroad.

Many colleges and universities have responded to the food insecurity of their students by creating on-campus food pantries.

©M.P. King /Wisconsin State Journal

EDUCATION, INDUSTRIALIZATION, AND THE "CREDENTIAL SOCIETY"

As societies change, so too does the role of **education**, *the transmission of society's norms, values, and knowledge base by means of direct instruction*. For much of human history, education occurred informally, within the family or the immediate community. Children often learned by doing—by working alongside their parents, siblings, and other relatives in the home, in the field, or on the hunt. With the emergence of industrial society, **formal education**, *education that occurs within academic institutions such as schools*, became increasingly common.

Education: The transmission of society's norms, values, and knowledge base by means of direct instruction.

Formal education: Education that occurs within academic institutions such as schools.

As schooling became important in industrial societies, it increasingly came to be seen as the birthright of all society members. Today, the norm is **mass education**, *the extension of formal schooling to wide segments of the population*. Consistent with the democratic ideals held in most economically advanced

Mass education: The extension of formal schooling to wide segments of the population.

or advancing societies, it is also the principal means by which people acquire the skills they need to participate effectively as workers and citizens in the midst of dramatic technological, cultural, and economic changes. In this section we provide a brief glance at historical educational institutional changes in the United States to provide some pretext for our discussion later in this chapter.

Contemporary society requires its members to master a large number of complex skills. People must know how to read and write and do basic math, but that is rarely enough. Societies need people to organize production, invent new products, and program computers; others engage in creating art or literature, curing diseases, resolving human conflicts, and addressing scientific challenges such as climate change. Building an educated population requires more than the on-the-job training of apprentices or helpers. It requires the transmission of more knowledge than most families are willing or able to pass on from one generation to the next.

The first formal educational institutions in the United States were created in the 17th century by the religious leaders of the New England Puritan communities. Their goal was to provide religious education; children were taught to read so they could study Scripture (Monroe, 1940; Vinovskis, 1995). In 1647, the Massachusetts Bay Colony passed a law requiring every community of 50 or more people to establish a town school. The law, named Ye Old Deluder Satan Act, was intended to protect New England's youth from acquiescing to the temptations of the devil.

Literacy: The ability to read and write at a basic level.

By the 18th century, the goal of education had shifted from religious training to cultivating practical and productive skills (Vinovskis, 1995). The emergence of industrial societies not only increased the need for people to be literate— **literacy** is defined as *the ability to read and write at a basic level*—but it also required that they learn skills, work habits, and discipline that would prepare them for jobs as industrial laborers, accountants, inventors, designers, merchandisers, lawyers, operators of complex machinery, and more (Bergen, 1996).

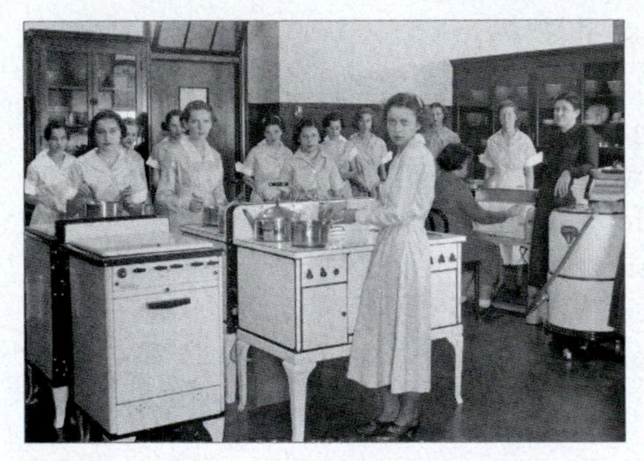

In the United States, girls and boys used to be educated separately and unequally in many public schools. Far fewer young women had the opportunity to go to college, and their high school education emphasized domestic skills such as meal preparation.

From the outset, schools in the United States were divided along social class lines. The sons of the middle and upper classes went to private schools that trained them for business and the professions. There were initially few public schools, and those that existed provided working- and lower-class children with the minimal education necessary for them to acquire the skills and obedience for factory work or farming (Bowles & Gintis, 1976; Wyman, 1997).

When workers began forming labor unions in the 19th century, one of their demands was for free **public education** for their children, *a universal education system provided by the government and funded by tax revenues rather than student fees* (fees served to exclude economically disadvantaged students from the classroom; Horan & Hargis, 1991). Political activists, philanthropic organizations, and newspapers joined the unions in their demand. By the late 19th century, public elementary schools had been established in most of the industrial centers of the United States, and mass public schooling soon spread throughout the country. In some states, school attendance was compulsory for at least the first 6 years.

Public education: A universal education system provided by the government and funded by tax revenues rather than student fees.

The concept of public education was soon expanded to include high schools, and by the end of the 19th century, the average U.S. student achieved 8 years of schooling, while 10% completed high school and 2% completed college or university (Bettelheim, 1982; Vinovskis, 1992; Walters & James, 1992).

Education in the United States has had a long history of segregation by race, ethnicity, and gender. For instance, African Americans brought forcibly to the United States through slavery were forbidden by law to be educated. Postslavery, African American schools were poorly funded, with fewer books, older buildings, and less-well-paid instructors (Virginia Museum of History and Culture, n.d.). Local Black communities often created their own schools (National Park Service, n.d.). For many Native American students, schooling in the late 19th and early 20th centuries meant being forcibly sent to boarding schools to be assimilated into white culture—to "kill the Indian and save the man" (Carlisle Indian School Project, n.d.). For Hispanic children, separate schools were maintained in large states like Texas and California (Wells, 1989). Importantly, schools for African Americans were segregated by law in the southern states until the U.S. Supreme Court ruled the practice unconstitutional in 1954;

elsewhere in the country, Blacks and whites attended different schools that offered unequal opportunities in education because they lived in different neighborhoods and schools were locally funded (Aviel, 1997; Bergen, 1996).

Girls and boys were also historically educated separately and unequally, with girls receiving training in cooking and homemaking skills and boys studying academic subjects such as literature and mathematics (Riordan, 1990; Tyack & Hansot, 1982). Institutions of higher education discriminated against women, prompting the founding of women's colleges, hospitals, and organizations like the American Association of University Women (AAUW), founded in 1881. AAUW brought women together to advocate for their rights to be educated; created a venue for published research "from an 1885 paper disproving a prevailing myth that college impairs a woman's fertility to, most recently, a study documenting the economic impact of workplace sexual harassment"; and "supported the academic achievements of many thousands of scholars" (https://www.aauw.org/about/history/).

Today, the United States is a **credential society**, one in which access to desirable jobs and social status depends on the possession of a certificate or diploma certifying the completion of formal education (Collins, 1979; Vinovskis, 1995). Since a person's job is a major determinant of income and social class, educational credentials play a major role in shaping opportunities for social and economic mobility. A socially validated credential such as a bachelor's degree or professional degree serves as a filter, determining the kinds of jobs and promotions for which a person is eligible. Regardless of intelligence or skill, those with only high school diplomas have a difficult time competing in the job market with those who have college degrees. For instance, if a position announcement indicates that a college degree is required, the candidate with only a high school diploma is unlikely to be considered at all.

Credential society: A society in which access to desirable work and social status depends on the possession of a certificate or diploma certifying the completion of formal education.

College credentials have become increasingly important over the past 80 years for both women and men in many fields. As Duffin (2020) shows in her comparison graduation rates between 1940 and 2019, over a third of men and women today are college graduates. Women's college graduation rates changed from only 3.8% to 36.6% of women during that time. Men's rates, on the other hand, grew from 5.5% to 35.4% in 2019 (Duffin, 2020). Today, women between the ages of 21–24 hold well over half (57.7%) of the college degrees for their age group (Gould et al., 2019).

The rates of minority students entering college also grew exponentially. In 2019, 59% of Asian students between the ages of 18–24 (or 1.3 million) were enrolled, joined by 37% of Black students (2.6 million), and 36% of Hispanic students (3.7 million) (National Center for Education Statistics, 2020a). To illustrate the significance of this change, consider the statistics cited by African American sociologist W. E. B. Du Bois in 1928 at the National Interracial Conference in Washington, D.C. As you may recall from earlier chapters, Du Bois was the first African American to graduate from Harvard with a PhD in 1895 and contributed significantly to the discipline of sociology:

> Today finds us with the educational part of our convention answered by facts. We have 19,000 [African American] college students, where we had less than 1,000 in 1900, and we are graduating annually 2,000 Bachelors of Arts, when in 1900 we sent out less than 150. It is admitted now without serious question that the American Negro can use modern education for his group development, in economic and spiritual life. (DuBois, 1930, retrieved from http://www.webdubois.org/dbTNCitizen.html)

THEORETICAL PERSPECTIVES ON EDUCATION

Sociologically speaking, what is the role of the educational system in society? For insight, we turn to functionalist, conflict, and symbolic interactionist theories. Functionalists highlight the ways in which the educational system is functional for society; conflict theorists point to its role in reinforcing and reproducing social stratification. Symbolic interactionists illuminate how relational processes in the classroom may contribute to educational success—or failure.

Functionalist Perspectives

As we have noted in other chapters, Émile Durkheim's work (1922/1956, 1922/1973) is foundational to American functionalist theory. Durkheim believed that education played a key role in maintaining a modern (organic) society where people were matched with their skills, worked together harmoniously, and understood their place in society.

According to Durkheim, modern "organic" societies are complex, specialized, and interdependent. This complexity creates a special problem for *social solidarity*—the bonds that unite the members of a social group. These organic societies are no longer characterized by high degrees of cultural, religious, or social likeness, so social ties have weakened. Durkheim believed that mass education addressed this problem by teaching the norms and values necessary to produce and maintain social solidarity. He advocated *moral education*, meaning that educational institutions should not only teach the academic knowledge and training necessary for members to fulfill their economic roles in modern society, but they should also teach a curriculum that would help students understand their role in society. Durkheim thought that education was particularly crucial to children's understanding of how society works. As he put it in his article *Education, Its Natural Role* (1911):

> We thus arrive at the following definition: education is the action exercised by the adult generations over those that are not yet ready for social life. Its purpose is to round and develop in the child a certain number of physical, intellectual and moral states which are demanded of him both by the political society as a whole and by the specific environment for which he is particularly destined . . . it emerges from the foregoing definition that education consists of a methodological socialization of the young generation. (Durkheim, 1911, p. 51, cited by Filoux, 1993)

As Émile Durkheim pointed out, school is a crucial agent of socialization. Among the lessons children learn in school are obedience to authority (such as raising one's hand) and conformity to schedules, imperatives that some sociologists say are rooted in early capitalism's need for compliant workers.

Compassionate Eye Foundation/Robert Daly/OJO Images

Twentieth century functionalist theories echo Durkheim's concerns about social solidarity, emphasizing the function of formal education in socializing people into the norms, values, and skills necessary for society to survive and thrive (Parsons & Mayhew, 1982). Functionalist theory also proposes that education has both manifest and latent functions (Bourdieu & Coleman, 1991; Merton, 1968). What are these functions?

The *manifest* (intended) functions of education include the transmission of the general knowledge and specific skills needed in society and the economy, such as literacy, *the ability to read and write*, and numeracy, *the ability to understand and work with numbers*. The *latent* (unintended) functions include the spread of societal norms and values that Durkheim argued should be explicit concerns of moral education. For example, beginning with kindergarten, children learn to organize their lives according to schedules, to sit at desks, to follow rules, and to show respect for authority. Sociologist Harry Gracey (1991) has argued that "the unique job of the kindergarten seems . . . to be teaching children the student role. The student role is the repertoire of behavior and attitudes regarded by educators as appropriate to children in school" (p. 448). Having "mastered" the student role, children internalize the external social norms and rules that govern the school day and their academic lives.

Consider other latent functions of the system of mass public education in the United States. For example, in keeping children occupied from about 8:00 a.m. until 3:00 p.m., schools serve as supervisors for a large population of children whose parents work to contribute to both the micro-level economies of their homes and the productivity of the macro-level economy. This became much more apparent during the spring of 2020, when schools closed to address the COVID-19 pandemic and parents scrambled to address homeschooling, daycare, and fears about their own job losses. Indeed, mothers were disproportionately impacted by school closures, with preliminary estimates being that women between the ages of 25–44 who were not working during the 2020 spring months of school closures were three times as likely as men of that age to claim they were not working due to childcare demands (Heggeness & Fields, 2020).

Conflict theorists have critiqued what they see as the functionalists' lack of emphasis on how schools may also have the latent function of reproducing social inequality. Functionalist theory assumes, for instance, that the educational system educates people in accordance with their abilities and potential, awarding credentials to those who deserve them and who can contribute most to society while withholding credentials from those incapable of doing the most demanding work. Conflict theorists, however, say that schools function to reproduce the existing class system, favoring those who are already the most advantaged and putting obstacles in the paths of those who are disadvantaged. They point to substantial differences in educational attainment across socioeconomic groups (a topic we take up later in this chapter) as evidence that one's socioeconomic class status is as important as intellectual capability in influencing educational attainment within the institution of education (Bowles & Gintis, 1976).

Notably, although education may socialize students into society's norms and values, it also potentially undermines societal authority by promoting a critical approach to dominant ideas. Education contributes to the development of a capacity for self-direction (Miller et al., 1986), and students often develop inquiring, critical spirits because they are exposed to views and ways of thinking that challenge their previously held ideas. For example, Phelan and McLaughlin (1995) found that people with higher levels of education are more sympathetic to and less likely to blame homeless people for their condition than are people with lower levels of education.

In keeping with the idea of education as having latent as well as manifest effects, consider that education provides *more* than just curricular learning. For instance, for some college students, co-curricular and peer experiences are a significant part of college. In 2016, Pew Research found that well over half (62%) of those who had graduated from 2- and 4-year colleges thought their time in college had helped them grow personally as well as intellectually. Again, this socialization may well support functionalist ideas of education as a moralizing force, but it may contradict them as well.

Based on your experiences in college are you more likely to see college as challenging your former ideas or reinforcing them? And would you agree that college has helped you grow both personally and intellectually?

Conflict Perspectives

Conflict theorists agree that education trains people in the dominant norms and values of society and the work skills and habits demanded by the economic system. Nevertheless, they reject the functionalist assertion that the system of education channels individuals into the positions for which they are best suited in terms of ambition, skills, and talents. Instead, they believe, it reproduces rather than reduces social stratification and, rather than ensuring that the best people train for and conscientiously perform the most socially important jobs (Davis & Moore, 1945), it ensures that the discovery of talent will be limited (Tumin, 1953).

Conflict theorists point out that poor and working-class children have fewer opportunities to demonstrate their talents and abilities because they lack equal access to educational opportunities. Moreover, part of the "hidden curriculum" of the classroom is to socialize members of the working class to accept their class position (Bowles & Gintis, 1976). Children are taught at an early age to define their academic ambitions and abilities in keeping with the social class of their parents. Conflict theorists argue that lowered educational ambitions are reinforced through inferior educational opportunities as well as labeling and discrimination in the classroom (Bowles & Gintis, 1976; Glazer, 1992; Kozol, 1991; Oakes, 1985; Willis, 1990).

Among the most prominent conflict theorists of education are Samuel Bowles and Herbert Gintis, who in their 1976 book, *Schooling in Capitalist America*, made three key points. First, they argued that schools not only impart cognitive skills but also "prepare people to function well and without complaint in the hierarchical structure of the modern corporation" (p. ix). Second, they used statistical data to support the argument that parental economic status is passed on to children, at least in part, through unequal educational opportunity, though the advantages conferred on children of higher-social-status families are not limited to their educational preparation. Third, they suggested that the modern school system was not the product of the evolutionary perfection of democracy applied to education, but that it instead served the interests of raising profits for capitalist enterprises such as factories by teaching pupils to be docile.

Frequently cited results of Bowles and Gintis's early work include Figure 12.1, which shows the powerful correlation between socioeconomic status (as measured by income) and educational attainment. The data, from a sample of men with similar childhood IQ scores in 1962, demonstrates an unmistakable relationship between the socioeconomic backgrounds of the subjects and the average number of years of education they completed.

A precise update of Bowles and Gintis's data is not available; however, other data on family income, demonstrated academic potential, and educational attainment show that they are still related. For example, Figure 12.2 shows bachelor's degree completion rates by both family income (socioeconomic status) and math achievement early in high school. Note that among the highest-scoring students (fourth quartile), 74% of high-income students had graduated from college a decade later, whereas only 41% of low-income students had completed a degree. The graduation gap is consistent across score categories: In every category, students in the middle- and top-income categories were more likely to graduate than were their lower-income counterparts.

Conflict theorists also point out that educational inequality has racialized and gendered dimensions. In terms of race, consider the experience of Malcolm X, a prominent champion of the rights of African Americans, who was assassinated in 1965. In his autobiography, Malcolm X recounts how, despite being a top student, his high school English teacher discouraged him from becoming a lawyer:

> Mr. Ostrowski looked surprised, I . . . remember he kind of half-smiled and said, "Malcolm, one of life's first needs is for us to be realistic. Don't misunderstand me, now. We all like you, you know that. . . . A lawyer—that's no realistic goal for a [Black man]. You need to think about something you can be. You're good with your hands—making things. . . . Why don't you plan on carpentry? People like you as a person—you'd get all kinds of work." (Haley & Malcolm, 1964, p. 41)

FIGURE 12.1 ■ Relationship Between U.S. Family Income and Years of Schooling for White Males Ages 35–44, 1962

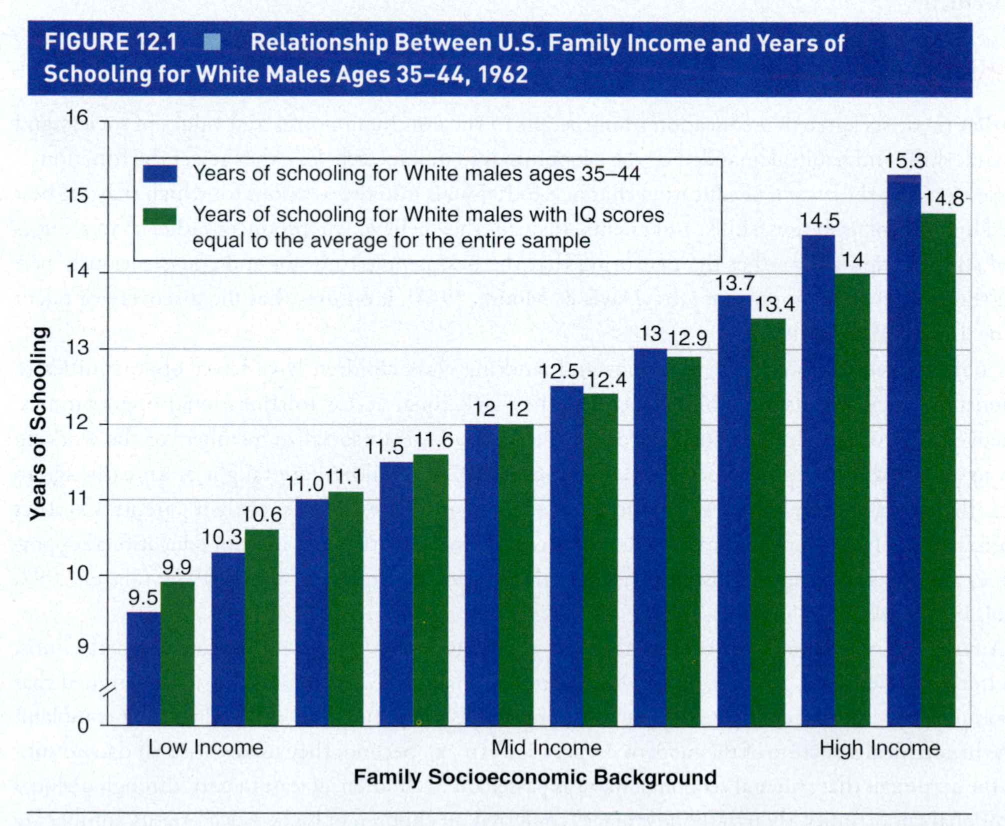

Source: Bowles, Samuel and Herbert Gintis. *Schooling in Capitalist America: Educational Reform and the Contradictions of Economic Life.* Copyright © 1977 by Samuel Bowles, Samuel and Herbert H. Gintis. Reprinted by permission of Basic Books, a member of the Perseus Books Group.

FIGURE 12.2 ■ Percentage of Spring 2002 High School Sophomores Who Earned a Bachelor's Degree or Higher by 2012, by Socioeconomic Status and Mathematics Achievement Quartile in 2002

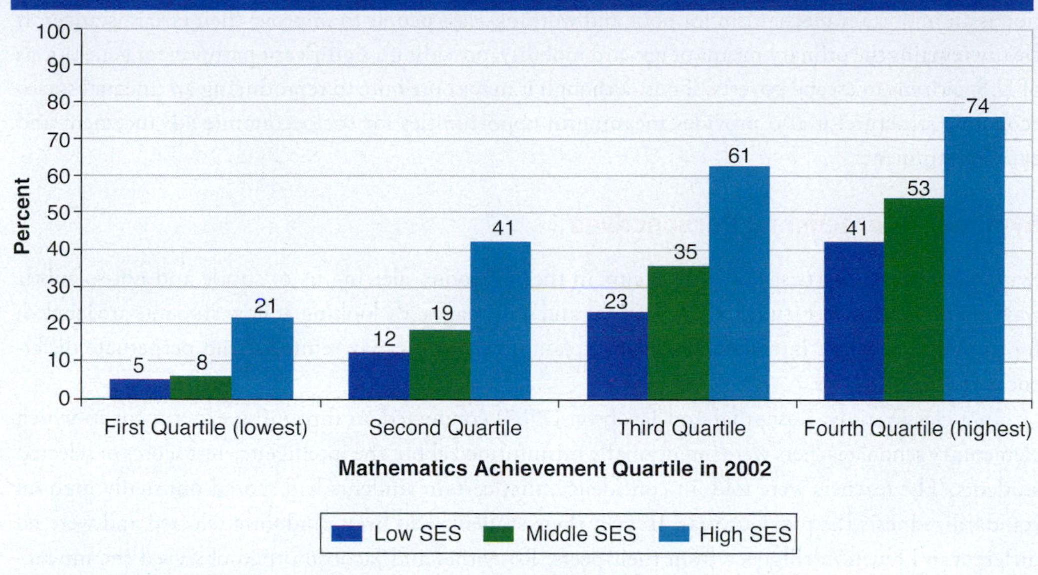

Source: "The Condition of Education." (2012). National Center for Education Statistics, U.S. Department of Education.

More than a half century later, author Jonathan Kozol (2000) documented the "savage inequalities" that poor and minority children continue to experience in terms of lower educational expectations and opportunities, including what they learn in the classroom not to prepare them for their futures, but also to encourage them to feel that they "deserve" different outcomes:

> Many people in Mott Haven [an impoverished neighborhood in the South Bronx] do a lot of work to make sure they are well-informed about the conditions in their children's public schools. Some also know a great deal more about the schools that serve the children of the privileged than many of the privileged themselves may recognize. They know that "business math" is not the same as calculus and that "job-readiness instruction" is not European history or English literature. They know that children of rich people do not often spend semesters of their teenage years in classes where they learn to type an application for an entry-level clerical position; they know that these wealthy children are too busy learning composition skills and polishing their French pronunciation and receiving preparation for the SATs. They come to understand the process by which a texture of enlightenment is stitched together for some children while it is denied to others. They also understand that, as the years go by, some of these children will appear to have deserved one kind of role in life, and some another. (pp. 100–101)

Today, significant disparities in education continue. Globally, for instance, while women around the world are more educated in every country than they were 50 years ago, two thirds of the world's women still have less education than do men (Evans et al., 2020). In the United States, one example of a disparity in secondary education that can influence higher education are Advanced Placement (AP) courses. AP courses provide important opportunities that help prepare students for college-level work and offer a chance to earn college credit even before high school graduation. According to the Education Commission of the States (2016), requiring that all high schools offer AP courses helps decrease the achievement gap between advantaged and disadvantaged schools by offering equality of opportunity. Currently, less than half of all states (20) require that all high schools or school districts offer at least one AP course (or another advanced opportunity course, such as International Baccalaureate [IB] or dual enrollment); this is an improvement since 2007, however, when it was mandated by only 12 states (Education Commission of the States, 2016). Although thousands of students across the country enroll in AP courses every year, not all students have equal access to these accelerated and challenging classes:

A ProPublica 2018 study, for instance, showed that nationwide, white students are almost twice as likely as Black students to be enrolled in AP courses (https://projects.propublica.org/miseducation).

Critics of conflict theories of education point out that even in highly stratified societies, education is an important mechanism for poor and working-class people to improve their circumstances. It clearly remains the primary means of upward mobility, providing a significant pathway for generations of U.S. citizens to escape poverty. Thus, although it may contribute to reproducing an unequal socioeconomic structure, it also provides meaningful opportunities for socioeconomic advancement and career attainment.

Symbolic Interactionist Perspectives

Symbolic interactionists study what occurs in the classroom, alerting us to subtle and not-so-subtle ways that schools affect students' interactions and self-images. By looking at how students are labeled, for instance, symbolic interactionists help explain how schools may reinforce and perpetuate differences among students.

In a classic study, Rosenthal and Jacobson (1968) conducted an intriguing experiment in which elementary school teachers were intentionally misinformed about the intelligence test scores of selected students. The teachers were told, in confidence, that certain students had scored unusually high on standardized tests the previous year. In fact, these students had been randomly selected and were no different in known intelligence from their peers. Rosenthal and Jacobson then observed the interactions between these students and their teachers and monitored the students' academic performance. The students labeled *exceptional* soon outperformed their peers, a difference that persisted for several years.

The teachers described the labeled students as more curious or more interested and communicated their heightened expectations of these learners through their voices, facial expressions, and use of praise. Enacting a self-fulfilling prophecy, the students came to see themselves through their teachers' eyes and began performing as if they were, in fact, more intelligent than their peers, earning still more positive attention from teachers. Younger students, whose self-images were more flexible, exhibited the greatest improvements in performance. Rosenthal and Jacobson concluded that the teachers behaved differently toward some students because the students had been labeled *exceptional*.

Subsequent studies have explored the impact of teachers' expectations and the labeling process on students' performance and potential. Some examples include:

- A study of student–teacher interaction in a predominantly African American kindergarten found that such labels as *fast* and *slow*, which the teacher assigned by the eighth day of class, tended to stay with the labeled students throughout the year (Rist, 1970). Further research supported the presence of harmful effects of teachers' stereotypical beliefs about minority students' competence on students' performance (Garcia & Guerra, 2004).
- A study of gender, race, and grades found that female and Asian American students frequently received classroom grades higher than their actual test scores, while Hispanic, Black, and white males received lower grades (Farkas et al., 1990a, 1990b). The researchers explained the differences by the teachers' perceptions of their students' attitudes. Those who appeared to be attentive and cooperative were judged to be hard workers and good students and were graded up; those who appeared indifferent or hostile were graded down (Rosenbloom & Way, 2004).
- A decade-long study of 10th graders (2002–2012), *The Education Longitudinal Study*, found teachers' expectations and students' college-going outcomes were significantly related, with teacher expectations predictive of college completion rates. In fact, after controlling for student demographics, teacher expectations were more predictive of college success than many major factors, including student motivation and student effort.
- A study of teachers' expectations of African American students' potential for college found that both non-Black and Black teachers had lower expectations for their Black students,

especially Black male students. However, in general, Black teachers' expectations were 30% to 40% higher than non-Black teachers' expectations (Gershenson, Holt, & Papageorge, 2015).

- A comparison study of students who had been diagnosed with mild and severe Attention Deficit Hyperactivity Disorder (ADHD) compared to those who had not been so diagnosed, supported that teachers were more likely to apply the labeling process to those diagnosed with mild ADHD, giving them poorer ratings on their social and academic behaviors than their students who were not so diagnosed. On the other hand, those with severe ADHD received teacher interventions that the researchers saw as beneficial (Owens, 2020).

Student labeling can also affect disciplinary actions taken when students misbehave. Even as early as preschool, Black students are four times as likely to be suspended as white students, and twice as likely to be expelled (Young, 2016). Black students have been found to receive harsher punishments for similar infractions when compared to other students, such as longer suspension (Harriot, 2017).

Classroom labeling has been studied in other countries as well. In one influential study, Paul Willis (1990) found that British boys from working-class families were systematically labeled as low academic achievers and socialized to think of themselves as capable of doing only working-class jobs. The boys understood quite well that this labeling process worked against them, and they resisted it by the use of humor and other challenges to authority. These behaviors reinforced their teachers' perception that the boys would never make it and would eventually drop out of school and assume their "rightful position" in the working class. The boys thus accepted their teachers' labeling, creating a self-fulfilling prophecy in which they wound up in working-class jobs.

How teachers interact with their students and their expectations of them can play an important role in a student's future success. Increasingly, these interactions may be mediated by technologies. How do you think these technologies influence the possibility of student labeling?

Ariel Skelley Digital Vision

Symbolic interactionism is well suited to studying the ways in which teachers and administrators consciously or unintentionally affect their students, but critics note that because it focuses on social interaction, it does not provide a picture of the role of the educational system in society as a whole or uncover and illustrate how structural problems such as public policies create unequal access of educational resources. Furthermore, many other factors are likely to impact students over the course of their education, and students, themselves, have some agency to accept or reject labels (Boronksi & Hassan, 2020). Given all of these insights on education, which approach do you see as the most applicable to your educational experiences?

EDUCATION, OPPORTUNITY, AND INEQUALITY

Functionalists argue that education is a vehicle for mobility and for filling the positions necessary for society to survive and thrive. Conflict theorists point out that the educational system reinforces existing inequalities by unequally according opportunities based on class, race, or gender. In fact, the U.S. educational system may operate to open avenues to mobility for all students *and* to create obstacles to achievement among the less privileged. That is, it may both reduce and reproduce inequality.

Sociologists continue to be interested in all facets of education. As the American Sociological Association (ASA) section on education notes: "Sociologists of education examine the ways in which formal schooling influences individuals and the ways society affects educational institutions." Importantly, the ASA educational section statement also points out that a "central concern within this research tradition is the process by which educational systems contribute to or alleviate social inequality in broader society" (American Sociological Association, n.d.). In the following sections, we look at issues of education and inequality, focusing on questions about childhood and adult literacy, school segregation by race and income, and the college dropout phenomenon. We conclude with a look at the relationship between education and income in the United States.

Word Poverty and Literacy

Researcher Louisa Cook Moats uses the term *word poverty* to characterize the impoverished language environments in which some children grow up. Word poverty is a particular problem in economically disadvantaged homes: Research on a community in California found that by age 5, children in impoverished language environments had heard 32 million fewer words spoken to them than the average middle-class child. Perhaps not surprisingly, the fewer words that were spoken to children, the fewer they could actively use themselves: In a study of how many words children could produce at age 3, "children from impoverished environments used less than half the number of words already spoken by their more advantaged peers" (Wolf, 2008, pp. 102–103).

Word poverty is also linked to a deficit of books in the homes of many children. Research conducted in three Los Angeles communities found that in the most economically impoverished community in the study, it was common to find no children's books in the home. In low- to middle-income homes, an average of three books could be found. By contrast, in the most affluent families, there was an average of 200 books in each home (Wolf, 2008). According to a global study, the deficit or wealth of books in a home is of significance in children's later schooling: Being raised in a home without books is as likely to affect children's educational attainment as having parents with very low educational attainment. The researchers conclude that "growing up in a home with 500 books would propel a child 3.2 years further in education, on average" (Evans et al., 2010, p. 179). In the United States, the advantage to having an expansive library is equal to an average of more than 2 years of education. Commenting on the study, an article on the website ScienceDaily noted,

> The researchers were struck by the strong effect having books in the home had on children's educational attainment even above and beyond such factors as education level of the parents, the country's GDP [gross domestic product], the father's occupation or the political system of the country. (University of Nevada, Reno, 2010, para. 10)

Why is word poverty significant? The answer is that it is linked to low literacy. A fundamental necessity for any country in the modern world is a population that is functionally literate.

Literacy is also necessary for individual success in education and the job market. The Program for the International Assessment of Adult Competencies (PIAAC) defines *literacy* as "understanding, evaluating, using and engaging with written text to participate in society, to achieve one's goals and to develop one's knowledge and potential" (Rampey et al., 2016, p. 2). PIAAC identifies six categories of literacy, using a numerical scale (Figure 12.3). The PIAAC study, which was based on a sample of 8,670 people (even though the number in the cited figure is slightly smaller, as it excludes those over age 65), found the following:

- In 2017, 19% of adults were at or below Level 1 on the literacy test. At Level 1, respondents performed simple tasks such as locating information in a text. At this level, "Only basic vocabulary knowledge is required, and the reader is not required to understand the structure of sentences or paragraphs or make use of other text features.
- Another 33% of respondents reached Level 2 on the test. "Tasks at this level require respondents to make matches between the text and information and may require paraphrasing or low-level inferences."
- About 35% of respondents were at Level 3. At this level, readers were expected to read "dense or lengthy" texts and to interpret, evaluate, and infer information. "Competing information is often present, but it is not more prominent than the correct information."
- About 14% of respondents were at Levels 4 and 5. At these levels, readers were examining, taking account of subtle rhetorical clues, evaluating evidence-based arguments, and applying abstract ideas (PIAAC Proficiency Levels for Literacy, 2020) (Rampey et al., 2016, p. B-3).

Wolf (2008) suggests that a strong foundation in literacy is a key to later educational success. Students who enter high school with shaky foundations in literacy have a higher probability of school

FIGURE 12.3 ■ Literacy Levels of U.S. Adults Ages 16-74 in Households and Prisons, 2017

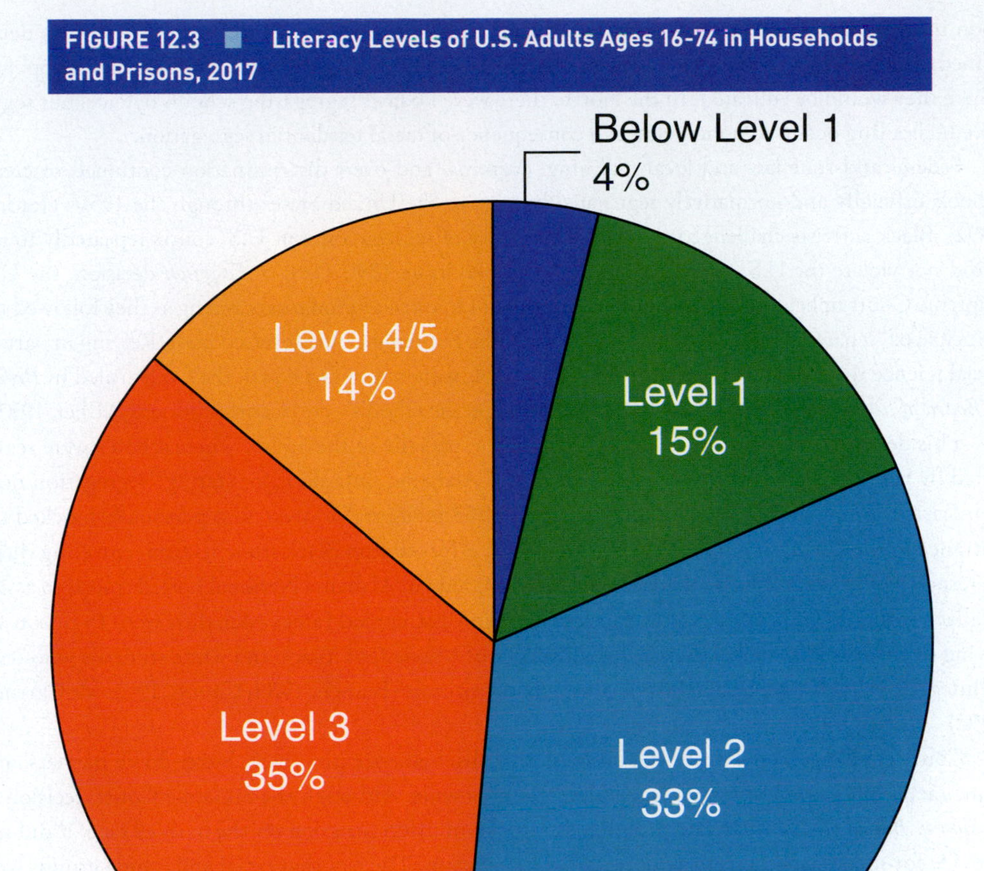

Source: U.S. Department of Education, National Center for Education Statistics, Statistics Canada, and Organisation for Economic Co-operation and Development, Program for the International Assessment of Adult Competencies. (2020). *PIAAC 2012/2014 and PIAAC 2017 literacy, numeracy, and problem-solving TRE assessments.* https://nces.ed.gov/surveys/piaac/ideuspiaac

failure than do their peers who grew up in homes with books and parents who read to them at an early age. At the same time, failure to complete high school is likely to affect someone's ability to complete literacy activities successfully. If you were to research the correlation between literacy and high school completion, how would you proceed? What variables would you choose to study and why? What kind of relationship would you hypothesize?

School Segregation

School segregation, the education of racial minorities in schools that are geographically, economically, and/or socially separated from those attended by the racial majority (usually whites), is a long-standing pattern that is worsening today, despite more than four decades of civil rights legislation intended to alleviate it and reduce its devastating effects. School segregation has long been linked to educational inequality in the United States.

School segregation: The education of racial minorities in schools that are geographically, economically, and/or socially separated from those attended by the racial majority.

Before slavery was abolished in the United States, it was, as we earlier noted, a crime to teach enslaved people to read and write, and formal education was reserved solely for whites. Following the abolition of slavery and the end of the Civil War, Black students could be educated, but Jim Crow laws soon initiated a century of discrimination against Black Americans in the South. These laws determined, among other things, where Black Americans could live, where they could eat and shop, and where they would be educated. In the North, there were no laws segregating schools by race, but segregated schooling occurred nonetheless as a consequence of racial residential segregation.

Federal and state law and local redlining, customs, and overt discrimination continued to create schools officially and normatively segregated by race in the United States through the 1950s (Jordan, 1992). Black activists challenged the constitutionality of segregation, but U.S. courts repeatedly found it did not violate the U.S. Constitution. For example, in its 1896 *Plessy v. Ferguson* decision, the U.S. Supreme Court upheld the states' rights to segregate public accommodations as long as they followed the principle of "separate but equal." In 1954, however, the Supreme Court reversed itself. Relying in part on social science research showing that segregated schools were not in fact equal, the Court ruled in *Brown v. Board of Education of Topeka* that laws segregating public schools were unconstitutional (Miller, 1995).

This decision met with considerable resistance, especially in the South, where schools were segregated by law. Indeed, Governor George Wallace of Alabama, infamous for stating "segregation now, segregation tomorrow, segregation forever" in his 1963 inaugural address, also personally blocked the entrance to the University of Alabama in 1963 in an effort to stop Black students from enrolling there. In response, President John F. Kennedy called in 100 federal Alabama National Guard troops to assist. Wallace stood down (Bell, 2013). A Black college student named James Meredith went to prison for trying to enroll and attend classes at the University of Mississippi. Black and white students who tried to integrate schools were beaten by police and fellow citizens (Branch, 1988; Chong, 1991; McCartney, 1992).

Court challenges, civil protests, and mass civil disobedience ultimately broke down barriers, and some racial integration of schools took place country-wide. Although the Supreme Court decision in *Brown v. Board of Education* had prohibited purposeful discrimination on the basis of race, it did not provide for specific methods to achieve school integration. The fact that racial and ethnic groups were residentially segregated meant that most Blacks would continue to attend schools that were predominantly Black, while whites would continue to attend mostly white schools.

Subsequent Court decisions provided one method of achieving integration: school busing, a Court-ordered program of transporting public school students to schools outside their neighborhoods. Mandated busing proved highly controversial, provoking criticism among some academics and hostility among many parents and policy makers. Controversy erupted in 1974 when Black students were bused into poor Irish neighborhoods in South Boston, whose schools were among the worst in the state. Instead of providing equal educational opportunity, busing worsened racial conflict in some of Boston's most economically disadvantaged neighborhoods. Violence resulted, and over the next 10 years, public school enrollment in the city plummeted (Frum, 2000).

Today, despite decades of civil rights activism and laws aimed at promoting integration, racial segregation persists in U.S. schools, and in some places, it has even worsened. An article on schools in Louisville, Kentucky, which have gained attention for their desegregation efforts, points out that

> [N]ationwide, in 1954, zero percent of black students attended majority-white schools. By 1972, that number was 36.4 percent. School integration reached its peak in 1988, when 43.5 percent of black students attended majority-white schools, but that number has declined since then, and in 2011, stood at just 23.2 percent. (Semuels, 2015, para. 64)

How is this possible? First, the movement over time of middle- and upper-class whites into largely white school districts in suburban or outlying areas left mostly poor minorities in U.S. inner-city schools, many of which are highly segregated (Coleman et al., 1982; Kozol, 2005; Orfield & Eaton, 1996). Because they are often located in low-income neighborhoods, highly segregated schools also tend to be the most poorly funded. In fact, a recent study by EdBuild found that "students who are enrolled in nonwhite school districts receive $23 billion less in funding than those enrolled in white districts despite serving the same number of students (https://edbuild.org/content/23-billion#CA, 2020). Importantly, while state formulas determining school funding vary, most U.S. school districts still depend most heavily on local property tax revenues. Although this system ensures that those who live in areas with high property values will generally accrue adequate—or even excellent—funds for the academic programs and physical maintenance of their schools, it also puts those who live in lower-income rural and urban areas at a distinct disadvantage, since even high property tax rates cannot bring in the level of resources that schools in middle- to upper-class areas enjoy (Ball et al., 1995; Kozol, 2005).

Second, U.S. Supreme Court decisions such as the 1991 *Board of Education of Oklahoma City v. Dowell*, the 1992 *Freeman v. Pitts*, and the 1995 *Missouri v. Jenkins* limited the scope of previous laws aimed at promoting racial integration of schools. The Court ruled that segregated schools resulting from "residential preferences" are a result of people making choices about where to reside and are therefore beyond the scope of the law. The Court declared that school districts that previously had made an effort to integrate schools could send students back to neighborhood schools even if those schools were segregated and inferior (Orfield & Eaton, 1996).

Hispanic and Black students are more likely to be in segregated schools today than in earlier decades. Public schools in some cities are much more likely to be attended by minority children than white children. In Chicago in 2019, for instance, 46.6% of students enrolled in public schools were Hispanic, 35.9% were Black, and only 10.8% were white (Masterson, 2019). In Washington, D.C., 2017–2018 public school enrollment was 60% African American, 20% Hispanic, 15% white, and 5% "some other race" (DC Public Schools, n.d.). According to the New York City Council, in 2019 "in New York City public schools, 74.6% of black and Hispanic students attend a school with less than 10% white students. Additionally, 34.3% of white students attend a school with more than 50% white students" (https://council.nyc.gov/data/school-diversity-in-nyc/). Figure 12.4 shows the concentration of students by race in public schools in the United States in 2017. Significantly, in 2017, the amount of white students enrolled in public schools had declined in all 50 states, with some, like Nevada, seeing a large drop (24%) (National Center for Education Statistics, 2019a).

Asian Americans, as we have noted in Chapter 9, are a very diverse group with family roots in countries as varied as, among others, China, India, the Philippines, Japan, Vietnam, and South Korea. Although these wide variations exist, in general Asian Americans are "the most likely of any major racial/ethnic group to live in mixed racial neighborhoods and marry across racial lines" (Tang, 2016). More segregated communities such as "Chinatowns" and "Little Saigons," which are populated with

FIGURE 12.4 ■ Racial/Ethnic Enrollment in Public Schools

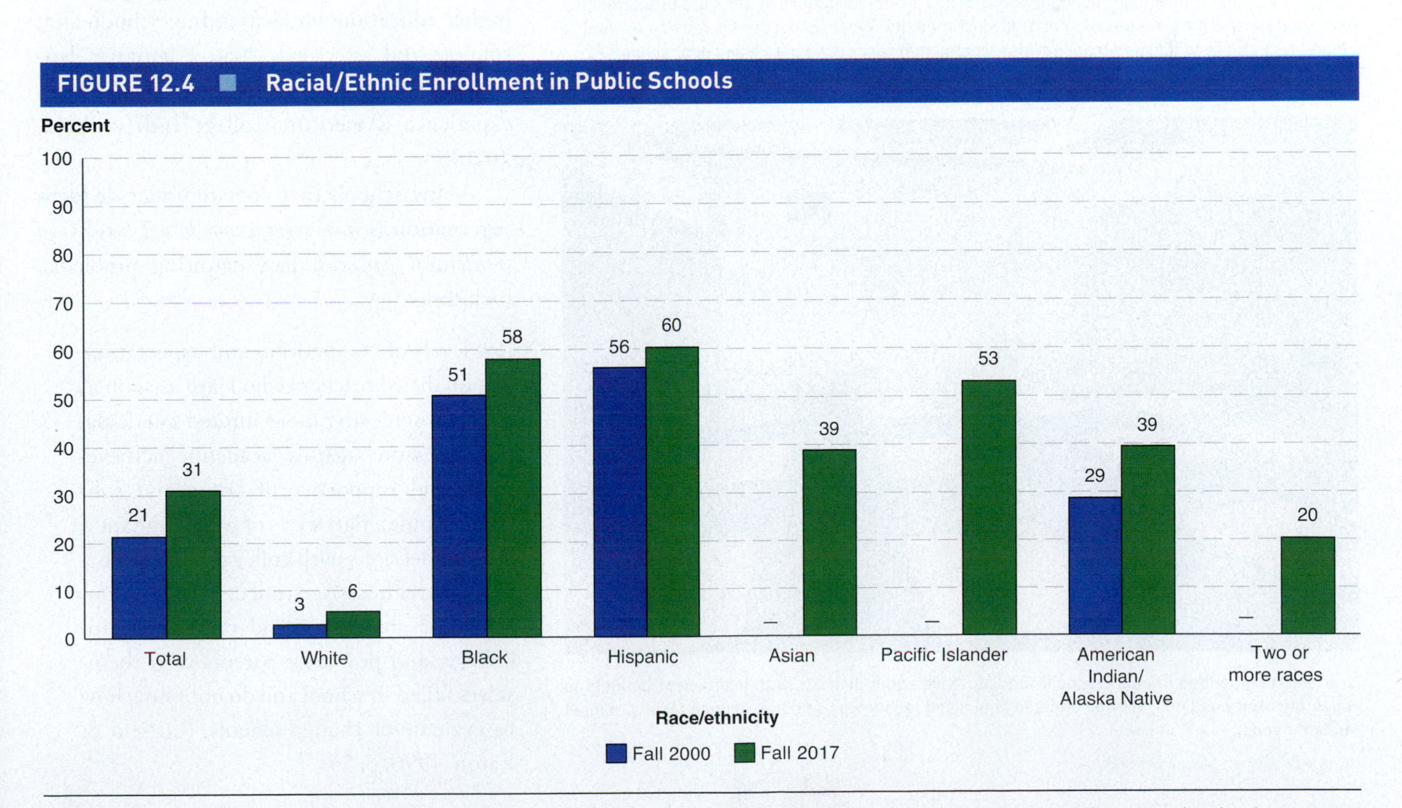

Source: Digest of Education Statistics. (2019). *Racial/Ethnic Enrollment in Public Schools.* Retrieved from https://nces.ed.gov/programs/coe/pdf/coe_cge.pdf

recent immigrants from China, Vietnam, South Korea, and other Asian countries, are an exception, and integration becomes more common in later generations (Chen, 1992; Loo, 1991; Zhou, 2009). By contrast, then, most Asian American students are integrated into schools with whites, and, as we show throughout this chapter, tend to have outcomes that mirror or exceed white students (with intragroup variation).

American Indian and Alaska Native high school students face unique educational attainment challenges influenced by factors such as high rates of poverty and disproportionately high rates of suicide (Camera, 2015; Mongeau, 2016). Many students attend schools on reservations. These schools are administered by the Bureau of Indian Education (BIE), which employs teachers and sets the curriculum for 183 elementary and secondary schools situated across 23 states (BIA.gov). Students in these schools have significantly lower graduation rates than other U.S. high school students (53% vs. 85%) (National Indian Education Center, n.d.). On the other hand, American Indian and Alaska Native high school students' public school graduation rate of 73% is also significantly lower than the overall nationwide high school graduation rate.

De facto segregation: School segregation based largely on residential patterns, which persists even though legal segregation is now outlawed in the United States.

Across the country, unequal funding of public schools and unequal resources are problems that perpetuate disparities in education and upward mobility. What resources do you see as vital to elementary and secondary educational classrooms that vary between rich and poor schools?

©Peter Noyce PLB/Alamy Stock Photo

Dramatic inequalities in public school funding leave some districts with insufficient budgets to repair decrepit buildings, provide updated books and technology, and hire enough faculty to meet student needs.

©David Grossman / Alamy Stock Photo

Treaties between the U.S. government and Indian nations have sought to ensure education that recognizes the value of Native American culture and tradition. Yet the teachers employed by the Bureau of Indian Affairs are ordinarily expected to cover the standard subjects of U.S. school curricula—and in English. Problems within the tribal communities have resulted, since Native Americans often see such instruction as failing to respect their linguistic and cultural differences. The BIE also funds 26 tribal colleges and universities (BIA.gov). Tribal colleges are often located on or adjacent to reservations and allow students to pursue higher education while attending schools that support and interweave Native heritage, language, and cultural practices into their college experience (American College Indian Fund, 2020).

Many schools that operate under **de facto segregation** (*school segregation based largely on residential patterns*) face daunting problems, including

> low levels of competition and expectation, less qualified teachers who leave as soon as they get seniority, more limited curricula, peer pressure against academic achievement and supportive of crime and substance abuse, high levels of teen pregnancy, few connections with colleges and employers who can assist students, less serious academic counseling and preparation for college, and powerless parents who themselves failed in school and do not know how to evaluate or change schools. (Orfield & Eaton, 1996, p. 54)

BEHIND THE NUMBERS: MINORITY STUDENTS' COLLEGE ENROLLMENT

As minority students in colleges nationwide grow, Black, Hispanic, and Native American students are still underrepresented, especially in elite colleges (Table 12.1).

TABLE 12.1 ■ College Enrollment Rates, 2017

College Participation Rates for 18- to 24-Year-Olds	Percentage
Total, all students	40
American Indian/Alaska Natives	20
Asian Americans	65
Black Americans	36
Hispanic Americans	36
Pacific Islander	33
White Americans	41

Source: National Center for Education Statistics. (2020a, May). *College Enrollment Rates.* Retrieved from https://nces.ed.gov/programs/coe/indicator_cpb.asp

Indeed, according to the *New York Times,* Black and Hispanic students are more underrepresented at these institutions than they were 35 years ago (Ashkenas et al., 2017).

What is behind the numbers of lower minority student enrollment at elite colleges?

- *Inadequate preparation for college due to underfunded, segregated schools.* Segregated schools tend to have fewer experienced teachers, a smaller selection of advanced courses, and inadequate resources (Ashkenas et al., 2017).
- *Undermatching.* High-achieving, economically disadvantaged minority students tend to lack sufficient guidance counselor assistance, awareness of opportunities, and college recruiter attention (Harris, 2018; Hoxby & Avery, 2013).
- *Admission practices of elite schools,* such as standardized test requirements (ACT and SAT), may make counselors more likely to steer minority students to less-selective schools (McGill, 2015). A perceived *lack of a culture of inclusion* may also decrease enrollment (McGill, 2015).
- *Legacy admissions* replicate old patterns of segregation. For instance, Harvard's records of 2009–2014 admissions showed 43% of white students admitted benefited from a status such as legacy, athlete, or connection to a donor, while only 16% of Black, Latinx, and Asian students received this consideration (Hu, 2020, para. 4).

Given what you have learned in this chapter, what other factors might influence enrollment differences? What methods would you use to understand enrollment differences between racial groups? Which theoretical perspectives would be most useful in framing a research study about such differences?

ISSUES IN U.S. HIGHER EDUCATION

Nearly three quarters of U.S. high school graduates will enroll in college believing that higher education will provide them jobs and higher earnings. Is this true? Next, we look at the relationship between education, employment, and earnings. We also examine college internships, a phenomenon that generates both praise and criticism. We end with a discussion of the college dropout phenomenon, its dimensions, probable causes, and possible cures.

Education, Employment, and Earnings

Does higher education translate into job stability? The short answer is that it indeed appears to make it more likely. Let's examine the correlation between educational attainment and the labor force participation rate, a measure of the proportion of those of working age (usually 16–64) who are either employed or unemployed and actively seeking work. The labor force participation rate does not include those who are institutionalized (in prison, for instance) or those who are serving in the armed forces; other groups who are not participating in the labor force include many full-time students and homemakers.

In 2019, labor force participation rates of adults varied significantly by education: 45% of those with less than a high school education were in the labor force, 57% of those with a high school education were employed, and 64% of those with some college or an associate degree were working. For adults with a bachelor's degree or higher, labor force participation rates rose to 73% (Current Population Survey, Annual Social and Economic Supplement, 2010–2019, https://www.census.gov/library/visualizations/2020/comm/labor-force-by-education.html). Figure 12.5 shows the labor force participation by level of education since 2010. What patterns do you see reflected in these data?

Furthermore, educational attainment and income and educational attainment and vulnerability to unemployment are both strongly correlated (see Figures 12.6 and 12.7). As we have noted, educational attainment grew more important in the postindustrial era, as the living wage jobs of the industrial era in sectors such as automobile manufacturing, steel, and textiles fell victim to outsourcing and automation. The foundation of the U.S. economy is now based on advanced professional occupations (which require higher education and often even graduate degrees) and service jobs (which are often part-time, low-pay, and low-benefit positions in sectors such as child and elder care, retail, and hospitality). Although some manufacturing still exists in the United States, new manufacturing jobs are far less likely to be unionized, are more likely to involve short-term contracts, and are characterized by a lower pay scale than were similar jobs in the past. Several new manufacturing jobs demand proficiency with high-technology equipment, which may require either a college education or other advanced training beyond high school.

In 2018, according to the Center for Education Statistics, about 35% of 25- to 29-year-olds had earned bachelor's degrees and had median earnings across all fields of $50,600 with variation depending on disciplinary field (Figure 12.7). Interestingly, though, their average median earnings,

FIGURE 12.5 ■ U.S. Labor Force Participation by Level of Education

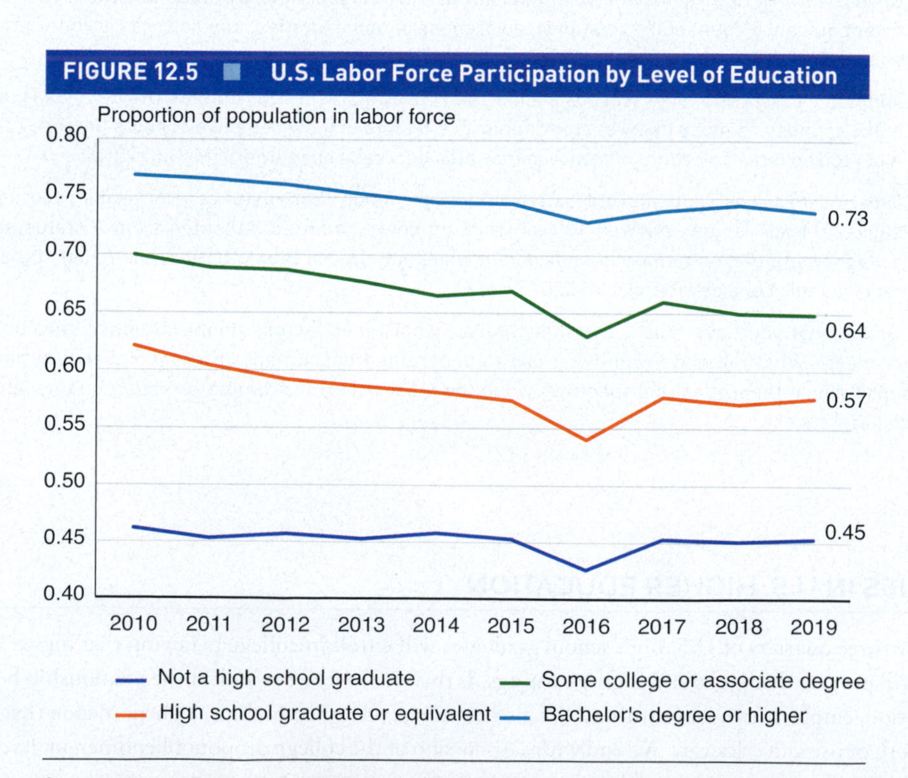

Source: United States Census Bureau. (2020). *Current Population Survey: Annual Social and Economic Supplement, 2010–2019.* Retrieved from https://www.census.gov/library/visualizations/2020/comm/labor-force-by-education.html

FIGURE 12.6 ■ Real Average Hourly Wages of Young College Graduates, 2019

Wages of young college graduates today are just above their 2000 levels

Real hourly wages for employed young college graduates (ages 21–24) not enrolled in further schooling, 1989–2019 (2018$)

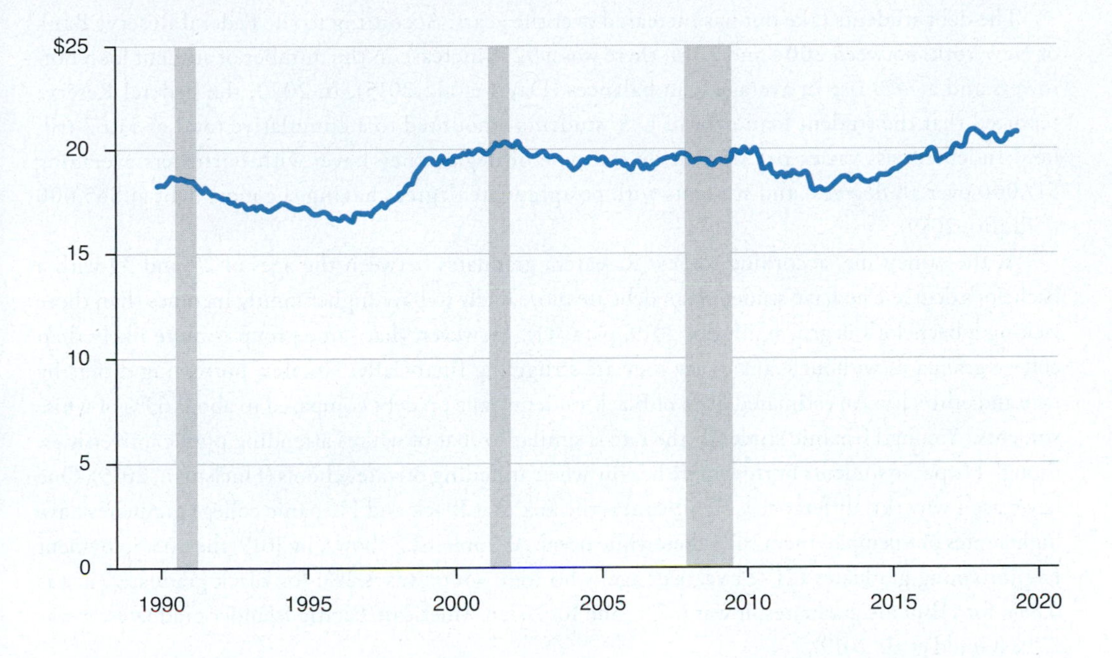

Source: Gould, E., Mokhiber, Z., & Wolfe, J. (2019, May 14). *Class of 2019*. Figure H. Washington, DC: Economic Policy Institute. Reprinted with permission.

Note: Data are for high school graduates ages 17–20 and college graduates ages 21–24 who are not enrolled in further schooling. Wages are in 2018 dollars.

FIGURE 12.7 ■ Median Annual Earnings of 25- to 29-Year-Old Bachelor's Degree Holders, by Selected Fields of Study, 2010 and 2018

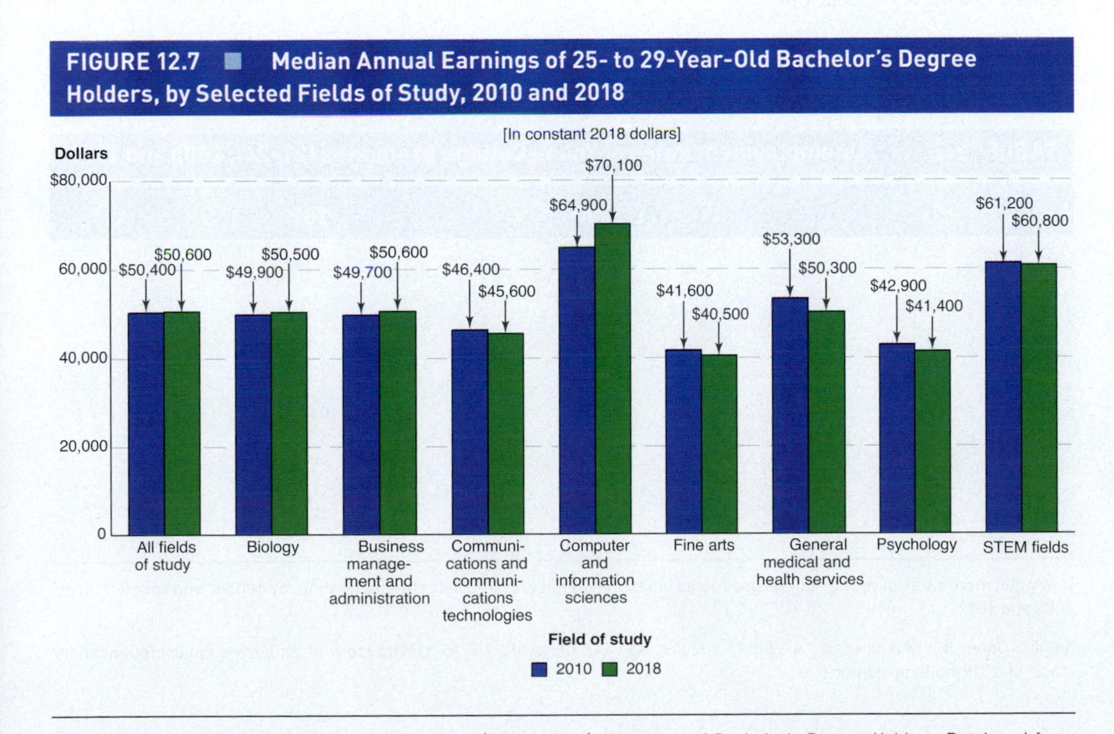

Source: National Center for Education Statistics. (2020b, May). *Outcomes of Bachelor's Degree Holders*. Retrieved from https://nces.ed.gov/programs/coe/indicator_sbc.asp

adjusted to 2018 dollars, had remained almost unchanged since 2010 (National Center for Education Statistics, 2020a). The average unemployment rate varied by gender and race (Table 12.2).

Importantly, though, while college graduates may have higher incomes, one common experience that at least a third of graduates share that can offset some of these gains is student loan debt (Cilluffo, 2019). According to the National Center for Education Statistics, the percentage of students taking out loan debt in 2016–2017 was close to 50% (McFarland et al., 2019).

The debt students take out has increased over the years: According to the Federal Reserve Bank of New York, between 2004 and 2014, there was a 92% increase in the number of student loan borrowers and a 74% rise in average loan balances (Davis et al., 2015). In 2020, the Federal Reserve reported that the student loan debt of U.S. students amounted to a cumulative total of $1.67 trillion! Indebtedness varies per student by the type of degree they have, with borrowers averaging $17,000 over all degrees, and students with postgraduate degrees having the most debt at $45,000 (Cilluffo, 2019).

At the same time, according to Pew Research, graduates between the ages of 25 and 39 with a bachelor's degree who have student loan debt are more likely to have higher family incomes than those lacking a bachelor's degree (Cilluffo, 2019, para. 11). However, that same group is more likely than college graduates without loans to say they are struggling financially. Notably, borrowing differs by race and ethnicity: An estimated 80% of Black students take on debt compared to about 63% of white students. Among Hispanic students, the rate is similar to that of whites attending public universities, though Hispanic students borrow more heavily when attending private schools (Huelsman, 2015). One key reason why this difference is significant is the fact that Black and Hispanic college graduates have higher rates of unemployment than their white peers: As Table 12.2 shows, in 2019, the unemployment rate for young graduates (21–24 years of age) who were white was 4.2%; for Black graduates, it was 6.8%; for Hispanic graduates, it was 6.7%; and for Asian American/Pacific Islander graduates, it was 7.7% (Gould et al., 2019).

Given the slow wage growth and the rise of student debt, is higher education worth it? The assertion that higher education opens doors to employment and higher lifetime earnings is an important counterbalance, and it remains essentially correct. According to the Bureau of Labor Statistics (2019), the higher the education, the higher the average level of income and the lower the rate of unemployment (Figure 12.8). What other patterns do you see in these data? What other benefits do you see in obtaining a higher education?

TABLE 12.2 ■ Unemployment Rates of Young College Graduates (21–24), 2000 and 2019

	2000	2019
Women	3.9%	4.1%
Men	4.7%	6.4%
White	4.0%	4.2%
Black	5.6%	6.8%
Hispanic	4.8%	6.4%
AAPI	5.9%	7.7%

Unemployment rates of young college graduates (ages 21–24) not enrolled in further schooling, by gender and race/ethnicity, 2000 and 2019

Source: Gould, E., Mokhiber, Z., & Wolfe, J. (2019, May 14). *Class of 2019.* Retrieved from https://www.epi.org/publication/class-of-2019-college-edition/

FIGURE 12.8 ■ 2019 Earnings and Employment Rate by Education Level

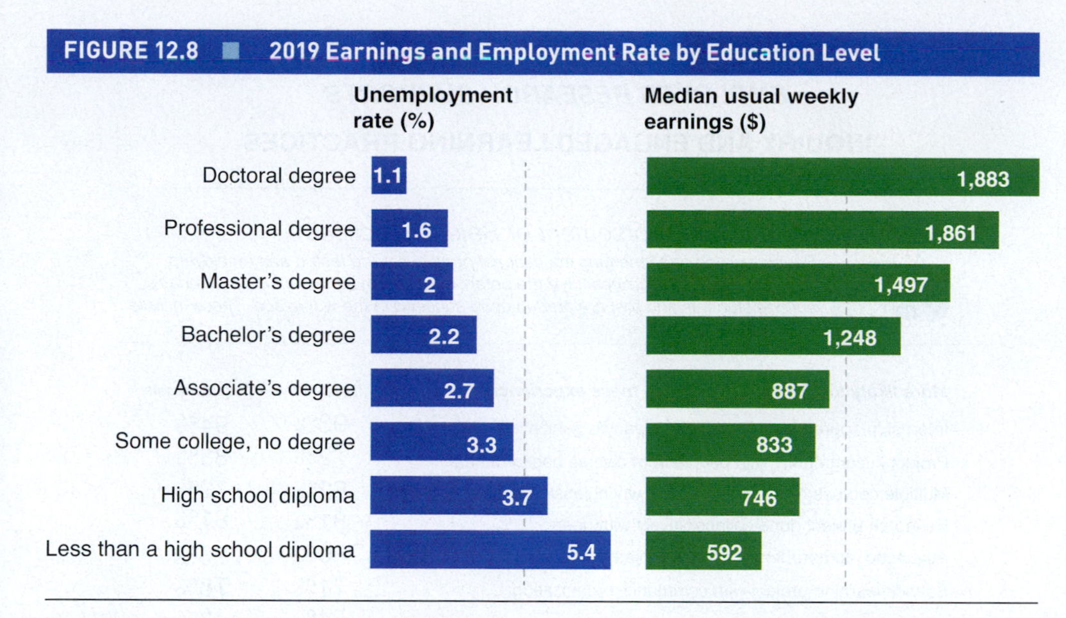

Source: U.S. Bureau of Labor Statistics. (2020, May). Learn More, Earn More: Education Leads to Higher Wages, Lower Unemployment. *Career Outlook*. Retrieved from https://www.bls.gov/careeroutlook/2020/data-on-display/education-pays.htm

Internships and Higher Education

Thousands of U.S. college students build their résumés and job experience with internships. A 2014 survey of 43,000 graduating seniors at about 700 universities found that 61% had an internship or co-op experience during college; most of those internships (over 53%) were unpaid (Venator & Reeves, 2015).

Many universities require students to complete an internship to earn credit toward a major or graduation. Internship sites have become increasingly flexible with internships. For instance, during the academic year 2020–2021, which was impacted by COVID-19, remote internships, including local, national, and international, became increasingly common (NACE staff, 2020b). During the preceding summer, 46% of employers who responded to a poll by the National Association of Colleges and Employers said that their internships would be moving internships online (NACE staff, 2020c).

While some internships, particularly in business, engineering, and financial services, offer stipends or pay, most do not. Additionally, more students than ever are also working in unpaid internships *after* graduation, having been unable to secure paid entry-level positions, even with their degrees. Both students and their universities often believe internships are a vital component of building human capital in preparation for the world of paid and professional work. Indeed, a 2018 American Colleges and Universities survey of U.S. employers found that internship experience was a significant factor in hiring a new employee (Figure 12.9).

Internships, both paid and unpaid, have become a staple of summer or a part of the regular academic year for students across the United States.

©Wendy Maeda/The Boston Globe via Getty Images

Racial differences are present in who gets an internship and whether or not that internship is paid or unpaid, according to the 2019 National Association of Colleges and Employers Student Survey Report (NACE staff, 2020a). According to NACE, white, Asian American, and international students on F1 visas are more likely to have paid internships while first-generation, women, African American, Hispanic, and multiracial students are more likely to have unpaid internships (NACE staff, 2020a). Additionally, first-generation students and Hispanic students are among those who are most likely *not* to have any internship opportunity (NACE staff, 2020b).

FIGURE 12.9 ■ Employer's Desired Characteristics in College Graduates

EMPLOYER RESEARCH SUPPORTS

INQUIRY AND ENGAGED LEARNING PRACTICES

Employer Endorsement of Select Practices

Seven existing and emerging educational practices were tested and employers believe that these practices have the potential to improve the education of today's college students and prepare graduates to succeed in the workplace. These include:

More likely to hire employees with these experiences:	Executives	Hiring Managers
Internship/apprenticeship with a company/organization	93%	94%
Project in community with people from diverse backgrounds	72%	83%
Multiple courses requiring significant writing assignments	82%	72%
Research project done collaboratively with peers	81%	81%
Advanced, comprehensive senior project/thesis	80%	76%
Service learning project with community organization	71%	78%
Study abroad program	54%	47%

*Employer-Related Civic Engagement**

(Company currently does this or is considering doing it)

	Executives	Hiring Managers
Organize opportunities for employees to volunteer	71%	72%
Give employees time off to volunteer	62%	63%
Provide in-kind donations of equipment/supplies to charitable organizations	62%	63%
Provide pro-bono services to charitable organizations	56%	49%

Source: Hart Research Associates. *Fulfilling the American Dream: Liberal Education and the Future of Work* (Washington, DC: AAC&U, 2018) www.aacu.org/leap/public-opinion-research.

*Hart Research Associates (unpublished data, 2018)

1818 R St. NW, Washington, DC 20009 202.387.3760 www.aacu.org

Source: American Association of Colleges & Universities. (2018). *Fulfilling the American Dream: Liberal Education and the Future of Work: Selected Findings from Online Surveys of Business Executives and Hiring Managers.* Retrieved from https://www.aacu.org/sites/default/files/files/LEAP/5EngagedLearnPractices.pdf

Critics raise questions about the legality and morality of employing young people, sometimes full-time, without paying them wages or salaries. Another way of looking at internships is to ask how an arrangement intended by law to benefit students and other prospective workers may benefit other entities—perhaps even to the detriment of the intern. In a sharp critique of internships in the book *Intern Nation: How to Earn Nothing and Learn Little in the Brave New Economy*, Ross Perlin (2011) raises the following critiques:

- Universities benefit from mandating student internships for credit, because they can require students to pay tuition dollars for an experience that—unlike the classroom experience—costs the university comparably little.
- Internships provide employers with the services of often well-educated and enthusiastic workers—for no pay.

- Lower-income students stand to lose more by interships because students from better-off families can afford (at least for a while) to work without pay in order to secure human and social capital; poor and working-class students need to work for pay, even if it's in a field unrelated to their area of career interest.
- All students are potentially disadvantaged where jobs that once were or might otherwise have been entry-level jobs are reinvented as unpaid internships. This shift removes an economically and professionally important stepping-stone for students leaving school with degrees.
- Internships constrict social and professional mobility, contribute to growing inequality, and support an economy in which those in the top tier are becoming less and less diverse. Even more seriously, a fundamental ethic in American life is under threat: the idea that a hard day's work demands a fair wage.

An article on unpaid internships in Washington, D.C., where thousands of students arrive every summer to work on Capitol Hill with advocacy organizations and in other political or cultural institutions, notes:

> Such interns may cost their employers nothing, but some economists worry such programs do carry a cost. Free labor could be depressing wages in Washington while turning the federal government and Capitol Hill into arenas where only wealthy students can afford to work. In turn, children of privilege get another leg up on their less fortunate classmates. (Shepherd, 2016, para. 6)

The U.S. Department of Labor, Wage and Hour Division (2018) recognizes legal internships as a means for preparing interns for work and has established a set of six criteria that most public and private institutions must meet:

- The internship, even though it includes actual operation of the facilities of the employer, is similar to training that would be given in an educational environment.
- The internship experience is for the benefit of the intern.
- The intern does not displace regular employees but works under close supervision of existing staff.
- The employer that provides the training derives no immediate advantage from the activities of the intern, and on occasion, its operations may actually be impeded.
- The intern is not necessarily entitled to a job at the conclusion of the internship.
- The employer and the intern understand that the intern is not entitled to wages for the time spent in the internship.

What, then, are we to conclude about internships, a growing and pervasive phenomenon that will be part of the experience of thousands of students this year and in years to come? Clearly, students benefit from spending time in a professional market that values both formal educational credentials and hands-on work experience. Universities benefit from giving their students opportunities to link classroom learning to experiential learning. And employers benefit from the creativity, skills, and enthusiasm of young workers. Finally, based on the evidence from employers in Figure 12.10, it is likely that students benefit significantly from their internship experiences being among the characteristics employers look for when hiring. On the other hand, who receives an internship, whether it is paid or unpaid, and the quality of the internship vary. As we have noted, minority and poorer students are more likely to be disadvantaged in obtaining internships, and in being reimbursed by them. How can students, their schools, the law, and employers ensure opportunity for all and contribute to making internships meaningful work experiences that provide foundations for careers? What do you think?

FIGURE 12.10 ■ Completion for 6-Year College by Race/Ethnicity for First-Time Students Starting in 2010 (4-year schools)

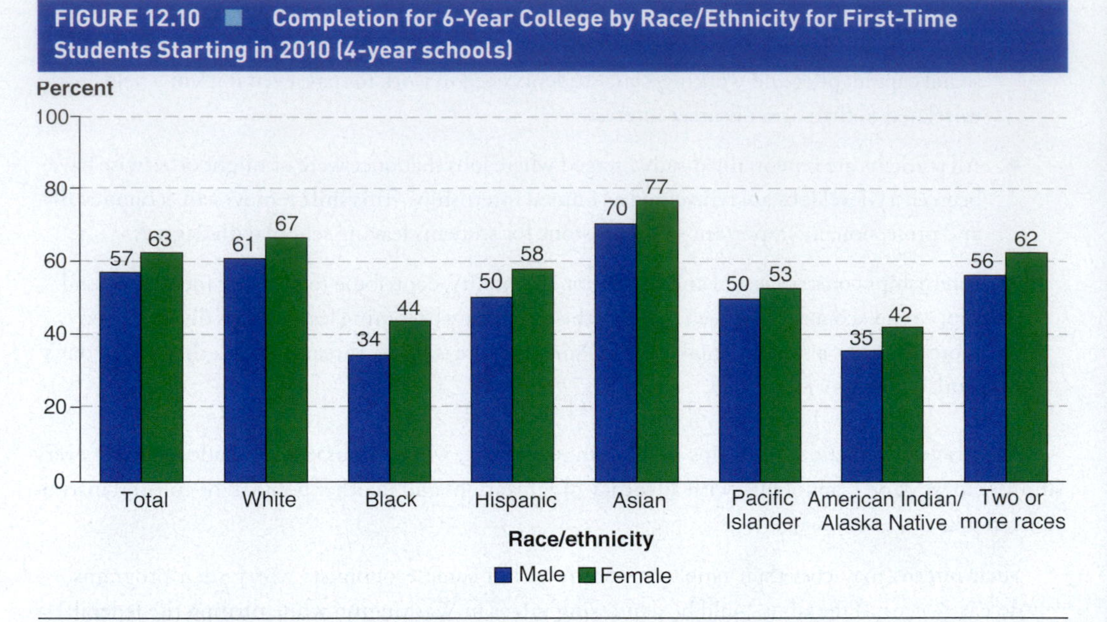

Source: U.S. Department of Education. (2020). *Status and Trends in the Education of Racial and Ethnic Groups, 2019*. Retrieved from https://nces.ed.gov/programs/raceindicators/indicator_red.asp

Notes: Data are run at the end of each month listed by federal fiscal year. Due to rounding and timing differences, the total figures may differ slightly from those in the Portfolio Summary report.

Dropping In, Dropping Out: Why Are College Dropout Rates So High?

Most high school graduates in the United States today go to college. According to data from the Bureau of Labor Statistics (2020), in 2019, 66.2% of high school completers enrolled. This figure has grown over time: In 1960, it was 45% and in 1990, 60%. Students of all racial and ethnic backgrounds are enrolling at high rates. Data suggest that education is more critical than ever for raising earning potential and strengthening competitiveness in the job market. In light of this, it is not surprising that most students are seeking to continue their education beyond high school.

Some observers, however, point out that all is not well in U.S. higher education. In fact, they note that the high rate of enrollment obscures a troubling reality: Many students leave college with no degree—and with debt. Journalist David Leonhardt (2009, para. 3) writes that "in terms of its core mission—turning teenagers into educated college graduates—much of the system is simply failing." Leonhardt argues that while the United States does an excellent job getting high school graduates to enroll in higher education, colleges have been far less successful in fostering timely graduation of students—or graduating them at all. Statistics suggest that many who enroll in college never finish with degrees, even though many end their college careers with substantial debt.

As we illustrate in Table 12.3, about 60% of students who enrolled in 4-year institutions went on to graduate within 6 years in both 2015/2016 and 2016/2017.

Significant differences, however, occurred in 6-year completion rates by race and ethnicity, with white and Asian students more likely to graduate in 4 years than Black, Native American, and Hispanic students (Figure 12.10). What other patterns do you see in these data?

There are also notable differences by institution: For example, although the 6-year completion rate at the country's most selective schools (those that accept 25% or fewer applicants) is about 87%, this falls to about 62% for somewhat selective schools (those accepting 50% to 74% of applicants), and drops to 31% for open-admission schools (National Center for Education Statistics, 2019b). Dropouts are costly, both to the nation as a whole, which loses potentially educated and productive workers, and to individuals, whose earning potential is diminished by their failure to obtain degrees and whose financial security may be compromised when they leave college with debt but no degree.

So, what is behind the college dropout phenomenon? Several factors contribute. First, high college costs drive many students out of higher education. The cost of a 4-year private college has outpaced

TABLE 12.3 ■ Undergraduate Retention and Graduation Rates

Undergraduate Retention and Graduation Rates	2015–2016	2016–2017
4-year institutions		
Retention-rate of first-time undergraduates	80.8%	81.0%
Graduation rate (within 6 years of starting program) of first-time, full-time undergraduates	59.7%	60.4%
2-year institutions		
Retention rate of first-time undergraduates	62.3%	62.5%
Graduation rate (within 150% of normal time for degree completion) of first-time, full-time undergraduates	30.3%	31.6%

Source: U.S. Department of Education, National Center for Education Statistics. (2019b). *The Condition of Education 2019 (NCES 2019-144). Undergraduate Retention and Graduation Rates.* Retrieved from https://nces.ed.gov/fastfacts/display.asp?id=40

inflation as well as wage growth, and student borrowing has risen. Costs are a particularly acute issue for lower-income students, who are more likely than their better-off peers to drop out before completing a degree. The financial burden of a college education falls more heavily on those with fewer resources, even when support (such as federal Pell Grants) defray some tuition costs.

Second, the rigors of college work lead some students to drop out. This factor is complex. With an increasing proportion of high school graduates enrolling in college, there may be more new students who are unready for the workload or the level of work. Half of students in associate degree programs and about a fifth of those in bachelor's programs are required by their institutions to enroll in remedial classes to address academic shortcomings. Some of these courses do not confer college credit, raising the cost of an education, lengthening the time needed to earn a degree, and increasing the likelihood that a student will leave without completing college.

Advocates for students suggest that colleges can do more to help students stay and succeed with coaching, scheduling that meets the needs of working students, and accelerated programs that speed the time to degree with rigorous work offered in concentrated time periods (Lewin, 2011a). At the same time, budget cuts, particularly at state institutions, have reduced rather than expanded opportunities to provide targeted services to struggling students.

Third, part-time attendance raises the probability of noncompletion. In fact, only about a quarter of students today fit the stereotypical model of full-time, on-campus attendees (Complete College America, 2011). According to Carnevale and Smith (2018), most college students (70%) work and attend school—a figure that amounts to about 14 million workers. Of those, a little under half (43%) are working in jobs that are low-income and are disproportionately Black, Hispanic, and women. Students who are attending college and working in low-income jobs, according to these authors, are less likely to graduate from college, even if they have high GPAs.

EDUCATION IN A GLOBAL PERSPECTIVE

Education enriches individuals, communities, and countries. Today, more people than ever worldwide are literate, and more are attending institutions of secondary and higher education, completing degrees, and sharing knowledge globally across new communication platforms. U.S. and international classrooms partner via the Internet so that students can meet other students to share interests and ideas. Many U.S. students study abroad for credit and to see places they might only have read about in books. Many international students annually come to the United States. In short, education can bring people and cultures together to foster greater understanding, innovation, and prosperity. Next, we take a look at how the United States lines up with its global peers in areas related to education such as college completion, job opportunities, and study abroad.

Higher Education and Job Opportunities

How do American students compare in their college graduation rates to international students? Some data from the Organisation for Economic Co-operation and Development (OECD), an international organization that includes the United States and 36 other member states across the globe, help us to answer that question. The OECD is dedicated to both fostering and tracking its member states' economic and policy development and changes. According to the Organisation for Economic Co-operation and Development *Education at a Glance* 2020 report, in 2018 the United States ranked 10th in college degree attainment (Figure 12.11). The OECD also found that higher education raises employment rates across all OECD countries from an average of 64% for 21- to 64-year-olds without a college degree to over 86% of those with a college education and up to 90% for those with a doctoral degree in all but three countries (Russian Federation, Slovak Republic, and Greece; p. 76). Work experience while studying in

FIGURE 12.11 ■ Percentage of the Population Who Had Attained Any Postsecondary Degree in Organisation for Economic Co-operation and Development (OECD) Countries, by Selected Age Groups, 2018

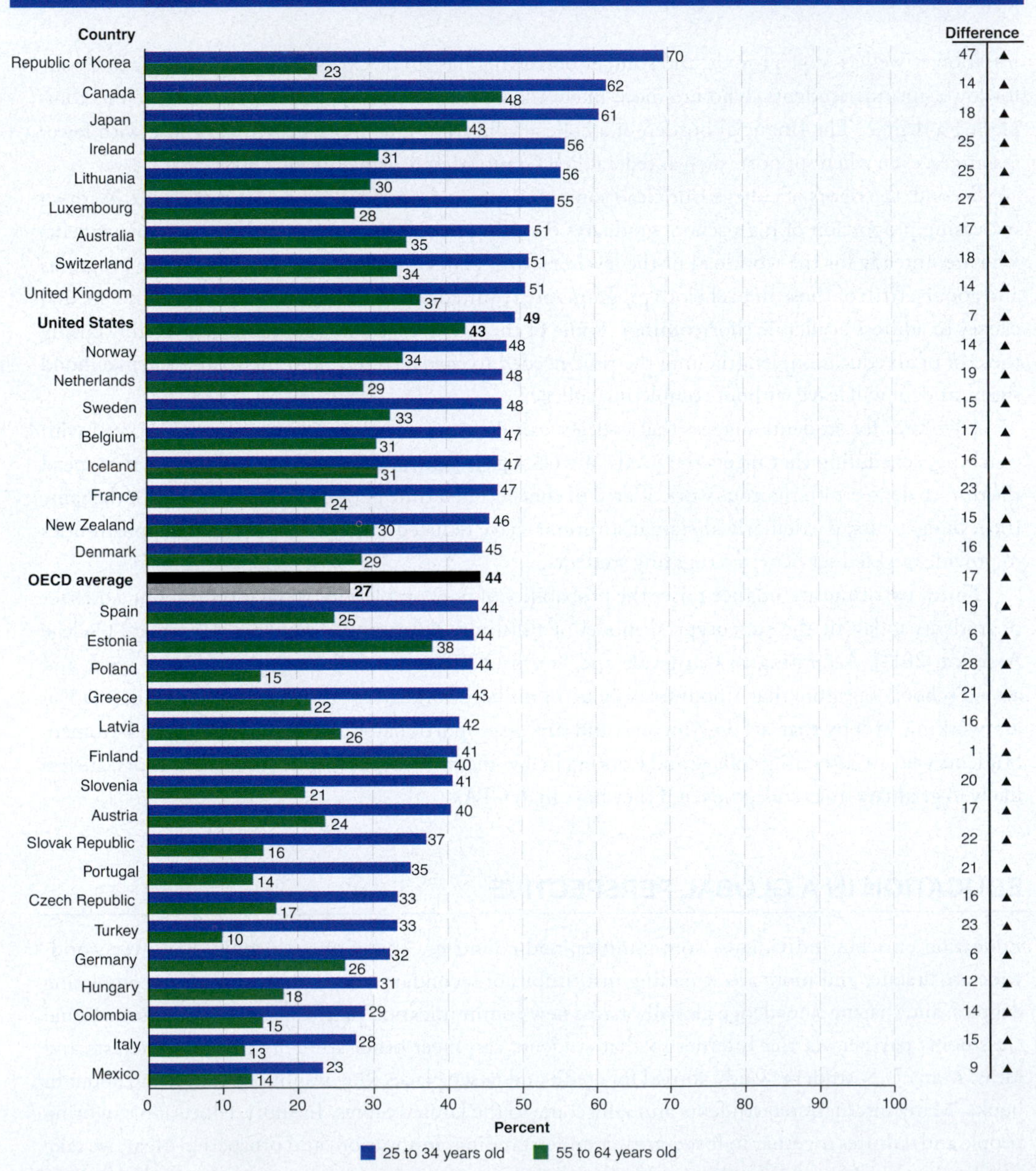

Source: National Center for Education Statistics. (2020a, May). Retrieved from https://nces.ed.gov/programs/coe/indicator_cac.asp

college also produced a higher employment rate across over half of the OECD countries (2019). Finally, across OECD countries, those with college degrees earn about 54% more than those with only secondary degrees, although their pay varies widely among educational fields (p. 88).

American and International Students Study Abroad

Will you, like an increasing number of college students, study abroad during college? In 2017–2018, a total of 341,751 American students studied abroad for college credit, a major growth from earlier generations. For example, in 1996–1997, only 100,000 American students studied abroad (Open Doors Data, 2019). At the same time, the number of international students coming to study in the United States far outnumbers the American students who study abroad who, in 2018–2019, numbered 1,095,0299 students. This figure had risen in sheer numbers but fallen by proportion of total U.S. higher education enrollees by 7% from the previous year (Open Doors Data, 2019).

Where do U.S. students go to study? Table 12.4 shows comparison data of the top 10 locations for U.S. students abroad and the top 10 locations for international students studying in the United States. As you can see, the top destinations for U.S. students include the United Kingdom, France, Spain, and Italy. China, Australia, Costa Rica, and Japan also draw over 8,000 U.S. students annually. What about international students who come to the United States? The U.S. states that host the most international students include California, New York, Texas, and Massachusetts (Open Doors Data, 2019).

Whether the study-abroad experience takes students from the United States to London, Rome, Beijing, Beirut, Rio de Janeiro, Havana, Cairo, or elsewhere, students say their worldviews are forever transformed by the opportunity to live in places they might otherwise know only from books or television. As one student said, "I will never again look at a story about the Middle East with such a one-sided perspective." Another added, "I genuinely enjoyed watching the bottom fall out of every one of my preconceived ideas about the Muslim world" (quoted in Conlin, 2010).

International students come to the United States for similar reasons that U.S. students go abroad. For instance, their top fields of study while in the United States are engineering, math, computer

Study abroad brings diverse students together around the shared goal of exploring new cultures. Thinking back to your earlier study of culture and socialization, what patterns of experiences do you think study-abroad students might share?

©Barry Iverson / Alamy Stock Photo

TABLE 12.4 ■ Top Destinations for U.S. Students Studying Abroad and Top Places of Origin for International Student Studying in the United States

RANK	Destination Country for U.S. Students	Country of Origin for International Students
1	United Kingdom	China
2	Italy	India
3	Spain	South Korea
4	France	Saudi Arabia
5	Germany	Canada
6	Ireland	Vietnam
7	China	Taiwan
8	Australia	Japan
9	Costa Rica	Brazil
10	Japan	Mexico

Source: Data from Institute of International Education. Open Doors Data (2019). *Fast Facts.*

sciences, business and management, and social sciences (Open Doors Data, 2019). These are also the top five fields of study for American students going abroad (Open Doors Data, 2019). Both sets of students are likely to experience culture shock, especially when it involves venturing to a new destination with a culture quite divergent from that of the student's own country.

Importantly, the experiences of international students worldwide "shifted" dramatically during the spring of 2020, when COVID-19 closed borders and sent international students the world over scrambling to return home, sometimes finding their entry barred on arrival (Halpern, 2020). The experiences for students and agencies assisting their travel has, at the time of this writing, led to some dramatic changes, including canceled and modified programs, and shorter hybrid experiences (online and face to face). The CEO of the Center for International Educational Exchange (CIEE), an organization that "operated 65 centers in 40 countries and was the largest State Department sponsor for inbound foreign students," put it this way in the fall of 2020:

> Hybridization is COVID-19's legacy on study abroad . . . there are currently five programs lined up for fall: London, Berlin, Rome, Lisbon, and Seoul. The plan is to start instruction virtually and then add a travel component, if and when it's safe. The pandemic is forcing this galvanization of online and in-person education and international experience. . . . Students can study art and architecture in Rome as they take a German class with someone in Berlin. They can be anywhere they want and still progress toward their college degrees. (James P. Pellow, president and CEO of CIEE, quoted in Halpern, 2020, para. 11)

One of COVID-19's other legacies, unfortunately, may be an increase in the social class gulf of those who can study abroad. Diversity offices, which helped connect first-generation and lower-income students who had been less likely to have access to these opportunities, saw cuts in budgets and programming as a result of the economic downturn in 2020 (Halpern, 2020).

In late October 2020, reports revealed some challenges that international undergraduate students were facing in the United States. While undergraduate international student enrollment in U.S. institutions had fallen 14%, 9 of 10 of the previous year's international students were still in the United States (Fischer, 2020). Some stayed by choice and some because they could not travel (Fischer, 2020). Many of these students were reported to be struggling for reasons associated with COVID-19 such as decreased social networks because of social distancing, loneliness, anxiety from isolation, financial worries, and prejudice against Asian students, one third of whom are from China (Fischer, 2020, para. 22).

WHAT'S NEXT? EDUCATION AFTER COVID-19

At the time of this writing, educators and students worldwide continue to adapt to the challenges of promoting and obtaining a quality education while maintaining appropriate social distancing protocols given the COVID-19 pandemic. To that end, many institutions that serve elementary, secondary, and higher education moved to learning online, using the Internet and programs like Zoom to continue to provide connection with students. What's next with this significant change in educational delivery?

As we earlier noted, sociologists who study education view educational inequality as a "central concern" (American Sociological Association, n.d.). Using your sociological imagination, you will likely realize that the change to online learning and its impact takes place within a context of already unequal living and learning situations.

For instance, a survey by the U.S. Bureau of the Census showed that the majority of the 56 million students enrolled in elementary and secondary schools nationwide had Internet access and connectivity in the spring of 2020. On the other hand, 4.4 million school-age children lacked access to a computer, and 3.7 million lacked Internet connectivity (USA Facts, 2020). The "homework gap" (the lack of Internet connectivity that is more likely to impact Black, Hispanic, and households with low income) has previously been documented (Auxier & Anderson, 2020). But that may become more significant with the expectation of attending classes online. In the fall of 2020, for example, nearly 6 in 10 parents with lower income told Pew Research that their school-age child would face at least one "digital obstacle" to online learning (Vogels,

2020). Students with disabilities are also an important group to bear in mind when considering the impact of learning from a distance. In the United States, students with disabilities, about two thirds of which are male, make up 14% of public school enrollment, and about a third have a specific learning disability (Schaeffer, 2020). Drawing on a previous study of adults with disabilities and lower use of technology, Schaeffer asserts that students with disabilities may face "unique barriers" to digital education (para. 4).

How do you think the changes in educational delivery will impact students? Which sociological paradigm would you find most useful to employ as a framework for research? What type of research would you do to understand the impact of online learning? What groups would you study?

WHAT CAN I DO WITH A SOCIOLOGY DEGREE?

Considering Graduate and Professional Education

Chapter 12 discusses the nature of our "credential society," where, in the contemporary world of work, your credentials—specifically, the higher education degree or degrees you earn—will have a big impact on your path to a career. While it is true that education pays with increased income opportunities and decreased risk of unemployment, the choice of whether to pursue advanced study is a very individual decision.

There are several types of graduate degrees. The Master of Arts (MA) and Master of Science (MS) academic degrees focus on intellectual development and the mastery of a core of knowledge in a particular discipline. Professional master's degrees include the Master of Business Administration (MBA) and the Master of Social Work (MSW), which are designed to advance careers or professions in their respective fields. The Doctor of Medicine (MD) and Juris Doctorate (JD) degrees focus on the acquisition of advanced skills and knowledge in the areas of medical practice and law, respectively. The Doctor of Philosophy degree (PhD), or *doctorate*, focuses on the broad mastery of disciplinary knowledge, the development of areas of specialization in the discipline, and rigorous, original research that contributes to the discipline and, perhaps, to policy and general societal knowledge of important topics.

A degree in sociology provides a strong foundation for graduate and professional education in any field. The core knowledge and diverse skills that you acquire as a sociology major prepare you for the rigors of graduate or professional school by helping you develop competencies in research, writing, communication, and critical thinking. The three main exams that are requirements for graduate admissions—the GRE (graduate school), MCAT (medical school), and LSAT (law school)—are designed to measure skills considered essential for success in those programs. These are skills inherent in sociology and have been described in the various career boxes throughout this book, including synthesizing information, problem solving using data analysis, making evidence-based arguments, and communicating complex ideas effectively in writing.

Andrew Barondess, JD Candidate, Class of 2020 at UCLA School of Law

The George Washington University, BA in Sociology

Obtaining a sociology degree enabled me to have a seamless transition into graduate school. Studying sociology affords one the opportunity to develop and hone a broad set of skills. While pursuing my degree, I gained invaluable experience not only in writing and researching but also in critical analysis of societal issues such as poverty and crime. Furthermore, my degree gave me a leg up in law school over others who may not have had firsthand experience researching, writing, and thinking critically. A sociology degree offers a better understanding of society and why people act the way they do. This is particularly important for tackling the complex legal problems faced in law school and afterward as a practicing attorney. Lastly, while I chose to pursue a legal education after undergrad, a sociology degree provides the building blocks for success in any graduate program.

Career Data: Lawyers

- 2019 Media Pay: $122,960 per year
- $59.11 per hour

- Typical Entry-Level Education: Doctoral or professional degree
- Job Outlook, 2019–2029: 4% (as fast as average)

Source: Bureau of Labor Statistics. (2019). *Occupational outlook handbook*.

SUMMARY

- **Education** is the transmission of society's norms, values, and knowledge base by means of direct instruction.
- **Mass education** spread with industrialization and the need for widespread **literacy**. Today, the need is not only for literacy but for specialized training as well. All industrial societies today, including the United States, have systems of **public education** that continue through the high school level and frequently the university level as well. Such societies are sometimes termed **credential societies**, in that access to desirable jobs and social status depends on the possession of a certificate or diploma.
- Functionalist theories of education emphasize the role of the school in serving the needs of society by socializing students and filling positions in the social order. Conflict theories emphasize education's role in reproducing rather than reducing social inequality. Symbolic interactionist theory, by focusing on interactions such as those that take part in the classroom itself, reveals how teachers' perceptions of students—as well as students' self-perceptions—are important in shaping students' performance.
- Early literacy and later educational attainment are powerfully correlated. Access to books and early reading experiences are among the strongest predictors of basic literacy at an early age. Researchers measure multiple levels of literacy.
- U.S. public schools are highly segregated by race and ethnicity. Before the 1954 U.S. Supreme Court decision in *Brown v. Board of Education*, segregation was legal. Since that time, schools have continued to show **de facto segregation** because segregated residential patterns still exist, because many white parents decide to send their children to private schools, and because the courts have recently limited the scope of previous laws aimed at promoting full integration.
- Differences in school funding by race, ethnicity, and class reinforce existing patterns of social inequality. In general, the higher someone's social class, the more likely they are to complete high school and college. Those with low income, in contrast, are often trapped in a cycle of low educational attainment and poverty.
- There are strong demonstrable relationships between educational attainment and employment prospects and between educational attainment and income both nationally and internationally. There is also a correlation between the socioeconomic status of a family and the probability of its members' further educational attainment.
- Paid and unpaid internships are increasingly required of students by their colleges. Racial, gender, and class inequalities are present in internship access. While required internships have been criticized, they are also typically well received by students and future employers.
- International comparisons of higher education achievement in OECD countries show that the United States ranks in the top 10 of OECD countries for higher educational degree attainment.
- Study abroad sends students from the United States around the world but also brings a significantly higher number of international students to the United States. Both groups tend to study in related fields.

KEY TERMS

credential society (p. 323)
de facto segregation (p. 334)
education (p. 321)
food insecurity (p. 320)
formal education (p. 321)
literacy (p. 322)
mass education (p. 321)
public education (p. 322)
school segregation (p. 331)

DISCUSSION QUESTIONS

1. Identify four large historical changes in the administration of education in the United States. What do you think was the most important of these changes?
2. How would each sociological theoretical perspective on education explain your experience in college? Which best describes it and why?
3. What is the current state of racial segregation in U.S. public schools? How has it changed since the civil rights era of the 1960s? What sociological factors help explain high levels of racial segregation in schools? How can the United States address high levels of racial segregation in our schools?
4. List key reasons why students drop out of college. How can identifying the sociological roots of the problem help us to develop effective policies to address it?
5. You have been asked to present a report on internships or studying abroad from a sociological perspective, incorporating a sociological theory and important trends. What will you share and why?

©Chris Martin/Alamy Stock Photo

13 RELIGION AND SOCIETY

WHAT DO YOU THINK?

1. Can something as personal as religion be studied by sociologists? Why or why not?
2. Is the influence of religion decreasing or increasing in American society?
3. How is the institution of religion both a source of social stability and of social conflict?
4. Why are young adults in the United States more likely than previous generations to claim no religious affiliation?

LEARNING OBJECTIVES

13.1 Explain why and how sociologists study religion.

13.2 Apply classical and contemporary sociological perspectives to analyze the place of religion in human societies.

13.3 Describe five different types of religious organizations.

13.4 Identify key characteristics of the six major global religions.

13.5 Discuss religion in the United States including trends in affiliation, the practice of civil religion, politics and religion, and the importance of considering gender, sexuality, and race.

13.6 Discuss globalization's effects on relations between religious groups.

A NEW RELIGION RISES: THE JEDI FAITHFUL

One of the most successful movie franchises in history is also the foundation of a small but growing religion. The original *Star Wars* film, which debuted in 1977, and the nine prequels and sequels that followed have spawned not only toys, books, and video games but also spiritual beliefs and values that have caught the attention of the film's legions of fans.

Jediism is the religion of adherents who embrace "a belief that collective thought can influence external change," a variation on the Force that forms a central theme in the *Star Wars* films (Rowen, 2018, p. 5). Among its component parts are entreaties to followers to engage in meditation, self-improvement, and service (Rowen, 2018). As one adherent noted in an interview, "We are absolutely looking to achieve the outcomes of any other religion.... A better life, and a better death" (Shea, 2017, para. 7).

Yoda is an important figure in the *Star Wars* universe.

The Jedi religion emerged out of early role-playing games based on the films, which offered the outlines of a moral code, including the ideas that "There is no emotion; there is peace" and "There is no ignorance; there is knowledge" (Rowen, 2018, para. 10). These games had a strong following, which may explain the moderate success of an e-mail circulated in 2001 that entreated those without a dominant religion to identify themselves as *Jedi*. The United States does not have a question about religion on the national census, but among those in countries that use the question, including the United Kingdom and Australia, over 550,000 people marked their religion as Jedi. According to the article, "In the United Kingdom, more people listed their religion as Jedi than Jewish or Buddhist" (Rowen, 2018, para. 11).

While some people provided the answer in jest, an article from the British Broadcasting Corporation suggests that at least some of the self-reported Jedi are earnest devotees. It notes,

> Beth Singler, a researcher in the Divinity Faculty of Cambridge University, estimates that there are about 2,000 people in the UK who are "very genuine" about being Jedi. That's roughly the same number as the Church of Scientology, she says. Jediism is not a joke for them but an inspiration. (DeCastella, 2014, para. 5)

Estimates of U.S. followers range from 5,000 to 10,000 (Rowen, 2018).

In 2015, the Temple of the Jedi Order, which operates in the United States, successfully petitioned the Internal Revenue Service for status as a tax-exempt ministry. This success has not been realized, however, in the United Kingdom, Australia, or New Zealand, where Jedi adherents have failed to gain official recognition as a religion. This raises some interesting sociological questions: What makes a religion? Who has the power to decide what belief systems do or do not constitute a religion?

Rowen (2018) writes,

> One common argument against the validity of Jediism is that, unlike traditional religions, it recognizes a canon it understands to be fictional. But throughout the 20th century, scientific advances that directly contradicted religious texts led to a demythologizing of Judeo-Christian religious books too. Many, including church leaders, came to understand the Bible's cosmology and myths not as literal, but as fable.... Once traditional religion takes some of its stories as fictional, its parallels to pop-culture inspired religions, like Jediism, become quite pronounced: Belief in the moral force of stories about saints is similar to belief in the moral force of stories about superheroes and masters of telekinesis. (p. 28)

So, is Jediism a religion? Does it fulfill the same functions as acknowledged religions? What makes it more or less real than the religions from which it draws, including the Judeo-Christian and Buddhist traditions? What do you think?

In this chapter, we examine the relationship between religion and society, beginning with how sociologists think about and study religion. We explore classical and contemporary theoretical perspectives on religion and society and examine the place of religion in people's private and public lives. We discuss five different ways of categorizing religions, and we explore six world religions. We look at several issues of religion in the United States, including changing religious affiliations, civil religion, and religion and intersectionality. We conclude by looking at expanding global contacts and relationships between religious groups.

WHY AND HOW DO SOCIOLOGISTS STUDY RELIGION?

Consider these facts:

- 9 in 10 Americans believe in a higher power (Pew Forum on Religion and Public Life, 2018).
- Globally, of 102 countries surveyed, 49% of the populations say they pray daily (Diamant, 2019b).
- Worldwide, 57 countries experienced some level of violence due to religious conflicts between groups in 2017 (Lipka & Majumdar, 2017).

These numbers show that domestically and worldwide, religion is a social phenomenon impacting individuals and groups, potentially even those who are unaffiliated (not officially part of a religious group). As such, it is an area of interest to sociologists who regard religion as an important source of beliefs, practices, and social action, organization, cohesion, and conflict. The American Sociological Association (2020) explains the scope, methods, and importance of religion to sociology as follows:

> Sociology uses the tools of social science to explore religious beliefs and practices, humanism and other secular approaches to understanding, and organizations rooted in shared belief

A daily prayer ritual, also known as Salah, is one of the five pillars of Islam. It is performed five times a day by Muslim worshippers facing the direction of the Kaaba in Mecca (in Saudi Arabia).

©Robertus Pudyanto/Getty Images

In Hinduism, cows symbolize wealth, strength, abundance, and a full earthly life. They roam freely in many Indian villages and cities.

©iStockphoto.com/flocu

systems. It seeks to understand the development of religious commitment and the impact of religious organizations on individuals, groups, and societies. Sociology also studies spirituality, which may be defined as individual and group efforts to find meaning for existence within or independent of organized religion.

Without assessing the validity of specific religious beliefs or practices, sociologists study the relationships among religion, the economy and other dimensions of social structure, culture, and social movements. According to the American Sociological Association (2020), sociology also examines conditions associated with the rise of religious fundamentalism.

As we show in this chapter, the sociological study of religion was built on the work of classical theorists featured in Chapter 1 such as Durkheim (2008). Following Durkheim's work, many sociologists came to define **religion** as *a system of common beliefs and rituals centered on sacred things that unites believers and provides a sense of meaning and purpose.* We use this definition, pointing out, in the paragraphs that follow, several ways that it informs this chapter, and several additional steps we take beyond this definition to connect it with current sociological thinking in the study of religion.

First, religion is a form of culture, providing people with a sense of purpose and meaning. Recall from Chapter 3 that culture is composed of *the beliefs, norms, behaviors, and products common to the members of a particular group.* Throughout history, humans have told stories about the world in which they live. Before the coming of Christianity to Northern Europe, for instance, people told stories about gods of nature or divinities who brought good fortune or ill fate to explain the tension between good and evil in human existence. Communities have always sought to construct logical frameworks to explain the world around them, and organized religion is only one way they engage in the quest for understanding and control.

Religion: A system of common beliefs and rituals centered on sacred things that unites believers and provides a sense of meaning and purpose.

Rituals: Routinized acts typically performed in collective ceremonies and holding shared meaning.

Religions commonly tell coherent and compelling stories about the forces that transcend everyday life, in ways that other aspects of culture, such as a belief in democracy, typically cannot (Geertz, 1973; Wuthnow, 1988). Where we cannot find empirical answers to fundamental questions about life, death, and fate, faith may stand in. Religion draws together those who identify with it, helping to create common worldviews and shared religious symbols and practices.

Second, religious beliefs, practices, and rituals *can function to bring people together, creating social cohesion.* **Rituals**, for example, *are routinized acts typically performed in collective ceremonies and holding shared meaning.* Examples include ritual washings; making the sign of the cross in Catholic ceremonies; or standing, facing the flag, and placing one's right hand over one's heart for the pledge of allegiance. Believers frequently draw together as a group to engage in rituals that also identify them as group members. Durkheim (2008) argued that religious rituals create and reinforce social cohesion. Robert Merton (1968) added to this idea, suggesting that while rituals have a manifest (or obvious) function, they also have the latent (secondary or unintentional) function of reinforcing group solidarity. For example, the manifest function of southwestern Hopi, Pueblo, Navajo, and Mojave rain dances was to bring rain; but their latent function was to bring people together, thus strengthening community ties.

When a single religion dominates a society, it may function as a source of social stability. If several religions compete for resources and power in a shared space, however, differences between religious groups may lead to conflict. This brings us to our third point: *Religion can serve as a source of social power that divides people and creates social conflict.* Outside of Durkheim, other classical theorists, such as Marx and McLellan (2000), W. E. B. Du Bois (1903/2003), and Charlotte Perkins Gilman (1922/2003), showed that religion can also be a source of social power and conflict and, as such, is an essential part of the study of society. Today, on a global scale, destabilizing worldwide religious conflicts include struggles among Hindus, Muslims, and Sikhs in India; Muslims and Christians in Bosnia and Kosovo; and Sunni and Shiite Muslims in Iraq and Pakistan. The recent influx of refugees from the Middle East and North Africa into predominantly secular European states such as France and Germany has also led to social tension, which stems from perceived and real differences in cultures and religious orientations.

Some online massive multiplayer online role-playing games like *World of Warcraft* are imbued with religious and supernatural narratives that allow people to encounter and experiment with diverse religious and supernatural ideas.

Sociologists also point out that religions and religious practices and beliefs vary over time, place, and space. Our fourth point is that *sociologists are interested in how religion is socially organized, "institutionalized," and enacted* (Edwards, 2019). Within Christianity and Judaism, for example, religious practice often occurs in formal organizations, such as churches or synagogues. Yet this is not necessarily true of Hinduism and Buddhism, whose rituals may be practiced in the home or in natural settings. In Islamic societies such as in Saudi Arabia and Iran, religious beliefs and practices are incorporated into daily life and guide political, cultural, and even economic practices. Mass media has also given rise to new ways of belonging, believing, participating, and encountering religious (or nonreligious) worldviews. Take the example of Massive Multiplayer Online Role-Playing games (MMORPGs) like *Assassin's Creed* and *World of Warcraft* (and *Star Wars* games, as we mentioned in the introduction). As Schaap and Aupers (2020) put it, such games are "full-fledged virtual fantasy worlds infused with religious narratives and tropes about supernatural deities, transcendent spirits, animated objects, mysticism, and magic":

> These games, then, offer players the opportunity to fully immerse themselves in a world of religion without necessarily embracing or converting to a particular tradition; through role-playing they are motivated to experience enchantment without belief and freely experiment with religious narratives, roles, and identities within the boundaries of the magic circle. (Schaap & Aupers, 2020, p. 891)

Fifth, sociologists study religion and religious behavior like they study other social institutions and interactions. While so doing, they are mindful that religious beliefs and practices are deeply personal, creating a profound sense of connection with forces that transcend everyday reality. Sociologists do not question the depth of religious feelings; nor do they question the truth or falseness of beliefs. Instead, they study religion as sociologists, not as believers or atheists. They examine religious behavior in all of its quantitative and qualitative dimensions, using surveys, in-depth interviews, focus groups, observations, and content and historical analysis. Sociologists can view religions from macro, meso, and micro levels. Those who study religion as a social institution might examine how religion functions (or is a source of conflict) on a worldwide or societal level. Or, they might study religion's interconnections with other social institutions like family and health. Those who focus on the interactions might study how religion is practiced, and how meaning is negotiated within religious groups. They might examine such phenomena as ways that "getting religion" coincides with community or individual experiences of loss, grief, poverty, or disaster.

If you were going to conduct a sociological study of religion, what would you study? Would you be more interested in the interactions and practices of individuals or the structure of groups? Would you be more interested in religious conflict or religion and its functions? How would these influence your research? Importantly, what types of facts and theories might you need to know before you even began your study?

THEORETICAL PERSPECTIVES ON RELIGION AND SOCIETY

In this section, we examine three classical and one contemporary sociological perspective on the role of religion in society. While no single theoretical perspective can capture the full sociological picture of religious behavior, these perspectives can help us begin to think about why religion exists and persists.

Classical Views: Religion, Society, and Secularization

Classical sociological theorists Émile Durkheim, Karl Marx, and Max Weber each highlighted different dimensions of religion—its functions, its ties to power and inequality, and its connection to social change. As you read each theory, consider which you find the most compelling and why.

Émile Durkheim

The functions of religion. Religion serves the important social function of "uniting believers in a moral community" (Durkheim, 1912/2008, p. 47). Such was one of the insights of Émile Durkheim's *The Elementary Forms of the Religious Life*, a study of early Central Australian Aborigines, small hunting and gathering tribes whom he believed had lived in much the same way for thousands of years. This work has had an enduring impact on sociological theories of religion.

Profane: A sphere of routine, everyday life.

Sacred: That which is set apart from the ordinary; the sphere endowed with spiritual meaning.

Totems: Within the sacred sphere, ordinary objects believed to have acquired transcendent or magical qualities connecting humans with the divine.

Durkheim said that the Australian Aborigines divided their world into two parts: the **profane**, *a sphere of routine, everyday life*, and the **sacred**, *the sphere endowed with spiritual meaning that is set apart from the ordinary everyday life.* Their sacred activities included rituals and ceremonies that provided heightened emotional awareness and a spiritual connection with divine forces and other members of the community. During such rituals, Durkheim suggested, the tribal members lost their sense of individuality, feeling at one with the larger group. The sacred sphere also included many ordinary objects that had become extraordinary and had come to represent the tribe and its connection to the divine. These Durkheim called **totems**, *ordinary objects believed to have acquired transcendent or magical qualities connecting humans with the divine.* Durkheim theorized that *totemism* was the earliest form of religion. According to Durkheim, the "god of the clan"—the object of worship—was "nothing else than the clan itself, personified and represented to the imagination" (p. 206). Durkheim's writings highlight the importance of the group in constructing and worshipping, through the objects of devotion, an image of itself.

Based on his research, Durkheim (2008) came to describe religion as a "unified system of beliefs and practices related to sacred things" that are held in awe and elevated above the "profane" elements of daily existence and that unite believers in a moral community (p. 47). Durkheim's key theoretical conclusion was that the sacred serves an important social function—it brings the community together, reaffirms the community's norms and values through ritual practices, and strengthens its collective social bonds. Society, to Durkheim, "sacralized" itself—that is to say it projected itself onto divine objects or beings. In short, the sacred *is* the group, endowed with divine powers and purpose. By worshipping the sacred, society worships itself, becoming stronger and more cohesive as a result. Durkheim's work suggests, then, that one of religion's key functions is to create and reinforce collective bonds.

Symbolic interactionist Herbert Blumer (1986) provides additional insight into how symbolic meaning is created through objects and collectively shared. Blumer noted that the meaning we give to objects grows out of the ways in which others respond to them and the actions we take toward them. A flag, as a physical object, is only a piece of colored fabric, but when citizens of a country endow it with a meaning it is then elevated to a sacred symbol. A wafer is only a wafer *unless* it is blessed in a Christian religious ceremony, when, for some groups, it becomes a representation of the body of Christ and its consumption becomes an act of worship.

Secularization: The rise in worldly thinking, particularly as seen in the rise of science, technology, and rational thought, and a simultaneous decline in the influence of religion.

Durkheim also theorized that modern industrial societies were experiencing a decline of the reach of the sacred and a widening of the increase of the profane. He thought that nonreligious forms of behavior would begin to provide the same functions previously provided by religion and that people were likely eventually to reject their earlier religiosity. This societal trend toward the decline of religiosity is called **secularization**, or *the rise in worldly thinking, particularly as seen in the rise of science,* *technology, and rational thought, and a simultaneous decline in the influence of religion.* We address this process throughout the chapter.

Durkheim's most important contributions to the sociology of religion were his observations that sacred rituals and objects embody the community itself, functioning to strengthen social solidarity. However, his perspective seems most applicable to highly homogeneous and cohesive societies such as the small Australian tribes that provided inspiration for his ideas. In modern, complex societies, which are racially, socially, and ethnically diverse, religion no longer serves such a clear purpose. Durkheim recognized the potential for *anomie* (a state of normlessness that occurs when people lose touch with their shared rules and values that give order and meaning to their lives) in societies where the role of religion had become less central. He also believed, however, that religious beliefs, practices, and institutions would decline in industrialized societies, but that the functions that religion served, such as providing moral and group connections, would instead be achieved through the division of labor in the workplace and socialization in schools.

Do you agree with Durkheim that society is becoming increasingly secular? If so, has that created the conditions that he predicted might arise? (For example, do the workplace and educational institutions serve as a substitute for social solidarity and integration?) These are questions that continue to interest sociologists today.

Karl Marx

Religion, power, and inequality. Karl Marx said very little about religion, but his insight of its role in a capitalist society and its ties to power has had an important influence on sociologists of religion, sensitizing them to the fact that religious beliefs are "rooted in economic, social, and political conditions and religious divisions falling along class lines" (Goldstein, 2020, p. 476). Marx and McLellan (2000) famously wrote that "religion is the sigh of the oppressed creature, the feeling of a heartless world, and the soul of soulless circumstances. It is the opium of the people" (p. 72).

Marx saw religion in his time as serving the interests of the ruling class. It provided an outlet for human misery that obscured the true source of suffering among the subordinate classes—their exploitation by the ruling class. In other words, by promising spiritual solutions—such as a better afterlife—as the answer to human suffering, religion discourages oppressed people from understanding and attempting to overcome oppression in their present lives. That is to say, if there are rewards in the afterlife, then people should strive for their reward in the afterlife, not the present life. Challenges to the unequal structure of society, then, were not only unnecessary, but also possibly counterproductive because they distracted people from their main purpose, which was to look to the afterlife for better conditions. To Marx, such a focus serves the interests of the powerful, who can then better maintain passivity among the economically deprived masses.

Marx also believed that once capitalism was overthrown, religion would no longer be necessary, since economic inequality and social stratification he saw as inevitable in the capitalist system would no longer exist. In this way, Karl Marx, like Durkheim, saw secularization as inevitable. While Durkheim was concerned about the fraying of social bonds that he believed would accompany secularization, Marx viewed secularization as a progressive trend, since the social solidarity and harmony promoted by religion were contrary to the interests of the working masses because it prevented their recognition of their own exploitation.

Marx's contribution to the sociological study of religion has been to sensitize researchers to the relationship of religious beliefs, practices, and organizations with the economic and social conditions that people experience. His insight that religion may distract people from the immediate problems of their daily lives is illuminating, particularly where the same elite circles hold both religious and political power.

On the other hand, Marx's idea that religion is purely a mystification enabling the ruling class to deceive the masses is open to challenge. First, while religions have supported ruling groups in many historical instances, they have also challenged such groups. Catholicism, along with the labor movement, was a powerful driving force in the social movements that overthrew the Polish communist state through nonviolent mass mobilization, replacing it with a democratic government in 1989. Second, for many people, religious beliefs fill a need that has little to do with political or economic power—a function Marx ignored. When countries such as Cuba and the former Soviet Union sought to follow Marx's ideas and marginalize religion, they were remarkably unsuccessful. Religious beliefs flourished underground, and in countries such as Poland, Ukraine, and Uzbekistan, they experienced a resurgence after the fall of communism.

Max Weber

Religious values and activities as a source of social change. Religion was a key subject for sociologist Max Weber, who wanted to explain, in particular, why the culture of modern capitalism had emerged first in England, France, and Germany rather than in India or China—which had once been more advanced than Northern and Western Europe in science, culture, and commerce. Weber concluded that capitalism first appeared where the Protestant Reformation had taken hold and that the driving force behind its development was Protestantism's religious tenets and the economic behaviors they fostered.

Weber (2002) found that the beliefs of early Protestantism provided fertile ground for capitalism's development. First was the idea that God places each person on Earth to fulfill a particular calling. Whether someone would be saved was predestined by God, but people could find evidence of God's plan in a life dedicated to hard work because economic success was an indicator of salvation. Second, since Protestantism held that consumption-centered lifestyles were sinful, believers were expected to live simple lives, work hard, and save and reinvest their earnings rather than enjoy the immediate gratifications of idleness or acquisition. A hardworking, frugal, sober population reinvesting its earnings in new economic activities was, suggested Weber, a fundamental foundation for economic growth and, significantly, for early capitalism.

Yet Weber also wrote that once capitalism took hold, it became institutionalized and shed the religious ethic that fueled its development. Capitalism, scientific ways of thinking, and bureaucratic forms of organization, he said, would marginalize religion in the drive for productivity and profit. Weber was concerned about the disenchantment he believed would result from secularization and bureaucratization. Modern society, in his well-known metaphor, would become an "iron cage," imprisoning people in rationalized but irrational bureaucratic structures, rote work, and lives bereft of spirituality or creativity.

While Weber's idea that a religious ethic of hard work and thrift contributes to economic growth has been used to explain examples of economic success around the world (Berger, 1986; Berger & Hsiao, 1988; Morishima, 1982), it has also been widely criticized. For example, his conclusions were based on the writings of Protestant theologians rather than on the actual practices of Protestants. During the colonial period he cited as the birthplace of the U.S. capitalist ethic, "Boston's taverns were probably fuller on Saturday night than were its churches on Sunday morning" (Finke & Stark, 1992, p. 23). Furthermore, some scholars argue that capitalism developed among Jews and Catholics—and, for that matter, among Hindus, Muslims, and Confucians—as well as among Protestants (Collins, 1980; Hunter, 1987).

The "Religious Economy" Perspective

While Durkheim, Marx, and Weber believed that the strength of religion, in general, weakens when challenged by competing religious or secular viewpoints, a very different theory of religion emerged in the late 20th century in the United States. This perspective, tailored to modern societies, which house a wide variety of religious faiths (including the United States), was the **religious economy** perspective. It suggests that *religions can be understood as organizations in competition with one another for followers* (Finke & Stark, 1988, 1992, 2005; Hammond, 1992; Moore, 1994; Stark & Bainbridge, 1996; Warner, 1993). In this view, religious competition, rather than religious monopoly, is the best way to ensure religious vitality.

Religious economy: An approach to the sociology of religion that suggests that religions can be understood as organizations in competition with one another for followers.

The religious economy perspective suggests that competition leads to *increased engagement* in religious organizations for two reasons. First, competition compels each religious group to exert more effort to win followers, reaching out to the masses in a variety of ways in order to capture their attention. Second, the presence of a multitude of religions means there is likely to be something for just about everyone. In a culturally diverse society such as the United States, a single religion may appeal to only a limited range of followers, but the diversity of religious choices—Indian *gurus*, fundamentalist preachers, Muslim *imams*, Catholic priests, Scientologists, Wiccan covens, and mainline Protestant churches—may encourage broad religious participation.

The culturally diverse United States offers Americans a variety of religious choices.

©Bob Daemmrich via Alamy Stock Photo

The religious economy analysis is adapted from the business world, in which competition (in theory) encourages the emergence of specialized products that appeal to specific markets. Sociologists who embrace this perspective even borrow the language of business. According to Finke and Stark (1992), a successful religious group must be well organized for competition, have eloquent preachers who are engaging "sales reps" in spreading the word, offer beliefs and rituals packaged as an appealing product, and develop effective marketing techniques. Religion, in this view, is a business much like any other.

Megachurch: A church with over 2,000 weekly attendees.

One modern example of corporate religion is the **megachurch**, *a church of over 2,000 weekly attendees*. Megachurches are typically Protestant, with many having evangelical or fundamentalist theologies. They exist primarily in the United States, numbering upwards of 1,750 locations (Sanders, 2020), but they are also seen across the globe, most notably in Seoul, South Korea, where the Yoido Full Gospel Church led by Young Hoon Lee has over 80,000 members. Cornelio (2020) points out that megachurches in Southeast Asia are particularly attractive to youth because "these spaces embody a message of prosperity and purpose" (p. 927).

The Hartford Institute for Religious Research (n.d.) indicates that most megachurches have these characteristics:

- A charismatic, authoritative senior minister
- A very active, 7-day-a-week congregational community
- A multitude of diverse social and outreach ministries
- An intentional small-group system or other structures of intimacy and accountability
- Innovative and often contemporary worship format
- A complex, differentiated organizational structure

Megachurches are also marked by their deep link historically and culturally to consumer capitalism (Sanders, 2020, p. 489). Megachurches have their own brands, rock bands, huge video screens, coffee shops, daycares, and theaters. Their clergy are sometimes well-known celebrities, and bestselling authors such as widely recognized U.S. megachurch pastors T. D. Jakes (bishop of the megachurch The Potter's House in Dallas, Texas), Rick Warren (founder and senior pastor of Saddleback Church in Lake Forest, California), and Joel and Victoria Osteen (co-pastors of the Houston, Texas, megachurch Lakewood).

The Osteens exemplify charisma and celebrity, as well as prosperity and consumption. They are frequently in the news for their guests or events. Their physical church is housed in the Compaq Center, Houston (a repurposed stadium formerly the home of the NBA Houston Rockets), but they televise their services weekly to millions of viewers, domestically and across the globe; have their own radio channel on Sirius XM "Joel Osteen Radio"; and also enlist Facebook and YouTube platforms. Several services with Kanye West in 2019 and 2020 drew viewers and media attention. In the 2020 event, for instance, West and Osteen both walked on water in a music video, and Osteen gave a sermon about the biblical meaning of walking on water (Willis, 2020). Furthermore, as the *Washington Post* noted

Joel Osteen, pastor of the Lakewood megachurch in Houston, Texas, presides over a service in the former Houston Rockets stadium. In 2016, Lakewood was the largest megachurch in the United States, weekly hosting about 52,000 people.

after the November 2019 event: "Kanye West and Joel Osteen are a match made in market-driven heaven," said Jonathan L. Walton, dean of Wake Forest Divinity School, who noted how West sold $170 "Jesus Is King" sweatshirts with his new album (quoted in Bailey, 2019, para. 2).

The religious economy perspective also points out that the consequence of having so many choices is that people may also *switch* from one religious group to another. While people's motivations are not easy to pinpoint, a study conducted by the Pew Research Center (2015a) found that about 28% of U.S. adults change their religious affiliations over the course of their lives. If we count those who change their affiliations *within* Protestant religious traditions (for instance, from Baptist to Lutheran), the figure rises to 42%. At the same time, some are leaving the religious group in which they were raised and not joining a new group. In fact, the proportion of the U.S. population claiming no religious affiliation has risen significantly in recent decades (Figure 13.1). According to a Pew Research Center (2019b) report, "The

FIGURE 13.1 ■ Growth of the Religiously Unaffiliated

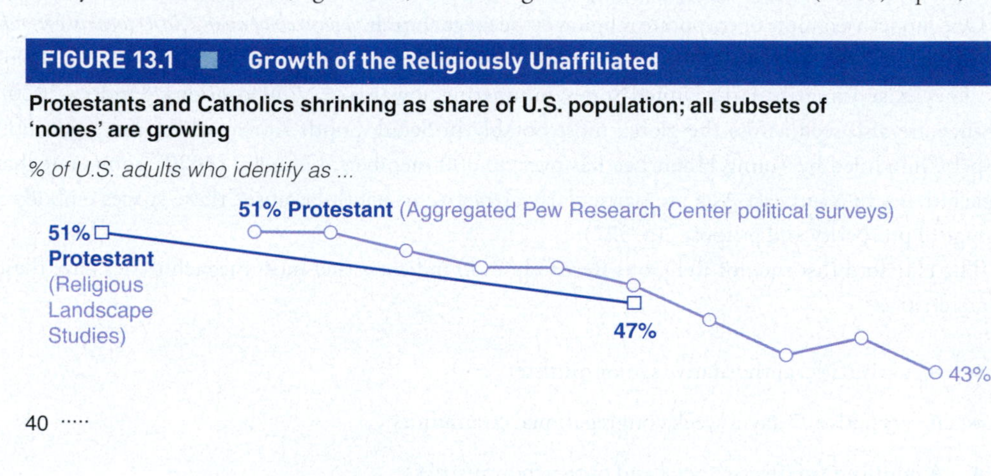

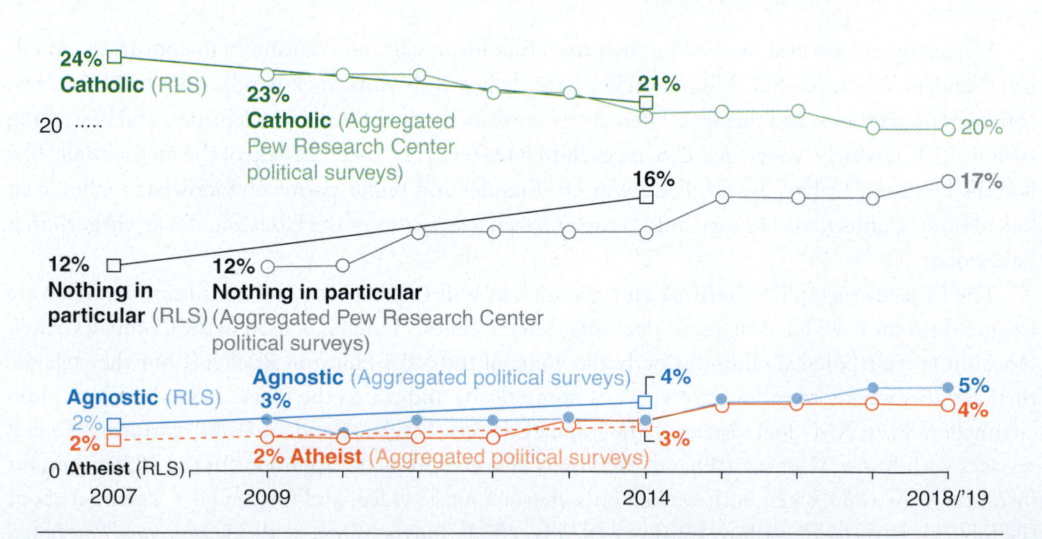

Source: Pew Forum on Religion and Public Life. (2019c). *Protestants and Catholics shrinking as share of U.S. population "nones" are growing.* https://www.pewforum.org/2019/10/17/in-u-s-decline-of-christianity-continues-at-rapid-pace/pf_10-17-19_rdd-update-new3/

religiously unaffiliated share of the population, consisting of people who describe their religious identity as atheist, agnostic or 'nothing in particular,' now stands at 26%, up from 17% in 2009" (para. 1).

One critique of the religious economy approach is that it likely overestimates the extent to which people rationally pick and choose among different religions, as if they were shopping for a new car or a pair of shoes. Among deeply committed believers, particularly in societies or communities that lack religious pluralism, it is not obvious that religion is a matter of rational choice. Even when people are allowed to choose among different religions, most are likely to practice their childhood religions without considering alternatives (Roof, 1993). And some religious groups are more likely to retain childhood members. According to the Pew Research Center:

> Hindus, Muslims and Jews are the three religious traditions that retain the largest shares of the adherents raised within their group. Among all U.S. adults who say they were raised as Hindus, fully 80% continue to identify with Hinduism as adults; most of those who no longer identify as Hindus now describe themselves as unaffiliated. Roughly three-quarters of those raised as Muslims (77%) and Jews (75%) also continue to identify with their childhood faiths. (Pew Forum on Religion and Public Life, 2015b, para. 15)

Furthermore, as the most common shift in affiliation is from a religious group to the nonaffiliated group (Pew Forum on Religion and Public Life, 2015b), it is not clear that U.S. believers are trading one set of beliefs for another that is better marketed or more enticing. They would, in fact, seem to be leaving organized religion behind. Do you think the religious economy perspective can account for this large and growing shift? Why or why not?

TYPES OF RELIGIOUS ORGANIZATIONS

Religions, as Barker (2020) puts it, are "notoriously difficult to classify" (p. 538). But early theorists of religion such as Max Weber, Ernst Troeltsch (1931), and Richard Niebuhr (1929) were interested in classifying religious organizations on a continuum based on the degree to which they are well established and conventional: Churches, for instance, reside at one end (they are conventional and well established), *cults* are at the other (they are neither), and *sects* fall somewhere in between. These distinctions were predominantly based on the study of European and U.S. religions, and there is some debate over how well they apply outside of these parameters. In the sections that follow we provide these theoretical distinctions between churches, sects, cults, and new religious movements. We find these distinctions useful for your consideration, and also your critique. While you read through each, consider how well you think they match religion as it is practiced today.

Church

A **church** is *a well-established religious organization that exists in a fairly harmonious relationship with the larger society* (Finke & Stark, 1992). Churches, as classified by Ernst Troeltsch (1931), were mainstream groups, formally and bureaucratically organized, that reflected their religious communities' prevailing values and beliefs. Troeltsch's model is heavily based on Christian historical tradition and has been criticized for its inability to apply to other religious groups (Richardson, 2020, p. 135). In the United States, sociological examples of "church" types of religious organizations today include, among many others, Presbyterians, United Methodists, Greek Orthodox, and Roman Catholics.

Church: A well-established religious organization that exists in a harmonious relationship with the larger society.

A church can take one of two forms. An **ecclesia** is *a church that is formally allied with the state and is the official religion of the society.* As such, it is likely to enjoy special rights and privileges that other churches lack. In Greece, for instance, while the practice of other religions is not prohibited or punished, the constitution holds that the Greek Orthodox Church is the prevailing religion of the country. Ecclesias, particularly those in Eastern and Central Europe, may also combine religion and national culture (Pew Forum on Religion and Public Life, 2017b). Returning to the example of Greece, in 2017 an overwhelming number of Greek citizens (90%) identified as Greek Orthodox and three quarters said that "being Orthodox is important to being 'truly Greek'" (Pew Forum on Religion and Public Life, 2017b). Their reasons for belonging are worth noting because they show a blend of faith, family

Ecclesia: A church that is formally allied with the state and is the official religion of the society.

traditions, and national identity: A little under half (41%) said that they identify with their religious group because of personal faith, while about a quarter (26%) said that they identify with their religious group because of national culture. On the other hand, about a third (32%) identified with the Greek Orthodox Church because of a blend of personal faith, family tradition, and national culture (Pew Forum on Religion and Public Life, 2017b).

Denomination: A church that is not formally allied with the state.

A **denomination**, in contrast, is *a church that is not formally allied with the state.* Since there is no established church in the United States, all U.S. churches are, by definition, denominations. Typically, however, sociologists of religion have identified the Protestant Mainline of "Episcopal, Presbyterian, Methodist, Lutheran, Reformed, and Baptist" with the term *denomination* (Bouma, 2020, p. 202). The existence of denominations allows for freedom of religious choice, so different denominations may compete with one another for membership, and none enjoys the special favor of the state.

Sect

Sect: A religious organization that has splintered off from an established church to restore perceived true beliefs and practices believed to have been lost by the established religious organization.

Unlike a church, which exists in relative harmony with the larger society, a **sect** is *a religious organization that has splintered off from an established church in an effort to restore perceived true beliefs and practices believed to have been lost by the established religious organization.* In the United States, sects draw followers from among lower-income households, racial and ethnic minorities, the rural poor, and those with lower rates of educational attainment (Lehman & Sherkat, 2018). Members of sects may hold some religious beliefs similar to other dominant religious groups, and yet these beliefs may also have important differences. For example, sects are marked by exclusivism in membership and beliefs, or as Lehman and Sherkat (2018) put it—they "claim to hold exclusive access to supernatural rewards and compensators" (p. 780). Because they are splinter groups, they may exist in *tension* with more established religious organizations and society in general (Barker, 2020). They may, for instance, ask their followers to reject worldliness and societal practices. As Lehman and Sherkat (2018) put it:

> Nearly 1 in 10 white Americans identify with one of the many other exclusivist sects—groups ranging from evangelical Pentecostals (like the Assembly of God, and Pentecostal Holiness) to the staid Churches of Christ, to insular exclusivist groups like the Amish and Mennonites. Sectarians as a group have the highest rates of biblical inerrancy and church attendance, and they are more opposed to homosexuality, abortion, and women's equality compared to all other religious identifications. Sectarians also have the second lowest rates of educational attainment, and the lowest incomes of any identification." (p. 785)

While churches tend to intellectualize religious practice, sects may emotionalize it, emphasizing heightened personal experience and religious conversion (Finke & Stark, 1992; Stark & Bainbridge, 1996).

Sects provide new sources of religious ideas and vitality outside mainstream faiths. Successful sects often grow in size and evolve into churches, becoming bureaucratized and losing their emotional appeal (Niebuhr, 1929). A new sect may then break off, seeking to return to its religious roots. This occurred within both U.S. Protestantism and Japanese Buddhism in the late 20th century. Disturbed by the increasingly intellectualized and liberal direction taken by mainstream Protestant churches, numerous sects broke off, seeking to return to what they viewed as the biblical roots of the Protestant faith (Finke & Stark, 1992). Similarly, Buddhist sects in Japan sought a return to original Buddhist beliefs in response to what they regarded as the social isolation and irrelevance of mainstream Buddhist groups (Davis, 1991). Adherents may form their own closed-off (endogamous) groups who separate physically from other people, living in communities and enclaves such as some groups of Amish and Mennonites, or Hassidic Jews, or they may simply have beliefs that separate them from others, such as Protestant fundamentalist and evangelical religious groups.

Cults and New Religious Movements

Cult: A religious organization that is thoroughly unconventional with regard to the larger society.

A **cult** is *a religious organization that is thoroughly unconventional with regard to the larger society* (Finke & Stark, 1992; Richardson, 2009). While sects often originate as offshoots of well-established religious organizations, cults tend to be new, with unique beliefs and practices that typically originate outside the religious mainstream. They may be led by charismatic figures who draw on a wide range of teachings to develop their novel ideas (Stark & Bainbridge, 1996). The presence of a powerful personality can thus define a cult, and the loss of the leader can spell its end.

Cults may have relationships with the larger society that are characterized by strife and distrust. For instance, "doomsday cults" that prophesize the end of the world have emerged from time to time. One example is Church Universal and Triumphant (CUT), led by the charismatic leader Elizabeth Clare Prophet. CUT's doctrine amalgamated bits of major religions such as Christianity and Buddhism along with mysticism, astrology, Western philosophy (Melton, 1996), and a belief in reincarnation (rebirth). Prophet herself, who led the cult after the death of her husband, Mark Prophet, claimed to have lived past lives as Marie Antoinette, Queen Guinevere of King Arthur's court, and the biblical figure Martha.

In 1990, Prophet predicted that the world would end, and hundreds of her followers fled to bomb shelters near Yellowstone National Park, where CUT had created a self-reliant community. Apparently, the group was also stockpiling weapons in its underground bunkers, and several members were later prosecuted on weapons charges. While CUT lost members after the failed doomsday prediction, it continued to operate and grow in later years, though Elizabeth Clare Prophet passed away in 2009.

Similar to sects, cults flourish when there is a breakdown in well-established societal belief systems or when segments of the population feel alienated from the mainstream and seek meaning elsewhere. Cults may originate within or outside a society. Interestingly, what is perceived as a cult in one country may be accepted as established religious practice in another. Christianity began as an indigenous cult in ancient Jerusalem, and in many Asian countries today, evangelical Protestantism is regarded as a cult imported from the United States.

The term *cult* has negative connotations today, and many sociologists now instead use the phrase *new religious movements* to characterize novel religious organizations that lack mainstream credibility (Barker 2020; Hadden, 1993; Hexham & Poewe, 1997). **New religious movements** (NRMs), considered *new spiritual groups or communities that occupy a peripheral place in a country's dominant religious landscape*, have most of the same characteristics as cults. They are comprised of "first generation" members who are enthusiastic about the new religion. They are typically led by a charismatic leader/founder and tend to attract "atypical sections of the general population" whose primary identity comes to be that of belonging to the group. These groups may also change quickly because their "prophecies may fail, experimental lifestyles may break down, charismatic leaders die, and the movement comes to depend on a more rational and/or traditional leadership" (Barker, 2020, pp. 536–539). Liogier (2020) argues that they are more likely to develop "in advanced industrial societies and with individuals and groups that are dominant in both economic and symbolic resources" (p. 322).

New religious movements: New spiritual groups or communities that occupy a peripheral place in a country's dominant religious landscape.

The growth of NRMs presented an ideal situation for sociological research into the development of new religions, helping to shed light on how the world's great religions came into being. There are many NRMs—by one estimate, between 1,500 and 2,000 in North America and perhaps 10,000 in Africa—but few are enduring, and most exist outside their countries' religious mainstreams (Hadden, 2006). NRMs may, however, evolve into sustained religious movements.

One well-known global NRM is Scientology. The Church of Scientology began in the United States in the 1950s, largely as a self-help movement. Today, it has members across the globe, and it benefits from the endorsement of a number of Hollywood figures such as Elisabeth Moss, Laura Prepon, Tom Cruise, and John Travolta. As it has grown, Scientology also has courted its fair share of controversy. Its founder, L. Ron Hubbard, was a charismatic leader revered by adherents and reviled by detractors. Scientologists believe that Hubbard's ideology provides a path to greater self-awareness and spiritual enlightenment, while critics counter that the secretive religion "brainwashes" members, convincing them to pay exorbitant sums of money to access the esoteric knowledge central to their faith. While the controversy continues, Scientology's high-profile followers and considerable financial resources make it likely that this religious movement is here to stay (Urban, 2011).

Pope Francis, the head of the Catholic Church, visits Dublin, Ireland, in August 2018, a predominantly Catholic nation.

THE GREAT WORLD RELIGIONS

Today, while thousands of different religions are followed across the world, three—Christianity, Islam, and Hinduism—are practiced by about 70% of its people (Hackett & McClendon, 2017). Next, we explore these and three other world religions (Judaism, Buddhism, and Confucianism) that have had powerful influence on global religious, political, and social practices.

Christianity

With an estimated 2.3 billion followers—nearly a third of the world's population—Christianity encompasses a broad spectrum of denominations, sects, and even new religious movements (Diamant, 2019a). It remains the world's largest religious group (see Figure 13.2). Common to all these is the belief that Jesus of Nazareth was the Messiah or savior foretold in the Hebrew Bible. While doctrinal differences separate the Christian faiths, almost all teach that at the beginning of time, humans fell from God's grace through their sinful acts and that acceptance of Christ and his teachings provides the key to salvation. Most Christians also believe in the New Testament account of the Resurrection, according to which Jesus rose from the dead on the third day after his crucifixion and then ascended to heaven. Christianity is a form of **monotheism**—*belief in a single all-knowing, all-powerful God*—although in most Christian faiths God is also regarded as a trinity made up of a Heavenly Father, His Son the Savior, and His sustaining Holy Spirit.

Monotheism: Belief in a single all-knowing, all-powerful God.

When Christianity emerged in Palestine some 2,000 years ago, it was a persecuted sect outside the mainstream of Jewish and Roman religious practices. Within four centuries, it had become an ecclesia

FIGURE 13.2 ■ Christians Are the World's Largest Religious Group

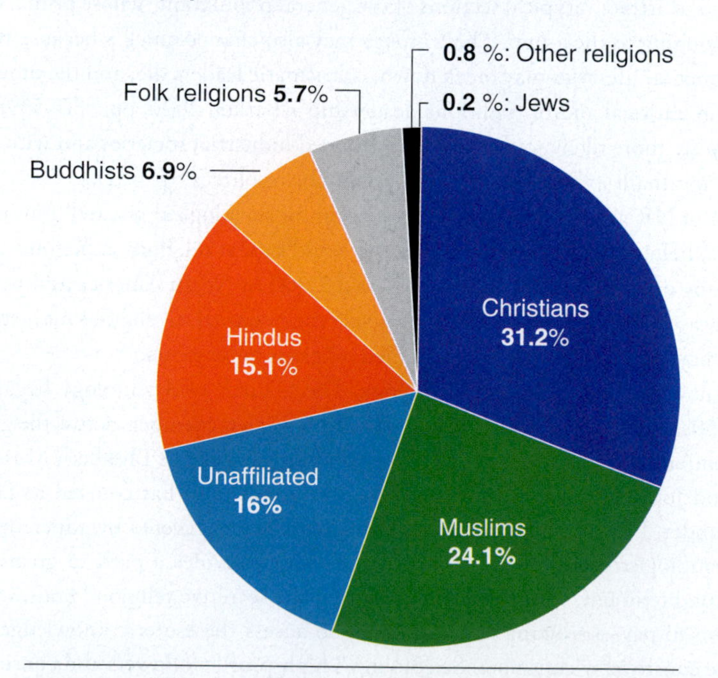

Source: Hackett, C., & McClendon, D. (2017, April 5). *Christians remain world's largest religious group, but they are declining in Europe*. https://www.pewresearch.org/fact-tank/2017/04/05/christians-remain-worlds-largest-religious-group-but-they-are-declining-in-europe/

(official religion) of the Roman Empire. In the 11th century, it divided into the Eastern Orthodox Church (based in Turkey) and the Catholic Church (based in Rome). A second great split occurred within the Catholic branch when the 16th-century Protestant Reformation gave rise to numerous Protestant denominations, sects, and new religious movements. Protestants tend to emphasize a direct relationship between the individual and God. Catholics, by contrast, emphasize the importance of the church hierarchy as intermediary between the individual and God, with the pope in Rome being the final earthly authority.

Christianity continues to gain adherents and grow around the world. Growth is particularly robust in parts of Africa: For instance, the number of Catholic converts is high in the Democratic Republic of Congo, Uganda, Nigeria, and Kenya (Rocca, Hong, & Ulick, n.d.). By 2060, the Pew Research Center estimates that there will be about 3 billion Christians and 3 billion Muslims worldwide (Diamant, 2019a). While the United States and Brazil are predicted to remain in the top two of Christian populations worldwide in 2060, the Pew Research Center predicts changes in other countries (Diamant, 2019a; Figure 13.3). What changes stand out to you? What do you think might cause these changes?

Islam

With 1.8 billion believers, or about 25% of the world's population, Islam is the second largest and the fastest-growing religion in the world today (Diamant, 2019a). *Muslim* is the term for those who practice *al-Islam*, an Arabic word meaning submission without reservation to God's will. About 60% of Muslims live in the Asia Pacific region, and about 20% in the Middle East and North Africa. Sub-Saharan Africa, Europe, and the Americas have growing Muslim populations as well (Figure 13.4; Pew Forum on Religion and Public Life, 2011). In the United States, Muslims constitute about 1% of the total population (Pew Forum on Religion and Public Life, 2015a). Importantly, African Americans constitute 20% of the total U.S. Muslim population (Mohamed & Diamant, 2019).

Although modern Islam dates to the seventh-century Arab prophet Muhammad, Muslims trace their religion to the ancient Hebrew prophet Abraham, also the founder of Judaism. The precepts of Islam as revealed to Muhammad are contained in a sacred book dictated to his followers and called the *Koran* (or *Qur'an*), which means "recitation." Muhammad's ideas were not initially accepted in his birthplace of Mecca, so in the year 622, he and his followers moved to Medina (both cities are in today's

FIGURE 13.3 ■ Ten Countries With the Largest Christian Populations, 2015 and 2060

2015	2015 CHRISTIAN POPULATION	% OF COUNTRY THAT IS CHRISTIAN	% OF WORLD'S CHRISTIAN POPULATION IN 2015	2060	2060 CHRISTIAN POPULATION	% OF COUNTRY PROJECTED TO BE CHRISTIAN	% OF WORLD'S CHRISTIAN POPULATION IN 2060
1 U.S.	248,180,000	76.9%	10.9%	1 U.S.	262,330,000	63.9%	8.6%
2 Brazil	179,910,000	88.5	7.9	2 Brazil	186,550,000	86.0	6.1
3 Mexico	113,620,000	94.6	5.0	3 Nigeria	174,170,000	37.2	5.7
4 Russia	103,490,000	73.1	4.5	4 D.R. Congo	160,070,000	95.7	5.2
5 Philippines	94,300,000	92.5	4.1	5 Philippines	152,320,000	91.9	5.0
6 Nigeria	86,650,000	48.1	3.8	6 Mexico	127,790,000	90.6	4.2
7 D.R. Congo	72,090,000	95.9	3.2	7 Tanzania	117,930,000	67.4	3.9
8 China	70,890,000	5.2	3.1	8 Uganda	96,200,000	83.8	3.1
9 Ethiopia	57,450,000	62.4	2.5	9 Kenya	91,780,000	81.4	3.0
10 Germany	54,880,000	67.3	2.4	10 Ethiopia	87,660,000	57.1	2.9
Subtotal	1,081,460,000		47.5	Subtotal	1,456,900,000		47.7
Subtotal for rest of world	1,194,790,000		52.5	Subtotal for rest of world	1,597,560,000		52.3
World total	2,278,250,000		100.0	World total	3,054,460,000		100.0

Source: Diamant, J. (2019a, April 1). *The countries with the 10 largest Christian populations and the 10 largest Muslim populations.* https://www.pewresearch.org/fact-tank/2019/04/01/the-countries-with-the-10-largest-christian-populations-and-the-10-largest-muslim-populations/

In the last month of the Muslim year, several million Muslims annually make a pilgrimage (a *hajj*) to the holy shrine of Kaaba in Mecca. The *hajj* is a once-in-a-lifetime requirement for all Muslims, and a journey many save for years. In 2020, the number was drastically curbed due to COVID-19 concerns, and no one over age 65 was allowed to make the pilgrimage.

©STR/AFP via Getty Images

The *Bar Mitzvah* (for boys) and *Bat Mitzvah* (for girls) rituals are an important rite of passage for Jewish youth.

©iStockphoto.com/3bugsmom

Saudi Arabia). This migration, called the *hijra*, marks the beginning of Islam, which later spread across Arabia. Muhammad is not worshipped by Muslims, who believe in positive devotion to Allah (God). Nor is Muhammad a messiah; rather, he is a teacher and prophet, the last in a line that includes Abraham, Noah, Moses, and Jesus.

Islam's sacred *sharia* (or "way") includes prescriptions for worship, daily life, ethics, and even government. Muslim life is governed by the Five Pillars of Islam, which are: (1) accepting Allah as God and Muhammad as Allah's messenger; (2) worshipping according to rituals, including facing toward Mecca and bowing in prayer at five set times each day; (3) observing Ramadan, a month of prayer and fasting during the daylight hours; (4) giving *alms* (donations) to those who are poor or in need; and (5) making a holy pilgrimage to Mecca at least once in a lifetime (Weeks, 1988). The Koran also invokes *jihad* as a spiritual, personal struggle for enlightenment. Terrorist groups, such as the Islamic State (or ISIS), interpret *jihad* as armed struggle against the West and Western values, but this violent, intolerant interpretation is rejected by the vast majority of the Muslim community and scholars of religion:

> French scholarship refers to this relationship as the Islamization of radicals versus radicalization of Islam. Others see Islam as a background cause or more important, as a justificatory instrument employed in the service of a modern extremist totalitarian ideology whose political agenda (e.g., a regional rebalancing of power) far exceeds any observance of Islam. (Ghanbarpour-Dizboni, 2020, p. 404)

Muslims around the world are divided into Sunnis and Shiites. Though both groups share much in common, there are also tensions between the communities in some countries; Iraq, for instance, is Shiite dominated but has a strong Sunni minority. Globally, there are more Sunnis than Shiites: Sunnis are estimated to comprise between 85% and 90% of all Muslims (Pew Research Center, 2011b).

Islam is expected to grow. By 2060, the Pew Research Center estimates there will be an equal number of Christians (3 billion) and Muslims (3 billion) worldwide (Diamant, 2019a). Consider Figure 13.4, which shows the changes in Muslim population by country between 2015 and 2060, predicted by the Pew Research Center (2019). What changes stand out to you? What do you think might cause these changes? How do these compare to changes Pew predicts for Christianity?

Judaism

With 14.3 million followers worldwide, Judaism is the *smallest* of the world's major religions—constituting about .20% of the world's population (Pew Forum on Religion and Public Life, 2015c). Most Jews live in two countries—Israel (home to about 42% of the global Jewish population) and the United States (39%)—with small populations in many other countries, numbering upwards of 100,000–200,000 in countries such as Russia, Argentina, Germany, and Australia to over 250,000–500,000 in countries such as the United Kingdom, Canada, and France (DellaPergola, 2010; Sawe, 2018).

Judaism has exerted a strong influence on the world, first as a key foundation of Islam and Christianity. Second, in European and U.S. culture, Jews have played a role disproportionate to their numbers in such diverse fields as music, literature, science, education, and business. Third, the existence of Israel as a Jewish state since 1948 has given the Jewish people and faith international prominence.

FIGURE 13.4 ■ Ten Countries With the Largest Muslim Populations, 2015 and 2060

2015	2015 MUSLIM POPULATION	% OF COUNTRY THAT IS MUSLIM	% OF WORLD'S MUSLIM POPULATION IN 2015	2060	2060 MUSLIM POPULATION	% OF COUNTRY PROJECTED TO BE MUSLIM	% OF WORLD'S MUSLIM POPULATION IN 2060
1 Indonesia	219,960,000	87.1%	12.6%	1 India	333,090,000	19.4%	11.1%
2 India	194,810,000	14.9	11.1	2 Pakistan	283,650,000	96.5	9.5
3 Pakistan	184,000,000	96.4	10.5	3 Nigeria	283,160,000	60.5	9.5
4 Bangladesh	144,020,000	90.6	8.2	4 Indonesia	253,450,000	88.1	8.5
5 Nigeria	90,020,000	50.0	5.1	5 Bangladesh	181,800,000	91.9	6.1
6 Egypt	83,870,000	95.1	4.8	6 Egypt	124,380,000	99.6	4.2
7 Iran	77,650,000	99.5	4.4	7 Iraq	94,000,000	99.3	3.1
8 Turkey	75,460,000	98.0	4.3	8 Turkey	813,410,000	97.9	3.0
9 Algeria	37,210,000	97.9	2.1	9 Iran	82,980,000	99.7	2.8
10 Iraq	36,200,000	99.0	2.1	10 Afghanistan	81,270,000	99.7	2.7
Subtotal	1,143,200,000		65.2	Subtotal	1,806,790,000		60.5
Subtotal for rest of world	609,420,000		34.8	Subtotal for rest of world	1,180,600,000		39.5
World total	1,752,620,000		100.0	World total	2,987,390,000		100.0

Source: Diamant, J. (2019a, April 1). *The countries with the 10 largest Christian populations and the 10 largest Muslim populations.* https://www.pewresearch.org/fact-tank/2019/04/01/the-countries-with-the-10-largest-christian-populations-and-the-10-largest-muslim-populations/

Judaism was one of the first religions to teach monotheism. Similar to many other religions, it teaches that its followers are God's chosen people, but unlike other religions, it does not teach that followers have a duty to convert others to their faith. The primary religious writing for the Jews is the *Torah* (or "law"), a scroll on which are inscribed the first five books of the Bible. Biblical tradition holds that Jewish law was given by God to Moses when he led the Jews out of slavery in Egypt about 3,500 years ago, and since then, it has been elaborated upon by *rabbis*, or teachers. Today, it is codified in books called the *Mishnah* and the *Talmud*.

Three principal divisions in Judaism reflect differing perspectives on the nature of biblical law. Orthodox Judaism believes that the Bible derives from God and that its teachings are absolutely binding, while Reform Judaism views the Bible as a historical document containing important ethical precepts, but it is not literally the word of God. Conservative Judaism occupies a middle ground, maintaining many traditional practices while adapting others to modern society.

Jews have often suffered persecution, and anti-Semitism has a long global history. From the 12th century on, European and Russian Jews were often forced to live in special districts termed *ghettos*, where they lacked full rights as citizens and were targets of harassment and violence (Tuchman, 1987). Partly in reaction to these conditions, and partly because the Torah identifies Jerusalem as the center of the Jewish homeland, some Jews embraced **Zionism**, *a movement calling for the return of Jews to Palestine and the creation of a Jewish state.* (Zion is a biblical name for the ancient city of Jerusalem.)

Zionism: A movement calling for the return of Jews to Palestine and the creation of a Jewish state.

Zionists established settlements in Palestine early in the 20th century, living for the most part peacefully with their Arab and Palestinian neighbors. Following World War II and the Nazi extermination of 6 million Jews, the state of Israel was created as a homeland for the survivors, an action that both gave refuge to Jews and ignited territorial tensions in the region that continue to this day.

Judaism is expected to grow from 14.3 million to 16.3 million in 2060 (15%) (Pew Research Center, 2017b).

Hinduism

Hinduism, about 2,000 years older than Christianity, is one of the oldest religions in the world and the source of Buddhism and Sikhism. In 2015, 15% of the world's population identified as Hindu

(Pew Research Center, 2015c). Hinduism is not based on the teachings of any single individual, and its followers do not trace their origins to a single deity. It is a broadly defined religion that calls for an ideal way of life. Hindus in 2015 numbered 1.1 billion throughout the world, primarily in India, where they make up the majority of the population.

Indian social structure is characterized by a caste system, officially abolished in 1949 but still powerful, in which people are believed to be born to a certain status that they must occupy for life. This system has its origins in the Hindu belief that one achieves an ideal life in part by performing the duties appropriate to one's caste. Hindus, similar to Buddhists, believe in *samsara*, the reincarnation of the soul according to a person's *karma*, or actions on Earth. Whether someone is reborn into a higher or lower caste depends on the degree to which the person is committed to *dharma*, or the ideal way of life. Although orthodox Hinduism requires observance of caste duties, for the past 500 years, religious societies have organized around *gurus* who break with caste conventions, emphasizing devotional love as the central spiritual act. This tradition influenced Mahatma Gandhi, the leader of India's independence movement, and others who have viewed Hinduism as a vehicle for social reform (Juergensmeyer, 1995).

Perhaps because Hinduism does not have a central organization or leader, its philosophy and practice are particularly diverse. Religious teachings touch all aspects of life, from the enjoyment of sensual pleasures to stark renunciation of earthly pursuits. Hindus believe in the God-like unity of all things, yet their religion also has aspects of **polytheism**, *the belief that there are different gods representing various categories of natural forces.* For example, Hindus worship different gods representing aspects of the whole, such as the divine dimension of a spiritual teacher (Basham, 1989).

Polytheism: The belief that there are different gods representing various categories of natural forces.

The population of Hindus is expected to rise apace with the global population, growing to about 1.4 billion by 2060 (Pew Research Center, 2017b).

Buddhism

Buddhism was founded in India by Siddhartha Gautama five centuries before Christ. According to legend, the young Siddhartha renounced an upper-caste life of material splendor in search of a more meaningful existence. A lifetime of wandering, occasional poverty, and different spiritual practices eventually taught him the way to achieve enlightenment, and he became the *Buddha*, the awakened or enlightened one. Buddhism is an example of a **nontheistic religion**, in that it involves *belief in the existence of divine spiritual forces rather than a god or gods.* It is more a set of rules for righteous living than a doctrine of belief in a particular god. Buddhists make up 7% of the world's population today, numbering 500 million people (Pew Research Center, 2017b).

Nontheistic religion: Belief in the existence of divine spiritual forces rather than a god or gods.

Gautama Buddha's philosophy is contained in the Four Noble Truths. First, all beings—gods, humans, and animals—are caught up in an endless round of suffering and rebirth, the result of their karma or actions. Second, suffering results from desire or attachment. To the extent that we depend on wealth or friends or family or even religious beliefs for satisfaction, we are condemned to suffer unending frustration and loss. Third, suffering can be overcome if we break the endless cycle of karma and rebirth and achieve nirvana—a blissful state of emptiness. Fourth, the means of achieving nirvana are contained in the Eightfold Path, which advocates ethical behavior, a simple lifestyle, renunciation of material pleasures, meditation, and (eventually) enlightenment.

Many people practice certain precepts of Buddhism without calling themselves Buddhists, so it is difficult to estimate accurately the number of Buddhists in the world today. Estimates put the figure at about 500 million, a figure that is, interestingly, expected to decline to 462 million in 2060 because of low fertility rates and aging populations in three countries where Buddhism is prevalent—China, Thailand, and Japan (Pew Forum on Religion and Public Life, 2017a, para. 17). Theravada Buddhism—the "Way of the Elders" or the "Lesser Vehicle"—predominates in Myanmar, Thailand, Laos, Cambodia, and Sri Lanka. It is strongly identified with local cultures, and traditional rulers in Thailand and Sri Lanka are religious figures as well. Mahayana Buddhism—the "Greater Vehicle"—is practiced in China, Korea, and Japan, where it is not the official religion but mixes with other cultural strands. In China, Buddhism is intertwined with Taoist folk religion and Confucian codes of ethical practice. In Japan, Shinto Buddhism combines emperor worship with elements of **animism**, *the belief,*

Animism: The belief that naturally occurring phenomena, such as mountains and animals, are possessed of indwelling spirits with supernatural powers.

common to many religions, *that naturally occurring phenomena, such as mountains and animals, are possessed of indwelling spirits with supernatural powers.*

Buddhism's meditative lifestyle may strike some people as incompatible with life in a modern industrial society, which emphasizes work, consumption, achievement, and all forms of karma that Buddhists regard as barriers to enlightenment and happiness. On the other hand, it is primarily Buddhist monks rather than ordinary practicing Buddhists who devote extended periods to meditation. Perhaps because of its emphasis on contemplation and meditation, Buddhism continues to attract followers in Western countries, including celebrities such as actor Orlando Bloom, singer Tina Turner, late Apple founder Steve Jobs, and golf star Tiger Woods (Lampman, 2006; MacLeod, 2011).

Confucianism

Confucianism, the name of which comes from the English pronunciation of the name of its founder, K'ung-Fu-tzu (551–479 B.C.), is more of a philosophical system for ethical living on Earth than a religion honoring a transcendental god (Fingarette, 1972). K'ung-Fu-tzu never wrote down his teachings, but his followers compiled many in a book called *The Analects* that became the foundation of official ethics and politics for some 2,000 years in China, until Confucianism was banned after the communist revolution in 1949. Although there are only about 6 million practitioners of Confucianism—almost all in Asia—this religion has had enormous influence in China, neighboring countries such as Korea and Vietnam, and Japan (Barrett, 2001).

Confucianism emphasizes harmony in social relations; respect for authority, hierarchy, and tradition; and the honoring of elders. Rulers are expected to be morally virtuous, setting an example for others to imitate. The group is more important than the individual. The key element in Confucian ethics is *jen*, meaning "love" or "goodness" and calling for faithfulness and altruism; we should never do anything to another person we would not want done to ourselves. Contemporary Confucianism, sometimes called "neo-Confucianism," has mystical elements as well as moral ones, including belief in the *Tao* (pronounced *dow*), or "way of being," determined by the natural harmony of the universe.

Confucianism teaches that opposites are not necessarily antagonistic; together, they make up a harmonious whole that is constantly in a creative state of tension or change. The two major principles in the universe, *yin* (the female principle) and *yang* (the male principle), are found in all things, and their dynamic interaction accounts for both harmony and change. The *I Ching* (Book of Changes) contains philosophical teachings based on this view and a technique for determining a wise course of action.

Scholars have argued that Confucian values such as respect for authority and a highly disciplined work ethic partly explain rapid economic growth in Singapore, Taiwan, China, and other Asian countries (Berger, 1986; Berger & Hsiao, 1988; MacFarquhar, 1980). On the one hand, this perspective supports Weber's argument that actions rooted in religious beliefs are linked to economic development. On the other hand, it challenges Weber's exclusive emphasis on Western religion, and Protestantism in particular, as the source of this development.

Establishment Clause: The passage in the First Amendment to the U.S. Constitution that states, "Congress shall make no law respecting an establishment of religion, or prohibiting the free exercise thereof."

RELIGION IN THE UNITED STATES

The United States is the most religiously diverse country in the world, with an estimated 1,500 distinct religions (Melton, 1996; Smith, 2002). Many of those groups are small, however, with most American adherents belonging to the world's major religions we previously explored. Compared to citizens of other post-industrial nations, Americans are unusually religious, practicing their religions in public forums, including houses of worship and also virtually, with about a fifth saying that they regularly share their faith online and half noting that they have seen someone else share their faith online (Pew Forum on Religion and Public Life, 2014). Many also practice it in private: 55% indicate that they pray daily (Lipka, 2015a). The First Amendment to the U.S. Constitution's **Establishment Clause** ("*Congress shall make no law respecting an*

COVID-19–related health concerns caused many religious groups to suspend gathering and move to online worship. For some religious groups, however, COVID-19 created freedom of religion concerns as states in the U.S. barred or restricted mass gatherings including religious ones. In July 2020 the U.S. Supreme Court refused to declare a 50-person ban on worship services as unconstitutional.

FIGURE 13.5 ■ If the United States Had 100 People: Charting Americans' Religious Affiliations

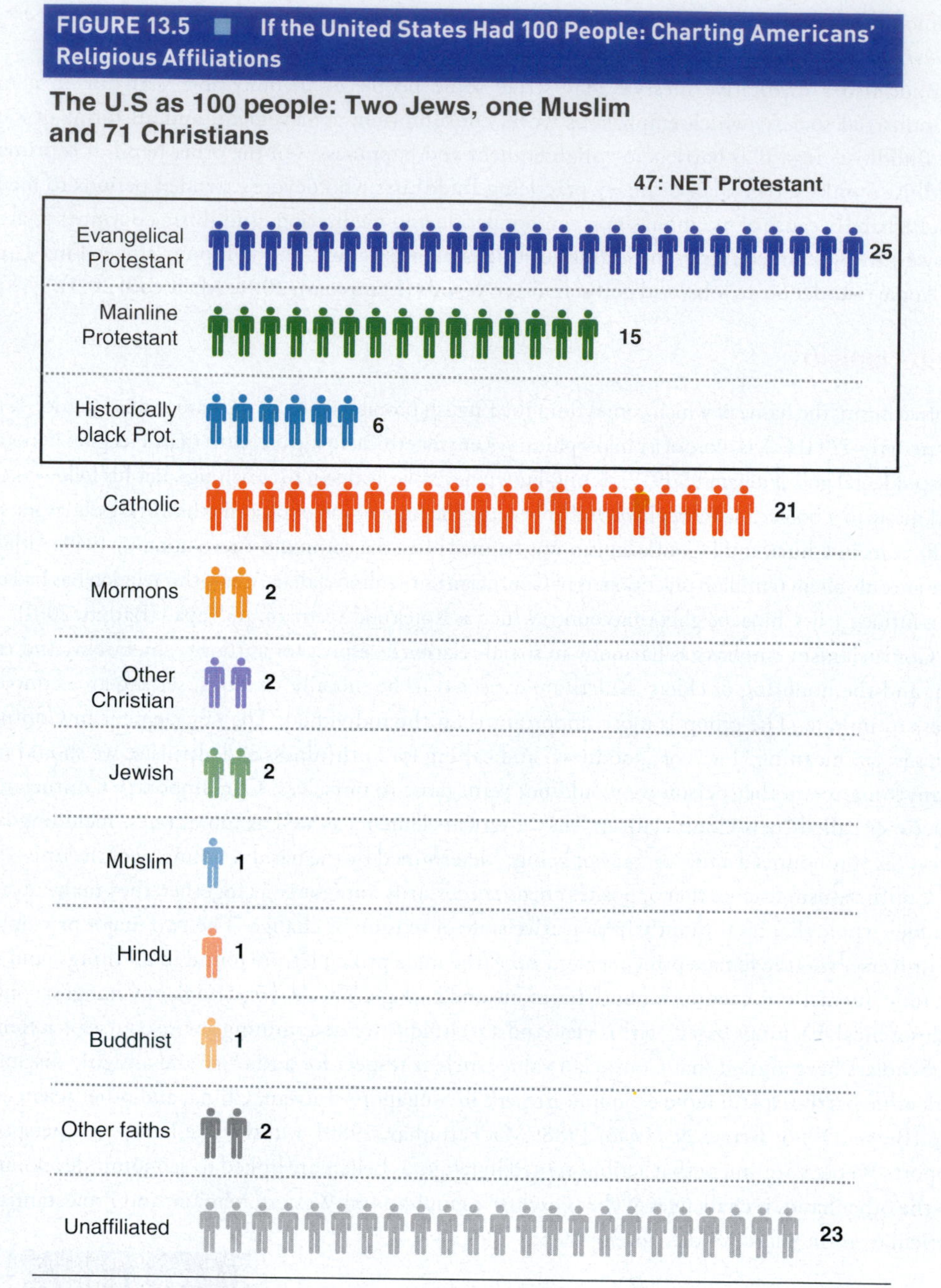

Source: Sandstrom, A. (2016, November 14). *If the U.S. Had 100 people: Charting Americans' religious affiliations.* Pew Research Center. https://www.pewresearch.org/fact-tank/2016/11/14/if-the-u-s-had-100-people-charting-americans-religious-affiliations/

establishment of religion, or prohibiting the free exercise thereof") codifies a separation of church and state. But, in America, politics have become increasingly intertwined with religion. Indeed, ideas about America can take on the look and feel of their own religion.

Most American adults profess an affiliation with a religious group: In interviews conducted between 2018 and 2019, 65% of Americans self-identified as Christian, with smaller numbers identifying as Jewish (2%), Muslim (0.9%), and Buddhist (0.7%) (Pew Forum on Religion and Public Life, 2019c). Figure 13.5 shows another way to consider religious diversity in the United States, and it importantly illustrates another consideration to which we will turn in this section: the fact that a portion of the American public, including an increasing proportion of youth, is also "unaffiliated."

In the sections that follow we take a closer look at all of these aspects of American religiosity, including trends in affiliation, civil religion, and religion and politics. We conclude with a brief discussion of the importance of considering American religiosity intersectionally.

Trends in Religious Affiliation

The continuing high levels of religious group membership among Americans can be traced, in part, to the religious socialization, and the social and friendship networks that churches, synagogues, and mosques provide to connect people who share the same beliefs and values. Additionally, as we explained in the religious economy perspective earlier, the United States provides an enormous number of such organizations to which to belong.

Importantly, religious affiliation has taken a dramatic downward turn in recent decades. The proportion of the U.S. population that describes itself as *unaffiliated* is rising. This change has been driven most fully by younger generations, who are more likely than young adults of earlier generations to indicate no religious affiliation (Figure 13.6; Lipka, 2015b); in studies done in the 1970s and 1980s, about half as many young adults identified themselves as unaffiliated with a religion. These trends are also true of European and Latin American youth (Cornelio, 2020). The Pew Forum on Religion and Public Life (2010) suggests that the rise of youth who say they are unaffiliated is "a result, in part, of the decision by many young people to leave the religion of their upbringing without becoming involved with a new faith" (p. 4). Additional reasons, according to Cornelio (2020), are overall societal secularization (young adults may not be exposed to beliefs and practices as they are growing up and thus may be unlikely to attach to a religious group), and religious individualization of expression (young adults may not need an institution to feel that they can relate to the sacred, saying they are "spiritual but not religious") (p. 927).

Data show that the unaffiliated are more likely than the affiliated to say that religious organizations are overly concerned with money and power, too focused on rules, and too closely involved in politics (Pew Forum on Religion and Public Life, 2012). Robert Wuthnow (2007) has argued that declining affiliation correlates with other demographic changes, including declining rates of marriage and parenthood among young adults. Research also suggests a link between the decline of religious affiliation and the decline of social capital as defined by Robert Putnam (2000), who writes that Americans are increasingly likely to live separate lives and disengage from community activities, a phenomenon he famously termed "bowling alone."

FIGURE 13.6 ■ Changing Religious Affiliations of U.S. Adults by Generation, 2014

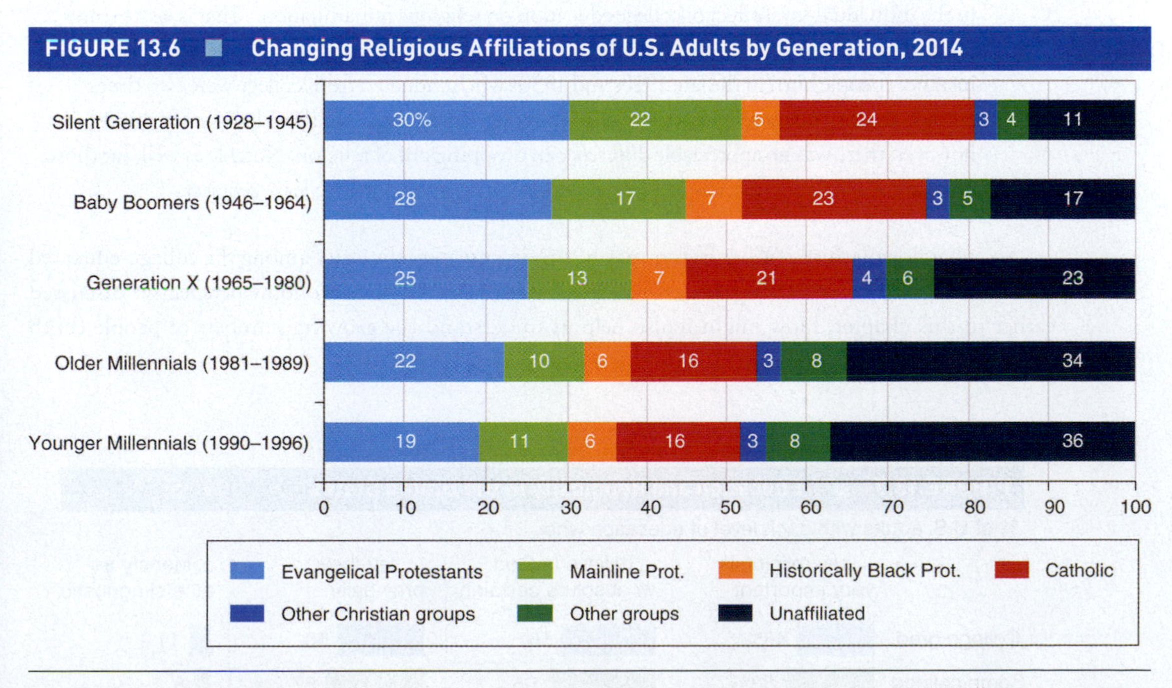

Source: Pew Research Center. (2015, May). *Millennials increasingly are driving growth of "nones."* http://www.pewresearch.org/fact-tank/2015/05/12/millennials-increasingly-are-driving-growth-of-nones/

Note: 2014 Religious Landscape Study conducted June 4–Sept. 30, Figures may not add to 100% because of rounding. Don't know/refused answers not shown. "Other Christian groups" includes Mormons, Orthodox Christians, Jehovah's Witnesses, and a number of smaller Christian groups.

Education is another possible factor to consider in the story of religious affiliation. Secularization theory posits that religiosity (strong religious beliefs, feelings, and practices) declines as societies modernize (Weber, 1963). Secularization has long been understood as a phenomenon that more fully affects the highly educated. Religiosity, by extension, is also widely held to correlate negatively with educational attainment. That is, more highly educated people are less likely to profess or be committed to organized religion. Data, including a Pew Research Center study (2017a), also seem to point to that conclusion: Among those with a high school education or less, fully 58% indicate that religion is "very important" to them, whereas about 46% of college graduates say the same. An "absolute certainty" about the existence of God is also more widely shared among those with some college or a high school education or less than among those with a college degree, and the more highly educated are far more likely to profess no religion (Figure 13.7).

At the same time, data also show considerable variation in this relationship based on membership in particular religions and generations. Consider the following:

- College-educated people who identify with Christianity are as likely or even more likely to say that they are religiously observant than are less-educated members of their religion. For instance, about 52% of college-educated Christians indicated that they attend church services weekly compared to 45% of those with some college education and 46% of those with a high school degree or less.
- "Religious commitment" for Christians (attendance at religious services, frequency of prayer, belief in God, and the self-reported importance of religion in one's life) seems to show little variation between those with a completed college education (70%), some college (73%), and no college (71%). By contrast, among Jews, there is considerably more variation, with just over half of Jews with some or no college saying that they believe in God with "absolute certainty" (54%), while only about 28% of college graduates saying the same (Pew Forum on Religion and Public Life, 2013, 2017a).
- Generational differences in the connection between education and religiosity are important. Sociologist Philip Schwadel (2020) argues that "increases in higher education have led to a decline in the individual-level effect of college education on religious nonaffiliation." That is, as a greater proportion of each generation has gained access to higher education, the effect has dissipated. For instance, people born in the late 1920s and 1930s who graduated from college were two times as likely to drop out of religion than were non–college graduates. For those born in the 1960s, however, there was no appreciable difference in dropping out of religion. Notable as well, for those born in the 1970s, those *without* a college education are more likely to leave religion.

So, what do you think? What factors may be driving greater religiosity among the college-educated and declining religiosity among the less-educated? Recall the religious economy perspective discussed earlier in this chapter. How might it also help us understand the growing numbers of people (both youth and adults) who do not affiliate with a religion?

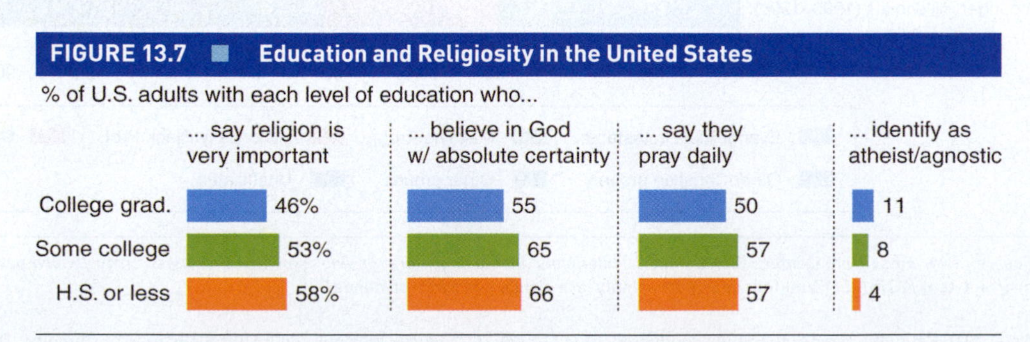

FIGURE 13.7 ■ Education and Religiosity in the United States

Source: Pew Research Center. (2017, April 26). *In America, does more education equal less religion?* http://www.pewforum.org/2017/04/26/in-america-does-more-education-equal-less-religion/

Religion and Politics

Protestantism in the United States can be divided into three strands: conservative, moderate, and liberal. Conservative Protestants emphasize a literal interpretation of the Bible, Christian morality in daily life as well as public politics, and conversion of others through evangelizing. Liberal Protestants generally adopt a more flexible, humanistic approach to their religious practices, and moderate Protestants are somewhere in between. All groups grew from the 1920s through the 1960s, but both liberal and moderate churches have since experienced a decline in membership, while the number of conservative Protestants has grown (Pew Forum on Religion and Public Life, 2008).

A key aspect of this growth is the rise of **evangelicalism**, *a belief in spiritual rebirth (conventionally denoted as being "born again")*. This often includes the admission of personal sin and salvation through acceptance of Christ, a literal interpretation of the Bible, an emphasis on highly emotional and personal spiritual piety, and a commitment to spreading "the Word" to others (Balmer, 1989). (The word *evangel* comes from the Greek for "bringing good news.") We can interpret the rise of evangelicalism as a response to growing U.S. secularism, religious diversity, and, in general, the decline of once-core Protestant values in U.S. life (Wuthnow, 1988). Megachurches, which we introduced previously, are often associated with evangelicalism.

Evangelicalism: A belief in spiritual rebirth (conventionally denoted as being "born again").

Conservative Protestants have been an active force in U.S. politics, and their growing numbers may be linked to a recent rise in their political influence, particularly within the Republican Party. Since the 1980s they have, in fact, become a "key voting bloc in presidential elections" (Black, 2016, p. 15). Data from the past decade confirm the emergence of a "worship attendance gap" in presidential politics; specifically,

> [T]he more observant members of religious communities tend to vote Republican while their less observant co-religionists tend to vote Democratic. This attendance gap has been largest among the white Christian traditions, but has appeared in a more modest form within nearly all religious affiliations. (Dionne & Green, 2008, p. 5)

Not only are strongly religious voters more likely to vote Republican, but they also seem increasingly more likely to vote overall. That is, they are active in politics to a degree that some other groups are not.

Many U.S. voters' political beliefs and actions are shaped by religious faith. Not surprisingly, then, some voters expect their political leaders to share their faith: A 2016 survey showed that 51% of

FIGURE 13.8 ■ American Support for an Atheist Presidential Candidate, 2016

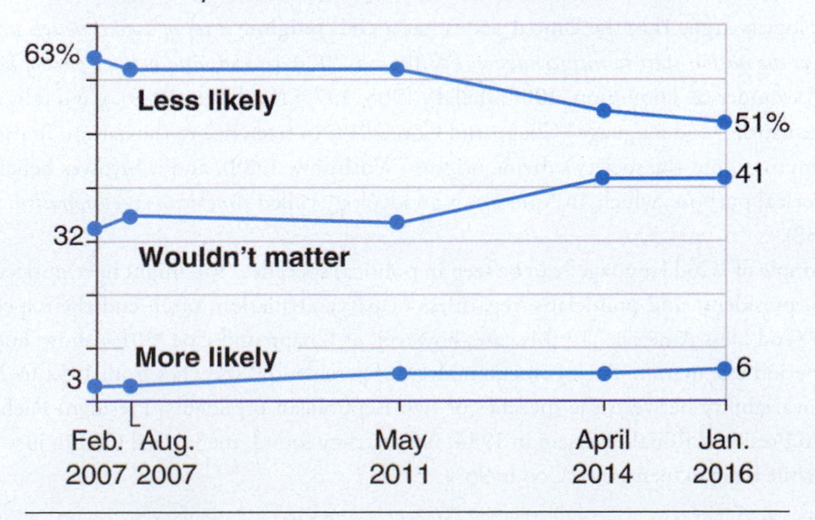

Source: Pew Research Center. (2016, June 1). *10 facts about atheists*. http://www.pewresearch.org/fact-tank/2016/06/01/10-facts-about-atheists/

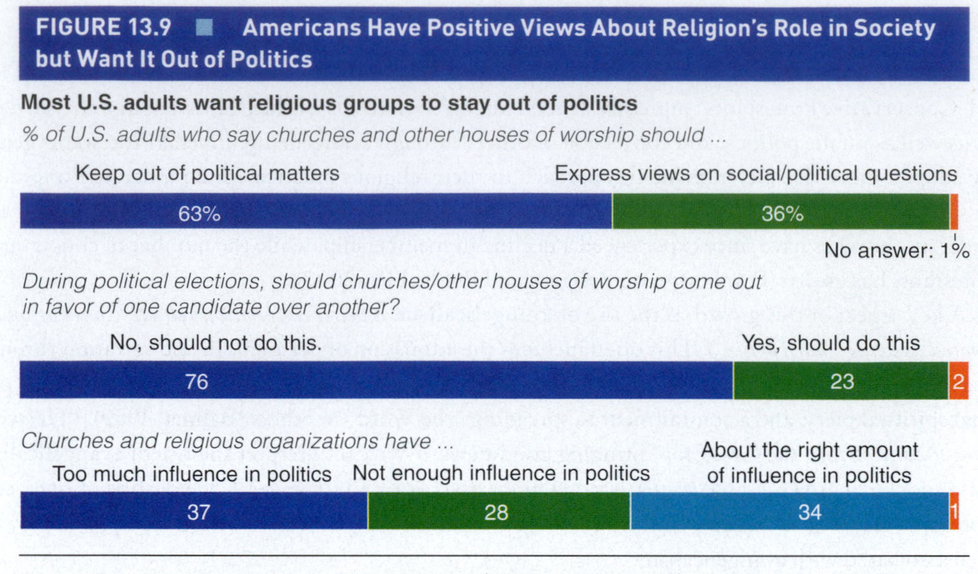

Source: Pew Forum on Religion and Public Life. (2019d, November 15). *Americans have positive views about religion's role in society but want it out of politics.* https://www.pewforum.org/2019/11/15/americans-have-positive-views-about-religions-role-in-society-but-want-it-out-of-politics/

respondents would be less likely to support a presidential candidate who "does not believe in God." At the same time, some belief systems do not confer an advantage: 42% of respondents indicated they would not vote for a Muslim candidate and 23% would reject a Mormon candidate. The influence of religion on presidential politics may, however, be on the wane: The same study showed that a growing proportion of Americans would not be influenced by a candidate's professed atheism (Figure 13.8; Pew Research Center, 2016b), and a more recent poll found that two thirds of Americans wanted religious groups to stay out of political matters (Figure 13.9; Pew Forum on Religion and Public Life, 2019d).

Finally, the emerging religious nones are, as sociologist Phil Schwadel (2020) points out, as large in numbers as evangelical Protestants and Catholics, and, as such, are also important to study as groups and individual actors in American politics (p. 180). Schwadel finds, however, that the "nones" can actually best be conceptualized as three groups: (1) those who view their religion as "nothing in particular (or NIPS); (2) atheists; and (3) agnostics. He finds that while atheists and agnostics are likely to be Democrat, atheists are much more likely to be politically active than agnostics, while NIPS are more likely to be Republican and "uninterested in politics" (p. 188).

"Civil Religion" in the United States

Civil religion: A set of sacred beliefs and practices that connect the nation-state to transcendence and become part of how a society sees itself.

Some sociologists argue that the United States has a **civil religion**, *a set of sacred beliefs and practices that "connect the nation-state to transcendence"* (Williams, 2020, p. 138) *and become part of how a society sees itself* (Alexander & Thompson, 2008; Bellah, 1968, 1975; Mathisen, 1989). Civil religion usually involves the use of "God language" Chapp and Coe (2019) in reference to the nation. It also draws on historical myths about the society's divine origins (Wuthnow, 1988), and it involves beliefs about its sacred historical purpose, which, in America, is an ideology called *American exceptionalism* (Williams, 2020, p. 138).

An example of "God language" can be seen in political speeches. You might have noticed that even today, U.S. presidents and politicians, regardless of party affiliation, often end their speeches with the words "God bless America." In this case, however, as Chapp and Coe (2019) show, audience and historical period can matter. Their content analysis of presidential speeches from 1952 to 2016 shows interesting variability between the speeches of two Republican presidents: President Richard Nixon in 1972 and President Ronald Reagan in 1984. Nixon, they found, mentioned God in just 16% of his speeches, while Reagan mentioned God in 95%:

> In 1972, Richard Nixon was seeking reelection as U.S. president. Facing South Dakota Senator George McGovern in what would eventually become a landslide victory for Nixon, the

> incumbent president talked about many things over the course of the campaign—especially his plan to end the Vietnam War and the opening of formal relations with China. One thing Nixon rarely talked about, however, was God. Over the course of his campaign, Nixon mentioned God in just 16 percent of his speeches. When all was said and done, Nixon had employed less religious language than any other television-era presidential candidate. Contrast these rhetorical choices with those made by another Republican president who, roughly a decade later, would also coast to reelection. Ronald Reagan, facing Minnesota Senator Walter Mondale in 1984, mentioned God in 95 percent of his speeches. Over the course of that campaign—often remembered for its famous "Morning in America" political ads—Reagan's use of religious language reached a high that is still unmatched in modern presidential campaign history. (p. 398)

The infusion of divine meaning into historical events is visible in President Abraham Lincoln's (1809–1865) oratory on the Civil War. In the Gettysburg Address, Lincoln, who was not highly religious himself, drew on religious imagery to attach meaning to the brutal destruction and enormous human cost of the war. He labeled his fellow citizens the "almost chosen people" and called the United States "the last, best hope on earth."

Civil religion is also associated with the Pledge of Allegiance. Consider its phrase "one nation under God," a phrase added by Congress only in 1954, 2 years after President Harry Truman declared May 7 the National Day of Prayer, a day when presidents proclaim that "the people of the United States may turn to God in prayer and meditation at churches, in groups, and as individuals" (Lipka, 2015a). Historically speaking, both of these decisions were taken, in part, to underscore the contrast between the "Christian" United States and the "atheistic" Soviet Union at a time when the rival countries were locked in a tense political relationship during the cold war and Americans feared the threat of "godless communism." The phrase "under God" is consistent with the construction of a civil religion and a nation worshipping itself and envisioning itself as fulfilling a divine destiny. Even though the First Amendment to the U.S. Constitution clearly calls for a separation of church and state, this profession of allegiance to "one nation, under God" has come to be seen as central to U.S. citizenship. Notably, however, the solicitor general of the United States has argued before the Supreme Court that "under God" is "descriptive" and "ceremonial" rather than a prayer or "religious invocation" (Pew Forum on Religion and Public Life, 2004), suggesting that it embodies a societal rather than an overtly religious value.

Civil religion in the United States is found in other dimensions of American life, such as the widespread societal veneration of sports and winning athletes. Popular sporting events such as the Olympic Games, the Super Bowl, and the World Series are infused with patriotic rituals that reinforce national pride and emphasize a sense of American exceptionalism. Finally, while we here explore civil religion as an American phenomenon, it is not unique to America. Sociologists of religion have studied it as a phenomenon in various global locations including "North Africa, Denmark, Australia, Korea, Chile, and Malaysia" (Williams, 2020, p. 138).

What do you think of the concept of civil religion? Do you think the content of the Pledge of Allegiance endorses religion, or is it a civil rather than a religious declaration? What do you think is the function of the phrase "under God" in the Pledge of Allegiance? Should we interpret it as patriotic or religious?

The Olympic Games might be considered a component of civil religion. The games, as well as ads that are shown during the Olympics, are suffused with patriotic content, and winning may reinforce a widespread sense of national pride.

©XIN LI/Getty Images Sport/Getty Images

Religion, Gender, Sexuality, and Race

As we noted earlier, sociologists see religions as social institutions that reflect the norms, roles, values, and practices embraced by communities and societies. As such, they also reflect societal inequalities and stereotypes around gender, sexuality, and race; influence behaviors; and can be a source of social inclusion and solace or exclusion and conflict.

Worldwide and in the United States, women are more likely than men to belong to, attend, and engage in religious activities (Pew Forum on Religion and Public Life, 2016). And yet, religion remains a "gendered institution," where gender is a fundamental way of organizing religious structures and practices (Acker, 2006) and the principal deities, prophets, leaders, and religious spaces are historically male. The prolific 20th-century feminist sociologist Charlotte Perkins Gilman spent much of her career pointing out the social and economic inequalities of gender inherent in patriarchal society. Her book *His Religion and Hers* (1922/2003) showed the importance of understanding how the messages of gender were different for men and women in religious texts and practices. Like Elizabeth Cady Stanton's two-volume work *The Women's Bible* (1898), Gilman sought to sensitize women and men of the 20th century to the fact that God is depicted as male, beliefs typically emphasize male religious and political superiority, and women are often excluded from positions of theological power.

In the 1960s–1990s an organized upsurge of second wave feminism influenced a feminist spirituality movement within and outside of organized religions. Women sought reform within Christianity and Judaism for the rights of women to be faith leaders; for the reimagination of God as ungendered; for nonsexist language in Scriptures, rituals, and services; and for the redesigning of traditions and rituals along nonsexist lines (Eller, 2000; Wallace, 1992; Weidman, 1984). Some women left organized groups to reestablish religious traditions that predated Judaism and Christianity, including the celebration of goddess-based religions such as Dianic Wicca. Today, more Christian Protestant denominations have female clergy, and even some Catholic churches offer women the opportunity to act as lay clergy, although the mainstream Catholic Church continues to oppose the practice. In Judaism women have become ordained rabbis in Conservative, Reform, and Reconstructionist Judaism (Hein, n.d.).

On the other hand, the same glass ceiling that keeps women from rising to the highest positions in the workforce is also operating as a *stained-glass ceiling* when it comes to religion. For instance, Church Clarity (a nonprofit organization that ranks the 100 largest American evangelical churches featured in the Christian publication *Outreach Magazine*) found that only 1 of the 100 churches had a woman senior pastor. Importantly, she did not serve alone but was instead co-pastoring with her husband (Church Clarity, 2017). Muslim women also encounter gender barriers. For instance, in the Islamic Society of North America's report on women-friendly mosques in the United States, researchers found that only 14% of mosques scored excellent, while 63% scored only fair or poor on four dimensions: providing women the ability to attend Friday services, physical barriers, women's programming, and ability to serve on the mosque boards. Interestingly, the report found that Shiite mosques are more women-friendly than Sunni mosques, and that African American mosques also tend to be more women-friendly (Sayeed et al., 2013).

Like gender, sexuality is also an important area of study in religion. Many religious groups have historically regulated their members' sexualities from proscriptions around virginity to strictures regarding heterosexuality and homosexuality. Page (2020) explains that the relationship between religion and sexuality is "complex," and that American religions have run the gamut from support of LGBTQ+ people to hostility. Indeed, a Pew 2013 study of LGBTQ+ Americans found that of the 52% who are religiously affiliated in the sample, "one-third say there is a conflict between their religious beliefs and their sexual orientation or gender identity," and just under one third (29%) say "they had been made to feel unwelcome in a place of worship" (Pew Research Center, 2013, para. 2). It is likely that the experience of inclusion or exclusion varies by religious group and even by region. For instance, Barton's (2012) auto-ethnographic research of 59 gay men, lesbians, and bisexuals in the Bible Belt reveals her LGBTQ+ respondents have experienced increased scrutiny, shame, and stigma in an area where conservative Christianity permeates schools, homes, and businesses. On the other hand, LGBTQ+ Christians, Jews, and Muslims have also experienced religious groups who have embraced them, creating LGBTQ+–friendly spaces and support networks (Page, 2020). Finally, Whitehead's 2013 study of Christian congregations shows that gender and sexuality are intersectional—that is, the acceptance of women to head their clergy, according to Whitehead (2013), is statistically associated with their acceptance of gay and lesbian leadership in churches. Even in groups that appear to be inclusive in leadership, however, "underlying assumptions of the congregation continue to limit them" (Ammerman, cited in Whitehead, 2013, p. 490).

Race and ethnicity are also important in the practice and landscape of American religion. In the early 20th century, sociologist William Edward Burghardt Du Bois (1868–1963) produced the first sociological insights on religion and African Americans in the United States. Enlisting statistics, observation, and his insights as a Black sociologist who had experienced the impact of living in a racist society, Du Bois discussed the essential spiritual, social, and economic role of the Black church (1899/1996; 1903/1989). He also brought attention to the sheer amount of Black churches in America, noting in *The Souls of Black Folk's Faith of Our Fathers* that *"the census of 1890 showed nearly 24,000 Black churches in the country with total enrolled membership of over 2.5 million."* Zooming forward 60

FIGURE 13.10 ■ The Most and Least Racially Diverse U.S. Religious Groups

How Racially Diverse Are U.S. Religious Groups?

% of each religious group in each racial/ethnic category, and each group's diversity score on the Herfindahl-Hirschman index

Group	Values as printed (White, Black, Asian, Mix/Other, Latino)	Index
Seventh-day Adventist	37% 32 8 8 15	9.1
Muslim	38 28 28 3 4	8.7
Jehovah's Witness	36 27 6 32	8.6
Buddhist	44 3 33 8 12	8.4
"Nothing in particular"	64 12 5 5 15	6.9
Catholic	59 3 3 2 34	6.7
All U.S. adults	***66 12 4 4 25***	***6.6***
Assemblies of God	66 3 5 25	6.2
Church of God (Cleveland, Tenn.)	65 3 3 28	6.2
Churches of Christ	69 16 4 10	6.1
American Baptist Churches USA	73 10 5 11	5.5
Atheist	78 3 7 2 10	4.7
Agnostic	79 3 4 4 9	4.5
Presbyterian Church in America	80 6 3 5 6	4.4
Orthodox Christian	81 8 3 2 6	4.2
Anglican Church	83 12 4	3.7
Church of God in Christ	5 84 4 8	3.5
Southern Baptist Convention	85 6 5 3	3.4
Mormon	85 5 8	3.4
Presbyterian Church (U.S.A.)	88 5 3 4	2.8
Church of the Nazarene	88 2 3 7	2.7
Unitarian	88 7 4	2.7
United Church of Christ	89 8 2	2.5
Jewish	90 2 2 2 4	2.3
Episcopal Church	90 4 3 2	2.3
Hindu	4 2 91 2	2.1
United Methodist Church	94 2 2	1.4
African Methodist Episcopal Church	2 94 3	1.4
Lutheran Church-Missouri Synod	95 2 2	1.2
Evang. Lutheran Church in America	96 2	1.0
National Baptist Convention	99	0.2

MORE DIVERSE ▲ ▼ LESS DIVERSE

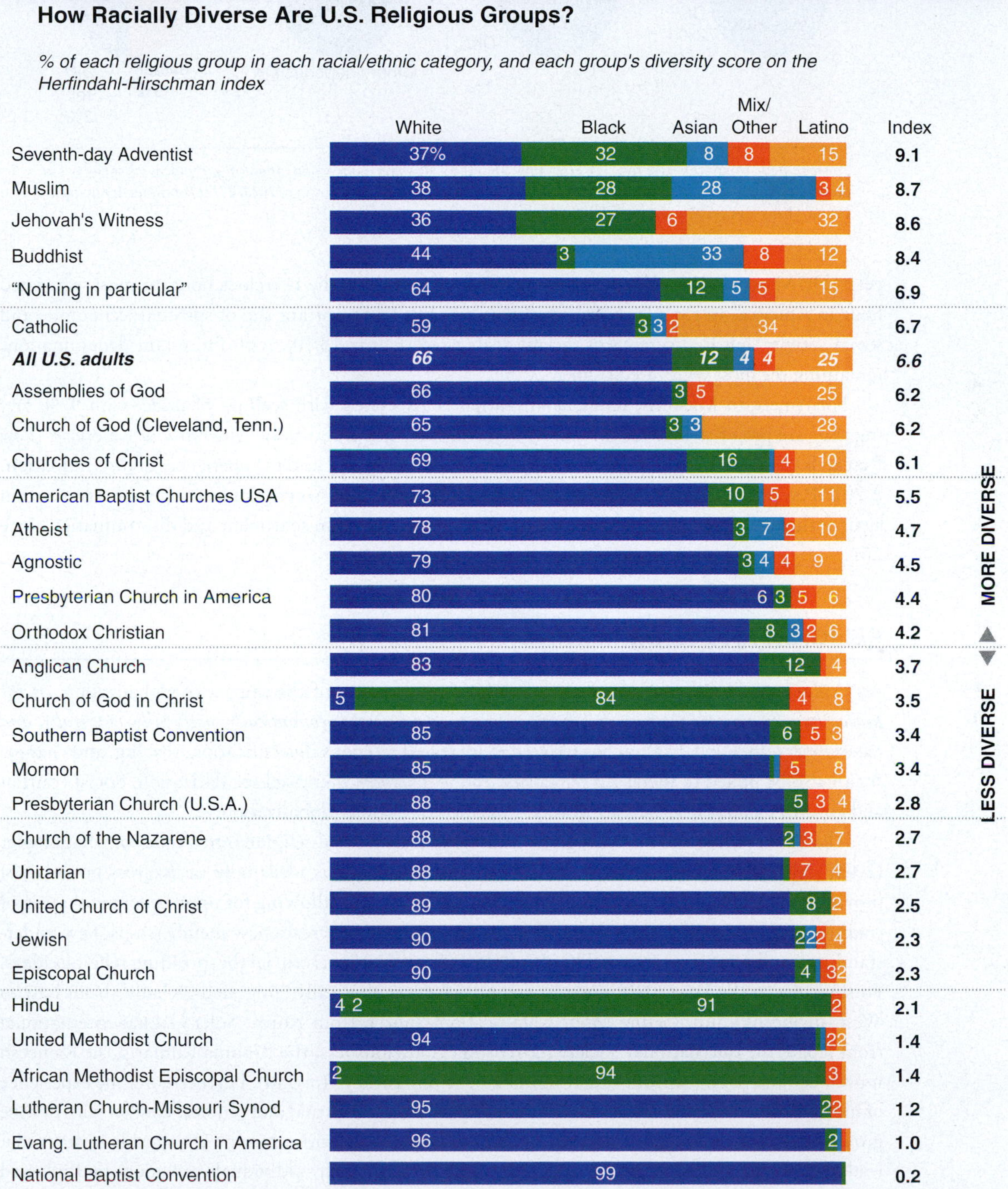

Source: Lipka, M. (2015c, July 27). *The most and least racially diverse U.S. religious groups.* https://www.pewresearch.org/fact-tank/2015/07/27/the-most-and-least-racially-diverse-u-s-religious-groups/

Note: Pew Research Center's 2015 analysis of the most and least racially diverse religious groups consists of five ethnic and racial group categories: Hispanics, as well as non-Hispanic whites, Blacks, Asians, and an umbrella category of other races and mixed-race Americans. If a religious group had exactly equal shares of each of the five racial and ethnic groups (20% each), it would get a 10.0 on the index; a religious group made up entirely of one racial group would get a 0.0. By comparison, U.S. adults' overall rate at 6.6 on the scale. And indeed, the purpose of this scale is to compare groups to each other, not to point to any ideal standard of diversity (Lipka, 2015c, para. 3).

FIGURE 13.11 ■ Among U.S. Muslims, Widespread Concerns About Their Place in Society

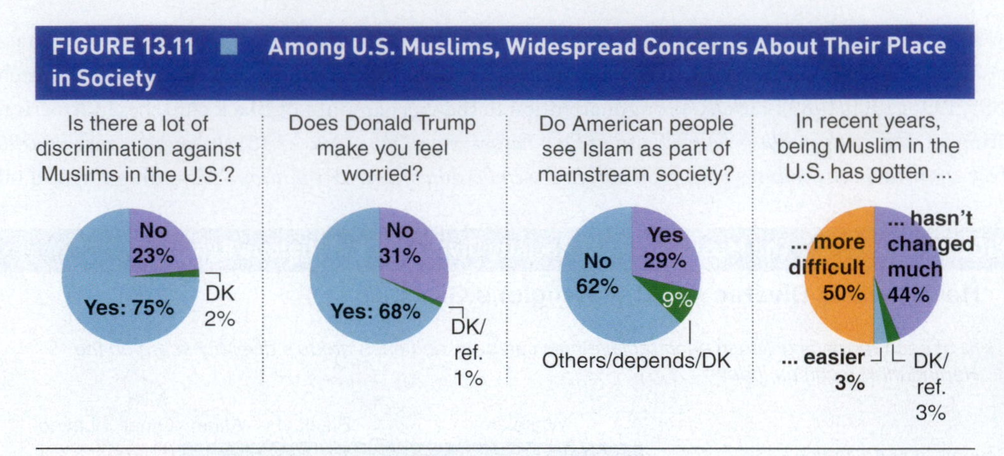

Source: Pew Research Center. (2017b, July 26). *U.S. Muslims concerned about their place in society, but continue to believe in the American dream.* https://www.pewforum.org/2017/07/26/findings-from-pew-research-centers-2017-survey-of-us-muslims/

years, Dr. Martin Luther King Jr. is famously quoted for calling the 11 o'clock hour the most segregated hour in America. In the 21st century, religion remains an important site of spiritual, economic, and social activity, but it also remains racially segregated (Figure 13.10), with Protestant denominations remaining the most segregated (Lipka, 2015c).

For American Muslims, religious affiliation is associated with feelings of pride, with 97% saying they are proud to be Muslim and 89% saying they are proud to be Muslim *and* American (Pew Research Center, 2017b). Being Muslim American, however, can lead to experiences of discrimination. A 2017 Pew poll (Figure 13.11) reveals that a majority of Muslim Americans feel there is discrimination against them, experiencing an increase of feelings of anti-Muslim sentiment and discrimination since 2011.

RELIGION AND GLOBALIZATION

An exploration of religion is incomplete without a discussion of the impact of globalization, or *the process by which people worldwide become increasingly connected economically, politically, culturally, and environmentally.* Globalization has resulted in increased religious diversification, sharing, and change. It can also be linked to increased religiosity, and, conversely, increased secularization. Social conflict and governmental restrictions are also associated with religious globalization.

Thanks in part to the Internet, globalization has transformed religion into a fluid, or what Liogier (2020) calls a "de-territorialized" form. That is to say, religious ideas now easily cross boundaries, providing unparalleled contact among religious traditions, and allowing for unprecedented sharing of traditions and ideas. The flow and exchange of new ideas may create new memberships, new understandings, and new connections. The global flow of people, too, is crucial for spreading religious ideas. For instance, as Barker (2020) notes, many NRMs have physically "migrated globally" from East to West, including "Sun Myung Moon's Unification Church from Korea, Soka Gakkai International from Japan, the International Society for Krishna Consciousness, the Brahma Kumaris, the Rajneesh movement and Sai Baba from India, and Hizmet from Turkey" (p. 536). Furthermore, the experience of individuals and identities is crucial to this story—increased virtual and physical contact with diverse groups and ideas also means that even *individual* religious identities may be changeable in ways that lead toward novel religious ideas; or, as Liogier (2020) points out, individuals can adopt multiple and seemingly conflicting identities given their potential reach with different types of religious communities simultaneously (e.g., online and real-time).

Globalization can also destabilize religions and promote secularization. For instance, the global emergence of a scientific/technological culture that influences people's views and behaviors also presents a powerful challenge to traditional religions. It means that, in some countries, many fewer people employ religious explanations for natural or social phenomena, as increasing numbers probe beyond explanations that refer to a divine plan. Religion may still enrich personal lives, but its role in public life—at least in the most economically advanced countries—has declined. Thus, the physical migration of religious groups to secularized nations can result in state or intergroup conflict and tensions. This is the case of Muslim immigrants in France where France "continues to enforce a national ban on full-face coverings in public, and local authorities also impose various restrictions that mostly affect Muslim women" (Pew Research Center, 2019a, para. 12).

The growing contact between large religions with mass followings that claim to possess exclusive accounts of history and the nature of reality can fuel "culture wars" as well as real wars that leaders cast as battles between religions and values or even between good and evil. In fact, rather than leading to a merging of religions (in the manner of a global culture of consumption or popular music), globalization may be fostering a backlash against blending that reveals itself a return to an imagined fundamental, basic religious state often marked by literalism of foundational texts and the belief in inerrancy and religious nationalism. **Religious nationalism** is *the linkage of religious convictions with beliefs about a nation's or ethnic group's social and political destiny*. This nationalism can emerge from religious groups or from societal responses against religious groups. It is on the rise in countries around the world where religious nationalist movements have revived traditional religious beliefs and rejected the separation of religion and the state (Beyer, 1994; Kinnvall, 2004). While the trend is most pronounced among Islamic fundamentalists in some Middle Eastern and North African countries, the United States has experienced the rising political influence of evangelical Protestantism and India has experienced the rise of Hindu nationalism.

Religious nationalism: The linkage of religious convictions with beliefs about a nation's or ethnic group's social and political destiny.

Christian, Islamic, and Hindu religious nationalist movements accept modern technology, politics, and economics; many use the Internet and social media to disseminate information and ideas. At the same time, they interpret religious values strictly and reject secularization, drawing selectively on traditions and past events that serve their current beliefs and interests. Benedict Anderson (1991) writes about "imagined communities," suggesting that nations are real but not natural—that is, they are unified by both real and invented traditions and heroes that legitimate their claims about territories, the past, and the present. Part of the Israeli and Palestinian conflict over land, for instance, is rooted in different narratives of history that tell diverging stories about who has true dominion over the territory.

Religious affiliations can bring communities together or tear them apart. India is the world's largest democracy, the world's second most populous country (more than 1.3 billion), and home to Hindus, Muslims, Christians, and believers in dozens of other religions. Diversity and democracy sometimes appear to be on shaky ground there, however, particularly when the extreme nationalist ideology of *Hindutva* (or "Hinduness") comes into contact with Muslim extremism. In the northern Indian town of Ayodhya, a mosque was built in the early 16th century on the site where Hindus believe Lord Ram, a sacred deity in Hinduism, was born. Since the 19th century, clashes between Hindus and Muslims have taken place on the site. In 1992, Hindu nationalists attacked the mosque, sparking riots across the country that killed more than 2,000 people. The site continues to be the object of contention, though the conflict appears to have shifted to the Indian courts: In 2010, a high court ruled that the site should be split among Muslims, Hindus, and a local sect (BBC, 2012).

Today, three quarters of the world's population live in countries with high or very high levels of social hostilities involving religion, which "run the gamut from vandalism of religious property and desecration of sacred texts to violent assaults resulting in deaths and injuries" (Pew Forum on Religion and Public Life, 2015a, para. 2). Millions around the world are subject to state restrictions and social hostilities based on religion, and the number of restrictions has risen over the past decade. The Pew Forum on Religion and Public Life in 2019 categorized 52 governments worldwide as having high or very high religious restrictions, a number that had grown from 40 in 2007. These ranged from discriminatory policies and practices to outright religious bans (Figure 13.12). The report notes that social hostilities related to religious norms have significantly risen, while moderate rises have been seen in harassment and religious violence organized by groups. At the same time, however, most of the world's religions actively embrace the values of peace, humanity, and charity. Many believe religion is the basis of morality. On the other hand, religion in various forms, including religious nationalism, has been associated with intolerance, discrimination, and violence. How are we to reconcile these two sides

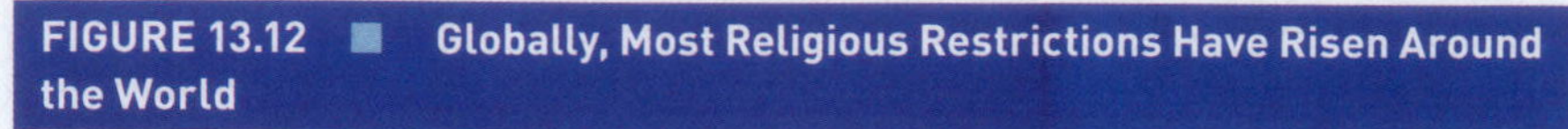

FIGURE 13.12 ■ Globally, Most Religious Restrictions Have Risen Around the World

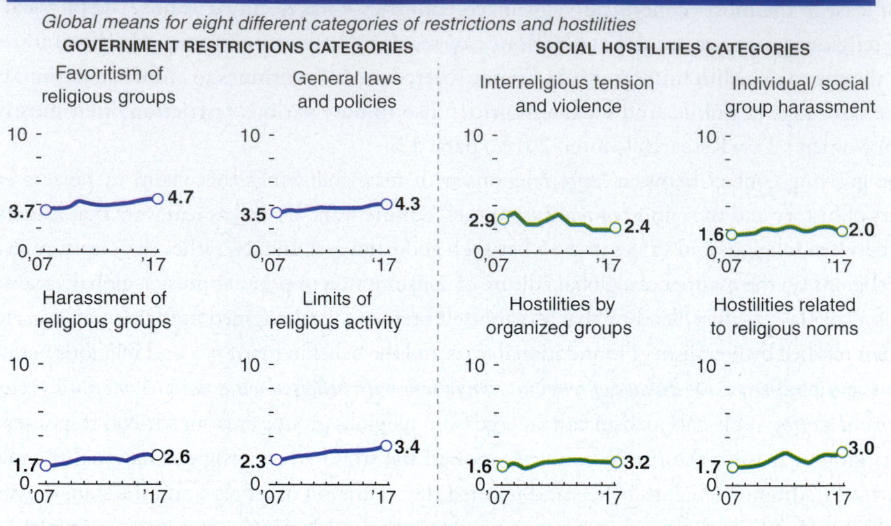

Source: Pew Research Center. (2019a, July 16). *A closer look at how religious restrictions have risen around the world.* https://www.pewforum.org/2019/07/15/a-closer-look-at-how-religious-restrictions-have-risen-around-the-world/

of religion as a local, national, and global institution? Is religion as an institution a force for positive change or a rationale for conflict? Might it be both? What do you think?

WHAT'S NEXT? RELIGION AND SOCIETY

In this chapter we have introduced the topic of religion showing that studying religion from a sociological perspective does not mean embracing or rejecting religious values; it means examining and analyzing an institution that is central to humans on individual, societal, and global levels. The global reach of religion is especially relevant as humans continue to use technology to expand their boundaries beyond physical spaces. Fascinating questions remain for you to explore: Will religiosity increase or decline in coming years? What about religious affiliations? What will American religiosity and practice really look like in the year 2060? Cultural and religious diversity is also important. Will some religious groups become more prevalent, or will we continue to see a profusion of groups like the religious economy theory predicts? How will gender, sexuality, and race and ethnicity continue to shape religious groups, and how will religious groups respond to the continuing growth of sexual, racial, and ethnic diversity? And though America's political foundations call for the separation of church and state, religious organizations seek to influence politics and public policy, and to request exemptions from local, state, and federal regulations. Can religious devotion mix comfortably with domestic and foreign policy, or are the two a volatile brew to be avoided? Sociologists will continue to use a variety of methods to explore these and other questions.

WHAT CAN I DO WITH A SOCIOLOGY DEGREE?

Community Resource and Service Skills

Community resource competencies link knowledge of nonprofit, government, and private community resources with the skills to access appropriate services and funding to best serve clients, organizations, and communities. Resources in communities take multiple forms, including individual donors, volunteers, politicians, business owners, religious leaders, schools, libraries and community centers, and public and private service agencies. Service skills may be developed through the study of and active participation in community organizations that engage with local populations and issues.

As a sociology major, you will develop important intercultural competencies and understandings of diversity. You will also learn important occupational professional skills, such as the ability to gather and summarize data in order to characterize community needs effectively and develop the habits of mind to be resourceful in addressing problems in ways that take different perspectives into account. Many educational institutions offer opportunities for service learning or volunteering that enable students to become familiar with the particular needs and resources of their own communities and to gain experience listening to and learning from both service providers and the people they serve.

Taylor Day, Second Harvest Heartland, Agency Compliance Specialist

BA, Sociology, University of Minnesota–Twin Cities

When I declared my sociology major, I knew I was interested in groups of people and their relationships. What I quickly discovered through my coursework was the strong connection between sociology and social justice, and this connection motivated my current interests in addressing food insecurity.

I have worked at a food bank since graduating with my sociology degree in 2016 from the University of Minnesota–Twin Cities. I have had several different positions throughout my time there—first working with large groups of volunteers as they repacked bulk produce into family-friendly sizes and then working with seniors on a USDA commodities program. I currently work directly with our partner food shelves and other community-based organizations. In these roles, I have worked very closely with both government agencies and nonprofits and have had the opportunity to engage in direct client service.

My ability to work well with those who are food insecure and with the agencies that serve those who are food insecure was greatly impacted by what I learned through my sociology degree. Sociology helped give me the background to understand social structures and how these strctures directly impact groups of people. Many systems affect those who are food insecure, along with the organizations that fuction in the hunger relief network, so to be able to use my sociology background to understand these social structures has been invaluable.

Also, my ability to think critically and understand social dynamics, which grew from my education in sociology, has helped me effectively problem solve with people to meet a common goal. Sociology pushes you to understand why people interact in the ways that they do and guides you to build relationships in an intentional way. When I graduated, I never though I would be doing capacity building and compliance work for a large nonprofit, but all of the knowledge and skills I learned through my sociology degree are the reason why I am passionate in what I do today.

Career Data: Social and Community Service Manager

- 2019 Median Pay: $67,150 per year
- $32.28 per hour
- Typical Entry-Level Education: Bachelor's degree
- Job Growth, 2019–2029: 17% (much faster than average)

Source: Bureau of Labor Statistics, *Occupation Outlook Handbook*, 2020.

SUMMARY

- **Religion** is a system of common beliefs and rituals centered on sacred things that unites believers and provides a sense of meaning and purpose. Religion may serve many functions in society, lending groups a common worldview, helping to ritualize or routinize behaviors or beliefs, and providing people with a sense of purpose.

- Sociologists study religion in a similar manner to other social institutions using methods and theories of the discipline of sociology. They study religion as a culture and examine how religion helps to organize and structure societies and group behavior, along with the functions religion serves for producing and maintaining group solidarity. They are also interested in religion as a source of power and conflict, between the state and between individuals and groups.
- Classical theorists have differing interpretations of religion's sociological function. Marx emphasized the role of religion in maintaining the capitalist power structure; Durkheim looked at the role of religion in reinforcing social solidarity; Weber highlighted the role of Protestantism in the development of capitalism.
- The **religious economy** perspective emphasizes the role of competition between groups as religions seek followers and potential adherents seek affiliations. The **megachurch** is one example of a group that appeals to believers as consumers, providing religion alongside of entertainment and services in a large setting.
- Religions manifest themselves in more than shared beliefs. They often take the form of organizations including **churches**, **sects**, **cults**, and **New Religious Movements**.
- Christianity, Islam, and Hinduism are the three largest and most practiced religions on Earth, but Judaism, Buddhism, and Confucianism are also influential.
- Americans have historically had high religiosity, although the number of the affiliated is declining. Youth and young adults in the United States today are less likely than older adults to claim a religious affiliation. At the same time, religious faith continues to be influential in U.S. life, including in politics.
- **Civil religion** involves the elevation of a nation as an object of worship; it involves a set of sacred beliefs and practices that become part of how a society sees itself. Civil religion, while not unique to the United States, can be seen in many facets of daily life there.
- Gender and sexuality are contested areas in religions, and groups provide both solace and discrimination to women and LGBTQ+ and racial and ethnic minoritized populations.
- Globalization can both increase religious diversity and promote secularization. It also can result in peace and religious conflict at the global level.

KEY TERMS

animism (p. 368)
church (p. 361)
civil religion (p. 374)
cult (p. 362)
denomination (p. 362)
ecclesia (p. 361)
Establishment Clause (p. 369)
evangelicalism (p. 373)
megachurch (p. 359)
monotheism (p. 364)
new religious movements (p. 363)
nontheistic religion (p. 368)
polytheism (p. 368)
profane (p. 356)
religion (p. 354)
religious economy (p. 358)
religious nationalism (p. 379)
rituals (p. 354)
sacred (p. 356)
sect (p. 362)
secularization (p. 356)
totems (p. 356)
Zionism (p. 367)

DISCUSSION QUESTIONS

1. Why do sociologists study religion? How do they study it? Which of their reasons for studying it is most interesting to you and why? Which method would you be most likely to use to study a sociologically informed religious question?

2. Compare and contrast how Durkheim, Marx, and Weber conceptualized the role of religion in society. What does the religious economy approach add to their theories? Which theory or approach most interests you and why?
3. Identify the five different types of religious organizations introduced in this chapter and compare their characteristics. Do you find the classifications useful? Why or why not?
4. Consider the six major religions introduced in this chapter. How have these religions operated in a globalizing world? How has globalization impacted their relations with each other?
5. Describe the role of religion in U.S. society today. How is it changing? With what groups is it changing? Considering affiliation, civil religion, politics, and race and gender, how might we expect it to change in the near future?
6. This chapter introduces the rise of people claiming no religious affiliation. It also explores the globalization of religiosity, and the ability of people to eclectically sift and sort through religious identities online. Drawing on these and other materials from this chapter, describe two major changes that you think a student of sociology might read about in this chapter 20 years from now.

©Kent Nishimura / Los Angeles Times via Getty Images

14 THE STATE, WAR, AND TERROR

WHAT DO YOU THINK?

1. What means do governments have to exercise control over their populations? What means do citizens have to exercise control over their own governance?
2. Why do states go to war? What are the functions and dysfunctions of war? Is war inevitable?
3. It is sometimes said that "one person's terrorist is another person's freedom fighter." What does this statement mean? Do you agree with this statement?

LEARNING OBJECTIVES

14.1 Describe the ideal-typical characteristics of the modern state.

14.2 Explain theories of state power.

14.3 Identify forms of authority and how they function.

14.4 Describe the major forms of state governance in the modern world.

14.5 Describe key characteristics of the U.S. political system.

14.6 Apply sociological perspectives to understand war and its historical and contemporary functions.

14.7 Take a sociological perspective on terrorists and terrorism in the modern world.

WHO CHOOSES AN AMERICAN PRESIDENT?

Did you participate in the 2020 presidential election? The U.S. election of 2020 saw historically high turnout across the country. This included longtime voters who have never missed an election in their adult years, as well as older adults who have never voted and young voters exercising their right to vote for the first time. In spite of the COVID-19 pandemic, Americans were eager to vote and did so in large numbers both in-person and by mail.

Following the election in November 2020, Americans took to the streets to protest against President Donald J. Trump's calls to suspend the counting of votes in some contested states.

©Eric McGregor/Getty Images

In the case of congressional representatives, state legislators, school board members, judges, and other candidates on the ballot, the winner of the most votes wins the election. Because of the Electoral College system, the United States has historically used to select a president, however, the winner of the popular vote in a presidential election may or may not ascend to the presidency. Since 2000, on two occasions, a president was elected who did not win the popular vote: George W. Bush in 2000 and Donald J. Trump in 2016.

Under the current electoral system, when you vote for president, you are actually voting for your state's electors. Each state has the same number of electors as it has members in its congressional delegation. California, the most populous state, has the most electors (54). Small states including Delaware, Montana, and Vermont have only 3. In all but two U.S. states, the election is winner-take-all, so a presidential candidate who gets only one vote more than his or her opponent in a given state gets all of that state's electoral votes.

The Electoral College system was written into the U.S. Constitution more than 200 years ago by our country's founders, who feared that direct popular vote for the presidency might lead to (unruly) rule by the masses, even though at the time only about 6% of the population could legally vote. As a recent *Atlantic Monthly* article on the Electoral College notes, "The delegates to the Philadelphia convention had scant conception of the American presidency—the duties, powers, and limits of the office. But they did have a handful of ideas about the method for selecting the chief executive. When the idea of a popular vote was raised, they griped openly that it could result in too much democracy" (Codrington, 2019, para. 5).

This was not, however, the only concern driving the founders' decision to forgo other options being considered, which included both the popular vote and congressional selection of the president (Tarr, 2019). Delegates from the slave-holding Southern states were not in favor of a popular vote because it held a clear disadvantage for them: While the sizes of the populations of the American North and South were similar, about a third of those living in the South were enslaved peoples, who had no rights and no political voice. The Southern delegates were wary of a system, like the popular vote for president, that would disadvantage them because they had fewer eligible voters. The Electoral College emerged as a compromise to distribute power broadly across the states in the Union, including those where more than 93% of the country's enslaved people lived (Codrington, 2019).

Should an election process rooted in a time when slavery was legal, when women were excluded from the vote, and when power was concentrated in the hands of a small group of white, educated, land-holding men continue to determine the outcome of the U.S. presidency? Two weeks after the 2020 election, *The Washington Post* published an editorial arguing that it is time to end the Electoral College: "Americans are not going to be satisfied with leaders who have been rejected by a majority of voters" (Editorial Board, 2020). How should Americans choose their president? Can an understanding of the conditions under which the Electoral College emerged change our perspective on the current system? What do you think?

We begin this chapter with a discussion of power and the modern nation-state and an examination of citizenship rights and their provision. We then look at theoretical perspectives on state power, its exercise, and its beneficiaries. A consideration of the types of authority and forms of governance in the modern world provides the background for an examination of the U.S. political system. We then turn to a discussion of war and society and an analysis of war from the functionalist and conflict perspectives. This is followed by a critical look at the issue of terrorism as well as the question of defining who is a terrorist. We conclude with a consideration of the question of why we study state power and its manifestations in phenomena that range from elections to making war.

THE MODERN STATE

For most of human history, people lived in small and homogeneous communities within which they shared languages, cultures, and customs. Today, however, the world's more than 7.5 billion people are distributed across 195 countries (the number of countries recognized by the United States, though other entities exist that claim statehood, including Palestine and Kurdistan). On the world stage, countries are key actors: They are responsible for war and peace, the economic and social welfare of their citizens, and the quality of our shared global environment and security.

Obtaining citizenship is a dream for many immigrants to the United States. Every year, thousands take the Naturalization Oath of Allegiance. Immigration has been a key building block of U.S. society and its cultural, economic, and political life.

©REUTERS/David Ryder

Social scientists commonly characterize the modern country as a **nation-state**—that is, *a single people* (a nation) *governed by a political authority* (a state) (Gellner, 1983). Very few countries neatly fit this model, however. Most are made up of many different peoples brought together through warfare, conquest, immigration, or boundaries drawn by colonial authorities without respect to ethnic or religious differences of the time. For instance, many Native Americans think of themselves as belonging to the Navajo, Lakota, Pawnee, or Iroquois Nation rather than only to the United States. In Nigeria, most people identify primarily with others who are Yoruba, Ibo, or Hausa rather than with a country called Nigeria. In Iraq, a shared Iraqi identity is less common than allegiance to the Sunni or Shiite Muslim or the Kurdish community. Because most political entities are not characterized by the homogeneity implied by the definition of *nation-state*, we will use the more familiar terms *country* and *state*.

Nation-state: A single people (a *nation*) governed by a political authority (a *state*); similar to the modern notion of *country*.

While not all countries possess them in equal measure, the characteristics we list below represent what Max Weber would term an *ideal-typical model*, a picture that approximates but does not perfectly represent reality. Modern countries emerged along with contemporary capitalism, which benefited from strong central governments and legal systems that regulated commerce and trade both within and across borders (Mann, 1986; Wallerstein, 1974; Weber, 1921/1979). This history accounts for several unique features of modern countries that distinguish them from earlier forms of political organization.

Law: A system of binding and recognized codified rules of behavior that regulate the actions of people pertaining to a given jurisdiction.

Underlying the social organization of the modern country is a system of **law**, the *codified rules of behavior that regulate the actions of people pertaining to a given jurisdiction* (Chambliss & Seidman, 1982). The rule of law is a critical aspect of democratic governance.

The governments of modern countries claim complete and final authority over the people who reside within the countries' borders (Hinsley, 1986).

Citizens: Legally recognized individuals who are part of a political community in which they are granted certain rights and privileges and, at the same time, have specified responsibilities.

Noncitizens: Individuals who reside in a given jurisdiction but do not possess the same rights and privileges as citizens; sometimes referred to as *residents, temporary workers*, or *aliens*.

People living within a country's borders are divided between **citizens**, *legally recognized individuals who are part of a political community in which they are granted certain rights and privileges and, at the same time, have specified responsibilities*, and **noncitizens**, *individuals who reside in a given jurisdiction but do not possess the same rights and privileges as citizens* (Held, 1989). (Noncitizens are *sometimes referred to as residents, temporary workers, or aliens.*)

Citizenship rights may take several forms. *Civil rights*, which protect citizens from injury by individuals and institutions, include the right to equal treatment in places such as the school or workplace regardless of race, gender, sexual orientation, or disabilities. *Political rights* ensure that citizens can participate in governance, whether by voting, running for office, or openly expressing political opinions. *Social rights*, which call for the governmental provision of various forms of economic and social security, include such things as retirement pensions and guaranteed income after losing a job or becoming disabled. Citizens are afforded legal protections from arbitrary rule and in turn are expected to pay taxes, to engage in their own governance through voting or other activities, and to perform military service (with specific expectations of such service varying by country). In reality, the extent to which all people enjoy the full rights of citizenship in any given country varies. Below, we discuss two specific aspects of citizenship rights—the first relates to the evolution of state provisions for ensuring social rights, the second to the degree to which citizens enjoy freedom in the form of political rights and civil liberties.

The Welfare State

In most modern countries, political and civil rights evolved and were institutionalized before social rights were realized. The category of *social rights* is broad and encompasses entitlements that include health care insurance, old-age pensions, unemployment benefits, a minimum wage floor for workers, and a spectrum of other benefits intended to ensure social and economic security for the citizenry. Social rights have largely been won by groups of citizens mobilizing (on the basis of the civil and political rights they enjoy) to realize their interests.

Welfare state: A government or country's system of providing for the financial and social well-being of its citizens, typically through government programs that provide funding or other resources to individuals who meet certain criteria.

Social rights are often embodied in what is termed the **welfare state**, *a government or country's system of providing for the financial and social well-being of its citizens, typically through government programs that provide funding or other resources to individuals who meet certain criteria.* The welfare state has been a part of Western systems of governance in the post–World War II period. Social Security, a social welfare program that ensures a stable income for retired workers, and Medicare, which provides for at least basic health care for the elderly, are examples of the U.S. welfare state. Social Security was created through the Social Security Act of 1935, signed by President Franklin D. Roosevelt, which established a social insurance program to provide continuing income for retired workers at age 65 or older (Social Security Administration, 2013). Although the provision of basic health care coverage was the vision of President Harry S. Truman, Medicare and Medicaid did not get signed into law until 1965, when they were endorsed by President Lyndon B. Johnson. The welfare state has long been a hallmark of advanced and wealthy countries; most developing states have much weaker social safety nets, and few have had the resources to provide for retirees or, often, for the unemployed or the sick.

While it has come to represent a culmination of the three key rights of citizenship in modern countries, the welfare state is shrinking rather than expanding. Factors such as global wage competition, the economic pressure of aging populations, and large budget deficits have hampered the expansion of social rights in many Western countries. While it is difficult to reduce benefits to already existing constituencies—for example, in the United States, discussion of reducing Social Security payments or raising the age of eligibility evokes protest from many retirees and soon-to-be retirees—the economic crises of recent years have led to attempts by lawmakers to curb benefits to less-influential constituencies, including the poor and immigrants.

The law bringing Social Security into being was signed in 1935 by President Franklin D. Roosevelt. Social Security benefits have contributed to the significant reduction of poverty among seniors in the United States. Seniors have a lower rate of poverty than children under age 18.

Political Rights and Civil Liberties

Every year, Freedom House, an organization dedicated to monitoring and promoting democratic change and human rights, publishes an evaluation of "freedom" in 195 countries, as well as 14 related and disputed territories (Figure 14.1). The report includes ratings that measure political rights (based on the electoral process, political pluralism, and participation) and civil liberties (based on freedom of expression and belief, rights of association and organization, rule of law, and individual rights).

Even in the United States, which earns a designation of "free" in Freedom House's survey, problems with voting procedures have raised the question of whether some voters, especially minorities, have been denied a political voice in elections. One way minorities have been disenfranchised is through felony convictions. African Americans are incarcerated at a significantly higher rate than other groups, and "all but two states suspend voting rights" while incarcerated for a felony. Most states restore suffrage upon release; however, 11 states require additional steps to regain the right to vote, and in certain cases financial penalties must be paid, which may have the effect of disenfranchising poor voters (Freedom House, 2020). Nationwide, 5.2 million Americans were disenfranchised due to a felony conviction, including 43% who have completed their sentences. Disenfranchisement is vastly disproportionate by race, as 1 of every 16 African Americans has been disenfranchised, a rate that is 3.7 times higher than that of other Americans (The Sentencing Project, 2020).

Disenfranchisement is also disproportionate by region, as Southeastern states have higher rates compared to Northeastern and Midwestern states. Three states—specifically Alabama, Mississippi, and Tennessee—have more than 8% of the adult population classified as disenfranchised. The most startling example of this phenomenon, however, is Florida: An estimated 900,000 Floridians have completed their sentences and have yet to regain their voting rights. This is true even after a 2018 ballot referendum in the state that "promised to restore their voting rights." Overall, Florida is home to over 1.1 million Americans barred from voting, many because they cannot afford to pay court-ordered sanctions to restore their suffrage, and even then the state is not obligated to tell them the total amount of their sanction (Uggen et al., 2020).

Recently, some U.S. state legislatures have sought to implement voter identification laws, which require voters to show identification when they vote in-person. As of 2020, six states had strict photo ID laws, while others had a variety of other laws ranging from required nonphoto ID to no required documents (Figure 14.2; Panetta & Liu, 2020). Critics argue that states with strict photo ID requirements disenfranchise millions of Americans. It is estimated that up to 11% of eligible voters do not have the kind of ID necessary to vote in those states. They point out that members of minority groups—in particular, those who are elderly, students, minorities, have disabilities, or are low-income voters—are among those least likely to have photo identification in the form of a driver's license or passport (Brennan Center for Justice, 2020). The practical effect of voter ID laws, they say, is not to combat voter fraud (which has been very infrequently documented in the United States; there were fewer than 31 credible voting fraud cases identified from 2000 to 2014, and only 4 in the 2016 election) but rather to disenfranchise minorities and the poor (Brennan Center for Justice, 2020; Cohen, 2012).

FIGURE 14.1 ■ Freedom Status Worldwide, 2019

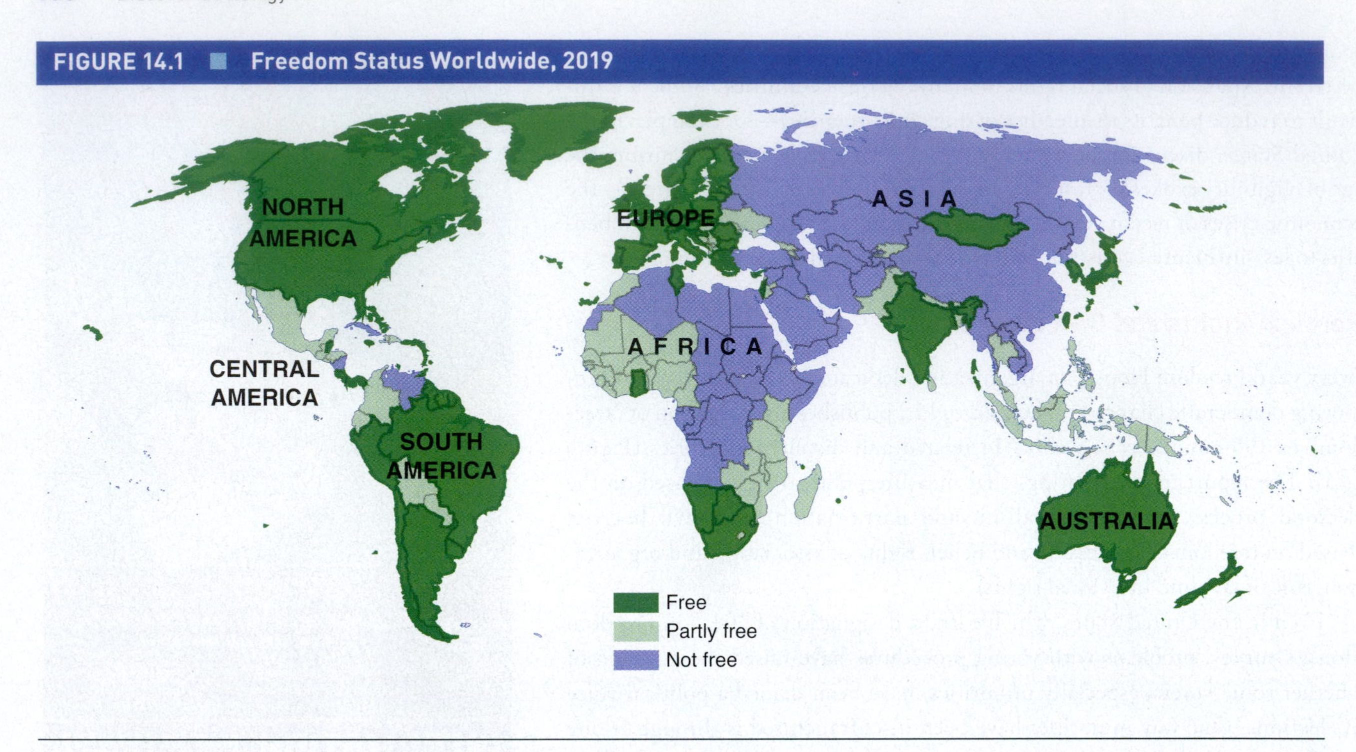

Source: Freedom House. (2020). 2019 Freedom in the World.

FIGURE 14.2 ■ Voter Identification Laws in the United States, November 2020

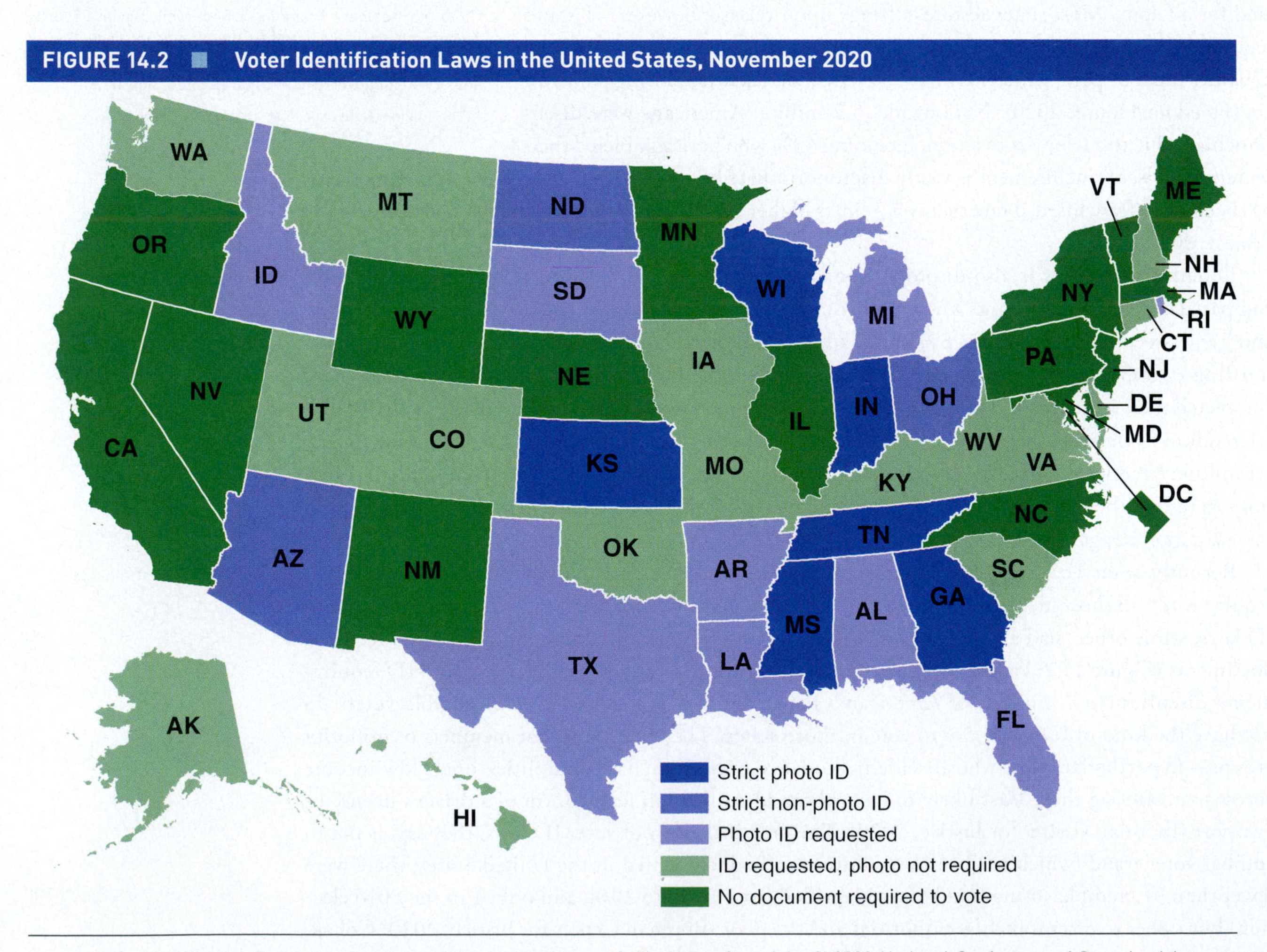

Source: Voter Identification Requirements: Voter ID Laws. Legislature & Elections.

Later in this chapter, we discuss more fully the issue of *political voice* (the representation of a group's interests in bodies such as state and national legislatures) and the ways in which political voice may vary. In the next section, we turn to broader sociological perspectives on state power, which can give us a fuller context for analyzing contemporary debates on power and politics.

THEORIES OF STATE POWER

Sociologists have developed different approaches to explaining how authority is exercised in modern states. Next, we highlight two theoretical approaches, which disagree about how expansively power is shared. These theories are based largely on governance in the United States today, although we can also apply them to other contemporary democratic states.

The Functionalist Perspective and Pluralist Theory

Classical sociologist Émile Durkheim (1922/1956, 1922/1973b) saw government as an institution that translates broadly shared values and interests into fair-minded laws and effective policies. Modern functionalist theorists recognize that societies are socially and culturally heterogeneous and likely to have a greater diversity of needs and perspectives. The government, they suggest, is a neutral umpire, balancing the conflicting values, norms, and interests of a variety of competing groups in its laws and actions.

In the United States, most people agree on such general values as liberty, democratic governance, and equality of opportunity, but there is vigorous public debate over issues such as abortion, the death penalty, the government's role in ensuring access to health care, and the degree to which the United States should take a leading role in global affairs. Recognizing the pluralistic—that is, diverse—nature of contemporary societies, sociologists and political scientists have developed theories of government that highlight state power and how it is exercised.

Pluralist theory tries to answer the question, "Given that modern societies are pluralistic, how do they resolve the inevitable conflicts?" To answer this question, Robert Dahl (1961, 1982, 1989) studied decision making in New Haven, Connecticut. Dahl (1961) concluded that power is exercised through the political process, which is dominated by different groups of leaders, each having access to a different amalgamation of political resources. Dahl argues that, in their efforts to exert political influence, individuals come together in **interest groups**—*groups made up of people who share the same concerns on particular issues who use their organizational and social resources to influence legislation and the functioning of social institutions*. An interest group may be short-lived, such as a local citizens' group that bands together to have a road repaved or a school built, or long-lasting, such as a labor union or a manufacturers' association.

Interest groups: Groups made up of people who share the same concerns on particular issues who use their organizational and social resources to influence legislation and the functioning of social institutions.

Dahl's theory asserts that interest groups serve the function of ensuring that everyone's perspectives (that is, their values, norms, and interests) are represented in the government. The influence of one group is offset by the power of another. For instance, if a group of investors bands together to seek government approval to clear-cut a forest to build homes, a group of citizens concerned about the environment may coalesce into an interest group to oppose the investors' plan. The ideal result, according to Dahl, would be a compromise: Perhaps cutting would be limited or some particularly sensitive areas would be preserved. Similarly, from this perspective, big businesses and organized labor may routinely face off over issues of pay, benefits, and worker voice in decision making, but neither exercises disproportionate influence on the political process.

When powerful interests oppose one another, pluralists see compromise as the likely and optimal outcome, and the role of the government is to broker solutions that benefit as many interest groups as possible. In this view, power is dynamic, passing from one stakeholder to another over time rather than becoming concentrated in the hands of a powerful few. Competition and the fluidity of power contribute to democratic governance and society.

A critical view of the pluralist characterization of political power points out that government is unlikely to represent or recognize all interests (Chambliss & Seidman, 1982; Domhoff, 2006). Critics also dispute the assumption that government is a neutral mediator between competing interests. They

Lobbying for special interests that range from multinational corporations and fossil fuel companies to workers' unions and immigrant rights organizations is common in the U.S. political system. What factors affect the influence that special interests exercise in the legislative and executive branches?

©Cyrus McCrimmon/The Denver Post/Getty Images

argue that laws may favor some groups over others. For example, when the U.S. Constitution was framed by white male property owners, only white male property owners could vote: People without property, women, Blacks, and American Indians were excluded from the political process. The disproportionate influence of some constituencies over others continues today. Consider the following example: President Trump's 2017 tax cut saw a large break for America's 400 wealthiest households. In fact, for the first time in history those households paid a lower tax rate than any other income group in the country. The rate came in at 23% of total income, compared to 47% in 1980 and 70% in 1950 (Leonhardt, 2019).

Indeed, political practices demonstrate definitively that some interest groups are more powerful than others. Furthermore, governments do not apply the rules neutrally, as the theory claims. Rather, political leaders interpret (or even bend) the rules to favor the most powerful groups in society, including big business and other moneyed interests that finance increasingly costly political campaigns for those who favor them (Domhoff, 2006, 2009; Friedman, 1975, 1990).

The Conflict Perspective and Class Dominance Theory

Conflict theory highlights power differences between social groups. This perspective recognizes that modern societies are pluralistic, but it argues that the interests of social groups are often incompatible with one another. Furthermore, conflict theory posits that some groups are more powerful than others and are therefore more likely to see their interests and values reflected in government policy making. Groups with greater resources use their power to create systems of law, economy, politics, and education that favor them, their children, and other group members.

Unlike pluralistic theory, which views competing interests as having relatively comparable and shifting opportunities and access to power, conflict theory sees power as being concentrated in the hands of a few privileged groups and individuals. The gains of the elite, conflict theorists suggest, come at the expense of those who have fewer resources, including economic, cultural, and social capital.

Conflict theory is rooted in the ideas of Karl Marx. You may recall from previous chapters that Marx believed the most important sources of social conflict are economic, and that, as a consequence, class conflict is fundamental to all other forms of conflict. Within a capitalist society, government represents and serves the interests of the capitalist class or *bourgeoisie*, the ruling class that exerts disproportionate influence on the government. Still, a well-organized working class may effectively press government for such economic reforms as a shorter working day or the end of child labor.

Class dominance theory: The theory that power is concentrated in the hands of a small group of elite or upper-class people who dominate and influence societal institutions; a variation of conflict theory.

Contemporary conflict theory extends Marx's concept of the *bourgeoisie* (which focused on those who own the means of production) to include other groups that wield considerable power. **Class dominance theory** argues that *power is concentrated in the hands of a small group of elite or upper-class people who dominate and influence societal institutions* (Domhoff, 1983, 1990, 2002; Mills, 1956/2000a). These individuals have often attended the same elite schools, belong to the same social organizations, and cycle in and out of top positions in government, business, and the military (the so-called *revolving door*). Class dominance theory complements Marx's original ideas with a focus on elite social networks as a center of political power rather than only on capitalism and the exercise of economic power.

G. William Domhoff (2002, 2006, 2009) posits that we can show the existence of a dominant class by examining the answers to several basic questions, which he calls *power indicators:* Who benefits? Who governs? Who wins?

In terms of "Who benefits?" Domhoff asks us to consider who gets the most of what is valued in society. Who gets money and disproportionate amounts of material goods? What about leisure and travel, or status and prestige? Domhoff (2006) asserts that "those who have the most of what people want are, by inference, the powerful" (p. 13).

In terms of "Who governs?" he asks who is positioned to make the important political and economic and legal decisions in the country or community. Are all demographic and economic groups relatively well represented? Are some disproportionately powerful? Domhoff (2006) suggests, "If a group or class is highly overrepresented or underrepresented in relation to its proportion of the population, it can be inferred that the group or class is relatively powerful or powerless" (p. 14).

Asking "Who wins?" entails inquiring about which group or groups have their interests realized most often. Domhoff concedes that movements with fewer resources—including, for instance, environmental groups—may win desired legislation sometimes. However, we need to look at who has their desires realized most consistently and most often. Is it small interest groups representing civil rights, environmental activists, or same-sex marriage advocates? Is it large corporate interests with friends in high places and the ability to write big campaign donation checks?

After examining the power indicators in his book *Who Rules America?*, Domhoff concludes that it is the upper class, particularly the owners and managers of large for-profit enterprises, that benefits, governs, and wins. This, he suggests, challenges the premise of pluralist theories that power is dynamic, passing between a variety of interests and groups. Domhoff argues that there exists a small but significant **power elite**, which is *a group of people with a disproportionately high level of influence and resources who utilize their status to influence the functioning of societal institutions.* Though the corporations, organizations, and individuals who make up the power elite may be divided on some issues, Domhoff contends that cooperation is stronger than competition among them: The members of the power elite are united by a common set of interests (including a probusiness and antiregulation economic environment) and common enemies (including environmentalists and progressive labor and consumer activists).

Power elite: A group of people with a disproportionately high level of influence and resources who utilize their status to influence the functioning of societal institutions.

From a critical perspective, class dominance theory may overemphasize the unified nature of the ruling class. For instance, Domhoff highlights the fact that many members of the power elite share similar social backgrounds. Often, they attend the same private high schools and colleges, spend their vacations in the same exclusive resorts, and marry into one another's families. They share a strong belief in the importance and value of capitalism and, as Domhoff argues, are steeped in a similar set of worldviews. However, it is difficult to show that they necessarily share the same political beliefs or even economic orientations (Chambliss & Seidman, 1982; Chambliss & Zatz, 1994).

Furthermore, government decisions sometimes appear to be in direct opposition to the expressed interests of powerful groups. For example, when faced with major conflicts between labor and management during the Depression years of the 1930s, the U.S. government passed laws legalizing trade unions and giving workers the right to bargain collectively with their corporate employers, even though both laws were strongly opposed by corporation executives and owners (Chambliss & Zatz, 1994; Skocpol, 1979; Tilly, 1975). Mark Smith (2000) found that when businesses act to influence public policy to support or oppose a given issue, they may experience backlash as labor and public interest groups organize in opposition to the perceived power seizure. As the influence of money in politics has grown, however, examples of powerful—and wealthy—groups having their legislative interests overridden are more challenging to find.

Does a power elite exercise disproportionate influence in the political sphere of the United States? Domhoff and other conflict theorists would answer in the affirmative. A pluralist perspective might see corporations, the upper class, and policy organizations as some among many players who compete in the political power game, balanced by other groups such as unions and environmentalists, and answer in the negative. What do you think?

POWER AND AUTHORITY

In the preceding section, we looked at different perspectives on how state power functions and who it serves. We now turn to the question of how countries exercise their power in practice, asking, "How do governments maintain control over their citizens?"

One way that states exercise power is through outright **coercion**—*the threat or use of physical force to ensure compliance.* Relying solely on coercion, however, is costly and difficult because it requires surveillance and sometimes suppression of the population, particularly those segments that might be inclined to dissent. Governments that ground their authority in coercion are vulnerable to instability,

Coercion: The threat or use of physical force to ensure compliance.

as they generally fail to earn the allegiance of their people. It is more efficient, and in the long run more enduring, if a government can establish legitimate authority, which, as you recall from Chapter 5, is *power that is recognized as deserved or earned.*

One of sociology's founders, Max Weber (1864–1920), was also one of the first social scientists to analyze the nature of legitimate authority, and his ideas have influenced our understanding of power and authority in the modern world. Weber sought to answer the question, "Why do people consent to give up power, allowing others to dominate them?" His examination of this question, which was based on detailed studies of societies throughout history, identified three key forms of legitimate authority: traditional, rational-legal, and charismatic.

Traditional Authority

Traditional authority: Power based on a belief in the sanctity of long-standing traditions and the legitimate right of rulers to exercise authority in accordance with those traditions.

For most of human history, state power relied on **traditional authority**, *power based on a belief in the sanctity of long-standing traditions and the legitimate right of rulers to exercise authority in accordance with these traditions* (Weber, 1921/1979). Traditional rulers claim power on the basis of age-old norms, beliefs, and practices. When the people being governed accept the legitimacy of traditional authority, it tends to be relatively stable over time. The monarchies of Europe, for example, ruled for hundreds of years based on traditional authority. Their people were considered the king's or queen's subjects, whose loyalty derived from their recognition of the fundamental legitimacy of monarchical rule, with its long-standing hierarchy and distribution of power on the basis of blood and birth. In modern Europe, however, monarchies such as those in Denmark and Sweden have little more than symbolic power—they have been largely stripped of political power. Traditional authority, in these instances, coexists with the rational-legal authority exercised by modern elected bodies, which we discuss next.

Traditional authority supports the exercise of power at both the macro and micro levels: Just as reverence for traditional norms and practices may give legitimacy to a state, a religion, or other government, it may also drive the decisions and actions of families. If a family marks a particular holiday or date with an obligatory ritual, then even if an individual questions the need for that ritual, the reasoning "We've *always* done that" is a micro-level exercise of traditional authority that ensures compliance and discourages challenges by any member of the group.

Rational-Legal Authority

Rational-legal authority: Power based on a belief in the lawfulness of enacted rules (laws) and the legitimate right of leaders to exercise authority under such rules.

Traditional authority, in Weber's view, was incompatible with the rise of modern capitalist states. Capitalism is based on forms of social organization that favor rational, rule-governed calculation rather than practices grounded in tradition. As capitalism evolved, traditional authority gave way to **rational-legal authority**, *power based on a belief in the lawfulness of enacted rules (laws) and the legitimate right of leaders to exercise authority under such rules* (Weber, 1921/1979). The legitimacy of rational-legal authority derives from a belief in the rule of law. We do something not simply because it has always been done that way but because it conforms to established rules and procedures.

In a system based on rational-legal authority, leaders are regarded as legitimate as long as they act according to law. Laws, in turn, are enacted and enforced through formal, bureaucratic procedures rather than reflecting custom and tradition or the whims of a ruler. Weber argued that rational-legal authority is compatible with modern economies, which are based on rational calculation of costs and benefits, profits, and other economic decisions. Rational-legal authority is commonly exercised in the ideal-typical modern state described earlier in this chapter. In practice, we could take the United States, Canada, Japan, or the countries of the European Union as specific examples of states governed by rational-legal authority.

Charismatic Authority

Charismatic authority: Power based on devotion inspired in followers by the personal qualities of a leader.

Weber's third form of authority can threaten both traditional and rational-legal authority. **Charismatic authority** is *power based on devotion inspired in followers by the personal qualities of a leader* (Weber, 1921/1979). It derives from widespread belief in a given community that an individual has a gift of great—even divine—powers, so it rests most significantly on an individual personality rather than on that individual's claim to authority on the basis of tradition, legal election, or appointment. Charismatic authority may also be the product of a *cult of personality*, an image of a leader that is carefully manipulated by the leader and other elites. In North Korea's long-standing dictatorship, power has passed through several generations of the same family, as has the government's careful construction of a cult of personality around each leader that elevates him as supremely intelligent, patriotic, and worthy of unquestioning loyalty.

Prominent charismatic leaders in religious history include Moses, Jesus Christ, the Prophet Muhammad, and Buddha. Some military and political rulers whose power was based in large part on charisma are Julius Caesar, Napoleon Bonaparte, Vladimir Lenin, and Adolf Hitler (clearly, not all charismatic leaders are charitable, ethical, or good).

Charismatic leaders have also emerged to lead communities and countries toward democratic development. Václav Havel, a dissident playwright in what was then communist Czechoslovakia, challenged the authority of a government that was not elected and that citizens despised; he spent years in prison and doing menial jobs because he was not permitted to work in his artistic field. Later, he helped lead the 1989 opposition movement against the government and, with its fall, was elected president of the newly democratic state. (Czechoslovakia no longer exists; its Czech and Slovak populations wanted to establish their own nation-states, and in 1993 they peacefully formed two separate republics.) Similarly, Nelson Mandela, despite spending 27 years in prison, was a key figure in the opposition movement against the racist policies of apartheid in South Africa. Mandela became South Africa's first democratically elected president in 1994, some 4 years after his release from prison. His death in 2013 marked the passing of a significant era in South African politics.

Notably, Weber also pointed to a phenomenon he termed the *routinization of charisma*. That is, with the decline, departure, or death of a charismatic leader, his or her authority may be transformed into legal-rational or even traditional authority. While those who follow may govern in the charismatic leader's name, the authority of successive leaders rests either on emulation (traditionalized authority) or on power that has been routinized and institutionalized (rationalized authority).

Authority exists in a larger context, and political authority is often sited in a government. Governance takes place in a variety of forms, which we discuss in the following section.

FORMS OF GOVERNANCE IN THE MODERN WORLD

In the modern world, the three principal forms of governance are authoritarianism, totalitarianism, and democracy. Next, we discuss each type, offering ideal-typical definitions as well as illustrative examples.

Authoritarianism

Under **authoritarianism**, *ordinary members of society are denied the right to participate in government, and political power is exercised by and for the benefit of a small political elite*. At the same time, authoritarianism is distinguished from totalitarianism (which we will discuss shortly) by the fact that at least some social, cultural, and economic institutions exist that are not under the control of the state. Two prominent types of authoritarianism are monarchies and dictatorships.

Authoritarianism: A form of governance in which ordinary members of society are denied the right to participate in government, and political power is exercised by and for the benefit of a small political elite.

Monarchy is *a form of governance in which power resides in an individual or a family and is passed from one generation to the next through hereditary lines*. Monarchies, which derive their legitimacy from traditional authority, were historically the primary form of governance in many parts of the world and in Europe until the 18th century. Today, the formerly powerful royal families of Europe have been either dethroned or relegated to peripheral and ceremonial roles. For example, the queens of England, Denmark, and the Netherlands and the kings of Sweden and Spain do not have any significant political power or formal authority to govern. A few countries in the modern world are still ruled by monarchies, including Saudi Arabia, Jordan, Qatar, and Kuwait. Even the monarchs of these nations, however, govern with the consent of powerful religious and social or economic groups.

Monarchy: A form of governance in which power resides in an individual or a family and is passed from one generation to the next through hereditary lines.

In the territory of Saudi Arabia, for instance, the royal family, also known as the House of Saud, has ruled for centuries, though the country of Saudi Arabia itself was established only in 1932. The Basic

Dictatorship: A form of governance in which power rests in a single individual.

Totalitarianism: A form of governance that denies popular political participation in government and also seeks to regulate and control all aspects of the public and private lives of citizens.

Law of 1992 declared Saudi Arabia to be a monarchy ruled by the sons and grandsons of King Abdul Aziz al-Saud, making the country the only one in the world named after a family. The constitution of the country is the Koran, and, consequently, *sharia* law (which is based on Islamic traditions and beliefs) is in effect. The country does not hold national elections, nor is the formation of independent political parties permitted. However, the royal rulers govern within the bounds of the constitution, tradition, and the consent of religious leaders, the *ulema.*

A more modern form of authoritarianism is **dictatorship**, *a form of governance in which power rests in a single individual.* An example of an authoritarian dictatorship is the government of Iraqi president Saddam Hussein, who ruled his country from 1979 to 2003, when he was deposed. As this case shows, the individual in power in a dictatorship is actually closely intertwined with an inner circle of governing elites. In Iraq, Saddam was linked to the inner circle of the Báath Party. Furthermore, because of the complexity of modern society, even the most heavy-handed authoritarian dictator requires some degree of support from military leaders and an intelligence apparatus. No less important to the dictator's power is the compliance of the masses, whether it is gained through coercion or consent.

Here, North Korean totalitarian leader Kim Jong-un visits the Korean People's Army. According to a United Nations report, North Korea's human rights violations include "violations of the freedoms of expression; discrimination; freedom of movement; right to food and right to life; arbitrary detention, torture, and execution; abductions and enforced disappearances from other countries" (United Nations Human Rights Commission, 2014).

©KNS/AFP/Getty Images

We might argue that today, it would be difficult for a single individual or even a handful of individuals to run a modern country effectively for any length of time. In recent years, many dictators have been deposed by foes or ousted in popular revolutions. Some have even allowed themselves to be turned out of office by relatively peaceful democratic movements, as happened in the former Soviet Union and the formerly communist states of Eastern Europe, such as Czechoslovakia and Hungary. China, which has become progressively more capitalistic while retaining an authoritarian communist government, remains an exception to this pattern.

Totalitarianism

When authoritarian dictatorships persist and become entrenched, the end result may be a totalitarian form of government. **Totalitarianism** *denies popular political participation in government and also seeks to regulate and control all aspects of the public and private lives of citizens.* In totalitarianism, there are no limits to the exercise of state power. All opposition is outlawed, access to information not provided by the state is stringently controlled, and citizens are required to demonstrate a high level of commitment and loyalty to the system. A totalitarian government depends more on coercion than on legitimacy in exercising power.

It thus requires a large intelligence apparatus to monitor the citizenry for antigovernment activities and to punish those who fail to conform. Members of the society are urged to inform on any of their fellow members who break the rules or criticize the leadership.

Nazi Germany, the Soviet Union under the leadership of Lenin and later Stalin, Chile under Augusto Pinochet, and the Spain of Francisco Franco are historical examples of totalitarian regimes. This image of "Big Brother," a symbol of totalitarianism's penetration of private as well as public life, comes from the film version of George Orwell's classic book, *1984*.

©AF archive/Alamy Stock Photo

One characteristic shared by totalitarian regimes of the 20th century was a ruthless commitment to power and coercion over the rule of law. Soviet leader Vladimir Lenin has been quoted as stating that "the dictatorship—and take this into account once and for all—means unrestricted power based on force, not on law" (Amis, 2002, p. 33). Another characteristic of these regimes was a willingness to destroy the opposition by any means necessary. Joseph Stalin's regime in the Soviet Union, which lasted from 1922 to 1953, purged millions of perceived, potential, or imagined enemies; Stalin's Great Terror tore apart the ranks of even the Soviet military apparatus. Martin Amis (2002) cites the following statistics in characterizing Stalin's war on his own military: From the late 1930s to about 1941, Stalin purged 3 of 5 marshals, 13 of 15 army commanders, 154 of 186 divisional commanders, and at least 43,000 officers lower down the chain of command (p. 175). An often-told story about Stalin cites him telling his political inner circle that each should find two replacements for himself.

Perhaps more than any other political system, totalitarianism is built on terror and the threat of terror—including genocide, imposed famine, purges, deportation, imprisonment, torture, and murder. Fear keeps the masses docile and the dictator in power. Torture has a long, brutal history in the dictatorships of the world, and, in trying to uncover its function, Amis (2002) makes the compelling observation that "torture, among its other applications, was part of Stalin's war against truth. He tortured, not to force you to reveal a fact, but to force you to collude in a fiction" (p. 61).

Today, few totalitarian states exist. Certainly, in the age of the Internet, control of information is an enormous challenge to states that seek full control of their populations. North Korea remains one of the last totalitarian regimes on the planet, where the citizenry is isolated from the rest of the world and few North Koreans outside the elite have Internet connections, smartphones, and computers—or even regular electricity and access to nutritious food.

Democracy

Democracy literally means "the rule of the people." (The word comes from the Greek *demos*, "the people," and *kratos*, "rule.") The concept of democracy originated in the Greek city-state of Athens during the fifth century BCE, where it took the form of **direct democracy**, *in which all citizens fully participate in their own governance.* This full participation was possible because Athens was a small community by today's standards and because the vast majority of its residents (including women and enslaved persons, on whose labor the economy relied) were excluded from citizenship (Sagan, 1992).

Direct democracy: A political system in which all citizens fully participate in their own governance.

Representative democracy: A political system in which citizens elect representatives to govern them.

Direct democracy is rarely possible today because of the sheer size of most countries and the complexity of their political affairs. One exception is the referendum process that exists in some U.S. states, including California and Oregon. In Oregon, for instance, the signatures of a specified percentage of registered voters can bring a referendum to the ballot. In 2014, for instance, Oregonians approved an effort to permit adults to purchase marijuana for recreational use, though the quantity is limited and the drug may not be used in public.

Democracy in the modern world more typically takes the form of **representative democracy**, *a political system in which citizens elect representatives to govern them.* In a representative democracy, elected officials are expected to make decisions that reflect the interests of their constituents. Representative democracy first took hold in the industrial capitalist countries of Europe. It is now the principal form of governance throughout the world, although some parts of the populations in democratic states may be disenfranchised. For instance, only in recent years have women been legally granted the right to vote in many countries (Figure 14.3). Some countries, such as China, the largest remaining authoritarian society, claim to have free elections for many government positions, but eligibility is limited to members of the Communist Party. Thus, even though voting is the hallmark of representative democracy, the mere option of voting does not ensure the existence of a true democracy.

The 2020 election in the United States took place in the context of a global pandemic. In spite of the health threat, over 152 million Americans voted.

©KAMIL KRZACZYNSKI/Contributor

THE U.S. POLITICAL SYSTEM

Politics in democratic societies is *the art or science of influencing public policy*; it is structured around competing political parties whose purpose is to gain control of the government by winning elections. Political parties serve this purpose by defining alternative policies and programs, building their membership, raising funds for their candidates, and helping to organize political campaigns. Not only must candidates win elections and retain their offices, but they must also, once in office, make decisions with far-reaching financial and social effects. These decisions ideally reflect the needs and desires of

Politics: The art or science of influencing public policy.

FIGURE 14.3 ■ When Women Won the Right to Vote in Selected Countries

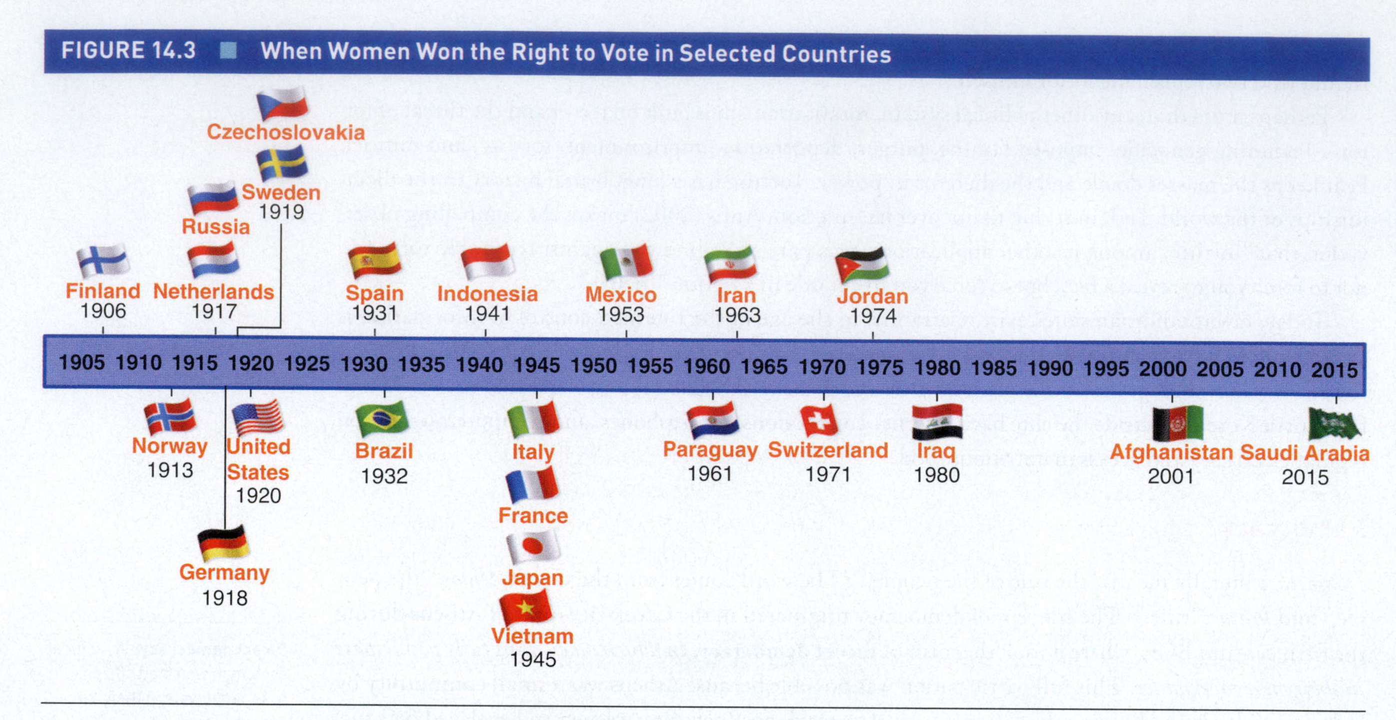

Source: Data from Women's Suffrage: When Did Women Vote? Interactive Map. Scholastic.com

their constituents as well as the interests of their parties and the entities that contribute to their campaigns. Some politicians argue that their constituents' issues take priority; other observers suggest that politicians are beholden to party or donor interests.

In the following section, we discuss electoral politics in the United States. Sociologists take an interest in electoral politics because it is an important site at which power in modern countries is exercised. Thus, key questions that sociologists ask—How is this functional for society? Who benefits from the existing social order? How do perceptions structure behaviors in the electoral process?—can be applied to electoral politics.

Electoral Politics and the Two-Party System

Most modern democracies are based on a parliamentary system, in which the chief of state (called a *prime minister*) is the head of the party that has the largest number of seats in the national legislature (typically called a *parliament*). Britain, for example, has a parliamentary system. This arrangement can give a significant degree of influence to minority parties (those that have relatively few representatives in parliament), since the majority party often requires minority party support to pass legislation or even to elect a prime minister.

In the United States, the president is chosen by voters—although, as happened in the 2000 and 2016 presidential elections, a candidate who wins the popular vote (in these cases, Al Gore and Hillary Clinton, respectively) cannot become president without also winning the requisite number of Electoral College votes.

The separate election of the president and Congress (rather than having the legislature choose a national leader, as is common in parliamentary democracies) is intended to help ensure a separation of powers between the executive and legislative branches of the government. At the same time, it weakens the power of minority or third parties, since—unlike in parliamentary systems—they are unlikely to have much impact on who will be selected chief of state. In Britain or Germany, by contrast, if a minority party stops voting with the majority party in parliament, its members can force a national election, which might result in a new prime minister. This gives minority parties potential power in parliament to broker deals that serve their interests. No such system exists in the United States and, as a consequence, third parties play only a minor role in national politics. No third-party candidate has won a presidential election since Abraham Lincoln was elected in 1860.

The domination of national elections and elected positions by the Republican and Democratic Parties is virtually ensured by the current political order. Parties representing well-defined interests are ordinarily eliminated from the national political process, since there are few avenues by which they can exert significant political power. Unlike in many other democracies, there are no political parties in the United States that effectively represent the exclusive interests of labor, environmentalists, or other constituencies at the national level. On the contrary, there is a strong incentive for political groups to support one of the two major political parties rather than to "waste" their votes on third parties that have no chance at all of winning the presidency; at most, such votes are generally offered as "protest" votes.

Third parties can occasionally play an important—even decisive—role in national politics, particularly when voters are unhappy with the two dominating parties. The presidential campaign of H. Ross Perot of the Reform Party in 1992 was probably significant in taking votes away from Republican George H. W. Bush and helping Democrat Bill Clinton to win the presidential election. Perot, running at the head of his own party organization, won nearly 19% of the popular vote. In 2000, the situation favored Republicans, as Democrat Al Gore probably lost votes to Green Party candidate Ralph Nader; in some states, George W. Bush had fewer votes than Gore and Nader combined, but more than Gore alone. Bush won the electoral votes of those states. Another candidate from the Green Party, Jill Stein, was also active in the 2016 presidential race, though she did not earn a significant number of votes (1.1%) and functioned primarily as an option for voters who were disaffected with both parties. Many voters who feel that the choices they have do not match up with their interests opt out of voting entirely, however. The issue of apathy in U.S. politics is one that we take up in the next section.

VOTER ACTIVISM AND APATHY IN U.S. POLITICS

One consequence of the lack of political choices in the entrenched two-party system in the United States may be a degree of political apathy, reflected in voter turnouts that are among the lowest in the industrialized world. Among democracies, the United States scores in the bottom fifth when it comes to voter participation. Whereas many European countries typically have voter turnouts between 70% and 90% of eligible voters, in 2000, about 55% of the voting-eligible U.S. population participated in the presidential election. The percentage has risen in subsequent presidential election years: Estimates put turnouts for 2008 at about 62%, 2012 around 59%, 2016 at 60%, and 2020 at a little more than 66% (Figure 14.4).

Historically, the proportion of eligible voters turning out for elections in the United States has varied by education (Figure 14.5), race and ethnicity (Figure 14.6), and age (Figure 14.7). Voters who are

FIGURE 14.4 ■ U.S. Voter Participation in Presidential Election Years, 1972–2020

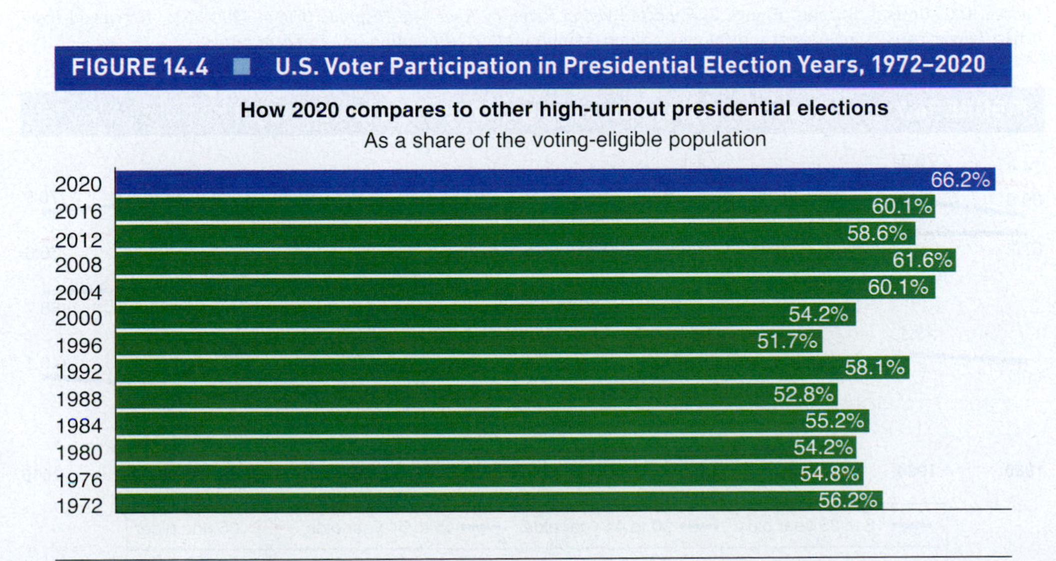

Source: Schaul, K., Rabinowitz, K., & Mellnik, T. (2020). 2020 Turnout Is the Highest in Over a Century. *Washington Post.* Retrieved from https://www.washingtonpost.com/graphics/2020/elections/voter-turnout/

FIGURE 14.5 ■ Reported Voting Rates by Educational Attainment, 1980–2016 (percentage)

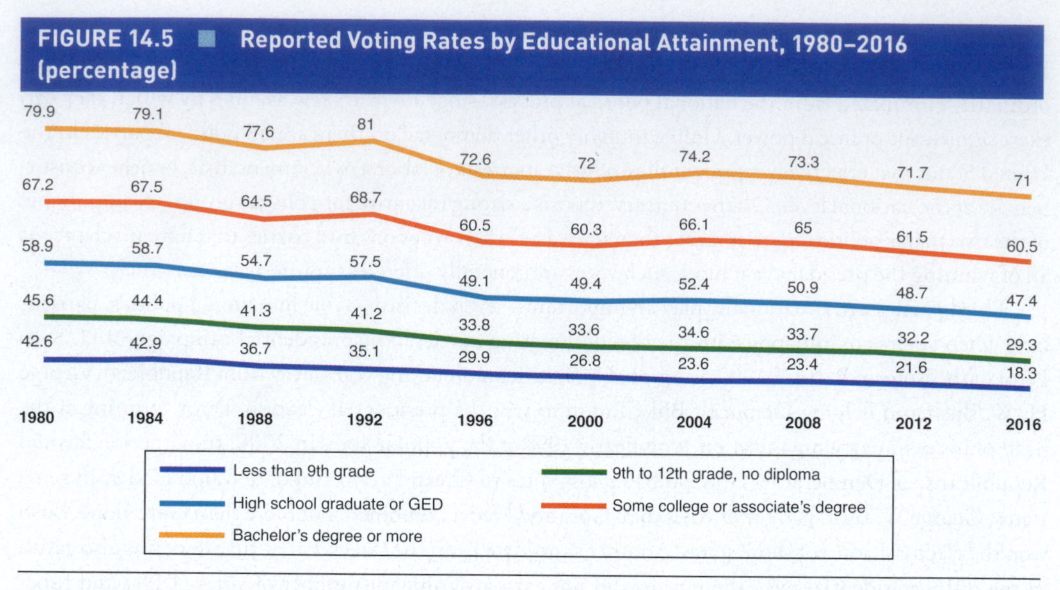

Source: U.S. Census`Bureau. *Table A-2. Reported Voting and Registration by Region, Educational Attainment, and Labor Force: November 1964 to 2016.* Retrieved from https://www. census.gov/data/tables/time-series/demo/voting-and-registration/voting-historical-time-series.html

FIGURE 14.6 ■ Reported Voting Rates by Race and Hispanic Origin, 1980–2016 (percentage)

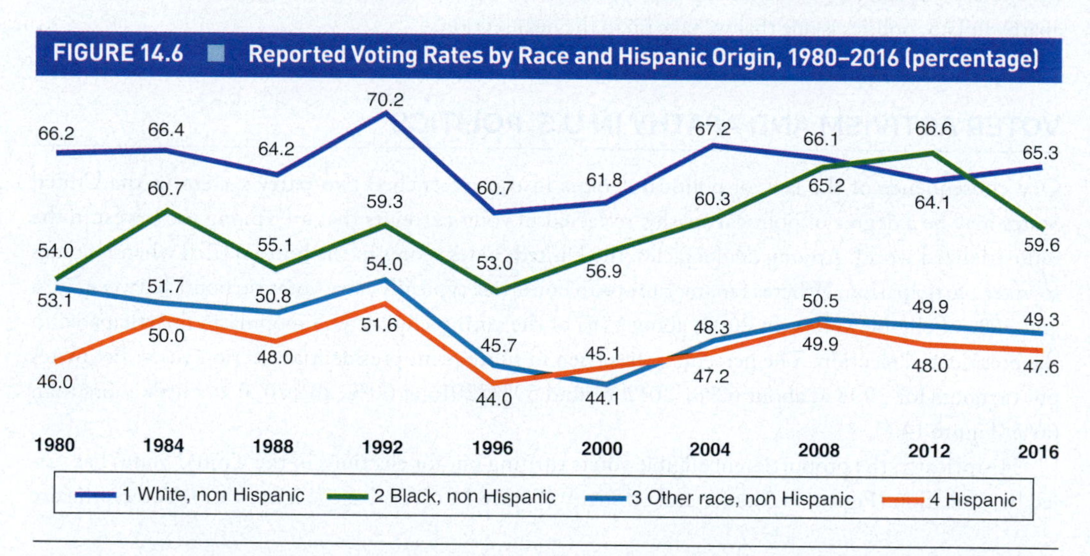

Source: U.S. Census Bureau. Figure 2. *Reported Voting Rates by Race and Hispanic Origin: 1980–2016.* Retrieved from https://www. census.gov/newsroom/blogs/random-samplings/2017/05/voting_in_america.html

FIGURE 14.7 ■ Reported Voting Rates by Age, 1980–2016 (percentage)

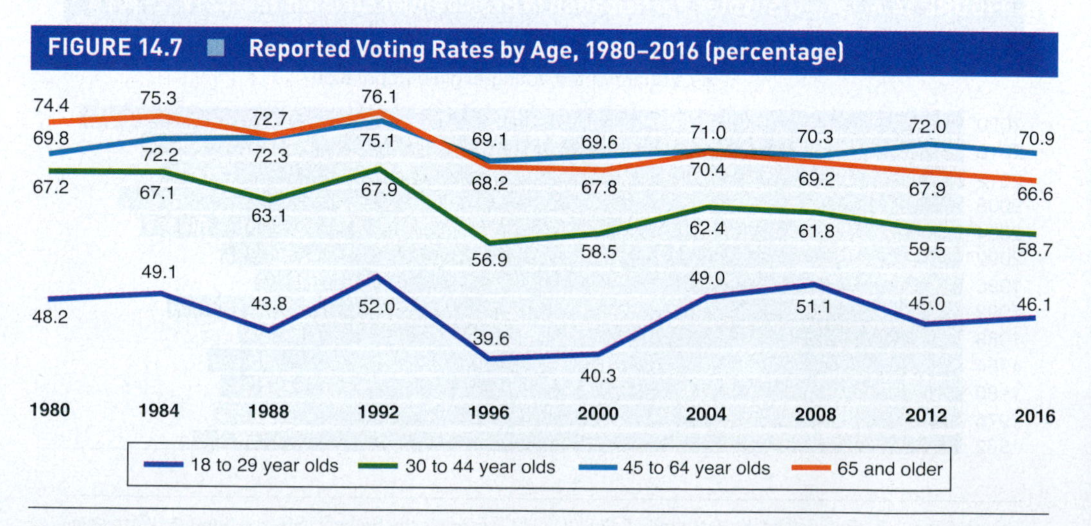

Source: U.S. Census Bureau. *Figure 4. Reported Voting Rates by Age: 1980–2016.* Retrieved from https://www.census.gov/newsroom/ blogs/random-samplings/2017/05/voting_in_america.html

white, older, and more educated have historically had greater influence than other demographic groups on the election of officials and, consequently, on government policies. Interestingly, however, data suggest that President Barack Obama's 2012 reelection was driven in part by the votes of young people (60% of voters age 18–29 cast a vote for Obama) and minorities (for example, about 70% of ethnic Latinx voters cast a ballot for Obama; Pew Research Center for the People and the Press, 2012). Indeed, turnout among younger Americans has been trending upward since the 2018 midterms. In 2018, 18- to 29-year-olds saw the largest increase in turnout for any age group, an estimated 79% increase compared to the midterm elections of 2014 (Misra, 2020).

According to the Center for Information and Research on Civic Learning and Engagement (CIRCLE) at Tufts University, voter turnout for those ages 18–29 rose by about 8% in 2020, compared to 2016. In total, 53% of these "youth voters" cast their ballot in 2020, and comprised 17% of the total vote. Some observers suggest that this increase was a response to issues widely perceived by young adults as urgent, including the COVID-19 pandemic, the consequences of systemic racism, and climate change. Additionally, 2020 saw a notable rise in social media campaigns, both by political groups and celebrities, encouraging young people to vote (Pike, 2020).

The historically lower proportions of young Americans and members of minority groups among voters prompt us to ask why people who are poor or working class, minority, and/or young are less likely to be active voters. One thesis is that voters do not turn out if they do not perceive that the political parties represent their interests (Delli Carpini & Keeter, 1996). Some of Europe's political parties represent relatively narrow and specific interests. If lower- to middle-class workers can choose a workers' party (for instance, the Labour Party in Britain or the Social Democratic Party in Germany), or environmentalists a Green Party (several European states, including Germany, have active Green Parties), or minority ethnic groups a party of their ethnic kin (in the non-Russian former Soviet states that are now democracies, Russians often have their own political parties), they may be more likely to participate in the process of voting. This is particularly likely if membership in the legislative body—say, a parliament—is proportionally allocated, in contrast to a winner-takes-all contest such as that in the United States.

In the United States, the legislative candidate with the greater number of votes wins the seat; the loser gets nothing. In several European countries, including Germany, parties offer lists of candidates, and the total proportion of votes received by each party determines how many members of the list are awarded seats in parliament. In *proportional voting*, small parties that can break a minimum barrier (in Germany, it is 5% of the total vote) are able to garner at least a small number of seats and enjoy a political voice through coalition building or by positioning themselves in the opposition.

Consider the winner-takes-all system and the proportional division of electoral votes. Is one more representative of the will of the people than the other? What do you think?

Some other reasons for low voter turnout among some demographic groups are practical: Low-wage workers may work two or more jobs and may not be able to visit polling places on the designated day of voting (for federal elections in the United States, the first Tuesday following the first Monday of November, usually between about 6:00–8:00 a.m. and 6:00–8:00 p.m.). In many European states, Election Day is a national holiday, and workers are given the day off to participate in the voting process. Recently, more U.S. states have offered early voting, extending the opportunity to vote by several days or even weeks at designated polling places. Oregon has allowed voting by mail since 1998, and other states have also begun to offer this alternative. Data suggest that these initiatives increase voter participation. Indeed, in 2020, over 101 million Americans voted early, with about 65 million of those votes coming via mailed ballots. The 2020 election saw the greatest voter turnout in over 100 years, and mail-in ballots in 2020 accounted for nearly 74% of the total turnout of voters in 2016 (McDonald, 2020).

Research suggests that young people are generally not apathetic about civic involvement; in fact, many volunteer and are eager to give back to their communities (*The Atlantic*, 2014). Voting, however, has not historically inspired the same commitment. The young are less likely to be courted by the parties and candidates, who tailor messages to attract the interests of groups such as the elderly, who vote in larger numbers, and other groups that are historically more likely to turn out at the polls or make campaign donations. Perhaps because of a perception that the candidates and parties do not seem to speak to or for them, younger voters have reciprocated with limited participation in the voting process. Other factors have also been identified as relevant in influencing the youth vote, including contact from organizations or campaigns and the accessibility of information about how, where, and when to vote (CIRCLE, n.d.).

The U.S. electorate is changing. Over the past few decades, Republicans and Democrats seem increasingly divided, though the changes we see are not only ideological. While, as we discuss in this chapter, young people are less likely to vote than older generations, their numbers make them a potentially powerful political force. In 2020, the voters of Gen Z, who were between about 18 and 24, made up about a tenth of eligible voters and their numbers will continue to grow: This is significant, not least because this is the most diverse generation in U.S. history. Baby Boomers and older generations (those over about 56), meanwhile, accounted for about four in ten eligible voters. Millennials, those between about 25 and 39, comprised about a quarter of the voting-eligible population, as did Gen X, which is sandwiched between Millennials and Baby Boomers (Figure 14.8).

Before we move on to the next section, which discusses the issue of political influence, let's pause to consider the following questions: How might our elected government and its policies be altered if people turned out to vote in greater numbers? Would higher turnouts among the poor, minorities, and young people make issues particularly pertinent to them higher priorities for decision makers? What do you think?

Power and Politics

While parts of the general public may show apathy about elections and politics, wealthy and powerful individuals, corporations, labor unions, and interest groups have a great deal at stake. Legislators have the power to make decisions about government contracts and regulations, taxes, federal labor and environmental and health standards, national security budgets and practices, and a spectrum of other important policies that affect profits, influence, and the division of power among those competing to have a voice in legislation.

The shortest route to political influence is through campaign contributions. The cost of campaigning has gone up dramatically in recent years, and candidates for public office must spend vast sums of money on getting elected. The presidential contest alone in the 2020 election was estimated to cost a staggering $6.6 billion, more than the cost of the presidential race and every

FIGURE 14.8 ■ Eligible Voters by Generation, 2020

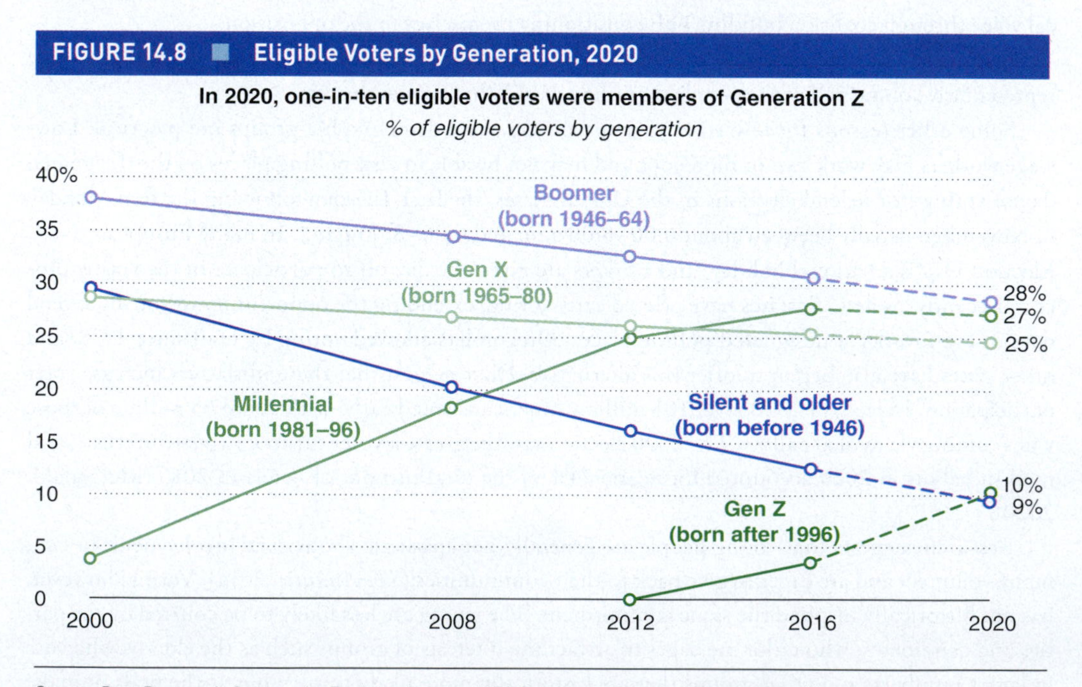

Source: Pew Research Center. *An Early Look at the 2020 Electorate.* Retrieved from https://www.pewsocialtrends.org/essay/an-early-look-at-the-2020-electorate/

congressional campaign combined in 2016 (Goldmacher, 2020). A large part of that spending was financed by donations, both large and small. The Biden campaign was the first to raise $1 billion in donations alone, nearly doubling the Trump campaign's $596 million raised (Center for Responsive Politics, 2020b). While large sums of money are given by wealthy individuals via super PACs, small-dollar donors (those who contribute $200 or less) made up 22% of total donations in the 2020 cycle. Further exploring these trends, in 2020, $1.8 billion worth of presidential campaign spending went to television ads. To put that in perspective, the entire cost of the 2016 presidential campaign, primaries included, was $2.4 billion (Goldmacher, 2020).

Donald Trump, one of the most divisive presidents in recent U.S. history, lost reelection in 2020. Given the growing number of Gen Z voters, how big of a role do you think this cohort played in the election's outcome?

©Gerardo Mora/Getty Images

It was not just the presidential race that broke spending records in 2020. Four separate Senate races crossed the $200 million mark. In South Carolina, the Democratic challenger to incumbent senator Lindsey Graham broke fund-raising records by being the first Senate candidate to raise over $100 million from outside donors. All told, the Democratic Party had had a sizable monetary advantage all the way down the ballot. Democratic candidates spent $5.5 billion, while Republican candidates spent about $3.8 billion. Significantly, those figures exclude an additional $1.3 billion in spending by self-funded Democratic presidential candidates Michael Bloomberg and Tom Steyer (Goldmacher, 2020). Does spending equal winning? The evidence in 2020 was mixed: While Democrat Joe Biden won the presidency with a majority of the popular vote and an Electoral College victory, several well-funded Democratic Senate candidates did not beat their Republican opponents. At the same time, winning opponents, while less well funded, often enjoyed the benefits of incumbency, which include greater media exposure and name recognition.

In November 2020, Democrat Joseph R. Biden was elected to the U.S. presidency. Unlike his predecessor, Donald J. Trump, Biden won both the popular vote and the Electoral College vote. Biden's vice president, Kamala Harris, is the first female vice president elected in the United States and the first woman of color to hold that position. In this photo, Biden and Harris address the country after their win.

©Pool/Pool

A substantial proportion of the money candidates and parties raise comes from corporate donors and well-funded interest groups. In many instances, companies and well-resourced interest groups donate money to candidates of both parties in order to ensure that they will have a voice and a hand in decision making, regardless of the electoral outcome. Clearly, while most politicians would deny that there is an explicit *quid pro quo* (a term that means "something for something") with big donors, most would also admit that money can buy the time and interest of a successful candidate and determine which issues are most likely to be heard.

The 2012 election had ushered a new player—the super PAC—into electoral politics. **Political action committees (PACs)** are *organizations created by groups such as corporations, unions, environmentalists, and other interest groups for the purpose of gathering money and contributing to political candidates who favor the groups' interests.* In 2010, the U.S. Supreme Court ruled in *Citizens United v. Federal Election Commission* that the government cannot restrict the monetary expenditures of such organizations in political campaigns, citing the First Amendment. This effectively means that these entities can contribute unlimited amounts of money to PACs. The term *super PAC* is used to describe the numerous well-funded PACs that have sprung up since the *Citizens United* decision.

Political action committees (PACs): Organizations created by groups such as corporations, unions, environmentalists, and other interest groups for the purpose of gathering money and contributing to political candidates who favor the groups' interests.

While super PACs cannot contribute directly to specific campaigns or parties, they can demonstrate support for particular candidates through, for instance, television advertising. Many critics of the *Citizens United* decision have been disturbed by the implications of unlimited corporate spending in politics. The primary drivers of this new political spending have been extremely wealthy individuals and (often) anonymous donors. According to the Center for Responsive Politics, during the 2020 election cycle, the top 100 individual donors to super PACs were responsible for over 69% of donations made (Center for Responsive Politics, 2020a, 2020b).

Lobbyists: Paid professionals whose job it is to influence legislation.

Efforts to influence legislation are not limited to direct or indirect campaign contributions. Special interest groups often hire **lobbyists**, *paid professionals whose job it is to influence legislation.* Lobbyists commonly maintain offices in Washington, D.C., or in state capitals, and the most powerful lobbies are staffed by full-time employees. Many of the best-funded lobbies represent foreign governments. Lobbying is especially intense when an industry or other interest group stands to gain or lose a great deal if proposed legislation is enacted. Oil companies, for instance, take special interest in legislation that would allow or limit drilling on U.S. territories, as do environmental groups. In an instance such as this, lobbyists from green groups and the oil and gas industry generally stand opposed to one another and seek to influence political decision makers to side with them.

Many lobbyists are former politicians or high-level government officials. Since lobbyists are often experts on matters that affect their organizations' interests, they may help in writing the laws that elected officials will introduce as legislation. Consider the following example described.

In an article titled "A Stealth Way a Bill Becomes a Law," *Bloomberg Businessweek* pointed out that several state-level bills rejecting cap-and-trade legislation (which is intended to reduce carbon dioxide emissions, believed by most scientists to contribute to climate change) used identical wording: "There has been no credible economic analysis of the costs associated with carbon mandates" (Fitzgerald, 2011). What was the source of this wording? It was supplied by the American Legislative Exchange Council (ALEC), an organization supported by companies including Walmart, Visa, Bayer, ExxonMobil, and Pfizer. More recently, ALEC worked with several large companies on a model bill focused on ensuring protection for companies from liability lawsuits during the COVID-19 pandemic. Such a law was supported by businesses seeking to protect themselves from lawsuits by employees who became ill on the job (Lacy, 2020). In exchange for a large membership fee, a corporation can buy itself a seat on the bill-writing task force, which prepares model legislation, primarily for Republican political decision makers.

Is the interaction of private-sector corporations and public-sector legislators an example of fruitful and appropriate cooperation on matters of mutual interest? When, if ever, is the writing of laws by corporate sponsors appropriate? When is it inappropriate?

Social Movements, Citizens, and Politics

Well-organized, popularly based social movements can also be important in shaping public policy. Among the most important social movements of the 19th and 20th centuries was the drive for women's suffrage, which invested half a century of activism to win U.S. women the right to vote. The movement's leaders fought to overcome the ideas that women ought not to vote because their votes were represented by their husbands; because the muddy world of politics would besmirch feminine purity; and because women, like adolescents and lunatics, were not fit to vote.

The women's suffrage movement was born in the United States in 1869, when Susan B. Anthony and Elizabeth Cady Stanton founded the American Woman Suffrage Association. It worked for decades to realize its goal: In 1920, the 19th Amendment to the Constitution was ratified and women were granted the right to vote on a national level (some states had granted this right earlier).

The second wave of the women's movements, which began in the 1960s, boasted other important achievements, including the passage of laws prohibiting gender-based job discrimination and rules easing women's ability to obtain credit independent of their husbands.

The temperance movement, symbolized by Carry Nation's pickaxe attacks on saloons in the early 1900s, sought to outlaw the use and sale of alcoholic beverages. This movement eventually resulted in the 1919 ratification of the 18th Amendment to the Constitution, which made it a crime to sell or distribute alcoholic beverages. The 21st Amendment eventually repealed Prohibition in 1933.

The labor movement grew throughout the first half of the 20th century, providing a powerful counterweight to the influence of business in U.S. politics. Labor unions were critical in getting federal and state laws passed to protect the rights of workers, including minimum wage guarantees, unemployment compensation, the right to strike, and the right to engage in collective bargaining. By midcentury, at the height of the unions' power, roughly 25% of all U.S. workers belonged to labor unions. Today, globalization and the flight of U.S. factories to low-wage areas have contributed to a decline in union membership, and just under 10% of workers belong to unions (U.S. Bureau of Labor Statistics, 2020). Furthermore, union membership is much more common among public-sector workers (33.6% in 2019) than those who work in the private sector (6.2%) (U.S. Bureau of Labor Statistics, 2020). At the same time, large unions (including the American Federation of Labor and Congress of Industrial Organizations [AFL-CIO]) have retained a good deal of political power, and candidates (particularly Democrats) vie for the unions' endorsements, which bring with them a large potential bloc of voters.

Social movements have long been a catalyst for policy change in the United States. In the last few decades, we have seen issues such as same-sex marriage and marijuana use advocated at the grassroots level, and eventually appear on legislative agendas and Election Day referendums. A current example of a social movement that has led to notable changes in state and local politics is gun control. Following the Parkland, Florida, school shooting in February of 2018, a movement led by student survivors mobilized activists and supporters. All over the country, including in states run by Republican governors, 50 new laws were enacted that restricted access to firearms in some capacity (Vasilogambros, 2020).

On the one hand, social movements provide a counterbalance to the power and influence in politics of large corporate donors, which we discussed previously. These movements offer a political voice to grassroots groups representing interests contrary to those of big business, such as labor rights and environmental protection. On the other hand, if we return to Domhoff's (2002) question, "Who wins?" we see that these groups rarely have more influence than large corporations and donors.

Constituents. Wealthy individuals, interest groups, PACs, and lobbyists exert considerable political influence through their campaign contributions. Still, these factors alone are not sufficient to explain political decisions. If they hope to be reelected, elected representatives must also serve their constituents. That is why politicians and their aides poll constituents, read their mail and e-mail, and look closely at the last election results.

One way politicians seek to win their constituents' support is by securing government spending on projects that provide jobs for or otherwise help their communities and constituents. If a new prison is to be built, for instance, legislators vie to have it placed in their district. Although a prison may seem like an undesirable neighbor, it can represent an economic windfall for a state or region. Among other things, prisons provide jobs to individuals who may not have the education or training to work in professional sectors of the economy and would otherwise be unemployed or working in the poorly paid service sector.

Projects that legislators push to bring to their home districts are sometimes labeled "pork." Pork may be superfluous or unnecessary for the macro-level economy but good for the legislator's home district. On the other hand, when a government commission proposes closing military bases, cost-conscious members of Congress will support the recommendation—*unless* any of the bases marked for closure are in their districts.

Politicians spend substantial amounts of time in their home districts. Over the course of the calendar year, the U.S. Congress is in session for an average of 103 days (Library of Congress, 2012). Congressional representatives spend much of their off-session time in their home districts because they are interested in hearing the views of their constituents—as well as in raising money and getting reelected.

Contradictions in Modern Politics: Democracy and Capitalism

Leaders in modern democratic capitalist societies such as the United States are caught between potentially contradictory demands. They seek widespread popular support, yet they must satisfy the demands of the elites whose financial backing is essential for electoral success. On the one hand, voters are likely to look to their political leaders to back benefits such as retirement income (in the form of Social Security, for instance), housing supports (affordable housing for low-income families or mortgage tax breaks for wealthier ones), and environmental protection. On the other hand, such programs are costly to implement and entail economic costs to corporations, developers, and other members of the elite. Some theorists argue that modern governments thus are caught in a conflict between their need to realize the interests of the capitalist class and their desire to win the support and loyalty of other classes (Held, 1989; Offe, 1984; Wolfe, 1977).

Jürgen Habermas (1976), a contemporary theorist with a conflict orientation, argues that modern countries have integrated their economic and political systems, reducing the likelihood of economic crisis while increasing the chances of a political crisis. He terms this the *legitimation crisis*. Governments have intervened in the market and, to some degree, solved the most acute contradictions of capitalism—including extreme income inequalities and tumultuous economic cycles—that Marx argued could be addressed only in a proletarian revolution. Governments often act to keep inflation and deflation in check, to regulate interest rates, and to provide social assistance to those who have lost jobs. Thus, economics is politicized, and the citizenry may come to expect that economic troubles will be solved through state structures and social welfare.

To understand Habermas's argument more fully, imagine a postindustrial U.S. city. The loss of jobs and industries manifests itself as a crisis—thousands of jobs in auto and other manufacturing industries move abroad, local businesses suffer as the amount of disposable income held by local people plummets, and economic pain is acute. How, in modern society, does our hypothetical city (which has hundreds of authentic counterparts in the United States) respond? Does it erupt in revolutionary fervor, with displaced laborers calling for class struggle? Or do people look to their local, state, and federal governments to provide relief in the form of tax cuts or credits, unemployment benefits, and plans for attracting new industries?

The citizenry of modern capitalism, says Habermas, does not widely question the legitimacy of capitalism. If there is a crisis, it is political, and it is solved with policies that may smooth capitalism's bumpy ride. In a sense, the state becomes the focus of discontent—in a democracy, political decision makers can be changed and a crisis averted. The economic system that brings many of these crises into being, however, remains in shadow, its legitimacy rarely questioned.

In the next part of the chapter, we look into some of the other challenges confronted by states and their populations, including war and terrorism. States are key players in modern warfare, and military conflict is an important domestic political issue and global concern. Terrorism has also become increasingly entwined with war today, as recent wars undertaken by the United States, for instance, have been part of an effort to combat the threat of terrorism.

WAR, STATE, AND SOCIETY

Conflict between ethnic or religious groups, states, and other social entities has a long history. War has been part of human societies, cultures, and practices in some form for millennia. In the 5th century BCE, the ancient Greeks created a game called *petteia*, the first board game known to have been modeled on war. In the 6th century CE, chess, another game of strategic battle, was born in northern India; it developed into its modern form by the 15th century. Military training in ancient Greece also gave birth to the first Olympic Games. In the 20th century, war games took on far more advanced forms, ranging from battlefield exercises used to prepare for defensive or offensive war to sophisticated computer simulations used for both popular entertainment and military readiness training (Homans, 2011).

Today, the countries of the world spend trillions of dollars preparing for war or fighting in wars. At the same time, armed conflict and associated casualties have declined (Figure 14.9). Goldstein (2011) suggests that the nature of armed conflict has changed, shifting from larger wars in which powerful state actors confronted one another directly (such as World War II or the Korean War of the 1950s) to asymmetrical guerrilla wars, such as those the United States has fought in Iraq and Afghanistan in the past decades. He notes,

> Worldwide, deaths caused directly by war-related violence in the new century have averaged about 55,000 per year, just over half of what they were in the 1990s (100,000 a year), a third of what they were during the Cold War (180,000 a year from 1950 to 1989), and a hundredth of what they were in World War II. (p. 53)

Whatever the forms war has taken, it has been a key part of the human experience throughout history. What explains its existence and persistence? Recall from earlier chapters that *manifest functions* are intended and obvious, while *latent functions* are hidden, unexpected, or "nonpurposive" (in Robert Merton's words). Functionalists look at a phenomenon or an institution that exists in society, assert that

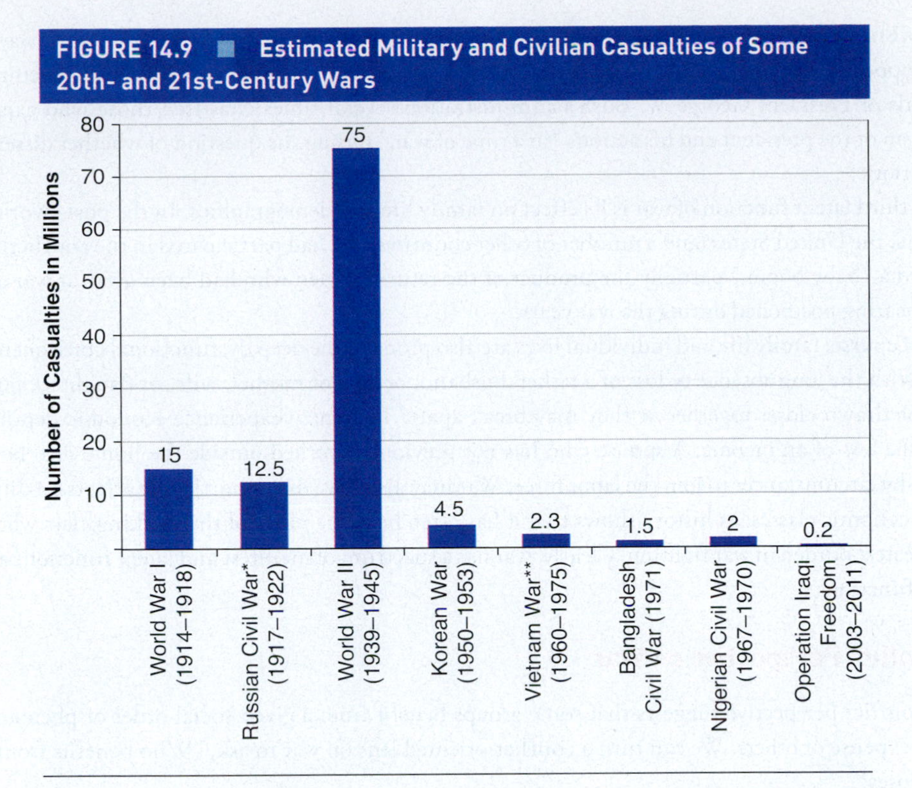

FIGURE 14.9 ■ Estimated Military and Civilian Casualties of Some 20th- and 21st-Century Wars

Sources: Leitenberg, M. (2006). *Deaths in Wars and Conflicts in the 20th Century*. Cornell University Peace Studies Program, Occasional Paper No. 29 (3rd ed.); Fischer, H. (2010). *Iraq Casualties: U.S. Military Forces and Iraqi Civilians, Police, and Security Forces*. Congressional Research Service.

**Including Soviet–Polish conflict.*

***Including North Vietnam versus South Vietnam.*

its existence presupposes a function, and ask what that function is. If something did not serve a function, it would cease to exist. Does war have a function? We begin with a functionalist perspective on war, considering its role and consequences at the macro and micro societal levels.

A Functionalist Perspective on War

What are the manifest functions of war? Historically, one function has been to gain territory. The Roman Empire (27 BCE–476 CE) waged war on surrounding territories, acquiring a substantial swath of the Middle East, including Cleopatra's Egypt, and then holding it with its massive armies. Another manifest function of war is to gain control of the natural resources of another state, while a third is to prevent the disintegration of a territorial unit. The American Civil War (1861–1865) sought to avert the secession of the South, which still favored slavery, from the North, which sought to abolish slavery.

What about the latent functions of war? First, war has historically operated as a stimulus to the economy: The term **war economy** refers to *the phenomenon of war boosting economic productivity and employment, particularly in capital- and labor-intensive sectors such as industrial production*. Notably, however, the wars in which the United States took part in the 20th century were fought outside its borders, and the benefits to the U.S. economy, especially in the World War II era, were not necessarily repeated elsewhere. The Soviet Union, France, Belgium, Poland, and many other European countries on whose territory World War II was waged emerged with shattered economies.

War economy: The phenomenon of war boosting economic productivity and employment, particularly in capital- and labor-intensive sectors such as industrial production.

As well, the first wars in which the United States has engaged in the 21st century—the conflict in Afghanistan that began in October 2001 and the military occupation of Iraq, which began in March 2003—have arguably had negative effects on the U.S. economy and have failed to benefit all but a few large corporations in the defense and energy sectors.

A second latent function of war is the fostering of patriotism and national pride. In times of war, governments implore their citizens to rally around the national cause, and citizens may display their patriotism with flags or demonstrations. Even those who oppose military action may shy away from open opposition in a climate of war-inspired patriotism: During the early years of the conflict in Iraq, officials of President George W. Bush's administration several times chastised those who expressed criticism of the president and his actions "in a time of war," raising the question of whether dissent was unpatriotic.

A third latent function of war is its effect on family life and demographics. In the post–World War II years, the United States (and a number of other countries that had participated in the conflict) experienced a "baby boom," partially the product of the return of men who had been away at war and of childbearing postponed during the war years.

Of course, family life and individual lives are also prone to the deep dysfunctional consequences of war. With the long absence or loss of a father, husband, or son (or mother, wife, or daughter), families may be drawn closer together or they may break apart. They may experience economic deprivation with the loss of an income. A spouse who has not previously worked outside the home may be compelled by circumstances to join the labor force. War may also have disproportionate effects on different socioeconomic classes, as history shows that it has often been members of the working class who bear the greatest burden in war fighting. Clearly, war has a spectrum of manifest and latent functions as well as dysfunctions.

A Conflict Perspective on War

The conflict perspective suggests that some groups benefit from a given social order or phenomenon at the expense of others. We can turn a conflict-oriented lens on war to ask, "Who benefits from war? Who loses?"

While we might be inclined to answer that the war's victor wins and the defeated state or social group loses, the conflict perspective offers us the opportunity to construct a more nuanced picture. Consider the conflict in Iraq that began with the U.S. occupation of that country in March 2003: The war commenced as an effort to topple the dictatorial regime of Saddam Hussein, which was believed by President George W. Bush and some of his political allies to possess weapons of mass destruction (though none were subsequently found in the country). While most troops were withdrawn in 2011, a small number reengaged in 2014 as the United States sought to support Iraq's fight for stability and battles against ISIS militants who had launched deadly terror attacks against soldiers and civilians. Who benefited from the war in Iraq that commenced in 2003?

Beneficiaries of any conflict include those who are freed from oppressive state policies or structures or from ongoing persecution by the defeat of a regime. Among the beneficiaries in Iraq, we might count the minority Kurdish population, who were victims of Saddam Hussein's genocidal attacks in 1987 and 1988 and were threatened by the Iraqi dictator's presence. Were the rest of the people of Iraq beneficiaries? To the degree that Saddam was an oppressive political tyrant, the answer may be yes; Saddam's ruling Báath Party, composed primarily of Sunni Muslims, persecuted the majority Shiite Muslim population of the country, particularly after some Shiites sought to foment an uprising at the end of the First Gulf War (1990–1991). Ordinary Sunni Muslims as well had no voice in the single-party state that Saddam ruled with a strong hand. At the same time, the long war has led to thousands of civilian and military casualties, fundamentally destabilized the country, and left it with a badly damaged economy and infrastructure. Today, Iraq continues to be plagued by *sectarian violence*—that is, violence between religious groups—that presents, together with ISIS violence, an existential threat to the country itself.

Other beneficiaries of the nearly decade-long Iraq War were corporations (mostly U.S.-based) that profited from lucrative government contracts to supply weapons and other military supplies. War generates casualties and destruction—it also generates profits. In assessing who benefits from war, we cannot overlook capitalist enterprises for which war is a business and investment opportunity. Among those that benefit directly from war and conflict are private military corporations (PMCs), which provide military services such as training, transportation, and the protection of human resources and infrastructure. The use of private contractors in war has a long history. The new U.S. government, fighting against the British in the American Revolution, paid private merchant ships to sink enemy ships and steal their cargo. The modern military term *company*, which refers to an organized formation of 200 soldiers, comes from the private "companies" of mercenaries who were hired to fight in conflicts during the Middle Ages in Europe.

At the same time, until recently, wars in the modern world were largely fought by the citizens of the nation-states involved. Sociologist Katherine McCoy (2009) writes,

> Scholars have long thought of fighting wars as something nation-states did through their citizens. Max Weber famously defined the modern state as holding a monopoly over the legitimate use of violence, meaning that only state agents—usually soldiers or police—were allowed to wield force. (p. 15)

In today's conflicts, governments, including the U.S. government, increasingly rely on PMCs to provide a vast array of services that used to be functions of the governments or their militaries. This situation raises critical questions about the accountability and control of these private armies, which are motivated by profit rather than patriotism (McCoy, 2009). The rise of PMCs has been driven by reductions in the size of armies since the end of the Cold War, the availability of smaller advanced weaponry, and a political-ideological trend toward the privatization and outsourcing of activities previously conducted by governments (Singer, 2003).

Syria's civil war began in 2011 and has continued largely unabated to this time. One of the war's key consequences has been mass displacement of populations as millions have fled Syria seeking shelter in countries across the globe. In this photo, Syrian refugees wait to cross at the Syrian–Turkish border.

©Anadolu Agency

The conflict perspective on war also asks, "Who loses?" Losers, of course, include those on both sides of a conflict who lose their lives, limbs, or livelihoods in war. Increasingly, according to some reports, they have been civilians, not soldiers. By one estimate, in World War I, about 5% of casualties were civilians (Swiss & Giller, 1993). In World War II, the figure has been estimated at 50% (Gutman & Rieff, 1999). While some researchers argue that the figure is much higher (Swiss & Giller, 1993), Goldstein (2011) suggests that the ratio of civilian to military casualties has remained at about 50:50 into the 21st century. Specific casualty figures also vary widely, often depending on the methodologies and motivations of the organizations or governments doing the counting. Apart from casualties, however, modern warfare has unleashed massive human flight in the form of refugees: According to the United Nations, at the end of 2018, armed conflict had displaced at least 30 million people, half of them children (United Nations, n.d.).

What about the less-apparent losers? Who else pays the costs of war? Modern military action has substantial financial costs, which are largely borne by taxpayers. In the decade between 2001, when the United States was the victim of terrorist attacks, and 2011, when the Iraq War ended and the war in Afghanistan began to wind down, the country spent an estimated $7.6 trillion on defense and homeland security. Even since the end of active U.S. engagement in Iraq and Afghanistan, the United States has continued to devote a substantial part of its federal budget to the Departments of Defense and Homeland Security. At the same time, many U.S. domestic programs in areas such as education, job training, and environmental conservation have lost funding as budgets have shrunk. Growth in the defense and security allocations of the federal budget has not been without costs.

In the next section, we consider the phenomena of terrorists and terrorism, which have been drivers—and consequences—of some of the world's most recent armed conflicts.

TERRORISTS AND TERRORISM

The "global war on terror" (GWOT) was initiated in 2001 after the September 11 attacks on the United States, which claimed 2,977 victims, as four commercial airliners were hijacked by Al-Qaeda terrorists. Two of the planes were crashed by the hijackers into New York City's World Trade Center towers; one into the Pentagon in Washington, D.C.; and one by desperate and heroic passengers into the ground in western Pennsylvania to thwart a fourth attack. The unprecedented events of September 2001 brought terrorism

A terrorist attack on the United States on September 11, 2001, brought down the World Trade Center Towers in New York City. The same group was also responsible for an attack on the Pentagon. A total of 2,977 people were killed by the September 11 terrorist attacks.

©Jose Jimenez/Primera Hora/Getty Images

and terrorists more fully than ever into the U.S. experience and consciousness. The political response was to refocus domestic priorities on the war on terror and homeland security, drawing resources and attention from other areas such as education and immigration reform. Terrorism became a key theme of U.S. politics, policies, and spending priorities and a subject of concern and discussion among diplomats, decision makers, and ordinary citizens.

The terms *terrorist* and *terrorism*, however, are broad and may not be defined or understood in the same ways across groups or countries. Acts of violence labeled by one group as terrorism may be embraced as heroic by another. The label of terrorist may be inconsistently applied depending on the ethnicity, religion, actions, and motivations of an individual or a group. Next, we examine these concepts and consider their usage with a critical eye.

Who Is a Terrorist?

Close your eyes and picture a terrorist. Why do you think that particular image appeared to you? The images we generate are culturally conditioned by the political environment, the mass media, and the experiences we have, and they differ across communities, countries, and cultures. The idea of a terrorist is not the same across communities because, as we will see, violent acts condemned by one community may be embraced by another as necessary sacrifices in the pursuit of political ends.

It has been said that one person's terrorist is another person's freedom fighter. Michael Collins was born in West Cork, Ireland, in 1890. Before he turned 20, he had sworn allegiance to the Irish Republican Brotherhood, a group of revolutionaries struggling for Irish independence from three centuries of British rule, and he worked and fought with them throughout the first decades of the 20th century. In Ireland and Northern Ireland today, Collins is widely regarded as a hero (Coogan, 2002). The 1996 film *Michael Collins*, starring Liam Neeson and Julia Roberts, cast him in a generally positive light: The film's tagline declared, "Ireland, 1916. His dreams inspired hope. His words inspired passion. His courage forged a nation's destiny."

In Britain, however, many consider Collins to be a terrorist. In 1920, while he was director of intelligence for the Irish Republican Army (IRA), his secret service squad assassinated 14 British officers (Coogan, 2002). The British responded to the IRA with violence as well. Notably, the Continuity Irish Republican Army continues to be on the U.S. Department of State's (2012a) global list of terrorist groups.

Was Michael Collins a terrorist or a hero? How do we judge Britain's violent military response? The label of *terrorist* is a subjective one, conditioned by whether one rejects or sympathizes with the motives and actions under discussion. As an expert on terrorism notes, "If one party can successfully attach the label *terrorist* to its opponent, then it has indirectly persuaded others to adopt its moral viewpoint" (Hoffman, 2006, p. 23).

What Is Terrorism?

Terrorism: "The unlawful use of violence or threat of violence to instill fear and coerce governments or societies. Terrorism is often motivated by religious, political, or other ideological beliefs and committed in the pursuit of goals that are usually political" (U.S. Department of Defense, 2011).

The question of *who is a terrorist* is closely tied to the question of *what is terrorism*. There is no single definition of **terrorism**. The U.S. Department of Defense (2011) defines it as "*the unlawful use of violence or threat of violence to instill fear and coerce governments or societies. Terrorism is often motivated by religious, political, or other ideological beliefs and committed in the pursuit of goals that are usually political.*" The Federal Bureau of Investigation (FBI) definition of terrorism adds another layer to the phenomenon. The FBI highlights *terrorism* as "the unlawful use of force or violence against persons or property to intimidate or coerce a government, the civilian population, or any segment thereof, in furtherance of political or social objectives" (National Institute of Justice, 2011). The years following the terror attacks of September 11, 2001, have seen a rise in the number of victims of terror attacks. Almost 20 years after these attacks, many of the threats are domestic—that is, they come from inside the United States, rather than from outside actors.

Domestic terrorism is not new. On April 20, 1995, 168 people perished in the bombing of a federal building in Oklahoma City, Oklahoma. While initial media suspicion pointed to foreign perpetrators, further investigation determined that American Timothy McVeigh, with the cooperation of a small group of antigovernment compatriots, was responsible for the crime. In the wake of the incident, the U.S. government increased its scrutiny of domestic threats. The terrorist incidents of September 11, 2001, which were perpetrated by Islamic radicals, shifted attention to the Middle East, including Afghanistan and, later, Iraq and Pakistan, among others.

Recent incidents of mass shootings in the United States have refocused attention on domestic threats to peace and security. Among a spate of incidents in recent years are the following: In June of 2015, Dylan Roof murdered nine African American churchgoers during a prayer service in South Carolina. Roof later indicated that he had hoped to ignite a "race war." In June of 2016, Omar Mateen killed 49 people at a gay nightclub in Florida. During a call to 911, Mateen pledged allegiance to the Middle Eastern terror group ISIS. In July of 2016, Micah Xavier Johnson shot to death five Dallas police officers who were working at a peaceful rally by Black Lives Matter activists. Johnson's online activities suggest he sought to target white officers in particular because he was angry about the killings of Black men by law enforcement. In February of 2018, a former student, Nikolas Cruz, opened fire in a Florida high school, killing 17. Cruz was expelled from the school the year prior for disciplinary problems (Martin et al., 2018).

The year 2019 brought no respite from violence. Based on a definition of *mass shooting* that classifies any incident with four or more shooting victims (excluding the shooter), there were 419 mass shootings in the United States in that year. The list of incidents includes the killing of 22 people and wounding of 24 in a Walmart in El Paso, Texas, and a drive-by shooting in Midland and Odessa, Texas, that left 7 victims dead and 24 wounded (Silverstein, 2020).

In 2014, the U.S. Department of Justice relaunched the work of a group focused on domestic threats. As a Council on Foreign Relations publication points out, however, there is inconsistency in understandings and legal approaches to what terrorism is and whether domestic and international incidents of violence both fall under that term (Masters, 2011). Are individuals or small groups in the United States who target government buildings, public events such as festivals or demonstrations, or other groups for violence *terrorists*? What is the significance of using that term rather than *criminal* or even *violent extremist*?

In 2019 and 2020, the most significant terror threat in the United States came from domestic terror groups, including white supremacists. In the state of Michigan (shown), a group of right-wing antigovernment extremists hatched a plot to kidnap Governor Gretchen Whitmer in late 2020. The plot was not successful, but threats to government officials have continued across the country.

©Jeff Kowalsky/Contributor

Recent events point to a growing consensus on the threat of *domestic terrorism* in the United States, though the term is still inconsistently applied. For example, a Homeland Threat Assessment released by the U.S. Department of Homeland Security in October 2020 suggested that "racially and ethnically motivated violent extremists—specifically white supremacist extremists (WSEs)—will remain the most persistent and lethal threat in the homeland" (quoted in Jones et al., 2020, para. 2). According to the report, white supremacists and similar groups were responsible for about 67% of terrorist plots and attacks in the United States in 2020—their targets included demonstrators at public protests, as well as individuals targeted on the basis of racial, ethnic, religious, and political identities (Jones et al., 2020).

Social Media and Modern Terrorism

According to the National Institute of Justice (2011), Title 22 of the U.S. Code, Section 2656f(d), defines *terrorism* as "premeditated, politically motivated violence perpetrated against noncombatant targets by subnational groups or clandestine agents, usually intended to influence an audience." Note the attention in this definition to the presence of an audience. This definition points more deliberately to terrorism as an instrument of horrific political theater whose direct victims are props on the stage of a larger political or ideological play. While such "media-oriented terrorism" does not, by one analysis, make up the majority of the terror acts of the past half century, it is widespread and has historical roots in the acts of 19th-century anarchists, who pioneered the concept of "propaganda of the deed" (Surette et al., 2009).

In some sense, the media offer a stage for acts of atrocity, not only functioning as reporters of terror events but also conditioning terrorist groups' selection of targets and actions. Media attention, which has expanded from the print media and television to include the Internet, the "Twitterverse," and other new media, has a powerful multiplier effect on modern terrorism, offering a broad platform of publicity even for small and relatively weak groups or "lone wolf" terrorists whose combat and political capabilities are otherwise very limited (Surette et al., 2009).

Social media has become a fundamental part of contemporary terrorism. If, as Timothy Furnish (2005) writes, "the purpose of terrorism is to strike fear into the hearts of opponents in order to win political concession," then social media has multiplied the effects of acts of terror, expanding the audience for horrific violence and transforming atrocities into media shows that can be played over and over again.

The terror group called the Islamic State (also known as ISIS [the Islamic State of Iraq and Syria] and ISIL [the Islamic State of Iraq and the Levant]) has used Twitter, Facebook, YouTube, and other sites to disseminate images of the killing of soldiers and civilians and the destruction of communities across their Middle Eastern battlegrounds. ISIS is an offshoot of Al-Qaeda, the terror group that claimed responsibility for the attacks in the United States on September 11, 2001. While both Al-Qaeda and ISIS are adherents of an extreme brand of Sunni Islam, the two groups apparently broke over ISIS's unfettered willingness to slaughter Muslim civilians. A writer on the *Vox Media* news website, writing of ISIS's use of images depicting atrocities against Iraqi soldiers, has pointed out that the multitude of graphic images is not merely ISIS bragging about its murderousness. ISIS has a well-developed social media presence, which has been used both to intimidate Iraqis who might oppose it, and win supporters in its battle with Al-Qaeda for influence over the international Islamist extremist movement (Beauchamp, 2014). Today, ISIS has lost considerable territory in the Middle East, including in Syria (Cordesman, 2020), but it remains an active online presence and control of virtual "territory" is also of value in, for instance, enticing new recruits. For this purpose, ISIS has used slickly produced videos glamorizing violence and touting its military successes; most videos, in an effort to appeal to foreign recruits, are produced in English, German, or French (NBC News, 2015).

Social media has also been an important platform for domestic terrorists in the United States. In their effort to elude public and government scrutiny, groups like white nationalists and other right-wing groups like the antigovernment Boogaloo Boys make use of the so-called dark web. But domestic terror groups have a presence on mainstream media platforms as well. While social media giants like Facebook and Twitter make some effort to police the presence of hate groups and violent extremists, their efforts are limited and deleted pages can be easily replaced. Social media has opened new avenues for recruitment and organizing that did not exist before. One writer has pointed out, for instance, that "Social media has lowered the collective-action problem that individuals who might want to be in a hate group would face. You can see that there are people out there like you. That's the dark side of social media" (Diep, 2017, para. 4).

How should social media platforms respond to the threats posed to individuals and society by international and domestic terrorists? Can social media provide remedies to terror threats as well? What do you think?

WHAT'S NEXT IN U.S. POLITICS AND GOVERNANCE?

In the modern world, politics and the state directly affect the lives of everyone. Understanding how politics and the state work is essential to our lives as informed, active local and global citizens. In the face of the apparently overwhelming power of the state and the seeming distance of political decision making from the lives of most people, it is easy for us to shrug our shoulders and feel powerless. Yet one of the lessons we learn from the sociological analysis of the state and politics is that both are subject to influence by ordinary citizens, especially when people are mobilized into social movements and interest groups, politically aware, and able to evaluate politics and policies critically. Public ignorance and apathy benefit those who use politics to ensure their own or their social groups' well-being; active citizenship is an authentic instrument of power, even where it faces significant obstacles.

The year 2020 was politically dramatic: It featured mass demonstrations against racism in the criminal justice system, the dangerous COVID-19 pandemic, and a contentious U.S. presidential election. Voting in the context of the pandemic was a new challenge—but it was one that was willingly taken up by millions of Americans. Indeed, the turnout for the 2020 vote was one of the largest in history. Will this tumultuous time kick off a new era of civic involvement in public activism and electoral participation? Will young adults become more involved, challenging the popular image of civic disengagement of Millennials and Gen Zers? Will the coming decade see more visible young leaders—and perhaps a generational shift in political leadership after a long period of domination by Baby Boomers? What do you think?

WHAT CAN I DO WITH A SOCIOLOGY DEGREE?

Written Communication Skills

Written communication is an essential skill for a broad spectrum of 21st-century careers. Sociology students have many opportunities to practice and sharpen written communication skills. Among others, sociologists learn to write *theoretically*, applying classical and contemporary theories to construct an analysis of social issues and phenomena, and to write *empirically*, preparing and communicating evidence-based arguments about the social world. Sociology majors write papers in a variety of forms and for a variety of audiences; these may include reaction papers, book reviews, theoretical analyses, research papers, quantitative analysis reports, field note write-ups, letters to decision makers or newspaper editors, and reflections on sociological activities or experiences. Excellent written communication is fundamental in many occupational fields, including politics, business and entrepreneurship, communications and marketing, law and criminal justice, community organizing and advocacy, journalism, higher education, law, and public relations.

Matthew Thibault, Social Media Trainer and Strategist in Social Business for Dell Technologies

University of Texas–Austin, BA in Sociology

As a recent graduate, I never could have imagined how a degree in sociology would kickstart my career and give me the writing and communication skills that I rely on as a professional.

Currently, my role is as a Social Media Strategist and Trainer for Dell Technologies in Round Rock, Texas, and the day-to-day work for my team often revolves around transforming events, strengthening partnerships, engaging with customers, and building an authentic presence for the Dell brand on social media. As a first job, social media has been a great opportunity to grow and acquire new hard skills in marketing while still feeling connected to those sociological roots.

On the content side, 280 characters can seem like a daunting limit for a Twitter post that needs to convey a lot of high-level information and yet, working within those restrictions connects back to word-count requirements practiced in the classroom. Building out social media training presentations that resonate with teams all over the globe for Dell is challenging for any trainer, and yet tasks like that are made much easier by having some insight into how social groups behave, how messages can be framed, what cultural barriers may arise, and how a single message may be perceived in different ways. For me, leaving college with experience and education regarding some of those deeper social situations was great preparation for working in an inherently communication-heavy industry and a globalized workforce.

The greatest barriers I have encountered in marketing and social media have often been rooted in poor communication, and in my experience, sociology has provided me with the chance to become a bridge: a bridge between teams, a bridge between regions, and a bridge between diverse peoples and communities. By filling in the gap, sociology has and will always be for me a way to transform our world and workforce for the better.

Career Data: Training and Development Manager

- 2019 Median Pay: $113,350
- $54.50 per hour
- Typical Entry-Level Education: Bachelor's degree
- Job Growth Outlook, 2019–2029: 7% (faster than average)

Source: Bureau of Labor Statistics, *Occupational outlook handbook*, 2019.

SUMMARY

- The world today is politically divided into 195 **nation-states**. Most countries are made up of many different peoples, brought together through warfare, conquest, or boundaries drawn by colonial authorities without respect to preexisting ethnic or religious differences.
- Modern countries are characterized by governments that claim complete and final authority over their **citizens**, systems of **law**, and notions of citizenship that contain obligations as well as civil, social, and political rights.
- State power is typically based on one of three kinds of legitimate authority: **traditional authority**, based on custom and tradition; **rational-legal authority**, based on a belief in the law; or **charismatic authority**, based on the perceived inspirational qualities of a leader.
- Functionalist theories of power argue that the role of the government is to mediate neutrally between competing interests; they assert that the influence of one group is usually offset by that of another group with an opposing view. Conflict theories of state power draw the opposite conclusion: The state serves the interests of the most powerful economic and political groups in society. Different versions of social conflict theories emphasize the importance of a **power elite**, structural contradictions, and the relative autonomy of state power from the economic elites.
- Governance in the modern world takes a number of forms, including **authoritarianism** (including **monarchies** and **dictatorships**), **totalitarianism**, and democracy.
- Democracy is one of the primary forms of governance in the world today, and most countries claim to be democratic in theory (if not in practice). Most democratic countries practice **representative democracy** rather than **direct democracy**.
- The U.S. political system is characterized by low voter turnouts. Voter participation varies, however, on the basis of demographic variables such as age and education.
- In the United States, elected officials depend heavily on financial support to get elected and to remain in office. Fund-raising is a major part of **politics**, and individuals and organizations that contribute heavily do so in hopes of influencing politicians. Special interests use **lobbyists** to exercise influence in U.S. politics. Politicians still depend on their constituents' votes to get elected, and so they must satisfy voters as well as special interests.
- We can examine war from various sociological perspectives. The functionalist perspective asks about the manifest (obvious) and latent (hidden) functions of war and conflict in society. The conflict perspective asks who benefits from war and conflict and who loses.
- The "global war on terror" was initiated in 2001 after the September 11 terrorist attacks on U.S. soil. The United States and its allies are fighting terror threats domestically, globally, and online.
- No single image of a terrorist is shared across communities and countries and cultures. Irishman Michael Collins is an example of someone regarded as a hero by some and a terrorist by others. **Terrorism** is a calculated use of violence to coerce or to inspire fear. It is also "theater"—intended to send a powerful message to a distinct or a global audience.

KEY TERMS

authoritarianism (p. 395)
charismatic authority (p. 394)
citizens (p. 388)
class dominance theory (p. 392)
coercion (p. 393)
dictatorship (p. 396)
direct democracy (p. 397)
interest groups (p. 391)
law (p. 388)
lobbyists (p. 403)

monarchy (p. 395)
nation-state (p. 387)
noncitizens (p. 388)
political action committees (PACs) (p. 403)
politics (p. 397)
power elite (p. 393)
rational-legal authority (p. 394)
representative democracy (p. 397)
terrorism (p. 410)
totalitarianism (p. 396)
traditional authority (p. 394)
war economy (p. 407)
welfare state (p. 388)

DISCUSSION QUESTIONS

1. In this chapter, you learned about theories of state power. Would you say that U.S. governance today is characterized more by pluralism or by the concentration of power in the hands of an elite? Cite evidence supporting your belief.
2. What is authoritarianism? What potential roles do modern technology and social media play in either supporting or challenging authoritarian governments around the world?
3. The chapter raised the issue of low voting rates for young people. Recall the reasons given in the chapter and then think about whether you can add others. Do most of the young people you know participate in elections? What kinds of factors might explain their participation or nonparticipation?
4. What are the manifest and latent functions and dysfunctions of war? Review the points made in the chapter. Can you add some of your own?
5. What is terrorism? How should this term be defined and by whom? When should domestic incidents of mass violence be labeled terrorism? Explain your reasoning.

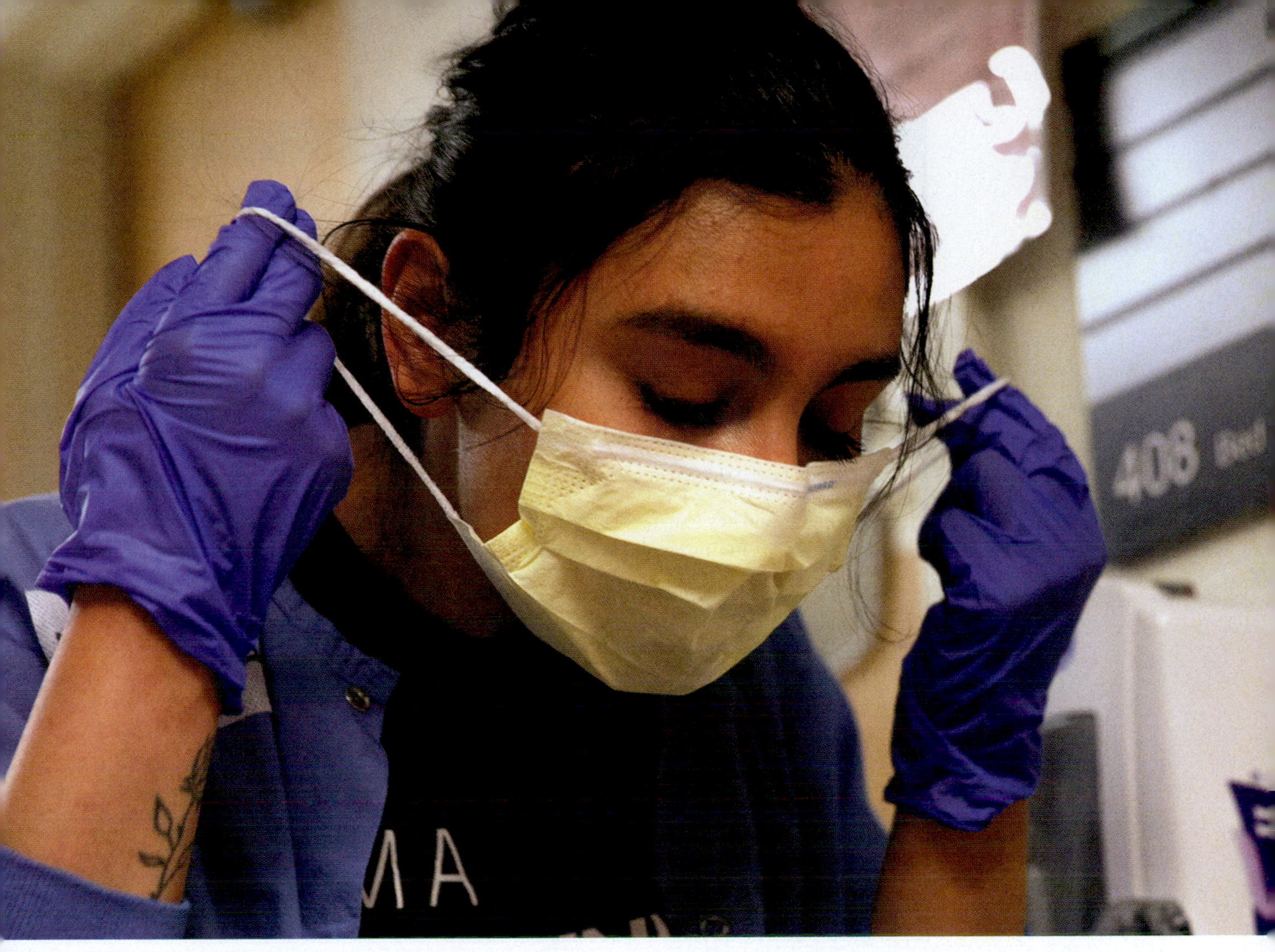

©Karen Ducey/Getty Images

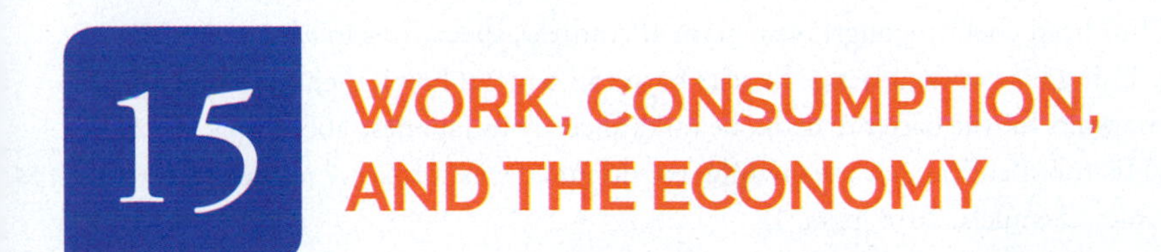

15 WORK, CONSUMPTION, AND THE ECONOMY

WHAT DO YOU THINK?

1. How did the global COVID-19 pandemic challenge and change the U.S. labor market? Were these challenges and changes experienced differently by men and women?
2. What will be the role of robots and artificial intelligence in the future economy? Will these technologies replace human workers? Or will they contribute to the creation of new jobs?
3. Why has average household debt grown in recent decades?

LEARNING OBJECTIVES

15.1 Describe the three major economic revolutions that have shaped the contemporary world.

15.2 Understand key social indictors used to describe the state of the economy and labor market, including employment and unemployment.

15.3 Discuss current and potential effects of automation and artificial intelligence on the labor market.

15.4 Compare the economic characteristics of capitalism and communism.

15.5 Distinguish between the formal and informal economy.

15.6 Describe historical and contemporary trends in U.S. consumption and consumer debt.

15.7 Discuss the relationship between globalization and the U.S. economy.

ROBOTS AND JOBS

An article on the automation of restaurant labor in Japan begins as follows:

> Visitors to Henn-na, a restaurant outside Naasaki, Japan, are greeted by a peculiar sight: their food being prepared by a row of humanoid robots that bear a passing resemblance to the Terminator. The "head chef," incongruously named Andrew, specializes in *okonomiyaki*, a Japanese pancake. Using his two long arms, he stirs batter in a metal bowl, then pours it onto a hot grill. While he waits for the batter to cook, he talks cheerily in Japanese about how much he enjoys his job. His robot colleagues, meanwhile, fry donuts, layer soft-serve ice cream into cones, and mix drinks. (Semuels, 2018, para. 1)

The author notes that in Japan, with its aging population and barely growing workforce, robots appear to fill a void in the labor market. But what about in the United States? The article points out that a broad spectrum of restaurants, including Wendy's, Panera, and McDonald's, are shifting to kiosk-based ordering. Some well-known chains such as Olive Garden and Red Robin are also bringing out tablet ordering that will reduce or eliminate the need for waitstaff and cashiers (Semuels, 2018). As in Japan, changes are also afoot in the kitchen: Economist Martin Ford (2015) quotes the co-founder of a robotics company that produced a machine that shapes burgers from ground meat and grills them precisely to order as saying, "Our device isn't meant to make employees more efficient . . . it's meant to completely obviate them" (p. 12). By one estimate, over half of the tasks workers do in U.S. hotels and restaurants could be automated using only the technologies that are already available (Semuels, 2018).

While some restaurant owners claim that their goal is to improve the customer experience by moving employees out from the kitchen and from behind cash registers (Semuels, 2018), living workers are a significant cost to employers: About 30% of business costs in the industry are comprised by labor (DePillis, 2015). While robots are also costly, the chief executive officer (CEO) of the company that owns Henn-na points out that "since you can work them 24 hours a day, and they don't need a vacation, eventually it's more cost-efficient to use the robot" (Semuels, 2018, para. 3).

The restaurant industry is, however, a vital part of the U.S. employment picture, and its significance has grown in the wake of manufacturing's decline. Indeed, by one estimate, food-service and accommodation jobs in the United States employ about 13.7 million workers (Semuels, 2018). These are jobs that have functioned for generations as stepping stones for young workers seeking a first job, as well as the primary occupations of breadwinners who have not completed high school or college.

While the shift to robotic workers has evolved slowly to this point, the effect of the global COVID-19 pandemic may accelerate the speed of changes. In an effort to protect workers and customers, companies around the world are looking to robots to take up tasks done by human workers, including cleaning, food preparation, and taking temperatures. For example, Walmart, the country's biggest private employer, is using robots to scrub its floors (Thomas, 2020). Will a return to normal life after the pandemic reverse these changes—or will they become permanent? The answer to those questions remains to be seen.

In this chapter, we discuss key issues in economic sociology and examine the implications of a new globalized economy—postindustrial, technologically sophisticated, and consumption oriented—for the world and for U.S. society in particular. We begin with a brief historical overview of the three great economic revolutions that have transformed human society. We then examine the characteristics and potential sociological implications of our evolving high-technology economy, focusing (as we did in the opening story) on the automation of jobs, robotics, artificial intelligence, and the future of work. We look at capitalism and communism, the two principal types of economic systems that dominated the 20th century and continue to influence the 21st century. Next, we examine work in the formal and informal economies. We also discuss social and economic issues of consumption, hyperconsumption, and debt. The chapter concludes with a consideration of the changes and challenges globalization has brought to our economic system and prospects.

The COVID-19 pandemic made many common occupations dangerous to workers, who risked exposure to the virus. The pandemic may have accelerated the shift to automation and robotic labor in many industries, including hospitality.

©piranka/Getty Images

THE ECONOMY IN HISTORICAL PERSPECTIVE

The **economy** is *the social institution that organizes the ways in which a society produces, distributes, and consumes goods and services*. By **goods**, we mean *objects that have an economic value to others, whether they are the basic necessities for survival* (a safe place to live, nutritious food to eat, weather-appropriate clothing) *or things that people simply want* (designer clothing, an iPhone, popcorn at the movies). **Services** are *economically productive activities that do not result directly in physical products; they can be relatively simple* (shining shoes, working a cash register, waiting tables at a restaurant) *or quite complex* (repairing an airplane engine or computer, conducting a medical procedure).

Economy: The social institution that organizes the ways in which a society produces, distributes, and consumes goods and services.

Goods: Objects that have an economic value to others, whether they are the basic necessities for survival or things that people simply want.

Services: Economically productive activities that do not result directly in physical products; they may be relatively simple or quite complex.

In human history, three technological revolutions have brought radically new forms of economic organization. The first led to the growth of agriculture several millennia ago, and the second led to modern industry some 250 years ago. We are now in the throes of the third revolution, which has carried us into a digital and postindustrial age.

The Agricultural Revolution and Agricultural Society

The agricultural revolution vastly increased human productivity over that of earlier hunting, gathering, and pastoral societies. This achievement was spurred by the development of innovations such as irrigation and crop rotation methods as well as by expanding knowledge about animal husbandry and the use of animals in agriculture. For example, the plow, which came into use about 5,000 years ago, had a transformational effect on agriculture when it was harnessed to a working animal. Greater productivity led to economic surplus. While the majority of people in agricultural societies still engaged in subsistence farming, an increasing number could produce surplus crops, which they could then barter or sell.

Eventually, specialized economic roles evolved. Some people were farmers; others were landowners who profited from farmers' labor. A number of families specialized in the making of handicrafts, working independently on items of their own design. This work gave rise to *cottage industries*—so called because the work was usually done at home.

The production of agricultural surpluses, as well as handicrafts, created an opportunity for yet another economic role to emerge—that of merchants, who specialized in trading surplus crops and crafted goods. Trading routes developed and permanent cities grew up along them, and the number and complexity of economic activities increased. By about the 15th century, early markets arose to serve as sites for the exchange of goods and services. Prices in markets were set (as they are in free markets today) at the point where *supply* (available goods and services) was balanced by *demand* (the degree to which those goods and services are wanted).

The Industrial Revolution and Industrial Society

The industrial revolution, which began in England with the harnessing of water and steam power to run machines such as looms, increased productivity still further. Cottage industries were replaced by factories—the hallmark of industrial society—and urban areas became centers of economic activity, attracting rural laborers seeking work and creating growing momentum for urbanization. Industrialization spread through Europe and the United States and then to the rest of the world. The change was massive. In 1810, about 84% of the U.S. workforce worked in agriculture and only 3% in manufacturing; by 1960, only 8% of all U.S. workers labored in agriculture and fully a quarter of the total workforce was engaged in manufacturing (Blinder, 2006).

Karl Marx saw industrial workers as instruments of labor tethered to an exploitive system. One 19th-century British mother described her 7-year-old son to a government commission: "He used to work 16 hours a day. . . . I have often knelt down to feed him, as he stood by the machine, for he could not leave it or stop" (quoted in Hochschild, 2003, p. 3).

Industrial society is characterized by the increased use of machinery and mass production, the centrality of the modern industrial laborer, and the development of a class society rooted in the modern division of labor.

Increased Use of Machinery and Mass Production

Machines increase the productive capacity of individual laborers by enabling them to efficiently produce more goods at lower cost. New machines have historically required new sources of energy as well: Waterwheels gave way to steam engines, then the internal combustion engine, and eventually, electricity and other modern forms of power.

Mass production: The large-scale, highly standardized manufacturing of identical commodities on a mechanical assembly line.

In 1913, automobile mogul Henry Ford introduced a new system of manufacturing in his factories. **Mass production** is *the large-scale, highly standardized manufacturing of identical commodities on a mechanical assembly line.* Under Ford's new system, a continuous conveyor belt moved unfinished automobiles past individual workers, each of whom performed a specific operation on each automobile: One worker would attach the door, another the windshield, another the wheels. (The term *Fordism* is sometimes used to describe this system.) Mass production resulted in the development of large numbers of identical components and products that could be produced efficiently at lower cost. This linked system of production became a foundation for the evolution and expansion of productive industries that went far beyond auto manufacturing.

The Birth of the Industrial Laborer

Reserve army of labor: A pool of job seekers whose numbers outpace the available positions and thus contribute to keeping wages low and conditions of work tenuous.

Scientific management: A practice that sought to use principles of engineering to reduce the physical movements of workers.

With the birth of industry came the rise of the industrial labor force, comprising mostly migrants from poorer rural areas or abroad seeking their fortunes in growing cities. Often the number of would-be workers competing for available jobs created a surplus of labor. Karl Marx described this as a **reserve army of labor**, *a pool of job seekers whose numbers outpace the available positions and thus contribute to keeping wages low and conditions of work tenuous* (those who do not like the conditions of work are easy to replace with those seeking work).

If it is possible to create an assembly line on which each worker performs a single, repetitive task, why not design those tasks to be as efficient as possible? This was the goal of **scientific management**, *a practice that sought to use principles of engineering to reduce the physical movements of workers.* Frederick Winslow Taylor's *Principles of Scientific Management,* published in 1911, gave factory managers the information they needed to greatly increase their control over the labor process by giving explicit instructions to workers regarding how they would perform their well-defined tasks. While Taylor was focused on the goal of efficiency, "Taylorism" also had the consequence of further deskilling work. Deskilling rendered workers more vulnerable to layoffs, since they—like the components they were making—were standardized and therefore easily replaced (Braverman, 1974/1988).

Classes in Industrial Capitalism

New economic classes developed along with the rise of industrial capitalist society. One important new class was composed of industrialists who owned what Marx called the *means of production*—for example, factories. Another was made up of wage laborers—workers who did not own land, property, or tools. They had only their labor power to sell at the factory gate. Work in early industrial capitalism was demanding, highly regimented, and even hazardous. Workers labored at tedious tasks for 14 to 16 hours a day, 6 or 7 days a week, and were at risk of losing their jobs if economic conditions turned unfavorable or if they raised too many objections (recall the concept of the *reserve army of labor*). The pool of exploitable labor was expanded by migrant workers from rural areas and abroad, and even children of poor families were sometimes forced to labor for wages.

Influenced by the poor conditions they saw in 19th-century English factories, Karl Marx and Friedrich Engels posited that these two classes—the *bourgeoisie* (capitalists) and the *proletariat* (working class)—would come into conflict. They argued in the *Manifesto of the Communist Party* (1848) that the bourgeoisie exploited the proletariat by appropriating the surplus value of their labor. That is, capitalists paid workers the minimum they could get away with and kept the remainder of the value generated by the finished products for themselves as profit or as a means to gather more productive capital in their own hands. The exploitation of wage labor by capitalists would, they believed, end in revolution and the end of private ownership of the means of production.

While some observers of early capitalism, including Marx and Engels, offered scathing critiques of the social and economic conditions of factory laborers, the early and middle decades of the 20th century (with the exception of the period of the Great Depression) witnessed improved conditions and opportunities for the blue-collar workforce in the United States. In the early 20th century, Henry Ford, the patriarch of Fordist production, took the audacious step of paying workers on his Model T assembly line in Michigan fully $5 for an 8-hour day, nearly three times the wage of a factory employee in 1914. Ford reasoned that workers who earned a solid wage would become consumers of products such as his Model T. Indeed, his workers bought, his profits grew, and industrial laborers (and, eventually, the workers of the unionized U.S. car industry) set off on a slow but steady path to the middle class (Reich, 2010).

The class structure that emerged from advanced industrial capitalism in the United States, Europe, Japan, Canada, and other modern states boasted substantial middle classes composed of workers who ranged from well-educated teachers and managers to industrial workers and mechanics with a high school or technical education. The fortunes of blue-collar and semiprofessional workers were boosted by a number of factors. Among these were extended periods of low unemployment in which workers had greater leverage in negotiating job conditions (Uchitelle, 2007). Unions supported auto workers, railroad workers, and workers in many other industries in the negotiation of contracts that ensured living wages, job security, and benefits. Unionization surged following the Great Depression and the 1935 passage of the Wagner Act, which "guaranteed the rights of workers to join unions and bargain collectively" (VanGiezen & Schwenk, 2001), growing to more than 27% of the labor force by 1940. At their peak in 1979, U.S. unions claimed 21 million members (Mayer, 2004).

Changes in the U.S. economy have shaken the relatively stable middle class that emerged around the middle of the 20th century and was, at least at its origin, largely rooted in workers without a college education but with access to jobs that provided security, benefits, and a living wage. Since the 1970s, mass layoffs have grown across industries, though manufacturing has been hit hardest (Uchitelle, 2007). As a result, today, the industrial laborer is less likely to belong to a union, less likely to have appreciable job security, more likely to have experienced a decline in wages and benefits—and less likely to be a member of the middle class (Table 15.1).

TABLE 15.1 ■ Selected Characteristics of Industrial and Postindustrial Societies

CHARACTERISTIC	INDUSTRIAL SOCIETY	POSTINDUSTRIAL SOCIETY
Principal technology	Industrial machinery	Advanced technologies including computers, automation of tasks
Key types of labor categories	Industrial workers and professionals	"Knowledge workers" and service workers
Type of production	Mass production	Flexible production
Labor control	"Scientific management"	Outsourcing (threatened and real), technological control of work environment and worker activities
Selected social stratification characteristics	Development of a modern class society with a dominant economic class, an expanding middle class that may integrate workers, and an economic underclass	Segmentation of the middle class by educational attainment, concentration of wealth and income at the top, and an expanding stratum of working poor

The Information Revolution and Postindustrial Society

During the past quarter century, the "information revolution," which began with Intel's invention of the microchip in 1971, has altered economic life, accelerating changes in the organization of work that were already under way. Pressured by global competition that intensified by the end of the 1970s, U.S. firms began to move away from the inflexible Fordist system of mass production, seeking ways to accommodate rapid changes in products and production processes and to reduce high labor costs that were making U.S. products less competitive. Postindustrial economic organization is complex, so the following sections focus on only some of the key aspects, including the growth of automation and flexible production, reliance on outsourcing and offshoring, and the growth of the service economy.

Automation and Flexible Production

Automation: The replacement of human labor by machines in the production process.

Postindustrial production relies on ever-expanding **automation**, *the replacement of human labor by machines in the production process.* Today, robots can perform tedious and dangerous work that once required the labor of hundreds of workers. While automation increases efficiency, it has also eliminated jobs.

Computer-driven assembly lines can be quickly reprogrammed, allowing manufacturers to shift to new products and designs rapidly and to shorten the time from factory to buyer. "Just-in-time" delivery systems also minimize the need for businesses to maintain warehouses full of parts and supplies; instead, parts suppliers ship components to factories on an as-needed basis so they move right to the production floor and into the products. Such reliance on more flexible, less standardized forms of production is sometimes termed *post-Fordism.*

Notably, while to this point, automation has had its most visible impact on manufacturing jobs, it is becoming significant (as we saw in the opening story) in the large U.S. service industry as well. Consider, for instance, the mass expansion of self-checkout lines at supermarkets, drug stores, and home improvement stores, among many others. In many cities, one can easily go into a retail store, find the items one needs, and check out—without ever speaking to another human being. While offering lower labor costs to employers and some convenience to consumers, self-ordering and self-checkout technologies are reducing the numbers of jobs available at dining and retail establishments. Further along in the chapter, we look at the potentially dramatic shifts in the future labor market as automation and artificial intelligence reach into other sectors of the labor market.

Reliance on Outsourcing and Offshoring

Businesses can perform activities associated with producing and marketing a product in-house or they can contract some of the work to outside firms, which in turn can do their own subcontracting to other firms. The term *outsourcing* often describes the use of low-cost foreign labor, but it can also mean contracting U.S. workers to do a job, typically for less pay than a regular company employee would earn.

In the United States, for instance, more foreign-owned car manufacturers are based in the South rather than in the industrial Midwest, which is the historical home of unionized auto manufacturing. By contrast to U.S.-owned companies, these companies employ a significant proportion of auto workers who are temporary rather than permanent workers—about 25%—which keeps their cost of employment lower (Burden, 2019).

Outsourcing has affected a wide spectrum of industries. For example, United Airlines used to rely on its own mechanics to service the company's planes. The mechanics were well paid and enjoyed benefits negotiated by their union. By the late 1990s, however, United increasingly turned to nonunionized mechanics operating from lower-cost, lower-wage hangars in the South. The terrorist attacks of September 11, 2001, which temporarily halted air travel, exacerbated the financial difficulties of airlines. In spite of billions in government aid and loans, airlines have continued to struggle. Major carriers have become even more reliant on outsourcing to cut costs (Uchitelle, 2007). The economic crisis precipitated by COVID-19, which significantly reduced demand for air travel, is likely to continue to whittle down the permanent workforces of air carriers across the globe.

The phenomenon of contracted work—that is, temporary work that minimizes the commitment of employer and employee to a long-term economic relationship and, in many cases, relieves the employer of the burden of providing benefits—is not limited to the blue-collar workforce. Computer giant Microsoft's use of "permatemps," initiated in the 1990s, is a striking example. During this time, 1,500 permatemps worked with the 17,000 regular domestic employees of the company. While they performed comparable tasks, the permatemps (some of whom had been in their jobs for 5 years or more) not only were denied the same vacation, health, and retirement benefits as other workers but also were denied discounts at the Microsoft store, opportunities for further job training, and even use of the company basketball court. A class-action suit was filed against Microsoft, and the company agreed to an out-of-court settlement of $97 million (FACE Intel, 2000). Consider as well the significant number of part-time and temporary instructors who fill the teaching ranks at many colleges and universities in the United States. Unlike their full-time peers in the academy, these part-time, temporary instructors, often referred to as *adjuncts*, rarely enjoy the benefits of having their own office, getting employer-provided health and retirement benefits, earning a secure and reliable salary, or even knowing if they will be employed in the coming semester. Today, just under half of all instructors in higher education are employed part-time (Lederman, 2019). This "adjunctification" of higher education has created a crisis of employment in many fields where students earning doctoral degrees have significant challenges finding full-time academic jobs in their field of study.

Offshoring refers more specifically to the practice among U.S. companies of contracting with businesses outside the country to perform services that would otherwise be done by U.S. workers. The movement of manufacturing jobs overseas to lower-wage countries, as noted earlier, has been taking place since the 1970s and 1980s. More recently, however, workers and policy makers have expressed concern about the offshoring of professional jobs, such as those in information technology. According to a Congressional Research Service paper on the topic, this trend has been fostered by the widespread adoption of technologies enabling rapid transmission of voice and data across the globe, economic crises in the United States that have created greater pressure to achieve economic efficiencies (such as lower labor costs), and the availability of a growing pool of well-educated and often English-speaking labor abroad (Levine, 2012).

Transformation of the Occupational and Class Structure

Among the most highly compensated workers in the modern economy are those who invent or design new products, engineer new technologies, and solve problems. They are creative people who "make things happen," organizers who bring people together, administrators who make firms run efficiently, legal and financial experts who help firms to be profitable, and computer scientists who are driving digital networking innovations (Bell, 1973; Reich, 2010). Workers in this category are sometimes called *symbolic analysts* (Reich, 1991) or *knowledge workers*. Most symbolic analysts are highly educated professionals who engage in mental labor and, in some way, the manipulation of symbols (numbers, computer codes, words). They include engineers, university professors, physicians, scientists, lawyers, financiers, and bankers, among others.

While the ranks of symbolic analysts have grown overall in recent decades and the ranks of routine production workers in manufacturing have been declining, most job growth over this period has been concentrated in the service sector. Services constitute a diverse sector of the labor market. As of 2019, the service sector employed more than 130 million U.S. workers and accounted for almost 80% of the U.S. workforce (U.S. Bureau of Labor Statistics, 2020a). Service occupations include a number of jobs that require higher education, including financial and private educational services, but also include retail sales, home health and nurses' aides, food service, and security services.

Many service positions do not require extensive education or training, and a growing fraction are part-time rather than full-time, are nonunionized, and have few or no benefits. Quite a few of these jobs require people skills that are stereotypically associated with females and are often viewed as "women's jobs" (but by no means invariably, since private security guards, a growing occupation, tend to be men). By contrast, many routine production jobs in the past were manufacturing jobs that commonly employed men. The decline in manufacturing employment opportunities, along with declines in educational attainment among men, has made unemployment and underemployment particularly acute for some demographic groups, including minority males (Autor, 2010).

Some manufacturing jobs have returned to the United States after the decline of this job sector beginning in the late 1970s and early 1980s. Many of today's industrial jobs, however, pay less and are temporary contract positions rather than permanent jobs.

©David Butow/Corbis via Getty Images

Together, these diverse labor market trends point to significant shifts in the U.S. class structure. Economist David Autor (2010) argues that a polarization of job opportunities has taken place, particularly in the past two decades. Autor sees a modern economy characterized by "expanding opportunities in both high-skill, high-wage occupations and low-skill, low-wage occupations, coupled with contracting opportunities in middle-wage, middle-skill, white-collar and blue-collar jobs." He views this as the basis of a split in the middle class: Specifically, the group that used to rely on manufacturing jobs to maintain their membership in the middle class is losing ground, while those who occupy the upper, professional rungs of the middle class have been able to maintain their status and access new opportunities.

However, not all economists agree with this assessment. Economist Alan Blinder (2006) argues that

> many people blithely assume that the critical labor-market distinction is, and will remain, between highly educated (or highly skilled) people and less-educated (or less-skilled) people—doctors versus call-center operators, for example. The supposed remedy for the rich countries, accordingly, is more education and a general "upskilling" of the work force. But this view may be mistaken. (p. 118)

Blinder suggests that the more critical social division in the future may not be between jobs that require high levels of education and those that do not but rather between work that can be wirelessly outsourced and work that cannot. Consider the growth of online university education. Whereas a college professor may be able to accommodate 100 or even 500 students in a massive lecture hall, an online instructor can have thousands of students and teach them at a considerable cost savings to the institution—and, in some instances, to the students. Some universities, such as the Massachusetts Institute of Technology (MIT), are offering free college course lectures online (though these are not normally available for credit). While this is not outsourcing as we typically define it, trends suggest that even many highly educated workers will be vulnerable to technological changes in the decades ahead.

The Service Economy and Emotional Labor

As discussed, recent decades have seen the expansion of the service sector of the U.S. economy. Many service jobs today require a substantial amount of emotional labor. According to sociologist Arlie Russell Hochschild (2003), **emotional labor** is *the commodification of emotions, including "the management of feeling to create a publicly observable facial and bodily display"* (p. 7). Like physical labor, the symbol of the industrial economy, emotional labor is also "sold for a wage and . . . has exchange value" (p. 7).

Emotional labor: "The commodification of emotions, including the management of feeling to create a publicly observable facial and bodily display" (Hochschild, 2003).

Hochschild (2003) uses the example of flight attendants, who do emotional labor in the management of airline passengers' comfort, good feelings, and sense of safety, but we could also use as examples customer service workers, retail sales associates, and restaurant servers. While these workers may enjoy their jobs, they are also forced to feign positive feelings even when such feelings are absent and to work relentlessly to evoke positive feelings in their customers. In addition to producing emotions, workers in these jobs may also be called upon to suppress feelings, particularly when poor behavior on the part of a customer or client evokes frustration or anger. The emotional laborer is, in a sense, compelled to "sell" his or her smile in exchange for a wage, just as the industrial laborer sells his or her physical labor. The emotional laborer's actions are programmed for profit and efficiency, as he or she is asked to perform emotions that maximize both.

Consider, for instance, the common use of scripts for employees in sales jobs or restaurants that closely instruct workers on how to maintain a message and close a sale, even if it causes the employee (or the consumer) distress or discomfort. The strain between real and performed feelings, notes Hochschild, leads to an emotive dissonance—a disconnect—between what the worker really feels and the emotions to be shown or suppressed. Hochschild posits that just as Marx's proletarian laboring in a mill was alienated from the work and from himself or herself, so too is the emotional laborer alienated from work and his or her emotional life.

In the next section of the chapter, we pause to consider the numbers that enable us to know what is happening in the economy—that is, to follow the good or ill fortunes of workers in the labor market. Specifically, we consider the figures that economists—and the rest of us—use to talk about employment, unemployment, and the state of the labor market.

UNDERSTANDING THE DATA: UNEMPLOYMENT, EMPLOYMENT, AND UNDEREMPLOYMENT IN THE UNITED STATES

Researchers, economists, and politicians—along with many U.S. workers—pay close attention to figures that tell us about the situation in the labor market. According to the U.S. Bureau of Labor Statistics (BLS), in October 2020, the labor force participation rate was 61.7% and over 149,800,000 U.S. residents were employed. At the same time, about 6.9% of U.S. workers were counted by the BLS as unemployed (U.S. Bureau of Labor Statistics, 2020). What do these numbers tell us? What do they illuminate, and what do they obscure?

Consider some of the most frequently cited BLS figures—unemployment in the United States. According to the BLS, the **unemployed** are *people who are jobless, have actively looked for work in the prior 4 weeks, and are available for work.* The BLS figures are based on the monthly Current Population Survey, which uses a representative sample of 60,000 households and has been conducted every month since 1940. While the BLS cannot count every U.S. household, the size of the sample and its configuration are believed to ensure a statistically accurate representation of the U.S. labor force.

Unemployed: People who are jobless, have actively looked for work in the prior 4 weeks, and are available for work.

Official unemployment figures (Figure 15.1), importantly, do not include those who, after a brief or extended period of joblessness, have given up looking for work or whose job seeking is *passive*—for instance, limited to scanning newspaper or online classified ads. That is, the unemployment rate does not tell us about everyone who is jobless but would like to work—rather, it tells us who is jobless and actively looking.

Those who are not employed and are not looking for work are categorized by the BLS as **not in the labor force**, because they are *neither officially employed nor officially unemployed. Persons who would like to work and have searched actively for a job in the past 12 months (but not in the prior 4 weeks)* are

Not in the labor force: Persons who are neither officially employed nor officially unemployed.

FIGURE 15.1 ■ U.S. Unemployment Rate, 2010–2020

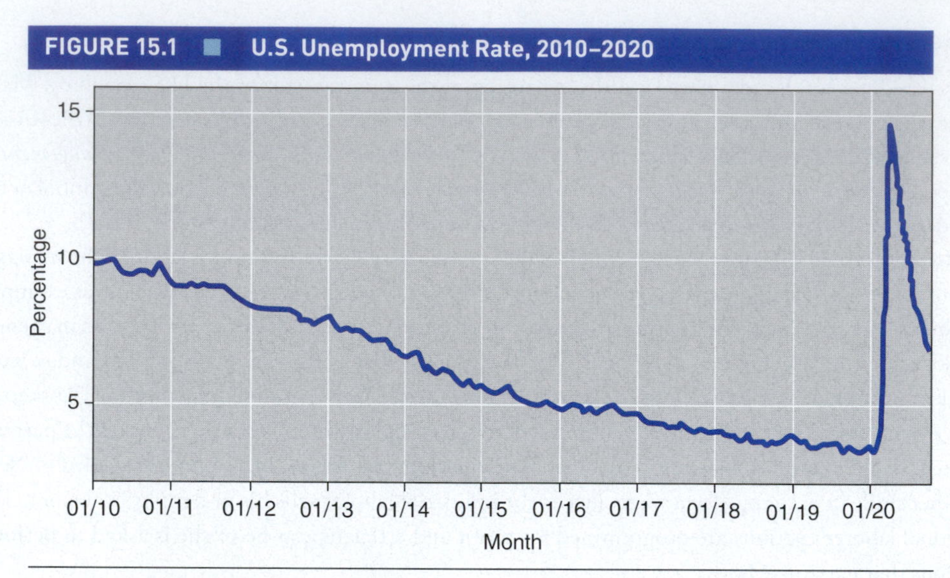

Source: U.S. Bureau of Labor Statistics. (2020, November 15). *Unemployment Rate.* Retrieved from https://data.bls.gov/timeseries/LNS14000000

Marginally attached to the labor force: Persons who would like to work and have searched actively for a job in the past 12 months (but not in the prior 4 weeks).

Discouraged workers: Those who would like to work but have given up searching, believing that no jobs are available for them.

Employed: Employed persons are those who are 16 years of age or older in the civilian, noninstitutional population who did any paid work—even as little as 1 hour—in the reference week or worked in their own businesses or farms.

Labor force participation rate: This figure represents the total of those who are counted by the Bureau of Labor Statistics as employed and those who are counted as unemployed, as a proportion of the civilian, noninstitutionalized population 16 years of age and over.

categorized as **marginally attached to the labor force**: In October 2020, there were about 1.9 million of such potential workers. The BLS also identifies a category it calls **discouraged workers**, which includes *those who would like to work but have given up searching, believing that no jobs are available for them*: Fall 2020 data put this figure at about 588,000 (U.S. Bureau of Labor Statistics, 2020). Widely cited official unemployment statistics omit these categories and may thus underestimate the numbers of those who need and want to work.

Who, then, according to BLS calculations, is **employed**? In BLS statistics, *employed persons are those who are 16 years of age or older in the civilian, noninstitutional population (that is, not in the military or in mental or penal institutions) who did any paid work—even as little as 1 hour—in the reference week or worked in their own businesses or farms.* This figure is a useful overview of economic activity and is valuable for examining labor market conditions over time, but it fails to capture the problem of underemployment.

Also significant is the labor force participation rate. As noted, as of October 2020, the **labor force participation rate** was just over 61%. This rate shows us the number of officially employed *and* unemployed persons as a percentage of the population 16 years of age or older (excepting those who are on active duty in the U.S. military or who are institutionalized, including being in prison; U.S. Bureau of Labor Statistics, 2020). As such, it gives an indication of the percentage of people who work or actively want to work. Notably, the labor force participation rate has been falling for several years, particularly among men. While some of the decline may be attributable to residents who have left the labor force due to discouragement, some of the decline in labor force participation is also due to the growing number of Baby Boomers aging into retirement, as well as the increase in students going to college after high school and, in some cases at least, not working during their studies.

Underemployment manifests in two key ways. First, the underemployed include workers forced to work part-time when they would like to work full-time. According to the BLS, in October 2020, there were about 6.6 million involuntary part-time workers in the United States; these are workers who are in part-time positions because their hours have been reduced or they are unable to find a full-time job (U.S. Bureau of Labor Statistics, 2020). Reduced hours in the 2020 data are in part an effect of the COVID-19 pandemic, which saw considerable reductions in demand in sectors like hospitality (including restaurants, shops, and hotels). Even those workers who were able to retain their jobs may have experienced reductions in their scheduled hours. Second, workers in jobs that are below their skill or credential level can also be considered underemployed. According to a recent report by the Federal Reserve Bank of New York, in September 2020, fully 43% of new college graduates and 33% of all college graduates were underemployed. These data gauge the percentage of workers who have a college degree but are working in a job that does not require a college degree (2020).

Unemployment and employment figures are important measures that help to track trends over time in the labor market and enable comparisons over time and space, and between different demographic groups. At the same time, while they illuminate some aspects of the complex labor market, they obscure others. Being critical consumers of labor market data entails understanding what these figures mean.

THE TECHNOLOGICAL REVOLUTION AND THE FUTURE OF WORK

In the previous section, we discussed key characteristics of the postindustrial economy of recent decades and today. In this section, we consider possible paths of economic development in the years ahead, focusing in particular on ways in which expanded automation and the rise of artificial intelligence may reshape the future labor market in the United States and the world.

Technological change has historically wrought both prosperity and pain in the economy and labor market. For example, while innovations such as the mechanization of agriculture in the 19th century dramatically improved productivity and enabled some landowners and farmers to grow a surplus that could be sold at a profit, it also created significant disruptions for agricultural laborers. At the same time, new technological innovations in production created a mass of new jobs in urban industry, spurring rural to urban migration and a new economic order heavily rooted in manufacturing. Put simply, in the past, technological innovations and accompanying economic shifts have put workers out of jobs, but in the words of economist Martin Ford (2015), "It never became systematic or permanent. New jobs were created and dispossessed workers found new opportunities" (p. x).

Emotional labor entails creating or suppressing feelings (whether positive or negative) in return for a wage. Have you ever worked as an emotional laborer? In what ways does Hochschild's concept capture or fail to capture your experience?

Today, we are experiencing a new era of change and, potentially, disruption as technology is changing how work and the workplace are structured. What are evolving trends in the economy and labor market? Next, we discuss two key points.

Big Names, Few Workers: Digital Networking Companies in the Contemporary Economy

The most visible and well-known companies in the United States today do not employ large numbers of workers. Consider an observation made by computer scientist Jaron Lanier (2013):

> At the height of its power, the photography company Kodak employed more than 140,000 people and was worth $28 billion. They even invented the first digital camera. But today Kodak is bankrupt, and the new face of digital photography has become Instagram. When Instagram was sold to Facebook for a billion dollars in 2012, it employed only thirteen people. (p. 2)

Lanier's example points to a significant contemporary trend: We are seeing the dramatic rise of companies whose fortunes are tied to digital networking—Instagram, Facebook, WhatsApp, and Snapchat are only a few. Companies engaged in digital networking are among the most financially valuable firms operating in the global economy. At the same time, the number of jobs they directly contribute to the labor market is relatively low. For example, in 2015, Snapchat had a valuation of $55 billion; at that time, it had 330 employees. When Facebook recently purchased WhatsApp at a cost of $22 billion, the company had a total of 55 employees. Figure 15.2 offers a visual representation of the highest valuation per employee. That is, it helps us see that some companies have both high valuations and few employees: Snapchat's valuation per employee is $48 million. As one article noted, "Ultimately, software has proven to be one of the most headcount-efficient businesses in the world" (Chen, 2015, para. 6).

FIGURE 15.2 ■ Visualization of Value of Tech Companies per Employee

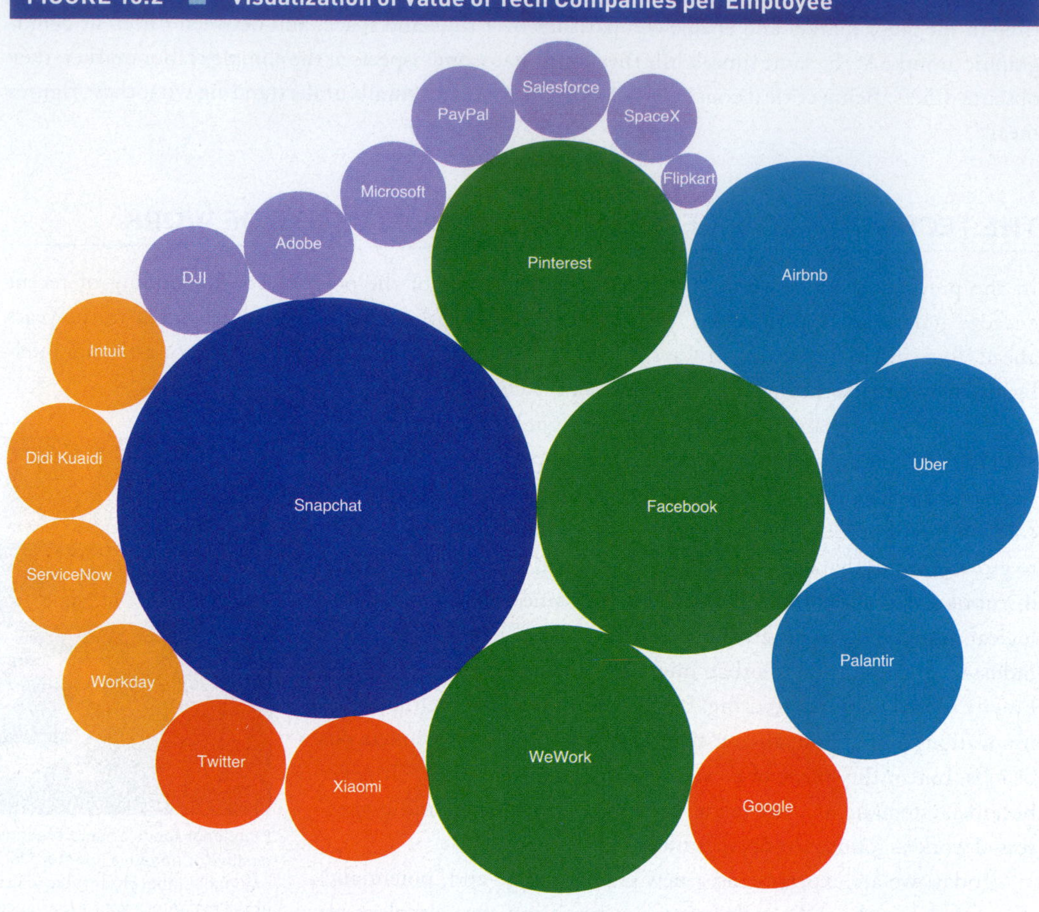

Source: Data from Chen, L. (2015, August 11). The Most Valuable Employees: Snapchat Doubles Facebook. *Forbes.*

While headcount efficiency is a boon for companies and shareholders, it is less beneficial for job seekers. In the heyday of U.S. manufacturing, prominent and technologically modern companies were providers of significant numbers of jobs: In 1955, for instance, General Motors had a workforce of over 576,000; U.S. Steel employed about 268,000 workers; and Chrysler had over 167,000 employees ("America's 5 Biggest Employers," 2010). The digitally networked economy is producing new, well-paying, and interesting jobs, but will there be enough of them to employ the workforce of the future? This is a question that remains to be fully answered.

Rise of the Robots?

Economist Martin Ford (2015) writes that we are in a moment when the role of technology in the workplace is changing. Ford notes that

> [this] shift will ultimately challenge one of our basic assumptions about technology: that machines are tools that increase the productivity of workers. Instead machines themselves are turning into workers, and the line between the capability of labor and capital is blurring as never before. (p. xii)

Ford (2015) suggests that the change is underpinned by "the relentless acceleration of computer technology" (p. xii). When Ford writes that "machines themselves are turning into workers," he is recognizing the myriad ways in which technology has the potential to disrupt the contemporary labor market—and, by extension, society.

For example, consider the seemingly imminent introduction of self-driving cars on our roads. An article in *The Washington Post* looks at one company's efforts to bring this goal to fruition in the city of Pittsburgh:

> Uber is about to build a vast, high-tech playground in one of this city's poorest areas.
>
> The ride-hailing giant wants a protected place to test driverless Ubers, part of its effort to replace costly human drivers.
>
> So on the site of an abandoned steel mill south of the Hot Metal Bridge, the company will carve out a 20-plus square block, Pac-Man-like maze lined with trapezoidal obstacles. It is the same place where thousands of workers once streamed in to take punishing jobs at the beehive-shaped ovens for baking coal and the furnaces that were fueled by it. (Laris, 2016, p. 1)

As self-driving cars become part of our world, "costly human drivers" may indeed be replaced. A key question, then, is what will those human drivers do? According to the U.S. Bureau of Labor Statistics (2020b), in 2019, the country had about 2 million heavy and tractor trailer truck drivers and about 1.5 million delivery drivers. These are only two of the many occupations that are based on driving, including taxi and Uber drivers.

Significantly, the advent of artificial intelligence is also changing the labor market for those with advanced education. Consider the case of journalists. The effect of the Internet on journalism has been widely discussed. On the one hand, as more people get their news for free on the Internet, paid subscriptions to newspapers and magazines have fallen and the availability of jobs for reporters has contracted. On the other hand, the Internet has seen the birth of new online news sites that have opened opportunities for writers. The effects of artificial intelligence on journalism, however, are still nascent and may be even more disruptive. New technology is permitting media outlets to produce narrative without employing reporters. According to Ford (2015),

Self-driving trucks are coming to a highway near you. What will be the benefits of self-driving trucks to society and the economy? What might be the costs?

©AP Photo/Tony Avelar

> Narrative Science's technology is used by top media outlets . . . to produce automated articles in a variety of areas, including sports, business, and politics. The company's software generates a news story every thirty seconds, and many of these are published on widely known websites that prefer not to acknowledge the use of their services. (pp. 84–85)

In 2015, the Associated Press, a major producer of news content, revealed that it has used a fully automated program for the production of some content since 2014: The "robot reporter" writes about 1,000 articles a month (Gleyo, 2015). When asked by a reporter to predict how many news articles would be written algorithmically within 15 years, the co-founder of Narrative Science estimated the proportion to be over 90% (Ford, 2015). In 2018, Bloomberg News, which is heavily oriented toward business and financial news, published thousands of stories written by artificial intelligence (Martin, 2019).

Significantly, some observers of artificial intelligence trends suggest that while human input is still needed in decision making, robots are *better* at many tasks than humans. An article on robot surgeons noted that "in experiments on pigs, surgical stitches made by autonomous robots were as good or better than stitches made by skilled surgeons" (Seaman, 2016). The author of the study is quoted as saying, "No matter how steady a surgeon's hands are, there is always some tremor." Hence, "Using autonomous robots in some of the 44.5 million soft tissue surgeries in the United States each year might reduce human errors and improve efficiency, surgical time and access to quality surgeons." Human medical providers will remain a central part of patient care and care decisions for the foreseeable future, but medicine is only one of many fields in which change is coming.

As artificial intelligence becomes capable of taking on both routine tasks (as it has been doing already) and more complex analytical and technological tasks (as it is beginning to do), it is worth asking how decision makers and society will respond. Will advances in artificial intelligence create new jobs and new job sectors that enable society and the economy to flourish? Or will computer intelligence push "costly" human workers out of jobs permanently? What do you think?

Tomorrow's Jobs?

When you graduate from college, you may go on to get further education or you may venture into the labor market to look for work. What job will you be doing? The answer to that question depends on a variety of factors: Of course, it will depend on your skills and interests, which will guide your search. It will also depend on the general health of the labor market, both nationally and in your local area. As we noted earlier in our chapter on gender, sexuality, and society (Chapter 10), the job one gets is, in many respects, an outcome of labor supply factors (what a prospective employee wants and brings to a job) and labor demand factors (what a prospective employer seeks).

There is, however, another significant variable to consider: Some jobs that currently exist may evolve out of existence, while other jobs—which we have perhaps yet to imagine—will come into being. Consider some of the jobs that are done by thousands of workers today that did not exist (at least in their present form) just a decade ago:

- Uber, Lyft, and other on-demand car service drivers: While drivers do jobs that are similar to those done for decades by taxi drivers, services such as Uber do not employ drivers and the company does not have a responsibility to provide benefits or job security. Rather, drivers use their personal vehicles and rely on the platform to link them up with customers, for which they pay a fee to the company. In 2015, Uber doubled the number of drivers worldwide who are registered with the company (Moore, 2017; World Economic Forum, 2016).
- Driverless car engineers: This even-newer job category, which demands advanced technical knowledge, could possibly render the previous one obsolete in the future (World Economic Forum, 2016).
- YouTube content creators: Early bloggers were around a decade ago, but today, vloggers (who post content on sites such as YouTube) can make considerable sums of money if they find sufficient followers. Related to this job is the podcast producer: Today, podcasts draw wide audiences who enjoy podcasts that appeal to mass audiences as well as more specific demographics, such as turkey hunters, banjo players, and comic book collectors.
- Drone operators: While still in its infancy, this job could take off as large retailers such as Amazon gain permission in local markets to do drone deliveries of goods. Across the globe, the number of unmanned aerial vehicles has risen dramatically. Piloting those drones in the interest of commerce and security, among others, presents new opportunities (World Economic Forum, 2017).
- Social media manager: This job is among the older jobs on our list. Social media managers work for the companies that dominate the social media market, such as Instagram, Twitter, and Facebook, but they are also increasingly in demand at smaller firms that seek to shape their online image and to reach out to customers using these platforms.
- Social media influencer: According to an article in *Adweek*, "a new report from #Hashoff, an influencer discovery and content marketing platform, found that 28 percent of creators

they surveyed consider being an influencer their main job. Six months ago, just 12 percent of respondents said it was their full-time job" (Main, 2017). Social media influencers use platforms such as Instagram to tag brands with the goal of enticing their followers to buy the fashion or other item they are showing. In return, they are paid by the companies. While in the past, this role was fulfilled by traditional advertising agencies and marketing campaigns, today, social media influencers are a key vehicle for disseminating marketing messages to potential consumers.

As we have discussed in this chapter, technological changes will likely bring dramatic shifts to the U.S. and global labor markets. Some jobs will disappear. Other new ones will be born. What characteristics do the new jobs we have described share? Can you think of other jobs we could add to this list? Does the list provide any clues about other future jobs and job sectors and who may occupy those—and who may be left behind?

TYPES OF ECONOMIC SYSTEMS

Two principal types of economic systems dominated the 20th century—and both continue to exist in the 21st century: capitalism and socialism. Industrialization occurred in capitalist economic systems in North America, Western Europe, Japan, and South Korea, among others, and under socialism in the Soviet Union, Eastern Europe, China, Vietnam, and Cuba. After 1989, the collapse of socialism in Eastern Europe and the (former) Soviet Union fostered the expansion of capitalist market systems. Orthodox socialism appears to be in decline elsewhere as well, notably in China, which remains politically tied to the Communist Party but is increasingly entrepreneurial and capitalist in its economic practices. Even Cuba, which remains steadfast in its socialist rhetoric, recently introduced some capitalist-style reforms.

Even though capitalism and socialism as ideologies share a common commitment to economic growth and increased living standards, they differ profoundly in their ideas about how the economy should be organized to achieve these goals and the degree to which the fruits of economic labor should be shared by everyone in a community. The following descriptions are of ideal-typical (that is, model) capitalist and socialist systems. Real economies often include some elements of both.

Capitalism

Capitalism is *an economic order characterized by the market allocation of goods and services, production for private profit, and private ownership of the means of producing wealth.* Workers sell their labor to the owners of capital in exchange for a wage, and capitalists are then free to make a profit on the goods and services their workers produce. Capitalism emphasizes free, unregulated (or minimally regulated) markets and private (rather than government) economic decision making about the goods and services that should be produced.

At the same time, governments in capitalist economies often play a key role in shaping economic life, even in countries such as the United States that have historically tended to keep the government's role to a minimum (which is sometimes referred to as *laissez-faire* capitalism—literally, *hands-off* capitalism). In a capitalist country, the labor market is comprised of both public-sector jobs and private-sector jobs. The **public sector** is *linked to the government (whether national, state, or local) and encompasses production or allocation of goods and services for the benefit of the government and its citizens.* The **private sector** also *provides goods and services to the economy and consumers, but its primary motive is gaining profit.*

Because capitalists compete with one another for customers, they experience persistent pressure to keep costs and therefore prices down. They can gain a competitive edge by adopting innovative processes such as mass production (think of early Fordism), reducing expensive inventories, and developing new products that either meet existing demands or create new demands (this is what sociologists call *manufactured needs*). One important process of innovation is minimizing the cost of labor, which capitalists have historically done by adopting technologies that increase productivity and keep wages

Capitalism: An economic order characterized by the market allocation of goods and services, production for private profit, and private ownership of the means of producing wealth.

Public sector: The sector of the labor market in which jobs are linked to the government (whether national, state, or local) and encompass production or allocation of goods and services for the benefit of the government and its citizens.

Private sector: The sector of the labor market that provides goods and services to the economy and consumers with the primary motive of gaining profit.

low. Automation has enabled a reduction in production costs, and technological advances such as artificial intelligence promise to further push down costs, though this has the potential to entail a significant loss of jobs in some sectors.

On the one hand, capitalism can create uneven development, inequality, and conflict between workers and employers, whose interests may be at odds. On the other hand, it has been successful in producing diverse and desirable products and services, encouraging invention and creativity by entrepreneurs who are willing to take risks in return for potential profit, and raising living standards in many countries across the globe.

The role of government varies widely among different capitalist economies. In the United States and England, for example, there is greater skepticism about government's role in the private sector and greater emphasis on the private sector as the means for allocating goods and services (though Britain, unlike the United States, has nationalized health care). In contrast, in many European economies, the government takes a strong role in individual lives. Sweden and France offer "cradle to grave" social supports, with paid parental leave, child allowances, national health insurance, and generous unemployment benefits. Japan, on the other hand, does not expect government to take such a major role but does expect businesses to assume almost family-like responsibility for the welfare of their employees.

A Case of Capitalism in Practice: A Critical Perspective

Profit is the driving motive of capitalist systems. While the desire for profit spawns creativity and productivity, it may also give rise to greed, corruption, and exploitation. Industries cut costs in order to increase profits; there is economic logic in such a decision. Cutting costs, however, can also compromise the health and safety of workers and consumers.

What do such compromises look like? A case study of profit over people in the meat industry in the United States offers one example. In the first decade of the 20th century, Upton Sinclair's novel *The Jungle* offered a powerful and frightening fictionalized account of the real-life problems of the meat industry. The novel chronicles the struggles of a Lithuanian immigrant family working and struggling in "Packingtown," Chicago's meat district. The following excerpt offers a snapshot of one family member's workplace experience in Packingtown:

> It was only when the whole ham was spoiled that it came into the department of Elzbieta. Cut up by the two-thousand-revolutions-a-minute flyers, and mixed with half a ton of other meat, no odor that ever was in a ham could make any difference. There was never the least attention paid to what was cut up for sausage. There would be meat stored in great piles in rooms, and the water from leaky roofs would drip over it, and thousands of rats would race about on it. . . .
>
> Such were the new surroundings in which Elzbieta was placed, and such was the work she was compelled to do. It was stupefying, brutalizing work; it left her no time to think, no strength for anything. She was part of the machine she tended, and every faculty that was not needed for the machine was doomed to be crushed out of existence. (Sinclair, 1906/1995, pp. 143–145)

Concerns have been raised about the use of chemicals in the poultry processing industry today. The chemicals may present a danger to both workers and consumers. What are the benefits of automating dangerous jobs? What are the potential costs?

©iStockphoto.com/roibu

Sinclair was critical of capitalism and the profit motive, which he felt underpinned the suffering of the workers and the stomach-turning products churned out in the filthy packinghouses. His work was a stirring piece of social criticism dressed as fiction, and it spurred change. President Theodore Roosevelt's inquiry into the conditions described by Sinclair brought about legislation requiring federal inspection of meat sold through interstate commerce and the accurate labeling of meat products and ingredients (Schlosser, 2012). The novel had little effect, however, on the conditions experienced by industrial laborers, to which Sinclair had sought to draw attention. As he later wryly remarked, "I aimed for the public's heart . . . and by accident I hit it in the stomach."

In the years since Sinclair's novel was published, capitalism has evolved (as has regulation), though profit and the need to cut costs have remained basic characteristics. What does cost cutting look like in today's more closely regulated meat industry? In *Fast Food Nation* (2002), writer Eric Schlosser describes his experience in a 21st-century meat processing plant:

> A man turns and smiles at me. He wears safety goggles and a hardhat. His face is splattered with gray matter and blood. He is the "knocker," the man who welcomes cattle to the building. Cattle walk down a narrow chute and pause in front of him, blocked by a gate, and then he shoots them in the head with a captive bolt stunner. . . .
>
> For eight and a half hours, he just shoots. . . .
>
> When a sanitation crew arrives at a meatpacking plant, usually around midnight, it faces a mess of monumental proportions. Workers climb ladders with hoses and spray the catwalks. They get under tables and conveyor belts, climbing right into the bloody muck, cleaning out grease, fat, manure, leftover scraps of meat. (pp. 170–171, 177)

The work is not only brutal; it is also dangerous. More recently, concerns have been raised about the use of chemicals in the poultry processing industry. U.S. demand for chicken and turkey is high and growing. In response, companies have sought to increase efficiency by allowing more chemical decontamination of bird carcasses on the production line:

> To keep speeds up, the new regulations "would allow visibly contaminated poultry carcasses to remain online for treatment"—rather than being discarded or removed for offline cleaning. . . . The heightened use of chemicals would follow a pattern that has already emerged in poultry plants. In a private report to the House Appropriations Committee, the USDA [United States Department of Agriculture] said that [in] plants that have already accelerated line speeds, workers have been exposed to larger amounts of cleaning agents.
>
> Exposure to chemicals has affected line workers, but also USDA and industry inspectors, who describe a variety of ailments ranging from respiratory problems and skin rashes to irritated eyes and nasal ulcers. (Kindy, 2013, paras. 6–7)

Critics of capitalism would suggest that conditions in the meat industry—past and present—illuminate a fundamental problem of capitalism: Capital accumulation and profit are based on driving down the costs of production. The case of the meat industry shows that capitalists sometimes drive down the costs by compromising worker and consumer safety. The potentially high human cost of industrial profits is a central point of the critique of capitalism.

Socialism and Communism

Modern ideas about communism and socialism originated in the theories of 19th-century philosophers and social scientists, especially those of Karl Marx. **Communism**, in its ideal-typical form, is *a type of economic system without private ownership of the means of production and, theoretically, without economic classes or economic inequality.* In an ideal-typical communist society, the capitalist class has been eliminated, leaving only workers, who manage their economic affairs cooperatively and distribute the fruits of their labor "to each according to his needs, from each according to his abilities." Since Marx believed that governments exist primarily to protect the interests of capitalists, he concluded that once the capitalist class and private property were eliminated, there would be no need for the state, which would, in his words, "wither away."

Communism: A type of economic system without private ownership of the means of production and, theoretically, without economic classes or economic inequality.

Marx recognized that there would most likely have to be a transitional form of economic organization between capitalism and communism, which he termed **socialism**. In a socialist system, *theoretically, the government manages the economy in the interests of the workers; it owns the businesses, factories, farms, hospitals, housing, and other means of producing wealth and redistributes that wealth to the population through wages and services.* The laborer works for a state-run industrial enterprise, the farmer works for a state-run farm, and the bureaucrat works in a state agency. Profit is not a driving economic imperative because there is no private property and no private profit.

Socialism: A type of economic system in which, theoretically, the government manages the economy in the interests of the workers; it owns the businesses, factories, farms, hospitals, housing, and other means of producing wealth and redistributes that wealth to the population through wages and services.

Before the collapse of the Soviet Union and its socialist allies in Eastern Europe, nearly a third of the world's population lived in socialist countries. Far from withering away as Marx predicted, these socialist governments remained firmly in place until 1989 (1991 in the Soviet Union), when popular revolutions ushered in transformations—not to the classless communist economies Marx envisioned, but to new capitalist states. The largest remaining socialist country in the world—China—has transitioned over the past 20 years into a market economy of a size and scale to nearly rival that of the United States, though the state still exercises control over large industries.

These transformations occurred in part because socialism proved too inflexible to manage a modern economy. Having the central government operate tens of thousands of factories, farms, and other enterprises was a deterrent to economic growth. Furthermore, though the capitalist class was eliminated, a new class emerged—the government bureaucrats and communist officials who managed the economy and who were often inefficient, corrupt, and more interested in self-enrichment than in public service (Djilas, 1957). Moreover, most socialist governments were intolerant of dissent, often persecuting, imprisoning, and exiling those who disagreed with their policies. At the same time, socialist governments were often successful in eliminating extreme poverty and providing their populations with housing, universal education, health care, and basic social services. Inequality was typically much lower in socialist states than in capitalist economies—although the overall standard of living was lower as well.

In the socialist period in the Soviet Union and allied countries of Eastern Europe, the state controlled all production, essentially eliminating competition in the marketplace for goods and services. Instead of advertisements in public spaces, political posters and propaganda elevated the achievements and builders of socialism and denigrated the capitalist way of life.

©Universal History Archive/Getty Images

The dramatic rise in economic inequality and poverty in the newly capitalist states of the former Soviet Union and Eastern Europe has created some nostalgia for the socialist past, particularly among the elderly, who have a threadbare social safety net in many states. While few miss the authoritarian political governments, there is some longing for the basic economic and social security that socialism offered.

A Case of Socialism in Practice: A Critical Perspective

In theory, a driving motive of socialist systems is achievement of a high degree of economic equality. This is realized in part through the creation of full-employment economies. In the Soviet Union, full employment gave all citizens the opportunity to earn a basic living, but it also brought about some socially undesirable results. For instance, inefficiency and waste flourished in enterprises that were rewarded for how much raw material they consumed rather than how much output they produced (Hanson, 2003). Human productivity was only partially utilized when work sites had to fill required numbers of positions but did not have meaningful work for all who occupied them. Disaffection and anger grew in workplaces where promotions were as likely to be based on political reliability, connections, and Communist Party membership as on merit. A system that theoretically ensured the use of resources for the good and equality of all workers was undermined by the realities of Soviet-style communism.

In practice, socialist systems such as that of postwar Hungary were characterized by both low wages and low productivity: A popular saying among workers was that "we pretend to work and the state pretends to pay us." Lacking a competitive labor market, workers may not have felt compelled to work particularly hard; unemployment was rare. At the same time, public-sector (government) jobs, which made up the bulk of the labor market, were poorly paid; many workers sought supplementary pay in the informal economy (Ledeneva, 1998). Furthermore, in the absence of a profit motive, there was limited entrepreneurial activity and the consumer market offered goods and services that were largely mediocre and often difficult to obtain.

Socialism in practice, according to sociologists Burawoy and Lukács (1992), was in part a performance:

> Painting over the sordid realities of socialism is simultaneously the painting of an appearance of brightness, efficiency, and justice. Socialism becomes an elaborate game of pretense which

> everyone sees through but which everyone is compelled to play. . . . The pretense becomes a basis against which to assess reality. If we have to paint a world of efficiency and equality—as we do in our [factory] production meetings, our brigade competitions, our elections—we become more sensitive to and outraged by inefficiency and inequality. (p. 129)

In the book he coauthored with Lukács, *The Radiant Past: Ideology and Reality in Hungary's Road to Capitalism* (1992), Burawoy, a U.S. sociologist who spent time working in socialist enterprises in Poland and Hungary as part of his study of socialist economies, recounts an instance of such a "painting ritual" when the Hungarian prime minister makes a visit to the Lenin Steel Works. Areas of the factory to be visited are literally painted over in bright hues, debris is swept up, and workers halt their productive tasks to create an appearance of productivity, for the prime minister "had to be convinced that the Lenin Steel Works was at the forefront of building socialism" (p. 127). For critics of socialism, the case of Hungarian steel in the socialist period highlights a fundamental problem of the system as it was practiced: Its weaknesses were made more rather than less apparent by the "painting rituals" that asked workers to pretend socialism was fundamentally efficient and equal when their own experience showed it was not. This was among the flaws that led to the collapse of socialism in Eastern Europe and the Soviet Union.

Work: Any human effort that adds something of value to the goods and services that are available to others.

Barter economy: An economy based on the exchange of goods and services rather than money.

Formal economy: All work-related activities that provide income and are regulated by government agencies.

WORKING ON AND OFF THE BOOKS

Work consists of *any human effort that adds something of value to the goods and services that are available to others.* By this definition, work includes paid labor in the factory or office, unpaid labor at home, and volunteer work in the community. Workers include rock stars and street musicians, corporate executives and prostitutes, nurses and babysitters. Almost the only activities excluded from this definition of work are those that individuals conduct purely for their own pleasure or benefit, such as pursuing a hobby or playing a musical instrument for fun.

Day laborers often work for low pay in unregulated conditions. Some of them are illegal migrants. Their status and language barriers make it challenging for them to report dangerous or abusive conditions of work.

©REUTERS/Lucas Jackson

The concept of work as consisting exclusively of labor that is sold for a wage is a relatively recent development of modern industrial society. Throughout most of human history, work was not done for a wage. In agricultural societies, subsistence farming was common: Families often worked their own plots of land and participated with others in the community in a **barter economy**, *an economy based on the exchange of goods and services rather than money.* With the advent of industrial society, however, work shifted largely to the economic setting of a formal, paid, and regulated job.

Today, work for pay occurs in two markets: the formal economy and the informal (or underground) economy. We look at each of these next.

The Formal Economy

The **formal economy** consists of *all work-related activities that provide income and are regulated by government agencies.* It includes work for wages and salaries as well as self-employment; it is what people ordinarily have in mind when they refer to *work*. It has grown in importance since the industrial revolution. Indeed, one of the chief functions of government in industrial society is regulation of the formal economy, which contributes to the shape and character of the labor market (Sassen, 1991; Tilly & Tilly, 1994).

In the United States, as in most countries of the world, private businesses are supposed to register with governmental entities ranging from

The term *underground economy* may evoke images of drugs, weapons, and stolen passports, but this type of economy also involves more mundane products, such as food. Los Angeles and other major cities have unlicensed and unregistered vendors providing food and services for residents, which has a mixture of positive and negative effects.

©REUTERS/Lucy Nicholson

tax bureaus (the Internal Revenue Service) to state and local licensing agencies. Whether they work in the private or the public sector, U.S. workers must pay income, Medicare, and Social Security taxes on their earnings, and employers are expected to withhold such taxes on their behalf and report employee earnings to the government. Numerous agencies regulate wages and working conditions, occupational health and safety, the environmental effects of business activities, product quality, and relationships among firms.

When statistical indicators such as employment and unemployment are tabulated by government entities such as the U.S. Bureau of Labor Statistics, they rely on data from the formal economy. Work is also done—and value produced—in the informal economy, but this is not included in BLS numbers, so some part of economic activity remains largely invisible in mainstream data.

The Informal (or Underground) Economy

Informal (or underground) economy: Those income-generating economic activities that are not regulated by the governmental institutions that ordinarily regulate similar activities.

A notable amount of income-generating work avoids formal regulation and is not organized around officially recognized jobs. This part of the economy is termed the **informal (or underground) economy**; it includes *all income-generating activities that are not regulated by the governmental institutions that ordinarily regulate similar activities.* Some of these sources of income are illegal, such as selling guns or pirated DVDs, drug dealing, and sex trafficking. Other work activities are not illegal but still operate under the government radar. These include selling goods at garage sales and on Internet auction sites such as eBay; housecleaning, gardening, and babysitting for employers who pay without reporting the transactions to the government; and informal catering of neighborhood events for unreported pay. In one way or another, most of us participate in the informal economy at some point in our lives.

Workers' reasons for participating in the informal economy are varied. A worker with a regular job may take up a second job "off the books" to make up a deficit in his or her budget, or someone may be compelled to work outside the legal economy because of his or her undocumented immigration status. Others may find the shadow economy more profitable (Schneider & Enste, 2002). Many of the underground economy's workers occupy the lowest economic rungs of society and are likely to be pursuing basic survival rather than untold riches. They are disproportionately low income, female, and immigrant, and the jobs they do lack the protections that come with many jobs in the formal economy, such as health care benefits, unemployment insurance, and job security. While "off the books" jobs enable more people to make a living, the work they do cannot typically be cited on a résumé, and "colleagues" in activities such as fixing cars or making and selling food informally cannot be used as references; hence, parlaying experience in the informal economy into a job in the formal economy is rarely tenable (Venkatesh, 2008).

Among the employers in the illegal U.S. underground economy are unlicensed "sweatshop" factories that make clothing, furniture, and other consumer goods (Castells & Portes, 1989; Sassen, 1991). Factories in the United States are competing with factories around the world where workers are paid a fraction of U.S. wages. To remain competitive (or to raise profits), U.S. firms sometimes subcontract their labor to low-cost sweatshop firms in the informal sector. Requirements to comply with environmental laws, meet health and safety standards, make contributions to Social Security and other social benefit programs, and pay taxes lead some businesses to seek to establish themselves off the books. Small businesses may operate without licenses, and large firms may illegally subcontract out some of their labor to smaller, unlicensed ones.

Shadow economies exist around the globe, though their size varies dramatically. According to one report, of 12 developed nations, the biggest shadow economies were to be found in Greece (21%), Italy (20%), and Spain (17%), while the smallest were in the Netherlands (8%), Switzerland (6%), and the United States (5%) (McCarthy, 2017). According to International Monetary Fund estimates, the largest shadow economies are found in developing states around the globe: For instance, shadow economies comprise about 65% of the gross domestic product (GDP) in Georgia, 62% in Bolivia, and 61% in Zimbabwe. Altogether, the average size of the shadow economies of all the countries is about 32% of the GDP (Medina & Schneider, 2018).

CONSUMERS, CONSUMPTION, AND THE U.S. ECONOMY

As we have seen in this chapter, production has been an important part of the rise of modern capitalist economies. In modern industrial countries, including the United States, however, production has receded in importance. The economies of advanced capitalist states today rely heavily on consumption to fuel their continued growth. Today, an estimated 70% of the U.S. economy is linked to consumption. In this section, we examine consumption and its relationship to the economy as well as to our lives as consumers.

Means of consumption: "Those things that make it possible for people to acquire goods and services and for the same people to be controlled and exploited as consumers" (Ritzer, 1999).

Theorizing the Means of Consumption

Karl Marx is well known for his concept of the *means of production* (defined in Chapter 1), which forms a basis for his theorizing on capitalism, exploitation, and class. While the early industrial era in which Marx wrote influenced his focus on production, he also sought to understand consumption in 19th-century capitalism. Marx defined the term *means of consumption* as "commodities that possess a form in which they enter individual consumption of the capitalist and working class" (quoted in Ritzer, 1999, p. 56). Marx distinguished between the levels of consumption of different classes, suggesting that *subsistence consumption* ("necessary means of consumption") characterizes the working class, whereas *luxury consumption* is the privilege of the exploiting capitalist class. In sum, Marx's definition focused on the consumption of the end products of the exploitative production process.

"Cathedrals of consumption," such as the casinos and hotels of Las Vegas, entice consumers to spend with bright, fun, and fantastical venues. The Luxor Hotel simulates the splendor of Ancient Egypt, featuring replicas of the Great Sphinx and Great Pyramid of Giza.

©Visions of America/UIG via Getty Images

Sociologist George Ritzer has expanded Marx's concept. He distinguishes between the end product (that is, a consumer good such as a pair of stylish dress shoes, a new car, or a gambling opportunity) and the means of consumption that allow us to obtain the good (for instance, the shopping mall, the car dealership, or the Las Vegas casino). For Ritzer (1999), the **means of consumption** are *"those things that make it possible for people to acquire goods and services and for the same people to be controlled and exploited as consumers"* (p. 57). For example, a venue such as a mall offers the consumer buying options and opportunities, but it is also part of a system of consumer control through which consumers are seduced into buying what they do not need, thinking they need what they merely want, and spending beyond their means.

One of Disney World's long-standing attractions is the Jungle Cruise, described on the park's website as an adventure cruise of the most "exotic and 'dangerous' rivers in Asia, Africa, and South America," although it is a virtually danger-free boat trip on a man-made waterway populated by plastic figures. Modern consumers, sociologist George Ritzer suggests, are buying the fantasy rather than the reality of such experiences.

©Jennifer Wright/Alamy

Ritzer's concept of the means of consumption also integrates German sociologist Max Weber's ideas about rationalization, enchantment, and disenchantment. Briefly, the Weberian perspective holds that premodern societies were more "enchanted" than modern societies. That is, societies or communities, which were often small and homogeneous, were grounded in ideas that he characterized as magical and mystical. Individuals and groups defined and pursued goals based on abstract teachings such as the ideals and ideas of a religion rather than on detailed, specific rules and regulations. Even early capitalism was linked to an enchanted world. Weber theorized that early Protestantism (and Calvinism in particular) embraced values of thrift, efficiency, and hard work and viewed economic success as an indicator of divine salvation. This so-called *Protestant ethic,* which he identified as characteristic in Northern Europe, laid foundations for the rise of capitalism, though capitalism eventually shed its religious aspects (Weber, 2002).

Modern capitalism lacks authentic enchantment: It is a highly rationalized system characterized by efficiency, predictability, and the pursuit of profit (rather than divine salvation!). This heavily bureaucratized and regulation-reliant environment is virtually devoid of spontaneity, spirituality, or surprise. Ritzer argues, however, that enchantment is important for controlling consumers, because consumption is, at least in part, a response to a fantasy about the item or service being consumed. Consequently, disenchanted structures must be "reenchanted" through spectacle and simulation (Baudrillard, 1981/1994), which draw in consumers. For instance, Disney simulates a kind of childhood dreamworld (think of the Magic Kingdom), Niketown is a sports fantasy, and Las Vegas aims to bring to a single city the dazzle of Egyptian pyramids, New York's towering urban structures, and Paris's Eiffel Tower. In such a context, the consumer is not merely buying sneakers (say, at Niketown) but embracing a broader fantasy about athletic achievement. In sum, from Ritzer's perspective, the means of consumption are a modern instrument of control not of the *worker* but of the *consumer,* who is enchanted, led to believe that he or she needs certain goods, and given optimal—sometimes nearly inescapable—avenues for acquisition, such as malls with long hallways and few exits to maximize the number of shops a consumer must pass before exiting.

A Historical Perspective on Consumption

Consumer society is a political, social, and economic creation. Consider, for instance, that during World War II, the U.S. government asked its citizen-consumers to serve the greater good by reducing consumption. In contrast, in the wake of the terror attacks on the United States in 2001 and the wars that followed, citizen-consumers were encouraged to spend more money to stimulate the economy. Former secretary of labor Robert Reich termed this appeal for consumption *market patriotism*. Taking a broader perspective, economist Juliet Schor (1998) argues that consumption patterns and the dramatic growth of consumption in the United States are heavily driven by Americans' reliance on reference groups. That is, consumers compare themselves and their consumption to the reference groups in their social environments. Significantly, says Schor, those reference groups have changed. In the 1950s, suburban middle-class consumers knew and emulated their neighbors. The substantial number of women outside the paid workforce meant that neighbors were more aware of what others were doing, wearing, and driving. By the 1970s, more women were moving into the workforce; consequently, fewer people knew their neighbors, and the workplace became an important source of reference groups. In contrast to the economically homogeneous neighborhood, however, the workplace is heterogeneous. Low five-figure wages coexist in the same space as high six-figure salaries, and coworkers may aspire upward and far beyond their means.

The 1980s, 1990s, and 2000s brought further upscaling of ambitions and spending, as television sold a powerful picture of consumer decadence masked as normal life. In the 1990s, young consumers embraced media referents such as the television sitcom *Friends,* about a group of young professionals living in lavish New York City apartments, wearing ever-changing stylish wardrobes, and casually consuming the pleasures around them. Lavish consumption came to seem normal rather than unreachable, even for people of average incomes (Schor, 1998). A new generation is now exposed to reality TV shows, including *Island Hunters, Keeping Up With the Kardashians,* and *Selling Sunset,* which highlight the benefits (but not the costs) of conspicuous consumption.

A somewhat different perspective on how the U.S. consumer economy has been built and sustained is offered by Robert Reich (2010), who believes the consumption-driven economy originated in a "basic bargain" between workers and employers that offered good pay in sectors such as manufacturing, creating a consumer class that could afford to spend (recall Henry Ford's decision to pay above-average wages so his auto workers could buy cars). Reich notes that, until about 1970, pay rose more quickly in the middle- and lower-income segments of the U.S. labor pool than it did at the top. Consumption grew with the standard of living. The real value of workers' pay stagnated in the 1970s, however, profoundly affected by forces that included globalization and automation. While income rose at the very top of the economic ladder, in the middle and lower strata, it stalled. Consumption continued to rise, however, driven not by gains in income but by the growing credit markets, which offered new ways to spend with or without cash on hand. Next, we review the consequences of that shift to credit-driven spending.

Credit: Debt and More Debt

Do you have a credit card? Do you carry debt? If you answered yes to either or both these questions, you are not alone. In early 2019, the U.S. Federal Reserve estimated that Americans' consumer debt reached $4 trillion. Credit card debt comprised about $1 trillion of this total, with the average household credit card debt at about $4,293 (Dickler, 2019). The average nonmortgage debt varies by generation: Gen Z averages a debt of about $10,900; Millennials hold about $27,200 in debt; Gen X has the highest debt at $32,800; and Baby Boomers have about $28,500 in debt. These debts encompass car loans, student loans, credit cards, and personal loans (DeMatteo, 2020).

In the early years of the 21st century, "Americans were more likely to go bankrupt than to get divorced" (Quiggin, 2010, p. 26). Consumption (or overconsumption) was not the direct cause of bankruptcies (most of which were precipitated by the loss of a job or unexpected health care costs), but the "culture of indebtedness," the widespread tendency to owe a great deal of money, left people less able to bear any added financial stress. While financial reforms passed into law in 2005 made the declaration of bankruptcy more onerous and less common, the financial crisis that began in 2007 saw another rise in bankruptcies: More than 1.5 million bankruptcy filings were made in 2010 (Administrative Office of the United States Courts, 2011). The number of filings has gone down since then, with bankruptcy filings in the United States falling to 766,698 for the 2017 calendar year (ACA International, 2018). While bankruptcies had not risen in mid-2020, even after the COVID-19 pandemic began, history suggests that bankruptcies peak not during an economic crisis, but in the years that follow: For example, according to a government website that records the activities in U.S. courts, bankruptcy filings escalated in the 2 years following the Great Recession before peaking in 2010 (United States Courts, 2020).

Humorist Will Rogers commented during the Great Depression that the United States was the first country to drive to the poorhouse in an automobile (cited in Sullivan, Warren, & Westbrook, 2000, p. 3). Car debt was a particularly notable burden during the late 1990s and early 2000s, when many U.S. drivers opted to purchase high-end cars, especially aggressively advertised sport utility vehicles. Rising gasoline prices through the 2000s, coupled with the economic recession that struck in 2007, reversed the trend, and consumers began opting for more fuel-efficient sedans, hybrid vehicles, and smaller crossover vehicles that combine features of cars and SUVs. Since the first hybrid cars came on the U.S. scene in late 1999, Americans have purchased over 4 million of these fuel savers (Cobb, 2016), though larger and less fuel-efficient vehicles continue to be popular in the United States. In fact, Americans are buying more expensive vehicles than in the past and, according to one report, taking out longer-term loans that result in greater interest payments. The lengthened loan period makes it appear to a consumer that the car or truck is a good deal because the monthly payments are modest—but the long-term cost is significant (Arnold, 2019). Automobile debt makes up about a tenth of Americans' consumer debt, with the average new car loan at about $32,000 and the average used car loan at about $25,000 (Jones, 2020).

Americans are avid consumers of trucks and cars and, taken together, auto loan debt is about .2 trillion. What drives the robust market for trucks and cars, particularly expensive vehicles? What are the manifest and latent functions of vehicles for consumers?

GLOBALIZATION AND THE NEW ECONOMIC ORDER

The U.S. economic order today has been powerfully affected by the emergence of a unified global economic system. In fact, some writers have argued that it no longer makes sense to think of the United States—or any other country—as an isolated economic society at all: In many respects, we can regard the world as a single economic unit (Friedman, 2005). We conclude this chapter by examining how global economic interdependence and the global labor market have affected work and economic life in the United States.

Global Economic Interdependence

The U.S. economy is interwoven with the economies of other countries. Many goods made in the United States are sold in foreign markets, while many goods bought by U.S. consumers are made by foreign workers. Economic integration is multidimensional and can be shallow or deep (Dicken, 1998). Shallow integration is more characteristic of the globalization of several decades past, when a single product (say, a German automobile) was made in a single country and that country's government would regulate its export as well as the import of other goods into the country. Countries did business with one another, but their ties were looser and less interdependent.

Deep integration is characteristic of the modern global economy in which corporations are often multinational rather than national, products are made of raw materials or parts from a spectrum of countries (Figure 15.3), and a corporation's management or engineering may be headquartered in one country while the sales force or customer service contingent may reside anywhere from Denver to Delhi. Familiar companies such as Nike, Apple, and Ford are among the many with globalized labor forces.

A Global Market for Labor

As a result of economic globalization, a growing number of U.S. workers are competing with workers all over the world. This trend may affect the job prospects of all workers, whether they hold high school diplomas or advanced degrees. There are substantial wage differences between countries. While the United States is intermediate among industrial countries, its wages are considerably higher than those in developing countries (Table 15.2).

FIGURE 15.3 ■ Global Origins of Boeing 787 Parts

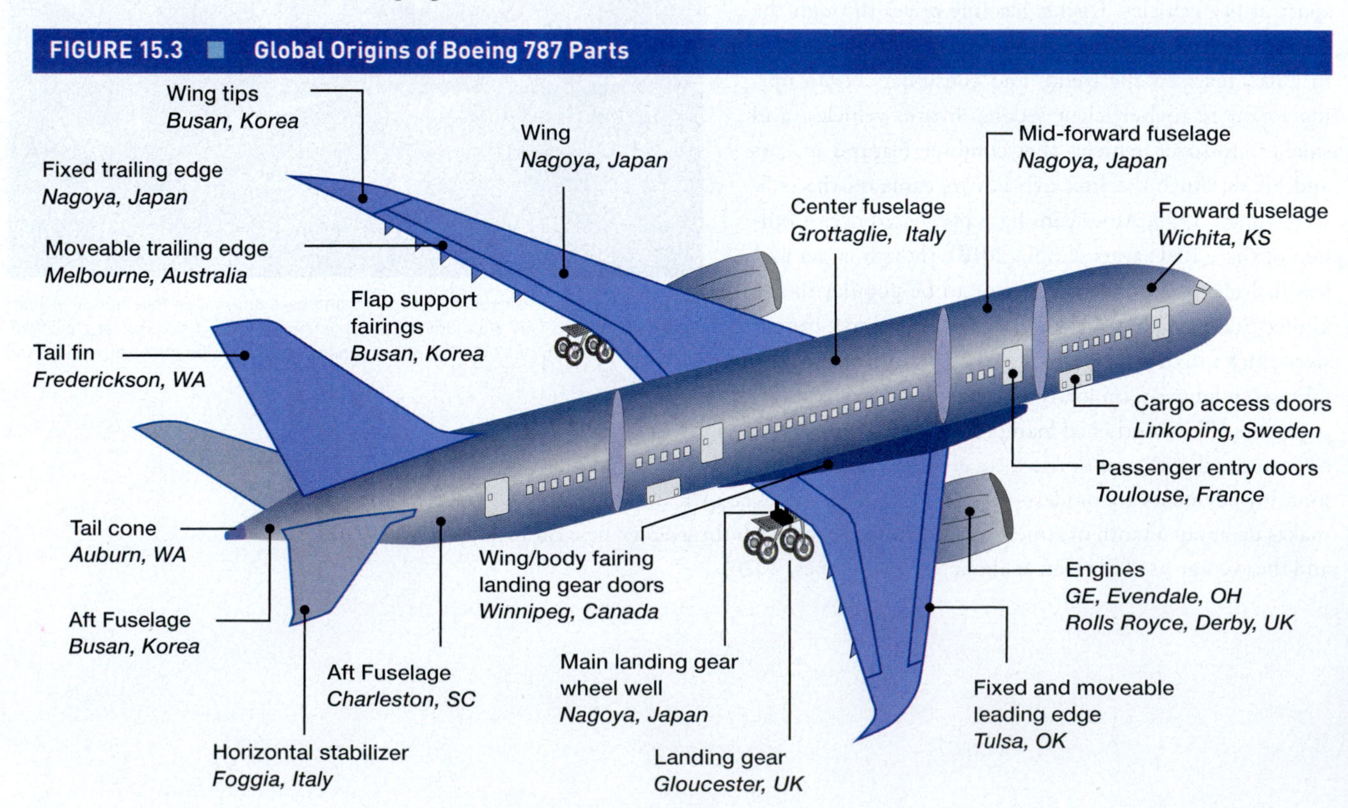

TABLE 15.2 ■ Average Wages in Select Countries, 2019

COUNTRY	AVERAGE YEARLY WAGES (USD)
1. Luxembourg	$68,681
2. United States	$65,836
3. Norway	$54,027
4. Australia	$54,401
5. Canada	$53,198
6. Germany	$53,638
7. United Kingdom	$47,226
8. Japan	$38,617
9. Spain	$38,758
10. Korea	$42,285
11. Greece	$27,459
12. Portugal	$26,634
13. Lithuania	$28,914
14. Hungary	$26,223
15. Mexico	$17,594

Source: Organisation for Economic Co-operation and Development. (2020). *Average Wages*. Retrieved from https://data.oecd.org/earnwage/average-wages.htm

Jobs will increasingly go wherever on the planet suitable workers can be found. Low labor costs, the decline or absence of labor unions, and governments that enforce worker compliance through repressive measures are all factors influencing the globalization of labor. Some sociologists call this trend a *race to the bottom* (Bonacich & Appelbaum, 2000), in which companies seeking to maximize profits chase opportunities to locate wherever conditions are most likely to result in the lowest costs. This has been the case in, for example, apparel manufacturing; much of the clothing we buy and wear today, including brands sold at popular shops such as H&M and Zara, is made by young workers abroad who labor for very low wages under poor working conditions. China, Bangladesh, Vietnam, and India are the world's leaders in garment manufacturing. China accounts for 21%; followed by Bangladesh and India, both at 14%; and Vietnam at 12%. Most garment companies look to locations with low labor costs to manufacture their goods, which makes Bangladesh, which has a monthly minimum wage of only $65, a popular site. This minimum wage is about $50 less than that of India and Vietnam, and $85 less than China (Knack, 2017). As one advocate for ethical manufacturing and wages notes, "If something is very cheap, you have to ask yourself, is it really possible to make it in a factory that is run properly, with a living wage?" (Butler, 2019, para. 17).

The minimum wage in Bangladesh was established in 2013 when a government-appointed panel voted to raise the minimum wage for garment workers; under the plan, it was scheduled to rise from a monthly minimum of about $38 to $66. However, recently Bangladesh unions and workers organized a conference to demand an increase in the monthly minimum wage to $192, due to the rising cost of living (BDnews24, 2018). Whether this will result in better conditions for workers or an abandonment of the country by low-wage-seeking manufacturers remains to be seen.

We have seen above that the emergence of a global labor market has resulted in job losses and declining wages in many U.S. industries, including auto and apparel manufacturing. We may be witnessing the emergence of a *global wage,* equivalent to the lowest worldwide cost of obtaining comparable labor

TABLE 15.3 ■ Annual Compensation of Selected CEOs, 2019

CORPORATION	CEO	COMPENSATION
Alphabet	Sundar Pichai	$280,621,552
Intel Corporation	Robert Swan	$66,935,100
Walt Disney Company	Robert A. Iger	$47,517,762
Pinterest, Inc.	Benjamin Silbermann	$46,222,113
Netflix, Inc.	Reed Hastings	$38,577,129
Visa	Alfred Kelly Jr.	$24,265,771
Facebook	Mark Zuckerberg	$23,415,973
UnitedHealth Group	David Wichmann	$18,886,989
Pfizer, Inc.	Albert Bourla	$17,928,963
Delta Air Lines, Inc.	Edward Bastian	$17,325,378

Source: Adapted from AFL-CIO. (2020). *100 Highest Paid CEOs.* Retrieved from https://aflcio.org/paywatch/highest-paid-ceos

Note: Annual compensation includes the value of executives' base salary, value of stock and option awards, and other financial compensation vehicles.

for a particular task once the costs of operating at a distance are considered. For virtually any job, this wage is far lower than U.S. workers are accustomed to receiving.

The global labor market is not limited to manufacturing. A global market is emerging for a wide range of professional and technical occupations as well. In fact, some of the "knowledge worker" jobs touted as the jobs of the future may be among the most vulnerable to globalization. Unlike cars or clothing, engineering designs and computer programs can move around the globe electronically at no cost, and transportation time is, for all practical purposes, nonexistent. Electronic engineering, computer programming, data entry, accounting, insurance claims processing, and other specialized services, such as medical image reading, can now be inexpensively purchased in such low-wage countries as India, Malaysia, South Korea, China, and the Philippines, where workers communicate digitally with employers in the United States.

Among those selling their labor on the new global market are highly educated professionals from Central and Eastern European countries such as Estonia and Hungary, both of which have full literacy, educated populations, and many individuals fluent in English and other world languages. While we often associate cheap labor with the low-wage factories of developing countries, these post-Soviet European states also advertise their educated workers (on their investment-promoting websites, for instance) as cheap labor. Indeed, much global labor is low in cost. As of 2018, the average monthly wage in well-educated Estonia was $1,445 and the average monthly wage for IT workers was $2,496, far below that of workers in the United States and many parts of Western and Northern Europe (workinestonia.com, 2018).

While globalization has had some negative effects on earnings for American workers in the lower- to middle-income ranges, corporate executive salaries have skyrocketed. Even in the midst of a plummeting economy (2007–2009), as the federal government was distributing billions of bailout dollars to corporations, banks, and investment firms to prevent them from failing, CEOs in the United States were bringing home multiple millions of dollars in compensation. Table 15.3 shows levels of compensation for CEOs of leading global companies—year to year, these figures have risen. It is notable that in the economic crisis precipitated by the COVID-19 pandemic, CEOs of companies seeking large government bailouts, including airlines and cruise lines, continued to earn millions in compensation (though some took voluntary pay cuts), even as workers were let go and taxpayer money was used to prevent company failures (Kelly, 2020).

WHAT'S NEXT? WORK AND THE ECONOMY

The labor market has experienced marked changes in recent decades. In the United States, it has experienced deindustrialization in the form of lost manufacturing capacity and jobs. It has also seen the influx of a rising number of workers with college and professional education. At the same time, the service industry has expanded, as consumers with cash and credit have opted to spend money on leisure, eating out, cars, and other consumer amenities.

Further changes in the labor market may be afoot. Consider the importance of technological innovations, like those discussed in this chapter, including the growing sophistication of artificial intelligence and robotics. Which sectors of the labor market are particularly vulnerable—and which will retain their need for human workers well into the future? It is clear that the lower labor costs of technologically enabled work will change the face of many occupations, not only in blue-collar manufacturing, but also in white-collar professional fields.

It is likely that the COVID-19 pandemic will accelerate some changes in the labor force, as changes made to protect workers and customers—including the replacement of humans by robots in service interactions and tasks like cleaning and food preparation—become permanent. While protecting the safety of workers is clearly a manifest function of these changes, a latent function—or dysfunction—may well be the loss of jobs in some sectors.

In our modern capitalistic economy, one's work is often closely tied not only to one's earning power, but to one's identity. As some jobs become unavailable to human workers, will they be replaced by other jobs that offer paid, sufficient, and meaningful work? What might the jobs of the future look like? Who will do them? What do you think?

WHAT CAN I DO WITH A SOCIOLOGY DEGREE?

Data and Information Literacy

Paradoxically, in the modern world, we are surrounded by information, but we are not always truly well informed. The abilities to distinguish between credible and questionable data, to seek out solid and reliable sources of information, and to use those sources wisely are critical skills in our information society and in today's job market. *Data and information literacy* encompasses the skills to identify the information needed to understand an issue or problem, to seek out credible and accurate sources of information, to recognize what a body of data illuminates and what it obscures, and to apply the information to a description and analysis of the issue at hand.

As a student, you need the skills of data and information literacy to complete tasks such as writing research papers and preparing class presentations. As a consumer, you employ these skills to guide your decisions about the purchase of a home or a vehicle. As a citizen, you need the tools of data and information literacy to make informed political choices about which candidates or causes to support. Data and information literacy is no less significant in the world of work.

As a sociology major, you will be asked to do research on social issues and problems. You are likely to encounter and utilize diverse information sources, such as databases of the U.S. government (for example, the U.S. Bureau of Labor Statistics or the U.S. Census Bureau) and international organizations (like the World Bank or the United Nations), academic books and research journals, and the mass media. You will have the opportunity to develop data and information literacy skills that enable you to be a solid researcher and a critical consumer of information.

Celisa Walker

University of Roehampton, London, MA in Media Communication and Culture; University of Nevada, Reno, BA in Sociology

When I first decided to study sociology, I did not have a clear idea of what my life would be beyond university. I only knew that a degree in sociology would provide me with important skills regardless of whatever career I chose. I found this to be true after graduating from the University of Nevada when I joined the team at Ildico Inc., a luxury timepiece retailer. In this role, I assisted the CFO in inventory and accounting duties. During my time in this position, I was able to exercise my strengthened time management and critical thinking skills on a daily basis. After spending a few years with the company, I returned to school to obtain a master's degree.

This past summer, I completed a master's degree at the University of Roehampton in London, England, in Media Communication and Culture. Media courses taken during my undergraduate studies were what I enjoyed most, so I turned to a more media-focused degree to study the subject further. Like sociology, this degree has further strengthened my written communication and enhanced my understanding of people. It has also given me the opportunity to gain international experience. Outside of university requirements, I served as marketing ambassador for the study abroad agency, Study Across the Pond. In this role, I advised students looking to obtain a degree in the UK on what they could expect from the experience. As a fellow international student, I took this role seriously as I now have an intimate understanding of what it is like to make such a life-changing yet rewarding move.

A degree in sociology has made me a better student, a better employee, and a more understanding and open-minded person as a whole. I feel that this is particularly important in today's increasingly connected, yet often divided, society. I am now well-equipped to tackle any role head on and to encourage others to be better versions of themselves.

Career Data: Market Research Analyst

- 2019 Median Pay: $63,790 per year
- $30.67 per hour
- Typical Entry-Level Education: Bachelor's degree
- Job Outlook, 2019–2029: 18% (much faster than average)

Source: Bureau of Labor Statistics, *Occupational outlook handbook*, 2020.

SUMMARY

- The **economy**—the social institution that organizes the ways in which a society produces, distributes, and consumes **goods** and services—is one of the most important institutions in society.
- Three major technological revolutions in human history have brought radically new forms of economic organization. The first led to agriculture, the second to modern industry, and the third to the postindustrial society that characterizes the modern United States.
- Industrial society is characterized by **automation**, the modern factory, **mass production**, **scientific management**, and modern social classes. Postindustrial society is characterized by the use of computers, the increased importance of higher education for well-paying jobs, flexible forms of production, increased reliance on outsourcing, and the growth of the service economy.
- Although postindustrial society holds the promise of prosperity for people who work with ideas and information, automation and globalization have also allowed for new forms of exploitation of the global workforce and job loss and declining wages for some workers in manufacturing and other sectors.

- Artificial intelligence has the potential to transform the labor market, as machines shift from the role of being an instrument for human workers to being workers themselves. This change may affect jobs for both less-educated and highly educated employees.
- **Capitalism** and **socialism** are the two principal types of political economic systems that emerged with industrial society. While both are committed to higher standards of living through economic growth, they differ on the desirability of private property ownership and the appropriate role of government. Both systems have theoretical and practical strengths and weaknesses.
- **Work** consists of any human effort that adds something of value to goods and services that are available to others. Economists consider three broad categories of work: the **formal economy**, the **informal (or underground) economy**, and unpaid labor.
- The informal economy is an important part of the U.S. economy, even though it does not appear in official labor statistics. Although in industrial societies, the informal economy tends to diminish in importance, in recent years, this process has reversed itself.
- In the modern economy, consumption replaces production as the most important economic process. The means of consumption, as defined by sociologist George Ritzer (1999), are "those things that make it possible for people to acquire goods and services and for the same people to be controlled and exploited as consumers" (p. 57). A shopping mall offers consumers buying options, but it also is part of a system of consumer control, as consumers are seduced into buying what they do not need.
- We acquire goods in part based on our consideration of reference groups. As consumption reference groups have changed in the past decades, U.S. consumers have increased spending and taken on a much larger debt load.
- Economic globalization is the result of many factors: technological advances that greatly increased the speed of communication and transportation while lowering their costs, increased educational attainment in low- and middle-income countries, and the opening of many national economies to the world capitalist market. Globalization has had profound effects on the U.S. economy.

KEY TERMS

automation (p. 422)
barter economy (p. 435)
capitalism (p. 431)
communism (p. 433)
discouraged workers (p. 426)
economy (p. 419)
emotional labor (p. 425)
employed (p. 426)
formal economy (p. 435)
goods (p. 419)
informal (or underground) economy (p. 436)
labor force participation rate (p. 426)
marginally attached to the labor force (p. 426)
mass production (p. 420)
means of consumption (p. 437)
not in the labor force (p. 425)
private sector (p. 431)
public sector (p. 431)
reserve army of labor (p. 420)
scientific management (p. 420)
services (p. 419)
socialism (p. 433)
unemployed (p. 425)
work (p. 435)

DISCUSSION QUESTIONS

1. How is unemployment in the United States measured? What aspects of this phenomenon does the unemployment rate measure and what aspects does it fail to capture?

2. What effects might the expansion of robotics and artificial intelligence have on the U.S. and global workforce? What evidence of the effect is available today? What sectors of the labor market may be affected in the future?

3. What are the main differences between the formal economy and the informal economy? What are the similarities? What sociological factors explain the existence of the informal economy in the United States?

4. What are the main characteristics of a socialist economic system? Where have such systems been found in recent history? What are their strengths and weaknesses?

5. What sociological factors explain the dramatic rise of consumer debt in the United States over the past three to four decades? Why should this be of concern to society and to policy makers?

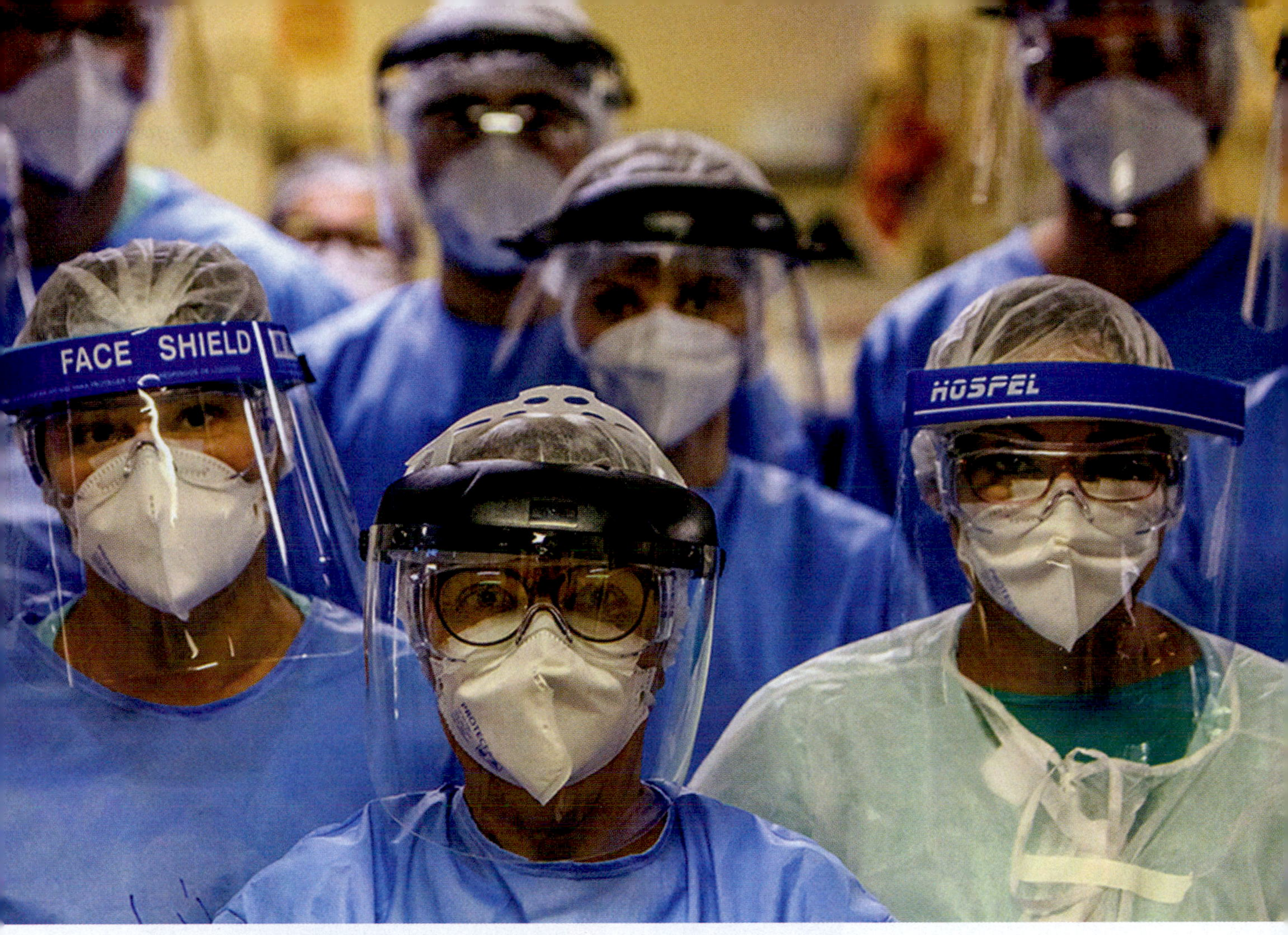

© Andy Cross/The Denver Post via Getty Images

16 HEALTH AND MEDICINE

WHAT DO YOU THINK?

1. How might your health be influenced by the society in which you live?
2. Why are the poor more likely than the middle class to be overweight or obese? What sociological factors might researchers look at to understand this correlation?
3. Why might HIV/AIDS be a concern of sociologists?

LEARNING OBJECTIVES

16.1 Explain sociological definitions of health and medicine.

16.2 Describe the sociological importance of the sick role, the social construction of illness, and stigma.

16.3 Recognize key issues in health care in the United States.

16.4 Identify four public health problems in the United States.

16.5 Discuss HIV/AIDS as a sociological concern.

16.6 Understand the relationship between globalization and public health.

Andrea Pattaro/AFP/Getty Images

THE (SOCIAL) POWER OF PANDEMICS

"It was a time when some people refused to wear masks and some who wore them failed to wear them effectively" (Little, 2020).

"A lot of cities did reopen too soon and had to close down a second time and, in some cases, a third time" (John Barry, from interview with Terry Gross on Fresh Air, *May 14, 2020).*

"Educators tried teaching from alternate classroom settings, including a frigid outdoor physics class at the University of Montana in January" (National Archives, 2020).

Can you identify the circumstances surrounding the quotes above? You might think that these arose during the 2020 COVID-19 pandemic, when the public grappled with masking, reopening, and schooling. A **pandemic** is an outbreak of disease that occurs over a wide area. COVID-19 was an infectious respiratory infection that traveled around the world, sickening millions and killing thousands. These quotes, however, actually address people's behavior over 100 years ago as they grappled with the 1918 Great Influenza. That virulent global pandemic infected 500 million people worldwide, killing over 50 million. During that pandemic, in the United States alone, over one fourth of all Americans were infected; 675,000 died; and the United States life expectancy in that year dropped by 12 years (National Archives, 2020)! Both the 1918 Great Influenza and COVID-19 are only several of a long history of global diseases that travel rapidly, creating both biological and social chaos.

Pandemic: An outbreak of disease that occurs over a wide area.

Diseases like influenza and COVID-19 are sociologically important because they occur in preexisting social contexts and social structures. As such, they bring with them social predictabilities—they reveal power and inequality at the same time that they change and challenge people's behaviors. They magnify social prejudices, revealing social fault lines and inadequacies, disproportionately impacting those without access to care. And some of them travel globally, changing cultures, practices, and societies in their wake.

The bubonic plague, known as the Black Death, is one such example. Arriving in the 14th century in Europe by way of Asia, it eliminated a third of the European population in only 20 years (Tuchman, 1978). It was, in short, a significant biological phenomenon. But it, too, had many social consequences. For instance, during the 14th century Black Death outbreak in Europe, according to Tuchman (1978), "the rich fled to their country places," "the urban poor died in their burrows," false information was widely spread and believed, and Jewish people were scapegoated as "well poisoners," resulting in purges and lynching of Jews.

People's behavior during COVID-19 fell into patterns that bore some similarity to both pandemics. In the United States in the early days of the pandemic, wealthy New Yorkers, alarmed by rising infection and deaths, fled to less-populated summer homes in the Hamptons and surrounding tourist areas, worrying the full-time residents of the areas who did not think their communities could medically handle them (Tully & Stowe, 2020).

China, where the virus first emerged, was scapegoated by politicians, most notably the president of the United States, Donald J. Trump, who called the virus the "China virus" repeatedly, most notably in his August 2020 presidential nomination speech (Elliot, 2020). On March 20, 2020, the House of Representatives of the U.S. Congress, noting the rise in hate crimes and anti-Asian sentiment due to the virus, introduced a resolution to condemn all forms of anti-Asian sentiment as it related to COVID-19 (https://www.congress.gov/bill/116th-congress/house-resolution/908/text).

And, disproportionately, subgroups of Americans, whose health access and quality of health were already stratified around race, class, and gender, suffered unequally during the pandemic. Especially hard hit were communities of color (Moore et al., 2020). Finally, struggles around appropriate public behavior were many, and educators at all levels grappled with how best to educate their students while maintaining everyone's health.

Given these social consequences, it might not surprise you, then, that sociologists are very interested in the interplay between health and illness and human behavior, social structure, and interaction. In this chapter, we focus on the role that social forces play in health and health care in the United States, and we address issues at the crossroads of medicine, health, public policy, and sociology. We begin by distinguishing *health* from *medicine*. We then look at the ways ideas about health and illness are socially constructed. We explore the relationship between social class, race, gender, and health care and outcomes in the United States, delving into the important issue of health care access and reform. Furthermore, we highlight sociological issues related to public health, including tobacco use, teenage pregnancy, obesity, and opioid addiction. We also discuss the sociology of HIV/AIDS, a global health challenge that continues to threaten lives, livelihoods, and entire communities and countries. We consider global issues of health and their sociological roots. We end with a consideration of what's next in health, medicine, and illness.

Research has shown that people of every age group benefit from regular exercise. In this photo, older women in South Africa participate in local soccer matches. Soccer is wildly popular in that country, and it is played by members of every age group.

©REUTERS/Siphiwe Sibeko

SOCIOLOGICAL DEFINITIONS OF HEALTH AND MEDICINE

Medical Sociology is one of the American Sociological Association's (ASA) largest interest sections with over 1,000 members (https://www.asanet.org/communities/sections/sites/medical-sociology). Sociological theories, concepts, and methods are increasingly incorporated in medical schools. Some of you may have noticed that the Medical College Admissions Test (MCAT) now contains a testing section on sociological knowledge.

Sociologists note that while the sociological studies of health and medicine are closely related, it is useful to distinguish between them. **Health** is *the extent to which a person experiences a state of mental, physical, and social well-being.* This definition, put forth by the World Health Organization (WHO, 2005), draws attention to the interplay of psychological, physiological, and sociological factors in a person's sense of well-being. It underscores that excellent health cannot be achieved if the body is disease-free but the mind is troubled or the social environment is harmful.

Health: The extent to which a person experiences a state of mental, physical, and social well-being.

Medicine: An institutionalized system for the scientific diagnosis, treatment, and prevention of illness.

Preventive medicine: Medicine that emphasizes a healthy lifestyle that will prevent poor health before it occurs.

Medicine is *an institutionalized system for the scientific diagnosis, treatment, and prevention of illness.* As such, it identifies and treats physiological and psychological conditions that prevent a person from achieving a state of normal health. In this effort, medicine typically applies scientific knowledge derived from physical sciences such as chemistry, biology, physics, psychology, and sociology. In the United States, we usually view medicine in terms of the failure of health: When people become ill, they seek medical advice to address the problem. Yet, as the definition suggests, medicine and health can go hand in hand.

The field of **preventive medicine**—*medicine that emphasizes a healthy lifestyle that will prevent poor health before it occurs*—is of key interest to health professionals, patients, and policy makers. Many national groups and agencies exist to promote the public's health, conduct research on the topic, and educate the public so that they can engage in proven preventative measures. For instance, the Centers for Disease Control's (CDC) National Center for Chronic Disease Prevention and Health Promotion (NCCDPHP) shows that while six in ten adult Americans have a chronic illness, four individual lifestyle behaviors (tobacco use, alcohol consumption, nutrition, and exercise) can make a significant impact in improving people's health, life expectancy, and costs of their care (https://www.cdc.gov/chronicdisease/resources/infographic/chronic-diseases.htm). The American College of Preventive Medicine (ACPM) links national and international organizations, communities, and providers to address public health problems that can be prevented or managed such as diabetes, hypertension, brain health, and violence (https://www.acpm.org/initiatives). Finally, the federal Office of Disease Prevention and Health Promotion (ODPHP) identifies five sociological *social determinants* of health that they define as "conditions in the environments in which people are born, live, learn, work, play, worship, and age that affect a wide range of health, functioning, and quality-of-life outcomes and risks." They identify the five as:

- economic stability
- education
- social and community context
- health and health care
- neighborhood and built environment. (Healthy People, Office of Disease Prevention and Health Promotion, 2020)

SOCIOLOGICAL APPROACHES TO ILLNESS

The Sick Role

Have you ever noticed that we expect those who are "sick" to behave in certain ways while we expect those who are not sick to respond to them appropriately? Sociologist Talcott Parsons (1975) certainly noticed that we have different expectations of those who are sick. He deemed this phenomenon the **sick role**, describing being ill as an actual social status marked by particular characteristics and expectations.

Sick role: A social role rooted in cultural definitions of the appropriate behavior of and response to people labeled as sick.

Parsons, a functionalist, believed that society operated through a series of agreed-upon roles. He thought the "sick role" allowed society to remain functional even when some of its members were ill and unable to fulfill their "normal" roles. This is partly because a "sick role" let sick people adopt an alternative role with behaviors that were still socially recognized, accepted, and allowed for continued role performances in a different capacity. In the United States, the "sick role" includes the right to be excused from social responsibilities. For instance, if you are ill you might be excused from classes (being a student), from work (being an employee), and from activities in which you normally engage (being a friend, for example). Furthermore, even if an illness is the result of a risky lifestyle, society does not usually hold the person who is ill accountable. On the other hand, the sick person is socially obligated to try to get well and to seek competent medical help in order to do so. Failure to seek help can lead others to refuse to confer on the suffering individual the benefits of the sick role. Can you think of examples where people have been denied the ability to assume the "sick role"?

The notion that a sick person is enacting a social role may also remind you of Erving Goffman's (1959) dramaturgical theory that we explored in Chapter 4. Recall that Goffman suggested that life is like a dramatic play where humans are actors on a social stage complete with front and back stages, scripts for certain settings, costumes, and props. In order to define situations in ways that are favorable to ourselves, he argued, we all play roles on the "front stage" that conform to what is expected and that will show us in the best light and contribute to a smooth social interaction.

Think about the last time you were at the doctor's office. How did you play the sick role? How did your doctor respond? To answer those questions, you might integrate Parson's ideas of the sick role with Erving Goffman's ideas of dramaturgy (1963). For example, Goffman would encourage you to conceptualize your doctor's office as a stage: Your doctor wears a "costume" (often a white lab coat and stethoscope) while you (playing the role of patient) are also asked to don a "costume" (a cloth or paper gown rather than street clothing). Parsons would say that your doctor's role is to greet you, ask questions about your illness, examine you, and offer advice. You, as a patient, are expected to play a more passive role, submitting to an examination, accepting the diagnosis, and taking advice rather than dispensing it. But you might also imagine what would happen in the event of some breakdown in the typical medical ritual: For instance, what if your doctor arrived dressed in evening attire, or if you gave your doctor medical counsel or refused to lie on the examining table, choosing instead to sit in the doctor's chair? The result would be failed expectations about the encounter, as well as an unsuccessful social and medical interaction.

Sociologist Talcott Parsons introduced the concept of the *sick role*, which offers sociologists the opportunity to think about the condition of being ill as a social status with particular characteristics and expectations.

©Tetra Images via Alamy Stock Photo, 2006

As Parsons pointed out, the sick person has an expected role, as do doctors, nurses, and others who are part of the "sick play." Both Parsons and Goffman, then, help us to understand that there are some important social norms and structures that underlie illness and interactions, and we are socialized to follow them. Importantly, though, cultural, class, and educational differences can have a significant impact on such interactions. Indeed, nursing and medical schools are increasingly aware of the need to have their students be "culturally competent"—that is, to understand differences in cultural norms and practices that accompany interactions and to respond appropriately in

Telemedicine, medical care delivered virtually, can allow people 24-hour access to medical assistance from their homes. It may also change the interactional dramaturgy of the traditional doctor's office.

©Jessica Rinaldi/The Boston Globe via Getty Images

this context. Describing the importance of culturally competent care for health practitioners, Lotz (2020) lists several useful scenarios that describe some cultural differences between groups that may impact health care interactions:

> [A] Chinese patient may be showing respect to a physician by avoiding eye contact. In American culture, a lack of eye contact can be considered rude or an indicator of depression. In both Muslim and Navajo cultures, eye contact has other distinct meanings. Understanding the nuances between cultures and how language has different meanings for different people is crucial when treating patients of backgrounds different from one's own. (para. 2)

The examples in this section relied heavily on face-to-face interaction. But, due to the dangers of face-to-face contact during the 2020 COVID-19 outbreak, many people gained access to virtual medical care known as *telemedicine* or *telehealth.* What, if any, impact do you think the technology of telemedicine has had on the sick role and the interactional dramaturgy of the doctor's office?

The Social Construction of Illness, Labeling, and Stigma

Talcott Parsons's "sick role" rests on the idea that societal members share definitions of what constitutes a "legitimate" illness. That is to say, illnesses that are culturally defined as legitimate, such as cancer and heart disease, entitle those so diagnosed to adopt the role of a sick person. The afflicted are forgiven for missing time at work, spending days in bed, and asking others for consideration and assistance. A seriously ill person who persists in leading a normal life is given credit for an extraordinary exertion of effort.

Importantly, as Berger and Luckman (1966) point out, reality is *socially constructed*—we daily create our definitions of reality in interaction, and these come to take on structure that guides our individual and societal behaviors. Understanding the concept of social construction is important when we consider the sociology of health and medicine. It allows us to grasp that the cultural definitions of sickness, health, and appropriate behavior that seem normal in our time or society may, indeed, vary widely across time, space, and place, and even between groups (Cockerham & Glasser, 2000; Foucault, 1988; Parsons, 1951, 1975; Sagan, 1987). Whose version of reality holds sway, however, is often associated with social power and can carry all of the blind spots and prejudices of the creator.

Sociologists point out that cultural definitions of mental illness and how it is viewed and treated are socially constructed and change over time and place. Here, a woman dresses her son, who is mentally ill, on his bed in a secure ward at the Institute of Human Behaviour and Allied Sciences, Delhi, India. Parents and spouses are encouraged to come and stay in the ward to offer help and stability. How does this differ from how mental illness is treated in the United States?

©In Pictures Ltd./Corbis via Getty Images

Take the case of mental illness. Explanations for, treatments of, and acceptance around mental illness clearly show that society plays a very solid role in creating circumstances that have punished, ostracized, or even rewarded individuals labeled as mentally ill. In different eras and societies people with mental illness have been labeled in a number of ways; for instance, some were seen through a religious lens to possess unusual spiritual qualities or, alternatively, to be demonically possessed. In modern societies, mental illness has been associated with a wide range of causes—having physiological origins, being a product of biological inferiority, indicative of a personal character or moral weakness, or inherent criminality (Foucault, 1988). As a result, treatments have widely varied.

The identification of what symptoms constitute a mental illness, *who* is labeled as mentally ill, who does the labeling, and the consequences of that labeling have real individual and social consequences. Erving Goffman addressed these consequences in two of his famous works: *Asylums* (1961) and *Stigma* (1963b). In *Asylums*, Goffman explored the confinement to mental hospitals (asylums) of those labeled mentally ill. He saw these places as total institutions, a concept we explored in Chapter 4. Recall that in total institutions our self is stripped away and remade. Goffman's work illustrated a number of those ways such as loss of possessions, loss of autonomy and privacy, and forced social relationships because of shared space. In *Stigma*, Goffman (1963b) showed that diseases and disfigurations can leave a social mark—a **stigma**—or *an attribute that is deeply discrediting to an individual or group because it overshadows other attributes.* Goffman identified three types of stigma:

Stigma: An attribute that is deeply discrediting to an individual or group because it overshadows other attributes.

- physical (those associated with bodies and "disfigurement");
- social (such as race or religion);
- moral ("blemishes of individual character" associated with mental illness or incarceration).

To Goffman, the stigmatized individual developed a "spoiled identity" or a loss of self and social control defined by the social identification and subsequent labeling. People who are stigmatized, then, can be deprived of individual agency, stereotyped, and discriminated against. Do you think this labeling and stigmatization is still experienced by individuals with mental illness?

Another example of the social construction of illness and the ensuing societal labeling can be seen in changes in U.S. society's response to alcoholism. In the middle of the 20th century, people addicted to alcohol were widely seen as weak and of questionable character. In 1956, however, the American Medical Association (2013) declared alcoholism an illness that required medical intervention. This medical model of alcoholism was widely accepted. We now view alcoholics as ill. This results in their ability to adopt the sick role and interact differently in family, work, and other settings. For instance, family members may offer sympathy, workplaces may offer leave, and federal programs fund research and programs to increase awareness and fight the disease of alcoholism. Importantly, a disease model of drug addiction also exists (Le Moal & Koob, 2007), but someone addicted to illegal drugs is more likely than an alcoholic to be denied the sick role. Those addicted to cocaine, heroin, and methamphetamine, for example, face the possibility of being sent to prison if they are found in possession of the drugs, and they may or may not be referred for treatment of their addiction.

Conflict theorists point out that the social construction of reality is also influenced by power relations. In the case of drug addiction, what is understood as *illness* and what is labeled as *deviance* not only transforms the status of the individual, it also leads to social stigma. It can also lead to criminal charges, loss of many social rights, and incarceration.

In 23 states and the District of Columbia, substance use during pregnancy is considered to be child abuse; another three states categorize it as grounds for civil commitment. Of particular note is the South Carolina Supreme Court's holding that a viable fetus is a "person" under the state's criminal child-endangerment statute: Thus, "maternal acts endangering or likely to endanger the life, comfort, or health of a viable fetus" are criminal acts of child abuse (quoted in Guttmacher Institute, 2020). This is significant because, according to a recent study on the issue of criminal charges for child harm in pregnancy, "the judicial decision depended on the disposition of the question of whether . . . a fetus is a child. The balance in the courts in favor of treating substance use during pregnancy depends on the definition of a child for the purposes of criminal statutes" (Angelotta & Appelbaum, 2017, p. 193).

While health concerns about the welfare of infants born to addicted mothers are important, it is also important to understand that criminalizing pregnant mothers keeps them from seeking treatment because they fear their child will be taken away and the criminal sanctions that will ensure. These policies are more likely to impact women of color and women in lower socioeconomic brackets who are unable to afford treatment in the first place or may fear the risk of surveillance and law enforcement if they reveal their addiction (Stone, 2015).

What examples come to mind of other conditions that have been historically defined (or not defined) as illness? What illnesses can you think of that have been stigmatized? How might their stigmatization show evidence of power in labeling?

SOCIOLOGICAL APPROACHES TO HEALTH CARE IN THE UNITED STATES

Health care: All those activities intended to sustain, promote, and enhance health.

Health care can be sociologically defined broadly as *all those activities intended to sustain, promote, and enhance health.* An adequate health care system includes more than the provision of medical services for all of those who need them—it also encompasses policies and organizations that minimize violence and the chance of accidents, whether on the highways, at work, or at home; policies that promote a clean, nontoxic environment; ecological protection; and the availability of clean water, fresh air, and sanitary living conditions. To those ends, in this section we explore issues related to the sociological definition of health care: health and public safety, health care access, and the causes and consequences of social inequality in health and medicine.

Health and Public Safety Issues

By the standards noted above, few societies come close to providing excellent health care for their citizens. Some, however, do much better than others. The record of the United States in this regard is mixed.

On the one hand, the U.S. government spends vast sums of money in its efforts to construct safe highways, provide clean drinking water, and eliminate or reduce air and ground pollution. Laws regulate working conditions with the aim of promoting healthy and safe workplace environments: The federal governmental agency OSHA (the Occupational Safety and Health Administration), an arm of the United States Department of Labor, is meant to guard worker health by enforcing regulations and maintaining public safety in a number of sectors including health care. OSHA responds in many ways, from providing basic workplace safety information and guidance to actively monitoring workplace injuries and fatalities. It also fields emerging issues. For instance, beginning on February 18, 2020, OSHA posted a list of daily and cumulative totals of whistleblower complaints of COVID-19–related workplace safety violations (https://www.whistleblowers.gov/covid-19-data#complaints_filed).

Many other state and federal agencies also provide for the public health. For instance, the U.S. Food and Drug Administration (FDA) is a federal organization that protects public health by researching, inspecting, and regulating medications, foods, devices, cosmetics, vaccines, devices that emit radiation (e.g., microwaves), products such as tobacco, and many other items.

FIGURE 16.1 ■ Rate of Violent Gun Deaths per 100,000 Population for Selected Countries, 2016

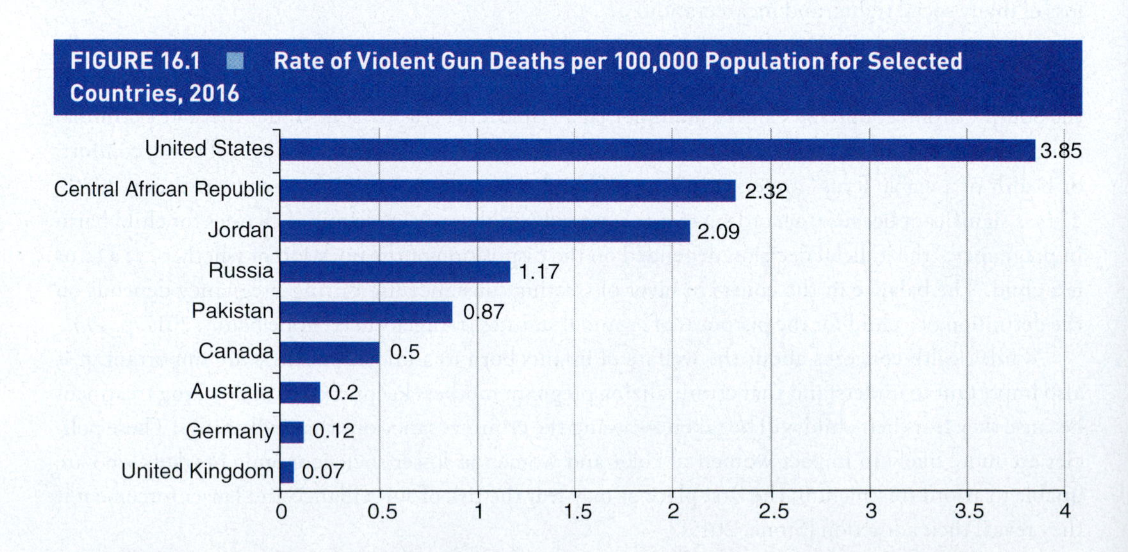

Source: Data from Institute for Health Metrics and Evaluation (IHME). (2018). GBD Compare. Seattle, WA: IHME, University of Washington.

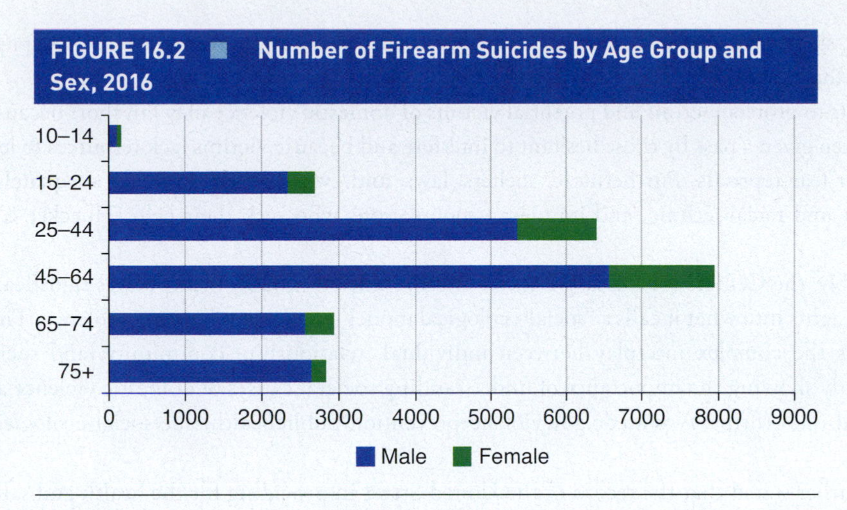

FIGURE 16.2 ■ Number of Firearm Suicides by Age Group and Sex, 2016

Source: CDC. *Suicide Rates in the United States Continue to Increase.* Retrieved from https://www.cdc.gov/nchs/products/databriefs/db309.htm

At the local level, society provides for public health by sending health inspectors to visit the premises of restaurants and grocery stores to check that food is handled in a sanitary manner, and agricultural inspectors check the quality of U.S. and imported food products. States require drivers to use seat belts, motorcyclists to wear helmets, and children to be strapped into car seats, all of which have been shown to reduce injuries and fatalities in road accidents. While these efforts do not guarantee the safety of life, work, food, or transport, they make an important cumulative contribution to public safety.

On the other hand, the safety of Americans is compromised on a number of dimensions. The deregulated U.S. gun industry results in a high level of gun ownership, gun accidents, and gun violence (Figure 16.1). The United States ranks behind 28 countries worldwide in the number of deaths caused by gun violence (Institute of Health Metrics, 2021). The incidence of homicide due to gun violence is 25 times higher in the United States than in 22 other developed countries (Aizenman, 2018; Grinshteyn & Hemenway, 2016). In 2018 alone, the United States experienced 39,740 firearm-related deaths, or, as the CDC puts it, "109 people dying from a firearm-related injury each day" (Centers for Disease Control and Prevention, 2020c).

Gun violence disproportionately impacts young African American males: Homicide is a leading cause of death among young African American males, and the majority of this violence is perpetrated using guns (Langley & Sugarmann, 2020). In fact, Black men are 13 times more likely to be killed by gun violence than are white men ("Gun Violence by the Numbers," n.d.; Kaiser Family Foundation, 2006; Violence Policy Center, 2010). Significantly as well, of the over 40,000 Americans who die by suicide each year, just over half are killed by a firearm (National Institute of Mental Health, 2018) (Figure 16.2).

Domestic violence is yet another significant public health concern in the United States. This is despite multiple federal, state, and local measures including the Violence Against Women Act (VAWA), shelters, programming, and public awareness campaigns. The Centers for Disease Control and Prevention (2020j) estimates that "1 in 4 women and nearly 1 in 10 men have experienced contact sexual violence, physical violence, and/or stalking by an intimate partner during their lifetime." Women are disproportionately impacted. At least 85% of victims of domestic violence are women, and an average of three women are murdered by a husband or boyfriend every day in the United States. Annually, about 1,500 people—disproportionately women—die from domestic violence (Huecker & Smock, 2020). In 2010, 38% of all female murder victims in the United States were killed by a husband or boyfriend (National Center for Victims of Crime, 2012).

Intimate partner violence also starts young, affecting, according to the CDC, "millions of teens" each year (Centers for Disease Control and Prevention, 2020j). Additionally, 9% of high school students report purposeful physical abuse by a partner within the past 12 months (National Center for Injury Prevention and Control, 2014). Children are also commonly victims of domestic violence. About 1 in 10 American children witness violence at home or experience it themselves (Huecker & Smock, 2020). Additionally, LGBTQ+ populations are vulnerable to domestic violence and sexual assault for many reasons, such as social stigma in reporting; fear of "outing" oneself or one's partner; and lack of services, providers, shelters, law enforcement, and legal assistance that fully incorporate LGBTQ+ populations (NCADV.org, 2018).

Efforts to protect victims and potential victims of domestic violence may fall short because batterers are often given a pass by those hesitant to interfere and because victims lack resources to leave their abusers or fear reprisals. Furthermore, shelters, laws, and even hotlines may not adequately protect LGBTQ+ and racial, ethnic, and immigrant populations who seek their help (Huecker & Smock, 2020).

Notably, the Centers for Disease Control and Prevention (2020l) incorporates significant sociological insights into what it calls a "social-ecological model" to prevent domestic violence. This model "considers the complex interplay between individual, relationship, community, and societal factors" clearly showing the importance of understanding societal causes of domestic violence as well as individual ones (https://www.cdc.gov/violenceprevention/publichealthissue/social-ecologicalmodel.html).

We earlier stated that the record of the United States in providing for the health and safety of its citizens is mixed. How did citizens feel about health and safety during the COVID-19 pandemic, and how did that compare with opinions of citizens from other countries? In August 2020, Pew Research conducted an international poll of 14 countries across Europe, North America, and Asia to ascertain how citizens felt that their countries were handling the virus. All told, Denmark's citizens were most positive, with 95% of its citizens agreeing that their country had done a good job handling the virus. Eighty-eight percent of Canadians also felt that their country had done a good job. On the other hand, the United States and the United Kingdom ranked lowest of all countries in citizen approval. In the United States, under half (47%) of Americans thought their country was doing a good job, while the majority (52%) thought their country was doing a bad job. In the UK, 46% of their citizens thought their country was doing a good job, while the majority (54%) thought it was doing a bad job (Devlin & Connaughton, 2020).

As we have illustrated in this section, Americans are annually impacted by a number of safety concerns that influence public health and require coordinated responses at local, state, and federal levels. We have also shown some patterns of differences by race, gender, and sexuality. What other important variables do you think influence violence and safety?

One important reason why the poor—as well as some families in the working and middle classes—in the United States are less likely to experience good health is that a notable proportion are unable to access regular care for prevention and treatment of disease. Here, a woman is able to engage in preventative care by doing yoga and checking her blood sugar. Both activities require patient education and access.

vgajic/getty images

Access to Health Care

Can you afford to go to the doctor if you are ill? If not, do you wait it out, see what happens, and hope for the best? These are, and continue to be, difficult issues for a significant number of Americans. Their ability to seek medical care has changed, to some extent, over the past 15 years, with changes in national and state policies and economic up- and downturns.

Some context is important to understand current access to health care in the United States. In the fall of 2010, 3 years after the start of the Great Recession and shortly before the Patient Protection and Affordable Care Act (which we discuss later) was signed into law, the U.S. Census Bureau reported that more than 16% of people in the United States were without health insurance, the highest figure in 23 years (Kaiser Family Foundation, 2010). Key sources of this decline were the economic crisis and the associated rise in unemployment; most U.S. adults get health insurance coverage from their employers. Workplace coverage is variable, however, and ranges from full benefits requiring little or no financial contribution from the employee to partial benefits paid for through shared employer and employee contributions. Cost-saving measures in U.S. workplaces in recent decades have shifted a greater share of the cost of these benefits from employers to employees.

As the economic picture improved in the years after 2010, many people went back to work, but millions of employees were still uninsured or underinsured. In fact, in 2017, the American Community Survey showed that 28 million Americans were uninsured, with working males with low-income jobs between the ages of 19 and 64 who had less than a high school education the most likely to lack coverage (Berchick, 2018, para. 10). Lack of insurance had also been worsened by a changing labor market and economic structure that favored part-time or contractual "gig" employment. These jobs had fewer benefits, such as the employer-based health insurance coverage that has traditionally applied to full-time employees.

In 2020, a substantial number of Americans accessed health care through government-funded programs such as Medicare and Medicaid. Medicare was created in 1965 to serve as a federal health insurance program for people age 65 and older and those with permanent disabilities, regardless of income or medical history (Kaiser Family Foundation, 2014). In 2016, it covered about 47.8 million over age 65 and about 9 million with permanent disabilities (NCPSSM, 2018).

Medicaid, on the other hand, is a shared federal and state health insurance program designed to assist low-income people of all ages with health care. It is not, however, available to everyone who needs long-term services; to be eligible for Medicaid, individuals must meet stringent financial qualifications (Kaiser Family Foundation, 2012). It reached an enrollment of 72 million in July 2019 (Kaiser Family Foundation State Health Facts, 2019).

A contemporary issue related to Medicare is the fact that members of the post–World War II baby-boomer generation (those born between about 1946 and 1964) have entered the 65+-year-old cohort. The massive numbers of baby boomers have a significant effect on the nation's need for health care dollars and resources. The U.S. Census Bureau reports that between 2000 and 2010, the 65+-age cohort grew at a faster rate than did the total population; the total population of the United States increased by less than 10%, while the population of those age 65 and older grew by more than 15% (Werner, 2011). "Boomers" in 2020 numbered an estimated 73 million. By 2030, all boomers will be over 65 (America Counts Staff, 2019). The increase in eligible Medicare recipients, medical advancements that extend the lives of the elderly, and a relatively smaller tax base are the ingredients of a debate over care and government spending that will grow more acute in the years to come (Antos, 2011).

At the opposite end of the age spectrum, the Children's Health Insurance Program (CHIP) was created in the late 1990s in an effort to cover more uninsured children. Because individual states administer CHIP in partnership with the federal government, state governments largely dictate its implementation, so the comprehensiveness of coverage and eligibility standards varies from state to state. At the start of 2018, however, there were about 9.6 million children enrolled in CHIP programs (Medicaid.gov, 2020).

While the care that the poorest U.S. adults can access through Medicaid is limited, it is often the working poor and other low-income employees who are shut out of insurance coverage altogether. They are most likely to be working in economic sectors such as the service industry (fast food restaurants, retail establishments, and the like) that provide few or no insurance benefits to employees—employees earn too little to afford self-coverage but too much to qualify for government health coverage. The fact, as noted earlier, that low-income people are more likely to have health problems also affected their ability to get insurance coverage before the passage of the 2010 Patient Protection and Affordable Care Act (known simply as the Affordable Care Act, or ACA, and also referred to as "Obamacare"), because insurers were allowed to exclude those with preexisting conditions such as diabetes, high blood pressure, and other illnesses and disabilities.

The Patient Protection and Affordable Care Act, signed into law by President Barack Obama in 2010, was created to expand insurance coverage to more people in the United States at a time when the numbers of the uninsured had been rising. The goal of this massive health insurance overhaul was to insure more people and make coverage more broadly accessible and affordable, in part by requiring that everyone buy insurance and that private insurance companies offer coverage under new terms that extend benefits to those who may have had difficulty purchasing insurance in the past, such as those with preexisting conditions. Among the ACA provisions was also the requirement that insurance companies permit young people up to age 26 to remain on their parents' health insurance policies if they lacked other coverage. Recent changes to the ACA by the Republican-dominated Congress have done away with some provisions of the law, including the individual mandate. While the ACA remains in force at the time of this writing, the effects of changes made in the last year are not yet clear. A number of Republican-led states refused to expand Medicaid coverage with state dollars; by August 2020 a remaining 12 states had not expanded coverage (Roubein, 2020).

Since its passage, the ACA has been the source of heated political debate. President Obama and other supporters of the act argued that the law has expanded insurance coverage to a broader swath of people, many of whom had been locked out of the insurance market due to preexisting conditions or unaffordability of individual insurance policies. They suggest that the law supported this expansion of coverage through the operation of new state-level insurance markets (or exchanges) that keep prices down by enabling purchasers to buy insurance as part of a group. Those with low incomes became eligible for federal subsidies to support their insurance purchases. Supporters also noted that greater coverage meant that more people could seek primary and preventive care, which helps to keep patients out of emergency rooms and hospitals, where care is far more costly.

Opponents have argued that the U.S. government overstepped its limits in requiring that people purchase health insurance or pay a penalty tax for failing to do so; thus, a Republican-dominated Congress, in 2017, eliminated the individual mandate. The effect of this change, however, may be to increase the cost of insurance, as those who are sicker opt to stay insured, while those who are in good health leave the marketplace. There have also been attempts to portray the ACA as a path to socialized medicine, though most people will receive their insurance through private insurance companies rather than through the government.

Both supporters and opponents of the ACA have expressed concerns about the costs of health care in the United States. Indeed, the United States spends three times more per capita on health care ($11,072 in 2019) than the average spending ($3,851) of eight other developed countries (Figure 16.3).

At the same time, many of the United States' health indicators compare poorly to those of its peers. Opponents of health care reform have argued that the ACA drives up costs by requiring insurers to cover those who have costly health conditions. Supporters of the law point out that having a large pool of uninsured contributes to higher costs when they fail to get preventive care and must resort to far more expensive emergency room care or hospitalization. Certainly, an aging U.S. population will likely need more, not fewer, health care services in the future. How the U.S. government and states will address these needs in the future remains to be seen.

Regardless of political struggles, the Kaiser Family Foundation indicates clear improvements in health coverage, access, and use for all racial and ethnic groups since it has been in place (Artiga & Orgera, 2019). Notably, however, health disparities remain. Post-ACA, in overall health, whites fare

FIGURE 16.3 ■ Per Capita Health Care Spending for Selected Developed Countries, 2019

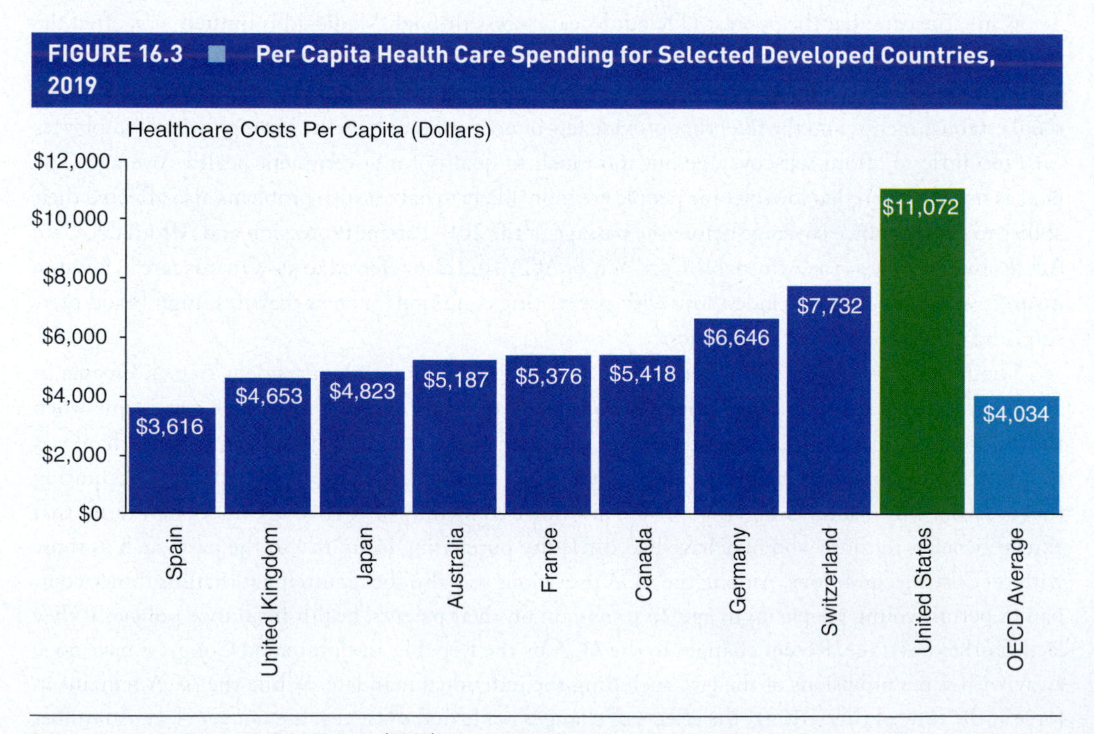

Source: Peter G. Peterson Foundation. (2020). *Per Capita Healthcare Costs International Comparison.* Retrieved from https://www.pgpf.org/chart-archive/0006_health-care-oecd

Note: Data are for 2019 or latest available date. Chart uses purchasing power parities to convert data into U.S. dollars.

better than nonelderly Blacks, Hispanics, and Native Americans, with particularly marked disparities in "teen birth rates, infant mortality rates, and HIV or AIDS diagnosis and death rates" (Artiga & Orgera, 2019). It is to these disparities that we now turn.

Intersecting Social Inequalities in Health and Medicine

As the preceding section showed, health—even with access to insurance—can differ by social class and race-ethnicity. So, too, gender and sexuality as well as other social categories are separately and intersectionally interconnected with health. In this section, we further explore health inequalities surrounding social class, race and ethnicity, and gender and sexuality.

Social class figures prominently as a barrier or a connector to access to health care, medicine, and a pathway to well-being. It can also help people set a foundation for a healthy life from conception through old age. This has been recognized by the U.S. Department of Health and Human Services. Its Healthy People 2020 campaign lists "addressing class inequality" as one of its overarching goals to improving health outcomes. Let's examine some of the links between social class and health.

Social class helps provide access to some basic building blocks of health: food, exercise, and preventive medical care. Among children, parental poverty affects their birthweight, food security, housing stability, and maltreatment, with the former two factors playing a significant role in the likelihood of developing chronic illnesses and other negative outcomes such as malnutrition, stunted growth, and suppressed immunity (Henry, 2010).

Poor children and adults may have less-healthy diets than do their higher-income counterparts: Inexpensive foods may be highly processed, fatty, and high in sugar. Fresh fruits and vegetables and lean meats may be out of financial reach for those who struggle to make ends meet. Full-time work and limited grocery stores in neighborhoods can limit opportunities to shop for and prepare healthy foods. Children in poor communities may also lack access to safe places for active outdoor play and exercise, contributing to higher rates of obesity.

Social class also intersects dramatically with race, ethnicity, and gender. Inequalities start even before birth, since poor mothers are less likely to have access to prenatal care. But the intersection between race and gender is also very important. An Annie E. Casey Foundation report suggests that "at any age, and at any income, education or socioeconomic level, an African American mother is more than twice as likely to lose her infant as a white woman" (Shore & Shore, 2009, p. 6). Indeed, the CDC's 2017 data show that Black mothers and American Indian/Alaska Native mothers (11.4 and 9.4 deaths per 1,000 live births, respectively) experience the highest rates (see Table 16.1; Centers for Disease Control and Prevention, 2019) while Asian and non-Hispanic white mothers experience the lowest rates of infant mortality (3.6 and 4.9 deaths per 1,000 live births, respectively).

Black mothers are also significantly more likely than mothers of all other racial/ethnic groups to give birth to infants with low birth weight (13.68%) or very low birth weight (3%): By comparison, only 7% of infants of white or Hispanic mothers have low birth weight and 1% are of very low birth weight (Martin et al., 2019). Low birth weight is a risk factor for short-term and long-term health issues early on such as jaundice, infections, respiratory and blood problems, and later such chronic illnesses as heart disease, diabetes, and high blood pressure (https://www.marchofdimes.org/complications/low-birthweight.aspx).

Why do racial differences in mortality and birth weight exist? Researchers point to factors such as income, education, psychosocial stress caused from discrimination, and, as we address below, environmental racism (Burris & Hacker, 2017), with some pointing to a combination of factors creating epigenetic modifications that result in unequal health outcomes (Vick & Burris, 2017).

Inequalities in health by race and ethnicity are compounded in childhood by disproportionate poverty rates in families of color. According to a report by the Annie E. Casey Foundation (2019), childhood poverty is a significant issue for African American and Native American families, with just over a third (33%) of children in both groups experiencing poverty. Conversely, about a quarter of Latinx children (26%), and just over 10% (11%) of white and Asian American/Pacific Islander children experience poverty.

TABLE 16.1 ■ Infant Mortality Rates by Race, 2018

Racial/Ethnic Group	Rate per 1,000 Live Births
Non-Hispanic Black	10.8
Native Hawaiian or Other Pacific Islander	9.4
American Indian/Alaska Native	8.2
Hispanic	4.9
Non-Hispanic White	4.6
Asian	3.6

Source: Centers for Disease Control and Prevention. (2019). *Infant Mortality.* Centers for Disease Control Division of Reproductive Health, National Center for Chronic Disease Prevention and Health Promotion. Retrieved from https://www.cdc.gov/reproductivehealth/MaternalInfantHealth/InfantMortality.htm

Recessions and economic slumps also hurt families, straining their ability to afford quality food, housing, and health care. Low-income Americans have a greater probability of exposure to violence and the mental and physical health problems that entails. Their work is also more likely to involve physical and health risks than is the work of middle- and upper-class people (Commission to Build a Healthier America, 2009). Lower-income people are more likely to live in areas that have high levels of air pollution, which raises their risks of asthma, heart disease, and cancer (Calderón-Garcidueñas & Torres-Jardón, 2012). The poor have a considerably higher risk of exposure to dangerous levels of lead from paint in older homes or aging public infrastructure, including lead pipes (Weitz, 2017).

Modern industry has generated wealth, convenience, and comfort. It has also wrought consequences that include waste and pollution. These pollutants exist in our air, water, and soil, many the by-products of factory waste or unsafe chemicals. They pose significant health issues and safety risks. The case of environmental pollution and its health consequences shows the intersection of race and social class.

The important fact that American neighborhoods and even whole cities remain segregated by race and social class (Massey, 2016) means that some populations are more vulnerable than others to these pollutants being located in their neighborhoods and impacting their health. This has led sociologists, most notably sociologist Robert Bullard (known as the father of environmental justice), to enlist the term **environmental racism**, or *policies and practices that impact the environment of communities where the population is disproportionately composed of ethnic and racial minorities.* For instance, factories are often situated in cities and communities comprised of poor and minority populations who may welcome them for the jobs they provide and may not have the political power or wealth to challenge them (Weitz, 2017).

Environmental racism: Policies and practices that impact the environment of communities where the population is disproportionately composed of ethnic and racial minorities.

Environmental pollutions such as particulate matter (a combination of liquid and solid particles in the air) are more likely to burden areas that are disproportionately inhabited by Black residents. The Environmental Protection Agency (EPA), in a study that examined disparities in the location of particulate matter–emitting facilities and the demographics of the residential population, found that at all levels, these facilities were most likely to be located by Black residents (Ihab et al., 2018).

The case of polluted water in Flint, Michigan, is illustrative of the concept of environmental racism, bringing together the confluence of region, pollution, and race. Flint, a formerly prosperous industrial city heavily reliant on its General Motors plant, experienced economic collapse in the wake of mass deindustrialization and plant closings in the 1970s. A substantial proportion of Flint's 99,800 residents today lack well-paying jobs, are African American, and live below the poverty line. It was, at least in part, the city's poor financial condition that led to the decision to switch away from the clean water supply coming from Detroit to the Flint River, which was less costly to access (Kennedy, 2016).

As a result, Flint residents experienced a number of illnesses and physical symptoms, including hair loss and skin rashes. They also experienced lead poisoning, a condition that can lead to impaired cognition, behavioral disorders, and delayed puberty, among other health consequences (CNN, 2016).

In some homes, lead levels showed as high as 13,200 parts per billion; by comparison, water contaminated with 5,000 parts per billion is classified by the EPA as hazardous waste (Kennedy, 2016). According to a later study by Virginia Tech researchers, Flint's "river water was found to be 19 times more corrosive than water from Detroit" (CNN, 2016). *The New York Times* writes that "residents and advocates have expressed outrage over the government's failure to protect Flint's children, something many of them say would not have happened if the city were largely white" (Goodnough, 2016). By 2019, 15 city officials had been indicted but no one had gone to jail. New businesses had begun to invest in Flint partly due to the investment made to fix Flint's water system. Still pending were the results of studies to understand the impact that the polluted water may have had on the cognitive development of thousands of Flint's children (Carmody, 2019).

FIGURE 16.4 ■ Life Expectancy in the United States, 2017

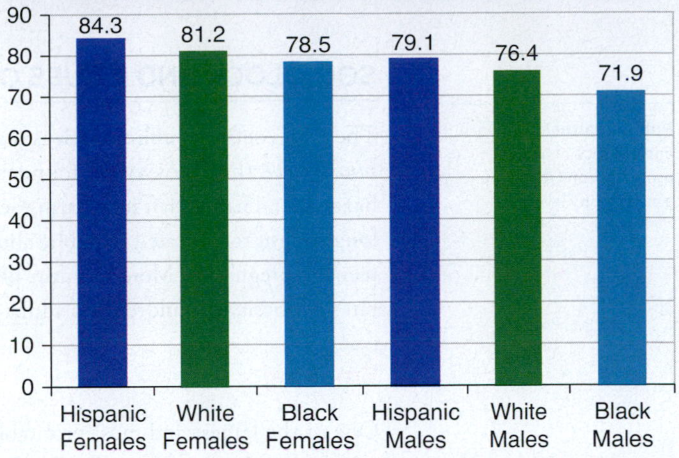

Hispanic Females	White Females	Black Females	Hispanic Males	White Males	Black Males
84.3	81.2	78.5	79.1	76.4	71.9

Source: Arias, E., & Jiaquan, X. (2019, June 24). *United States Life Expectancy 2017.* National Center for Health Statistics, *United States Life Tables, 2017. National Vital Statistics Reports 68*(7). Retrieved from https://www.cdc.gov/nchs/data/nvsr/nvsr68/nvsr68_07-508.pdf

Finally, life expectancy in the United States also reveals intersecting racial and gender patterns (Figure 16.4). In 2017, the overall life expectancy at birth in the United States was 78.6 years (Arias & Jiaquan, 2019), a number that has declined in recent years. Hispanic women's life expectancy was highest at 84.3, while white women's life expectancy was second highest at 81.2. Black women have a life expectancy of 78.5 years. Hispanic males had the highest life expectancy at 79.1 years, while white men had a life expectancy of 76.4 years. Black men had the lowest life expectancy of all groups. At 71.9 years, their life expectancy was nearly 13 years lower than Hispanic women and nearly 5 years lower than white men (Arias & Jiaquan, 2019).

You might notice that Hispanic women and men have longer average life expectancy than do their white and Black peers. Given that Hispanic families and communities are more likely to be poor than are white ones, and that Hispanics also suffer negative effects from prejudice and discrimination, how can this be explained? Among the hypotheses advanced to explain it is the healthy migrant effect, which suggests that immigrants are less likely to be drawn from the fraction of a population that is in ill health or of advanced age. As well, researchers have posited a "salmon bias," which posits that immigrant Hispanic residents in the United States may return to their home country when ill, particularly if they lack access to good health care options in the United States (Murphy et al., 2017; Scommegna, 2017). Some research also shows that Hispanic immigrants are less likely to be smokers than are their nonimmigrant counterparts (Scommegna, 2017). Recent research by Boen and Hummer (2019), however, points out the importance of understanding that, regardless of difference in mortality rates, African American and Hispanic populations continue to experience significant health disparities relative to whites. These include "higher disability, and elevated depressive, metabolic, and inflammatory risks," with Hispanics more likely than whites to have lives characterized by "more socioeconomic hardship, stress, and health risks than whites, and similar health risks to blacks" (Boen & Hummer, 2019, p. 434).

A "worrisome lack of progress on health equity" was found in a recent study of health data of 5.4 million Americans over 25 years (1993–2017) showing significant health decreases in the general population (Zimmerman & Anderson, 2019). Using the CDC's Behavioral Risk Factor Surveillance System, the authors found that while Black/white disparities had improved slightly, health differences by income disparity had grown. According to the lead author on the study:

> What these results show is that there was a big decline in the 1990s in terms of health. . . . So, what we're seeing now, what we're calling deaths of despair—it's not just about the opioid

crisis—there's something that's been adversely affecting Americans' health for, you know, a couple of decades at least. (Newman, 2019)

SOCIOLOGY AND ISSUES OF PUBLIC HEALTH IN THE UNITED STATES

Public health: The science and practice of health protection and maintenance at a community level.

The CDC defines **public health** as *the science and practice of health protection and maintenance at a community level* (https://www.cdcfoundation.org/what-public-health). Public health officials try to control hazards and habits that may harm the health and well-being of the population. For instance, they have long sought to educate the public about the hazards of tobacco use and sought to decrease the rate of teenage pregnancy. More recently, public health officials have warned about the challenges of rising rates of obesity in children and adults and opioid addiction.

Smoking

One of the largest and most profitable industries in the United States is the manufacture and sale of tobacco products, estimated to be a $47.1 billion industry. Over 34 million adults are current smokers (U.S. Department of Health and Human Services, 2020). Tobacco use is also the number one cause of preventable death and disease, resulting in 480,000 deaths each year, 41,000 of which are due to secondhand smoke exposure (Centers for Disease Control and Prevention, 2018d). Fully 9 in 10 lung cancers are due to smoking, and mortality among smokers is three times higher than among nonsmokers. Smoking costs the United States health care system $170 billion annually (U.S. Department of Health and Human Services, 2020).

Importantly, as Figure 16.5 illustrates, the smoking rate continues to drop, falling from a high of 52.5% in 1965 (U.S. Department of Health and Human Services, 2020) to about 13.7% of the population today (Centers for Disease Control and Prevention, 2018d). Like alcohol use, smoking has come to be treated as a "health condition that can benefit from treatment," with declines in numbers being attributed to an increasing number of available "behavioral" and "pharmacological" therapies and smoking cessation interventions such as call centers and apps (U.S. Department of Health and Human Services, 2020). At the same time, however, tobacco products such as e-cigarettes (vapes) have continued to diversify and target new audiences such as young adults (U.S. Department of Health and Human Services, 2020). Figure 16.6 illustrates tobacco product use among high school students. This use has been increasing in the past 10 years, with 1 in 4 high school students and 1 in 10 middle school students now saying that they had vaped in the past month (Centers for Disease Control and Prevention, 2020k).

FIGURE 16.5 ■ Cigarette Smoking in the United States, 2018

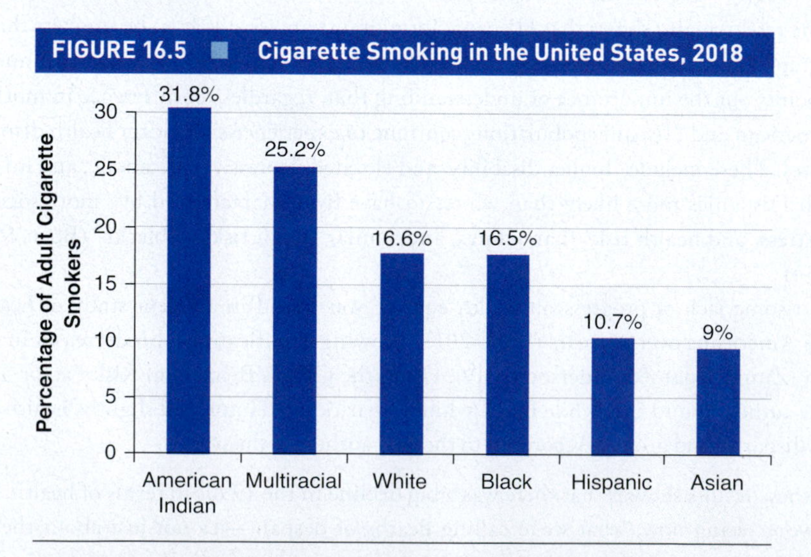

Source: Centers for Disease Control and Prevention. (2018b). *Current Cigarette Smoking Among Adults in the United States.* Retrieved from https://www.cdc.gov/tobacco/data_statistics/fact_sheets/adult_data/cig_smoking/index.htm#

FIGURE 16.6 ■ High School Use of Tobacco in the United States, 2019

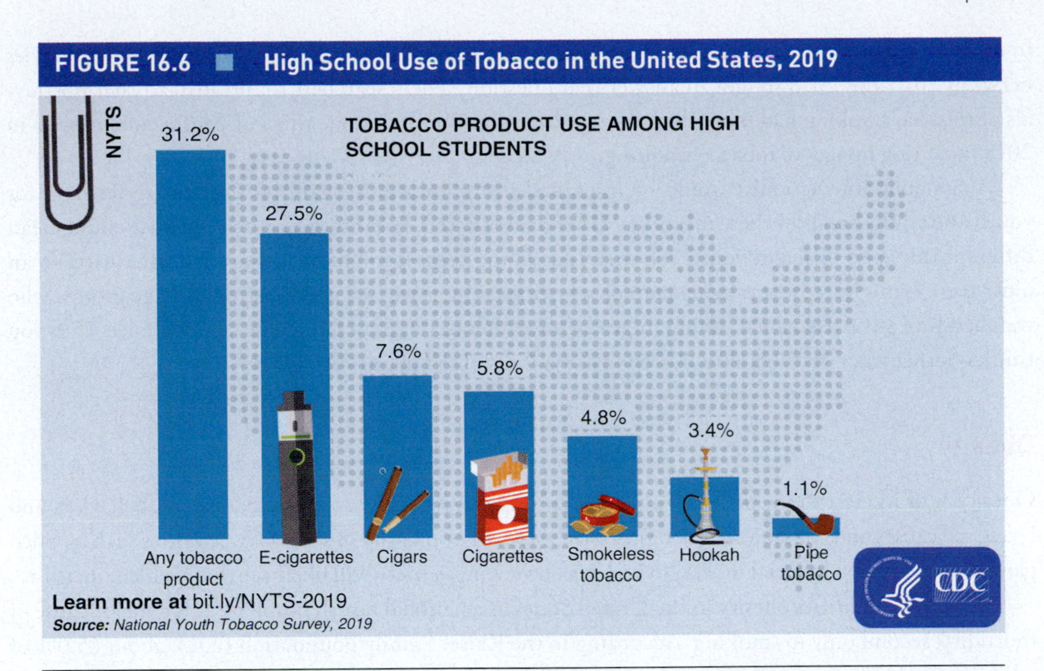

Source: Centers for Disease Control and Prevention. *Current Tobacco Use Among Youth in the United States.* Retrieved from https://www.cdc.gov/tobacco/infographics/youth/index.htm#youth-tobacco

Morbidity: The rate of illness in a particular population.

Mortality: The rate of death in a particular population.

While statistics on **morbidity**, *the rate of illness in a particular population*, and **mortality**, *the rate of death in a particular population*, highlight important medical aspects of cigarette smoking, we can also use sociological analysis to illuminate this public health issue. Why do people continue to smoke and so many young people take up smoking despite the evidence of its ill effects? Why do smoking rates vary between groups? For instance, why do more men than women smoke? Why does the government not regulate the production and sale of such an addictive and dangerous product more stringently?

Several sociological theoretical frameworks (critical theory and symbolic interactionism) might offer us some insight into these questions. Critical theorists, for instance, might first point out that tobacco companies have spent considerable money to keep their products on the market. For instance, in 2018, they spent over $23 million lobbying Congress for legislation preferable to them, an amount that has been typical of annual lobbying expenditures over the past few years (Open Secrets.org). Critical theorists would also encourage us to focus on choices people make to smoke by pointing to the mass media and advertising and how images and media messages might impact consumption.

Although cigarette advertising has been banned from television for over 50 years, advertisers have been able to share images in many other venues including, most recently, social media (Andrews, 2019). Additionally, e-cigarettes are not explicitly addressed in the laws and, in 2019, companies like Juul launched social media and television campaigns advertising the *health* features of their products in a $10 million campaign called "Make the Switch" where they aimed to "help adults find a healthier alternative to smoking cigarettes" (Andrews, 2019, para. 4).

Vaping, like cigarette smoking, can symbolize fun and leisure to those who vape. Vape.com, an annual international conference, brings together vapers and e-cigarette users for consumerism and competitions. In this photo we see a vaping competitor in Pretoria, South Africa.

©GUILLEM SARTORIO/AFP via Getty Images

Mass media has also historically enlisted gender to promote, construct, and reinforce gender stereotypes (Kilbourne, 1999). Adult male smoking, for instance, was often portrayed as an act of independence, ruggedness, and machismo exemplified in the iconic image of the Marlboro Man. On the other hand, adult female smoking has been associated with images that are elegant, chic, and playful or carefree.

What messages do youth receive about smoking in popular culture? The U.S. surgeon general has established a causal link between youth smoking and cigarette smoking portrayed in movies (Tynan et al., 2019). Have these images decreased over time? A study of youth-rated movies between 2010 and 2018 sought to answer that question. The researchers found instead that portrayals of tobacco smoking had more than *doubled* over that time, with one third of youth-rated movies in 2018 including images of tobacco smoking.

A symbolic interactionist would be interested in the meaning cigarettes or e-cigarettes have for youth and adults and how these meanings differ or are shared. For instance, e-cigarette use could mean different things to different youth cohorts: To a young teen, it might be a symbol of maturity; to an older teen, it might represent being cool or rebellious. Or, as we saw previously, to all age groups who watched Juul advertisements, vaping might be a representation of making a healthier choice. Can you think of other ways that cigarettes function as symbols of self in our society?

Obesity

Obesity is of interest to sociologists because it corresponds to many social activities and behaviors and has social causes and consequences. For instance, with the popularity of sedentary activities such as video games, participation in social media, and television viewing, society will likely see this problem increase.

The CDC identifies obesity in the United States as a national health problem: It is a major cause of mortality, second only to smoking. According to the Kaiser Family Foundation (n.d.), about 65.0% of adults in the United States between the ages of 20 and 74 self-report that they are overweight or obese, and the CDC put the 2015–2016 prevalence rate of obesity at 39.8% of the population for adults (Hales et al., 2017).

The CDC map (Figure 16.7) shows the prevalence of self-reported obesity in U.S. states and territories in 2018. Notably, the south and midwest have the highest prevalence of obesity (over 33%), while the northeast and west have the lowest (28% and 26.9%, respectively) (Centers for Disease Control and Prevention, 2018c).

The rate of obesity in American children is a significant concern: It is twice what it was in the late 1970s. Today, 1 in 6 children and adolescents are obese (Hales et al., 2017). Children with obesity may experience social and physical difficulties such as bullying, high blood pressure, joint problems, high

FIGURE 16.7 ■ Prevalence of Self-Reported Obesity Among U.S. Adults by State and Territory, 2018

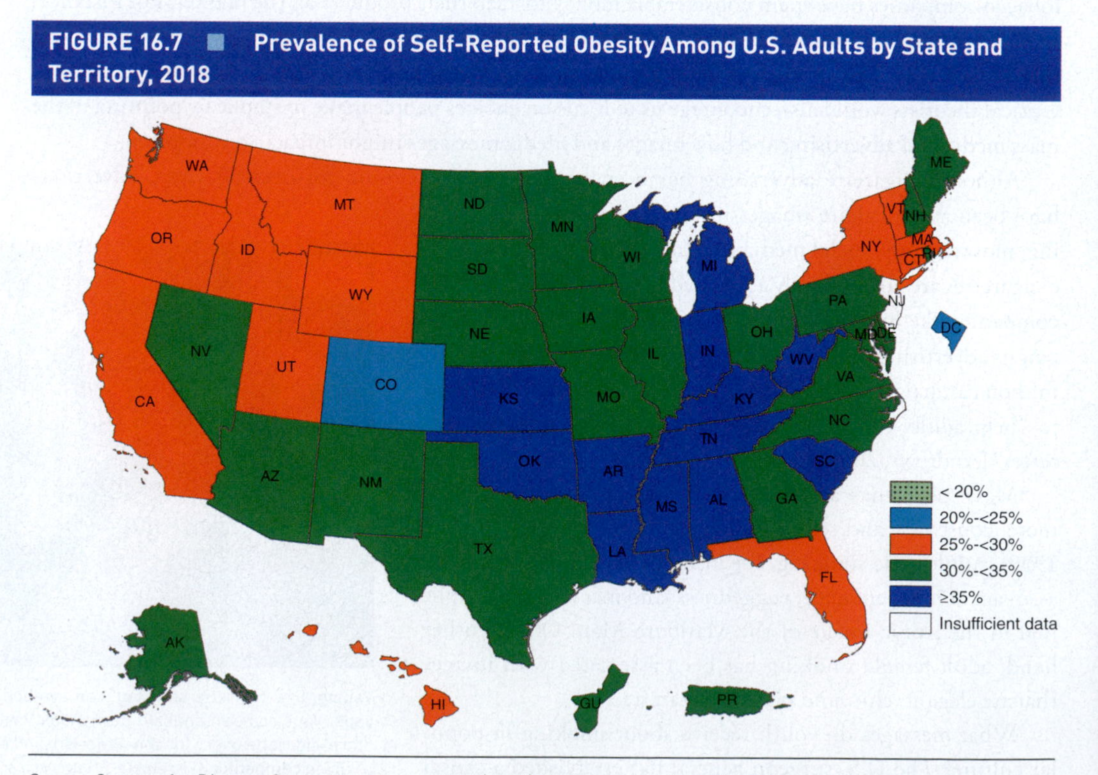

Source: Centers for Disease Control and Prevention. (2018). Based on data from the Behavioral Risk Survey.

cholesterol, and fatty liver disease. Very obese children have been observed to suffer health problems once believed to affect only older adults, including heart attacks and Type 2 diabetes.

Factors contributing to Americans' rise in size have been linked to food, exercise, and social changes. Families today eat more meals outside the home than in the past, and many of these meals are consumed at fast food establishments. As well, the portions diners are offered in restaurants are growing because many ingredients have become very inexpensive. In *Fast Food Nation*, Eric Schlosser (2012) notes that "commodity prices have fallen so low that the fast food industry has greatly increased its portion sizes, without reducing profits, in order to attract customers" (p. 243), a point supported by mathematician and physicist Carson C. Chow, who argues that the obesity epidemic in the United States is an outcome of the overproduction of food since the 1970s (cited in Dreifus, 2012). Federal subsidies for food production favor meat and dairy, which soak up almost three quarters of these funds. Just over 10% support the production of sugar, oils, starches, and alcohol, and less than a third of 1% support the growing of vegetables and fruits. These data show that the U.S. Congress has opted to subsidize the production of foods that contribute to obesity rather than those, including fruits and vegetables, recommended in the government's own nutrition guidelines (Rampell, 2010).

Physician and scientist Deborah A. Cohen (2014) argues in her book, *A Big Fat Crisis*, that "obesity is primarily the result of exposure to an obesogenic environment" (p. 191), and she points to three key components of that environment. First, she notes (consistent with Chow) that factors such as agricultural advances have led to an abundance of cheap food. Second, she suggests that the availability of food, particularly junk food, has grown: More than 41% of retail stores, including hardware stores, furniture stores, and drugstores, offer food. Third, food advertising has vastly expanded. Cohen notes that grocery stores today earn more from companies paying for prime display locations than from consumers buying groceries.

Obesity is also linked to social class. Poor access to nutritious food in the United States is more likely to result in obesity rather than emaciation. Consider, for example, that some of the country's poorest states have the highest obesity rates (see Figure 16.7). In West Virginia and Mississippi, 39.5% of adults are obese; in Louisiana and Alabama, over 36% are obese (Centers for Disease Control and Prevention, 2018c). Among the demographic groups most likely to be poor are also those most at risk of obesity; over half of African American women are obese (55%), as are 51% of Hispanic women

FIGURE 16.8 ■ Self-Reported Overweight and Obesity Rates in the United States by Race and Ethnicity, 2018

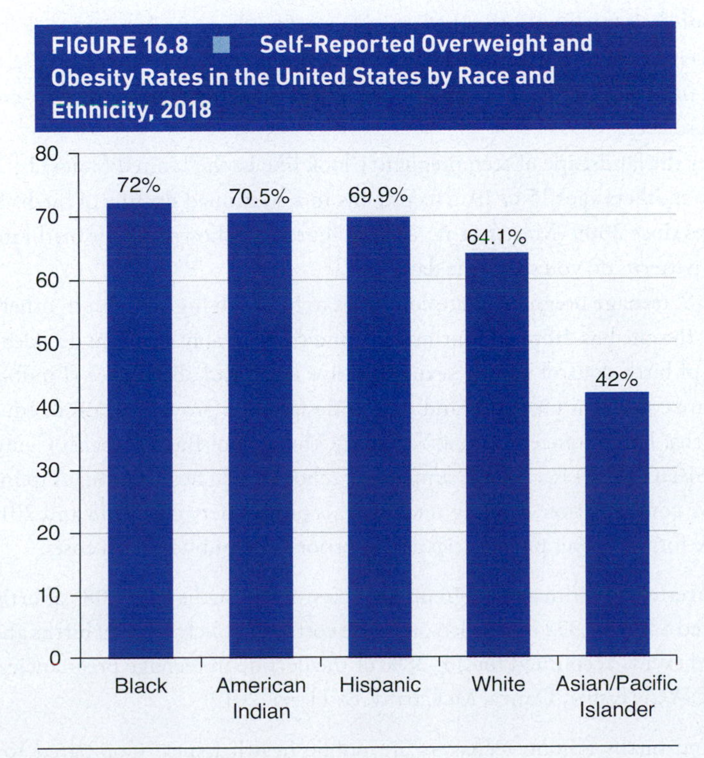

Source: Kaiser Family Foundation. (2018).

(Hales, Carroll, Fryar, & Ogden, 2017). Figure 16.8 shows differences in self-reported obesity rates by race and ethnicity in the United States in 2018.

Those without a high school education are more likely to be obese (35.5%) than those who complete high school (over 32%) or college (just under 22%) (Centers for Disease Control and Prevention, 2018c). According to the *Handbook on Obesity*, "In heterogeneous and affluent societies like the United States, there is a strong inverse correlation of social class and obesity" (quoted in Critser, 2003, p. 117).

Clearly, obesity is a complex phenomenon driven by a variety of factors—biological, genetic, environmental, social, and economic. As you will see below, poverty is also an important factor in the prevalence of obesity. From a sociological perspective, we consider the connection between the personal trouble and the public issue of obesity and being overweight. That is, if one individual or a handful in a community are obese, that may be a personal trouble, attributable to genetics, illness, food availability, ability to exercise, or other factors. However, when 42% of the population is obese (Hales et al., 2017), including majorities in some communities, this is a public issue and one that, to paraphrase C. Wright Mills, we may not explain by focusing only on individual cases. Rather, we need to seek out its sociological roots.

Consider how this issue might look through the conflict lens. Who benefits, and who loses? While "losers" in this instance are surely those whose health is compromised by excessive weight, there are also macro-level effects such as lost productivity when employees miss work due to obesity-linked illnesses (such as diabetes). In fact, the CDC has estimated that the medical care costs associated with obesity in the United States total about $209 billion annually (Cawley & Meyerhoefer, 2012).

Who benefits? The food industry, particularly fast food companies, arguably benefits when consumers prioritize quantity over quality. By offering bigger portions (which cost only a bit more to provide), restaurants draw bigger crowds and bigger profits. The massive U.S. weight loss industry also benefits, since the rise in obesity exists in the presence of widespread societal obsession with thinness. Often, the same companies that market high-fat, unhealthy foods also peddle "lite" versions (Lemonnier, 2008).

Teen Pregnancy

Sociologists are interested in teen pregnancy because, as Émile Durkheim might point out, it is a *social fact*, or phenomenon that varies in different times and places. Durkheim might also note that teen pregnancy can only be explained with other social facts. That is, to understand sociologically both the rise and fall of the rates of teen pregnancies and births, we must recognize that these are not only personal troubles or individual issues, but that they are fundamentally tied to other economic, social, and cultural issues in society.

So, what does the landscape of teen pregnancy look like in the United States? In 2018, there were 179,871 births to mothers ages 15 to 19, a record low in a continued decline of births for teen mothers in all racial groups since 2009 (Martin et al., 2019). Figure 16.9 shows teenage birthrates between 2017 and 2018. What patterns do you see in the data?

While the U.S. teenage pregnancy rate continues to be much higher than in other Western industrialized nations, the rate has dropped continuously and considerably in recent decades for reasons such as increased use of birth control among sexually active teens and abstinence (Lindberg et al., 2016). Also important are consistent programs and centers designed to provide access to education and contraception for teens. For instance, in New York City, the School-Based Health Center Reproductive Health Project (SBHC RHP) is a project supporting school-based health projects to increase the availability of effective contraception. A study tracking this project between 2008 and 2017 found significant results in the form of fewer pregnancies and abortions and public health costs:

> The project averted an estimated 5,376 pregnancies, 2,104 births and 3,085 abortions, leading to an estimated $30,360,352 in avoided one-time costs of publicly funded births and abortions. These averted events accounted for 26–28% of the decline in teenage pregnancies, births and abortions in NYC. (Fisher, Danza, McCarthy, & Tiezzi, 2019)

Pregnancy and births among teenagers are public health issues. Compared to older mothers, teen mothers have worse health, more pregnancy complications, and more stillborn, low-weight, or

FIGURE 16.9 ■ Birthrates for Females Aged 15–19, by Race and Hispanic Origin of Mother: United States, 2017 and 2018

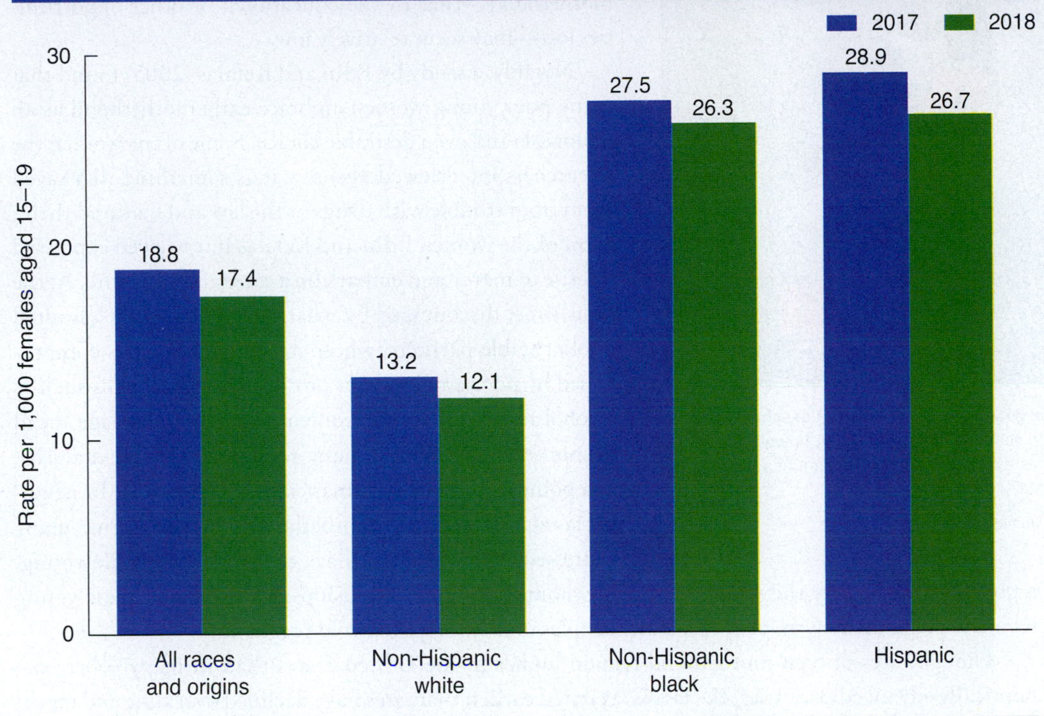

Source: Martin, J., Hamilton, B., Osterman, M., & Driscoll, A. (2019, November 27). Births: Final Data for 2018 *National Vital Statistics Report, 68*(13).

medically fragile infants. In fact, in 2018, just under 12% of teen mothers had late or no prenatal care (Martin et al., 2019). Teen pregnancy and birth are also associated with another public health problem: poverty.

Giving birth early compounds the risk that young women will be solely responsible for the raising and supporting of their children. It also puts them at higher risk for becoming or remaining poor; In 2018, 26.8% of female-headed households in the United States lived below the poverty line, compared with 5.4% of married-couple families (Semega et al., 2019). Parenthood is a leading cause of dropping out of school among teenage women; teen mothers are at greater risk than their peers of not completing high school—only 50% of teenage mothers earn a high school diploma before age 22, and fewer than 2% earn a college degree by age 30 (Centers for Disease Control and Prevention, 2017c; National Campaign to Prevent Teen Pregnancy, 2010).

The relationship between teen pregnancy and birth and poverty is complicated. On the one hand, as noted, early motherhood compounds the risk of poverty. On the other hand, poverty is itself a risk factor for teenage motherhood: An estimated 80% of teen mothers grew up in low-income households (Shore & Shore, 2009), and poor teens have a higher incidence of early sexual activity, pregnancy, and birth than do their better-off peers (National Campaign to Prevent Teen Pregnancy, 2010).

In her book *Dubious Conceptions: The Politics of Teenage Pregnancy* (1996), sociologist Kristin Luker suggests that poverty is a *cause* as well as a *consequence* of teen pregnancy and birth. She argues that poor young women's probability of early motherhood is powerfully affected by "disadvantage and discouragement" (p. 111). By *disadvantage* she means the social effects of poverty, which reduce opportunities for a solid education and the realization of professional aspirations. Consider, for instance, a high school senior from an affluent household: She may spend her 18th birthday contemplating whether to begin college immediately or take a year off for travel abroad. A young woman who hails from a poor household in rural Louisiana or the Bronx's depressed Mott Haven neighborhood may have received an inferior education in her underfunded school and, having little money, has no hope for college. Travel beyond her own state or even city is unthinkable. Local jobs in the service industry are an option, as is motherhood. *Discouragement*, according to Luker, is the effect of poverty that may prevent poor young

Early parenthood is a leading reason why teen women drop out of school. About a third cite this reason for leaving high school. Staying in school, however, is key to job prospects that enable families to stay out of poverty. What might schools do to encourage young mothers to graduate?

©Fairfax Media/Fairfax Media via Getty Images

women from exercising agency in confronting obstacles. In an impoverished situation, the *opportunity costs* of early motherhood—that is, the educational or other opportunities lost—may seem relatively low.

Notably, a study by Edin and Kefalas (2005) found that many poor young women embrace early motherhood as an honorable and even desirable choice. Some of the women the researchers interviewed also saw it as something that saved them from trouble with drugs or the law and matured them. Most of the women Edin and Kefalas interviewed expressed a desire to marry and embark on a career in the future. At the same time, discouraged by what they perceived as a limited pool of stable partners, whose marriageability was compromised by poor employment prospects and problems such as alcohol and drug use, the women did not put marriage ahead of motherhood, though many retained hopes for marriage at a point when they felt financially independent. In neighborhoods where early motherhood was the norm, many expressed a preference to have their children while young, a preference shared by the young men with whom they had relationships. While few of these young women's pregnancies were planned, many couples took no steps to avoid pregnancy.

Though rates of teen motherhood remain higher in the United States than in many other economically advanced countries, the fact is, as stated earlier, that rates have declined over time and repeat births for teen mothers are also declining. Awareness campaigns, the access to and proper use of condoms, and counseling have proven to be effective measures (Dee et al., 2017; U.S. Department of Health and Human Services, 2013). There have also been small drops in the numbers of teenagers approving of and engaging in premarital sexual activity, and the rate of births among teenage women has dropped compared to the rate in earlier decades (Centers for Disease Control and Prevention, 2019; Ventura & Hamilton, 2011).

Opioid Addiction

In 2017, the United States Department of Health and Human Services and President Donald J. Trump declared opioid use a public health emergency (USDHHS, n.d.; Gramlich, 2018). This may not surprise you, as Pew Research also reveals that a stunning percentage of Americans (90% in rural areas) identify "drug addiction" as a problem in their community (Gramlich, 2018). Images of opioid overdoses feature prominently in news stories, and Narcan (a prescription emergency nasal spray designed to reverse the effects of opioid overdose) has become a household name. In some locations it is sold over the counter in drug stores. In other words, this private trouble of opioid addiction has certainly, and obviously, become a prominent social issue.

The opioid emergency did not just happen. It was precipitated by the deceptive marketing of opioid painkillers by drug companies as nonaddictive, the widespread availability of those drugs, and the subsequent overprescription by health care providers (Quinones, 2015; Venkataramani et al., 2019). Its impact was significant. In 2018, 10.3 million Americans over the age of 12 (or 3.7% of the population) were counted as misusing prescription opioids (SAMHSA, 2019). Another 808,000 used heroin, an illegal opioid. Notably, heroin use involves "primarily white men and women in their late 20s living outside of large urban areas" (Cicero, Ellis, Surratt, & Kurtz, 2014). The accompanying issue of heroin use is associated with opioids because many addicted to prescription opioids—in particular, painkillers such as OxyContin and Vicodin—were no longer able to find ways to obtain these drugs (Cicero, Ellis, Surratt, & Kurtz, 2014). The authors point out that "the factors driving this shift may be related to the fact that heroin is cheaper and more accessible than prescription opioids, and there seems to be widespread acceptance of heroin use among those who abuse opioid products."

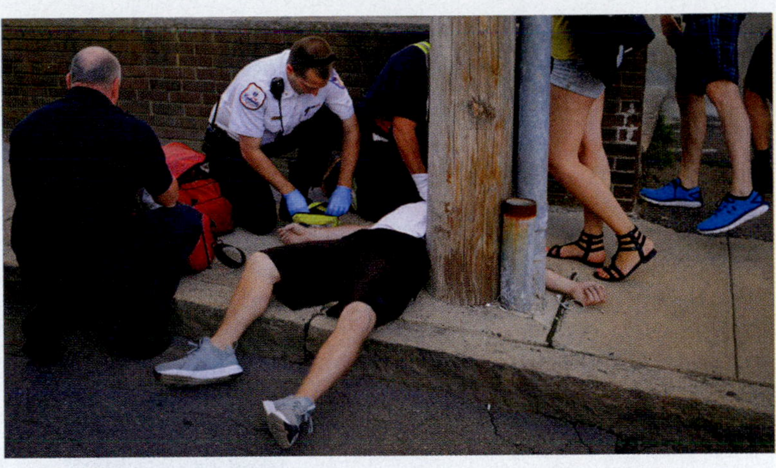

Opioid addiction and overdose is both a private trouble and increasingly a social issue. During the first 6 months of the COVID-19 pandemic in 2020, *The New York Times* reported a 13% increase in opioid deaths in the United States (Katz et al., 2020).

©REUTERS/Brian Snyder

Opioids have been deadly for many Americans: in 2017, over 42,000 people died from an opioid overdose, an average of 116 each day (U.S. Department of Health and Human Services, n.d.). The number of opioid overdose deaths has grown five-fold since 1999 (Centers for Disease Control and Prevention, 2017). Opioid use and deaths are also concentrated geographically, as illustrated in Figure 16.10. In 2018, the states with the highest age-adjusted drug overdose death rates were West Virginia, Delaware, Maryland, Pennsylvania, Ohio, and New Hampshire.

Notably, several researchers have identified opioid deaths as "deaths of despair." An interesting study by Venkataramani et al. (2019) suggests a link between these and declining economic conditions. They examined 112 manufacturing counties' mortality rates from opioid overdose between 1999 and 2016, comparing those with closures of U.S. automotive plants and those without. They found opioid deaths, 5 years after plant closures, to be associated with "increased mortality" with the overdose rates 85% higher in counties with plant closures. They also found that the victims were "primarily non-Hispanic white men." This is not to say, however, that Black and Hispanic communities are untouched by this phenomenon, and, indeed, those communities have suffered significant losses around opiate use as well (Tiger, 2017). At the same time, Tiger (2017) argues, it is important to understand the significant impact of the confluence of race and class when it comes to lower-class white users, who she finds are "caught at the intersection of the criminal justice, drug treatment, and child protection systems."

Social stigma has been identified as a significant "hindrance" in the identification, treatment, and structural challenges associated with opioid use (Tsai et al., 2019). For instance, stereotypes can influence public attitudes to see those who are addicted as being "moral failures"; such attitudes and actions may transfer to all types of providers who work with users; lawmakers and law enforcers can encode and enact laws and services influenced by stigma; and those who need treatment may internalize this stigma, which then can prevent them from seeking treatment (Tsai et al., 2019). Tiger (2017) writes of the lower social class of predominantly white users in Vermont and the social distance between their lives and those of their doctors, which she credits for the "punitive medicalization" of opioid addiction:

> Even the most sympathetic physicians I spoke with endorsed monitoring and coerced treatment. One said that there was "no high-level thinking in Vermont" and "no one understands the medical piece." When I asked about the best way to treat addiction, this doctor told me it was suboxone (buprenorphine) combined with "tight control . . . put an ankle bracelet on them and tightly monitor them. . . . If you mess up, you go to jail. Folks do best when there are consequences." Other medical providers were frustrated with their patients, viewing their poverty-related struggles such as lack of transportation or difficulty finding employment as "excuses" for not succeeding in recovery. Their patients' continued smoking and poor eating habits are also a regular source of frustration. One doctor who called addiction a "disease" insists that the criminal justice oversight of a sick person is not a contradiction, but a mechanism to ensure sorely needed "accountability."

The ongoing opioid epidemic in the United States will require continued research, programming, and action. Efforts are being made by families, public officials, law enforcement, and health professionals to address the problem across the country, though the problem is still acute. The medicalization or criminalization of addiction can have a significant effect on how users are viewed and treated. What images have you seen of opioid addiction in the mass media? How have the stories been told? Would

FIGURE 16.10 ■ 2018 Drug Overdose Death Rates by State

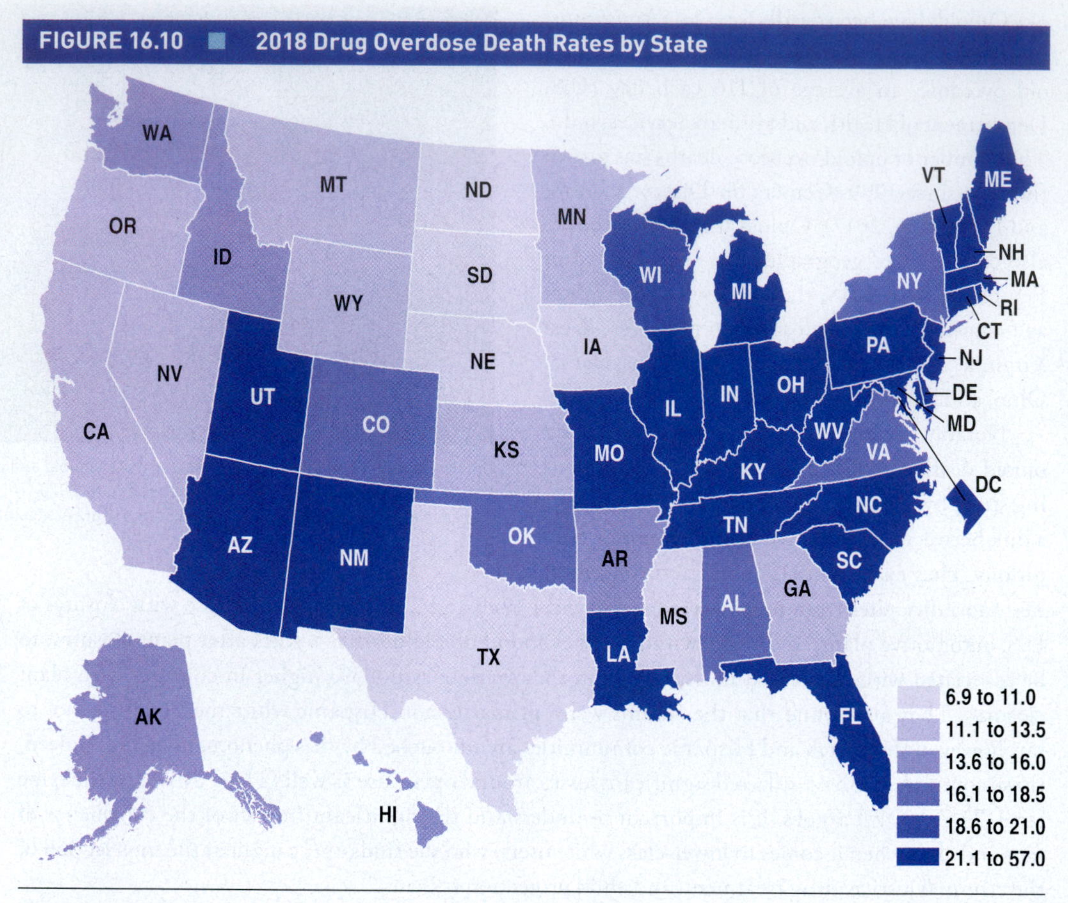

Source: CDC (2018). *Number and Age-Adjusted Rates of Drug Overdose Deaths by State, US 2018*. https://www.cdc.gov/drugoverdose/data/statedeaths/drug-overdose-death-2018.html

you agree that opioid addiction is marked by stigma? What, if any, impact do you think this has on a public health response?

THE SOCIOLOGY OF HIV/AIDS

The case of acquired immunodeficiency syndrome (AIDS) and the virus that causes it, human immunodeficiency virus (HIV), exemplifies the many ways in which the study of sociology can intersect with health. For instance, HIV/AIDS, recognized as a global pandemic in the 1980s, has widely impacted many people worldwide socially and economically, as well as at the level of their personal health (Nall, 2020). Figure 16.12 shows the worldwide prevalence of HIV/AIDS. This disease shares commonalities with many issues we have already discussed: It illustrates how diseases are socially constructed; how stigma can accompany these constructions and impact care; how existing social inequalities around race, class, gender, sexuality, and nationality, and region or nation can act as barriers or roadblocks; and the importance of community, state, federal, and global political responses and cooperation. This disease is also a story of the pursuit of both medical breakthroughs and profits in a globalizing world.

In 2019, about 38 million people around the globe were living with HIV/AIDS (Centers for Disease Control and Prevention, 2020a). While most were adults (36.2 million), a significant number—1.8 million—were children under the age of 15 (Centers for Disease Control and Prevention, 2020a). Some have HIV without knowing it. In fact, according to the CDC, in 2019, about 7.1 million people (or 19% of people globally) still needed to be tested (Centers for Disease Control and Prevention, 2020a). Importantly, while globally, well over half (67%, or 25.4 million) of people with HIV were able to access antiretroviral therapy (ART) globally, 12.6 million were not.

According to UNAIDS, there has been an 11% decrease in new infections globally since 2010 and, since their peak in 2005, AIDS-related deaths have dropped by 48%. The situation, however, varies by region, and HIV/AIDS continues to present substantial medical and societal challenges. For instance, while the number of new infections in Eastern and Southern Africa declined, in the regions of Eastern and Central Europe and Central Asia, as well as in the Middle East and North Africa, new infections and AIDS-related deaths increased (UNAIDS, 2017). In 2020, COVID-19 "seriously interrupted" responses to AIDS, with COVID-19 lockdowns and border closures that threatened medical supply chains, costs, and stock (UNAIDS, 2020a). As the UNAIDS June 2020 press release put it:

> Since 24.5 million people were on antiretroviral therapy at the end of June 2019, millions of people could be at risk of harm—both to themselves and others owing to an increased risk of HIV transmission—if they cannot continue to access their treatment. A recent modelling exercise estimated that a six-month disruption of antiretroviral therapy in sub-Saharan Africa alone could lead to 500,000 additional AIDS-related deaths. (para. 4)

HIV/AIDS and Gender and Sexuality

The spread of sexually transmitted diseases, including HIV/AIDS, can be better understood if we examine how its spread connects with gender, sexuality, and inequality. Globally speaking, in 2019, women and girls accounted for just under half—48%—of the new HIV/AIDS infections (UNAIDS, 2020c). The number of women with HIV/AIDS has continued to rise: In sub-Saharan Africa around 5,550 young women are infected weekly; these same women ages 15 to 24 years old are twice as likely to be living with HIV than are men of the same age (UNAIDS, 2020c).

Norms and traditions in many regions reinforce women's lower status in society. In some traditional communities in Africa, for example, it is socially acceptable—or even desirable—for men to have multiple sexual partners both before and after marriage (Thobejane, 2014). In this case, marriage itself becomes a risk factor for women. Many women also still lack accurate knowledge regarding sexually transmitted diseases, a problem made more acute by widespread female illiteracy in poor regions. Women who are not infected may not know how to protect themselves, and women who are infected may not know how to protect their partners.

We earlier addressed the risk of domestic violence and sexual assault and indicated that women are at higher risk of these phenomena. This holds true worldwide and has an effect on the conditions of transmission of HIV/AIDS. Data gathered by the United Nations (UNAIDS, UNFPA, & UNIFEM, 2004; UNAIDS, 2010) suggest that up to half the women in the world may experience violence from a domestic partner at some point; this includes rape, which is not likely to take place with a condom. Compared to men, women are 1.5 times more likely to acquire HIV through physical and sexual violence including rape (UNAIDS, 2020c).

Vulnerability to HIV/AIDS also varies depending on sexual orientation and gender identity. Women who are most at risk for contracting HIV are marginalized groups like sex workers, drug injection users, transgender women, and women in prison (UNAIDS, 2020b). In the case of transgender populations, data gathered between 2009 and 2014 reveal that transgender women had higher rates of HIV infection (84%) than did transgender men (15.4%). Transgender women of color are at higher risk, with HIV rates the highest for Black and African American women (44.2%) and Hispanic/Latinx women (25.8%), compared to white transgender women (6.7%) (Clark et al., 2017).

In the United States, gay and bisexual men remain the most affected populations (see Figure 16.11). In 2018, adult and adolescent gay and bisexual men from the ages of 13–34 made up 64% of new HIV infections. Black and African American men are the highest population by race and ethnicity to be diagnosed at 37%, Hispanic and Latino at 29%, and white at 28%.

Finally, the previously discussed challenges around COVID-19 and HIV/AIDS are significant to all of the populations discussed in this section. One important group that the UNAIDS project addresses in its June 2020 release is pregnant women, who can transmit HIV to their unborn children. As the UNAIDS June 2020 press release put it:

> If services to prevent mother-to-child transmission of HIV were similarly halted for six months, the estimated increases in new child HIV infections would be 162% in Malawi, 139% in Uganda, 106% in Zimbabwe and 83% in Mozambique.

FIGURE 16.11 ■ HIV Diagnoses in the United States for the Most-Affected Subpopulations, 2018

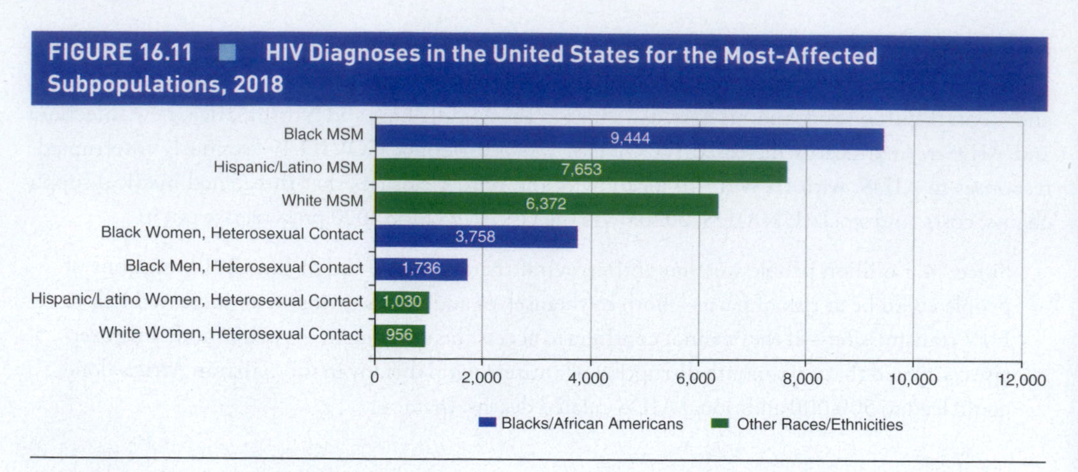

Source: Centers for Disease Control and Prevention Surveillance Report. (2020a, May). *Diagnoses of HIV Infection in the United States and Dependent Areas 2018.* Retrieved from http://www.cdc.gov/hiv/library/reports/hiv-surveillance.html

Note: MSM stands for "men who have sex with men."

Poverty and HIV/AIDS

Across the globe, there is a powerful relationship between the risk of HIV/AIDS, poverty, and sex work. In developing countries, economic insecurity and the lack of gainful employment sometimes drive workers (particularly men) to seek work far from home. For example, migrant workers from surrounding countries toil in the mines of South Africa. Away from their families and communities, some seek out the services of sex workers, who may be infected (UNAIDS, 2010). The sex workers, themselves, are often victims of dire and desperate economic circumstances. Women in the sex trade, some of whom have been trafficked and enslaved, are highly vulnerable to HIV/AIDS. They have little protection from robbery or rape and limited power to negotiate safe sex with paying customers, though some countries, such as Thailand, have sought to empower sex workers to demand condom use (Avert, 2017).

Poor states, as well as poor individuals, are vulnerable to the ravages of disease. Consider the cases of these southern African states: HIV prevalence among young women ages 15–24 is about 10% in Botswana, 14% in Lesotho, and 18% in Swaziland (PRB, 2017). The high rates of infection and death among young

FIGURE 16.12 ■ HIV Prevalence Worldwide, 2020

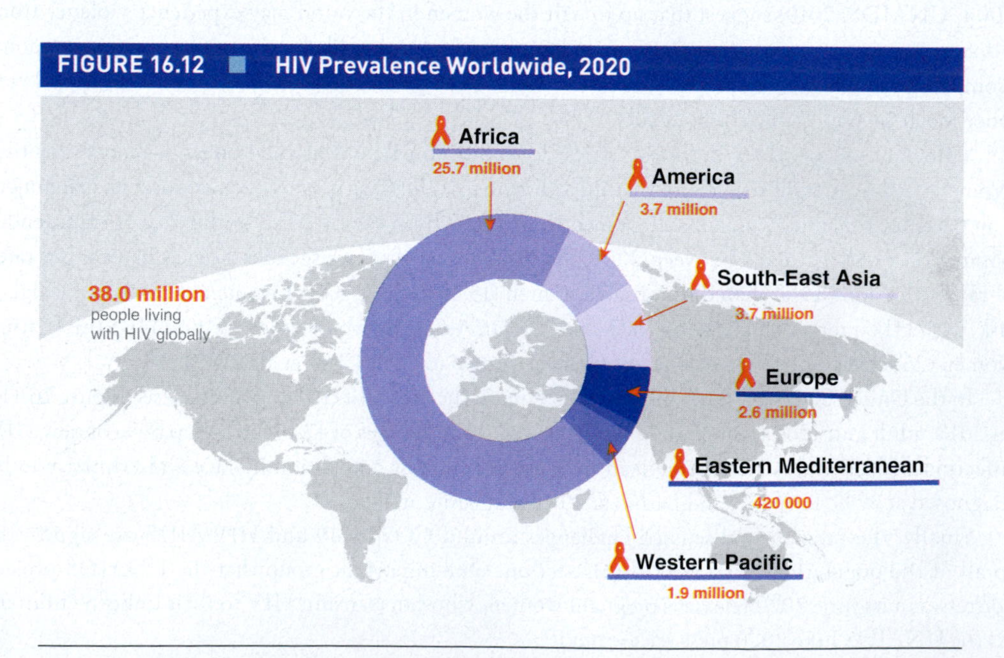

Source: World Health Organization. (2020). *Global Health Observatory Data: HIV/AIDS.* Retrieved from http://www.who.int/gho/hiv/epidemic_status/cases_all/en/

and middle-aged adults also mean that countries are left with diminished workforces. Without productive citizens, the state of a country's economy declines, further reducing the resources that might be put into HIV/AIDS prevention or treatment. While HIV/AIDS is far from limited to poor victims or poor countries, poverty clearly increases the risk of disease at both the individual and the national levels.

What other challenges of poverty do you think might contribute to the spread of HIV/AIDS and the inability to address it adequately?

Stigma, Violence, and HIV/AIDS

Social stigma from HIV/AIDS was and continues to be an important issue. Returning to an earlier discussion of the social construction of illness, those among the first to contract HIV/AIDS in the United States encountered significant social stigma. In the early days of the emergence of HIV/AIDS in the United States, gay male communities in large American cities such as Los Angeles, New York City, and San Francisco were disproportionately impacted. Associated with gay men, in a time of significant cultural discrimination and prejudice toward the gay community, the American government put little funding into crucial research and treatment (Shilts, 1987). Indeed, in the early days of this pandemic, AIDS stigma prompted former surgeon general C. Everett Koop to note:

> We are fighting a disease, not people. Those who are already afflicted are sick people and need our care as do all sick patients. The country must face this epidemic as a unified society. We must prevent the spread of AIDS while at the same time preserving our humanity and intimacy. (Koop, 1986, quoted in Fassbender, Byrnes, & Levine, 2019)

Importantly, this stigma is still a very real impediment to treatment for those with HIV/AIDS, with UNAIDS data showing that worldwide "stigma and discrimination at healthcare facilities . . . interferes with their ability to live with HIV successfully and which discourages others from seeking care . . . (including) denial of care, dismissive attitudes, coerced procedures or breach of confidentiality" (UNAIDS, 2020b, p. 13). And in the United States, as the Centers for Disease Control and Prevention (2020d) puts it:

> Gay, bisexual, and other men who have sex with men (MSM) are the population most affected by HIV in the United States. Stigma, homophobia, and discrimination put MSM of all races/ethnicities at risk for multiple physical and mental health problems and can affect whether they seek and receive high-quality health services, including HIV testing, treatment, and other prevention services.

Finally, violence can spread HIV/AIDS. The rape of men by other males, not uncommon in prison settings, can also be implicated in the spread of the infection. In many countries, the incidence of HIV/AIDS in prisons is significantly higher than the incidence of the disease in the noninstitutionalized population. Part of this phenomenon is linked to the sharing of needles among drug-injecting prisoners, tattooing with unsterile equipment, or consensual male sexual activity, but part is also linked to the underreported sexual violence behind bars (Avert, 2017).

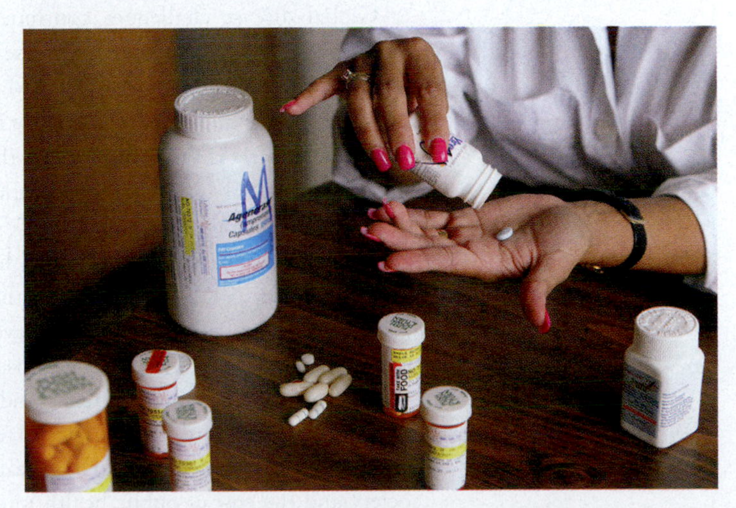

Much progress has been made in developing medicine that helps keep HIV/AIDS under control and maintains one's quality of life, and even better advancements have been made in prevention, awareness, and education on how to avoid contracting HIV. Nevertheless, it continues to be a global epidemic that claims millions of lives.

HIV/AIDS is a medical issue. It is also a sociological issue. Vulnerability to infection is compounded by factors such as gender stereotypes and poverty. At a time when hope of new treatments and prevention strategies has materialized but the pandemic continues to ravage communities and countries, a sociological perspective can help us to identify the social roots of HIV/AIDS and to seek the most fruitful paths for combating its spread.

GLOBAL ISSUES IN HEALTH AND MEDICINE

In terms of global health and medicine, public health agencies tend to agree that we are collectively "in this together." As became clear during the Great Influenza of 1918 and the COVID-19 Pandemic of 2020, a cough can quickly spread around the world. As we have shown in this chapter, however, health is much broader than communicable disease. Thanks to global material changes in people's daily lives worldwide and individual and collective efforts, NGOs, and numerous organizations dedicated to public health and well-being, health outcomes have improved in many areas of the world and life expectancies have increased. Access to clean water, sewage systems, and medicines has broadened. Globalization has also encouraged access to more educated populations who are increasingly knowledgeable about the importance of diet and changing lifestyle practices. Of crucial importance in improving global health is the ability to access medical care, medicines, and vaccines. For example, the WHO's plan for "Health for All by the Year 2000" successfully immunized half the world's children against measles, polio, and smallpox (Steinbrook, 1988).

Widespread vaccinations have been responsible for many landmark gains in global public health. In 2004, an estimated 78% of children in the world were vaccinated against diphtheria, tetanus, and whooping cough due to the efforts of the Bill and Melinda Gates Foundation, working with the Global Alliance for Vaccines and Immunization (Bill and Melinda Gates Foundation, 2006). In 2020, the incidence of polio, a disease that in 1988 was noted to have paralyzed 1,000 children per day, had decreased by 99%, an accomplishment achieved through the efforts of the Global Polio Eradication Initiative (GPEI) (Bill and Melinda Gates Foundation, n.d.). These achievements have produced a sharp decline in death rates in most of the world's countries (Andre et al., 2008).

At the same time, however, political unrest, destabilization, poverty, and other issues can impact the maintenance of public health in one country and influence flows to other locations. For instance, Venezuela's health care system's ability to address vaccine-preventable diseases and ensure the public health has been significantly impaired by the confluence of authoritarianism, political unrest, and economic freefall. Hyperinflation rates there have soared to over 45,000%, and the average daily wage of $1.79 has forced people to travel beyond borders for food and jobs (Paniz-Mondolfi et al., 2019). Added to this was the erosion since 2010 of Venezuela's public health infrastructure and the suspension of its national immunization program (Paniz-Mondolfi et al., 2019). Subsequently, diphtheria and measles outbreaks have occurred within and beyond Venezuela (for instance, in Brazil), and new concerns have emerged about the potential for a re-emergence of polio (Paniz-Mondolfi et al., 2019).

Global disease challenges continue with mosquito-borne illnesses such as the Zika virus, malaria, and West Nile virus—diseases that, while endemic to certain locales, can also spread through travel (Laurent et al., 2019). Malaria, for instance, has been called "the most severe public health problem worldwide" (Centers for Disease Control and Prevention, 2020i). According to some estimates, malaria is a threat to no less than half the global population. In 2018 alone, according to the CDC, it killed 405,000 people worldwide, and infected an estimated 228 million (Centers for Disease Control and Prevention, 2020i). The most vulnerable populations are children and pregnant women in sub-Saharan Africa, which has the most malaria deaths (Centers for Disease Control and Prevention, 2018b). According to the CDC, direct costs of malaria amount to over $12 billion worldwide. Global efforts, however, have resulted in an estimated 25% decrease in malaria deaths between 2010 and 2016 (Centers for Disease Control and Prevention, 2020i).

The toll taken by malaria is felt by individuals, communities, and nations. For individual families, malaria is costly in terms of drugs, travel to clinics, lost time at work or school, and burial, among other expenses. For governments, malaria means the potential loss of tourism and productive members of society and the cost of public health interventions, including treatments and mosquito nets, which many individuals are unable to pay for themselves (Centers for Disease Control and Prevention, 2012). At the same time, the fight against malaria, as well as fights against HIV/AIDS and tuberculosis, have broadly benefited from substantial proportions of available funding from international and national donors and governments seeking to improve the health of populations in the developing world.

The 2020 COVID-19 pandemic is the most recent example of the global spread of a potentially fatal disease. Like HIV/AIDS, it was marked by rapid spread around the world to post-industrial and industrializing nations alike. Like HIV/AIDS, COVID-19 is also a global issue in terms of treatment and prevention. As Dr. William Schaffner, an infectious disease expert, noted:

> This virus doesn't need passports. In a few short months it has travelled to all of the continents of the world except Antarctica. If there were ever an event that showed us how we need to work tougher as a global community, this is it. (Schaffner, quoted in Reuters, 2020)

Globalization, in the case of COVID-19, is both functional and dysfunctional for real and potential victims of the infection. On the one hand, COVID-19 was global in its path of spread, and it appears likely that its defeat will also be global, as it was for other once-deadly and widespread diseases such as smallpox, polio, and malaria (Steinbrook, 1988). At the same time, like HIV/AIDS, world governments are collaborating to combat the disease, although, in the United States, this collaborative effort was deeply impaired by the Trump administration, which cut funding to the World Health Organization (WHO) in April 2020 during the COVID-19 pandemic. According to Reuters, the United States is one of WHO's biggest contributors, giving $400 million in 2019 to the organization focused on improving global health. Additionally, the Trump administration declared in August 2020 that it would not participate with a broad alliance of the World Health Organization and 170 countries participating in the Global COVAX Facility (COVID-19 Vaccine Global Access). According to White House spokesperson Judd Deere:

> The United States will continue to engage our international partners to ensure we defeat this virus, but we will not be constrained by multilateral organizations influenced by the corrupt World Health Organization and China. (Rauhala & Abutaleb, 2020)

Importantly, one underaddressed global health threat worldwide is chronic disease. Heart disease, stroke, and cancer have long been chronic maladies associated with the habits of the populations of developed countries, such as overeating, lack of exercise, and smoking. One scientist notes that while 80% of global deaths from chronic diseases take place in low- and middle-income countries, those illnesses receive the smallest fraction of donor assistance for health. Of the nearly $26 billion allocated for health in 2009, only 1% targeted chronic disease (Lomborg, 2012).

Chronic disease also increasingly threatens poor countries, driven by a dramatic rise in both obesity and smoking. According to the World Health Organization, global obesity rates "nearly tripled" between 1975 and 2016. The WHO estimates that about half the adult populations of Brazil, Russia, and South Africa are overweight. In Africa, around 8% of adults are obese. While these figures are low compared to those in the United States, where two thirds of adults are estimated to be overweight and one third are obese, the numbers are rising. Children are also an important group with increasing obesity rates:

> In 2019, an estimated 38.2 million children under the age of 5 years were overweight or obese. Once considered a high-income country problem, overweight and obesity are now on the rise in low- and middle-income countries, particularly in urban settings. In Africa, the number of overweight children under 5 has increased by nearly 24% since 2000. Almost half of the children under 5 who were overweight or obese in 2019 lived in Asia. (World Health Organization, 2020)

A variety of factors contribute to this phenomenon, including growing incomes in many parts of the developing world, which enable more consumption, economic changes that shift work from physical labor to indoor and sedentary labor, and the movement of fast-food restaurants into new regions where people can now afford to splurge on burgers and soda (Kenny, 2012).

While smoking has decreased in many developed countries in recent decades, it has grown dramatically in some parts of the world. Today, about 80% of smokers live in the developing world (Qian et al., 2010). China, for instance, is home to about 40% of smokers worldwide (Liu et al., 2017). Three hundred million Chinese are smokers and 740 million Chinese are exposed to second-hand smoke (WHO, 2020). An estimated 62.5% of men and 3.4% of Chinese women have smoked and, importantly, generations born after 1970 have smoked exclusively manufactured cigarettes, exposing them to greater risks of carcinogens (Liu et al., 2017). Tobacco use has grown fourfold in China since the 1970s and has become a key component of the nation's growing prosperity. Cigarettes, particularly expensive brands, are given as gifts to friends and family; red cigarettes are special presents for weddings, bringing "double happiness." China also has its own tobacco manufacturing industry, which is run by the government. This creates a conflict of interest, since the same entity that regulates tobacco and might be interested in promoting better public health is profiting from the large number of tobacco users (PBS, 2010). Since 2001, when China joined the World Trade Organization and its markets opened to new goods, Western cigarette makers have also been aggressively marketing their products there, targeting relatively untapped consumer categories such as women, who are otherwise less likely than men to smoke (Qian et al., 2010).

Globalization as applied to health and medicine can be seen through a functional lens as allowing countries to improve overall health and well-being by sharing many positive technological and medicinal innovations, evidence, practices, and resources. Their shared motivations to do so could be considered through Ulrich Beck's (1992) framework of **risk society**—*a society where people are joined globally in their efforts to decrease danger and risk from large-scale biological, environmental, or military (nuclear) threats.* Or, as Beck, quoted in Tooze, says, "What yesterday was still far away will be found today and, in the future, 'at the front door'" (Ulrich Beck, 1992, quoted in Tooze, 2020, p. 2). That is to say, in a globalizing world, no one is isolated from diseases or unhealthy products from distant places; we are all part of the same community, linked by travel, commerce, and communications. Neither are we isolated from the far-reaching consequences of health dangers that threaten to destabilize regions far from our own. Finally, we are also part of a global consumer society where global capitalism influences companies to pursue medical profits that can result in advantageous innovations that raise the standards of many (such as many vaccines) or products used in ways that work against public health (in the United States, the manufacture and profit-driven overprescription of opioids). Globalization can also be seen through the lens of numerous conflict approaches: It can enable the flow of unsafe products and problematic practices, and it can create worldwide winners and losers, increasing social class and health inequalities. What are some of the ways that you see globalization connected to public health? Which perspective best captures your examples?

Risk society: A society where people are joined globally in their efforts to decrease danger and risk from large-scale biological, environmental, or military (nuclear) threats.

WHAT'S NEXT? TECHNOLOGY AND HEALTH AND ILLNESS

We began this chapter by examining behaviors and inequalities that magnified health impacts during several past and one ongoing global pandemic. Throughout the chapter we illustrated the importance of patterns of human behavior in defining, creating, and fighting disease and the importance of taking a sociological approach to health, medicine, and illness. While COVID-19 had specific biological impacts on the population, it also underscored the importance of understanding social roots in imagining creative, constructive responses to a significant global health problem that threatened lives, livelihoods, and world economies. The impact of this disease has reverberated through all social institutions, as families, educational and religious systems, world economies, and, of course, health systems tried to grapple with safety, well-being, and daily life.

So, what's next? For one thing, technologies that were present before COVID-19 may continue to provide new services, reach new populations, and create new research opportunities for students of health, medicine, and sociology. Take telemedicine (or telehealth). This service received widespread usage in many more homes in early 2020, when the need to provide safe and widespread access to primary mental and physical health care medical services was expanded with Congress's passing of the CARES Act (Coronavirus Aid, Relief, and Economic Security). While some form of telehealth had existed for over 20 years with platforms in place in over 50 large U.S. medical systems, it was vastly underutilized because insurance companies placed many restrictions on its use and reimbursement structures (Allen, 2017; Avadhanula, 2020).

The CARES Act lifted these restrictions, new platforms for telehealth care delivery grew, and the number of patients accessing online health care dramatically expanded (Avadhanula, 2020). Previously, this type of delivery system was especially important in areas where lack of medical personnel, clinics, and hospitals required people to travel long distances for care. While patients who have acute or urgent needs are still best served by personal visits to a health care provider, and while those who do not own computers or smartphones or cannot pay the fees for online consultations may still be locked out of these opportunities, telehealth technology may offer a potential vehicle for bringing medical advice to both advantaged and underserved communities. Telemedicine also offers new avenues for those needing addiction treatment (Mosely, 2020).

©Michael S. Williamson/The Washington Post via Getty Images

Thinking sociologically, what populations do you see as most benefiting from this type of medical care? Are there potential losers? Are there potential pitfalls to the use of these technologies as well? Can you think of other ways that technological innovations could be used to address medical needs across the income spectrum?

WHAT CAN I DO WITH A SOCIOLOGY DEGREE?

Qualitative Research Skills

Sociologists use qualitative research skills to gather rigorous, in-depth information on social behaviors such as those surrounding health and institutions such as medicine. Qualitative research highlights data that cannot be *quantified* (that is, cannot be converted into numbers). It relies on the gathering of data through methods such as focus groups, participant and nonparticipant observation, interviews, and archival research. Generally, population samples are small in qualitative research because the aim of the research is to gain deep understanding.

Throughout this book, you have encountered qualitative research studies that contribute to our knowledge of the social world. As you advance in your sociological studies, you will have the opportunity to learn how to do qualitative sociology. For example, you may learn to prepare interview questions that will allow you to accurately assess respondents' attitudes toward a particular social trend, or you may learn to take detailed field notes on observations you make of a practice or population you seek to study.

Knowledge of qualitative research methods is a beneficial skill in medical sociology and in today's job market. Learning to collect data through observation, interviews, and focus groups, for instance, prepares you to do a wide variety of job tasks, including survey development, questionnaire design, data collection and reporting, and market research. Furthermore, qualitative research experience fosters communication competencies through the processes of small-group management and rapport building, as well as negotiation with study participants.

Dakota Z Ross-Cabrera, Research Associate, National Research Center on Hispanic Children and Families

BA, American Studies, Columbia University

MA, Sociology, Columbia University

I became interested in pursuing an MA in sociology while completing a Fullbright Teaching Assistantship in La Rioja, Spain. During my year in the small wine region, I observed my young students' changing racial and ethnic identities and saw a unique intersection of religion, class, race, and ethnicity. My Spanish and Moroccan students divided along ethnic and religious lines, despite living in small villages with constant contact. This experience piqued my interest in sociology, specifically, intersectional Latinx identity and the factors that shape self-identification.

After completing my MA in sociology, I decided to test my interest in immigration, identity, race, and poverty, by working for Morningside Heights Legal Services. The pro bono law clinic provides clients with high-quality legal assistance from practicing lawyers, law students, and law professors. I gained anew legal perspective on issues such as immigration, mass incarceration, and adolescent representation. My sociological knowledge allowed me to bring a level of analysis and critical understanding to intersectional themes such as poverty, sexuality, and immigration status present in each case. Painting a legal picture of the clients involved understanding racial and ethnic identities, social boundaries, and discriminatory practices faced by clients. Sociology gave me the tools to assist in conducting qualitative interviews on sensitive topics and underscored themes and trends relevant to the legal argument.

Eventually, I shifted jobs to work for the National Research Center on Hispanic Children and Families and am working on social policy. I am currently conducting a scan on social-insurance policies (such as food stamps and cash assistance) in the 13 states with the largest Hispanic populations in the U.S. The goal of this study is to examine ways in which social-insurance policies fail to reach Latinxs and provide guidance for restructuring these policies so that low-income individuals can effectively access and use these services. I work with a team of economists, developmental psychologists, and quantitative-method sociologists. Sociology has helped me paint a national picture of the diversity of Latinxs, with a focus on country of origin, state of residence, educational attainment, and level of education, and to match my team's analysis of social-insurance policies and practices with the profile and needs of the Latinx community in the U.S. to create viable solutions for the community.

Career Data: Operations Research Analyst

- 2019 Median Pay: $84,810 per year
- $40.78 per hour
- Typical Entry-Level Education: Bachelor's degree
- Project Job Growth by 2019–2029: 26% (much faster than average)

Source: Bureau of Labor Statistics, Occupational outlook handbook 2019.

SUMMARY

- Health and medicine, while clearly related, are not the same thing. **Health** is the degree to which a person experiences a generalized state of wellness, while **medicine** is an institutionalized approach to the prevention of illness.
- Notions of illness are socially constructed, as are the social roles that correspond to them. The sociological concept of the **sick role** is important to an understanding of societal expectations and perceptions of the ill individual. The dramaturgical perspective allows us to consider the components of interaction necessary to carry out the sick role on the part of the performer and the audience.
- Not all forms of addiction are treated the same in society. Some (including alcoholism) are medicalized, while others (including drug use) are criminalized.
- The U.S. health care system does not serve all segments of the population equally. Good health and good **health care** are still often privileges of class and race and intersect with gender and sexuality.
- **Public health** issues such as smoking, obesity, teen pregnancy, and opioid use can be examined sociologically. The sociological imagination gives us the opportunity to see the relationship between private troubles (such as being addicted to tobacco, being obese, or becoming a teen mother) and public issues ranging from the relentless drive for profits in a capitalist country to the persistent poverty of generations.
- The global pandemic of HIV/AIDS demands a sociological approach as well as a medical approach. The mass spread of the infection is closely intertwined with sociological issues. Gender inequality makes some populations more vulnerable to infection. Poverty renders both individuals and countries more vulnerable to the disease. Continued **stigma** and violence can interfere with diagnosis, treatment, and even transmission.

- Rising standards of living in many parts of the developing world and international cooperation around diseases such as vaccinations have had many positive effects, but global interconnections can also spread problematic practices, products, and diseases. Chronic illness, exacerbated by diets predisposing people worldwide to obesity, are often underexamined in considering global health and illness.
- Global interconnections can be functional, but they can also be conceptualized as increasing individual and collective risk.

KEY TERMS

environmental racism (p. 460)
health (p. 450)
health care (p. 454)
medicine (p. 450)
morbidity (p. 463)
mortality (p. 463)
pandemic (p. 449)
preventive medicine (p. 450)
public health (p. 462)
risk society (p. 476)
sick role (p. 451)
stigma (p. 453)

DISCUSSION QUESTIONS

1. What is the sociological difference between health and medicine? Pick a topic relevant to each field and explain how a sociologist might study it using a qualitative or quantitative approach.
2. Some illnesses are accepted while others are stigmatized. Why? Identify and compare the characteristics of an illness (or of drug use) that is stigmatized versus one that is not.
3. Provide three specific examples of the intersectionality of health and race, class, gender, and sexuality.
4. The chapter looked at cigarettes and smoking through a sociological lens. Recall how we applied the functionalist, symbolic interactionist, and conflict perspectives to this topic, and try applying those perspectives to junk food, such as soda, candy, and fast food.
5. How is HIV/AIDS a sociological issue as well as a medical one? What are key sociological roots of the spread of this disease in communities and countries?

©CHANDAN KHANNA/AFP via Getty Images

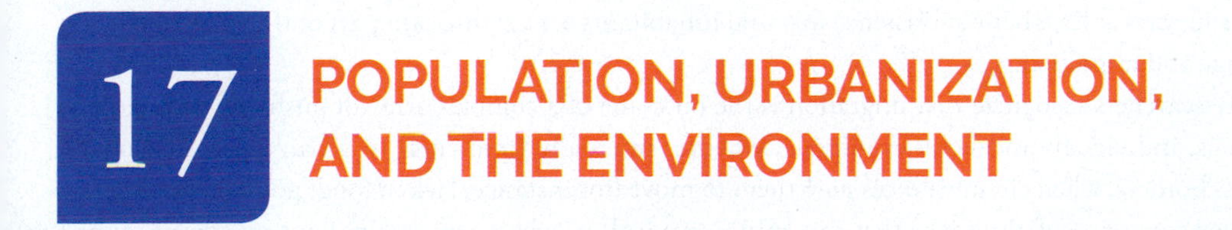

17 POPULATION, URBANIZATION, AND THE ENVIRONMENT

WHAT DO YOU THINK?

1. Should we be concerned about global population growth? What are the benefits and costs of a growing global population?
2. What factors drive rural to urban migration? What are factors that *push* people out of rural areas? What are factors that *pull* them toward urban life?
3. Is it possible to balance the imperatives of economic growth and environmental protection?

LEARNING OBJECTIVES

17.1 Identify key global population trends, including regional differences and birth, death, and growth rates.

17.2 Understand the debate over population growth and consumption.

17.3 Describe U.S. historical and contemporary trends in urbanization, the rise of global cities, and the rapid growth of cities in the developing world.

17.4 Take a sociological perspective on attitudes toward and causes and consequences of climate change.

CLIMATE MIGRANTS

What is the relationship between climate change and migration? A recent *New York Times Magazine* article asks this question and explores recently developed models that seek to use historical and recent data to predict some future scenarios (Lustgarten, 2020).

The article points out that "For most of human history, people have lived within a surprisingly narrow range of temperatures, in places where the climate supported abundant food production. But as the planet warms, that band is suddenly shifting north" (para. 7). Geographic shifts in fruitful land that supports at least basic subsistence for rural inhabitants are a significant part of the story of climate change and migration.

Researchers recognize that migration is the outcome of a complex array of push and pull factors. That is, individuals and families migrate, whether internally from rural to urban areas, or globally across borders, when circumstances *push* them to move (for instance, lack of food, political or religious persecution, or a deficit of jobs that can ensure survival) or when new destinations *pull* them toward mobility (for instance, when they have other family abroad or if job opportunities are perceived to be abundant (Solimano, 2010). The goal of recent research focused on climate change and migration is to sort out and recognize the specific effect of climate as a push factor in migration decisions.

Consider the case of Jorge A., a Guatemalan farmer from the Alta Verapaz region. The region has lush mountains with coffee plantations, as well as dry forests and expansive valleys. For many years, its residents were largely stable. Of late, however, Alta Verapaz has been wracked by floods and drought. After losing his corn crops to devastating floods, Jorge A. borrowed $1,500 to purchase okra seeds, but drought prevented the crop from taking root. While weather has always been a source of risk for farmers, climate change has rendered some storm patterns like El Niño more acute. In March of this year, Jorge A. paid $2,000 to a people smuggler (sometimes called a coyote) and set out north with his 7-year-old son, hoping to find the means to survive—something he was no longer able to do in Alta Verapaz (Lustgarten, 2020).

Jorge A. is one man, but some researchers suggest that the number of climate migrants fleeing destructive weather, poverty, and hunger is likely to rise significantly if the worst effects of climate change cannot be mitigated. One study, based on past patterns of the effect of droughts on movement, posited that by the year 2080, climate change effects in Mexico could drive 6.7 million migrants from that country alone north to the United States. The United States will not be spared either, however. A two-degree (Celsius) rise in the global temperature is predicted to lead to the loss of the West Antarctic ice sheet. The resulting rise of sea levels could displace as many as 79 million people, many of them on the Eastern Seaboard of the United States (Lynas, 2020).

Can the damaging effects of climate change on productive land be mitigated with serious, sustained action on the part of communities and government? What is needed to spur significant public and private action on climate change? If that fails because change is perceived as economically costly or politically unpalatable, will countries in more temperate zones, like the United States, be willing to accept desperate migrants seeking new means to survive—and what will happen to those left behind?

This chapter takes on a variety of critical contemporary concerns. First, we discuss global population growth and look at the debate on rising populations. Next, we examine urbanization, gentrification, and the growth of megacities around the world. Finally, we turn to the problem of climate change and discuss the value of taking a sociological perspective on its causes and consequences. While the three key topics in this chapter are diverse, they share a common thread: *Individual choices add up to phenomena that can have powerful wider impacts.* A family may choose to buy an SUV to transport their children, while another chooses a fuel-efficient vehicle. A family may choose to have six children rather than one, or it may opt to leave a rural village to move to a city or to cross a border in search of better opportunities. What appears at the micro level as an individual decision with direct effects only on a particular family can add up to macro-level phenomena with national or global effects and consequences. At the same time, these individual *choices* are, as we saw in the opening story, often themselves a consequence of larger social, economic, political, or environmental phenomena that leave families and communities little choice in their pursuit of survival.

GLOBAL POPULATION GROWTH

The world's population is growing at a rapid rate, expanding as much since 1950 as it did in the preceding 4 million years. By 1850, the global population reached 1 billion; by 1950, it was 2.3 billion. As of 2012, it had reached more than 7 billion; and in mid-2020, there were an estimated 7.8 billion inhabitants on our planet. Based on 2020 data, the Population Research Bureau estimates that the global population will reach 9.9 billion by 2050 (Population Reference Bureau, 2020). The study of population, including the rise in number of the Earth's inhabitants, is called *demography*. **Demography** is the *science of population size, distribution, and composition* (Keyfitz, 1993).

Demography: The science of population size, distribution, and composition.

Population growth is highly uneven around the world, with the greatest expansion taking place in developing countries. Consider that about half of the increase in global population between 2010 and 2050 is projected to take place in nine countries, all but one of which are in the developing world: India, Pakistan, Nigeria, Ethiopia, the United States, the Democratic Republic of Congo, Tanzania, China, and Bangladesh (Figure 17.1). In the United States, most population growth will take place as a result of immigration; elsewhere, it will be the product of *natural population increase*—that is, it will result from births outpacing deaths.

Jakarta, the capital of Indonesia, is one of Asia's rapidly growing cities. While urbanization is a key characteristic of modernity, it also brings new health, cultural, political, and economic challenges.

While the global total fertility rate (TFR)—the average number of births per woman, as noted in Chapter 8—was 2.3 in mid-2020, TFR differed substantially across countries and regions (see Table 17.1). In reporting TFR, the Population Reference Bureau distinguishes among *more developed countries, less developed countries*, and *least developed countries*. In the more developed countries in 2017, the TFR was 1.6; in less developed countries, it was 2.5 (2.8 when China is excluded); and in the least developed countries, it was 4.1 (Population Reference Bureau, 2020). *Replacement rate fertility*—the rate at which two parents are only replacing themselves—is represented by a TFR of 2.1 (with an allowance of risk for mortality in the fraction above 2.0). Below this rate, populations decline; above it, they grow.

FIGURE 17.1 ■ Estimated World Population Growth, 1950–2100

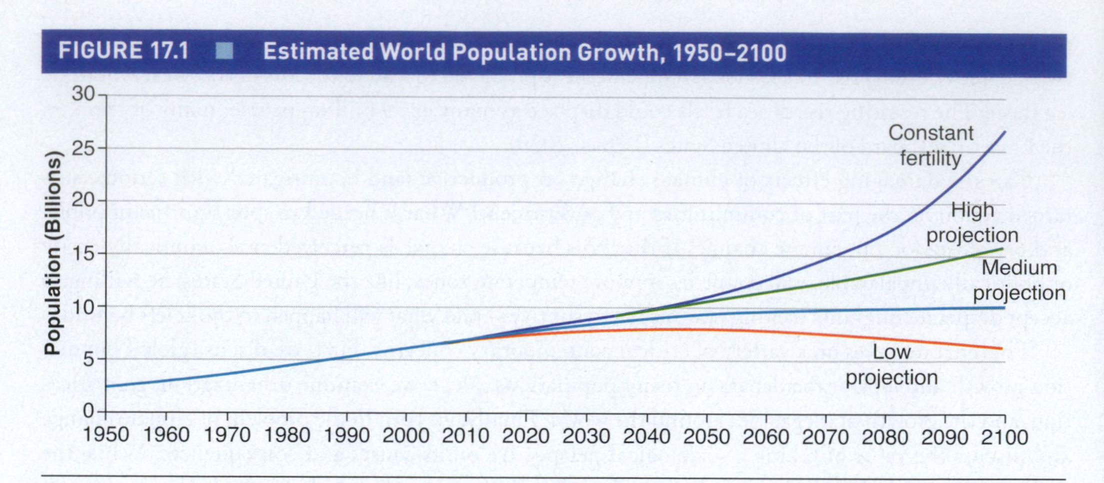

Source: United Nations, Department of Economic and Social Affairs, Population Division. (2011). *World Population Prospects: The 2010 Revision, Highlights and Advance Tables* (Working Paper ESA/P/WP.220). New York: Author.

TABLE 17.1 ■ Total Fertility Rates for Selected Countries, 2020

COUNTRY	TOTAL FERTILITY RATE
South Korea	0.9
Spain	1.3
Germany	1.6
Russia	1.6
United States	1.7
Turkey	2.3
Mexico	2.1
Israel	3.0
Egypt	2.9
Senegal	4.6
Mozambique	4.9
Niger	7.1

Source: Population Reference Bureau. (2020). 2020 World Population Data Sheet. Washington, DC: Author. Retrieved from https://www.prb.org/wp-content/uploads/2020/07/letter-booklet-2020-world-population.pdf

Central Africa, which comprises mainly less developed and least developed countries, is the fastest-growing region in the world. Its population is expected to grow from 280 million to 413 million between 2020 and 2050, whereas South America's population is estimated to rise from 429 million to 497 million in the same period, and Western Europe's will grow only modestly, from 195 million to 201 million. Other regions will lose population, including Eastern Europe, where the population is expected to decrease from about 291 million in 2020 to 263 million in 2050, due to below-replacement-level fertility rates (Population Reference Bureau, 2020).

Population momentum: The tendency of population growth to continue beyond the point when replacement rate fertility has been achieved because of the high concentration of people of childbearing age.

China and India, the world's most populous countries, present interesting cases for a discussion of population growth. China and India both have more than 1 billion inhabitants and high **population momentum**, which is *the tendency of population growth to continue beyond the point when replacement rate fertility has been achieved because of the high concentration of people of childbearing age.* China long sought to check its population growth with a one-child policy, which slowed its rate of growth and put its total fertility rate below replacement rate (1.8), though a recent easing of the policy has led to a small rise in the TFR. India's population growth has declined markedly over time (today, the TFR is 2.2), but the nation's population momentum remains high and its current population of 1.3 billion is expected to grow to about 1.5 billion by 2030 and 1.6 billion by 2050 (Table 17.2; Population Reference Bureau, 2020).

To the east of China is Japan, which, like its close neighbor Russia and distant neighbors in Eastern and Western Europe, is also experiencing population decline: It has a population of about 126 million people and a TFR of 1.3; by current projections, its population will fall to 109.9 million by 2050. Russia, currently one of the world's most populous countries, with 146 million inhabitants, has a TFR of 1.6 and is also expected to lose population by mid-century, declining to about 136 million people (Population Reference Bureau, 2020).

While some developing countries struggle with rapid population growth because it puts a strain on basic services such as sanitation and education, as well as on the conservation of natural resources, many modern industrialized countries are lamenting a "birth dearth" that leaves aging populations

TABLE 17.2 ■ Current and Projected Population Growth for Global Regions, 2020, 2030, and 2050

REGION	2020 POPULATION (MILLIONS)	MID-2030 PROJECTED POPULATION (MILLIONS)	MID-2050 PROJECTED POPULATION (MILLIONS)	BIRTHS PER 1,000 POPULATION
World	7,773	8,937	9,876	19
Sub-Saharan Africa	1,094	1,591	2,192	36
Northern Africa	244	306	367	24
Western Africa	401	587	818	37
Eastern Africa	445	645	874	35
Middle Africa	180	281	413	42
Southern Africa	68	79	87	20
Northern America	368	406	435	11
Central America	179	203	216	18
Caribbean	43	45	46	16
South America	429	476	497	15
Western Asia	281	344	389	20
South Central Asia	1,956	2,238	2,510	2
Central Asia	75	89	101	23
South Asia	1,967	2,269	2,456	21
Southeast Asia	662	749	800	18
East Asia	1,641	1,662	1,585	10
Northern Europe	106	112	115	11
Western Europe	195	201	202	10
Eastern Europe	292	279	264	10
Southern Europe	153	153	148	8
Oceania	43	53	63	17

Source: Population Reference Bureau. (2020). 2020 World Population Data Sheet. Washington, DC: Author. Retrieved from https://www.prb.org/wp-content/uploads/2020/07/letter-booklet-2020-world-population.pdf

Imagine a pond with a single water lily that doubles in size each day. On the 30th day, it covers the entire pond. On what day does it cover half the pond? The answer gives us a way of thinking about how population momentum drives rising population sizes, even given a constant growth rate. The answer can be found at the end of the chapter (Edward O. Wilson, "Is Humanity Suicidal?" *New York Times Magazine*, May 30, 1993).

dependent for their social welfare (in the form of public retirement benefits, for instance) on the financial contributions of fewer young workers.

Demography and Demographic Analysis

Demographers have developed statistical techniques for predicting future population levels on the basis of current characteristics. Annual population growth or decline in a country is the result of four factors: (1) the number of people born in the country during the year, (2) the number who die, (3) the number who immigrate into the country, and (4) the number who emigrate out. In the language of demographers, population changes are based on **fertility** (*the number of live births in a given population*), mortality (*the number of deaths in a given population*), and **net migration** (*in-migration minus out-migration*).

Let's look at fertility first. Demographers estimate future fertility on the basis of past fertility patterns of women of childbearing age. Although it is possible to make a rough estimate of population growth on the basis of **crude birthrate**—*the number of births each year per 1,000 women*—a far more accurate measure is **age-specific fertility rate**, *the number of births typical for women of a specific age in a particular population.* Demographers divide women into 5-year cohorts—for example, women ages 15 to 19, 20 to 24, 25 to 29, and so on. If the current average number of live births per 1,000 women is known for each of these age groups, it is relatively easy to project future fertility. Five years from now, for example, today's 15- to 19-year-old women will be 20 to 24, which means that the fertility rates of today's 20- to 24-year-old women can be applied to them. In this way, each successive cohort of women can be "aged" at 5-year intervals, and the result is an estimate of total live births. Since fertility rates in most cultures peak during women's late teens and 20s, the largest number of babies will be born to women in these age groups; thereafter, as the cohort ages into the 30s and 40s, the total number of babies born to the group will decline, dropping to zero as the cohort ages out of childbearing altogether.

Fertility: The number of live births in a given population.

Net migration: In-migration minus out-migration.

Crude birthrate: The number of births each year per 1,000 women.

Age-specific fertility rate: The number of births typical for women of a specific age in a particular population.

Crude death rate: The number of deaths each year per 1,000 people.

Age-specific mortality rate: An estimate of the number of deaths typical in men and women of specific ages in a particular population.

Life expectancy: The average number of years a newborn is expected to live based on existing health conditions in the country.

The second source of population change is mortality. Again, although **crude death rate**—the *number of deaths each year per 1,000 people*—yields a rough measure, demographers prefer to rely on **age-specific mortality rate**, or *an estimate of the number of deaths typical in men and women of specific ages in a particular population.* Similar to age-specific fertility rates, these rates are then applied to successive cohorts of men and women as they age. As you might guess, female mortality rates also affect the number of babies born, since as a cohort of females ages, some of its members will die, leaving fewer women of childbearing age. While in economically advanced regions (such as Western Europe) most women live well beyond their childbearing years, female mortality rates in younger cohorts may have notable effects on birthrates in regions with higher rates of early mortality such as Southern Africa (which includes countries such as South Africa, Botswana, and Namibia), in which an estimated 10.5% of women ages 18 to 24 are infected with HIV/AIDS (Population Reference Bureau, 2017).

One measure of the overall mortality of a society is its **life expectancy**, *the average number of years a newborn is expected to live based on existing health conditions in the country.* In almost all societies, the life expectancy at birth is higher for females than for males. In the United States in 2017, for example, the average life expectancy for females was 81 years, while it was 76 for males. In Eastern Europe, the gap between male and female life expectancy averages 10 years (Population Reference Bureau, 2020). Some of this gap is attributable to the fact that males are more likely than females to die in early childhood, to die from accidents or violence in young adulthood, and to experience early death from poor health in middle to later life.

Life expectancy varies significantly among countries of the world. It also varies by gender (Table 17.3). Women in Hong Kong rank at the top in terms of life expectancy (88 years), followed by women in San Marino and Japan (both 87 years). However, men in the Central African Republic rank

TABLE 17.3 ■ Life Expectancy at Birth by Gender for Selected Countries, 2020

	Life Expectancy at Birth	
Country	**Males**	**Females**
Democratic Republic of the Congo	58	62
Nigeria	54	56
South Africa	62	68
India	68	70
Pakistan	67	71
Senegal	66	70
Egypt	73	75
China	75	79
Mexico	72	78
United States	76	81
Denmark	79	83
Israel	81	85

Source: Population Reference Bureau. (2020). *2020 World Population Data Sheet.* Author. Retrieved from https://www.prb.org/wp-content/uploads/2020/07/letter-booklet-2020-world-population.pdf

very low, with a life expectancy at birth of only 51 (Population Reference Bureau, 2020). By region, life expectancy is lowest for men and women (57 and 59, respectively) in Western Africa and is highest in Western and Southern Europe, with life expectancies for men at 79 and women at 84.

It is difficult to make predictions about population growth with precision because predictions depend on assumptions about human behavior. If a government effectively implements family planning programs, for example, the fertility of the population may differ substantially after a decade or two. An unforeseen epidemic (on a scale such as that of HIV/AIDS or COVID-19) may increase mortality; conversely, the development of new drugs (such as new antibiotics or widely distributed immunizations) could greatly reduce it. For this reason, demographers typically offer a range of estimates of future populations: a low estimate that assumes high mortality and low fertility; a high estimate that assumes low mortality and high fertility; and an intermediate estimate that represents an informed figure somewhere in between.

Population forecasting also depends on fertility, but again, momentum is critical. In 1979, China decided to limit fertility by rewarding one-child families with additional income and preferential treatment in terms of jobs, housing, health care, and education, while threatening punishment for those who refused to keep their families small. This policy reduced China's TFR well below replacement rate fertility. Yet this reduction does not mean China's population will decline markedly from the 1.4 billion people it reached in 2020. With about 17% of its population under the age of 15—nearing childbearing years—China has substantial population momentum. According to the Population Reference Bureau (2020), its population will grow to about 1.42 billion in 2030 before falling back to 1.36 billion by 2050.

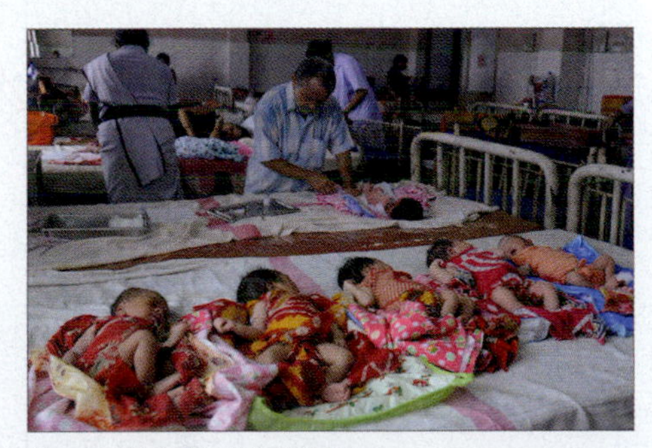

About a third of India's 1.3 billion people are under age 15, representing momentum for future population growth, even if family planning leads to smaller cohorts in the future. But son preference skews sex ratios at birth in some regions, which may affect the number of future mothers.

©ARINDAM DEY/AFP/Getty Images

Theory of the First Demographic Transition

Extrapolating from the Western model of population change, demographers have argued that many societies go through roughly the same stages of population transition. They suggest that societies begin with an extended stage of low or no growth resulting from high fertility and equally high mortality, pass through a transitional stage of explosive growth resulting from high fertility and low mortality, and end up in a final stage of slow or no growth resulting from low fertility and low mortality (Figure 17.2). This perspective on population change is called the *theory of the first demographic transition*. We outline this transition in more detail next.

Early agricultural societies had high fertility and mortality rates that generally counterbalanced one another. Consequently, the population was either stable or grew very slowly. Crude death rates as high as 50 per 1,000 were caused by harsh living conditions, unstable food supplies, poor medical care, and lack of disease control. Epidemics, famines, and wars produced high rates of death, periodically reducing the population even more dramatically. Such societies had to develop strong norms and institutions in support of high fertility to prevent a decline in population. Children were valued for a variety of cultural and economic reasons, particularly where they made contributions to hunting, farming, herding, weaving, and the other work necessary in a household-based economy (Simon, 1981).

As societies modernized (that is, became more urban and industrial), their birthrates initially remained high. At the same time, mortality rates plummeted due to improved food supplies brought about by growing trade links, sanitation control that accompanied greater public health knowledge and prosperity, and, eventually, modern medicine. Consider, for instance, the discovery that hand washing is important for doctors attending women giving birth. While this does not sound revolutionary today, it had a critical impact on maternal and child survival. In the 19th century, women in Europe and the United States had stunning rates of maternal mortality: Up to 25% of women delivering babies at hospitals died from puerperal sepsis, also known as *childbed fever*. Dr. Ignaz Semmelweis, a Viennese physician practicing in the 1840s, observed this phenomenon and recommended that attending physicians wash their hands with a chlorinated solution before assisting in childbirth. Semmelweis and his findings were harshly criticized, and he was ostracized for speaking out; his medical colleagues, nearly all of whom hailed from the upper classes, did not believe that gentlemen (even those attending a birth after dissecting a cadaver) could have dirty hands. As support for germ theory spread, however, Semmelweis's discovery proved to be critically important for the reduction of maternal mortality (Nuland, 2003).

FIGURE 17.2 ■ The First Demographic Transition Model in Four Stages

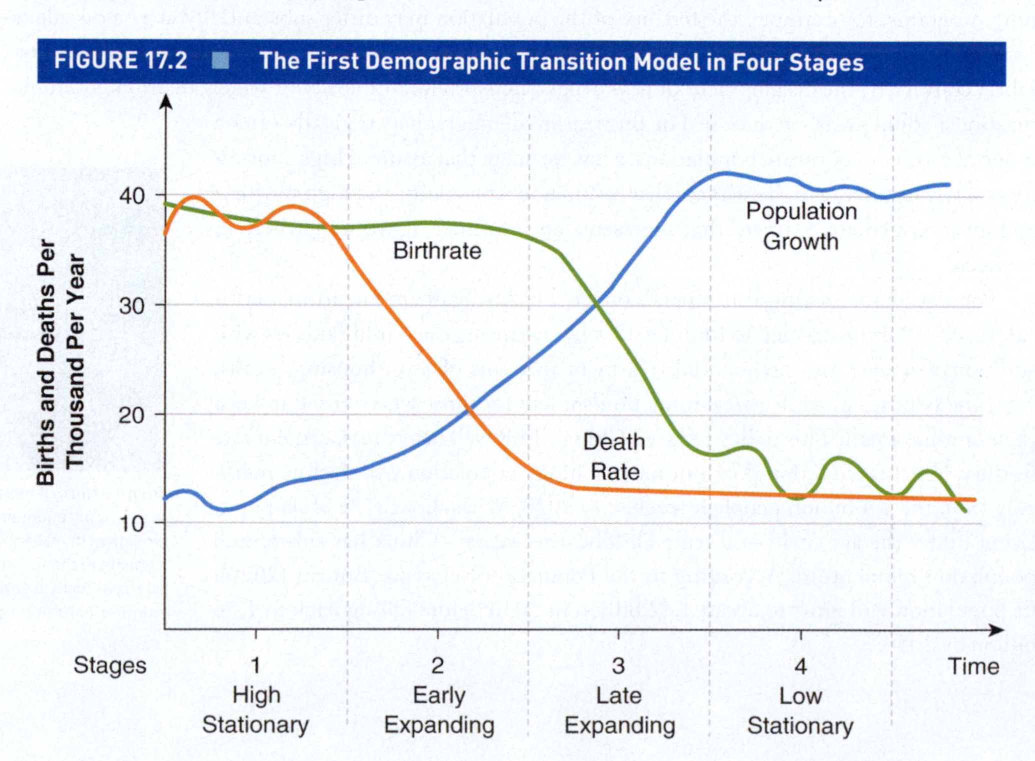

Because people did not initially change their fertility behavior—specifically, they did not stop having many children—as death rates dropped, the result was rapid population growth. Eventually, however, expanding industrialization and urbanization brought about a fall in fertility. Although during the early stages of industrialization, children often worked in factories and contributed to family income, the hardships of child labor eventually led to its legal prohibition. Rather than being an economic necessity, then, children became an expense. Urban living was also less amenable to large families than rural life had been. Population growth stabilized as low mortality came to be matched by falling fertility. At this point, the first demographic transition in the industrialized societies of the West was complete.

The theory of the first demographic transition offers a useful perspective based on a pattern observed in Western countries. We can critique it, however, for the same reason; that is, it describes the historical experience of today's modern Western societies. It does not describe nearly as well the experience of the developing world, which accounts for most global population growth today. In newly industrializing low-income countries, families do not always drastically reduce their fertility. And many low-income nations have not industrialized, yet their mortality rates have declined because populations have access to food, medicine (particularly antibiotics), agricultural and sanitation technologies, and pesticides, which contribute to better health and longer lives. Drops in mortality, however, have not been accompanied by drops in fertility as in modern industrialized states, where decreased fertility stemmed from economic growth, urbanization, industrialization, and expanded educational opportunities.

High fertility combined with low (or relatively low) mortality underlies much of the population explosion in the developing world. Fertility remains high in poor countries for a number of reasons, including health and culture. For instance, families in regions that still experience high child mortality are more likely to have "extra" children to ensure that some survive, and notions about the ideal size of families are culturally variable.

Dr. Ignaz Semmelweis observed higher maternal mortality in Clinic 1 of his hospital, staffed by male obstetricians who performed autopsies, than in Clinic 2, staffed by female midwives who did not. Chlorine washing of male birth attendants' hands reduced maternal mortality until 1850, when Semmelweis left the hospital and old practices resumed.

©Bettmann/Getty Images

Economic factors also shape fertility decisions. Consider the relationship between economic rationality and childbearing. The saying "Children are a poor man's riches" highlights the fact that in agricultural societies in particular, children (specifically male children) are a value more than a cost. In rich and industrialized countries, children, while emotionally valued, are economically costly and contribute little or nothing to the household in terms of economic value. Economic rationality, then, is present in both the decision of a poor household in the developing world to have many children and the decision of a rich household in the developed world to have few.

One of the most important factors contributing to both reduced fertility and improved child survival is the education of women (Figure 17.3). According to one study, for every 1-year increase in the average education of women in their childbearing years, a country experienced a 9.5% fall in child mortality (Gakidou, Cowling, Lozano, & Murray, 2010). Strikingly, a child born to a literate woman is 50% more likely to survive to age 5 than a child born to a woman who cannot read (United Nations Educational, Scientific and Cultural Organization, 2010).

Women who are more educated are more likely to be familiar with family planning methods and more likely to use them. Better-educated women are also more likely to have jobs in the formal labor force and, consequently, to limit their fertility in order to bring in economic resources. Finally, women with some economic resources of their own also tend to have more decision-making power in the family, allowing them to participate in making choices about birth control and family size (Pradhan, 2015; United Nations, 1995).

Better health services for children and declining infant and child mortality also have important effects on fertility decisions. When women have reason to believe that all or most of their offspring will survive into adulthood, they are less likely to have "extra" children to ensure that a few survive (Kibirige, 1997).

FIGURE 17.3 ■ Relationship Between Fertility and Female Education: Total Fertility Rate by Years of Schooling

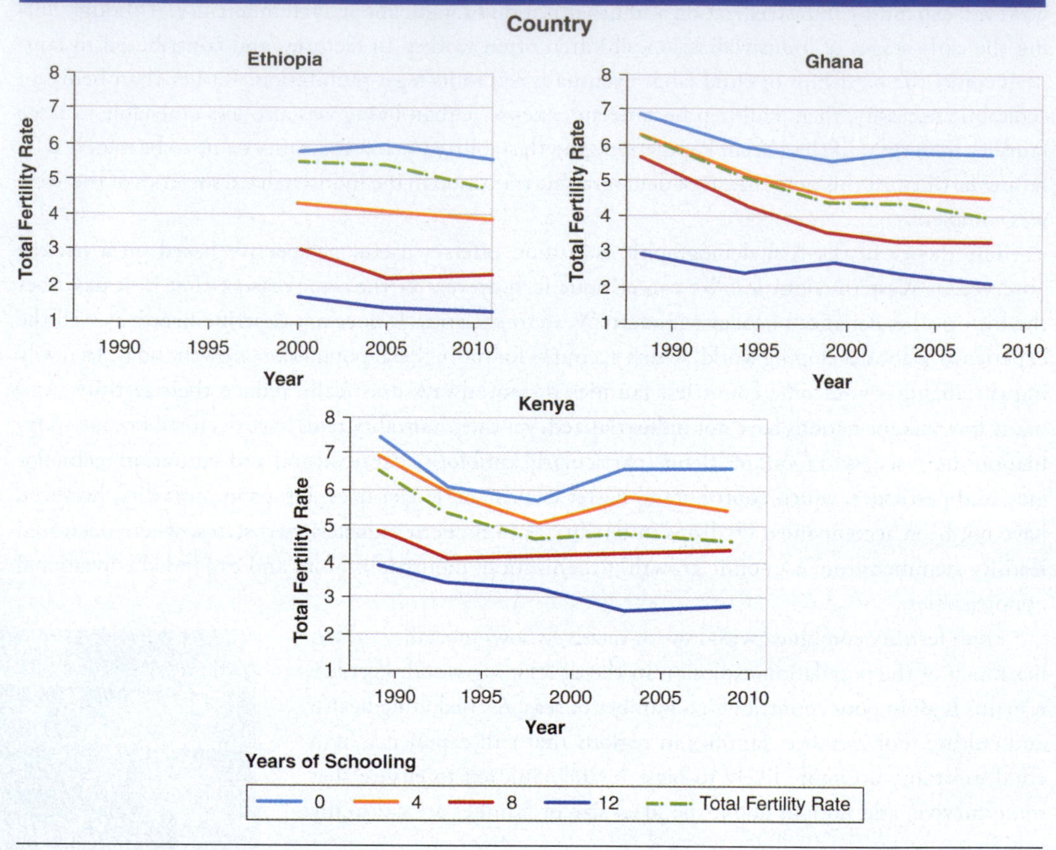

Source: Pradhan, E. (2015, November 27). *The Relationship Between Women's Education and Fertility.* World Economic Forum.

In the West African country of Mali, women with a secondary education or more bear an average of three children, while women without education have an average of seven children. Global demographic data show a strong correlation between women's educational attainment and fertility.

©REUTERS/Joe Penne

Developed countries and the governments of some high-fertility countries have helped finance family planning programs in order to educate people about birth control. In addition, they have provided residents with condoms, birth control pills, and other means of reducing fertility. While such programs have met with some success, they often run up against deep-seated religious and other cultural values, some of which are found in the developed states as well. For instance, during Ronald Reagan's presidency and the presidencies of both George H. W. Bush and George W. Bush, U.S. governmental funding of family planning programs abroad was cut significantly because birth control (and especially abortion) violated the beliefs of conservative supporters of those administrations. It is clear that global efforts in family planning are fraught with problems. It remains to be seen whether they will ultimately succeed in reducing fertility (and hence global population growth).

Is a Second Demographic Transition Occurring in the West?

In the theory of the first demographic transition, population reaches a point of stabilization in Stage 4. What happens after that? Some demographers argue that at least in the most developed countries, stabilization has been followed by a *second demographic transition* (Lesthaeghe, 1995; McNicoll, 2001) characterized by broad changes in family patterns. Countries experiencing a second demographic transition may be characterized by increased rates of divorce and cohabitation, for instance, as well as decreased rates of marriage and fertility and a rise of nonmarital births as a proportion of all births. According to some demographers, Germany, France, and Sweden are among the countries experiencing a second demographic transition. Because changes in family patterns are associated with smaller families, the second demographic transition often includes a decline in the population's **rate of natural increase (RNI)**—that is, *the crude birthrate minus the crude death rate.*

Rate of natural increase (RNI): The crude birthrate minus the crude death rate.

If a country has a negative RNI, this can lead to population declines. Russia, for example, is experiencing relatively rapid loss of population: People are dying at a faster rate than they are being born, and Russia can expect to see a population loss of about 10% through the middle of the century (Population Reference Bureau, 2020). The result is an inverted age pyramid, wider at the top and narrower at the bottom (Figure 17.4). Some countries with a negative RNI, however, still experience growing populations as a result of immigration. By contrast, countries with above-replacement-level fertility rates and growing populations have normal pyramids, such as that for Kenya shown in Figure 17.5.

What accounts for the second demographic transition? Demographer Ron Lesthaeghe (1995) argues that "the motivations underlying the 'second transition' are clearly different from those supporting the 'first transition,' with individual autonomy and female emancipation more central to the second than to the first" (p. 18). According to this perspective, more people associate personal satisfaction with consumption and personal fulfillment and are not as likely to seek fulfillment through family relationships. Female emancipation has also been broadened by the medical evolution of the "perfectly contracepting society" (Westoff & Ryder, 1977), such that women have unprecedented control over their fertility, and many have chosen smaller families or have opted not to have children.

Critics suggest that the second transition describes only a fraction of the world's population. However, the number of countries experiencing below-replacement birthrates is rising, and a sociological and demographic perspective such as the theory of the second transition offers sociologists analytical tools for understanding these key trends in advanced industrial states.

Falling Fertility in Developed Countries

In economically developed countries such as England, Germany, and Japan, growing proportions of young adults are choosing to forgo having children. According to a recent article on the topic,

FIGURE 17.4 ■ Population Pyramid for Russia, 2020

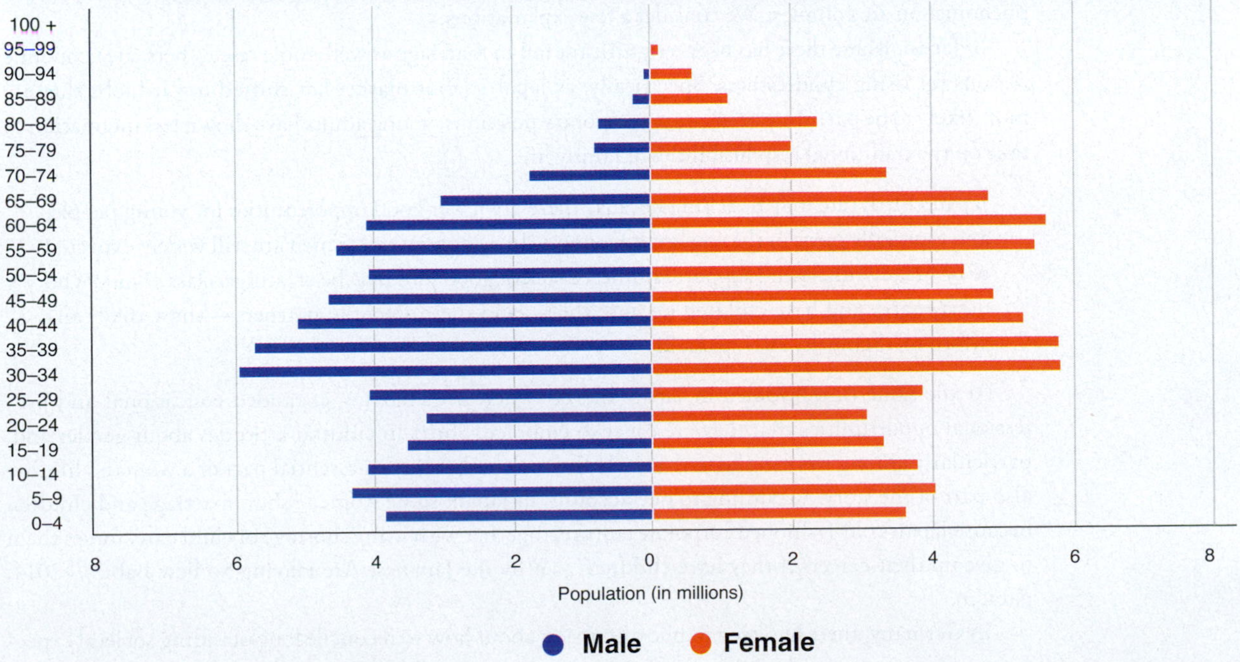

Source: Central Intelligence Agency. *The World Fact Book.*

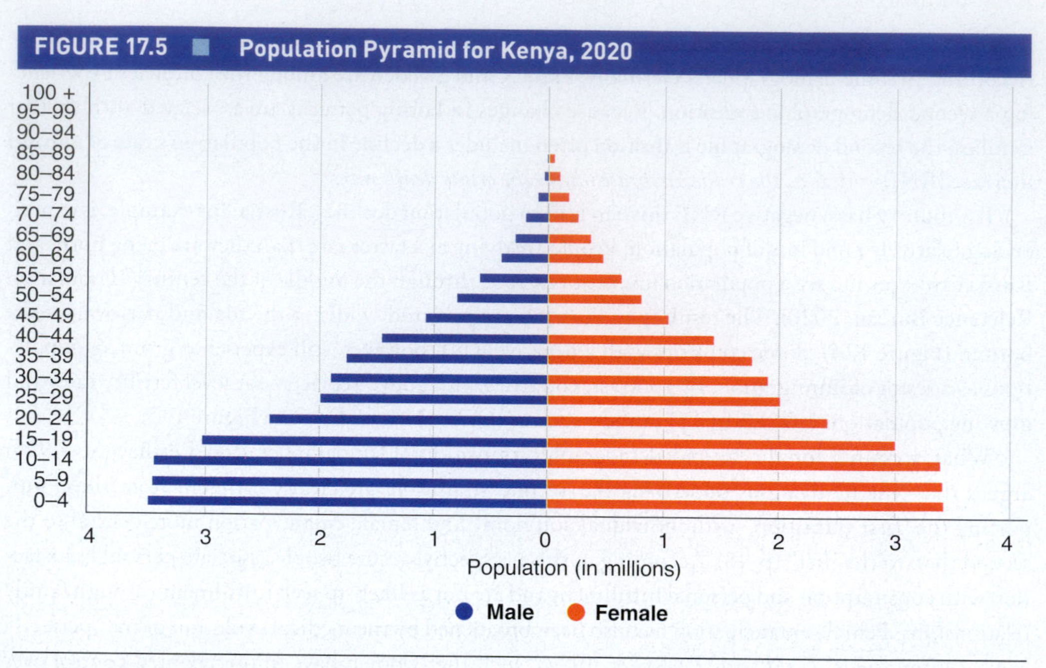

FIGURE 17.5 ■ Population Pyramid for Kenya, 2020

Source: Central Intelligence Agency. *The World Fact Book*.

Just 9% of English and Welsh women born in 1946 had no children. For the cohort born in 1970—who, barring a few late surprises, can be assumed to be done with babies—the proportion is 17%. In Germany, 22% of women reach their early 40s without children; in Hamburg, 32% do. ("The Rise of Childlessness," 2017, para. 2)

Declines in childbearing are being felt in demographic indicators:

> Since Japan began counting its newborns more than a century ago, more than a million infants have been added to its population each year.... No longer, in the latest discomforting milestone for a country facing a steep population decline. Last year, the number of births in Japan dropped below one million for the first time. (Soble, 2017, paras. 1–2)

What explains the rise in childlessness among young adults? The sociological roots of this growing phenomenon are complex. We consider a few explanations.

In Japan, where there has been a significant fall in marriage as well, some researchers cite economic reasons for rising childlessness. Specifically, as Japan's labor market has shifted toward jobs that are more likely to be part-time or insecure temporary positions, young adults have shown less inclination to take on the traditional responsibilities of family life:

> Japan's birth rate may be falling because there are fewer good opportunities for young people, and especially men, in the country's economy. In a country where men are still widely expected to be breadwinners and support families, a lack of good jobs may be creating a class of men who don't marry and have children because they—and their potential partners—know they can't afford to. (Semuels, 2017, para. 2)

At the same time, women in Japan and elsewhere are enjoying expanded educational and professional opportunities that they are eager to embrace. Shifts in cultural attitudes about gender and, particularly among women themselves, whether motherhood is an essential part of a woman's life, are also part of the story. According to one account, in Japan, some women "shun marriage and children because Japan's old-fashioned corporate culture, together with a dire shortage of child care, forces them to give up their careers if they have children" ("Why the Japanese Are Having So Few Babies," 2014, para. 3).

In Germany, there has been public discussion about how to reconcile long-standing societal expectations that women with children remain at home to raise the children rather than returning to the workforce with the country's concern about rising childlessness. Women who wish to have children but also to continue their careers have faced challenges in Germany (Nicholson, 2013).

Interestingly, German demographer Michaela Kreyenfeld points to the possibility of some explanatory variation for childlessness by gender:

> Women often have no children because they have prioritised education or work in their 20s and 30s. Men are more likely to remain childless because women do not view them as good boyfriend material—let alone good husband or father material. They have a problem finding partners. ("The Rise of Childlessness," 2017, para. 15)

Decisions about having a child or children are highly personal. At the same time, they are made in cultural, economic, and political contexts that influence the decisions of young adults. The sociological imagination helps us to see how individual choices add up to societal-level outcomes and how societal conditions structure those choices.

MALTHUS AND MARX: HOW MANY PEOPLE ARE TOO MANY?

High rates of population growth in some parts of the world may seem daunting, even alarming to some, but numbers alone do not tell the entire story. There is still a great deal of sparsely settled land in the world, and if the planet's 7.8 billion people were all somehow transplanted to the territory of the United States, the resulting crowding would be no greater than currently exists in the country of Taiwan.

Is the world overpopulated? This question is the subject of debate. On one side are those who predict catastrophe if population growth is not slowed or stopped altogether. Activists who fear that a population doomsday is just around the corner often conclude that drastic measures are required, including stringent public policies that promote small families. On the other side are those who argue that while population growth should be slowed, extreme measures are unwarranted. They tend to favor expanded female education, voluntary family planning programs (though some groups object to contraception as well as abortion), and economic policies that raise living standards, making smaller families a more rational economic choice.

Malthus: Overpopulation and Natural Limits

The argument that the world is overpopulated was first made two centuries ago by British social philosopher Thomas Malthus (1766–1834). Malthus (1798/1926) developed the theory of exponential population growth: the belief that, similar to compound interest, a constant rate of population growth produces a population that grows by an increasing amount with each passing year. **Exponential population growth**, thus, refers to *a constant growth rate that is applied to a base that is continuously growing in size, producing a population that grows by an increasing amount with each passing year.* Malthus also posited that although population grows exponentially, the food supply does not; the Earth's resources are finite. Consequently, though the population may continue to double, the amount of food is more likely to grow at a constant rate. The result, according to Malthus's dire warning, is a growing mismatch between population and food resources. Unless we take steps to control our population growth, Malthus predicted, nature will do it for us: Wars fought over scarce resources, epidemics, and famine will keep population in check. Back when only a billion people occupied the entire planet, Malthus believed doomsday was already on the horizon.

Exponential population growth: A constant rate of population growth applied to a base that is continuously growing in size, producing a population that grows by an increasing amount with each passing year.

How accurate was Malthus's prediction? Although war, epidemic, and famine have in fact been sadly evident throughout human history, and population has continued to grow exponentially, the food supply has grown along with it. Malthus failed to recognize that modern technology can also be applied to agriculture, yielding an exponential growth in food supplies—at least for a time. A report by the Food and Agriculture Organization of the United Nations (2009) concludes that although the world's population may reach over 9 billion by 2050, "the required increase in food production can be achieved if the necessary investment is undertaken and policies conducive to agricultural production are put in place," along with "policies to enhance access by fighting poverty, especially in rural areas, as well as effective safety net programmes" (p. 2).

Although Malthus's predictions of global catastrophe have not been borne out, there may be a limit to the carrying capacity of the planet. World population cannot continue rapid growth indefinitely without consequences. Yet to point out that population growth presents serious challenges is not the same as concluding that the limits have been reached. Despite Malthus's pessimistic prophecies and Paul Ehrlich's dire warnings in *The Population Bomb* (1968), the long-predicted population doomsday has yet to arrive. In fact, the connection between population growth and human misery appears to be more complicated than many analyses have suggested. The issue is not simply how much additional food will be required to feed more mouths, but also whether the food that is produced will reach those who need it. Mass hunger in many countries is as much a product of politics as of true lack of food; in civil conflicts, for instance, hunger may be used as a weapon, with opposing forces blocking food shipments to enemy areas.

Simon: A Modern Critic Takes on Malthus

Economist Julian Simon (1977, 2000) became well known for his pointedly anti-Malthusian perspective on population. Simon not only rejected the notion that unchecked population growth would lead humanity down a path to hunger, deprivation, and poverty but also posited precisely the opposite. He argued that population growth has positive economic effects. In his most recent work, published posthumously, Simon (2000) suggested that *sudden modern progress* (SMP)—the rapid rise in living standards and technology—is the result of population growth and density. Put another way, the great population growth of the modern period was a causal factor of "the great breakthrough" (which is also the title of his book)—that is, population growth brings about technological progress. More people means more minds and more innovation, so human-generated resources, Simon argued, can overcome limitations on natural resources. Unlike Malthus, Simon was encouraged rather than daunted by the prospect of rising populations.

Indeed, the question of whether more minds can balance the pressure caused by more bodies is an engaging and imperative one. Can population growth be both problematic and powerful? What do you think?

Marx: Overpopulation or Maldistribution of Wealth?

Malthus forecast misery and inevitable overpopulation under conditions of growth. Simon saw population growth and density as necessary conditions for economic progress. Karl Marx focused on the unequal distribution of resources across populations.

Marx was concerned about the dominance of an economic system that enables the wealthy few to consume the world's resources at the expense of the impoverished masses.

He was critical of Malthus for claiming that overpopulation is the central cause of human starvation and misery. In Marx's (1992a) view, the central problem is not a mismatch between population size and resource availability, but rather the unequal distribution of resources; in most societies, as well as in the world as a whole, he argued, the members of a small elite enjoy the lion's share of the wealth and resources while the majority are left to take up the crumbs that fall from the richly endowed tables of the few.

Parente (2008) notes that even after adjustments are made for differences in relative prices and gross domestic product per capita, the living standards in the wealthiest industrial countries are about 50 to 60 times greater than those in the world's poorest countries. Most of the world's resources and goods are consumed by the West: Western Europe and North America, with 12% of the world's population, account for 60% of private consumption expenditures. The United States alone, home to about 5% of the world's population, burns about a quarter of the globe's coal, oil, and natural gas. In contrast, sub-Saharan Africa and South Asia, which together are home to more than one third of humanity, account for just over 3% of private consumer expenditures. At the beginning of the new millennium, about two fifths of the Earth's inhabitants lived on less than $2 per day (Worldwatch Institute, 2011). Marx argued that such maldistribution is the result of a capitalist economic system that divides people into unequal social classes.

Both Malthus and Marx were writing when the world's population was around 1 billion people. Marx's criticism of Malthus has stood the test of time, since world population has doubled and more than doubled again since Malthus's predictions were made, but global resources have not run dry. Looking critically at Marx's ideas, we might note that although the maldistribution of wealth is an important factor in understanding poverty and human misery, Marx underestimated the importance of population growth itself as a variable.

Malthus, Marx, and Modernity

Consider this question: What is the greater threat to the health and survival of our global environment—the rapid growth of the populations of the developing world or the overconsumption of resources by the small stratum of the wealthy? While some policy makers in the West express concern about the threats posed by unchecked population growth or the use of "dirty" technologies in developing states or the decimation of rainforests in the Amazon and elsewhere, there is little vigorous mainstream debate over the global threat presented by the recklessly wasteful consumption of resources by Western consumers. To cite only one example of the way Western consumption is masked by a focus on the global poor: Broad media attention has been given to the millions of acres of rainforest lost to clear-cutting in poor states, not least because of the immense biodiversity that has been sacrificed. Many in the United States have donated money to campaigns aimed at saving these precious resources. At the same time, heedless U.S. consumers of steaks, burgers, and other beef products may not recognize that some of the clear-cutting is done by ranchers seeking land on which to farm cattle, the meat from which will be sent to our supermarkets, restaurants, and dinner tables.

Like many cities, the capital of the Philippines, Manila, is a study in contrasts, with a nascent professional middle class and a persistent problem of poverty. In this photo, homeless children sleep beneath a bustling city bridge.

The question of whether overpopulation or overconsumption is the greater threat is a provocative one. Neither phenomenon is without consequences. Malthus feared that population would outpace food production; Marx posited that elites would consume far more than their share. A modern take on this debate points us to the conflict between underdevelopment in some states and "overdevelopment" in others.

As we shall see later in this chapter, environmental stresses have grown substantially since Malthus's and Marx's time, and some scientists believe we are approaching a point of no return in inflicting environmental damage on the planet. While overconsumption presents threats to our future, the consequences of the population explosion must also be faced. Since much of the world's population increase is currently concentrated in urban areas, we will next examine the impact of urbanization on modern life before turning to the environmental effects of urbanization and population growth combined.

URBANIZATION

What is a *city*? Early urban sociologist Louis Wirth (1938) wrote that "a sociologically significant definition of the city seeks to select those elements of urbanism which mark it as a distinctive mode of human group life" (p. 4). For sociological purposes, and thus for ours too, Wirth's definition is useful: A **city** *is "a relatively large, dense, and permanent settlement of socially heterogeneous individuals"* (p. 8).

City: A relatively large, dense, and permanent settlement of socially heterogeneous individuals.

In the eyes of some literary writers, cities are grim places of human degradation and misery. In the 19th-century poem "The City of Dreadful Night," James Thomson (1874) describes such a place:

That city's atmosphere is dark and dense,

Although not many exiles wander there,

With many a potent evil influence,

Each adding poison to the poisoned air;

Infections of unalterable sadness,

Infections of incalculable madness,

Infections of incurable despair.

Some writers have praised the modernity, power, and culture of the city, while others have recognized the dramatic contradictions of cities and their ability to both attract and repel. Honoré de Balzac wrote of the inhabitants of Paris in 1833:

> By dint of taking in everything, the Parisian ends by being interested in nothing. No emotion dominating his face, which friction has rubbed away, it turns gray like the faces of those houses upon which all kinds of dust and smoke have blown.... [The Parisian] grumbles at everything, consoles himself for everything, jests at everything, forgets, desires, and tastes everything, seizes all with passion, quits all with indifference—his kings, his conquests, his glory, his idols of bronze or glass—as he throws away his stockings, his hats, and his fortune. In Paris, no sentiment can withstand the drift of things.

Cities have become part of our lives and lore, but they were not always so. Cities have a long history, but until the Industrial Revolution, most people were rural dwellers. Today, most of the world's people live in cities and embody the beauties, miseries, and contradictions of those places. Below, we turn a sociological lens on the cities of the United States and the world.

The Rise of Industry and Early Cities

Agricultural surplus: Food beyond the amount required for immediate survival.

Preindustrial cities, based on both agriculture and trade, first appeared 10,000 to 12,000 years ago. The development of settled agricultural areas enabled farmers to produce an **agricultural surplus**, *food beyond the amount required for immediate survival.* This surplus in turn made it possible for cities to sustain populations in which residents were not engaged primarily in farming. The first known cities were small, their populations seldom exceeding a few thousand, since the surplus production of 10 or more farmers was required to support a single nonfarming city dweller. The need for access to transportation routes and rich soil for farming figured prominently in the siting of the earliest cities along major river systems (Hosken, 1993). Early city residents included government officials, priests, handicraft workers, and others specializing in nonagricultural occupations, although many city dwellers engaged in some farming as well. Until modern times, very few cities in the world surpassed 100,000 people. More than 2,000 years ago, Rome was considered an enormous metropolis, with 800,000 people; today, it would be comparable in population to a U.S. city such as San Francisco (Mumford, 1961).

Urbanization: The concentration of people in urban areas.

The Industrial Revolution of the 18th century radically changed the nature of cities. While cities of the past had served primarily as centers of trade, industrial cities emerged as centers of manufacturing. Although some of the earliest English factories were in small cities, by the 19th century, industrialization was moving hand in hand with **urbanization**, *the concentration of people in urban areas.* At the beginning of the 19th century, there were barely 100 places in England with more than 5,000 inhabitants; by the end of the century, there were more than 600 that, together, contained more than 20 million people. The city of London grew from 1.1 million to 7.3 million people between 1800 and 1910 (Hosken, 1993), becoming a center of industry as well as an ever more important hub of commerce and culture. In the United States as well, the explosion of cities coincided with the onset of industrialization at the end of the 19th century. By the early 20th century, most U.S. citizens could be classified as urban, and today, as in Britain and other Western European states, the vast majority live in metropolitan areas.

Sociologists and the City

The early industrial cities were grimy places in which people lived in shanties and shacks, often in the shadows of the factories where they spent more than a dozen working hours each day. In the absence of proper sanitation and sewage systems, illness and epidemics were common, and many people died of typhoid, cholera, dysentery, and tuberculosis. Writers such as Charles Dickens captured the miseries of early urban industrial life, where men, women, and children toiled in dank, squalid conditions and lived lives of deprivation and degradation. Early sociologists also turned their lenses on the city, some viewing urban life as bordering on the pathological. At the same time, most recognized that cities provided opportunities for individuality and creativity.

During the rapid urbanization of the 19th century, some of the earliest sociologists worried about the differences between a presumably serene country life and the "death and decay" of city life (Toennies, 1963). The rural community (*Gemeinschaft*, in the original German) was contrasted with urban society (*Gesellschaft*), much to the disadvantage of the latter. Rural community life was said to be characterized by intimate relationships, a strong sense of family, powerful folkways and mores, and stabilizing religious foundations. Urban life, by contrast, was believed to be characterized by impersonal and materially based relationships, family breakdown, and the erosion of traditional beliefs and religious values. The behavior of city dwellers was viewed as governed no longer by long-standing social norms, but by cold cost-benefit calculations, individual preferences rather than a sense of collective good, and ever-changing public opinion.

While alienation was assumed to be a product of this grimly efficient urban world, Émile Durkheim put forth the notion that the *mechanical solidarity* of traditional community life (based on homogeneity) could be replaced by the *organic solidarity* of modern societies, with complex divisions of labor in which people were heterogeneous but interdependent for survival and prosperity.

During the 1920s and 1930s, researchers at the University of Chicago turned their city into a vast laboratory for urban studies, pioneering urban sociology as a field. Early 20th-century U.S. sociology centered on the study of "social problems" such as hoboes, the mentally ill, juvenile delinquents, criminals, prostitutes, and others who were seen as casualties of urban living. Urbanism was believed to be a specific way of life that resulted from the geographic concentration of large numbers of socially diverse people. Sociologists believed that one feature of this life was a good deal of mutual mistrust, leading city residents to segregate themselves on the basis of race, ethnicity, class, and even lifestyle into neighborhoods of like-minded people (Wirth, 1938). Of course, as we saw in our discussion of race in Chapter 9, self-segregation should be distinguished from imposed segregation that is the result of individual or institutional racism.

Although early sociologists often linked city life with pathology, even in supposedly impersonal cities, people maintain intense social networks and close personal ties. Numerous sociological studies have found that significant community relationships persist within even the largest cities (Duneier, 1992; Fischer, 1982, 1984; Gans, 1962a, 1962b; Liebow, 1967; Whyte, 1943; Wirth, 1928). Physical characteristics such as size and density do not by themselves account for urban problems. What is important is the way a particular society organizes itself in cities. Next, we look at cities in the United States.

Cities in the United States

Transportation and communication technologies have played important roles in helping to shape U.S. cities. Highways and urban rail and subway systems enabled cities to expand outward by allowing people to travel greater distances between home and work. Similarly, modern construction technologies permitted cities to expand skyward, with massive skyscrapers that take up little land space but create living or working space for thousands.

The Social Dynamics of U.S. Cities and Suburbs

Political and economic forces are also key to shaping modern cities. Sociologists John Logan and Harvey Molotch (1987; Molotch, 1976; Warner, Molotch, & Lategola, 1992) have argued that cities

Urban growth machine: Those persons and institutions that have a stake in an increase in the value of urban land and that constitute a power elite in cities.

are shaped by what they call the **urban growth machine**, *those persons and institutions that have a stake in an increase in the value of urban land and that constitute a power elite in most cities.* These people and institutions are said to include downtown businesses, real estate owners (particularly those who own commercial and rental property), land developers and builders, newspapers (whose advertising revenues are often tied to the size of the local population), and the lawyers, accountants, architects, real estate agents, construction workers, and others whose income is tied to serving those who own land. Logan and Molotch view the urban growth machine as dominating local politics in most U.S. cities, with the result that cities often compete with one another to house factories, office buildings, shopping malls, and other economic activities that will increase the value of the land owned by the members of the growth machine.

The great highway building projects of the postwar period literally paved the roads to America's suburbanized future. Robert Fishman suggests that "every true suburb is the outcome of two opposing forces, an attraction toward the opportunities of the great city and a simultaneous repulsion against urban life" (quoted in Gainsborough, 2001, p. 33). Is this an accurate characterization? Why or why not?

G. William Domhoff (2002) expands this notion with his concept of *growth coalitions:* groups "whose members share a common interest in intensifying land use in their geographic locale" (p. 39). Growth coalitions are a powerful force in local politics, though they encounter conflicts in their pursuit of growth and profit. Domhoff notes that

> neighborhoods are something to be used and enjoyed in the eyes of those who live in them, but they are often seen as sites for further development by growth coalitions, who justify new developments with a doctrine claiming the highest and best use for land. (p. 40)

Consider Domhoff's conflict-oriented analysis of U.S. politics and decision making, which we examined in Chapter 14. Recall that Domhoff asks us to ponder the questions *Who benefits? Who governs? Who wins?* Looking at growth politics—the increasing development of already crowded suburbs or the gentrification of urban neighborhoods at the expense of historical residents and businesses—we may ask who has the power and the resources to realize their interests. Are those who favor *use value* (the enjoyment of the neighborhood) or those who favor *exchange value* (the economic development of an area) more powerful? Is the shape of modern cities and suburbs a product primarily of those who inhabit them or those who profit from them? Can these values be balanced?

The post–World War II development of U.S. cities and suburbs illustrates how the growth machine operates at the national level. The rapid growth of suburbs during the 1950s and the 1960s is often attributed to people's preference for suburban living, but it was also a product of economic forces and government policies designed to stimulate the postwar economy (Jackson, 1985; Mollenkopf, 1977). The government's influence is illustrated by the 1956 National Interstate and Defense Highways Act, which established a highway trust fund paid for by a federal tax on gasoline. This legislation ensured a self-renewing source of funding to construct high-speed freeways connecting cities and suburbs across the country. The legislation eventually financed nearly 100,000 miles of highway, characterized by President Dwight Eisenhower at the time as "the greatest public works program in history" and enough to build a "Great Wall" around the world 50 feet wide and 9 feet high. The law originated in the planning efforts of a powerful consortium of bankers, corporations, and unions connected with the automobile, petroleum, and construction industries (Mollenkopf, 1977).

The Highways Act, in combination with other government programs, spurred the growth of the U.S. economy, both by improving transportation and by promoting the automobile and construction industries. The existence of freeways encouraged people to buy more cars and to drive additional miles, generating still more gasoline tax revenues for highway construction. Large numbers of people bought houses in the suburbs with federally insured and subsidized loans, commuted on federally financed highways to work in federally subsidized downtown office buildings, and even shopped in suburban shopping centers built in part with federal dollars. All this development ushered in a quarter century of growth and relative prosperity for working- and middle-class residents of the suburbs.

In fact, the rapid growth of suburbs in this period represents one of the most dramatic population shifts in U.S. history. In 1950, cities were home to about 33% of the U.S. population, considerably more than their small but growing suburbs (23%). During the 1960s the suburbs overtook the cities, and by 1990, the suburban population had reached 46%, while the urban population had declined slightly to 31% (Frey & Speare, 1991). Between 2012 and 2016, 55% of Americans lived in the suburbs, while the proportion in central cities dropped slightly to 31% (Pew Research Center, 2018b). In 2018, 80.6% of the U.S. population inhabited metropolitan areas (U.S. Census Bureau, 2018).

The emergence of postindustrial society has encouraged further development beyond the economic activity of the urban core. Modern information technology has made it easier for corporations to locate high-tech factories and office parks in once-remote suburban or even rural locations, where they find relatively inexpensive land, modern industrial facilities, fewer environmental problems, and incentive packages that can improve their profitability (Maidenberg, 2016). Since 1980, more than two thirds of employment growth in the United States has taken place outside central cities, and even manufacturing, while in decline nationally, has found a home in the suburbs; today, more than 70% of manufacturing is suburban (Wilson, 2010). According to *Crain's Chicago Business*, for example,

> Overall industrial employment... fell 21 percent between 2003 and 2013, as companies became more productive, and some firms shuttered while others moved factory work elsewhere. But the suburbs' decline was lighter than the city's: from 2003–2013, suburban manufacturing jobs fell 18 percent to about 292,000 positions, according to data provided by CMAP. By contrast, Chicago lost 33 percent of its manufacturing jobs, ending 2013 with 64,439 positions. (Maidenberg, 2016, para. 6)

While those who benefit from such relocations tend to be well-educated managerial, technical, and professional specialists, the suburbs are also home to a substantial proportion of entry-level positions. This is significant because it contributes to the loss of employment opportunities in central cities. Sociologist William Julius Wilson (2010) writes of the *spatial mismatch* between urban job seekers and suburban jobs, noting that "opportunities for employment are geographically disconnected from the people who need the jobs" (p. 41). He offers the example of Cleveland, where, "although entry-level workers are concentrated in inner-city neighborhoods, 80 percent of the entry-level jobs are located in the suburbs" (p. 41). Where public transportation is limited or few residents own cars, reaching suburban jobs can be a significant obstacle for urban dwellers.

The migration of middle- and upper-income families from the cities begun in the post–World War II era evolved into a critical socioeconomic disparity between the suburbs and central cities. Analyzing the concentrated poverty of urban core areas such as South Chicago, Wilson (2010) points out that political forces created a new urban poverty that plagues inner-city neighborhoods deeply segregated by both race and class. These forces included government support for highway building, a postwar mortgage lending boom that benefited predominantly white veterans and their families, and the decline of the industrial base that long provided economic sustenance to U.S. cities and less-educated workers.

Gentrification and U.S. Cities

The decline of U.S. cities, most visible in the economically distressed urban cores of cities, including Chicago, Washington, D.C., and Baltimore, has fostered efforts toward **urban renewal**, *the transformation of old neighborhoods with new buildings, businesses, and residences.* Urban renewal is linked to **gentrification**, a process characterized by *change in the socioeconomic composition of older and poorer neighborhoods with the remodeling of old structures and building of new residences and shops to attract new middle- and high-income residents.* Gentrification may have a variety of demographic effects, reducing the number of racial and ethnic minority residents and lowering the average household size, as families are replaced by young singles and couples with more robust stores of disposable income (Beauregard, 1986).

Urban renewal: The transformation of old neighborhoods with new buildings, businesses, and residences.

Gentrification: The change in the socioeconomic composition of older and poorer neighborhoods with the remodeling of old structures and building of new residences and shops to attract new middle- and high-income residents.

Gentrification may transform struggling neighborhoods into flourishing and economically viable urban spaces that offer cultural and business opportunities to residents and visitors. Gentrification may also have the effect of reducing crime, as abandoned buildings return to active use and the number of residents in a neighborhood rebounds: Some studies have found a relationship between gentrification and decreases in violent crime in cities (Barton, 2014; Macdonald & Stokes, 2019; Kirk & Laub, 2010); one study correlated a rise in urban coffee shops in gentrifying neighborhoods to a decline in homicides. Interestingly, the effect on robberies was mixed: While they declined in predominantly white and Latino neighborhoods, they rose in predominantly Black neighborhoods (Papachristos, Smith, Scherer, & Fugiero, 2011). Cities also benefit from gentrification as the process rebuilds a middle-class and upper-middle-class base of residents who pay city taxes and pushes up the value of taxable property and, consequently, revenues for city governments. Some of this is the result of the building of new luxury housing units and some derives from the transformation of rental units into higher-priced condominiums.

Gentrification also heralds a rise in rents and other costs of living, which may push out longtime low-income residents who cannot afford to be part of the boom in condominiums or the luxury amenities intended to make gentrified spaces inviting to upwardly mobile new residents. For older residents, gentrification may entail the loss of local cultural spaces and businesses. In 2015, the *Washington Post* reported on the closing of a small hair salon in the newly gentrified Bloomingdale neighborhood in the District of Columbia:

> [Latosha] Jackson-Martin's father, William Jackson, opened Jak & Co. Hairdressers downtown 50 years ago, and moved it to now-trendy Bloomingdale in 1988. For much of its past quarter-century, the store has been surrounded by a laundromat, a uniform business and liquor stores with Plexiglass windows.
>
> Now, it's nestled between a pub with an extensive whiskey and scotch menu, a gourmet bakery and a Mexican restaurant that sells cucumber margaritas.
>
> "I want people in the community to know, especially young people, that the community is filled with people who have and people who don't have," Jackson-Martin said. "I want people to know that we put up a fight to stay where we are, but we are in the 'has not'... I can't afford to pay double the rent like the other folks." (Stein, 2015)

Jackson-Martin's experience mirrors that of a spectrum of older businesses that not only cannot afford rising retail rents but may also be facing a loss of their traditional customer base as new residents move in and demand grows for different shops and restaurants. Her experience also highlights some of the social tensions that underlie gentrification, as the interests of old and new residents come into conflict.

What, then, are we to conclude about gentrification? How can urban renewal and efforts to bring upwardly mobile residents back to U.S. city centers and to revitalize economically distressed neighborhoods be balanced with the needs and aspirations of low-income residents and long-term businesses at risk of displacement as the cost of living in the gentrified neighborhood climbs?

The Emergence of Global Cities

Global cities: Metropolitan areas that are highly interconnected with one another in their role as centers of global political and economic decision making, finance, and culture.

We live in an age of urban dominance. Today, more than half the world's population resides in cities, many of which are massive global centers such as London (population 9.3 million), New York (18.3 million), and Tokyo (37.4 million) (World Atlas, 2020). **Global cities** are *metropolitan areas that are highly interconnected with one another in their role as centers of global political and economic decision making, finance, and culture* (Sassen, 1991). Global cities are the terrain where a spectrum of globalization processes assume concrete, localized forms; their economic role is defined as much by the role they play in the global economy as by their influence in their immediate geographic regions (Sassen, 2005).

Saskia Sassen (1991, 2005) identifies four principal functions of global cities. First, they are command posts in the organization of the world economy. Second, they serve as key locations for businesses related to finance, accounting, marketing, design, and other highly specialized (and profitable) services that are replacing manufacturing as the leading economic sectors. Third, they are the most important sites of innovation and new product development. Finally, they serve as the principal markets for global businesses. Sassen (2000) also notes that "whether at the global or regional level, these cities must inevitably engage each other in fulfilling their functions.... There is no such entity as a single global city" (p. 4).

In global cities, multinational corporations and international bankers maintain their headquarters and oversee the operation of diverse production and management operations that are spread across the globe. Global cities are sites for the creation and concentration of enormous economic wealth: By one estimate, 100 cities account for 38% of the world's economy (Dobbs et al., 2011), a figure that continues to rise. New York City's $1.5 trillion gross domestic product (GDP) ranks it among the 20 largest economies in the world. Dominant global cities are on par economically with many countries (Figure 17.6). When looking at the interplay of globalization and emerging technology, global cities are a major economic unit.

At the same time, as we saw in our opening story, cities have long been magnets for those who cannot make an adequate living elsewhere, whether in other countries or in the impoverished rural areas or small towns of the cities' own countries. Low-wage services, low-skill factory production, and sweatshops coexist with the most profitable activities of international businesses in global cities; we find dire poverty and spectacular wealth side by side. In Los Angeles, for example, hundreds of thousands of immigrants from Mexico and Central America work in the shadows of the downtown skyscrapers that house the world's largest banks and corporations. These immigrants labor as janitors, domestics, or workers in small clothing factories that sew apparel for global garment manufacturers (Milkman, 2006; Singer, 2012). Cities are places of intense contrasts and contradictions: The inequalities that permeate relationships, institutions, and countries are on vivid display in the global cities of the world.

World Urbanization Today

Some of the most highly urbanized countries in the world today are those that, only a century ago, were almost entirely rural. As recently as 1950, only 18% of the inhabitants of developing countries lived in urban areas. In 2018, 55% did. For the first time in history, there are more people living in urban areas than in rural areas throughout the world, and the United Nations projects that this figure will grow to 68% by 2050 (United Nations Department of Economic and Social Affairs, 2018). Today, more than 30% of the world's poor inhabit cities—some in developed countries, most in developing countries. A UNICEF report (2012) identifies urban poverty as a critical and growing problem, particularly for children, noting that

> hundreds of millions of children today live in urban slums, many without access to basic services. They are vulnerable to dangers ranging from violence and exploitation to the injuries,

FIGURE 17.6 ■ Top 10 Metros by GDP With Comparable Nations (PPP-Adjusted $ Billions)

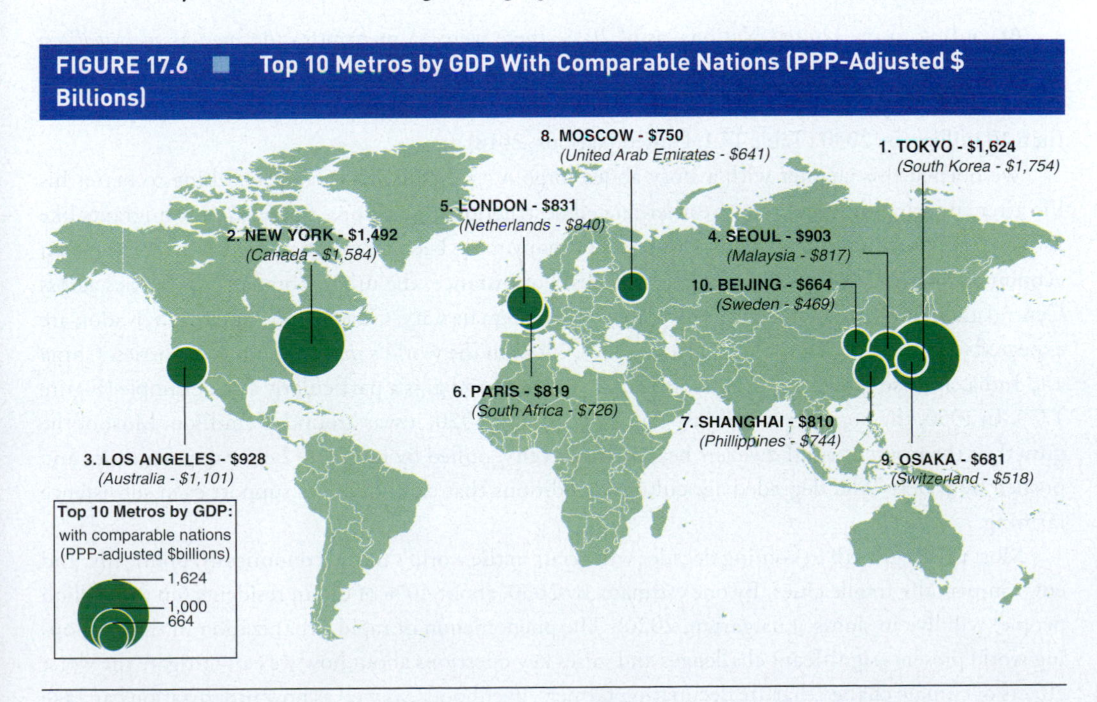

Source: Taylor Blake/Martin Prosperity Institute.

TABLE 17.4 ■ The World's 15 Largest Megacities, 2018 and 2030 (Estimate)

Rank	City, Country	Population In 2018 (thousands)	City, Country	Population in 2030 (thousands)
1	Tokyo, Japan	37,468	Delhi, India	38,939
2	Delhi, India	28,514	Tokyo, Japan	36,574
3	Shanghai, China	25,582	Shanghai, China	32,869
4	São Paulo, Brazil	21,650	Dhaka, Bangladesh	28,076
5	Ciudad de México (Mexico City), Mexico	21,581	Al-Qahirah (Cairo), Egypt	25,517
6	Al-Qahirah (Cairo), Egypt	20,056	Mumbai (Bombay), India	24,572
7	Mumbai (Bombay), India	19,980	Beijing, China	24,282
8	Beijing, China	19,618	Ciudad de México (Mexico City), Mexico	24,111
9	Dhaka, Bangladesh	19,578	São Paulo, Brazil	23,824
10	Kinki M.M.A. (Osaka), Japan	19,281	Kinshasa, Democratic Republic of the Congo	21,914

Source: United Nations. (2018). *The World's Cities in 2018.* New York: Author. Retrieved from *Annex Table: The World's Cities With 1 Million Inhabitants or More in 2018* (pp. 10–29).

illnesses and death that result from living in crowded settlements atop hazardous rubbish dumps or alongside railroad tracks. (p. v)

Megacities: Metropolitan areas or cities with a total population of 10 million or more.

According to the United Nations, as of 2018, there were 33 **megacities**, defined as *metropolitan areas or cities with a total population of 10 million or more.* About 6.9% of people worldwide currently reside in a megacity, and projections indicate that about 8.8% of people will reside in a city of more than 10 million by 2030 (Table 17.4; United Nations, 2018).

We opened this chapter with a story about Jorge A., a Guatemalan farmer seeking to better his life after enduring the persistent, weather-related destruction of his crops. Some climate migrants like Jorge seek opportunities by crossing borders, but many more become internal migrants, moving from economically distressed rural areas to large cities. For instance, the urban populations of cities across Central and South America, such as Mexico City, Guatemala City, Guadalajara, and San Salvador, are expected to rise dramatically (Lustgarten, 2020). Cities in the world's most populous countries, China and India, are also growing rapidly. The Indian city of Mumbai is a particularly stark example (Figure 17.7). In 1950, the population was just over 3 million. in 2020, it was around 20 million: Most of the growth is the result of rural dwellers heading to the city, pulled by hopes for better opportunities and pushed by poverty and degraded agricultural conditions that can no longer support even subsistence farming.

Most urban growth in coming decades will occur in the world's most economically, politically, and environmentally fragile cities. By one estimate, by 2030, about 40% of urban residents (up to 2 billion people) will live in slums (Lustgarten, 2020). The phenomenon of rapid urbanization in the developing world presents significant challenges and raises key questions about how we can mitigate the worst effects of climate change that are decimating farmers' livelihoods, as well as how urbanization can take place in ways that are equitable and sustainable.

FIGURE 17.7 ■ Mumbai Population, 2020

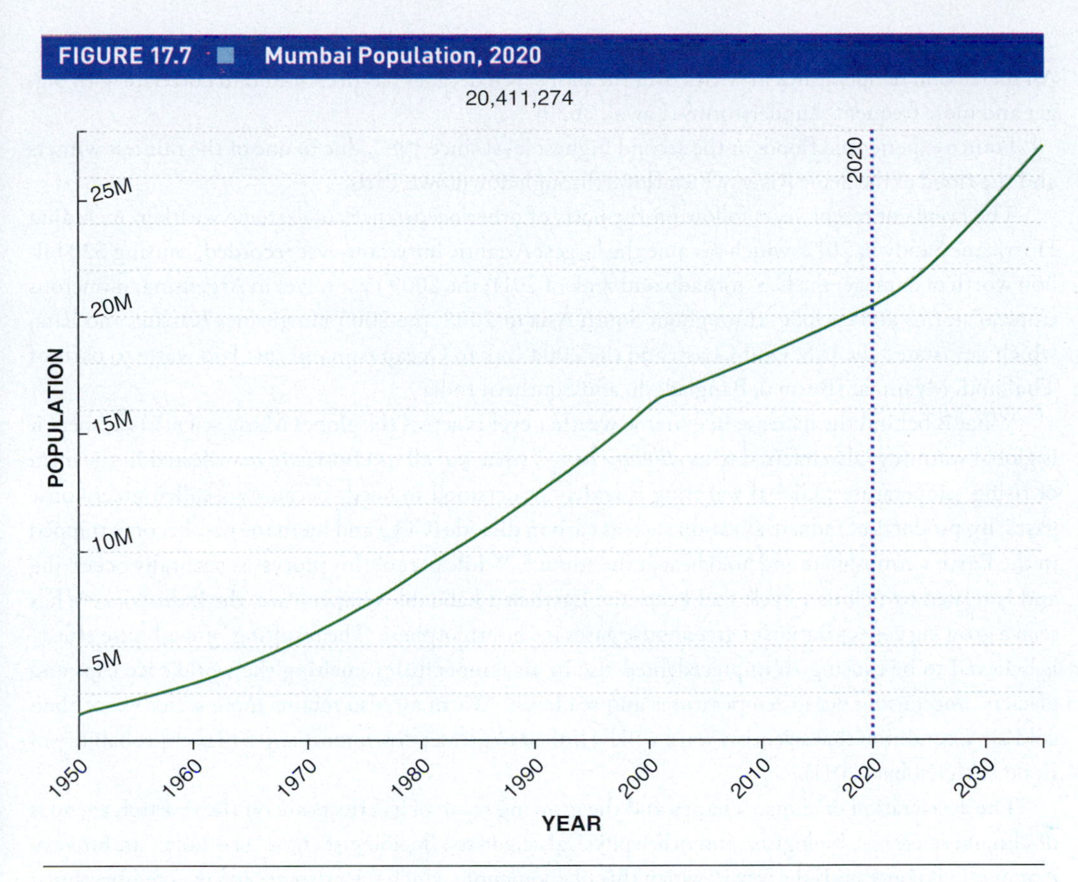

Source: World Population Review, https://worldpopulationreview.com/world-cities/mumbai-population.

THE LOCAL AND GLOBAL ENVIRONMENT

According to the National Oceanic and Atmospheric Administration (NOAA), "Earth's globally averaged temperature for 2019 made it the third warmest year in NOAA's 138-year climate record, behind 2018 (warmest) and 2017 (second warmest)" (2020, para. 2). Across the globe, a spectrum of devastating weather events have taken place. The U.S. government's National Climatic Data Center, part of NOAA, has reported the following:

In the slums of Mumbai, India, the juxtaposition between wealth and poverty is substantial.

©iStockphoto.com/Adrian Catalin Lazar

Global carbon dioxide levels in 2018 were higher than any point in the past 800,000 years. Excess carbon dioxide negatively affects the greenhouse effect, which helps keep the Earth at its normal temperature.

In Cape Town, South Africa, a multiyear drought has led to water use restrictions and a prediction of "Day Zero," a day when most of the city's taps will be turned off and active water rationings will commence.

Southwest America has been mired in drought, with conditions from 2000 to 2018 comparable to several megadroughts since 800 C.E. Global warming caused by human emissions of greenhouse gases was a major contributor to the droughts (Fountain, 2020).

In California, the 10 largest fires have occurred since 2000, including one in 2020 that burned over 400,000 acres of land. With warmer temperatures drying out California's forests and brush, the link between climate change and bigger fires is impossible to escape (Pierre-Louis & Schwartz, 2020).

Greenland's ice sheet has melted to a point of no return, and efforts to slow global warming will not stop it from disintegrating. The ice sheet dumps more than 280 billion metric tons of melting ice into the ocean each year, making it the greatest single contributor to global sea level rise (Claypool & Miller, 2020).

Africa has experienced bigger and more frequent thunderstorms as global temperatures have risen. An increase in temperatures in Africa over the past seven decades has been found to correlate with bigger and more frequent thunderstorms (Lawal, 2020).

France experienced floods at the second highest level since 1982, due to one of the rainiest winters and the flood of the Seine River, which flows through downtown Paris.

The events of recent years follow on the heels of other occurrences of extreme weather, including Hurricane Sandy in 2012, which became the largest Atlantic hurricane ever recorded, causing $20 billion worth of damage; the U.S. tornado outbreak of 2011; the 2009 heat waves in Argentina; numerous tropical storms and cyclones throughout South Asia in 2008; the 2005 Hurricanes Katrina and Rita, which devastated the U.S. Gulf Coast; and the 2004 Indian Ocean tsunami that laid waste to parts of Thailand, Myanmar (Burma), Bangladesh, and Southern India.

What is behind the increase in extreme weather events across the globe? Many scientists suspect it is global warming, also referred to as *climate change*, since not all its effects are manifested in the form of rising temperatures. Global warming is widely understood to occur because so-called greenhouse gases, by-products of industrialization such as carbon dioxide (CO_2) and methane gas, become trapped in the Earth's atmosphere and hold heat at the surface. While part of this process is naturally occurring and is related to a climate cycle that keeps the Earth at a habitable temperature, the industrial era has seen a growing concentration of greenhouse gases in the atmosphere. The resulting "greenhouse effect" is believed to be causing an unprecedented rise in air temperatures, melting the world's ice caps and glaciers, and raising ocean temperatures and sea levels. Warm air also retains more water vapor than cold air, a condition that scientists warn will be linked to greater downpours and a higher probability of floods (McKibben, 2011).

The acceleration of climate change and the growing reach of its effects are on the research agendas of climate scientists, biologists, and other physical scientists. Sociologists have also taken an interest. Among their concerns is the way in which this phenomenon, which scientists accept as a credible threat to our planet and its inhabitants, has been framed in the mainstream social and political discourse as a societal problem. Some sociologists have argued that public attention to problems such as climate change depends in part on the presence of a "social scare" (Ungar, 1992)—that is, an event (such as extreme weather) that draws attention to a phenomenon by allowing it to "piggyback on dramatic real-world events" (Ungar, 1992, p. 483).

Recent extreme weather events may be having an effect on public concern about climate change in the United States. A recent Pew Research survey found heightened levels of worry, noting that 60% of Americans indicated that global climate change is a major threat to the country, and 67% of U.S. adults say the federal government is doing too little to reduce the effects of global climate change (Funk & Kennedy, 2020).

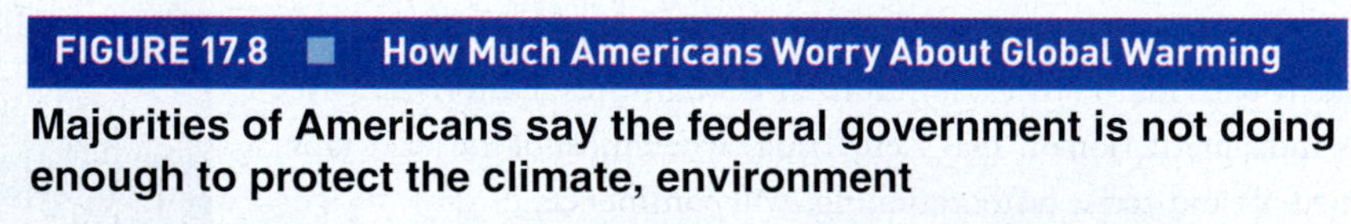

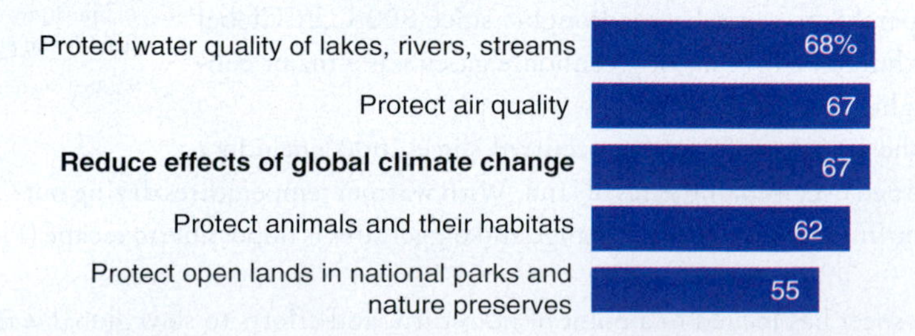

Source: Pew Research Center. (2020, April 21). *How Americans See Climate Change and the Environment in 7 Charts.* Retrieved from https://www.pewresearch.org/fact-tank/2020/04/21/how-americans-see-climate-change-and-the-environment-in-7-charts/

Figure 17.8 shows Americans' concern about the threat of global warming, a worry that has fluctuated over time. What accounts for changes in levels of concern? Research suggests that the creeping nature of climate change—the fact that effects are not constantly apparent and are perceived by some to be far in the future—leads some people to discount it as a potential problem (Moser & Dilling, 2004). Some sociologists also point to the **treadmill of production**, *the constant and aggressive growth needed to sustain the modern economy* (Schnaiberg & Gould, 1994). On the political and economic agenda, this growth has historically taken precedence over environmental concerns.

Climate change is leading to ice melt at the poles that could raise sea level significantly. The photo shows the disintegration of the Greenland ice sheet.

Governments—as well as companies that drill for oil, mine coal, manufacture products, or engage in other energy-intensive endeavors—are concerned about maintaining profitability in the private sector, which may lead them to ignore, minimize, or even deny the problem. Are these two imperatives—the need for a vigorous economy and the need for a clean and sustainable environment—irreconcilable? How can we build an economy that can both grow and be green?

Treadmill of production: The constant and aggressive growth needed to sustain the modern economy.

Population Growth, Modernization, and the Environment

Industrialization and urbanization, combined with rapid population growth, have taken a toll on the global environment and its resources. Threats to the environment exist in both underdevelopment and overdevelopment. On the one hand, the world's population, approaching 8 billion people in the third decade of this millennium, needs basic resources: water, food, shelter, and energy. As humans seek to meet these needs, they tax the Earth by impinging on the habitats of unique animal and plant species to make room for human habitation. More important, perhaps, the rise of the **global consumer class**—*those who actively use technology, purchase consumer goods, and embrace the culture of consumption*—has meant that more individuals are using more resources per person than ever before.

Global consumer class: Those who actively use technology, purchase consumer goods, and embrace the culture of consumption.

By some estimates, a quarter of the world's population falls within the global consumer class. A large percentage of the members of this class live in developed countries; according to the Worldwatch Institute, the United States and Canada, home to just over 5% of the world's population, account for about 31% of private consumption expenditures, and Western Europe, with around 6% of the globe's people, accounts for more than 28% of expenditures. However, the consumer class is expanding. More than half of its members now live in areas of advancing prosperity in the developing world, such as China and India. In 2009, China surpassed the United States to become the biggest market for automobiles on the planet (Langfitt, 2013), and half of the world's new shopping malls are being built in China, where luxury consumption is growing rapidly ("Chinese Consumers," 2014). The future consumer markets of the world are in developing countries rather than developed countries, which are characterized by sagging population growth and already high consumption rates.

In many respects, urbanization and industrialization are important achievements in the developing world, where more people have the opportunity to meet their needs and realize their dreams, leaving behind the travails of deprivation. Though the gulf between rich and poor remains wide, prosperity has advanced around the globe. Development's darker underbelly, however, becomes visible when we look at the problems presented by the consumer class and its growing ranks. Arguably, among the most critical problems of development is overconsumption.

Overconsumption may be understood as a recklessly wasteful use of resources, from fuel to food and consumer goods. Overconsumption is a symptom of development. Among its health consequences are *obesity* (having a body mass index [BMI] of 30 or higher) and *being overweight* (having a BMI between 25 and 29.9); a healthy BMI is between 18.5 and 24.9, according to the Centers for Disease Control and Prevention. More than two thirds of U.S. adults are overweight or obese, conditions that can lead to heart disease and other dangerous maladies (Centers for Disease Control and Prevention, 2015). Globally, about 13% of the world's adult population is obese and 39% are overweight (World Health Organization, 2020). Notably, more people in developing countries are joining the ranks of the overweight and obese as they move from more traditional diets to modern foods (Popkin, Adair, & Ng, 2012). High-fat and high-sugar foods are widely consumed: These foods are aggressively advertised, cheap, and popular, especially among young people seeking the symbols of a Western lifestyle.

Rising consumption also threatens the environment, as more people buy, use, and toss away ever more "stuff." Globally, 80 billion pieces of new clothing are purchased each year, and about 85% of the clothing Americans consume—3.8 billion pounds—is sent to landfills as solid waste each year, amounting to 80 pounds per American (Bick, Halsey, & Ekenga, 2018).

Consider as well the environmental effect of our ubiquitous electronic devices: Discarded computers, mobile phones, and the like are creating a growing toxic waste problem in all countries. Future obsolescence is built into many of the goods we buy: Having acquired an iPhone 8 or other premium device, many people would fully expect to graduate in a couple of years to an upgrade. Short-lived ownership of any device is both assumed and promoted by manufacturers. The nascent expansion of 5G networks, which promise unprecedented data speed, will further exacerbate the problem of e-waste, as older devices are not likely to be compatible with 5G. E-waste is a particularly pernicious part of the global waste stream: While recycling is available in some areas, a significant proportion ends up bundled with regular trash. This can release toxic chemicals like mercury into the environment (Semuels, 2019).

As more people across the globe strive, understandably, for prosperity and modernity, it may be wise for us to ask: Can such consumption be sustained? Those who live in the prosperous countries might also ask: Is it hypocritical to raise concerns about the global destruction wrought by overconsumption when the West has been the primary consumer of the world's resources for the past century? To give you the opportunity to consider these important questions, we discuss the environmental impacts of population growth in developing states and of consumption growth in developed states and newly prosperous areas.

Underdevelopment and Overdevelopment in the Modern World

Environmental problems that appear in one part of the globe often result from actions taken elsewhere and may have far-reaching effects for all of the planet's inhabitants. They are indeed global. However, many immediate environmental effects are local.

The most degraded local environments tend to be those inhabited by the poorest people. Air pollution is far worse in Mexico City, Mumbai, and Karachi than in the most polluted U.S. cities such as Los Angeles. China's rapid development has brought new prosperity to many Chinese, but the rising number of factories, power stations, and cars has also brought serious pollution. While China's situation has improved somewhat in recent years, it is still dire for many residents: China is home to 47 of the world's 100 most polluted cities, according to the *2019 World Air Quality Report* (Regan, 2020).

Disease runs rampant in the large cities of India, where the infrastructure (including sewage systems, which should carry away waste and ensure the flow of potable water) cannot meet the needs of growing populations. In some coastal areas of India, toxic pollution streams flow from shipbreaking yards, where laborers take apart ships no longer fit to sail the seas. Local environmental problems have also arisen from deforestation in the tropical rainforests of Central America and the destruction of mangrove ponds and wetlands in Asia, as these areas are transformed into agricultural production sites for growing numbers of farmers.

Population pressures in developing countries are important contributors to these problems. However, the world's developed states are also implicated. Consider, for instance, deforestation and wetlands destruction. Some of the rainforest land in the Amazon has been cleared to make room for cattle ranching that produces beef to be consumed not locally but in the United States and other developed nations. Similarly, the wetlands of some coastal areas of Asia have been transformed into shrimp farms to grow the delicacy enjoyed in wealthy countries. The ships in India's coastal shipbreaking yards are largely the corpses of Western fleets. The consequences of the developed world's consumption are felt in the furthest corners of the developing world (Eglitis, 2010).

Industrialization in both the developed and developing worlds is a critical aspect of the environmental equation. On the one hand, some developing countries, such as China, are aggressively pursuing industrialization and multisector economic growth. Because of its enormous size and rapid move toward industrialization, China is fast becoming the world's major contributor to greenhouse gases. Coal, a highly polluting source of energy, provides the country with three quarters of its energy, creating so much pollution in some urban areas that residents must wear surgical masks for protection. In the southern and eastern parts of China, where urbanization and industrialization are proceeding at historically unprecedented rates, the environmental damage has been considerable. On the other hand, just as it did in the West, industrialization is bringing greater prosperity and prospects to many Chinese. This raises the question of whether we must choose between a good life and a good Earth. What would it mean to achieve a balance between the needs and desires of people around the world and the needs of the planet?

THE GLOBAL ENVIRONMENT AND CLIMATE CHANGE: WHAT'S NEXT?

In this chapter, we looked at three key phenomena: demographic shifts, urbanization, and human-produced climate change. While different, they are not unrelated: For example, demographic stresses introduced by rapidly rising population can overtax community or family resources, and these in turn, exacerbated by acute climate-change-induced weather events, can drive people to cities and over borders seeking refuge and opportunities. At the same time, while developed countries have low or no natural population growth, they account for a substantial proportion of global consumption of electronics, energy, automobiles, and meat, which contributes disproportionately to environmental pollution and climate change. As sociologists, we seek to understand relationships between disparate and diverse phenomena and to understand how change in one part of a social system spurs change, sometimes expected and sometimes unexpected, in another part.

For decades, climate scientists have endeavored to draw the attention of decision makers and the global community to the actual and potential hazards of climate change. Have they been successful? From one perspective, they have been successful: In 2015, the Paris Agreement was signed by the United Nations Framework Convention on Climate Change. The agreement brings all nations together to undertake efforts to fight climate change and adapt to its effects, with enhanced support to assist developing countries in doing so (United Nations, 2020). In the United States, more people are expressing conern: 67% of American adults think that the federal government isn't doing enough to reduce the effects of global climate change (Pew Research Center, 2020). On the other hand, the rise of populist politics has been accompanied by a rejection of progressive climate policies and, in some cases, outright climate change denial. For example, Brazilian president Jair Bolsonaro called the fires that have been surging in the Amazon rainforest for years a "lie," despite photographic and videographic evidence of the fires (Lewis, 2020). In the United States, Republican senator of Georgia David Perdue said that there is "an active debate going on" regarding whether humans have had an impact on the environment (Cranley, 2019). Perdue is one of many American political figures who seek to obscure the reality of climate change to their constituents.

What's next? The societal, economic, and environmental costs of failure to recognize and act on climate change could be significant. As a recent book posits, "If we stay on the current business-as-usual trajectory, we could see two degrees [Celsius rise in Earth's temperature] as soon as the early 2030s, three degrees around mid-century, and four degrees by 2075 or so. If we're unlucky, with positive feedbacks . . . from thawing permafrost in the Arctic or collapsing tropical rainforests, then we could be in for five or even six degrees by century's end (Lynas, 2020). We know, however, that the path to a more sustainable future for families, communities, and countries is available. The price of renewable energy, such as wind and solar power, has fallen considerably, and awareness of climate change has grown in part thanks to the efforts of young environmental activists like Sweden's Greta Thunberg, who was *Time* magazine's 2019 Person of the Year and a nominee for the Nobel Peace Prize.

Sociology gives us a useful and unique perspective from which to look at the challenges posed by climate change. What sociological factors shape one's acceptance or denial of the science of climate change? What obstacles stand in the way of fundamental structural change to respond to climate change? Are they economic? Political? Cultural? Can social movements like those led by young climate activists foster change?

WHAT CAN I DO WITH A SOCIOLOGY DEGREE?

The Global Perspective

The study of sociology helps you develop a broad understanding of the social world, which includes relationships between cultures and countries over time and space and the ability to see the world from a variety of perspectives. A global perspective encompasses knowledge and skills. A global perspective evolves through study and experiences that lead to a strong understanding and appreciation of the significance and effects of global cultural, economic, political, and social connections on individuals, communities, and countries. It also encompasses the development of skills for working effectively in intercultural environments.

Understanding different cultures, recognizing a diverse spectrum of legitimate political and economic interests, and having the ability to see issues from multiple perspectives are key to global efforts to deal cooperatively with environmental and other threats to the planet. As a sociology student, you will have the opportunity to develop the kind of thoughtful global perspective that will enable you to make critical connections between decisions about, for instance, economic consumption or production, which are made at the individual or community or country level, and effects that are experienced globally.

Hannah Gdalman, former Peace Corps volunteer in Guatemala and current Peace Corps Fellow, Illinois State University, and Sociology Graduate Fellow, Financial Health Network

BA, Sociology, Knox College

MS, Sociology, Illinois State University, August 2020

One of the great things about studying sociology is its incredible range of applications. When I graduated with a BA in sociology/anthropology in 2009, I was mostly focused on the ways in which my coursework would help me in my upcoming service with the Peace Corps in rural Guatemala. Having a background in cultural studies and social science certainly contributed to my Peace Corps work in Youth Development; however, it also related to my future travels, nonprofit work, and more.

As a current graduate fellow in sociology and applied community and economic development, I have the luxury of dabbling in multiple corners of this far-reaching discipline. As a graduate fellow at the Financial Health Network in Chicago, I'll be working to promote consumer financial health through research, market analysis, and advisory services that, in turn, influence business leaders and policy makers to better meet the needs of financially vulnerable populations. The Financial Health Network has reached over 165 million customers through its focus on market innovation.

Simultaneously, I'm working on exciting research that furthers my passion for Central America and sustainable development. My fieldwork brought me back to Guatemala to undertake ethnographic research in volunteer tourism (that is, when people choose to travel in order to volunteer, usually in developing countries and communities). It is my hope that my research will better shed light on best practices in volunteer tourism by focusing on the interplay between tourists, host communities, and the local environment.

Sociology has given me the unique opportunity to balance qualitative and quantitative, domestic and international—one day, I can be analyzing big data and the next doing semistructured interviews in the Guatemalan countryside. I'm very grateful for the wide array of formative experiences, both academic and professional, that sociology has afforded me.

Sociology prepares students to succeed in a variety of postgraduation settings that offer the opportunity to serve individuals and communities domestically and globally. Positions are usually competitive and offer modest pay and/or stipends. You may want to learn about opportunities in organizations like the following:

Peace Corps

City Year

Teach for America

NYC Teaching Fellowships *

Americorps

* *This is one of a variety of city-based programs that are offered by Chicago, Boston, and Washington, D.C., among others.*

SUMMARY

- The world's population is growing at a rapid rate, having increased as much since 1950 as it did in the preceding 4 million years. Growth is highly uneven around the world, with most taking place in developing countries. Other regions, including most in Europe, are losing population.
- Annual population growth or decline in a country is the result of four factors: (1) the number of people born in the country during the year, (2) the number who die, (3) the number who immigrate into the country, and (4) the number who emigrate out. In the language of demographers, population changes are based on **fertility**, mortality, and **net migration**.
- The theory of the first demographic transition proposes that many societies go through roughly the same stages of population growth: low growth resulting from high fertility and equally high mortality, a transitional stage of explosive growth resulting from high fertility and low mortality, and a final stage of slow or no growth resulting from low fertility and low mortality.
- Advanced industrial states may be undergoing a second demographic transition, seen as changes in family patterns that affect population. For instance, in the world's industrialized states, divorce has increased, cohabitation has increased, marriage has declined, fertility has fallen, and nonmarital births as a proportion of all births have increased.
- Thomas Malthus developed the theory of **exponential population growth**: the belief that, similar to compound interest, a constant rate of population growth produces a population that grows by an increasing amount with each passing year. Malthus claimed that while population grows exponentially, the food supply does not; the Earth's resources are finite. Others, such as economist Julian Simon, have suggested that population growth increases humanity's potential for uncovering talent and innovation.
- Karl Marx was critical of Malthus and felt the central problem was not a mismatch between population size and resource availability, but rather the inequitable distribution of resources between the wealthy and the disadvantaged.
- Sociologist Louis Wirth (1938) defined the *city* as a relatively large, densely populated, and permanent settlement that brought together heterogeneous populations. While cities of the past served primarily as centers of trade, in the 18th century, industrial cities emerged as centers of manufacturing. By the 19th century, industrialization was advancing hand in hand with **urbanization**.
- Some of the most highly urbanized countries in the world today are those that, only a century ago, were almost entirely rural. As recently as 1950, only 18% of the inhabitants of developing countries lived in urban areas. Today, more than half do. For the first time in history, there are more people living in urban areas than in rural areas throughout the world (Population Reference

Bureau, 2012). More than 30% of the world's poor inhabit cities—some in the developed countries, most in developing countries.

- With the emergence of post-industrial society, **global cities** have appeared. These metropolitan areas are highly interconnected with one another and serve as centers of global political and economic decision making, finance, and culture. Examples include New York, London, Tokyo, Hong Kong, Los Angeles, Mexico City, and Singapore.
- The combination of rapid population growth and modernization, in the form of industrialization and urbanization, takes a toll on the global environment and its resources. Both underdevelopment and overdevelopment threaten the environment.
- The study of **demography** and population growth helps us gain a fuller understanding of the ways that micro-level events, such as childbearing decisions in a family, are linked to macro-level issues, such as population growth or decline, threats of mortality, and challenges to the sustainability of resources and development.

KEY TERMS

age-specific fertility rate (p. 486)
age-specific mortality rate (p. 486)
agricultural surplus (p. 496)
city (p. 495)
crude birthrate (p. 486)
crude death rate (p. 486)
demography (p. 483)
exponential population growth (p. 493)
fertility (p. 486)
gentrification (p. 499)
global cities (p. 500)
global consumer class (p. 505)
life expectancy (p. 486)
megacities (p. 502)
net migration (p. 486)
population momentum (p. 484)
rate of natural increase (RNI) (p. 491)
treadmill of production (p. 505)
urban growth machine (p. 498)
urban renewal (p. 499)
urbanization (p. 496)

DISCUSSION QUESTIONS

1. What factors have contributed to the decline of population growth in many modern countries? What are the benefits and consequences of fertility declines?
2. Do populations stop growing when fertility declines to replacement rate fertility (a total fertility rate of 2.1)? Explain your answer.
3. As you saw in the chapter, urbanization continues to increase across the globe. What draws people to cities? What sociological factors point to this trend continuing domestically and globally?
4. What is gentrification? What are some of the key costs and benefits of gentrification in U.S. cities?
5. Are economic growth and environmental protection irreconcilable values? Consider what you have read both in earlier chapters about economic growth and employment and in this chapter about environmental challenges such as climate change, and respond thoughtfully to the question.

Sarah M. Golonka / Alamy Stock Photo

18 SOCIAL MOVEMENTS AND SOCIAL CHANGE

WHAT DO YOU THINK?

1. Has your level of political activism changed since you came to college? What are the issues driving campus activism today?
2. Do people behave differently in crowds than they do individually or in small groups? What sociological factors explain crowd behavior?
3. Does social media contribute positively to activism—or does it function only to make people think they are making a difference? Might it do both?

LEARNING OBJECTIVES

18.1 Apply sociological perspectives to understand characteristics and paths of social change.

18.2 Describe key sources of social change in society.

18.3 Identify different types of social movements.

ACTIVIST AMERICA?

In 2014, Americans seemed sadly lacking in participation in civic activities, such as voting, volunteering, calling elected officials, donating to a campaign or cause, or attending political meetings. A study that year lamented that "forty-one percent of Americans do not participate very often in any of 10 bedrock activities of American civic and political life," and just 1% could be classified as very politically active ("Only One Percent," 2014, para. 1). The midterm election voter turnout that year was an underwhelming 37%.

By 2018, however, that had changed. A joint survey by the *Washington Post* and the Kaiser Family Foundation found a growing number of Americans actively engaging in social and political activism. According to the survey, "One in 5 Americans have protested in the streets or participated in political rallies since the beginning of 2016. Of those, 19% said they had never before joined a march or a political gathering" (Jordan & Clement, 2018, p. A1). Furthermore, in 2018, more than 47% of the voting-eligible population (almost half of all eligible voters) cast a ballot during the midterm elections—the highest midterm turnout since 1966 (Domonoske, 2018).

By 2020, images of massive protests against police violence, both in the United States and across the world, quickly became iconic. Indeed, according to *The New York Times*, four polls conducted in June 2020 indicated that between 15 million and 26 million people had marched in protest against the death of George Floyd, a Black man who died in March 2020 when Derek Chauvin, a police officer, held his knee on Floyd's neck, suffocating him (Buchanan, Bui, & Patel, 2020).

Young people are at the forefront of many of the biggest social movements in the United States today.

©Dinendra Haria/SOPA Images/LightRocket via Getty Images

Who is engaging in this wave of activism? The data, collected from a random representative sample of 1,850 adults ages 18 years and older, suggest that about a fifth of Americans attended a political rally of some kind in the past 2 years. Of those who attended a rally, most categorized themselves as Democrats (40%) or independents (36%). Another 20% reported that they were Republicans. Rallygoers were also broken down by educational attainment: About 21% reported having a high school education or less, while 29% reported some college, and 50% reported having a college degree of some kind (Jordan & Clement, 2018).

Significantly, however, this statistical portrait of protest may be missing a robust new wave of activism among those who are under age 18. That is, recent years have seen a dramatic rise in public activism by young people, particularly high school students. Among the most high profile are survivors of the February 2018 mass shooting at Marjorie Stoneman Douglas High School in Parkland, Florida. Just over a month after the tragedy, a group of students who had lost 17 of their classmates organized to lead a massive rally in Washington, D.C., called "March for Our Lives." Highlighting issues of school safety and gun control, the march drew an estimated 800,000 demonstrators to Washington's streets. Rallies were also held in dozens of other cities and towns around the country (Shabad, Baily, & McCausland, 2018). The students continued to take their activism to other well-off suburban schools similar to their own, as well as to urban schools and neighborhoods that have long been threatened by deadly violence (Zornick, 2018). The group also mobilized to register young voters in a cross-country tour across America called "Road to Change," registering over 50,000 new voters in 2018 (March for Change.org).

In 2020, adopting the slogan "Our Power," the group also mobilized to support #BlackLivesMatter and the fight for racial justice across groups, continued to work for voting rights, and planned an October march to the steps of the Supreme Court to protest the nomination of a new Supreme Court justice before the 2020 presidential election.

Teen activism to promote action on global climate change is also on the rise, particularly in Europe. Greta Thunberg, the Swedish teen who inspired climate strikes around the world, has become a global phenomenon, speaking at the United Nations and the World Economic Forum; inspiring thousands of young people around the globe to carry out protests and school walkouts of their own; leading 4 million protesters in a global climate strike on September 20, 2019, that involved all seven continents; and being recognized as *Time* magazine's Person of the Year in 2019 (Felsenthal, 2019; Watts, 2019).

Sociologists are keenly interested in social change and social movements. We begin this chapter with an overview of sociological theorizing on social change. We continue with an examination of key sources of social change, focusing in particular on collective behavior and resources. Next, we provide an overview of forms that social movements take, and we conclude with reflections on the nature of social change going forward in a rapidly changing and globalized world.

SOCIOLOGICAL PERSPECTIVES ON SOCIAL CHANGE

The concept of *social change* is broad-based. It refers to small-group changes, such as a social club changing a long-standing policy against admitting women or minorities, and to global-level and national-level transformations, like the outsourcing of jobs to low-wage countries and the rise of social movements that seek to address the threat of climate change.

When sociologists speak of social change, they are generally referring to changes that occur throughout the social structure of an entire society. *Societies* are understood sociologically as entities comprising those people who share a common culture and common institutions. *Social change* may refer to changes within small, relatively isolated communities, such as those of the Amish or the small, culturally homogeneous tribes that dot the Amazon basin; changes across complex and modern societies, such as the United States, Japan, or Germany; or changes common across similar societies, such as the economically advanced states of the West or the Arab countries of North Africa and the Middle East.

Sociological perspectives on social change begin with particular assumptions about both the social world and basic processes of change. Three key social-change theories in sociology are functionalist theories, conflict theories, and cyclical theories. We now briefly consider each theoretical perspective and discuss its utility for helping us understand the nature of social change in the world today.

The Functionalist Perspective

How can society change when, according to functionalists, social stability is essential to its smooth functioning? As we discuss in this section, functionalist theories of social change assume that as societies develop, they become more complex and interdependent, and as a result, the changes that occur encourage the healthy integration of emergent forms of interdependence.

Differentiation: The development of increasing societal complexity through the creation of specialized social roles and institutions.

For instance, Herbert Spencer (1892) argued that modern societies are marked by **differentiation**—*the development of increasing societal complexity through the creation of specialized social roles and institutions.* Spencer was referring to what Émile Durkheim conceptualized as the division of labor, which is characterized by the sorting of people into interdependent occupational and task categories (and, by extension, class categories). Think of medieval England, when craftspeople worked at homemade tools and shoes that they exchanged for food or clothing, using a broad range of skills to act relatively independently of one another. Compare this with modern society, where factory workers each produce parts of an automobile, managers sell completed cars to dealerships, and salespeople sell them to customers. Today, people master a narrow range of tasks within a large number of highly specialized (differentiated) institutional roles and thus are highly interdependent. (Note the similarity here to Durkheim's notion that societies evolve over time from *mechanical* to *organic solidarity*—the former being characteristic of traditional, homogeneous societies and the latter characteristic of diverse, modern societies.)

The earliest functionalist theories of social change were *evolutionary theories*, which assumed that all societies begin as simple or primitive and eventually develop into more complicated and civilized forms along a single, unidirectional evolutionary path. During the 20th century, however, this notion of unilinear development became increasingly shaky, as anthropologists came to believe that societies evolve in many different ways. More recent evolutionary theories (sometimes termed *multilinear*) argue that multiple paths to social change exist, depending on the particular circumstances of the society (J. D. Moore, 2004;Sahlins & Service, 1960). Technology, environment, population size, and social organization are among the factors that play roles in determining the path a society takes.

Some evolutionary theorists viewed societies as eventually reaching an equilibrium state in which no further change would occur unless an external force set it in motion. For example, Durkheim believed that primitive or less developed societies were largely unmarked by change unless population growth resulted in such a differentiation of social relationships that organic solidarity replaced mechanical solidarity. Talcott Parsons (1951) viewed societies as equilibrium systems that constantly seek to maintain balance—the status quo—unless something external disrupts their equilibrium, such as changes in technology or economic relationships with other societies. Parsons later came to argue, however, that societies do change by becoming more complicated systems that are better adapted to their external environments (Parsons & Shils, 2001).

Up to the early 20th century, it was commonly believed that allowing women to vote was "unnatural" and would disrupt society. Elizabeth Cady Stanton, Susan B. Anthony, Angelina Weld Grimke, and many other women challenged both public beliefs and legal practices that prevented women from casting ballots.

Undeniably, contemporary societies contain many more specialized roles and institutions than have earlier ones, but evolutionary theories also assume that social changes are progressive and that "modern" societies are more evolved than earlier "primitive" ones. Such beliefs appealed to countries whose soldiers, missionaries, and merchants were conquering or colonizing much of the rest of the world, since these beliefs helped justify those imperialist actions as part of the "civilizing" mission of a more advanced people. Anthropologists and sociologists eventually rejected these ideas (Nolan & Lenski, 2009).

How do you think social change happens in our increasingly complex societies? Do you, like Parsons, believe that society is becoming better able to deal with changes? Do you, like evolutionary theorists, see social change as progressive?

The Conflict Perspective

In contrast to functionalists who emphasize interdependence and differentiation as drivers of change, conflict perspectives emphasis that change is created by social inequality and the resulting competition for scarce resources that they believe cause conflict. Unlike functionalists, they do not see social stability as natural. Responding to the conflicts and contradictions can potentially bring a society to the brink of sharp and sometimes violent breaks with the past.

Karl Marx focused on the contradictions and conflicts built into capitalist societies, where the world is divided between owners of the means of production and workers who own only their own labor power and must sell it under conditions not of their own making. In Marx's view, the revolutionary transformation of a society into a new type—from feudalism to capitalism or from capitalism to socialism, for example—would occur when the concentration of power in one social class was sufficient to transform the consciousness of those oppressed to rise up and create a social movement able to transform political and economic institutions into new sets of social relationships. As we have seen throughout this text, Marx's conflict theory adhered to its own evolutionary view of social change, in which all societies would advance to the same final destination: a classless, stateless society. We have earlier noted several weaknesses in this theory. Of particular importance is Marx's tendency to overemphasize economic conflict while underestimating cultural conflict and other noneconomic and significant factors, such as gender, ethnicity, race, and nationalism.

Later conflict theorists have addressed key questions about processes of social change, such as how groups come to want and pursue social change. Italian Marxist Antonio Gramsci (1971), for instance, highlighted the importance of ideas in maintaining order and oppression in society. He observed that the ruling class is often able to create *ideological hegemony*, a generally accepted view of what is of value and how people should relate to their economic and social status in society. Ideological hegemony may lead people to consent to their own domination by, for instance, socializing them to believe that the existing hierarchy of power is the best or only way to organize society. Consider, for example, that in the past, women were socialized by schools, families, and religious institutions to believe they should not have jobs outside the home or vote. The idea that women should not hold positions outside the home could be considered a *hegemonic idea* of this period.

Gramsci also spoke of *organic intellectuals*—those who emerge from oppressed groups to create counterhegemonies that challenge dominant (and dominating) ideas. In the mid-19th century, women's suffrage activists—including Lucretia Mott, Susan B. Anthony, Elizabeth Cady Stanton, and Angelina Weld Grimke—were organic intellectuals, challenging powerful beliefs that women should be excluded from politics. Over time and through the efforts of activists, the counterhegemonic idea that women should have a voice in politics became the hegemonic, or dominant, belief in Western society.

In the 1950s, in response to the dominant functionalist paradigm, sociologist Ralf Dahrendorf (1958) published an influential article titled "Out of Utopia." Dahrendorf argued that functionalist theory, with its emphasis on how social institutions exist to maintain the status quo, overlooks critically important characteristics of society that lead to social conflict, such as the role of power, social change, and the unequal distribution of resources. The distribution of authority in society, said Dahrendorf, is a means of determining the probability of conflict. Where hierarchical structures, such as states; private economic entities, such as manufacturing firms; and even religious organizations are all dominated by the same elite, the potential for conflict is higher than in societies where authority is more dispersed. Put another way, if Group A dominates all or most key hierarchical authority structures and Group B is nearly always subordinate, conflict is likely because Group B has little stake in the existing social order. Nevertheless, if Group B has authority in some hierarchical structures and Group A has authority in others, neither group has great incentive to challenge the status quo.

Marx emphasized control of the means of production as a source of power and conflict, Gramsci highlighted control of dominant ideas in society as an important source of power and change, and Dahrendorf put authority and its concentration or distribution at the center of his work. Conflict theorists differ in their beliefs about what sources are most likely to underlie social conflict and social change, but all agree that social conflict and social change are both inevitable and desirable components of society and progress.

Rise-and-Fall Theories of Social Change

Rise-and-fall theories of social change: Theories that argue that social change reflects a cycle of growth and decline.

In contrast to theories that propose that societies progress, **rise-and-fall theories of social change** *argue that social change reflects a cycle of growth and decline.* Rise-and-fall or cyclical theories are common in the religious myths of many cultures, which view social life as a reflection of the life cycle of living creatures or the seasons of the year, with the end representing some form of return to the beginning. Sociology, emerging in an era that equated scientific and technological advancement with progress, at first tended to reject such cyclical metaphors in favor of more evolutionary or revolutionary ones that emphasized the forward motion of progress.

There have been several significant exceptions, however, among historically oriented social theorists. Pitirim Sorokin (1957/1970, 1962), a historical sociologist of the mid-20th century, argued that societies alternate among three different kinds of mentalities: those that give primacy to the senses, those that emphasize religiosity, and those that celebrate logic and reason. Societies that value hedonism and the satisfaction of immediate pleasures more highly than the achievement of long-term goals give primacy to the senses; religiosity occurs in societies that value following the tenets of a religion over enjoying the senses or solving problems through logic and reason. We tend to think of modern societies as defined largely by the emphasis on logic and reason.

All societies contain a mixture of religiosity, an emphasis on the senses, and the celebration of logic and reason. Sorokin's ideal types may nonetheless be useful for describing the *relative* emphasis of each of these modes of adaptation in different societies. For example, we might say that Europe and the United States today put greater emphasis on logic and reason than on religion or giving primacy to the senses; it would be a mistake, however, to say that there is no emphasis on the senses or religion, because these traits also play important roles in shaping Western societies.

In *The Rise and Fall of the Great Powers*, historian Paul Kennedy (1987) traces the conditions associated with national power and decline during the past five centuries. As nations grow in economic power, he argues, they often seek to become world military powers as well, a goal that proves to be their undoing in the long run. Wielding global military power eventually weakens a nation's domestic economy, undermining the prosperity that once fueled it. Kennedy forecasts that this might well be the fate of the United States.

More recently, writer Cullen Murphy (2007) has pointed to parallels between the Roman Empire and the United States, noting that Rome was also characterized by an overburdened and costly military, a deep sense of exceptionalism, and a tendency to denigrate and misunderstand other cultures. He also notes the Roman pattern of shifting the onus for providing services to citizens away from the public sector to the private sector, seeing this as a form of enrichment for the few but a disadvantage for the many. A key point in rise-and-fall narratives is that social change can be both progressive and regressive—power does not invariably beget more power; it may also beget decline.

Dr. Martin Luther King Jr.'s words and deeds inspired and continue to inspire social change. Sociologists such as Max Weber noted that the actions of a single "charismatic" person could be truly significant.

The most renowned sociologist considered by some to be a cyclical theorist is Max Weber. Although he took an evolutionary view of society as increasingly moving toward a politically and economically legal-rational society governed by rules and regulations, Weber (1919/1946) also emphasized the role of irrational elements in shaping human behavior. For example, although he wrote about the growing formal rationality of the modern world, he also recognized the possibility that a society's path could be altered by the appearance of a charismatic figure whose singular personal authority transcended institutionalized authority structures. Leaders who have been credited for drastically changing a nation's trajectory include Haile Selassie, who governed Ethiopia for half a century; Adolf Hitler in Germany; Mao Zedong in China; and Fidel Castro in Cuba. In the United States, Martin Luther King Jr. led the civil rights movement in the 1960s that fundamentally changed race relations.

Sociologists do not generally enlist cyclical theories. Even Weber's theory is not truly cyclical; his idea of charismatic authority is a sort of wild card, providing an unpredictable twist in an otherwise predictable march of social change from one form of authority to another. The more far-reaching versions of cyclical theory, such as Sorokin's theory that society swings among three different worldviews, are framed in such broad terms that it is challenging to test them.

SOURCES OF SOCIAL CHANGE

Social change ultimately results from human action. Sociologists studying how change occurs often analyze the mass action of large numbers of people and the institutionalized behaviors of organizations. In this section, we examine social change within the context of mass action by groups of people, focusing on theories of collective behavior and the role played by social movements.

Collective Behavior

Collective behavior is *voluntary, goal-oriented action that occurs in relatively disorganized situations in which society's predominant social norms and values cease to govern individual behavior* (Oberschall, 1973; Turner & Killian, 1987). Although collective behavior is usually associated with disorganized aggregates of people, it may also occur in highly regimented social contexts when order and discipline break down.

Collective behavior: Voluntary, goal-oriented action that occurs in relatively disorganized situations in which society's predominant social norms and values cease to govern individual behavior.

Beginning with the writings of the 19th-century French sociologist Gustave Le Bon, the sociological study of collective behavior has been particularly concerned with the behavior of people in **crowds**—*temporary gatherings of closely interacting people with a common focus.* People in crowds were traditionally seen as prone to being swept up in group emotions, losing their ability to make rational decisions as individuals. The "group mind" of the crowd has long been viewed as an irrational and dangerous aspect of modern societies, with crowds believed to consist of rootless, isolated individuals prone to herdlike behavior (Arendt, 1951; Fromm, 1941; Gaskell & Smith, 1981; Kornhauser, 1959).

Crowds: Temporary gatherings of closely interacting people with a common focus.

It has become clear that there can be a fair degree of social organization in crowds. For example, the Occupy Wall Street movement of 2011–2012 and the Arab Spring revolutions, which began in late 2010, and the #MeToo movement, which began in 2006 from a post on Tarana Burke's MySpace account, may have begun spontaneously. But they quickly developed a degree of predictability and organization and, in turn, became social movements.

It is important to note that crowds alone do not constitute social movements, but they are a critical ingredient in most cases. In a social-media age, sociologists may need to rethink the very notion of *spontaneity*, as collective action today is often rooted in activist social media that contributes to informing and organizing collective behavior. For instance, in mid-2020, over one third of social media users told Pew Research Center that they had used social media to "show their support for a cause, look up information about rallies or protests happening in their area, or encourage others to take action on issues they regard as important" through Facebook, Twitter, and other sites (Figure 18.1; Auxier, 2020, para. 1). The survey also found racial differences, with Hispanic and Black users more likely than whites to have looked up a rally or protest in their area (para. 2). Age and political party also makes a notable difference in social media activism. On the other hand, users were also skeptical about the perception of doing something on social media as adequate to change making. The authors of the survey note that while "most Americans believe these platforms are an effective tool for raising awareness and creating sustained movements, majorities also believe they are a distraction and lull people into believing they are making a difference when they're not" (Auxier, 2020, para. 2).

Sociologists seek to explain the conditions that may lead a group of people to engage in collective behavior, whether violent or peaceful. Next, we examine three principal sociological approaches: contagion theories, which emphasize nonsocial factors such as instincts; emergent norm theories, which seek out some kind of underlying social organization that leads a group to generate norms governing collective action; and value-added theories, which combine elements of personal, organizational, and social conditions to explain collective behavior.

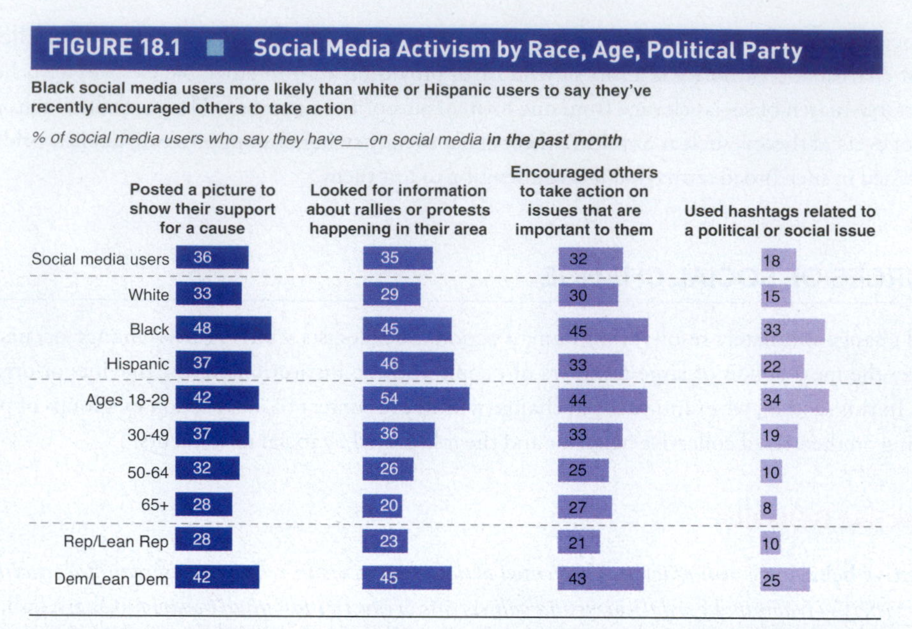

FIGURE 18.1 ■ Social Media Activism by Race, Age, Political Party

Source: Auxier, B. (2020, July 13). *Activism on Social Media Varies by Race and Ethnicity, Age, Political Party.* Washington, DC: Pew Research Center. Retrieved from https://www.pewresearch.org/fact-tank/2020/07/13/activism-on-social-media-varies-by-race-and-ethnicity-age-political-party/

Contagion Theories

Contagion theories assume that human beings can revert to herdlike behavior when they come together in large crowds. Herbert Blumer (1951), drawing on symbolic interactionism, emphasized the role of raw imitation, which leads people in crowds to mill about much like a group of animals, stimulating and goading one another into movement actions, whether peaceful or violent. Individual acts, therefore, become contagious; they are unconsciously copied until they eventually explode into collective action. A skilled leader can effectively manipulate such behavior, working the crowd until it reaches a fever pitch.

Sociologists have used contagion theories to explain the panicked flights of crowds; epidemics of bizarre collective behaviors, such as uncontrollable dancing or fainting; and reports of satanic child abuse. As we noted in Chapter 3, in 1983, a local moral panic erupted in a small California city after a parent of a preschool child accused teachers at her child's school of raping and sodomizing dozens of students. The trial in the case stretched on for years, but no wrongdoing was ever proven and no defendant convicted. Accusations in the case, which drew on allegations from children and parents, included stories about teachers chopping up animals at the school, clubbing to death a horse, and sacrificing a baby. Public accounts of the trial unleashed a national panic about abuse and satanism in childcare facilities, even though there was no serious documentation of such activities (Haberman, 2014). Some sociologists believe that a few well-publicized cases of deviant behavior—including wild accusations such as those just described—can trigger imitative behavior until a virtual epidemic emerges that then feeds on itself (Goode, 2009).

Take, for example, the phenomenon of suicide. Contagion has been linked in some research to its risk. For instance, data suggest that high levels of media coverage of celebrity suicides are followed by increased incidences of death by suicide and suicidal ideation, particularly among those who are demographically similar to the person who died. A controversy emerged in 2017 over the Netflix adaptation of a popular young adult novel, *13 Reasons Why*, which tells the story of a young women who kills herself and leaves behind 13 tapes meant for each of the people she believes led her to the decision to take her own life (Devitt, 2017). Research suggests that suicide can occur in clusters. That is, multiple suicides can occur in close temporal and geographic proximity: "It happens on average in at least five communities a year in this country. . . . Up to 5 percent of suicides among adolescents occur close to others, a higher rate than found in adults" (Carey, 2018, paras. 7–8). More recently, the deaths by suicide of celebrity chef Anthony Bourdain and designer Kate Spade led to a debate over how much media coverage is appropriate and how much may be harmful to those who might be at risk of contagion (Carey, 2018).

Although copycat behavior may occur in a group, community, or society, an explanation limited to this factor is unlikely to account fully for collective behavior. Furthermore, such explanations are sometimes used to discredit particular instances of collective behavior as resulting from an irrational (and therefore dangerous) tendency of people to jump on the bandwagon. In the 1960s, some people dismissed antiwar and civil rights protesters as misled "flower children" rather than recognizing them as people concerned about injustice and war. In 2020, #BlackLivesMatter peaceful protesters were sometimes conflated with those who engaged in looting and rioting (Wamsley, 2020). Sociologists, however, seek ways to determine *why* collective behavior occurs and to understand the rational and organizational basis for its emergence (Chafetz & Dworkin, 1983; L. Wright, 1993). We look next at what some other theories suggest.

Emergent Norm Theories

What role do norms and values play in the way people engage collectively? Some sociologists look to "emergent norms" to explain collective behavior. We can define **emergent norms** as *norms that are situationally created to support a collective action*. For example, Ralph H. Turner and Lewis M. Killian (1987) argue that even when crowd behavior appears chaotic and disorganized, norms emerge that explain the crowd's actions. Crowd members take stock of what is going on around them, are mindful of their personal motivations, and in general, collectively define the situation in which they find themselves. In this respect, crowd behavior is not very different from ordinary behavior; there is no need to fall back on instincts or contagion to explain it. For instance, some attendees in a crowd at a political rally may not agree with a candidate's position on, for instance, immigration, but the influence of the candidate and the crowd can function to create an environment where disagreeable positions on immigration come to seem rational, normative, and desirable. Hence, a cheering crowd of political supporters willing to work to elect a candidate can emerge even where there was not initially widespread embrace of a candidate's position on a given issue.

Emergent norms: Norms that are situationally created to support a collective action.

The emergent norm approach offers only a partial explanation of collective behavior. First, all crowds do not develop norms that govern their actions; crowds often emerge out of shared sets of norms among the participants. Second, purely spontaneous emotional outbursts may also occur as people act on their immediate impulses. Furthermore, when norms governing crowd behavior do emerge, they are unlikely, by themselves, to account fully for collective behavior.

For instance, when crowds gathered in cities around the world during the summer of 2020 to protest after George Floyd's murder by police officer Derek Chauvin, who kept his knee on Floyd's neck for 7 minutes and 46 seconds while Floyd repeatedly said "I can't breathe" (BBC, 2020), they were not only responding to the horror of his death. They were also responding to fear of and frustration with police violence against minorities, but their demonstrations were not merely a result of emergent norms, even though some actions were spontaneous. The grievances being expressed were rooted in long-standing disaffection with the treatment of poor minority communities and systemic institutionalized racism across the United States and abroad.

Value-Added Theory

Both contagion and emergent norm theories focus primarily on the micro level of individual action and thought, largely ignoring macrolevel factors—poverty, unemployment, governmental abuses of authority, and so on—that may explain the emergence of collective behavior. More than 50 years ago, Neil Smelser (1962) sought to develop what he termed a *value-added approach* to understanding collective behavior. He identified several micro- and macrolevel factors that each contribute to forming a foundation for collective behavior.

1. *Structural conduciveness* is a condition where the existing social structure favors the emergence of collective behavior.
2. *Structural strain* occurs when the social system breaks down.

3. *Generalized beliefs* are shared explanations of the conditions that are troubling people. People must define the problem, identify its causes, and—to use C. Wright Mills's phrase—come to see their personal troubles as public issues.
4. *Precipitating factors* are dramatic events that confirm the generalized beliefs of the group, thereby triggering action.
5. *Mobilization for action* occurs when leaders arise who encourage action.
6. *The failure of social control* leaves those charged with maintaining law and order unable to do so in the face of mounting pressures for collective action.

Smelser's approach has been used to analyze collective behavior in a variety of settings, including self-help groups (D. H. Smith & Pillemer, 1983), social welfare organizations (M. J. Smith & Moses, 1980), nuclear weapons freeze activism (Tygart, 1987), militias on the Internet (Weeber & Rodheaver, 2003), and "Trump's America" (Fuist & Williams, 2019). The theory's strength is that it combines societal, organizational, and individual-level factors into one comprehensive theory. Yet it has also been criticized for emphasizing the part that people's *reactions* play in collective behavior more than the fact that people themselves are conscious agents creating the conditions that can bring about significant social change.

Parts of Smelser's approach could easily be applied to the #BlackLivesMatter movement as it emerged in late May and early June 2020 around the death of George Floyd. Let's look at the first three dimensions: For instance, COVID-19 had provided "structural conduciveness" because many people's lives had changed precipitously due to the social distancing, loss of loved ones, and loss of jobs. "Structural strain" was evident as health care systems were under stress from the pandemic, people were being evicted from their homes because of the pandemic, and politicians were unable to work in a bipartisan fashion to create assistance to address it. "Generalized beliefs" that supported the awareness of racism and police brutality were already in place, based on the multiple prominent shootings of Black men and women in America. As Edwards, Lee, and Esposito noted in 2019:

> The killings of Oscar Grant, Michael Brown, Charleena Lyles, Stephon Clark, Tamir Rice, among many others, and the protests that followed have brought sustained national attention to the racialized character of police violence against civilians. (p. 1)

How would you apply the additional components of Smelser's approach to the summer protests of 2020? In total, do you believe these construct a reasonable explanation of what Buchanan et al. (2020) dubbed the "largest movement in American history"?

How Do Crowds Act?

All of us have participated in some form of collective "spontaneous" behavior: from rumors to fads and fashions, to panics and crazes, and riots. We discuss each of these forms of collective behavior next.

Rumors

Rumors: Unverified forms of information that are transmitted informally, usually originating in unknown sources.

Rumors are *unverified forms of information that are transmitted informally, usually originating in unknown sources.* The classic study on rumors was conducted more than 70 years ago by Gordon W. Allport and Leo Postman (1947). In one version of this research, a white student was asked to study a photograph depicting an urban scene: two men on a subway car, one menacing the other. The student was then asked to describe the picture to a second white student, who in turn was asked to pass the information along to a third, and so on. Eventually, after numerous retellings, the information changed completely to reflect the students' previously held beliefs. For example, as the "rumor" in the study took shape, the person engaging in the menacing act was described as Black and the victim as white—even though in the actual photograph the reverse was true.

Allport and Postman's (1947) research revealed several features unique to rumors. The information they contain is continually reorganized according to the belief systems of those who are passing them along. Some information is forgotten, and some is altered to fit into more familiar frameworks, such as the racist preconceptions offered in the earlier example. Furthermore, the degree of alteration varies according to the nature of the rumor; it is greatest for rumors that trigger strong emotions or that pass through large numbers of people. For a rumor to have an effect, it must tap into collectively held beliefs, fears, or hopes. For some, the rumor that the world will be ending imminently is a hopeful message; for others, it is a source of great fear. Rumors often reinforce subcultural beliefs. The rumor that the Central Intelligence Agency and the National Security Agency are planting listening devices in everyone's homes feeds into the belief that the government is out to control us. Political campaigns are infamous for starting and perpetuating rumors: In the 2016 presidential campaign, the Internet was rife with rumors about the health of candidate Hillary Clinton, who was said to be hiding a serious illness. While an abundance of evidence was available to contradict the rumors, they continued to be widely embraced, particularly on the far right of the political spectrum.

The acceleration of rumors through social media groups is an important phenomenon for sociologists to study. In 2020, a number of instances of social media rumor spreading were reported in the press, such as false reporting about extremists setting fires in Oregon. Many of these rumors had been traced to QAnon followers who

> believe there is a "deep state" within the US government that is controlled by a cabal of Satan-worshiping pedophiles. According to the baseless conspiracy theory, the cabal is largely run by Democratic politicians and liberal celebrities—and Trump is trying to take them down. (O'Sullivan & Toropon, 2020)

What rumors have recently received attention? From where did they come? How were they amplified? What if any effect do you think these rumors have on society?

Fads and Fashions

The desire to join others in being different (itself perhaps something of an irony) continually feeds the rise of new looks and sounds. **Fads**—*temporary, highly imitated outbreaks of mildly unconventional behavior*—are particularly common responses to popular entertainment, such as music, movies, and books, and require social networks (electronic or otherwise) to spread (Iribarren & Moro, 2007). The fads of piercing body parts to wear ornaments and extensive body tattooing continue. Other fads have included wearing blue jeans with holes, staging "panty raids" on sorority houses popularized by the 1970s movie *Animal House*, and being a hipster, among many others.

Fads: Temporary, highly imitated outbreaks of mildly unconventional behavior.

As fads become popular, they sometimes cease being fads and instead become **fashions**—that is, *somewhat long-lasting styles of imitative behavior or appearance*. Georg Simmel first examined the sociological implications of fashions more than a century ago. He pointed out that fashions reflect a tension between people's desire to be different and their desire to conform. By adopting a fashion, a person initially appears to stand out from the group, yet the fashion itself reflects group norms. As the fashion catches on, more and more people adopt it, and it eventually ceases to express any degree of individuality. Its very success undermines its attractiveness, so the eventual fate of all fashions is to become unfashionable.

Fashions: Somewhat long-lasting styles of imitative behavior or appearance.

Simmel's observations offer another insight into fashions: Unlike fads, they often grow out of the continuous and well-organized efforts of those who work in design, manufacturing technology, marketing, and media to define what is in style. As grunge music became popular in the 1990s, it spawned a profitable clothing industry, and highly paid fashion designers created clothing that was grungy in everything but price. Today, there are a variety of fashion trends that resonate with different audiences and subcultures. Whether it means wearing skinny or torn jeans, oversize sunglasses or owl-like reading spectacles, Nikes or Chucks or Adidas, Gucci or Prada, manufacturers and marketers will spend millions of dollars to convince consumers that they must buy particular products to be fashionable and popular.

As masking during COVID-19 became normalized, some began adopting fashion-oriented types of masks, and designers began to develop "fashionable masks" for weddings and other events.

Political activism swept the country in the 1960s and 1970s with widespread demonstrations focused on civil rights, women's rights, and the Vietnam War. The dramatic protests and social transformations of this period helped fuel the reformulation of theories on social change.

Panics and Crazes

A **panic** is *a massive flight from something feared*. The most celebrated example was created by an infamous radio broadcast on the night before Halloween in 1938: Orson Welles's Mercury Theatre rendition of H. G. Wells's science fiction novel *War of the Worlds*. The broadcast managed to convince thousands that Martians had landed in Grover's Mill, New Jersey, and were wreaking havoc with deadly laser beams. People panicked, flooded the telephone lines with calls, and fled to "safer" ground.

Panics are often ignited by the belief that something is awry in the corporate world or in consumer technology. As the year 2000 approached, panic over the Y2K problem, also known as the "millennium bug," gripped many people who believed reports that computer systems worldwide would crash when the year 2000 began. (Supposedly computers would be unable to distinguish the year 2000 from 1900, because they used only 2 digits to designate the year.) Another panic involved the Mayan calendar, which was projected to end during our calendar equivalent of December 2012. The fact that the structure of the Mayan calendar and the Mayan system of counting and noting dates did not pass December 2012 led many to believe that the Mayans had predicted the end of the world. Some panics, similar to that involving Y2K, reflect the fear that, in modern industrial society, we are highly dependent on products and technological processes about which we have little knowledge and over which we have no control.

A **craze** is *an intense attraction to an object, a person, or an activity*. Crazes are similar to fads but are more intense. Body disfigurement has been a periodic craze, ranging from nose piercing to putting rings through nipples, navels, lips, and tongues. The fact that these practices instill horror in some people probably accounts, in part, for the attraction they hold for others. In many cultures, body disfigurement is considered a necessary condition of beauty or attractiveness. Although such practices would be regarded as crazes in the West, they are normal enhancements of beauty in other cultures (D. E. Brown, Edwards, & Moore, 1988).

Panic: A massive flight from something that is feared.

Craze: An intense attraction to an object, a person, or an activity.

Riot: An illegal, prolonged outbreak of violent behavior by a sizable group of people directed against individuals or property.

Riots

A **riot** is *an illegal, prolonged outbreak of violent behavior by a sizable group of people directed against individuals or property*. Riots represent a form of crowd behavior; often, they are spontaneous, although sometimes they are motivated by a conscious set of concerns. Prison and urban riots are common examples. During a riot, conventional norms, including respect for the private property of others, are suspended and replaced with other norms developed within the group. For example, inmates may destroy property to force prison officials to adopt more humane practices, and the theft of property during an urban riot may reflect the participants' desire for a more equitable distribution of resources.

The very use of the term *riot* to characterize a particular action is often highly political. In 1773, a crowd of Bostonians protesting British taxation of the American colonies seized a shipment of tea from a British vessel and dumped it into Boston Harbor to protest taxation without representation. Although the British Crown roundly condemned this action as the illegal act of a rioting mob, U.S. history books celebrate the Boston Tea Party as the noble act of inspired patriots and an opening salvo in the Revolutionary War.

SOCIAL MOVEMENTS

Social movement: A large number of people who come together in a continuing and organized effort to bring about (or resist) social change and who rely at least partially on noninstitutionalized forms of political action.

Theories of collective behavior generally emphasize the passive, reactive side of human behavior. Social movement theory, in contrast, regards human beings as agents of their own history—actors who have visions and goals, analyze existing conditions, weigh alternative courses of action, and organize themselves as best they can to achieve success.

A **social movement** is *a large number of people who come together in a continuing and organized effort to bring about (or resist) social change and who rely at least partially on noninstitutionalized forms of political action*. Social movements thus have one foot outside the political establishment, and this is what distinguishes them from other efforts aimed at bringing about social change. Their political activities are not limited to such routine efforts as lobbying or campaigning; they include noninstitutionalized political actions such as boycotts, marches, and other demonstrations and civil disobedience.

Social movements often include some degree of formal organization oriented toward achieving longer-term goals, along with supporting sets of beliefs and opinions, but their strength often derives from their ability to disrupt the status quo by means of spontaneous, relatively unorganized political actions. As part of its support for the civil rights movement in the 1960s, the National Association for the Advancement of Colored People (NAACP) advocated the disruption of normal business activities, such as boycotting buses and restaurants, to force integration. The people who participate in social movements typically are outside the existing set of power relationships in society; such movements provide one of the few forms of political voice available to the relatively powerless (McAdam, McCarthy, & Zald, 1988; Tarrow, 1994).

The modern-day Tea Party movement emerged in the early 21st century as a protest against big government. While its membership declined over the years, some of Donald J. Trump's strongest support came from Tea Party members.

©Darren McCollester/Getty Images

An ongoing example is the Dreamer movement of Dreamers and their supporters, which has supported passage of the Dream Initiative (DACA). The Dreamers are undocumented individuals, arriving before the age of 16, mostly from Mexico, El Salvador, Guatemala, and Honduras. They were "younger than 31 on 15 June 2012, when the DACA program began" (*The Guardian*, n.d.). This immigration reform legislation allows undocumented young people who migrated to the United States with their families when they were children to have access to higher education and, over time, permanent residency or citizenship. An executive order signed by President Barack Obama in 2012 allowed Dreamers to apply for deferred action permits and avoid deportation under certain conditions. In 2017, the Donald J. Trump administration rescinded the 2012 memo that allowed for the Dream Initiative, seeking to wind down the program. This move was challenged by U.S. district courts in California, New York, Maryland, and the District of Columbia. In June 2020, the U.S. Supreme Court ruled that the Trump administration had unlawfully attempted to terminate the program but "also recognized that the federal government ultimately retains the legal authority to end the DACA initiative if it were to do so in compliance with the Administrative Procedure Act (APA)" (American Immigration Council, 2020).

The body of research on social movements in the United States is partially the result of movements that began in the late 1950s and gained attention and support in the 1960s and early 1970s. Theories of collective behavior, with their emphasis on the seemingly irrational actions of unorganized crowds, were ill equipped to explain the rise of well-organized efforts by hundreds of thousands of people to change government policies toward the Vietnam War and civil rights for African Americans. As these two social movements spawned others, including the second-wave feminist movement, which saw women demanding greater rights and opportunities in the workplace, sociologists had to rethink their basic assumptions and develop new theoretical perspectives.

Next, we examine different types of social movements, looking especially at sociological theories about why they arise.

The Dreamers and supporters protest in front of the Supreme Court in November 2019. This ruling, which ultimately upheld the original DACA temporarily, has an impact on over 700,000 young immigrants.

©Jahi Chikwendiu/The Washington Post via Getty Images

Types of Social Movements

Social movements are typically classified according to the direction and degree of change they seek. For the purposes of our discussion, we will distinguish five different kinds: reformist, revolutionary, rebellious, reactionary, and utopian (Table 18.1). In fact, these distinctions are not clear-cut, and the categories are not mutually exclusive. Rather, they represent ideal types. In the final section of the chapter, we will also consider some examples of a new, sixth category: new social movements that aim to change values and beliefs.

Reformist Movements

Reformist social movements: Movements seeking to bring about social change within the existing economic and political system.

Reformist social movements *seek to bring about social change within the existing economic and political system* and usually address institutions such as the courts and lawmaking bodies and/or public officials. They are most often found in societies where democratic institutions make it possible to achieve social change within the established political processes. Yet even reformist social movements can include factions that advocate more sweeping, revolutionary social changes. Sometimes, the government fails to respond, or it responds very slowly, raising frustrations. At other times, the government may actively repress a movement, arresting its leaders, breaking up its demonstrations, and even outlawing its activities.

The American Woman Suffrage Association, formed in 1869 by Susan B. Anthony and Elizabeth Cady Stanton, was a reformist organization that resulted in significant social changes. During the latter part of the 19th century, it became one of the most powerful political forces in the United States,

TABLE 18.1 ■ Principal Types of Social Movements

TYPE	PRINCIPAL AIMS	EXAMPLES
Reformist	To bring about change within the existing economic and political system	U.S. civil rights movement March for Our Lives, the student-led gun control movement Climate change activism, like that practiced by Greta Thunberg Teacher activism seeking better pay and funding for public education
Revolutionary	To fundamentally change the existing social, political, and/or economic system in light of an alternative vision	1776 U.S. Revolutionary War 1905 and 1917 Russian Revolutions 1991 South African antiapartheid movement 2010 Arab Spring
Rebellious	To fundamentally alter the existing political and/or economic system without a detailed alternative vision	Nat Turner slave rebellion Urban riots following the assassination of Martin Luther King Jr.
Reactionary	To restore an earlier social system—often based on a mythical vision of the past	White supremacist organizations in the United States European skinheads and neofascist movements
Utopian	To withdraw from society and create a utopian community	Religious communities such as the Quakers, Shakers, and Mennonites 1960s communes in the United States
New social movements	To make fundamental changes in values, culture, and private life	Gay, lesbian, bisexual, and transgender rights movements Disability rights movement Body acceptance movement Mindfulness movements

seeking to liberate women from oppression and ensure them the right to vote (Vellacott, 1993), which precipitated the first wave of the women's movement. In 1872, Victoria Woodhull helped to organize the Equal Rights Party, which nominated her for the U.S. presidency (even though, by law, no woman could vote for her); she campaigned on the issues of voting rights for women, the right of women to earn and control their own money, and free love (Underhill, 1995). After a half century of struggle by numerous social movement activists, women finally won the right to vote with the ratification of the 19th Amendment to the U.S. Constitution in 1920.

In June 2019, hundreds of thousands of demonstrators marched through the streets of former British colony Hong Kong's streets to protest legislation that would permit extradition to mainland China. Many citizens of Hong Kong, which today is a semiautonomous territory of China, fear that extradited prisoners could face politically charged trials in communist China.

The civil rights movement of the late 1950s and the 1960s called for social changes that would enforce the constitutionally mandated civil rights of African Americans; it often included nonviolent civil disobedience directed at breaking unjust laws. The ultimate aim of the civil rights movement was to change those laws, rather than to change society as a whole. Thus, for example, when Rosa Parks violated the laws of Montgomery, Alabama, by refusing to give up her seat on a city bus to a white person, she was challenging the city ordinance, not the government itself.

Much early civil rights activism was oriented toward registering southern Blacks to vote, so that by exercising their legal franchise, they could achieve a measure of political power. Within the civil rights movement, however, there were activists who concluded that the rights of Black Americans would never be achieved through reformist activities alone. Like many social movements, the civil rights movement was marked by internal struggles and debates regarding the degree to which purely reformist activities were adequate to the movement's objectives (Branch, 1988). The Black Panther Party, for example, argued for more radical changes in U.S. society, advocating for "Black Power" instead of merely fighting for an end to racial segregation. The Black Panthers often engaged in reformist activities, such as establishing community centers and calling for the establishment and support of more Black-owned businesses. At the same time, they also engaged in revolutionary activities, such as arming themselves against what they viewed to be a hostile police presence within Black neighborhoods.

The experience of U.S. labor unions, another example of a reformist social movement, shows the limits of the reformist approach to social change. Organized labor's principal demands have been for fewer hours, higher wages and benefits, job security, and safer working conditions. (In Europe, similar demands have been made, although workers there have sought political power as well.) Labor unions within the United States seldom appeal to a broad constituency beyond the workers themselves, and as a result, their success has depended largely on workers' economic power. U.S. workers have lost much of that power since the early 1970s, as economic globalization has meant the loss of many jobs to low-wage areas. Threats of strikes are no longer quite as menacing, as corporations can close factories down and reopen them elsewhere in the world.

Revolutionary Movements

Revolutionary social movements *seek to fundamentally alter the existing social, political, and economic system, in keeping with a vision of a new social order.* They frequently result from the belief that reformist approaches are unlikely to succeed because the political or economic system is too resistant. In fact, whether a social movement becomes predominantly reformist or revolutionary may well hinge on the degree to which its objectives can be achieved within the system.

Revolutionary social movements: Movements seeking to fundamentally alter the existing social, political, and economic system, in keeping with a vision of a new social order.

Revolutionary movements call for basic changes in economics, politics, norms, and values, offering a blueprint for a new social order that can be achieved only through mass action, usually by fostering conflict between those who favor change and those who favor the status quo. They are directed at clearly identifiable targets, such as a system of government believed to be unjust or an economy believed to be based on exploitation. Yet even the most revolutionary of social movements are likely to have reformist elements: members or factions who believe some change is possible within the established institutions. In most social movements, members debate the relative importance of reformist and revolutionary activities. Although the rhetoric may favor revolution, most day-to-day activities are likely to support reform. Only when a social movement is suppressed and avenues to reform are closed off will its methods call for outright revolution.

Revolutionary social movements sometimes, although by no means always, include violence. In South Africa, for example, the movements that were most successful in bringing about an end to apartheid were largely nonviolent. Those that defeated socialism in the former Soviet Union and Eastern Europe did so with a minimal amount of bloodshed. Nevertheless, revolutionary movements associated with the Arab Spring, which began in 2010 in countries such as Egypt, Tunisia, and Libya, resulted in considerable violence, most often perpetrated against the protesters by those already in power or their allies. It is unclear, however, whether the Arab Spring movements were truly revolutionary; most new governments are not radically more democratic than their predecessors. It takes time for political and economic conditions to change within any given country, however, and although some dictators have been removed from power, it remains to be seen whether these changes in political office will result in the changes desired by constituents.

Rebellions

Rebellions: Movements seeking to overthrow the existing social, political, and economic systems but lacking detailed plans for a new social order.

Rebellions *seek to overthrow the existing social, political, and economic systems but lack detailed plans for a new social order.* They are particularly common in societies where effective mobilization against existing structures is difficult or impossible because of the structures' repressive nature. The histories of European feudalism and U.S. slavery are punctuated by examples of rebellions. Nat Turner, an enslaved Black American, led other enslaved people in an 1831 uprising against their white enslavers in the state of Virginia. Before the uprising was suppressed, 55 whites were killed, and subsequently, Turner and 16 of his followers were hanged (K. S. Greenberg, 2003).

Reactionary Movements

Reactionary social movements: Movements seeking to restore an earlier social system—often based on a mythical past—along with the traditional norms and values that once presumably accompanied it.

Reactionary social movements *seek to restore an earlier social system—often based on a mythical past—along with the traditional norms and values that once presumably accompanied it.* These movements are termed *reactionary* because they arise in reaction to recent social changes that threaten or have replaced the old order. They are also sometimes referred to as *countermovements* or *resistance movements* for the same reason.

For these groups, a mythical past is often the starting point for pursuing goals aimed at transforming the present. The Ku Klux Klan (KKK), the White Aryan Resistance, and other white supremacist organizations have long sought to return to a United States where whites held exclusive political and

In August 2017, about 600 white nationalist demonstrators marched in Charlottesville, Virginia. Their rally, "Unite the Right," featured symbols of the Confederacy, the KKK, and Nazism, and was met with vigorous protests. It ended in deadly violence with the killing of Heather Heyer, a participant in the counterprotests. How would you characterize these movements?

©AP Photo/Steve Helber

economic power. Their methods have ranged from spreading discredited social and biological theories that expound the superiority of the "white race" to acts of violence against Black, Asian, Hispanic/Latinx, Jewish, gay and lesbian, and other Americans deemed to be inferior or otherwise a threat to the "American way of life" (Gerhardt, 1989; J. W. Moore, 1991).

Whether a social movement is viewed as reactionary or revolutionary depends, to some extent, on the observer's perspective. In Iran, for example, a social movement led by the Ayatollah Khomeini overthrew the nation's pro-U.S. leader in 1979 and created an Islamic republic that quickly reestablished traditional Muslim laws. In the pronouncements of U.S. policy makers and the mass media, the new Iranian regime was reactionary: It required women to be veiled, turned its back on democratic institutions, and levied death sentences on those who violated key Islamic values or otherwise threatened the Islamic state. Yet from the point of view of the clerics who led the upheaval, the movement overthrew a corrupt and brutal dictator who had enriched his family at the expense of the Iranian people and had fostered an alien way of life offensive to traditional Iranian values. From this standpoint, the movement claimed to be revolutionary, promising to provide a better life for Iranians.

As globalization threatens traditional ways of life around the world, we might expect to see an increase in reactionary social movements. This is especially likely to be the case if threats to long-standing traditional values are accompanied by declines in standards of living. In Germany, for example, a decline in living standards for many working-class people has spawned a small but significant resurgence of Nazi ideology and a group named the Alternative for Germany party (AFG), which has also worked its way into the state military (Bennhold, 2020). White supremacist groups often blame immigrants for their economic woes. The result has been a vocal campaign by skinhead groups against immigrants, particularly in the states that made up the former East Germany, which have seen greater economic upheavals than other parts of the country. It has also resulted in violence including the assassination of a politician, attack on a synagogue, and killing of nine immigrants (Bennhold, 2020).

Utopian Movements

Utopian social movements *seek to withdraw from the dominant society by creating their own ideal communities.* The youth movements of the 1960s had a strong utopian impulse; many young (and a few older) people "dropped out" of conventional society and formed their own communities, starting alternative newspapers, health clinics, and schools, and in general, seeking to live according to their own value systems outside the established social institutions. Some sought to live communally as well, pooling their resources and sharing tasks and responsibilities. They saw these efforts to create intentional communities, based on cooperation rather than competition, as the seeds of a revolutionary new society.

Utopian social movements: Movements seeking to withdraw from the dominant society by creating their own ideal communities.

Although some religious utopian movements have endured, those based on social philosophy have not. Some utopian socialist communities were founded in the United States during the 19th century; some provided models for the socialist collectives of the 1960s. Few lasted for any length of time. The old ways of thinking and acting proved remarkably tenacious, and the presence of the larger society—which remained largely unchanged by the experimentation—was a constant temptation. Alternative institutions such as communally run newspapers and health clinics found they had to contend with well-funded mainstream competitors. Most folded or reverted to mainstream forms (Fairfield, 1972; Nordhoff, 1875/1975; Rothschild & Whitt, 1987).

BEHIND THE NUMBERS: THERE WERE MILLIONS . . . OR NOT

How Many Attended?

- *How many people attended the Million Man March, a massive 1995 grassroots gathering to highlight concerns of Black men, their families, and their communities?*
- *Was the 2017 presidential inauguration audience, as noted by incoming president Donald J. Trump, "the biggest ever inauguration audience"? (Swaine, 2018).*
- *Was the Women's March on Washington in 2017 "likely the largest single-day demonstration in recorded U.S. history"?*

Who Decides?

Answering these questions is, perhaps, not surprisingly contentious. Consider the Million Man March numbers (Figure 18.2). As Ira Flatow noted on NPR's *Science Friday*:

> It depends on whom you ask. According to the U.S. Park Police, about 400,000. But the organizers of the march took issue with that number and asked for a recount. And using different images . . . a crowd counting expert at Boston University estimated the crowd to be closer to 800,000—almost twice the number that the Park Service had. (Counting Crowds, 2010)

How Do They Get Those Numbers?

Media accounts often cite the number of demonstrators in a unit of space at a moment in time (e.g., "200 protesters at 10 a.m. in Trafalgar Square"). In taking a look behind the numbers, we find that the National Park Service does not conduct official head counts of demonstrators in public spaces such as the National Mall in Washington, D.C., a popular venue for large gatherings. Journalists often rely on best-guess estimates for crowds. Official tools for counting include:

- Crowd-counting experts
- Satellite pictures that capture a moment in time
- Social media and crowdsourcing such as Google Docs

The January 21, 2017, Women's March on Washington, D.C., together with marches in other U.S. cities, may have gathered from 3 to 4 million people, figures assembled in part through a Google spreadsheet tweeted by a political science professor. The public was invited to share information and sources to the spreadsheet: According to an account of the effort, the spreadsheet currently has entries for nearly 550 cities and towns in the United States, from the march in D.C. (470,000 to 680,000 participants) to a protest in Show Low, Arizona (1 participant). The spreadsheet also tallies attendance at rallies in more than 100 cities around the world (Waddell, 2017).

- Crowd-counting programs

Researchers at the University of Central Florida introduced the world's first mass crowd-counting program software in 2015 to count the number of demonstrators in Barcelona calling for the independence of the region of Catalonia. The computer program scanned 67 aerial images of the demonstrators, whose protest action stretched for 3.2 miles. The data, ready in only 30 minutes, led researchers to conclude that just over half a million people were in attendance, a figure significantly below that offered by the protest organizers (*Science Daily*, 2015).

FIGURE 18.2 ■ Protest Turnout for the Million Man March

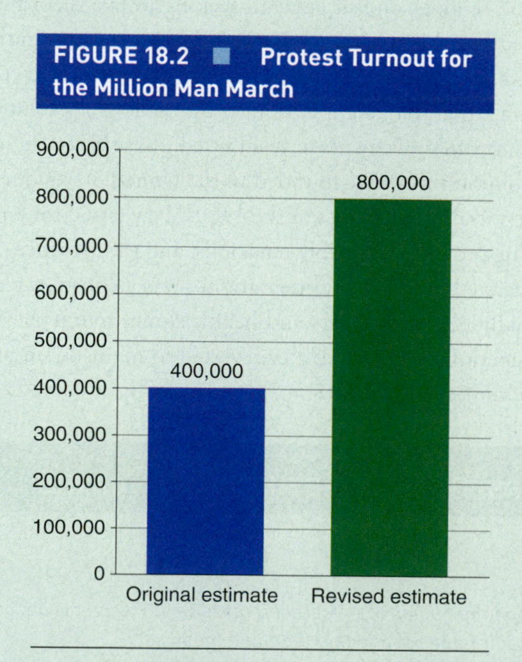

Source: Data complied by author.

- Other proxy measures

  Researchers can also use measures that help to replace counting actual people such as the number of trips taken on the metro or the number of people going through a turnstile.

What Are the Sociological Implications for Counting Crowds?

As Chapter 2 shows, the methods to obtain information are often as important as the information itself. All three of our examples show that numbers are also political. Crowd size matters—it can symbolize power, politics, relevance, and devotion. If you were going to measure crowd size, what measures would you use and why? How would you enlist your sociological imagination and what you know about research methods to help you?

Why Do Social Movements Arise?

Although social movements have existed throughout history, modern society has created conditions in which they thrive and multiply. The rise of the modern democratic nation-state, along with the development of capitalism, has fueled their growth. Democratic forms of governance, which emphasize social equality and rights of political participation, legitimate the belief that people should organize themselves politically to achieve their goals. Democratic nation-states give rise to social movements—and often protect them as well. Capitalism, which raises universal economic expectations while producing some inequality, further spurs the formation of such movements.

Given these general historical circumstances, sociologists have advanced several theories to explain why people sometimes come together to create or resist social change. Some focus on the micro level, looking at the characteristics and motivations of the people who join social movements. Some focus on the organizational level, looking at the characteristics that result in successful social-movement organizations. Some focus on the macro level, examining the societal conditions that give rise to social movements. More recently, theories have emphasized cultural dimensions of social movements, stressing the extent to which social movements reflect—as well as shape—larger cultural understandings. An ideal theory would bridge all these levels, and some efforts have been made to develop one.

Microlevel Approaches

Much research has focused on what motivates individuals to become active members of social movements. Psychological factors turn out to be poor predictors. Neither personality nor personal alienation adequately accounts for activist leanings. Rather, participation seems motivated more by psychological identification with others who are similarly afflicted (Marwell & Oliver, 1993; McAdam, 1982).

Sociology generally explains activism in such terms as having had prior contact with movement members, belonging to social networks that support movement activity, and having a history of activism (McAdam, 1986; Snow, Zurcher, & Ekland-Olson, 1980). Coming from a family background of social activism may also be important. One study, for example, found that many white male activists during the early 1960s social movements had parents who themselves had been activists 30 years earlier (Flacks, 1971). A lack of personal constraints may also be a partial explanation; it is obviously easier for individuals to engage in political activity if work or family circumstances afford them the necessary time and resources (McCarthy & Zald, 1973). Finally, a sense of moral rightness may provide a powerful motivation to become active, even when the work is difficult and the monetary rewards are small or nonexistent (Jenkins, 1983).

Social movements always suffer from the **free-rider problem**, however—that is, *many people avoid the costs of social-movement activism (such as time, energy, and other personal resources) and still benefit from its success* (Marwell & Oliver, 1993). Why not let others join the social movement and do the hard work, since if the movement succeeds then everyone will benefit, regardless of degree of participation? Clearly, it takes a great deal of motivation and commitment, as well as a conviction that their efforts may make a difference, for people to devote their time to mailing leaflets or organizing marches; building such motivation and commitment is a major challenge faced by all social movements.

Free-rider problem: The problem that many people avoid the costs of social-movement activism (such as time, energy, and other personal resources) and still benefit from its success.

The free-rider problem can also be connected to "activism" on social media. For instance, Christensen (2011) identifies a phenomenon he calls *slacktivism* or "political activities that have no impact on real-life political outcomes, but only serve to increase the feel-good factor of the participants" (cited in Cabrera et al., 2017, p. 1). We can apply the idea of slacktivism to social media and free-riding, because social media can allow users to click from the distance of their homes while avoiding other in-person costs including, as mentioned above, time, energy, and resources. However, in a thoughtful article about student activism, Cabrera et al. show that activism is not a zero-sum game, writing of the importance of social media along with in-person social interactions. As they put it:

> There is no substitute for in-person social activism to create the "tensions" described by Dr. King (1964); however, this does not preclude online space from being able to generate new tensions (Karp, 2010, 2012). If social justice is the overall goal, then online activity can and needs to be one component of this strategy. Yes, Scott-Heron was correct that the revolution will be live, but online engagement can be conducted in real time. Struggling with the tension between in-person and online engagement defines this generation of student activists. . . . (pp. 12–13)

Organizational-Level Approaches

Social-movement organizations (SMOs): Formal organizations that seek to achieve social change through noninstitutionalized forms of political action.

Some recent research has been devoted to understanding how social movements are consciously and deliberately organized to create social change. This research focuses on **social-movement organizations (SMOs)**, *formal organizations that seek to achieve social change through noninstitutionalized forms of political action.* The study of SMOs represents a major sociological step away from regarding social change as resulting from unorganized individuals and crowds. Instead, it places the study of social change within the framework of the sociology of organizations.

Because SMOs constitute a type of formal organization, sociologists use the same concepts and tools to study civil rights organizations and revolutionary groups as they do to study business firms and government bureaucracies. Researchers conceptualize SMOs' actions as rational, their goals as more or less clearly defined, and their organizational structures as bureaucratically oriented toward specific measurable goals (Jenkins, 1983; McCarthy & Zald, 1977).

SMOs range from informal volunteer groups to professional organizations with full-time leadership and staff. A single social movement may sustain numerous such organizations: A partial list associated with the 1960s civil rights movement includes the NAACP, the Student Nonviolent Coordinating Committee (SNCC), the Congress of Racial Equality (CORE), the Southern Christian Leadership Conference (SCLC), Students for a Democratic Society (SDS), and the Black Panther Party. As social movements grow, so too do the number of SMOs associated with them, each vying for members, financial support, and media attention.

Resource mobilization theory: A theory that focuses on the ability of social-movement organizations to generate money, membership, and political support to achieve their objectives.

One influential approach to the study of SMOs is **resource mobilization theory**, *a theory that focuses on the ability of social movement organizations to generate money, membership, and political support to achieve their objectives.* This approach argues that since discontent and social strain are always present among some members of any society, these factors cannot explain the rise or the relative success of social movements. Rather, what matters are differences in the resources available to different groups and how effectively they use them. The task for sociologists, then, is to explain why some SMOs are better able to deploy scarce resources than others (Jenkins, 1983; McAdam, 1988). Among the most important resources are tangible assets, such as money, facilities, and means of communication, as well as such intangibles as a central core of dedicated, skilled, hardworking members (Jenkins, 1983).

Much like businesses, then, SMOs rise or fall on their ability to be competitive in a resource-scarce environment. Some scholars have even written of social-movement "industries," with competing organizations engaging in "social marketing" to promote their particular "brands" of social change (Jenkins, 1983; Zald & McCarthy, 1980).

Governmental policies are important determinants of the success or failure of SMOs. The government may repress an organization, driving it underground so that it has difficulty operating. Or the government may favor more moderate organizations (for example, Martin Luther King Jr.'s SCLC) over other, more radical ones (such as the Black Panther Party). Other ways in which the government affects SMOs are through regulating them, providing favorable tax treatment for those that qualify, and refraining from excessive surveillance or harassment (McAdam et al., 1988).

The success or failure of SMOs also depends on their ability to influence the mass media. During the 1960s, the anti–Vietnam War organizations became very effective in commanding the television spotlight, although this effectiveness proved a mixed blessing: Media coverage frequently sensationalized demonstrations rather than presenting the underlying issues, thus contributing to rivalries and tensions within the antiwar movement (Gitlin, 1980). Today, arguably, social media exercises even greater influence on movement success.

This iconic image shows one man standing in opposition to four tanks in Tiananmen Square, China, on June 5, 1989. Thousands of pro-democracy demonstrators sought political and economic changes in a weekslong occupation of the square. Many were injured or killed in a government crackdown.

©AP Photo/Jeff Widener

Although some scholars have argued that larger, more bureaucratic SMOs are likely to be successful in the long run (W. Gamson, 1975), others have claimed that mass defiance, rather than formal organization, is the key to success (Piven & Cloward, 1977). Paradoxically, too much success may undermine social movements, since their strength derives partly from their being outside society's power structures as they make highly visible demands for social change. Once a group's demands are met, the participants are often drawn inside the very power structures they once sought to change. Movement leaders become bureaucrats, their fights are conducted by lawyers and government officials, and rank-and-file members disappear; the militant thrust of the organization is then blunted (Piven & Cloward, 1977).

A related problem is *goal displacement*, which occurs when an SMO's original goals become redirected toward enhancing the organization and its leadership (McCarthy & Zald, 1973). The U.S. labor movement is an example: Once labor unions became successful, many of them became large and prosperous bureaucracies that were perceived as distanced from the needs of their rank-and-file members.

In the end, SMOs have to motivate people to support their causes, often with dollars as well as votes. Many groups engage in **grassroots organizing**, *attempts to mobilize support among the ordinary members of a community*. This organizing may range from door-to-door canvassing to leafleting to get people to attend massive demonstrations. Most social movements emerge from a group that has some grievance, and their active members consist largely of people who will directly benefit from any social change that occurs.

Grassroots organizing: Attempts to mobilize support among the ordinary members of a community.

Some SMOs also depend on **conscience constituents**, *people who provide resources for a social-movement organization but who are not themselves members of the aggrieved group that the organization champions* (McCarthy & Zald, 1973). Such supporters are motivated by strong ethical convictions rather than by direct self-interest in achieving the social movement's goals. The National Coalition for the Homeless, for example, consists primarily of public interest lawyers, shelter operators, and others who advocate on behalf of homeless people; only a relatively small number of homeless people are directly involved in the organization. Homeless advocacy groups raise money from numerous sources, including media celebrities and direct mailings to ordinary citizens.

Conscience constituents: People who provide resources for a social-movement organization but who are not themselves members of the aggrieved group that the organization champions.

Macrolevel Approaches

Regardless of the efforts particular SMOs make, large-scale economic, political, and cultural conditions ultimately determine a movement's success or failure. For a social movement to arise and succeed, conditions must be such that people feel it is necessary and are willing to support it. Therefore, social movements emerge and flourish in times of other social change, particularly if people experience that change as disruptive of their daily lives (McAdam et al., 1988; Tilly, 1978). For example, the labor movement arose with the emergence of industrial capitalism, which brought harsh conditions to the lives of many people, and the women's movement reemerged in the 1960s, when expanded educational opportunities for women left many female college graduates feeling marginalized and alienated as full-time homemakers and workplace discrimination threw obstacles in the way of their workplace aspirations.

Some political systems encourage social movements, while others repress them (Gale, 1986). When a government is in crisis, it may respond by becoming more repressive, or it may create a space for social movements to flourish. The former action occurred in China in 1989, when thousands of students and workers, frustrated by deteriorating economic conditions and rigid government controls, took to the streets to demand greater economic and political freedom. The brutal crackdown at Beijing's Tiananmen Square, televised live to a global audience, ended the nascent social movement for democracy. Likewise, government crackdowns on demonstrators and social-movement participants have been widespread throughout the Middle East and North Africa as a second wave of Arab Spring movements continued into 2020 in countries such as Jordan, Sudan, Tunisia, Syria, Egypt, Lebanon, Algeria, Iraq, and Morocco (Tisdall, 2019). According to an analyst of these conflicts, Indira Lakshmanan:

> The problems that brewed in a cauldron of discontent from the early 2000s, sparking the Arab uprisings . . . a massive youth bulge, high unemployment, low wages, education systems mired in the past, a lack of innovation and absence of freedoms—are still stewing, and getting worse. . . . The strongmen haven't delivered a system to address the underlying problems. (Lakshmanan, quotes in Tisdall, 2019, para. 11)

Just as economic and political collapse may facilitate the rise of social movements, so too may prosperity. Resources for social activism are more abundant, mass media and other means of communication are more likely to be readily available, and activists are more likely to have independent means of supporting themselves. Prosperous societies are also more likely to have large classes of well-educated people, a group that has historically provided the leadership in many social movements (McAdam et al., 1988; McCarthy & Zald, 1973; Zald & McCarthy, 1980).

Finally, even the spatial organization of society may have an impact on social movements. Dense, concentrated neighborhoods or workplaces facilitate social interaction and spur the growth of social movements. A century and a half ago, Karl Marx recognized that cities and factories were powerful breeding grounds for revolutionary insurgency against capitalism, since they brought previously isolated workers together in single locations. Subsequent research has sustained his conclusion (Marx & Engels, 1848/1998; Tilly, 1975). The concentration of students on college campuses contributed to the rise of student activism in the 1960s (Lofland, 1985). As we saw in our opening story, some students (and this time at the level of high school as well as college) are actively engaged in activism today.

Finally, the sociologically significant processes of globalization and technological change provide both resources and constraints for people everywhere. A globalized economy has both positive and negative effects. It opens up the possibility of an increase in the standard of living for people worldwide as assisted by its increases in global productive capacity, technological advances, and global cooperation. It may also lead to lowered wages and to job losses in high-wage industrial countries, as well as to exploitative labor conditions in the low-wage countries of the world. Concerns about such problems have given rise to labor and environmental groups that operate across national borders (T. Barry & Sims, 1994), and, as a result, social movements themselves are increasingly internationalized (G. T. Marx & McAdam, 1994).

The LGBTQ+ Pride movement is a global movement for tolerance, equality, and pride. In this photo, participants pose at a Pride celebration in London. The year 2019 marked 50 years since the Stonewall riots in New York City, which are considered by many to be the origin of the LGBTQ+ rights movement.

Cultural-Level Studies and Frame Alignment

Much of the research we have discussed emphasizes the political, economic, and organizational conditions that either help or hinder the rise of social movements. Sociologists have often regarded social movements as by-products of favorable social circumstances rather than as the active accomplishments of their members. Today, however, instead of stressing how important it is for conditions to be ripe for social movements to thrive, many sociologists are thinking about how SMOs themselves are continually interpreting events so as to align themselves better with the cultural understanding of the wider society. The 21st-century Tea Party movement is a good example. Although it had a long list of political goals it wanted to achieve, it succeeded in rallying people around the idea that "big government" and taxation were threats to freedom and liberty (Berman, 2010). These are ideas that resonate with those who have some suspicion of intrusive government. The movement sought to create a good fit between itself and people who were its likely constituents.

Sociologists think of that fit in terms of **frame alignment**, *the process by which the interests, understandings, and values of a social-movement organization are shaped to match those in the wider society.* If their members' understandings align with the understandings of others in a community or society, social movements are likely to be successful; otherwise, they are likely to fail. SMOs achieve frame alignment in a variety of ways, ranging from modifying the beliefs of members to attempting to change the beliefs of the entire society (Snow, Rochford, Worden, & Benford, 1986).

Frame alignment: The process by which the interests, understandings, and values of a social-movement organization are shaped to match those in the wider society.

In one common situation, people already share the social movement's concerns and understandings but lack the means to bring about the desired changes. In this case, there is no need for the SMO to get people to change their thinking about the problem; rather, the task is to get people to support the movement's efforts to do something about it. The SMO must get the word out, whether through informal networks, direct-mail campaigns, or (more recently) social media.

The #MeToo movement in the United States is an awareness campaign against sexual harassment and assault, first started in 2006. In 2017, following the surfacing of sexual-violence accusations of prominent film producer Harvey Weinstein, actress Alyssa Milano took to Twitter encouraging survivors of sexual violence to spread the hashtag #MeToo, to give the world "a sense of the magnitude of the problem" (Gilbert, 2017). Since then, #MeToo has become an international movement, bringing awareness to the fact that millions of women (and men) have been victims of sexual assault and/or harassment in many spheres (MeToo, n.d.). The movement calls for action against perpetrators and has shown survivors that they have an entire community to which to turn for support. The #MeToo movement has manifested as both online activism and live protest actions.

Finally, an SMO may seek to build support by attempting to change the way people think entirely. Revolutionary SMOs, for example, urge people to stop thinking of themselves as victims of bad luck, focusing attention instead on the faults of the political or economic system, which presumably requires a drastic overhaul.

In sum, SMOs are competing for the hearts and minds of their constituents, with whom they must somehow bring their own beliefs and analyses into alignment if they are to succeed.

New Social Movements

New social movements: Movements that have arisen since the 1960s and are fundamentally concerned with the quality of private life, often advocating large-scale cultural changes in how people think and act.

Social movements have often served as a means to an end: People come together to achieve specific objectives, such as improving the conditions of workers, gaining equality for the disadvantaged, or protesting a war. In the past, participation in such social movements was often separate from members' personal lives. Since the 1960s, however, many social movements have sought to break the boundary between politics and personal life. In addition to serving as a means for changing the world, the SMO has come to be seen as a vehicle for personal change and growth (Giugni & Passy, 1998). Social movements that have embraced this perspective have been labeled **new social movements**. Although they often address political and economic issues, they *are fundamentally concerned with the quality of private life, often advocating large-scale cultural changes in the way people think and act*. In a sense, the new social movements reflect the *sociological imagination*, which calls for us to understand the relationship between our personal experiences and larger social forces.

The disability rights movement can be seen as a new social movement. Here Denise and Neil Jacobson, who met in 1968 at Camp Jened, a camp for teens with disabilities, await the premier of the documentary about that experience, Crip Camp. As Denise Jacobson noted about disability rights: "You can pass a law but until you change society's attitude the law won't change much" (*Crip Camp*, 2020).

©George Pimentel/Getty Images

The Disability Rights Movement, a local, national, and global movement, could be seen as an example of a new social movement. In the United States, members of the movement fought for civil rights for people with disabilities for many years, finally achieving the passage of the Americans with Disabilities Act of 1990 signed into law by President George

H. W. Bush. The law put into place safeguards regarding employment and accommodations in public places, removing legal discrimination against those with disabilities. Such discriminations were profound, including denial of services and educational opportunities, lack of accessibility for wheelchairs, lack of housing, medical discrimination, and many more. The 2020 documentary *Crip Camp* reveals how Camp Jened, a summer camp for those with disabilities located in the Catskill Mountains in New York, "fostered a sense of community and creativity that fed directly into the American disability rights movement in the 1970s" (Kenigsberg, 2020, para. 1). The movement continues today, targeting perceptions about people with disabilities, as well as drawing attention to and addressing continuing gaps in accessibility such as inadequacies in public school disability accommodations (Shapiro & Bowman, 2020).

New social movements may be formally organized, with clearly defined roles (leadership, recruiting, and so on), or they may be informal and loosely organized, preferring spontaneous and confrontational methods to more bureaucratic approaches. Part of the purpose of new social movements in protesting, in fact, is not to force a distinction between "them" and "us" but to draw attention to the movement's own right to exist as equals with other groups in society (J. Gamson, 1991; Omvedt, 1992; Tucker, 1991).

The new social movements aim to improve life in a wide range of areas subject to governmental, business, or other large-scale institutional control, from the workplace to sexuality, health, education, and interpersonal relationships. Four characteristics set these movements apart from earlier ones (Melucci, 1989):

1. The new social movements focus not only on the distribution of material goods but also on the control of symbols and information—an appropriate goal for an "information society" in which the production and ownership of knowledge are increasingly valuable.

2. People join new SMOs not purely to achieve specific goals but also because they value participation for its own sake. For instance, LGBTQ+ (lesbian, gay, bisexual, transgender, queer, and more) movements have provided safe havens for members in addition to pursuing social change.

3. Rather than large, bureaucratically run, top-down organizations, the new social movements are often networks of people engaged in routine daily activities. For example, a small online movement was begun by a woman who objected to an unannounced $5 charge on her credit card bill; her protest was joined by thousands of others, and the bank rescinded the charge. Groups trying to raise awareness of climate change and threats to the environment often are loosely organized and register their concerns online and through other media, such as newspapers and television talk shows.

4. The new social movements strongly emphasize the interconnectedness of planetary life and may see their actions as tied to a vision of the planet as a whole, rather than centering on narrow self-interest. "Think globally, act locally" is the watchword, and includes, but is not limited to, an awareness of environmental issues.

WHAT'S NEXT? SOCIAL CHANGE AND SOCIAL MOVEMENTS

Human beings make their own history, but they do not make it out of thin air. Every generation inherits certain *constraints*—characteristics of the society that limit their vision, choices, and *resources* that they can mobilize in new and creative ways. People are constrained by existing institutions and social relationships. Social structures provide the resources for human action, even as the actions themselves are oriented toward changing those structures (Giddens, 1985).

Sociologists study all aspects of social movements and social change, enlisting both qualitative and quantitative methods to study them. Notably, the American Sociological Association's section on the study of collective behavior and social movements has over 800 members and describes itself as "one of ASA's largest sections" (https://www.asanet.org/asa-communities/sections/collective-behavior-and-social-movements).

So, what's next? In an era of rapid social change, widespread immigration, and threats to democratic institutions across the world, one emerging issue is to understand and illustrate the sociological mechanisms through which democracies repress the dissent of their citizens, effectively undermining the democratic process of protest, communication, collective voices, and collective action (Subramaniam, 2020). For instance, after the world's largest democracy, India, passed the Citizenship Amendment Act of December 2019, a number of significant protests broke out across the large country. The act created a pathway to citizenship for Hindu, Sikh, Buddhist, Jain, Parsi, and Christian religious minorities that had fled to India seeking religious freedom. But it intentionally did not allow Muslims this same freedom. Police force was broadly enlisted against citizens, including students and professors at universities in India. Curfews and police detentions were common. Shutdowns of the Internet were also part of the state response to silence millions of protesters in some of the provinces. In this case, unlike others we have shown, the state intervened in the use of *social media* to quell protests in addition to enlisting other measures.

Why would a sociologist be interested in studying state power? How have other democratic societies around the world, including the United States, repressed dissent? What sociological methods could you enlist to study this issue at micro, meso, and macro levels? What challenges might you encounter in attempting to do such research?

WHAT CAN I DO WITH A SOCIOLOGY DEGREE?

Conflict Dynamics and Resolution Skills

Conflict resolution skills are of value when two or more entities—individuals, employees and employers, social groups, businesses, or countries, among others—engage in a process to resolve a disagreement that may be related to values or perceptions or to social, economic, environmental, or political interests. Sociological study offers students the opportunity to develop competencies in the analysis of conflict dynamics. Understanding the dynamics of conflict involves the ability to research and recognize the fundamental issues at the root of conflicts, to take the perspectives of different parties in a conflict, and to communicate effectively with groups in conflict. Understanding conflict dynamics and using this knowledge to resolve conflicts are related to sociology's general orientation toward problem analysis and problem solving.

It is in our interest as sociologists and citizens of the world to think in informed and creative ways about averting and addressing conflicts over resources as well as other sources of tension between states, societies, and groups. As a sociology student, you will have the opportunity to study different theoretical perspectives on conflict and its sources and to become familiar with research studies that delve into the roots of conflicts from the community to the global level. Understanding conflict dynamics and utilizing that knowledge to address and prevent or resolve conflicts are skills of great value in a heterogeneous and complex social and global environment.

Epiphany Summers, Organizing Director, Dream Defenders

Ursinus College, BS in Sociology and Psychology

The George Washington University, MA in Sociology

I initially became interested in sociology in college because it helped me to make sense of the community I grew up in and the systems at play around me. I grew up poor in Philadelphia with a hard-working family. It was sociologists like Michelle Alexander who helped me link the impacts of the 1994 Crime Bill to the overcriminalization of Black men. It was reading Karl Marx that gave me context for divisions of class that I was seeing for the first time in college. And it was diving into ideas like intersectionality that helped me to understand why race, class, and gender must be incorporated into the way we see the differences between one another. With all of this information, I went on to study sociology at the graduate level. I was passionate about making sense of the community around me, so I focused my Master's thesis on a qualitative research project where I conducted focus groups with people like me: young Black women. In my program, I learned about how to create generative questions, organize data, and, overall, a lot about Black culture. While in graduate school, I also did community service working with at-risk youth. Because I wanted to make a greater impact on people's lives and opportunities, I decided that I wanted to do social justice work. This would allow me to apply the many concepts and ideologies I learned in sociology and find solutions to the many community problems that existed around me. After completing my Master's degree in sociology, I started my career in community organizing. Today, I am the statewide Organizing Director for an organization called Dream Defenders in Florida, a youth racial justice organization. Much of what I studied in sociology helps me to be successful. I train people to collect qualitative data from the communities we organize in via individual conversations, house meetings, and surveys. I then train people on how to use that data to organize people to act. Many of the conversations I had in sociology classrooms are things I get to take action on every day!

Career Data: Training and Development Manager

- 2019 Median Pay: $113,350
- $54.50 per hour
- Typical Entry-Level Education: Bachelor's degree
- Job Growth Outlook, 2019–2029: 7% (faster than average)

Source: U.S. Bureau of Labor Statistics (2020).

SUMMARY

- Sociologists disagree about whether social change is gradual or abrupt and about whether all societies are changing in roughly the same direction. The evolutionary, revolutionary, and **rise-and-fall theories of social change** are three approaches to these questions.
- Some early sociologists viewed **collective behavior** as a form of group contagion in which the veneer of civilization gave way to more instinctive, herdlike forms of behavior.
- A more sociological approach, **emergent norm theory**, examines the ways in which **crowds** and other forms of collective behavior develop their own rules and shared understandings.
- The most comprehensive theory of collective behavior—value-added theory—attempts to consider the necessary conditions for collective behavior at the individual, organizational, and even societal levels.
- Social movements have been important historical vehicles for bringing about social change. They are usually achieved through **social-movement organizations (SMOs)**, which we study using the tools and understandings of organizational sociology.
- We can classify social movements as **reformist**, **revolutionary**, **rebellious**, **reactionary**, or **utopian**, depending on their vision of social change.
- **Resource mobilization theory** argues that we can explain the success or failure of SMOs not by the degree of social strain that may explain their origins but by their organizational ability to marshal the financial and personal resources they need.
- In recent years, sociologists have sought to explain how **social movements** align their own beliefs and values with those of their potential constituents in the wider society. **Frame alignment** activities range from modifying the beliefs of the SMO to attempting to change the beliefs of the entire society.

- Many social movements depend heavily on **conscience constituents** for their support. Micromobilization contexts are also important incubators of social movements.
- Globalization has created an opportunity for the formation of global social movements, since many of the problems in the world today are global and require global solutions.
- **New social movements**, organized around issues of personal identity and values, differ from earlier social movements in that they focus on symbols and information, as well as material issues; participation is frequently seen as an end in itself; the movements are organized as networks rather than bureaucratically; and they emphasize the interconnectedness of social groups and larger social entities.

KEY TERMS

collective behavior (p. 517)
conscience constituents (p. 531)
craze (p. 522)
crowds (p. 517)
differentiation (p. 514)
emergent norms (p. 519)
fads (p. 521)
fashions (p. 521)
frame alignment (p. 533)
free-rider problem (p. 529)
grassroots organizing (p. 531)
new social movements (p. 533)
panic (p. 522)
reactionary social movements (p. 526)
rebellions (p. 526)
reformist social movements (p. 524)
resource mobilization theory (p. 530)
revolutionary social movements (p. 525)
riot (p. 522)
rise-and-fall theories of social change (p. 516)
rumors (p. 520)
social movement (p. 522)
social-movement organizations (SMOs) (p. 530)
utopian social movements (p. 527)

DISCUSSION QUESTIONS

1. Research and describe one current social movement. Explain it using functionalist, conflict, and rise and fall approaches. Which do you think most accurately helps you understand this movement and why?
2. Identify what you see as two current fads and two current fashions that you think are sociologically relevant. Explain why you think this and be sure to explain why you have identified them as either a fad or fashion (that is, how do they differ?) If you were going to study these sociologically, what methods would you choose and why? What would you hope to learn from studying these phenomena?
3. Identify the different types of social movements. Then, design your own social movement. What problem or issue would you want to address? How would you disseminate information and engage other participants? How would you overcome the problems of social movements that were identified in this chapter?

GLOSSARY

Achieved status: Social position linked to an individual's acquisition of socially valued credentials or skills.

Agency: The ability of individuals and groups to exercise free will and to make social changes on a small or large scale.

Age-specific fertility rate: The number of births typical for women of a specific age in a particular population.

Age-specific mortality rate: An estimate of the number of deaths typical in men and women of specific ages in a particular population.

Agricultural surplus: Food beyond the amount required for immediate survival.

Alliance (or coalition): A subgroup that forms between group members, enabling them to dominate the group in their own interest.

Animism: The belief that naturally occurring phenomena, such as mountains and animals, are possessed of indwelling spirits with supernatural powers.

Anomie: A state of normlessness that occurs when people lose touch with the shared rules and values that give order and meaning to their lives.

Anticipatory socialization: Adoption of the behaviors or standards of a group one emulates or hopes to join.

Antimiscegenation laws: Laws prohibiting interracial sexual relations and marriage.

Ascribed status: Social position linked to characteristics that are socially significant but cannot generally be altered (such as race or sex).

Assimilation: The absorption of a minority group into the dominant culture.

Authoritarianism: A form of governance in which ordinary members of society are denied the right to participate in government, and political power is exercised by and for the benefit of a small political elite.

Automation: The replacement of human labor by machines in the production process.

Barter economy: An economy based on the exchange of goods and services rather than money.

Behaviorism: A psychological perspective that emphasizes the effect of rewards and punishments on human behavior.

Beliefs: Particular ideas that people accept as true.

Bias: A characteristic of results that systematically misrepresent the true nature of what is being studied.

Bourgeoisie: The capitalist (or property-owning) class.

Bureaucracies: Formal organizations characterized by written rules, hierarchical authority, and paid staff, intended to promote organizational efficiency.

Capital crime: A crime, such as murder, that is severe enough to merit the death penalty.

Capital offense: Crimes punishable by death.

Capitalism: An economic order characterized by the market allocation of goods and services, production for private profit, and private ownership of the means of producing wealth.

Caste society: A system in which social positions are closed, so that individuals remain at the social level of their birth throughout life.

Causal relationship: A relationship between two variables in which one variable is the cause of the other.

Charismatic authority: Power based on devotion inspired in followers by the personal qualities of a leader.

Church: A well-established religious organization that exists in a harmonious relationship with the larger society.

Cisgender: Those whose gender identity matches their sex assigned at birth.

Citizens: Legally recognized individuals who are part of a political community in which they are granted certain rights and privileges and, at the same time, have specified responsibilities.

City: A relatively large, dense, and permanent settlement of socially heterogeneous individuals.

Civil religion: A set of sacred beliefs and practices that connect the nation-state to transcendence and become part of how a society sees itself.

Class: A person's economic position in society, which is usually associated with income, wealth, and occupation (and sometimes associated with political voice).

Class conflict: Competition between social classes over the distribution of wealth, power, and other valued resources in society.

Class dominance theory: The theory that power is concentrated in the hands of a small group of elite or upper-class people who dominate and influence societal institutions; a variation of conflict theory.

Class society: A system in which social mobility allows an individual to change their socioeconomic position.

Class-dominant theories: Theories that propose that what is labeled deviant or criminal—and therefore who gets punished—is determined by the interests of the dominant class.

Coercion: The threat or use of physical force to ensure compliance.

Coercive organizations: Organizations in which members are forced to give unquestioned obedience to authority.

Cognitive development: The theory, developed by Jean Piaget, that an individual's ability to make logical decisions increases as the person grows older.

Cohabitation: Living together as a couple without being legally married.

Collective behavior: Voluntary, goal-oriented action that occurs in relatively disorganized situations in which society's predominant social norms and values cease to govern individual behavior.

Collective conscience: The common beliefs and values that bind a society together.

Common-law marriage: A type of relationship in which partners live as if married but without the formal legal framework of traditional marriage.

Communism: A type of economic system without private ownership of the means of production and, theoretically, without economic classes or economic inequality.

Concepts: Ideas that summarize a set of phenomena.

Conscience constituents: People who provide resources for a social-movement organization but who are not themselves members of the aggrieved group that the organization champions.

Conversation analysis: The study of how participants in social interaction recognize and produce coherent conversation.

Correlation: The degree to which two or more variables are associated with one another.

Craze: An intense attraction to an object, a person, or an activity.

Credential society: A society in which access to desirable work and social status depends on the possession of a certificate or diploma certifying the completion of formal education.

Crime: Any act defined in the law as punishable by fines, imprisonment, or both.

Critical race theories (CRTs): Theories that focus on the relationship between race, power, and racism.

Critical thinking: The ability to evaluate claims about truth by using reason and evidence.

Crowds: Temporary gatherings of closely interacting people with a common focus.

Crude birthrate: The number of births each year per 1,000 women.

Crude death rate: The number of deaths each year per 1,000 people.

Cult: A religious organization that is thoroughly unconventional with regard to the larger society.

Cultural capital: Wealth in the form of knowledge, ideas, verbal skills, and ways of thinking and behaving.

Cultural inconsistency: A contradiction between the goals of ideal culture and the practices of real culture.

Cultural pluralism: The coexistence of different racial and ethnic groups as characterized by the acceptance of one another's differences.

Cultural relativism: A worldview whereby the practices of a society are understood sociologically in terms of that society's norms and values, and not the norms and values of another society.

Culture: The beliefs, norms, behaviors, languages, and products common to the members of a particular group that bring meaning to their social worlds.

De facto segregation: School segregation based largely on residential patterns, which persists even though legal segregation is now outlawed in the United States.

Deductive reasoning: Starts from broad theories about the social world but proceeds to break them down into more specific and testable hypotheses.

Demography: The science of population size, distribution, and composition.

Denomination: A church that is not formally allied with the state.

Dependency theory: The theory that the poverty of some countries is a consequence of their exploitation by wealthy states, which control the global capitalist system.

Dependent variables: Variables that change as a result of changes in other variables.

Deviance: Any attitude, behavior, belief, or condition that violates cultural norms or societal laws and results in disapproval, hostility, or sanction if it becomes known.

Dictatorship: A form of governance in which power rests in a single individual.

Differential association theory: The theory that deviant and criminal behavior is learned and results from regular exposure to attitudes favorable to acting in ways that are deviant or criminal.

Differential opportunity theory: The theory that people differ not only in their motivations to engage in deviant acts but also in their opportunities to do so.

Differentiation: The development of increasing societal complexity through the creation of specialized social roles and institutions.

Direct democracy: A political system in which all citizens fully participate in their own governance.

Discouraged workers: Those who would like to work but have given up searching, believing that no jobs are available for them.

Discrimination: The unequal treatment of individuals on the basis of their membership in a group.

Document analysis: The examination of written materials or cultural products: previous studies, newspaper reports, court records, campaign posters, digital reports, films, pamphlets, and other forms of text or images produced by individuals, government agencies, private organizations, or others.

Domestic (or family) violence: Physical or sexual abuse committed by one family member against another.

Double consciousness: Among African Americans, an awareness of themselves as both American and Black, and never free of racial stigma.

Dramaturgical approach: Developed by Erving Goffman, the study of social interaction as if it were governed by the practices of theatrical performance.

Dyad: A group consisting of two persons.

Ecclesia: A church that is formally allied with the state and is the official religion of the society.

Economic capital: Money and material that can be used to access valued goods and services.

Economy: The social institution that organizes the ways in which a society produces, distributes, and consumes goods and services.

Education: The transmission of society's norms, values, and knowledge base by means of direct instruction.

Ego: According to Sigmund Freud, the part of the mind that is the "self," the core of what is regarded as a person's unique personality.

Egocentric: Experiencing the world as if it were centered entirely on oneself.

Emergent norms: Norms that are situationally created to support a collective action.

Emic perspective: The perspective of the insider, the one belonging to the cultural group in question.

Emotional labor: "The commodification of emotions, including the management of feeling to create a publicly observable facial and bodily display" (Hochschild, 2003).

Employed: Employed persons are those who are 16 years of age or older in the civilian, noninstitutional population who did any paid work—even as little as 1 hour—in the reference week or worked in their own businesses or farms.

Endogamous: A characteristic of marriages in which partners are limited to members of the same social group or caste.

Environmental racism: Policies and practices that impact the environment of communities where the population is disproportionately composed of ethnic and racial minorities.

Establishment Clause: The passage in the First Amendment to the U.S. Constitution that states, "Congress shall make no law respecting an establishment of religion, or prohibiting the free exercise thereof."

Ethnicity: Characteristics of groups associated with national origins, languages, and cultural and religious practices.

Ethnocentrism: A worldview whereby one judges other cultures by the standards of one's own culture and regards one's own way of life as normal and often superior to others.

Ethnomethodology: A sociological method used to study the body of common-sense knowledge and procedures by which ordinary members of a society make sense of their social circumstances and interaction.

Etic perspective: The perspective of the outside observer.

Evangelicalism: A belief in spiritual rebirth (conventionally denoted as being "born again").

Experiments: Research techniques for investigating cause and effect under controlled conditions.

Exponential population growth: A constant rate of population growth applied to a base that is continuously growing in size, producing a population that grows by an increasing amount with each passing year.

Expulsion: The process of forcibly removing a population from a particular area.

Extended families: Social groups consisting of one or more parents, children, and other kin, often spanning several generations, living in the same household.

Fads: Temporary, highly imitated outbreaks of mildly unconventional behavior.

Family: Two or more individuals who identify themselves as being related to one another, usually by blood, marriage, or adoption, and who share intimate relationships and dependency.

Fashions: Somewhat long-lasting styles of imitative behavior or appearance.

Feminism: The belief that social equality should exist between the sexes; also, the social movements aimed at achieving that goal.

Feminist perspectives on deviance: Perspectives that suggest that studies of deviance have been biased because almost all the research has been done by, and about, males, largely ignoring female perspectives on deviant behavior as well as analyses of differences in the types and causes of female deviance.

Fertility: The number of live births in a given population.

Fieldwork: A research method that uses in-depth and often extended study to describe and analyze a group or community; also called *ethnography*.

Folkways: Fairly weak norms that are passed down from the past, the violation of which is generally not considered serious within a particular culture.

Food deserts: Areas (often urban neighborhoods or rural towns) characterized by poor access to healthy and affordable food.

Food insecurity: A lack of consistent access to enough food for an active, healthy life.

Formal economy: All work-related activities that provide income and are regulated by government agencies.

Formal education: Education that occurs within academic institutions such as schools.

Formal organization: An organization that is rationally designed to achieve particular objectives, often by means of explicit rules, regulations, and procedures.

Formal rationality: A context in which people's pursuit of goals is shaped by rules, regulations, and larger social structures.

Formal social control: Official attempts to discourage certain behaviors and visibly punish others; most often exercised by the state.

Foster care: A situation in which a child is cared for by people who are not his or her parents for either a brief or extended period of time.

Frame alignment: The process by which the interests, understandings, and values of a social-movement organization are shaped to match those in the wider society.

Free-rider problem: The problem that many people avoid the costs of social-movement activism (such as time, energy, and other personal resources) and still benefit from its success.

Gender: The norms, roles, and behavioral characteristics associated in a given society with being a man or a woman.

Gender identity: One's sense of being a woman, man, or outside of the gender binary.

Gender roles: The actions, attitudes, and behaviors that are considered appropriately masculine or feminine in a particular culture.

Gender socialization: The process of learning social norms and values around gender.

Gender wage gap: The difference between the earnings of women who work full-time year-round as a group and those of men who work full-time year-round as a group.

Generalized other: The abstract sense of society's norms and values by which people evaluate themselves.

Genocide: The mass, systematic destruction of a people or a nation.

Gentrification: The change in the socioeconomic composition of older and poorer neighborhoods with the remodeling of old structures and building of new residences and shops to attract new middle- and high-income residents.

Glass ceiling: An artificial boundary that allows women to see the next occupational or salary level even as structural obstacles keep them from reaching it.

Glass escalator: The nearly invisible promotional boost that men gain in female-dominated occupations.

Global cities: Metropolitan areas that are highly interconnected with one another in their role as centers of global political and economic decision making, finance, and culture.

Global consumer class: Those who actively use technology, purchase consumer goods, and embrace the culture of consumption.

Global culture: A type of culture—some would say U.S. culture—that has spread across the world in the form of Hollywood films, fast-food restaurants, and popular music heard in virtually every country.

Global elite: A transglobal class of professionals who exercise considerable economic and political power that is not limited by national borders.

Global inequality: The systematic disparities in income, wealth, health, education, access to technology, opportunity, and power among countries, communities, and households around the world.

Globalization: The process by which people worldwide become increasingly connected economically, politically, socially, culturally, and environmentally.

Goods: Objects that have an economic value to others, whether they are the basic necessities for survival or things that people simply want.

Grassroots organizing: Attempts to mobilize support among the ordinary members of a community.

Gross national income–purchasing power parity per capita (GNI-PPP): A comparative economic measure that uses international dollars to indicate the amount of goods and services someone could buy in the United States with a given amount of money.

Groupthink: A process by which the members of a group ignore ways of thinking and plans of action that go against the group consensus.

Habitus: The internalization of objective probabilities and subsequent expression of those probabilities as choice.

Health care: All those activities intended to sustain, promote, and enhance health.

Health: The extent to which a person experiences a state of mental, physical, and social well-being.

Hegemonic masculinity: The culturally normative idea of male behavior that often emphasizes strength, control, and aggression.

Hidden curriculum: The classroom socialization into the norms, values, and roles of a society that schools provide along with the "official" curriculum.

High culture: Music, theater, literature, and other cultural products that are held in particularly high regard in society.

Human capital: The skills, knowledge, and credentials a person possesses that make them valuable in a particular workplace.

Human trafficking: The recruitment, transport, transfer, harboring, or receipt of a person by such means as threat or use of force or other forms of coercion, abduction, fraud, or deception for the purpose of exploitation.

Hyperreality: Mass reproduction of images and realities leading to a culture where language and images are separated completely, and reproductions of reality are more real than the reality they intend to produce.

Hypotheses: Ideas about the world, derived from theories, that describe possible relationships between social phenomena.

I: According to George Herbert Mead, the part of the self that creatively responds to a social situation. It is the impulse to act; it is creative, innovative, unthinking, and largely unpredictable.

Id: According to Sigmund Freud, the part of the mind that is the repository of basic biological drives and needs.

Ideal culture: The values, norms, and behaviors that people in a society profess to embrace.

Income: The amount of money a person or household earns in a given period of time.

Independent or experimental variables: Variables the researcher changes intentionally.

Indirect labor costs: The time, training, or money spent when an employee takes time off to care for sick family members, opts for parental leave, arrives at work late, or leaves a position after receiving employer-provided training.

Individual discrimination: Overt and intentional unequal treatment, often based on prejudicial beliefs.

Inductive reasoning: Starts from specific data, such as interviews, observations, or field notes, that may focus on a single community or event and endeavors to identify larger patterns from which to derive more general theories.

Inequality: Differences in wealth, power, political voice, educational opportunities, and other valued resources.

Infant mortality rate: The number of deaths of infants under age 1 per 1,000 live births per year.

Informal (or underground) economy: Those income-generating economic activities that are not regulated by the governmental institutions that ordinarily regulate similar activities.

Informal social control: The unofficial means through which deviance and deviant behaviors are discouraged in society; most often occurs among ordinary people during their everyday interactions.

Institutionalized discrimination: Discrimination enshrined in law, public policy, or common practice; it is unequal treatment that has become a part of the routine operation of such major social institutions as businesses, schools, hospitals, and the government.

Interest groups: Groups made up of people who share the same concerns on particular issues who use their organizational and social resources to influence legislation and the functioning of social institutions.

International families: Families that result from globalization.

International governmental organization (IGO): An international organization established by treaties between governments to facilitate and regulate trade between the member countries, promote national security, protect social welfare and human rights, or ensure environmental protection.

International nongovernmental organization (INGO): An international organization established by agreements between the individuals or private organizations making up the membership and existing to fulfill an explicit mission.

Intersectionality: The confluence of social statuses that shape people's lives, access to resources and power, justice, health, and well-being.

Intersex: Those born with characteristics associated with sex (e.g., genitals, gonads, and chromosome patterns) that do not fit typical binary notions of male or female bodies.

Interview: A detailed conversation designed to obtain in-depth information about a person and his or her activities.

Iron law of oligarchy: Robert Michels's theory that there is an inevitable tendency for a large-scale bureaucratic organization to become ruled undemocratically by a handful of people.

Labeling theory: A symbolic interactionist approach holding that deviant behavior is a product of the labels people attach to certain types of behavior.

Labor demand factors: Factors that highlight the needs and preferences of the employer.

Labor force participation rate: This figure represents the total of those who are counted by the Bureau of Labor Statistics as employed and those who are counted as unemployed, as a proportion of the civilian, noninstitutionalized population 16 years of age and over.

Labor supply factors: Factors that highlight reasons why women or men may prefer particular occupations.

Language: A symbolic system composed of verbal, nonverbal, and written representations that are vehicles for conveying meaning.

Latent functions: Functions of a phenomenon or institution that are not recognized or expected.

Law: A system of binding and recognized codified rules of behavior that regulate the actions of people pertaining to a given jurisdiction.

Laws: Codified norms or rules of behavior.

Leading questions: Questions that tend to elicit particular responses.

Legitimate authority: A type of power that is recognized as deserved or earned.

Liberal feminism: The belief that women's inequality is primarily the result of imperfect institutions, which can be corrected by reforms that do not fundamentally alter society itself.

Life chances: The opportunities and obstacles a person encounters in education, social life, work, and other areas critical to social mobility.

Life expectancy: The average number of years a newborn is expected to live based on existing health conditions in the country.

Literacy: The ability to read and write at a basic level.

Lobbyists: Paid professionals whose job it is to influence legislation.

Looking-glass self: The concept developed by Charles Horton Cooley that our self-image results from how we interpret other people's views of us.

Macrolevel paradigms: Theories of the social world that are concerned with large-scale patterns and institutions.

Mandatory minimum sentences: Legislation stipulating that a person found guilty of a particular crime must be sentenced to a minimum numbers of years in prison.

Manifest functions: The obvious and intended functions of a phenomenon or institution.

Marginally attached to the labor force: Persons who would like to work and have searched actively for a job in the past 12 months (but not in the prior 4 weeks).

Marriage: A culturally approved relationship, usually between two individuals, that provides a framework for economic cooperation, emotional intimacy, and sexual relations.

Mass education: The extension of formal schooling to wide segments of the population.

Mass media: Media of public communication intended to reach and influence a mass audience.

Mass production: The large-scale, highly standardized manufacturing of identical commodities on a mechanical assembly line.

Material culture: The physical objects that are created, embraced, or consumed by society that help shape people's lives.

Matrix of domination: A system of social positions in which any individual may concurrently occupy a status (for example, gender, race, class, or sexual orientation) as a member of a dominated group and a status as a member of a dominating group.

Me: According to George Herbert Mead, the part of the self through which we see ourselves as others see us.

Means of consumption: "Those things that make it possible for people to acquire goods and services and for the same people to be controlled and exploited as consumers" (Ritzer, 1999).

Means of production: The sites and technology that produce the goods we need and use.

Medicine: An institutionalized system for the scientific diagnosis, treatment, and prevention of illness.

Megachurch: A church with over 2,000 weekly attendees.

Megacities: Metropolitan areas or cities with a total population of 10 million or more.

Meritocracy: A society in which personal success is based on talent and individual effort.

Microlevel paradigm: A theory of the social world that is concerned with small-group social relations and interactions.

Minorities: Less powerful groups that are dominated politically, economically, and socially by a more powerful group and, often, discriminated against on the basis of characteristics deemed by the majority to be socially significant.

Modernization theory: A market-oriented development theory that envisions development as evolutionary and guided by modern institutions, practices, and cultures.

Monarchy: A form of governance in which power resides in an individual or a family and is passed from one generation to the next through hereditary lines.

Monogamy: A form of marriage in which a person may have only one spouse at a time.

Monotheism: Belief in a single all-knowing, all-powerful God.

Morbidity: The rate of illness in a particular population.

Mores: Strongly held norms, the violation of which seriously offends the standards of acceptable conduct of most people within a particular culture.

Mortality: The rate of death in a particular population.

Multicultural feminism: The belief that inequality must be understood—and ended—for all women, regardless of race, class, nationality, age, sexual orientation, physical ability, or other characteristics.

Multiculturalism: A commitment to respecting cultural differences rather than trying to submerge them into a larger, dominant culture.

Nation-state: A single people (a *nation*) governed by a political authority (a *state*); similar to the modern notion of *country*.

Negative correlation: A relationship showing that as one variable increases, the other decreases.

Net financial assets: A measure of wealth that excludes illiquid personal assets, such as the home and vehicles.

Net migration: In-migration minus out-migration.

New religious movements: New spiritual groups or communities that occupy a peripheral place in a country's dominant religious landscape.

New social movements: Movements that have arisen since the 1960s and are fundamentally concerned with the quality of private life, often advocating large-scale cultural changes in how people think and act.

Noncitizens: Individuals who reside in a given jurisdiction but do not possess the same rights and privileges as citizens; sometimes referred to as *residents, temporary workers*, or *aliens*.

Nonmaterial culture: The abstract creations of human cultures, including language, ideas about behavior, and social practices.

Nontheistic religion: Belief in the existence of divine spiritual forces rather than a god or gods.

Normative organizations (or voluntary associations): Organizations that people join of their own will to pursue morally worthwhile goals without expectation of material reward.

Norms: Accepted social behaviors and beliefs.

Not in the labor force: Persons who are neither officially employed nor officially unemployed.

Nuclear families: Families characterized by one or two parents living with their biological, dependent children in a household with no other kin.

Objectivity: The ability to represent the object of study accurately.

Occupation: A person's main vocation or paid employment.

Occupational gender segregation: The concentration of men and women in different occupations.

Official poverty line: The dollar amount set by the government as the minimum necessary to meet the basic needs of a family.

Operational definition: Describes the concept in such a way that it can be observed and measured.

Organization: A group with an identifiable membership that engages in concerted collective actions to achieve a common purpose.

Organized crime: Crime committed by criminal groups that provide illegal goods and services.

Pandemic: An outbreak of disease that occurs over a wide area.

Panic: A massive flight from something that is feared.

Patriarchy: Any set of social relationships in which men dominate women.

Personal power: Power that depends on the ability to persuade rather than the ability to command.

Phrenology: A theory that the skull configurations of deviant individuals differ from those of nondeviants.

Pluralistic societies: Societies made up of many diverse groups with different norms and values.

Political action committees: Organizations created by groups such as corporations, unions, environmentalists, and other interest groups for the purpose of gathering money and contributing to political candidates who favor the groups' interests.

Political power: The ability to exercise influence on political institutions and/or political actors to realize personal or group interests.

Politics: The art or science of influencing public policy.

Polyandry: A form of marriage in which a woman may have multiple husbands.

Polygamy: A form of marriage in which a person may have more than one spouse at a time.

Polygyny: A form of marriage in which a man may have multiple wives.

Polytheism: The belief that there are different gods representing various categories of natural forces.

Popular culture: The entertainment, culinary, and athletic tastes shared by the masses.

Population: The whole group of people studied.

Population momentum: The tendency of population growth to continue beyond the point when replacement rate fertility has been achieved because of the high concentration of people of childbearing age.

Positional power: Power that depends on the leader's role in the group.

Positive correlation: A relationship showing that as one variable rises or falls, the other does as well.

Positivism: An approach to research that is based on scientific evidence.

Power: The ability to mobilize resources and achieve goals despite the resistance of others.

Power elite: A group of people with a disproportionately high level of influence and resources who utilize their status to influence the functioning of societal institutions.

Prejudice: A belief about an individual or a group that is not subject to change on the basis of evidence.

Presentation of self: The creation of impressions in the minds of others to define and control social situations.

Preventive medicine: Medicine that emphasizes a healthy lifestyle that will prevent poor health before it occurs.

Primary deviance: The first step in the labeling of deviance, it occurs at the moment an activity is labeled deviant.

Primary groups: Small groups characterized by intense emotional ties, face-to-face interaction, intimacy, and a strong, enduring sense of commitment.

Principle of falsification (or falsifiability): The principle, advanced by philosopher Karl Popper, that to be scientific, a theory must lead to testable hypotheses that can be disproved if they are wrong.

Private sector: The sector of the labor market that provides goods and services to the economy and consumers with the primary motive of gaining profit.

Profane: A sphere of routine, everyday life.

Proletariat: The working class; wage workers.

Property crimes: Crimes that involve the violation of individuals' ownership rights, including burglary, larceny/theft, motor vehicle theft, and arson.

Psychoanalysis: A psychological perspective that emphasizes the complex reasoning processes of the conscious and unconscious mind.

Public education: A universal education system provided by the government and funded by tax revenues rather than student fees.

Public health: The science and practice of health protection and maintenance at a community level.

Public sector: The sector of the labor market in which jobs are linked to the government (whether national, state, or local) and encompass production or allocation of goods and services for the benefit of the government and its citizens.

Qualitative research: Research that is characterized by data that cannot be quantified (or converted into numbers), focusing instead on generating in-depth knowledge of social life, institutions, and processes.

Qualitative variables: Variables that express qualities and do not have numerical values.

Quantitative research: Research that gathers data that can be quantified and offers insight into broad patterns of social behavior and social attitudes.

Quantitative variables: Factors that can be counted.

Queer theory: A theory that describes how societies develop ideas about sexuality and gender, rejecting claims that gender identities and sexuality are fixed or stable.

Race: A group of people who share a set of characteristics (usually physical characteristics) deemed by society to be socially significant.

Racism: The idea that one racial or ethnic group is inherently superior to another; often results in institutionalized relationships between dominant and minority groups that create a structure of economic, social, and political inequality based on socially constructed racial or ethnic categories.

Radical feminism: The belief that women's inequality underlies all other forms of inequality, including economic inequality.

Random sampling: Sampling in which everyone in the population of interest has an equal chance of being chosen for the study.

Rape culture: A social culture that provides an environment conducive to rape that blames victims for their victimization.

Rate of natural increase (RNI): The crude birthrate minus the crude death rate.

Rational-legal authority: Power based on a belief in the lawfulness of enacted rules (laws) and the legitimate right of leaders to exercise authority under such rules.

Reactionary social movements: Movements seeking to restore an earlier social system—often based on a mythical past—along with the traditional norms and values that once presumably accompanied it.

Real culture: The values, norms, and behaviors that people in a society actually embrace and exhibit.

Rebellions: Movements seeking to overthrow the existing social, political, and economic systems but lacking detailed plans for a new social order.

Reference groups: Groups that provide standards for judging our attitudes or behaviors.

Reformist social movements: Movements seeking to bring about social change within the existing economic and political system.

Refugees: People who have fled war, violence, conflict, or persecution and have crossed an international border to find safety in another country.

Reliability: The extent to which researchers' findings are consistent with the findings of different studies of the same thing or with the findings of the same study over time.

Religion: A system of common beliefs and rituals centered on sacred things that unites believers and provides a sense of meaning and purpose.

Religious economy: An approach to the sociology of religion that suggests that religions can be understood as organizations in competition with one another for followers.

Religious nationalism: The linkage of religious convictions with beliefs about a nation's or ethnic group's social and political destiny.

Replication: The repetition of a previous study using a different sample or population to verify or refute the original findings.

Representative democracy: A political system in which citizens elect representatives to govern them.

Research methods: Specific techniques for systematically gathering data.

Reserve army of labor: A pool of job seekers whose numbers outpace the available positions and thus contribute to keeping wages low and conditions of work tenuous.

Resocialization: The process of altering an individual's behavior through control of his or her environment, for example, within a total institution.

Resource mobilization theory: A theory that focuses on the ability of social-movement organizations to generate money, membership, and political support to achieve their objectives.

Restrictive covenants: Contractual agreements that restrict the use of land, ostensibly in order to preserve the value of adjacent land or a neighborhood.

Revolutionary social movements: Movements seeking to fundamentally alter the existing social, political, and economic system, in keeping with a vision of a new social order.

Riot: An illegal, prolonged outbreak of violent behavior by a sizable group of people directed against individuals or property.

Rise-and-fall theories of social change: Theories that argue that social change reflects a cycle of growth and decline.

Risk society: A society where people are joined globally in their efforts to decrease danger and risk from large-scale biological, environmental, or military (nuclear) threats.

Rituals: Routinized acts typically performed in collective ceremonies and holding shared meaning.

Role-taking: The ability to take the roles of others in interaction.

Rumors: Unverified forms of information that are transmitted informally, usually originating in unknown sources.

Sacred: That which is set apart from the ordinary; the sphere endowed with spiritual meaning.

Sample: A small number of people; a portion of the larger population selected to represent the whole.

Scapegoating: Assigning misplaced blame to a person or group based on an alleged characteristic or action.

School segregation: The education of racial minorities in schools that are geographically, economically, and/or socially separated from those attended by the racial majority.

School-to-prison pipeline: The policies and practices that push students, particularly at-risk youth, out of schools and into the juvenile and criminal justice system.

Scientific: A way of learning about the world that combines logically constructed theory and systematic observation.

Scientific management: A practice that sought to use principles of engineering to reduce the physical movements of workers.

Scientific method: A process of gathering empirical (scientific and specific) data, creating theories, and rigorously testing theories.

Scientific theories: Explanations of how and why scientific observations are as they are.

Second shift: The unpaid housework that women typically do after they come home from their paid employment.

Secondary deviance: The second step in the labeling of deviance, it occurs when a person labeled deviant accepts the label as part of his or her identity and, as a result, begins to act in conformity with the label.

Secondary groups: Groups that are impersonal and characterized by functional or fleeting relationships.

Sect: A religious organization that has splintered off from an established church to restore perceived true beliefs and practices believed to have been lost by the established religious organization.

Secularization: The rise in worldly thinking, particularly as seen in the rise of science, technology, and rational thought, and a simultaneous decline in the influence of religion.

Segregation: The practice of separating people spatially or socially on the basis of race or ethnicity.

Self-fulfilling prophecy: As a consequence of labeling, people develop deviant self-concepts, living up to the label imposed on them.

Serial monogamy: The practice of having more than one wife or husband—but only one at a time.

Services: Economically productive activities that do not result directly in physical products; they may be relatively simple or quite complex.

Sex: The anatomical and other biological characteristics that differ between males and females and that originate in human genes.

Sexism: The belief that one sex is innately superior to the other and is therefore justified in having a dominant social position.

Sexual division of labor: The phenomenon of dividing production functions by gender and designating different spheres of activity, the "private" to women and the "public" to men.

Sexuality: A term used to encompass sexual identity, attraction, and relationships.

Sick role: A social role rooted in cultural definitions of the appropriate behavior of and response to people labeled as sick.

Significant others: According to George Herbert Mead, the specific people who are important in children's lives and whose views have the greatest impact on the children's self-evaluations.

Simulacra: Copies of objects and things that have no reality in the first place.

Social bonds: Individuals' connections to others.

Social capital: The personal connections and networks that enable people to accomplish their goals and extend their influence.

Social categories: Categories of people who share common characteristics without necessarily interacting or identifying with one another.

Social class reproduction: The way in which class status is reproduced from generation to generation, with parents passing on a class position to their offspring.

Social closure: The ability of a group to strategically and consciously exclude outsiders or those deemed "undesirable" from participating in the group or enjoying the group's resources.

Social conflict paradigm: A theory that seeks to explain social organization and change in terms of the conflict that is built into social relations; also known as conflict theory.

Social control: The attempts by certain people or groups in society.

Social control theory: The theory that explains that the probability of delinquency or deviance among children and teenagers is rooted in social control.

Social desirability bias: A response bias based on the tendency of respondents to answer a question in a way that they perceive will be favorably received.

Social diversity: The social and cultural mixture of different groups in society and the societal recognition of difference as significant.

Social dynamics: The laws that govern social change.

Social embeddedness: The idea that economic, political, and other forms of human behavior are fundamentally shaped by social relations.

Social epidemiology: The study of communities and their social statuses, practices, and problems with the aim of understanding patterns of health and disease.

Social facts: Qualities of groups that are external to individual members yet constrain their thinking and behavior.

Social inequality: A high degree of disparity in income, wealth, power, prestige, and other resources.

Social learning: A perspective that emphasizes the way people adapt their behavior in response to social rewards and punishments.

Social mobility: The upward or downward status movement of individuals or groups over time.

Social movement: A large number of people who come together in a continuing and organized effort to bring about (or resist) social change and who rely at least partially on noninstitutionalized forms of political action.

Social power: The ability to exercise social control.

Social solidarity: The bonds that unite the members of a social group.

Social statics: The way society is held together.

Social stratification: The systematic ranking of different groups of people in a hierarchy of inequality.

Socialism: A type of economic system in which, theoretically, the government manages the economy in the interests of the workers; it owns the businesses, factories, farms, hospitals, housing, and other means of producing wealth and redistributes that wealth to the population through wages and services.

Socialist feminism: The belief that women's inequality results from the combination of capitalistic economic relations and male domination (patriarchy), arguing that both must be fundamentally transformed before women can achieve equality.

Socialization: The process by which people learn the culture of their society.

Social-movement organizations (SMOs): Formal organizations that seek to achieve social change through noninstitutionalized forms of political action.

Sociological imagination: The ability to grasp the relationship between individual lives and the larger social forces that help to shape them.

Sociological theories: Logical, rigorous frameworks for the interpretation of social life that make particular assumptions and ask particular questions about the social world.

Sociology: The scientific study of human social relations, groups, and societies.

Spurious relationship: A correlation between two or more variables caused by another factor that is not being measured rather than a causal link between the variables themselves.

Standpoint epistemology: A philosophical perspective that argues that what we can know is affected by the position we occupy in society.

Standpoint theory: A perspective that says the knowledge we create is conditioned by where we stand or by our subjective social position in our daily lives.

State crimes: Criminal or other harmful acts of commission or omission perpetrated by state officials in the pursuit of their jobs as representatives of the government.

Statistical data: Quantitative information obtained from government agencies, businesses, research studies, and other entities that collect data for their own or others' use.

Status: The prestige associated with a social position.

Stereotyping: The generalization of a set of characteristics to all members of a group.

Stigma: An attribute that is deeply discrediting to an individual or a group because it overshadows other attributes and merits the individual or group may possess.

Strain theory: The theory that when there is a discrepancy between the cultural goals for success and the means available to achieve those goals, rates of deviance will be high.

Stratified sampling: Dividing a population into a series of subgroups and taking random samples from within each group.

Structural contradiction theory: The theory that conflicts generated by fundamental contradictions in the structure of society produce laws defining certain acts as deviant or criminal.

Structural functionalism: A theory that seeks to explain social organization and change in terms of the roles performed by different social structures, phenomena, and institutions; also known as functionalism.

Structural strain: In Merton's reformulation of Durkheim's functionalist theory, a form of anomie that occurs when a gap exists between the culturally defined goals of a society and the means available in society to achieve those goals.

Structuralism: The idea that an overarching structure exists within which culture and other aspects of society must be understood.

Structure: Patterned social arrangements that have effects on agency and are, in turn, affected by agency.

Subcultural theories: Theories that explain deviance in terms of the conflicting interests of different segments of the population.

Subcultures: Cultures that exist together with a dominant culture but differ from it in some important respects.

Superego: According to Sigmund Freud, the part of the mind that consists of the values and norms of society insofar as they are internalized, or taken in, by the individual.

Survey: A research method that uses a questionnaire or interviews administered to a group of people in-person or by telephone or e-mail to determine their characteristics, opinions, and behaviors.

Symbolic interactionism: A microsociological perspective that posits that both the individual self and society as a whole are the products of social interactions based on language and other symbols.

Symbols: Representations of things that are not immediately present to our senses.

Taboos: Powerful mores, the violation of which is considered serious and even unthinkable within a particular culture.

Terrorism: "The unlawful use of violence or threat of violence to instill fear and coerce governments or societies. Terrorism is often motivated by religious, political, or other ideological beliefs and committed in the pursuit of goals that are usually political" (U.S. Department of Defense, 2011).

"Three strikes" laws: State and federal laws that sentence an individual to life in prison who has been found guilty of committing three felonies or serious crimes punishable by a minimum of a year in prison.

Total fertility rate (TFR): The average number of children a woman in a given country will have in her lifetime if age-specific fertility rates hold throughout her childbearing years (ages 15–49).

Total institutions: Institutions that isolate individuals from the rest of society to achieve administrative control over most aspects of their lives.

Totalitarianism: A form of governance that denies popular political participation in government and also seeks to regulate and control all aspects of the public and private lives of citizens.

Totems: Within the sacred sphere, ordinary objects believed to have acquired transcendent or magical qualities connecting humans with the divine.

Traditional authority: Power based on a belief in the sanctity of long-standing traditions and the legitimate right of rulers to exercise authority in accordance with those traditions.

Transactional leader: A leader who is concerned with accomplishing the group's tasks, getting group members to do their jobs, and making certain that the group achieves its goals.

Transformational leader: A leader who is able to instill in group members a sense of mission or higher purpose, thereby changing (transforming) the nature of the group itself.

Transgender: An umbrella term used to describe those whose gender identity, expression, or behavior differs from their assigned sex at birth or is outside the gender binary.

Treadmill of production: The constant and aggressive growth needed to sustain the modern economy.

Triad: A group consisting of three persons.

Underemployed: Working in jobs that do not make full use of one's skills or working part time when one would like to be working full time.

Unemployed: People who are jobless, have actively looked for work in the prior 4 weeks, and are available for work.

Urban growth machine: Those persons and institutions that have a stake in an increase in the value of urban land and that constitute a power elite in cities.

Urban renewal: The transformation of old neighborhoods with new buildings, businesses, and residences.

Urbanization: The concentration of people in urban areas.

Utilitarian organizations: Organizations that people join primarily because of some material benefit they expect to receive in return for membership.

Utopian social movements: Movements seeking to withdraw from the dominant society by creating their own ideal communities.

Validity: The degree to which concepts and their measurements accurately represent what they claim to represent.

Value neutrality: The characteristic of being free of personal beliefs and opinions that would influence the course of research.

Values: The abstract and general standards in society that define ideal principles, such as those governing notions of right and wrong.

Variable: A concept or its empirical measure that can take on two or more possible values.

Verstehen: The German word for interpretive understanding; Weber's proposed methodology for explaining social relationships by having the sociologist imagine how subjects might perceive a situation.

Violent crimes: Crimes that involve force or threat of force, including murder and negligent manslaughter, rape, robbery, and aggravated assault.

War economy: The phenomenon of war boosting economic productivity and employment, particularly in capital- and labor-intensive sectors such as industrial production.

"War on drugs:" Actions taken by U.S. state and federal governments to curb the illegal drug trade and reduce drug use by punishing drug possession, use, and trafficking more harshly.

Wealth (or net worth): The value of everything a person owns minus the value of everything he or she owes.

Welfare state: A government or country's system of providing for the financial and social well-being of its citizens, typically through government programs that provide funding or other resources to individuals who meet certain criteria.

White-collar crime: Crime committed by people of high social status in connection with their work.

Work: Any human effort that adds something of value to the goods and services that are available to others.

World systems theory: The theory that the global capitalist economic system has long been shaped by a few powerful economic actors, who have constructed it in a way that favors their class interests.

Zero-tolerance policies: School or district policies that set predetermined punishments for certain misbehaviors and punish the same way, no matter the severity or the context of the behavior.

Zionism: A movement calling for the return of Jews to Palestine and the creation of a Jewish state.

REFERENCES

Abdullah, H. (2016, April 5). HUD seeks to end housing discrimination against ex-offenders. *NBC News*. Retrieved from http://www.nbcnews.com/news/us-news/hud-seeks-end-housing-discrimination-against-ex-offenders-n550471

Abramson, C. M. (2015). *The end game: How inequality shapes our final years*. Cambridge, MA: Harvard University Press.

Achen, A. C., & Stafford, F. P. (2005). *Data quality of housework hours in the Panel Study of Income Dynamics: Who really does the dishes?* (PSID Technical Series Paper 05-04). Institute for Social Research, University of Michigan. Retrieved from http://psidonline.isr.umich.edu/Publications/Papers/tsp/2005-04_Data_Qual_of_Household_Hours-Dishes.pdf

Acierno, R., Hernandez-Tejada, M., Muzzy, W., & Steve, K. (2009). *Final report: The National Elder Mistreatment Study*. Report submitted to the U.S. Department of Justice, National Institute of Justice. Retrieved from https://www.ncjrs.gov/pdffiles1/nij/grants/226456.pdf

Ackbar, S. (2011). *Constructions and socialization of gender and sexuality in lesbian-/gay-headed families*. Doctoral dissertation, University of Windsor. Retrieved from ProQuest (NR77959.

Acker, J. (1992). From sex roles to gendered institutions. *Contemporary Sociology, 21*(5), 565–569. Retrieved from http://www.jstor.org/stable/2075528

Acker, J. (2006). Inequality regimes: Gender, class, and race in organizations. *Gender and Society, 20*(4), 441–464.

Adachi, P. J. C., & Willoughby, T. (2011). The effect of video game competition and violence on aggressive behavior: Which characteristic has the greatest influence? *Psychology of Violence, 1*, 259–274.

Adely, H. (2020, September 28). *"We don't have the white experience": Arab Americans grapple with race categories on census*. Retrieved from https://www.northjersey.com/story/news/new-jersey/2020/09/28/census-2020-arab-americans-grapple-race-categories-census/3525270001/

Administration for Children and Families. (2020, January 15). *Child abuse, neglect data released*. Retrieved from https://www.acf.hhs.gov/media/press/2020/child-abuse-neglect-data-released

Administrative Office of the United States Courts. (2011). *Report of statistics required by the Bankruptcy Abuse Prevention and Consumer Protection Act of 2005*. Washington, DC: Government Printing Office. Retrieved from http://www.uscourts.gov/uscourts/Statistics/BankruptcyStatistics/BAPCPA/2010/2010BAPCPA.pdf

Adorno, T. (1975). The culture industry reconsidered. *New German Critique, 6*(Fall), 12–19.

Advancement Project. (2010). *Test, punish, and push out: How "zero tolerance" and high-stakes testing funnel youth into the school-to-prison pipeline*. Retrieved from http://b.3cdn.net/advancement/d05cb2181a 4545db07_r2im6caqe.pdf

Ahola, A. S., Christianson, S., & Hellstrom, A. (2009). Justice needs a blindfold: Effects of gender and attractiveness on prison sentences and attributions of personal characteristics in a judicial process. *Psychiatry, Psychology, and Law, 16*, S90–S100.

Ahuja, M., Barnes, R., Chow, E., & Rivero, C. (2014, September 4). The changing landscape on same-sex marriage. *Washington Post*. Retrieved from http://www.washingtonpost.com/wp-srv/special/politics/same-sex-marriage

Aina, O. E., & Cameron, P. A. (2011). Why does gender matter? Counteracting stereotypes with young children. *Dimensions of Early Childhood, 39*(1), 11–19.

Aizenman, N., & Silver, M. (2019). *How the U.S. compares with other countries in deaths from gun violence*. Retrieved from https://www.npr.org/sections/goatsandsoda/2019/08/05/743579605/how-the-u-s-compares-to-other-countries-in-deaths-from-gun-violence

Akbulut, Y., Dönmez, O., & Dursun Ö. Ö. (2017). Cyberloafing and social desirability bias among students and employees. *Computers in Human Behavior*, 728–795.

Albanese, J. S. (1989). *Organized crime in America*. New York: Anderson.

Aldrich, H. E., & Marsden, P. V. (1988). Environments and organizations. In N. J. Smelser (Ed.), *Handbook of Sociology* (pp. 361–392). Newbury Park, CA: Sage.

Alegria, S. (2019). Escalator or step stool? Gendered labor and token processes in tech work. *Gender & Society, 33*(5), 722–745. Retrieved from https://doi.org/10.1177/0891243219835737

Alexander, J. C., & Thompson, K. (2008). *A contemporary introduction to sociology: Culture and society in transition*. Boulder, CO: Paradigm.

Alexander, M. (2010). *The new Jim Crow: Mass incarceration in the age of colorblindness*. New York: New Press.

Alfano, S. (2009, February 11). Poll: Women's movement worthwhile. *CBS News*. Retrieved from http://www.cbsnews.com/2100-500160_162-965224.html

Alianza for Youth Justice and UCLA's Latino Policy and Politics Initiative. (2020, September 8). *The Latinx data gap in the youth justice system*. Annie E. Casey Foundation. Retrieved from https://www.aecf.org/resources/the-latinx-data-gap-in-the-youth-justice-system/

Allen, A. (2017, November 8). A hospital without patients. *Politico*. Retrieved from https://www.politico.com/agenda/story/2017/11/08/virtual-hospital-mercy-st-louis-000573/

Allen, V. L., & Levine, J. M. (1968). Social support, dissent and conformity. *Sociometry, 31*, 138–149.

Allport, G. W., & Postman, L. (1947). *The psychology of rumor*. New York, NY: Holt.

Alper, M., Durose, M. R., & Markman, J. (2018, May). *2018 update on prisoner recidivism: A nine-year follow-up period (2005–2014)*. Retrieved from https://www.bjs.gov/content/pub/pdf/18upr9yfup0514.pdf

Alter, C., Haynes, S., & Worland, J. (2019, December 23–30). Time 2019 person of the year: Greta Thunberg. *Time Magazine*. Retrieved from https://time.com/person-of-the-year-2019-greta-thunberg/

America Counts Staff. (2019, December 10). *2020 census will help policymakers prepare for the incoming wave of aging boomers*. Retrieved from https://www.census.gov/library/stories/2019/12/by-2030-all-baby-boomers-will-be-age-65-or-older.html

American Association of Colleges & Universities. (2018). *Fulfilling the American Dream: Liberal education and the future of work. Selected findings from online surveys of business executives and hiring managers*. Retrieved from https://www.aacu.org/sites/default/files/files/LEAP/2018EmployerResearchReport.pdf

American Association of University Women. (2016, Spring). *The simple truth about the gender pay gap*. Washington, DC: Author. Retrieved from http://www.aauw.org/research/the-simple-truth-about-the-gender-pay-gap/

American Bar Association. (2019). *Commission on Women in the Profession*. Retrieved from https://www.americanbar.org/content/dam/aba/administrative/women/current_glance_2019.pdf

American College Health Association. (2020). *American College Health Association–National College Health Assessment III: Undergraduate student reference group executive summary spring 2020*. Silver Spring, MD: American College Health Association.

American College Indian Fund. (2020). *Tribal colleges*. Retrieved from https://collegefund.org/students/tribal-colleges/

American Immigration Council. (2020, August 27). *The Dream Act, DACA, and other policies designed to protect Dreamers. Fact sheet on DACA*. Retrieved from https://www.americanimmigrationcouncil.org/research/dream-act-daca-and-other-policies-designed-protect-dreamers

American Medical Association. (2013). *AMA history timeline 1941–1960*. Retrieved from http://www.ama-assn.org/ama/pub/about-ama/our-history/ama-history-timeline.page

American Sociological Association. (2020). *Religion and spirituality*. Retrieved from https://www.asanet.org/topics/religion-and-spirituality

American Sociological Association. (n.d). *About the section*. Retrieved from https://www.asanet.org/asa-communities/sections/sites/sociology-education/about-section

Amis, M. (2002). *Koba the dread*. New York: Miramax.

Amnesty International. (2015, March 19). *Nigeria: Hundreds of oil spills continue to blight Niger delta*. Retrieved from https://www.amnesty.org/en/latest/news/2015/03/hundreds-of-oil-spills-continue-to-blight-niger-delta/

Amnesty International. (2020). *Death penalty in 2019: Facts and figures*. Retrieved from https://www.amnesty.org/en/latest/news/2020/04/death-penalty-in-2019-facts-and-figures/

Andersen, M. L., & Collins, P. H. (Eds.). (1992). *Race, class, and gender: An anthology*. Stamford, CT: Wadsworth.

Anderson, B. (1991). *Imagined communities: Reflections on the origin and spread of nationalism*. London: Verso.

Anderson, M., & Jiang, J. (2018, May 31). *Teens, social media & technology 2018*. Pew Research Center. Retrieved from https://www.pewresearch.org/internet/2018/05/31/teens-social-media-technology-2018/

Anderson, M., & Perrin, A. (2017, May 17). *Technology use among seniors*. Washington, DC: Pew Research Center. Retrieved from https://www.pewinternet.org/2017/05/17/technology-use-among-seniors

Anderson, N. (1940). *Men on the move*. Chicago: University of Chicago Press.

Anderson, T. D. (2000). Sex-role orientation and care-oriented moral reasoning: An online test of Carol Gilligan's theory. *Dissertation Abstracts International, 61*, 1618–A. Retrieved from http://search.proquest.com.proxygw.wrlc.org/socabs/docview/60390650/13C8E238511536831E/1?accountid=11243

Andre, F. E., Boy, R., Bock, H. L., Clemens, J., Datta, S. K., John, T. J., ... Schmitt, H. J. (2008). Vaccination greatly reduces disease, disability, death and inequity worldwide. *Bulletin of the World Health Organization, 86*. Retrieved from http://www.who.int/bulletin/volumes/86/2/07-040089/en

Andrews, M. (2019, August 20). *Cigarettes can't be advertised on TV. Should Juul ads be permitted?* National Public Radio. Retrieved from https://www.npr.org/sections/health-shots/2019/08/20/752553108/cigarettes-cant-be-advertised-on-tv-should-juul-ads-be-permitted

Angier, N. (2000, August 22). Do races differ? Not really, genes show. *New York Times*. Retrieved from http://www.nytimes.com/2000/08/22/science/do-races-differ-not-really-genes-show.html?pagewanted=all&src=pm

Annie E. Casey Foundation. (2019, June 16). *2019 KIDS COUNT data book*. Retrieved from https://www.aecf.org/resources/2019-kids-count-data-book/?msclkid=e2d008be3db9102f5c40

Antos, J. (2011). Medicare reform and fiscal reality. *Journal of Policy Analysis and Management, 30*, 934–942.

Anupam, B., Olenski, A. R., & Blumenthal, D. M. (2016, July 11). Sex differences in physician salary in U.S. public medical schools. *The JAMA Network*. Retrieved from http://archinte.jamanetwork.com/article.aspx?articleid=2532788#Results

Anzaldúa, G. (Ed.). (1990). *Making face, making soul: Haciendo caras*. San Francisco: Aunt Lute Foundation.

APM Research Lab Staff. (2020, October 15). *The color of coronavirus: Covid-19 deaths by race and ethnicity in the United States*. Retrieved from https://www.apmresearchlab.org/covid/deaths-by-race

Appadurai, A. (1996). *Modernity at large: Cultural dimensions of globalization*. Minneapolis, MN: University of Minnesota Press.

Appadurai, A. (2014). Arjun Appadurai. *Globalizations 11*(4), 481–490. Retrieved from https://doi-org.thor.nebrwesleyan.edu/10.1080/14747731.2014.951209

Appelbaum, B. (2015). Out of trouble, but criminal records keep men out of work. *New York Times*. Retrieved from http://www.nytimes.com/2015/03/01/business/out-of-trouble-but-criminal-records-keep-men-out-of-work.html?_r=0

Appelo, T. (2012). THR poll: 'Glee' and 'Modern Family' drive voters to favor gay marriage—Even many Romney voters. *The Hollywood Reporter*. Retrieved from http://www.hollywoodreporter.com/news/thr-poll-glee-modern-family-386225

Archer, D., & Gartner, R. (1984). *Violence and crime in cross-national perspective*. New Haven, CT: Yale University Press.

Arends, B. (2020, March 6). *White-collar criminal prosecutions fall to their lowest level on record, study says.* Retrieved from https://www.marketwatch.com/story/its-disturbing-us-justice-department-white-collar-criminal-prosecutions-fall-to-their-lowest-level-on-record-2020-03-06

Arendt, H. (1951). *The origins of totalitarianism.* New York, NY: Harcourt, Brace.

Arias, E., & Jiaquan, X. (2019, June 24). United States life tables, 2017. *National Vital Statistics Reports 68*(7). Retrieved from https://www.cdc.gov/nchs/data/nvsr/nvsr68/nvsr68_07-508.pdf

Armstrong, L., Phillips, J. G., & Saling, L. L. (2000). Potential determinants of heavier Internet use. *International Journal of Human-Computer Studies, 53,* 537–550.

Arnold, C. (2019, October 31). The 7-year car loan: Watch your wallet. *National Public Radio.* Retrieved from https://www.npr.org/2019/10/31/773409100/the-7-year-car-loan-watch-your-wallet

Artiga, S., & Orgera, K. (2019, November 12). *Key facts on health and health care by race and ethnicity.* Kaiser Family Foundation. Retrieved from https://www.kff.org/disparities-policy/report/key-facts-on-health-and-health-care-by-race-and-ethnicity/

Asch, S. (1952). *Social psychology.* Englewood Cliffs, NJ: Prentice Hall.

Asencio, E. K., & Burke, P. J. (2011). Does incarceration change the criminal identity? A synthesis of labeling and identity theory perspectives on identity change. *Sociological Perspectives, 54*(20), 163–182.

Ashkenas, J., Park, H., & Pearce, A. (2017, August 24). Even with affirmative action, Blacks and Hispanics are more underrepresented at top colleges than 25 years ago. *New York Times* Interactive. Retrieved fromhttps://www.nytimes.com/interactive/2017/08/24/us/affirmative-action.html

The Aspen Institute Project Play. (2019). Retrieved from https://assets.aspeninstitute.org/content/uploads/2019/10/2019_SOP_National_Final.pdf?_ga=2.159918073.59867992.1591033615-1484098388.1591033615

Association of American Medical Colleges. (2019, December 9). *The majority of U.S. medical students are women, new data show.* Retrieved from https://www.aamc.org/news-insights/press-releases/majority-us-medical-students-are-women-new-data-show

The Atlantic. (2014). *When it comes to politics, do millennials care about anything?* Retrieved from https://www.theatlantic.com/sponsored/allstate/when-it-comes-to-politics-do-millennials-care-about-anything/255/

Aubrey, J. S., & Frisby, C. M. (2011). Sexual objectification in music videos: A content analysis comparing gender and genre. *Mass Communication and Society, 14,* 475–501.

Aubrey, J. S., & Harrison, K. (2004). The gender-role content of children's favorite television programs and its links to their gender-related perceptions. *Media Psychology, 6,* 111–146.

Autor, D. (2010, April). *The polarization of job opportunities in the U.S. labor market: Implications for employment and earnings.* Washington, DC: Center for American Progress and the Hamilton Project. Retrieved from http://www.scribd.com/doc/52779456/The-Polarization-of-Job-Opportunities-in-the-U-S-Labor-Market

Auxier, B. (2020, July 13). *Activism on social media varies by race and ethnicity, age, political party.* Pew Research Center. Retrieved from https://www.pewresearch.org/fact-tank/2020/07/13/activism-on-social-media-varies-by-race-and-ethnicity-age-political-party/

Auxier, B., & Anderson, M. (2020, March 16). *As schools close due to the coronavirus, some U.S. students face a digital "homework gap."* Pew Research Center Fact Tank. Retrieved from https://www.pewresearch.org/fact-tank/2020/03/16/as-schools-close-due-to-the-coronavirus-some-u-s-students-face-a-digital-homework-gap/

Auxier, B., & McClain, C. (2020, September 9). *Americans think social media can help build movements, but can also be a distraction.* Pew Research Center. Retrieved from https://www.pewresearch.org/fact-tank/2020/09/09/americans-think-social-media-can-help-build-movements-but-can-also-be-a-distraction/

Avadhanula, S. (2020, July 2). *Telemedicine during the COVID-19 pandemic—and beyond.* Retrieved from https://health.usnews.com/health-care/for-better/articles/telemedicine-during-the-covid-19-pandemic-and-beyond

Aviel, D. (1997). Issues in education: A closer examination of American education. *Childhood Education, 73,* 130–132.

Avrahami-Winaver, A., Regev, D., & Reiter, S. (2020). Pictorial phenomena depicting the family climate of deaf/hard of hearing children and their hearing families. *Frontiers in Psychology, 11,* 2221. doi:10.3389/fpsyg.2020.02221

Babbie, E. R. (1998). *The practice of social research.* Belmont, CA: Wadsworth.

Babcock, P., & Marks, M. (2010, August). *Leisure college, USA: The decline in student study time* (Research Education Outlook No. 7). Washington, DC: American Enterprise Institute for Public Policy. Retrieved from http://www.aei.org/files/2010/08/05/07-EduO-Aug-2010-g-new.pdf

Badger, E. (2014, April 15). *Pollution is segregated too.* Retrieved from https://www.washingtonpost.com/news/wonk/wp/2014/04/15/pollution-is-substantially-worse-in-minority-neighborhoods-across-the-u-s/

Badger, E. (2016, May 22). This can't happen by accident. *Washington Post.* Retrieved from https://www.washingtonpost.com/graphics/business/wonk/housing/atlanta/

Badger, E., & Eilperin J. (2016, March 14). The cruelest thing about buying diapers. *Washington Post.* Retrieved from https://www.washingtonpost.com/news/wonk/wp/2016/03/14/the-cruelest-thing-about-buying-diapers/

Bahler, K. (2020, April 27). The class of 2020 desperately wants jobs but is graduating into record unemployment. Can they ever catch up? *Newsweek.* Retrieved from https://www.newsweek.com/2020/05/15/class-2020-all-dressed-nowhere-go-1500142.html

Bailey, K., West, R., & Anderson, C. A. (2011). The influence of video games on social, cognitive, and affective information processing. *Handbook of Social Neuroscience,* 1001–1011.

Bailey, S. P. (2019, November 19). How Kanye West put Joel Osteen's prosperity gospel back under the spotlight. *Washington Post.* Retrieved from https://www.washingtonpost.com/religion/2019/11/19/how-kanye-west-put-joel-osteens-prosperity-gospel-back-under-spotlight/

Baker-Smith, C., Coca, V., Goldrick-Rab, S., Looker, E., Richardson, B., & Williams, T. (2020). *#RealCollege 2020: Five years of evidence on campus basic needs insecurity*. Retrieved from https://hope4college.com/wp-content/uploads/2020/02/2019_RealCollege_Survey_Report.pdf

Baldwin, J. D., & Baldwin, J. I. (1986). *Behavior principles in everyday life*. Englewood Cliffs, NJ: Prentice Hall.

Baldwin, J. D., & Baldwin, J. I. (1988). Factors affecting AIDS-related sexual risk-taking behavior among college students. *Journal of Sex Research, 25*, 181–196.

Ball, S. J., Bowe, R., & Gewirtz, S. (1995). Circuits of schooling: A sociological exploration of parental choice of school in social class contexts. *Sociological Review, 43*, 52–77.

Balmer, Z. (1989). *Mine eyes have seen the glory: A journey into the evangelical subculture in America*. New York: Oxford University Press.

Bandura, A. (1977). *Social learning theory*. Englewood Cliffs, NJ: Prentice Hall.

Bandura, A., & Walters, R. H. (1963). *Social learning and personality development*. New York: Holt, Rinehart & Winston.

Banerjee, B. (2014, June 3). Indian gang rape case highlights lack of toilets. *Associated Press*. Retrieved from http://bigstory.ap.org/article/india-gang-rape-case-highlights-lack-toilets

Barber, B. K. (1992). Family, personality, and adolescent problem behaviors. *Journal of Marriage and the Family, 54*, 69–79.

Bare Branches, Redundant Males. (2015, April 18). *The Economist*. Retrieved from http://www.economis.com/news/asia/21648715-distorted-sex-ratios-birth-generation-ago-are-changing-marriage-and-damaging-societies-asias

Barker, E. (2020). New religious movements. *The SAGE encyclopedia of the sociology of religion* (pp. 5–14). Thousand Oaks, CA: SAGE.

Barner, M. (1999). Sex-role stereotyping in FCC- mandated children's educational television. *Journal of Broadcasting & Electronic Media, 43*, 551–564.

Barnes, R. (2012, November 26). Justices decline to consider whether Constitution requires insanity defense. *Washington Post*. Retrieved from https://www.washingtonpost.com/politics/justices-decline-to-consider-whether-constitution-requires-insanity-defense/2012/11/26/d7a3cc62-3816-11e2-8a97-363b0f9a0ab3_story.html?utm_term=.4c83389ec92e

Baron, E. (2020, September 11). Theranos founder Elizabeth Holmes, charged with fraud over her defunct blood-testing startup, to claim mental condition affecting "issue of guilt." *Orlando Sentinel*. Retrieved from https://www.orlandosentinel.com/news/nationworld/ct-nw-theranos-elizabeth-holmes-20200911-cruhyo3unvfjnmmzk5jngexot4-story.html

Barrera, M. (1979). *Race and class in the Southwest*. Notre Dame, IN: University of Notre Dame Press.

Barrett, D. B. (2001). *World Christian encyclopedia*. New York: Oxford University Press.

Barrett, J. (2020, June 26). *Nascar president on Bubba Wallace: Noose, concern "was real," but I should have said "alleged"*. Retrieved from https://www.dailywire.com/news/nascar-president-on-bubba-wallace-investigation-noose-concern-was-real-but-i-should-have-said-alleged

Barroso, A. (2020, July 7). *61% of U.S. women say "feminist" describes them well; many see feminism as empowering, polarizing. Pew Research*. Retrieved from https://www.pewresearch.org/fact-tank/2020/07/07/61-of-u-s-women-say-feminist-describes-them-well-many-see-feminism-as-empowering-polarizing/#more-369224

Barroso, A., Parker, K., & Bennett, J. (2020, May 27). *As millennials near 40, they're approaching family life differently than previous generations*. Pew Research Center. Retrieved from https://www.pewsocialtrends.org/2020/05/27/as-millennials-near-40-theyre-approaching-family-life-differently-than-previous-generations/

Barry, K. (1979). *Female sexual slavery*. Englewood Cliffs, NJ: Prentice Hall.

Barry, T., & Sims, B. (1994). *The challenge of cross-border environmentalism: The U.S.–Mexico case*. Albuquerque, NM: Resource Center Press.

Barthel, M. (2016, April 20). How to stop cheating in college. *The Atlantic Monthly*. Retrieved from https://www.theatlantic.com/education/archive/2016/04/how-to-stop-cheating-in-college/479037/

Barthel, M., & Stocking, G. (2020, April 6). Older people account for large shares of poll workers and voters in U.S. general elections. *Pew Research*. Retrieved from https://www.pewresearch.org/fact-tank/2020/04/06/older-people-account-for-large-shares-of-poll-workers-and-voters-in-u-s-general-elections/

Barton, B. (2012). *Pray the gay away: The extraordinary lives of Bible Belt gays*. New York: New York University Press.

Barton, M. (2014, September 17). Gentrification and violent crime in New York City. *Crime and Delinquency*. Retrieved from https://doi.org/10.1177/0011128714549652

Basham, A. L. (1989). *The origins and development of classical Hinduism*. New York: Oxford University Press.

Basow, S. (2004). The hidden curriculum: Gender in the classroom. In M. Paludi (Ed.), *Praeger guide to the psychology of gender* (pp. 117–132). Westport, CT: Praeger.

Batalova, J. (2020, March 4). *Immigrant women and girls in the United States. Migrant Policy Institute*. Retrieved from https://www.migrationpolicy.org/article/immigrant-women-and-girls-united-states-2018#Fertility

Baudrillard, J. (1981/1994). *Simulacra and simulation* (Glaser, Trans.). Ann Arbor: University of Michigan Press.

Baudrillard, J. (1988). *America*. London, England: Verso.

Bauman, Z. (1998). *Globalization: The human consequences*. New York: Columbia University Press.

Bauman, Z. (2001). *Modernity and the Holocaust*. Ithaca, NY: Cornell University Press.

BBC. (2012, December 6). *Timeline: Ayodhya holy site crisis*. Retrieved from http://www.bbc.com/news/world-south-asia-11436552

BBC. (2014a, February 17). *Former Barclays employees named in Libor criminal case*. Retrieved from http://www.bbc.co.uk/news/business-26228635

BBC. (2017, November 30). *Mulan: Disney casts Chinese actress Liu Yifei in lead role*. Retrieved from https://www.bbc.com/news/world-asia-china-42177082

BBC. (2020, July 16). *George Floyd: What happened in the final moments of his life*. Retrieved from https://www.bbc.com/news/world-us-canada-52861726

Beaman, J., & Taylor, C. J. (2020, June 12). #CourageIsBeautiful but PPE is better: White supremacy, racial capitalism, and COVID-19. *Contexts Magazine*. Retrieved from https://contexts.org/blog/courageisbeautiful-but-ppe-is-better-white-supremacy-racial-capitalism-and-covid-19/

Bearak, M. (2016, June 27). *Young Brits are angry about older people deciding their future, but most didn't vote*. Retrieved from https://www.washingtonpost.com/news/worldviews/wp/2016/06/27/young-brits-are-angry-about-older-people-deciding-their-future-but-most-didnt-vote/

Beauchamp, Z. (2014, June 16). *What ISIS has to gain from tweeting these photos of a massacre*. Vox Media. Retrieved from http://www.vox.com/2014/6/16/5814900/isis-photos-horrifying-iraq

Beauregard, R. (1986). The chaos and complexity of gentrification. In N. Smith & P. Williams (Eds.), *Gentrification of the city*. London: Unwin Hyman.

Beck, J. (2016, March 1). The Instagrams of food deserts. *The Atlantic*. Retrieved from http://www.theatlantic.com/health/archive/2016/03/the-instagrams-of-food-deserts/471540/

Beck, R. (2015). *We believe the children: A moral panic in the 1980s*. New York: Perseus.

Beck, U. (1992). *Risk society: Towards a new modernity* (Ritter, Trans.). Thousand Oaks, CA: SAGE.

Bell, D. (1973). *The coming of post-industrial society: A venture in social forecasting*. New York: Basic Books.

Bell, D. (2013, June 13). George Wallace stood in a doorway at the University of Alabama 50 years ago today. *U.S. News and World Report*. Retrieved from https://www.usnews.com/news/blogs/press-past/2013/06/11/george-wallace-stood-in-a-doorway-at-the-university-of-alabama-50-years-ago-today

Bellah, R. N. (1968). Meaning and modernization. *Religious Studies*, *4*, 37–45.

Bellah, R. N. (1975). *The broken covenant: American civil religion in time of trial*. New York: Seabury Press.

Bellstrom, K. (2015, June 29). GM's Mary Barra sets a *Fortune* 500 record for female CEOs. *Fortune*. Retrieved from http://fortune.com/2015/06/29/female-ceos-fortune-500-barra/?iid=sr-link9

Bennett, J. (2015, August 8). A master's degree in... masculinity? *New York Times*. Retrieved from http://www.nytimes.com/2015/08/09/fashion/masculinities-studies-stonybrook-michael-kimmel.html

Bennhold, K. (2020, July 3). As Neo-Nazis seed military ranks, Germany confronts "an enemy within." *New York Times*. Retrieved from https://www.nytimes.com/2020/07/03/world/europe/germany-military-neo-nazis-ksk.html

Benveniste, A. (2020, August 4). *The Fortune 500 now has a record number of female CEOs: A whopping 38*. Retrieved from https://www.cnn.com/2020/08/04/business/fortune-500-women-ceos/index.html

Benzow, A., & Fikri, K. (2020). *The expanded geography of high-poverty neighborhoods. Economic Innovation Group*. Retrieved from https://eig.org/wp-content/uploads/2020/04/Expanded-Geography-High-Poverty-Neighborhoods.pdf

Berchick, E. (2018, September 12). *Most uninsured were working age adults*. Retrieved from https://www.census.gov/library/stories/2018/09/who-are-the-uninsured.html

Bergen, T. J., Jr. (1996). The social philosophy of public education. *School Business Affairs*, *62*, 22–27.

Berger, J. M. (2015, March 6). *The ISIS Twitter census: Making sense of ISIS's use of Twitter*. Retrieved from http://www.brookings.edu/blogs/order-from-chaos/posts/2015/03/06-isis-twitter-census-berger

Berger, P. L. (1986). *The capitalist revolutions: Fifty propositions about prosperity, equality, and liberty*. New York: Basic Books.

Berger, P. L., & Hsiao, H.-H. M. (Eds.). (1988). *In search of an East Asian development model*. Piscataway, NJ: Transaction Books.

Berger, P., & Luckman, T. (1966). *The social construction of reality: A treatise in the sociology of knowledge*. Garden City, NY: Anchor Books.

Berkley, S. (2013, September 12). How cell phones are transforming health care in Africa. *MIT Technology Review*, guest blog. Retrieved from http://www.technologyreview.com/view/519041/how-cell-phones-are-transforming-health-care-in-africa

Berkowitz, B., Gamio, L., Lu, D., Uhrmacher, K., & Lindeman, T. (2016, July 27). The math of mass shootings. *Washington Post*. Retrieved from https://www.washingtonpost.com/graphics/national/mass-shootings-in-america/

Berman, R. (2010, July 5). *Gallup: Tea Party's top concerns are debt, size of government*. Retrieved from https://thehill.com/blogs/blog-briefing-room/news/107193-gallup-tea-partys-top-concerns-are-debt-size-of-government

Bernard, J. (1981). *The female world*. New York: Free Press.

Bernard, J. (1982). *The future of marriage*. New Haven, CT: Yale University Press.

Berns, R. (1989). *Child, family, community: Socialization and support*. New York: Holt, Rinehart & Winston.

Berry, T., & Wilkins, J. (2017). The gendered portrayal of inanimate characters in children's books. *Journal of Children's Literature*, *43*(2), 4–15. Retrieved from https://search-proquest-com.thor.nebrwesleyan.edu/docview/1964434636?accountid=28187

Bettelheim, B. (1979). *Surviving, and other essays*. New York: Knopf.

Bettelheim, B. (1982). Difficulties between parents and children: Their causes and how to prevent them. In N. Stinnett et al. (Eds.), *Family strengths 4: Positive support systems* (pp. 5–14). Lincoln: University of Nebraska Press.

Beyer, P. (1994). *Religion and globalization*. London: Sage.

Bhatia, S. (2013, February 3). *Women, rape, and lack of toilets*. Feminist Wire. Retrieved from http://www.thefeministwire.com/2013/02/op-ed-women-rape-and-lack-of-toilets

Bhattacharjee, P. (2017, November 15). How does your child's screen time measure up? *CNN*. Retrieved from https://www.cnn.com/2017/11/15/health/screen-time-averages-parenting/index.html

Bhutta, N., Chang, A. C., Dettling, L. J., & Hsu, J. W. (2020, September 28). *Disparities in wealth by race and ethnicity in the 2019 survey of consumer finances*. Retrieved from https://www.federalreserve.gov/econres/notes/feds-notes/disparities-in-wealth-by-race-and-ethnicity-in-the-2019-survey-of-consumer-finances-20200928.htm

Bick, R., Halsey, E., & Ekenga, C. (2018, December 27). *The global environmental injustice of fast fashion*. Retrieved from https://ehjournal.biomedcentral.com/articles/10.1186/s12940-018-0433-7

Bilal, M., Zia-ur-Rehman, M., & Raza, I. (2010). Impact of family friendly policies on employees' job satisfaction and turnover intention (A study on work-life balance at workplace). *Interdisciplinary Journal of Contemporary Research in Business, 2*, 378–395.

Bill and Melinda Gates Foundation. (2006). *Ensuring the world's poorest children benefit from lifesaving vaccines*. Retrieved from http://www.gatesfoundation.org/learning/Documents/GAVI.pdf

Bill and Melinda Gates Foundation. (2013). *What we do: Vaccine delivery strategy*. Retrieved from http://www.gatesfoundation.org/vaccines/Documents/vaccines-fact-sheet.pdf

Bill and Melinda Gates Foundation. (n.d.). *Polio*. Retrieved from https://www.gatesfoundation.org/our-work/programs/global-development/polio

Bishaw, A. (2014). *Changes in areas with concentrated poverty: 2000 to 2010 (American Community Survey Report 27)*. Washington, DC: U.S. Census Bureau. Retrieved from http://www.census.gov/content/dam/Census/library/publications/2014/acs/acs-27.pdf

Bishop, C. J., Kiss, M., Morrison, T. G., Rushe, D. M., & Specht, J. (2014). The association between gay men's stereotypic beliefs about drag queens and their endorsement of hypermasculinity. *Journal of Homosexuality, 62*, 554–567.

Bishop, K. (2009). Dead man still walking: Explaining the zombie renaissance. *Journal of Popular Film and Television, 33*, 196–205.

Bishop, K. (2010). *American zombie gothic: The rise and fall (and rise) of the walking dead in popular culture*. Jefferson, NC: McFarland.

Bishop, M., & Hicks, S. L. (Eds.). (2009). *Hearing, mother father deaf*. Washington, DC: Gallaudet University Press.

Black, A. (2016). Evangelicals and politics: Where we've been and where we're headed. *Evangelicals*(Fall 2016). National Association of Evangelicals.

Blackburn, G., & Scharrer, E. (2019). Video game playing and beliefs about masculinity among male and female emerging adults. *Sex Roles: A Journal of Research, 80*(5–6), 310–324. https://doi.org/10.1007/s11199-018-0934-4

Blalock, H. (1967). *Toward a theory of minority group relations*. New York: Wiley.

Blau, P. M. (1964). *Exchange and power in social life*. New York: Wiley.

Blau, P. M. (1977). *Inequality and heterogeneity: A primitive theory of social structure*. New York: Free Press.

Blau, P. M., & Meyer, M. (1987). *Bureaucracy in modern society* (3rd ed.). New York: Random House.

Blinder, A. S. (2006). Offshoring: The next industrial revolution? *Foreign Affairs, 85*, 113–128.

Block, A., & Weaver, A. (2004). *All is clouded by desire: Global banking, money laundering, and international organized crime*. Praeger.

Block, J., & Subramanian, S. (2015). Moving beyond "food deserts": Reorienting United States policies to reduce disparities in diet quality. *PLoS Med, 12*(12): e1001914. Retrieved from https://doi.org/10.1371/journal.pmed.1001914

Block, M., Cox, A., & Giratiknon, T. (2015, July 8). Mapping segregation. *New York Times*. Retrieved from http://www.nytimes.com/interactive/2015/07/08/us/census-race-map.html

Blow, C. M. (2015, July 30). The shooting of Samuel Du Bose. *New York Times*. Retrieved from http://www.nytimes.com/2015/07/30/opinion/charles-blow-the-shooting-of-samuel-dubose.html

Blumberg, J. (2007, October 23). A brief history of the Salem witch trials. *Smithsonian*. Retrieved from http://www.smithsonianmag.com/history-archaeology/brief-salem.html

Blumer, H. (1951). Collective behavior. In A. M. Lee (Ed.), *Principles of sociology* (pp. 166–222). New York, NY: Barnes & Noble.

Blumer, H. (1969). *Symbolic interactionism: Perspective and method*. Englewood Cliffs, NJ: Prentice Hall.

Blumer, H. (1970). *Movies and conduct*. New York: Arno Press.

Blumer, H. (1986). *Symbolic interactionism: Perspective and method*. University of California Press.

Boen, C. E., & Hummer, R. A. (2019, December). Longer—but harder—lives? The Hispanic health paradox and the social determinants of racial, ethnic, and immigrant-native health disparities from midlife through late life. *Journal of Health and Social Behavior 60*(4), 434–445.

Boers, E., Afzali, M. H., & Conrod, P. (2020). Temporal associations of screen time and anxiety symptoms among adolescents. *The Canadian Journal of Psychiatry, 65*(3), 206–208. doi:10.1177/0706743719885486

Bohnert, D., & Ross, W. H. (2010). The influence of social networking Web sites on the evaluation of job candidates. *Cyberpsychology, Behavior, and Social Networking, 13*, 341–347.

Bonacich, E., & Appelbaum, R. P. (2000). *Behind the label: Inequality in the Los Angeles apparel industry*. Berkeley: University of California Press.

Bond, S. (2020, October 12). *Facebook bans Holocaust denial, reversing earlier policy*. Retrieved from https://www.npr.org/2020/10/12/923002012/facebook-bans-holocaust-denial-reversing-earlier-policy

Booker, B. (2020). *After mounting pressure, Washington's NFL franchise drops its team name*. Retrieved from https://www.npr.org/sections/live-updates-protests-for-racial-justice/2020/07/13/890359987/after-mounting-pressure-washingtons-nfl-franchise-drops-its-team-name

Booker, M. K. (2001). *Monsters, mushroom clouds, and the Cold War: American science fiction and the roots of post-modernism, 1946–1964*. Westport, CT: Greenwood.

Boronksi, T., & Hassan, N. (2020). *Sociology of education* (2nd ed.). London, England: SAGE.

Bos, H., & Sandfort, T. G. M. (2010). Children's gender identity in lesbian and heterosexual two-parent families. *Sex Roles, 62*, 114–126.

Bouma, G. (2020). Denomination. *The SAGE encyclopedia of the sociology of religion* (pp. 202–204). Thousand Oaks, CA: SAGE.

Bourdieu, P. (1977). *Outline of a theory of practice*. New York: Cambridge University Press.

Bourdieu, P. (1984). *Distinction: A social critique of the judgment of taste*. Cambridge, MA: Harvard University Press.

Bourdieu, p. (1991). *Language and symbolic power*. New York, NY: Cambridge University Press.

Bourdieu, P., & Coleman, J. S. (1991). *Social theory for a changing society*. Boulder, CO: Westview Press.

Bourdieu, P., & Wacquant, L. J. D. (1992). *An invitation to reflexive sociology*. Chicago, IL: University of Chicago Press.

Bowles, S., & Gintis, H. (1976). *Schooling in capitalist America: Educational reform and the contradictions of economic life*. New York: Basic Books.

Bowling Green State University. (2012, February 11). Finding love has no expiration date: People over 60 are fastest growing demographic in online dating. *Science Daily*. Retrieved from www.sciencedaily.com/releases/2012/02/120211095051.htm

Boyce, M., & Hollingsworth, A. (2015). *Sudanese refugees in Chad: Passing the baton to no one*. Retrieved from http://www.refugeesinternational.org/reports/2015/9/29/sudanese-refugees-in-chad-passing-the-baton-to-no-one

Braga, A. A., & Dusseault, D. (2018). Can homicide detectives improve homicide clearance rates? *Crime & Delinquency*, *64*(3), 283–315.

Branch, T. (1988). *Parting the waters: America in the King years, 1954–1963*. New York: Simon & Schuster.

Brandon, S. G. (1973). *Ancient empires*. New York: Newsweek Books.

Brandt, A. M. (1983). Racism and research: The case of the Tuskegee syphilis study. In J. W. Leavitt & R. L. Numbers (Eds.), *Sickness and health in America* (pp. 392–404). Madison: University of Wisconsin Press.

Branson, C. E., & Cornell, D. G. (2009). A comparison of self and peer reports in the assessment of middle school bullying. *Journal of Applied School Psychology*, *25*(1), 5–27.

Braverman, H. (1988). *Labor and monopoly capital: The degradation of work in the 20th century*. New York: Monthly Review Press. (Original work published 1974)

Bremer, C. (2012, March). *Economic commentary* (T. Vaughn, managing director). Retrieved from http://www.tom-vaughn.com/Economic-Commentary-March-2012.c3472.htm

Bremer, J., & Rauch, P. K. (1998). Children and computers: Risks and benefits. *Journal of the American Academy of Child and Adolescent Psychiatry*, *37*, 559–560.

Brennan Center for Justice. (2020). Voter ID. Retrieved from https://www.brennancenter.org/issues/ensure-every-american-can-vote/vote-suppression/voter-id

Brenner, P. S., & DeLamater, J. D. (2014). Social desirability bias in self-reports of physical activity: Is an exercise identity the culprit? *Social Indicators Research*, *117*(2), pp. 489–504.

Brewster, M. E. (2017). Lesbian women and household labor division: A systematic review of scholarly research from 2000 to 2015. *Journal of Lesbian Studies*, *21*(1), 47–69. doi:10.1080/10894160.2016.114235

Brittany, P. B. (2019). "They look at you like you're nothing": Stigma and shame in the child support system. *Symbolic Interaction 42*(4), 640–668. doi:http://dx.doi.org.thor.nebrwesleyan.edu/10.1002/symb.427

Brock, J., & Cocks, T. (2012, March 8). *Insight: Nigeria oil corruption highlighted by audits*. Reuters. Retrieved from http://www.reuters.com/article/2012/03/08/us-nigeria-corruption-oil-idUSBRE8270GF20120308

Bronson, J., & Carson, E. A. (2019, April). *Prisoners in 2017*. Retrieved January 18, 2020, from https://www.bjs.gov/content/pub/pdf/p17.pdf

Brown, C. S. (2020, September 3). *Media messages to young girls: Does "sexy girl" trump "girl power"?* Briefing paper. Council on Contemporary Families. Retrieved from https://contemporaryfamilies.org/girls-media-messaging-brief-report/

Brown, D. (2010, September 16). A mother's education has a huge effect on a child's health. *Washington Post*. Retrieved from http://www.washingtonpost.com/wp-dyn/content/article/2010/09/16/AR2010091606384.html

Brown, D. E., Edwards, J. W., & Moore, R. B. (1988). *The penis inserts of Southeast Asia*. Berkeley, CA: Center for South and Southeast Asian Studies.

Brumberg, J. J. (1997). *The body project: An intimate history of American girls*. New York: Vintage Books.

Buchanan, L., Bui, Q., & Patel, J. K. (2020, July 3). Black Lives Matter may be the largest movement in U.S. history. *New York Times* Interactive. Retrieved from https://www.nytimes.com/interactive/2020/07/03/us/george-floyd-protests-crowd-size.html

Budiansky, S., Goode, E. E., & Gest, T. (1994, January 16). The Cold War experiments. *U.S. News & World Report*. Retrieved from http://www.usnews.com/usnews/news/articles/940124/archive_012286.htm

Budiman, A. (2020a, August 20). Key findings about U.S. immigrants. *Fact Tank*. Pew Research Center. Retrieved from https://www.pewresearch.org/fact-tank/2020/08/20/key-findings-about-u-s-immigrants/

Budiman, A. (2020b, October 1). Americans are more positive about the long-term rise in U.S. racial and ethnic diversity than in 2016. *Fact Tank*. Pew Research Center. Retrieved from https://www.pewresearch.org/fact-tank/2020/10/01/americans-are-more-positive-about-the-long-term-rise-in-u-s-racial-and-ethnic-diversity-than-in-2016/

Budiman, A., Cilluffo, A., & Ruiz, N. G. (2019, May 22). Key facts about Asian origin groups in the U.S. *Fact Tank*. Retrieved from https://www.pewresearch.org/fact-tank/2019/05/22/key-facts-about-asian-origin-groups-in-the-u-s/

Buechler, S. M. (1990). *Women's movements in the United States: Women's suffrage, equal rights, and beyond*. New Brunswick, NJ: Rutgers University Press.

Bukszpan, D. (2019, August 4). Why fans keep watching "The Walking Dead" even though they don't like it anymore. *CNBC Entertainment*. Retrieved from https://www.cnbc.com/2019/08/02/why-fans-keep-watching-the-walking-dead-even-though-they-dont-like-it-anymore.html

Burawoy, M., & Lukács, J. (1992). *The radiant past: Ideology and reality in Hungary's road to capitalism*. Chicago: University of Chicago Press.

Burden, M. (2019, July 15). Growing use of temps irks UAW. *Automotive News*. Retrieved from https://www.autonews.com/automakers-suppliers/growing-use-temps-irks-uaw

Bureau of Labor Statistics. (2017). Unemployment rates and earnings by educational attainment. *Employment projections*. Washington, DC: U.S. Government Printing Office. Retrieved from https://www.bls.gov/emp/chart-unemployment-earnings-education.htm

Bureau of Labor Statistics. (2019). Unemployment rates and earnings by educational attainment. *Employment projections*. Washington, DC: U.S. Government Printing Office.

Bureau of Labor Statistics. (2020, July 17). News release. Retrieved from https://www.bls.gov/news.release/pdf/wkyeng.pdf

Bureau of Labor Statistics Reports. (2019, December). *Women in the labor force: A data book*. Retrieved from https://www.bls.gov/opub/reports/womens-databook/2019/home.htm

Bureau of Labor Statistics, U.S. Department of Labor. (2020). *Occupational outlook handbook*. Social and Human Service Assistants. Retrieved May 19, 2020, from https://www.bls.gov/ooh/community-and-social-service/social-and-human-service-assistants.htm

Burgess, E. W. (1925). The growth of the city. In R. E. Park, E. W. Burgess, & R. McKenzie (Eds.), *The city*. Chicago: University of Chicago Press.

Burgess, R. L., & Akers, R. L. (1966). A differential association-reinforcement theory of criminal behavior. *Social problems*, *14*(2), 128–147.

Burns, C. (2019, January 4). The gay and transgender wage gap. *American Progress*.

Burns, J. M. (1978). *Leadership*. New York: Harper & Row.

Burris, H. H., & Hacker, M. R. (2017). Birth outcome racial disparities: A result of intersecting social and environmental factors. *Semin Perinatol 41*(6), 360–366. doi:10.1053/j.semperi.2017.07.002

Butler, S. (2019, January 21). Why are wages so low for garment workers in Bangladesh? *The Guardian*. Retrieved from https://www.theguardian.com/business/2019/jan/21/low-wages-garment-workers-bangladesh-analysis

Cabeza, M. F., Johnson, J. B., & Tyner, L. J. (2011). Glass ceiling and maternity leave as important contributors to the gender wage gap. *Southern Journal of Business and Ethics*, *3*, 73–85.

Cabrera, N., Matias, C., & Montoya, R. (2017). Activism or slacktivism? The potential and pitfalls of social media in contemporary student activism. *Journal of Diversity in Higher Education*. 10.1037/dhe0000061

Calderón-Garcidueñas, L., & Torres-Jardón, R. (2012). Air pollution, socioeconomic status and children's cognition in megacities: The Mexico City scenario. *Frontiers in Developmental Psychology, 3*. Retrieved from http://www.frontiersin.org/Developmental_Psychology/10.3389/fpsyg.2012.00217/full

Camarota, S. (2005). *Birth rates among immigrants in America: Comparing fertility in the U.S. and home countries*. Washington, DC: Center for Immigration Studies. Retrieved from http://www.cis.org/articles/2005/back1105.pdf

Camera, L. (2015, November 6). Native American students left behind. *U.S. News & World Report*. Retrieved from http://www.usnews.com/news/articles/2015/11/06/native-american-students-left-behind

Campbell, A. (1984). *The girls in the gang*. New Brunswick, NJ: Rutgers University Press.

Campbell, A., & Muncer, S. (Eds.). (1998). *The social child*. East Sussex, England: Psychology Press.

Campbell, K., Klein, D. M., & Olson, K. (1992). Conversation activity and interruptions among men and women. *Journal of Social Psychology*, *132*, 419–421.

CareerBuilder. (2015). *Employers reveal biggest resume blunders in annual CareerBuilder survey*. Retrieved from http://www.careerbuilder.com/share/aboutus/pressreleasesdetail.aspx?sd=8/13/2015&id=pr909&ed=12/31/2015

CareerBuilder. (2018). *Employers share their most outrageous resume mistakes and instant deal breakers in a new career builder survey*. Retrieved from http://press.careerbuilder.com/2018-08-24-Employers-Share-Their-Most-Outrageous-Resume-Mistakes-and-Instant-Deal-Breakers-in-a-New-CareerBuilder-Study

Carey, B. (2018, June 8). Can one suicide lead to others? *New York Times*. Retrieved from https://www.nytimes.com/2018/06/08/health/suicide-bourdain-spade.html

Carlisle Indian School Project. (n.d.). https://carlisleindianschoolproject.com/

Carlson, D. L., Petts, R. J., & Pepin, J. R. (2020). *Men and women agree: During the COVID-19 pandemic men are doing more at home*. Retrieved from https://contemporaryfamilies.org/covid-couples-division-of-labor

Carlson, M. W., & Hans, J. D. (2020). Maximizing benefits and minimizing impacts: Dual-earner couples' perceived division of household labor decision-making process. *Journal of Family Studies*, *26*(2), 208–255. doi:10.1080/13229400.2017.1367712

Carmody, S. (2019, April). *5 years after Flint's crisis began, is the water safe?* Michigan's news hour. Retrieved from https://www.npr.org/2019/04/25/717104335/5-years-after-flints-crisis-began-is-the-water-safe

Carnevale, A., & Smith, N. (2018). *Balancing work and learning. Implications for low-income students*. Retrieved from https://1gyhoq479ufd3yna29x7ubjn-wpengine.netdna-ssl.com/wp-content/uploads/Low-Income-Working-Learners-ES.pdf

Carpenter, Z. (2015, November 4). Think it's hard finding a place to live? Try doing so with a criminal record. *The Nation*. Retrieved from https://www.thenation.com/article/public-housing-criminal-record

Carreyrou, J. (2019). *Bad blood*. Picador.

Carson, E. A. (2015). *Prisoners in 2014*. Washington, DC: U.S. Department of Justice, Bureau of Justice Statistics. Retrieved from http://www.bjs.gov/content/pub/pdf/p14.pdf

Carson, E. A. (2018). *Prisoners in 2016*. Washington, DC: U.S. Department of Justice, Bureau of Justice Statistics. Retrieved from http:/www. bjs.gov/index.cfm?ty=pbdetail&iid=6187

Carson, E. A., & Anderson, E. (2016, December). *Prisoners in 2015*. Washington, DC: Bureau of Justice Statistics. Retrieved from https://www.bjs.gov/content/pub/pdf/p15.pdf

Cassidy, C. (2016, June 5). Patchy reporting undercuts national hate crimes count. *Associated Press*. Retrieved from http://bigstory.ap.org/article/8247a1d2f76b4baea2a121186dedf768/ap-patchy-reporting-undercuts-national-hate-crimes-count

Cassidy, L., & Hurrell, R. M. (1995). The influence of victim's attire on adolescents' judgments of date rape. *Adolescence*, *30*(118), 319.

Castells, M., & Portes, A. (1989). World underneath: The origins, dynamics, and effects of the informal economy. In A. Portes, M. Castells, & L. A. Benton (Eds.), *The informal economy: Studies in advanced and less developed countries* (pp. 11–37). Baltimore: Johns Hopkins University Press.

Cavalli-Sforza, L. L., Menozzi, P., & Piazza, A. (1994). *The history and geography of human genes*. Princeton, NJ: Princeton University Press.

Cawley, J. (2001). The impact of obesity on wages. *Journal of Human Resources*, *39*, 451–474.

Center for Responsive Politics. (2016). *Tobacco industry profile: Summary, 2016*. Retrieved from https://www.opensecrets.org/lobby/indusclient.php?id=A02

Center for Responsive Politics. (2020a). *Super PACs: How many donors give*. Retrieved from https://www.opensecrets.org/outside-spending/donor-stats

Center for Responsive Politics. (2020b). *2020 election to cost $14 billion, blowing away spending records*. Retrieved from https://www.opensecrets.org/news/2020/10/cost-of-2020-election-14billion-update

Centers for Disease Control and Prevention. (2012). *Impact of malaria*. Retrieved from http://www.cdc.gov/malaria/malaria_world-wide/impact.html

Centers for Disease Control and Prevention. (2014). *HIV surveillance report: Diagnoses of HIV infection in the United States and dependent areas, 2014*. Retrieved from http://www.cdc.gov/hiv/pdf/library/reports/surveillance/cdc-hiv-surveillance-report-us.pdf

Centers for Disease Control and Prevention. (2015). *Adult obesity facts*. Retrieved https://www.cdc.gov/obesity/data/adult.html

Centers for Disease Control and Prevention. (2017a). *Marriage and divorce*. Retrieved from https://www.cdc.gov/nchs/fastats/marriage-divorce.htm

Centers for Disease Control and Prevention. (2017b). *Tobacco use: Extinguishing the epidemic*. Retrieved from https://www.cdc.gov/chronicdisease/resources/publications/aag/pdf/2017/tobacco-aag-H.pdf

Centers for Disease Control and Prevention. (2018a). *Adult obesity prevalence maps*. Retrieved from https://www.cdc.gov/obesity/data/prevalence-maps.html

Centers for Disease Control and Prevention. (2018b). *Current cigarette smoking among adults in the United States*. Retrieved from http://www.cdc.gov/tobacco/data_statistics/fact_sheets/adult_data/cig_smoking/

Centers for Disease Control and Prevention. (2018c). *Malaria's impact worldwide*. Retrieved from https://www.cdc.gov/malaria/malaria_worldwide/impact.html

Centers for Disease Control and Prevention. (2018d). *National Intimate Partner and Sexual Violence Survey*. Retrieved from https://www.cdc.gov/violenceprevention/nisvs/index.html

Centers for Disease Control and Prevention. (2018e). *Prevalence of self-reported obesity among U.S. adults by race/ethnicity, state and territory, BRFSS, 2018*. Retrieved from https://www.cdc.gov/obesity/data/prevalence-maps.html

Centers for Disease Control and Prevention. (2019). *Infant mortality*. Centers for Disease Control Division of Reproductive Health, National Center for Chronic Disease Prevention and Health Promotion. Retrieved from https://www.cdc.gov/reproductivehealth/MaternalInfantHealth/InfantMortality.htm

Centers for Disease Control and Prevention. (2020a). *2018 drug overdose rates*. Retrieved from https://www.cdc.gov/drugoverdose/data/statedeaths/drug-overdose-death-2018.html

Centers for Disease Control and Prevention. (2020b, July 24). *Health equity considerations and racial and ethnic minority groups*. Retrieved from https://www.cdc.gov/coronavirus/2019-ncov/community/health-equity/race-ethnicity.html

Centers for Disease Control and Prevention. (2020c). *HIV and gay and bisexual men*. Retrieved from https://www.cdc.gov/hiv/group/msm/index.html

Centers for Disease Control and Prevention. (2020d). *HIV surveillance report, 2018 (updated) (Vol. 31). Diagnoses of HIV infection in the United States and dependent areas, 2018*. Retrieved from http://www.cdc.gov/hiv/library/reports/hiv-surveillance.html

Centers for Disease Control and Prevention. (2020e, May). *How common are firearm injuries?* https://www.cdc.gov/violenceprevention/firearms/fastfact.html

Centers for Disease Control and Prevention. (2020f). *Increase in suicide mortality rate in the United States, 1999–2018*. Retrieved from https://www.cdc.gov/nchs/products/databriefs/db362.htm.

Centers for Disease Control and Prevention. (2020g). *Malaria*. Retrieved from https://www.cdc.gov/parasites/malaria/index.html

Centers for Disease Control and Prevention. (2020h). *Malaria's impact worldwide*. Retrieved from https://www.cdc.gov/malaria/malaria_worldwide/impact.html

Centers for Disease Control and Prevention. (2020i). *Preventing sexual violence*. Retrieved from https://www.cdc.gov/violenceprevention/sexualviolence/fastfact.html

Centers for Disease Control and Prevention. (2020j, May). Surveillance Report. *Diagnoses of HIV infection in the United States and dependent areas 2018*. Retrieved from http://www.cdc.gov/hiv/library/reports/hiv-surveillance.html

Centers for Disease Control and Prevention. (2020k, July 7). *The global HIV/AIDS epidemic*. Retrieved from https://www.hiv.gov/hiv-basics/overview/data-and-trends/global-statistics

Centers for Disease Control and Prevention. (2020l). *What is intimate partner violence?* Retrieved from https://www.cdc.gov/violenceprevention/intimatepartnerviolence/fastfact.html

Centers for Disease Control and Prevention. (2020m). *Youth and tobacco use*. Retrieved from https://www.cdc.gov/tobacco/data_statistics/fact_sheets/youth_data/tobacco_use/index.htm

Centers for Disease Control and Prevention. (n.d). *HIV and African Americans*. Retrieved from https://www.cdc.gov/hiv/group/racialethnic/africanamericans/

Chafetz, J. S. (1984). *Sex and advantage: A comparative, macro-structural theory of sex stratification*. Totowa, NJ: Rowman & Allanheld.

Chafetz, J. S. (1997). Feminist theory and sociology: Underutilized contributions for mainstream theory. *Annual Review of Sociology*, *23*, 97–120.

Chafetz, J. S., & Dworkin, A. G. (1983). Macro and micro process in the emergence of feminist movements: Toward a unified theory. *Western Sociological Review*, *14*(1), 27–45.

Chaffey, D. (2015, April 27). *Social network popularity by country*. Retrieved from http://www.smartinsights.com/social-media-marketing/social-media-strategy/new-global-social-media-research/attachment/2015-social-network-popularity-by-country/

Chaffey, D. (2016). *Global social media research summary 2016*. Retrieved from http://www.smartinsights.com/social-media-marketing/social-media-strategy/new-global-social-media-research/

Chalk, F., & Jonassohn, K. (1990). *The history and sociology of genocide: Analyses and case studies*. New Haven, CT: Yale University Press.

Chambliss, W. J. (1973). The Saints and the Roughnecks. *Society, 11*, 24–31.

Chambliss, W. J. (1988a). *Exploring criminology*. New York: Macmillan.

Chambliss, W. J. (1988b). *On the take: From petty crooks to presidents*. Bloomington: Indiana University Press.

Chambliss, W. J. (2001). *Power, politics, and crime*. Boulder, CO: Westview Press.

Chambliss, W. J., & Hass, A. (2011). *Criminology: Connecting theory, research, and practice*. New York: McGraw-Hill.

Chambliss, W. J., & King, H. (1984). *Boxman: A professional thief's journey*. New York: Macmillan.

Chambliss, W. J., Michalowski, R., & Kramer, R. C. (Eds.). (2010). *State crime in the global age*. London: Willan.

Chambliss, W. J., & Seidman, R. B. (1982). *Law, order, and power*. Reading, MA: Addison-Wesley.

Chambliss, W. J., & Zatz, M. S. (1994). *Making law: The state, the law, and structural contradictions*. Bloomington: Indiana University Press.

Chandler, A. (2014, June 15). Should Twitter have suspended the violent ISIS Twitter account? *The Wire*. Retrieved from http://www.thewire.com/global/2014/06/should-twitter-have-suspended-the-violent-isis-twitter-account/372805

Chapp, C. B., & Coe, K. (2019). Religion in American presidential campaigns, 1952–2016: Applying a new framework for understanding candidate communication. *Journal for the Scientific Study of Religion*, 58(2), 398–414. Retrieved from https://doi.org/10.1111/jssr.12590

Chaves, M. (1993). Denominations as dual structures: An organizational analysis. *Sociology of Religion, 54*, 147–169.

Chaves, M. (1994). Secularization as declining religious authority. *Social Forces, 72*, 749–774.

Chen, H. S. (1992). *Chinatown no more: Taiwan immigrants in contemporary New York*. Ithaca, NY: Cornell University Press.

Chen, L. (2015, August 11). The most valuable employees: Snapchat doubles Facebook. *Forbes*. Retrieved from http://www.forbes.com/sites/liyanchen/2015/08/11/the-most-valuable-employees-snapchat-doubles-facebook/#1ba06be3f754

Cheng, E., & Wilkinson, T. (2020, April 23). Agonizing over screen time? Follow the three C's. *New York Times Magazine*. Retrieved from https://www.nytimes.com/2020/04/13/parenting/manage-screen-time-coronavirus.html

Chenoweth, E., & Pressman, J. (2017, February 7). This is what we learned by counting the women's marches. *Washington Post*. Retrieved from https://www.washingtonpost.com/news/monkey-cage/wp/2017/02/07/this-is-what-we-learned-by-counting-the-womens-marches/

Cherlin, A., Cumberworth, E., Morgan, S. P., & Wimer, C. (2013). The effects of the Great Recession on family structure and fertility. *The ANNALS of the American Academy of Political and Social Science, 650*(1), 214–231. Retrieved from https://doi.org/10.1177/0002716213500643

Chesney-Lind, M. (1989). Girls' crime and woman's place: Toward a feminist model of female delinquency. *Crime & Delinquency, 33*(1), 5–29.

Chesney-Lind, M. (2004). Beyond bad girls: Feminist perspectives on female offending. In C. Sumner (Ed.), *The Blackwell companion in criminology*. Oxford, UK: Blackwell.

Chesney-Lind, M., & Morash, M. (2013). Transformative feminist criminology: A critical rethinking of a discipline. *Journal of Critical Criminology, 21*(3), 287–304.

Chia, R. C., Allred, L. J., Grossnickle, W. F., & Lee, G. W. (1998). Effects of attractiveness and gender on the perception of achievement-related variables. *Journal of Social Psychology, 138*, 471–477.

Chiang, S. (2009). Personal power and positional power in a power-full "I": A discourse analysis of doctoral dissertation supervision. *Discourse and Communication, 3*, 255–271.

Childhelp. (2010). *National child abuse statistics: Child abuse in America*. Retrieved from http://www.childhelp.org/pages/statistics#gen-stats

Childhelp National Child Abuse Hotline. (2020). *Child abuse statistics and facts*. Retrieved from https://www.childhelp.org/child-abuse-statistics/

ChildStats. (2013). *America's children: Key national indicators of well-being, 2013*. Retrieved from http://www.childstats.gov/americaschildren/famsoc3.asp

ChildTrends. (2018). *Late or no prenatal care*. Retrieved from https://www.childtrends.org/indicators/late-or-no-prenatal-care

Chimere-dan, O. (1992). Apartheid and demography in South Africa. *Etude Population Africaine, 7*, 26–36. https://doi.org/doi:10.11564/7-0-419. PMID:12321499

Chodorow, N. (1999). *The reproduction of mothering: Psychoanalysis and the sociology of gender*. Berkeley: University of California Press.

Chong, D. (1991). *Collective action and the civil rights movement*. Chicago: University of Chicago Press.

Christakis, N. A. (2019, April). How AI will rewire us. *Atlantic Monthly*. Retrieved from https://www.theatlantic.com/magazine/archive/2019/04/robots-human-relationships/583204/

Christie, L. (2014, June 5). *America's homes are bigger than ever*. Retrieved from http://money.cnn.com/2014/06/04/real_estate/american-home-size/index.html

Church Clarity. (2017). *Scoring America's 100 largest churches for clarity*. Retrieved from https://www.churchclarity.org/resources/scoring-americas-100-largest-churches-for-clarity

Cicero, T. J., Ellis, M. S., Surratt, H. L., & Kurtz, S. P. (2014). The changing face of heroin use in the United States. *JAMA Psychiatry, 71*, 821–826. doi:10.1001/jamapsychiatry.2014.366

Cichowski, L., & Nance, W. E. (2004). *More marriages among the deaf may have led to doubling of common form of genetic deafness in the U.S.* Virginia Commonwealth University News Center. Retrieved from http://www.news.vcu.edu/news/More_marriages_among_the_deaf_may_have_led_to_doubling_of_common

Cilluffo, A. (2019, August 13). *5 facts about student loans*. Retrieved from https://www.pewresearch.org/fact-tank/2019/08/13/facts-about-student-loans/

Clarion Project. (2014, September 10). *The Islamic State's magazine*. Retrieved from http://www.clarionproject.org/news/islamic-state-isis-isil-propaganda-magazine-dabiq

Clark, H. (2006, March 8). Are Women Happy Under the Glass Ceiling? *Forbes*. Accessed at https://www.forbes.com/2006/03/07/glass-ceiling-opportunities--cx_hc_0308glass.html#7a44d3b73e39

Clark, H., Babu, A. S., Wiewel, E. W., Opoku, J., & Crepaz, N. (2017). *Diagnosed HIV infection in transgender adults and adolescents: Results from the National HIV Surveillance System, 2009–2014*. Retrieved from https://pubmed.ncbi.nlm.nih.gov/28035497/

Claypool, M., & Miller, B. (2020, August 14). *Greenland's ice sheet has melted to a point of no return, according to new study*. CNN. Retrieved from https://www.cnn.com/2020/08/14/weather/greenland-ice-sheet/index.html

Cliff, G., & Wall-Parker, A. (2017). Statistical analysis of white collar crime. *Oxford research encyclopedias:* Criminology and criminal Justice. Retrieved from https://oxfordre.com/criminology/view/10.1093/acrefore/9780190264079.001.0001/acrefore-9780190264079-e-267#acrefore-9780190264079-e-267-bibItem-0058

Cline, E. L. (2013). *Overdressed: The shockingly high cost of cheap fashion*. New York, NY: Portfolio.

Cloward, R. A., & Ohlin, L. E. (1960). *Delinquency and opportunity: A theory of delinquent gangs*. Glencoe, IL: Free Press.

CNN. (2016, May 5). *Flint water crisis fast facts*. Retrieved from http://www.cnn.com/2016/03/04/us/flint-water-crisis-fast-facts/

Coates, T. (2015, April). The Black family in the age of mass incarceration. *The Atlantic*. Retrieved from https://www.theatlantic.com/magazine/archive/2015/10/the-black-family-in-the-age-of-mass-incarceration/403246

Cobb, J. (2016). *Americans buy their four-millionth hybrid car.* Retrieved from http://www.hybridcars.com/americans-buy-their-four-millionth-hybrid-car/

Cockerham, W. C., & Glasser, M. (2000). *Readings in medical sociology* (2nd ed.). Englewood Cliffs, NJ: Prentice Hall.

Codrington, W. (2019, November 17). The Electoral College's racist origins. *The Atlantic Monthly*. Retrieved from https://www.theatlantic.com/ideas/archive/2019/11/electoral-college-racist-origins/601918/

Cohen, A. (2012, March 16). How voter ID laws are being used to disenfranchise minorities and the poor. *Atlantic*. Retrieved from http://www.theatlantic.com/politics/archive/2012/03/how-voter-id-laws-are-being-used-to-disenfranchise-minorities-and-the-poor/254572

Cohen, A. K. (1955). *Delinquent boys: The culture of the gang*. Glencoe, IL: Free Press.

Cohen, D. A. (2014). *A big fat crisis: The hidden forces behind the obesity crisis and how we can end it*. New York: Nation Books.

Cohen, P. (2014, December 12). Fewer births and divorces, more violence: How the recession affected the American family. *The Conversation*. Retrieved from https://theconversation.com/fewer-births-and-divorces-more-violence-how-the-recession-affected-the-american-family-34272

Cohn, D., Livingston, G., & Wang, W. (2014). *After decades of decline, a rise in stay-at-home mothers*. Washington, DC: Pew Research Center. Retrieved from http://www.pewsocialtrends.org/2014/04/08/after-decades-of-decline-a-rise-in-stay-at-home-mothers

Coleman, J. M., & Hong, Y.-Y. (2008). Beyond nature and nurture: The influence of lay gender theories on self-stereotyping. *Self and Identity, 7*, 34–53.

Coleman, J. S. (1990). *The foundations of social theory*. Cambridge, MA: Harvard University Press.

Coleman, J. S., Hoffer, T., & Kilgore, S. (1982). *High school achievement: Public, Catholic, and private schools compared*. New York: Basic Books.

College Board. (2020). *SAT suite of assessments annual report*. Retrieved from https://reports.collegeboard.org/pdf/2020-total-group-sat-suite-assessments-annual-report.pdf

Collins, P. H. (1990). *Black feminist thought: Knowledge, consciousness and the politics of empowerment*. New York: Routledge.

Collins, R. (1979). *The credential society: An historical sociology of education and stratification*. New York: Academic Press.

Collins, R. (1980). Weber's last theory of capitalism: A systematization. *American Sociological Review, 45*, 925–942.

Commission to Build a Healthier America. (2009, April). *Race and socioeconomic factors affect opportunities for better health* (Issue brief 5). Princeton, NJ: Robert Wood Johnson Foundation. Retrieved from http://www.commissiononhealth.org/PDF/506edea1-f160-4728-9539-aba2357047e3/Issue%20Brief%205%20April%2009%20-%20Race%20and%20Socioeconomic%20Factors.pdf

Common Sense Media. (2015). *The common sense census: Media use by tweens and teens*. Retrieved from https://www.commonsensemedia.org/research/the-common-sense-census-media-use-by-tweens-and-teens

Complete College America. (2011, September). *Time is the enemy*. Retrieved from https://www.luminafoundation.org/files/resources/time-is-the-enemy.pdf

Condron, D. J. (2009). Social class, school and non-school environments, and Black/White inequalities in children's learning. *American Sociological Review, 74*, 685–708.

Condry, J. C. (1989). *The psychology of television*. Hillsdale, NJ: Erlbaum.

Conley, D. (1999). *Being Black, living in the red: Race, wealth, and social policy in America*. Berkeley: University of California Press.

Connell, R. W. (2005). Change among the gatekeepers: Men, masculinities, and gender equality in the global arena. *Signs, 30*, 1801–1826.

Connell, R. W., & Messerschmidt, J. W. (2005). Hegemonic masculinity: Rethinking the concept. *Gender & Society, 19*, 829–859.

Connor, P. (2018). *Most displaced Syrians are in the Middle East, and about a million are in Europe*. Pew Research. Retrieved from https://www.pewresearch.org/fact-tank/2018/01/29/where-displaced-syrians-have-resettled/

Consumer Financial Protection Bureau. (2015). *CFPB orders Citibank to pay $700 million in consumer relief for illegal credit card practices*. Retrieved from http://www.consumerfinance.gov/about-us/news-room/cfpb-orders-citibank-to-pay-700-million-in-consumer-relief-for-illegal-credit-card-practices/

Coogan, T. P. (2002). *Michael Collins: The man who made Ireland*. New York: Palgrave Macmillan.

Cook, K.J. (2016). Has criminology awakened from its "Androcentric Slumber"? *Feminist Criminology, 11*(4), 334–353.

Cooley, C. H. (1909). *Social organization: A study of the larger mind*. New York: Charles Scribner's Sons.

Cooley, C. H. (1964). *Human nature and the social order*. New York: Schocken Books. (Original work published 1902)

Coontz, S. (2000). Historical perspectives on family studies. *Journal of Marriage and the Family, 62*, 283–297.

Coontz, S. (2005). *Marriage, a history: From obedience to intimacy, or how love conquered marriage*. New York: Penguin Books.

Cordesman, A. H. (2020, September 9). *The real world capabilities of ISIS: The threat continues. Center for Strategic and International Studies*. Retrieved from https://www.csis.org/analysis/real-world-capabilities-isis-threat-continues

Cornelio, J. (2020). Youth. *The SAGE encyclopedia of the sociology of religion* (pp. 927–928). Thousand Oaks, CA: SAGE.

Counting crowds: Results may vary. (2010, November 5). *NPR*. Retrieved from http://www.npr.org/templates/story/story.php?storyId=131099075

Country Music Association. (2019, February 1). *CMA: Millennials remain main driver of country music growth*. Retrieved from http://www.insideradio.com/cma-millennials-remain-main-driver-of-country-music-growth/article_fd9154e2-25f1-11e9-9b3d-c7034b8f54b6.html

Cranley, E. (2019, April 29). These are the 130 current members of Congress who have doubted or denied climate change. *Business Insider*. Retrieved from https://www.businessinsider.com/climate-change-and-republicans-congress-global-warming-2019-2

Critser, G. (2003). *Fatland: How Americans became the fattest people in the world*. New York: Houghton Mifflin.

Crome, A. (2015). Religion and the pathologization of fandom: Religion, reason, and controversy in *My Little Pony* fandom. *Journal of Religion and Popular Culture, 27*(2), 130–147.

Crowe, N., & Hoskins, K. (2019). Researching transgression: Ana as a youth subculture in the age of digital ethnography. *Societies* 9(3). doi:http://dx.doi.org.thor.nebrwesleyan.edu/10.3390/soc9030053

Culhane, D. (2010, July 11). Five myths about America's homeless. *The Washington Post*. Retrieved from http://www.washingtonpost.com/wp-dyn/content/article/2010/07/09/AR2010070902357.html

Culhane, D., Treglia, D., Byrne, T., Metraux, S., Kuhn, R., Doran, K., Johns, E., & Schretzman, M. (2019). *The emerging crisis of aged homelessness*. Actionable Intelligence for Social Policy, University of Pennsylvania. Retrieved from https://www.aisp.upenn.edu/wp-content/uploads/2019/01/Emerging-Crisis-of-Aged-Homelessness-1.pdf

Curtiss, S. (1977). *Genie: A psycholinguistic study of a modern-day "wild child."*. New York, NY: Academic Press.

Cuthbert, K. (2019). "When we talk about gender we talk about sex": (A)sexuality and (a)gendered subjectivities. *Gender & Society, 33*(6), 841–864. Retrieved from https://doi.org/10.1177/0891243219867916

Cutright, P., & Fernquist, R. M. (2000). Effects of societal integration, period, region, and culture of suicide on male age-specific suicide rates: Twenty developed countries, 1955–1989. *Social Science Research, 29*, 148–172.

Dahl, R. A. (1961). *Who governs?* New Haven, CT: Yale University Press.

Dahl, R. A. (1982). *Dilemmas of a pluralist democracy: Autonomy vs. control*. New Haven, CT: Yale University Press.

Dahl, R. A. (1989). *Democracy and its critics*. New Haven, CT: Yale University Press.

Data USA. (2020). *Data USA: Flint Michigan*. Retrieved from https://datausa.io/profile/geo/flint-mi/#economy

Davey, M., & Smith, M. (2016, April 8). Hastert molested at least four boys, prosecutors say. *New York Times*. Retrieved from http://www.nytimes.com/2016/04/09/us/dennis-hastert-molested-at-least-four-boys-prosecutors-say.html

David, D. S., & Brannon, R. (1976). *The forty-nine percent majority: The male sex role*. New York, NY: Random House.

Davis, A., Kimball, W., & Gould, E. (2015, May 27). *The class of 2015*. Retrieved from http://www.epi.org/publication/the-class-of-2015/

Davis, K., & Moore, W. (1945). Some principles of stratification. *American Sociological Review, 10*, 242–249.

Davis, S. N., Greenstein, T. N., & Marks, J. P. (2007). Effects of union type on division of household labor: Do cohabitating men really perform more housework? *Journal of Family Issues, 28*(9), 1246–1272. http://dx.doi.org/10.1177/0192513x07300968

Davis, W. (1991). *Fundamentalism in Japan: Religious and political*. Chicago: University of Chicago Press.

DC Public Schools. (n.d). *At a glance enrollment*. Retrieved from https://dcps.dc.gov/page/dcps-glance-enrollment

De Wild, L., & Augers, S. (2019). Playing the other: Role-playing religion in video games. *European Journal of Cultural Studies, 22*(5–6), 867–884. Retrieved from https://doi.org/10.1177/1367549418790454

De Wild, L., Augers, S., Kasen, C., & Canada, I. (2018). "Things greater than thou": Post-apocalyptic religion in games. *Religions, 9*(6).

Death Penalty Information Center. (2018). *Facts about the death penalty.* Retrieved from https://deathpenalty-info.org/documents/FactSheet.pdf

Death Penalty Information Center. (2020a, August 31). *Coronavirus prison fatalities surpass two decades of executions; COVID-19 has killed more California death row prisoners than the state has executed in 27 years.* Retrieved from https://deathpenaltyinfo.org/news/coronavirus-prison-fatalities-surpass-two-decades-of-executions-covid-19-has-killed-more-california-death-row-prisoners-than-the-state-has-executed-in-27-years

Death Penalty Information Center. (2020b, September 8). *Facts about the death penalty.* Retrieved from https://files.deathpenaltyinfo.org/documents/pdf/FactSheet.f1599571382.pdf

Death Penalty Information Center. (n.d). *Execution of juveniles in the U.S. and other countries.* Retrieved from https://deathpenaltyinfo.org/execution-juveniles-us-and-other-countries

DeBeaumont, R. (2009). Occupational differences in the wage penalty for obese women. *Journal of Socio-Economics, 38*, 344–349.

DeChoudhury, M., Sharma, S., & Kiciman, E. (2016, February). *Characterizing dietary choices, nutrition, and language in food deserts via social media.* Paper presented at Computer-Supported Cooperative Work and Social Computing 2016, San Francisco, California. http://dx.doi.org/10.1145/2818048.2819956

Dee, D. L., Pazol, K., & Cox, S. (2017). Trends in repeat births and use of postpartum contraception among teens: United States, 2004–2015. Retrieved from https://www.cdc.gov/mmwr/volumes/66/wr/mm6616a3.htm?s_cid=mm6616a3_e#suggested citation

Delgado, R., & Stefancic, J. (2007). Critical race theory and criminal justice. *Humanity & Society*, 31(2–3), 133–145. doi:http://doi.org/10.1177/016059760703100201

DellaPergola, S. (2010). *World Jewish population 2010* (Current Jewish Population Reports 2). New York: Berman Jewish DataBank. Retrieved from http://www.jewishdatabank.org/Reports/World_Jewish_Population_2010.pdf

Delli Carpini, M. X., & Keeter, S. (1996). *What Americans know about politics and why it matters.* New Haven, CT: Yale University Press.

DeMatteo, M. (2020, November 12). The average Gen X-er has $32,878 in non-mortgage debt—here's how they compare to other generations. *CNBC.* Retrieved from https://www.cnbc.com/select/average-american-debt-in-40s/

DeNavas-Walt, C., & Proctor, B. D. (2014). *Income and poverty in the United States: 2013*(Current Population Reports P60-249). Washington, DC: U.S. Census Bureau. Retrieved from http://www.census.gov/content/dam/Census/library/publications/2014/demo/p60-249.pdf

Denison, E., & Kitchen, A. (2015). *Out on the fields.* Retrieved from https://outonthefields.com/wp-content/uploads/2020/11/Out-on-the-Fields-Final-Report-1.pdf

Denizet-Lewis, B. (2003, August 3). Double lives on the down low. *New York Times Magazine.* Retrieved from http://www.nytimes.com/2003/08/03/magazine/double-lives-on-the-down-low.html?pagewanted=all&src=pm

Denno, B. W. (1990). *Biology and violence from birth to adulthood.* Cambridge: Cambridge University Press.

Department of Labor. (2018). *Fact sheet #71: Internship programs under the Fair Labor Standards Act.* Retrieved from https://www.dol.gov/agencies/whd/fact-sheets/71-flsa-internships

DePillis, L. (2015). Minimum-wage offensive could speed arrival of robot-powered restaurants. *Washington Post.* Retrieved from https://www.washingtonpost.com/business/capitalbusiness/minimum-wage-offensive-could-speed-arrival-of-robot-powered-restaurants/2015/08/16/35f284ea-3f6f-11e5-8d45-d815146f81fa_story.html?wprss=rss_homepage

DeSantis, A., & Kayson, W. A. (1997). Defendants' characteristics of attractiveness, race, sex and sentencing decisions. *Psychological Reports, 81*, 679–683.

Desilver, D. (2019, August 29). Ten facts about American workers. *Pew Research.* Retrieved from https://www.pewresearch.org/fact-tank/2019/08/29/facts-about-american-workers/

Desilver, D., Lipka, M., & Fahmy, D. (2020, June 3). *10 things we know about race and policing in the U.S.* Retrieved from https://www.pewresearch.org/fact-tank/2020/06/03/10-things-we-know-about-race-and-policing-in-the-u-s/

Desmond, M. (2015, March). *Unaffordable America: Poverty, housing, and eviction.* Fast Forward, No.22. Madison: University of Wisconsin, Institute for Research on Poverty. Retrieved from http://www.irp.wisc.edu/publications/fastfocus/pdfs/FF22-2015.pdf

Desmond, M. (2016a). *Evicted: Poverty and profit in the American city.* New York, NY: Crown.

Desmond, M. (2016b, June 9). *Evicted: Housing, poverty, and policy.* Presentation at Georgetown Law School, Washington, DC.

Dettling, L. J., Hsu, J. W., Jacobs, L., Moore, K. B., Thompson, J. P., & Llanes, E. (2017). *Recent trends in wealth-holding by race and ethnicity: Evidence from the survey of consumer finances.* Board of Governors of the Federal Reserve System. Retrieved from https://www.federalreserve.gov/econres/notes/feds-notes/recent-trends-in-wealth-holding-by-race-and-ethnicity-evidence-from-the-survey-of-consumer-finances-20170927.htm

de Vise, D. (2012, May 21). Is college too easy? As study time falls, debate rises. *Washington Post.* Retrieved from http://www.washingtonpost.com/local/education/is-college-too-easy-as-study-time-falls-debate-rises/2012/05/21/gIQAp7uUgU_story.html

Devitt, P. (2017, May 8). 13 Reasons Why and suicide contagion. *Scientific American.* Retrieved from https://www.scientificamerican.com/article/13-reasons-why-and-suicide-contagion1

Devlin, K. (2020, April 9). *Most European students learn English in school.* Retrieved from https://www.pewresearch.org/fact-tank/2020/04/09/most-european-students-learn-english-in-school/

Devlin, K., & Connaughton, A. (2020, August 27). *Most approve of national response to COVID-19 in 14 advanced economies.* Retrieved from https://www.pewresearch.org/global/2020/08/27/most-approve-of-national-response-to-covid-19-in-14-advanced-economies/

Diamant, J. (2019a, April 1). *The countries with the 10 largest Christian populations and the 10 largest Muslim populations.* Retrieved from https://www.pewresearch.org/fact-tank/2019/04/01/the-countries-with-the-10-largest-christian-populations-and-the-10-largest-muslim-populations

Diamant, J. (2019b). *With high levels of prayer, U.S. is an outlier among wealthy nations*. Retrieved from https://www.pewresearch.org/fact-tank/2019/05/01/with-high-levels-of-prayer-u-s-is-an-outlier-among-wealthy-nations/

Diaz, J. D. (1999). *Suicide in the Las Vegas homeless population: Applying Durkheim's theory of suicide*. (Doctoral dissertation, University of Nevada, Las Vegas.

Dicken, P. (1998). *The global shift: Transforming the world economy* (3rd ed.). New York: Guilford Press.

Dickler, J. (2019). *Consumer debt hits trillion*. CNBC. Retrieved from https://www.cnbc.com/2019/02/21/consumer-debt-hits-4-trillion.html

Diep, F. (2017, August 15). How social media helped organize and radicalize America's white supremacists. *Pacific Standard*. Retrieved from https://psmag.com/social-justice/how-social-media-helped-organize-and-radicalize-americas-newest-white-supremacists

Digest of Education Statistics. (2019). *Racial/ethnic enrollment in public schools*. Retrieved from https://nces.ed.gov/programs/coe/pdf/coe_cge.pdf

Dilmac, B. (2009). Psychological needs as a predictor of cyber-bullying: A preliminary report on college students. *Educational Sciences: Theory and Practice, 9*, 1308–1325.

DiMaggio, P. J., & Powell, W. (1983). The iron cage revisited: Institutional isomorphism and collective rationality in organizational fields. *American Sociological Review, 48*, 147–160.

Dion, K., Berscheid, E., & Walster, E. (1972). What is beautiful is good. *Journal of Personality and Social Psychology, 24*, 285–290.

Dionne, E. J., Jr., & Green, J. C. (2008). *Religion and American politics: More secular, more evangelical... or both?* Washington, DC: Brookings Institution. Retrieved from http://www.brookings.edu/~/media/research/files/papers/2008/2/religion%20green%20dionne/02_religion_green_dionne.pdf

Dishion, T. J., McCord, J., & Poulin, F. (1999). When interventions harm: Peer groups and problem behavior. *American Psychologist, 54*, 755–764.

Djilas, M. (1957). *The new class: An analysis of the communist system*. New York: Harvest Books.

Dobbs, R., Smit, S., Remes, J., Manyika, J., Roxburgh, C., & Restrepo, A. (2011). *Urban world: Mapping the economic power of cities*. McKinsey Global Institute. Retrieved from https://www.mckinsey.com/featured-insights/urbanization/urban-world-mapping-the-economic-power-of-cities#

Doey, L., Coplan, R. J., & Kingsbury, M. (2013). Bashful boys and coy girls: A review of gender differences in childhood shyness. *Sex Roles, 70*, 255–266.

Dokoupil, T. (2012a, July 9). Is the Web driving us mad? *Newsweek*. Retrieved from newsweek/2012/07/08/is-the-internet-making-us-crazy-what-the-new-research-says.html">http://www.thedailybeast.com/newsweek/2012/07/08/is-the-internet-making-us-crazy-what-the-new-research-says.html

Dokoupil, T. (2012b, July 16). Tweets, texts, email, posts: Is the onslaught making us crazy? *Newsweek*, pp. 24–30.

Dolan, K. (2020, August 25). World's billionaires list: The richest in 2020. *Forbes*. Retrieved from https://www.forbes.com/billionaires/

Doleac, J. L., & Hansen, B. (2018). *The unintended consequences of "ban the box," statistical discrimination and employment outcomes when criminal histories are hidden*. Retrieved from http://jenniferdoleac.com/wp-content/uploads/2017/03/Doleac_Hansen_PP2017.pdf

Dollard, J. (1957). *Caste and class in a Southern town* (3rd ed.). New York: Anchor Books.

Dolnick, E. (1993, September). Deafness as culture. *Atlantic Monthly*, pp. 37–53).

Domhoff, G. W. (1983). *Who rules America now?* Simon and Schuster.

Domhoff, G. W. (1990). *The power elite and the state: How policy is made in America*. New York: Aldine de Gruyter.

Domhoff, G. W. (2002). *Who rules America? Power and politics* (4th ed.). New York: McGraw-Hill.

Domhoff, G. W. (2006). *Who rules America? Power, politics, and social change* (5th ed.). New York: McGraw-Hill.

Domhoff, G. W. (2009). *Who rules America? Challenges to corporate and class dominance* (6th ed.). New York: McGraw-Hill.

Dominus, S. (2017, May 11). Is an open marriage a happier marriage? *New York Times Magazine*. Retrieved from https://www.nytimes.com/2017/05/11/magazine/is-an-open-marriage-a-happier-marriage.html.

Douglas, D. (2013, March 6). Attorney general says big banks' size may inhibit prosecution, *Washington Post*, A12.

Downs, E., & Smith, S. (2010). Keeping abreast of hypersexuality: A video game character content analysis. *Sex Roles, 62*(11), 721–733.

Drake, B. (2014, January 7). *Number of older Americans in the workforce is on the rise*. Washington, DC: Pew Research Center. Retrieved from http://www.pew research.org/fact-tank/2014/01/07/number-of-older-americans-in-the-workforce-is-on-the-rise/

Dreifus, C. (2012, May 14). A mathematical challenge to obesity. *New York Times*. Retrieved from http://www.nytimes.com/2012/05/15/science/a-mathemat-ical-challenge-to-obesity.html

Drew, J. (2015, April 6). *A list of tribal laws prohibiting gay marriage*. Retrieved from https://www.yahoo.com/news/list-tribal-laws-prohibiting-gay-marriage-162248831.html?ref=gs

Droke, A. (2017). *Top 10 schools to find a husband*. Retrieved from https://www.collegemagazine.com/top-10-schools-find-husband-mrs-degree/

Du Bois, W. E. B. (1899/1996). *Philadelphia negro: A social study*. Philadelphia: University of Philadelphia Press.

Du Bois, W. E. B. (1920/2003). *Dark water: Voices from within the veil*. Amherst, NY: Prometheus Books.

Du Bois, W. E. B. (1930). The negro citizen. In C. S. Johnson (Ed.), *The negro in civilization*. New York: Henry Holt and Company. Retrieved from http://www.webdubois.org/dbTNCitizen.html

Du Bois, W. E. B. (2008). *The souls of Black folk*. Rockville, MD: Arc Manor. (Original work published 1903)

Duffin, E. (2020, March 31). Percentage of the U.S. population with a college degree, by gender 1940–2019. *Statistica*. Retrieved from https://www.statista.com/statistics/184272/educational-attainment-of-college-diploma-or-higher-by-gender/

Duggan, M. (2015, December 15). *Who plays video games and identifies as a "gamer". Internet &Technology*. Retrieved from http://www.pewinternet.org/2015/12/15/who-plays-video-games-and-identifies-as-a-gamer/

Duggan, M., Ellison, N. B., Lampe, C., Lenhart, A., & Madden, M. (2015, January 9). *Social media update 2014*. Washington, DC: Pew Research Center. Retrieved from http://www.pewinternet.org/2015/01/09/social-media-update-2014/

Duhigg, C., & Barboza, D. (2012, January 25). In China, human costs are built into an iPad. *New York Times*. Retrieved from http://www.nytimes.com/2012/01/26/business/ieconomy-apples-ipad-and-the-human-costs-for-workers-in-china.html?_r=2&pagewanted=print

Duneier, M. (1992). *Slim's table: Race, respectability, and masculinity*. Chicago: University of Chicago Press.

Dunn, A. (2019, September 24). *Younger, college-educated black Americans are most likely to feel need to "code-switch"*. Retrieved from https://www.pewresearch.org/fact-tank/2019/09/24/younger-college-educated-black-americans-are-most-likely-to-feel-need-to-code-switch/

Durkheim, É. (1951). *Suicide*. New York: Free Press. (Original work published 1897)

Durkheim, É. (1956). *Education and sociology* (S. L. Fox, Trans.). New York: Free Press. (Original work published 1922)

Durkheim, É. (1973a). *Émile Durkheim on morality and society*. Chicago: University of Chicago Press. (Original work published 1922)

Durkheim, É. (1973b). *Moral education: A study in the theory and application of the sociology of education*. New York: Free Press. (Original work published 1922)

Durkheim, É. (1997). *The division of labor in society*. New York: Free Press. (Original work published 1893)

Durkheim, É. (2008). *The elementary forms of the religious life*. New York: Dover. (Original work published 1912)

Dvorak, P. (2020, October 1). Domestic abuse has risen during the pandemic: Groups like the House of Ruth are ready. *Washington Post*. Retrieved from https://www.washingtonpost.com/local/domestic-abuse-has-risen-during-the-pandemic-groups-like-the-house-of-ruth-are-ready/2020/10/01/e46922c8-03ea-11eb-b7ed-141dd88560ea_story.html?utm_campaign=wp_afternoon_buzz&utm_medium=email&utm_source=newsletter&wpisrc=nl_buzz&carta-url=https%3A%2F%2Fs2.washingtonpost.com%2Fcar-lntr%2F2be398a%2F5f7639299d2fda0efb3afe6b%2F596b97819bbc0f403f93ab29%2F10%2F57%2F2f924ecd260ec7b3849c30443cbe99a8

Dworkin, A. (1981). *Pornography: Men possessing women*. New York: Pedigree.

Dworkin, A. (1987). *Intercourse*. New York: Free Press.

Dworkin, A. (1989). *Letters from the war zone: Writings, 1976–1987*. New York: Dutton.

Dwoskin, E. (2012, March 28). Will you marry me (after I pay off my student loans)? *Bloomberg Businessweek*. Retrieved from http://www.businessweek.com/articles/2012-03-28/will-you-marry-me-after-i-pay-off-my-student-loans

Dwyer, R. E., Hodson, R., & McCloud, L. (2013). Gender, debt, and dropping out of college. *Gender & Society, 27*(1), 30–55. http://dx.doi.org/10.1177/0891243212464906

Dyer, J., Pettyjohn, H., & Saladin, S. (2020). Academic dishonesty and testing: How student beliefs and test settings impact decisions to cheat. *Journal of the National College Testing Association* 4(1).

Eccles, J. S., & Barber, B. L. (1999). Student council, volunteering, basketball, or marching band: What kind of extracurricular involvement matters? *Journal of Adolescent Research, 14*, 10–43.

The Economist. (2019, February 16). Men still pick "blue" jobs and women "pink" jobs. Retrieved from https://www.economist.com/finance-and-economics/2019/02/16/men-still-pick-blue-jobs-and-women-pink-jobs

Ed Build.org. (n.d). *23 billion*. Retrieved from https://edbuild.org/content/23-billion#CA

Eddy, M. B. (1999). *Christian science: No and yes*. Boston: Author. (Original work published 1887)

Edin, K., & Kefalas, M. (2005). *Promises I can keep: Why poor women put motherhood before marriage*. Berkeley: University of California Press.

Editorial Board. (2020, November 15). Opinion: Abolish the electoral college. *Washington Post*. Retrieved from https://www.washingtonpost.com/opinions/abolish-the-electoral-college/2020/11/15/c40367d8-2441-11eb-a688-5298ad5d580a_story.html

Education Commission of the States. (2016). *50-state comparison: Advanced placement policies*. Retrieved from https://www.ecs.org/advanced-placement-policies/

Edwards, F. R., Lee, H., & Esposito, M. (2019). Risk of being killed by police use of force in the United States by age, race–ethnicity, and sex. *Proceedings of the National Academy of Sciences, 116*, 16793–16798. Retrieved from https://www.pnas.org/content/116/34/16793

Edwards, J. (2012, October 7). This video of Haitians reading "#FirstWorldProblems" from Twitter is making people really angry. *Business Insider*. Retrieved from http://www.businessinsider.com/anger-over-haitians-reading-firstworldproblems-from-twitter-2012-10#ixzz33sB1IhdT

Edwards, K. E. (2007). *"Putting my man face on": A grounded theory of college men's gender identity development*. (Doctoral dissertation, University of Maryland-College Park. Retrieved from ProQuest (3260431)

Edwards, K. L. (2019). Presidential address: Religion and power—A return to the roots of social scientific scholarship. *Journal for the Scientific Study of Religion 58*(1), 5–19. Retrieved from https://doi-org.thor.nebrwesleyan.edu/10.1111/jssr.12583

Effinger, A., & Burton, K. (2014, April 9). Trailer parks lure Wall Street investors looking for double-wide returns. *Bloomberg*. Retrieved from http://www.bloomberg.com/news/2014-04-10/trailer-parks-lure-investors-pursuing-double-wide-returns.html

Eglitis, D. S. (2010). The uses of global poverty: How economic inequality benefits the West. In J. J. Macionis & N. V. Benokraitis (Eds.), *Seeing ourselves: Classic, contemporary, and cross-cultural readings in sociology* (8th ed. ed., pp. 199–206). New York: Pearson.

Ehrenreich, B. (2001). *Nickel and dimed: On (not) getting by in America*. New York: Metropolitan Books.

Ehrenreich, B., & Hochschild, A. R. (2002). Introduction. In B. Ehrenreich & A. R. Hochschild (Eds.), *Global woman: Nannies, maids, and sex workers in the new economy* (pp. 1–14). New York: Metropolitan Books.

Eichler, A. (2012, May 30). *Unpaid overtime: Wage and hour lawsuits have skyrocketed in the last decade. Huffington Post*. Retrieved from http://www.huffingtonpost.com/2012/05/30/wage-hour-lawsuits_n_1556484.html

Eisenbrey, R. (2012, March 2). Pushing back against illegal unpaid internships. *Economic Policy Institute Blog*. Retrieved from http://www.epi.org/blog/pushing-back-illegal-unpaid-internships

Eligon, J. A. (2016, January 21). Question of environmental racism in Flint. *New York Times*. Retrieved from http://www.nytimes.com/2016/01/22/us/a-question-of-environmental-racism-in-flint.html

Eller, C. (2000). *The myth of matriarchal prehistory: Why an inventive past won't give women a future*. Boston: Beacon Press.

Elliot, S. (2020, August 28). *Donald Trump slams "China virus" in nomination speech as crowd ignores distancing*. Retrieved from https://www.mirror.co.uk/news/us-news/donald-trump-slams-china-virus-22590714

Elliott, S. K. (2016, October 18). *How American Indian reservations came to be*. Retrieved from https://www.pbs.org/wgbh/roadshow/stories/articles/2015/5/25/how-american-indian-reservations-came-be/

Elsesser, K. (2019, December 11). Lawsuit says capital ACT and SAT are biased: Here's what the research says. *Forbes*. Retrieved from https://www.forbes.com/sites/kimelsesser/2019/12/11/lawsuit-claims-sat-and-act-are-biased-heres-what-research-says/#d8c2db83c429

EM (Equal Measures) 2030. (2019a). *SDG Gender Index, tracking gender equality*. Retrieved from https://data.em2030.org/2019-sdg-gender-index/explore-the-2019-index-data/

EM (Equal Measures) 2030. (2019b). *Global report 2019*. Retrieved from https://www.equalmeasures2030.org/wp-content/uploads/2019/07/EM2030_2019_Global_Report_English_WEB.pdf

EM (Equal Measures) 2030. (2019c). *Bending the curve towards gender equality by 2030*. Retrieved from https://www.equalmeasures2030.org/products/bending-the-curve-towards-gender-equality-by-2030/

Emerson, R. M. (1962). Power-dependence relations. *American Sociological Review*, *27*, 31–41.

Emmanuel, A. (1972). *Unequal exchange: A study of the imperialism of trade*. New York: Monthly Review Press.

Engels, F. (1942). *The origins of family, private property, and the state*. New York, NY: International. (Original work published 1884)

Environmental Protection Agency. (2011). *Electronics waste management in the United States through 2009*. Washington, DC: Author.

Epstein, D. M. (1993). *Sister Aimee: The life of Aimee Semple McPherson*. New York: Harcourt Brace Jovanovich.

Erber, G., & Sayed-Ahmed, A. (2005). Offshore outsourcing: A global shift in the present IT industry. *Intereconomics*, *40*, 100–112.

Erikson, E. H. (1950). *Childhood and society*. New York: Norton.

Erle, S., & Hendry, H. (2020). Monsters: Interdisciplinary explorations in monstrosity. *Palgrave Communications*, *6*(1), 1–7.

Ethnologue Languages of the World. (2021). *Endangered languages*. Retrieved from https://www.ethnologue.com/guides/how-many-languages-endangered

Etter, G. (1998). Common characteristics of gangs: Examining the cultures of the new urban tribes. *Journal of Gang Research*, *5*, 19–33.

Etzioni, A. (1975). *A comparative analysis of complex organizations: On power, involvement, and their correlates*. New York: Free Press.

Evans, D. K., Akmal, M., & Jakilea, P. (2020). *Gaps in education: The long view*. The Center for Global Development. Retrieved from https://www.edu-links.org/sites/default/files/media/file/gender-gaps-education-long-view.pdf

Evans, M. D. R., Kelley, J., Sikora, J., & Treiman, D. J. (2010). Family scholarly culture and educational success: Books and schooling in 27 nations. *Research in Social Stratification and Mobility*, *28*, 171–197.

Evon, D. (2020, July 15). *Is Wayfair trafficking children via overpriced items?* Retrieved from https://www.snopes.com/fact-check/wayfair-trafficking-children/

FACE Intel (Former and Current Employees of Intel). (2000). *Related class action lawsuits: A huge victory for the worker*. Retrieved from http://www.faceintel.com/relatedclassactions.htm

Fahmy, D. (2018). *Key findings about Americans' belief in God*. Washington, DC: Pew Research Center. Retrieved from https://www.pewresearch.org/fact-tank/2018/04/25/key-findings-about-americans-belief-in-god/

Fainaru-Wada, M., & Fainaru, S. (2014). *League of denial: The NFL, concussions, and the battle for truth*. New York: Three Rivers Press.

Faiola, A. (2015, April 21). A global surge of refugees leaves Europe struggling to cope. *Washington Post*. Retrieved from https://www.washingtonpost.com/world/europe/new-migration-crisis-overwhelms-european-refugee-system/2015/04/21/3ab83470-e45c-11e4-ae0f-f8c46aa8c3a4_story.html

Faiola, A., & Herrero, A. V. (2020, September 7). Women and children confront dual threat of domestic violence and illness. *Washington Post*. Retrieved from https://search-proquest-com.thor.nebrwesleyan.edu/docview/2440413678?accountid=28187

Fairfield, R. (1972). *Communes USA*. Baltimore, MD: Penguin Books.

Faiths unite for day of dignity to help homeless. (2015, October 18). Retrieved from http://news3lv.com/archive/faiths-unite-for-day-of-dignity-to-help-homeless

Faludi, S. (1991). *Backlash: The undeclared war against American women*. New York: Crown.

Fantasy Sports and Gaming Association (FSGA). (2019). *Industry demographics*. Retrieved from https://thefsga.org/industry-demographics/

Farkas, G., Grobe, R. P., Sheehan, D., & Shuan, Y. (1990). Cultural resources and school success: Gender, ethnicity, and poverty groups within an urban school district. *American Sociological Review*, *55*, 127–142.

Farkas, G., Sheehan, D., & Grobe, R. P. (1990). Coursework mastery and school success: Gender, ethnicity, and poverty groups within an urban school district. *American Educational Research Journal, 27*, 807–827.

Fassbender, L., Byrnes, C., & Levine, R. (2019). Looking back at past epidemics. *Pennsylvania Legacies, 19*(1) 40–41. Retrieved August 27, 2020, from https://doi.org/10.5215/pennlega.19.1.0040

Fausto-Sterling, A. (2000). *Sexing the body: Gender politics and the construction of sexuality*. New York, NY: Basic Books.

Federal Bureau of Investigation. (2017a). Violent crime. In *Crime in the United States 2017 (Uniform Crime Reports)*. Washington, DC: Author. Retrieved from https://ucr.fbi.gov/crime-in-the-u.s/2017/crime-in-the-u.s.-2017/topic-pages/violent-crime

Federal Bureau of Investigation. (2017b). Property crime. In *Crime in the United States 2017 (Uniform Crime Reports)*. Washington, DC: Author. Retrieved from https://ucr.fbi.gov/crime-in-the-u.s/2017/crime-in-the-u.s.-2017/topic-pages/property-crime

Federal Bureau of Investigation. (2019). *Federal hate crime statistics 2019*. Retrieved from https://www.fbi.gov/news/pressrel/press-releases/fbi-releases-2019-hate-crime-statistics

Federal Bureau of Investigation (FBI). (2020a). *2019 preliminar y semiannual crime statistics overview*. Retrieved from

Federal Bureau of Investigation (FBI). (2020b). *FBI warns of money mule schemes exploiting the COVID-19 pandemic*. Retrieved from https://www.fbi.gov/news/pressrel/press-releases/fbi-warns-of-money-mule-schemes-exploiting-the-covid-19-pandemic

Federal Bureau of Investigation (FBI). (n.d.). *Transnational organized crime*. Retrieved from https://www.fbi.gov/investigate/organized-crime

Federal Reserve. (2020, September 8). *Consumer credit*. Retrieved from https://www.federalreserve.gov/releases/g19/current/default.htm

Federal Reserve Bank of New York. (2018, July). *The labor market for recent college graduates*. Retrieved from https://www.newyorkfed.org/research/college-la-bor-market/college-labor-market_underemploy-ment_rates.html

Federal Reserve Bank of New York. (2020, October 22). *The labor market for recent college graduates*. Retrieved from https://www.newyorkfed.org/research/college-labor-market/college-labor-market_underemployment_rates.html

Federal Reserve System & Brookings Institution. (2008). *The enduring challenge of concentrated poverty in America: Case studies from communities across the U.S.* Washington, DC: Authors. Retrieved from http://www.frbsf.org/community-development/files/cp_fullreport.pdf

Felsenthal, E. (2019, December 30). The choice. *Time Magazine person of the year edition*.

Fenstermaker Berk, S. (1985). *The gender factory: The apportionment of work in American households*. New York: Plenum Press.

Fenstermaker, S., & West, C. (2002). *Doing gender, doing difference: Inequality, power, and institutional change*. New York: Routledge.

Ferment, C. A. (1989, January). Political practice and the rise of an ethnic enclave: The Cuban American case. *Theory and Society, 18*, 47–48.

Fieldstadt, E., & Kaplan, E. (2019, October 2). *Felicity Huffman released from prison on 11th day of 14-day sentence*. Retrieved from https://www.nbcnews.com/pop-culture/celebrity/felicity-huffman-released-prison-end-14-day-sentence-n1071921

Figlio, D. N. (2005). *Testing, crime and punishment* (NBER Working Paper 11194). Cambridge, MA: National Bureau of Economic Research.

Filoux, J.-C. (1993). Émile Durkheim. *Prospects: The Quarterly Journal of Comparative Education, 23*(1/2), 303–320.

Financial Crimes Enforcement Network. (2014). JPMorgan admits violation of the Bank Secrecy Act (press release). Retrieved from http://www.fincen.gov/news_room/nr/pdf/20140107.pdf

Finckenauer, J. O., & Waring, E. (1996). Russian emigre crime in the U.S.: Organized crime or crime that is organized? *Transnational Organized Crime, 2*, 139–155.

Fine, L. (2012). *Sexual identity and postsecondary education: Outcomes, institutional factors, and narratives* (Electronic thesis or dissertation). Ohio State University, Ohio.

Fingarette, H. (1972). *Confucius: The secular as sacred*. Long Grove, IL: Waveland Press.

Finke, R., & Stark, R. (1988). Religious economies and sacred canopies: Religious mobilization in American cities, 1906. *American Sociological Review, 53*, 41–49.

Finke, R., & Stark, R. (1992). *The churching of America, 1776–1980: Winners and losers in our religious economy*. New Brunswick, NY: Rutgers University Press.

Finke, R., & Stark, R. (2005). *The churching of America, 1776–2005: Winners and losers in our religious economy*. New Brunswick, NY: Rutgers University Press.

Firestone, S. (1971). *The dialectic of sex*. London: Paladin.

Fischer, C. (1982). *To dwell among friends: Personal networks in town and city*. Chicago: University of Chicago Press.

Fischer, C. (1984). *The urban experience* (2nd ed.). New York: Harcourt Brace Jovanovich.

Fischer, K. (2020, October 20). Where are most international students? Stranded here, needing colleges' help. *Chronicle of Higher Education*. Retrieved from https://www.chronicle.com/article/where-are-most-international-students-stranded-here-needing-colleges-help

Fisher, R., Danza, P., McCarthy, J., & Tiezzi, L. (2019). Provision of contraception in New York City school-based health centers: Impact on teenage pregnancy and avoided costs, 2008–2017. *Perspectives on Sexual and Reproductive Health, 51*, 201–209. doi:10.1363/psrh.12126

Fishman, P. (1978). Women's work in interaction. *Social Problems, 25*(4), 397–406. http://dx.doi.org/10.1525/sp.1978.25.4.03a00050

Fitzgerald, A. (2011, July 28). A stealth way a bill becomes a law. *Bloomberg Businessweek*. Retrieved from http://www.businessweek.com/magazine/a-stealth-way-a-bill-becomes-a-law-07282011.html

Flacks, R. (1971). *Youth and social change*. Chicago, IL: Markham.

Flores, A. R., Herman, J. L., Gates, G. J., & Brown, T. N. T. (2016). *How many adults identify as transgender in the United States?* Los Angeles, CA: The Williams Institute GLAAD. Retrieved from https://www.glaad.org/

Flores, A. R., Langton, L., Meyer, H. I., & Romero, A. P. (2020, October). Victimization rates and traits of sexual and gender minorities in the United States: Results from the National Crime Victimization Survey, 2017. *Science Advances 6*. Retrieved from https://advances.sciencemag.org/content/advances/6/40/eaba6910.full.pdf

Fontenot, K., Semega, J., & Kollar, M. (2018, September 12). *Income and poverty in the United States: 2017.* United States Census Bureau. Retrieved from https://www.census.gov/library/publications/2018/demo/p60-263.html

Food and Agriculture Organization of the United Nations. (2009). *How to feed the world in 2050.* Rome: Author. Retrieved from http://www.fao.org/fileadmin/templates/wsfs/docs/expert_paper/How_to_Feed_the_World_in_2050.pdf

Food and Agriculture Organization of the United Nations. (2020, October 5). *Sustainable development goals.* Retrieved from http://www.fao.org/sustainable-development-goals/indicators/211/en/

Ford, D. (2015, July 24). *Who commits mass shootings?* Retrieved from http://www.cnn.com/2015/06/27/us/mass-shootings/

Ford, J., & Gomez-Lanier, L. (2017). Are tiny homes here to stay? A review of literature on the tiny house movement. *Family and Consumer Sciences Research Journal, 45*(4), 394–405. Retrieved from https://doi.org/10.1111/fcsr.12205

Fortune. (2016, March 30). The world's 19 most disappointing leaders. Archived from the original on November 23, 2016. Retrieved December 2, 2016.

Foucault, M. (1988). *Madness and civilization: A history of insanity in the age of reason.* New York: Vintage Books.

Foucault, M. (2005). *The order of things: An archaeology of the human sciences* (Sheridan, Trans.). London, UK.

Fountain, H. (2020, July 8). In parched southwest, warm spring renews threat of "megadrought." *New York Times.* Retrieved from https://www.nytimes.com/2020/07/08/climate/southwest-megadrought-climate-change.html

Fox, R., Corretjer, O., & Webb, K. (2019). Benefits of foreign language learning and bilingualism: An analysis of published empirical research 2012–2019. *Foreign Language Annals, 52*(4), 699–726. Retrieved from https://doi.org/10.1111/flan.12424

Frank, A. G. (1966). The development of underdevelopment. *Monthly Review, 18*(4), 17–31.

Frank, A. G. (1979). *Dependent accumulation and underdevelopment.* Macmillan.

Frank, R. (2015, June 15). *Millionaires control 41% of world's wealth, expected to take more.* Retrieved from http://www.cnbc.com/2015/06/15/millionaires-control-41-of-worlds-wealth.html

Freedom House. (2020). *Freedom in the world 2020.* Retrieved from https://freedomhouse.org/country/united-states/freedom-world/2020

Freeland, C. (2012). *Plutocrats: The rise of the new global super-rich and the fall of everyone else.* New York, NY: Penguin Books.

Freeman, D. W. (2012, January 12). *Video game-obsessed mom neglects kids, starves dogs.* Retrieved from http://www.cbsnews.com/news/video-game-obsessed-mom-neglects-kids-starves-dogs/

Freeman, T., & Posner, M. (n.d). *17 facts everyone should know about Hasidic Jews.* Retrieved from https://www.chabad.org/library/article_cdo/aid/4079238/jewish/17-Facts-Everyone-Should-Know-About-Hasidic-Jews.htm

Freire, P. (1972). *Pedagogy of the oppressed.* New York: Herder & Herder.

Freud, S. (1905). *Standard Edition* (Vol. 7). London: Hogarth.

Freud, S. (1929). *Standard Edition* (Vol. 21). London: Hogarth.

Freud, S. (1933). *New introductory lectures on psychoanalysis.* New York: Norton.

Frey, W. H., & Speare, A. (1991). *U.S. metropolitan area population growth, 1960–1990: Census trends and explanations* (Population Studies Center Research Report No. 91-212). Ann Arbor: Institute for Social Research, University of Michigan.

Friedan, B. (1963). *The feminine mystique.* New York: Norton.

Friedan, B. (1981). *The second stage.* New York: Summit.

Friedersdorf, C. (2017, June 27). How people like you fuel extremism. *The Atlantic Monthly.* Retrieved from https://www.theatlantic.com/politics/archive/2017/06/together-people-like-you-fuel-extremism/531702/

Friedman, H. L. (2013). Tiger girls on the soccer field. *Contexts, 12,* 30–35.

Friedman, L. M. (1975). *The legal system: A social science perspective.* New York: Russell Sage Foundation.

Friedman, L. M. (1990). *The republic of choice: Law, authority, and culture.* Cambridge, MA: Harvard University Press.

Friedman, S., Squires, G. D., & Galvan, C. (2010). *Cybersegregation in Dallas and Boston: Is Neil a more desirable tenant than Tyrone or Jorge?* Paper presented at the annual meeting of the Population Association of America, Dallas, TX.

Friedman, T. L. (2000). *The Lexus and the olive tree: Understanding globalization.* New York, NY: Anchor Books.

Friedman, T. L. (2005). *The world is flat: A brief history of the twenty-first century.* New York: Farrar, Straus and Giroux.

Fromm, E. (1941). *Escape from freedom.* New York, NY: Farrar & Rinehart.

Frum, D. (2000). *How we got here: The '70s.* New York: Basic Books.

Fry, R. (2016). *Millennials match baby boomers as largest generation in U.S. electorate, but will they vote?* Washington, DC: Pew Research Center. Retrieved from http://www.pewresearch.org/fact-tank/2016/05/16/millennials-match-baby-boomers-as-largest-generation-in-u-s-electorate-but-will-they-vote/

Fry, R. (2019, June 20). *U.S. women near milestone in the college-educated labor force.* Retrieved from https://pewrsr.ch/2ZEVQB3

Fry, R., Passel, J., & Cohn, D. (2020, September 4). A majority of young adults in the U.S. live with their parents for the first time since the Great Depression. *Pew Research Center.* Retrieved from https://www.pewresearch.org/fact-tank/2020/09/04/a-majority-of-young-adults-in-the-u-s-live-with-their-parents-for-the-first-time-since-the-great-depression/

Frye, N. K., & Breaugh, J. A. (2004). Family-friendly policies, supervisor support, work–family conflict, family–work conflict, and satisfaction: A test of a conceptual model. *Journal of Business and Psychology, 19,* 197–220.

Fuchs, H. (2020, July 16). Government executes second federal death row prisoner in a week. *New York Times.* Retrieved from https://www.nytimes.com/2020/07/16/us/politics/wesley-ira-purkey-executed.html

Fuist, T. N., & Williams, R. H. (2019). "Let's call ourselves the super elite": Using the collective behavior tradition to analyze Trump's America. *Sociological Forum, 34,* 1132–1152.

Funk, C., & Kennedy, B. (2020). *How Americans see climate change and the environment in 7 charts*. Retrieved from https://www.pewresearch.org/fact-tank/2020/04/21/how-americans-see-climate-change-and-the-environment-in-7-charts/

Furnish, T. (2005). Beheading in the name of Islam. *Middle East Quarterly, 12*, 51–55. Retrieved from http://www.meforum.org/713/beheading-in-the-name-of-islam

Gainsborough, J. F. (2001). *Fenced off: The suburbanization of American politics*. Washington, DC: Georgetown University Press.

Gakidou, E., Cowling, K., Lozano, R., & Murray, C. J. L. (2010). Increased educational attainment and its impact on child mortality in 175 countries between 1970 and 2009: A systematic analysis. *The Lancet, 376*, 959–974.

Gale, R. P. (1986). Social movements and the state: The environmental movement, counter movement, and governmental agencies. *Sociological Perspectives, 29*, 202–240.

Gallaudet Research Institute. (2005). *A brief summary of estimates for the size of the deaf population in the USA based on available federal data and published research*. Retrieved from http://research.gallaudet.edu/Demographics/deaf-US.php

Gallaudet University. (2012). *Local and regional deaf populations*. Retrieved from http://libguides.gallaudet.edu/content.php?pid=119476&sid=1029190

Gamble, J. L., & Hess, J. J. (2012). Temperature and violent crime in Dallas, Texas: Relationships and implications of climate change. *Western Journal of Emergency Medicine, 13*, 239–246.

Gamson, J. (1991). Silence, death, and the invisible enemy: AIDS activism and social movement "newness". *Social Problems, 36*(4), 351–367.

Gamson, W. (1975). *The strategy of social protest*. Homewood, IL: Dorsey Press.

Gans, H. J. (1962a). Urbanism and suburbanism as ways of life. In A. Rose (Ed.), *Human behavior and social processes*. Boston: Houghton Mifflin.

Gans, H. J. (1962b). *The urban villagers: Group and class in the life of Italian-Americans*. New York: Free Press.

Gans, H. J. (1972). The positive functions of poverty. *American Journal of Sociology, 78*, 275–289.

Gao, G. (2016, July 7). *Biggest share of whites in U.S. are Boomers, but for minority groups it's Millennials or younger*. Washington, DC: Pew Research. Retrieved from http://www.pewresearch.org/fact-tank/2016/07/07/biggest-share-of-whites-in-u-s-are-boomers-but-for-minority-groups-its-millennials-or-younger/

Garcia, C., & Vanek Smith, S. (2020, February 24). *For richer or... richer* [Interview with Branko Milanovic]. *National Public Radio*. Retrieved from https://www.npr.org/2020/02/24/809050430/for-richer-or-richer

Garcia, S. B., & Guerra, P. L. (2004). Deconstructing deficit thinking: Working with educators to create more equitable learning environments. *Education and Urban Society, 36*, 150–168.

Garfinkel, H. (1963). A conception of, and experiments with, "trust" as a condition of stable concerted actions. In O. J. Harvey (Ed.), *Motivation and social interaction* (pp. 187–238). New York: Ronald Press.

Garfinkel, H. (1985). *Studies in ethnomethodology*. New York: Blackwell.

Gaskell, G., & Smith, P. (1981, August 20). The crowd in history. *New Society*, pp. 303–304.

Gates, G. (2017, January 11). *In U.S., More Adults Identifying as LGBT*. Retrieved from https://news.gallup.com/poll/201731/lgbt-identification-rises.aspx

Gates, G. J. (2011). *How many people are lesbian, gay, bisexual, and transgender?* Los Angeles, CA: The Williams Institute, University of California–Los Angeles. Retrieved from http://williamsinstitute.law.ucla.edu/wp-content/uploads/Gates-How-Many-People-LGBT-Apr-2011.pdf

Gatsiounis, I. (2008, March 20). In Thailand, pollution from shrimp farms threatens a fragile environment. *New York Times*. Retrieved from http://www.nytimes.com/2008/03/20/business/worldbusiness/20iht-rbogcoast.1.11278833.html?_r=0

Gauntlett, D. (2008). *Media, gender, and identity: An introduction* (2nd ed.). New York: Taylor & Francis.

Geertz, C. (1973). *The interpretation of cultures*. New York: Basic Books.

Geiger, A. W., & Gramlich, J. (2019, November 22). *6 facts about marijuana*. Pew Research Forum. Retrieved from https://www.pewresearch.org/fact-tank/2019/11/22/facts-about-marijuana/

Geiger, A. W., & Livingston, G. (2019, February 13). *8 facts about love and marriage in America*. Retrieved from https://www.pewresearch.org/fact-tank/2019/02/13/8-facts-about-love-and-marriage/

Gelb, A., & Denney, J. (2018, January 16). National prison rate continues to decline amid sentencing, re-entry reforms. *Pew Charitable Trusts*. Retrieved from https://www.pewtrusts.org/en/research-and-analysis/articles/2018/01/16/national-prison-rate-continues-to-decline-amid-sentencing-re-entry-reforms

Gellner, E. (1983). *Nations and nationalism*. Ithaca, NY: Cornell University Press.

Gerding, A., & Signorielli, N. (2014). Gender roles in tween television programming: A content analysis of two genres. *Sex Roles, 70*, 43–56.

Gerhardt, K. F. G. (1989). *The silent brotherhood: Inside America's racist underground*. New York, NY: Free Press.

Geronimus, A. (1992). The weathering hypothesis and the health of African-American women and infants: Evidence and speculations. *Ethnicity and Disease, 2*, 207–221.

Gershoff, E. T., & Grogan-Kaylor, A. (2016, June). Spanking and child outcomes: Old controversies and new meta-analyses. *Journal of Family Psychology, 30*, 453–469 http://dx.doi.org/https://doi.org/10.1037/fam0000191

Gerstel, N., & Gallagher, S. (1994). Caring for kith and kin: Gender, employment, and the privatization of care. *Social Problems, 41*, 519–539.

Ghanbarpour-Dizboni, A. (2020). ISIS. *The SAGE encyclopedia of the sociology of religion* (pp. 403–404). Thousand Oaks, CA: SAGE.

Ghosh, B. N. (2001). *Dependency theory revisited*. London: Ashgate.

Gibbs, N. (2009, October 14). The state of the American woman: What women want now. *Time*. Retrieved from http://www.time.com/time/specials/packages/article/0,28804,1930277_1930145_1930309,00.html

Giedd, J. N. (2004). Structural magnetic resonance imaging of the adolescent brain. *Annals of the New York Academy of Sciences, 1021*, 77–85.

Gieseler, C. (2018). Gender-reveal parties: Performing community identity in pink and blue. *Journal of Gender Studies, 27*(6), 661–671. https://doi.org/10.1080/09589236.2017.1287066

Giffords Law Center. (2020, August 1). *Gun violence statistics*. Retrieved from https://lawcenter.giffords.org/facts/gun-violence-statistics/

Gilbert, D. L. (2011). *The American class structure in an age of growing inequality* (8th ed.) Thousand Oaks, CA: Pine Forge.

Gilbert, S. (2017). The Movement of #metoo. *The Atlantic Monthly*. Retrieved from https://www.theatlantic.com/entertainment/archive/2017/10/the-movement-of-metoo/542979/

Gilbertson, G. A., & Gurak, D. T. (1993, June). Broadening the enclave debate: The labor market experiences of Dominican and Colombian men in New York City. *Sociological Forum, 8*, 205–220.

Gilchrist, Tracy E. (2020, January 30). Schitt's Creek's Emily Hampshire: Stevie, Dan Levy "changed my life.". Retrieved from https://www.advocate.com/television/2020/1/30/schitts-creeks-emily-hampshire-stevie-dan-levy-changed-my-life

Gilligan, C. (1982). *In a different voice: Psychological theory and women's development*. Cambridge, MA: Harvard University Press.

Gilligan, C., Ward, J. V., & Taylor, J. M. (Eds.). (1989). *Mapping the moral domain: A contribution of women's thinking to psychological theory and education*. Cambridge, MA: Harvard University Press.

Gilman, C. P. (1998/1898). *Women and economics: A study of the economic relation between men and women as a factor in social evolution*. Berkeley: University of California Press. Retrieved from http://ark.cdlib.org/ark:/13030/ft896nb5rd/

Gilman, C. P. (2006). *Women and economics: A study of the economic relation between men and women as a factor in social evolution*. New York, NY: Cosimo. (Original work published 1898)

Gitlin, T. (1980). *The whole world is watching: Mass media in the making and unmaking of the new left*. Berkeley: University of California Press.

GLAAD Accelerating Acceptance. (2020). Retrieved from GLADD and Harris Poll. (2018). *Accelerating acceptance 2018* [Survey report]. Retrieved from https://www.glaad.org/files/aa/2017_GLAAD_Accelerating_Acceptance.pdf

GLAAD Media Institute. (2020). *GLAAD studio responsibility index*. Retrieved from https://www.glaad.org/sites/default/files/GLAAD%202020%20Studio%20Responsibility%20Index.pdf

GLAAD Media Reference Guide. (2016, October). Retrieved from https://www.glaad.org/reference

GLAAD Studio Responsibility Index. (2020). Retrieved from https://www.glaad.org/sites/default/files/GLAAD%202020%20Studio%20Responsibility%20Index.pdf

Glass, A. (2019, October 21). FIFA Women's World Cup breaks viewership records. *Forbes*. Retrieved from https://www.forbes.com/sites/alanaglass/2019/10/21/fifa-womens-world-cup-breaks-viewership-records/#5ae94dc21884

Glassdoor. (2015). *Here's how much more CEOs earn than their employees*. Retrieved from https://www.glassdoor.com/blog/heres-ceos-earn-employee/

Glaze, L. E., & Herberman, E. J. (2013, December). *Correctional populations in the United States, 2012* (BJS Bulletin NCJ 243936). Washington, DC: U.S. Department of Justice, Bureau of Justice Statistics. Retrieved from http://www.bjs.gov/content/pub/pdf/cpus12.pdf

Glazer, N. (1992). The real world of education. *The Public Interest* (Winter), 57–75.

Glazer, N. (1997). *We are all multiculturalists now*. Cambridge, MA: Harvard University Press.

Glewwe, P. (1999). Why does mother's schooling raise child health in developing countries? Evidence from Morocco. *Journal of Human Resources, 34*, 124–159.

Gleyo, F. (2015, October 10). *AP has a robot journalist that writes a thousand articles per month*. Retrieved from http://www.techtimes.com/articles/93473/20151010/ap-has-a-robot-journalist-that-writes-a-thousand-articles-per-month.htm

Glock, C. Y., & Bellah, R. N. (1976). *The new religious consciousness*. Berkeley: University of California Press.

Glynn, S. (2019, May 10). *Breadwinning mothers continue to be the U.S. norm. Center for American Progress*. Retrieved from https://www.americanprogress.org/issues/women/reports/2019/05/10/469739/breadwinning-mothers-continue-u-s-norm/

Goffman, E. (1959). *The presentation of self in everyday life*. New York: Doubleday.

Goffman, E. (1961). *Asylums: Essays on the social situation of mental patients and other inmates*. Garden City, NY: Anchor Books.

Goffman, E. (1963a). *Behavior in public place*. New York: Free Press.

Goffman, E. (1963b). *Stigma: Notes on the management of spoiled identity*. Englewood Cliffs, NJ: Prentice Hall.

Goffman, E. (1967). *Interaction ritual: Essays on face to face behavior*. Garden City, NY: Anchor.

Goffman, E. (1972). *Relations in public: Microstudies of the public order*. New York: Harper & Row.

Goldberg, C. A. (2008). Introduction to Émile Durkheim's "Anti-Semitism and Social Crisis.". *Sociological Theory, 26*(4), 299–321. https://doi.org/10.1111/j.1467-9558.2008.00331_1.x

Goldhagen, D. J. (1997). *Hitler's willing executioners*. New York, NY: Vintage Books.

Goldin, C. (2014). A grand gender convergence: Its last chapter. *American Economic Review, 104*, 1091–1119.

Goldin, C., Katz, M. F., & Kuziemko, I. (2006). *The homecoming of American college women: The reversal of the college gender gap (Working Paper 12130)* . Cambridge, MA: National Bureau of Economic Research. Retrieved from http://faculty.smu.edu/millimet/classes/eco7321/papers/goldin%20et%20al.pdf

Goldmacher, S. (2020, October 28). The 2020 campaign is the most expensive ever (by a lot). *New York Times*. Retrieved from https://www.nytimes.com/2020/10/28/us/politics/2020-race-money.html

Goldrich-Rab, S. (2020, February 20). Housing and food insecurity affecting many college students, new data says. *Washington Post*. Retrieved from https://www.washingtonpost.com/education/2020/02/20/housing-food-insecurity-affecting-many-college-students-new-data-says/

Goldscheider, F. K., & Waite, L. J. (1991). *New families, no families? The transformation of the American home*. Berkeley: University of California Press.

Goldstein, J. S. (2011, September/October). World peace could be closer than you think. *Foreign Policy*, 53–56.

Goldstein, W. S. (2020). Marx, Karl and Engels, Friedrich. *The SAGE encyclopedia of the sociology of religion* (pp. 473–476). Thousand Oaks, CA: SAGE.

Goldstone, J. A. (2001). Towards a fourth generation of revolutionary theory. *Annual Review of Political Science*, *4*, 139–187.

Gonzalez, I. (2020, March 6). *The price Cuban women pay in caring for others*. Retrieved from https://havanatimes.org/features/the-price-cuban-women-pay-in-caring-for-others/

Goode, E. (2009). *Moral panics: The social construction of deviance*. Chichester, England: Wiley-Blackwell.

Gooden, A. M., & Gooden, M. A. (2001). Gender representation in notable children's picture books: 1995–1999. *Sex Roles*, *45*, 89–101.

Goodnough, A. (2016, January 29). Flint weighs scope of harm to children caused by lead in water. *New York Times*. Retrieved from http://www.nytimes.com/2016/01/30/us/flint-weighs-scope-of-harm-to-children-caused-by-lead-in-water.html

Goody, J. (1983). *The development of the family and marriage in Europe*. Cambridge: Cambridge University Press.

Gottfredson, M. R., & Hirschi, T. (2004). *A general theory of crime*. Stanford, CA: Stanford University Press. (Original work published 1990)

Gould, E., Mokhiber, Z., & Wolfe, J. (2017). *The class of 2018*. Economic Policy Institute. Retrieved from https://www.epi.org/files/pdf/147514.pdf

Gould, E., Mokhiber, Z., & Wolfe, J. (2018). *Class of 2018: College edition*. Economic Policy Institute. Retrieved from https://www.epi.org/publication/class-of-2018-college-edition/

Gould, E., Mokhiber, Z., & Wolfe, J. (2019). *Class of 2019: College edition*. Economic Policy Institute. Economic Policy Institute Retrieved from https://www.epi.org/publication/class-of-2019-college-edition/

Gourevitch, P. (1999). *We wish to inform you that tomorrow we will be killed with our families: Stories from Rwanda*. New York, NY: Farrar, Straus and Giroux.

GovTrack. (2018, August 16). *A bill would make English the official language*. Retrieved from https://govtrackinsider.com/a-bill-would-make-english-the-official-language-975fa22ea0f2

Gracey, H. L. (1991). Learning the student role: Kindergarten as academic boot camp. In J. M. Henslin (Ed.), *Down to earth sociology: Introductory readings* (6th ed.). New York: Free Press.

Graf, N. (2018, April 18). *A majority of U.S. teens fear a shooting could happen at their school and most parents share their concern*. Retrieved from https://www.pewresearch.org/fact-tank/2018/04/18/a-majority-of-u-s-teens-fear-a-shooting-could-happen-at-their-school-and-most-parents-share-their-concern/

Graf, N. (2019, November 6). Key findings on marriage and cohabitation in the U.S. Pew Research Center. Retrieved from https://www.pewresearch.org/fact-tank/2019/11/06/key-findings-on-marriage-and-cohabitation-in-the-u-s/

Graham, R. (2020, August 31). Liberty will investigate university's operations under Jerry Falwell Jr. *New York Times (online)*. Retrieved from https://www.nytimes.com/2020/08/31/us/liberty-jerry-falwell-investigation.html

Gramlich, J. (2017, March 1). *Most violent and property crimes in the U.S. go unsolved*. Retrieved from https://www.pewresearch.org/fact-tank/2017/03/01/most-violent-and-property-crimes-in-the-u-s-go-unsolved/

Gramlich, J. (2018, May 30). *As fatal overdoses rise, many Americans see drug addiction as a major problem in their community*. Pew Research Center. Retrieved from https://www.pewresearch.org/fact-tank/2018/05/30/as-fatal-overdoses-rise-many-americans-see-drug-addiction-as-a-major-problem-in-their-community/

Gramlich, J. (2019, October 17). *5 facts about crime in the U.S.* Retrieved from https://www.pewresearch.org/fact-tank/2019/10/17/facts-about-crime-in-the-u-s/

Gramlich, J. (2020, May 6). *Black imprisonment rate in the U.S. has fallen by a third since 2006*. Retrieved from https://www.pewresearch.org/fact-tank/2020/05/06/share-of-black-white-hispanic-americans-in-prison-2018-vs-2006/

Gramsci, A. (1971). *Selections from the prison notebooks* (Q. Hoare & G. N. Smith, Trans.). London, England: Lawrence & Wishart.

Great Nonprofits. (2016). *Islamic Relief USA*. Retrieved from http://greatnonprofits.org/org/islamic-relief-usa

Green, A., & Robeson, W. W. (2020, July 31). *Closing university child care centers hurts both student parents and future educators*. Retrieved from https://iwpr.org/media/press-hits/closing-university-child-care-centers-hurts-both-student-parents-and-future-educators

Greenberg, A. (2015, June 2). *The dark web drug lords who got away*. Retrieved from https://www.wired.com/2015/06/dark-web-drug-lords-got-away/

Greenberg, K. S. (2003). *Nat Turner: A slave rebellion in history and memory*. New York, NY: Oxford University Press.

Greenburg, C. A. (2008). Introduction to Émile Durkheim's "Anti-Semitism and social crisis.". *Sociological Theory*, *26*(4), 299–323.

Greenwood, J., Guner, N., Kocharkov, G., & Santos, C. (2014). Marry your like: Assortative mating and income inequality. *American Economic Review*, *104*(5), 348–353.

Gregg, M. (2014, September 25). *Five ways cyberterrorists could target the U.S.* Retrieved from http://www.huffingtonpost.com/michael-gregg/five-ways-cyber terrorists_b_5874860.html

Gressel, C. M., Rashed, T., Maciuika, L. A., Sheshadri, S., Coley, C., Kongeseri, S., & Bhavani, R. R. (2020). Vulnerability mapping: A conceptual framework towards a context-based approach to women's empowerment. *World Development Perspectives*, *20*, 100245. https://doi.org/10.1016/j.wdp.2020.100245

Grey, S. (2006). *Ghost plane: The true story of the CIA torture program*. New York, NY: Macmillan.

Griffin, S. (1978). *Woman and nature: The roaring inside her*. New York: Harper & Row.

Griffin, S. (1979). *Rape, the power of consciousness*. New York: Harper & Row.

Griffin, S. (1981). *Pornography as silence: Culture's revenge against nature*. New York: Harper & Row.

Groom, N., & Dobuzinskis, A. (2019, February 25). *Oscars not so white? Academy Awards winners see big shift.* Retrieved from https://www.reuters.com/article/us-awards-oscars-diversity-idUSKCN1QE0N4

Gross, T. (2019, August 14). On "pose," Janet Mock tells the stories she craved as a young trans person. *Fresh Air.* Retrieved from https://www.npr.org/2019/08/14/750931179/on-pose-janet-mock-tells-the-stories-she-craved-as-a-young-trans-person

Gross, T. (2020, May 14). *What the 1918 flu pandemic can tell us about the COVID-19 crisis.* Retrieved from https://www.npr.org/2020/05/14/855986938/what-the-1918-flu-pandemic-can-tell-us-about-the-covid-19-crisis

Grusky, O., Bonacich, P., & Webster, C. (1995). The coalition structure of the four person family. *Current Research in Social Psychology, 2,* 16–28.

Gubrium, J. F. (1975/1995). *Living and dying at Murray Manor.* Charlottesville: University of Virginia Press.

Guilmoto, C. Z. (2011). *Skewed sex ratios at birth and future marriage squeeze in China and India, 2005–2100* (Working Paper 15). Paris: Centre Population & Développement.

Gump, L. S., Baker, R. C., & Roll, S. (2000). Cultural and gender differences in moral judgment: A study of Mexican Americans and Anglo-Americans. *Hispanic Journal of Behavioral Sciences, 22,* 78–93.

Gunnell, J. J., & Ceci, S. J. (2010). When emotionality trumps reason: A study of individual processing style and juror bias. *Behavioral Sciences & the Law, 28,* 850–877.

Gunnoe, M. L. (1997). Toward a developmental contextual model of the effects of parental spanking on children's aggression. *Archives of Pediatrics and Adolescence, 151,* 768–775.

Guo, J. (2016, January 25). Researchers have found a major problem with 'The Little Mermaid' and other Disney movies. *Washington Post.* Retrieved from https://www.washingtonpost.com/news/wonk/wp/2016/01/25/researchers-have-dis-covered-a-major-problem-with-the-little-mermaid-and-other-disney-movies/

Gurdus, L. (2019, March 9). *Female CEOs are scarce, but history shows they can produce huge returns.* Retrieved from https://www.cnbc.com/2019/03/09/female-ceos-are-scarce-but-history-shows-they-can-produce-big-returns.html

Gutman, R., & Rieff, D. (1999). *Crimes of war: What the public should know.* New York: Norton.

Guttmacher Institute. (2020, July 1). *Substance use during pregnancy.* Retrieved from https://www.guttmacher.org/state-policy/explore/substance-use-during-pregnancy

Haberman, C. (2014, March 9). The trial that unleashed hysteria over child abuse. *New York Times.* Retrieved from http://www.nytimes.com/2014/03/10/us/the-trial-that-unleashed-hysteria-over-child-abuse.html?_r=0

Habermas, J. (1976). *Legitimation crisis.* London: Heinemann.

Habermas, J. (1989). *The structural transformation of the public sphere: An inquiry into a category of bourgeois society.* Cambridge, MA: MIT Press. (Original work published 1962)

Hackett, C. (2017, November 29). *5 facts about the Muslim population.* Retrieved from https://www.pewresearch.org/fact-tank/2017/11/29/5-facts-about-the-muslim-population-in-europe/

Hackett, C., & McClendon, D. (2017, April 5). *Christians remain world's largest religious group, but they are declining in Europe.* Retrieved from https://www.pewresearch.org/fact-tank/2017/04/05/christians-remain-worlds-largest-religious-group-but-they-are-declining-in-europe/

Hadden, J. K. (1993). *Religion and the social order: The handbook on cults and sects in America.* Bingley, England: Emerald Group.

Hadden, J. K. (2006). *New religious movements.* Hartford Institute for Religion Research. Retrieved from http://hirr.hartsem.edu/denom/new_religious_movements.html

Haggerty, R. A. (1991). *Dominican Republic and Haiti: Country studies* (2nd ed.). Washington, DC: Federal Research Division, Library of Congress.

Haj-Yahia, M. M., & Cohen, H. C. (2009). On the lived experience of battered women residing in shelters. *Journal of Family Violence, 24,* 95–109.

Hales, C. M., Carroll, M. D., Fryar, C. D., & Ogden, C. L. (2017). *Prevalence of obesity among adults and youth: United States, 2015–2016.* NCHS data brief, no 288. Hyattsville, MD: National Center for Health Statistics. Retrieved from https://www.cdc.gov/nchs/data/databriefs/db288.pdf

Haley, A., & Malcolm, X. (1964). *The autobiography of Malcolm X.* New York: Ballantine Books.

Hall, E. (1966). *The hidden dimension.* Garden City, NY: Doubleday.

Hall, E. (1973). *The silent language.* New York: Doubleday.

Hall, P. M. (2003). Interactionism, social organization, and social processes: Looking back there, reflecting now here, and moving ahead then. *Symbolic Interaction, 26,* 33–55.

Halpern, A. (2020, September 2). The pandemic is shifting how students study abroad. *Conde Naste Traveler online.* Retrieved from https://www.cntraveler.com/story/the-pandemic-is-shifting-how-students-study-abroad

Halpern, H. P., & Perry-Jenkins, M. (2016). Parents' gender ideology and gendered behavior as predictors of children's gender-role attitudes: A longitudinal exploration. *Sex Roles, 74*(11–12), 527–542.

Hamblin, J. (2016, June 16). Toxic masculinity and murder. *The Atlantic.* Retrieved from http://www.theatlantic.com/health/archive/2016/06/toxic-masculinity-and-mass-murder/486983/

Hamel, L., Rao, M., Levitt, L., Claxton, G., Cox, C., Pollitz, K., & Brodie, M. (2014). *Survey of non-group health insurance enrollees: A first look at people buying their own health insurance following implementation of the Affordable Care Act.* Menlo Park, CA: Kaiser Family Foundation. Retrieved from http://kaiserfamilyfoundation.files.wordpress.com/2014/06/survey-of-non-group-health-insurance-enrollees-findings-final1.pdf

Hamermesh, D. S. (2011). *Beauty pays: Why attractive people are more successful.* Princeton, NJ: Princeton University Press.

Hamermesh, D. S., & Parker, A. (2005). Beauty in the classroom: Professorial pulchritude and putative pedagogical productivity. *Economics of Education Review, 24,* 369–376.

Hamm, M. S. (2002). Apocalyptic violence: The seduction of terrorist subcultures. *Theoretical Criminology, 8*(3), 323–339. http://dx.doi.org/10.1177/1362480604044612

Hammond, P. E. (1992). *Religion and personal autonomy: The third disestablishment in America.* Columbia: University of South Carolina Press.

Handlin, O. (1991). *Boston's immigrants, 1790–1880: A study in acculturation.* Cambridge, MA: Belknap Press.

Haney, C., Banks, W. C., & Zimbardo, P. G. (1973). Interpersonal dynamics in a simulated prison. *International Journal of Criminology and Penology, 1,* 69–97.

Haninger, K., & Thompson, K. M. (2004). Content and ratings of teen-rated video games. *Journal of the American Medical Association, 291*(7), 856–865. http://dx.doi.org/10.1001/jama.291.7.856

Hannon, E. (2012, April 8). India's census: Lots of cellphones, too few toilets. *Weekend Edition,* NPR. Retrieved from http://www.npr.org/2012/04/08/150133880/indias-census-lots-of-cellphones-too-few-toilets

Hanson, P. (2003). *An economic history of the USSR, 1945–1991.* New York: Longman.

Hardoon, D., Fuentes-Nieva, R., & Ayele, S. (2016, January 18). *An economy for the 1%: How privilege and power in the economy drive extreme inequality and how this can be stopped. Oxfam International.* Retrieved from http://policy-practice.oxfam.org.uk/publications/an-economy-for-the-1-how-privilege-and-power-in-the-economy-drive-extreme-inequ-592643

Harlan, C., & Nakashima, E. (2011, August 29). Suspected North Korean cyber attack on a bank raises fears for S. Korea, allies. *Washington Post.* Retrieved from https://www.washington-post.com/world/national-security/suspected-north-korean-cyber-attack-on-a-bank-raises-fears-for-s-korea-allies/2011/08/07/gIQAvWwIoJ_story.html

Harper, B. (2000). Beauty, stature and the labour market: A British cohort study. *Oxford Bulletin of Economics and Statistics, 62,* 771–800.

Harring, H. A., Montgomery, K., & Hardin, J. (2011). Perceptions of body weight, weight, weight management strategies, and depressive symptoms among US college students. *Journal of American College Health, 59,* 43–50.

Harrington, M. (1963). *The other America: Poverty in the United States.* New York: Simon & Schuster.

Harris, J. R. (2009). *The nurture assumption: Why children turn out the way they do* (2nd ed.). New York: Free Press.

The Harris Poll. (2019). *Accelerating acceptance. A survey of American acceptance and attitudes toward LGBTQ Americans.* Executive Summary. Retrieved from https://www.glaad.org/sites/default/files/Accelerating%20Acceptance%202019.pdf

Hartford Institute for Religious Research. (n.d.). *Mega churches.* Retrieved from http://hirr.hartsem.edu/megachurch/megachurches.html

Harth, E. (2001). *Last witnesses: Reflections on the wartime internment of Japanese Americans.* London, England: Palgrave Macmillan.

Hartmann, H. (1984). The unhappy union of Marxism and feminism: Toward a more progressive union. In A. M. Jaggar & P. S. Rothenberg (Eds.), *Feminist frameworks: Alternative theoretical accounts of the relations between women and men* (2nd ed., pp. 172–188). New York: McGraw- Hill.

Harvard Medical School. (2010, July). Marriage and men's health. *Harvard Men's Health Watch Newsletter.* Retrieved from http://www.health.harvard.edu/newsletters/Harvard_Mens_Health_Watch/2010/July/marriage-and-mens-health

Harvard Medical School. (2019, June 5). *Marriage and men's health. Harvard Men's Health Watch.* Retrieved from https://www.health.harvard.edu/mens-health/marriage-and-mens-health

Hattery, A. J. (2001). *Families in crisis: Men and women's perceptions of violence in partner relationships.* Blacksburg, VA: Southern Sociological Society.

Hatzenbuehler, P. L., Gillespie, J. M., & O'Neil, C. E. (2012, April). Does healthy food cost more in poor neighborhoods? An analysis of retail food cost and spatial competition. *Agricultural and Resource Economics Review, 41,* 43–56.

Hausmann, R., Tyson, L. D., & Zahidi, S. (2011). *The global gender gap report 2011.* Geneva, Switzerland: The World Economic Forum.

Haveman, H. A., & Wetts, R. (2019). Contemporary organizational theory. *Sociology Compass, 13*(3). https://doi.org/10.1111/soc4.12664

Healthy People, Office of Disease Prevention and Health Promotion. (2020). *Social determinants of health.* Retrieved from https://www.healthypeople.gov/2020/topics-objectives/topic/social-determinants-of-health

Healy, M. (2020, August 18). Domestic violence rose during lockdown—and injuries are dramatically more severe, study finds. *Los Angeles Times.* Retrieved from https://www.latimes.com/science/story/2020-08-18/intimate-partner-violence-spiked-80-after-pandemic-lockdown-began

Hedwig, L. (2011). Inequality as an explanation for obesity in the United States. *Sociology Compass, 5,* 215–232.

Hegewisch, A., & Liepmann, H. (2012). *Fact sheet: The gender wage gap by occupation.* Washington, DC: Institute for Women's Policy Research. Retrieved from http://www.iwpr.org/publications/pubs/the-gender-wage-gap-by-occupation

Hegewisch, A., & Tesfaselassie, A. (2019, April 2). *Fact sheet: The gender wage gap by occupation 2018 and by race and ethnicity.* Institute for Women's Policy Research. Retrieved from https://iwpr.org/iwpr-issues/employment-and-earnings/the-gender-wage-gap-by-occupation-2018/

Heggeness, M. L., & Fields, J. M. (2020, August 18). *Working moms bear brunt of home schooling while working during COVID-19.* United States Census Bureau. Retrieved from https://www.census.gov/library/stories/2020/08/parents-juggle-work-and-child-care-during-pandemic.html

Hein, Avi. (n.d.). *Women in Judaism: A history of women's ordination as rabbis. The Jewish Virtual Library.* Retrieved from https://www.jewishvirtuallibrary.org/a-history-of-women-s-ordination-as-rabbis

Held, D. (1989). *Political theory and the modern state.* Stanford, CA: Stanford University Press.

Heller, K. (2016, December 28). Meet the elite group of authors who sell 100 million books—or 350 million. *Independent.* Retrieved from https://www.independent.co.uk/arts-entertainment/books/meet-the-elite-group-of-authors-who-sell-100-million-books-or-350-million-paolo-coelho-stephen-king-a7499096.html

Henry, T. (2010, November 15). Even short-term poverty can hurt kids' health. *CNNHealth.* Retrieved from http://thechart.blogs.cnn.com/2010/11/15/even-short-term-poverty-can-hurt-kids-health

Heppner, C. M. (1992). *Seeds of disquiet: One deaf woman's experience.* Washington, DC: Gallaudet University Press.

Herbert, B. (2010, March 5). Cops vs. kids. *New York Times*. Retrieved from http://www.nytimes.com/2010/03/06/opinion/06herbert.html?_r=0

Heritage, J., & Greatbatch, D. (1991). On the institutional character of institutional talk: The case of news interviews. In D. H. Zimmerman & D. Boden (Eds.), *Talk and social structure* (pp. 93–137). Cambridge, England: Polity Press.

Hersey, P., Blanchard, K., & Natemeyer, W. (1987). *Situational leadership, perception, and the use of power*. Escondido, CA: Leadership Studies.

Hershberg, R. M. (2018). Consejos as a family process in transnational and mixed-status Mayan families. *Journal of Marriage and Family 80*(2), 334–348. Retrieved from https://doi.org/10.1111/jomf.12452

Hess, H. (1973). *Mafia and mafiosi: The structure of power*. Farnborough, England: Saxon House.

Hesse-Biber, S. (1997). *Am I thin enough yet? The cult of thinness and the commercialization of identity*. New York: Oxford University Press.

Hexham, I., & Poewe, K. (1997). *New religions as global cultures: Making the human sacred*. Boulder, CO: Westview Press.

Hill, L. E., & Johnson, H. P. (2002). *How fertility changes across immigrant generations*. San Francisco: Public Policy Institute of California. Retrieved from http://www.ppic.org/content/pubs/rb/RB_402LHRB.pdf

Hillin, T. (2016, May 24). *Facebook apologized after fat-shaming a model? but the damage was already done*. Retrieved from http://fusion.net/story/306275/facebook-apologized-after-fat-shaming-model-tess-holliday/

Hinchliffe, E. (2020, May 18). The number of female CEOs in the Fortune 500 hits an all-time record. *Fortune*. Retrieved from https://fortune.com/2020/05/18/women-ceos-fortune-500-2020/

Hine, T. (2000). *The rise and fall of the American teenager*. New York: Bard/Avon.

Hinsley, F. H. (1986). *Sovereignty* (2nd ed.). Cambridge, England: Cambridge University Press.

Hirschi, T. (1969). *Causes of delinquency*. Berkeley: University of California Press.

Hirschi, T. (2004). Self-control and crime. In R. F. Baumeister & K. D. Vohs (Eds.), *Handbook of self-regulation: Research, theory, and applications* (pp. 537–552). New York: Guilford.

Ho, C. (1993). The internationalization of kinship and the feminization of Caribbean migration: The case of Afro-Trinidadian immigrants in Los Angeles. *Human Organization*, *52*, 32–40.

Hobbs, F., & Stoops, N. (2002). *Demographic trends in the 20th century* (Census 2000 Special Report CENSR-4). Washington, DC: U.S. Census Bureau. Retrieved from http://www.census.gov/prod/2002pubs/censr-4.pdf

Hochschild, A. (2001a). The nanny chain. *The American Prospect*. Retrieved from http://prospect.org/article/nanny-chain.

Hochschild, A. (2001b). *The time bind: When work becomes home and home becomes work*. New York: Holt.

Hochschild, A. R. (2003). *The managed heart: Commercialization of human feeling*. Berkeley: University of California Press.

Hoffman, B. (2006). *Inside terrorism*. New York: Columbia University Press.

Holbrook, A. L., & Krosnick, J. A., (2009). Social desirability bias in voter turnout reports: Tests using the item count technique. *Public Opinion Quarterly*, *74*(1), pp. 37–67.

Hollander, M. M. (2015). The repertoire of resistance: Non-compliance with directives in Milgram's "obedience" experiments. *British Journal of Social Psychology*, *54*(3), 425–444.

Hollin, C. (2019). Opportunity theory. In E. McLaughlin (Ed.), *The SAGE dictionary of criminology* (4th ed.). Thousand Oaks, CA: SAGE.

Homans, C. (2011, September/October). Anthropology of an idea: War games. *Foreign Policy*, pp. 30–31.

Hong, S. (2016). Representative bureaucracy, organizational integrity, and citizen coproduction: Does an increase in police ethnic representativeness reduce crime? *Journal of Policy Analysis and Management*, *35*(1), 11–33.

Hoops, J. F. (2020), De-essentializing race through dialectic analysis of *The Big Sick*. In J. D. Hamlet (Ed.), *Films as rhetorical texts: Cultivating discussion about race, racism, and race relations*. Lanham, MD: Lexington Books.

Hooton, E. A. (1939). *The American criminal: An anthropological study*. Cambridge, MA: Harvard University Press.

Hopkins, D. J. (2009). No more Wilder effect, never a Whitman effect: When and why polls mislead about black and female candidates. *Journal of Politics*, *71*, 769–781.

Hopper, R. (1991). Hold the phone. In D. H. Zimmerman & D. Boden (Eds.), *Talk and social structure* (pp. 217–231). Cambridge: Polity Press.

Horan, P. M., & Hargis, P. G. (1991). Children's work and schooling in the late nineteenth century family economy. *American Sociological Review*, *56*, 583–596.

Horkheimer, M. (1947). *The eclipse of reason*. Oxford: Oxford University Press.

Horkheimer, M., & Adorno, T. (2007). *Dialectic of Enlightenment*. Stanford, CA: Stanford University Press. (Original work published 1944)

Horowitz, J. M., & Fetterolf, J. (2020, April 30). *Worldwide optimism about future of gender equality, even as many see advantages for men*. Retrieved from https://www.pewresearch.org/global/2020/04/30/worldwide-optimism-about-future-of-gender-equality-even-as-many-see-advantages-for-men/

Horowitz, J. M., Graf, N., & Livingston, G. (2019, November 6). *Marriage and cohabitation in the U.S. Pew Research*. Retrieved from https://www.pewsocialtrends.org/2019/11/06/marriage-and-cohabitation-in-the-u-s/

Horowitz, J. M., & Igielnik, R. (2020, July 7). *A century after women gained the right to vote, majority of Americans see work to do on gender equality*. Retrieved from https://www.pewsocialtrends.org/2020/07/07/a-century-after-women-gained-the-right-to-vote-majority-of-americans-see-work-to-do-on-gender-equality/

Hosken, F. (1993). City. In *Academic American encyclopedia*. Danbury, CT: Grolier.

Houghton, S., Hunter, S. C., Rosenberg, M., Wood, L., Zadow, C., Martin, K., & Shilton, T. (2015). Virtually impossible: Limiting Australian children's and adolescents' daily screen based media use. *BMC Public Health*. doi:10.1186/1471-2458-15-5

Howarth, C. (2006). Race as stigma: Positioning the stigmatized as agents, not objects. *Journal of Community and Applied Social Psychology, 16*(6), 442–451.

Hoxby, C., & Avery, C. (2013). The missing "one-offs": The hidden supply of high-achieving, low-income students. *Brookings Papers on Economic Activity*, pp. 1–50.

Hu, K. (2020, August 27). *Op-ed: The real affirmative action dominating admission to elite colleges benefits privileged white kids*. Retrieved from https://www.latimes.com/opinion/story/2020-08-27/affirmative-action-yale-harvard-admissions-legacies

Huber, J. (2006). Comparative gender stratification. In J. S. Chafetz (Ed.), *Handbook of the sociology of gender* (pp. 65–80). New York, NY: Springer.

Huddleston, T., Jr. (2020, July 21). *Jeff Bezos added $13 billion to his net worth in one day?that's a record*. Retrieved from https://www.cnbc.com/2020/07/21/bezos-record-multibillion-dollar-net-worth-gain-bloomberg.html

Huecker, M., & Smock, W. (2020, June 26). *Domestic violence*. Retrieved from https://www.ncbi.nlm.nih.gov/books/NBK499891/

Huelsman, M. (2015, May 19). *The debt divide: The racial and class bias behind the "new normal" of student borrowing*. Retrieved from http://www.demos.org/publication/debt-divide-racial-and-class-bias-behind-new-normal-student-borrowing

Huffman, M. L., & Torres, L. (2002). It's not only "who you know" that matters: Gender, personal contacts, and job lead quality. *Gender & Society, 16*, 793–813.

Hughes, C., & Haynes, D. (2020, August 21). *Lori Loughlin, Mossimo Giannulli get prison in college scandal*. Retrieved from https://www.upi.com/Top_News/US/2020/08/21/Lori-Loughlin-Mossimo-Giannulli-get-prison-in-college-scandal/2361598032024/

Human Rights Campaign. (2020, August 27). *How HIV impacts LGBTQ people*. Retrieved from https://www.hrc.org/resources/hrc-issue-brief-hiv-aids-and-the-lgbt-community?utm_source=GS&utm_medium=AD&utm_campaign=BPI-HRC-Grant&utm_content=441762699074&utm_term=hiv%20stigma&gclid=EAIaIQobChMI0eO_7Ki86wIVMP_jBx2DhgBhEAAYAyAAEgKgPPD_BwE

Human Rights Campaign Foundation. (2020). *LGBTQ youth are living in crisis: Key findings from HRC foundation analysis of CDC data*. Retrieved from https://hrc-prod-requests.s3-us-west-2.amazonaws.com/ProjectThrive-YRBSData-Statement-100920.pdf?mtime=20201009165824&focal=none

Human Rights Watch. (n.d.). *Syria. Events of 2019. Human Rights Watch World report*. Retrieved from https://www.hrw.org/world-report/2020/country-chapters/syria#

Hunt, D., & Ramón, A-C. (2020, February 6). *UCLA Hollywood diversity report*. Retrieved from https://socialsciences.ucla.edu/wp-content/uploads/2020/02/UCLA-Hollywood-Diversity-Report-2020-Film-2-6-2020.pdf

Hunter, J. D. (1987). *Evangelicalism: The coming generation*. Hutchinson, KS: de Wit Books.

Hunter, J. D. (1992). *Culture wars: The struggle to control the family, art, education, law, and politics in America*. New York: Basic Books.

Hutcheon, D. (1999). *Building character and structure*. Westport, CT: Praeger.

Hutchinson, A. (2016, March 18). *Here's why Twitter is so important, to everyone*. Retrieved from http://www.socialmediatoday.com/social-networks/heres-why-twitter-so-important-everyone

Hvistendahl, M. (2011). *Unnatural selection: Choosing boys over girls, and the consequences of a world full of men*. New York: Public Affairs.

Hyman, H. H. (1942). The psychology of status. *Archives of Psychology, 38*, 147–165.

Ihab, M., Benson, A. F., Luben, T. J., Sacks, J. D., & Richmond-Bryant, J. (2018, April 1). Disparities in distribution of particulate matter emission sources by race and poverty status. *American Journal of Public Health, 108*(4), 480–485.

Immerwahr, D. (2007). Caste or colony? Indianizing race in the United States. *Modern Intellectual History, 4*, 275–301.

Ingraham, C. (1999). *White weddings: Romancing heterosexuality in popular culture*. New York: Routledge.

Institute for Economics and Peace. (2015). *Global terrorism index*. Retrieved from http://static.visionofhumanity.org/sites/default/files/2015%20Global%20Terrorism%20Index%20Report_2.pdf

Institute for Health Metrics and Evaluation. (2021, March 25). *On gun violence the U.S. is an outlier*. Retrieved from http://www.healthdata.org/acting-data/gun-violence-united-states-outlier

Institute for Women's Policy Research. (2014, January). *Pay secrecy and wage discrimination*. Retrieved from https://iwpr.org/wp-content/uploads/2020/09/Q016.pdf

Institute for Women's Policy Research. (2020a, September). *Same gap, different year. The gender wage gap: 2019 earnings differences by gender, race, and ethnicity*. Retrieved from https://iwpr.org/wp-content/uploads/2020/09/Gender-Wage-Gap-Fact-Sheet-2.pdf

Institute for Women's Policy Research. (2020b, October). *Latinas projected to reach equal pay in 2220*. Retrieved from https://iwpr.org/wp-content/uploads/2020/10/Latina-Women-Equal-Pay-Day-Policy-Brief.pdf

Institute of International Education. (2010, November). *Study abroad by U.S. students slowed in 2008/09 with more students going to less traditional destinations* (press release). Retrieved from http://www.iie.org/Who-We-Are/News-and-Events/Press-Center/Press-Releases/2010/2010-11-15-Open-Doors-US-Study-Abroad

Institute of International Education. (2017). *International students*. Retrieved from https://www.iie.org/Research-and-Insights/Open-Doors/Data/International-Students.

International Labour Organization. (2014). Domestic workers: Making decent work a reality for domestic workers world wide. Retrieved from http://www.ilo.org/global/topics/domestic-workers/lang-en/index.htm

International Telecommunication Union. (n.d.). *World Telecommunication/ICT development report and database*. Retrieved from https://data.worldbank.org/indicator/IT.NET.USER.ZS

Isidore, C. (2012, September 6). 3 answers to the auto bailout debate. *CNNMoney*. Retrieved from http://money.cnn.com/2012/09/06/autos/auto-bailout

Jackson, K. T. (1985). *Crabgrass frontier: The suburbanization of America*. New York: Oxford University Press.

Jaffee, S., & Hyde, J. (2000). Gender differences in moral orientation: A meta analysis. *Psychological Bulletin, 126*, 703–726.

Jaggar, A. M. (1983). *Feminist politics and human nature.* Totowa, NJ: Rowman & Allanheld.

Jain, R. (2020, February 19). *India to have 966 million mobile users by 2023, but less than 5% will have 5G: Report. Business Insider.* Retrieved from https://www.businessinsider.in/tech/mobile/news/india-to-have-966-million-mobile-users-by-2023-but-less-than-5-will-have-5g-report/articleshow/74202680.cms

Jancelewicz, C. (2020, February 9). *Oscars 2020 winners: "Parasite" makes Academy Awards history with big sweep.* Retrieved from https://globalnews.ca/news/6522353/2020-oscars-winners-academy-awards/

Janis, I. L. (1972). *Victims of groupthink.* Boston: Houghton Mifflin.

Janis, I. L. (1989). *Crucial decisions: Leadership in policy making and crisis management.* New York: Free Press.

Janis, I. L., & Mann, L. (1977). *Decision making: A psychological analysis of conflict, choice, and commitment.* New York: Free Press.

Jargowsky, P. (2015, August 7). *Architecture of segregation: Civil unrest, the concentration of poverty, and public policy.* Retrieved from https://tcf.org/content/report/architecture-of-segregation/

Jenkins, C. D. (1983). Social environment and cancer mortality in men. *New England Journal of Medicine, 308,* 395–408.

Jerrim, J., & Macmillan, L. (2015). Income inequality, intergenerational mobility, and the Great Gatsby curve: Is education the key? *Social Forces, 94*(2), 505–505. Retrieved from https://doi.org/10.1093/sf/sov075

Johnson, J. (2011, November 27). College administrators worry that use of prescription stimulants is increasing. *Washington Post.* Retrieved from http://articles.washingtonpost.com/2011-11-27/local/35281941_1_prescription-drugs-study-drugs-prescription-stimulants

Johnson, J. M., & Ferraro, K. J. (1984). The victimized self: The case of battered women. In J. A. Kotarba & A. Fontana (Eds.), *The existential self in society* (pp. 119–130). Chicago: University of Chicago Press.

Johnson, M. (2013, January 23). *The history of Twitter.* Retrieved from http://socialnomics.net/2013/01/23/the-history-of-twitter/

Jones, J. (2020, January 8). *Average car payment/loan statistics.* Lending Tree. Retrieved from https://www.lendingtree.com/auto/debt-statistics/

Jones, J. M. (2017, June 22). In U.S., 10.2% of LGBT adults now married to same-sex spouse. *Gallup.* Accessed at https://news.gallup.com/poll/212702/lgbt-adults-married-sex-spouse.aspx?utm_source=alert&utm_medium=email&utm_content=morelink&utm_campaign=syndication.

Jones, N. A., & Bullock, J. (2012). *The two or more races population: 2010 (Census Brief C2010BR-13).* Washington, DC: U.S. Census Bureau. Retrieved from http://www.census.gov/prod/cen2010/briefs/c2010br-13.pdf

Jones, R. (2019, April 26). Apartheid ended 29 years ago. How has South Africa changed? *National Geographic* Online. Retrieved from https://www.nationalgeographic.com/culture/2019/04/how-south-africa-changed-since-apartheid-born-free-generation/#close

Jones, R. P., & Cox D. (2015). *How race and religion shape millennial attitudes on sexuality and reproductive health.* Washington, DC: Public Religion Research Institute. Retrieved from http://www.prri.org/wp-content/uploads/2015/03/PRRI-Millennials-Web-FINAL.pdf

Jones, S. G., Doxsee, C., Harrington, N., Hwang, G., & Suber, J. (2020, October). *The war comes home: The evolution of domestic terrorism in the United States.* Center for Strategic and International Studies. Retrieved from https://www.csis.org/analysis/war-comes-home-evolution-domestic-terrorism-united-states

Jordan, M. (1992, January 9). Big city schools become more segregated in the 1980s, a study says. *Washington Post,* p. A3.

Jordan, M., & Clement, S. (2018, April 6). Rallying nation: In reaction to Trump, millions of Americans are joining protests and getting political. *Washington Post.* Retrieved from https://www.washingtonpost.com/news/national/wp/2018/04/06/feature/in-reaction-to-trump-millions-of-americans-are-joining-protests-and-getting-political/?utm_term=.4450c306495e

Josephson Institute Center for Youth Ethics. (2012). *Report card on the ethics of American youth.* Retrieved from http://charactercounts.org/programs/reportcard/2012/index.html

Josephson Institute of Ethics. (2009). *A study of values and behavior concerning integrity: The impact of age, cynicism and high school character.* Los Angeles: Author.

Juergensmeyer, M. (1995). The social significance of Radhasoami. In D. Lorenzen (Ed.), *Bhakti religion in North India: Community identity and political action* (pp. 67–89). Albany: State University of New York Press.

Junco, R. (2012). Too much face and not enough books: The relationship between multiple indices of Facebook use and academic performance. *Computers in Human Behavior, 28,* 187–198.

Junn, J., Wolbrecht, C., et al. (2020, August 18). *Justice demands the vote: 100 years of women's suffrage.* Retrieved from https://genderpolicyreport.umn.edu/justice-demands-the-vote-100-years-of-womens-suffrage/

Jurjevich, J. (2019, May 23). *Race/ethnicity and the 2020 Census.* Retrieved from https://www.census2020now.org/faces-blog/same-sex-households-2020-census-r3976

Juujärvi, S., Ronkainen, K., & Silvennoinen, P. (2019). The ethics of care and justice in primary nursing of older patients. *Clinical Ethics 14*(4), 187–194. Retrieved from https://doi.org/10.1177/1477750919876250

Kagay, M. R. (1994, July 8). Poll on doubt of Holocaust is corrected: Roper says 91% are sure it occurred. *New York Times.*

Kahlenberg, S. G., & Hein, M. M. (2010). Progression on Nickelodeon? Gender-role stereotypes in toy commercials. *Sex Roles, 62,* 830–847.

Kaiser Family Foundation. (2006, July). *Race, ethnicity, and health care: Fact sheet.* Menlo Park, CA: Author. Retrieved from http://kaiserfamilyfoundation.files.wordpress.com/2013/01/7541.pdf

Kaiser Family Foundation. (2010, September 17). Census Bureau: Recession fuels record number of uninsured Americans. *Kaiser Health News.* Retrieved from http://www.kaiserhealthnews.org/daily-reports/2010/september/16/uninsured-census-statistics.aspx

Kaiser Family Foundation. (2012). *Medicaid and long-term care services and supports*. Retrieved from http://www.kff.org/medicaid/upload/2186-09.pdf

Kaiser Family Foundation. (2014). *Medicare at a glance*. Retrieved from http://kff.org/medicare/fact-sheet/medicare-at-a-glance-fact-sheet

Kaiser Family Foundation. (2017). *Distribution of medical school graduates by gender*. Retrieved from https://www.kff.org/other/state-indicator/medical-school-graduates-by-gender/?dataView=1¤tTimeframe=0&sortModel=%7B%22colId%22:%22Location%22,%22sort%22:%22asc%22%7D

Kaiser Family Foundation. (2018). *33 states and DC have adopted Medicaid expansion as of May 2018*. Retrieved from https://www.kff.org/medicaid/slide/33-states-and-dc-have-adopted-medicaid-ex-pansion-as-of-may-2018/

Kaiser Family Foundation. (2020). *Status of state Medicaid expansion decisions: Interactive map*. Retrieved from https://www.kff.org/medicaid/issue-brief/status-of-state-medicaid-expansion-decisions-interactive-map/

Kaiser Family Foundation State Health Facts. (2019, July). *Total monthly Medicaid/CHIP enrollment and pre-ACA enrollment*. Retrieved from https://www.kff.org/health-reform/state-indicator/total-monthly-medicaid-and-chip-enrollment/?currentTimeframe=18&sortModel=%7B%22colId%22:%22Location%22,%22sort%22:%22asc%22%7D

Kallen, H. M. (1915, February 25). Democracy versus the melting pot: A study of American nationality. *The Nation*.

Kambhampati, S., & Luna, J. (2020, November 18). *2021's Congress will feature the most women of color ever*. Retrieved from https://www.msn.com/en-us/news/politics/2021s-congress-will-feature-the-most-women-of-color-ever/ar-BB1b8EFK?ocid=uxbndlbing

Kanazawa, S., & Still, M. C. (2000). Parental investment as a game of chicken. *Politics and the Life Sciences, 19*, 17–26.

Kandal, T. R. (1988). *The woman question in classical sociological theory*. Gainesville: University of Florida Press.

Kanter, R. M. (1983). *The change masters: Innovation for productivity in the American corporation*. New York: Simon & Schuster.

Kappeler, V. E., Sluder, R. D., & Alpert, G. P. (1998). *Forces of deviance: Understanding the dark side of policing*. Prospect Heights, IL: Waveland Press.

Kapuscinski, R. (2001). *The shadow of the sun*. New York, NY: Vintage Books.

Kara, S. (2009). *Sex trafficking: Inside the business of modern slavery*. New York: Columbia University Press.

Karpinski, A. C., & Duberstein, A. (2009). *A description of Facebook use and academic performance among undergraduate and graduate students*. Columbus: Ohio State University, College of Education and Human Ecology. Retrieved from http://researchnews.osu.edu/archive/facebook2009.jpg

Kasarda, J. (1993). Urban industrial transition and the underclass. In W. J. Wilson (Ed.), *The ghetto underclass* (pp. 43–64). Newbury Park, CA: Sage.

Kate, H. A. (2016). The gender buffet: LGBTQ parents resisting heteronormativity. *Gender & Society, 30*(2), 189–212. https://doi.org/10.1177/0891243215611370

Katz, J., & Chambliss, W. J. (1995). Biology and crime. In J. F. Sheley (Ed.), *Criminology: A contemporary handbook* (2nd ed.). Belmont, CA: Wadsworth.

Katz, J., Goodnough, A., & Sanger-Katz, M. (2020, July 15). In shadow of pandemic, U.S. drug overdose deaths resurge to record. *New York Times*. Retrieved from https://www.nytimes.com/interactive/2020/07/15/upshot/drug-overdose-deaths.html

Kaufman, J. M. (2009). Gendered responses to serious strain: The argument for a general strain of deviance. *Justice Quarterly, 26*, 410–444.

Kavner, L. (2012, August 15). Compliance, a low budget indie, might be the most disturbing movie ever made. *Huffington Post*. Retrieved from http://www.huffingtonpost.com/2012/08/15/compliance-movie-film_n_1779123.html

Kelley, B., & Carchia, C. (2013, July 11). "Hey, data data—swing!": The hidden demographics of youth sports. *ESPN The Magazine*. Retrieved from http://espn.go.com/espn/story/_/id/9469252/hidden-demographics-youth-sports-espn-magazine

Kelly, J. (2020, March 11). Airline CEOs take pay cuts as people avoid flying due to COVID-19. *Forbes*. Retrieved from https://www.forbes.com/sites/jackkelly/2020/03/11/airline-ceos-take-pay-cuts-as-people-avoid-flying-due-to-covid-19/?sh=3f3ed99c2ad4

Kendi, I. (2020, June 1). The American nightmare. Retrieved from https://www.theatlantic.com/ideas/archive/2020/06/american-nightmare/612457/

Kenigsberg, B. (2020, March 24). "Crip Camp" review: After those summers, nothing was the same. *New York Times*. Retrieved from https://www.nytimes.com/2020/03/24/movies/crip-camp-review.html?auth=login-google

Kennedy, M. (2016, April 20). *Lead-laced water in Flint: A step-by-step look at the makings of a crisis*. Retrieved from http://www.npr.org/sections/thetwo-way/2016/04/20/465545378/lead-laced-water-in-flint-a-step-by-step-look-at-the-makings-of-a-crisis

Kennedy, P. (1987). *The rise and fall of the great powers: Economic change and military conflict from 1500 to 2000*. New York, NY: Random House.

Kenning, C., & Halladay, J. (2008, January 25). Cities study dearth of healthy food. *USA Today*. Retrieved from http://usatoday30.usatoday.com/news/health/2008-01-24-fooddesert_N.htm

Kenny, C. (2012, June 4). The global obesity bomb. *Bloomberg Businessweek*. Retrieved from http://www.businessweek.com/articles/2012-06-04/the-global-obesity-bomb

Kessler, E.-M., Racoczy, K., & Staudinger, U. (2004). The portrayal of older people in prime time television series: The match with gerontological evidence. *Ageing and Society, 24*, 531–552.

Keyfitz, N. (1993). Thirty years of demography and *Demography*. *Demography, 30*, 533–549.

Khanna, P. (2010, August 16). Beyond city limits: The age of nations is over. The new urban era has begun. *Foreign Policy*. Retrieved from http://www.foreignpolicy.com/articles/2010/08/16/beyond_city_limits?page=full

Khullar, A. (2014, March 25). *WHO: Air pollution caused one in eight deaths*. Retrieved from http://www.cnn.com/2014/03/25/health/who-air-pollution-deaths/

Kibirige, J. S. (1997). Population growth, poverty, and health. *Social Science & Medicine, 45*, 247–259.

Kilbourne, J. (1999). *Deadly persuasion: Why women and girls must fight the addictive power of advertising*. New York: Free Press.

Kilman, C. (2016, January 17). Small house, big impact: The effect of tiny houses on community and environment. *Undergraduate Journal of Humanistic Studies* (Carleton College). Retrieved from https://apps.carleton.edu/ujhs/assets/charlie_kilman_tinyhouses__4_.pdf

Kimmel, M. S. (1986). Toward men's studies. *American Behavioral Scientist, 29*(5), 517–529. http://dx.doi.org/10.1177/000276486029005002

Kimmel, M. S. (1996). *Manhood in America: A cultural history*. New York: Free Press.

Kimmel, M. S. (2013). *Angry white men: Masculinity in America at the end of an era*. New York: Nation Books.

Kimmel, M., & Wade, L. (2018). *Ask a feminist: Michael Kimmel and Lisa Wade discuss toxic masculinity*. Retrieved from http://signsjournal.org/kimmel-wade-toxic-masculinity/

Kindy, K. (2013, April 25). *At chicken plants, chemicals blamed for health ailments are poised to proliferate*. Retrieved from https://www.washingtonpost.com/politics/at-chicken-plants-chemicals-blamed-for-health-ailments-are-poised-to-proliferate/2013/04/25/d2a65ec8-97b1-11e 2-97cd-3d8c1afe4f0f_story.html

King, H., & Chambliss, W. J. (1984). *Harry King: A professional thief's journey*. New York: Macmillan.

Kinnvall, C. (2004). Globalization and religious nationalism: Self, identity, and the search for ontological security. *Political Psychology, 25*, 741–767.

Kirk, A. (2019, January 23). The Oscars is bad at representing ethnic minorities and women—but just how bad? *Telegraph UK*. Retrieved from https://www.telegraph.co.uk/films/0/oscars-bad-representing-ethnic-minorities-women-just-bad

Kirk, D., & Laub, J. (2010). *Neighborhood change and crime in the modern metropolis*. Retrieved from https://pdfs.semanticscholar.org/9e2b/397d8a85ce6c1a93abc827f2a0064cb6b566.pdf

Kissane, R. J., & Winslow, S. (2016). "'You're underestimating me and you shouldn't': Women's agency in fantasy sports. *Gender and Society, 30*(5), 819–841. Retrieved from https://doi.org/10.1177/0891243216632205

Kissinger, H. A. (2015). *Elizabeth Holmes. Time*. Retrieved from https://time.com/collection-post/3822734/elizabeth-holmes-2015-time-100/

Kluckhohn, F. R., & Strodtbeck, F. L. (1961). *Variations in value orientations*. Evanston, IL: Row, Peterson.

Knight Frank Research. (2019). *The wealth report* (13th ed.). Retrieved from https://content.knightfrank.com/resources/knightfrank.com/wealthreport/2019/the-wealth-report-2019.pdf

Kochhar, R. (2020, June 11). *Unemployment rose higher in three months of COVID-19 than it did in two years of the Great Recession*. Pew Research. Retrieved from https://www.pewresearch.org/fact-tank/2020/06/11/unemployment-rose-higher-in-three-months-of-covid-19-than-it-did-in-two-years-of-the-great-recession/

Kocchar, R., & Cilluffo, A. (2018, July 12). *Income inequality in the U.S. is rising most rapidly among Asians*. Retrieved from https://www.pewsocialtrends.org/2018/07/12/income-inequality-in-the-u-s-is-rising-most-rapidly-among-asians/?utm_source=AdaptiveMailer&utm_medium=email&utm_campaign=18-7-12%20Income%20Inequality&org=982&lvl=100&ite=2866&lea=634931&ctr=0&par=1&trk=

Kochhar, R., & Fry, R. (2014, December 12). *Wealth inequality has widened along racial, ethnic lines since end of Great Recession*. Washington, DC: Pew Research Center. Retrieved from http://www.pewresearch.org/fact-tank/2014/12/12/racial-wealth-gaps-great-recession/

Kochhar, R., Fry, R., & Taylor, P. (2011). *Wealth gaps rise to record highs between Whites, Blacks, Hispanics*. Washington, DC: Pew Research Center. Retrieved from http://www.pewsocialtrends.org/files/2011/07/SDT-Wealth-Report_7-26-11_FINAL.pdf

Koeze, E., & Popper, N. (2020, April 7). The virus changed the way we internet. *New York Times*. Retrieved from https://www.nytimes.com/interactive/2020/04/07/technology/coronavirus-internet-use.html

Kohlberg, L. (1969). Stage and sequence: The cognitive-developmental approach to socialization. In A. Goslin (Ed.), *Handbook of socialization theory and research* (pp. 347–480). Chicago: Rand McNally.

Kohlberg, L. (1983). *The philosophy of moral development*. New York: Harper & Row.

Kohlberg, L. (1984). *The psychology of moral development*. New York: Harper & Row.

Kohn, M. L. (1989). *Class and conformity: A study in values* (2nd ed.). Chicago: University of Chicago Press.

Kollmayer, M., Schultes, M.-T., Schober, B., Hodosi, T., & Spiel, C. (2018). Parents' judgments about the desirability of toys for their children: Associations with gender role attitudes, gender-typing of toys, and demographics. *Sex Roles, 79*(5–6), 329–341. Retrieved from https://doi-org.thor.nebrwesleyan.edu/10.1007/s11199-017-0882-4

Kolowich, S. (2011, August 22). What students don't know. *Inside Higher Ed*. Retrieved from http://www.insidehighered.com/news/2011/08/22/erial_study_of_student_research_habits_at_illinois_university_libraries_reveals_alarmingly_poor_information_literacy_and_skills

Kornhauser, W. (1959). *The politics of mass society*. Glencoe, IL: Free Press.

Kornrich, S., & Furstenberg, F. (2013). Investing in children: Changes in parental spending on children, 1972–2007. *Demography, 50*, 1–23.

Koss, M. D. (2015). Diversity in contemporary picturebooks: A content analysis. *Journal of Children's Literature, 41*(1), 32.

Kozol, J. (1991). *Savage inequalities: Children in American schools*. New York: HarperCollins.

Kozol, J. (1995). *Amazing grace: Lives of our children and the conscience of a nation*. New York: Crown.

Kozol, J. (2000). *Ordinary resurrections: Children in the years of hope*. New York: Crown.

Kozol, J. (2005). *The shame of the nation: The restoration of apartheid schooling in America*. New York: Three Rivers Press.

Kraft, A. (2016, April 18). *Chinese restaurant fires subpar robot waiters*. Retrieved from http://www.cbsnews.com/news/chinese-restaurant-fires-subpar-robot-waiters/

Kramer, A. (2013, May 13). *How are savings groups changing lives?* Oxfam America. Retrieved from http://firstperson.oxfamamerica.org/2013/05/how-are-savings-groups-changing-lives/

Kramer, R. C., & Michalowski, R. J. (2005). War, aggression and state crime: A criminological analysis of the invasion and occupation of Iraq. *British Journal of Criminology*, *45*, 446–469.

Kramer, S. (2019, December 12). *U.S. has world's highest rate of children living in single-parent households*. Retrieved from https://pewrsr.ch/2LLvbxW

Kraut, R., Patterson, M., Lundmark, V., Kiesler, S., Mukopadhayay, T., & Scherlis, W. (1998). Internet paradox: A social technology that reduces social involvement and psychological well-being? *American Psychologist*, *53*, 1017–1032.

Kristof, N., & WuDunn, S. (2009). *Half the sky: Turning oppression into opportunity for women worldwide*. New York, NY: Knopf.

Kristofferson, K., White, K., & Peloza, J. (2013). The nature of slacktivism: How the social observability of an initial act of token support affects subsequent prosocial action. *Journal of Consumer Research*, *40*(6), 1149–1166.

Kroeger, T., & Gould, E. (2017) *The class of 2017*. Economic Policy Institute. Retrieved from http://www.edpi.org/publication/the-class-of-2017/

Krogstad, J. (2016). *2016 electorate will be the most diverse in U.S. history*. Washington, DC: Pew Research Center. Retrieved from http://www.pewresearch.org/fact-tank/2016/02/03/2016-electorate-will-be-the-most-diverse-in-u-s-history/

Krogstad, J-M., & Noe-Bustamante, L. (2020, September 10). *Key facts about U.S. Latinos for National Hispanic Heritage Month*. Retrieved from https://www.pewresearch.org/fact-tank/2020/09/10/key-facts-about-u-s-latinos-for-national-hispanic-heritage-month/

Kronk, E. A. (2013, April 16). One statute for two spirits: Same-sex marriage in Indian country. *Jurist Forum*. Retrieved from http://jurist.org/forum/2013/04/elizabeth-kronk-two-spritis.php

Kubrin, C. E. (2005). Gangstas, thugs, and hustlas: Identity and the code of the street in rap music. *Social Problems*, *52*, 360–378.

Labropoulou, E., Liakos, C., Halasz, S., & Qiblawi, T. (2020, September 10). *Fire ravages Europe's largest migrant camp on Lesbos*. *CNN*. Retrieved from https://www.cnn.com/2020/09/09/europe/greece-lesbos-fires-intl/index.html

Lacarte, V., & Hayes, J. (2020, September 28). *Projections for pay equity by race/ethnicity*. Institute for Women's Policy Research. Retrieved from https://iwpr.org/iwpr-publications/quick-figure/pay-equity-projection-race-ethnicity-2020/

Lacy, A. (2020, July 21). ALEC is close to passing model bill that would protect companies from coronavirus-related lawsuits. *The Intercept*. Retrieved from https://theintercept.com/2020/07/21/alec-coronavirus-corporate-liability-bill/

Lafrance, A. (2016, March 10). The human–robot trust paradox. *Atlantic Monthly*. Retrieved from https://www.theatlantic.com/technology/archive/2016/03/humans-robots-future/472749

Lamerichs, N. (2014). Costuming as subculture: The multiple bodies in cosplay. *Scene 2*(1), 113–125.

Lampman, J. (2006, September 14). American Buddhism on the rise. *Christian Science Monitor*. Retrieved from http://www.csmonitor.com/2006/0914/p14s01-lire.html

Lane, H. (1992). *The mask of benevolence: Disabling the deaf*. New York: Random House.

Lane, H. (2005). Ethnicity, ethics, and the deaf world. *Journal of Deaf Studies and Deaf Education*, *10*, 291–310.

Langfitt, F. (2013, April 29). *As the car market moves east, an extravaganza in Shanghai*. National Public Radio. Retrieved from http://www.npr.org/blogs/thetwo-way/2013/04/27/179025891/as-the-car-market-moves-east-an-extravaganza-in-shanghai

Langley, M., & Sugarmann, J. (2020, June). *Black homicide victimization in the United States: An analysis of 2017 homicide data*. Violence Policy Center. Retrieved from https://vpc.org/studies/blackhomicide20.pdf

Langton, L., Planty, M., & Sandholtz, N. (2013). *Hate crime victimization, 2003–2011* (NCJ 241291). Washington, DC: U.S. Department of Justice, Bureau of Justice Statistics. Retrieved from http://www.bjs.gov/index.cfm?ty=pbdetail&iid=4614

Lanier, J. (2013). *Who owns the future?* New York: Simon & Schuster.

Lareau, A. (2002). Invisible inequality: Social class and childrearing in Black families and White families. *American Sociological Review*, *67*, 747–776.

Lareau, A. (2011). *Unequal childhoods: Class, race and family life*. Berkeley and Los Angeles, CA: University of California Press.

Laris, M. (2016, June 9). This government competition could completely change the American city. *Washington Post*. Retrieved from https://www.washingtonpost.com/local/trafficandcommuting/can-a-wonked-out-reality-competition-help-save-the-american-city/2016/06/08/f5f0b3d8-112f-11e6-8967-7ac733c56f12_story.html

Laub, J., & Sampson, R. J. (2003). *Shared beginnings, divergent lives*. Cambridge, MA: Harvard University Press.

Laurent, S. A., & Lo, C. C. (2019). Risk society online: Zika virus, social media and distrust in the Centers for Disease Control and Prevention. *Sociology of Health & Illness 41*(7), 1270–1288. Retrieved from https://doi-org.thor.nebrwesleyan.edu/10.1111/1467-9566.12924

Lauzen, M., Dozier, D., & Horan, N. (2008). Constructing gender stereotypes through social roles in prime-time television. *Journal of Broadcasting & Electronic Media*, *52*, 200–214.

Lawal, S. (2020, February 10). Africa, a thunder and lightning hot spot, may see even more storms. *New York Times*. Retrieved from https://www.nytimes.com/2020/02/10/climate/lightning-africa-climate-change.html?auth=login-email&login=email

Lawson, K. M., Crouter A. C., & McHale. S. M. (2015, October). Links between family gender socialization experiences in childhood and gendered occupational attainment in young adulthood. *Journal of Vocational Behavior*, *90*, 26–35.

Le Moal, M., & Koob, G. F. (2007). Drug addiction: Pathways to the disease and pathophysiological perspectives. *European Neuropsychopharmacology*, *17*, 377–393.

Ledeneva, A. V. (1998). *Russia's economy of favours: Blat, networking, and informal exchange*. Cambridge, England: Cambridge University Press.

Lederman, D. (2019, November 29). The faculty shrinks, but tilts to full time. *Inside Higher Education*. Retrieved from https://www.insidehighered.com/news/2019/11/27/federal-data-show-proportion-instructors-who-work-full-time-rising

Lee, M. M., Carpenter, B., & Meyers, L. S. (2007). Representations of older adults in television advertisements. *Journal of Aging Studies, 21*, 23–30.

Lehman, D., & Sherkat, D. E. (2018). Measuring religious identification in the United States. *Journal for the Scientific Study of Religion 57*(4), 779–794. Retrieved from https://doi-org.thor.nebrwesleyan.edu/10.1111/jssr.12543/

Lemert, E. (1951). *Social pathology*. New York, NY: McGraw-Hill.

Lemonnier, J. (2008, February 18). Big players in diet industry shift focus to online presences. *Consumer Lab*. Retrieved from http://consumerlab.wordpress.com/2008/02/18/big-players-in-diet-industry-shift-focus-to-online-presences

Lempert, D. (2007). *Women's increasing wage penalties from being overweight and obese*. Washington, DC: U.S. Bureau of Labor Statistics. Retrieved from http://www.bls.gov/osmr/abstract/ec/ec070130.htm

Leonhardt, D. (2009, September 8). Colleges are failing in graduation rates. *New York Times*. Retrieved from http://www.nytimes.com/2009/09/09/business/econ-omy/09leonhardt.html

Leonhardt, D. (2019, October 6). The rich really do pay lower taxes than you. *New York Times*. Retrieved from https://www.nytimes.com/interactive/2019/10/06/opinion/income-tax-rate-wealthy.html

Lester, D. (Ed.). (2000). *Suicide prevention: Resources for the millennium*. Philadelphia: Brunner-Routledge.

Lesthaeghe, R. (1995). The second demographic transition in Western countries: An interpretation. In O. Mason & A. Jensen (Eds.), *Gender and family change in industrialized countries* (pp. 17–62). Oxford: Clarendon Press.

Levin, D. (2020, October 12). No home, no wi-fi: Pandemic adds to strain on poor college students. *New York Times*. Retrieved from https://www.nytimes.com/2020/10/12/us/covid-poor-college-students.html?login=smartlock&auth=login-smartlock

Levine, L. (2012, December 17). *Offshoring (or offshore outsourcing) and job loss among U.S. workers* (CRS 7-5700; RL32292). Washington, DC: Congressional Research Service. Retrieved from http://fas.org/sgp/crs/misc/RL32292.pdf

Levine, M., & Crowther, S. (2008). The responsive bystander: How social group membership and group size can encourage as well as inhibit bystander intervention. *Journal of Interpersonal Psychology, 95*, 1429–1439.

LeVine, R. A., LeVine, S., Schnell-Anzola, B., Rowe, M. L., & Dexter, E. (2012). *Literacy and mothering: How women's schooling changes the lives of the world's children*. New York: Oxford University Press.

Levine, S., & Laurie, N. O. (Eds.). (1974). *The American Indian today*. Penguin Books.

Levintova, H. (2015, October 1). Girls are the fastest-growing group in the juvenile justice system. *Mother Jones*. Retrieved from http://www.motherjones.com/politics/2015/09/girls-make-ever-growing-proportion-kids-juvenile-justice-system

Levitt, P. (2004, October 1). *Transnational migrants: When "home" means more than one country*. Migration Policy Institute. Retrieved from http://www.migrationpolicy.org/article/transnational-migrants-when-home-means-more-one-country

Lewin, T. (2011a, September 27). College graduation rates are stagnant even as enrollment rises, a study finds. *New York Times*. Retrieved from http://www.nytimes.com/2011/09/27/education/27remediation.html

Lewin, T. (2011b, October 25). Screen time higher than ever for children. *New York Times*. Retrieved from http://www.nytimes.com/2011/10/25/us/screen-time-higher-than-ever-for-children-study-finds.html

Lewis, S. (2020, August 12). Brazil's Jair Bolsanaro calls new Amazon fires a "lie"—as videos show the rainforest burning. *CBS News*. Retrieved from https://www.cbsnews.com/news/brazil-jair-bolsanaro-amazon-fires-lie-rainforest-burn

Liasson, M. (2020, June 20). *As the culture wars shift, President Trump struggles to adapt*. Retrieved from https://www.npr.org/2020/06/20/881096897/as-the-culture-wars-shift-president-trump-struggles-to-adapt

Library of Congress. (2012). *Days in session calendars: 112th Congress 2nd session*. Retrieved from http://thomas.loc.gov/home/ds/h1122.html

Liebow, E. (1967). *Tally's corner: A study of Negro streetcorner men*. Boston: Little, Brown.

Lienert, P., & Thompson, M. (2014, April 2). *GM avoided defective switch redesign in 2005 to save a dollar each*. Reuters. Retrieved from http://www.reuters.com/article/2014/04/02/us-gm-recall-delphi-idUSBREA3105R20140402

Light, H. K., & Martin, R. E. (1986). American Indian families. *Journal of American Indian Education*, 1–5.

Lindberg, L. D., Santelli, J. S., & Desai, S. (2016). Understanding the decline in adolescent fertility in the United States, 2007–2012. *J Adolescent Health*, 1–7. Retrieved from https://www.cdc.gov/teenpregnancy/about/

Liogier, R. (2020). Globalization. *The SAGE encyclopedia of the sociology of religion* (pp. 320–325). Thousand Oaks, CA: SAGE.

Lipka, M. (2015a, May 6). *5 facts about prayer*. Washington, DC: Pew Research Center. Retrieved from http://www.pewresearch.org/fact-tank/2015/05/06/5-facts-about-prayer/

Lipka, M. (2015b). *Millennials increasingly are driving growth of 'nones.'* Pew Research Center, Fact Tank. Retrieved from http://www.pewresearch.org/fact-tank/2015/05/12/millennials-increasingly-are-driving-growth-of-nones/

Lipka, M. (2015c, July 27). *The most and least racially diverse U.S. religious groups*. Retrieved from https://www.pewresearch.org/fact-tank/2015/07/27/the-most-and-least-racially-diverse-u-s-religious-groups/

Lipka, M., & Majumdar, S. (2017). *How religious restrictions around the world have changed over a decade*. Pew Research Center. Retrieved from https://www.pewresearch.org/fact-tank/2019/07/16/how-religious-restrictions-around-the-world-have-changed-over-a-decade/

Little, B. (2020, May 6). *When mask-wearing rules in the 1918 pandemic faced resistance*. Retrieved from https://www.history.com/news/1918-spanish-flu-mask-wearing-resistance

Liu, S., Zhang, M., Yang, L., Li, Y., Wang, L., Huang, Z., Wang, L., Chen, Z., & Zhou, M. (2017). Prevalence and patterns of tobacco smoking among Chinese adult men and women: Findings of the 2010 national smoking survey. *Journal of Epidemiology and Community Health, 71*(2), 154–161. Retrieved from https://doi.org/10.1136/jech-2016-207805

Livingston, G. (2014). *Growing number of dads home with the kids.* Washington, DC: Pew Research Center. Retrieved from http://www.pewsocialtrends.org/2014/06/05/growing-number-of-dads-home-with-the-kids

Livingston, G., & Brown, A. (2017, May 18). *Intermarriage in the U.S. 50 years after Loving vs. Virginia.* Pew Research Center. Retrieved from https://www.pewsocialtrends.org/2017/05/18/intermarriage-in-the-u-s-50-years-after-loving-v-virginia

Livingston, G., & Thomas, D. (2019, December 16). *Among 41 countries, only U.S. lacks paid parental leave.* Pew Research Center. Retrieved from https://www.pewresearch.org/fact-tank/2019/12/16/u-s-lacks-mandated-paid-parental-leave/

Livingstone, S., & Brake, D. R. (2010). On the rapid rise of social networking sites: New findings and policy implications. *Children and Society, 24*, 75–83.

Logan, J. R., Minca, E., & Adar, S. (2012). The geography of inequality: Why separate means unequal in American public schools. *Sociology of Education, 85*, 287–301.

Logan, J. R., & Molotch, H. L. (1987). *Urban fortunes: The political economy of place.* Berkeley: University of California Press.

Lomborg, B. (2012, April 30). The high cost of heart disease and cancer. *Slate.* Retrieved from http://www.slate.com/articles/technology/copenhagen_consen-sus_2012/2012/04/copenhagen_consensus_ideas_for_reducing_cancer_and_heart_disease.html

Lombroso, C. (1896). *L'homme criminel.* Paris: F. Alcan.

Longshore, C. T. (2017, October 1). Mental illness and homeless baby-boomers: What can be done? *Behavioral Health News.* Retrieved from https://behavioralhealthnews.org/mental-illness-and-homeless-baby-boomers-what-can-be-done/

Lonsdorf, K. (2017, August 22). *From Rolls-Royce to Grey Poupon, A look at brand mentions in chart-topping songs.* NPR. Retrieved from https://www.npr.org/2017/08/22/545314024/rolls-royce-tops-list-as-musics-most-popular-brand

Loo, C. M. (1991). *Chinatown: Most time, hard time.* New York: Praeger.

López, G., Ruiz, N. G., & Patten, E. (2017, September 8). Key facts about Asian Americans, a diverse and growing population. *Fact Tank.* Pew Research Center. Retrieved from https://www.pewresearch.org/fact-tank/2017/09/08/key-facts-about-asian-americans/

Lopez, M. H., Gonzalez-Barrera, A., & Krogstad, J. M. (2018, September 11). *Latinos are more likely to believe in the American Dream, but most say it is hard to achieve.* Washington, DC: Pew Research Center. Retrieved from https://www.pewresearch.org/fact-tank/2018/09/11/latinos-are-more-likely-to-believe-in-the-american-dream-but-most-say-it-is-hard-to-achieve/

Lopez, M. H., Krogstad, J. M., & Passel, J. S. (2020, September 15). *Who is Hispanic?* Retrieved from https://www.pewresearch.org/fact-tank/2020/09/15/who-is-hispanic/

Lotz, L. (2020). *What is culturally competent care?* Retrieved from https://www.nursepractitionerschools.com/faq/what-is-culturally-competent-care

Lucas, J. W., & Lovaglia, M. J. (1998). Leadership status, group size, and emotion in face-to-face groups. *Sociological Perspectives, 41*, 617–637.

Lustgarten, A. (2020, July 26). Refugees from the Earth. *New York Times Magazine*, 8–23, 43.

Lynas, M. (2020). *Our final warning: Six degrees of climate emergency.* London, England: 4th Estate

Maas, P. (1997). *Serpico.* New York: HarperTorch.

Macdonald, J., & Stokes, R. (2019, September 30). *Gentrification, land use, and crime.* Retrieved from https://doi.org/10.1146/annurev-criminol-011419-041505

MacFarquhar, R. (1980, February 9). The post-Confucian challenge. *The Economist* (pp. 67–72).

MacKinnon, C. A. (1982). Feminism, Marxism, method and the state: An agenda for theory. *Signs, 7*, 515–544.

Mackun, P., & Wilson, S. (2011). *Population distribution and change: 2000 to 2010* (Census Brief CS2020BR-01). Washington, DC: U.S. Census Bureau. Retrieved from http://www.census.gov/prod/cen2010/briefs/c2010br-01.pdf

MacLeod, C. (2011, November 2). In China, tensions rising over Buddhism's quiet resurgence. *USA Today.* Retrieved from http://www.usatoday.com/news/religion/story/2011-11-01/tibetan-buddhism-china-communist-tension/51034604/1

Madlock, P. E., & Westerman, D. (2011). Hurtful cyber-teasing and violence: Who's laughing out loud? *Journal of Interpersonal Violence, 26*, 3542–3560.

Madrigal, A. (2011 November). What's wrong with #FirstWorldProblems? *Atlantic Monthly.* Retrieved from http://www.theatlantic.com/technology/archive/2011/11/whats-wrong-with-firstworldproblems/248829

Mahdi, S. (2020, May 11). *War in Yemen forces more girls into child marriage.* Retrieved from https://www.dw.com/en/child-marriage-on-the-rise-in-yemen/a-53390598

Maher, J. K., Herbst, K. C., Childs, N. M., & Finn, S. (2008). Racial stereotypes in children's television commercials. *Journal of Advertising Research, 48*, 80–93.

Maher, L. (1997). *Sexed work: Gender, race, and resistance in a Brooklyn drug market.* New York: Oxford University Press.

Maidenberg, M. (2016, April 29). *Where do Chicago manufacturing, transportation employees live and work?* Retrieved from http://www.chicagobusiness.com/article/20160429/NEWS05/160429806/where-do-chicago-manufacturing-transportation-employees-live-and-work

Malacrida, C. (2005). Discipline and dehumanization in a total institution: Institutional survivors' descriptions of time-out rooms. *Disability & Society, 20*, 523–537.

Malik, R. (2019, June 20). *Working families are spending big money on childcare.* Center for American Progress. Retrieved from https://www.americanprogress.org/issues/early-childhood/reports/2019/06/20/471141/working-families-spending-big-money-child-care/

Malinauskas, B. M., Raedeke, T. D., Aeby, V. G., Smith, J. L., & Dallas, M. B. (2006). Dieting practices, weight perceptions, and body composition: A comparison of normal weight, overweight, and obese college females. *Nutrition Journal, 5*.

Mallicoat, S. L. (2019). *Women, gender, and crime* (3rd ed.). Thousand Oaks, CA: SAGE.

Malone Gonzalez, S. (2019). Making it home: An intersectional analysis of the police talk. *Gender & Society, 33*(3), 363–386. http://dx.doi.org/10.1177/0891243219828340

Malthus, T. (1926). *First essay on population*. London: Macmillan. (Original work published 1798)

Mann, C. C. (2011, June). The birth of religion. *National Geographic*, 34–59.

Mann, M. (1986). *The sources of social power: Vol. 1. A history of power from beginning until 1760*. New York: Cambridge University Press.

Mapping Prejudice. (2020). *What are covenants?* Retrieved from https://mappingprejudice.umn.edu/s

Marcuse, H. (1964). *One-dimensional man*. Boston, MA: Beacon Press.

Margolis, E. (2001). *The hidden curriculum in higher education*. New York: Routledge.

Margonelli, L. (2010, February 7). Eternal life. *New York Times*. Retrieved from https://www.nytimes.com/2010/02/07/books/review/Margonelli-t.html

Marini, M. M. (1990). Sex and gender: What do we know? *Sociological Forum, 5*, 95–120.

Markert, J. (2010). The changing face of racial discrimination: Hispanics as the dominant minority in the United States—A new application of power-threat theory. *Critical Sociology, 36*, 307–327.

Markusen, E. (2002). Mechanisms of genocide. In C. Rittner, J. K. Roth, & J. M. Smith (Eds.), *Will genocide ever end?* (pp. 83–90). St. Paul, MN: Paragon House.

Marlowe, C. M., Schneider, S. L., & Nelson, C. E. (1996). Gender and attractiveness biases in hiring decisions: Are more experienced managers less biased? *Journal of Applied Psychology, 81*, 11–21.

Maroto, M. E., Snelling, A., & Linck, H. (2015). Food insecurity among community college students: Prevalence and association with grade point average. *Community College Journal of Research and Practice, 39*(6), 515–526.

Martin, C. L., & Fabes, R. A. (2001). The stability and consequences of young children's same-sex peer interactions. *Developmental Psychology, 37*, 431–446.

Martin, D. S. (2012, March 1). *Vets feel abandoned after secret drug experiments. CNN*. Retrieved from http://edition.cnn.com/2012/03/01/health/human-test-subjects

Martin, J. A., Hamilton, B. E., Osterman, M. J. K., & Driscoll, A. K. (2019, November 27). Births: Final data for 2018. *National Vital Statistics Reports. 68*(13). Retrieved from https://www.cdc.gov/nchs/fastats/unmarried-childbearing.htm

Martin, N. (2019, February 8). Did a robot write this? How AI is impacting journalism. *Forbes*. Retrieved from https://www.forbes.com/sites/nicolemartin1/2019/02/08/did-a-robot-write-this-how-ai-is-impacting-journalism/?sh=4c8d91bd7795

Martin, N. L., & Kposowa, A. J. (2019). Race and consequences: An examination of police abuse in America. *Journal of Social Sciences, 15*, 1–10.

Martineau, H. (1837). *Society in America*. New York: Saunders & Otley.

Maruschak, L. M., & Minton, T. D. (2020). *Correctional populations in the United States 2017–2018*. Bureau of Justice Statistics. Retrieved from https://www.bjs.gov/content/pub/pdf/cpus1718.pdf

Marwell, G., & Oliver, P. (1993). *The critical mass in collective action: A micro-social theory*. New York, NY: Cambridge University Press.

Marx, G. T., & McAdam, D. (1994). *Collective behavior and social movements: Process and structure*. Englewood Cliffs, NJ: Prentice Hall.

Marx, K. (1992a). *Capital: A critique of political economy*. (Vol. 1). New York: Penguin Classics. (Original work published 1867)

Marx, K. (1992b). *Capital: A critique of political economy*. (Vol. 2). New York: Penguin Classics. (Original work published 1885)

Marx, K. (1992c). *Capital: A critique of political economy*. (Vol. 3). New York: Penguin Classics. (Original work published 1894)

Marx, K. (2000). Towards a critique of Hegel's *Philosophy of Right*: Introduction. In D. McLellan (Ed.), *Karl Marx: Selected writings* (Rev. ed.) New York: Classic Books International. (Original work published 1844)

Marx, K., & Engels, F. (1998). *The communist manifesto*. New York: Verso. (Original work published 1848)

Masci, D. (2018, August 2). *5 facts about the death penalty*. Pew Research Forum. Retrieved from https://www.pewresearch.org/fact-tank/2018/08/02/5-facts-about-the-death-penalty/

Masci, D., Brown, A., & Kiley, J. (2019, June 24). *5 facts about same-sex marriage*. Washington, DC: Pew Research Center. Retrieved from https://www.pewresearch.org/fact-tank/2019/06/24/same-sex-marriage/

Mason, M., McDowell, R., Mendoza, M., & Htusan, E. (2015, December 14). *Global supermarkets selling shrimp peeled by slaves*. Associated Press. Retrieved from http://bigstory.ap.org/article/8f64f-b25931242a985bc30e3f5a9a0b2/ap-global-supermarkets-selling-shrimp-peeled-slaves

Massey, D. S. (2011). Epilogue: The past and future of Mexico–U.S. migration. In O.-V. Mark (Ed.), *Beyond la frontera: The history of Mexico–U.S. migration* (pp. 241–265). New York: Oxford University Press.

Massey, D. S. (2016). Residential segregation is the linchpin of racial stratification. *City & Community, 15*(1), 4–7.

Massey, D. S., & Denton, N. A. (1993). *American apartheid: Segregation and the making of the underclass*. Boston: Harvard University Press.

Masters, J. (2011, February 7). *Militant extremists in the United States*. Council on Foreign Relations. Retrieved from http://www.cfr.org/terrorist-organizations-and-networks/militant-extremists-united-states/p9236

Masterson, M. (2019). *CPS enrollment declines by 6,000 students*. Retrieved from https://news.wttw.com/2019/11/08/cps-enrollment-declines-6000-students

Mathisen, J. A. (1989). Twenty years after Bellah: Whatever happened to American civil religion? *Sociological Analysis, 50*, 129–146.

Matthew, W. C., & Jason, D. H. (2020). Maximizing benefits and minimizing impacts: Dual-earner couples' perceived division of household labor decision-making process. *Journal of Family Studies*, *26*(2), 208–225. doi:10.1080/13229400.2017.1367712

Maxouris, C. (2020, June 26). *Bubba Wallace responds to FBI findings: "Whether tied in 2019, or whatever, it was a noose."* CNN. Retrieved from https://edition.cnn.com/2020/06/24/us/bubba-wallace-response-fbi-hate-crime-investigation/index.html

Mayer, G. (2004). *Union membership trends in the United States*. Washington, DC: Congressional Research Service. Retrieved from http://digitalcommons.ilr.cornell.edu/cgi/viewcontent.cgi?article=1176&context=key_workplace

Mazur, E., & Richards, L. (2011). Adolescents' and emerging adults' social networking online: Homophily or diversity? *Journal of Applied Developmental Psychology*, *32*, 180–188.

Mazzella, R., & Feingold, A. (1994). The effects of physical attractiveness, race, socioeconomic status, and gender of defendant and victims on judgments of mock jurors: A meta-analysis. *Journal of Applied Social Psychology*, *24*, 1315–1344.

McAdam, D. (1982). *Political process and the development of Black insurgency, 1930–1970*. Chicago, IL: University of Chicago Press.

McAdam, D. (1986). Recruitment to high-risk activism: The case of freedom summer. *American Journal of Sociology*, *92*, 64–90.

McAdam, D., McCarthy, J. D., & Zald, M. N. (1988). Social movements. In N. J. Smelser (Ed.), *Handbook of sociology* (pp. 695–737). Newbury Park, CA: Sage.

McCarthy, J. (2014, June 2). *Double rape, lynching in India exposes caste fault lines. National Public Radio*. Retrieved from http://www.npr.org/blogs/parallels/2014/06/02/318259419/double-rape-lynching-in-india-exposes-caste-fault-lines

McCarthy, J. D., & Zald, M. N. (1973). *The trend of social movements in America: Professionalization and resource mobilization*. Morristown, NJ: General Learning.

McCarthy, J. D., & Zald, M. N. (1977). Resource mobilization and social movements: A partial theory. *American Journal of Sociology*, *82*, 1212–1241.

McCartney, J. T. (1992). *Black power ideologies: An essay in African American political thought*. Philadelphia: Temple University Press.

McCoy, K. (2009). Uncle Sam wants them. *Contexts*, *8*, 14–19.

McDonald, M. (2020). *2020 general election early vote statistics*. Retrieved from https://electproject.github.io/Early-Vote-2020G/index.html

McDonald, S., & Day, J. C. (2010). Race, gender, and the invisible hand of social capital. *Sociology Compass*, *4*, 532–543.

McDonald, S., Lin, N., & Ao, D. (2009). Networks of opportunity: Gender, race, and job leads. *Social Problems*, *56*, 385–402.

McDonald, S., & Mair, C. A. (2010). Social capital across the life course: Age and gendered patterns of network resources. *Sociological Forum*, *25*, 335–359.

McFarland, J., Hussar, B., Zhang, J., Wang, X., Wang, K., Hein, S., Diliberti, M., Forrest Cataldi, E., Bullock Mann, F., & Barmer, A. (2019). *The condition of education 2019* (NCES 2019-144). U.S. Department of Education. National Center for Education Statistics. Retrieved from https://nces.ed.gov/pubsearch/pubsinfo.asp?pubid=2019144

McGregor, J. (2014, January 3). Zappos says goodbye to bosses. *Washington Post*. Retrieved from http://www.washingtonpost.com/blogs/on-leadership/wp/2014/01/03/zappos-gets-rid-of-all-managers

McGregor, J. (2020). Average CEO earnings soared to $21.3 million last year and could rise again in 2020 despite the coronavirus recession. *Washington Post*. Retrieved from https://www.washingtonpost.com/business/2020/08/18/corporate-executive-pay-increase/

McGuire, L. C., Okoro, C. A., Goins, R. T., & Anderson, A. (2008). Characteristics of American Indian and Alaska native adult caregivers: Behavioral Risk Factor Surveillance System, *Ethnicity & Disease*, *18*, 520. 2000

McIntosh, K., Moss, E., Nunn, R., & Shambaugh, J. (2020). *Examining the Black–white wealth gap*. Brookings Institution. Retrieved from https://www.brookings.edu/blog/up-front/2020/02/27/examining-the-black-white-wealth-gap/

McIntosh, P. (1990). White privilege: Unpacking the invisible knapsack. *Independent School*, *49*(2), 31–36.

McKenna, K. Y. A., & Bargh, J. A. (1998). Coming out in the age of the Internet: Identity "demarginalization" through virtual group participation. *Journal of Personality and Social Psychology*, *75*, 681–694.

McKibben, B. (2011, April 7). Resisting climate reality. *New York Review of Books*. Retrieved from http://www.nybooks.com/articles/archives/2011/apr/07/resisting-climate-reality/?pagination=false

McKibben, B. (2020, August 20). 130 degrees. *The New York Review of Books*, 8–10.

McLaughlin, E. (2019). State crime. In E. McLaughlin (Ed.), *The SAGE dictionary of criminology* (4th ed.) Thousand Oaks, CA: SAGE.

McLean, B., & Elkind, P. (2003). *The smartest guys in the room: The amazing rise and scandalous fall of Enron*. New York: Penguin/Portfolio.

McLean, B., & Nocera, J. (2010). *All the devils are here: The hidden history of the financial crisis*. New York: Penguin/Portfolio.

McLoyd, V. C., & Smith, J. (2002). Physical discipline and behavior problems in African American, European American, and Hispanic children: Emotional support as a moderator. *Journal of Marriage and Family*, *64*, 40–53.

McLuhan, M. (1964). *Understanding media: The extensions of man*. New York, NY: McGraw-Hill.

McNeely, C. L. (1995). *Constructing the nation-state: International organization and prescriptive action*. Westport, CT: Greenwood.

McNicoll, G. (2001). Government and fertility in transitional and post-transitional societies. *Population and Development Review*, *27*, 129–159.

Mead, G. H. (1934). *Mind, self, and society*. Chicago: University of Chicago Press.

Mead, G. H. (1934). *Mind, self, and society from the standpoint of a social behaviorist*. Chicago, IL: University of Chicago Press.

Mead, G. H. (1938). *The philosophy of the act*. Chicago: University of Chicago Press.

Medicaid.gov. (2020). https://www.medicaid.gov/.

Mednick, S. A., Jr. Gabrielli, W. F., & Hutchings, B. (1987). Genetic factors in the etiology of criminal behavior. In S. A. Mednick, T. E. Moffitt, & S. A. Stack (Eds.), *The causes of crime: New biological approaches*. Cambridge: Cambridge University Press.

Mehra, A., Dixon, A. L., Brass, D. J., & Robertson, B. (2006). The social network ties of group leaders: Implications for group performance and leader reputation. *Organization Science, 17*, 64–79.

Melton, J. G. (Ed.). (1996). *Encyclopedia of American religions* (5th ed.) New York: Gale Research.

Melucci, A. (1989). *Nomads of the present: Social movements and individual needs in contemporary society*. Philadelphia, PA: Temple University Press.

Melvin, D., Walsh, N. P., & Hume, T. (2016, January 15). *Starvation in Syria a 'War Crime,' U.N. chief says*. Retrieved from http://www.cnn.com/2016/01/15/middleeast/syria-madaya-starvation/

Mendelson, S. (2017). Box office: 'Fate of the Furious' joins 'Furious 7' in the $1B club. *Forbes*, April, *30*. Retrieved from https://www.forbes.com/sites/scottmendelson/2017/04/30/box-office-fate-of-the-furious-joins-furious-7-in-the-1-billion-club/#7514f82f2f54

Merino, S. (2021). Irreligious socialization? The adult religious preferences of individuals raised with no religion. *Secularism and Nonreligion, 1*, 1–16.

Merken, S., & James, V. (2020). Perpetrating the myth: Exploring media accounts of rape myths on "women's" networks. *Deviant Behavior, 41*(9), 1176–1191. Retrieved from https://doi-org.thor.nebrwesleyan.edu/10.1080/01639625.2019.1603531

Merton, R. K. (1938). Social structure and anomie. *American Sociological Review, 3*, 672–682.

Merton, R. K. (1948). The self-fulfilling prophecy. *The Antioch Review, 8*(2), 193–210. http://dx.doi.org/10.2307/4609267

Merton, R. K. (1968). *Social theory and social structure*. New York: Free Press.

Merton, R. K. (1996). *On social structure and science*. Chicago: University of Chicago Press.

Messerschmidt, J. W., & Rohde, A. (2018). Osama bin Laden and his jihadist global hegemonic masculinity. *Gender & Society, 32*(5), 663–685. Retrieved from https://doi.org/10.1177/0891243218770358

Metzl, J. M. (2019). *Dying of whiteness: How the politics of racial resentment is killing America's heartland*. New York: Basic Books.

Meyer, C. G., & Chen, H.-M. (2019). Vanilla and kink: Power and communication in marriages with a BDSM-identifying partner. *Sexuality & Culture: An Interdisciplinary Quarterly, 23*(3), 774–792. Retrieved from https://doi.org/10.1007/s12119-019-09590-x

Meyrowitz, J. (1985). *No sense of place: The impact of electronic media on social behavior*. Oxford, England: Oxford University Press.

Michalowski, R., & Dubisch, J. (2001). *Run for the wall: Remembering Vietnam on a motorcycle pilgrimage*. New Brunswick, NJ: Rutgers University Press.

Milanovic, B. (2019). *Capitalism, alone: The future of the system that rules the world*. Cambridge, MA: Harvard University Press.

Milgram, S. (1963). Behavioral study of obedience. *The Journal of Abnormal and Social Psychology, 67*(4), 371–378. https://doi.org/10.1037/h0040525

Milkman, R. (2006). *L.A. story: Immigrant workers and the future of the U.S. labor movement*. New York: Russell Sage Foundation.

Miller, K. A., Kohn, M. A., & Schooler, C. (1986). Educational self-direction and personality. *American Sociological Review, 5*, 372–390.

Miller, K. E., Melnick, M. J., Barnes, G. M., Farrell, M. P., & Sabo, D. F. (2005). Untangling the links among athletic involvement, gender, race, and adolescent academic outcomes. *Sociology of Sport Journal, 22*, 178–193.

Miller, L. P. (1995). Tracking the progress of *Brown*. *Teachers College Record, 96*, 609–613.

Millett, K. (1970). *Sexual politics*. Garden City: Doubleday.

Mills, C. W. (2000a). *The power elite*. New York: Oxford University Press. (Original work published 1956)

Mills, C. W. (2000b). *The sociological imagination* (40th anniversary ed.). New York: Oxford University Press. (Original work published 1959)

Miner, H. (1956). Body ritual among the Nacirema. *American Anthropologist, 58*, 503–507.

Mirrlees, T. (2013). *Global entertainment media: Between cultural imperialism and cultural globalization*. London, England: Routledge.

Misachi, J. (2018, May 9). What is the official language of the United States? World facts. *World Atlas* online. Retrieved from https://www.worldatlas.com/articles/what-is-the-official-language-of-the-united-states.html

Mishel, L., & Kandra, J. (2020, August 18). *CEO compensation surged 14% in 2019 to $21.3 million*. Economic Policy Institute. Retrieved from https://www.epi.org/publication/ceo-compensation-surged-14-in-2019-to-21-3-million-ceos-now-earn-320-times-as-much-as-a-typical-worker/

Misra, J. (2020). *Voter turnout rates among all voting age and major racial and ethnic groups were higher than in 2014*. U.S. Bureau of the Census. Retrieved from https://www.census.gov/library/stories/2019/04/behind-2018-united-states-midterm-election-turnout.html

Mitchiner, J., & Sass-Lehrer, M. (2011). My child can have more choices: Reflections of deaf mothers on cochlear implants for their children. In R. Paludneviciene & I. W. Leigh (Eds.), *Cochlear implants: Evolving perspectives*. Washington, DC: Gallaudet University Press.

Mitter, S. (2020, July 1). *Corporate leaders Jeff Bezos, Mukesh Ambani, Elon Musk, and other billionaire wealth gainers of the pandemic*. *Yourstory.com*. Retrieved from https://yourstory.com/2020/06/jeff-bezos-mukesh-ambani-elon-musk-billionaire-wealth-gainers

Mizruchi, M. S., & Potts, B. B. (1998). Centrality and power revisited: Actor success in group decision making. *Social Networks, 20*, 353–387.

Mohamed, B., & Diamant, J. (2019, January 17). *Black Muslims account for a fifth of all US Muslims and about half are converts to Islam*. Retrieved from https://www.pewresearch.org/fact-tank/2019/01/17/black-muslims-account-for-a-fifth-of-all-u-s-muslims-and-about-half-are-converts-to-islam/

Mohamed, H., & Chughtai, A. (2019, February 9). What you need to know about Africa's refugees. *Al Jazeera*. Retrieved from https://www.aljazeera.com/indepth/interactive/2019/02/africa-refugees-190209130248319.html

Mollenkopf, J. (1977). The postwar politics of urban development. In J. Walton & D. E. Carns (Eds.), *Cities in change* (2nd ed., pp. 549–579). Boston: Allyn & Bacon.

Moloney, C. J. (2012). *The buffalo slaughter and the conquest of the West* (Master's thesis, George Washington University.

Moloney, C. J., & Chambliss, W. J. (2014). Slaughtering the bison, controlling Native Americans: A state crime and green criminology synthesis. *Critical Criminology*, *22*, 319–338.

Molotch, H. L. (1976). The city as a growth machine. *American Journal of Sociology*, *82*, 309–333.

Molotch, H. L., & Boden, D. (1985). Talking social structure: Discourse, domination and the Watergate hearings. *American Sociological Review*, 273–288.

Monaghan, A. (2014, November 13). *US wealth inequality—top 0.1% worth as much as the bottom 90%.* Retrieved from https://www.theguardian.com/business/2014/nov/13/us-wealth-inequality-top-01-worth-as-much-as-the-bottom-90

Mongeau, L. (2016). Pulling reservation schools back from the brink. *The Hechinger Report.* Retrieved from http://hechingerreport.org/pulling-reservation-schools-back-brink

Monroe, P. (1940). *Founding of the American public school system.* New York: Macmillan.

Moody, C. (2020). Most brown and Black Americans are exposing themselves to coronavirus for a paycheck. *Vice Online.* Retrieved from https://www.vice.com/en_ca/article/xgqpyq/most-brown-and-black-americans-are-exposing-themselves-to-coronavirus-for-a-paycheck?utm_source=stylizedembed_vice.com&utm_campaign=z3bdmx&site=vice

Moon, B., & Morash., M. (2017). Gender and general strain theory: A comparison of strains, mediating, and moderating effects explaining three types of delinquency. *Youth and Society*, *49*(4), 484–504.

Moon, D., Tobin, T. W., & Sumerau, J. E. (2019). Alpha, omega, and the letters in between: LGBTQI conservative Christians undoing gender. *Gender & Society*, *33*(4), 583–606. Retrieved from https://doi.org/10.1177/0891243219846592

Moore, J., & Pinderhughes, R. (2001). The Latino population: The importance of economic restructuring. In M. L. Andersen & P. H. Collins (Eds.), *Race, class, and gender: An anthology* (4th ed., pp. 251–258). Belmont, CA: Wadsworth.

Moore, J. T., Ricaldi, J. N., & Rose, C. E,. et al. (2020, August 21). *Disparities in incidence of COVID-19 among underrepresented racial/ethnic groups in counties identified as hotspots during June 5–18, 2020–22 states. February–June* 2020. Retrieved from https://doi.org/10.15585/mmwr.mm6933e1

Moore, J. W. (1991). *Going down to the barrio: Homeboys and homegirls in change.* Philadelphia, PA: Temple University Press.

Moore, L. R. (1994). *Selling God: American religion in the marketplace culture.* New York: Oxford University Press.

Moore, P. (2016, April 26). Most Americans think released felons should have the vote. *YouGov.* Retrieved from https://today.yougov.com/news/2016/04/26/most-americans-think-released-felons-should-have-v

Moreno, E. J. (2020, August 21). Zuckerberg questioned by FTC in Facebook antitrust probe. *The Hill.* Retrieved from https://thehill.com/policy/technology/513069-zuckerberg-questioned-by-ftc-in-facebook-antitrust-probe

Morgan, R. E., & Kena, G. (2017, December). *Crime victimization, 2016.* Washington, DC: Bureau of Justice Statistics. Retrieved from https://www.bjs.gov/content/pub/pdf/cv16.pdf

Morgan, R. E., & Oudekerk, B. A. (2019). *Criminal victimization 2018.* Bureau of Justice Statistics. Retrieved from https://www.bjs.gov/content/pub/pdf/cv18.pdf

Morishima, M. (1982). *Why has Japan "succeeded"? Western technology and the Japanese ethos.* New York: Cambridge University Press.

Mosely, T. (2020, May 21). *Addiction experts turn to telemedicine to help patients. National Public Radio.* Retrieved from https://www.wbur.org/hereandnow/2020/05/21/addiction-treatment-telemedicine

Moser, S., & Dilling, L. (2004). Making climate hot: Communicating the urgency and challenge of global climate change. *Environment*, *46*, 32–46.

Motlagh, J. (2012, September 19). In a world hungry for cheap shrimp, migrants labor overtime in Thai sheds. *Washington Post.* Retrieved from http://www.washingtonpost.com/world/asia_pacific/in-a-world-hungry-for-cheap-shrimp-migrants-labor-overtime-in-thai-sheds/2012/09/19/3435a90e-01a4-11e2-b257-e1c2b3548a4a_story.html

Mueller, A. S., Abrutyn, S., & Osborne, M. (2017). Durkheim's "Suicide" in the zombie apocalypse. *Contexts*, *16*(2), 44–49.

Muller, J. (2014, February 15). UAW's loss and what it means for your paycheck. *Forbes.* Retrieved from http://www.forbes.com/sites/joannmuller/2014/02/15/uaws-loss-and-what-it-means-for-your-paycheck

Muller, T., & Espenshade, T. J. (1985). *The fourth wave: California's newest immigrants.* Washington, DC: Urban Institute Press.

Mumford, L. (1961). *The city in history: Its origins, its transformations, and its prospects.* New York: Harcourt.

Muncer, S. J., & Campbell, A. (2000). Comments on "Sex differences in beliefs about aggression: Opponent's sex and the form of aggression" by. J. Archer and A. Haigh. *British Journal of Social Psychology*, *39*, 309–311.

Münchener, Rückversicherungs-Gesellschaft. (2016). *Natural loss events worldwide 2015.* Retrieved from www.munichre.com/site/wrap/get/documents_E1656163460/mram/assetpool.munichreamerica.wrap/PDF/07Press/2015_World_map_of_nat_cats.pdf

Muncie, J. (2019). Biological criminology. In E. McLaughlin (Ed.), *The SAGE dictionary of criminology* (4th ed.). Thousand Oaks, CA: SAGE.

Murdock, G. P. (1949). *Social structure.* New York: Macmillan.

Murphy, C. (2007). *Are we Rome? The fall of an empire and the fate of America.* New York, NY: Houghton Mifflin Harcourt.

Murphy, M. (2012, December 18). *But what about the men? Masculinity and mass shootings.* Retrieved from http://www.feministcurrent.com/2012/12/18/but-what-about-the-men-on-masculinity-and-mass-shootings/

Murphy, M. (2019, April 19). Cellphones now outnumber the world's population. *Quartz.* Retrieved from https://qz.com/1608103/there-are-now-more-cellphones-than-people-in-the-world/

Murphy, R. (1988). *Social closure: The theory of monopolization and exclusion.* Oxford: Clarendon.

Mutchler, J. E., Baker, L. E., & Lee, S. (2007). Grandparents responsible for grandchildren in Native-American families. *Social Science Quarterly, 88*, 990–1009.

Mutharayappa, R., Choe, M. K., Arnold, F., & Roy, T. K. (1997, March). *Son preference and its effect on fertility in India* (National Family Survey Subject Reports No. 3). Honolulu: East-West Center Program on Population. Retrieved from http://scholarspace.manoa.hawaii.edu/bitstream/handle/10125/3475/NFHSsubjrpt003.pdf?sequence=1

Myrdal, G. (1963). *Challenge to affluence*. New York: Random House.

NACE staff. (2020a, July 24). *Racial disproportionalities exist in terms of intern representation.* National Association of Colleges and Employers (NACE). Retrieved from https://www.naceweb.org/diversity-equity-and-inclusion/trends-and-predictions/racial-disproportionalities-exist-in-terms-of-intern-representation/

NACE staff. (2020b, August 7). *Women are underrepresented among paid interns*. National Association of Colleges and Employers (NACE). Retrieved from https://www.naceweb.org/diversity-equity-and-inclusion/trends-and-predictions/women-are-underrepresented-among-paid-interns/

NACE staff. (2020c, August 21). *First generation students are underrepresented in internships.* National Association of Colleges and Employers (NACE). Retrieved from https://www.naceweb.org/diversity-equity-and-inclusion/trends-and-predictions/first-generation-students-underrepresented-in-internships/

Nadworny, E., & Kamenetz, A. (2019, March 13). *Does it matter where you go to college? Some context for the admissions scandal*. National Public Radio. Retrieved from https://www.npr.org/2019/03/13/702973336/does-it-matter-where-you-go-to-college-some-context-for-the-admissions-scandal

Nall, R. (2020, April 24). The history of HIV and AIDS in the United States. *Healthline*. Retrieved from https://www.healthline.com/health/hiv-aids/history

Narayan, U., & Harding, S. (2000). *Decentering the center: Philosophy for a multicultural, postcolonial, and feminist world.* Bloomington: Indiana University Press.

Nash, B. (n.d.a). *The numbers:* Black Panther. Nash Information Services. Retrieved from https://www.the-numbers.com/movie/Black-Panther

Nash, B. (n.d.b). *Top 2019 movies at the worldwide box office: The numbers*. Nash Information Services. Retrieved from https://www.the-numbers.com/box-office-records/worldwide/all-movies/cumulative/released-in-2019

National Archives. (2020). *Influenza 1918 to COVID-19*. Retrieved from https://usnatarchives.tumblr.com/post/615386175560105984/influenza-1918-to-covid-19.

National Association of Realtors. (2012). *The digital house hunt: Consumer and market trends in real estate*. Retrieved from http://www.realtor.org/sites/default/files/Study-Digital-House-Hunt-2013-01_1.pdf

National Association of the Deaf. (2000). *NAD position statement on cochlear implants*. Retrieved from http://www.nad.org/issues/technology/assistive-listening/cochlear-implants

National Campaign to Prevent Teen Pregnancy. (2010). *Why it matters: Teen pregnancy, poverty, and income disparity.* Retrieved from http://www.thenationalcampaign.org

National Center for Charitable Statistics. (2018). *The non-profit sector in brief 2018*. Washington, DC: Urban Institute. Retrieved from https://nccs.urban.org/publication/nonprofit-sector-brief-2018#the-nonprofit-sector-in-brief-2018-public-charites-giving-and-volunteering

National Center for Education Statistics. (2016a, May). *Immediate college enrollment rate*. Retrieved from http://nces.ed.gov/programs/coe/indicator_cpa.asp

National Center for Education Statistics. (2016b). *Annual earnings of young adults*. Retrieved from http://nces.ed.gov/programs/coe/indicator_cba.asp

National Center for Education Statistics. (2018). *Number of instructional days and hours in the school year, by state: 2018*. Retrieved from https://nces.ed.gov/programs/statereform/tab5_14.asp

National Center for Education Statistics. (2019a). *Racial/ethnic enrollment in public schools*. Retrieved from https://nces.ed.gov/programs/coe/indicator_cge.asp

National Center for Education Statistics. (2019b). *The condition of education 2019 (NCES 2019-144). Undergraduate retention and graduation rates*. Retrieved from https://nces.ed.gov/programs/coe/indicator_ctr.asp

National Center for Education Statistics. (2020a, May). *College enrollment rates*. Retrieved from https://nces.ed.gov/programs/coe/indicator_cpb.asp

National Center for Education Statistics. (2020b, May). *Employment outcomes of bachelor's degree holders*. Retrieved from https://nces.ed.gov/programs/coe/indicator_sbc.asp

National Center for Injury Prevention and Control. (2014). *Understanding teen dating violence: Fact sheet*. Washington, DC: Centers for Disease Control and Prevention. Retrieved from http://www.cdc.gov/violenceprevention/pdf/teen-dating-violence-2014-a.pdf

National Center for Victims of Crime. (2012). *Intimate partner violence*. Retrieved from http://www.victimsofcrime.org/library/crime-information-and-statistics/intimate-partner-violence

National Center on Elder Abuse. (2020). *What is known about the incidence and prevalence of elder abuse in the community setting?* Retrieved from https://ncea.acl.gov/What-We-Do/Research/Statistics-and-Data.aspx

National Coalition Against Domestic Violence. (2018, June 6). *Domestic violence and the LGBTQ community*. Retrieved from https://ncadv.org/blog/posts/domestic-violence-and-the-lgbtq-community

National Conference of State Legislatures. (2020). *Felon voting rights*. Retrieved from https://www.ncsl.org/research/elections-and-campaigns/felon-voting-rights.aspx

National Employment Law Project. (2014, April). *The low-wage recovery: Industry employment and wages four years into the recovery* (Data Brief). New York: Author. Retrieved from http://www.nelp.org/page/-/reports/low-wage-recovery-industry-employment-wages-2014-report.pdf?nocdn=1

National Institute of Drug Abuse. (2019). *Opioid summaries by state*. Retrieved from https://www.drugabuse.gov/drugs-abuse/opioids/opioid-summaries-by-state

National Institute of Justice. (2011, September 13). *Terrorism*. Retrieved from http://www.nij.gov/topics/ crime/terrorism/welcome.htm

National Institute of Mental Health. (2020). *Eating disorders.* https://www.nimh.nih.gov/health/topics/eating-disorders/index.shtml

National Institute on Drug Abuse. (2019). *Monitoring the future study: Trends in prevalence of various drugs for 8th graders, 10th graders and 12th graders; 2016–2019 (in percent).* Rockville, MD. Retrieved from https://www.drugabuse.gov/drug-topics/trends-statistics/monitoring-future/monitoring-future-study-trends-in-prevalence-various-drugs

National Oceanic and Atmospheric Administration (NOAA). (2020). *2019 was the second warmest year on record.* Retrieved from https://www.climate.gov/news-features/featured-images/2019-was-second-warmest-year-record

National Park Service. (n.d.). *African Americans and education during Reconstruction: The Tolson's Chapel schools.* Retrieved from https://www.nps.gov/articles/african-americans-and-education-during-reconstruction-the-tolson-s-chapel-schools.htm

National Partnership for Women and Families. (2020, September). *America's women and the wage gap.* Retrieved from https://www.nationalpartnership.org/our-work/resources/economic-justice/fair-pay/americas-women-and-the-wage-gap.pdf

National Public Radio. (2010, November 5). Counting crowds: Results may vary. *Science Friday.* Retrieved from http://www.npr.org/templates/story/story.php?storyId=131099075

National Retail Federation. (2019, September 25). *Social media influencing near-record Halloween spending.* Retrieved from https://nrf.com/media-center/press-releases/social-media-influencing-near-record-halloween-spending

National Student Clearinghouse Research Center. (2019, October 7). *High school benchmarks—2019.* Retrieved from https://nscresearchcenter.org/high-school-benchmarks-2019/

National Survey of Student Engagement. (2012). *Fostering student engagement campuswide: Annual results 2012.* Bloomington: Indiana University Center for Postsecondary Research. Retrieved from http://nsse.iub.edu/html/annual_results.cfm

Naylor, N. T. (2002). *Wages of crime: Black markets, illegal finance, and the underworld economy.* Ithaca, NY: Cornell University Press.

NBC News. (2015, April 22). *ISIS using social media and violence to recruit* [Video]. Retrieved from http://www.nbcnews.com/watch/long-story-short/isis-using-social-media-and-violence-to-recruit-432161347692

NCPSSM. (2018). *Medicare fast facts.* Retrieved from https://www.ncpssm.org/our-issues/medicare/medicare-fast-facts/

Neate, R. (2015, May 3). *America's trailer parks: The residents may be poor but the owners are getting rich.* Retrieved from http://www.theguardian.com/lifeandstyle/2015/may/03/owning-trailer-parks-mobile-home-university-investment

Neuman, W. L. (2000). *Social research methods: Qualitative and quantitative approaches.* Toronto: Allyn & Bacon.

Neus, N. (2020, July 29). Major children's hospital apologizes for performing cosmetic genital surgeries on intersex infants. *CNN Health.* Retrieved from https://www.cnn.com/2020/07/29/health/intersex-surgeries-chicago-hospital/index.html

New America Foundation. (2012). *Federal education budget project.* Retrieved from http://febp.newamerica.net/k12

Newcomb T. C. (2008). *Parameters of parenting in Native American families* (Doctoral dissertation, Oklahoma State University) Retrieved from ProQuest (3320882)

Newman, K. (2019, June 28). America has a health equity problem. *U.S. News & World Report Online.* https://www.usnews.com/news/healthiest-communities/articles/2019-06-28/health-equity-in-america-still-a-problem.

Newport, F. (2018, May 22). *In U.S., estimate of LGBT population rises to 4.5%.* Retrieved from https://news.gallup.com/poll/234863/estimate-lgbt-population-rises.aspx

Newsom, J. S. (Director). (2015). *The mask you live in* (DVD). The Representation Project.

Neyazi, T. A. (2010). Cultural imperialism or vernacular modernity? Hindi newspapers in a globalizing India. *Media, Culture & Society, 32,* 907–924.

Ng, G. (2019, February 27). In 2019, Oscars isn't so White. But let's aim for a colourful Oscars. *EOnline.* Retrieved from https://www.eonline.com/ap/news/1018585/in-2019-oscars-isn-t-so-white-but-let-s-aim-for-a-colourful-oscars

Ng, S., & Deng, F. (2017, August 22). Language and power. *Oxford Research Encyclopedia of Communication.* Retrieved from https://oxfordre.com/communication/view/10.1093/acrefore/9780190228613.001.0001/acrefore-9780190228613-e-436

Nicholas, S. E. (2009). "I live Hopi, I just don't speak it": The critical intersection of language, culture and identity in the lives of contemporary Hopi youth. *Journal of Language, Identity & Education, 8,* 321–334.

Nichols, M. (2020, October 1). *U.N. to tackle gender equality, chief calls it "greatest" rights challenge. Reuters.* Retrieved from https://www.reuters.com/article/women-un-idUSL1N2GR1LN

Niebuhr, H. R. (1929). *The social sources of denominationalism.* New York: Meridian Books.

Nielsen. (2017). *The Nielsen total audience report: Q2 2017.* Retrieved from http://www.nielsen.com/us/en/insights/reports/2017/the-nielsen-total-audience-q2-2017.html

Nielsen. (2018). *The Nielsen total audience report Q3 2018.* www.nielsen.com

Nielsen. (2020). *Playback time: Which consumer attitudes will shape the streaming wars?* Retrieved from https://www.nielsen.com/us/en/insights/article/2020/playback-time-which-consumer-attitudes-will-shape-the-streaming-wars/

Nisbet, R. (1970). *The social bond: An introduction to the study of society.* New York: Knopf.

Noe-Bustamante, L., Flores, A., & Shah, S. (2019, September 16). *Facts on Hispanics of Cuban origin in the United States, 2017.* Retrieved from https://www.pewresearch.org/hispanic/fact-sheet/u-s-hispanics-facts-on-cuban-origin-latinos/

Nolan, P., & Lenski, G. (2009). *Human societies: An introduction to macrosociology.* Boulder, CO: Paradigm.

Nomaguchi, K., & Milkie, M. A. (2020). Parenthood and well—being: A decade in review. *Journal of Marriage & Family, 82*(1), 198–223. Retrieved from https://doi-org.thor.nebrwesleyan.edu/10.1111/jomf.12646

Nordhoff, C. (1975). *The communistic societies of the United States.* New York, NY: Harper & Row. (Original work published 1875)

Nuland, S. B. (2003). *The doctors' plague: Germs, childbed fever, and the strange story of Ignac Semmelweis*. New York: Norton.

Oakes, J. (1985). *Keeping track: How schools structure inequality*. New Haven, CT: Yale University Press.

Oakley, M., Farr, R. H., & Scherer, D. G. (2017). Same-sex parent socialization: Understanding gay and lesbian parenting practices as cultural socialization. *Journal of GLBT Family Studies, 13*(1), 56–75. doi:10.1080/1550428X.2016.1158685

Obama administration must do more to protect children harvesting tobacco. (2014, May 18). *Washington Post*. Retrieved from http://www.washingtonpost.com/opinions/obama-administration-must-do-more-to-protect-children-harvesting-tobacco/2014/05/18/23b8a7c4-dd36-11e3-b745-87d39690c5c0_story.html

Oberschall, A. (1973). *Social conflict and social movements*. Englewood Cliffs, NJ: Prentice Hall.

O'Dea, R. E., Lagisz, M., & Jennions, M. D., & Nakagawa, S. (2018). Gender differences in individual variation in academic grades fail to fit expected patterns for STEM. *Nat Commun 9*, 3777. Retrieved from https://doi.org/10.1038/s41467-018-06292-0

O'Dea, T. (1966). *The sociology of religion*. Upper Saddle River, NJ: Prentice Hall.

Offe, C. (1984). *Contradictions of the welfare state*. Cambridge: MIT Press.

Ogden, C. L., Carroll, M. D., Kit, B. K., & Flegal, K. M. (2012). *Prevalence of obesity in the United States, 2009–2010* (NCHS Data Brief 82). Hyattsville, MD: National Center for Health Statistics. Retrieved from http://www.cdc.gov/nchs/data/databriefs/db82.pdf

Ogunlesi, T., & Busari, S. (2012, September 14). *Seven ways mobile phones have changed lives in Africa. CNN*. Retrieved from http://www.cnn.com/2012/09/13/world/africa/mobile-phones-change-africa

O'Kane, C. (2020, June 17). *Aunt Jemima to change name and image due to origins based on a racial stereotype*. Retrieved from https://www.cbsnews.com/news/aunt-jemima-change-name-image-racial-stereotype/

Okun, A. (2013, October 9). *Some terrible people on Twitter have decided that it's "Fat Shaming Week."* Retrieved from https://www.buzzfeed.com/alannaokun/some-terrible-people-on-twitter-have-decided-that-its-fat-sh?utm_term=.cxw62AABWz#.ogZvdDDLjw

Olivieri, E. (2014). *Occupational choice and the college gender gap* [Working paper]. Retrieved from https://docs.google.com/viewer?a=v&pid=sites&srcid=ZGVmYXVsdGRvbWFpbnxlbGlzYW9saXZpZXJpfGd4OjhkNmI5NTg2NTgyOTYx

Omvedt, G. (1992). "Green earth, women's power, human liberation": Women in peasant movements in India. *Development Dialogue, 1*(2), 116–130.

O'Neill, T. (2007, February). Curse of the black gold: Hope and betrayal on the Niger Delta. *National Geographic*. Retrieved from http://ngm.nationalgeographic.com/2007/02/nigerian-oil/oneill-text

Open Doors Data. (2019). *Fast facts*. Retrieved from https://opendoorsdata.org/fast_facts/

Orfield, G., & Eaton, S. E. (1996). *Dismantling desegregation: The quiet reversal of* Brown v. Board of Education. New York: Norton.

Organisation for Economic Co-operation and Development. (2011). Chart A1.1. Percentage of population that has attained tertiary education, by age group (2009). In *Education at a glance*. Paris: Author. Retrieved from http://www.oecd.org/education/highereducationandadultlearn-ing/48630299.pdf

Ortiz, E., & Lubell, B. (2017, May 9). Penn State fraternity death: Why did no one call 911 after Timothy Piazza got hurt. *NBC News*. Retrieved from https://www.nbcnews.com/news/us-news/penn-state-fraternity-death-why-did-no-one-call-911-n756951

O'Sullivan, D., & Toropon, K. (2020, September 11). *QAnon fans spread fake claims about real fires in Oregon*. Retrieved from https://www.cnn.com/2020/09/11/tech/qanon-oregon-fire-conspiracy-theory/index.html

Owens, D. C. (2018). *Medical bondage, race, gender, and the origins of medical gynecology*. Athens: University of Georgia Press.

Owens, J. (2020). Relationships between an ADHD diagnosis and future school behaviors among children with mild behavioral problems. *Sociology of Education, 93*(3), 191–214. Retrieved from https://doi.org/10.1177/0038040720909296

Oxfam. (2015). A decade of saving for change [Brochure]. Retrieved from https://www.oxfamamerica.org/static/media/files/SFC timeline-final-AA.pdf

Oxfam. (2020, October 5). *5 shocking facts about extreme global inequality and how to even it up*. Retrieved from https://www.oxfam.org/en/5-shocking-facts-about-extreme-global-inequality-and-how-even-it

Oxner, R. (2020, October 16). *Disney warns viewers of racism in some classic movies with strengthened label*. Retrieved from https://www.npr.org/sections/live-updates-protests-for-racial-justice/2020/10/16/924540535/disney-warns-viewers-of-racism-in-some-classic-movies-with-strengthened-label

Page, S. J. (2020). Sexuality. *The SAGE encyclopedia of the sociology of religion* (pp. 745–747). Thousand Oaks, CA: SAGE.

Pager, D. (2003). The mark of a criminal record. *American Journal of Sociology, 108*(5), 937–975.

Paglen, T., & Thompson, A. C. (2006). *Torture taxi: On the trail of the CIA's rendition flights*. Hoboken, NJ: Melville House.

Pal, A. (2019, January 19). *From pariah to demi-god: Transgender leader a star at massive Indian festival*. Retrieved from https://www.reuters.com/article/us-india-religion-kumbh-transgender-idUSKCN1PE04D

Panetta, G., & Liu, Y. (2020, August). In 34 states, you'll need to show ID to vote on Election Day. See what the law requires in each state. *Business Insider*. Retrieved from https://www.businessinsider.com/voter-identification-requirements-in-each-state-2020-8

Paniz-Mondolfi, A. E., Tami, A., Grillet, M. E., Marquez, M., Hernandez-Villena, J., Escalona-Rodriguez, M. A., . . . & Oletta, J. (2019). Resurgence of vaccine-preventable diseases in Venezuela as a regional public health threat in the Americas. *Emerging Infectious Diseases, 25*(4), 625–632.

Papachristos, A. V., Smith, C. M., Scherer, M. L., & Fugiero, M. A. (2011). More coffee, less crime? The relationship between gentrification and neighborhood crime rates in Chicago, 1991 to 2005. *City & Community, 10*, 215–240.

Pape, R. (2005). *Dying to win: The strategic logic of suicide terrorism*. New York: Random House.

Paquette, D. (2016, March 10). Pay doesn't look the same for men and women at top newspapers. *Washington Post*. Retrieved from https://www.washingtonpost.com/news/wonk/wp/2016/03/10/pay-doesnt-look-the-same-for-men-and-women-at-top-newspapers

Parente, S. L. (2008). Narrowing the economic gap in the 21st century. K. R. Holmes, E. J. Feulner, & M. A. O'Grady (Eds.), *2008 index of economic freedom*. Washington, DC: Heritage Foundation.

Parker, K. (2012, September 20). *Where the public stands on government assistance, taxes, and presidential candidates*. Pew Research. Retrieved from http://www.pewsocialtrends.org/2012/09/20/where-the-public-stands-on-government-assistance-taxes-and-the-presidential-candidates/

Parker, K. (2018, March 7). Women in majority-male workplaces report higher rates of gender discrimination. Retrieved from https://www.pewresearch.org/fact-tank/2018/03/07/women-in-majority-male-workplaces-report-higher-rates-of-gender-discrimination/

Parker, K., Graf, N., & Igielnik, R. (2019, January 17). Generation Z looks a lot like millennials on key social and political issues. Pew Research Center. Retrieved from https://www.pewsocialtrends.org/2019/01/17/generation-z-looks-a-lot-like-millennials-on-key-social-and-political-issues/

Parker, K., & Igielnik, R. (2020, March 7). On the cusp of adulthood and facing an uncertain future what we know about Gen Z so far. Retrieved from https://www.pewsocialtrends.org/essay/on-the-cusp-of-adulthood-and-facing-an-uncertain-future-what-we-know-about-gen-z-so-far/

Parker, K., Morin, R., & Horowitz, J. (2019, March 19). *Looking to the future, public sees an America in decline on many fronts*. Retrieved from https://www.pewsocialtrends.org/2019/03/21/public-sees-an-america-in-decline-on-many-fronts/

Parker-Pope, T. (2010, April 14). Is marriage good for your health? *New York Times Magazine*. Retrieved from http://www.nytimes.com/2010/04/18/magazine/18marriage-t.html?pagewanted=all

Parkin, F. (1979). Social closure and class formation. In A. Giddens & D. Held (Eds.), *Classes, power, and conflict* (pp.175–184). Los Angeles: University of California Press.

Parrado, E. A., & Morgan, S. P. (2008). Intergeneration fertility among Hispanic women: New evidence of immigrant assimilation. *Demography, 45*, 651–671.

Parsons, T. (1951). *The social system*. New York: Free Press.

Parsons, T. (1954). The kinship system of the contemporary United States. In *Essays in sociological theory*. (pp. 189–194). New York: Free Press.

Parsons, T. (1960). Some principle characteristics of industrial societies. In T. Parson (Ed.), *Structure and process in modern societies*. New York: Free Press. (pp. 132–168)

Parsons, T. (1966). *Societies: Evolutionary and comparative perspectives*. Upper Saddle River, NJ: Prentice Hall.

Parsons, T. (1967). *The structure of social action*. New York: Free Press.

Parsons, T. (1975). The sick role and the role of the physician reconsidered. *Milbank Memorial Fund Quarterly, Health and Society, 53*, 257–278.

Parsons, T. (2007). *Social structure and personality*. New York: Free Press. (Original work published 1964)

Parsons, T., & Bales, R. F. (1955). *Family, socialization and interaction process*. Glencoe, IL: Free Press.

Parsons, T., & Mayhew, H. D. (1982). *On institutions and social evolution: Selected writings*. Chicago: University of Chicago Press.

Parsons, T., & Smelser, N. J. (1956). *Economy and society*. New York: Free Press.

Pascoe, C. J. (2007). *Dude, you're a fag: Masculinity and sexuality in high school*. Berkeley: University of California Press.

Patel, A. (2013, November 23). Horrors of India's brothels documented. *BBC*. Retrieved from http://www.bbc.com/news/world-asia-india-24530198

Patterson, D. (1989). *Power in law enforcement: Subordinate preference and actual use of power base in special weapons teams (SWAT)* (Doctoral dissertation, Fielding Institute Santa Barbara, CA).

Pauly, M. (2019, November 21). It's 2019, and states are still making exceptions for spousal rape. *Mother Jones*. Retrieved from https://www.motherjones.com/crime-justice/2019/11/deval-patrick-spousal-rape-laws

PBS. (2010, May 31). China faces growing health threat from prevalent tobacco use. *NewsHour*. Retrieved from http://www.pbs.org/newshour/bb/health/jan-june10/tobacco_05–31.html

PBS. (2011). Caught in the crossfire: Arab Americans. *9/11 stories*. Retrieved from http://www.pbs.org/itvs/caughtinthecrossfire/arab_americans.html

Pearlstein, S. (2010, October 6). The costs of rising economic inequality. *Washington Post*. Retrieved from http://www.washingtonpost.com/wp-dyn/content/article/2010/10/05/AR2010100505535.html

Pecanha, S., & Wallace, T. (2015, June 20). The f light of refugees around the globe. *New York Times*. Retrieved from http://www.nytimes.com/interactive/2015/06/21/world/map-flow-desperate-migration-refugee-crisis.html?_r=1

Peek, K. L. (1999). *The good, the bad, and the "misunderstood": A study of the cognitive moral development theory and ethics in the public sector*. Fort Lauderdale, FL: Nova Southeastern University.

Peguero, A. A., Varela, K. S., Marchbanks, M. P., Blake, J., & Eason, J. M. (2018, October). School punishment and education: Racial/ethnic disparities with grade retention and the role of urbanicity. Urban Education. doi:10.1177/0042085918801433

Perlin, R. (2011). *Intern nation: How to earn nothing and learn little in the brave new economy*. New York: Verso.

Perlin, R. (2012, February 6). These are not your father's internships. *New York Times*. Retrieved from http://www.nytimes.com/roomfordebate/2012/02/04/do-unpaid-internships-exploit-college-students/todays-internships-are-a-racket-not-an-opportunity

Perlman, M. (2019, January 22). *How the word "queer" was adopted by the LGBTQ community*. Retrieved from https://www.cjr.org/language_corner/queer.php

Perrin, A. (2018). *5 facts about Americans and video games*. Washington, DC: Pew Research Center. Retrieved from https://pewrsr.ch/2vbbMxD

Perrin, A., & Anderson, M. (2019, April 10). *Share of U.S. adults using social media, including Facebook, is mostly unchanged since 2018*. Retrieved from https://pewrsr.ch/2VxJuJ3

Perrin, A., & Kumar, M. (2019, July 25). *About three-in-ten U.S. adults say they are "almost constantly" online*. Retrieved from https://www.pewresearch.org/fact-tank/2019/07/25/americans-going-online-almost-constantly/

Pettit, K. L. S., & Reuben, K. (2010). *Investor-owners in the boom and bust*. Urban Institute. Retrieved from http://metrotrends.org/mortgagelending.html

Pew Forum on Religion and Public Life. (2004, June 14). Supreme Court upholds "under God" in Pledge of Allegiance: Court overturns lower court ruling on legal technicality. Retrieved from http://www.pewforum.org/Press-Room/Press-Releases/Supreme-Court-Upholds-Under-God-in-Pledge-of-Allegiance.aspx

Pew Forum on Religion and Public Life. (2008). *Summary of key findings: U.S. Religious Landscape Survey*. Washington, DC: Pew Research Center.

Pew Forum on Religion and Public Life. (2009). *Mapping the global Muslim population: A report on the size and distribution of the world's Muslim population*. Washington, DC: Pew Research Center. Retrieved from http://www.pewforum.org/2009/10/07/mapping-the-global-muslim-population/

Pew Forum on Religion and Public Life. (2010). *Religion among the millennials*. Washington, DC: Pew Research Center. Retrieved from http://www.pewforum.org/files/2010/02/millennials-report.pdf

Pew Forum on Religion and Public Life. (2011). *The future of the global Muslim population: Projections for 2010–n.d.* Washington, DC: *Pew Research Center*. Retrieved from http://www.pewforum.org/files/2011/01/FutureGlobalMuslimPopulation-WebPDF-Feb10.pdf

Pew Forum on Religion and Public Life. (2012). *'Nones' on the rise*. Retrieved from http://www.pewforum.org/2012/10/09/nones-on-the-rise/

Pew Forum on Religion and Public Life. (2014). *Religious landscape study*. Retrieved from http://www.pewforum.org/religious-landscape-study/

Pew Forum on Religion and Public Life. (2015a). *America's changing religious landscape*. Washington, DC: Pew Research Center. Retrieved from http://www.pewforum.org/2015/05/12/americas-changing-religious-landscape/

Pew Forum on Religion and Public Life. (2015b). *Chapter 2: Religious switching and intermarriage*. Retrieved from https://www.pewforum.org/2015/05/12/chapter-2-religious-switching-and-intermarriage/

Pew Forum on Religion and Public Life. (2015c). *The future of world religions: Population growth projections, 2010–2050*. Retrieved from https://www.pewforum.org/2015/04/02/religious-projections-2010-2050

Pew Forum on Religion and Public Life. (2016, March 26). *The gender gap in religion around the world*. Retrieved from https://www.pewforum.org/2016/03/22/the-gender-gap-in-religion-around-the-world/

Pew Forum on Religion and Public Life. (2017a, April 6). *The changing global religious landscape*. Retrieved from https://www.pewforum.org/2017/04/05/the-changing-global-religious-landscape/

Pew Forum on Religion and Public Life. (2017b, May 10). *Religious belief and belonging in Central and Eastern Europe*. Retrieved from https://www.pewforum.org/2017/05/10/religious-affiliation/

Pew Forum on Religion and Public Life. (2018a, April 25). *When Americans say they believe in God, what do they mean?* Retrieved from https://www.pewforum.org/2018/04/25/when-americans-say-they-believe-in-god-what-do-they-mean/

Pew Forum on Religion and Public Life. (2018b, August 1). *Why Americans go (and don't go) to religious services. Washington, DC: Pew Research Center*. Retrieved from https://www.pewforum.org/2018/08/01/why-americans-go-to-religious-services/

Pew Forum on Religion and Public Life. (2019a, July 16). *A closer look at how religious restrictions have risen around the world*. Retrieved from https://www.pewforum.org/2019/07/15/a-closer-look-at-how-religious-restrictions-have-risen-around-the-world/

Pew Forum on Religion and Public Life. (2019b, October 17). *In U.S., decline of Christianity continues at rapid pace*. Washington, DC: Pew Research Center. Retrieved from https://www.pewforum.org/2019/10/17/in-u-s-decline-of-christianity-continues-at-rapid-pace/

Pew Forum on Religion and Public Life. (2019c, October 17). *Protestants and Catholics shrinking as share of US population "nones" are growing*. Retrieved from https://www.pewforum.org/2019/10/17/in-u-s-decline-of-christianity-continues-at-rapid-pace/pf_10-17-19_rdd-update-new3

Pew Forum on Religion and Public Life. (2019d, November 15). *Americans have positive views about religion's role in society, but want it out of politics*. Retrieved from https://www.pewforum.org/2019/11/15/americans-have-positive-views-about-religions-role-in-society-but-want-it-out-of-politics/

Pew Research Center. (2011a). *Republican candidates stir little enthusiasm*. Washington, DC: Author. Retrieved from http://www.people-press.org/files/legacy-pdf/06-02-11%202012%20Campaign%20Release.pdf

Pew Research Center. (2011b). *Sunni and Shia Muslims*. Retrieved from http://www.pewforum.org/2011/01/27/future-of-the-global-muslim-population-sunni-and-shia/

Pew Research Center. (2012, March). *Faith on the move – the religious affiliation of international migrants. Washington, DC*. Retrieved from http://www.pewforum.org/2012/03/08/religious-migration-exec/

Pew Research Center. (2013). *A survey of LGBT Americans*. Retrieved from https://www.pewsocialtrends.org/2013/06/13/a-survey-of-lgbt-americans/

Pew Research Center. (2014a, January 27). *Climate change: Key data points from Pew Research*. Retrieved from http://www.pewresearch.org/key-data-points/climate-change-key-data-points-from-pew-research

Pew Research Center. (2014b). *Older adults and technology use*. Washington, DC: Author. Retrieved from http://www.pewinternet.org/files/2014/04/PIP_Seniors-and-Tech-Use_040314.pdf

Pew Research Center. (2015a). *Latest trends in religious restrictions and hostilities*. Retrieved from http://www.pewforum.org/2015/02/26/religious-hostilities

Pew Research Center. (2015b). *Parenting in America*. Washington, DC. Retrieved from http://www.pewsocialtrends.org/2015/12/17/parenting-in-america/

Pew Research Center. (2015c). *Projected changes in the global Hindu population*. Retrieved from https://www.pewforum.org/2015/04/02/hindus/

Pew Research Center. (2016). *Faith and the 2016 campaign*. Retrieved from http://www.pewforum.org/2016/01/27/faith-and-the-2016-campaign/

Pew Research Center. (2017a, April). *Global restrictions on religion rise modestly in 2015, reversing downward trend*. Retrieved from http://www.pewforum.org/2017/04/11/=global-restrictions-on-religion-rise-modestly-in-2015-reversing-downward-trend/

Pew Research Center. (2017b, July 26). *U.S. Muslims concerned about their place in society, but continue to believe in the American Dream*. Retrieved from https://www.pewforum.org/2017/07/26/findings-from-pew-research-centers-2017-survey-of-us-muslims

Pew Research Center. (2018a, February 5). *Social media fact sheet*. Retrieved from http://www.pewinternet.org/fact-sheet/social-media/

Pew Research Center. (2018b, May 22). *Demographic and economic trends in urban, suburban, and rural communities*. Retrieved from https://www.pewsocialtrends.org/2018/05/22/demographic-and-economic-trends-in-urban-suburban-and-rural-communities/

Pew Research Center. (2019a, May 16). *Trump's staunch GOP supporters have roots in the Tea Party*. Retrieved from https://www.pewresearch.org/politics/2019/05/16/trumps-staunch-gop-supporters-have-roots-in-the-tea-party/

Pew Research Center. (2019b). *Newspapers fact sheet*. Retrieved from https://www.journalism.org/fact-sheet/newspapers/

Pew Research Center. (2020, August). *Important issues in the 2020 election*. Retrieved from https://www.pewresearch.org/politics/2020/08/13/important-issues-in-the-2020-election/

Pew Research Center for the People and the Press. (2012, November 26). *Young voters supported Obama less, but may have mattered more*. Retrieved from http://www.people-press.org/2012/11/26/young-voters-supported-obama-less-but-may-have-mattered-more

Phelan, A. M., & McLaughlin, H. J. (1995). Educational discoveries: The nature of the child and practices of new teachers. *Journal of Teacher Education*, *46*, 165–174.

Piaget, J. (1926). *The language and thought of the child*. New York: Harcourt, Brace.

Piaget, J. (1928). *Judgment and reasoning in the child*. New York: Harcourt, Brace.

Piaget, J. (1930). *The child's conception of physical causality*. New York: Harcourt, Brace.

Piaget, J. (1932). *The moral judgment of the child*. New York: Harcourt, Brace.

Pierce, A. (2015, July 22). *Hey feminism, what's new?* Retrieved from https://now.org/blog/hey-feminism-whats-new/

Pierre-Louis, K., & Schwartz, J. (2020, August 22). Why does California have so many wildfires? *New York Times*. Retrieved from https://www.nytimes.com/article/why-does-california-have-wildfires.html

Pike, L. (2020, November 7). Why so many young people showed up on Election Day. *Vox*. Retrieved from https://www.vox.com/2020/11/7/21552248/youth-vote-2020-georgia-biden-covid-19-racism-climate-change

Pipes, D., & Durán, K. (2002, August). *Muslim immigrants in the United States*. Washington, DC: Center for Immigration Studies. Retrieved from http://www.cis.org/sites/cis.org/files/articles/2002/back802.pdf

Piven, F. F., & Cloward, R. A. (1977). *Poor people's movements: Why they succeed, how they fail*. New York, NY: Random House.

Podsakoff, P., & Schriesheim, C. (1985). Field studies of French and Raven's bases of power: Critique, reanalysis, and suggestions for future research. *Psychological Bulletin*, *97*, 387–411.

Ponton, L. (2000). *The sex lives of teenagers*. New York: Dutton.

Popkin, B. M., Adair, L. S., & Ng, S. W. (2012). Global nutrition transition and the pandemic of obesity in developing countries. *Nutrition Reviews*, *70*, 3–21.

Popper, K. (1959). *The logic of scientific discovery*. New York: Basic Books.

Population Reference Bureau. (2010). *2010 world population data sheet*. Washington, DC: Author. Retrieved from http://www.prb.org/pdf10/10wpds_eng.pdf

Population Reference Bureau. (2012). *2012 world population data sheet*. Author. Washington, DC: Retrieved from http://www.prb.org/pdf12/2012-population-data-sheet_eng.pdf

Population Reference Bureau. (2013). *2013 world population data sheet*. Washington, DC: Author. Retrieved from http://www.prb.org/pdf13/2013-population-data-sheet_eng.pdf

Population Reference Bureau. (2015). *2015 world population data sheet*. Washington, DC: Author. Retrieved from http://www.prb.org/pdf15/2015-world-popula-tion-data-sheet_eng.pdf

Population Reference Bureau. (2017). *2017 world population data sheet*. Washington, DC: Author. Retrieved from http://www.prb.org/Publications/Datasheets/2017/2017-world-population-data-sheet.aspx

Population Reference Bureau. (2018). *World population data sheet*. Retrieved from https://www.prb.org/2018-world-population-data-sheet-with-focus-on-changing-age-structures/

Population Reference Bureau. (2020). *2020 world population data sheet*. Washington, DC: Author. Retrieved from https://www.prb.org/wp-content/uploads/2020/07/letter-booklet-2020-world-population.pdf

Poteet, G. A. (2007). *Perceptions of pretty people: An experimental study of interpersonal attractiveness* (Master's thesis. Washington State University). Retrieved from http://www.dissertations.wsu.edu/Thesis/Spring2007/a_poteet_050307.pdf

Potok, M. (2015, November 16). *FBI: Reported hate crimes down nationally except against Muslims*. Retrieved from https://www.splcenter.org/hatewatch/2015/11/16/fbi-reported-hate-crimes-down-nationally-except-against-muslims

Potter, H., Boggs, B., & Dunbar, C. (2017). Discipline and punishment: How schools are building the school-to-prison pipeline. In *Advances in race and ethnicity in education*, Vol. 4 (pp. 65–90). Somerville, MA: Emerald Publishing Limited.

Poushter, J., & Kent, N. (2020, June 25). *The global divide on homosexuality persists*. Retrieved from https://www.pewresearch.org/global/2020/06/25/global-divide-on-homosexuality-persists/

Powell, R. (2013). Social desirability bias in polling on same-sex marriage ballot initiatives. *American Politics Research*, *41*, 1052–1070.

Power, C., Reed, A., & Lee, A. (2019, October 7). *Education matters: Keeping girls in school can improve economic development in Burkina Faso's Sahel region.* Retrieved from https://www.prb.org/keeping-girls-in-school-can-improve-economic-development-in-burkina-fasos-sahel-region/

Power, S. (2002). *"A problem from hell": America and the age of genocide.* New York, NY: Basic Books.

Pradhan, E. (2015, November 27). *The relationship between women's education and fertility.* Retrieved from https://www.weforum.org/agenda/2015/11/the-relationship-between-womens-education-and-fertility/

Prell, R. (1999). *Fighting to become Americans: Jews, gender, and the anxiety of assimilation.* Boston, MA: Beacon Press.

Presser, S. (1990). Can changes in context reduce vote overreporting in surveys? *Public Opinion Quarterly, 54,* 586–593.

Preston, P. (1994). *Mother father deaf: Living between sound and silence.* Cambridge, MA: Harvard University Press.

Propublica. (2018). *Miseducation.* Retrieved from https://projects.propublica.org/miseducation

Proulx, C. M., & Snyder-Rivas, L. A. (2013). The longitudinal associations between marital happiness, problems, and self-rated health. *Journal of Family Psychology, 27,* 194–202. http://dx.doi.org/10.1037/a0031877

Putnam, R. (2000). *Bowling alone: The collapse and revival of American community.* New York: Simon & Schuster.

Qian, J., Cai, M., Gao, J., Tang, S., Xu, L., & Critchley, J. A. (2010). Trends in smoking and quitting in China from 1993 to 2003: National Health Service survey data. *Bulletin of the World Health Organization, 88.* Retrieved from http://www.who.int/bulletin/volumes/88/10/09-064709/en/index.tml

Queen, S. A., Habenstein, R. W., & Adams, J. B. (1961). *The family in various cultures* (2nd ed.). Philadelphia: B. Lippincott.

Quiggin, J. (2010). *Zombie economics: How dead ideas still walk among us.* Princeton, NJ: Princeton University Press.

Quinney, R. (1970). *Crime and justice in America.* New York, NY: Little, Brown.

Quinones, S. (2015). *Dreamland: The true tale of America's opiate epidemic.* New York: Bloomsbury Publishing.

Raghavan, S. (2016, July 7). In Yemen, child brides are part of the ravages of civil war. *Washington Post,* pp. A1, A14.

Rampell, C. (2010, March 9). Why a Big Mac costs less than a salad. *New York Times.* Retrieved from http://economix.blogs.nytimes.com/2010/03/09/why-a-big-mac-costs-less-than-a-salad

Rampey, B. D., Finnegan, R., Goodman, M., Mohadjer, L., Krenzke, T., Hogan, J., & Provasnik, S. (2016, March). *Skills of U.S. unemployed, young, and older adults in sharper focus: Results from the Program for the International Assessment of Adult Competencies (PIAAC) 2012/2014: First look.* Washington, DC: National Center for Education Statistics. Retrieved from http://nces.ed.gov/pubsearch/pubsinfo.asp?pubid=2016039

Randall, V. R. (1999). *History of tobacco. Boston University Medical Center.* Retrieved from http://academic.udayton.edu/health/syllabi/tobacco/history.htm#industry

Rankin, J., & Kernsmith, R. (2019). Social control theories. In E. McLaughlin (Ed.), *The SAGE dictionary of criminology* (4th ed.). Thousand Oaks, CA: SAGE.

Rashbaum, W. K., & Goldstein, J. (2016, June 20). Three N.Y.P.D. commanders are arrested in vast corruption case. *New York Times.* Retrieved from https://www.nytimes.com/2016/06/21/nyregion/nypd-arrests.html?mcubz=1

Rasmussen, E. E., & Densley, R. L. (2017). Girl in a country song: Gender roles and objectification of women in popular country music across 1990 to 2014. *Sex Roles, 76,* 188–201. Retrieved from https://doi.org/10.1007/s11199-016-0670-6

Ratner, M., & Ray, E. (2004). *Guantánamo: What the world should know.* New York, NY: Chelsea Green.

Rauhala, E., & Abutaleb, Y. (2020, September 1). Trump administration says U.S. won't join World Health Organization–linked effort to develop, distribute coronavirus vaccine. *Washington Post.* Retrieved from https://www.inquirer.com/health/coronavirus/trump-world-health-organization-coronavirus-covid-19-vaccine-20200901.html?ocid=uxbndlbing

Raven, B., & Kruglianski, W. (1975). Conflict and power. In P. Swingle (Ed.), *Structure of conflict* (pp. 177–219). New York: Academic Press.

Raymond, N. (2020, July 16). *Canadian businessman, California mom get prison terms in U.S. college scandal.* Retrieved from https://www.msn.com/en-us/news/us/canadian-businessman-california-mom-get-prison-terms-in-us-college-scandal/ar-BB16MkiK?ocid=uxcntrlbingb4load

Reaney, P., & Goldsmith, B. (2008, April 4). *Husbands create 7 hours of extra housework a week: Study.* Reuters. Retrieved from http://www.reuters.com/article/2008/04/04/us-housework-husbands-idUSN 0441782220080404

Reczek, C. (2020). Sexual and gender-minority families: A 2010 to 2020 decade in review. *Journal of Marriage & Family, 82*(1), 300–325. Retrieved from https://doi-org.thor.nebrwesleyan.edu/10.1111/jomf.12607

Reese, P. (2019, October 7). *When masculinity turns "toxic": A gender profile of mass shootings.* Retrieved from https://www.latimes.com/science/story/2019-10-07/mass-shootings-toxic-masculinity

Regan, H. (2020). *21 of the world's cities with the worst air pollution are in India. CNN.* Retrieved from https://www.cnn.com/2020/02/25/health/most-polluted-cities-india-pakistan-intl-hnk/index.html

Reich, R. (1991). *The work of nations: Preparing ourselves for 21st century capitalism.* New York: First Vintage Books.

Reich, R. (2001, April 9). The case (once again) for universal health insurance. *American Prospect.* Retrieved from http://prospect.org/article/case-once-again-universal-health-insurance

Reich, R. (2010). *Aftershock: The next economy and America's future.* New York: Knopf.

Reiman, J., & Leighton, P. (2012). *The rich get richer and the poor get prison.* New York, NY: Prentice Hall.

Reingold, J. (2016). How a radical shift left Zappos reeling. *Fortune.* Retrieved from http://fortune.com/zappos-tony-hsieh-holacracy/

Reiter, S. (2020). Pictorial phenomena depicting the family climate of deaf/hard of hearing children and their hearing families. *Frontiers in Psychology, 11.* Retrieved from https://doi.org/10.3389/fpsyg.2020.02221

Renjini, D. (2000). *Nayar women today: Disintegration of matrilineal system and the status of Nayar women in Kerala*. India: India Classical.

Renzetti, C. M., & Curran, D. J. (1992). *Women, men, and society* (2nd ed.). Boston: Allyn & Bacon.

Reskin, B., & Padavic, I. (2002). *Women and men at work* (2nd ed.). Thousand Oaks, CA: Sage.

Reuters. (2020, April 14). *Factbox: Global reaction to Trump withdrawing WHO funding*. Retrieved from https://www.reuters.com/article/us-health-coronavirus-trump-who-reaction-idUSKCN21X0CN

Reynolds, D. (2019). Meet the 10 gay, lesbian, and bisexual members of the 116th Congress. *The Advocate* Online. Retrieved from https://www.advocate.com/politicians/2019/1/03/meet-10-gay-lesbian-and-bisexual-members-116th-congress

Rhodes, A. (2016, June 28). *Young people—if you're so upset by the outcome of the EU referendum, then why didn't you get out and vote?* Retrieved from http://www.independent.co.uk/voices/eu-referendum-brex-it-young-people-upset-by-the-outcome-of-the-eu-ref-erendum-why-didnt-you-vote-a7105396.html

Richards, C. (2012). Playing under surveillance: Gender performance and the conduct of the self in a primary school playground. *British Journal of Sociology of Education, 33*, 373–390.

Richardson, J. (2009). Satanism in America: An update. *Social Compass, 56*, 552–563.

Richardson, J. T. (2020). Church–sect theory. *The SAGE encyclopedia of the sociology of religion* (pp. 135–136). Thousand Oaks, CA: SAGE.

Rico, B., Kreider, R. M., & Anderson, L. (2018, July 9). *Growth in interracial and interethnic married-couple households*. Retrieved from https://www.census.gov/library/stories/2018/07/interracial-marriages.html

Ridgeway, C. L., & Correll, S. J. (2004). Unpacking the gender system: A theoretical perspective on gender beliefs and social relations. *Gender & Society, 18*, 510–531.

Ridgeway, C. L., & Smith-Lovin, L. (1999). The gender system and interaction. *Annual Review of Sociology, 25*, 191–217.

Ridley, M. (1998). *The origins of virtue: Human instincts and the evolution of cooperation*. New York: Viking Press.

Riley, C. (2012, March 26). Can 46 rich dudes buy an election? *CNNMoney*. Retrieved from http://money.cnn.com/2012/03/26/news/economy/super-pac-donors/index.htm

Riordan, C. (1990). *Girls and boys in school: Together or separate?* New York: Teachers College Press.

Rios, V. (2012). Stealing a bag of potato chips and other crimes of resistance. *Contexts Magazine* (Winter), 49–53.

Risman, B. J. (1998). *Gender vertigo: American families in transition*. New Haven, CT: Yale University Press.

Rist, R. S. (1970). Student, social class, and teacher expectations: The self-fulfilling prophecy in ghetto education. *Harvard Educational Review, 40*, 411–451.

Ritzer, G. (1999). *Enchanting a disenchanted world: Revolutionizing the means of consumption*. Thousand Oaks, CA: Pine Forge.

Ritzer, G. (2007). *The globalization of nothing*. Thousand Oaks, CA: Pine Forge.

Robertson, C. (2018, April 25). A lynching memorial is opening. The country has never seen anything like it. *New York Times*. Retrieved from https://www.nytimes.com/2018/04/25/us/lynching-memorial-alabama.html

Rodenhizer, K., & Edwards, K. (2017). *The impacts of sexual media exposure on adolescent and emerging adults' dating and sexual violence attitudes and behaviors: A critical review of the literature*. Retrieved from http://journals.sagepub.com/doi/abs/10.1177/1524838017717745.

Romano, A. (2016, October 30). *The history of satanic panic in the US—and why it's not over yet*. Retrieved from https://www.vox.com/2016/10/30/13413864/satanic-panic-ritual-abuse-history-explained

Romer, D., Jamieson, P. E., & Jamieson, K. H. (2017). The continuing rise of gun violence in PG-13 movies, 1985 to 2015. *Pediatrics*, January. Retrieved from http://pediatrics.aappublications.org/content/early/2017/01/09/peds.2016-2891

Romm, C. (2015, January 28). Rethinking one of psychology's most famous experiments. *The Atlantic Monthly*. Retrieved from https://www.theatlantic.com/health/archive/2015/01/rethinking-one-of-psychologys-most-infamous-experiments/384913/

Romo, V. (2020, June 5). *NFL on kneeling players' protests: "We were wrong," Commissioner says. National Public Radio. Updates: The Fight Against Racial Injustice*. Retrieved from https://www.npr.org/sections/live-updates-protests-for-racial-justice/2020/06/05/871290906/nfl-on-kneeling-players-protests-we-were-wrong-commissioner-says

Roof, W. C. (1993). *A generation of seekers: Spiritual journeys of the baby boom generation. San Francisco: HarperSanFrancisco*.

Roof, W. C., & McKinney, W. (1990). *American main- line religion: Its changing shape and future*. New Brunswick, NJ: Rutgers University Press.

Roper, W. (2020, June 2). *Black Americans 2.5X more likely than whites to be killed by police* (digital image). Retrieved from https://www-statista-com.thor.nebrwesleyan.edu/chart/21872/map-of-police-violence-against-black-americansPicture

Roschelle, A. R., & Kaufman, P. (2004). Fitting in and fighting back: Stigma management strategies among homeless kids. *Symbolic Interaction, 27*, 23–46. Retrieved from http://onlinelibrary.wiley.com/doi/10.1525/si.2004.27.1.23/abstract

Rose, E. K. (2018). The rise and fall of female labor force participation during World War II in the United States. *The Journal of Economic History, 78*(3), 673–711. Retrieved from https://doi.org/10.1017/S0022050718000323

Rosen, J. (2014, September 5). Animal Traffic. *New York Times*. Retrieved from https://www.nytimes.com/ 2014/09/05/t-magazine/animal-trafficking-black-market.html

Rosenbaum, S. (2013, September 4). Pledge of Allegiance challenged in Massachusetts Supreme Court. *NBC News*. Retrieved from http://www.nbcnews.com/news/us-news/pledge-allegiance-challenged-massachusetts-supreme-court-v20327848

Rosenbloom, S. R., & Way, N. (2004). Experiences of discrimination among African American, Asian American, and Latino adolescents in an urban high school. *Youth & Society, 35*, 420–451.

Rosenfeld, M. J., & Thomas, R. J. (2012). Searching for a mate: The rise of the Internet as a social intermediary. *American Sociological Review, 77*(4), 523–547.

Rosenthal, R., & Jacobson, L. (1968). *Pygmalion in the classroom*. New York: Holt, Rinehart & Winston.

Rosenwald, M. S. (2016, May 17). Youth sports participation is up slightly, but many kids are still left behind. *The Washington Post*. Retrieved from https://www.washingtonpost.com/news/local/wp/2016/05/17/youth-sports-participation-is-up-slightly-but-many-kids-are-still-left-behind/?utm_term=.d9bca4eae1cb

Rosoff, S., Pontell, H., & Tillman, R. (2010). *Profit without honor: White-collar crime and the looting of America*. Upper Saddle River, NJ: Prentice Hall.

Rostow, W. W. (1961). *The stages of economic growth*. Cambridge: Cambridge University Press.

Rothe, D. L. (2009). *State criminality: The crime of all crimes*. Lanham, MD: Lexington Books.

Rothkopf, D. (2008). *Superclass: The global power elite and the world they are making*. New York: Farrar, Straus and Giroux.

Rothschild, J., & Whitt, A. (1987). *The cooperative workplace: Potentials and dilemmas of organizational democracy and participation*. New York, NY: Cambridge University Press.

Rothschild-Whitt, J. (1979). The collectivist organization: An alternative to rational-bureaucratic models. *American Sociological Review, 44*, 509–527.

Roubein, R. (2020, August 5). Missouri voters latest to approve Medicaid expansion. *Politico*. Retrieved from https://www.politico.com/news/2020/08/05/missouri-approves-medicaid-expansion-391678

Rowbotham, S. (1973). *Woman's consciousness, man's world*. Middlesex, England: Pelican.

Rowen, B. (2018, March 16). The Jedi Faithful. *Pacific Standard*. Retrieved from https://psmag.com/economics/the-jedi-faithful

Ruberg, B. (2018). Queerness and video games: Queer game studies and new perspectives through play. *Glq, 24*(4), 543–543.

Rubin, B. (1996). *Shifts in the social contract: Understanding change in American society*. Thousand Oaks, CA: Pine Forge.

Rubin, L. B. (2006). What am I going to do with the rest of my life? *Dissent, 53*, 88–94.

Rugh, J. S., & Massey, D. S. (2014). Segregation in post-civil rights America: Stalled integration or the end of the segregated century? *Du Bois Review: Social Science Research on Race, 11*, 205–232.

Ruppanner, L., & Maume, D. J. (2016). The state of domestic affairs: Housework, gender, and state-level institutional logics. *Social Science Research, 60*, 15–28.

Rwanda: How the genocide happened. (2008, December 18). BBC News. Retrieved from http://news.bbc.co.uk/2/hi/1288230.stm

Ryan, R. A. (1981). Strengths of the American Indian family: State of the art. In F. Hoffman (Ed.), *The American Indian family: Strengths and stresses*. Isleta, NM: American Indian Social Research and Development Associates.

Rymer, R. (1993). *Genie: A scientific tragedy*. New York: HarperCollins.

Rymer, R. (2012, July). Vanishing voices. *National Geographic*, pp. 60–93.

Saad, L., & Jones, J. M. (2016, March 16). *U.S. concern about global warming at eight-year high*. Retrieved from http://www.gallup.com/poll/190010/concern-global-warming-eight-year-high.aspx

Sabo, D. F., Miller, K. E., Farrell, M. P., Melnick, M. J., & Barnes, G. M. (1999). High school athletic participation, sexual behavior, and adolescent pregnancy: A regional study. *Journal of Adolescent Health, 25*, 597–613.

Sadker, D. M., & Sadker, M. P. (1997). *Failing at fairness: How our schools cheat girls*. New York: Scribner.

Sadker, D. M., & Zittleman, K. (2009). *Still failing at fairness: How gender bias cheats girls and boys in school and what we can do about it*. New York: Scribner.

Sadker, D. M., Zittleman, K., & Sadker, M. P. (2003). *Teachers, schools, and society. New York: McGraw-Hill.*

Saez, E. (2010, July 17). *Striking it richer: The evolution of top incomes in the United States (updated with 2008 estimates)*. Berkeley: Econometrics Laboratory, University of California–Berkeley. Retrieved from http://elsa.berkeley.edu/~saez/saez-UStopincomes-2008.pdf

Saez, E., & Zucman, G. (2014). *Wealth inequality in the United States since 1913: Evidence from capitalized income tax data*. (NBER Working Paper 201625). Cambridge, MA: National Bureau of Economic Research. Retrieved from http://gabriel-zucman.eu/files/SaezZucman2014.pdf

Sagan, E. (1992). *The honey and the hemlock: Democracy and paranoia in ancient Athens and modern America*. New York: Basic Books.

Sagan, L. L. (1987). *The health of nations: True causes of sickness and well-being*. New York: Basic Books.

Sahlins, M. D., & Service, E. R. (1960). *Evolution and culture*. Ann Arbor, MI: Ann Arbor Paperbacks.

Salaita, S. (2005). Ethnic identity and imperative patriotism: Arab Americans before and after 9/11. *College Literature, 32*(2), 146–168. http://dx.doi.org/10.1353/lit.2005.0033

Salami, B., Duggleby, W., & Rajani, F. (2017). The perspective of employers/families and care recipients of migrant live—in caregivers: A scoping review. *Health Soc Care Community, 25*, 1667–1678. doi:https://doi.org/10.1111/hsc.12330

Samari, G., & Coleman-Minahan, K. (2018). Parental gender expectations by socioeconomic status and nativity: Implications for contraceptive use. *Sex Roles, 78*(9–10), 669–684.

Sanders, G. (2020). Megachurch. *The SAGE encyclopedia of the sociology of religion* (pp. 489–491). Thousand Oaks, CA: SAGE.

Sandoz, M. (1961). *These were the Sioux*. New York: Dell.

Sandstrom, A. (2016, November 14). *If the U.S. had 100 people: Charting Americans' religious affiliations*. Retrieved from https://www.pewresearch.org/fact-tank/2016/11/14/if-the-u-s-had-100-people-charting-americans-religious-affiliations/

Sangweni, Y. (2017, February 17). The way-too-short list of Black Oscar winners. *Essence Magazine*. Retrieved from https://www.essence.com/celebrity/way-too-short-list-black-oscar-winners/

Santos, F. (2020, September 30). Elderly and homeless: America's next housing crisis. *New York Times Magazine*. Retrieved from https://www.nytimes.com/2020/09/30/magazine/homeless-seniors-elderly.html

Sassen, S. (1991). *The global city: New York, London, Tokyo*. Princeton, NJ: Princeton University Press.

Sassen, S. (2000). *Cities in a world economy* (2nd ed.). Thousand Oaks, CA: Pine Forge.

Sassen, S. (2005). *The global city: Introducing a concept*. Retrieved from http://www.saskiasassen.com/pdfs/publications/the-global-city-brown.pdf

Savage, R. (2020, January 8). *Walking in two worlds: Canada's "two-spirit" doctor guiding trans teenagers*. Retrieved from https://www.reuters.com/article/us-canada-lgbt-indigenous-feature-trfn-idUSKBN1Z7252

Sawe, B. E. (2018, May 23). *Countries with the largest Jewish populations*. Retrieved from https://www.worldatlas.com/society/

Sayeed, S., al-Adawlya, A., & Bagby, I. (2013). Women and the American mosque. *The U.S. mosque study 2011* (report 3). Islamic Society of North America. Retrieved from http://www.hartfordinstitute.org/The-American-Mosque-Report-3.pdf

Schaap, J., & Aupers, S. (2020). Video and role-playing games. *The SAGE encyclopedia of the sociology of religion* (pp. 890–891). Thousand Oaks, CA: SAGE.

Schaefer, D. R. (2011). Resource characteristics in social exchange networks: Implications for positional advantage. *Social Networks, 33*, 143–151.

Schaefer, R. T. (2009). *Race and ethnicity in the United States* (5th ed.). Upper Saddle River, NJ: Pearson Prentice Hall.

Schaeffer, K. (2020, April 23). *As schools shift to online learning amid pandemic here's what we need to know about disabled students in the U.S.* Pew Research Fact Tank. Retrieved from https://www.pewresearch.org/fact-tank/2020/04/23/as-schools-shift-to-online-learning-amid-pandemic-heres-what-we-know-about-disabled-students-in-the-u-s/

Schafft, K. A., Jensen, E. B., & Hinrichs, C. C. (2009). Food deserts and overweight schoolchildren: Evidence from Pennsylvania. *Rural Sociology, 74*, 153–177.

Schaul, K., Rabinowitz, K., & Mellnik, T. (2020, November 19). 2020 turnout is the highest in over a century. *Washington Post*. Retrieved from https://www.washingtonpost.com/graphics/2020/elections/voter-turnout/

Scheff, T. J. (1966). *Being mentally ill: A sociological theory*. Chicago, IL: Aldine.

Schegloff, E. (1990). *On the organization of sequences as a source of "coherence" in talk-in-interaction*. In B. Dorval (Ed.), *Conversational organization and its development* (pp. 55–77). Norwood, NJ: Ablex.

Schegloff, E. (1991). Reflections on talk and social structure. In D. H. Zimmerman & D. Boden (Eds.), *Talk and social structure* (pp. 44–70). Cambridge: Polity Press.

Schlosser, E. (2012). *Fast food nation: The dark side of the all-American meal*. Boston: Houghton Mifflin Harcourt.

Schmidt, S. (2017). *The wedding industry in 2017 and beyond*. May 16. Marketresearch.com. Retrieved from https://blog.marketresearch.com/the-wedding-industry-in-2017-and-beyond

Schnabel, L., Hackett, C., & McClendon, D. (2018). Where men appear more religious than women: Turning a gender lens on religion in Israel. *Journal for the Scientific Study of Religion, 57*(1), 80–94. Retrieved from https://doi-org.thor.nebrwesleyan.edu/10.1111/jssr.12498

Schnaiberg, A., & Gould, K. A. (1994). *Environment and society: The enduring conflict*. New York: St. Martin's Press.

Schneider, D. M., & Gough, K. (1974). *Matrilineal kinship*. Berkeley: University of California Press.

Schneider, F., & Enste, D. (2002, March). Hiding in the shadows: The growth of the underground economy. *Journal of Economic Issues, 30*. Retrieved from http://www.imf.org/external/pubs/ft/issues/issues30/index.htm

Schneider, G. S., & Vozzella, L. (2018, June 4). How a reshaped Virginia legislature learned to love Medicaid. *The Washington Post*. Retrieved from https://www.washingtonpost.com/local/virginia-politics/how-a-reshaped-virginia-legislature-learned-to-love-medicaid-expansion/2018/06/04/aeac0f50-65b9-11e8-99d2-0d678ec08c2f_story.html?utm_term=.d7e59e52d0dc

Schor, J. B. (1998). *The overspent American: Why we want what we don't need*. New York: HarperPerennial.

Schulte, B. (2015). *Why parents should stop hoping their kids will get married*. Retrieved from https://www.washingtonpost.com/news/wonk/wp/2015/05/17/why-parents-should-stop-expecting-their-kids-to-get-married/

Schuman, H., & Presser, S. (1981). *Questions and answers in attitude surveys: Experiments on question form, wording, and context*. New York: Academic Press.

Schuppe, J. (2016, June 19). 30 years after basketball star Len Bias' death, its drug war impact endures. *NBC News*. Retrieved from http://www.nbcnews.com/news/us-news/30-years-after-basketball-star-len-bias-death-its-drug-n593731

Schuster, H., & Bullard, G. (2020, September 30). *Tenant activism in D.C. has surged during the pandemic*. Retrieved from https://www.npr.org/local/305/2020/09/30/918652622/tenant-activism-in-d-c-has-surged-during-the-pandemic

Schwadel, P. (2020). The politics of religious nones. *Journal for the Scientific Study of Religion, 59*(1), 180–189.

Schwarz, H. (2015, April 28). *There are 390,000 gay marriages in the U.S. The Supreme Court could quickly make it half a million*. Retrieved from https://www.washing tonpost.com/news/the-fix/wp/2015/04/28/heres-how-many-gay-marriages-the-supreme-court-could-make-way-for/

Scott, B. (2020, October 16). *Pandemic forces more women to leave the workforce*. National Public Radio. Retrieved from https://www.npr.org/2020/10/16/924648105/pandemic-forces-more-women-to-leave-the-workforce

Scott, J. (2005, May 16). Life at the top in America isn't just better, it's longer. *New York Times*. Retrieved from http://www.nytimes.com/2005/05/16/national/class/HEALTH-FINAL.html?pagewanted=all

Scott, W. R., & Meyer, J. W. (1994). *Institutional environments and organizations: Structural complexity and individualism*. Thousand Oaks, CA: Sage.

Seager, J. (2003). *The Penguin atlas of women in the world*. New York: Penguin Books.

Seaman, A. M. (2016, June 14). Robots may push surgeons to the sidelines – but not soon. *Washington Post*, E6.

Sebald, H. (2000). *Adolescence: A social psychological approach* (4th ed.). Englewood Cliffs, NJ: Prentice Hall.

Sellin, T. (1938). *Culture, conflict and crime*. New York: Social Science Research Council.

Semega, J., Kollar, M., Creamer, J., & Mohanty, A. (2019). *Income and poverty in the United States: 2018*. U.S. Census Bureau, Current Population Reports, P60–266. Washington, DC: U.S. Government Printing Office.

Semega, J., Kollar, M., Shrider, E. A., & Creamer, J. F. (2020). *Income and poverty in the United States*. U.S. Bureau of the Census. Retrieved from https://www.census.gov/content/dam/Census/library/publications/2020/demo/p60-270.pdf

Semuels, A. (2015, March 27). *The city that believed in desegregation*. Retrieved from http://www.theatlantic.com/business/archive/2015/03/the-city-that-believed-in-desegregation/388532/

Semuels, A. (2017, July 20). The mystery of why Japanese people are having so few babies. *The Atlantic*. Retrieved from https://www.theatlantic.com/business/archive/2017/07/japan-mystery-low-birth-rate/534291/

Semuels, A. (2018, January/February). Robots will transform fast food. *Atlantic Monthly*. Retrieved from https://www.theatlantic.com/magazine/archive/2018/01/iron-chefs/546581

Semuels, A. (2019, March 5). Is this the end of recycling? *The Atlantic*. Retrieved from https://www.theatlantic.com/technology/archive/2019/03/china-has-stopped-accepting-our-trash/584131/

Senior, J. (2016, February 21). Review: In 'Evicted,' home is an elusive goal for America's poor. *New York Times*. Retrieved from http://www.nytimes.com/2016/02/22/books/evicted-book-review-mat-thew-desmond.html

Sennett, R. (1998). *The corrosion of character: The personal consequences of work in the new capitalism*. New York: Norton.

The Sentencing Project. (2011). *Felony disenfranchisement*. Retrieved from http://www.sentencingproject.org/template/page.cfm?id=133

The Sentencing Project. (2020a). *Fact sheet: Trends in U.S. corrections*. Retrieved from https://www.sentencingproject.org/wp-content/uploads/2020/08/Trends-in-US-Corrections.pdf

The Sentencing Project. (2020b). *Felony disenfranchisement*. Retrieved from https://www.sentencingproject.org/issues/felony-disenfranchisement/

The Sentencing Project. (2020c, October). *Locked out 2020: Estimates of people denied voting rights due to a felony conviction*. Retrieved from https://www.sentencingproject.org/publications/locked-out-2020-estimates-of-people-denied-voting-rights-due-to-a-felony-conviction/

The Sentencing Project. (n.d.). *Criminal justice facts*. Retrieved from https://www.sentencingproject.org/criminal-justice-facts/

Serajuddin, U., & Hamadeh, N. (2020, July 1). New World Bank country classifications by income level: 2020–2021. *World Bank Blogs*. Retrieved from https://blogs.worldbank.org/opendata/new-world-bank-country-classifications-income-level-2020-2021

Serota, K. B., & Levine, T. R. (2015). A few prolific liars: Variation in the prevalence of lying. *Journal of Language and Social Psychology, 34*(2), 138–157. Retrieved from https://doi-org.thor.nebrwesleyan.edu/10.1177/0261927X14528804

Setoodeh, R. (2016, January 19). George Clooney on white Oscars: "We're moving in the wrong direction". *Variety*. Retrieved from http://variety.com/2016/film/news/george-clooney-white-oscars-1201682504

Shabad, R., Baily, C., & McCausland, P. (2018, March 24). At March for Our Lives, survivors lead hundreds of thousands in call for change. *NBC News*. Retrieved from https://www.nbcnews.com/news/us-news/march-our-lives-draws-hundreds-thousands-washington-around-nation-n859716

Shahani-Denning, C. (2003). *Physical attractiveness bias in hiring: What is beautiful is good*. Hempstead, NY: Hofstra University Office for Research and Sponsored Programs. Retrieved from http://www.hofstra.edu/pdf/orsp_shahani-denning_spring03.pdf

Shamir, R. (2011). Mind the gap: The commodification of corporate social responsibility. *Symbolic Interaction, 28*, 229–253. Retrieved from http://onlinelibrary.wiley.com/doi/10.1525/si.2005.28.2.229/abstract

Shanmugasundaram, S. (2018, April 15). *Hate crimes explained*. *Southern Poverty Law Center*. Retrieved from https://www.splcenter.org/20180415/hate-crimes-explained#collection

Shao, R., & Wang, Y. (2019). The relation of violent video games to adolescent aggression: An examination of moderated mediation effect. *Frontiers in Psychology*. Retrieved from https://doi.org/https://doi.org/10.3389/fpsyg.2019.00384

Shapiro, J., & Bowman, E. (2020, July 27). *One laid groundwork for the ADA; the other grew up under its promises*. Retrieved from https://radio.wpsu.org/post/one-laid-groundwork-ada-other-grew-under-its-promises

Shattuck, R. (1980). *The forbidden experiment*. New York: Farrar, Straus and Giroux.

Shaw, M. (2010). Sociology and genocide. In D. Bloxham & A. D. Moses (Eds.), *The Oxford handbook of genocide studies* (pp. 142–161). New York, NY: Oxford University Press.

Sheinin, D., Thompson, K., McDonald, S. N., & Clement, S. (2016, January 31). New wave feminism. *Washington Post*, pp. A1–A17.

Sheldon, W. H. (1949). *Varieties of delinquent youth: An introduction to constitutional psychiatry*. New York: Harper.

Shepherd, K. (2016, July 5). Part time jobs and thrift: How unpaid interns in DC get by. *New York Times*. Retrieved from http://www.nytimes.com/2016/07/06/us/part-time-jobs-and-thrift-how-unpaid-interns-in-dc-get-by.html?_r=0

Shilts, Ry. (1987). *And the band played on: Politics, people, and the AIDS epidemic*. New York: St. Martin's Press.

Shipler, D. K. (2005). *The working poor: Invisible in America*. New York: Vintage Books.

Shmerling, R. (2016, November 30). The health advantages of marriage. *Harvard Health Blog*. Retrieved from https://www.health.harvard.edu/blog/the-health-advantages-of-marriage-2016113010667

Shore, R., & Shore, B. (2009). *Reducing infant mortality* (KIDS COUNT Indicator Brief). Baltimore, MD: Annie E. Casey Foundation. Retrieved from http://www.aecf.org/m/resourcedoc/AECF-KCReducingInfantMortality-2009.pdf

Silva, E. B. (2001). *White supremacy and racism in the post-civil rights era*. Boulder, CO: Lynne Rienner.

Silver, L., & Johnson, C. (2018, October 9). *Majorities in sub-Saharan Africa own mobile phones, but smartphone ownership is modest*. *Pew Research*. Retrieved from https://www.pewresearch.org/global/2018/10/09/majorities-in-sub-saharan-africa-own-mobile-phones-but-smartphone-adoption-is-modest/

Silver, N. (2015, May 1). *The most diverse cities are often the most segregated*. Retrieved from http://fivethirtyeight.com/features/the-most-diverse-cities-are-often-the-most-segregated/

Silverman, R. M. (2005). Community socioeconomic status and disparities in mortgage lending: An analysis of metropolitical Detroit. *Social Science Journal, 42*, 479–486.

Silverstein, J. (2020, January 2). There were more mass shootings than days in 2019. *CBS News*. Retrieved from https://www.cbsnews.com/news/mass-shootings-2019-more-than-days-365/

Simmel, G. (1955). *Conflict and the web of group affiliations* (K. Wolf, Trans.). Glencoe, IL: Free Press.

Simon, J. L. (1977). *The economics of population growth*. Princeton, NJ: Princeton University Press.

Simon, J. L. (1981). *The ultimate resource*. Princeton, NJ: Princeton University Press.

Simon, J. L. (2000). *The great breakthrough and its cause*. *Ann Arbor:*. University of Michigan Press.

Simon, M. (2017, August 2). Companion robots are here: Just don't fall in love with them. *Wired*. Retrieved from https://www.wired.com/story/companion-robots-are-here

Simon-Thomas, E. (2019, June 17). Is marriage really bad for women's happiness? *Greater Good Magazine*. Retrieved from https://greatergood.berkeley.edu/article/item/is_marriage_really_bad_for_womens_happiness

Simpson, M. E., & Conklin, G. H. (1989). Socioeconomic development, suicide, and religion: A test of Durkheim's theory of religion and suicide. *Social Forces*, *67*, 945–964.

Sinclair, U. (1995). *The jungle*. New York: Doubleday, Page. (Original work published 1906)

Singer, A. (2012). *Immigrant workers in the U.S. labor force*. Washington, DC: Brookings Institution. Retrieved from http://www.brookings.edu/research/papers/2012/03/15-immigrant-workers-singer#8

Singer, P. W. (2003). *Corporate warriors: The rise of the privatized military industry*. Ithaca, NY: Cornell University Press.

Skinner, B. F. (1938). *The behavior of organisms. Cambridge, MA: B. F. Skinner Foundation*.

Skinner, B. F. (1953). *Science and human behavior*. Cambridge, MA: B. F. Skinner Foundation.

Skinner, O. D., & McHale, S. M. (2018). The development and correlates of gender role orientations in African-American youth. *Child Development*, *89*(5), 1704–1719. Retrieved from https://doi-org.thor.nebrwesleyan.edu/10.1111/cdev.12828

Sklair, L. (2002). *Globalization: Capitalism and its alternatives*. Oxford: Oxford University Press.

Sklar, J. (2020, April 24). Zoom fatigue is taxing the brain. Here is why that happens. *National Geographic*. Retrieved from https://www.nationalgeographic.com/science/2020/04/coronavirus-zoom-fatigue-is-taxing-the-brain-here-is-why-that-happens/

Skloot, R. (2017). *The immortal life of Henrietta Lacks*. New York: Broadway Books.

Skocpol, T. (1979). *States and social revolutions*. New York: Cambridge University Press.

Slaughter, A. M. (2015). *Unfinished business: Women men work family*. New York: Random House.

Slovak, K., & Singer, J. B. (2011). School social workers' perceptions of cyberbullying. *Children & Schools*, *33*, 1–16.

Smelser, N. J. (1962). *The theory of collective behavior*. New York: Free Press.

Smith, B. (1990). Racism and women's studies. In G. Anzaldúa (Ed.), *Making face, making soul: Haciendo caras*. San Francisco: Aunt Lute Foundation.

Smith, D. (1987). *The everyday world as problematic: A feminist sociology*. Boston: Northeastern University Press.

Smith, D. (1990). *The conceptual practices of power: A feminist sociology of knowledge*. Boston: Northeastern University Press.

Smith, D. (2005). *Institutional ethnography: A sociology for people*. Walnut Creek, CA: AltaMira Press.

Smith, D. (2018, April 27). Ida B Wells: The unsung heroine of the civil rights movement. *The Guardian* Online. Retrieved from https://www.theguardian.com/world/2018/apr/27/ida-b-wells-civil-rights-movement-reporter

Smith, M. (2000). *American business and political power: Public opinions, elections, and democracy*. Chicago: University of Chicago.

Smith, P. K. (2009). *Obesity among poor Americans: Is public assistance the problem?* Nashville. TN: Vanderbilt University Press.

Smith, P. K., Mahdavi, J., Carvalho, M., Fisher, S., Russell, S., & Tippett, N. (2008). Cyberbullying: Its nature and its impact in secondary school pupils. *Journal of Child Psychology and Psychiatry*, *49*, 376–385.

Smith, R. W. (2002). As old as history. In C. Rittner, J. K. Roth, & J. M. Smith (Eds.), *Will genocide ever end?* (pp. 31–34). St. Paul, MN: Paragon House.

Smith, S. (2001). *Allah's mountains: The battle for Chechnya*. London: I. B. Tauris.

Smith, S. C., Choueiti, M., & Pieper, K. (2014). *Race/Ethnicity in 600 popular films: Examining on screen portrayals and behind the camera diversity*. Los Angeles, CA: Annenberg School for Communication and Journalism, University of Southern California. Retrieved from http://annenberg.usc.edu/pages/~/media/MDSCI/Racial%20Inequality%20in%20Film%202007-2013%20Final.ashx

Smith, S. L., Choueiti, M., Pieper, K., Yao, K., Case, A., & Choi, A. (2019, September). *Inequality in 1,200 popular films: Examining portrayals of gender, race/ethnicity, LGBTQ & disability from 2007 to 2018*. Annenberg Foundation. Retrieved from http://assets.uscannenberg.org/docs/aii-inequality-report-2019-09-03.pdf

Smith, S. L., Pieper, K., & Choueiti, M., with assistance from Tofan, A., DePauw, A., & Case, A. (2018, January). *Still rare, still ridiculed: Portrayals of senior characters on screen in popular films from 2015 and 2016*. Inclusion Initiative. Retrieved from https://annenberg.usc.edu/sites/default/files/2018/01/22/Still%20Rare%20Still%20Ridiculed%20Final%20Report%20January%202018.pdf

Smith, T. W. (2002). Religious diversity in America: The emergence of Muslims, Buddhists, Hindus, and others. *Journal for the Scientific Study of Religion*, *41*, 577–585.

Smith-Greenaway, E. (2013). Maternal reading skills and child mortality in Nigeria: A reassessment of why education matters. *Demography*, *50*, 1551–1561.

Smits, D. (1994). The frontier army and the destruction of the buffalo: 1865–1883. *Western Historical Quarterly*, *25*(3), 312–338.

Smock, P. J., Manning, W. D., & Porter, M. (2005). "Everything's there except money": How money shapes decisions to marry among cohabitating adults. *Journal of Marriage and Family*, *67*, 680–696.

Snow, D. A., Rochford, E. B., Jr., Worden, S. K., & Benford, R. D. (1986). Frame alignment processes, micromobilization, and movement participation. *American Sociological Review, 51*, 464–481.

Snow, D. A., Zurcher, L. A., Jr., & Ekland-Olson, S. (1980). Social networks and social movements: A microstructural approach to differential recruitment. *American Sociological Review, 45*, 787–801.

Social Media Fact Sheet. (2019, June 12). Washington, DC: Pew Research Center. Retrieved from https://www.pewresearch.org/internet/fact-sheet/social-media/

Social Security Administration. (2013). *Social Security history*. Retrieved from http://www.ssa.gov/history

Solimano, A. (2010). *International migration in the age of crisis and globalization: Historical and recent experience*. Cambridge, England: Cambridge University Press.

Somashekhar, S., Lowery, W., Alexander, K. L., Kindy, K., & Tate, J. (2015, August 8). Black and unarmed. *Washington Post*. Retrieved from http://www.washingtonpost.com/sf/national/2015/08/08/black-and-unarmed

Soria, K. M., & Horgos, B. (2020). Social class differences in students' experiences during the COVID-19 pandemic. *SERU COVID-19 survey policy briefs*. Retrieved from https://cshe.berkeley.edu/seru-covid-survey-reports

Sorokin, P. (1962). *Society, culture, and personality: Their structure and dynamics*. New York, NY: Cooper Square.

Sorokin, P. (1970). *Social and cultural dynamics: A study of change in major systems of art, truth, ethics, law and social relationships*. Boston, MA: Extending Horizons Books, Porter Sargent Publishers. (Original work published 1957)

Sorokowska, A., Sorokowski, P., Hilpert, P., Cantarero, K., Frackowiak, T., Ahmadi, K., & John, D. P, Jr., et al. (2017). Preferred interpersonal distances: A global comparison. *Journal of Cross-Cultural Psychology, 48*(4), 577–592. Retrieved from https://doi.org/10.1177/0022022117698039

Spark, C., Porter, L., & de Kleyn, L. (2019). "We're not very good at soccer": Gender, space and competence in a Victorian primary school. *Children's Geographies, 17*(2), 190–203. doi:https://doi.org/10.1080/14733285.2018.1479513

Sparrow, R. (2005). Defending deaf culture: The case of cochlear implants. *Journal of Political Philosophy, 13*, 135–152.

Speed, B. (2016, June). *How did different demographic groups vote in the EU referendum?* Retrieved from http://www.newstatesman.com/politics/staggers/2016/06/how-did-different-demographic-groups-vote-eu-referendum

Spitzer, S. (1975). Toward a Marxian theory of deviance. *Social Problems, 22*(5), 641–651. http://dx.doi.org/10.2307/799696

Squires, G. D. (2003). Racial profiling, insurance style: Insurance redlining and the uneven development of metropolitan America. *Journal of Urban Affairs, 24*, 391–410.

Squires, G. D., Friedman, S., & Saidat, C. E. (2002). Experiencing residential segregation: A contemporary study of Washington, DC. *Urban Affairs Review, 38*, 155–183.

Srivastava, R. (2019, August 23). *Sign of the times: Mumbai green lights women figures on traffic signals*. Retrieved from https://www.reuters.com/article/us-india-cities-women-trfn-idUSKCN24Z1W9

Stark, R., & Bainbridge, W. S. (1996). *A theory of religion*. New York: Peter Lang. (Original work published 1987)

Statistical Research Department. (2020, August 31). *People shot to death by U.S. police, by race 2017–2020*. Retrieved from https://www.statista.com/statistics/585152/people-shot-to-death-by-us-police-by-race/

Stearns, E., & Glennie, E. J. (2006). When and why dropouts leave high school. *Youth and Society, 38*(1), 29–57. http://dx.doi.org/10.1177/0044118x05282764

Stein, P. (2015, April 10). *After 50 years, a D.C. store will close 'due to gentrification.'* Retrieved from https://www.washingtonpost.com/news/local/wp/2015/04/10/after-50-years-a-d-c-store-will-close-due-to-gentrification/?tid=pm_local_pop_b

Steinbrook, R. (1988, January 29). AIDS summit delegates adopt a unanimous call for action. *Los Angeles Times*. Retrieved from http://articles.latimes.com/1988-01-29/news/mn-26467_1_aids-control

Stephens, T., Kamimura, A., Yamawaki, N., Bhattacharya, H., Mo, W., Birkholz, R., Makomenaw, A., & Olson, L. M. (2016). Rape myth acceptance among college students in the United States, Japan, and India. *SAGE Open, 6*(4), https://doi.org/10.1177/2158244016675015.

Stephenson, W. (2014, June 3). Welcome to West Port Arthur, Texas, ground zero in the fight for climate justice. *The Nation*. Retrieved from https://www.thenation.com/ article/welcome-west-port-arthur-texas-ground-zero-fight-climate-justice/

Stepick, A., III., & Grenier, G. (1993). Cubans in Miami. In J. Moore & R. Pinderhughes (Eds.), *In the barrios: Latinos and the underclass debate* (pp. 79–100). New York, NY: Russell Sage Foundation.

Stevenson, B. (2010). Beyond the classroom: Using Title IX to measure the return to high school sports. *Review of Economics & Statistics, 92*, 284–301.

Stevenson, B., & Zlotnik, H. (2018, May). Representations of men and women in introductory economics textbooks. *AEA Papers and Proceedings, 108*, 180–185.

Stiglitz, J. E. (2012). *The price of inequality: How today's divided society endangers our future*. New York: Norton.

Stokes, R., & Chevan, A. (1996). Female-headed families: Social and economic context of racial differences. *Journal of Urban Affairs, 8*, 245–268.

Stokoe, E., & Sikveland, R. (2019). The backstage work negotiators do when communicating with persons in crisis. *Journal of Sociolinguistics*. doi:https://doi.org/10.1111/josl.12347

Stolle, D. (1998). Why do bowling and singing matter? Group characteristics, membership, and generalized trust. *Political Psychology, 19*, 497–525.

Stone, J. (2016). *Brexit protest: Thousands march against Leave vote in London*. Retrieved http://www.independent.co.uk/news/uk/politics/brexit-eu-referendum-protest-march-london-saturday-2-july-anti-result-live-a7111581.html

Stone, R. (2015, December). Pregnant women and substance use: Fear, stigma, and barriers to care. *Journal of Health Justice 3*(2). Retrieved from https://www.ncbi.nlm.nih.gov/pmc/articles/PMC5151516/

Straus, M. A., & Gelles, R. J. (Eds.). (1990). *Physical violence in American families: Risk factors and adaptations to violence in 8,145 families*. New Brunswick, NJ: Transaction.

Straus, M. A., Gelles, R. J., & Steinmetz, S. K. (1988). *Behind closed doors: Violence in the American family.* Newbury Park, CA: Sage.

Straus, M. A., Sugarman, D. B., & Giles-Sims, J. (1997). Spanking by parents and subsequent antisocial behavior of children. *Archives of Pediatrics and Adolescence, 151*, 761–767.

Subramaniam, M. (2020, June 15). The 2019 Citizenship Amendment Act in India. *Critical Mass, 45*(1). Retrieved from http://cbsm-asa.org/2020/06/protesting-the-2019-citizenship-amendment-act-in-india/#more-1071

Substance Abuse and Mental Health Services Administration. (2019). In *Key substance use and mental health indicators in the United States: Results from the 2018 National Survey on Drug Use and Health* (HHS Publication No. PEP19-5068, NSDUH Series H-54). Rockville, MD: Center for Behavioral Health Statistics and Quality, Substance Abuse and Mental Health Services Administration. Retrieved from https://www.samhsa.gov/data/

Sullivan, T. A., Warren, E., & Westbrook, J. (2000). *The fragile middle class: Americans in debt.* Binghamton, NY: Vail-Ballou Press.

Sumter, S. R., Vandenbosch, L., & Ligtenberg, L. (2017). Love me Tinder: Untangling emerging adults' motivations for using the dating application Tinder. *Telematics and Informatics, 34*(1), 67–78.

Sunstein, C. R., & Hastie, R. (2014). Making dumb groups smarter. *Harvard Business Review, 92*(12), 90–98.

Supple, A. J., Ghazarian, S. R., Frabutt, J. M., Plunkett, S. W., & Sands, T. (2006). Contextual influences on Latino adolescent ethnic identity and academic outcomes. *Child Development, 77*(5), 1427–1433.

Surette, R., Hansen, K., & Noble, G. (2009). Measuring media oriented terrorism. *Journal of Criminal Justice, 37*, 360–370.

Sutherland, E. H. (1929). The person v. the act in criminology. *Cornell Law Quarterly, 14*, 159–167.

Sutherland, E. H. (1983). *White collar crime: The uncut version.* New Haven, CT: Yale University Press. (Original work published 1949)

Swaine, J. (2018, September 6). *Trump inauguration crowd photos were edited after he intervened.* Retrieved from https://www.theguardian.com/world/2018/sep/06/donald-trump-inauguration-crowd-size-photos-edited

Swiss, S., & Giller, J. E. (1993). Rape as a crime of war. *Journal of the American Medical Association, 270*, 619–622.

Tang, S. (2016 October 31). *Understanding Asian-American assimilation outcomes, 1940–2000.* Retrieved from https://journeys.dartmouth.edu/censushistory/2016/10/31/understanding-asian-american-assimilation-outcomes-1940-2000/

Tankersley, J. (2016, January 6). What top researchers discovered when they re-ran the numbers of income inequality. *Washington Post.* Retrieved from https://www.washingtonpost.com/news/wonk/wp/2016/01/06/what-top-researchers-discovered-when-they-re-ran-the-numbers-of-income-inequality/

Tannenbaum, F. (1938). *Crime and the community.* New York, NY: Columbia University Press.

Tarr, G. A. (2019, November 29). Five common misconceptions about the electoral college. *The Atlantic Monthly.* Retrieved from https://www.theatlantic.com/ideas/archive/2019/11/five-common-misconceptions-about-electoral-college/602596/

Tarrow, S. G. (1994). *Power in movement: Social movements, collective action, and politics.* New York: Cambridge University Press.

Taub, A. (2016, June 21). Brexit, explained: 7 questions about what it means and why it matters. *New York Times.* Retrieved from http://www.nytimes.com/2016/06/21/world/europe/brexit-britain-eu-explained.html?r=0

Taylor, P., & Lopez, M. H. (2013, May 8). *Six take-aways from the Census Bureau's voting report.* Washington, DC: Pew Research Center. Retrieved from http://www.pewresearch.org/fact-tank/2013/05/08/six-take-aways-from-the-census-bureaus-voting-report/

Taylor, S., & Butcher, M. (2007). *Extra-legal defendant characteristics and mock juror ethnicity re-examined.* Paper presented at the annual conference of the British Psychological Society, York Conference Park, York, England.

Tews, M. J., Stafford, K., & Zhu, J. (2009). Beauty revisited: The impact of attractiveness, ability, and personality in the assessment of employment suitability. *International Journal of Selection and Assessment, 17*, 92–100.

Thobejane, T. D. (2014). An exploration of polygamous marriages: A worldview. *Mediterranean J Soc Sci, 5*(Suppl. 7), 1058–1066.

Thomas, G. M., Meyer, J. W., Ramirez, F. O., & Boli, J. (1987). *Institutional structure: Constituting state, society, and the individual.* Newbury Park. CA: Sage.

Thomas, W. I., & Thomas, D. S. (1928). *The child in America: Behavior problems and programs.* New York: Knopf.

Thomas, Z. (2020, April 18). Coronavirus: Will covid-19 speed up the use of robots to replace human workers? *BBC.* Retrieved from https://www.bbc.com/news/technology-52340651

Thomson, J. (1874). The city of dreadful night. *National Reformer.*

Thorne, B. (1993). *Gender play: Girls and boys in school.* New Brunswick, NJ: Rutgers University Press.

Tiger, R. (2017). Race, class, and the framing of drug epidemics. *Contexts Magazine.* 16(4).

Tilly, C. (1975). *The formation of national states in Europe.* Princeton, NJ: Princeton University Press.

Tilly, C. (1978). *From mobilization to revelation.* Reading, MA: Addison-Wesley.

Tilly, C., & Tilly, L. (1994). Capitalist work and labor markets. In N. J. Smelser & R. Swedberg (Eds.), *The handbook of economic sociology.* Princeton, NJ: Princeton University Press.

Tisdall, S. (2019, January 26). Will corruption, cuts and protest produce a new Arab spring? *The Guardian. Retrieved from https://www.theguardian.com/world/2019/jan/26/sudan-egypt-corruption-arab-spring*

Toennies, F. (1963). *Gemeinschaft and Gesellschaft.* New York: Harper & Row. (Original work published 1887)

Tolan, P., Gorman-Smith, D., & Henry, D. (2005). Family violence. *Annual Review of Psychology, 57*, 557–583.

Tolnay, S. E., & Beck, E. M. (1995). *A festival of violence: An analysis of southern lynchings, 1882–1930.* Urbana-Champaign: University of Illinois Press.

Toossi, M., & Torpey, E. (2017, May). Older workers: Labor force trends and career options. In *Career Outlook.* U.S. Bureau of Labor Statistics. Retrieved from https://www.bls.gov/careeroutlook/2017/article/older-workers.htm

Tooze, A. (2020, August 1). The sociologist who could save us from coronavirus. *Foreign Policy.* Retrieved from https://foreignpolicy.com/2020/08/01/the-sociologist-who-could-save-us-from-coronavirus/

Trac Reports. (2020). *Corporate and white-collar prosecutions at all-time lows.* Retrieved from https://trac.syr.edu/tracreports/crim/597/

Transparency International. (2011). *Corruption perceptions index 2011* . Retrieved from http://cpi.transparency.org/cpi2011/results

Trent, S., Williams, J., Thornton, C., & Shanahan, M. (2004). *Farming the sea, costing the earth: Why we must green the blue revolution.* London: Environmental Justice Foundation. Retrieved from http://www.ejfoundation.org/pdf/farming_the_sea_costing_the_earth.pdf

Trimble, L. B., & Kmec, J. A. (2011). The role of social networks in the job attainment process. *Sociology Compass*, *5*, 165–178.

Tripodi, S., Pettus-Davis, C., Bender, K., Fitzgerald, M., Renn, T., & Kennedy, S. (2019). Pathways to recidivism for women released from prison: A mediated model of abuse, mental health issues, and substance use. *Criminal Justice and Behavior*, *46*(9), 1219–1236.

Troeltsch, E. (1931). *The social teaching of the Christian churches* (Vol. 1). New York: Macmillan.

Trudeau, M. (2009, February 5). *More students turning illegally to "smart" drugs.* National Public Radio. Retrieved from http://www.npr.org/templates/story/story.php?storyId=100254163

Truman, J. L., & Morgan, R. E. (2014, April). *Nonfatal domestic violence, 2003–2012.* U.S. Department of Justice, Office of Justice Programs, Bureau of Justice Statistics. Retrieved from http://www.bjs.gov/content/pub/pdf/ndv0312.pdf

Tsai, A. C., Kiang, M. V., Barnett, M. L., Beletsky, L., Keyes, K. M., McGinty, E. E., & Venkataramani, A. S. (2019). Stigma as a fundamental hindrance to the United States opioid overdose crisis response. *PLoS Medicine*, *16*(11), e1002969. https://doi.org/10.1371/journal.pmed.1002969

Tuchman, B. W. (1978). *A distant mirror: The calamitous 14th century.* Knopf.

Tucker, I. (2018, November 17). Elizabeth Stokoe: "We all talk, but we don't really know how." *The Guardian.* Retrieved from https://www.theguardian.com/science/2018/nov/17/elizabeth-stokoe-conversation-analyst-talk-interview

Tucker, K. H. (1991). How new are the new social movements? *Theory, Culture and Society*, *8*(2), 75–98. http://dx.doi.org/10.1177/026327691008002004

Tully, T., & Stowe, S. (2020). *The wealthy flee coronavirus. Vacation towns respond: "Stay away."* New York: New York Times Company.

Tumin, M. M. (1953). Some principles of stratification: A critical analysis. *American Sociological Review*, *18*, 387–393.

Tumin, M. M. (1963). On inequality. *American Sociological Review*, *28*. 19–26.

Tumin, M. M. (1985). *Social stratification: The forms and functions of inequality* (2nd ed.). Englewood Cliffs, NJ: Prentice Hall.

Turanovic, J., Reisig, M., & Pratt, T. (2015). Risky lifestyles, low self-control, and violent victimization across gendered pathways to crime. *Journal of Quantitative Criminology*, *31*(2), 183–206.

Turner, M. A., Popkin, S. J., & Rawlings, L. (2009). *Public housing and the legacy of segregation.* Washington, DC: Urban Institute Press.

Turner, R. H., & Killian, L. M. (1987). *Collective behavior* (3rd ed.). Englewood Cliffs, NJ: Prentice Hall.

Twenge, J. M., Spitzberg, B. H., & Campbell, W. K. (2019). Less in-person social interaction with peers among U.S. adolescents in the 21st century and links to loneliness. *Journal of Social and Personal Relationships*, *36*(6), 1892–1913. Retrieved from https://doi.org/10.1177/0265407519836170

Tyack, D., & Hansot, E. (1982). *Managers of virtue: Public school leadership in America, 1820–1980.* New York: Basic Books.

Tygart, C. E. (1987). Social structure linkages among social movement participants: Toward a synthesis of micro and macro paradigms. *Sociological Viewpoints*, *3*(1), 71–84.

Tyler, I. E. (2018). Resituating Erving Goffman: From stigma power to Black power. *The Sociological Review*, *66*(4), 744–765. doi:https://doi.org/10.1177/0038026118777450

Tyman, K., Saylor, C., Taylor, L. A., & Comeaux, C. (2010). Comparing children and adolescents engaged in cyberbullying with matched peers. *Cyberpsychology, Behavior, and Social Networking*, *13*, 195–199.

Tynan, M. A., Polansky, J. R., Driscoll, D., Garcia, C., & Glantz, S. A. (2019). Tobacco use in top-grossing movies—United States, 2010–2018. *MMWR Morbidity and Mortality Weekly Report 68*, 974–978. Retrieved from https://www.cdc.gov/mmwr/volumes/68/wr/mm6843a4.htm?s_cid=mm6843a4_w

Tyson, A. (2020, July 22). *Republicans remain far less likely than Democrats to view COVID-19 as a major threat to public health.* Fact Tank. Pew Research Center. Retrieved from https://www.pewresearch.org/fact-tank/2020/07/22/republicans-remain-far-less-likely-than-democrats-to-view-covid-19-as-a-major-threat-to-public-health/

Uchitelle, L. (2007). *The disposable American: Layoffs and their consequences.* New York: Vintage Books.

Uggen, C., Larson, R., & Shannon, S. (2016, October 6). 6 million lost voters: State-level estimates of felony disenfranchisement, 2016. *The Sentencing Project.* https://www.sentencingproject.org/publications/6-million-lost-voters-state-level-estimates-felony-disenfranchisement-2016

Uggen, C., Larson, R., Shannon, S., & Pulido-Nava, A. (2020). *Locked out 2020: Estimates of people denied voting rights due to a felony conviction.* The Sentencing Project. Retrieved from https://www.sentencingproject.org/publications/locked-out-2020-estimates-of-people-denied-voting-rights-due-to-a-felony-conviction/

Umaña-Taylor, A. J., & Hill, N. E. (2020). Ethnic–racial socialization in the family: A decade's advance on precursors and outcomes. *Journal of Marriage and Family*, *82*(1), 244–271. https://doi.org/10.1111/jomf.12622

UNAIDS. (2017). *UNAIDS data 2017.* Joint United Nations Programme on HIV/AIDS (UNAIDS). Retrieved from http://www.unaids.org/sites/default/files/media_asset/20170720_Data_book_2017_en.pdf

UNAIDS. (2020a, June 22). *Press release: COVID-19 could affect the availability and cost of antiretroviral medicines, but the risks can be mitigated.* Retrieved from https://www.unaids.org/en/resources/presscentre/pressreleaseandstatementarchive/2020/june/20200622_availability-and-cost-of-antiretroviral-medicines

UNAIDS. (2020b, July 6). *UNAIDS data 2020.* Retrieved from https://www.unaids.org/sites/default/files/media_asset/2020_aids-data-book_en.pdf

UNAIDS. (2020c, August 27). *Global HIV and AIDS statistics 2020 fact sheet.* Retrieved from https://www.unaids.org/en/resources/fact-sheet

Underhill, B. (1995). *The woman who ran for president: The many lives of Victoria Woodhull.* New York, NY: Bridge Works.

UNESCO. (2019). *Out-of-school children and youth.* Retrieved from http://uis.unesco.org/en/topic/out-school-children-and-youth

Ungar, S. (1992). The rise and (relative) decline of global warming as a social problem. *Sociological Quarterly, 33,* 483–501.

UNICEF. (2016, March). *The AIDS epidemic continues to take a staggering toll, especially in sub-Saharan Africa.* Retrieved from http://data.unicef.org/hiv-aids/global-trends.html

UNICEF. (2019). Fast facts: 10 facts illustrating why we must #EndChildMarriage. Retrieved from https://www.unicef.org/press-releases/fast-facts-10-facts-illustrating-why-we-must-endchildmarriage

UNICEF. (2020a, March 1). *Malnutrition. UNICEF data.* Retrieved from https://data.unicef.org/topic/nutrition/malnutrition/

UNICEF. (2020b). *Girls' education.* Retrieved from https://www.unicef.org/education/girls-education

Union of International Associations. (2011). Historical overview of number of international organizations by type, 1909–2011. In *Yearbook of international organizations, 2011/2012 edition.* Herndon, VA: Brill.

United Nations. (1995). *Women's education and fertility behavior: Recent evidence from the demographic and health surveys.* New York: Author.

United Nations. (2015, December 8). *Infographic: Human rights of women.* Retrieved from https://www.unwomen.org/en/digital-library/multimedia/2015/12/infographic-human-rights-women

United Nations. (2020). *What is the Paris Agreement?* Retrieved from https://unfccc.int/process-and-meetings/the-paris-agreement/what-is-the-paris-agreement

United Nations. (n.d.). *Refugees.* Retrieved from https://www.un.org/en/sections/issues-depth/refugees/

United Nations Department of Economic and Social Affairs. (2018). *68% of the world's population projected to live in urban areas by 2050, says UN.* Retrieved from https://www.un.org/development/desa/en/news/population/2018-revision-of-world-urbanization-prospects.html

United Nations Educational, Scientific and Cultural Organization. (2010). *Education counts: Towards the millennium development goals.* Paris: Author. Retrieved from http://unesdoc.unesco.org/images/0019/001902/190214e.pdf

United Nations Environment Programme. (2009). *Recycling: From e-waste to resources.* Berlin: Author. Retrieved from http://www.unep.org/PDF/PressReleases/E-Waste_publication_screen_FINALVER SION-sml.pdf

United Nations Fact Sheet. (n.d.). *Fact sheet intersex.* Retrieved from https://docs.google.com/viewerng/viewer?url=http://interactadvocates.org/wp-content/uploads/2016/02/United-Nations_FactSheet_Intersex.pdf

United Nations High Commissioner for Refugees. (2020a). *UNHCR global trends: Forced displacement in 2019.* Retrieved from https://www.unhcr.org/en-us/statistics/unhcrstats/5ee200e37/unhcr-global-trends-2019.html?query=45.7%20million%20people%20internally%20displaced,%20with%2079.5%20million%20of%20them%20refugees

United Nations High Commissioner for Refugees. (2020b). *What is a refugee?* Retrieved from https://www.unhcr.org/what-is-a-refugee.html

United Nations High Commissioner for Refugees (UNHCR). (2020c, October 10). *COVID-19 inducing "widespread despair" among refugees, UNHCR appeals for urgent support for mental health.* Retrieved from https://www.unhcr.org/en-us/news/press/2020/10/5f80502c4/covid-19-inducing-widespread-despair-among-refugees-unhcr-appeals-urgent.html

United Nations Human Rights. (2020, October 28). *Children, women, migrants all at increased risk of exploitation and trafficking during second COVID wave, UN expert warns.* Retrieved from https://www.ohchr.org/EN/NewsEvents/Pages/DisplayNews.aspx?NewsID=26443&LangID=E

United Nations Population Fund (UNFPA). (2012). *Sex imbalances at birth: Current trends, consequences, and policy implications.* Retrieved from https://www.unfpa.org/publications/sex-imbalances-birth

United Nations Women. (n.d.). *Economic empowerment of women.* Retrieved from https://www.unwomen.org/-/media/headquarters/attachments/sections/library/publications/2013/12/un%20women_ee-thematic-brief_us-web%20pdf.pdf?la=en

United States Advisory Council on Human Trafficking. (2020). *Annual report 2020.* Retrieved from https://www.state.gov/united-states-advisory-council-on-human-trafficking-annual-report-2020/

United States Census Bureau. (2018). *Poverty: 2017 and 2018.* Retrieved from https://www.census.gov/content/dam/Census/library/publications/2019/acs/acsbr18-02.pdf

United States Courts. (2020, July 18). *Bankruptcy filings fall 11.8 percent for year ending June 30.* Retrieved from https://www.uscourts.gov/news/2020/07/29/bankruptcy-filings-fall-118-percent-year-ending-june-30

United States Department of Health and Human Services. (n.d.). *What is the U.S. opioid epidemic?* Retrieved from https://www.hhs.gov/opioids/about-the-epidemic/index.html

United States Department of Labor. (2020). *Usual weekly earnings of wage and salary workers: Second quarter 2020.* Washington, DC: Bureau of Labor Statistics. Retrieved from https://www.bls.gov/news.release/pdf/wkyeng.pdf

United States Drug Enforcement Administration. (2020). *Drug scheduling*. Retrieved from https://www.dea.gov/drug-scheduling

University of Chicago Law School Faculty Blog. (2006, February 2). *Deliberation day and political extremism*. Retrieved from https://uchicagolaw.typepad.com/faculty/2006/02/deliberation_da.html

University of Nevada, Reno. (2010, May 21). Books in home as important as parents' education in determining children's education level. *ScienceDaily*. Retrieved from http://www.sciencedaily.com/releases/2010/05/100520213116.htm

Urban, H. B. (2011). *The Church of Scientology: A history of a new religion*. Princeton, NJ: Princeton University Press.

Urban Institute. (2017). *Nine charts about wealth inequality in America*. Retrieved from http://apps.urban.org/features/wealth-inequality-charts/

USA Facts. (2020a, January 20). *Native Americans and the US census: How the count has changed*. Retrieved from https://usafacts.org/articles/native-americans-and-us-census-how-count-has-changed/

USA Facts. (2020b, September 28). *4.4 million households with children don't have consistent access to computers for online learning during the pandemic*. Retrieved from https://usafacts.org/articles/internet-access-students-at-home/

U.S. Bureau of Labor Statistics. (2013a, March 8). *Table A-1. Employment status of the civilian population by sex and age*. Retrieved from http://www.bls.gov/news.release/empsit.t01.htm

U.S. Bureau of Labor Statistics. (2013b). *Union members—2012* (USDL-13-0105). Washington, DC: U.S. Department of Labor. Retrieved from http://www.bls.gov/news.release/pdf/union2.pdf

U.S. Bureau of Labor Statistics. (2015c). *May 2015 national occupational employment and wage estimates*. Retrieved from http://www.bls.gov/oes/current/oes_nat.htm

U.S. Bureau of Labor Statistics. (2015d, December 15). *Employment by major industry sector*. Retrieved from http://www.bls.gov/emp/ep_table_201.htm

U.S. Bureau of Labor Statistics. (2020a, September 1). *Employment by major industry sector*. Retrieved from https://www.bls.gov/emp/tables/employment-by-major-industry-sector.htm

U.S. Bureau of Labor Statistics. (2020b, November 15). *Labor force characteristics*. Retrieved from https://www.bls.gov/cps/lfcharacteristics.htm#emp

U.S. Bureau of Labor Statistics. (2020c). Learn more, earn more: Education leads to higher wages, lower unemployment. *Career Outlook*. Retrieved fromhttps://www.bls.gov/careeroutlook/2020/data-on-display/education-pays.htm

U.S. Bureau of Labor Statistics. (2020d, September 1). *Occupational outlook handbook*. Retrieved from https://www.bls.gov/ooh/

U.S. Bureau of Labor Statistics. (2020e). *Union members summary*. Retrieved from https://www.bls.gov/news.release/union2.nr0.htm

U.S. Census Bureau. (2011). *American Indian and Alaska Native Heritage Month: November 2011*. Retrieved from https://www.census.gov/newsroom/releases/archives/facts_for_features_special_editions/cb11-ff22.html

U.S. Census Bureau. (2012). *Characteristics of same-sex couple households: 2012*. Retrieved from http://www.census.gov/hhes/samesex

U.S. Census Bureau. (2015). *Families in poverty by type of family: 2013 and 2014*. Retrieved from http://www.census.gov/library/publications/2015/demo/p60-252.html

U.S. Census Bureau. (2016, November 17). *The majority of children live with two parents, Census Bureau reports*. Retrieved from https://www.census.gov/newsroom/press-releases/2016/cb16-192.html

U.S. Census Bureau. (2017). *America's families and living arrangements: 2017: Adults* (A table series): Table A1. Retrieved from https://www.census.gov/data/tables/2017/demo/families/cps-2017.html

U.S. Census Bureau, (2018, August 9). *One in five Americans live in rural areas*. Retrieved from https://www.census.gov/library/stories/2017/08/rural-america.html

U.S. Census Bureau. (2019, June 19). Population estimates. *Quick Facts*. Retrieved from https://www.census.gov/quickfacts/fact/table/US/PST045219

U.S. Centers for Disease Control and Prevention. (2015). *National intimate partner and sexual violence survey: 2015 data brief*. Retrieved from https://www.cdc.gov/violenceprevention/datasources/nisvs/2015NISVSdatabrief.html

U.S. Centers for Disease Control and Prevention. (2019). *Preventing intimate partner violence*. Retrieved from https://www.cdc.gov/violenceprevention/pdf/ipv-factsheet508.pdf

U.S. Centers for Disease Control and Prevention. (2020). *Marriage and divorce*. National Center for Health Statistics. Retrieved from https://www.cdc.gov/nchs/fastats/marriage-divorce.htm

U.S. Department of Commerce, Economics and Statistics Administration. (2010, January). *Middle class in America* (prepared for the Office of the Vice President of the United States Middle Class Task Force). Washington, DC: Author. Retrieved from http://www.commerce.gov/sites/default/files/docu-ments/migrated/Middle%20Class%20Report.pdf

U.S. Department of Defense. (2011). *Dictionary of military terms*. Retrieved from http://www.dtic.mil/doctrine/dod_dictionary

U.S. Department of Education. National Center for Education Statistics. (2020a). Status dropout rates. *The Condition of Education 2020* (NCES 2020-144). Retrieved from https://nces.ed.gov/fastfacts/display.asp?id=16

U.S. Department of Education. National Center for Education Statistics. (2020b). *College enrollment rates*. Retrieved from https://nces.ed.gov/programs/coe/indicator_cpb.asp

U.S. Department of Education. National Center for Education Statistics, Statistics Canada and Organization for Economic Cooperation and Development, Program for the International Assessment of Adult Competencies (PIAAC). (2020). *PIAAC 2012/2014 and PIAAC 2017 literacy, numeracy, and problem-solving TRE assessments*. Retrieved from https://nces.ed.gov/surveys/piaac/ideuspiaac

U.S. Department of Health and Human Services. (2010). *Statistics and research: Child maltreatment 2010*. Retrieved from http://www.acf.hhs.gov/programs/cb/research-data-technology/statistics-research/child-maltreatment

U.S. Department of Health and Human Services. (2020). *Smoking cessation: A report of the surgeon general—executive summary*. Atlanta, GA: U.S. Department of Health and Human Services. Centers for Disease Control and Prevention, National Center for Chronic Disease Prevention and Health Promotion, Office on Smoking and Health.

U.S. Department of Health and Human Services Office of Minority Health. (n.d.). *Profile: American Indian/Alaska Native*. Retrieved from https://www.minorityhealth.hhs.gov/omh/browse.aspx?lvl=3&lvlid=62

U.S. Department of Justice Office of Public Affairs. (2017). *Volkswagen AG agrees to plead guilty and pay $4.3 billion in criminal and civil penalties*. Retrieved from https://www.justice.gov/opa/pr/volkswagen-ag-agrees-plead-guilty-and-pay-43-billion-criminal-and-civil-penalties-six

U.S. Department of Labor, Wage and Hour Division. (2018). *Internship programs under the Fair Labor Standards Act*. Washington, DC: Author. Retrieved from http://www.dol.gov/whd/regs/compliance/whdfs71.pdf

U.S. Department of State. (2012, September 28). *Foreign terrorist organizations*. Retrieved from http://www.state.gov/j/ct/rls/other/des/123085.htm

U.S. Elections Project. (2013, February 9). *2012 general election turnout rates*. Retrieved from http://elections.gmu.edu/Turnout_2012G.html

U.S. Government Accountability Office. (2011). *Child maltreatment: Strengthening national data on child fatalities could aid in prevention* (GAO-11-599). Washington, DC: Government Printing Office. Retrieved from http://www.gao.gov/new.items/d11599.pdf

U.S. Government Accountability Office. (2016). *K–12 Education: Better use of information could help agencies identify disparities and address racial discrimination* (GAO-16-345). Washington, DC: Government Printing Office. Retrieved from http://www.gao.gov/products/GAO-16-345

U.S. Securities and Exchange Commission. (2015). *SEC adopts rule for pay ratio disclosure*. Washington, DC: Author. Retrieved from http://www.sec.gov/news/pressrelease/2015-160.html

Valens, A. (2018, March 29). *How big is the transgender population, really?* Retrieved from https://www.dailydot.com/irl/transgender-population-in-us/

Valentine, G. (2006). Globalizing intimacy: The role of information and communication technologies in maintaining and creating relationships. *Women's Studies Quarterly, 34*, 365–393.

Van DeBosch, H., & Van Cleemput, K. (2008). Defining cyberbullying: A qualitative research into the perceptions of youngsters. *CyberPsychology & Behavior, 11*, 499–503.

van Dijke, M., & Poppe, M. (2006). Striving for personal power as a basis for social power dynamics. *European Journal of Social Psychology, 36*(4), 537–566. http://dx.doi.org/10.1002/ejsp.351

Van Kessel, P., Toor, S. , & Smith, A. (2019, July 25). *A week in the life of popular YouTube channels*. Retrieved from https://www.pewresearch.org/internet/2019/07/25/a-week-in-the-life-of-popular-youtube-channels

VanGiezen, R., & Schwenk, A. E. (2001, Fall). Compensation before World War I through the Great Depression. *Compensation and Working Conditions* (U.S. Department of Labor, Bureau of Labor Statistics). Retrieved from http://www.bls.gov/opub/mlr/cwc/compensation-from-before-world-war-i-through-the-great-depression.pdf

Vasel, K. (2017, February 2). Couples are spending a record amount to get married. *CNN*. Retrieved from https://money.cnn.com/2017/02/02/pf/cost-of-wedding-budget-2016-the-knot/index.html

Vasilogambros, M. (2018). *After Parkland, states pass 50 new gun-control laws*. Pew Trusts. Retrieved from https://www.pewtrusts.org/en/research-and-analysis/blogs/stateline/2018/08/02/after-parkland-states-pass-50-new-gun-control-laws

Vatz, S. (2013, May 24). *Why America stopped making its own clothes*. Retrieved from http://ww2.kqed.org/lowdown/2013/05/24/madeinamerica/

Veblen, T. (1899). *The theory of the leisure class*. New York: Macmillan.

Vedder, R. (2020, April 7). Why the coronavirus will kill 500–1,000 colleges. *Forbes Magazine* online. Retrieved from https://www.forbes.com/sites/richardvedder/2020/04/07/500-1000-colleges-to-disappear-survival-of-the-fittest/#4c4a078e11a1

Vega, T. (2015, October 30). Out of prison and out of work: Jobs out of reach for former inmates. *CNN Money*. Retrieved from http://money.cnn.com/2015/10/30/news/economy/former-inmates-unemployed

Vellacott, J. (1993). *From Liberal to Labour with women's suffrage: The story of Catherine Marshall*. Montreal, Canada: McGill-Queen's University Press.

Venator, J., & Reeves, R. V. (2015, July 7). *Unpaid internships: Support beams for the glass floor*. Washington, DC: Brookings Institution. Retrieved from http://www.brookings.edu/blogs/social-mobility-memos/posts/2015/07/07-unpaid-internships-reeves

Venkataramani, A. S., Bair, E. F., O'Brien, R. L., & Tsai, A. C. (2019). Association between automotive assembly plant closures and opioid overdose mortality in the United States: A difference-in-differences analysis. *JAMA Intern Med, 180*(2), 254–262. Retrieved from doi:https://doi.org/10.1001/jamainternmed.2019.5686

Venkatesh, S. A. (2008). *Off the books*. Cambridge, MA: Harvard University Press.

Ventura, S. J., & Hamilton, B. E. (2011). *U.S. teenage birth rate resumes decline* (NCHS Data Brief 58). Hyattsville, MD: National Center for Health Statistics. Retrieved from http://www.cdc.gov/nchs/data/databriefs/db58.pdf

Vick, A. D., & Burris, H. H. (2017). Epigenetics and health disparities. *Current Epidemiological Rep, 4*(1), 31–37.

Villa, M. (2017, January 11). *Women own less than 20 percent of the world's land: It's time to give them equal property rights*. World Economic Forum. Retrieved from https://www.weforum.org/agenda/2017/01/women-own-less-than-20-of-the-worlds-land-its-time-to-give-them-equal-property-rights

Villarreal, Y. (2020, July 17). Inside Netflix's eye-opening look at arranged marriage, your next reality TV obsession. *Los Angeles Times*. Retrieved from https://www.latimes.com/entertainment-arts/tv/story/2020-07-17/indian-matchmaking-netflix-explained

Vinovskis, M. A. (1992). Schooling and poor children in 19th-century America. *American Behavioral Scientist, 35*, 313–331.

Vinovskis, M. A. (1995). *Education, society, and economic opportunity: A historical perspective on persistent problems*. New Haven, CT: Yale University Press.

Vinson, A. H. (2019). Short white coats: Knowledge, identity, and status negotiations of first-year medical students. *Symbolic Interaction, 42*(3), 395–411. doi:http://dx.doi.org.thor.nebrwesleyan.edu/10.1002/symb.400

Violence Policy Center. (2010). *Black homicide victimization in the United States: An analysis of 2007 homicide data*. Washington, DC: Author. Retrieved from http://www.vpc.org/studies/blackhomicide10.pdf

Virginia Museum of History and Culture. (n.d.). *Beginnings of Black education*. Retrieved from https://www.virginiahistory.org/collections-and-resources/virginia-history-explorer/civil-rights-movement-virginia/beginnings-black

Vogels, E. A. (2020). *59% of U.S. parents with lower incomes say their child may face digital obstacles in schoolwork*. Pew Research Fact Tank. Retrieved from https://www.pewresearch.org/fact-tank/2020/09/10/59-of-u-s-parents-with-lower-incomes-say-their-child-may-face-digital-obstacles-in-schoolwork/

Vollet, J. W., George, M. J., Burnell, K., & Underwood, M. K. (2020). Exploring text messaging as a platform for peer socialization of social aggression. *Developmental Psychology* 56(1), 138–152. Retrieved from https://doi.org/10.1037/dev0000848

Wacquant, L. (2002). From slavery to mass incarceration. *New Left Review, 13*, 40–61.

Waddell, K. (2017, January 23). The exhausting work of tallying America's largest protest. *The Atlantic Monthly*. Retrieved from https://www.theatlantic.com/technology/archive/2017/01/womens-march-protest-count/514166/

Wade, C., & Tavris, C. (1997). *Psychology*. New York: Longman.

Wagner, L. (2016, January 22). Film academy votes to increase diversity. *NPR*. Retrieved from http://www.npr.org/sections/thetwo-way/2016/01/22/464016379/film-academy-votes-to-increase-diversity

Wald, M. L. (2014, March30). U.S. agency knew about G.M. flaw but did not act. *New York Times*. Retrieved from http://www.nytimes.com/2014/03/31/business/ us-regulators-declined-full-inquiry-into-gm-ignition-flaws-memo-shows.html?_r=0

Wallace, R. (1992). *They call him pastor: Married men in charge of Catholic parishes*. Mahwah, NJ: Paulist Press.

Wallenstein, A. (2014, February 10). How *The Walking Dead* breaks every rule we know about TV hits. *Variety*. Retrieved from http://variety.com/2014/tv/news/how-the-walking-dead-breaks-every-rule-we-know-about-tv-hits-1201089433

Wallerstein, I. (1974). *The modern world-system*. New York: Academic Press.

Wallerstein, I. (2011a). *The modern world-system I: Capitalist agriculture and the origins of the European world-economy in the sixteenth century*. Berkeley: University of California Press. (Original work published 1974)

Wallerstein, I. (2011b). *The modern world-system II: Mercantilism and the consolidation of the European world-economy, 1600–1750*. Berkeley: University of California Press. (Original work published 1980)

Wallerstein, I. (2011c). *The modern world-system III: The second era of great expansion of the capitalist world-economy, 1730–1840s*. Berkeley: University of California Press. (Original work published 1989)

Wallerstein, I. (2011d). *The modern world system IV: Centrist liberalism triumphant, 1789–1914*. Berkeley: University of California Press.

Walsh, A., & Hemmens, C. (2019). *Law, justice and society: A socio-legal introduction* (5th ed.). Oxford University Press.

Walters, P. B., & James, R. J. (1992). Schooling for some: Child labor and school enrollment of Black and White children in the early 20th century South. *American Sociological Review, 57*, 635–650.

Wamsley, L. (2020, August 31). *D.C. mayor Bowser blames outside agitators as cause of weekend violence. NPR*. Retrieved from https://www.npr.org/sections/live-updates-protests-for-racial-justice/2020/08/31/908022325/d-c-mayor-bowser-blames-outside-agitators-as-cause-of-weekend-violence

Wang, W., & Parker, K. (2011). *Women see value and benefits of college: Men lag on both fronts, survey finds*. Washington, DC: Pew Research Center. Retrieved from http://www.pewsocialtrends.org/files/2011/08/Gender-and-higher-ed-FNL-RPT.pdf

Wang, W., & Parker, K. (2014, September 24). *Record number of Americans have never married*. Washington, DC: *Pew Research Center*. Retrieved from http://www.pewsocialtrends.org/2014/09/24/record-share-of-americans-have-never-married/

Wang, W., & Taylor, P. (2011). *For millennials, parenthood trumps marriage. Pew Research*. Retrieved from https://www.pewresearch.org/social-trends/2011/03/09/for-millennials-parenthood-trumps-marriage/

Ward, L. M. (2016). *Media and sexualization: State of empirical research, 1995–2015*. Retrieved from https://www.tandfonline.com/doi/abs/10.1080/00224499.2016.1142496

Warner, G., Beardsley, E., & Hadid, D. (2020, May 27). From Niqab to N-95. *NPR Rough Translation*. Retrieved from https://www.npr.org/2020/04/28/847433454/from-niqab-to-n95

Warner, K., Molotch, H. L., & Lategola, A. (1992). *Growth of control: Inner workings and external effects*. Berkeley: University of California Press.

Warner, R. S. (1993). A work in progress toward a new paradigm for the sociological study of religion in the United States. *American Journal of Sociology, 98*, 1044–1093.

Warren, K., Mitten, D., D'Amore, C., & Lotz, E. (2019). The gendered hidden curriculum of adventure education. *Journal of Experiential Education, 42*(2), 140–154. Retrieved from https://doi-org.thor.nebrwesleyan.edu/10.1177/1053825918813398

Warrick, J. (2016, May 28). ISIS fighters seem to be trying to sell sex slaves online. *Washington Post*. Retrieved from https://www.washingtonpost.com/world/national-security/isis-fighters-appear-to-be-trying-to-sell-their-sex-slaves-on-the-internet/2016/05/28/b3d1edea-24fe-11e6-9e7f-57890b612299_story.html

Washington, H. (2007). *Medical apartheid: The dark history of medical experimentation on Black Americans from colonial times to the present*. New York: Doubleday.

Washington Post. (2020). In 1,267 days, President Trump has made 20,055 false or misleading claims. Retrieved from https://www.washingtonpost.com/graphics/politics/trump-claims-database/?utm_term=.b04b1667f114&tid=lk_inline_manual_2&itid=lk_inline_manual_3&itid=lk_inline_manual_3

Wasserman, I. M. (1999). *African Americans and the criminal justice system: An explanation for changing patterns of Black male suicide*. Paper presented at the annual conference of the Midwest Sociological Society, Minneapolis.

Watson, A. (2019, February 28). *Highest grossing movies at the North American box office 2018*. Retrieved from https://www.statista.com/statistics/825900/box-office-revenue-of-the-top-grossing-movies-2018

Watson, A. (2020, May 23). *Media-use statistics & facts. Statistica*. Retrieved from https://www.statista.com/topics/1536/media-use/

Watson, J. B. (1924). *Behaviorism*. New York: People's Institute.

Watts, J. (2019, March 11). Greta Thunberg, schoolgirl climate change warrior: 'Some people can let things go. I can't'. *The Guardian*.

Watts, L. K., Wagner, J., Velasquez, B., & Behrens, P. I. (2017). Cyberbullying in higher education: A literature review. *Computers in Human Behavior, 69*, 268–274.

Weber, M. (1946). *From Max Weber: Essays in sociology* (H.Gerth & C. W. Mills, Eds. & Trans.). New York: Oxford University Press. (Original work published 1919)

Weber, M. (1963). *The sociology of religion*. Boston: Beacon Press. (Original work published 1921)

Weber, M. (1979). *Economy and society: An outline of interpretive sociology* (2 vols.). Berkeley: University of California Press. (Original work published 1921)

Weber, M. (2002). *The Protestant ethic and the spirit of capitalism, and other writings*. New York: Penguin Books. (Original work published 1904–1905)

Weber, M. (2012). *The theory of social and economic organization*. Eastford, CT: Martino Fine Books. (Original work published 1921)

Webster, H. (2014, July 14). What parents and kids should know about selfies. *U.S. News & World Report*. Retrieved from http://health.usnews.com/health-news/health-wellness/articles/2014/07/14/what-parents-and-kids-should-know-about-selfies

The Wedding Report. (2019). Retrieved from https://wedding.report/index.cfm/action/wedding_statistics/view/market/id/00/idtype/s/location/united_states/

Weeber, S. C., & Rodheaver, D. G. (2003). Militias at the millennium: A test of Smelser's theory of collective behavior. *Sociological Quarterly, 44*(2), 181–204.

Weed News marijuana legality map. (2020). Retrieved from https://www.weednews.co/marijuana-legality-states-map/

Weeks, J. R. (1988). The demography of Islamic nations. *Population Bulletin, 43*, 5–54.

Weidman, J. L. (Ed.). (1984). *Christian feminism: Visions of a new humanity*. New York: Harper & Row.

Weimann, G. (2004). *Cyberterrorism? How real is the threat?* Washington, DC: United States Institute of Peace. Retrieved from https://www.usip.org/sites/default/files/sr119.pdf

Weiss-Wendt, A. (2010). The state and genocide. In D. Bloxham & A. D. Moses (Eds.), *The Oxford handbook of genocide studies* (pp. 81–101). New York, NY: Oxford University Press.

Weitz, R. (2012). *The sociology of health, illness, and health care: A critical approach* (6th ed.). Boston: Wadsworth Cengage Learning.

Weitz, R. (2017). *The sociology of health, illness, and health care: A critical approach* (7th ed.). Boston, MA: Cengage.

Weitzer, R., & Kubrin, C. E. (2009). Misogyny in rap music: A content analysis of prevalence and meanings. *Men and Masculinities, 12*, 3–29.

Wellman, B., & Hampton, K. (1999). Living networked on and offline. *Contemporary Sociology, 28*, 648–654.

Wells, A. S. (1989). Hispanic education in America: Separate and unequal. *Eric/Cue Digest 59*. Retrieved from https://www.ericdigests.org/pre-9214/hispanic.htm

Wells-Barnett, I. B. (1892). *Southern horrors: Lynch law in all its phases*. Retrieved from http://www.gutenberg.org/files/14975/14975-h/14975-h.htm

Welsh, R. (1998). Severe parental punishment and aggression: The link between corporal punishment and delinquency. In I. A. Hyman & J. H. Wise (Eds.), *Corporal punishment in American education: Readings in history, practice and alternatives* (pp. 126–142). Temple University Press.

Werdigier, J. (2010, June 4). J. P. Morgan penalized by regulator in Britain. *New York Times*, B3.

Werner, C. A. (2011). *The older population: 2010* (Census Brief C2010BR-09). Washington, DC: U.S. Census Bureau. Retrieved from http://www.census.gov/prod/cen2010/briefs/c2010br-09.pdf

Wertheimer, B. (1977). *We were there: The story of working women in America*. New York: Pantheon.

West, C. (1995). In K. Crenshaw, N. Gotanda, G. Peller, & K. Thomas (Eds.), *Critical race theory: The key writings that formed the movement*. New York: New Press.

West, C., & Zimmerman, D. H. (1977). Woman's place in everyday talk: Reflections on parent–child interactions. *Social Problems, 24*, 521–529.

West, C., & Zimmerman, D. H. (1983). Small insults: A study of interruptions in conversations between unacquainted persons. In B. Thorne, C. Kramarae, & N. Henley (Eds.), *Language, gender, and society* (pp. 102–117). Rowley, MA: Newbury House.

West, C., & Zimmerman, D. H. (1987). Doing gender. *Gender & Society, 1*(2), 125–151. http://dx.doi.org/10.1177/0891243287001002002

West, M., & Chew, H. E. (2014). *Reading in the mobile era: A study of mobile reading in developing countries. UNESCO*. Retrieved from https://unesdoc.unesco.org/ark:/48223/pf0000227436

Western, B., & Pettit, B. (2010). Incarceration and social inequality. *Daedalus, 139*, 8–19.

Westoff, C. F., & Ryder, N. B. (1977). *The contraceptive revolution*. Princeton, NJ: Princeton University Press.

Weston, L. (2014, September 9). OECD: The U.S. has fallen behind other countries in college completion. *Business Insider.* Retrieved from https://www.busi nessinsider.com/r-us-falls-behind-in-college-competition-oecd-2014-9

Whalen, J., Zimmerman, D. H., & Whalen, M. R. (1990). When words fail: A single case analysis. *Social Problems, 35*, 335–362.

Whelan, A., Wrigley, N., Warm, D., & Cannings, E. (2002). Life in a 'food desert.'. *Urban Studies, 39*, 2083–2100.

Whitehead, A. L. (2013). Gendered organizations and inequality regimes: Gender, homosexuality, and inequality within religious congregations. *Journal for the Scientific Study of Religion, 52*(3), 476–493. Retrieved from https://doi-org.thor.nebrwesleyan.edu/10.1111/jssr.12051

WHO,UNICEF, &UNFPA,World Bank Group, &the United Nations Population Division. (2019). *Trends in maternal mortality: 2000 to 2017.* Geneva: World Health Organization. Retrieved from https://data.worldbank.org/indicator/SH.STA.MMRT?locations=YE

WHO/UNICEF Joint Monitoring Programme (JMP) for Water Supply, Sanitation and Hygiene. (2020, October 5). *People using safely managed sanitation services (% of population).* The World Bank. Retrieved from https://data.worldbank.org/indicator/SH.STA.SMSS.ZS

Whyte, W. F. (1943). *Street corner society: The social structure of an Italian slum.* Chicago: University of Chicago Press.

Whyte, W. F. (1991). *Participatory action research.* Newbury Park, CA: Sage.

Williams, C. (1995). *Still a man's world: Men who do women's work.* Berkeley: University of California Press.

Williams, J. P. (2018 April 20). Segregation's legacy. *U.S. News and World Report* Online. Retrieved from https://www.usnews.com/news/the-report/articles/2018-04-20/us-is-still-segregated-even-after-fair-housing-act

Williams, J. P., & Kirschner, D. (2012). Coordinated action in the massively multiplayer online game *World of Warcraft. Symbolic Interaction.* Advance online publication. Retrieved from http://onlinelibrary.wiley.com/doi/10.1002/j.1533-8665.2012.00022.x/abstract

Williams, R. (2020). Civil religion. *The SAGE encyclopedia of the sociology of religion* (pp. 138–140). Thousand Oaks, CA: SAGE.

Williams, R. M., Jr. (1970). *American society: A sociological interpretation* (3rd ed.). New York: Knopf.

Willis, H., Hill, E., Stein, R., Triebert, C., Laffin, B., & Jordan, D. (2020, August 11). New footage shows delayed medical response to George Floyd. *New York Times.* Retrieved from https://www.nytimes.com/2020/08/11/us/george-floyd-body-cam-full-video.html

Willis, J. (2020, September 7). *Kanye West walks on water during Sunday service with Joel Osteen: Watch.* Retrieved from https://www.etonline.com/kanye-west-walks-on-water-during-sunday-service-with-joel-olsteen-watch-152613

Willis, P. (1990). *Common culture: Symbolic work at play in the everyday cultures of the young.* Boulder, CO: Westview Press.

Willoughby, T., Adachi, P. J. C., & Good, M. (2012). A longitudinal study of the association between violent video game play and aggression among adolescents. *Developmental Psychology, 48*, 1044–1057.

Wilson, B. J. (2008). Media and children's aggression, fear, and altruism. *The Future of Children*, pp. 87–118.

Wilson, D. (2019). Social control. In E. McLaughlin (Ed.), *The SAGE dictionary of criminology* (4th ed.). Thousand Oaks, CA: SAGE.

Wilson, D. C., Moore, D. W., McKay, P. F., & Avery, D. R. (2008). Affirmative action programs for women and minorities: Expressed support affected by question order. *Public Opinion Quarterly, 72*, 514–522.

Wilson, T. P. (1991). Social structure and the sequential organization of interaction. In D. H. Zimmerman & D. Boden (Eds.), *Talk and social structure* (pp. 22–43). Cambridge, England: Polity Press.

Wilson, W. J. (1978). *The declining significance of race: Blacks and changing American institutions.* Chicago: University of Chicago Press.

Wilson, W. J. (1996). *When work disappears: The world of the new urban poor.* New York: Vintage Books.

Wilson, W. J. (2010). *More than just race: Being Black and poor in the inner city.* New York: Norton.

Wingfield, A. H. (2008). Racializing the glass escalator: Reconsidering men's experiences with women's work. *Gender & Society, 23*(1), 5–26. http://dx.doi.org/10.1177/0891243208323054

Wingfield, A. H. (2019). "Reclaiming our time": Black women, resistance, and rising inequality: SWS presidential lecture. *Gender & Society, 33*(3), 345–362.

Wirth, L. (1928). *The ghetto.* Chicago: University of Chicago Press.

Wirth, L. (1938). Urbanism as a way of life. *American Journal of Sociology, 44*, 1–24.

Wirth, L. (1945). The problem of minority groups. In R. Linton (Ed.), *The science of man in the world crisis* (pp. 347–372). New York: Columbia University Press.

Woldoff, R. A. (2011). *White flight/Black flight: The dynamics of racial change in an American neighborhood.* Ithaca, NY: Cornell University Press.

Wolf, M. (2008). *Proust and the squid: The story and science of the reading brain.* New York: HarperPerennial.

Wolfe, A. (1977). *The limits of legitimacy.* New York: Free Press.

Wolff, E. N. (2017). *Household wealth trends in the United States, 1962 to 2016: Has middle class wealth recovered?* (No. w24085). National Bureau of Economic Research.

Wolfson, A. (2005, October 9). A hoax most cruel: Caller coaxed McDonald's managers into strip-searching a worker. *Courier Journal.* Retrieved from http://www.courier-journal.com/apps/pbcs.dll/article?AID=/20051009/NEWS01/510090392&loc=interstitialskip&nclick_check=1

Women in the U.S. Congress. (2019). Center for American Women and Politics, Eagleton Institute of Politics, Rutgers University. Retrieved from https://www.cawp.rutgers.edu/women-us-congress-2019

Wonacott, M. E. (2002). *Gold-collar workers* (ERIC Digest EDO-CE-02-234). Retrieved from http://www.calpro-online.org/ERIC/docs/dig234.pdf

Wong, A. (2019, February 20). The U.S. teaching population is getting bigger, and more female. *The Atlantic*. Retrieved from https://www.theatlantic.com/education/archive/2019/02/the-explosion-of-women-teachers/582622/

Wood, G. S. (1993). *The radicalism of the American Revolution*. New York: Vintage Books.

Wood, R. G., Goesling, B., & Avellar, S. (2007). *The effects of marriage on health: A synthesis of recent research evidence*. Princeton, NJ: Mathematica Policy Research. Retrieved from http://www.mathematica mpr.com/publications/PDFs/marriagehealth.pdf

Woodiwiss, M. (2000). *Organized crime: The dumbing of discourse*. In G. Mair & R. Tarling (Eds.), *British Criminology Conference: Selected proceedings* (Vol. 3). London: British Society of Criminology. Retrieved from http://www.britsoccrim.org/volume1/017.pdf

Work Estonia. (2018, September). *Salary levels and everyday costs in Estonia*. Retrieved from https://www.workinestonia.com/salary-levels-and-everydays-costs-in-estonia/

World Atlas. (2020). *The 150 largest cities in the world*. Retrieved from https://www.worldatlas.com/citypops.htm

The World Bank. (2015a). *WDI 2016 maps*. Retrieved from http://data.worldbank.org/products/wdi-maps

The World Bank. (2015b). *Improved sanitation facilities (% of population with access)*. Retrieved from http://data.worldbank.org/indicator/SH.STA.ACSN

The World Bank. (2017). *People using safely managed sanitation services, rural*. Retrieved from https://data.worldbank.org/indicator/SH.STA.SMSS.RU.ZS

The World Bank. (2018). *Many governments take steps to improve women's economic inclusion, although legal barriers remain widespread*. Retrieved from https://www.worldbank.org/en/news/press-release/2018/03/29/many-governments-take-steps-to-improve-womens-economic-inclusion-although-legal-barriers-remain-widespread

The World Bank. (2019). *Individuals using the internet (% of population)*. Retrieved from https://data.worldbank.org/indicator/IT.NET.USER.ZS

The World Bank. (2020a, April 16). *Poverty: Overview*. Retrieved from https://www.worldbank.org/en/topic/poverty/overview

The World Bank. (2020b, October 5). *Poverty*. The World Bank Group. Retrieved from https://www.worldbank.org/en/topic/poverty/overview

World Economic Forum. (2017). *More efficient and safer: How drones are changing the workforce*. Retrieved from https://www.weforum.org/agenda/2017/06/more-efficient-and-safer-how-drones-are-changing-the-workplace/

World Health Organization. (2005). *Widespread misunderstandings about chronic disease—and the reality*. Retrieved from http://www.who.int/chp/chronic_disease_report/media/Factsheet2.pdf

World Health Organization. (2020). *Obesity and overweight*. Retrieved from https://www.who.int/en/news-room/fact-sheets/detail/obesity-and-overweight

World Prison Studies. (2020, January 18). Retrieved from https://www.prisonstudies.org/

Worldwatch Institute. (2004). *State of the world: Consumption by the numbers*. Washington, DC: Author.

Worldwatch Institute. (2010). *State of the world: Transforming cultures from consumerism to sustainability*. Washington, DC: Author.

Worldwatch Institute. (2011). *State of the world: Innovations that nourish the planet*. Washington, DC: Author.

Wortmann, S. L. (2020, February 14). *Journeys to and beyond prison walls: Gendered pathways and programming in six United States women's prisons [Conference Presentation]*. Future of Women Conference, Bangalore, India.

Wright, E. O. (1994). *Interrogating inequality: Essays on class analysis, socialism and Marxism*. New York: Verso.

Wright, E. O. (1998). *Classes* (2nd ed.). New York: Verso.

Wright, J., & Cullen, F. (2012). The future of biosocial criminology: Beyond scholars' professional ideology. *Journal of Contemporary Criminal Justice*, *28*, 237–253.

Wright, J. P., Tibbetts, S. G., & Daigle, L. E. (2008). *Criminals in the making: Criminality across the life course*. Thousand Oaks, CA: Sage.

Wright, L. (1993, May 24). Remember Satan: Part II. *New Yorker*, pp. 54–76.

Wrong, D. H. (1959). The functional theory of stratification: Some neglected considerations. *American Sociological Review*, *24*, 772–782.

Wuthnow, R. (1976). *The consciousness reformation*. Berkeley: University of California Press.

Wuthnow, R. (1978). *Experimentation in American religion: The new mysticisms and their implications for churches*. Berkeley: University of California Press.

Wuthnow, R. (1988). *The restructuring of American religion: Society and faith since World War II*. Princeton, NJ: Princeton University Press.

Wuthnow, R. (1989). *Communities of discourse: Ideology and social structure in the Reformation, the Enlightenment, and European socialism*. Cambridge, MA: Harvard University Press.

Wuthnow, R. (2007). *After the baby boomers*. Princeton, NJ: Princeton University.

Wyman, A. (1997). *Rural women teachers in the United States: A sourcebook*. Lanham, MD: Scarecrow Press.

Wyss, S. (2007). "This was my hell": The violence experienced by gender non-conforming youth in US high schools. *International Journal of Qualitative Studies in Education*, *17*(5), 709–730.

Xu, J. Q., Murphy, S. L., Kochanek, K. D., & Arias, E. (2018). *Mortality in the United States, 2017*. NCHS data brief, no 355. Hyattsville, MD: National Center for Health Statistics. Retrieved from https://www.cdc.gov/nchs/products/databriefs/db328.htm

Yang, C. (2017). Local labor markets and criminal recidivism. *Journal of Public Economics*, *147*, 16–29.

Ybarra, M., Strasburger, V., & Mitchell, K. (2014). Sexual media exposure, sexual behavior, and sexual violence victimization in adolescence. *Clinical Pediatrics*. Retrieved from http://journals.sagepub.com/doi/abs/10.1177/0009922814538700

Young, S., & Martin, D. S. (2012 March 9). *CNN readers share stories about secret army drug testing program. CNN*. Retrieved from http://edition.cnn.com/2012/03/09/health/soldier-guinea-pigs/index.html

Yuhas, A. (2020, April 8). Don't expect a coronavirus baby boom. *New York Times*. Retrieved from https://www.nytimes.com/2020/04/08/us/coronavirus-baby-boom.html

Zahidi, S. (2019, June 18). *Accelerating gender parity in Globalization 4.0*. Retrieved from https://www.weforum.org/agenda/2019/06/accelerating-gender-gap-parity-equality-globalization-4/

Zald, M., & McCarthy, J. D. (1980). Social movement industries: Competition and cooperation among movement organizations. In L. Kriesberg (Ed.), *Research in social movements, conflicts and change* (Vol. 3). Greenwich, CT: JAI Press.

Zaleski, K. L., Gundersen, K. K., Baes, J., Estupinian, E., & Vergara, A. (2016). Exploring rape culture in social media forums. *Computers in Human Behavior*, *63*, 922–927. Retrieved from https://doi.org/10.1016/j.chb.2016.06.036

Zhou, M. (2009). *Contemporary Chinese America: Immigration, ethnicity, and community transformation*. Philadelphia: Temple University Press.

Zimbardo, P. G. (1974). On "obedience to authority". *American Psychologist*, *29*, 566–567.

Zimmerman, D. H. (1984). Talk and its occasion: The case of calling the police. In D. Schiffrin (Ed.), *Meaning, form, and use in context: Linguistic applications* (pp. 210–228). Washington, DC: Georgetown University Press.

Zimmerman, D. H. (1992). The interactional organization of calls for emergency assistance. In P. Drew & J. Heritage (Eds.), *Talk at work: Interaction in institutional settings* (pp. 418–469). New York: Cambridge University Press.

Zimmerman, D. H., & West, C. (1975). Sex roles, interruptions, and silences in conversations. In B. Thorne & N. Henley (Eds.), *Language and sex: Difference and dominance* (pp. 105–129). Rowley, MA: Newbury House.

Zimmerman, F. J., & Anderson, N. W. (2019). Trends in health equity in the United States by race/ethnicity, sex, and income, 1993–2017. *JAMA Netw Open*, *2*(6), e196386. doi:https://doi.org/10.1001/jamanetworkopen.2019.6386

Zinn, M. B., Weber, L., Higginbotham, E., & Dill, B. T. (1986). The costs of exclusionary practices in women's studies. *Signs*, *11*, 290–303.

Zornick, G. (2018, April 3). How the #Neveragain Movement Is Disrupting Gun Politics. *The Nation*. Retrieved from https://www.thenation.com/article/how-the-neveragain-movement-is-disrupting-gun-politics

INDEX